REVIEWS in

and GEOCHEMISTRY

Volume 50 2002

BERYLLIUM:

MINERALOGY, PETROLOGY, AND GEOCHEMISTRY

EDITOR: **EDWARD S. GREW**
University of Maine, Orono, Maine

Series Editor: ***Paul H. Ribbe***

Virginia Polytechnic Institute and State University
Blacksburg, Virginia

MINERALOGICAL SOCIETY of AMERICA

Washington, DC

REVIEWS IN MINERALOGY AND GEOCHEMISTRY

(Formerly: REVIEWS IN MINERALOGY)

ISSN 1529-6466

Volume 50

Beryllium:
Mineralogy, Petrology, and Geochemistry

ISBN 0-93995062-6

Additional copies of this volume as well as others in this series may be obtained at moderate cost from:

THE MINERALOGICAL SOCIETY OF AMERICA
1015 EIGHTEENTH STREET, NW, SUITE 601
WASHINGTON, DC 20036 U.S.A.

BERYLLIUM: Mineralogy, Petrology, and Geochemistry

FOREWORD

The book has been several years in the making, under the experienced and careful oversight of Ed Grew (University of Maine), who edited (with Larry Anovitz) a similar, even larger volume in 1996: *BORON: Mineralogy, Petrology, and Geochemistry* (*RiMG* Vol. 33, reprinted with updates and corrections, 2002). There are 14 chapters in *BERYLLIUM*, and an overview of their contents may be found in the Introduction (Chapter 1). Addenda and corrections to *BERYLLIUM* (and other volumes in this series) are posted at http://www.minsocam.org .

Paul H Ribbe, Series Editor
Blacksburg, Virginia
November 6, 2002

COVER

The cover illustrates the entry for beryllium in the Periodic Table. From upper left to right, it gives the atomic number, valence state (the only one known in minerals), chemical symbol, atomic weight and electronic configuration. Unlike boron, beryllium has no diagnostic flame color, so I selected for the block a color of the Be mineral euclase, $BeAlSiO_4(OH)$. Beryllium minerals have found use as colorful gemstones, the best known which are aquamarine (blue to green from Fe), emerald (green from Cr^{3+}), and alexandrite (red and green from Cr^{3+}), others are chrysoberyl (green-yellow from Fe^{3+}), morganite (pink from Mn) and Euclase. See George Rossman's site:

http://minerals.gps.caltech.edu/FILES/Visible/index.htm .

The blue color in the cover is from a zoned euclase from Zimbabwe, which corresponds to the absorption spectra below kindly provided by Rossman. This mineral is blue because the Fe^{2+}/Fe^{3+} intervalance band centered near 670 nm absorbs the red end of the visible spectrum but allows the blue end to pass. Fe^{2+} bands are seen near 850 and 1250 nm and a weak Fe^{3+} feature is near 440 nm. Sharp OH overtones appear near 1430 nm. The three spectral curves in the figure are for linearly polarized light vibrating in the directions in which the alpha, beta and gamma indices of refraction would be measured.

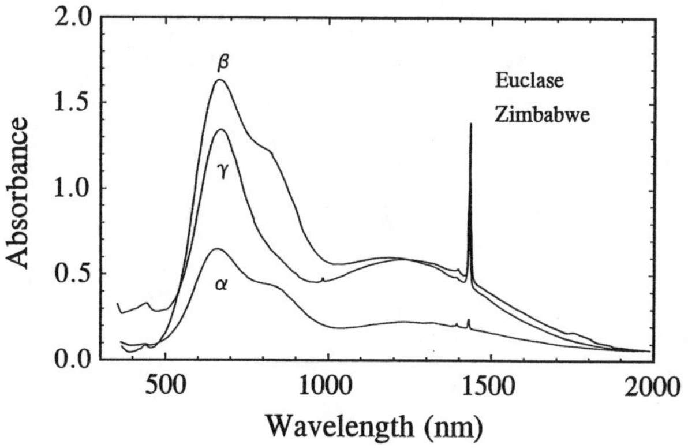

Table of Contents

1 Mineralogy, Petrology and Geochemistry of Beryllium: An Introduction and List of Beryllium Minerals

Edward S. Grew

2 Behavior of Beryllium During Solar System and Planetary Evolution: Evidence from Planetary Materials

Charles K. Shearer

3 Trace-Element Systematics of Beryllium in Terrestrial Materials

Jeffrey G. Ryan

4 Rates and Timing of Earth Surface Processes From *In Situ*-Produced Cosmogenic Be-10

Paul R. Bierman, Marc W. Caffee, P. Thompson Davis, Kim Marsella, Milan Pavich, Patrick Colgan, David Mickelson

5 Cosmogenic Be-10 and the Solid Earth: Studies in Geomagnetism, Subduction Zone Processes, and Active Tectonics

Julie D. Morris, John Gosse, Stefanie Brachfeld, Fouad Tera

6 Environmental Chemistry of Beryllium-7

James M. Kaste, Stephen A. Norton, Charles T. Hess

7 Environmental Chemistry of Beryllium

J. Veselý, S.A. Norton, P. Skřivan, V. Majer, P. Krám, T. Navrátil, J.M. Kaste

8 Beryllium Analyses by Secondary Ion Mass Spectrometry

Richard L. Hervig

9 The Crystal Chemistry of Beryllium

Frank C. Hawthorne, Danielle M.C. Huminicki

10 Mineralogy of Beryllium in Granitic Pegmatites

Petr Černý

11 Beryllium in Silicic Magmas and the Origin of Beryl-Bearing Pegmatites

David London, Joseph M. Evensen

12 Beryllium in Metamorphic Environments (emphasis on aluminous compositions)

Edward S. Grew

13 Be-Minerals: Synthesis, Stability, and Occurrence in Metamorphic Rocks

Gerhard Franz, Giulio Morteani

14 Non-pegmatitic Deposits of Beryllium: Mineralogy, Geology, Phase Equilibria and Origin

Mark D. Barton, Steven Young

1 Mineralogy, Petrology and Geochemistry of Beryllium: An Introduction and List of Beryllium Minerals

Edward S. Grew

Department of Geological Sciences
5790 Bryand Center, University of Maine
Orono, Maine 04469
esgrew@maine.edu

RATIONALE FOR VOLUME

In 1996, in collaboration with Lawrence Anovitz, I edited *Boron Mineralogy, Petrology and Geochemistry*, volume 33 in the *Reviews in Mineralogy* series, a book that has been reprinted with addenda in 2002 (further addenda and corrections are posted at http://www.minsocam.org, where you may also find corrections to this volume). Many of the same reasons for inviting investigators to contribute to a volume on boron apply equally to a volume on beryllium. Like B, Be poses analytical difficulties, and it has been neglected in many geochemical, mineralogical and petrological studies. However, with the development of instruments to measure cosmogenic isotopes, greater availability and refinement of the ion microprobe, and with overall improvement in analytical technology, interest in Be and its cosmogenic isotopes has increased, and more studies are being published. Thus, I decided that it was an appropriate time to invite those actively involved in research on Be to contribute to this volume, which is intended to be a companion to *Boron Mineralogy, Petrology and Geochemistry*. NOTE: In this chapter, individual review papers are referred to by author name(s) and chapter number.

BRIEF HISTORY OF BERYLLIUM

The Be mineral beryl and its colored variants emerald, aquamarine, and "chrysoberyl" (= golden beryl, not the present chrysoberyl) were known to the ancients, and Pliny the Elder had noted that many persons considered emerald and beryl "to be of the same nature" (Sinkankas 1981, p. 20; also Dana 1892; Weeks and Leicester 1968). However, not until 1798 was it realized that beryl contained a previously unknown constituent; analyses before then yielded only silica, alumina, lime and minor iron oxide (Vauquelin 1798; Anonymous 1930; Weeks and Leicester 1968; Greenwood and Earnshaw 1997). The mineralogist René Just Haüy asked Nicolas Louis Vauquelin (Fig. 1) to analyze beryl and emerald in order to test his supposition that these two minerals were identical as his measurements of crystals, hardness and density implied. In the process, Vauquelin (1798) not only demonstrated the near chemical identity of beryl and emerald, the latter differing in its tenor of chrome, but also isolated a new earth from both, his *terre du béril*, similar yet distinct from alumina. The editors of the *Annales de Chimie et de Physique* proposed to name it *glucine* (Greek, sweet) because its water-soluble salts had a sweet taste, and because this name would be easy to remember. However, Klaproth (1802, p. 79) thought it would be more sensible to call the new earth "Beryllerde (Beryllina)" because yttrium salts are also sweet-tasting, and the name beryllium is now generally accepted for the element.

Because Be is so similar to Al, chemists were confused about the valency and atomic weight of Be. Avdeyev, Brauner and Mendeleyev argued that the periodic law required Be to be divalent (Mendeleyev 1897). Divalency was confirmed in 1884 when the vapor density of beryllium chloride was found to correspond to the formula $BeCl_2$, a confirmation that Mendeleyev considered "as important in the history of the periodic law as the discovery of scandium" (Mendeleyev 1897, v. 2, p. 485; Stock 1932; Everest 1973).

1529-6466/00/0048-0001$05.00

Figure 1. Postage stamp honoring N.-L. Vauquelin issued by the French government in 1963. For more information on the stamp see "Phil – Ouest: Timbres de l'Ouest de la France" <http://www.le-lann.com/PhilOuest/DO/index.htm?FenD= http://www.le-lann.com/Phil-Ouest/DO/Album.html>. Reproduction courtesy of Roy Kristiansen.

ECONOMICS OF BERYLLIUM

Beryllium is a strategic material because of its very low absorption cross-section with respect to thermal neutrons, which, together with other special properties, makes Be a good neutron moderator material (Everest 1973; Greenwood and Earnshaw 1997; Emsley 2001). It is also the best reflector of neutrons, so that overall, it finds wide application in nuclear reactors. Indeed, the neutron was discovered in 1932 when Be was bombarded by alpha-particles. Beryllium is transparent to X-rays. For such a light metal, Be is strong and refractory (Table 1), and resistant to corrosion because of an oxide film. Beryllium metal finds extensive applications in alloys with copper, aluminum and other metals, and BeO is an excellent refractory finding some use in ceramics.

Table 1. Selected information on elemental beryllium

PHYSICAL PROPERTIES		ISOTOPES	
Property	*Value*	*Isotope*	*Half-life*
Atomic No.	4	^{6}Be	5.0×10^{-21} sec
Atomic Wt.	9.012182(3)	^{7}Be	53.28 days
Configuration	$[He]2s^2$	^{8}Be	$\sim 7 \times 10^{-17}$ sec
Density, 20 °C	1.848 g/cm^3	^{9}Be	stable
Melting point	1287 °C	^{10}Be	1.52×10^6 years
Boiling point	2471 °C	^{11}Be	13.8 sec
Valence	2	^{12}Be	24. ms
		^{14}Be	4.3 ms

Source of information: CRC Handbook of Chemistry and Physics, 79th Edition (1998-1999); ms = millisecond

Although the metal was first isolated in 1828, independently by Wöhler (1828) and Bussy (1828), commercially feasible processes took over 100 years to develop as demand for beryllium, its alloys and its compounds increased. Isolating high-purity beryllium metal and working it have both proved to be technologically difficult. Beginning in 1898 chemists have attempted to apply electrolysis of Be fluorides for commercial production (e.g., Stock, 1932), but current production of Be metal depends largely on reduction of

BeF_2 with Mg or electrolysis of Be chloride-alkali chloride electrolytes (Schwenzfeier, 1955; Windecker, 1955; Everest, 1973; Greenwood and Earnshaw 1997).

Until about 1960, pegmatitic beryl was the primary source of beryllium and the U.S. was forced to import over 90% of the beryl that its beryllium industry needed during the period 1936-1952 (Griffith 1955). However, in the 1960s, the situation changed radically with the exploitation of epithermal deposits such as Spor Mountain, Utah (Staatz and Griffits 1961; Staatz 1963; Barton and Young, Chapter 14), which, together with hydrothermal deposits, became the primary source of economic beryllium (Everest 1973; Wood 1992). A similar shift to the plutonic equivalents of Spor Mountain, i.e., hydrothermal bertrandite-phenakite-fluorite deposits such as Yermakovskoye, Buratiya in Transbaikalia, took place in the former Soviet Union (Ginzburg 1975; Novikova et al. 1994; Kupriyanova et al. 1996). The beryllium ore in the epithermal deposits, commonly bertrandite, is less refractory than beryl, and the larger size of the deposits result in their being more economic to process. Table 2 summarizes some present-day production and consumption statistics on beryllium. Kupriyanova et al. (1996) recommended increased exploration for and exploitation of Be deposits in Russia in order to meet the increased demand for Be as Russia's economy stabilizes, implying that the demand for beryllium is an indicator of a country's technological and industrial development.

Table 2. A few economic facts on beryllium.

Year(s)	1997-2000*	2001*
PRODUCTION IN METRIC TONS		
Worldwide	280-344	280
U.S.	180-243	180
China	55	55
Russia	40	40
Kazakhstan	4	4
Other countries	1-2	2
1998 APPARENT CONSUMPTION IN METRIC TONS		
U.S.	300-385	295†
PRICE (U.S. DOLLARS) PER POUND OF METAL		
Domestic, vacuum-cast ingot	327-421	338†
USE		
Electronic components plus aerospace & defense applications	estimated 80% of total consumption	

Source: U.S. Geol. Survey Minerals Information, Mineral Commodity Summary for Beryllium , January, 2002, at
http://minerals.usgs.gov/minerals/pubs/commodity/beryllium/100302.
* Production amounts for 2000 and 2001 are estimated
† Estimated

This brief overview of Be economics would not be complete without mention of the wide use of Be minerals as gemstones, notably emerald and aquamarine (varieties of beryl) and alexandrite (chrysoberyl); taaffeite-group minerals, euclase, phenakite and tugtupite are also fashioned into gemstones.

TOXICITY OF BERYLLIUM

Beryllium is highly toxic, and the hazard to humans is greatest in industrial settings; toxicity of Be compounds has also discouraged experimental investigations. However, officials were slow to realize this toxicity until it was shown in the late 1940s that lung beryllium disease ("berylliosis") was caused by Zn-Be silicate (synthetic beryllian willemite) used as an ingredient in phosphors (Skilleter 1990). A resurgence of beryllium disease in the 1990s led to regulations arbitrarily cutting the occupational exposure limit value from $2\mu g/m^3$ set in the 1940s to $0.2\mu g/m^3$ pending research to identify the real threshold limit below which disease will not occur (Ratney 2001). Veselý et al. (Chapter 7) review the literature on the environmental chemistry of Be and the spread of this anthropogenic contaminant, which can be toxic to aquatic plants and animals.

BERYLLIUM ABUNDANCE

Despite its low atomic number, Be is a rare element in meteorites and the stellar atmospheres, although less so in interstellar space, a paradox realized as early as 1926, and attributed to destruction of Be at high stellar temperatures (Goldschmidt 1954; Shearer, Chapter 2; see below). Beryllium is also a rare element on earth, but just how rare, required more effort. Goldschmidt and Peters (1932) reckoned that Be would be enriched in the Earth's crust because they expected Be to be lithophile like Li, and that given the similarity of Be to Al in chemical behavior, Be could have been readily overlooked as Washington (1931) conjectured was the case in igneous rocks. This situation stimulated Goldschmidt and Peters (1932) to conduct the first geochemical survey of Be contents in terrestrial rocks and minerals, as well as in meteorites and confirmed thereby the rarity of Be in meteorites and in all but a few rocks.

Current estimates are that the Earth's upper crust averages 3 ppm Be, which represents a 50-fold enrichment over the inferred content in primitive mantle, 60 ppb Be (Taylor and McLennan 1985, 1995; Wedepohl 1995); for the crust as a whole, Be is the 47th most abundant element (Emsley 2001). Like boron, Be is a quintessential crustal element.

ANALYZING BERYLLIUM

The analysis of the stable isotope ^{9}Be as a trace element in bulk samples has come a long ways since Goldschmidt and Peters (1932) carried out their survey using a flame-emission technique (Ryan, Chapter 3). The advent of plasma emission spectrometry (either inductively coupled or direct current-generated) around 1980 pushed detection limits down to the 0.1 ppm range. As a result, mantle and basaltic rocks could be analyzed with precision and accuracy, which was not possible with flame photometry, atomic absorption and fluorimetry, the methods available at that time. Major advances in the analysis of the cosmogenic isotopes ^{10}Be and ^{7}Be has opened up many new opportunities for geochemical studies of Be (reviewed in this volume by Bierman et al., Chapter 4; Morris et al., Chapter 5; and Kaste et al., Chapter 6).

Beryllium as a major constituent in minerals presents an even greater challenge to the analyst than boron (Anovitz and Grew 1996). Its chemical behavior is similar to Al, so separation of Be from Al "demands a somewhat complicated procedure" (Washington 1931). As a result, it has been overlooked in minerals, and not only early in the history of mineralogy, e.g., chrysoberyl (Seybert 1824). In the late 19th and early 20th centuries many Be minerals were introduced as new without evidence that Be was an essential component, e.g., bavenite, milarite (Palache 1931) and barylite (Aminoff 1923). Even in more recent characterizations Be was not often sought, e.g., roggianite; compare Passaglia (1969a,b) and Galli (1980) with Passaglia and Vezzalini (1988). In contrast, kolbeckite was originally reported to contain Be (Palache et al. 1944) and subsequently was found not to have it (Hey et al. 1982).

Single-crystal structure refinement (SREF) is a powerful tool for measuring Be independently of classical wet chemistry and microbeam technology (Hawthorne and Grice 1990; Ottolini and Oberti 2000).

Microbeam methods

Electron microprobe. Over the last ten years, several microbeam technologies have been developed for in situ analysis of Be. The great advantage of in situ analysis is that much less sample is needed and spatial variation in Be content can be assessed; petrology was revolutionized in the 1960s when compositional zoning of elements with $Z \geq 11$ (Na) in individual minerals could be mapped and quantified with the electron microprobe. For lighter elements, particularly Be, such capability seemed out-of-reach, and relatively recently Raudsepp (1995, p. 215) characterized the prospect of quantitative electron microprobe analysis as "presently dim." Nonetheless, the outlook has been improving. Meeker and Evensen (1999; also Evensen and Meeker 1997) assess the difficulties of Be analysis with the electron microprobe. One is crystallographic orientation, which is not normally a problem for heavier elements ($Z > 10$). Dyar et al. (2001) analyzed Be in phenakite and hambergite and reported BeO contents within ±1-2 wt % of the ideal BeO contents of 45.4-53.3 %, whereas Kleykamp (1999) reported successful analysis of Be in the synthetic compound $Li_2Be_2O_3$. Current top-of-the-line microprobes (e.g., JEOL 8200, Cameca SX 100) feature Be-analytical capability.

Ion microprobe. A widely used microbeam method for Be analysis is secondary ion-mass spectroscopy (SIMS) with the ion microprobe (Ottolini and Oberti 2000; Hervig, Chapter 8).

Inductively coupled plasma mass spectrometry. Another promising technology is laser-ablation inductively coupled plasma mass spectrometry (LA-ICP-MS), which is finding increasing application in the earth sciences (e.g., Durrant 1999; Sylvester 2001). However, relatively few investigators have applied it specifically to Be analyses, whether in earth sciences (Bea et al. 1994; Zack 2000; Flem et al. 2002), or in other sciences (Crain and Gallimore 1992). Beryllium can be measured as one of many elements in a trace-element package using Ar as a carrier gas (e.g., Bea et al. 1994; see Grew, Chapter 12). However, by restricting menus to 6-12 elements, conditions can be optimized, i.e., the ICP-MS is tuned for maximum sensitivity for a certain mass range and, in the case of Li, Be and B, sensitivity is markedly increased using He as carrier gas (Zack 2000).

Nuclear microprobe. Beryllium can also be analyzed with a nuclear microprobe using particle-induced gamma-ray emission (PIGE) spectrometry (Volfinger and Robert 1994; Rio et al. 1995; Lahlafi 1997; Calligaro et al. 2000; Kim et al. 2000). Detection limits depend on the nuclear reaction used and appear to be very sensitive to experimental conditions. This limit is reported to be 700-1800 ppm Be in some studies and thus only minerals containing substantial Be can be analyzed, e.g., emerald (Calligaro et al. 2000). In others, the reported limit was only 20 ppm, so that Be at trace levels in Li mica could be monitored (Volfinger and Robert 1994); Lahlafi (1997) obtained comparable (but not specified) detection limits in a study of trace Be in synthetic micas.

MINERALOGY OF BERYLLIUM

Minerals containing essential beryllium

About 110 mineral species, i.e., naturally occurring, inorganic solid compounds, have been reported to contain essential Be as of 2002. By essential, I mean that the mineral cannot form in the absence of Be. These minerals have been tabulated in Appendices 1 and 2; names not yet approved by the International Mineralogical Association (IMA) Commission on New Minerals and Mineral Names (CNMMN) are given in quotation marks in both appendices: The 98 valid and potentially valid species include 8 oxides and hydroxides, 4 borates, 25 phosphates and arsenates lacking silicate, and 61 silicates.

Valid species. A total of 89 minerals are considered valid species (Appendix 1), which includes species approved by the IMA CNMMN, beginning in 1959. Although absent from some lists of Be minerals (e.g., Ross 1964; Beus 1966), hyalotekite is included in my list because Be is reported in all analysis in which it was sought (cf. the Y-dominant, Be-free analogue kapitsaite, Pautov et al. 2000; Sokolova et al. 2000). Hawthorne and Huminicki (Chapter 9) review hyalotekite compositions and conclude that two end-member compositions are needed to describe reported compositions even if substitutions involving Y, Pb and K are ignored, and that one of these two end members contains essential Be (also Hawthorne 2002).

Potentially valid species. Nine minerals are considered potentially valid species (Appendix 1). Although not confirmed by IMA CNMMN, available data suggest that they meet the criteria for being distinct species (Nickel and Grice 1998), i.e., in terms of composition (e.g., several minerals in the gadolinite group, Y- and Sc-dominant milarite) or crystal structure (triclinic Fe- and Mn-dominant analogues of roscherite).

Problematic species. The validity of 12 minerals is considered unproven (Appendix 2). These are incompletely studied minerals (e.g., "glucine"), minerals whose status as distinct species requires confirmation (triclinic modification of leucophanite, Li-dominant analogue of odintsovite), or minerals that are likely a variety of another species ("gelbertrandite"), as well as minerals of purely historical interest (e.g., "muromontite"). I include these minerals in order to draw attention to the need for further mineralogical research, particularly in the gadolinite group, to which six of the problematic species belong.

How the appendices were compiled:

Column 1 in Appendices 1 and 2 gives the name and relationships, e.g., the group to which the mineral belongs (as defined in Fleischer's Glossary of Mineral Species, Mandarino 1999) and any analogues, e.g. bazzite (Sc-dominant analogue of beryl and stoppaniite), the crystal system and a generalized formula in which major substitutions are indicated. In a few cases, a formula with specific coefficients is given instead; this seems appropriate when only one complete analysis has been reported, e.g., minasgeraisite. Hawthorne and Huminicki (Chapter 9) give idealized end-member formulae based on crystal structure refinements in their Appendix A.

Column 2 gives in broad terms the environments where the mineral occurs.

Column 3 gives the type locality in italics (if known) together with other localities. I have added "and others" to indicate that the list is not complete. Locality information for minerals found in the former Soviet Union is largely taken from Pekov (1994, 1998); the original authors generally provided few specifics on localities. Areas unusually prolific in numbers of Be mineral species are listed in Table 3.

Column 4 gives sources for the information provided in Appendix 1, including any reviews of the mineral. The original description is listed first, whether consulted or not; references not consulted are so indicated in the bibliography. In some cases, partial descriptions of a given mineral were published before the mineral was introduced as a distinct species. If a cited paper has been abstracted in American Mineralogist, e.g., in

Table 3. Areas unusually prolific in number of beryllium mineral species

Species		Locality	Reference(s) in addition to Appendix 1
Type[1]	Total[2]		
10	12	Långban and Harstigen Fe-Mn deposits, Bergslagen ore region, Värmland Co., Sweden	Holtstam & Langhof (1999)
6.5[3]	22	Nepheline-syenite pegmatites, Oslo Region, Norway	Raade et al. (1980); Engvoldsen et al. (1991); Andersen et al. (1996)
6	17	Granite pegmatites, Minas Gerais state, Brazil (18 species in Brazil overall)	Bhaskara Rao & Silva (1968); Atencio (2000a,b)
5.5[4]	19	Ilímaussaq and Igaliko peralkaline complexes, southwest Greenland	Semenov (1969); Engell et al. (1971); Petersen & Secher (1993); Markl (2001b)
5	9	Granite pegmatite, Tip Top Mine, Custer Co., South Dakota, USA	Campbell & Roberts (1986)
4.5[4]	15	Lovozero and Khibiny peralkaline complexes, Kola Peninsula, Russia	Vlasov et al. (1966); Semenov (1972); Khomyakov (1995), Men'shikov et al. (1999); Yakovenchuk et al. (1999); Pekov (2000)
4	14	Granite pegmatites, Oxford County, Maine, U.S.A. (15 for state as a whole)	King & Foord (1994, 2000)
4	15	Granite pegmatites and greisens, Middle Urals (Yekaterinburg area), Russia	Yarosh (1980); Pekov (1994, 1998)
2	12	Hsianghualing tin deposit, Hunan, China	Huang et al. (1988)
2	8/10	Granite pegmatites, Baveno, Novara, Italy	Albertini (1983); Nova (1987); Pezzotta et al. (1999)
2	8	Granite pegmatites of central Madagascar, e.g., Sahatany River area, Anjanabonoina pegmatite	Lacroix (1922); Behier (1960); Wilson (1989); Lefevre & Thomas (1998); Pezzotta (1999)
2	4	Metamorphosed Fe-Mn-Zn deposit, Franklin, New Jersey, USA	Palache (1935); Frondel & Baum (1974); Dunn (1995)
1	14/16	Skarn, Lupikko, Pitkäranta, Karelia, Russia (including beryl in pegmatite)	Eskola (1951); Nefedov (1967); Nikol'skaya & Larin (1972); Nefedov et al. (1977); Schmetzer (1983); Pekov (1994, 1998)
1	11/12	Heftetjern granite pegmatite, Tørdal, Telemark, Norway	Juve & Bergstøl (1990); Raade & Kristiansen (2000), R. Kristiansen (pers. comm.)
1	9/10	Granite pegmatites, Pike's Peak batholith, Colorado, USA	Raines (2001)
1	8	Granite pegmatite, Foote Mine, Kings Mountain, Cleveland Co., North Carolina, USA	Marble & Hanahan (1978)
1	8	Li pegmatites, Eräjärvi area, Finland (7 in the Viitaniemi pegmatite)	Volborth (1954b); Mrose & von Knorring (1959); Lahti (1981)
1	5	Pegmatites, Khmara Bay, Enderby Land, Antarctica	Grew (1981, 1998); Grew et al. (2000)
1	5	Be-enriched gneiss, Høgtuva near Mo i Rana, Nordland County, Norway	Lindahl & Grauch (1988); Grauch et al. (1994)
0	15/17	Peralkaline pegmatites, Mont Saint-Hilaire, Quebec, Canada	Mandarino & Anderson (1989); Horváth & Gault (1990); Chao et al. (1990); Horváth & Pfenninger-Horváth (2000a,b)

Note: [1]Number of species for which area is the type locality. [2]Number of species from Appendix 1 confirmed in the given area/possible species, including unconfirmed identifications. [3]Chiavennite was discovered simultaneously at Chiavenna and Tvedalen; the former is considered the type locality. [4]Tugtupite was discovered almost simultaneously at Ilímaussaq and Lovozero; the former is considered the type locality.

that journal's section "New Mineral Names", then the volume number and page of the abstract is given in square brackets following the citation. Chemical Abstracts is cited if no abstract were published in American Mineralogist. If the original description were incomplete or erroneous, papers reporting the correction are also cited. A common error in original descriptions is overlooking the presence of Be in the mineral. Information has also been obtained from the review papers listed in Table 3. Crystal structure refinements are cited if the cited formula is based on them. Hawthorne and Huminicki (Chapter 9) discuss all minerals for which the crystal structure has been refined and cite the papers reporting the latest refinements. Černý (Chapter 10) considers every mineral found (or reported) in granitic pegmatites. Other chapters are cited if the mineral is considered in detail.

Information on the Be minerals listed in Appendix 1 also can be found in recently published compendia such as *Dana's New Mineralogy* (Gaines et al. 1997), *Handbook of Mineralogy* (Anthony et al. 1995, 1997, 2000), *Encyclopedia of Mineral Names* (Blackburn and Dennen 1997), *Strunz Mineralogical Tables* (Strunz and Nickel 2001), and the *International Encyclopedia of Minerals* (Mandarino, Editor-in-Chief, in preparation). Chemical analyses of each mineral are given in the *Handbook of Mineralogy*. Detailed descriptions, with hard-to-obtain information on minerals found in the former Soviet Union, are given in the Russian-language *Mineraly* handbooks (volumes 3 and 4, three issues each, 1972-1996, Chukhrov et al., 1972, 1981a,b; 1992a,b; Bokiy et al., 1996). These compendia (except the *International Encyclopedia of Minerals*, which was not available at the time) have been invaluable in assembling the information in Appendix 1. Older compendia providing useful information on Be minerals are Dana's *System of Mineralogy*, 6th (Dana 1892) and 7th editions (Palache et al. 1944, 1951); Beus (1966), Vlasov (1966), Ginzburg (1976), Burt (1982) and Clark (1993); Beus and Vlasov provide coverage in English of papers published in Russian and Chinese. Zeolite minerals, including tvedalite, a possible zeolite, are briefly described in the report of the subcommittee on zeolites (e.g., Coombs et al. 1997). Country-wide overviews of Be mineral occurrences (see also Appendices 1 and 2) include Pekov (1994, 1998) for the former USSR (14 new species, 46 overall), Kristiansen (1999a,b) and Nijland et al. (1998) for Norway (9 new species, 31 overall), Atencio (2000a,b) for Brazil (6 new species; 18 overall, all but one in Minas Gerais), and Hügi et al. (1968) and Hügi and Röwe (1970) for Switzerland (2 new species, 11 overall). Information on type minerals is also provided for Norway (Raade 1996) and Australia (Sutherland et al. 2000).

I have tried to be consistent in my two chapters in the transcription of geographic names from other writing systems; most importantly, Russian and Chinese. However, I have not insisted other contributors adhere to the same transliterations because there are often two or more variants in the literature itself. Examples include Yermakovskoye/ Ermakovskoe (Be deposit in Buryatiya, Russia), Hsianghualing/Xianghualing (Be deposit in Hunan, China).

Information that is too involved, incomplete or contradictory and consequently in need of further comments that could not be accommodated in Appendices 1 and 2 is given below under the name of the mineral in question.

Appendix 3 lists mineral synonyms of recent origin, the accepted equivalents, the original description, and the source of the renaming. A complete listing of mineral synonyms may be found in Clark (1993), de Fourestier (1999) and Bayliss (2000).

Dubious reports. In compiling Appendix 1, I have accepted published information as reported in good faith. However, there are a few cases where the reported information could be fraudulent or misleading, and these are not included in the list, but are mentioned here instead. Behoite has also been reported from Clear Creek, Burnet Co., Texas (Crook 1977), but given Crook's record of falsification with regard to other minerals (Peacor et al. 1982), this report of behoite should be viewed with caution. Kingsbury (1961, 1964) reported a detailed study of danalite and helvite and an occurrence of rhodizite from Meldon dyke, Okehampton, Devon, UK, respectively. By rights, both papers should be cited in Appendix 1. However, the findings by Ryback et al. (1998, 2001) that Kingsbury falsified localities of other minerals he reported casts serious doubts on the value of Kingsbury's (1961, 1964) studies of Be minerals.

Aminoffite. The original analysis (Hurlbut 1937) and crystal structure refinement (Coda et al. 1967; Hawthorne and Huminicki, Chapter 9) disagree on the Be/Si ratio and presence of Al. The mineral reported from Tuva, Russia by Kapustin (1973a,b) does not appear to be aminoffite: (1) the chemical analyses do not agree with those reported by Hurlbut (1937) or by Coda et al. (1967); (2) powder X-ray data differ from those on the type material (Mandarino 1964).

Bavenite. This mineral varies considerably in composition and most available analyses show excess Be relative to the idealized formula given in most references, $Ca_4Be_2Al_2Si_9O_{26}(OH)_2$. Beryllium varies inversely with Al according to the formula

$Ca_4(BeOH)_{2+x}Al_{2-x}Si_9O_{26-x}$, where $0 < x < 1$,

or $Ca_4Be_2Al_2Si_9O_{26}(OH)_2 - Ca_4Be_3AlSi_9O_{26}(OH)_3$

(Switzer and Reichen 1960; Berry 1963; Beus 1966; Cannillo et al. 1966; Petersen et al. 1995). The status of H is less clear; available chemical data show little evidence for an increase of H with Be. Petersen et al. (1995) reported a broad absorption band in their infrared spectrum that they attribute to molecular H_2O, whereas Komarova et al. (1967) found no such absorption features. Mark Cooper and Frank Hawthorne are currently carrying out a systematic study of bavenite crystal chemistry.

Bertrandite and "gelbertrandite" (cf. sphaerobertrandite). "Gelbertrandite" appears to be a colloidal or cryptocrystalline bertrandite containing variable amounts of molecular water that is lost easily on heating, and is thus not a valid species (Semenov, 1957; Pekov 1994, 2000). The variety of bertrandite in the Be-bearing tuff at Spor Mountain, Utah has chemical and physical properties intermediate between Lovozero "gelbertrandite" (Semenov 1957) and bertrandite proper (Montoya et al. 1962).

Beryllite. Beryllite is recognized as a distinct species yet shows certain similarities to sphaerobertrandite and bertrandite. For example, beryllite Be/Si ratio of 2.8 to 3.1 is variable from specimen to specimen and is close to sphaerobertrandite Be/Si, now shown to be 3.0. A bigger compositional difference is the presence of substantial H_2O^- in beryllite analyses. Although most low angle reflections in the powder X-ray diffraction patterns for beryllite and bertrandite are different, higher angle reflections ($d \leq 2.2$ Å) are remarkably similar.

Beryllonite. Nysten and Gustafsson (1993) and Charoy (1999) reported small Na deficits in electron microprobe analyses of beryllonite; Charoy (1999) also reported substantial F (0.64-1.19 wt %). Formulae calculated from Charoy's (1999, Table 2) analyses are $Na_{0.93}Be_{0.99}P_{1.00}O_{3.92}F_{0.08}$ and $Na_{0.92}BeP_{1.01}O_{3.96}F_{0.04}$ (assuming ideal Be content in second formula); these suggest that F (+ OH?) incorporation and Na deficit could be related through $(F,OH)\square O_{-1}Na_{-1}$.

Bityite. The Mica Subcommittee of the IMA CNMMN (e.g., Rieder et al. 1999) defined bityite, end-member composition $CaLiAl_2(AlBeSi_2)O_{10}(OH)_2$, as a trioctahedral brittle mica with $^{[6]}Li > ^{[6]}\square$. However, this definition overlooks both the variable Be content of bityite and solid solution between margarite and ephesite, ideally $NaLiAl_2(Al_2Si_2)O_{10}(OH)_2$. A definition citing Be as well as Li contents might have been preferable, e.g., that a Ca-rich mica with Be > 0.5 apfu as well as $^{[6]}Li > ^{[6]}\square$ would qualify as bityite (Lahti and Saikkonen 1985). Grew (Chapter 12) considers bityite-margarite solid solution in more detail.

Faheyite. This phosphate is one of the few Be minerals for which a crystal structure has not been reported (Hawthorne and Huminicki, Chapter 9). The idealized formula originally given for the mineral, $(Mn^{2+},Mg,Na)Fe^{3+}{}_2Be_2(PO_4)_4{\cdot}6H_2O$, is not balanced, so most compendia simplify it to $(Mn^{2+},Mg)Fe^{3+}{}_2Be_2(PO_4)_4{\cdot}6H_2O$ (e.g., Mandarino 1999). However, this simplification ignores the substantial Na reported in the original analysis, 0.93 wt % Na_2O, corresponding to 0.20 Na per formula unit in a formula (Appendix 1) calculated from the original analysis corrected for insoluble quartz and muscovite (Lindberg and Murata 1953). Several tenths wt % Ca are also reported in a semi-quantitative spectrogram. These findings suggest the possibility of a partially occupied (Na,Ca) site distinct from an M site occupied by divalent cations, but impurities could also be contributing to the difficulty of deducing a proper formula.

Gadolinite group. The mineral gadolinite-(Y) was first described by Geyer (Geijer) in1788 as "schwarzer Zeolith" (black zeolite) in material collected by C.A. Arrhenius and analyzed by Johan Gadolin in 1794 as "ytterbite", but Klaproth (1802) was the first to apply the name "gadolinite" (Dana 1892). Klaproth (1802) named the mineral in honor of Gadolin for his discovery of "yttria" (Y oxide mixed with Er, Tb and other REE oxides) in it (Weeks and Leicester 1968). Because gadolinite was the original source of the Y and related rare earths, Flink (1917c) opined that maybe it played a larger role in the history of inorganic chemistry than any other mineral.

The gadolinite group (Table 4) includes the valid Be silicates gadolinite-(Y), gadolinite-(Ce), hingganite-(Y), hingganite-(Yb) and minasgeraisite-(Y), as well as the borosilicates calcybeborosilite-(Y), datolite, and homilite (Mandarino 1999; DeMartin et al. 2001; Hawthorne and Huminicki, Chapter 9); the status of bakerite is under study. In addition, there are several minerals that have not been approved by the IMA CNMMN as valid species (Appendices 1, 2), namely, "calcio-gadolinite", calcium-rich gadolinite (Oftedahl 1972 reported Ca replaces Fe and not rare earths and Y as in calcio-gadolinite), Engström's "erdmannite" (Dana 1892; Ito and Hafner 1974; Raade 1996), "gadolinite-(Nd)" (Kartashov and Lapina 1994); "hingganite-(Ce)", "hingganite-(Nd)" (Kartashov and Lapina 1994), and the unnamed minerals reported by Miyawaki et al. (1987), Chao et al. (1990); Della Ventura et al. (1990), Wight and Chao (1995), and Jambor et al. (1998). Several of these minerals are potentially valid species. R. Miyawaki (pers. comm.) is preparing proposals to IMA CNMMN for Ce- and Nd-dominant analogues of hingganite-(Y) and hingganite-(Yb), which may be "hingganite-(Ce)" and "hingganite-(Nd)" according to the Levinson modifier, and the Ca-rich hingganite-like mineral with solid solution towards $YCaFe^{3+}{}_{0.33}Be_2Si_2O_8(OH)_2$ that was described by Miyawaki et al. (1987). The Ca-rich hingganite-like mineral is compositionally calcian, ferroan hingganite-(Y) because (Y, REE) > Ca and $\square$ > Fe, but has lower symmetry than hingganite-(Y), i.e., $P2_1$ vs. $P2_1/a$, and thus could be a distinct species. Several of the other unnamed minerals are similarly intermediate in composition between valid or potentially valid species for which a distinct end-member formula can be written (bold font in Table 4); space groups of these other intermediate phases have not been reported.

The end member compositions are related by both homovalent substitutions on one site, most importantly, CeY_{-1} (and other rare earths) and $CaFe_{-1}$, and coupled heterovalent substitutions involving different sites, most importantly, $CaB(YBe)_{-1}$, $CaFe^{3+}(YFe^{2+})_{-1}$, and $\square(OH)_2Fe_{-1}O_{-2}$ (Burt 1989; Miyawaki and Nakai 1996; Pezzotta et al. 1999). DeMartin et al. (2001) noted that no convincing evidence has been reported for the

Table 4. Idealized formulae for minerals of the gadolinite group.

Bakerite[1]	$Ca_4B_4(BO_4)(SiO_4)_3(OH)_3{\cdot}H_2O$(?)
Calcybeborosilite-(Y)	$Y_2\square B_2(Si_2O_8)O_2$
"Calciogadolinite-(Ce)" (see Appendix 2)	$CaCeFe^{2+}Be_2(Si_2O_8)O(OH)$
"Calciogadolinite-(Y)" (Ito 1967; Povarennykh 1972)	$CaYFe^{3+}Be_2(Si_2O_8)O_2$
Datolite	$Ca_2\square B_2(Si_2O_8)(OH)_2$
Gadolinite-(Ce)	$Ce_2Fe^{2+}Be_2(Si_2O_8)O_2$
"Gadolinite-(Nd)"	$Nd_2Fe^{2+}Be_2(Si_2O_8)O_2$
Gadolinite-(Y)	$Y_2Fe^{2+}Be_2(Si_2O_8)O_2$
"Hingganite-(Ce)"	$Ce_2\square Be_2(Si_2O_8)(OH)_2$
"Hingganite-(Nd)"	$Nd_2\square Be_2(Si_2O_8)(OH)_2$
Hingganite-(Y)	$Y_2\square Be_2(Si_2O_8)(OH)_2$
Hingganite-(Yb)	$Yb_2\square Be_2(Si_2O_8)(OH)_2$
Homilite	$Ca_2Fe^{2+}B_2(Si_2O_8)O_2$
Minasgeraisite-(Y)	$Y_2CaBe_2(Si_2O_8)O_2$
Unnamed	$YCaFe^{3+}_{0.33}\square_{0.67}Be_2Si_2O_8(OH)_2$

Note: **Bold** indicates mineral (valid species or not) for which an end-member composition is given. [1]The status of bakerite as a distinct species is currently under study. It probably corresponds to a B-rich datolite. (N. Perchiazzi & A. Kampf, pers. comm.). The vacancies indicated by □ mark the absence of Fe^{2+} or Ca (as in minasgeraisite), but are not true voids due to the presence of two protons (Demartin et al. 2001).

presence of substantial Fe^{3+} in fresh, unoxidized samples or for Ca substitution of Fe^{2+}. Thus, only the substitutions $CaB(YBe)_{-1}$ and $\square(OH)_2Fe_{-1}O_{-2}$ are important in describing compositional variations among Y-dominant gadolinite-group minerals. In addition, DeMartin et al. (2001) noted that the latter substitution should be written $(OH)_2Fe_{-1}O_{-2}$ because there are no vacancies on the site occupied by Fe; when Fe is absent, two protons take its place. Demartin et al. (2001) called into question the existence of minasgeraisite-(Y) in which Fe^{2+} is largely replaced by Ca.

Study of this mineral group is complicated by difficulties in analyzing individual rare earth elements, distinguishing Fe^{3+} and Fe^{2+}, complications due to the presence of other constituents (e.g., Bi, Th, U), and in detecting the presence of small amounts of hydroxyl, as well as by the common metamictization and alteration.

Gainesite group (gainesite, mccrillisite, selwynite). The analyses yield formulae deficient in Be relative to the ideal formulation $Na(Na, K, Cs)BeZr_2(PO_4)_4{\cdot}1\text{-}2H_2O$. In gainesite and mccrillisite, this deficiency is largely made up by Li, whereas in selwynite, Al + Be = 0.54 and Li was not detected. Neither the question of charge balance and stoichiometry nor the location of the water molecules has been addressed in a systematic crystallographic study of these minerals.

Herderite. Haidinger (1828a,b) did not analyze the type herderite, and subsequently, no one has conclusively demonstrated that the type material is F dominant. There were several attempts to determine F and OH in herderite from Ehrenfriedersdorf in the late 19th century, but the results were contradictory and resulted in considerable controversy, no doubt due to the difficulty of analyzing for F by the methods then in use (Weisbach 1884; Genth 1884; Winkler 1885; Leavens et al. 1978; Dunn et al. 1979; King and Foord 1994; Jahn 1998). Leavens et al. (1978) analyzed 41 samples by electron microprobe and found all to be OH dominant, assuming ideal F+OH content, but material from the type locality, Ehrenfriedersdorf, was not available for this survey. No chemical data are available for the herderite reported from Italy and Pitkäranta (see below), and Dunn and Wight (1976) analyzed the Brazilian gem only for F. In summary, a complete chemical analysis of herderite has not yet been published.

Haidinger (1828a,b) also provided no information on the paragenesis for the type herderite, which was part of A.G. Werner's collection, now housed in the museum of the Bergakademie Freiberg (Massanek et al. 1999). The holotype specimen was recently found anew after having been lost. This and other specimens from the Sauberg district near Ehrenfriedersdorf show herderite with apatite, "gilbertite" (mixture of muscovite with kaolinite), quartz, cassiterite, wolframite and fluorite in vugs (Jahn 1998; Massanek et al. 1999; Massanek, pers. comm.), purportedly from veins cutting mica schist (Franke 1934; Schneiderhöhn 1941, p. 131). The provenance of the herderite gemstone studied by Dunn and Wight (1976) is unknown except that it is from Brazil. There are also reports of herderite from Baveno and Cuasso al Monte, Italy (respectively, Guelfi and Orlandi 1987; Gramaccioli and Gentile 1991), but these reports need confirmation (F. Pezzotta, pers. comm.). Herderite is on the list of Be minerals for Mokrusha, Murzinka Region, Ural Mountains and for Pitkäranta, Karelia, Russia (Fersman 1940; Nefedov 1967; Nikol'skaya and Larin 1972; Pekov 1994), but in neither case is the identification as herderite backed by F analysis. Compared to hydroxylherderite, herderite remains a poorly characterized mineral, particularly in regard to the physico-chemical conditions that are conducive to incorporation of F in excess of OH.

Hingganite subgroup. Minerals of this subdivision of the gadolinite group (see above) have been introduced under a variety of names and compositions depending on whether Y or one of the rare earth elements were dominant (Table 4, Appendix 3). It should be noted that the crystal structure reported by Yakubovich et al. (1983) is for hingganite-(Y), not hingganite-(Yb), the title of this paper notwithstanding. The analysis given for hingganite-(Y) in Table 1 of Voloshin et al. (1983) corresponds to a formula very close to the "chemical formula" cited by Yakubovich et al. (1983).

Hsianghualite. The mineral was originally described in an abstract (Huang et al. 1958), which resulted in some confusion, e.g., in the SiO_2 content and in the origin of the name. Beus (1966, under the name "syanchualite", which is the transliteration of the Russian rendition of the original Chinese name) and Vlasov (1966) give the most complete descriptions in English. The name is from the locality, which means "fragrant flower." Three early analyses reported Li < 2 per formula unit, 0.03-0.60 wt % Na_2O, 0-0.54 wt % Al_2O_3 and 0.41 wt % H_2O^+ (Vlasov 1966; Chukhrov et al. 1972), which suggested the possibility of limited solid solution with analcime, which has been reported to contain 0.33 wt % BeO (Semenov 1972). However, two new analyses reported by Huang et al. (1988) gave Li > 2 per formula unit, 0.09 wt % Al_2O_3, 0.22 wt % Na_2O and no H_2O^+. The structure was refined independently by Academy (1973) and by Rastsvetaeva et al. (1991), who reported somewhat different coordinates for the atoms, but agree that hsianghualite has symmetry $I2_13$ and is closely related to analcime. F and Ca occupy positions very close to those occupied in analcime by H_2O and Na, respectively, whereas the Li position has no equivalent in analcime, i.e.,

$Li_2Ca_3[Be_3Si_3O_{12}]F_2$ and $\square_2(Na_2\square)[Al_2Si_4O_{12}]\cdot 2H_2O$. Hsianghualite has remarkably high density (D_{meas} = 2.97-3.00, D_{calc} = 2.95-2.98 g/cm^3) and refractive index (n = 1.613) for a zeolite, especially one containing over 20% (Li_2O + BeO). Calculation of the Gladstone-Dale relationship (Mandarino 1979, 1981) using D_{calc} yields -0.002 to +0.009, values well within the "superior" category (±0 to ±0.019), i.e., the density and refractive index are consistent with the chemical composition despite the abundance of light elements. Compared to cubic analcime (a = 13.725 Å), hsianghualite packs 108% more molecular weight into 82% the volume, so no wonder it is significantly denser and has a higher refractive index.

Hurlbutite. Charoy (1999) reported 0.53-0.79 wt % F, 0.52-0.80 SrO and 0.09-0.25 wt % Na_2O using an electron microprobe, suggesting possible solid solution in a mineral generally found to be free of impurities above the trace level (Mrose 1952; Staněk 1966; Nysten and Gustafsson 1993). Volborth (1954b) reported 1.51 wt % Na_2O, 0.95 wt % K_2O and 1.31 wt % SiO_2 in a wet-chemical analysis, but these constituents could be due to impurities.

Leifite group. This group includes leifite, its Cs-dominant analogue telyushenkoite, and, potentially, a K-dominant analogue (Sokolova et al. 2002). New refinements of the crystal structures of leifite and telyushenkoite by Sokolova et al. (2002) are the latest attempts to derive a viable formula for leifite, which is often given as $Na_2(Si,Al,Be)_7(O,OH,F)_{14}$ (e.g., Mandarino 1999; Blackburn and Dennen 1997), evidently derived from $(Na,H_3O)_2(Si,Al,Be)_7(O,OH,F)_{14}$ (Micheelsen and Petersen 1970), although Coda et al. (1974) deduced $Na_6[Si_{16}Al_2(BeOH)_2O_{39}]\cdot 1.5H_2O$ from their refinement of the crystal structure. Most compositional data do not fit either formula. The formula given by Hochleitner et al. (2000), $Na_{6-7}K_{<1}[(F,OH)_2|Be_2Al_{2-3}Si_{16-15}O_{39}]$, and a formula consistent with the structure refined by Sokolova et al. (2002), $^{A}(Na,\square)^{B}(H_2O,\square)Na_6Be_2(Al,Si,Zn)_3Si_{15}O_{39}F_2$, are in better accord with more recent analyses, most of which gave total alkalis ≅ 7, total F ≤ 2, Al ranging from 2 to 3 per 41 (O,OH,F), and Be ≅ 2 per formula unit (Khomyakov et al., 1979; Petersen et al., 1994b; Larsen and Åsheim, 1995; Men'shikov et al. 1999). However, Khomyakov et al. (1979) and Men'shikov et al. (1999) reported F approaching 3 per formula unit and ambiguity remains concerning the presence of molecular water or hydroxyl. In their structural refinement of leifite, Sokolova et al. (2002) reported electron density at the B site, which they concluded must be due to molecular H_2O, but could not find a position for OH. In contrast, Larsen and Åsheim (1995) found evidence for hydroxyl but not for molecular H_2O in the infrared spectrum of leifite from Norway. However, the leifite from Norway is K-dominant, and by analogy with telyushenkoite, it could lack H_2O at the B site (Sokolova et al. 2002).

Although Miller (1996) indicated that leifite is a potential ore of Be in the Strange Lake granitic pegmatite (Quebec-Labrador, Canada), he did not describe any unaltered leifite, but only pseudomorphs containing a Ca-Y silicate (gerenite, Jambor et al. 1998) replacing leifite. However, John Jambor (pers. comm.) has confirmed the presence of unaltered leifite in the Strange Lake intrusive.

Leucophanite and meliphanite. These minerals are compositionally distinct and Al is an essential constituent of meliphanite: the ideal formulae are $Ca_4Na_4Be_4Si_8O_{24}F_4$ and $Ca_4(Na,Ca)_4Be_4AlSi_7O_{24}(F,O)_4$, respectively (Grice and Hawthorne 2002). Analytical data available in the literature (Appendix 1; also Dana 1892; Portnov 1964; Chao 1967; Shatskaya and Zhdanov 1969; Khomyakov 1972; Ganzeyeva et al. 1973; Zubkov and Galadzheva 1974; Novikova 1976e) show that substitution of Na by Ca is overall more extensive in meliphanite than in leucophanite and there is little overlap between the two minerals in terms of Na, Ca, Al and Si; a few leucophanite samples have compositions approaching the ideal end-member composition $CaNaBeSi_2O_6F$. Most wet chemical analyses of leucophanite and meliphanite gave H_2O^+, often 0.5-2 wt%, but infrared spectra do not support the presence of a hydrous component. Grice and Hawthorne (2002) found no evidence for OH in a sample each of meliphanite and a leucophanite studied by infrared spectroscopy, confirming the results of most previous infrared studies of these minerals (Novikova et al. 1975; Novikova 1976e; Povarennykh and Nefedov 1971), whereas Shatskaya and Zhdanov (1969) reported absorption features for both OH and H_2O in an infrared spectrum of leucophanite. Of 30 leucophanite samples analyzed for rare earth elements (REE) and Y, total REE and Y ranged from 0.1 to 8 wt% (as oxide) in about 25 of them, i.e., several samples contained more than the total REE reported in a possible triclinic polymorph, 3 wt% (Cannillo et al. 1992; Appendix 2). Grice and Hawthorne (2002) were not able to detect triclinic symmetry in one of these REE-rich samples. Rare earths and Y were sought in 12 meliphanite samples, yielding up to 0.55 wt% REE + Y oxide. Strontium can replace Ca, e.g., 1.1 wt% SrO in leucophanite from Tajikistan (Grew et al. 1993).

Lovdarite. Petersen et al. (2002a) reported that the formula for lovdarite from Ilímaussaq, Greenland deviated in stoichiometry from that deduced by Merlino (1990) material from Lovozero and given in Appendix 1; i.e., Ilímaussaq lovdarite is variably deficient in K and Na, somewhat deficient in Be, but has a small excess of Si.

Magnesiotaaffeite-2N'2S. Kozhevnikov et al. (1975) gave their locality only as "eastern Siberia", and Pekov (1994) gave no information on this occurrence. N.N. Pertsev (pers. comm.) informed me that the locality is Sakhir-Shulutynskiy pluton, eastern Sayan, Russia.

"Makarochkinite". Although Grauch et al. (1994) suggested that "makarochkinite" was the same as høgtuvaite, Hawthorne and Huminicki (Chapter 9) note that dominance of Ti at one site in "makarochkinite" would make it a distinct species because Ti is not dominant at any site in høgtuvaite (see also Hawthorne 2002), i.e., an end member for "makarochkinite" could be $Ca_2(Fe^{2+}{}_5Ti)O_2[Si_5BeO_{18}]$ and that for høgtuvaite could be $Ca_2(Fe^{2+}{}_4Fe^{3+}{}_2)O_2[Si_5BeO_{18}]$. However, presently available samples of "makarochkinite" do not contain sufficient Ti to justify recognizing "makarochkinite" as a distinct species on this basis.

Meliphanite. This mineral was originally described under the name "melinophane" (Scheerer 1852), a spelling that was often used in the Russian-language literature until recently and occasionally used elsewhere. Scheerer (1852) named this mineral for both its yellow color and similarity to leucophane, but he did not specify whether he based his name on the Greek root μελι (meli – honey) or μηλινος (melinos – quince yellow) for the yellow color. Dana (1867, p. 405) criticized the name "melinophane" as "wrong and bad" presuming Scheerer confused "meli" with roots based on μελιν- (melin-), and thus proposed revising Scheerer's name to meliphane, the presently accepted form (with the suffix "ite" added). However, could Dana (1867) have neglected the possibility that Scheerer meant quince-yellow instead, as Raade (1996) presumed? Scheerer studied material from near Stavern, which is east of Langesundsfjord, but did not further specify the type locality. One possibility is Båtbukta on the island Stokkøya between Langesund and Helgeroa, where Brøgger and Reusch collected meliphanite in 1873 (Larsen 2000; see also Andersen et al. 1996). Presently meliphanite is regarded as the most abundant Be mineral in the alkaline pegmatites of the Langesundsfjord–Tvedalen area (Andersen et al. 1996).

"Muromontite" and "berylliumorthite". The minerals analyzed in the cited papers are metamict and often partially decomposed, so interpretation of the analytical data is difficult. Dana (1892, p. 526) wrote that "muromontite" is "apparently related to allanite", but Kerndt (1848), who originally described "muromontite", did not suggest this. He compared "muromontite" with allanite and gadolinite; "muromontite" has a morphology like that of allanite, but it was also found in grains lacking crystal faces. Kerndt noted that these three minerals did differ mainly in composition, and that they also differ in physical properties, but he realized that better euhedral crystals and more reliable chemical data would be needed to confirm whether the three were distinct species. The composition of "muromontite" has some features of allanite, most notably SiO_2 content, but the dominance of Y over Ce, presence of Be and low Ca is suggestive of gadolinite-(Y), i.e., Kerndt's (1848) analysis gave SiO_2 31.09, Al_2O_3 2.23, Y_2O_3 37.14 (includes all Y earths), Ce_2O_3 5.54, La_2O_3 3.54, FeO 11.23, MnO 0.91, MgO 0.42, BeO 5.52, CaO 0.71 Na_2O 0.65, K_2O 0.17, H_2O and loss on ignition 0.85, Sum 100 wt %, although the BeO content is little more than half that expected in gadolinite. Beryllium was subsequently reported in allanite from other localities, e.g., Forbes and Dahll (1855) reported 3.71 wt % BeO in an allanite from southern Norway. However, the composition of this mineral deviates from that expected in allanite by its high H_2O and total Fe (as FeO) contents, respectively 12.24 and 22.98 wt %. Quensel and Alvfeldt (1945) described their "berylliumorthite", i.e., beryllium allanite, with 3.83 wt % BeO from Skuleboda, Sweden as "decomposed muromontite". Although crystallographic measurements gave angles consistent with those for allanite, its composition deviates markedly from that of allanite (Hasegawa 1960), most notably the deficiency in SiO_2 (23.96 wt %), excessive H_2O^+ (7.98 wt %) and presence of substantial CO_2 (8.84 wt %). Iimori's (1939) analysis of allanite lacking crystal faces and containing 2.49 wt % BeO from Iisaka, Japan gives

$$(REE_{0.76}Ca_{0.86}Mn_{0.17}Th_{0.01})_{\Sigma=1.80}(Fe^{2+}{}_{0.88}Zr_{0.03})_{\Sigma=0.91}(Al_{1.46}Fe^{3+}{}_{0.42})_{\Sigma=1.88}Be_{0.58}Si_{2.99}O_{12}(OH),$$

a formula calculated assuming ideal OH content (Iimori's 3.33 wt % H_2O corresponds to 2.17 H per formula unit). The formula has an excess of tetrahedral cations (Be + Si = 3.57 vs. Si = 3 in allanite) and a deficit of ions in octahedral and higher coordination (4.59 vs. 5 in allanite). Iimori (1939) also reported an increase in Be content in weathered material and concluded that this was further evidence that Be was present in the unaltered allanite, which became enriched in Be during weathering. In summary, the existence of a Be-rich mineral related to allanite remains unproven, and it is doubtful whether the many reports of minor Be in allanite are really due to incorporation of Be in the allanite structure rather than to impurities and alteration, especially alteration associated with metamictization (see below).

Odintsovite. Petersen et al. (2001) cited low analytical totals and low Na contents as suggestive that the mineral from Ilímaussaq is the Li-dominant analogue (Appendix 2), $K_2(Na,Ca)_4(Li,Na)Ca_2(Ti,Fe^{3+},Nb)_2O_2[Be_4Si_{12}O_{36}]$, of the type odintsovite, but Li was not analyzed in the Ilímaussaq mineral.

Phenakite. The lion's share of phenakite localities are in granite pegmatites, and the reader is referred to Černý (Chapter 10) for a general discussion of these and to Gaines et al. (1997) and Anthony et al. (1995) for listings of the best known localities, though neither compendium gives the background literature on the localities. The localities and references given in Appendix 1 emphasize occurrences other than granitic pegmatites.

Roscherite, greifensteinite and zanazziite. "What exactly is roscherite?" (J. Jambor, 1991, *Am Mineral* 76, p. 1732). The question has yet to be answered (Hawthorne and Huminicki, Chapter 9). The confusion began early: Be was overlooked in the type roscherite (Lindberg 1958), which was reported to contain only Al (Slavík 1914), and the "roscherite" studied crystallographically by Fanfani et al. (1975) was later shown to be zanazziite (Leavens et al. 1990). On the basis of the dominant cation in the M1 site as defined by Leavens et al. (1990; Mg,Fe^{2+} site of Fanfani et al. 1975), monoclinic members of the roscherite group include an Mn-dominant species ("roscherite-1*M*" of Strunz and Nickel 2001), an Fe-dominant species (greifensteinite) and an Mg-dominant species (zanazziite)(Appendix 1). Triclinic modifications of Mn-dominant roscherite, i.e., "roscherite-1*A*" of Strunz and Nickel (2001), and of Fe-dominant roscherite were reported by Fanfani et al. (1977) and Leavens et al. (1990), respectively, thereby implying the existence of a triclinic modification of zanazziite as well. Nonetheless, this by no means exhausts all the possibilities. The M2 site as defined by Leavens et al. (1990; Al site of Fanfani et al. 1975) is occupied by variable amounts of Al, Fe^{3+}, Mn^{3+}, Fe^{2+}, Mn^{2+}, Mg and/or is partially vacant (□), and charge is balanced by variations in the OH/H_2O ratio (Hawthorne and Huminicki, Chapter 9). Consequently the number of possible species in a roscherite group is large. In addition to the localities listed in Appendix 1, unanalyzed minerals identified simply as roscherite have been described in granitic pegmatites from the Charles Davis mine, North Groton, New Hampshire and three other pegmatite mines in New Hampshire (Henderson et al. 1967); the Tip Top mine, Custer Co., South Dakota (Peacor et al. 1983; Kampf et al. 1992; cf. analyzed Mn-dominant material reported by Campbell and Roberts 1986); Weinebene, Carinthia, Austria (Walter et al. 1990; Sabor 1990); and Bendada, Beira Alta, Portugal (Schnorrer-Köhler and Rewitzer 1991).

Sphaerobertrandite. Until recently, sphaerobertrandite had been considered a Be- and H_2O-rich variety of bertrandite (Pekov 1994, 2000). However, a new study of material from Lovozero, Kola Peninsula and from Tvedalen, Norway (UK-10 of Andersen et al. 1996) shows that sphaerobertrandite is a distinct species; its validity was approved by a special decision of the CNMMN IMA in 2001 (A.O. Larsen and I. V. Pekov, pers. comm.; Pekov et al. in press). It turns out that sphaerobertrandite is markedly distinct from bertrandite in composition, symmetry and cell size (Table 5). Bukin (1967, 1969) reported synthesis of sphaerobertrandite that he distinguished from his synthetic bertrandite.

Table 5. Comparison of bertrandite and sphaerobertrandite.

	*Bertrandite**	*Sphaerobertrandite***
Ideal formula	$Be_4Si_2O_7(OH)_2$	$Be_3SiO_4(OH)_2$
Symmetry – space group	Orthorhombic – $Cmc2_1$	Monoclinic – $P2_1/c$
a (Å)	8.716(3)	5.081(3)
b (Å)	15.255(3)	4.639(1)
c (Å)	4.565(1)	17.664(9)
β (°)	90	106.09(5)
V (Å^3)	607.0†	400.0

Source of data: *Giuseppetti et al. (1992). **A.O. Larsen & I.V. Pekov (pers. comm.); Pekov et al. (in press). †Cell volume calculated from the individual dimensions; Giuseppetti et al. (1992) gave 592.22 Å^3

Tvedalite. The crystal structure has not been solved (Hawthorne and Huminicki, Chapter 9). S. Merlino (pers. comm., 1992, to Strunz and Nickel 2001, and 2002 to me) developed a possible model for the structure based on the structural features of chiavennite whereby tvedalite had monoclinic symmetry (space group *C2/m*) and a formula $Ca_2Mn_2Be_3Si_6O_{16}(OH)_6{\cdot}2H_2O$. However, Merlino was not able to confirm this model with diffraction data: single crystal diffraction data are unavailable and powder diffraction data did not sufficiently support the model.

Uralolite. Reports of this phosphate from Taquaral and Galiléia, Minas Gerais, Brazil were mentioned by Gaines et al. (1997), King and Foord (1994) and Anthony et al. (2000), but no details or citations were given. Specimens of uralolite on beryllonite from Linópolis (~30 km north of Galiléia) have been available commercially.

Väyrynenite. Electron microprobe analyses of väyrynenite (Ni and Yang 1992; Nysten and Gustafsson 1993; Roda et al. 1996; Huminicki and Hawthorne 2000), plus wet-chemical analysis of gem-quality material (Meixner and Paar 1976), gave much lower Ca, Na, K, and Al contents (except Al in one case) than the relatively high CaO (0.53-1.82 wt %), Na_2O (0.20-1.42 wt %), K_2O (0.03-1.18 wt %) and Al_2O_3 (0.40-2.78 wt %) contents reported in other wet chemical analyses (Volborth 1954b,c; Volborth and Stadner 1954; Mrose and von Knorring 1959; Gordiyenko et al. 1973). Impurities such as beryllonite, herderite, hurlbutite, apatite or muscovite, which are associated with väyrynenite, could be responsible for these high values. Constituents reported in significant amounts in electron microprobe analyses include F, i.e., 0.86 wt % F, equivalent to 0.08 F per formula unit, in gem-quality material from Pakistan (Huminicki and Hawthorne 2000), and Al, i.e., 0.90 wt % Al_2O_3, equivalent to 0.03 Al per formula unit, in a sample from Spain (Roda et al. 1996). The latter could be substituting for Be (Gordiyenko et al. 1973).

Welshite. Grew et al. (2001) published two empirical formulae based on new data, one of which is given in Appendix 1, the other is $Ca_2Mg_{3.8}Mn^{2+}_{0.1}Fe^{2+}_{0.1}Fe^{3+}_{0.8}Sb^{5+}_{1.2}O_2$–$[Si_{2.8}Be_{1.8}Fe^{3+}_{0.65}Al_{0.5}As_{0.25}O_{18}]$. Hawthorne and Huminicki (Chapter 9) discuss possible welshite end members and concludes that a mixture of $Ca_2Mg_4Sb_2O_2[Si_3Be_3O_{18}]$ and $Ca_2Mg_5SbO_2[Si_3Al_3O_{18}]$ best describes the empirical formula in Appendix 1 if certain simplifications are assumed.

Minerals containing non-essential beryllium

Beryllium contents in the great majority of minerals, including important rock-forming minerals, are mostly 10 ppm or less and rarely exceed 100 ppm Be (e.g., London and Evensen, Chapter 11; Grew, Chapter 12), even in Be-rich environments. Beryllium is thus an incompatible element in most geologic systems (Ryan, Chapter 3). However, several minerals have been reported to incorporate Be in substantial amounts (i.e., >0.1 wt % BeO or 350 ppm Be) or form series with isostructural Be minerals, although Be is not essential for their formation (e.g., Beus, 1956, 1966; this paper, Table 6).

Wet chemical analyses. Assessing the validity of the Be determinations is not easy. Older studies reporting significant Be by wet chemical analysis of bulk samples are suspect given the overall difficulty of Be analyses and its confusion with Al, and given the probability of impurities, particularly in secondary or metamict minerals. An example where impurities have contaminated analyzed material is the supposedly beryllian secondary mineral tengerite, $Y_2(CO_3)_3 \cdot 2\text{-}3H_2O$. Beryllium reported by Hidden (1905) and Iimori (1938) was attributed to admixed Be minerals by Miyawaki et al. (1993). The situation with "foresite" is not clear: 0.71 wt % BeO was reported in two of the four analyses cited by Dana (1892, p. 585). "Foresite" was later shown to be a mixture of stilbite and cookeite (Cocco and Garavelli 1958), but this finding leaves the source of Be unexplained; it could be either contamination or confusion with Al.

One example of possible confusion with Al (or another constituent) is the humite-group in which substantial BeO was reported in two studies (Jannasch and Locke 1894; Zambonini 1919). Neither of the analyzed Be-bearing humite-group mineral is from a Be-enriched environment. Although both Jannasch and Locke (1894) and Zambonini (1919) checked specifically that Be was present and Al absent or nearly so, these reports need confirmation. It should be noted that Ross (1964) suggested the possibility of Be incorporation in olivine- and humite-group minerals. Given that the presence of significant B in olivine and several humite-group minerals has been well documented (e.g., Grew 1996), Ross's (1964) suggestion cannot be dismissed out of hand.

Table 6. Minerals containing non-essential beryllium in amounts exceeding 0.1 wt % BeO or 350 ppm Be.

MINERAL (AND RELATED BERYLLIUM MINERAL)	FORMULA (INCLUDING BE WHERE KNOWN)	SUBSTITUTION BY WHICH BE IS INCORPORATED	MAXIMUM BE		REFERENCES FOR SUBSTITUTIONS AND MAXIMUM CONTENTS
			Formula units	*Wt % oxide*	
Allanite-(Ce), allanite-(Y)	$(Ce\ or\ Y,Ca)_2(Al,Fe^{2+},Fe^{3+})_3$-$O[Si_2O_7][SiO_4](OH)$	Not known	—	5.52	*Cf.* "Muromontite"
Analcime	$NaAlSi_2O_6 \cdot H_2O$	Not known	—	0.33	Semenov (1972)
Boralsilite	$Al_{16}B_6Si_2O_{37}$	Not known	0.126#	0.27#	Grew et al. (1998a); Grew (Ch 12)#
Calcybeborosilite-(Y)	$(Y,Ca)_2(\square,Fe)(B,Be)_2$-$(Si_2O_8)(OH,O)_2$	$YBe(CaB)_{-1}$	0.62-0.87	3.72-5.09	Semenov et al. (1963); Pekov et al. (2000)
Chevkinite-(Ce)	$(REE,Ca)_4(Fe^{2+},Mg)_2$-$(Ti,Fe^{3+})_3Si_4O_{22}$	Not known	—	2.15	Price (1888)
Ciprianiite (isostructural with hellandite)	$Ca_4(Th,REE,Ca)_2$-$(Al,Fe^{3+},Ti)(\square,Be,Li)_2$-$[Si_4B_4O_{22}](OH,F,O)_2$	Similar to substitution for hellandite?	0.82	1.95	Oberti et al. (2002); Della Ventura et al. (2002)
Clinohumite	$(Mg,Fe)_9(SiO_4)_4(F,OH)_2$	Not known	—	1.30	Zambonini (1919)
Cordierite (sodian orthorhombic)	$(\square_{1-y},Na,K)(Mg,Fe)_2$-$(Al_{4-x}Be_x)Si_5O_{18} \cdot nH_2O$ where $Na+K = y$ and $x \leq y$	$(Na,K)Be(\square Al)_{-1}$	0.437 *0.65*	1.77 —	Černý & Povondra (1966), Povondra & Langer (1971); Grew (Ch 12)††
Cordierite/indialite–BeMg-cordierite (synth. hexagonal)	$Mg_2Al_4Si_5O_{18} \cdot nH_2O$ – $Mg_2Al_2BeSi_6O_{18} \cdot nH_2O$	$BeSiAl_{-2}$	*1.0**	*4.40** (theoretically)	Hölscher & Schreyer (1989)
Cordierite/indialite–beryl (synth. orthorhombic and hexagonal)	$(\square_{1-y},Na,K)(Mg,Fe)_2$-$(Al_{4-x}Be_x)Si_5O_{18} \cdot nH_2O$ – $Al_2Be_3Si_6O_{18} \cdot nH_2O$ where $Na+K = y$ and $x \leq y$	$(Na,K)Be(\square Al)_{-1}$ $Be_3SiAl_{-2}(Mg,Fe)_{-2}$	*3.00*	—	Evensen & London (in press), London & Evensen (Ch 11)

Datolite (?)(un-named mineral B)	$(Ca,Y)_2(\square,Fe)(B,Be)_2$-$(Si_2O_8)(OH,O)_2$	$YBe(CaB)_{-1}$	0.442	2.75	Della Ventura et al. (1990)
Epistolite	$Na_2(Nb,Ti)_2Si_2O_9{\cdot}nH_2O^{**}$	Not known	—	0.29	Semenov (1969)
Gibbsite	$Al(OH)_3$	Not known	—	0.3	Semenov (1972)
Grossular	$Ca_3Al_2Si_3O_{12}$	Not known	—	0.19	Glass et al. (1944)
Hellandite-(Y)	$(Ca,REE)_4(Y,REE)_2$-$(Al,Fe^{3+},Ti)(\square,Be,Li)_2$-$[Si_4B_4O_{22}]\ (OH,F,O)_2$	~$(BeCaO)(\square REEOH)_{-1}$	0.61	1.40	Ma et al. (1986); Oberti et al. (1999, 2002); Della Ventura et al. (2002); Hawthorne & Huminicki (Ch 9)
Hellandite-(Ce) – mottanaite-(Ce)	$(Ca,REE)_4(Ce,REE,Th)_2$-$(Al,Fe^{3+},Ti)(\square,Be,Li)_2$-$[Si_4B_4O_{22}]\ (OH,F,O)_2$		1.18*	2.94*	
Högbomite group	Fe-Mg-Ti-Al oxide		—	0.10	Wilson (1977)
Humite†	$(Mg,Fe)_7(SiO_4)_3(F,OH)_2$	Not known	—	1.06	Jannasch & Locke (1894)
Kornerupine series	$(\square,Fe,Mg)(Mg,Al,Fe)_5Al_4$-$(Si,B,Al)_5O_{21}(OH,F)$	$BeSi(Al,B)_{-2}$ (?)	0.219#	0.74#	Grew et al. (1998b); Grew (Ch 12)#
Margarite–bityite	$(Ca,Na)(\square,Li)Al_2$-$[Si_2(Al,Be)_2]O_{10}(OH)_2$	~$BeLi(Al\square)_{-1}$	1.23*	8.05*	Rowledge & Hayton (1948); Grew (Ch 12)
Montmorillonite, zincian	$(Na,Ca)_{0.3}(Al,Mg)_2Si_4O_{10}$-$(OH)_2{\cdot}nH_2O^{**}$	Not known	0.07	0.40	Semenov (1972)
			—	0.93	
Mullite (synth)	$Al_6Si_2O_{13}$	Not known	—	*1.5*	Gelsdorf et al. (1958); Grew (Ch 12)
Rhodochrosite	$MnCO_3$	Not known	—	0.25	Grigor'yev (1967)
Saponite	$(Ca/2,Na)_{0.3}(Mg,Fe^{2+})_2$-$(Si,Al)_4O_{10}(OH)_2{\cdot}4H_2O^{**}$	Not known	0.5	2.8	Montoya et al. (1962)

Sapphirine–khmaralite	$(Al,Mg,Fe)_8(Al,Si,Be)_6O_{20}$	$BeSiAl_{-2}$	0.78*#; *1.0*	2.77*# *3.72*	Barbier et al. (1999, 2002); Hölscher (1987); Christy et al. (2002); Grew (Ch 12)#
Sillimanite	Al_2SiO_5	Not known	0.009#	0.13#	Grew et al. (1998c); Grew (Ch 12)#
Spessartine	$(Mn, Fe, Ca)_3(Al,Fe)_2Si_3O_{12}$	Not known	0.08	0.39	Iimori (1938)
Steenstrupine	$Na_{14}Ce_6Mn^{2+}Mn^{3+}Fe^{2+}{}_2(Zr,Th)(Si_6O_{18})(PO_4)_7\cdot 3H_2O$	Not known		1.93	Moberg (1898, 1999)
Tadzhikite-(Ce). (isostructural with hellandite)	$Ca_4(Ce,Y,REE)_2(Ti,Al,Fe^{3+})(\square,Be,Li)_2[Si_4B_4O_{22}](OH,F,O)_2$	Similar to substitution for hellandite?	0.08	0.20	Yefimov et al. (1970); Hawthorne et al. (1998); Oberti et al. (1999, 2002)
Vesuvianite	$(Ca,Na,REE)_{19}(Al,Mg,Fe,Ti)_{13}Si_{18}O_{68}(OH,F,O)_{10}$	Not known	—	0.15	Groat (1988), Groat et al. (1992)
			—	3.95	Hurlbut (1955)
Vinogradovite	$Na_8(Ti,Nb,Fe)_8O_8(Si_2O_6)_4[(Si,Al,Be)_4O_{10}]_2\cdot n(H_2O,Na,K)$	$NbBe(TiAl)_{-1}$	0.85‡	3±1.5§	Rønsbo et al. (1990); Kalsbeek & Rønsbo (1992)
Wavellite	$^{[6]}Al_3(PO_4)_2(OH,F)_3\cdot 5H_2O$	Admixed Be phosphate? (Beus 1966). Be → $^{[6]}Al$ unlikely	—	1.0	Preuss & von Gliszczynski (1950)
Werdingite	$(Mg,Fe)_2Al_{12}(Al,Fe)_2Si_4B_2(B,Al)_2O_{37}$	Not known	0.296#	0.61#	Grew et al. (1998c); Grew (Ch 12)#
Willemite (isostructural with phenakite)	$(Zn,Be)_2SiO_4$	$BeZn_{-1}$	— *0.30* (at 1300°C)	0.53 *7.9*$	Gurvich (1965); Hahn & Eysel (1970; Sharma & Ganguli (1973); Chatterjee & Ganguli (1975), Franz, Morteani (Ch 13)

Notes for Table 6: Italics indicate values obtained on synthetic materials. #Maximum Be contents are revised from original reports. ††In addition to the references cited in Chapter 12, Visser et al. (1994) reported Be data on cordierite. *Maxima are for the corresponding beryllium mineral. †Schäfer (1895) reported the humite to be monoclinic, and thus the analyzed mineral is either clinohumite or chondrodite although the stoichiometry corresponds to humite. **Formulae from Mandarino (1999). ‡Based on crystal structure refinement and does not correspond exactly to the laser probe analysis. §Based on semiquantitative laser probe analysis. $Calculated from the formula.

Clay minerals. A virtually unexplored, but potentially major, carrier of Be is the smectite group in which up to 2.8 wt % BeO (0.5 Be per 10 O, 2OH, $2H_2O$, Montoya et al. 1962, cf. Table 6). The crystallographic position of Be in these minerals is not known.

Metamict minerals. Allanite is most widespread of the metamict minerals reported to incorporate Be; and a distinct species "muromontite" (up to 5.52 wt % BeO, Appendix 2) was thought to be related to allanite (see above). In a compilation of 126 wet-chemical analyses of allanite, Hasegawa (1960) cited 15 (plus two Be-rich varieties discussed under "muromontite" above) with a BeO determination (trace to 1.35 wt %), including the analyses reported by Dana (1892), Nagashima (1951) and Hasegawa (1957, 1958). Significant amounts of BeO (0.28-0.47 wt %) were reported by Labuntsov (1939), Kimura (1960), and Khvostova (1962) in four samples, whereas Khvostova (1962) found only 3.2-252 ppm Be (≤0.07 wt % BeO) in six others. Machatschki (1948) suggested that Be replaces Si and $^{[4]}Al$ and Beus (1956, 1966) specified two substitutions for achieving charge balance:

$$2Ca^{2+} + [SiO_4]^{4-} = 2REE^{3+} + [BeO_4]^{6-}$$

$$[SiO_4]^{4-} = [BeO_2(F,OH)_2]^{4-}.$$

Machatschki (1948) concluded that incorporation of Be in allanite implied that the epidote-group as a whole had $[(Si,Be,Al)(O,OH)_3]$ chains. However, epidote-group minerals are not chain silicates, and there is little tetrahedral Al in allanite (Deer et al. 1986). In general, Be substitution for Si is difficult (Hawthorne and Huminicki, Chapter 9), and substantial Be substitution of Si in allanite does not appear likely.

Another rare-earth mineral reported to contain substantial beryllium is steenstrupine from Ilímaussaq, South Greenland (Moberg 1898, 1899). Three full analyses by Blomstrand were given; one of crystal fragments and two of more altered, massive and less dense material containing more H_2O, but only in the latter two analyses were BeO and Al_2O_3 determined separately, giving 1.22-1.93 wt % BeO (Table 6). Semenov (1969) was unable to confirm the presence of Be in his samples of steenstrupine from Ilímaussaq, but he found 2.65 wt % BeO in a "brown Ce-silicate" differing from steenstrupine in containing no P.

Analyses of the zircon-like mineral "alvite" included (BeO + Al_2O_3) contents of 4.40-14.11 wt %, but the presence of significant BeO was not demonstrated because it was not specifically separated from Al_2O_3 (Forbes and Dahll 1855; Bedr-Chan 1925).

The relatively high Be contents in these metamict minerals, as well as in chevkinite, fergusonite, homilite, microlite, samarskite, and yttrotantalite cited by Warner et al. (1959), as well as "hydro-thorite" (Semenov 1972), are most likely to have resulted from impurities or analytical difficulties or both, with the possible exception of homilite, a gadolinite-group mineral (Table 4). Beus (1966) suggested admixed gadolinite and adsorption of Be in a colloidal system.

Vesuvianite—a case study. This mineral is an example in which authoritative mineralogists reported a significant Be content and subsequent investigators have searched especially for it, particularly as vesuvianite appeared to be a possible commercial source of Be. Palache and Bauer (1930) reported 9.20 wt % BeO in vesuvianite from Franklin, New Jersey and speculated that Be could be generally present in vesuvianite but overlooked because of its being determined as Al. Washington (1931) went further in suggesting that Be-bearing vesuvianite (or another mineral) could explain the high Al contents of some igneous rocks because Be was analyzed as Al. Re-analysis (spectrographic) of the Palache and Bauer (1930) material gave only 0.17 wt % BeO, leading Hurlbut (1955) to conclude that the original analysis was in error. Hurlbut (1955) also reported wet-chemical and spectrographic analyses of 1.56-3.95 and 1.1 wt %, respectively, on another portion of the Palache and Bauer (1930) Franklin sample and cited several hundred unpublished analyses using a portable laboratory of vesuvianite from the western U.S. yielding BeO contents up to 1.5 wt %. Other investigators (Meen 1939; Glass et al. 1944; Eskola 1951; Kuchukova 1955; Beus 1957; Kapustin 1973b; Shpanov et al. 1976) have reported relatively high Be contents ranging from 0.5 to 1.5 wt % BeO. Beus (1957) noted the close association of Be with OH and F in Hurlbut's (1955) analyses of Franklin vesuvianite and in an analysis of a sample from Kazakhstan and suggested that Be substitutes for Si with charge balanced by (F,OH) substitution for O, i.e., Si + 2O = Be + 2(F,OH).

However, other studies have failed to confirm significant Be in vesuvianite. In their crystallographic study of the vesuvianite that Meen (1939) reported to contain 1.07 wt % BeO (equivalent to 1.27 Be per 18 Si), Rucklidge et al. (1975) failed to find Be at a possible tetrahedral site. Allen (1985) refined the occupancy of a T site of vesuvianite from Luning, Nevada assuming the presence of Be and obtained about 2 Be per formula unit, but no Be was found by ion microprobe. Groat et al. (1994) found B, Al and Fe at this site (their T1) in vesuvianite from other localities, a result that implies B, Al and/or Fe could be occupying Allen's (1985) T site in the Luning sample. Other studies have turned up no more than 0.3 wt % BeO (e.g., Goldschmidt and Peters 1932; Silbermintz and Roschkowa 1933; Neumann and Svinndal 1955; Henriques 1964; Eskola 1951; Warner et al. 1959; Serdyuchenko and Ganzeyeva 1976), even in vesuvianites from Be deposits (0.00064-0.034 wt % BeO, Shpanov et al. 1976; Novikova et al. 1994; 0.2 wt % BeO, Holser 1953). BeO contents ranging from 0.002 to 0.15 wt % that Groat (1988) and Groat et al. (1992) obtained by atomic absorption on 16 vesuvianite samples are probably the most reliable values reported for this mineral to date. These results suggest that significant Be in vesuvianite is a will-o'-the-wisp. Nonetheless, it remains entirely possible that under certain conditions Be could be a major constituent in vesuvianite; this has been demonstrated for B, which can be present in amounts sufficient to form the B analogue, wiluite (Groat et al. 1994, 1998).

Modern technology. More recent studies based on microbeam techniques such as secondary ion mass spectroscopy (ion microprobe, see Hervig, Chapter 8) or laser-ablation inductively-coupled plasma emission mass spectroscopy appear to be more reliable because the problem of impurities in bulk samples is avoided. These methods allow the analyses of very small samples where impurities pose the greatest problem or preclude wet chemical analysis altogether. Examples include boralsilite, werdingite and sillimanite (Table 6; Grew, Chapter 12). Analysis of these relatively fine-grained minerals, which are often intergrown with other phases, was greatly facilitated by use of the ion microprobe.

Best of all is supplementing chemical data with crystallographic or experimental studies that both confirm chemical analysis and elucidate the mechanism by which Be is incorporated. For example, the substitution by which Be is incorporated in cordierite, hellandite-group minerals, indialite and sapphirine-khmaralite has been reasonably well established by crystallographic or experimental studies (Table 6; Hawthorne and

Huminicki, Chapter 9; London and Evensen, Chapter 11; Grew, Chapter 12). On the basis of a semiqualitative "laser probe" analysis and deficiency in Si + Al in the formula, Rønsbo et al. (1990) calculated that significant Be was present in the Na titanosilicate vinogradovite from the Ilímaussaq complex, and this calculation was confirmed in a single crystal refinement (Kalsbeek and Rønsbo 1992). Vlasov et al. (1966) reported 0.08 wt % Be in vinogradovite from Lovozero. Beryllium is inferred to substitute for Si and Al occupying the Si(2) site in vinogradovite with charge balance achieved by the substitution $NbBe(TiAl)_{-1}$ (Table 6).

Hellandite group. The hellandite group differs from cordierite, sapphirine and other Be-bearing minerals in that Be does not substitute for B or Si, the only tetrahedral cations in this structure. Instead Be occupies a largely vacant site displacing H, its only occupant in Be- and Li-free hellandite (Oberti et al. 1999, 2002; Ottolini and Oberti 2000; Della Ventura et al. 2002; Hawthorne and Huminicki, Chapter 9).

Calcybeborosilite-(Y). This gadolinite-group mineral from a peralkaline pegmatite at Dara-i-Pioz, Tajikistan (Semenov et al. 1963; Povarennykh 1972; Rastsvetaeva et al. 1996; Pekov et al. 2000) is a valid species through "grandfathering" (*Am. Mineral.* 86:1537; J. Jambor, pers. comm.). However, Be is not an essential constituent, and thus it is not included in Appendix 1, but in Table 6. Its composition, e.g., $Ca_{0.96}(Y_{0.74}Ce_{0.10}REE_{0.18})_{\Sigma=1.02}U_{0.04}Th_{0.01}Fe^{2+}_{0.28}Mn_{0.05}Be_{0.62}B_{1.18}[Si_2O_8](OH_{1.35}O_{0.52}F_{0.13})$, ideally $(Y,Ca)_2(\square,Fe)(B,Be)_2(Si_2O_8)(OH,O)_2$, meets the criteria for a valid species, i.e., (Y,REE) > Ca distinguishes it from datolite and B > Be distinguishes it from gadolinite-(Y) and hingganite-(Y). The theoretical end-member is $Y_2\square B_2(Si_2O_8)O_2$.

Datolite (?). The composition of unnamed mineral B, e.g., $Ca_{1.07}(Y_{0.34}Ce_{0.24}Nd_{0.22}REE_{0.15})_{\Sigma=0.95}Th_{0.11}Fe^{2+}_{0.48}Be_{0.54}B_{1.24}Si_{2.22}O_8(OH)_2$, from a sanidinitic ejectum in pyroclastic rocks at Lake Vico, Latium, Italy (Della Ventura et al. 1990; Della Ventura, pers. comm., 1999) has Ca > (Y,REE), and thus appears to be beryllian, REE-rich datolite (Table 6), i.e., Be is not an essential constituent.

BERYLLIUM STUDIES

Beryllium in extraterrestrial systems. Why is beryllium so rare?

Beryllium has been a rare element since the very beginning of time: the calculated primordial Be abundance is negligibly small and the standard model of Big-Bang nucleosynthesis is "hopelessly ineffective in generating ^{6}Li, ^{9}Be, ^{10}B and ^{11}B" (Vangioni-Flam et al. 2000). These three light elements are "fragile" as a result of "selection principle at the nuclear level" (Vangioni-Flam et al. 2000), and are destroyed in stellar interiors. Beryllium and boron "are thought to be secondary elements formed from supernovae by spallation reactions between cosmic rays, alpha particles, and protons and heavier nuclei such as carbon, oxygen, and nitrogen in the interstellar medium" ("Editors' Choice", *Science*, v. 290, p. 1263, 2000, in reference to Primas et al. 2000a; also Cameron 1995). However, data on Be abundances obtained with the Hubble Space Telescope (Pagel 1991), Very Large Telescope in Chile ("Editor's Choice", *op. cit.*) and other new instrumentation show that Be is present in unexpected amounts in primitive metal-poor stars in the halo of our Galaxy; suggesting primary Be nucleosynthesis in the early galaxy, the so-called halo phase (e.g., Fields et al. 2000; Vangioni-Flam and Cassé 2001). "Editors' Choice" (*op. cit.*) even suggested the possibility of "primordial beryllium," i.e., revision of the standard model of Big Bang nucleosynthesis allowing for some synthesis of Be within a few minutes after the Big Bang (also Greenwood and Earnshaw 1997). Despite attempts to reconcile data obtained from increasingly sophisticated and sensitive instrumentation with theory on light element (LiBeB) abundances in the cosmos, investigators disagree, questions remain unresolved, and more data are called for (e.g., Thorburn and Hobbs 1996; Ramaty et al. 1998; Vangioni-Flam et al. 1999, 2000; Parizot and Drury 1999; Fields et al. 2000; Primas et al. 2000a,b; King 2002).

Shearer (Chapter 2) reviews the behavior of Be in our corner of the cosmos, the Solar System, with an emphasis on meteorites, the Moon and Mars, and the implications of this behavior for the evolution of the solar system.

Beryllium in terrestrial systems

Beryllium is a lithophile element concentrated in the residual phases of magmatic systems (e.g., Goldschmidt and Peters 1932; Goldschmidt 1958). Be contents of MORBs, arc basalts and ultramafic rocks are mostly <1 ppm (e.g., most recently, Santos et al. 2002; see Ryan, Chapter 3). Residual phases include acidic plutonic and volcanic rocks, whose geochemistry and evolution are covered, respectively, by London and Evensen (Chapter 11) and by Barton and Young (Chapter 14), while granitic pegmatites, which are well-known for their remarkable, if localized, Be enrichments and a wide variety of Be mineral assemblages, are reviewed by Černý (Chapter 10).

Alkaline rocks tend to be more enriched in Be than acid rocks (e.g., some evolved alkaline volcanic systems, Ryan, Chapter 3), and some Be deposits are associated with alkaline rocks (Barton and Young, Chapter 14). As regards alkaline plutonic systems, Goldschmidt (1958) noted that nepheline syenites contain more Be than other igneous rocks. Nonetheless, average Be contents of a given complex can vary considerably, e.g., in Russia, from 2 ppm in the Botogol' massif, eastern Sayan region to 12 ppm in the Lovozero complex, Kola Peninsula (Beus 1966, Table 118), and from one rock type to another in a given complex, e.g., from 4 to 24 ppm Be in the Lovozero complex (Beus 1966, Table 117; Vlasov et al. 1966; Hörmann 1978). Extreme enrichments are encountered locally, most notably in the Ilímaussaq complex, South Greenland, where peralkaline nepheline syenites (naujaite and lujavrite) range 7-44 ppm Be (Engell et al. 1971); Markl (2001a) reported 330-1970 ppm Be on 4 lujavrite samples. Regrettably alkaline plutonic rocks are not reviewed in this volume as acid rocks are, an unfortunate omission.

Not all Be concentrations have obvious magmatic affinities, either granitic or alkaline, e.g., one class of emerald deposits results from Be being introduced by heated brines (Franz and Morteani, Chapter 13; Barton and Young, Chapter 14). Pelitic rocks are an important reservoir of Be in the Earth's crust and their metamorphism plays an critical role in recycling of Be in subduction zones (Ryan, Chapter 3); eventually, anatectic processes complete the cycle, providing a source of Be for granitic rocks (London and Evensen, Chapter 11; Grew, Chapter 12).

In summary, I should emphasize that despite presentation in different chapters the systems and processes ranging from sedimentary through metamorphic and hydrothermal to magmatic (including pegmatitic) really constitute a continuum (and not a linear one) and should be considered holistically. That is, there are transitions from one system to another, and the boundaries staked out by authors for the material covered in individual chapters are to some degree rather arbitrary.

Cosmogenic isotopes

The cosmogenic isotopes ^{7}Be and ^{10}Be (Table 1) have found increasing applications in the geological sciences. The shorter-lived isotope ^{7}Be has wide applications in environmental studies (Kaste et al. Chapter 6). The use of the longer lived ^{10}Be to assess erosion rates was first proposed in 1985 and has rapidly expanded because of advances in analytical technology, leading to studies as diverse as soil production and erosion accelerated by agriculture (Greensfelder 2002; Bierman et al., Chapter 4). Beryllium-10 can yield independent temporal records of geomagnetic field variations for comparison with records obtained by measuring natural remnant magnetization, be a chemical tracer for processes in convergent margins, and can date events in Cenozoic tectonics (Morris et al., Chapter 5).

Spectroscopy of beryllium in minerals

Infrared and Raman. Powder infrared spectra of more than 35 Be minerals have been illustrated or discussed by Plyusnina (1963), Povarennykh and Nefedov (1971), Ross (1974a,b), Strens (1974), Moenke (1974), and Povarennykh et al. (1982) or by investigators cited in these papers; additional spectra are included in the descriptions of new minerals or in reports of new finds. A comprehensive investigation of infra-red (including single crystal spectra of beryl) and Raman spectroscopic properties of Be in bromellite, chrysoberyl, phenakite, bertrandite, beryl and euclase, was carried out by Hofmeister et al. (1987), who used the results to calculate heat capacities for comparison with experimental heat capacities of these phases. Pilati et al. (1998) obtained single-crystal Raman and infrared spectra on phenakite in order to make lattice-dynamical calculations.

Nuclear magnetic resonance. Beryllium-9 ($I = 3/2$) has favorable nuclear magnetic resonance (NMR) properties such as a small quadrupole moment ($Q = 5.288 \times 10^{-30} m^2$) and good receptivity (78.7 relative to ^{13}C), but Be toxicity has discouraged research (Bryce and Wasylishen 1999). Another unfavorable factor cited by Bryce and Wasylishen (1999) is the narrow chemical shift range (<50 ppm) for ^{9}Be, where ppm refers to parts per million of the energy of the applied magnetic field (typically given in hertz). Minerals studied by NMR of ^{9}Be include beryl (Sherriff et al. 1991) and tugtupite (Xu and Sherriff 1994) as well as synthetic chrysoberyl (Yeom et al. 1995), synthetic bearsite (Harrison et al. 1993), synthetic zeolites (e.g., Han et al. 1993), and synthetic materials unrelated to minerals (papers on "minerals" cited by Bryce and Wasylishen 1999).

Electron energy-loss spectroscopy. Parallel electron energy-loss spectroscopy (parallel EELS) in conjunction with transmission electron microscopy provides information on elemental composition, coordination, site symmetry, and bonding of atoms at very high spatial resolution (Garvie and Buseck 1996). Beryllium features have been studied in rhodizite, chrysoberyl, phenakite, bromellite, beryl, weinebeinite, Be-Te alloys and in synthetic Be-B-bearing materials (Brydson et al. 1988a,b 1989; Engel et al. 1988; Liu and Williams 1989; Hofer and Kothleitner 1993; Garvie et al. 1997; Garvie and Rez in prep.). EELS is also a valuable microanalytical tool for light elements. For example, Engel et al. (1988) quantified the cell contents of rhodizite and obtained 4.9 ± 0.5 Be per 28 oxygen, in good agreement with the 4.55 Be obtained by crystal structure refinement of a sample from the same locality (Pring et al. 1986), and Hofer and Kothleitner (1993) reported Be/Si = 0.49 ± 0.05 in beryl (vs. 0.5 nominal). While the K-edges of Be and B are well resolved in EELS spectra, interferences between the Be K edge and the $L_{2,3}$ edges of Mg, Al, Si and P introduces complexities for interpreting and quantifying spectra of silicate and phosphate minerals (Hofer and Kothleitner 1993; Garvie and Buseck 1999). For example, the addition of Mg made it impossible to quantify spectra of Mg-bearing beryl although Mg-poor beryl was amenable to quantification (Hofer and Kothleitner 1993).

CONCLUSION

As was the case with the boron volume, the present volume can be considered a progress report on the current state of research on Be and on applications of Be in the earth, planetary and cosmic sciences. It is intended also to be a guide and stimulus to further research on Be in these sciences. With this objective in mind I summarize directions for future research noted by contributors to the volume:

Planetary science. Shearer (Chapter 2) notes that early solar system processes, volatile elements in the lunar and martian mantles and surficial geochemical processes on Mars are three of many areas where more Be data could be beneficial.

Geochemistry. Ryan (Chapter 3) notes the absence of reliable data on the Be content of the mantle-derived rocks, so mantle Be content must estimated indirectly. This is not merely a question of sample availability; the main difficulty is analytical: mantle Be contents are too low to analyze by currently available technology. Surprisingly, the published data available for sedimentary rocks is relatively limited and old compared to data available on their metamorphic equivalents. This complicates assessing variations in Be content with metamorphic grade.

Cosmogenic isotopes. Bierman et al. (Chapter 4) notes that the accuracy of measuring ^{10}Be far outstrips the accuracy of interpreting the measurements, i.e., the science has not kept up with technology. Moreover, measurements using ^{10}Be and ^{26}Al, a complementary cosmogenic nuclide, provide evidence for erosion rates faster than those suggested by stratigraphic and geomorphic observations, a discrepancy than needs resolution if evolution of the Earth's surface is to be properly understood. Pairing of dating using ^{10}Be and the currently developing (U-Th)/He method has potential for proving insight into late Cenozoic tectonic evolution (Morris et al., Chapter 5). Application of the shorter-lived isotope ^{7}Be to various environmental studies is also relatively new, and understanding its distribution and residence times in various terrestrial ecosystems is far from complete (Kaste et al., Chapter 6).

Beryllium on Earth's surface. Veselý et al. (Chapter 7) note the limited information on Be distribution and mobility in soils and on diagenetic processes affecting Be in sediments, two critical links in the Be cycle on the Earth's surface. There is little data on Be concentrations in lake water.

Mineralogy. Although beryl is the most abundant Be mineral and by far the most studied (see Černý, Chapter 10, London and Evensen, Chapter 11, Franz and Morteani, Chapter 13), it is also the most complex chemically, a situation belied by the conventional rendering of its formula as $Be_3Al_2Si_6O_{18}$. Beryl incorporates significant variable amounts of Na, Cs, and H_2O in large channels (e.g., most recently, Pankrath and Langer 2002), as well as Li for Be and Mg, Fe and Sc for Al, so modeling its stability relations is a daunting task. Thus, many questions concerning beryl stability remain unanswered. This stability has practical as well as academic implications, most importantly in understanding the origin of emerald (Franz and Morteani, Chapter 13; Barton and Young, Chapter 14).

While compiling the list of Be minerals, I encountered problem areas in the characterization of less common Be minerals. Several of these are currently being addressed, e.g., bavenite and aminoffite, but others, e.g., beryllite and "glucine," are also in need of more thorough characterization. A framework for defining new species in the gadolinite group is much to be desired before more individual species are proposed to the IMA CNMMN. The vector representation constructed by Burt (1989) could provide such a framework; Miyawaki and Nakai (1996) used a somewhat different representation. DeMartin et al. (2001) urged crystallographic study of minasgeraisite-(Y) because the presence of significant Ca on the site normally occupied by Fe^{2+} "would require a considerable structural re-arrangement" in their opinion.

The incorporation of variable amounts of Be in minerals in which Be is either essential or non-essential is another problem area needing detailed and systematic crystallographic study. The conditions favoring substitution of Be for Al, Si and other constituents are not fully understood, for example, why do sapphirine, høgtuvaite and welshite incorporate significant amounts of Be, whereas the closely related mineral aenigmatite apparently does not? The incorporation of Be in otherwise vacant sites could be important in more cases than is generally recognized. This mechanism clearly applies to the hellandite group (Oberti et al. 1999, 2002) and could apply to other minerals such as vesuvianite, in which B is incorporated on a normally vacant site (e.g., Groat et al. 1994). Incorporation of Be even at trace levels raises questions in the case of olivine and clinopyroxene, in which a suitable site is not obvious (Ryan, Chapter 3).

Petrology. An area needing much work is the distribution of Be among rock-forming minerals and between these minerals and melt. London and Evensen (Chapter 11) and Grew (Chapter 12) have noted the gross inadequacy of available data, a significant portion of which are contradictory; Evensen and London (2002) have made a major contribution to rectifying the situation on Be partitioning in experimental granitic systems. Understanding the Be budget in crustal and mantle rocks depends on knowing which minerals are primary hosts of Be, particularly when obvious sinks such as cordierite and sapphirine are absent. Is Be lost from rocks as they are buried, heated and melted? Determining he extent to which Be is incorporated in melts (or lost to vapor) depends on knowing the fractionation between solid phases, melt and fluid. Overall, systems including Be and Be minerals are less well studied experimentally than corresponding systems without Be and common rock-forming minerals (Franz and Morteani, Chapter 13). In addition, the role of halogens has received relatively little attention from experimentalists although these elements, particularly F, play a major role in natural systems where Be is abundant. Experimental studies of Be in various fluid systems are needed to gain a better understanding of transport mechanisms for Be in hydrothermal environments, where many economic deposits of Be metal and minerals are formed (Barton and Young, Chapter 14), and deep in the Earth's crust, the source of most granites and associated Be-enriched pegmatites.

ACKNOWLEDGMENTS

I thank the following individuals who provided references that were not available from Interlibrary Loan or valuable information: Daniel Antencio, Peter Buseck, Herman Cho, Giancarlo Della Ventura, Pete Dunn, Gerhard Franz, Laurence Garvie, Joel Grice, Detlef Günther, Ulf Hålenius, George Harlow, Elsa Horváth, László Horváth, John Jambor, Vandall King, Roy Kristiansen, Alf Olav Larsen, Henry Longerich, Andreas Massanek, Olaf Medenbach, Stefano Merlino, Ritsuro Miyawaki, Per Nysten, Nikolay Pertsev, Federico Pezzotta, and Stanislav Vrána. George Rossman kindly provided the optical absorption spectrum for euclase and background information on it, and helped to determine the corresponding color for use on the cover of the volume. Roy Kristiansen very generously loaned me several critical publications from his personal library and made every effort to keep me abreast of the vast literature on Be minerals; he also supplied the image reproduced in Figure 1. Irina I. Kupryanova provided information on the names of specific localities for Be minerals in Russia. Nikita Chukanov and Igor Pekov kindly provided information on greifensteinite and sphaerobertrandite, respectively, in advance of publication. Staff of the Interlibrary Loan Department of Fogler Library, University of Maine, is thanked for their willingness to search for both obscure and old articles in a variety of languages and scripts, and their patience with my continual requests. Richard Bideaux, Carl Francis, Roy Kristiansen, Joseph Mandarino and Igor Pekov are thanked for their thoughtful and thorough comments on this introductory chapter. Financial support was provided by the U.S. National Science

Foundation, grants OPP-9813569 and OPP0087235, to the University of Maine.

The following are thanked for reviewing other chapters of this volume: Mark Barton, Edward Brook, Donald M. Burt, Petr Černý, Jim Channell, Bernard Charoy, Andrew Christy, Jack Dibb, Don Dingwell, Carl Francis, Gerhard Franz, Vanessa Gale, Sergey Krivovichev, Jana Kubizňáková, Devendra Lal, William P. Leeman, David Lindsey, David London, Steve Ludington, Roger Mitchell, Peter Nabelek, Colin Neal, Curtis Olsen, Luisa Ottolini, Lee Riciputi, Kevin Righter, Jeff Ryan, Denis M. Shaw, Steve Simon, Peter Wallbrink, Steve Young, and one reviewer who wished to remain anonymous.

REFERENCES

Academy of Geological Science and Academia Sinica, Section of Crystal Structure Analysis (1973) The crystal structure of hsianghualite. Acta Geol Sinica 1973(2):226-242 (in Chinese with an English summary) [Chem Abstracts 80:88177y]

Agakhanov AA, Pautov LA, Belakovskiy DI, Sokolova E, Hawthorne FC (2002) Telyushenkoite, Cs Na_6 $[Be_2Al_3Si_{15}O_{39}F_2]$, a new mineral. Zapiski Vseross Mineral Obshch (in press) (in Russian) [not consulted]

Albertini C (1983) Baveno, Italy. Mineral Rec 14(3):157-168

Allen FM (1985) Structural and Chemical Variations in Vesuvianite. PhD dissertation, Harvard Univ, Cambridge, Mass

Aminoff G (1923) An association of barylite and hedyphane from Långban. Geol Fören Förhandl 45:124-143 (in Swedish, not consulted) [Chem Abstracts 17:2254]

Aminoff G (1924) Über ein neues Mineral von Långban. Z Kristallogr 60:262-274 [Chem Abstracts 20:30]

Aminoff G (1925) Über Berylliumoxyd als Mineral und dessen Kristallstruktur. Z Kristallogr 62:113-122

Aminoff G (1926) Zur Kristallographie des Trimerits. Geol Fören Stockholm Förhandl 48(1):19-43 [Am Mineral 12:381]

Aminoff G (1933) On the structure and chemical composition of swedenborgite. Kungliga Svenska Vetenskapsakademiens Handlingar, 3rd Ser, 11:3-13 [Chem Abstracts 29:4702]

Andersen S, Sørensen I (1967) On beryllite and bertrandite from the Ilímaussaq alkaline intrusion, South Greenland. Contribution to the mineralogy of Ilímaussaq, No. 5. Meddel Grønland 181(4):11-27.

Andersen F, Berge SA, Burvald I (1996) Die Mineralien des Langesundsfjords und des umgebenden Larvikit-Gebietes, Oslo-Region, Norwegen. Mineralien-Welt 7(4):21-100

Anderson BW, Payne CJ, Claringbull GF, Hey MH (1951) Taaffeite, a new beryllium mineral, found as a cut gemstone. Mineral Mag 29:765-772 [Am Mineral 37:300-301]

Anonymous (1930) Discovering the sweet element "A classic of science". Science News Letter 18:346-347 [A translation of the essential parts of Vauquelin 1798]

Anovitz LM, Grew ES (1996) Mineralogy, petrology, and geochemistry of boron: An introduction. Rev Mineral 33:1-40

Anthony JW, Bideaux RA, Bladh KW, Nichols, MC (1995) Handbook of Mineralogy, vol. 2, Silica, Silicates. Mineral Data Publishing, Tucson, Arizona

Anthony JW, Bideaux RA, Bladh KW, Nichols, MC (1997) Handbook of Mineralogy, vol. 3, Halides, Hydoxides, Oxides. Mineral Data Publishing, Tucson, Arizona

Anthony JW, Bideaux RA, Bladh KW, Nichols, MC (2000) Handbook of Mineralogy, vol. 4, Arsenates, Phosphates, Vanadates. Mineral Data Publishing, Tucson, Arizona

Apollonov VN (1968) Euclase from Central Asia. Trudy Mineral Muz Akad Nauk SSSR 18:168-171 (in Russian)

Argamakov IG, Gordienko VV, Kotrly M, Zukova IA (1995) Die Pegmatit-"Schmuckdose" im Lovozero-Massiv, Halbinsel Kola, Rußland. Mineralien-Welt 6(1):37-43

Armbruster T (2002) Revised nomenclature of högbomite, nigerite, and taaffeite minerals. Eur J Mineral 14:389-395

Armbruster T, Libowitzky E, Diamond L, Auernhammer M, Bauerhansl P, Hoffmann C, Irran E, Kurka, A, Rosenstingl H (1995) Crystal chemistry and optics of bazzite from Furkabasistunnel (Switzerland). Mineral Petrol 52:113-126

Artini E (1901) A new mineral species found in the granite of Baveno. Atti Reale Accad Lincei Rendiconti Classe Sci Fis Matemat Nat 10:139-145 (in Italian)

Artini E (1915) Two minerals from Baveno containing rare earths: weibyeite and bazzite. Atti Reale Accad Lincei Rendiconti Classe Sci Fis Matemat Nat 24:313-319 (in Italian) [Chem Abstracts 9:2365-2366]

Atencio D (2000a) Minerals for which Brazil is the type locality. Rocks and Minerals 75:44-46

Atencio D (2000b) Type Mineralogy of Brazil (Preliminary Edition). Universidade de São Paulo, Instituto de Geociências and Museu de Geociências

Aurisicchio C, Fioravanti G, Grubessi O, Zanazzi PF (1988) Reappraisal of the crystal chemistry of beryl. Am Mineral 73:826-837

Baba S, Grew ES, Shearer CK, Sheraton JW (2000) Surinamite: A high-temperature metamorphic beryllosilicate from Lewisian sapphirine-bearing kyanite-orthopyroxene-quartz-potassium feldspar gneiss at South Harris, N.W. Scotland. Am Mineral 85:1474-1484

Barbier J, Grew ES, Moore PB, Su S-C (1999) Khmaralite, a new beryllium-bearing mineral related to sapphirine: A superstructure resulting from partial ordering of Be, Al and Si on tetrahedral sites. Am Mineral 84:1650-1660

Barbier J, Grew ES, Yates MG, Shearer CK (2001) Beryllium minerals related to aenigmatite. Geol Assoc Canada-Mineral Assoc Canada, Joint Ann Meet Abstr 26:7

Barbier J, Grew ES, Hålenius E, Hålenius U, Yates MG (2002) The role of Fe and cation order in the crystal chemistry of surinamite, $(Mg,Fe^{2+})_3(Al,Fe^{3+})_3O[AlBeSi_3O_{15}]$: A crystal structure, Mössbauer spectroscopic, and optical spectroscopic study. Am Mineral 87:501-513

Bayliss P (2000) Glossary of Obsolete Mineral Names. The Mineralogical Record, Tucson, Arizona

Bea F, Pereira MD, Stroh A (1994) Mineral/leucosome trace-element partitioning in a peraluminous migmatite (a laser ablation–ICP–MS study). Chem Geol 117:291-312

Bedr-Chan S (1925) Analyse des Alvits. Z Anorg Allgem Chem 144:304-306

Behier J (1960) Contribution à la minéralogie de Madagascar. Ann Géol Madagascar 29:1-78

Bel'kov IV, Denisov AP (1968) Melinophane from the Sakharyok alkaline massif. Materialy po Mineralogii Kol'skogo Poluostrova 6:221-224 (in Russian)

Berry LG (1963) The composition of bavenite. Am Mineral 48:1166-1168

Bertrand E (1880) Nouveau minéral des environs de Nantes. Bull Soc Minéral France 3:96-97

Bertrand E (1883) Nouveau minéral des environs de Nantes. Bull Soc Minéral France 6:248-252

Beus AA (1956) Characteristics of the isomorphous entry of beryllium into crystalline mineral structures. Geochemistry 1956(1):62-77

Beus AA (1957) Beryllian vesuvianite. Trudy Mineral Muz Akad Nauk SSSR 8:25-28 (in Russian)

Beus AA (1966) Geochemistry of Beryllium and Genetic Types of Beryllium Deposits. Freeman, San Francisco

Bhaskara Rao A. Silva JC (1968) Phosphate minerals of Brazilian pegmatites—A mineralogical review. Report 22nd Sess Int'l Geol Congress 1964 Delhi, India:157-192

Birch WD, Pring A, Foord EE (1995) Selwynite, $NaK(Be,Al)Zr_2(PO_4)_4 \cdot 2H_2O$, a new gainesite-like mineral from Wycheproof, Victoria, Australia. Can Mineral 33:55-58 [Am Mineral 80:1075]

Blackburn WH, Dennen WH (1997) Encyclopedia of Mineral Names. Can Mineral Spec Pub 1

Blomstrand CW (1876) Contribution to the knowledge of minerals of the Långban mines. B) Barylite, a new mineral from Långban. Geol Fören Förhandl 3:128-133 (in Swedish, not consulted)

Bøggild OB (1915) Leifite, a new mineral from Narsarsuk. Meddel Grønland 51: 427-433 (in Danish)[Chemical Abstracts 11:2651-2652]

Böggild OB (1920) Leifit, ein neues Mineral von Narsarsuk, Grönland. Z Kristallogr Mineral 55:425-429 [Böggild is the German spelling of Bøggild]

Bokiy GB, Mozgova NN, Sokolova MN, eds (1996) Minerals: Handbook. Vol IV, part 3: Silicates. Additions to vol 3 and 4. Nauka, Moscow (in Russian)

Bondi M, Griffin WL, Mattioli V, Mottana A (1983) Chiavennite, $CaMnBe_2Si_5O_{13}(OH)_2 \cdot 2H_2O$, a new mineral from Chiavenna (Italy). Am Mineral 68:623-627

Brøgger WC (1887) On "eudidymite", a new Norwegian mineral. Preliminary report. Nyt Magazin for Naturvidenskaberne 31:196-199 (in Norwegian) [Often given as 1890, although "printed" in 1887, cf. Raade 1996]

Brøgger WC (1890) Die Mineralien der Syenitpegmatitgänge der südnorwegischen Augit- und Nephelinsyenite. Z Kristallogr Mineral 16 [p 65-67 for hambergite; not consulted, see Raade 1996]

Brugger J, Gieré R (1999) As, Sb, Be and Ce enrichment in minerals from a metamorphosed Fe-Mn deposit, Val Ferrera, eastern Swiss Alps. Can Mineral 37:37-52

Bryce DL, Wasylishen RE (1999) Beryllium-9 NMR study of solid bis(2,4-pentanedionato-*O,O'*)beryllium and theoretical studies of ^{9}Be electric field gradient and chemical shielding tensors. First evidence for anisotropic beryllium shielding. J Phys Chem A 103:7364-7372

Brydson R, Vvedensky DD, Engel W, Sauer H, Williams BG, Zeitler E, Thomas JM (1988a) Chemical information from electron-energy-loss near-edge structure. Core hole effects in the beryllium and boron K-edges in rhodizite. J Phys Chem 92:962-966

Brydson R, Williams BG, Engel W, Lindner T, Muhler M, Schlögl R, Zeitler E, Thomas JM (1988b) Electron energy-loss spectroscopy and the crystal chemistry of rhodizite. Part 2. – Near-edge structure. J Chem Soc Faraday Trans 1, 84:631-646

Brydson R, Sauer H, Engel W, Thomas JM, Zeitler E (1989) Co-ordination fingerprints in electron loss near-edge structures: Determination of the local site symmetry of aluminium and beryllium in ultrafine minerals. J Chem Soc Chem Comm 15:1010-1012
Bukin GV (1967) Conditions of crystallization of phenakite-bertrandite-quartz association (experimental data). Doklady Akad Nauk SSSR 176:664-667 (in Russian)
Bukin GV (1969): Synthesis of epididymite under hydrothermal conditions. Trudy Mineral Muz Akad. Nauk SSSR 19:131-133 (in Russian)
Bulnaev KB (1996) Origin of the fluorite–bertrandite–phenakite deposits. Geology of Ore Deposits 38(2):128-136
Burke EAJ, Lustenhouwer WJ (1981) Pehrmanite, a new beryllium mineral from Rosendal pegmatite, Kemiö Island, southwestern Finland. Can Mineral 19:311-314 [Am Mineral 67:859]
Burns PC, Novák M, Hawthorne FC (1995) Fluorine-hydroxyl variation in hambergite: A crystal-structure study. Can Mineral 33:1205-1213
Burt DM (1975) Beryllium mineral stabilities in the model system CaO-BeO-SiO_2-P_2O_5-F_2O_{-1} and the breakdown of beryl. Econ Geol 70:1279-1292.
Burt DM (1978) Multisystems analysis of beryllium mineral stabilities: the system BeO-Al_2O_3-SiO_2-H_2O. Am Mineral 63:664-676
Burt DM (1980) The stability of danalite, $Fe_4Be_3(SiO_4)_3S$. Am Mineral 65:355-360
Burt DM (1982) Minerals of beryllium. *In* Černý P (ed) Granitic Pegmatites in Science and Industry, Mineral Assoc Can Short Course Handbook 8, p 135-148
Burt DM (1988) Stability of genthelvite, $Zn_4(BeSiO_4)_3S$: An exercise in chalcophilicity using exchange operators. Am Mineral 73:1384-1394
Burt DM (1989) Compositional and phase relations among rare-earth element minerals. Rev Mineral 21:259-307
Burt DM (1994) Vector representation of some mineral compositions in the aenigmatite group, with special reference to høgtuvaite. Can Mineral 32:449-457 [Am Mineral 80:405]
Bussy A-B (1828) Reported at the Séance du 16 août 1828 in the Section de Pharmacie. J Chimie Médicale Pharmacie Toxicologie 4:455-456
Calligaro T, Dran J-C, Poirot J-P, Querré G, Salomon J, Zwaan JC (2000) PIXE/PIGE characterization of emeralds using an external micro-beam. Nucl Instrum Methods Phys Research B 161-163:769-774
Cameron AGW (1995) Accounting for light-element abundances. Nature 373:286
Campbell TJ, Roberts WL (1986) Phosphate minerals from the Tip Top Mine, Black Hills, South Dakota. Mineral Rec 17:237-254
Cannillo E, Coda A, Fagnani G (1966) The crystal structure of bavenite. Acta Crystallogr 20:301-309
Cannillo E, Giuseppetti G, Tadini C (1969) The crystal structure of asbecasite. Atti Accad Nazionale Lincei Rendiconti Classe Sci Fis Matemat Nat 46:457-467 [Am Mineral 55:1818]
Cannillo E, Giuseppetti G, Mazzi F, Tazzoli V (1992) The crystal structure of a rare earth bearing leucophanite: $(Ca,RE)CaNa_2Be_2Si_4O_{12}(F,O)_2$. Z Kristallogr 202:71-79
Cassedanne JP (1989) The Ouro Preto topaz mines. Mineral Rec 20(3):221-233
Cassedanne JP, Alves JN (1994) The Jaguaraçu Pegmatite, Minas Gerais, Brazil. Mineral Rec 25:165-170
Cassedanne JP, Baptista A (1999) The Sapucaia Pegmatite, Minas Gerais, Brazil. Mineral Rec 30:347-360, 365-366
Cassedanne JP, Cassedanne JO (1987) La moraesite de la mine de tourmaline de Humaita, Minas Gerais, Brésil. Can Mineral 25:419-424
Černý P, Hawthorne FC, Jarosevich E (1980) Crystal chemistry of milarite. Can Mineral 18:41-57
Černý P, Hawthorne FC, Jambor JL, Grice JD (1991) Yttrian milarite. Can Mineral 29:533-541.
Černý P, Povondra P (1966) Beryllian cordierite from Věžná: (Na,K) + Be → Al. N Jahrb Mineral Monatsh 1966(2):36-44
Chadwick B, Friend CRL, George MC, Perkins WT (1993) A new occurrence of musgravite, a rare beryllium oxide, in Caledonides of North-East Greenland. Mineral Mag 57:121-129
Chang H-C (1969) Structural analysis of liberite. Int'l Geol Review 11:778-786
Chao C-L (1964) Liberite (Li_2BeSiO_4), a new lithium-beryllium silicate mineral from the Nanling Ranges, South China. Acta Geol Sinica 44(3):334-342 (in Chinese with English abstract)[Am Mineral 50: 519]
Chao GY (1967) Leucophanite, elpidite, and narsarsukite from the Desourdy quarry, Mont St. Hilaire, Quebec. Can Mineral 9:286-287
Chao GY, Conlon RP, Van Velthuizen J (1990) Mont Saint-Hilaire unknowns. Mineral Rec 21:363-368
Charoy B (1999) Beryllium speciation in evolved granitic magmas: phosphates versus silicates. Eur J Mineral 11:135-148
Chatterjee M, Ganguli D (1975) Phase relationships in the system BeO-ZnO-SiO_2. N Jahrb Mineral Monatsh 1975:518-526

Chistyakova MB, Moleva VA, Razmanova, ZP (1966) The first find of bazzite in the USSR. Doklady Akad Nauk SSSR 169:1421-1424 (in Russian) [Am Mineral 52:563-564].

Christy AG, Grew ES, Mayo SC, Yates MG, Belakovskiy DI (1998) Hyalotekite, $(Ba,Pb,K)_4(Ca,Y)_2Si_8(B,Be)_2(Si,B)_2O_{28}F$, a tectosilicate related to scapolite: new structure refinement, phase transitions and a short-range ordered 3*b* superstructure. Mineral Mag 62:77-92

Christy AG, Tabira Y, Hölscher A, Grew ES, Schreyer W (2002) Synthesis of beryllian sapphirine in the system MgO-BeO-Al_2O_3-SiO_2-H_2O and comparison with naturally occurring beryllian sapphirine and khmaralite. Part 1: Experiments, TEM and XRD. Am Mineral 87:1104-1112

Chukanov NV, Möckel S, Rastsvetayeva RK, Zadov AE (in press) Greifensteinite $Ca_2Be_4(Fe^{2+},Mn)_5(PO_4)_6(OH)_4{\bullet}6H_2O$—a new mineral from Greifenstein, Saxony. Zapiski Vseross Mineral Obshch (in Russian)

Chukhrov FV (1960) Mineralogy and zoning in the Eastern Kounrad. Trudy Inst Geol Rudnykh Mestorozhdeniy Petrogr Mineral Geokhim 50:5-237 (in Russian) [not consulted; see Chem Abstracts 55:15240d]

Chukhrov FV, Bonshtedt-Kupletskaya EM, Smol'yaninova NN, eds (1972) Minerals, Handbook, vol III, issue 1: Silicates with Single and Doubled Silicon-Oxygen Tetrahedra. Nauka, Moscow (in Russian)

Chukhrov FV, Smol'yaninova NN, eds (1981a) Minerals, Handbook, vol III, issue 2: Silicates with Linear Three-Membered Groups, Rings and Chains of Silicon-Oxygen Tetrahedra. Nauka, Moscow (in Russian)

Chukhrov FV, Smol'yaninova NN, eds (1981b) Minerals, Handbook, vol III, issue 3: Silicates with ribbons of silicon-oxygen tetrahedra. Nauka, Moscow (in Russian)

Chukhrov FV, Smol'yaninova NN, eds (1992a) Minerals, Handbook, vol IV, issue 1: Silicates with a Structure Transitional from Chain to Layered. Layered silicates. Nauka, Moscow (in Russian)

Chukhrov FV, Smol'yaninova NN, eds (1992b) Minerals, Handbook, vol IV, issue 2: Layered silicates. Layered Silicates with Complex Tetrahedral Radicals. Nauka, Moscow (in Russian)

Clark AM (1993) Hey's Mineral Index: Mineral Species, Varieties and Synonyms, 3rd edn. Chapman and Hall, London.

Clark AM, Fejer EE, Couper AG, von Knorring O, Turner RW, Barstow RW (1983) Iron-rich roscherite from Gunnislake, Cornwall. Mineral Mag 47:81-83

Cocco G, Garavelli C (1958) Re-examination of some zeolites from Elba. Atti Soc Toscana Sci Nat Mem Ser A 65:262-283 (in Italian) [Am Mineral 45:1135-1136]

Coda A, Rossi G, Ungaretti L (1967) The crystal structure of aminoffite. Atti Accad Nazionale Lincei Rendiconti Classe Sci Fis Matemat Nat 43(3-4):225-232

Coda A, Ungaretti L, Della Giusta A (1974) The crystal structure of leifite, $Na_6[Si_{16}Al_2(BeOH)_2O_{39}]{\cdot}1{\cdot}5H_2O$. Acta Crystallogr B30:396-401

Cooke JP Jr (1866) On danalite, a new mineral species from the granite of Rockport, Mass. Am J Sci 92:73-79

Coombs DS et al. (1997) Recommended nomenclature for zeolite minerals: report of the subcommittee on zeolites of the International Mineralogical Association, Commission on New Minerals and Mineral Names. Can Mineral 35:1571-1606

Correia Neves JM; Dutra CV; Karfunkel J; Karfunkel B; Schmidt JC; Quemeneur JG; Pedrosa Soares AC (1980) Mineralogy and geochemistry of the Ênio pegmatite, (Galiléia-Minas Gerais). An Acad Brasil Ciênc 52 (3):603-616 (in Portuguese with English summary)

Crain JS, Gallimore DL (1992) Determination of trace impurities in uranium oxides by laser ablation inductively coupled plasma mass spectrometry. J Anal Atomic Spectrometry 7:605-610

Crook WW III (1977) The Clear Creek pegmatite: A rare earth pegmatite in Burnet County, Texas. Mineral Rec 8(2):88-90

Damour A (1883) Note et analyse sur le nouveau minéral des environs de Nantes. Bull Soc Minéral France 6:252-254

Dana JD (1867) Chemical formulas of the feldspars, and of some other silicates. Am J Sci 94:398-409

Dana ES (1888) Preliminary notice of beryllonite, a new mineral. Am J Sci 136:290-291

Dana ES (1892) The System of Mineralogy, Sixth Edn. Wiley, New York

Dana ES, Wells HL (1889) Description of the new mineral, beryllonite. Am J Sci 137:23-32

Danø M (1966) The crystal structure of tugtupite—a new mineral, $Na_8Al_2Be_2Si_8O_{24}(Cl,S)_2$. Acta Crystallogr 20:812-816

Deer WA, Howie RA, Zussman J (1986) Rock-Forming Minerals, vol. 1B, Disilicates and Ring Silicates, 2nd edn. Longman, London

de Fourestier J (1999) Glossary of mineral synonyms. Can Mineral Spec Pub 2

Delamétherie JC (1792) De l'euclase. J Phys 41:155-156 [without proper credit to RJ Haüy (Dana 1892, p xlii)]

Della Ventura G, Parodi GC, Mottana A (1990) New rare earth minerals in the sandinitic ejecta within pyroclastic rocks of the Roman Potassic Province. Atti Accad Nazionale Lincei Rendiconti Classe Sci Fis Matemat Nat Ser 9, 1:159-163

Della Ventura G, Maras A, Mottana A, Parodi GC, Sacerdoti M, Stoppani FS (1991) Antimonian asbecasite in a syenitic ejectum within the Vico pyroclastic rocks (Roman potassic province). Atti Accad Nazionale Lincei Rendiconti Classe Sci Fis Matemat Nat Ser 9, 2:371-378

Della Ventura G, Rossi P, Parodi GC, Mottana A, Raudsepp M, Prencipe M (2000) Stoppaniite, $(Fe,Al,Mg)_4(Be_6Si_{12}O_{36})*(H_2O)_2(Na,\square)$, a new mineral of the beryl group from Latium (Italy). Eur J Mineral 12:121-127

Della Ventura G, Bonazzi P, Oberti R, Ottolini L (2002) Ciprianiite and mottanaite-(Ce), two new minerals of the hellandite group from Latium (Italy). Am Mineral 87:739-744

DeMark RS (1984) Minerals of Point of Rocks, New Mexico. Mineral Rec 15(3):149-156

DeMark RS (1989) Micromounting in New Mexico. Mineral Rec 20(1):57-64

Demartin F, Gramaccioli CM, Pilati T (1992) A first occurrence of euclase in the Swiss Alps: Discovery and refinement of the crystal structure. Schweiz Mineral Petrogr Mitt 72:159-165

Demartin F, Pilati T, Diella V, Gentile P, Gramaccioli CM (1993) A crystal-chemical investigation of Alpine gadolinite. Can Mineral 30:127-136

Demartin F, Gramaccioli CM, Pilati T (2000) Structure refinement of bazzite from pegmatitic and miarolitic occurrences. Can Mineral 38:1419-1424

Demartin F, Minaglia A, Gramaccioli CM (2001) Characterization of gadolinite-group minerals using crystallographic data only: the case of hingganite-(Y) from Cuasso al Monte, Italy. Can Mineral 39:1105-1114

de Parseval P, Fontan F, Aigouy T (1997) Composition chimique des minéraux de terres rares de Trimouns (Ariège, France). C R Acad Sci Paris Série II a, 324:625-630

de Roever EWF, Kieft C, Murray E, Klein E, Drucker WH (1976) Surinamite, a new Mg-Al silicate from the Bakhuis Mountains, western Surinam. I. Description, occurrence, and conditions of formation. Am Mineral 61:193-197

de Roever EWF, Lattard D, Schreyer W (1981) Surinamite: A beryllium-bearing mineral. Contrib Mineral Petrol 76:472-473 [Am Mineral 67:418]

Devouard B, Raith M, Rakotondrazafy R, El-Ghozzi M, Nicollet C (2002) Occurrence of musgravite in anorthite-corundum-spinel-sapphirine rocks ("sakenites") from South Madagascar: Evidence for a high-grade metasomatic event. 18th Gen Meeting Intl Mineral Assoc, Mineralogy for the New Millennium, Programme with Abstracts, p 207

Ding X, Bai G, Yuan Z, Sun L (1981) Yttroceberysite, a new Ce-Be-rich silicate. Geol Rev China 27:459-466 (in Chinese with English summary)[Am Mineral 73:442,935]

Ding X, Bai G, Yuan Z, Liu J (1984) Hingganite $[(Y,Ce)BeSiO_5(OH)]$(sic): New Data. Acta Petrol Mineral Anal 3(1):46-48 (in Chinese with English summary)[Chem Abstracts 101:233307z]

Ding X, Bai G, Yuan Z (1985) Mineralogical characteristics of some rare-metal alkaline granites in the Inner Mongolian Autonomous Region [China]. Zhonggue Dizhi Kexueyuan Kuangchan Dizhi Yanjiuso Sokan 14:71-88 (in Chinese)[not consulted; from Chem Abstracts 105:9441p]

Dunn PJ (1976) Genthelvite and the helvine group. Mineral Mag 40:627-636

Dunn PJ (1995) Franklin and Sterling Hill, New Jersey: the World's Most Magnificent Mineral Deposits. The Franklin-Ogdensburg Mineralogical Society, Franklin, New Jersey

Dunn PJ, Gaines RV (1978) Uralolite from the Dunton Gem Mine, Newry, Maine: A second occurrence. Mineral Rec 9:99-100

Dunn PJ, Wight W (1976) Green gem herderite from Brazil. J Gemmology 15(1):27-28

Dunn PJ, Wolfe CW, Leavens PB, Wilson WE (1979) Hydroxyl-herderite from Brazil and a guide to species nomenclature for the herderite/hydroxyl-herderite series. Mineral Rec 10(1):5-11

Dunn PJ, Peacor DR, Simmons WB, Gaines RV (1984) Sverigeite, a new tin beryllium silicate from Långban, Värmland, Sweden. Geol Fören Stockholm Förhandl 106:175-177 [Am Mineral 70:1332]

Dunn PJ, Peacor DR, Grice JD, Wicks FJ, Chi PH (1990) Wawayandaite, a new calcium manganese beryllium boron silicate from Franklin, New Jersey. Am Mineral 75:405-408

Durrant SF (1999) Laser ablation inductively coupled plasma mass spectrometry: achievements, problems, prospects. J Anal Atomic Spectrometry 14:1385-1403

Dyar MD, Wiedenbeck M, Robertson D, Cross LR, Delaney JS, Ferguson K, Francis CA, Grew ES, Guidotti CV, Hervig RL, Hughes JM, Husler J, Leeman W, McGuire AV, Rhede D, Rothe H, Paul RL, Richards I, Yates M (2001) Reference minerals for the microanalysis of light elements. Geostandards Newsletter 25:441-463

Dzhurayev ZT, Zolotarev AA, Pekov IV, Frolova LV (1998) Hambergite from pegmatite veins of the eastern Pamirs. Zapiski Vseross Mineral Obshch 127(4):132-139 (in Russian)

Ehlmann AJ, Mitchell RS (1970) Behoite, beta-$Be(OH)_2$, from the Rode Ranch pegmatite, Llano County, Texas. Am Mineral 55:1-9

Ellingsen HV, Haugen A, Raade G, Eldjarn K, Berg H-J (1995) Rare minerals in an amazonite pegmatite at Tennvatn in Nordland. Norsk Bergverksmuseum Skrifter 9:35-37 (in Norwegian)

Emsley J (2001) Nature's building blocks. An A-Z guide to the elements. Oxford University Press, Oxford

Engel W, Sauer H, Zeitler E, Brydson R, Williams BG, Thomas JM (1988) Electron energy-loss spectroscopy and the crystal chemistry of rhodizite. Part 1. Instrumentation and chemical analysis. J Chem Soc Faraday Trans 1, 84:617-629

Engell J, Hansen J, Jensen M, Kunzendorf H, Løvborg L (1971) Beryllium mineralization in the Ilímaussaq intrusion, South Greenland, with description of a field beryllometer and chemical methods. Grønlands Geol Undersøgelse Rapport 33:1-40

Engvoldsen T, Berge SA, Andersen F, Burvald I (1991) Pegmatite minerals from the Larvik ring complex. Stein (Nordic Magazine for Popular Geology) 18(1):15-71 (in Norwegian)

Erdmann A (1840) Investigation of leucophane, a new mineral from the Brevig area in Norway. Kongl Svensk Vetenskaps-Akad Handl:191-200 [In Swedish. Not consulted, see Raade, 1996]

Eskola P (1951) Around Pitkäranta. Ann Acad Scient Fennicæ Series A3 Geol-Geograph 27:1-90

Evensen JM, London D (2002) Experimental silicate mineral/melt partition coefficients for beryllium and the crustal Be cycle from migmatite to pegmatite. Geochim Cosmochim Acta 66:2239-2265

Evensen JM, London D (in press) Complete cordierite-beryl solid solutions in granitic systems: phase relations and complex crystal chemistry. Am Mineral

Evensen JM, Meeker GP (1997) Feasibility of Be analysis for geological materials using EPMA. Microscopy Soc Am Proceedings Microscopy and Microanalysis 1997 3(2):893-894 (abstract)

Evensen JM, London D, Hughes JM, Rakovan JF, Hervig RL, Kaszuba JP (2002) Crystal chemistry, crystallography, and petrogenesis of the beryllium micas. 18th Gen Meeting Intl Mineral Assoc, Mineralogy for the New Millennium, Programme with Abstracts, p 207

Everest DA (1973) 9. Beryllium. *In* Bailar JC, Emeléus HJ, Nyholm R, Trotman-Dickenson AF (eds) Comprehensive Inorganic Chemistry, vol. 1. Pergamon, Oxford, New York, p 531-590

Falster AU (1994) The mineralogy and geochemistry of the Animikie Red Ace Pegmatite, Florence County, Wisconsin. Master's thesis, University of New Orleans, New Orleans, Louisiana [not consulted; abstract available from GeoRef 98-08873]

Falster AU, Simmons WB (1989) Rhodizite formation in a Cs-Rb-B-Mn enriched pegmatite in Wisconsin. Geol Soc Am Abstr Progr 21(6):A119

Falster AU, Simmons WB, Webber KL (1996) The mineralogy and geochemistry of the Animikie Red Ace Pegmatite, Florence County, Wisconsin. Recent Research Developments in Mineralogy 1:7-67

Fanfani L, Nunzi A, Zanazzi PF, Zanzari AR (1975) The crystal structure of roscherite. Tschermaks Mineral Petrograph Mitt 22:266-277

Fanfani L, Zanazzi PF, Zanzari AR (1977) The crystal structure of a triclinic roscherite. Tschermaks Mineral Petrograph Mitt 24:169-178 [Am. Mineral 63:427]

Ferraris G, Prencipe M, Rossi P (1998) Stoppaniite, a new member of the beryl group: crystal structure and crystal-chemical implications. Eur J Mineral 10:491-496

Fersman AE (1940) Pegmatites; vol. I, Granitic Pegmatites, Third Edn. Akad Nauk SSSR (in Russian)

Fields BD, Olive KA, Vangioni-Flam E, Cassé M (2000) Testing spallation processes with beryllium and boron. Astron J 540:930-945

Fleischer M, Switzer G (1953) The bavenite problem. Am Mineral 38:988-993

Flem B, Larsen RB, Grimstvedt A, Mansfeld J (2002) In situ analysis of trace elements in quartz by using laser ablation inductively coupled mass spectrometry. Chem Geol 182:237-247

Flink G (1886) Mineralogical Notices. I. (1-16.). 14. Harstigite from Pajsberg. Bihang Kongl Svenska Vetensk-Akad Handl 12, Afd 2, No 2:59-63, Table II (in Swedish)

Flink G (1890) Ueber Pinakiolith und Trimerit, zwei neue Mineralien aus den Mangangruben Schwedens. Z Kristallogr 18:361-376

Flink G (1893) On some minerals from Greenland. Geol Fören Stockholm Förhandl 15:195-208, 467-470 [in Swedish]

Flink G (1917a) Contribution to Swedish mineralogy. 104. Trimerite from Jakobsberg. Arkiv Kemi Mineral Geol 6:46-48 (in Swedish)

Flink G (1917b) Contribution to Swedish mineralogy. 105. Harstigite from Harstigsgrufvan. Arkiv Kemi Mineral Geol 6:49-51 (in Swedish)

Flink G (1917c) Contribution to Swedish mineralogy. 116. Gadolinite. Arkiv Kemi Mineral Geol 6:82-83 (in Swedish)

Fontan F, Fransolet A-M (1982) Le béryl bleu riche en Mg, Fe et Na de la mine de Lassur, Ariège, France. Bull Minéral 105:615-620

Foord EE, Gaines RV, Crock JG, Simmons WB Jr, Barbosa CP (1986) Minasgeraisite, a new member of the gadolinite group from Minas Gerais, Brazil. Am Mineral 71:603-607

Foord EE, Brownfield ME, Lichte FE, Davis AM, Sutley SJ (1994) Mccrillisite, $NaCs(Be,Li)Zr_2(PO_4)_4$•1-$2H_2O$, a new mineral species from Mount Mica, Oxford County, Maine, and new data for gainesite. Can Mineral 32:839-842 [Am Mineral 80:1074]

Forbes D, Dahll T (1855) Mineralogical observations around Arendal and Kragerö. Nyt Magazin for Naturvidenskaberne 8(3):213-229 (in Norwegian)

Franke E (1934) Beitrag zur Kenntnis der Zinn-Wolframerzlagerstätte im Sauberg bei Ehrenfriedersdorf. Z prakt Geol 42(3):33-48

Franz G, Morteani G (1984) The formation of chrysoberyl in metamorphosed pegmatites. J Petrol 25:27-52

Franz G, Grundmann G, Ackermand D (1986) Rock forming beryl from a regional metamorphic terrain (Tauern Window, Austria): Parageneses and crystal chemistry. Tschermaks Mineral Petrogr Mitt 35:167-192

Frondel C, Baum JL (1974) Structure and mineralogy of the Franklin zinc-iron-manganese deposit, New Jersey. Econ Geol 69:157-180

Frondel C, Ito J (1965) Composition of rhodizite. Tschermaks Mineral Petrogr Mitt 10:409-412

Gadomski M, Wiewióra A, Szpila K (1971) Bavenite from Strzegom (Lower Silesia). Archiwum Mineral 29(1-2):5-28 (in Polish with abridged English version)

Gaines RV (1976) Beryl—A review. Mineral Rec 7:211-223

Gaines RV, Skinner HCW, Foord EE, Mason B, Rozenzweig A (1997) Dana's New Mineralogy: the System of Mineralogy of James Dwight Dana and Edward Salisbury Dana, 8th edn. Wiley, NewYork

Gallagher MJ, Hawkes JR (1966) Beryllium minerals from Rhodesia and Uganda. Bull Geol Surv Great Britain 25:59-75

Galli E (1980) The crystal structure of roggianite, a zeolite-like silicate. Proc 5th Int'l Conf on Zeolites, p 205-213 [not consulted. Am Mineral 68:852]

Ganzeyeva LV, Bedrzhitskaya KV, Shumkova NG (1973) Leucophane from alkaline metasomatites of the Russian Platform. *In* Mineralogicheskiye Issledovaniya 3. IMGRE, Moscow, p 25-28 (in Russian)

Gard JA (1969) An electron microscope and diffraction study of roggianite. Clay Minerals 8:112-113 [Am Mineral 55:322-323]

Garvie LAJ, Buseck PR (1996) Parallel electron energy-loss spectroscopy of boron in minerals. Rev Mineral 33:821-843

Garvie LAJ, Buseck PR (1999) Bonding in silicates: Investigation of the Si $L_{2,3}$ edge by parallel electron energy-loss spectroscopy. Am Mineral 84:946-964

Garvie LAJ, Buseck PR, Rez P (1997) Characterization of beryllium-boron-bearing materials by parallel electron energy-loss spectroscopy (PEELS). J Solid State Chem 133:347-355.

Gelsdorf G, Müller-Hesse H, Schwiete H-E (1958) Einlagerungsversuche an synthetischem Mullit und Substitutionsversuche mit Galliumoxyd und Germaniumdioxyd Teil II. Archiv Eisenhüttenwesen 29:513-519

Genth FA (1884) On herderite. Proc Am Phil Soc 21:694-699

Genth FA (1892) Contributions to mineralogy, No. 54; with crystallographic notes by S.L. Penfield. 6. Danalite. Am J Sci 144:385-386

Gerasimovsky VI (1939) Chkalovite. C R Acad Sci URSS 22:259-263 [Am Mineral 25:380]

Ginzburg AI (1957) Bityite-lithium-beryllium margarite. Trudy Mineral Muz Akad Nauk SSSR 8:128-131

Ginzburg AI, ed (1975) The genetic types of hydrothermal beryllium deposits. Nedra, Moscow (in Russian)

Ginzburg AI, ed (1976) The mineralogy of hydrothermal beryllium deposits. Nedra, Moscow (in Russian)

Ginzburg AI, Shatskaya VT (1966) Hypergene phosphates of beryllium and conditions of their formation. *In* Geologiya Mestorozhdeniy Redkikh Elementov 30. Nedra, Moscow, p 101-117 (in Russian)

Ginzburg AI, Novikova MI, Gal'chenko VI (1979) Eudidymite in fluorite-bertrandite-phenakite deposits. Doklady Acad Sci USSR Earth Sci Sections, 246:151-154

Giuseppetti G, Mazzi F, Tadini C, Larsen AO, Åsheim A, Raade G (1990) Berborite polytypes. N Jahrb Miner Abh 162:101-116 [Am Mineral 76:1734]

Giuseppetti G, Mazzi F, Tadini C, Galli E (1991) The revised crystal structure of roggianite: $Ca_2[Be(OH)_2Al_2Si_4O_{13}] <2.5H_2O$. N Jahrb Miner Monatsh 1991(7):307-314 [Am Mineral 77:452]

Giuseppetti G, Tadini C, Mattioli V (1992) Bertrandite: $Be_4Si_2O_7(OH)_2$, from Val Vigezzo (NO) Italy: the X-ray structural refinement. N Jahrb Miner Monatsh 1992(1):13-19

Glass JJ, Jahns RH, Stevens RE (1944) Helvite and danalite from New Mexico and the helvite group. Am Mineral 29:163-191

Goldschmidt VM (1958) Geochemistry. Oxford University Press, Orford, UK

Goldschmidt VM, Peters C (1932) Zur Geochemie des Berylliums. Nachrichten Akad Wissenschaften Göttingen, II. Math-Phys Klasse, p 360-376

Gordiyenko VV, Fedchenko VF, Zorina ML, Novikova YuN, Chernysheva VF (1973) First find of väyrynenite—$MnBe[PO_4](OH)$—in the USSR. Zapiski Vsesoyuz Mineral Obshch 102:432-435 (in Russian)
Graeser S (1966) Asbecasit und Cafarsit, zwei neue Mineralien aus dem Binnatal (Kt. Wallis) Schweiz mineral petrogr Mitt 46:367-375 [Am. Mineral. 52:1583-1584]
Graeser S (1995) Bergslagit aus den Schweizer Alpen. Aufschluss 46:15-22
Graeser S, Albertini C (1995) Arsenmineralien und ihre Begleiter aus Gneisklüften der Monte-Leone-Decke (II) Wannigletscher und Conca Cervandone. Lapis 20(7/8):41-64
Gramaccioli CM, Demartin F (2001) Gadolinit aus den Alpen. Lapis 26(3):24-31,50
Gramaccioli CM, Gentile P (1991) Description of the minerals of Lombardia. *In* Natura in Lombardia, i Minerali. Regione Lombardia, Giunta Regionale Ecologia, p 59-177 (in Italian) [not consulted]
Gramaccioli CM, Diella V, Demartin F, Orlandi P, Campostrini I (2000) Cesian bazzite and thortveitite from Cuasso al Monte, Varese, Italy: A comparison with the material from Baveno, and inferred origin. Can Mineral 38:1409-1418
Grauch RI, Lindahl I, Evans HT Jr, Burt DM, Fitzpatrick JJ, Foord EE, Graff P-R, Hysingjord J (1994) Høgtuvaite, a new beryllian member of the aenigmatite group from Norway, with new X-ray data on aenigmatite. Can Mineral 32:439-448 [Am Mineral 80:405]
Graziani G, Guidi G (1980) Euclase from Santa do Encoberto, Minas Gerais, Brazil. Am Mineral 65: 183-187
Greensfelder L (2002) Subtleties of sand reveal how mountains crumble. Science 295:256-258
Greenwood NN, Earnshaw A (1997) Chemistry of the Elements, 2nd edn. Butterworth-Heinemann, Oxford, Woburn
Grew ES (1981) Surinamite, taaffeite, and beryllian sapphirine from pegmatites in granulite-facies rocks in Casey Bay, Enderby Land, Antarctica: Am Mineral 66:1022-1033
Grew ES (1996) Borosilicates (exclusive of tourmaline) and boron in rock-forming minerals in metamorphic environments. Rev Mineral 33:387-502
Grew ES (1998) Boron and beryllium minerals in granulite-facies pegmatites and implications of beryllium pegmatites for the origin and evolution of the Archean Napier Complex of East Antarctica. Mem Nat Inst Polar Res Spec Issue 53:74-92
Grew ES, Belakovskiy DI, Fleet ME, Yates MG, McGee JJ, Marquez N (1993) Reedmergnerite and associated minerals from peralkaline pegmatite, Dara-i-Pioz, southern Tien Shan, Tajikistan. Eur J Mineral 5: 971-984
Grew ES, Yates MG, Belakovskiy DI, Rouse RC, Su S-C, Marquez N (1994) Hyalotekite from reedmergnerite-bearing peralkaline pegmatite, Dara-i-Pioz, Tajikistan and from Mn skarn, Långban, Värmland, Sweden: a new look at an old mineral. Mineral Mag 58:285-297
Grew ES, McGee JJ, Yates MG, Peacor DR, Rouse RC Huijsmans JPP, Shearer CK, Wiedenbeck M, Thost DE, and Su S-C (1998a) Boralsilite ($Al_{16}B_6Si_2O_{37}$): A new mineral related to sillimanite from pegmatites in granulite-facies rocks. Am Mineral 83:638-651
Grew ES, Pertsev NN, Vrána S, Yates MG, Shearer CK, Wiedenbeck M (1998b) Kornerupine parageneses in whiteschists and other magnesian rocks: is kornerupine + talc a high-pressure assemblage equivalent to tourmaline + orthoamphibole? Contrib Mineral Petrol 131:22-38
Grew ES, Yates MG, Huijsmans JPP, McGee JJ, Shearer CK, Wiedenbeck M, Rouse RC (1998c) Werdingite, a borosilicate new to granitic pegmatites. Can Mineral 36:399-414
Grew ES, Yates MG, Barbier J, Shearer CK, Sheraton JW, Shiraishi K, Motoyoshi Y (2000) Granulite-facies beryllium pegmatites in the Napier Complex in Khmara and Amundsen Bays, western Enderby Land, East Antarctica. Polar Geoscience 13:1-40
Grew ES, Hålenius U, Kritikos M, Shearer CK (2001) New data on welshite, e.g. $Ca_2Mg_{3.8}Mn^{2+}{}_{0.6}Fe^{2+}{}_{0.1}Sb^{5+}{}_{1.5}O_2[Si_{2.8}Be_{1.7}Fe^{3+}{}_{0.65}Al_{0.7}As_{0.17}O_{18}]$, an aenigmatite-group mineral. Mineral Mag 65:665-674
Grice JD, Hawthorne FC (2002) New data on meliphanite, $Ca_4(Na,Ca)_4Be_4AlSi_7O_{24}(F,O)_4$. Can Mineral 40:971-980
Grice JD, Robinson GW (1984) Jeffreyite, $(Ca,Na)_2(Be,Al)Si_2(O,OH)_7$, a new mineral species and its relation to the melilite group. Can Mineral 22:443-446 [Am Mineral 70:872]
Grice JD, Peacor DR, Robinson GW, Van Velthuizen J, Roberts WL, Campbell TJ, Dunn PJ (1985) Tiptopite $(Li,K,Na,Ca,\square)Be_6(PO_4)_6(OH)_4$, a new mineral species from the Black Hills, South Dakota. Can Mineral 23:43-46 [Am Mineral 71:230]
Griffith RF (1955) Historical note on sources and uses of beryllium. *In* White DW Jr, Burke JE (eds) The Metal Beryllium. The American Society for Metals, Cleveland, p 5-13
Grigor'yev NA (1963) Glucine—a new beryllium mineral. Zap Vsesoyuz Mineral Obshch 92: 691-696 (in Russian) [Am Mineral 49:1152]

Grigor'yev NA (1964) Uralolite—a new mineral. Zapiski Vsesoyuz Mineral Obshch 93:156-162 (in Russian) [Am Mineral 49:1776]
Grigor'yev NA (1967) Co-deposition of beryllium with manganese during formation of rhodochrosite under hypergene conditions. Doklady Akad Nauk SSSR 173:1411-1413 (in Russian)
Groat LA (1988) The Crystal Chemistry of Vesuvianite. PhD dissertation, Univ Manitoba, Winnipeg, Manitoba
Groat LA, Hawthorne FC, Ercit TS (1992) The chemistry of vesuvianite. Can Mineral 30:19-48
Groat LA, Hawthorne FC, Ercit TS (1994) The incorporation of boron into the vesuvianite structure. Can Mineral 32:505-523
Groat LA, Hawthorne FC, Ercit TS, Grice JD (1998) Wiluite, $Ca_{19}(Al,Mg,Fe,Ti)_{13}(B,Al,\square)_5Si_{18}O_{68}(O,OH)_{10}$, a new mineral species isostructural with vesuvianite, from the Sakha Republic, Russian Federation. Can Mineral 36:1301-1304
Guelfi F, Orlandi P (1987) Brief report from several Italian mineral localities. Rivista Mineralogica Italiana 10:59-61 [not consulted]
Gurvich SI (1965) Discovery of beryllium-bearing willemite in the USSR. Doklady Acad Sci USSR Earth Sci Sections 153:136-138
Haapala I, Ojanperä P (1972) Genthelvite-bearing greisens in southern Finland. Geol Surv Finland Bull 259:1-22
Habel M (2000) Neufunde aus dem östlichen Bayerischen Wald (IV). Mineralien-Welt 11(3):33-38
Hahn T, Eysel W (1970) Solid solubility in the system Zn_2SiO_4-Zn_2GeO_4-Be_2SiO_4-Be_2GeO_4. N Jahrb Mineral Monatsh 1970:263-276
Haidinger W (1828a) On herderite, a new mineral species. Philosophical Magazine and Annals Philosophy (London) New Series 4:1-3
Haidinger W (1828b) Ueber den Herderit, eine neue Mineralspecies. Pogg Ann Phys Chem 13:502-505 [German version of Haidinger 1828a]
Han S, Schmitt KD, Shihabi DS, Chang CD (1993) Isomorphous substitution of Be^{2+} into ZSM-5 zeolite with ammonium tetrafluorberyllate. J Chem Soc Chem Comm 1993:1287-1288
Hansen S, Fälth L, Johnsen O (1984a) Bergslagite, a mineral with tetrahedral berylloarsenate sheet anions. Z Kristallogr 166:73-80 [Am Mineral 70:436]
Hansen S, Fälth L, Petersen OV, Johnsen O (1984b) Bergslagite, a new mineral species from Långban, Sweden. N Jahrb Mineral Monatsh 1984(6):257-262 [Am Mineral 70:436]
Harrison WTA, Nenoff TM, Gier TE, Stucky GD (1993) Tetrahedral-atom 3-ring groupings in 1-dimensional inorganic chains: $Be_2AsO_4OH{\cdot}4H_2O$ and $Na_2ZnPO_4OH{\cdot}7H_2O$. Inorg Chem 32: 2437-2441
Hasegawa S (1957) Chemical studies of allanites and their associated minerals from the pegmatites in the northern part of the Abukuma Massif. Sci Reports Tohoku Univ 3rd Ser (Mineral Petrol Econ Geol) 5:345-371
Hasegawa S (1958) Chemical studies of allanites from the new localities in Fukushima and Kagawa Prefectures. Sci Reports Tohoku Univ 3rd Ser (Mineral Petrol Econ Geol) 6:39-56
Hasegawa S (1960) Chemical composition of allanite. Sci Reports Tohoku Univ 3rd Ser (Mineral Petrol Econ Geol) 6:331-387
Hassan I, Grundy HD (1991) The crystal structure and thermal expansion of tugtupite, $Na_8[Al_2Be_2Si_8O_{24}]Cl_2$. Can Mineral 29:385-390
Haüy RJ (1799) 10. Euclase (N.N.), c'est-à-dire, *facile à briser*. J Mines 5:258
Hawthorne FC (2002) The use of end-member charge-arrangements in defining new mineral species and heterovalent substitutions in complex minerals. Can Mineral 40:699-710
Hawthorne FC, Grice JD (1987) The crystal structure of ehrleite, a tetrahedral sheet structure. Can Mineral 25:767-774 [Am Mineral 74:504-505]
Hawthorne FC, Grice JD (1990) Crystal structure analysis as a chemical analytical method: application to light elements. Can Mineral 28:693-702
Hawthorne FC, Kimata M, Černý P, Ball N, Rossman GR, Grice JD (1991) The crystal chemistry of the milarite-group minerals. Am Mineral 76:1836-1856
Hawthorne FC, Cooper MA, Taylor MC (1998) Refinement of the crystal structure of tadzhikite. Can Mineral 36:817-822
Heinrich EW. Deane RW (1962) An occurrence of barylite near Seal Lake, Labrador. Am Mineral 47: 758-763
Henderson WA, Weber CH Mr, Weber CH Mrs (1967) Roscherite from North Groton, N. H. Rocks and Minerals 42:763
Henriques Å (1964) Geology and ores of the Åmmeberg District (Zinkgruvan), Sweden. Arkiv Mineral Geol 4(1):1-246

Hesse K-F, Stümpel G (1986) Crystal structure of harstigite, $MnCa_6Be_4[SiO_4]_2[Si_2O_7]_2(OH)_2$. Z Kristallogr 177:143-148
Hey MH, Milton C, Dwornik EJ (1982) Eggonite (kolbeckite, sterrettite), $ScPO_4 \cdot 2H_2O$. Mineral Mag 46:493-497
Hidden WE (1905) Some results of late minerals research in Llano County, Texas. Am J Sci 89:425-433
Hochleitner R, Weiß S, Horváth L (2000) Steckbrief Leifit. Lapis 25(7/8):9-11,86
Hofer F, Kothleitner G (1993) Quantitative microanalysis using electron energy-loss spectrometry. I. Li and Be in oxides. Microscopy Microanal Microstruct 4:539-560
Hofmeister AM, Hoering TC, Virgo D (1987) Vibrational spectroscopy of beryllium aluminosilicates: heat capacity calculations from band assignments. Phys Chem Minerals 14:205-224
Hölscher, A (1987) Experimentelle Untersuchungen im System $MgO-BeO-Al_2O_3-SiO_2-H_2O$: MgAl-Surinamit und Be-Einbau in Cordierit und Sapphirin. Unpubl PhD dissertation, Ruhr-Universität Bochum, Germany
Hölscher A, Schreyer W (1989) A new synthetic hexagonal BeMg-cordierite, $Mg_2[Al_2BeSi_6O_{18}]$, and its relationship to Mg-cordierite. Eur J Mineral 1:21-37
Holser WT (1953) Beryllium minerals in the Victorio Mountains, Luna County, New Mexico. Am Mineral 38:599-611
Holtstam D, Langhof J, eds (1999) Långban. The mines, their minerals, geology and explorers. Swedish Museum of Natural History, Raster Förlag, Stockholm
Holtstam D, Wingren N (1991) Zincian helvite, a pegmatite mineral from Stora Vika, Nynäshamn, Sweden. Geol Fören Stockholm Förhandl 113(2-3):183-184
Hörmann PK (1978) Beryllium. *In* Wedepohl KH (ed) Handbook of Geochemistry II/1. Springer, Berlin, p 4-B-1 to 4-O-1, 1-6
Horváth L, Gault RA (1990) The mineralogy of Mont Saint-Hilaire, Quebec. Mineral Rec 21(4):284-359
Horváth L, Pfenninger-Horváth E (2000a) The minerals of Mont Saint-Hilaire (Québec, Canada) Rivista Mineral Italiana 3:140-202 (In Italian with condensed version in English)
Horváth L, Pfenninger-Horváth E (2000b) Die Mineralien des Mont Saint-Hilaire. Lapis 25(7/8):23-61
Huang Y, Du S, Wang K, Chao C, Yu C (1958) Hsiang-hua-shih, a new beryllium mineral. Ti-chih-yueh-kan 7:35 (in Chinese) [not consulted; cf. "syanchualite", Beus 1966, p. 69-71; Am Mineral 44:1327-1328, 46:244; Chemical Abstracts 53:18766e]
Huang Y, Du S, Zhou X (1988) Hsianghualing rocks, mineral deposits and minerals. Beijing Sci Techn Publ Bur, Beijing (in Chinese with English summary)
Hudson DR, Wilson AF, Threadgold IM (1967) A new polytype of taaffeite—a rare beryllium mineral from the granulites of central Australia. Mineral Mag 36:305-310
Hügi T, Röwe D (1970) Berylliummineralien und Berylliumgehalte granitischer Gesteine der Alpen. Schweiz Mineral Petrograph Mitt 50:445-480
Hügi T, Saheurs J-P, Spycher E (1968) Distribution of Be in granitic rocks of the Swiss Alps. *In* Ahrens LH (ed) Origin and Distribution of the Elements. Pergamon, Oxford, p 749-760
Huminicki DC, Hawthorne FC (2000) Refinement of the crystal structure of väyrynenite. Can Mineral 38:1425-1432
Huminicki DC, Hawthorne FC (2001) Refinement of the crystal structure of swedenborgite. Can Mineral 39:153-158
Hurlbut CS Jr (1937) Aminoffite, a new mineral from Långban. Geol Fören Förhandl 59: 290-292 [Am Mineral 23: 293]
Hurlbut CS Jr (1955) Beryllian idocrase from Franklin, New Jersey. Am Mineral 40:118-120
Hyslop EK, Gillanders RJ, Hill PG, Fakes RD (1999) Rare-earth-bearing minerals fergusonite and gadolinite from the Arran granite. Scot J Geol 35(1):65-69
Iimori T (1938) Tengerite found in Iisaki, and its chemical composition. Scientific Papers Inst Phys Chem Research (Tokyo) 34:832-841
Iimori T (1939) A beryllium-bearing variety of allanite. Scientific Papers Inst Phys Chem Research (Tokyo) 36:53-55
Ito J (1967) Synthesis of calciogadolinite. Am Mineral 52:1523-1527
Ito J, Hafner SS (1974) Synthesis and study of gadolinites. Am Mineral 59:700-708
Jahn S (1998) Der Herderit von Ehrenfriedersdorf—ein sächsischer Zankapfel. Mineralien Welt 9(3): 24-28, 64
Jambor JL, Roberts AC, Grice JD, Birkett TC, Groat LA, Zajac S (1998) Gerenite-(Y), $(Ca,Na)_2(Y,\mathit{REE})_3Si_6O_{18} \cdot 2H_2O$, a new mineral species, and an associated Y-bearing gadolinite-group mineral, from the Strange Lake peralkaline complex, Quebec-Labrador. Can Mineral 36:793-800
Jannasch P, Locke J (1894) Über einen fluorfreien Humit. Z Anorgan Chem 7:92-95
Jensen A, Petersen OV (1982) Tugtupite: A gemstone from Greenland. Gems Gemol 18:90-94

Jermolenko V (2002) Kara-Oba: Mineralogische Perle der Betpak-Ebene bei Dzhambul, Kasachstan. Lapis 27(4):13-34, 50

Juve G, Bergstøl S (1990) Caesian bazzite in granite pegmatite in Tørdal, Telemark, Norway. Mineral Petrol 43:131-136

Kalsbeek N, Rønsbo JG (1992) Refinement of the vinogradovite structure, positioning of Be and excess Na. Z Kristallogr 200:237-245

Kampf AR (1991) Taaffeite crystals. Mineral Rec 22:343-347

Kampf AR (1992) Beryllophosphate chains in the structures of fransoletite, parafransoletite, and ehrleite and some general comments on beryllophosphate linkages. Am Mineral 77:848-856

Kampf AR, Dunn PJ, Foord EE (1992) Parafransoletite, a new dimorph of fransoletite from the Tip Top Pegmatite, Custer, South Dakota. Am Mineral 77:843-847

Kapustin YuL (1973a) First find of aminoffite in the USSR and relationships of minerals in the leucophanite-aminoffite group. Doklady Akad Nauk SSSR Earth Sci Sect 209:100-103

Kapustin YuL (1973b) Accessory beryllium mineralization in alkalic rocks of Tuva. *In* Borodin LS (ed) New Data on the Geology, Mineralogy and Geochemistry of Alkalic Rocks. Nauka, Moscow, p 66-90 (in Russian)[Chemical Abstracts 80:50458c]

Kartashov PM, Lapina MI (1994) New varieties of hingganite-gadolinite group minerals from alkaline-granit (sic) pegmatites of western Mongolia. Int'l Mineral Assoc 16th Gen Meeting Abstracts:195-196

Kartashov PM, Voloshin AV, Pakhomovskiy YaA (1993) Zoned crystalline gadolinite from the alkaline granitic pegmatites of Haldzan-Buragtag (Mongolian Altai). Zapiski Vseross Mineral Obshch 122(3):65-79 (in Russian)

Kenngott A (1870) Mittheilungen an Professor G. Leonhard. N Jahrb Mineral Geol Paläontol 1870:80-81 [Original description of milarite]

Kerndt T (1848) Chemische Untersuchung des Muromontits, eines neuen Cerminerals aus der Gegend von Mauersberg bei Marienberg im sächsischen Erzgebirge. J Prakt Chem 43:228-241

Khomyakov AP (1972) Loparite and rinkite from the Burpala alkaline massif (northern Baikal Region). *In* Mineralogicheskiye Issledovaniya 2. IMGRE, Moscow, p 25-30 (in Russian)

Khomyakov AP (1995) Mineralogy of Hyperagpaitic Alkaline Rocks. Clarendon Press, Oxford

Khomyakov AP, Stepanov VI (1979) The first find of chkalovite in Khibiny and its paragenesis. Doklady Akad Nauk SSSR 248:727-730 (in Russian)

Khomyakov AP, Semenov YeI, Bykova AV, Voronkov AA, Smol'yaninova NN (1975) New data on lovdarite. Doklady Akad Nauk Earth Sci Sections 221:154-157

Khomyakov AP, Bykova AV, Kaptsov VV (1979) New data on Lovozero leifite. *In* Semenov YeI (ed) New Data on the Mineralogy of Deposits in Alkaline Formations. Institut Mineralogii, Geokhimii i Kristallokhimii Redkikh Elementov, Moscow, p 12-15 (in Russian)

Khvostova VA (1962) Mineralogy of Orthite [allanite]. Trudy Inst Mineral Geokhim Kristallokhim Redkikh Elementov Akad Nauk SSSR 11:1-119 (in Russian)

Kim YS, Choi HW, Kim DK, Woo HJ, Kim NB, Park KS (2000) Analysis of light elements by PIGE. Anal Sci Technol 13(1):12-21 (in Korean)[not consulted; Chem Abstracts 133:98757]

Kimura K (1960) Chemical investigations of minerals containing rare elements from the Far East. No. 56. Allanite from Daibosatsu Pass, Yamanashi Prefecture. Nippon Kagaku Zasshi (J Chem Soc Japan, Pure Chem Sect) 81:1238-1239 (in Japanese) [Chem Abstracts 55:7173f]

King JR (2002) The evolution of galactic beryllium and boron traced by magnesium and calcium. Pub Astron Soc Pacific 114:25-28

King VT, Foord EE (1994) Mineralogy of Maine. Volume 1: Descriptive Mineralogy. Maine Geol Surv, Augusta, Maine

King VT, Foord EE (2000) Mineralogy of Maine—addenda to Volume 1. *In* King VT (ed) Mineralogy of Maine. Volume 2: Mining History, Gems, and Geology. Maine Geol Surv, Augusta, p 427-512

Kingsbury AWG (1961) Beryllium minerals in Cornwall and Devon: helvine, genthelvite, and danalite. Mineral Mag 32:921-940

Kingsbury AWG (1964) Some minerals of special interest in south-west England. *In* Hosking KFG, Shrimpton GJ (eds) Present Views of Some Aspects of the Geology of Cornwall and Devon. Royal Geological Society of Cornwall, Penzance, p 247-266

Klaproth MH (1802) Chemische Untersuchung des Gadolinits. *In* Klaproths Beiträge zur Chemische Kenntniss der Mineralkörper 3. HA Rottman, Berlin, p 52-79

Klaska KH, Jarchow O (1977) Die Bestimmung der Kristallstruktur von Trimerit $CaMn_2(BeSiO_4)_3$ und das Trimeritgesetz der Verzwillingung. Z Kristallogr 145:46-65

Klement'yeva LV (1969) A find of bromellite in the USSR. Doklady Akad Nauk SSSR Earth Sci Sections 188:152-154

Kleykamp H (1999) Quantitative X-ray microanalysis of beryllium using a multilayer diffracting device. J Anal Atom Spectrom 14:377-380

Kolitsch U (1996a) Bavenit aus dem Kalksilikatfels der Hohen Waid bei Schriesheim. Aufschluss 47:91-93

Kolitsch U (1996b) Bergslagit aus dem Rhyolithsteinbruch bei Sailauf im Spessart. Mineralien-Welt 7(5):45-46

Komarova GN, Moleva VA, Rudnitskaya ES, Dmitriyeva MT (1967) Bavenite from Transbaikaliya. Izv Akad Nauk SSSR Ser Geol 1967(7):57-66 (in Russian)

Konev AA, Vorob'yev YeI, Sapozhnikov AN, Piskunova LF, Ushchapovskaya ZF (1995) Odintso-vite—$K_2Na_4Ca_3Ti_2Be_4Si_{12}O_{38}$—a new mineral (Murunskiy Massif) Zapiski Vseross Mineral Obshch 124(5):92-96 (in Russian) [Am Mineral 81:1014]

Konev AA, Vorob'yev YeI, Lazebnik KA (1996) Mineralogy of the Murun Alkalic Massif. Izdatel'stvo Siber Otdel Rossiyskoy Akad Nauk, Novosibirsk, Russia

Kopchenova YeV, Sidorenko GA (1962) Bearsite—the arsenic analogue of moraesite. Zap Vsesoyuz Mineral Obshch 91:442-446 (in Russian) [Am Mineral 48:210-211]

Kozhevnikov OK, Dashkevich LM, Zakharov AA, Kashayev AA, Kukhrinkova NV, Sinkevich TP (1975) First taafeite [sic] find in the USSR. Doklady Acad Sci USSR, Earth Sciences Sections, 224:120-121

Kozlova PS (1962) Accessory epididymite and eudidymite in alkaline syenites of the south slope of the Talass ridge. Trudy Mineral Muz Akad Nauk SSSR 13:205-209 (in Russian)

Kristiansen R (1994) Two new minerals for Norway—manganocolumbite and hingganite-(Yb). Stein (Nordic Magazine for Popular Geology) 21(2):88-93 (in Norwegian)

Kristiansen R (1999a) Beryllium minerals in Norway. Norsk Bergverksmuseum Skr 15:34-46 (in Norwegian)

Kristiansen R (1999b) Beryllium minerals in Norway. Stein (Nordic Magazine for Popular Geology) 26(2):8-23 (in Norwegian)

Kuchukova MS (1955) Vesuvianite from the Kara-Tyube Mountains. Zapiski Uzbek Otdel Vsesoyuz Mineral Obshch 8:173-179 (in Russian)

Kupriyanova II, Novikova MI (1976a) Milarite. *In* Ginzburg AI (ed) The Mineralogy of Hydrothermal Beryllium Deposits. Nedra, Moscow, p 105-113 (in Russian)

Kupriyanova II, Novikova MI (1976b) Bavenite. *In* Ginzburg AI (ed) The Mineralogy of Hydrothermal Beryllium Deposits. Nedra, Moscow, p 131-139 (in Russian)

Kupriyanova II, Shpanov YeP, Novikova MI, Zhurkova ZA (1996) Beryllium of Russia: situation, problems of exploration and of industrial development of a raw-materials base. Komitet Rossiyskoy Federatsii po Geologii i Ispol'zovaniyu Nedr, Geologiya, Metody Poiskov, Razvedki i Otsenki Mestorozhdeniy Tverdykh Poleznykh Iskopayemykh: Obsor/AOZT "Geoinformmark", Moscow (in Russian)

Kuschel H (1877) Mittheilungen an Professor G. Leonhard. Milarite. N Jahrb Mineral Geol Paläontol 1877:925-926

Kuz'menko MV (1954) Beryllite—a new mineral. Doklady Akad Nauk SSSR 99:451-454 (in Russian) [Am Mineral 40:787-788]

Kwak TAP, Jackson PG (1986) The compositional variation and genesis of danalite in Sn-F-W skarns, NW Tasmania, Australia. N Jahrb Mineral Monatsh 1986:452-462

Labuntsov AN (1939) Pegmatites of the USSR II. Pegmatites of Northern Karelia and their Minerals. Izdatel'stvo Akad Nauk SSSR, Moscow, Leningrad (in Russian)

Lacroix A (1908a) Sur une nouvelle espèce minérale et sur les minéraux qu'elle accompagne dans les gisements tourmalinifères de Madagascar. C R Acad Sci Paris 146:1367-1371

Lacroix A (1908b) Les minéraux des filons de pegmatite à tourmaline lithique de Madagascar. Bull Soc Franç Minéral 31:218-247

Lacroix A (1922) Minéralogie de Madagascar, vol 1 and 2. Challamel, Paris

Lahlafi M (1997) Rôle des micas dans la concentration des éléments légers (Li, Be et F) dans les granites crustaux: étude expérimentale et cristallochimique. Unpublished Thèse Grade de Docteur, Université d'Orléans

Lahti SI (1981) On the granitic pegmatites of the Eräjärvi area in Orivesi, southern Finland. Geol Surv Finland Bull 314

Lahti SI, Saikkonen R (1985) Bityite $2M_1$ from Eräjärvi compared with related Li-Be brittle micas. Bull Geol Soc Finland 57:207-215

Langhof J, Holtstam D, Gustafsson L (2000) Chiavennite and zoned genthelvite-helvite as late-stage minerals of the Proterozoic LCT pegmatites at Utö, Stockholm, Sweden. GFF 122(2):207-212

Larsen AO (1988) Helvite group minerals from syenite pegmatites in the Oslo Region, Norway, Norway. Contribution to the mineralogy of Norway, No. 68. Norsk Geol Tidsskrift 68:119-124

Larsen AO (1996) Rare earth minerals from the syenite pegmatites in the Oslo Region, Norway. *In* Jones AP, Wall F, Williams CT (eds) Rare Earth Minerals. Chemistry, Origin and Ore Deposits. Mineral Soc Series 7. Chapman & Hall, London, p 151-166

Larsen AO (2000) Brøgger and Reusch; the first tour to Langesundsfjorden. Norsk Bergverksmuseum Skrift 17:35-38 (in Norwegian)

Larsen AO, Åsheim A (1995) Leifite from a nepheline syenite pegmatite on Vesle Arøya in the Langesundsfjord district, Oslo Region, Norway. Norsk Geol Tidsskrift 75:243-246

Larsen AO, Åsheim A, Berge SA (1987) Bromellite from syenite pegmatite, southern Oslo region, Norway. Can Mineral 25:425-428

Larsen AO, Åsheim A, Raade G, Taftø J (1992) Tvedalite, $(Ca,Mn)_4Be_3Si_6O_{17}(OH)_4 \cdot 3H_2O$, a new mineral from syenite pegmatite in the Oslo Region, Norway. Am Mineral 77:438-443

Leavens PB, Dunn PJ, Gaines RV (1978) Compositional and refractive index variations of the herderite–hydroxyl-herderite series. Am Mineral 63:913-917

Leavens PB, White JS, Nelen JA (1990) Zanazziite, a new mineral from Minas Gerais, Brazil. Mineral Rec 21:413-417 [Am Mineral 76:1732]

Lefevre M, Thomas L (1998) Les pegmatites de la vallée de la Sahatany Madagascar. Le Règne Minéral 19:15-28

Levinson AA (1962) Beryllium-fluorine mineralization at Aguachile Mountain. Coahuila, Mexico. Am Mineral 47:67-74

Lindahl I, Grauch RI (1988) Be-REE-U-Sn-mineralization in Precambrian granitic gneisses, Nordland County, Norway. *In* Zachrisson E (ed) Proc 7th Quadrennial IAGOD Symp (Luleå, Sweden). E Schweizerbart'sche Verlagsbuchhandlung, Stuttgart, p 583-594

Lindberg ML (1958) The beryllium content of roscherite from the Sapucaia pegmatite mine. Minas Gerais, Brazil, and from other localities. Am Mineral 43:824-838

Lindberg ML (1964) Crystallography of faheyite, Sapucaia pegmatite mine, Minas Gerais, Brazil. Am Mineral 49:395-398

Lindberg ML, Murata, KJ (1953) Faheyite, a new phosphate mineral from the Sapucaia pegmatite mine, Minas Gerais, Brazil. Am Mineral 38:263-270

Lindberg ML, Pecora WT, Barbosa AL de M (1953) Moraesite, a new hydrous beryllium phosphate from Minas Gerais, Brazil. Am Mineral 38:1126-1133

Liu Y (1990) A synthetic hsianhualite (sic). Int'l Mineral Assoc 15th Gen Meeting 28 June-3 July 1990 Beijing China Abstracts 2:549-550

Liu DR, Williams DB (1989) Accurate composition determination of Be-Ti alloys by electron energy loss spectroscopy. J Microscopy 156:201-210

Liu J, Zeng Y (1998) Preliminary study on fluid inclusions in hsianghualite. Huanan Dizhi Yu Kuangchan (Geology and Mineral Resources of South China) 1(53):56-63 (in Chinese with English abstract, which is reproduced in GeoRef 1999-057072]

Ma R, Zhang J, Yang F (1986) Hellandite of Quyang, in Hebei, China. Dizhi Xuebao [Acta Geol Sinica] 60(1):68-77 (in Chinese with English abstract) [Chem Abstracts 104:227840p]

Machatschki F (1948) Welche Schlüsse sind aus der Existenz von Berylliumorthiten auf die Struktur der Epidote zu ziehen? Tschermaks Mineral Petrogr Mitt 1(1):19-23

Maksimova NV, Ilyukhin VV, Belov NV (1973) Crystal structure of sorensenite. Doklady Akad Nauk SSSR 213:91-93 (in Russian)

Mandarino JA (1964) X-ray powder data for aminoffite Am Mineral 49:212-214

Mandarino JA (1979) The Gladstone-Dale relationship. Part III: Some general applications. Can Mineral 17:71-76

Mandarino JA (1981) The Gladstone-Dale relationship: Part IV. The compatibility concept and its application. Can Mineral 19:441-450

Mandarino JA (1999) Fleischer's Glossary of Mineral Species 1999. The Mineralogical Rec, Tucson, Arizona

Mandarino JA, Anderson V (1989) Monteregian treasures; the minerals of Mont Saint-Hilaire, Quebec. Cambridge University Press, Cambridge

Mandarino JA, Harris DC (1969) Epididymite from Mont St. Hilaire, Quebec. Can Mineral 9:706-709

Marble L, Hanahan J Jr (1978) The Foote minerals. Rocks and Minerals 53(4):158-173

Markl G (2001a) A new type of silicate liquid immiscibility in peralkaline nepheline syenites (lujavrites) of the Ilimaussaq alkaline complex, South Greenland. Contrib Mineral Petrol 141:458-472

Markl G (2001b) Stability of Na-Be minerals in late-magmatic fluids of the Ilímaussaq alkaline complex, South Greenland. Geol Greenland Surv Bull 190:145-158

Markl G, Schumacher JC (1997) Beryl stability in local hydrothermal and chemical environments in a mineralized granite. Am Mineral 82:194-202

Massanek A, Rank K, Weber W (1999) Die mineralogischen Sammlungen des Abraham Gottlob Werner. Lapis 24(9):21-31

Mazzi F, Ungaretti L, Dal Negro A, Petersen OV, Rönsbo JG (1979) The crystal structure of semenovite. Am Mineral 64: 202-210

Meeker GP, Evensen JM (1999) The analysis of Be in geological materials using the electron microprobe. Geol Soc Am Abstr Progr 31(7):A305-306
Meen VB (1939) Vesuvianite from Great Slave Lake region, Canada. Univ Toronto Studies Geol Ser 42:69-74
Meixner H, Paar W (1976) Ein Vorkommen von Väyrynenit-Kristallen aus „Pakistan". Z Kristallogr 143:309-318
Mendeléeff [Mendeleyev] DI (1897) The principles of chemistry (in 2 volumes). Translation of 6th Russian edn. Longmans, Green and Company, London
Men'shikov YuP, Denisov AP, Uspenskaya YeI, Lipatova EA (1973) Lovdarite—a new hydrous beryllosilicate of alkalies. Doklady Akad Nauk SSSR 213:429-432 (in Russian) [Am Mineral 59:874]
Men'shikov YuP, Pakhomovskiy YaA, Yakovenchuk VN (1999) Beryllium mineralization in veins of the Khibiny Massif. Zapiski Vseross Mineral Obshch 128:3-14 (in Russian)
Mereiter K, Niedermayr G, Walter F (1994) Uralolite, $Ca_2Be_4(PO_4)_3(OH)_3{\cdot}5H_2O$: new data and crystal structure. Eur J Mineral 6:887-896 [Am Mineral 80:1333]
Merlino S (1990) Lovdarite, $K_4Na_{12}(Be_8Si_{28}O_{72}){\cdot}18H_2O$, a zeolite-like mineral: structural features and OD character. Eur J Mineral 2:809-817 [Am Mineral 77: 212]
Merlino S, Pasero M (1992) Crystal chemistry of beryllophosphates: The crystal structure of moraesite, $Be_2PO_4(OH){\bullet}4H_2O$. Z Kristallogr 201:253-262
Metalidi VS (1988) Phenakite in basaltoid rocks of the Ovruch graben (Ukrainian Shield). Dopovidi Akad Nauk Ukrain RSR Ser B Geol Khim Biol Nauki 1988(1):12-14 (in Ukrainian) [Chem Abstracts 109:234261g]
Metcalf-Johansen J, Hazell RG (1976) The crystal structure of sorensenite, $Na_4SnBe_2(Si_3O_9)_2{\cdot}2H_2O$. Acta Crystallogr B32:2553-2556
Micheelsen H, Petersen OV (1970) Leifite, revised, and karpinskyite, discredited. Bull Geol Soc Denmark 20:134-151 [Am Mineral 57:1006]
Miller RR (1996) Structural and textural evolution of the Strange Lake peralkaline rare-element (NYF) granitic pegmatite, Quebec-Labrador. Can Mineral 34:349-371
Minakawa T, Yoshimoto Y (1998) Gugiaite from the Yuge Island, Ehime Prefecture. Abstr Annual Meeting Mineral Soc Japan 1997, Mineral J 20(1):27-28
Miyawaki R, Nakai I (1996) Crystal chemical aspects of rare earth minerals. *In* Jones AP, Wall F, Williams CT (eds) Rare Earth Minerals Chemistry, Origin and Ore Deposits. Mineral Soc Series 7. Chapman & Hall, London, p 21-40
Miyawaki R, Nakai I, Nagashima K, Okamoto A, Isobe T (1987) The first occurrences of hingganite, hellandite and wodginite in Japan. Kobutsugaku Zasshi 18(1):17-30 (in Japanese) [Am Mineral 75:432]
Miyawaki R, Kuriyama J, Nakai I (1993) The redefinition of tengerite-(Y), $Y_2(CO_3)_3{\cdot}2–3H_2O$, and its crystal structure. Am Mineral 78:425-432
Moberg JC (1898) Zur Kenntniss des Steenstrupins. Z Kristallogr Mineral 29:386-398
Moberg JC (1899) Contribution to the knowledge of steenstrupine. Meddel Grønland 20:245-263 (in Danish; nearly identical to Moberg 1898)
Moenke HHW (1974) Silica, the three-dimensional silicates, borosilicates and beryllium silicates. *In* Farmer VC (ed) The Infrared Spectra of Minerals, Monograph 4, Mineralogical Society, London, p 365-382
Montoya JW, Havens R, Bridges DW (1962) Beryllium-bearing tuff from Spor Mountain, Utah: its chemical, mineralogical and physical properties. U S Bur Mines Rep Invest 6084:1-15
Montoya JW, Baur GS, Wilson SR (1964) Mineralogical investigation of beryllium-bearing tuff, Honeycomb Hills, Juab County, Utah. U S Bur Mines Rep Invest 6408:1-11
Moor R, Oberholzer WF, Gübelin E (1981) Taprobanite, a new mineral of the taaffeite-group. Schweiz mineral petrogr Mitt 61:13-21 [Am Mineral 67:1076]
Moore PB (1967) Eleven new minerals from Långban, Sweden. Can Mineral 9:301 (abstract)
Moore PB (1968a) Relations of the manganese-calcium silicates, gageite and harstigite. Am Mineral 53:309-315
Moore PB (1968b) Relations of the manganese-calcium silicates, gageite and harstigite: A correction. Am Mineral 53:1418-1420
Moore PB (1969a) Joesmithite, a new amphibole-like mineral from Långban. Arkiv Mineral Geol 4: 487-492
Moore PB (1969b) Joesmithite: A novel amphibole crystal chemistry. Mineral Soc Am Spec Paper 2: 111-115
Moore PB (1971) Mineralogy & chemistry of Långban-type deposits in Bergslagen, Sweden. Mineral Rec 1(4):154-172

Moore PB (1978) Welshite, $Ca_2Mg_4Fe^{3+}Sb^{5+}O_2[Si_4Be_2O_{18}]$, a new member of the aenigmatite group. Mineral Mag 42:129-132

Moore PB (1988) The joesmithite enigma: Note on the $6s^2Pb^{2+}$ lone pair. Am Mineral 73:843-844

Moore PB, Araki T, Ghose S (1982) Hyalotekite, a complex lead borosilicate: its crystal structure and the lone-pair effect of Pb(II). Am Mineral 67:1012-1020

Moore PB, Araki T, Steele IM, Swihart GH, Kampf AR (1983) Gainesite, sodium zirconium beryllophosphate: a new mineral and its crystal structure. Am Mineral 68:1022-1028

Moore PB, Davis AM, Van Derveer DG, Sen Gupta PK (1993) Joesmithite, a plumbous amphibole revisited and comments on bond valences. Mineral Petrol 48:97-113

Mrose ME (1952) Hurlbutite, $CaBe_2(PO_4)_2$, a new mineral. Am Mineral 37:931-940

Mrose ME, von Knorring O (1959) The mineralogy of väyrynenite, $(Mn,Fe)Be(PO_4)(OH)$. Z Kristallogr 112:275-288

Nagashima K (1951) Chemical investigations of Japanese minerals containing rarer elements. 42. Be-bearing allanite. Nippon Kagaku Zasshi (J Chem Soc Japan, Pure Chem Sect) 72:52-53 (in Japanese) [Chem Abstracts 46: 3464e]

Nakai T (1938) On calcio-gadolinite, a new variety of gadolinite found in Tadati Village, Nagano Prefecture. Bull Chem Soc Japan 13(9):591-594 [Am Mineral 25:312-313]

Nazarova AS, Kuznetsova NN, Shashkin DP (1966) Babefphite—a barium-beryllium fluoride-phosphate. Doklady Akad Nauk SSSR 167:895-897 (in Russian) [Am Mineral 51:1547]

Nazarova AS, Solntseva LS, Yurkina KV (1975) Babefphite. Trudy Mineral Muz Akad Nauk SSSR 24:191-195 (in Russian)

Nedashkovskiy PG (1983) Phenakite-bearing alkaline pegmatites—a new genetic type of beryllium mineralization. Doklady Akad Nauk SSSR 271(1):157-158

Nefedov YeI (1967) Berborite, a new mineral. Doklady Akad Nauk SSSR Earth Sci Sect 174:114-117 [Am Mineral 53:348-349]

Nefedov EI, Griffin WL, Kristiansen R (1977) Minerals of the schoenfliesite-wickmanite series from Pitkäranta, Karelia, U.S.S.R. Can Mineral 15:437-445

Neumann H, Svinndal S (1955) The cyprin-thulite deposit at Øvstebø, near Kleppan in Sauland, Telemark, Norway. Norsk Geol Tidsskrift 34:139-156

Ni Y, Yang Y (1992) Väyrynenite, a rare beryllium mineral. Acta Petrol Mineral 11(3):252-258 (in Chinese with English abstract)

Nickel EH (1963) Eudidymite from Seal Lake, Labrador, Newfoundland. Can Mineral 7:643-649

Nickel EH, Grice JD (1998) The IMA Commission on New Minerals and Mineral Names: Procedures and guidelines on mineral nomenclature, 1998. Can Mineral 36:913-926

Nijland TG, Zwaan JC, Touret L (1998) Topographical mineralogy of the Bamble sector, south Norway. Scripta Geologica 118:1-46

Nikol'skaya ZhD, Larin AM (1972) Greisens of the Pitkäranta ore field. Zapiski Vsesoyuz Mineral Obshch 101:290-297 (in Russian)

Nimis P, Molin G, Visonà D (1996) Crystal chemistry of danalite from Daba Shabeli Complex (N Somalia). Mineral Mag 60:375-379

Nordenskiöld AE (1877) New minerals from Långban. Geol Fören Stockholm Förhandl 3:376-384 (in Swedish)

Nordenskiöld AE (1887) Mineralogical Contribution. 17. The correct composition of eudidymite. Geol Fören Stockholm Förhandl 9:434-436 (in Swedish)

Nordenskjöld N (1834) Description of phenakite, a new mineral from the Urals. Kongl Vetenkaps-Academiens Hand för år 1833:160-165, Tab VI (in Swedish)

Nova G (1987) Atlas of the Minerals of Baveno. Gruppo Mineralogico Lombardo, Milan (in Italian)

Novák M, Korbel P, Odehnal F (1991) Pseudomorphs of bertrandite and epididymite after beryl from Věžná, western Moravia, Czechoslovakia. N Jahr Mineral Monatsh 1991:473-480

Novák M, Burns PC, Morgan GB VI (1998) Fluorine variation in hambergite from granitic pegmatites. Can Mineral 36:441-446

Novikova MI (1967) Processes of alteration of phenakite. Zapiski Vsesoyuz Mineral Obshch 96:418-424 (in Russian)

Novikova MI (1976a) Bertrandite. *In* Ginzburg AI (ed) The Mineralogy of Hydrothermal Beryllium Deposits. Nedra, Moscow, p 78-88 (in Russian)

Novikova MI (1976b) Beryl. *In* Ginzburg AI (ed) The Mineralogy of Hydrothermal Beryllium Deposits. Nedra, Moscow, p 88-105 (in Russian)

Novikova MI (1976c) Euclase. *In* Ginzburg AI (ed) The Mineralogy of Hydrothermal Beryllium Deposits. Nedra, Moscow, p 70-74 (in Russian)

Novikova MI (1976d) Phenakite. *In* Ginzburg AI (ed) The Mineralogy of Hydrothermal Beryllium Deposits. Nedra, Moscow, p 57-69 (in Russian)

Novikova MI (1976e) Characteristics of the development of meliphanite, milarite and bavenite in a fluorite-phenakite-bertrandite deposit. *In* Ginzburg AI (ed) Ocherki po Geneticheskoy Mineralogii. Nauka, Moscow, p 59-65 (in Russian)

Novikova MI (1984) Leucophanite, meliphanite and eudidymite in a deposit of the fluorite-phenakite-bertrandite association. Mineral Zhurnal 6(5):84-90 (in Russian)

Novikova MI, Sidorenko GA, Shatskaya VT (1975) Composition and structural features of melinophane. *In* Mineraly i Paragenezisy Mineralov Endogennykh Mestorozhdenii (Collection of papers). Nauka, Leningrad, p 49-53 (in Russian)

Novikova MI, Shpanov YeP, Kupriyanova II (1994) Petrology of the Yermakovskoye beryllium deposit, western Transbaikalia. Petrologiya 2(1):114-127 (in Russian)

Nowacki W, Phan KD (1964) Composition quantitative de la bazzite de Val Strem (Suisse) détérminée par la microsonde électronique de Castaing. Bull Soc Franç Minéral Cristallogr 87:453 [Am Mineral 52:563-564]

Nysten P (1997) Paragenetic setting and crystal chemistry of milarites from Proterozoic granitic pegmatites in Sweden. N Jahrb Mineral Monatsh 1996:564-576

Nysten P (2000) Swedish mineral (4) Trimerite—the salmon-pink triplets. Geol Forum 28:22-24 (in Swedish)

Nysten P, Gustafsson L (1993) Beryllium phosphates from the Proterozoic granitic pegmatite at Norrö, southern Stockholm archipelago, Sweden. Geol Fören Stockholm Förhandl 115(2):159-164

Nysten P, Holtstam D, Jonsson E (1999) The Långban minerals. *In* Holtstam D, Langhof J (eds) Långban. The Mines, Their Minerals, Geology and Explorers, p 89-183 Swedish Museum of Natural History, Raster Förlag, Stockholm

Oberti R, Ottolini L, Camara F, Della Ventura G (1999) Crystal structure of non-metamict Th-rich hellandite-(Ce) from Latium (Italy) and crystal chemistry of the hellandite-group minerals. Am Mineral 84:913-921

Oberti R, Della Ventura G, Ottolini L, Hawthorne FC, Bonazzi P (2002) Re-definition, nomenclature and crystal-chemistry of the hellandite group. Am Mineral 87:745-752

Oftedal I (1972) Calcium-rich gadolinite from Kragerø. Norsk Geol Tidsskrift 52:197-200

Okrusch M (1971) Zur Genese von Chrysoberyll- und Alexandrit-Lagerstätten. Eine Literaturübersicht. Z Deutsch Gemmol Gesellschaft 20(3):114-124

Ottolini L, Oberti R (2000) Accurate quantification of H, Li, Be, B, F, Ba, REE, Y, Th, and U in complex matrixes: A combined approach based on SIMS and single-crystal structure refinement. Anal Chem 72:3731-3738

Pagel BEJ (1991) Beryllium and the Big Bang. Nature 354:267-268

Palache C (1931) On the presence of beryllium in milarite. Am Mineral 16:469-470

Palache C (1935) The minerals of Franklin and Sterling Hill, Sussex County, New Jersey. U.S. Geol Survey Prof Paper 180:1-135

Palache C, Bauer LH (1930) On the occurrence of beryllium in the zinc deposits of Franklin, New Jersey. Am Mineral 15:30-33.

Palache C, Berman H, Frondel C (1944) The System of Mineralogy of James Dwight Dana and Edward Salisbury Dana, Yale University. 7th ed, v. I. Elements, Sulfides, Sulfosalts, Oxides. Wiley, New York

Palache C, Berman H, Frondel C (1951) The System of Mineralogy of James Dwight Dana and Edward Salisbury Dana, Yale University. 7th ed, v. II. Halides, Nitrates, Borates, Carbonates, Sulfates, Phosphates, Arsenates, Tungstates, Molybdates, etc. Wiley, New York

Pankrath R, Langer K (2002) Molecular water in beryl, $^{VI}Al_2$ $[Be_3Si_6O_{18}]\cdot nH_2O$, as a function of pressure and temperature: An experimental study. Am Mineral 87:238-244

Parizot E, Drury L (1999) Spallative nucleosynthesis in supernova remnants. I. Analytical estimates. Astron Astrophys 346:329-339

Parker RL, Indergand P (1957) Kurze Mitteilungen zur Mineralogie der Schweiz 8. Ein neues schweizerisches Vorkommen von Bertrandit. Schweiz mineral petrogr Mitt 37:554-558

Passaglia E (1969a) Roggianite, nuovo minerale silicato. Rendiconti Soc Ital Mineral Petrol 25:105-106 (in Italian)

Passaglia E (1969b) Roggianite, a new silicate mineral. Clay Minerals 8:107-111 [Am Mineral 55:322-323]

Passaglia E, Vezzalini G (1988) Roggianite: revised chemical formula and zeolitic properties. Mineral Mag 52:201-206 [Am Mineral 74:505]

Pautov LA, Khvorov PV, Sokolova YeV, Ferraris G, Ivaldi G, Bazhenova LF (2000) Kapitsaite-(Y) $(Ba, K)_4(Y, Ca)_2Si_8(B, Si)_4O_{28}F$—A new mineral. Zapiski Vseross Mineral Obshch 129(6):42-49 [Am Mineral 86:1535]

Peacor DR, Simmons WB Jr, Essene EJ, Heinrich EW (1982) New data on and discreditation of "texasite," "albrittonite," "cuproartinite," "cuprohydromagnesite," and "yttromicrolite," with corrected data on nickelbischofite, rowlandite, and yttrocrasite. Am Mineral 67:156-169

Peacor DR, Dunn PJ, Roberts WL, Campbell TJ, Newbury D (1983) Fransoletite, a new calcium beryllium phosphate from the Tip Top pegmatite, Custer, South Dakota. Bull Minéral 106: 499-503 [Am Mineral 70: 215]

Peacor DR, Rouse RC, Ahn J-H (1987) Crystal structure of tiptopite, a framework beryllophosphate isotypic with basic cancrinite. Am Mineral 72:816-820

Pekov IV (1994) Remarkable finds of minerals of beryllium: from the Kola Peninsula to Primorie. World of Stones (Mir Kamnya) 4:10-26 (English), 3-12 (Russian)

Pekov IV (1995) Moraesite from an alkalic pegmatite of the Lovozero Massif. Ural'skiy Mineral Sbornik Ross Akad Nauk Ural'skoye Otdeleniye 5:256-260 (in Russian)

Pekov IV (1998) Minerals first discovered on the territory of the former Soviet Union. Ocean Pictures, Moscow

Pekov IV (2000) Lovozero Massif: History, Pegmatites, Minerals. Ocean Pictures, Moscow

Pekov IV, Voloshin AV, Pushcharovskiy Dyu, Rastsvetayeva RK, Chukanov NV, Belakovskiy DI (2000) New data on calcybeborosilite-(Y) (REE, Ca)$_2$(B, Be)$_2$[SiO$_4$]$_2$(OH,O)$_2$. Moscow Univ Geol Bull 55(2):62-70 [Am Mineral 86:1537]

Pekov IV, Chukanov NV, Larsen AO, Merlino S, Pasero M, Pushcharovsky DYu, Ivaldi G, Zadov AE, Grishin VG, Åsheim A, Taftø J, Chistyakova NI (in press) Sphaerobertrandite, $Be_3SiO_4(OH)_2$: new data, crystal structure and genesis. Eur J Mineral

Penfield SL (1894) On the crystallization of herderite. Am J Sci 147:329-339

Peng C-J, Tsao R-L, Zou Z-R (1962) Gugiaite, $Ca_2BeSi_2O_7$, a new beryllium mineral and its relation to the melilite group. Scientia Sinica 11:977-988 [Am Mineral 48:211-212]

Peng C, Chang Y, Chang K (1964) Crystallographic features of hsianghualite. Int'l Geol Rev 10:900-904

Perez J-B, Dusausoy Y, Babkine J, Pagel M (1990) Mn zonation and fluid inclusions in genthelvite from the Taghouaji complex (Aïr Mountains, Niger). Am Mineral 75:909-914

Petersen OV, Rønsbo JG (1972) Semenovite—a new mineral from the Ilímaussaq alkaline intrusion, south Greenland. Lithos 5:163-173 [Am Mineral 58:1114]

Petersen OV, Secher K (1993) The minerals of Greenland. Mineral Rec 24(2): 4-67

Petersen OV, Randløv J, Leonardsen ES, Rønsbo JG (1991) Barylite from the Ilímaussaq alkaline complex and associated fenites, South Greenland. N Jahrb Mineral Monatsh 1991:212-216

Petersen OV, Rønsbo JG, Leonardsen ES (1994a) Hingganite-(Y) from the Zomba-Malosa complex, Malawi. N Jahrb Mineral Monatsh 1994(4):185-192

Petersen OV, Rønsbo JG, Leonardsen ES, Johnsen O, Bollingberg H, Rose-Hansen J (1994b) Leifite from the Ilímaussaq alkaline complex, South Greenland. N Jahrb Mineral Monatsh 1994(2):83-90

Petersen OV, Micheelsen HI, Leonardsen ES (1995) Bavenite, $Ca_4Be_3Al[Si_9O_{25}(OH)_3]$, from the Ilímaussaq alkaline complex, South Greenland. N Jahrb Mineral Monatsh 1995(7):321-335

Petersen OV, Medenbach O, Bollhorn J (1997) Epididymite twins. Contribution to the mineralogy of Ilímaussaq, No. 99. N Jahrb Mineral Monatsh 1997(5):221-228

Petersen OV, Gault RA, Balić-Žunić T (2001) Odintsovite from the Ilímaussaq alkaline complex, South Greenland. N Jahrb Mineral Monatsh 2001(5):235-240

Petersen OV, Niedermayr G, Johnsen O, Gault RA, Brandstätter F (2002a) Lovdarite from the Ilímaussaq alkaline complex, South Greenland. N Jahrb Mineral Monatsh 2002(1):23-30

Petersen OV, Giester G, Brandstätter F, Niedermayr G (2002b) Nabesite, $Na_2BeSi_4O_{10}{\cdot}4H_2O$, a new mineral from the Ilímaussaq alkaline complex, South Greenland. Can Mineral 40:173-181

Pezzotta F (1999) Madagaskar. Ein Paradies voll mit Mineralien und Edelsteinen. ExtraLapis 17:1-96

Pezzotta F, Diella V, Guastoni A (1999) Chemical and paragenetic data on gadolinite-group minerals from Baveno and Cuasso al Monte, southern Alps, Italy. Am Mineral 84:782-789

Pilati T, Gramaccioli CM, Pezzotta F, Fermo P, Bruni S (1998) Single-crystal vibrational spectrum of phenakite, Be_2SiO_4, and its interpretation using a transferable empirical force field. J Phys Chem A 102:4990-4996

Plyusnina II (1963) Infrared absorption spectra of beryllium minerals. Geochemistry 2:174-190

Pokrovskiy PV, Grigor'yev NA, Potashko KA, Ayzikovich AN (1963) Moraesite from the Urals. Zapiski Vsesoyuz Mineral Obshch 92:232-239 (in Russian)

Pokrovskiy PV, Grigor'yev NA, Potashko KA (1965) Secondary beryllium phosphates and their distribution in the weathering crust on mica-fluorite greisens. Trudy Inst Geol Ural'sk Filial Akad Nauk SSSR 70:205-209 (in Russian)

Polyakov VO, Cherepivskaya Gye, Shcherbakova YeP (1986) Makarochkinite—a new beryllosilicate. *In* New and Little-Studied Minerals and Mineral Associations of the Urals. Akad Nauk SSSR Ural'skiy Nauchnyy Tsentr, Sverdlovsk, p 108-110 (in Russian)

Portnov AM (1964) Leucophanite from the northern Baikal Region. Trudy Mineral Muz Akad Nauk SSSR 15:229-231 (in Russian)

Povarennykh AS (1972) Crystal Chemical Classification of Minerals. Plenum, New York

Povarennykh AS, Nefedov YeI (1971) Infrared spectra of some beryllium and boron minerals. Geol Zhurn 31(5):13-27 (in Russian)

Povarennykh AS, Keller P, Kristiansen R (1982) Infrared absorption spectra of swedenborgite and queitite. Can Mineral 20:601-603

Povondra P, Langer K (1971) Synthesis and some properties of sodium-beryllium-bearing cordierite, $Na_xMg_2(Al_{4-x}Be_xSi_5O_{18})$. N Jahrb Mineral Abh 116:1-19

Preuss E, von Gliszczynski S (1950) Über den Berylliumgehalt einiger Wavellite. Geochim Cosmochim Acta 1:86-88

Price RC (1888) Analysis of tscheffkinite [chevkinite] from Nelson County, Virginia. Am Chem J 10:38-39

Primas F, Asplund M, Nissen PE, Hill V (2000a) The beryllium abundance in the very metal-poor halo star G 64-12 from VLT/UVES observations. Astron Astrophys 364:L42-L46

Primas F, Molaro P, Bonifacio P, Hill V (2000b) First UVES observations of beryllium in very metal-poor stars. Astron Astrophys 362:666-672

Pring A, Din VK, Jefferson DA, Thomas JM (1986) The crystal chemistry of rhodizite: a re-examination. Mineral Mag 50:163-172 [Am Mineral 72:1028; 73:194]

Quensel P, Alvfeldt O (1945) Berylliumorthite (muromontite) from Skuleboda feldspar quarry. Arkiv Kemi Mineralogi Geol 18A(22):1-17 (in Swedish with English summary)

Raade G (1996) Minerals originally described from Norway. Including notes on type material. Norsk Bergverksmuseum Skriftserie Skrift 11

Raade G, Kristiansen R (2000) Mineralogy and geochemistry of the Heftetjern granite pegmatite, Tørdal: a progress report. Norsk Bergverksmuseum Skr 17:19-25

Raade G, Haug J, Kristiansen R, Larsen AO (1980) Langesundsfjord. Lapis 5(10):22-28 (in German)

Raade G, Åmli R, Mladeck MH, Din VK, Larsen AO, Åsheim A (1983) Chiavennite from syenite pegmatites in the Oslo Region, Norway. Am Mineral 68:628-633

Ragu A (1994) Helvite from the French Pyrénées as evidence for granite-related hydrothermal activity. Can Mineral 32:111-120

Raimbault L, Bilal E (1993) Trace-element contents of helvite-group minerals from metasomatic albitites and hydrothermal veins at Sucuri, Brazil and Dajishan, China. Can Mineral 31:119-127

Raines E (2001) A brief summary of the mineral deposits of the Pikes Peak Batholith, Colorado. Rocks and Minerals 76:298-325

Ramaty R, Vangioni-Flam E, Cassé M, Olive KA, eds (1998) LiBeB, cosmic rays and related X- and gamma-rays. Astron Soc Pacific Conference Series 171

Rastsvetaeva RK, Rekhlova OYu, Andrianov VI, Malinovskii YuA (1991) Crystal structure of hsianghualite. Sov Phys Doklady 36(1):11-13)

Rastsvetaeva RK, Evsyunin VG, Kashaev AA (1995) Crystal structure of a new natural K,Na,Ca-titanoberyllosilicate. Crystallogr Reports 40(2):228-232 [Am Mineral 80:1332]

Rastsvetaeva RK, Pushcharovskii DYu, Pekov IV, Voloshin AV (1996) Crystal structure of "calcybeborosilite" and its place in the datolite-gadolinite isomorphous series. Crystallogr Reports 41(2):217-221

Ratney RS (2001) Is beryllium disease a fossil? – Not yet. Int'l Archives Occup Environ Health 74:159-161

Raudsepp M (1995) Recent advances in the electron-probe micro-analysis of minerals for the light elements. Can Mineral 33:203-218

Rieder M, Cavazzini G, D'yakonov YuS, Frank-Kamenetskii VA, Gottardi G, Guggenheim S, Koval' PV, Müller G, Neiva AMR, Radoslovich EW, Robert J-L, Sassi FP, Takeda H, Weiss Z, Wones DR (1999) Nomenclature of the micas. Mineral Mag 63:267-279

Rio S, Métrich N, Mosbah M, Massiot P (1995) Lithium, boron and beryllium in volcanic glasses and minerals studied by nuclear microprobe. Nucl Instrum Methods Phys Research B 100:141-148

Robinson GW, Grice JD, Van Velthuizen J (1985) Ehrleite, a new calcium beryllium zinc phosphate hydrate from the Tip Top pegmatite, Custer, South Dakota. Can Mineral 23:507-510 [Am Mineral 71:1544]

Robinson GW, King VT, Asselborn E, Cureton F, Tschernich R, Sielecki R (1992) What's new in minerals? Annual world summary of mineral discoveries covering April 1991 through April 1992. Mineral Rec 23:423-437

Roda E, Fontan F, Pesquera A, Velasco F (1996) The phosphate mineral association of the granitic pegmatites of Fregeneda area (Salamanca, Spain). Mineral Mag 60:767-778

Rønsbo JG, Petersen OV, Leonardsen ES (1990) Vinogradovite from the Ilímaussaq alkaline complex, South Greenland, a beryllium bearing mineral. N Jahrb Mineral Monatsh 1990:481-492.

Rose G (1834) Ueber den Rhodizit, eine neue Mineralgattung. Poggendorff Ann Phys Chem 33:253-256

Rose G (1836) Fernere Bemerkungen über den Rhodizit. Poggendorff Ann Phys Chem 39:321-323

Ross M (1964) Crystal chemistry of beryllium. U S Geol Survey Prof Paper 468:1-30

Ross SD (1974) Borates. *In* Farmer VC (ed) The Infrared Spectra of Minerals, Monograph 4, Mineralogical Society, London, p 205-226

Ross SD (1974) Phosphates and other oxy-anions of group V. *In* Farmer VC (ed) The Infrared Spectra of Minerals, Monograph 4, Mineralogical Society, London, p 383-422

Rossi P, Bellatreccia F, Caprilli E, Parodi GC, Della Ventura G, Mottana A (1995) A new occurrence of rare minerals in an ejectum in the pyroclastics of Vico Volcano, Roman Comagmatic Region, Italy. Atti Accad Nazionale Lincei Rendiconti Classe Sci Fis Matemat Nat Ser 9, 6:147-156

Rouse RC, Peacor DR, Dunn PJ, Campbell TJ, Roberts WL, Wicks FJ, Newbury D (1987) Pahasapaite, a beryllophosphate zeolite related to synthetic zeolite rho, from the Tip Top Pegmatite of South Dakota. N Jahrb Mineral Monatsh 1987:433-440 [Am Mineral 73:1496]

Rouse RC, Peacor DR, Merlino S (1989a) Crystal structure of pahasapaite, a beryllophosphate mineral with a distorted zeolite rho framework. Am Mineral 74:1195-1202

Rouse RC, Peacor DR, Metz GW (1989b) Sverigeite, a structure containing planar NaO_4 groups and chains of 3- and 4-membered beryllosilicate rings. Am Mineral 74:1343-1350

Rouse RC, Peacor DR, Dunn PJ, Su S-C, Chi PH, Yeates H (1994) Samfowlerite, a new Ca Mn Zn beryllosilicate mineral from Franklin, New Jersey: Its characterization and crystal structure. Can Mineral 32:43-53 [Am Mineral 80:185]

Rowledge HP, Hayton, JD (1948) 2. – Two new beryllium minerals from Londonderry. J Royal Soc West Austral 33:45-52 ["Bowleyite", Am Mineral 35:1091]

Rucklidge JC, Kocman V, Whitlow SH, Gabe EJ (1975) The crystal structures of three Canadian vesuvianites. Can Mineral 13:15-21

Ryback G, Clarke AM, Stanley CJ (1998) Re-examination of the A.W.G. Kingsbury collection of British minerals at the Natural History Museum, London. Geol Curator 6(9):317-322

Ryback G, Hart AD, Stanley CJ (2001) A.W.G. Kingsbury's specimens of British minerals. Part 1. Some examples of falsified localities. J Russell Soc 7(2):51-69

Sabor M (1990) Seltene Mineralien vom Brandrücken auf der Koralpe in Österreich. Lapis 15(11): 27-31, 74

Sacerdoti M, Parodi GC, Mottana A, Maras A, della Ventura G (1993) Asbecasite: crystal structure refinement and crystal chemistry. Mineral Mag 57:315-322

Sæbö PC (1966) Contributions to the mineralogy of Norway No. 35. The first occurrences of the rare mineral barylite, $Be_2BaSi_2O_7$, in Norway. Norsk Geol Tidsskrift 46:335-348

Santos JF, Schärer U, Gil Ibarguchi JI, Girardeau J (2002) Genesis of pyroxenite-rich peridotite at Cabo Ortegal (NW Spain): Geochemical and Pb-Sr-Nd isotope data. J Petrol 43:17-43

Schäfer RW (1895) Ueber die metamorphen Gabbrogesteine des Allalingebietes im Wallis zwischen Zermatt- und Saasthal. Tschermaks Mineral Petrograph Mitt 15: 91-134

Scheerer T (1852) Melinophan, eine neue Mineralspecies. J Prakt Chemie 55:449-451

Schetelig J (1922) Gadolinit. *In* Brøgger WC, Vogt T, Schetelig J (eds) Die Mineralien der südnorwegischen Granitpegmatitgänge. II. Silikate der Seltenen Erden (Y-Reihe und Ce-Reihe). Videnskapsselskapets Skrifter I. Mat-Naturv Klasse 1:88-123 (+ figures 12-23 and plates XIV-XV)

Schlatti M (1967) Synthese und Strukturtyp des Berylliumborates $Be_2BO_3(OH)\cdot H_2O$. Naturwiss 54:587

Schlatti M (1968) Hydrothermalsynthese und Strukturtyp des Berylliumborates $^2_\infty$ $Be_2BO_3OH.H_2O$. Tschermaks Mineral Petrogr Mitt 12:463-469

Schmetzer K (1983) Crystal chemistry of natural Be-Mg-Al oxides: taaffeite, taprobanite, musgravite. N Jahrbuch Mineral Abh 146:15-28 [Am Mineral 69:215]

Schmetzer K, Bernhardt H-J, Biehler R (1991) Emeralds from the Ural Mountains, USSR. Gems Gemol 27(2):86-99

Schneiderhöhn H (1941) Lehrbuch der Erzlagerstättenkunde. Erster Band. Die Lagerstätten magmatischen Abfolge. Gustav Fischer, Jena

Schnorrer-Köhler G, Rewitzer C (1991) Bendada—ein Phosphatpegmatit im Mittelteil Portugals. Lapis 16(5):21-33, 50

Schwenzfeier CW Jr (1955) The sulfate extraction of beryllium from beryl. *In* White DW Jr, Burke JE (eds) The Metal Beryllium. The American Society for Metals, Cleveland, p 71-101

Segalstad TV, Larsen AO (1978) Gadolinite-(Ce) from Skien, southwestern Oslo region, Norway. Am Mineral 63:188-195

Semenov YeI (1957) New hydrous silicates of beryllium-gelbertrandite and sphaerobertrandite. Trudy Inst Mineral Geokhim Kristallokhim Redkikh Elementov 1:64-69 (in Russian) [Am Mineral 43:1219-1220]

Semenov YeI (1969) Mineralogy of the Ilímaussaq Alkaline Massif (South Greenland). Moscow, Nauka (in Russian)

Semenov YeI (1972) Mineralogy of the Lovozero Alkaline Massif. Moscow, Nauka (in Russian)

Semenov YeI, Bykova AV (1960) Beryllosodalite. Doklady Akad Nauk SSSR 133:1191-1193 (in Russian) [Am Mineral 46:241]

Semenov EI, Sørensen H (1966) The mineralogy of Ilímaussaq. II. Eudidymite and epididymite from the Ilímaussaq intrusion, south Greenland. Meddel Grønland 181(2):1-21

Semenov YeI, Dusmatov VD, Samsonova NS (1963) Yttrium-beryllium minerals of the datolite group. Kristallografia 8:677-679 (in Russian; English translation Sov Phys Crystal 8: 539-541)

Semenov YeI, Gerassimovsky VI, Maksimova NV, Andersen S, Petersen OV (1965) Sorensenite, a new sodium-beryllium-tin-silicate from the Ilímaussaq Intrusion, South Greenland. Meddel Grønland 181(1):1-19 [Am Mineral 51:1547-1548]

Semenov YeI, Bose H, Sørensen H, Katayeva ZT (1987) Leucophane in alkaline pegmatites of Ilímaussaq. Mineral Zhurnal 9(2):84-85 (in Russian)

Serdyuchenko DP, Ganzeyeva LV (1976) Idocrase from Pitkyaranta, composition and isomorphism. Doklady Akad Nauk SSSR Earth Sci Sections 231:128-130

Sergeyev AD (1973) Bromellite find in metasomatic rocks of a tin deposit in Transbaikal. Doklady Acad Sci USSR Earth Sci Sections 213:151-153

Seybert H (1824) Analyses of the chrysoberyls from Haddam and Brazil. Am J Sci 8:105-112

Sharma KK, Ganguli D (1973) Crystallization of phenakite structures with Zn-Be and Mg-Be substitutions. Ceramurgia 3(3):155-158 (in Italian and English)

Sharp WN (1961) Euclase in greisen pipes and associated deposits, Park County, Colorado. Am Mineral 46:1505-1508

Shatskaya VT (1976a) Gadolinite. *In* Ginzburg AI (ed) The Mineralogy of Hydrothermal Beryllium Deposits. Nedra, Moscow, p 74-78 (in Russian)

Shatskaya VT (1976b) Leucophanite-meliphanite group. *In* Ginzburg AI (ed) The Mineralogy of Hydrothermal Beryllium Deposits. Nedra, Moscow, p 113-119 (in Russian)

Shatskaya VT, Zhdanov RG (1969) New data on leucophanite. Trudy Mineral Muz Akad Nauk SSSR 19:239-241 (in Russian)

Sherriff BL, Grundy HD, Hartman JS, Hawthorne FC, Černý P (1991) The incorporation of alkalis in beryl: multi-nuclear MAS NMR and crystal-structure study. Can Mineral 29:271-285

Shilin LL (1956) Karpinskyite—a new mineral. Doklady Akad Nauk SSSR 107:737-739 (in Russian) [Am Mineral 42:119-120]

Shilin LL, Semenov YeI (1957) The beryllium minerals epididymite and eudidymite in alkalic pegmatites of the Kola Peninsula. Doklady Akad Nauk SSSR 112:325-328 (in Russian)

Shpanov YeP (1976) Chrysoberyl. *In* Ginzburg AI (ed) The Mineralogy of Hydrothermal Beryllium Deposits. Nedra, Moscow, p 49-55 (in Russian)

Shpanov YeP, Kupriyanova II, Novikova MI (1976) Vesuvianite. *In* Ginzburg AI (ed) The Mineralogy of Hydrothermal Beryllium Deposits. Nedra, Moscow, p 170-173 (in Russian)

Silbermintz BA, Roschkowa EW (1933) Zur Frage des Vorkommens von Beryllium in Vesuvianen. Centralblatt Mineral Geol Paläont Abt A Mineral Petrogr 1933:249-254

Simmons WB, Pezzotta F, Falster AU, Webber KL (2001) Londonite, a new mineral species: the Cs-dominant analogue of rhodizite from the Antandrokomby granitic pegmatite, Madagascar. Can Mineral 39:747-755 [Am Mineral 87:356]

Simonov MA, Egorov-Tismenko YuK, Belov NV (1980) Use of modern x-ray equipment to solve fine problems of structural mineralogy by the example of the crystal structure of babefphite $BaBe(PO_4)F$. Sov Phys Crystallogr 25(1):28-30

Sinkankas J (1981) Emerald and other beryls. Chilton Book Company, Radnor, Pennsylvania

Sinkankas J (1988) Beryl: a summary. Rocks and Minerals 63:10-22

Sinkankas J, Read PG (1986) Beryl. Butterworths Gem Books, London

Skilleter DN (1990) To Be or not to Be—the story of beryllium toxicity. Chemistry in Britain 26(1):26-30

Slavík F (1914) New phosphates from Greifenstein in Saxony. Rozpravy České Akademie Císaře Františka Josefa pro Vědy, Slovesnost a Uměni, Třída II (mathematicko-přírodnická), 23(4):1-19 (in Czech) [a short version is abstracted in Chem Abstracts 10:31-32]

Sokolova EV, Ferraris G, Ivaldi G, Pautov LA, Khvorov PV (2000) Crystal structure of kapitsaite-(Y), a new borosilicate isotypic with hyalotekite—Crystal chemistry of the related isomorphous series. N Jahrb Mineral. Monatsh 2000(2):74-84

Sokolova E, Huminicki DMC, Hawthorne FC, Agakhanov AA, Pautov LA, Grew ES (2002) The crystal chemistry of telyushenkoite and leifite, $A\ Na_6\,[Be_2\,Al_3\,Si_{15}\,O_{39}\,F_2]$, A = Cs, Na. Can Mineral 40:183-192

Soman K, Druzhinin AV (1987) Petrology and geochemistry of chrysoberyl pegmatites of south Kerala, India. N Jahrb Mineral Abh 157: 167-183

Sørensen H (1960) Beryllium minerals in a pegmatite in the nepheline syenites of Ilímaussaq, South West Greenland. Report 21st Int'l Geol Congress (Norden) 17:31-35 [Am Mineral 46:241]

Sørensen H (1962) On the occurrence of steenstrupine in the Ilímaussaq massif, southwest Greenland. Meddel Grønland 167(1):1-251 [Not consulted. Am Mineral 48:1178]

Sørensen H (1963) Beryllium minerals in a pegmatite in the nepheline syenites of the Ilímaussaq, South West Greenland. Report 21st Int'l Geol Congress (Norden) 27:157-159.
Sørensen H, Danø M, Petersen OV (1971) On the mineralogy and paragenesis of tugtupite, $Na_8Al_2Be_2Si_8O_{24}(Cl,S)_2$, from the Ilímaussaq alkaline intrusion, south Greenland. Meddel Grønland 181(12):1-38
Staatz MH (1963) Geology of the beryllium deposits in the Thomas Range, Juab County, Utah. U S Geol Surv Bull 1142-M
Staatz MH, Griffitts WR (1963) Beryllium-bearing tuff in the Thomas Range, Juab County, Utah. Econ Geol 56:941-950
Staněk J (1966) Scholzite and hurlbutite from the pegmatites at Otov near Domažlice, SW Bohemia. Časopis pro Mineral Geol 11(1):21-26 (in Czech with English summary)
Stock A (1932) Beryllium. Trans Electrochem Soc 61:255-274
Strens RGJ (1974) The common chain, ribbon and ring silicates. *In* Farmer VC (ed) The Infrared Spectra of Minerals, Monograph 4, Mineralogical Society, London, p 305-330
Strunz H (1956) Bityit, ein Berylliumglimmer. Z Kristallogr 107:325-330
Strunz H, Nickel EH (2001) Strunz Mineralogical Tables. Chemical-Structural Mineral Classification System, 9th edn. Schweizerbart, Stuttgart
Sutherland FL, Pogson RE, Birch WD, Henry DA, Pring A, Bevan AWR, Stalder HA, Graham IT (2000) Mineral species first described from Australia and their type specimens. Mineralogy in Australian Museums 6:105-128
Switzer G, Reichen LE (1960) Re-examination of pilinite and its identification with bavenite. Am Mineral 45:757-762
Switzer G, Clarke RS Jr, Sinkankas J, Worthing HW (1965) Fluorine in hambergite. Am Mineral 50:85-95
Sylvester P, ed (2001) Laser-ablation-ICPMS in the Earth sciences. Principles and applications. Mineral Assoc Canada Short Course Series 29
Taylor SR, McLennan SM (1985): The continental crust: its composition and evolution. An examination of the geochemical record preserved in sedimentary rocks. Blackwell, Oxford
Taylor SR, McLennan SM (1995): The geochemical evolution of the continental crust. Rev Geophys 33:241-265
Teale GS (1980) The occurrence of högbomite and taaffeite in a spinel-phlogopite schist from the Mount Painter Province of South Australia. Mineral. Mag 43:575-577
Thorburn JA, Hobbs LM (1996) Beryllium abundances of six halo stars. Astron J 111:2106-2114
Vainshtein EE, Aleksandrova IT, Turanskaya NV (1960) Composition of the rare earths in gadolinites from deposits of different genetic types. Geochemistry 6:596-603
Vangioni-Flam E, Cassé M (2001) Evolution of lithium-beryllium-boron and oxygen in the early Galaxy. New Astronomy Rev 45:583-586
Vangioni-Flam E, Ramaty R, Olive KA, Cassé M (1998) Testing the primary origin of Be and B in the early galaxy. Astron Astrophys 337:714-720
Vangioni-Flam E, Cassé M, Audouze J (2000) Lithium-beryllium-boron: origin and evolution. Phys Rep 333-334:365-387
Vauquelin N-L (1798) De l'aigue marine, ou béril; et découverte d'une terre nouvelle dans cette pierre. Ann Chim Phys 26:155-177 [for a partial translation, see Anonymous 1930]
Vezzalini, G, Mattioli V (1979) Second finding of roggianite. Periodico Mineral 48(1):15-20 (in Italian)
Visser D, Kloprogge JT, Maijer C (1994) An infrared spectroscopic (IR) and light element (Li, Be, Na) study of cordierites from the Bamble Sector, South Norway. Lithos 32:95-107
Vlasov KA, ed. (1966) Geochemistry and mineralogy of rare elements and genetic types of their deposits, v. II, Mineralogy of rare elements. Israel Program for Scientific Translations, Jerusalem
Vlasov KA, Kuz'menko MZ, Es'kova EM (1966) The Lovozero Alkali Massif. Hafner, New York
Volborth A (1954a) Väyryneniiti, $[BeMn(PO_4)(OH,F)]$. Geologi (Finland) 6:7 [not consulted; Am Mineral 38:848]
Volborth A (1954b) Phosphatminerale aus dem Lithiumpegmatit von Viitaniemi, Eräjärvi, Zentral-Finnland. Ann Acad Scientiarum Fennicæ, Ser A, III Geol-Geograph 39:1-90 [Am Mineral 41:371]
Volborth A (1954c) Eine neue, die Phosphatanalyse verkürzende Methode und ihre Anwendung in der Analyse der Beryllium-Phospate. Z Anorg Allgem Chem 276:159-168
Volborth A (1956) Die Mineralparagenese im Lithiumpegmatit von Viitaniemi (Zentralfinnland) vom geochemischen Standpunkt, Tschermaks mineral petrogr Mitt (Ser 3) 5(4):273-283
Volborth A, Stradner E (1954) Väyrynenit, $BeMn[PO_4](OH)$, ein neues Mineral. Anz Österreich Akad Wiss (Wien) Math-Naturwiss Kl 91(2):21-23
Volfinger M, Robert J-L (1994) Particle-induced-gamma-ray-emission spectrometry applied to the determination of light elements in individual grains of granite minerals. J Radioanal Nuclear Chem Articles 185(2):273-291

Voloshin AV, Pakhomovskiy YaA, Men'shikov YuP, Povarennykh AS, Matviyenko YeN, Yakubovich OV (1983) Ytterbium hingganite—a new mineral from amazonite pegmatites of the Kola Peninsula. Doklady Akad Nauk SSSR 270:1188-1192 (in Russian. English translation: Int'l Geol Rev 26: 60-63) [Am Mineral 69:811].

Voloshin AV, Pakhomovskiy YaA, Rogachev DL, Tyusheva FN, Shishkin NM (1986) Ginzburgite—a new calcium-beryllium silicate from desilicated pegmatites. Mineral. Zhurnal 8(4):85-90 (in Russian with English abstract) [Am Mineral 73:439-440].

Voloshin AV, Pakhomovskiy YaA, Rogachev DL, Nadezhnina TN, Pushcharovskiy Dyu, Bakhchisaraytsev Ayu (1989) Clinobehoite—a new natural modification of $Be(OH)_2$ from desilicated pegmatites. Mineral Zhurnal 11(5):88-95 (in Russian with English abstract) [Am Mineral 76:666-667].

Von Knorring O (1985) Some mineralogical, geochemical and economic aspects of lithium pegmatites from the Karibib-Cape Cross pegmatite field in South West Africa/Namibia. Communications Geol Surv SW Africa/Namibia 1:79-84

Walter F (1992) Weinebeneite, $CaBe_3(PO_4)_2(OH)_2 \cdot 4H_2O$, a new mineral species: mineral data and crystal structure. Eur J Mineral 4:1275-1283 [Am Mineral 78:847-848]

Walter F, Postl W, Taucher J (1990) Weinebeneit: Paragenese und Morphologie eines neuen Ca-Be-Phosphates von der Spodumenpegmatitlagerstätte Weinebene, Koralpe, Kärnten. Mitt Abt Miner Landesmuseum Joanneum 58:37-43 [Am Mineral 78:847-848]

Warner LA, Holser WT, Wilmarth VR, Cameron EN (1959) Occurrence of nonpegmatite beryllium in the United States. U S Geol Surv Prof Paper 318

Washington HS (1931) Beryllium in minerals and igneous rocks. Am Mineral 16:37-41

Wedepohl KH (1995): The composition of the continental crust. Geochim Cosmochim Acta 59:1217-1232

Weeks ME, Leicester HM (1968) Discovery of the elements, 7th ed. Journal of Chemical Education, Easton, Pennsylvania

Weisbach A (1884) Über Herderit. N Jahrb Mineral Geol Paläont 1884(2):134-136

Werner AG (1789) Mineralsystem des Herrn Inspektor Werners mit dessen Erlaubnis herausgegeben von C.A.S. Hoffmann. Bergmännisches J 1:369-398

Werner AG (1790). Aeussere Beschreibungung des Olivins, Krisoliths, Berils, und Krisoberils, nebst noch einigen über diese Steine, besonders den erstern hinzugefügten Bemerkungen. Bergmännisches J 2: 54-94

Werner AG (1817) Letztes Mineral-System 2:29 [Not consulted; see Dana 1892]

West AR (1975) Crystal chemistry of liberite, Li_2BeSiO_4 and Li_2BeGeO_4. Bull Soc Franç Minéral Cristallogr 98:6-10

Wilson WE (1989) The Anjanabonoina pegmatite, Madagascar. Mineral Rec 20:191-200

Wight Q, Chao GC (1995) Mont Saint-Hilaire Revisited, Part 2. Rocks and Minerals 70:90-103, 131-138

Wilson AF (1977) A zincian högbomite and some other högbomites from the Strangways Range, Central Australia. Mineral Mag 41:337-344

Windecker CE (1955) The production of beryllium by the electrolysis of beryllium chloride. *In* White DW Jr, Burke JE (eds) The Metal Beryllium. The American Society for Metals, Cleveland, p 102-123

Winkler C (1885) Über Herderit. N Jahrb Mineral Geol Paläont 1885(1):172-174

Wöhler F (1828) Sur le glucinium et l'yttrium. Ann Chim Phys 39:77-84

Wood SA (1992) Theoretical prediction of speciation and solubility of beryllium in hydrothermal solution to 300 °C at saturated vapor pressure: Application to bertrandite/phenakite deposits. Ore Geology Rev 7:249-278

Ximen L, Peng Z (1985) Crystal structure of xinganite. Acta Mineral Sinica 5(4):289-293 (in Chinese with English abstract) [Am Mineral 73:441-442;935]

Xu Z, Sherriff BL (1994) ^{23}Na ^{27}Al ^{9}Be ^{29}Si solid state NMR study of tugtupite. Can Mineral 32:935-943

Yakovenchuk VN, Ivanyuk Gyu, Pakhomovskiy YaA, Men'shikov YuP (1999) Minerals of the Khibiny Massif. Izdatel'stvo "Zemlya", Moscow, Russia (in Russian, English translation in preparation)

Yakubovich OV, Matvienko EN, Voloshin AV, Simonov MA (1983) The crystal structure of hingganite-(Yb), $(Y_{0.51}TR_{0.36}Ca_{0.13}) \cdot Fe_{0.065}Be[SiO_4](OH)$. Sov Phys Crystallogr 28(3):269-271

Yakubovich OV, Malinovskii YuA, Polyakov VO (1990) Crystal structure of makarochkinite. Sov Phys Crystallogr 35(6):818-822 [Am Mineral 77: 448]

Yang Z, Fleck M, Pertlik F, Tillmanns, E, Tao K (2001) The crystal structure of natural gugiaite, $Ca_2BeSi_2O_7$. N Jahrb Mineral Monatsh 2001(4):186-192

Yarosh PYa, ed (1980) Minerals of the rocks and ores of the Urals. Akad Nauk SSSR Ural Nauchn Tsentr, Sverdlovsk [Yekaterinburg], Russia (in Russian)

Yefimov AF, Dusmatov VD, Alkhazov Vyu, Pudovkina ZG, Kazakova Mye (1970) Tadzhikite—a new rare-earth borosilicate from the hellandite group. Doklady Akad Nauk SSSR 195:1190-1193 (in Russian)

Yefimov AF, Yes'kova YeM, Loskutova LI, Shumkova NG (1971) A new find of barylite in the USSR. Trudy Mineral Muz Akad Nauk SSSR 20:198-201 (in Russian)

Yeom TH, Lim AR, Choh SH, Hong KS, Yu YM (1995) Temperature-dependent nuclear magnetic resonance study of ^{9}Be in an alexandrite single crystal. J Phys Condensed Matter 7:6117-6123

Yeremenko GK, Ryabtsev VV, Butulandi M, Goncharova YeI (1988) New data on rhodizite from pegmatites of Madagascar. Mineral Sbornik 42(2):68-71 (in Russian)

Zachariasen WH (1931) Meliphanite, leucophanite and their relation to melilite. Norsk Geol Tiddskrift 12:577-582

Zack T (2000) Trace element mineral analysis in high pressure metamorphic rocks from Trescolmen, Central Alps. Dissertation, Universität Göttingen

Zambonini F (1919) Sur la véritable nature du titanolivine [clinohumite] de la vallée d'Ala (Piémont). Bull Soc Franç Minéral 42:250-279

Zhang P, Yang Z, Tao K, Yang X (1995) Mineralogy and Geology of Rare Earths in China. Science Press, Beijing, China

Zubkov LB (1976) The helvine-danalite-genthelvite group. *In* Ginzburg AI (ed) The Mineralogy of Hydrothermal Beryllium Deposits. Nedra, Moscow, p 122-131 (in Russian)

Zubkov LB, Galadzheva NI (1974) Accessory leucophanite from zones of albitization in diabase in the Precambrian of the Russian Platform. Mineral Sbornik L'vov Gosudar Univ 28(2):72-76 (in Russian)

APPENDIX 1. LIST OF VALID AND POTENTIALLY VALID MINERAL SPECIES (IN QUOTATION MARKS) CONTAINING ESSENTIAL BERYLLIUM

MINERAL, CRYSTAL SYSTEM AND FORMULA	ENVIRONMENT	LOCALITIES	REFERENCES
Aminoffite (see note) Tetragonal $Ca_3Be_2Si_3O_{10}(OH)_2$	Veins and cavities in magnetite ore; fluorite veins associated with alkalic rocks	*Långban deposit, Filipstad, Värmland Co., Sweden;* Dugdinsk pluton and Bayan-Kol dike field, Tuva, Siberia, Russia	Hurlbut (1937) Coda et al. (1967); Kapustin (1973a,b); Chukhrov & Smol'yaninova (1981a); Nysten et al. (1999)
Asbecasite Trigonal $Ca_3(Ti,Sn^{4+},Fe)(As^{3+},Sb^{3+})_6Si_2(Be,B)_2O_{20}$	Alpine fissures, sandinitic ejectum, amazonite pegmatite	*Cherbadung (Cervandone), Binntal area, Switzerland;* Tre Croci, Viterbo Prov., Latium, Italy; Tennvatn, Nordland, Norway	Graeser (1966); Cannillo et al. (1969); A.O. Larsen (pers. comm., 1990), Della Ventura et al. (1991); Sacerdoti et al. (1993); Ellingsen et al. (1995); Graeser & Albertini (1995)
Babefphite Triclinic $BaBe(PO_4)F$	Fluorite-rare-metal deposit in skarn	*Aunik, Buratiya, Transbaikal (Siberia), Russia*	Nazarova et al. (1966, 1975); Simonov et al. (1980)
Barylite Orthorhombic $BaBe_2Si_2O_7$	Skarn and veins and vugs in Fe and Mn ores; nepheline syenite pegmatites; fenites and veins in alkalic and fenitized rocks	*Långban deposit, Filipstad, Värmland Co., Sweden;* Seal Lake, Labrador, Canada; Bratthagen, Lågendalen, Oslo region, Norway; Moskal' Ridge, west slope, Urals, Russia; Ilímaussaq, South Greenland; and others	Blomstrand (1876); Aminoff (1923); Heinrich & Deane (1962); Sæbö (1966); Yefimov et al. (1971); Petersen et al. (1991); Nysten et al. (1999)

Bavenite (see note) Orthorhombic $Ca_4Be_2Al_2Si_9O_{26}(OH)_2$–$Ca_4Be_3AlSi_9O_{25}(OH)_3$	Miarolitic cavities in granite and granite pegmatites; Alpine fissures; hydrothermal fluorite-phenakite mineralization; metasomatized peralkaline pegmatite; skarns	*Baveno, Novara Prov., Italy;* Strzegom, Lower Silesia, Poland; Yermakovskoye deposit, Buratiya, Russia; Ilímaussaq, South Greenland; Hohe Waid, Odenwald, Germany; and others	Artini (1901); Fleischer & Switzer (1953); Komarova et al. (1967); Gadomski et al. (1971); Kupriyanova & Novikova (1976b); Petersen et al. (1995); Kolitsch (1996a)
Bazzite (Sc-dominant analogue of beryl and stoppaniite) Hexagonal $(\square,Na,Cs)(Sc,Fe^{3+},Fe^{2+},Mg,Al)_2Be_3Si_6O_{18}\cdot nH_2O$	Miarolitic cavities in granite and granite pegmatites; alpine fissures	*Baveno, Novara Prov., Italy;* Kent granite massif, Kazakhstan; Heftetjern, Tørdal, Telemark, Norway; Furka-Basistunnel, Oberwald–Realp, Switzerland; and others	Artini (1915); Nowacki & Phan (1964); Chistyakova et al. (1966); Juve & Bergstøl (1990); Armbruster et al. (1995); Gramaccioli et al. (2000); Demartin et al. (2000)
Bearsite (structure very similar to that of moreasite) Monoclinic $Be_2AsO_4OH\cdot 4H_2O$	Oxidized zone of U deposit	*Bota-Burum deposit, SW Balkhash Region, Kazakhstan*	Kopchenova & Sidorenko (1962)

Behoite (dimorphous with clinobehoite) Orthorhombic β-$Be(OH)_2$	Granite pegmatite; Be-bearing tuff; miarolitic cavities in nepheline syenite and nepheline-syenite pegmatites	*Rode Ranch, Llano Co., Texas, USA*; Honeycomb Hills, Utah, USA; Hsianghualing, Nanling Ranges, Hunan, China; Mont Saint-Hilaire, Quebec, Canada; Tvedalen and Lågendalen, Oslo region, Norway.	Ehlmann & Mitchell (1970); Montoya et al. (1964); Crook (1977); Huang et al. (1988); Horváth & Gault (1990); Andersen et al. (1996)
Berborite (-1*T*, -2*T*, -2*H* polytypes identified) Trigonal, hexagonal $Be_2(BO_3)(OH)\cdot H_2O$	Skarn, nepheline syenite pegmatites	*Lupikko, Pitkäranta, Karelia, Russia;* Tvedalen and Siktesøya, Oslo region, Norway	Nefedov (1967), Schlatti (1967, 1968); Giuseppetti et al. (1990); Andersen et al. (1996)
Bergslagite (isostructural with hydroxylherderite) Monoclinic $CaBe[(As,P,Si)(O,OH)_4](OH)$	Veins, vugs and quartz-mica schist associated with Mn-Fe deposits; vug in rhyolite	*Långban deposit, Filipstad, Värmland Co., Sweden*; Falotta, Cavradi and Val Ferrera, Graubünden, Switzerland; Sailauf, Spessart, Germany	Hansen et al. (1984a,b); Graeser (1995); Kolitsch (1996b); Brugger & Gieré (1999)
Bertrandite (see note; cf. "gelbertrandite", sphaerobertrandite) Orthorhombic $Be_4Si_2O_7(OH)_2$	Granite pegmatites; hydrothermal fluorite mineralization; W-Mo deposits; alpine fissures; nepheline-syenite pegmatites	*Petit Port and Barbin, Nantes, France;* Aguachile Mtn., Coahuila, Mexico; Yermakovskoye deposit, Buratiya, Russia; Eastern Kounrad and Kara Oba, central Kazakhstan; Gotthard massif, Switzerland; Tvedalen, Oslo region, Norway; and others	Bertrand (1880, 1883), Damour (1883); Parker & Indergand (1957); Chukhrov (1960); Levinson (1962); Hügi & Röwe (1970); Novikova (1976a); Novikova et al. (1994); Andersen et al. (1996); Jermolenko (2002)

Beryl (Al-dominant analogue of bazzite and stoppaniite). Varietal names: emerald, aquamarine, heliodor, morganite, goshenite, vorobyevite, bixbite, Maxixe-type Hexagonal $(\square,Na,Cs)(Al,Fe^{3+},Fe^{2+},Mg)_2(Be,Li)_3Si_6O_{18}\bullet nH_2O$	Granite and granitic pegmatites; rhyolite; blackwall skarns associated with serpentine; hydrothermal fluorite mineralization; metamorphic-hydrothermal activity in carbonaceous shale and carbonate (emerald); hydrothermal in Fe-rich dolomite	"Izumrudnyye Kopi' (Emerald Pits), Central Urals, Russia; Muzo, Colombia; Habatachtal, Tauern Window, Austrian Alps; Minas Gerais, Brazil; and many others	Known from antiquity; Gaines (1976); Novikova (1976b); Sinkankas (1981, 1986, 1988); Fontan & Fransolet (1982); Deer et al. (1986); Franz et al. (1986); Aurisicchio et al. (1988); Schmetzer et al. (1991); Markl & Schumacher (1997); Franz & Morteani (Ch 13)
Beryllite (see note) Unknown $\sim Be_6Si_2O_9(OH)_2\cdot nH_2O$	Peralkaline pegmatites	*Lovozero (several localities), Kola Penin., Russia;* Ilímaussaq, South Greenland	Kuz'menko (1954); Andersen & Sørensen (1967)
Beryllonite (see note; topologically isostructural with trimerite) Monoclinic $NaBePO_4$	Granite pegmatite and granite; fractures in hornfels	*Stoneham, Oxford Co., Maine, USA*; Viitaniemi pegmatite, Eräjärvi, Finland; Mont Saint-Hilaire, Quebec, Canada; Norrö, near Stockholm, Sweden; Beauvoir granite, Massif Central, France; and others	Dana (1888); Dana & Wells (1889); Volborth (1954b); Lahti (1981); Horváth & Gault (1993); Nysten & Gustafsson (1993); King & Foord (1994); Charoy (1999)

Bityite (Mica group; see note) Monoclinic $(Ca,Na)(Li,\square)Al_2(Si,Be,Al)_4O_{10}(OH)_2$	Granitic pegmatites; aureole of a granite pegmatite (J. Evensen, pers. comm.)	*Maharitra, near Mont Bity (Ibity), Sahatany valley, Madagascar*; Londonderry, Western Australia; No Beer pegmatite, Bikita, Zimbabwe; Maantienvarsi pegmatite, Eräjärvi, Finland; and others	Lacroix (1908a,b); Rowledge & Hayton (1948); Strunz (1956); Ginzburg (1957); Gallagher & Hawkes (1966); Lahti & Saikkonen (1985); Evensen et al. (2002); (Grew, Ch 12)
Bromellite (isostructural with zincite) Hexagonal BeO	Vein in hematite-rich skarn; blackwall skarns associated with serpentine; metasomatized dolomite; nepheline syenite pegmatite	*Långban deposit, Filipstad, Värmland Co., Sweden*; Lupikko, Pitkäranta, Karelia, Russia; "Izumrudnyye Kopi" (Emerald Pits), Central Urals, Russia; Arkiya, Transbaikalia, Russia; Saga quarry, Tvedalen, Oslo region, Norway; Hsianghualing, Nanling Ranges, Hunan, China; Khibiny, Kola, Russia.	Aminoff (1925); Nefedov (1967); Klement'yeva (1969); Sergeyev (1973); Larsen et al. (1987); Huang et al. (1988); Men'shikov et al. (1999)
Chiavennite (a zeolite mineral) Orthorhombic. $(Ca,Na)(Mn,Fe)[Be,B(OH)]_2(Si,Al)_5O_{13}\cdot 2H_2O$	Granite and nepheline syenite pegmatites	*Chiavenna, Italian Alps, Lombardy*; Tvedalen, Oslo region, Norway; Utö, Stockholm, Sweden	Bondi et al. (1983); Raade et al. (1983); Langhof et al. (2000)
Chkalovite Orthorhombic $Na_2BeSi_2O_6$	Pegmatites associated with peralkaline rocks; sodalite xenoliths in peralkaline rock	*Lovozero, Kola Penin., Russia,* also Khibiny; Ilímaussaq, Greenland; Mont Saint-Hilaire, Quebec, Canada	Gerasimovsky (1939); Sørensen (1960); Khomyakov & Stepanov (1979); Horváth & Gault (1993); Argamakov et al. (1995); Markl (2001b)

Chrysoberyl (isostructural with forsterite Orthorhombic $BeAl_2O_4$	Granitic pegmatites; blackwall skarns; metasomatized carbonate rocks; hornfels xenoliths in peralkaline rocks	*Araçuaí(?), Minas Gerais, Brazil*; Maršíkov pegmatite, Moravia, Czech Republic; Trivandrum district, Kerala, India; Khibiny Massif, Kola Penin., Russia; and others	Werner (1789, 1790); Okrusch (1971); Shpanov (1976); Franz & Morteani (1984), Soman & Druzhinin (1987); Men'shikov et al. (1999); Antencio (2000b); Franz & Morteani (Ch 13)
Clinobehoite (dimorphous with behoite) Monoclinic $Be(OH)_2$	Hydrothermally altered pegmatite	*Emerald Pits, Middle Urals, Russia;* possibly also Lupikko, Pitkäranta, Karelia, Russia	Voloshin et al. (1989)
Danalite (Fe-dominant analogue of helvite and genthelvite) Isometric End member is $Fe_4Be_3Si_3O_{12}S$	Granite, granite pegmatite, albitized granite; skarns from carbonate rocks	*Rockport, Mass., USA*; St. Dizier and 3 other localities, NW Tasmania, Australia; Sucuri granite, Goias, Brazil; Daba Shabeli Complex, N. Somalia; and others	Cooke (1866); Dunn (1976); Burt (1980); Kwak & Jackson (1986); Raimbault & Bilal (1993); Nimis et al. (1996)
Ehrleite Triclinic $Ca_2ZnBe(PO_4)_2(PO_3OH)\cdot 4H_2O$	Granite pegmatite	*Tip Top mine, Custer Co., South Dakota, USA*	Robinson et al. (1985); Hawthorne & Grice (1987)
Epididymite (dimorphous with eudidymite) Orthorhombic $Na_2Be_2Si_6O_{15}\cdot H_2O$	Peralkaline and granite pegmatites; hornfels (rare)	*Narssârssuk, South Greenland,* also Ilímaussaq; Mont Saint-Hilaire, Quebec, Canada; Věžná, Moravia, Czech Republic; Khibiny and Lovozero, Kola Penin., Russia; and others	Flink (1893); Shilin & Semenov (1957); Semenov & Sørensen (1966); Mandarino & Harris (1969); Kapustin (1973b); Novák et al. (1991); Petersen et al. (1997); Men'shikov et al. (1999)

Euclase Monoclinic $BeAlSiO_4(OH)$	Granite pegmatites; mica-fluorite veins in metasomatized carbonate rocks; metamorphic-hydrothermal activity in carbonaceous shale and carbonate (with emerald); Alpine fissures; low-grade metamorphic rocks	*Probably Ouro Preto, Minas Gerais, Brazil,* also Santa do Encoberto; Pikes Peak granite, Park Co., Colorado, USA; Sargardon, Uzbekistan; Pizzo Giubine, Ticino, Switzerland; and others	Haüy (1799); Delamétherie (1792); Sharp (1961); Apollonov (1968); Novikova (1976c); Graziani & Guidi (1980); Cassedanne (1989); Demartin et al. (1992); Atencio (2000b); Grew (Ch 12)
Eudidymite (dimorphous with epididymite) Monoclinic $Na_2Be_2Si_6O_{15} \cdot H_2O$	Peralkaline pegmatites and syenite; metamorphosed peralkaline complex; hydrothermal fluorite-bertrandite-phenakite deposit	*Lille Arøy, Langesundsfjord, Norway* and nearby localities in the Oslo region; Seal Lake, Labrador, Canada; Ilímaussaq, South Greenland; Yermakovskoye deposit, Buratiya, Russia; and others	Brøgger (1887); Nordenskiöld (1887), Kozlova (1962); Nickel (1963); Semenov & Sørensen (1966); Ginzburg et al. (1979); Novikova (1984)

Faheyite (see note) Hexagonal $Na_{0.20}Mn^{2+}_{0.62}Mg_{0.21}Fe^{3+}_{1.98}Al_{0.01}Be_{2.15}P_{3.97}O_{16}$-$\cdot 6.11H_2O$	Granite pegmatites	*Sapucaia mine, Minas Gerais, Brazil;* Noumas, Steinkopf, Namibia; Roosevelt mine, near Custer, South Dakota, USA	Lindberg & Murata (1953); Lindberg (1964); Cassedanne & Baptista (1999); von Knorring (1985 and pers. comm. to R. Kristiansen); Robinson et al. (1992)
Ferrotaaffeite-6*N*'3*S* (formerly "pehrmanite"; Fe-dominant analogue of magnesiotaaffeite-6*N*'3*S*) Hexagonal $(Fe,Zn,Mg)_2Al_6BeO_{12}$	Granite pegmatite	*Rosendal pegmatite, Kemiö Island, Finland*	Burke & Lustenhouwer (1981); Armbruster (2002); Grew (Ch 12)
Fransoletite (dimorphous with parafransoletite) Monoclinic $Ca_3Be_2(PO_4)_2(PO_3OH)_2 \bullet 4H_2O$	Granite pegmatite	*Tip Top mine, Custer Co., South Dakota, USA*	Peacor et al. (1983); Kampf (1992)
Gadolinite-(Ce) (Gadolinite group, see note) Monoclinic $(Ce,La,Nd,Y)_2Fe^{2+}Be_2Si_2O_{10}$	Syenite and nepheline syenite pegmatites, pegmatoidal alaskite, granite; sodalite veins in hornfels	*Skien, SW Oslo region, Norway,* also Tvedalen, Larvik; Xihuashan granite, Huailai, Hebei Prov., China; Khibiny massif, Kola Penin., Russia; and others	Segalstad & Larsen (1978); Vainshtein et al. (1960); Zhang et al. (1995); Larsen (1996); Men'shikov et al. (1999)
"Gadolinite-(Nd)" (Gadolinite group, see note) Monoclinic $(Nd,Ce,Y)_2Fe^{2+}Be_2Si_2O_{10}$	Alkali-granite pegmatite	"Tatyana" pegmatite, Mt. Ulin-Khuren, Altai Mountains, western Mongolia	Kartashov & Lapina (1994)

Gadolinite-(Y) (Gadolinite group, see note) Monoclinic. $(Y,REE,Ca)_2(Fe^{2+},\square)(Be,B)_2Si_2(O,OH)_{10}$	Granite, alkali-granite and other pegmatites; metasomatized rocks, e.g., fluorite-leucophanite type; Alpine fissures	*Ytterby, Stockholm Archipelago, Sweden*; Hitterø (now Hidra), near Flekkefjord, Vest-Adger, Norway; Moos, Böckstein, Rauris Valley, Austria; "Tatyana" pegmatite, Mt. Ulin-Khuren, Altai Mountains, western Mongolia; and others	Klaproth (1802); Schetelig (1922); Ito & Hafner (1974), Shatskaya (1976a); Demartin et al. (1993); Kartashov et al. (1993), Pezzotta et al. (1999); Hyslop et al. (1999); Gramaccioli & Demartin (2001)
Gainesite (see note; Na-dominant analogue of mccrillisite and selwynite) Tetragonal ~$Na(Na,K)(Be,Li)Zr_2(PO_4)_4 \cdot 1.5H_2O$	Granite pegmatite	*Nevel quarry, Newry, Oxford Co., Maine, USA*	Moore et al. (1983; Foord et al. (1994)
Genthelvite (Zn-dominant analogue of helvite) Isometric End member is $Zn_4Be_3Si_3O_{12}S$	Alkaline granite and syenite and their pegmatites; greisen; metasomatites associated with alkaline granite; calc-silicate skarn	*West Cheyenne Canyon (Pikes Peak batholith), El Paso Co., Colorado, USA;* Kymi granite, Viipuri rapakivi massif, Finland; Perga deposit, Ukraine; Taghouaji complex, Aïr Mtns., Niger; and others	Genth (1892); Glass et al. (1944); Haapala & Ojanperä (1972); Dunn (1976); Zubkov (1976); Burt (1988); Perez et al. (1990); Raimbault & Bilal (1993)
Greifensteinite (Fe-dominant analogue of monoclinic roscherite and zanazzite; unknown mineral #3 of King & Foord) Monoclinic. $Ca_2(\square,Mn^{2+},Fe^{2+})_2(Fe^{2+},Mn^{2+})_4Be_4(PO_4)_6(OH)_4\text{-}\cdot 6H_2O$	Granite pegmatites, cavities in granite	*Greifenstein, near Ehrenfriedersdorf, Saxony, Germany;* Newry, Paris and Rumford, Oxford Co., Maine; Gunnislake, Cornwall, U.K.; Ênio pegmatite, Galiléia, Minas Gerais, Brazil	Chukanov (pers. comm.); Chukanov et al. (in press); Lindberg (1958); Correia Neves et al. (1980); Clark et al. (1983); King & Foord (1994, 2000)

Gugiaite (isostructural with the åkermanite, melilite group) Tetragonal $Ca_2BeSi_2O_7$	Skarn, e.g., in carbonate at contact with alkaline syenite; miarolitic cavity in granite	*Gugia, Liaoning Province, China*; Baveno, Novara Prov., Italy; Yuge Island, Ehime Pref., Japan	Peng et al. (1962); Nova (1987); Minakawa & Yoshimoto (1998); Yang et al. (2001)
Hambergite Orthorhombic $Be_2BO_3(OH,F)$	Nepheline syenite and granite pegmatites; skarn	*Helgeroa, Langesundsfjord, Oslo region, Norway*, also Tvedalen; Little Three Mine, San Diego Co., Calif., USA; Lupikko, Pitkäranta, Karelia, Russia; Kracovice, Czech Republic; eastern Pamirs, Tajikistan; and others	Brøgger (1890); Switzer et al. (1965), Nefedov (1967), Burns et al. (1995); Novák et al. (1998), Dzhurayev et al. (1998)
Harstigite Orthorhombic $Ca_6(Mn^{2+},Mg)Be_4(SiO_4)_2(Si_2O_7)_2(OH)_2$	Open fissure in manganese skarn	*Harstigen, Pajsberg, Sweden*	Flink (1886, 1917b); Moore (1968a,b); Hesse & Stümpel (1986)

Helvite (Mn-dominant analogue of danalite and genthelvite; isotypic with other members of the sodalite group and with tugtupite) Cubic End member is $Mn_4Be_3Si_3O_{12}S$	Pegmatites associated with granite, syenite, nepheline syenite, etc.; skarn; W-bearing quartz veins, greisen; hydrothermal veinlets in Mn deposits and metasomatized Mn skarns; sandinitic ejecta	*Schwarzenberg, Saxony, Germany*; Iron Mtn., New Mexico, USA; numerous localities in Langesundsfjord area, Oslo region, Norway; Vielle Aure and Loudervielle, Pyrenees, France, and others	A.G. Werner, e.g., Werner (1817), Glass et al. (1944); Dunn (1976); Larsen (1988), Holtstam & Wingren (1991); Raimbault & Bilal (1993); Ragu (1994); Rossi et al. (1995)
Herderite (see note; isostructural with hydroxyl-herderite, bergslagite) Monoclinic $CaBePO_4(F,OH)$	Tin vein; cut gem-stone	*Ehrenfriedersdorf, Saxony, Germany*; Brazil (specific locality unknown); Mogok Stone Tract, Myanmar (Burma)	Haidinger (1828a,b); Schneiderhöhn (1941); Burt (1975); Dunn & Wight (1976); Leavens et al. (1978); Dunn et al. (1979); Jahn (1998); G. Harlow (pers. comm.)
"Hingganite-(Ce)" (Gadolinite group, see note) Monoclinic $(Ce,Y,REE)_2(\square,Fe^{2+})Be_2Si_2O_8(OH,O)_2$	Granite and alkali-granite pegmatites	*Tahara, near Ena City, Gifu Pref., Japan*; "Tatyana" pegmatite, Mt. Ulin-Khuren, Altai Mountains, Mongolia	Miyawaki et al. (1987); Kartashov & Lapina (1994)
"Hingganite-(Nd)" (Gadolinite group, see note) Monoclinic. $(Nd,Ce,Y,REE)_2(\square,Fe^{2+})Be_2Si_2O_8(OH,O)_2$	Granite and alkali-granite pegmatites	*Tahara, near Ena City, Gifu Pref., Japan*; "Tatyana" pegmatite, Mt. Ulin-Khuren, Altai Mountains, Mongolia	Miyawaki et al. (in prep.); Kartashov & Lapina (1994)

Hingganite-(Y) (Gadolinite group, see note) Monoclinic $(Y,REE)_2(\square,Fe)\ Be_2Si_2O_8(OH,O)_2$	Peralkaline granite and associated pegmatite; dolomite associated with talc-chlorite deposit	*Xing'anling (Hingganling), Inner Mongolian Automonous Region, China*; Tastyg, Tuva, Russia; Malosi complex, Malawi; Trimouns, Ariège, France; Cuasso al Monte, Varese, Italy; and others	Ding et al. (1981, 1984, 1985); Semenov et al. (1963); Yakubovich et al. (1983); Ximen & Peng (1985); Petersen et al. (1994a); de Parseval et al. (1997); Pezzotta et al. (1999); Demartin et al. (2001)
Hingganite-(Yb) (Gadolinite group, see note) Monoclinic $(Yb,Y)_2(\square)\ Be_2Si_2O_8(OH)_2$	Amazonite pegmatite	*Ploskaya Mt., western Keivy, Kola Penin., Russia*; Tangen, Kragerø, Norway	Voloshin et al. (1983); Kristiansen (1994)
Høgtuvaite (cf. "makarochkinite", Appendix 2; aenigmatite group) Triclinic $(Ca_{1.8}Na_{0.2})(Fe^{2+}{}_{3.55}Fe^{3+}{}_{2.2}Ti_{0.25})O_2[Si_{4.5}BeAl_{0.5}O_{18}]$	Metamorphic, in gneiss and pegmatite	*Høgtuva Window, Mo i Rana, Nordland Co., Norway*	Grauch et al. (1994); Burt (1994); Grew (Ch 12)
Hsianghualite (see note; a zeolite mineral with an analcime-type structure) Cubic $Ca_3Li_2(Be_3Si_3O_{12})F_2$	Li-mica veins in "ribbon rock" (skarn)	*Hsianghualing, Nanling Ranges, Hunan Prov., China*	Huang et al. (1958, 1988); Peng et al. (1964); Beus (1966); Vlasov (1966); Academy (1973); Liu (1990); Rastsvetsaeva et al. (1991); Liu & Zeng (1998)
Hurlbutite (see note) Monoclinic $CaBe_2(PO_4)_2$	Granite pegmatites & granite	*Newport, Sullivan Co., New Hampshire, USA*; Viitaniemi pegmatite, Eräjärvi, Finland; Norrö, near Stockholm, Sweden; Beauvoir granite, Massif Central, France; etc	Mrose (1952); Volborth (1954b, 1956); Burt (1975); Nysten & Gustafsson (1993); Charoy (1999)

Hyalotekite Triclinic $(Ba,Pb,K)_4(Ca,Y)_2Si_8(B,Be)_2(Si,B)_2O_{28}F$	Mn-Fe rich skarn; peralkaline pegmatite	*Långban deposit, Filipstad, Värmland Co., Sweden*; Dara-i-Pioz, Tajikistan	Nordenskiöld (1877); Moore et al. (1982); Grew et al. (1994); Christy et al. (1998)
Hydroxylherderite (isostructural with herderite, bergslagite) Monoclinic $CaBePO_4(OH,F)$	Granite pegmatites and granites; greisens; hypergene in weathering crust over a fluorite-muscovite greisen	Paris, Oxford Co., Maine,USA, *also Stoneham and Buckfield; Virgem da Lapa and Golconda, Minas Gerais, Brazil; Beauvoir Granite, Massif Central, France; and others*	Penfield (1894); Burt (1975), Leavens et al. (1978); Dunn et al. (1979); Nysten & Gustafsson (1993); Charoy (1999); King & Foord (1994, 2000)
Jeffreyite (structurally related to melilite group) Orthorhombic $(Ca,Na)_2(Be,Al)Si_2(O,OH)_7$	Rodingitized granite dike	*Jeffrey mine, Asbestos, Quebec, Canada*	Grice & Robinson (1984)
Joesmithite (very closely related to clino-amphiboles, but not strictly isomorphic) Monoclinic. $Pb(Ca,Na)_2(Mg,Fe^{3+,2+},Mn)_5Si_{6.22}Be_{1.78}O_{22}(OH,F)_2$	Hematite ore, calcite-filled cavities	*Långban deposit, Filipstad, Värmland Co., Sweden*	Moore (1967; 1969a,b; 1988); Moore et al. (1993)
Khmaralite (forms a series with sapphirine) Monoclinic $Ca_{0.04}Mg_{5.44}Fe^{3+}{}_{0.25}Fe^{2+}{}_{1.73}Al_{14.19}Be_{1.56}B_{0.02}Si_{4.77}O_{40}$	Granulite-facies meta-pegmatite	*Khmara Bay, Enderby Land, Antarctica*	Grew (1981, 1998); Barbier et al. (1999); Grew et al. (2000); Barbier et al. (2002)

Leifite (see note; Na-dominant analogue of telyushenkoite) Trigonal $(Na,\square)(H_2O,\square)Na_6[Be_2(Al,Si,Zn)_3Si_{15}O_{39}F_2]$	Pegmatite and albitite associated with peralkaline intrusives	*Narssârssuk, South Greenland*; Mont Saint-Hilaire, Quebec, and Strange Lake, Quebec-Labrador, Canada; Lovozero and Khibiny, Kola Penin., Russia;	Bøggild (1915, 1920); Micheelson & Petersen (1970); Khomyakov et al. (1979); Petersen et al. (1994b); Miller (1996); Men'shikov et al. (1999); Hochleitner et al. (2000); Sokolova et al. (2002)
Leifite, K-dominant analogue Trigonal $(K,Na,\square)\square Na_6[Be_2(Al,Si,Zn)_3Si_{15}O_{39}F_2]$	Pegmatite and albitite associated with peralkaline intrusives	Vesle Arøya, Langesundsfjord, Oslo region, Norway; Ilímaussaq, S. Greenland; Mont Saint-Hilaire, Quebec, Canada	Larsen & Åsheim (1995); Petersen et al. (1994b); Sokolova et al. (2002)
Leucophanite (related to melilite, see note; cf. meliphanite below and reported triclinic modification, Appendix 2) Orthorhombic $CaNaBeSi_2O_6F$	Pegmatites and veins associated with peralkaline rocks; metasomatites associated with alkaline rocks; fluorite-phenakite-bertrandite metasomatite	*Låven Island, Langesundsfjord, Oslo region, Norway;* Yermakovskoye deposit, Buratiya, Russia; Khibiny Massif, Kola Penin., Russia; Dara-i-Pioz, Tajikistan; and others	H.M.T. Esmark. Erdmann (1840); Zachariasen (1931); Shatskaya (1976b); Novikova (1984); Grew et al. (1993); Semenov et al. (1987); Men'shikov et al. (1999); Grice and Hawthorne (2002)
Liberite Monoclinic β-Li_2BeSiO_4	Li-mica veins in "ribbon rock" (skarn)	*Hsianghualing, Nanling Ranges, Hunan, China*	Chao (1964); Chang (1969); West (1975)

Londonite (Cs-dominant analogue of rhodizite) Cubic $(Cs,K,Rb)Al_4Be_5B_{11}O_{28}$	Granitic pegmatite	*Antandrokomby, Manandona Valley, Madagascar;* also at 3 other localities in central Madagascar	Simmons et al. (2001)
Lovdarite (a zeolite mineral; see note) Orthorhombic $K_2Na_6(Be_4Si_{14}O_{36})\cdot 9H_2O$	Pegmatites associated with peralkaline complexes; vugs in phonolite	*Lovozero, Kola Penin., Russia*; Point of Rocks, near Springer, New Mexico, USA; Ilímaussaq, South Greenland	Mens'shikov et al. (1973); Khomyakov et al. (1975); DeMark (1984, 1989); Merlino (1990); Petersen et al. (2002a)
Magnesiotaaffeite-2*N*'2*S* (see note; formerly "taaffeite") Hexagonal $(Mg,Fe,Zn)_3Al_8BeO_{16}$	Gem gravels; skarns and other metasomatized rocks, including "ribbon rock"; spinel-phlogopite schist	*Sri Lanka;* Sakhir-Shulutynskiy pluton, East Sayan, Russia; Hsianghualing, Nanling Ranges, Hunan, China; Mount Painter, Australia; Lupikko, Pitkäranta, Karelia, Russia; and others	Anderson et al. (1951); Kozhevnikov et al. (1975); Teale (1980); Schmetzer (1983); Huang et al. (1988); Kampf (1991); Armbruster (2002); Grew (Ch 12)
Magnesiotaaffeite-6*N*'3*S* (formerly "musgravite"; the Mg-dominant analogue of ferrotaaffeite-6*N*'3*S*) Hexagonal $(Mg,Fe,Zn)_2Al_6BeO_{12}$	Phlogopite zone in meta-pyroxenite; peraluminous meta-pegmatite, skarn; gem gravels (?)	*Ernabella Mission, Musgrave Ranges, Australia;* Khmara Bay, Enderby Land, Antarctica; Dove Bugt, East Greenland; Sakeny, south Madagascar; and others	Hudson et al. (1967); Grew (1981); Schmetzer (1983), Chadwick et al. (1993) ; Armbruster (2002); Devouard et al. (2002); Grew (Ch 12)

Mccrillisite (see note, Cs-dominant analogue of gainesite and selwynite) Tetragonal. ~$Na(Cs,Li)(Be,Li)Zr_2(PO_4)_4 \cdot 1.2H_2O$	Granite pegmatite	*Mt. Mica, Paris, Oxford Co., Maine, USA*	Foord et al. (1994)
Meliphanite ("melinophane", see notes; cf. leucophanite) Tetragonal $Ca_4(Na,Ca)_4Be_4AlSi_7O_{24}(F,O)_4$	Pegmatites and veins associated with alkalic rocks; metasomatites associated with alkaline rocks; fluorite-phenakite-bertrandite metasomatite	*Langesundsfjord area, Oslo region, Norway,* also Tvedalen, Oslo region; Sakharyok massif, Kola Penin., Russia; Yermakovskoye deposit, Buratiya, Russia; and others	Scheerer (1852); Brøgger (1890); Bel'kov & Denisov (1968); Novikova et al. (1975); Novikova (1984); Grice and Hawthorne (2002)
Milarite (Osumilite group) Hexagonal. $K(\square,H_2O,Na)_2(Ca,Y,REE)_2(Be,Al)_3Si_{12}O_{30}$ with Be ≈ 1.5 to 3 apfu and H_2O and Na partially occupying the B site	Alpine fissures; alkalic rocks; granite pegmatites; hydrothermal veins; fluorite-phenakite-bertrandite metasomatite	*Val Giuf* (Val Milar only later), *Graubünden, Switzerland*; Věžná, Czech Republic; Stora Vika and Utö, south of Stockholm, Sweden; Yermakovskoye deposit, Buratiya, Russia; and others	Kenngott (1870); Kuschel (1877); Palache (1931); Kupriyanova & Novikova (1976a); Černý et al. (1980, 1991); Hawthorne et al. (1991); Nysten (1997); Cassedanne & Alves (1994); Hawthorne (2002)
Milarite, Sc-dominant analogue. Hexagonal	Granite pegmatite	*Heftetjern pegmatite, Tørdal, Telemark, Norway*	Raade & Kristiansen (2000); Černý (Ch 10)
Milarite, Y-dominant analogue Hexagonal	Granite pegmatite; peralkaline granite	Jaguaraçu, Minas Gerais, Brazil; Strange Lake, Quebec-Labrador, Canada; Heftetjern pegmatite, Tørdal, Telemark, Norway	Černý et al. (1991); Raade & Kristiansen (2000); Hawthorne (2002); Černý (Ch 10)

Minasgeraisite-(Y) (Gadolinite group, see note) Monoclinic $(Y_{0.72}REE_{0.41}Ca_{0.56}Bi_{0.31})_{\Sigma=2.00}$- $(Ca_{0.45}Mn_{0.20}Mg_{0.08}Fe_{0.05}Zn_{0.02}Cu_{0.01}\square_{0.19})_{\Sigma=1.00}$- $(Be_{1.55}Si_{0.24}B_{0.21})_{\Sigma=2.00}(Si_{1.95}P_{0.08})_{\Sigma=2.03}O_{10}$ with zones up to 0.65 Bi pfu.	Granite pegmatite	*Jaguaruçu pegmatite, Minas Gerais, Brazil*	Foord et al. (1986)
Moraesite (almost isostructural with bearsite) Monoclinic $Be_2PO_4(OH)\cdot4H_2O$ (varieties reported to contain Ca, Mg, Mn and Fe are probably mixtures of moraesite with other phosphates)	Granite pegmatites; alkalic pegmatite; in weathering crust over a fluorite-muscovite greisen	*Sapucaia mine, Minas Gerais, Brazil,* also Humaita and other localities in Minas Gerais; Boyevskoye deposit, Middle Urals, Russia; Lovozero, Kola Penin., Russia; and others	Lindberg et al. (1953); Pokrovskiy et al. (1963); Ginzburg & Shatskaya (1966); Cassedanne & Cassedanne (1987); King & Foord (1994); Merlino & Pasero (1992); Pekov (1995)
Mottanaite-(Ce) (Hellandite group) Monoclinic Ideally $Ca_4(CeCa)AlBe_2(Si_4B_4O_{22})O_2$	Alkali-syenitic ejectum	*Monte Cavalluccio, Sacrofano, Latium, Italy*	Oberti et al. (2002); Della Ventura et al. (2002)
Nabesite (a zeolite mineral) Orthorhombic $Na_2BeSi_4O_{10}\cdot4H_2O$	Veins associated with alkaline intrusive	*Ilímaussaq, South Greenland*	Petersen et al. (2002b)
Odintsovite (see note and possible Li-dominant analogue, Appendix 2) Orthorhombic $K_2(Na,Ca,Sr)_4(Na,Li)Ca_2$- $(Ti,Fe^{3+},Nb)_2O_2[Be_4Si_{12}O_{36}]$	Veins associated with peralkaline intrusive	*Malomurun alkaline complex, Irkutsk Oblast'-Sakha Republic (Yakutiya), Russia*	Konev et al. (1995, 1996); Rastsvetaeva et al. (1995); Petersen et al. (2001)
Pahasapaite (a zeolite mineral) Cubic. $(Ca_{5.5}Li_{3.6}K_{1.2}Na_{0.2}\square_{13.5})Li_8Be_{24}P_{24}O_{96}\cdot38H_2O$	Granitic pegmatite	*Tip Top mine, Custer Co., South Dakota, USA*	Rouse et al. (1987, 1989a)
Parafransoletite (dimorphous with fransoletite) Triclinic $Ca_3Be_2(PO_4)_2(PO_3OH)_2\cdot4H_2O$	Granite pegmatite	*Tip Top mine, Custer Co., South Dakota, USA*	Kampf et al. (1992); Kampf (1992)

Phenakite (see note; isostructural with willemite) Hexagonal Be_2SiO_4	Pegmatites associated with granites, including alkaline granites; nepheline-syenite pegmatite; trachyte; alpine fissures; greisens, hydrothermal fluorite mineralization; metasomatic rocks; fissures in hematite skarn	*"Izumrudnye" Kopi (Emerald Pits), Tokovaya River, middle Urals, Russia*; Yermakovskoye deposit, Buratiya, Russia; Håkestad-Larvikite quarry near Larvik, Oslo region, Norway; Ovruch graben, Ukraine; Långban deposit, Filipstad, Värmland Co., Sweden; and others	Nordenskjöld (1834); Novikova (1967, 1976d); Hügi & Röwe (1970); Burt (1978); Nedashkovskiy (1983); Metalidi (1988); Novikova et al. (1994); Andersen et al. (1996); Bulnaev (1996); Nysten et al. (1999); Franz & Morteani (Ch 13)
Rhodizite (K-dominant analogue of londonite) Cubic $(K,Cs,Rb)Al_4Be_5B_{11}O_{28}$	Granitic pegmatites	*Shaitanka (Rezh district) and Sarapulka (Murzinka), middle Urals; Russia*; Sahatany River area, Madagascar; Red Ace pegmatite, Florence Co, Wisconsin, USA	Rose (1834, 1836); Lacroix (1922); Frondel & Ito (1965); Pring et al. (1986); Yeremenko et al. (1988); Falster and Simmons (1989); Falster (1994); Falster et al. (1996); Simmons et al. (2001)

Roggianite (a zeolite mineral) Tetragonal $Ca_2[Be(OH)_2Al_2Si_4O_{13}]\cdot<2.5H_2O$	albitite dikes; hydrothermally altered pegmatite ("ginzburgite")	*Alpe Rosso, Val Vigezzo, Novara, Italy*, also Pizzo Marcio, 1km distant; Izumrudnye Kopi (Emerald Pits), middle Urals, Russia ("ginzburgite")	Passaglia (1969a,b); Gard (1969), Vezzalini & Mattioli (1979); Galli (1980); Voloshin et al. (1986); Passaglia & Vezzalini (1988); Giuseppetti et al. (1991)
Roscherite ("roscherite-1*M*" of Strunz & Nickel 2001; Mn-dominant analogue of zanazziite and greifensteinite; see note) Monoclinic $(Fe^{2+},Mg,Al,Fe^{3+},\square)Ca_2(Mn^{2+},Fe^{2+},Mg,Ca)_4Be_4$-$(PO_4)_6(OH)_{6-n}\cdot(4+n)H_2O$, where $0 \leq n \leq 2$	Miarolitic cavities in Sn-bearing Li-mica granite; granite pegmatites	*Greifenstein, near Ehrenfriedersdorf, Saxony, Germany,* Sapucaia pegmatite, Minas Gerais, Brazil; Tip Top mine, Custer Co., South Dakota, USA	Slavík (1914); Lindberg (1958); Campbell & Roberts (1986)
Roscherite ('roscherite-1*A*' of Strunz & Nickel 2001) Triclinic $(\square_{1/3}Fe^{3+}{}_{0.42}Mn^{3+}{}_{0.12}Al_{0.13})\,\square$-$Ca_2(Mn^{2+}{}_{3.91}Mg_{0.04}Ca_{0.05})Be_4(PO_4)_6(OH)_4\cdot6H_2O$	Granite pegmatite	Foote Mine, Kings Mountain, Cleveland Co., North Carolina, USA	Fanfani et al. (1977)
Roscherite (Fe-dominant analogue of "roscherite-1*A*" of Strunz & Nickel 2001; cf. greifensteinite) Triclinic. Only partial analysis available	Granite pegmatite	Lavra da Ilha pegmatite, Taquaral, Minas Gerais, Brazil	Leavens et al. (1990)
Samfowlerite Monoclinic $Ca_{14}Mn_3Zn_2(Zn,Be)_2Be_6(SiO_4)_6(Si_2O_7)_4(OH,F)_6$	Vugs in zinc ore	*Franklin, Sussex Co., New Jersey, USA*	Rouse et al. (1994); Dunn (1995)
Selwynite (see note, K-dominant analogue of gainesite and mccrillisite) Tetragonal $\sim NaK(Be,Al)Zr_2(PO_4)_4\cdot2H_2O$	Granite pegmatite	*Wycheproof, NW Victoria, Australia*	Birch et al. (1995)
Semenovite-(Ce) Orthorhombic $(Fe^{2+},Mn,Zn,Ti)(Ce,La,REE,Y)_2Na_{0-2}$-$(Ca,Na)_8(Si,Be)_{20}(O,OH,F)_{48}$	Albitite vein in peralkaline intrusive	*Ilímaussaq, South Greenland*	Petersen & Rønsbo (1972); Mazzi et al. (1979)

Sorensenite Monoclinic $Na_4SnBe_2Si_6O_{18}\cdot 2H_2O$ (with 0.75–1.36 wt % Nb_2O_5 *or* Nb/Sn = 4.5-8%)	Hydrothermal veins and pegmatitic bodies in sodalite nepheline syenite (peralkaline intrusive)	*Ilímaussaq, South Greenland*	Semenov et al. (1965), Maksimova et al. (1973), Metcalf-Johansen & Hazell (1976)
Sphaerobertrandite (see note; cf. bertrandite); UK-10, Tvedalen Monoclinic $Be_3SiO_4(OH)_2$	Peralkaline and nepheline syenite pegmatites; vein in "ribbon rock" (skarn)	*Mannepakhk, Lovozero, Kola Peninsula, Russia,* also Kuftn'yun, Sengischorr, Lepkhe-Nel'm Mountains, Lovozero, and Yukspor, Khibiny; Hsianghualing, Hunan Prov., China; Tvedalen, Oslo region, Norway; Ilímaussaq, South Greenland	Semenov (1957, 1969, 1972); Huang et al. (1988); Andersen et al. (1996); Pekov (2000); A.O. Larsen & I.V. Pekov (pers. comm.); Pekov et al. (in press)
Stoppaniite (Fe^{3+}-dominant analogue of beryl and bazzite) Hexagonal $(Na,\square)(Fe^{3+},Al,Mg)_2Be_3Si_6O_{18}\cdot H_2O$	Miarolitic cavities in sanidinite ejectum	*Vico Volcanic, District, Capranica (Viterbo), Italy*	Ferraris et al. (1998); Della Ventura et al. (2000)
Surinamite Monoclinic $(Mg,Fe^{2+})_3(Al,Fe^{3+})_3O[AlBeSi_3O_{15}]$	Metapelites and metamorphosed granite pegmatites	*Bakhuis Mtns., Surinam;* Khmara Bay, Enderby Land, Antarctica; South Harris, Scotland; and others	de Roever et al. (1976, 1981); Grew (1981); Baba et al. (2000); Grew et al. (2000); Barbier et al. (2002); Grew (Ch 12)

Sverigeite Orthorhombic $Na(Mn^{2+},Mg)_2Sn^{4+}[Be_2Si_3O_{12}(OH)]$	In calcite vein cutting franklinite ore	*Långban deposit, Filipstad, Värmland Co., Sweden*	Dunn et al. (1984); Rouse et al. (1989b); Nysten et al. (1999)
Swedenborgite Hexagonal $NaBe_4SbO_7$	Calcite veins in hematite-ore	*Långban deposit, Filipstad, Värmland Co., Sweden*	Aminoff (1924, 1933); Huminicki & Hawthorne (2001)
Telyushenkoite (Cs-dominant analogue of leifite) Trigonal $(Cs,Na,\square)\square Na_6Be_2(Al,Si,Zn)_3Si_{15}O_{39}F_2$	Peralkaline granite pegmatite	*Dara-i-Pioz, Tajikistan*	Agakhanov et al. (2002); Sokolova et al. (2002)
Tiptopite (isotypic with cancrinite group) Hexagonal $K_2(Li_{2.9}Na_{1.7}Ca_{0.7}\square_{0.7})Be_6P_6O_{24}(OH)_2\cdot 1.3H_2O$	Granite pegmatites	*Tip Top mine, Custer Co., South Dakota, USA*	Grice et al. (1985); Peacor et al. (1987)
Trimerite (topologically isostructural with beryllonite) Monoclinic $CaMn^{2+}_2(BeSiO_4)_3$	In veins and breccia fillings of calcite and other minerals that cut Fe and Mn ores and skarn	*Harstigen mine, near Filipstad, Värmland Co., Sweden*, also the nearby Långban and Jakobsberg mines	Flink (1890, 1917a); Aminoff (1926); Klaska & Jarchow (1977); Nysten et al. (1999); Nysten (2000)
Tugtupite (isotypic with sodalite and helvite groups) Tetragonal $Na_4AlBeSi_4O_{12}Cl$	Hydrothermal veins and pegmatites in peralkaline rocks; sodalite xenoliths in nepheline syenite	*Ilímaussaq, South Greenland;* Lovozero and Khibiny, Kola Penin., Russia; Mont Saint-Hilaire, Quebec, Canada	Semenov & Bykova (1960); Sørensen (1960, 1962, 1963); Danø (1966); Sørensen et al. (1971); Jensen & Petersen (1982); Hassan & Grundy (1991); Horváth & Gault (1990); Men'shikov et al. (1999); Markl (2001b)

Tvedalite (see note, a possible zeolite mineral) Orthorhombic $(Ca,Mn)_4Be_3Si_6O_{17}(OH)_4{\cdot}3H_2O$ *or* Monoclinic $Ca_2Mn_2Be_3Si_6O_{16}(OH)_6{\cdot}2H_2O$	Nepheline syenite pegmatite	*Vevja quarry, Tvedalen, Oslo region, Norway*	Larsen et al. (1992); S. Merlino (pers. comm.. to Strunz & Nickel 2001)
Unnamed (Gadolinite group; cf. calcio-gadolinite) Monoclinic $Ca_{0.90}(Y_{0.87}REE_{0.19})_{\Sigma=1.06}Fe^{3+}_{0.28}$-$Be_{2.01}Si_{2.01}O_{7.87}(OH)_{2.13}$	Pegmatite	Tahara, near Ena City, Gifu Pref., Japan	Miyawaki et al. (1987)
Uralolite (see note) Monoclinic $Ca_2Be_4(PO_4)_3(OH)_3{\cdot}5H_2O$	In weathering crust over a fluorite-muscovite greisen; granite pegmatites	*Boyevskoye deposit, Middle Urals, Russia*; Newry, Oxford Co., Maine, USA; Foote Mine, Kings Mountain, Cleveland Co., North Carolina, USA; Weinebene, Koralpe, Carintha, Austria; Eitzing, Oberfrauenwald, Bavaria, Germany	Grigor'yev (1964); Dunn & Gaines (1978); Marble & Hanahan (1978); Walter et al. (1990); Sabor (1990); Mereiter et al. (1994); Habel (2000)
Väyrynenite (see note) Monoclinic $Be(Mn^{2+},Fe^{2+},Mg)PO_4(OH,F)$	Granite pegmatites	*Viitaniemi pegmatite, Eräjärvi, Finland;* Ognev-Baken pegmatites, Kalba, Kazakhstan; Bendada, Beira Alta Prov., Portugal; Norrö, Stockholm archipelago, Sweden; Fregeneda area, Salamanca, Spain; and others	Volborth (1954a,b,c); Volborth & Stradner (1954); Mrose & von Knorring (1959), Gordiyenko et al. (1973); Schnorrer-Köhler & Rewitzer (1991); Nysten & Gustafsson (1993); Roda et al. (1996)

Wawayandaite Monoclinic $Ca_{12}Mn^{2+}{}_4B_2Be_{18}Si_{12}O_{46}(OH,Cl)_{30}$	Vein in zinc ore	*Franklin, Sussex Co., New Jersey, USA*	Dunn et al. (1990); Dunn (1995)
Weinebeneite (a zeolite mineral) Monoclinic $CaBe_3(PO_4)_2(OH)_2 \cdot 4H_2O$	Granite pegmatite	*Weinebene, Koralpe, Carintha, Austria*	Walter et al. (1990); Walter (1992)
Welshite (see note, aenigmatite group) Triclinic $Ca_2Mg_{3.8}Mn^{2+}{}_{0.6}Fe^{2+}{}_{0.1}Sb^{5+}{}_{1.5}O_2$-$[Si_{2.8}Be_{1.7}Fe^{3+}{}_{0.65}Al_{0.7}As_{0.17}O_{18}]$	In dolomite filling fractures in hematite ore, calcite veins	*Långban deposit, Filipstad, Värmland Co., Sweden*	Moore (1967, 1971, 1978); Nysten et al. (1999); Grew et al. (2001); Grew (Ch 12)
Zanazziite (see note; Mg-dominant analogue of monoclinic roscherite and greifensteinite) Monoclinic $(Mg,Fe^{2+})Ca_2(Mg,Fe^{2+},Mn^{2+}$,-$Al,Fe^{3+})_4Be_4(PO_4)_6(OH)_4 \cdot 6H_2O$	Granite pegmatites	*Lavra da Ilha pegmatite, Taquaral, Minas Gerais, Brazil;* Newry, Oxford Co., Maine., USA	Leavens et al. (1990); Fanfani et al. (1975, 1977); King & Foord (1994)

APPENDIX 2. LIST OF PROBLEMATIC SPECIES

MINERAL, CRYSTAL SYSTEM, FORMULA	ENVIRONMENT	LOCALITIES	REFERENCES
"Calcio-gadolinite" Not known $CaFe^{2+}(Ce,La)Be_2Si_2O_9(OH)$, a formula calculated and simplified from Nakai's analysis	Granite pegmatite	Tadati (Tadachi), Nagano Pref., Japan; Tatyana pegmatite, Mt. Ulin-Khuren, Altai, Mongolia	Nakai (1938); Ito (1967); Kartashov & Lapina (1994)
Calcium-rich gadolinite Monoclinic? $(Y,REE)_2(Ca,Fe)Be_2Si_2O_{10}$?	Granite pegmatite	Lindvikskollen, Kragerø, Norway	Oftedal (1972)
"Gelbertrandite". Probably is colloidal or cryptocrystalline bertrandite Orthorhombic. $Be_4Si_2O_7(OH)_2 \cdot nH_2O$	Peralkaline pegmatites; Be-bearing tuff	*Karnasurt and Mannepakht, Lovozero, Kola Penin., Russia;* Spor Mtn., Juab Co., Utah, USA	Semenov (1957); Montoya et al. (1962)
"Glucine" Not known $CaBe_4(PO_4)_2(OH)_4 \cdot 0.5H_2O$	In weathering crust over a fluorite-muscovite greisen; granite pegmatite (?)	*Boyevskoye deposit, Middle Urals, Russia;* Mt. Mica, Paris, Maine, USA (this report needs further substantiation)	Grigor'yev (1963), Pokrovskiy et al. (1965); Ginzburg & Shatskaya (1966); King & Foord (1994)
Leucophanite, cerian (potentially a distinct species because of its triclinic symmetry) Triclinic $(Ca,Ce,Nd,La,Pr)CaNa_2Be_2Si_4O_{12}(F,O)_2$	Pegmatite associated with nepheline syenite? (paragenesis not specified)	*Mont Saint-Hilaire, Quebec, Canada*	Cannillo et al. (1992)
"Makarochkinite" (see note; cf. høgtuvaite. Aenigmatite group) Triclinic $(Ca_{1.76}Na_{0.19}Mn_{0.05})(Fe^{2+}_{3.66}Fe^{3+}_{1.36}Ti_{0.54}$-$Mg_{0.25}Mn_{0.10}Nb_{0.06}Sn_{0.02}Ta_{0.01})O_2[Si_{4.45}BeAl_{0.49}$-$Fe^{3+}_{0.06}B_{0.01}O_{18}]$	Granite pegmatite	*Ishkul' Mtn., Ilmeny Natural Reserve, south Urals, Russia*	Polyakov et al. (1986); Yakubovich et al. (1990); Grauch et al. (1994); Barbier et al. (2001); Grew (Ch 12)

Appendix 2, continued. List of problematic species

"Muromontite" ("berylliumorthite", i.e., beryllium allanite; epidote group) Not known	Pegmatites	*Mauersberg, Saxony, Germany;* Næs grube, Arendal, S. Norway; Iisaka, Fukushima Pref., Japan; Skuleboda quarry, Uddevalla, Bohuslän, Sweden	Kerndt (1848), Forbes & Dahll (1855); Iimori (1939), Quensel & Alvfeldt (1945); Machatschki (1948); Hasegawa (1960)
Odintsovite: possible Li-dominant analogue Orthorhombic $K_2(Na,Ca,Sr)_4(Li,Na)Ca_2(Ti,Fe^{3+},Nb)_2O_2[Be_4Si_{12}O_{36}]$	Veins associated with peralkaline intrusive	*Ilímaussaq, South Greenland*	Petersen et al. (2001)
Unnamed (Gadolinite group) Presumably monoclinic $(Ca,Y,REE)_2(\square,Fe)\ Be_2Si_2O_8(O,OH)_2$	Aplite-pegmatite in peralkaline complex	Strange Lake, Quebec-Labrador, Canada	Jambor et al. (1998)
Unnamed (UK-48) (Gadolinite group) Monoclinic $CaY_{2-x}(Si,Be,B)_4(O,OH)_{10}\cdot 2H_2O$	Peralkaline syenite	Mont Saint-Hilaire, Que., Canada	Chao et al. (1990); Wight & Chao (1995)
Unnamed (UK48A) (Gadolinite group) Monoclinic $CaCe_{2-x}(Si,Be,B)_4(O,OH)_{10}\cdot 2H_2O$ (?)	Peralkaline syenite	Mont Saint-Hilaire, Que., Canada	Wight & Chao (1995)
Unnamed brown Ce-silicate possibly related to steenstrupine. Metamict. Suggested formula: $CaMn(Ce,La,Nd)_3Al_2Be_2Si_6O_{23}(OH)\cdot 6H_2O$	Peralkaline pegmatite	Ilímaussaq, South Greenland	Semenov (1969)

APPENDIX 3. MINERAL SYNONYMS OF RECENT ORIGIN

SYNONYM	EQUIVALENT	ORIGINAL DESCRIPTION	EQUIVALENCY EXPLAINED
"Yberisilite" (Tuva mineral of Semenov et al. 1963)	Hingganite-(Y)	Povarennykh (1972)	Ding et al. (1984); *Am. Mineral.* 73:935; Petersen et al. (1994a); Bokiy et al. (1996)
"Yttroceberysite"		Ding et al. (1981)	
"Xinganite"		Ximen & Peng (1985)	
"Karpinskyite"	Leifite	Shilin (1956)	Micheelsen & Petersen (1970)
"Beryllosodalite"	Tugtupite	Semenov & Bykova (1960)	Danø (1966)
"Ginzburgite"	Roggianite	Voloshin et al. (1986)	Passaglia & Vezzalini (1988), Giuseppetti et al. (1991)
"Taprobanite"*	Magnesiotaaffeite-2*N*'2*S*	Moor et al. (1981)	Fleischer (1982, *Am. Mineral.* 67:1076); Schmetzer (1983); Armbruster (2002)
Taaffeite	Magnesiotaaffeite-2*N*'2*S*	Anderson et al. (1951)	Armbruster (2002)
"Musgravite"*	Magnesiotaaffeite-6*N*'3*S*	Hudson et al. (1967); Schmetzer (1983)	
"Pehrmanite"*	Ferrotaaffeite-6*N*'3*S*	Burke & Lustenhouwer (1981)	

*Originally had been approved by IMA CNMMN; taaffeite is still accepted as a group name.

NB. Transliteration of the Cyrillic letter И/и gives either Roman letters I/i or Y/y, e.g., yberisilite or iberisilite. The various spellings of hingganite, e.g., "xinganite", result from the difficulty of rendering the Chinese name of the type locality, Xing'an/Hinggan, into English.

2 Behavior of Beryllium During Solar System and Planetary Evolution: Evidence from Planetary Materials

Charles K. Shearer

Institute of Meteoritics
Department of Earth and Planetary Sciences
University of New Mexico
Albuquerque, New Mexico 87131
cshearer@unm.edu

INTRODUCTION

A majority of this volume of *Reviews in Mineralogy and Geochemistry* is dedicated to the distribution and behavior of Be in terrestrial environments, yet the Earth makes up less than 0.02% the mass of the solar system. Therefore, it seems sensible to explore the behavior of Be outside our terrestrial home. In addition to remotely sensed data from planetary and solar surfaces, there is a large suite of planetary materials that have been used for exploring the geochemistry and mineralogy of the solar system (Shearer et al. 1998). The first planetary samples in our collection were delivered to Earth as meteorites. These samples include material from numerous asteroids, the Moon, and Mars. This is an ever-growing collection of over 7000 well-documented samples with a total mass greater than 16,000 kg. In the second half of the 20th century, humans have been far less passive and much more systematic in collecting materials from other planetary bodies. Robotic and human sample return missions have been made to the Earth's Moon, where over 2200 individual samples with a total weight of 384 kg have been collected. United States, Japanese and European survey teams have recovered over 16,000 meteorites in Antarctica. Interplanetary dust particles (<1 gram, ≈ 10,000 samples) are actively being collected in the stratosphere, from terrestrial polar ice, deep-sea sediments, and within impact features on spacecraft. Exciting new missions are underway or being planned to return samples from Mars, asteroids, comets, and other moons.

Since the pioneering work that laid the foundation of cosmochemistry (Oddo 1914; Harkins 1917; Goldschmidt 1926; Russell 1929; Atkinson and Houtermans 1929), significant inroads have been made into understanding the synthesis of Be in the cosmos and its behavior in extraterrestrial environments. Early studies from the 1960s and 1970s primarily emphasized the bulk Be contents of different meteorite groups. Much more recently (1990s), ion microprobe studies of individual phases in extraterrestrial materials have improved our understanding of its behavior in a variety of environments in the solar system. With this additional data from these studies, Be in planetary materials has been used to better understand stellar evolution, reconstruct early solar system and planetary processes, reconstruct planetary compositions and evaluate magmatic and impact processes.

THE ABUNDANCE OF BE IN THE SOLAR SYSTEM

The combination of data from the Sun and meteorites permits an estimate to be made of the solar system abundances of the elements. These abundances are shown in Figure 1 as the relative numbers of atoms (normalized to the value of 10^6 for silicon) as a function of atomic number. This abundance diagram has several noteworthy characteristics. First, the most abundant elements are the lightest ones, hydrogen and helium. They make up over 99% of the total mass of the solar system. The atomic ratio of H:He is about 10:1.

1529-6466/00/0050-0002$05.00

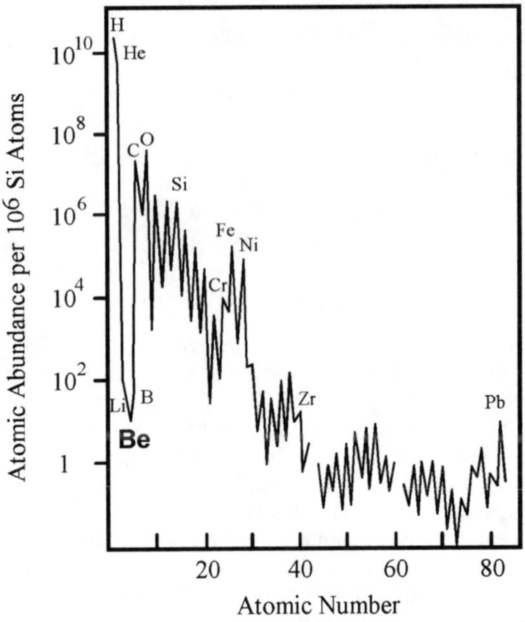

Figure 1. Solar system abundances of the elements showing the relative number of atoms normalized to the value of 10^6 for silicon.

Second, there is a general decrease in abundance with atomic number. Third, superimposed on this general decline is a conspicuous alternation of abundances in which elements with even atomic numbers are more abundant than the two adjacent elements with odd atomic numbers. This feature, first noted by Oddo (1914) and Harkins (1917), is a consequence of the nuclear processes by which the elements are synthesized. It is not the intent of this discussion to delve into a summary of nucleosynthesis, as many excellent introductions to this topic are available (e.g., Burbridge et al. 1957; Trimble 1975; Fowler 1984; Woolum 1988; Clayton 1988; Cox 1989; Reeves 1994). In addition to these overall features in elemental abundances, there are a number of peaks and valleys in the plot, most notably the peak at masses including and adjacent to iron (Z = 25 to 30) and the valley defined by Li, Be, and B (Z = 3, 4, 5, respectively). The questions as to what processes are reflected in the depletion of Be and how and where Be was synthesized are critical to our understanding of the early evolution of the solar system.

In his seminal geochemistry studies, Goldschmidt (1926, summarized in 1954) pointed out that the light elements Li, Be, and B are deficient in the solar system and that this deficiency must be the result of their instability during certain nuclear processes. Gamow (1928) and Gurney and Condon (1928) investigated the process of alpha-decay based on the wave mechanics in order to explain the experimentally found relationship between the decay constant and the energy of the alpha particles (Geiger and Nuttall 1912). Using these studies, Atkinson and Houtermans (1929) calculated the probability for the penetration of the protons into the nucleus of Be at stellar temperatures (1 to 4×10^7 degrees K). Based on their results, they concluded that the isotopes of Be are destroyed as a result of collisions with thermally accelerated protons in stellar atmospheres.

Several nucleosynthetic reactions have been proposed to produce and then consume

Be in a solar environment. During the early stages of solar evolution, exothermic hydrogen burning reactions produced a central core of helium. A first stage in He burning would be the reaction:

$$^{4}He + ^{4}He \Rightarrow ^{8}Be$$

This reaction is endothermic and the ^{8}Be nucleus is unstable by less than 0.1 MeV. In the stellar environments, the sustained conditions of high temperature and density give rise to a small equilibrium concentration of ^{8}Be. The small amount of ^{8}Be produced by this reaction can then capture an additional ^{4}He to produce ^{12}C:

$$^{8}Be + ^{4}He \Rightarrow ^{12}C + \gamma \ (\gamma = \text{photon}).$$

Be formation and destruction in stars is complex and a number of additional chains of thermonuclear reactions have been proposed by Gamow (1938):

$$^{3}He + ^{4}He \Rightarrow ^{7}Be + \gamma$$
$$^{7}Be + e^{-} \Rightarrow ^{7}Li + \nu \ (e^{-} = \text{electron}, \nu = \text{neutrino})$$
$$^{7}Li + ^{1}H \Rightarrow ^{4}He + ^{4}He,$$

$$^{3}He + ^{4}He \Rightarrow ^{7}Be + \gamma$$
$$^{7}Be + ^{1}H \Rightarrow ^{8}B + \gamma$$
$$^{8}B \Rightarrow ^{4}He + ^{4}He + e^{+} + \nu \ (e^{+} = \text{positron}),$$

$$^{4}He + ^{4}He \Rightarrow ^{8}Be + \gamma$$
$$^{8}Be + ^{1}H \Rightarrow ^{9}B + \gamma$$
$$^{9}B \Rightarrow ^{9}Be + e^{+}$$
$$^{9}Be + ^{1}H \Rightarrow ^{4}He + ^{4}He + ^{2}H$$

Because isotopes of Be are very unstable at the temperatures of stellar interiors, it appears probable that they were produced in regions of low density and temperature. Burbridge et al. (1957) referred to the non-stellar processes that were responsible for the synthesis of Be, in addition to deuterium, Li and B, as x-processes. They have also been referred to as l-processes in the literature (Woolum 1988). Since 1957, the nature of the x-processes has been defined in substantial detail. A variety of mechanisms have been suggested for the production of Be, primarily involving the breakup of heavier elements (i.e., C,N,O) through nuclear reactions with protons and alpha particles. Proposed processes include (1) reactions in regions near the stellar surface (Bernas et al. 1967; Hayakawa 1968); (2) reactions caused by the interaction of high-energy charged particles derived from the surface of the early Sun with the outer layers of planetesimals (Fowler et al. 1955, 1962); (3) reactions between galactic cosmic rays and nuclei in interstellar matter (Reeves et al. 1970; Meneguzzi et al. 1971; Mitler 1972; Reeves et al. 1973; Cassé et al. 1995; Cameron 1995; Ramaty et al. 1997, 1998); (4) reactions in supernova waves (Cameron 1973; Audouze and Truran 1975); and (5) cosmological nucleosynthesis (Wagoner et al. 1967).

Process 1 appears to be unlikely simply because even at the stellar surface, it is expected that Be will be consumed as a result of collisions with thermally accelerated protons (Atkinson and Houtermans 1929). The variation of Be and Li with spectral class of star is best explained in terms of the destruction mechanisms at the stellar surface (Reeves et al. 1970). In addition, ratios of Be to Li and B have been interpreted by Reeves et al. (1970) to indicate that their generation involved high energy processes such as spallation, rather than thermonuclear reactions. Process 2 calls upon an early irradiation of the planetesimals to produce Be. If this model is correct, this process should be reflected in the difference between the isotopic composition of meteoritic and terrestrial gadolinium (Murthy and Schmitt 1963; Eugster et al. 1970). These studies concluded that the meteoritic and terrestrial gadolinium had the same isotopic composition and therefore eliminates the model for early irradiation of small bodies as a process for producing Be.

Numerous lines of evidence indicate Be is produced by spallation reactions in galactic cosmic rays. Galactic cosmic rays consist mostly of particles (i.e., C,N,O, protons, alpha particles) traveling at very high velocities. They come from high-energy sources outside our solar system and were generated from supernovae explosions or accelerated out of the ambient interstellar medium. The debate concerning these models was discussed by Ramaty et al. (1998) and Webber (1997). Laboratory experiments have demonstrated that spallation processes analogous to high-energy cosmic ray protons hitting heavier nuclei would produce nuclei fragments in the mass range of 6 to 11. Jacobs et al. (1974) reported that analysis of comparative production cross-sections from carbon-nitrogen-oxygen targets indicated that it is possible to match the solar system abundance of Be with spallation between interstellar C,N, and O and cosmic ray protons (and alpha particles) of energies below about 25 MeV and spallation of cosmic ray C, N, and O in collision with interstellar H and He. Additional evidence for a spallation origin for Be comes from the elemental abundances found in cosmic rays (Trimble 1975). Although these abundances generally follow the pattern of solar system abundances, the peaks and valleys are considerably smoothed out (Fig. 2). In particular, the elements Li, Be and B are relatively more abundant in cosmic rays. This observation is consistent with the concept that the cosmic rays started with a "normal" pattern of elemental abundances that was modified by spallation processes on their journey through space.

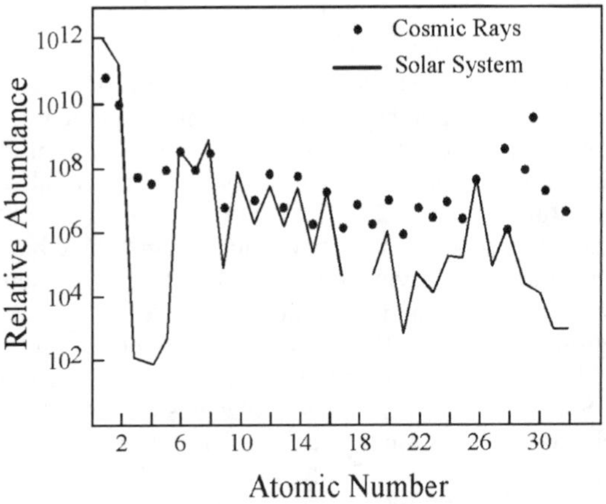

Figure 2. Comparison between the elemental composition of cosmic rays and solar system abundances. The two sets of data have been normalized to carbon. Although many elemental abundances in cosmic rays tend to reflect their solar system abundances, Be, Li, and B are substantially enriched in cosmic rays (after Trimble 1975; Cox 1989).

A lingering problem for producing the light elements only by cosmic ray spallation is that the B and Li isotopic ratios and the B/Be ratio in meteorites and interstellar medium are substantially different from that calculated for production by cosmic rays (Ramaty et al. 1997, 1998). For example, the estimates for the B and Li isotopic ratios for the solar system are $^{11}B/^{10}B = 4.0 \pm 0.1$ and $^{7}Li/^{6}Li = 12.1 \pm 0.1$, whereas that inferred from interstellar production is $^{11}B/^{10}B = 2.5$ and $^{7}Li/^{6}Li = 1.5$ (McKeegan et al. 2000). In addition, the observations of B abundance made with the Hubble Space Telescope show that the B/Be abundance ratio remains essentially constant, implying a common origin for these two elements (Ramaty et al. 1998). Meneguzzi et al (1971) proposed that differences in these ratios could be accounted for if an extra source of low energy cosmic rays could exist above that predicted by the power law spectrum. Based on the recent detection of an excess of γ-rays in the direction of the star-forming region in the Orion cloud, Cassé et al. (1995) suggested that Orion-like low-energy cosmic rays could make a significant contribution to the total Galactic light element inventory. They calculated that these low-energy cosmic rays have the ability to yield by spallation $^{11}B/^{10}B$ and B/Be ratios that were more similar to meteorites and interstellar medium. Alternatively, Woosley and Weaver (1995) and Field et al. (1995) suggested that the excess ^{11}B observed in meteorites and interstellar medium could be due to ^{12}C spallation by neutrinos in core-collapse supernovae.

BEHAVIOR OF BE DURING EVOLUTION OF THE EARLY SOLAR SYSTEM: EVIDENCE FROM METEORITES

Introduction

Meteorites represent a range of planetary environments and record a number of processes from presolar events, solar system condensation and early planetesimal assemblage (chondrites), early planetesimal-planetary differentiation and magmatism (iron meteorites, stony iron meteorites and achondrites), to large-scale planetary processes (lunar meteorites, SNC meteorites). Although there are abundant whole-rock analyses of Be in ordinary chondrites, there are limited bulk Be data for many meteorite groups. More recently, ion microprobe analyses have been used to better understand the evolution of the components making up the more primitive meteorites such as the various chondrite groups.

Classification of meteorites

Meteorites are divided into three major classes: iron, stony irons, and stones (Table 1). A typical iron meteorite consists mostly of metallic iron with 5 to 20% Ni. Iron meteorites most likely represent cores of disrupted asteroids or impact melts. Stony iron meteorites are divided into two groups: pallasites and mesosiderites. Pallasites consist of metal and silicates (mostly olivine). The metal is similar to that found in fractionated iron meteorites. Pallasites may represent cumulates from core-mantle boundaries of disrupted asteroids. Mesosiderites are composed of metal mixed in with basaltic, gabbroic, and pyroxenitic lithologies. Olivine is a minor component. The metal is distinct from that in the iron meteorites and more closely resembles metal found in unfractionated chondritic meteorites. The mesosiderites appear to represent surface-derived mixtures of crustal lithologies and perhaps an unfractionated metal derived from either the core or the surface.

There are many different types of stony meteorites. They are classified into two groups: chondrites and achondrites. Chondrites are agglomeritic rocks consisting of a wide variety of components (interstellar grains, refractory inclusions, chondrules, matrix, sulfides, silicates, Fe-Ni metal) with very different formational histories. These arrays of components provide information about events that predate solar system formation, primary processes that occurred during the earliest stages of solar system formation, and secondary processes that occurred following the assembly of asteroids. In contrast, achondrites are of asteroidal or planetary origin and reflect more extensive differentiation processes (high temperature metamorphism, melting, core formation). Achondrites such

Table 1. Classification of meteorites and potential planetary origins.

Meteorite groups			*Origin*
Irons			Cores of disrupted asteroids or impact melts
	Ataxites		
	Octahedrites		
	Hexahedrites		
Stony Irons			
	Pallasites		Cumulates from asteroidal core-mantle boundary
	Mesosiderites		Impacted melted crustal igneous lithologies
Stones			
	Achondrites		
		HED	Basaltic melts and cumulates, impact mixtures
		Aubrites	Melted and crystallized residue from partial melting
		Acapulcoites	metamorphism, partial melting and melt migration
		Lodranites	
		Winonaites	
		Urelilites	Partial melt residue
		Angrites	Basaltic melts and cumulates
		Brachinites	Metamorphosed chondrite, partial melt residue
		SNC	Basaltic melts and cumulates from Mars
		Lunar	Basalts, highland lithologies, impact mixtures from the Moon
	Chondrites		Agglomeritic rocks consisting of a wide variety of early solar system components that have been altered and metamorphosed to varying degrees.
		Carbonaceous	
		Ordinary	
		Enstatite	
		R	

as HED (Howardites, Eucrites, Diogenites), Aubrites, Acapulcoites, Lodranites, Winonaites, Urelites, Angrites and Brachinites, formed as a result of differentiation processes on an asteroidal scale. SNC (Shergottites, Nakhlites, Chassignites) and lunar meteorites represent differentiation on a planetary scale (Mars and the Moon, respectively). An extensive review of the mineralogy and petrology of these various types of meteorites can be found in *Reviews in Mineralogy*, Volume 36, "Planetary Materials" (1998).

Be abundance in meteorites

Beryllium (along with Li and B) is more abundant in meteorites than in the photosphere of the Sun (Fig. 3) because it is consumed in the solar interior by nuclear reactions (Fig. 3). Based on a available data set (Sill and Willis 1962; Quandt and Herr 1974; Newsom 1995), Be abundances in bulk meteorites exhibit a rather wide variation within and among meteorite groups (Fig. 4). The studies of Sill and Willis (1962) and Quandt and Herr (1974) analyzed Be in the bulk meteorites by extraction-fluorometrically techniques and irradiation-photoneutron counting techniques, respectively.

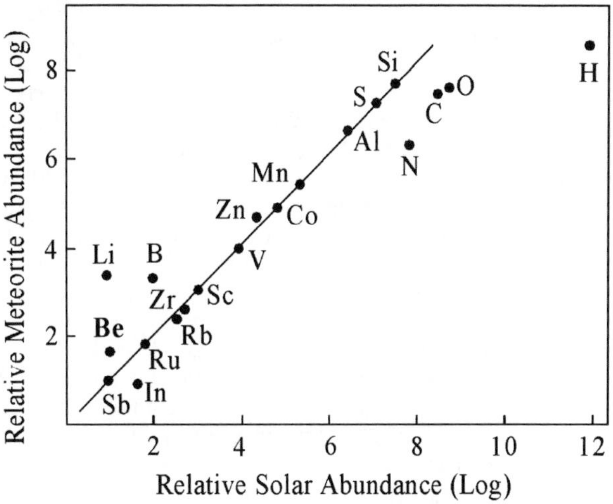

Figure 3. Comparison of elemental abundances (atomic numbers between 1 and 51) between carbonaceous chondrite meteorites and the Sun (after Cox 1989). Line is a 1:1 correspondence.

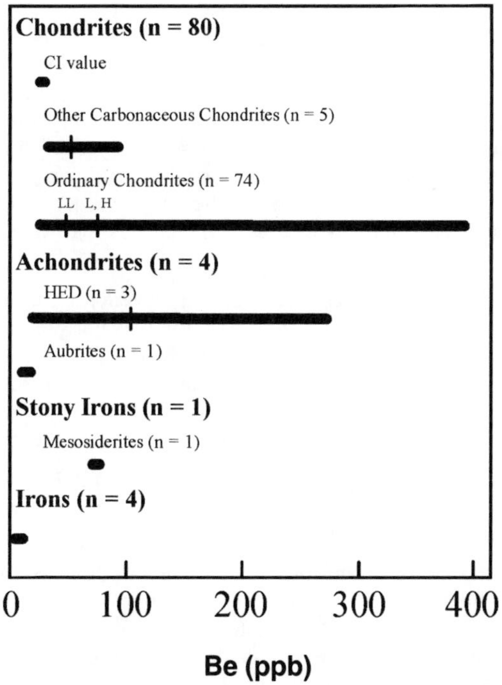

Figure 4. Concentration of Be in a variety of different meteorites. Vertical marks crossing the horizontal bars are arithmetic means for each group. Data compiled from Sill and Willis (1962), Quandt and Herr (1974) and Newsom (1995).

The CI (C = carbonaceous chondrite, I = Ivuna meteorite) carbonaceous chondrites are volatile-rich meteorites that have the most solar-like chemical compositions of all the primitive meteorite types. By virtue of this chemical characteristic, the elemental abundances in the CI carbonaceous chondrites have been used as elemental normalization factors to illustrate how far other planetary materials deviate from the bulk composition of the solar system. Although chemically primitive, the primary mineralogy of the CI carbonaceous chondrites has been altered at low temperature (<200°C) by aqueous solutions. As a result of this low temperature alteration, these meteorites are the most friable and delicate of the carbonaceous chondrites and are dominated by low temperature phases such as hydrous silicates (serpentine, smectite), sulfates, and carbonates. Only five specimens are known. The abundance of Be in CI carbonaceous chondrites has been compiled by Anders and Grevesse (1989) and Wasson and Kallemeyn (1988). The average CI abundance estimated by these two studies is 24.9 ppb and 27 ppb, respectively. The Li/Be of CI based on these two studies is approximately 60, whereas B/Be ranges from 35 to 44. Even in this fairly primitive sample of the solar system, Be is more abundant than in the photosphere. However, Be in CI does not deviate as much from solar abundances as do B or Li (Fig. 3).

The Be content in other carbonaceous chondrites (CM, CV, CO) ranges from 34 to 93 ppb. Numerous ordinary chondrites (H, L, LL) that have been analyzed for Be. They exhibit a variation in Be from 20 to 386 ppb. The range in the Be content for the LL ordinary chondrites (41-62 ppb) is far less than that of the H and L. Among the H and L ordinary chondrites, those that are the least affected by metamorphism (petrologic type 3) have limited variability, whereas many of those affected by thermal metamorphism (petrologic type 4-6) have a wider range in Be. The most common textural and mineralogical changes in ordinary chondrites resulting from thermal metamorphism, are (1) chondrule textures and outlines become increasingly obscured; (2) chondrules become increasingly devitrified; (3) matrix become increasingly recrystallized and coarser-grained; and (4) mineral compositions become increasingly equilibrated. It is unlikely that the elevated Be contents observed in several of these ordinary chondrites are due to extensive terrestrial contamination, because the elevated Be occurs in meteorites recovered soon after observed falls as well as in meteorites that were found many years after they fell.

Only four achondritic meteorites have been analyzed for Be: an aubrite (Norton County, highly reduced pyroxenite) and three HED meteorites (Johnstown, Pasamonte, and Sioux County). Of the HED meteorites, Johnstown is diogenite (orthopyroxenite), whereas Pasamonte, and Sioux County are unequilibrated and equilibrated eucrites (basalts), respectively. The two pyroxenites have low Be (6-13 ppb), whereas the two eucrites have elevated Be (37 and 276 ppb). The eucrite with the elevated Be was an observed fall. The only mesosiderite that was analyzed has a Be concentration of 76 ppb, while the iron meteorites have less than 8 ppb Be.

Behavior of Be during solar system condensation

Most models for the origin of the solar system are variations of the nebular models, in which the sun and then the planets condensed from an interstellar cloud of gas and dust. Observational conformation of this inceptive condition of our solar system is found in the star-forming regions of the Orion nebula and the Taurus molecular cloud (O'Dell and Beckwith 1997) . The behavior of Be during condensation of the solar nebula has been predicted by both mineralogical studies of refractory or calcium-aluminum-rich inclusions (CAIs) in chondritic meteorites and thermodynamic calculations of a nebular condensation sequence.

CAIs are a mineralogically and chemically diverse group of objects that occur mainly in carbonaceous chondrites. The CAIs are <1mm to ~10 cm in diameter and have mineralogies that are dominated by compounds having very high vaporization temperatures (~1300 K; Grossman and Larimer 1974; Ebel and Grossman 2000). Among the refractory phases present in the CAIs are corundum, hibonite, grossite, perovskite, anorthite, spinel, melilite, and "fassaite" (in the meteorite literature the obsolete term "fassaite" is used often incorrectly to refer to an aluminian titanian diopside or augite) (Fig. 5). Their formation in a high temperature space environment predates that of chondrites and chondrules. Therefore, they provide a record of some of the earliest events in the formation and evolution of the solar system. Of the carbonaceous chondrites, the CV chondrites contain the highest abundance and the most mineralogically diverse CAIs. Although far less common, Ca-, Al-rich objects which appear to be CAIs have been found in ordinary (Noonan 1975; Noonan and Nelen 1976; Bischoff and Keil 1983, 1984) and enstatite chondrites (Bischoff et al. 1983). The classification of CAIs has been based largely on the studies of Allende, a CV chondrite, and extended to other chondrites. These classification schemes have been based on either bulk trace element composition (e.g., Mason and Taylor 1982; MacPherson et al. 1988) or petrographic properties (Grossman 1975). The bulk composition classification scheme is based on the shapes of the chondrite-normalized REE patterns for the CAIs. REE pattern-shape reflects REE behavior during both condensation and post-condensation (evaporation, melting, aqueous alteration) processes (Fig. 6). Alternatively, Grossman (1975) divided CAIs in Allende into "coarse grained" and "fine-grained" inclusions. The "coarse-grained" inclusions were further divided into 3 groups (A-, B-, and C-types) based on the modal abundances of primary minerals "fassaite", melilite and the remaining phases (anorthite, perovskite, hibonite, and spinel) (Fig. 7). Type B1 CAIs have melilite-rich mantles and "fassaite"-rich cores. In type B2, these phases are evenly distributed. Melilite is primarily a simple solid solution between gehlenite ($Ca_2Al_2SiO_7$) and åkermanite ($Ca_2MgSi_2O_7$). In some cases Na_2O is present (< 0.1 to 0.5 wt %) and increases systematically with åkermanite content (MacPherson and Davis 1994). The "fine-grained" CAIs were further divided into spinel-rich inclusions and amoeboid olivine aggregates (Grossman and Ganapathy 1976). The abundance and distribution of Be and other trace elements in CAIs are important because they provide us with information concerning the behavior of refractory trace elements in the early solar system, can help us distinguish stellar from solar system processes, may yield information on light element nucleosynthesis, and may be a useful chronometer for the early solar system (Weller et al. 1978; Spivack et al. 1988, 1989; Beckett et al. 1990). In addition, the role of volatilization and fractional condensation in the origin of CAIs, temperatures and oxygen fugacities of nebular gas reservoirs, and the nature of trace element carriers have all been inferred from trace element studies of bulk inclusions (Boynton 1989; Fegley and Palme 1985; Wark 1987). Trace-element distribution in individual phases in CAIs provides clues into the nature of these processes. For example, complimentary REE patterns of coexisting pyroxene and melilite suggests crystallization from a melt (Mason and Martin 1974)

Spivack et al., (1989) measured Be contents of bulk and individual phases for three type-B and four fine-grained CAIs. The bulk Be concentrations for the coarse-grained CAI was 510 ppb. The fine-grained CAIs had bulk Be concentrations between 230 and 560 ppb. All of these CAIs were significantly enriched in Be relative to CI by factors of between 8.5 and 21. The data of Spivack et al. (1989) also showed that Be is systematically distributed among phases in the CAIs. Beryllium is concentrated in melilite relative to coexisting pyroxene, plagioclase, and spinel. In the coarse-grained, type-B CAIs, Be in melilite ranges from 0.39 to 0.57 ppm and is 2.3 to 3.8 times higher than the associated pyroxene. In melilite, there is also a strong positive correlation between åkermanite component and Be abundance. In the melilite mantles of one of the

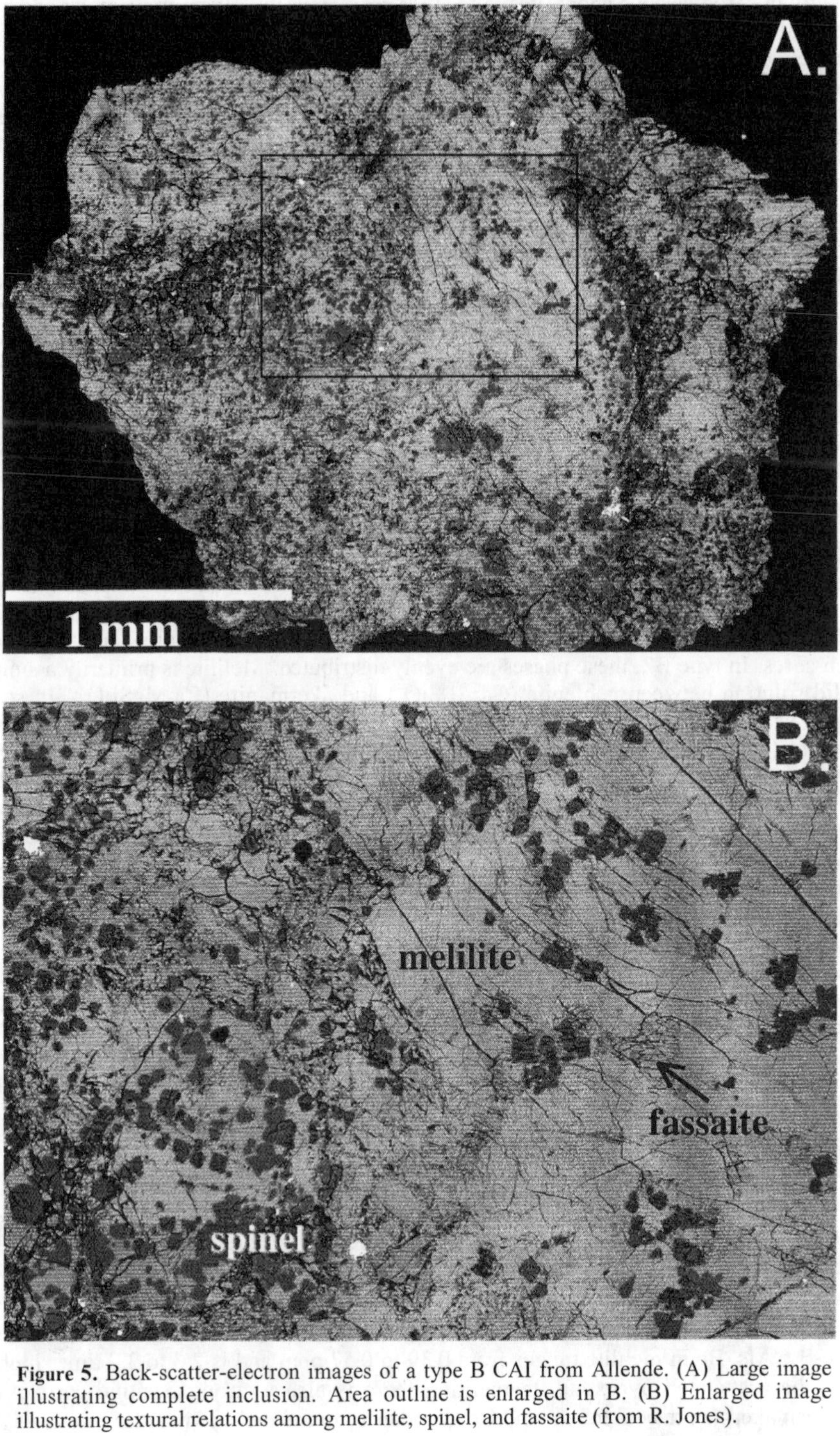

Figure 5. Back-scatter-electron images of a type B CAI from Allende. (A) Large image illustrating complete inclusion. Area outline is enlarged in B. (B) Enlarged image illustrating textural relations among melilite, spinel, and fassaite (from R. Jones).

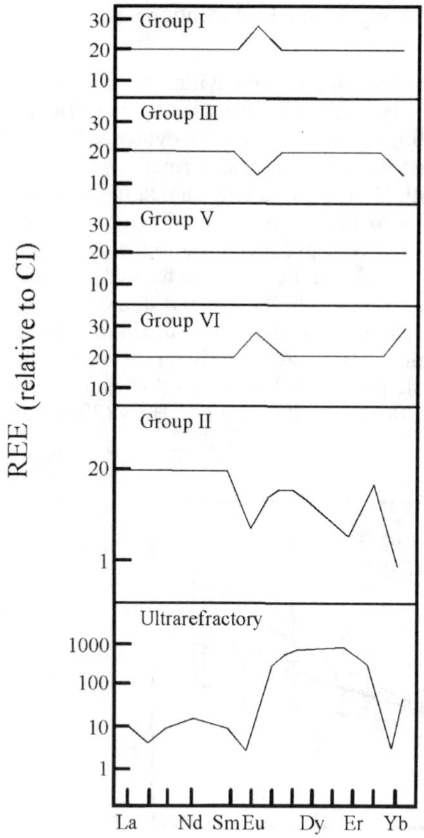

Figure 6 (left). Chondrite-normalized schematic REE patterns observed in refractory inclusions from different compositional groups (after MacPherson et al. 1988).

Figure 7 (below). Ternary diagram that illustrates the mineralogical compositions of Type A, B, and C refractory inclusions in terms of their abundance of melilite, "fassaite", and other phases (anorthite, perovskite, hibonite, and spinel) (after Wark 1987).

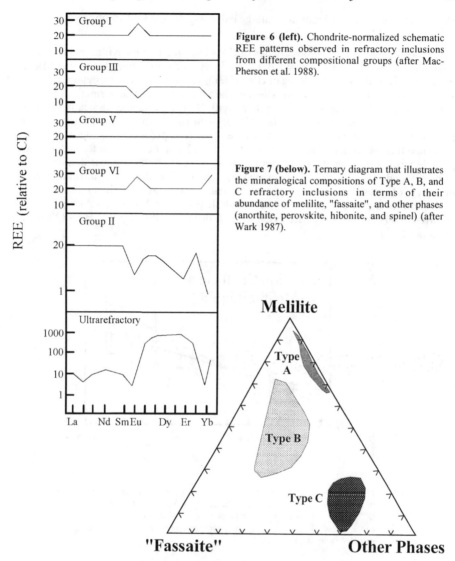

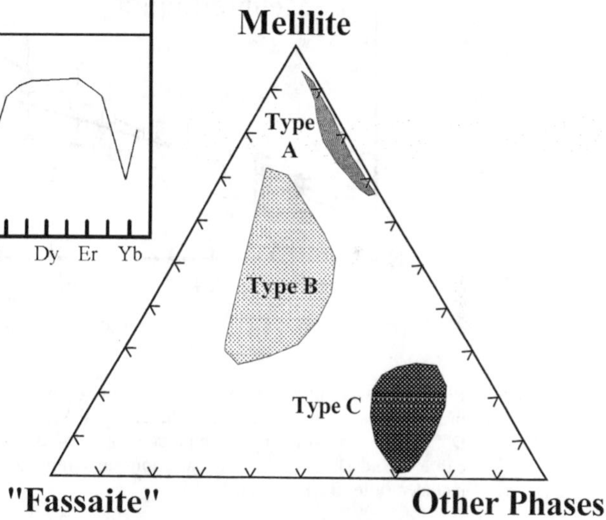

type-B CAIs studies by Spivack et al. (1989), the Be abundance increases with distance from the rim of the CAI (20 mol % åkermanite), toward the center of the inclusion. The melilite with the highest Be abundance (≈ 5 ppm) and åkermanite content (60 mol %) occurs near the "fassaite" core of the CAI. Spivack et al. (1989) suggested that the relationship between melilite composition and Be content may be due to either the higher distribution coefficient of Be for åkermanite relative to gehlenite or simply the incompatible behavior of Be. Their crystal chemical rational for the higher distribution coefficient for åkermanite is based on the preference of Be for the Mg tetrahedral site rather than the Al tetrahedral site in melilite. Åkermanite and gugiaite ($Ca_2BeSi_2O_7$) are isostructural, with Be^{2+} and Mg^{2+} occupying equivalent tetrahedral sites. In gehlenite, there are no tetrahedrally coordinated sites occupied by divalent cations, and therefore

Be^{2+} substitutions for Al^{3+} require a charge balancing coupled substitution such as $Be^{2+} + Si^{4+} \Leftrightarrow Al^{3+} + Al^{3+}$.

More recent ion microprobe studies of CAIs have focused on evidence for the extinct radioactive isotope of Be ($^{10}Be \Rightarrow {}^{10}B$, $t_{1/2} = 1.5$ My). The fingerprint for extinct ^{10}Be is an excess of ^{10}B in the CAI. McKeegan et al. (2000) reported persuasive evidence that ^{10}Be was present in a type B CAI from Allende when the inclusion was formed. In one of the CAIs analyzed, ^{10}B excesses are correlated with Be/B in a manner indicating that short lived ^{10}Be was incorporated into the CAI prior to (or during) crystallization and was preferentially partitioned into melilite where it decayed producing radiogenic ^{10}B. They estimated that the initial ($^{10}Be/^{9}Be$) was ~ 1×10^{-3} and that the ^{10}Be was formed by intense particle irradiation near the early sun (McKeegan et al. 2000). Chaussidon and Robert (1995) found variable $^{11}B/^{10}B$ in meteoritic chondrules, that they attributed to the presence of presolor grains that were nucleosynthesized in different interstellar environments and incorporated into the chondrule precursor. Follow-up studies have been reported in abstract form by McKeegan et al. (2001), MacPherson and Huss (2001), and Sugiura (2001).

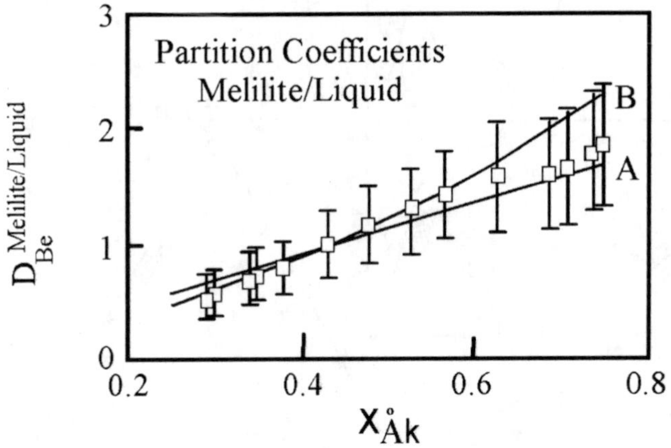

Figure 8. Distribution coefficients for Be between melilite and liquid in a CAI bulk composition as a function of $X_{åkermanite}$. Curve A is $D^{melilite/liquid}$ with variations in temperature and liquid ignored, whereas curve B and $D^{melilite/liquid}$ with variations in temperature and liquid included (after Beckett et al. 1990).

Further experimental and theoretical investigations of the behavior of Be in phases in CAIs were carried out by Beckett et al. (1990). The study of Beckett et al. (1990) determined distribution coefficients for Be between melilite and liquid over the entire range of melilite + spinel crystallization for a bulk composition corresponding to that of an average Type B inclusion from the Allende carbonaceous chondrite (Fig. 8). In this study, zoned crystals of melilite were grown during controlled cooling experiments to observe trace element zoning trends in the melilite (Fig. 9). They found that divalent cations (i.e., Be) substituting into the tetrahedral coordinated T_1 site have partition coefficients proportional to $X_{åkermanite}$. Beryllium is incompatible in gehlenitic melilite ($X_{åkermanite} < 0.43$), but compatible in åkermanitic melilite ($X_{åkermanite} > 0.43$). Beckett et al. (1990) calculated equations for melilite/liquid partition coefficients that included and ignored the effect of temperature and melt composition:

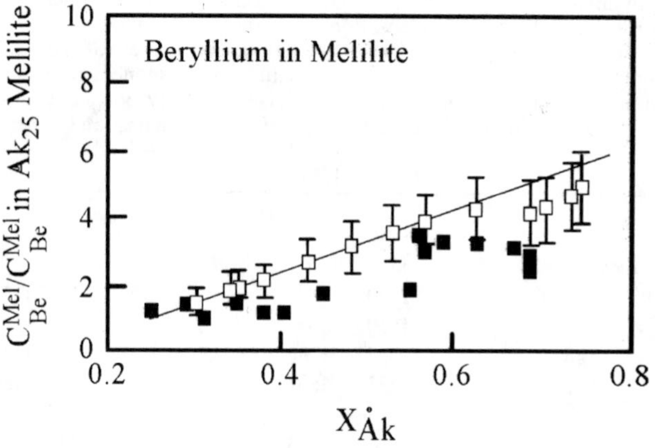

Figure 9. Ratio of Be concentration in melilite to the concentration at the onset of melilite crystallization for a synthetic CAI (□) and a naturally occurring type B1 inclusion (■) (after Beckett et al. 1990).

…with T and melt composition included

D_{Be} (± 0.09) = (0.15 ± 0.21) (9.95) $X_{Åk}$ exp {(0.89 ± 0.87) $X_{Åk}$};

…with T and melt composition ignored

D_{Be} (± 0.20) = (2.24 ± 0.21) $X_{Åk}$.

Two models for the origin of melilite mantles in Type B1 inclusions are (1) that they were formed by fractional crystallization from a melt (MacPherson and Grossman 1981; Stolper 1982; MacPherson et al. 1984) and (2) that they were formed by replacement of primary pyroxene during metamorphism (Meeker et al. 1983). Based on their experimental data, Beckett et al. (1990) were able to predict fractional crystallization behavior recorded in zoned melilite and experimentally-reproduced zoning observed in melilite from type B CAIs (e.g., Fig. 9). Therefore, they concluded that the melilite mantles in Type B1 inclusions were produced by fractional crystallization in a closed system.

Even though in most CAIs (e.g., Type B and C inclusions), the phases did not necessarily form by condensation processes, the interpretation of the refractory composition of CAIs is widely based on calculated equilibrium and non-equilibrium condensation sequences (e.g., Grossman 1972; Lattimer et al. 1978). Enrichment of CAIs in Be (230 to 560 ppb) relative to CI abundance (Be = 26.7 ppb) was interpreted to suggest that Be behaves as a refractory element, much like aluminum, and therefore condensed into solid phases at high temperatures during early stages of solar nebula condensation. The condensation sequence originated with the concept of a hot nebula in which all the constituents were at CI abundances in gaseous form at 10^{-3} bar. Thermodynamic calculations predicted the sequence of minerals condensing from the vapor. Condensation sequence was first calculated assuming equilibrium condensation at total pressures of 10^{-3} bar and an O/H ratio of 41 times the cosmic ratio (e.g., Grossman 1972; Lattimer et al. 1978). Under these conditions, the first solid to condense from the gas phase is corundum, followed by hibonite ($CaAl_{12}O_{19}$), perovskite ($CaTiO_3$), grossite ($CaAl_4O_7$), melilite, and spinel (Table 2). Subsequent sequences were calculated at nebular conditions thought to be more similar to regions of chondrite formation (P = 10^{-5} bar).

Table 2. Equilibrium condensation temperature (K) of major minerals that would condense from a gas phase with an enhanced O/H ratio and nebular pressures of 10^{-3} (Latimer et al. 1978) or 10^{-5} bar (as compiled by Wood 1988). Condensation temperatures for melilite, spinel, and forsterite calculated by Lauretta and Lodders (1997) at nebular pressures of 10^{-3} bar are in bold. Temperatures represent conditions of initial condensation or subsequent reaction. Temperatures in parentheses indicate conditions under which a phase is totally consumed by reactions that convert it to another mineral. Also included is the approximate condensation temperature of BeO.

Mineral	T (at P = 10^{-3})		T (at P = 10^{-5})
Corundum	1742		1758 (1513)
Hibonite	~1735		
Perovskite	1680		1647 (1393)
Grossite	~1650		
Melilite	1625	**~1580**	1625 (1450)
Spinel	1535	**1501**	1513 (1362)
(Fe,Ni) metal			1471
Forsterite	1430	**1443**	1444
Anorthite	1385		1362
Enstatite			1349
BeO			980-1250
Alkali-bearing feldspar			<1000
Ferrous olivines and pyroxenes			<1000
Troilite			700
Magnetite			405

Using thermodynamic approaches to calculate element behavior during condensation, Cameron et al. (1973), Spivack et al. (1989), and Lauretta and Lodders (1997) calculated the behavior of Be. In a gas of solar composition with total pressure between 10^{-5} and 10^{-3} bar, Spivack et al. (1989) concluded that monoatomic Be is the stable gaseous species. Lauretta and Lodders (1997) concurred that the dominant Be gas at high temperature (>1650 K) is monatomic Be, but also demonstrated that at lower temperatures (<1650 K), Be hydroxides such as BeOH and $Be(OH)_2$ supersede monatomic Be as the dominant gas species (Fig. 10). Other Be gas species such as BeH, BeH_2, and BeS account for less than 0.4% of Be.

Spivack et al. (1989) suggested that BeO is the dominant form in condensed solids. They calculated that the 50% condensation temperature for Be ranges from 1250 to 980 K. As shown in Table 2, this is significantly below the condensation temperature of many of the refractory major and trace elements in CAI. Therefore, they concluded that Be is not intrinsically refractory and that the Be enrichment in CAI reflects condensation as a trace constituent of a major phase. Earlier, Cameron et al (1973) had suggested that Be completely condenses in solid solution with spinel ($MgAl_2O_4$–$BeAl_2O_4$) at 1400 K. This calculation is somewhat compromised because spinel and chrysoberyl are not isostructural and solid solution between spinel and chrysoberyl is limited to less than 25% of the $BeAl_2O_4$ component in spinel (Franz and Morteani, this volume). More recent condensation calculations by Lauretta and Lodders (1997) indicate that Be condenses into melilite solid solution as a gugiaite ($Ca_2BeSi_2O_7$) component soon after melilite appears. If melilite is the only host phase for Be, then the 50% condensation temperature for Be is 1490±30 K and 100% of the Be is condensed by 1470 K (Fig. 10). The observations that Be is concentrated in melilite in CAIs (Spivack et al. 1989; Beckett et al. 1990) is consistent with these calculations. It is also possible that Be condenses into spinel and forsterite (Mg_2SiO_4–Be_2SiO_4) following its incorporation into melilite. Given that there is a limited solid solution of $BeAl_2O_4$ in spinel (Franz and Morteani, this volume), additional Be could condense into spinel solid solution at 1501 K. This will change the 50% condensation temperature for Be. Just prior to the appearance of spinel, ~34% of the Be is condensed into melilite. With the appearance of spinel the 50% condensation temperature of Be becomes equal to the condensation temperature of spinel. Lauretta and Lodders (1997) further suggested that once melilite disappears from the nebula, all the remaining Be will go into forsterite as it becomes stable at 1443 K. This is not likely because Be_2SiO_4 (phenakite) is not isostructural with forsterite and the incorporation of Be in olivine is very limited (Brenan et al. 1998).

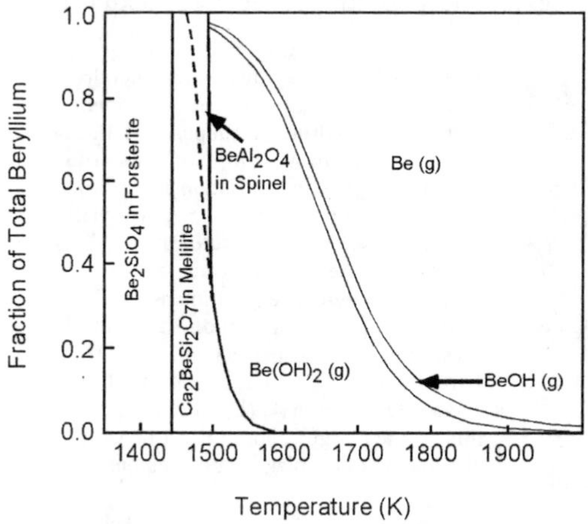

Figure 10. Plot illustrating the behavior of Be at a 10^{-3} bar total pressure. Dominant gas species prior to and during condensation are shown. The condensation of Be first occurs in melilite and then melilite + spinel. The dashed curve indicates the condensation of Be in melilite solid solution if the condensation of spinel is ignored. As melilite disappears from the solar nebula the remaining Be may go into forsterite as a phenakite component. However, the substitution of Be into olivine is very limited because Be_2SiO_4 (phenakite) is not isostructural with forsterite (after Lauretta and Lodders 1997).

Behavior of Be in chondrites and chondrules

Chondrules are the most abundant components of chondrites, composing up to 80% by volume of the ordinary and enstatite chondrites and greater than 15% of most other chondrite groups, except for the CI group in which chondrules are absent (Brearley and Jones 1998). A majority of chondrules are submillimeter, spherical objects consisting primarily of feldspathic glass and ferromagnesian silicates such as olivine and pyroxene (Fig. 11). Chondrule textures are suggestive of rapidly crystallized molten or partially molten droplets. Numerous textural classes of chondrules have been established, some of which are illustrated in Figure 11. The textures have been reproduced under experimental conditions that included initial temperatures of 1556 to 1700°C, heating times of 12 to 30 minutes, rapid cooling (~500°C/hour), and quenching temperatures below 1100°C (Hewins 1988; Lofgren and Russell 1986). Summaries of chondrule properties such as detailed textural characteristics, bulk and mineral chemistries and oxygen isotope data may be found in Grossman et al. (1988a) and Brearley and Jones (1998).

Relative to CI abundances, chondrules are depleted in siderophile and chalcophile elements. The chondrite matrix contains most of the siderophile and chalcophile elements, although the differences between matrix and chondrules vary among the chondrite groups. The greatest depletions are in the most reduced chondrite groups. These depletions are thought to be due to selective melting of silicate precursor material that was already depleted in siderophile and chalcophile elements (Grossman et al. 1988b). Perhaps these depletions were additional exacerbated by the loss of metal and sulfides as immiscible melts during the chondrule-forming process. Also evident are small depletions in volatile elements in all chondrules except in the CO3 chondrules. The amount of evaporative loss of an element from a molten droplet should be a function of its volatility and chondrule characteristics. Yet, correlations have not been found between depletions of volatile lithophile elements (i.e., Na) and chondrule size (Dodd and Walter 1972; Grossman and Wasson 1983), texture (i.e., thermal history), composition, the oxidation state of Fe, or the depletion of volatile siderophile elements (Grossman et al. 1988b; Grossman and Wasson 1983). Dodd and Walter (1972) observed that some chondrules containing abundant metallic Fe and low-Fe olivine tended to be lower in Na than chondrules with less abundant Fe and high-Fe olivine. Chondrules in EH3 and CV3 chondrites are also depleted in refractory lithophile elements (Grossman et al. 1988a,b). This evidence leads to the conclusion that the chondrule precursor materials were already depleted in volatiles and that in most cases chondrules did not undergo extensive

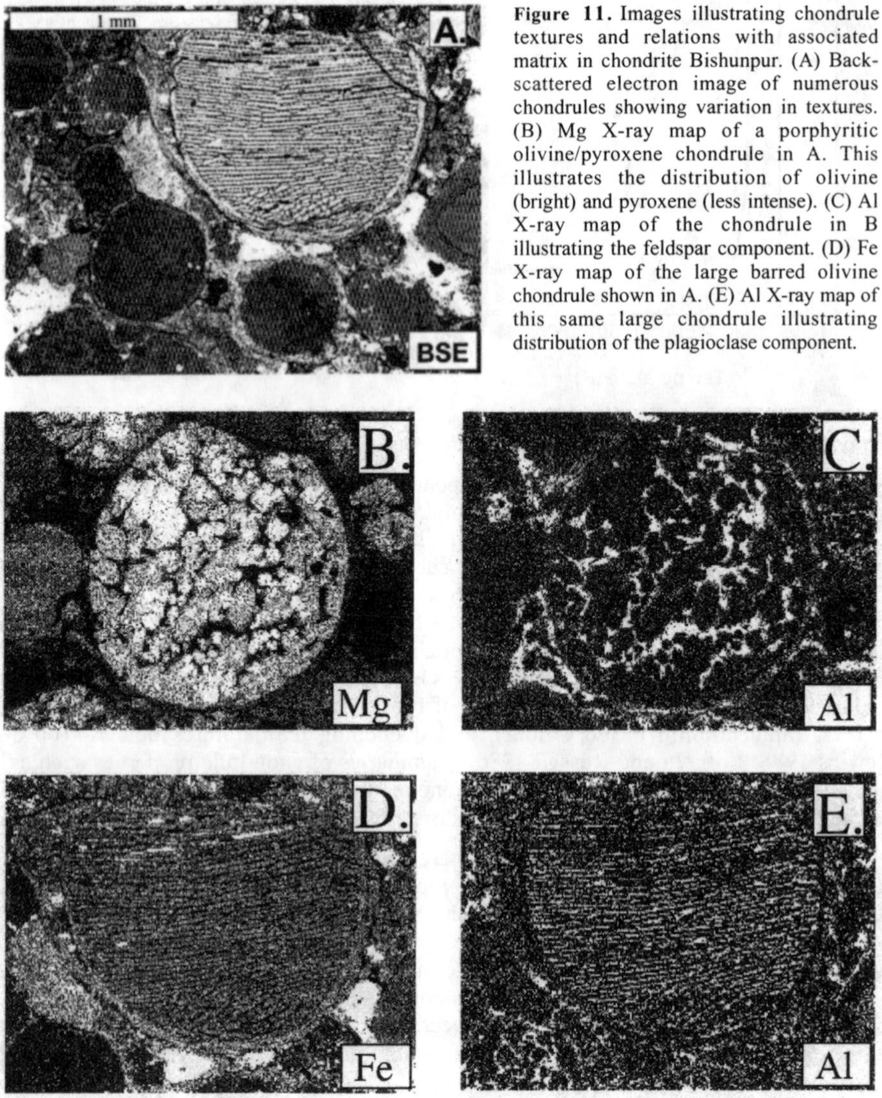

Figure 11. Images illustrating chondrule textures and relations with associated matrix in chondrite Bishunpur. (A) Back-scattered electron image of numerous chondrules showing variation in textures. (B) Mg X-ray map of a porphyritic olivine/pyroxene chondrule in A. This illustrates the distribution of olivine (bright) and pyroxene (less intense). (C) Al X-ray map of the chondrule in B illustrating the feldspar component. (D) Fe X-ray map of the large barred olivine chondrule shown in A. (E) Al X-ray map of this same large chondrule illustrating distribution of the plagioclase component.

fractionation by volatile loss (Grossman et al. 1988b; Taylor 1992).

Within the precision of U-Pb and Rb-Sr techniques, individual chondrules seem to be contemporaneous with one another and with their host chondrites (Swindle et al. 1983). Iodine-xenon relative ages for chondrules in Allende and Bjurbole show that the range in ages appears to be 5 million years. Argon-40/argon-39 ages show a spread of 200 million years, but this wider range most certainly reflects alteration on chondrite parent bodies. Chondrule formation in the solar system may have extended over a period of less than 10 million years. What is the link between the CAIs and chondrule formation? Two chondrules from Felix CO3 and few aluminum-rich chondrules from the CV chondrites Allende and Mokoia appear to have highly fractionated REE patterns similar to group II CAIs (Rubin and Wasson 1987; Misawa and Nakamura 1988; Jones et al. 2001). This may be interpreted as indicating that CAI production preceded chondrule formation. Many CAIs yield an initial $^{26}Al/^{27}Al$ ratio of $\sim 5 \times 10^{-5}$, whereas typical values for most chondrules are $<1 \times 10^{-5}$. If this difference is attributed solely to decay of ^{26}Al, it implies an age difference between the formation of CAIs and chondrules of at least 2 million years.

Although numerous planetary models have been proposed for the origin of chondrules (early impacts, collisions, volcanism), currently the most popular models involve a nebular origin. Most of these nebular models propose that precursor materials for the chondrules were dustballs entrained in the nebular gas in the inner nebula. These dustballs consisted of aggregates of silicates, whereas metal and sulfide particles remained dispersed. A transient, local heating event melted the dustballs, they were subsequently cooled rapidly (100s of degrees per hour), and then accreted into their parent bodies. The nature of the heating event(s) is not well understood (Boss 1996) and has been attributed by various workers to shock waves, nebular flares, bipolar outflows, x-wind and lightning. A detailed discussion of these models within the context of astronomical, mineralogical and textural observations is presented by Jones et al. (2000). Not only did the chondrule-forming events post-dated CAI formation, but it presumably occurred at a lower temperature. Minor reduction and volatilization occurred during the formation of chondrules, but many of the chemical depletions observed in chondrules reflect nebular processes that occurred prior to melting, Parent-body processes affecting chondrites and their chondrules include aqueous alteration, thermal metamorphism, and shock metamorphism. Distinguishing between the effects of nebular and parent body processes on chemical, mineralogical and textural of chondrites and chondrules has been a point of much debate.

Several studies have focused upon the distribution and behavior of Be (in addition to Li and B) in chondrules. These studies used the distribution of these elements in order to better understand light element nucleosynthesis, the role of volatilization during chondrule formation, and chondrule alteration on the chondrite parent bodies. The study of Brearley and Layne (1996) emphasized the behavior of Be, B, and Li in microchondrules in the CH carbonaceous chondrites ACFER 182, ACFER 214, and PAT 91546. Studies by Chaussidon and Robert (1995) and Hanon et al (1999) emphasized the behavior of these elements in chondrules in the carbonaceous chondrite Allende (CV) and ordinary chondrites Bishunpur (LL3.1), Semarkona (LL3.0), Hedjaz (L3.7), and Clovis (H3.6). Using the data of Hanon et al. (1999), Robert et al. (1999) emphasized not only these light lithophile elements, but also the isotopic systematics in chondrules in carbonaceous and ordinary chondrites.

The study by Brearley and Layne (1996) focused on glassy to cryptocrystalline chondrules in CH chondrites. These chondrules exhibit little evidence of microcrystals or chemical heterogeneities, except for rare SiO_2-rich rims. The majority of the chondrules are SiO_2-rich, pyroxene-normative and have extremely low FeO contents (<2.2 wt % FeO). The MgO contents of these chondrules range from 38 to 43 wt %. The CaO and Al_2O_3 contents of these chondrules range from ~0.25 wt % to ~5.5 wt % and have a chondritic Ca/Al ratio. A few of the chondrule analyses by Brearley and Layne (1996) have significantly higher FeO (>8%). The FeO-rich chondrules have much lower refractory element abundances than the FeO-poor ones.

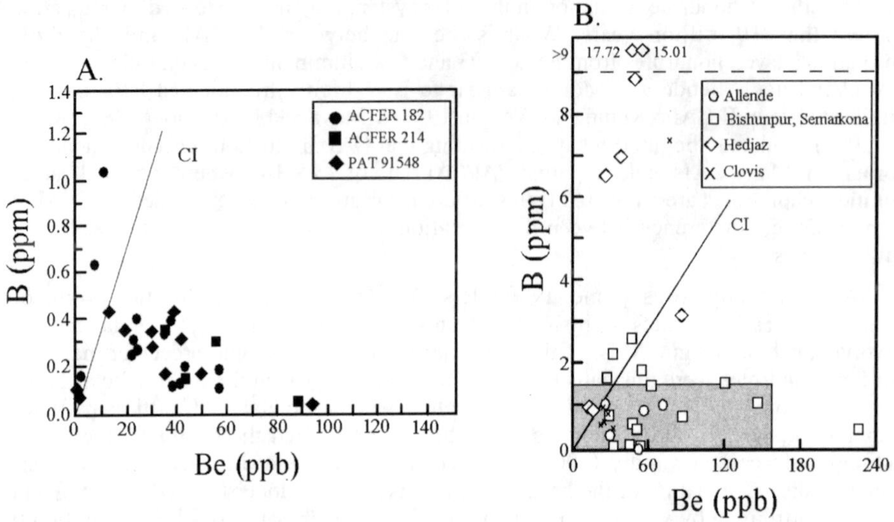

Figure 12. Plot of Be vs. B contents of individual chondrules. Data are ion microprobe analyses from (A) Brearley and Layne (1996) and (B) Hanon et al. (1999). The shaded area in B represents the range in A.

Beryllium abundances in the chondrules exhibit considerable variation from one chondrule to another (<0.3 to 3 × CI). The variability exceeds the range in Be observed in bulk carbonaceous chondrites. This chondrule population primarily defines a negative correlation between B and Be, with some scatter (Fig. 12A). A smaller population of chondrules plot off of this trajectory at exceedingly low abundances of Be. There is no apparent relationship between Li and Be. The Be/Al ratios (Fig. 13A) range from greater than chondritic to subchondritic. The chondrules from ACFER 214 and most of the chondrules from PAT 91548 define a group with Be/Al higher than CI and a positive correlation between Be and Al.

The inverse correlation between Be and B was interpreted by Brearley and Layne (1996) as indicating that the volatile content of this group of chondrules was controlled by the composition of the precursor materials rather than volatile loss during chondrule formation. Random loss of B following chondrule formation would have resulted in limited correlation between the volatile B and the refractory Be. Perhaps, only those chondrules with exceedingly low Be experienced B loss (Fig. 12). The positive correlation between Be and Al suggests that two distinct populations of glassy chondrules exist which have distinct Be/Al ratios resulting in differences in their precursor refractory element components. Brearley and Layne (1996) concluded that these two populations may represent two distinct reservoirs with different Be/Al that experienced mixing.

Unlike the glassy chondrules studied by Brearley and Layne (1996), the chondrules analyzed by Chaussidon and Robert (1995) and Hanon et al. (1999) are much more crystalline and consist of olivine (0 to 88 vol %), orthopyroxene (0 to 91 vol %) and mesostasis (0 to 63 vol %). These chondrules exhibit varying textures (Fig. 11), olivine compositions, and degrees of alteration and metamorphism. Chondrules in Semarkona and Bishunpur suffered the least alteration through the minor incorporation of H_2O (Grossman and Brearley 1997; Deloule and Robert 1996; Deloule et al. 1997). Chondrules in Allende, Hedjaz, and Clovis #1 exhibit much more alteration, reflected in partial recrystallization of mesostasis and reequilibration of Fe/(Fe+Mg) in olivine and orthopyroxene (Jones 1990; Krot et al. 1995).

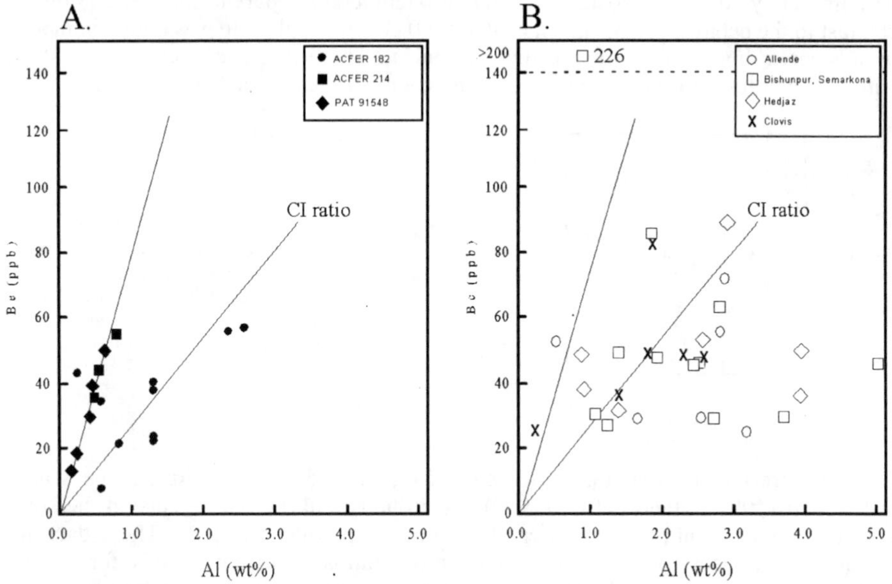

Figure 13. Plot of Al vs. Be contents of individual chondrules. Data are combined electron and ion microprobe analyses from (A) Brearley and Layne (1996) and (B) Hanon et al. (1999). Diagonal lines in A and B represent the CI ratio and the best-fit-line through ACFER 214 and PAT 91548 data.

Within a single chondrule, Be concentrations vary by an order of magnitude. For example, in one chondrule in Allende Be concentrations range from <1 ppb to 140 ppb. This variability is a result of mineralogical heterogeneity on the scale of the ion beam that was used to analyze the chondrules. As illustrated in Figure 14, there is a linear relationship between Be concentrations and the glass fraction of the ion probe spot. This indicates that Be is favorably partitioned into the mesostasis relative to olivine or orthopyroxene. This incompatible partitioning behavior of Be between olivine/melt and orthopyroxene/melt in chondrules is in agreement with its incompatible behavior in terrestrial (Ryan and Langmuir 1988) and lunar basalts (Shearer et al. 1994).

Bulk Be concentrations in these chondrules range from 25 to 239 ppm. The variations in concentration are present both among different chondrules within a given chondrite and among chondrules from different chondrites. Compared to the results of Brearley and Layne (1996), the relationships between Be and Al and Be and B are far more complex. Compared to the glassy chondrules in CH chondrites, these more crystalline chondrites show much more variability in Al, B and Be (Figs. 12B and 13B). Most plot at Be/Al values near or substantially lower than CI (Fig. 13B). A much smaller number of chondrules plot near the Be-enriched line defined by Brearley and Layne

(1996). Unlike the glassy CH chondrules, neither the whole chondrule population or sub-populations (textural type, meteorite class) shows a negative correlation between B and Be. Also a majority of these crystalline chondrules have B concentrations that are substantially higher than the CH glassy chondrules. The lack of B-Be correlation, the positive correlation between B-Al only in Allende (Hanon et al. 1999), and the wide range in B concentrations was interpreted as indicating that the volatile contents of this group of chondrules was not exclusively controlled by the composition of the precursor materials. Instead, the volatile content was influenced substantially by volatile addition predominately to the mesostasis during low-temperature parent-body alteration. In contrast to the behavior of B, Hanon et al. (1999) concluded that Be was highly immobile during parent-body alteration and metamorphism. Therefore, the heterogeneity of Be within chondrites reflects the chemical heterogeneity of their precursor materials.

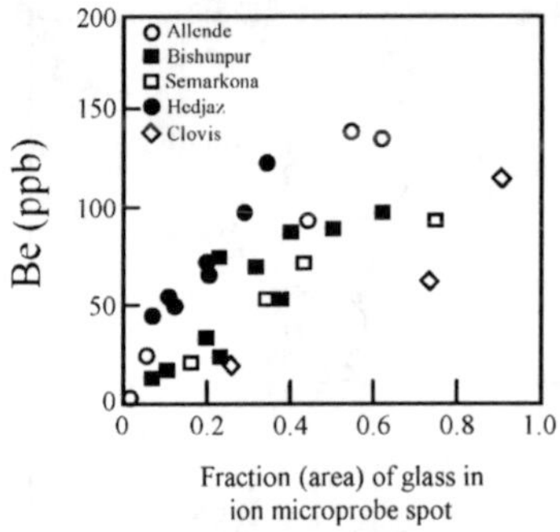

Figure 14. Plot of glass fraction of ion probe spot vs. Be concentration. Illustrates that Be preferentially partitions into the mesostasis relative to crystals (olivine and pyroxene) during chondrule crystallization (from Hanon et al. 1999).

Using trace element and isotopic ratios of Li, Be, and B in chondrules, Chaussidon and Robert (1995), Hanon et al. (1999) and Robert et al. (1999) calculated the solar system abundance and the character of the nuclear production of Be. These data are useful because theory predicts the production rate ratios of B/Be and Li/Be for various nucleosynthetic processes. Modeling of spallogenic reactions in the interstellar medium predicts B/Be ratios ranging from ~17 for nuclear collisions involving high-energy galactic cosmic rays and interstellar C and N (Meneguzzi et al. 1971) to 50-100 for collisions between C and O at low energy (10 to 100 MeV). Hanon et al. (1999) normalized the average Be concentrations in chondrules that they analyzed to a solar system H/Si ratio of 2.8×10^4 to calculate a hypothetical solar system Be/H ratio of 2.1 (+1.3, -0.8) $\times 10^{-11}$. This value is indistinguishable from the average galactic composition proposed by Reeves (1994) and the solar system abundance calculated by Anders and Grevesse (1989) from the Orgueil carbonaceous chondrite (1.3×10^{-11} and 2.6×10^{-11}, respectively).

The B/Be ratio in chondrules varies considerably. The strong positive correlation between B/Li and B/Be (Fig. 15) was interpreted by Hanon et al. (1999) as indicating that these ratios are primarily controlled by the heterogeneity of B in chondrules. The B/Be is less than solar in the least altered chondrites and greater than solar in chondrules from the highly metamorphosed chondrites. The average B/Be value for the chondrules is 15 and corresponds to the solar ratio and could be interpreted to imply that there was no fractionation of B from Be during condensation in the solar nebula (Chaussidon and Robert 1995; Lauretta and Lodders 1997). However, most of the least altered chondrules analyzed by Hanon et al (1999) as well as most of the glassy chondrules analyzed by Brearley and Layne (1996) plot at B/Be values less than solar (Fig. 15). Both Brearley and Layne (1996) and Hanon et al. (1999) interpreted the B/Be values for the chondrules that had experienced lower degrees of metamorphism to represent the ratios in chondrule precursor material. Lauretta and Lodders (1997) showed that chondrule precursor material may have experienced significant fractionation of B from Be during condensation from the solar nebula at temperatures above 900 K. This would have had the effect of lowering the B/Be ratio from both solar (~29) or CI chondrite (35-44) values.

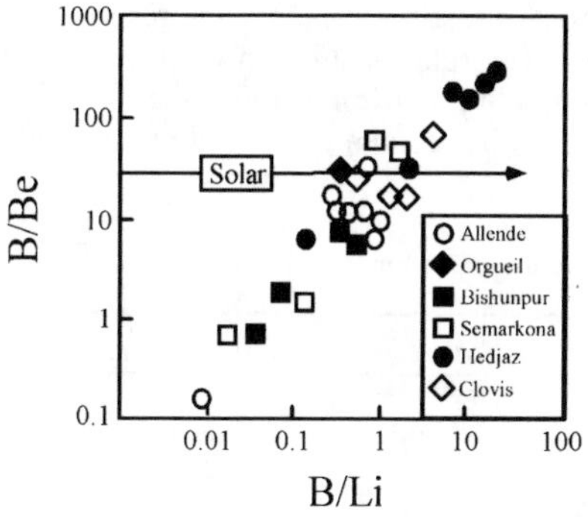

Figure 15. Ratios of B/Be vs. B/Li in chondrules from the study of Hanon et al. (1999). This high correlation results from the high-B heterogeneity among the chondrules compared to Li and Be. Compared to Orgueil, low B concentrations most likely reflect the heterogeneities in the chondrule precursors, whereas the high B concentrations are only observed in metamorphosed chondrites and perhaps reflect post-assembly, parent body processes (from Hanon et al. 1999).

BULK BE OF PLANETS AND BEHAVIOR OF BE DURING PLANETARY ACCRETION

According to the solar nebula theory, planetary development initiated with the condensation and formation of primitive aggregates of undifferentiated solar nebula material followed by accretion and differentiation of this primitive material. As discussed above, the memory of this primitive material is preserved in undifferentiated meteorites such as chondrites. As the terrestrial planets have undergone substantial differentiation since their assembly, determining the composition of their proto-material and understanding processes instrumental to their formation is a formidable task. In the early- to middle-twentieth century, a widely accepted category of theories of terrestrial planetary accretion contended that the planets accreted directly from the fine-grained primitive aggregates resembling chondritic meteorites. The time scale over which this process was calculated to occur was on the order of 10^7 to 10^8 years. More contemporary models advocate that the planets are the end products of a hierarchical accretionary process that first assembled a large number of kilometer-sized planetesimals from an initial protoplanetary disk of gas and dust. Growth during this stage was controlled by surface, electromagnetic and sticking forces. These small bodies then coalesced into protoplanets-planets whose evolution was controlled primarily by gravitational interactions. Some of the observations that support these hierarchical accretionary models include (1) left-over planetesimals in the solar system (= asteroids); (2) ubiquity of ancient (3.9 to 4.5 billion years old) cratered terrains across the solar system; and (3) the tilt of axes of rotation relative to the ecliptic of most of the terrestrial planets.

There are numerous permutations of these contemporary accretion models involving variables such as accretional mechanisms, accretional rates and composition-characteristics (size, differentiation) of accreting bodies. Accretionary mechanisms such as massive gravitational instabilities and sweeping-up processes whereby larger bodies collide with smaller ones have been proposed and modeled (i.e., Wetherill 1988; Weidenschilling and Cuzzi 1993; Lissauer and Stewart 1993). Numerous studies of planet formation suggest that the formation by dynamic friction of Mars size planetary embryos occurred in as little as 10^5 years (Greenberg et al. 1978; Wetherill and Stewart 1993; Canup and Agnor 1998, 2000; Stewart and Canup 1998). The final stages of planetary accretion, which is believed to persist for ~10^8 years, are characterized by large, discrete accretion events, such as the impact that formed the Moon, and subsequent events recorded on the lunar surface (Wetherill 1992; Canup and Agnor 1998, 2000). Geochemical and isotopic signatures attributed to these events have been interpreted in many different ways (e.g., Stewart and Canup 1998; Shearer and Newsom 2000). Proposed planetesimal compositions include chondrite-like bodies, highly differentiated bodies, to bodies with compositions representative of distinct solar system reservoirs (i.e., lower volatile content of the inner solar system). The relationship between the estimated composition of a planet and the solar system composition provides clues to their formation. The Be contents of many of the planets have been estimated and models for their assembly have been predicted.

Table 3. Calculated bulk planetary abundances for Be, Li, and B.

Planet:	*Mercury*	*Venus*	*Moon*		*Earth*	*Silicate Earth*		*Continental crust*	*Mars*
Be (ppb)	34	47	186	180	45	80	60	1500	88
Li (ppm)	0.87	1.94	8.7	0.83	1.85	1.6	0.83	13	1.94
B (ppb)	0.11	10	13	540	9.6	500	600	10000	2.26
Reference	1	1	2	3	1	4	5	5	6

References:
1 = Morgan and Anders (1980), 2 = Anders (1977), 3 = Taylor (1982), 4 = Ringwood (1991), 5 = Taylor & McLennan (1985), 6 = Morgan and Anders (1979)

Estimates for the bulk Be contents of most of the terrestrial planets have been made based on samples collected from planetary surfaces either by man (Moon, Earth), machine (Moon), or impact (Moon, Mars), in situ chemical measurements (Venus, Mars), remotely sensed geochemical and geophysical measurements (Mercury, Venus, Earth, Moon) and cosmochemical models for planetary assembly. A summary of the calculated bulk planetary contents of Be, Li, and B for the terrestrial planets is presented in Table 3. The bulk compositions of Mercury, Venus, Earth, and Mars were approximated by Morgan and Anders (1979, 1980). Their calculations for Mercury were based on very limited spectroscopic surface information suggesting a low FeO content (<5.5 wt %), the high density of Mercury (5.43 g cm^{-3}) implying an iron core making up greater than 65 wt % of the planet, and a cosmochemical model which assumes that planets and chondrites underwent similar fractionation processes in the solar nebula (Ganapathy and Anders 1979). Morgan and Anders (1980) used the same cosmochemical model (Ganapathy and Anders 1974) to calculate the bulk compositions of Venus. For Mars, Morgan and Anders (1979) combined a similar cosmochemical model with abundances of index elements obtained from orbital gamma-ray observations, Viking surface observations, and geophysical constraints. Unlike Morgan and Anders (1979), Wänke and Dreibus (1988) based their estimate of the bulk composition of Mars on elemental ratios in Martian meteorites (SNC). Unfortunately, they did not calculate Be abundances. Both Anders (1977) and Taylor (1982) calculated Be, Li, and B abundances for the silicate portion of the Moon. The Anders (1977) values for the Moon are based on a cosmochemical approach that involved mixing of 6 components. The composition of Taylor (1982) is based on geophysical constraints (heat flow, density), the composition of the lunar highland, and elemental ratios in lunar samples. Both Anders (1977) and Taylor

(1982) estimated a similar bulk Be for the silicate Moon (186 and 180 ppb, respectively), whereas their estimates of Li and B are dramatically different. The differences in LLE abundances in these two calculations is tied to differences in the behavior of the LLE assumed in each model. Both models assumed that Be would behave as a refractory element such as Al, whereas they differ on the relative volatile behavior of Li and B. The estimates of Be for bulk Earth (Morgan and Anders 1980), bulk silicate Earth (Ringwood 1991; Taylor and McLennan 1985), and bulk continental crust (Taylor and McLennan 1985) used a variety of methods. Morgan and Anders (1980) used seven cosmochemical components constrained by the mass of the Earth's core, U and Fe abundances, and the ratios K/U, Th/U, and FeO/MnO. The bulk Earth composition calculated by Ringwood (1991) is based upon his "pyrolite" primitive mantle model. Taylor and McLennan (1985) used a mixture of cosmochemical components for refractory elements, crustal data for volatile elements, and mantle nodule data for the siderophile elements.

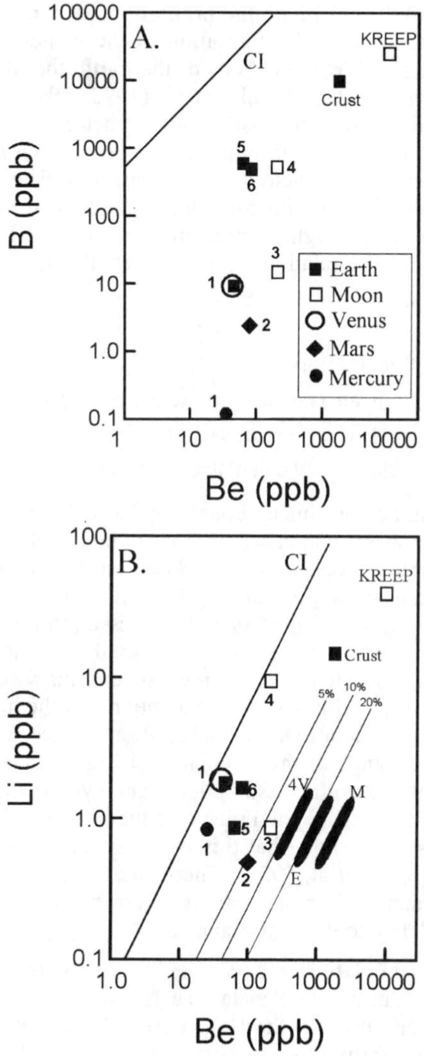

Figure 16. (A) Plot of B vs. Be for estimates of planetary bulk compositions (1,2) and individual planetary components such as bulk silicate planet (3,4,5,6), terrestrial crust, and KREEP. Altogether, planetary bulk compositions plot to the volatile-poor, refractory-rich side of chondritic abundances. (B) Plot of Li vs. Be for estimates of planetary bulk compositions. Diagram is also contoured with the percent of high-temperature condensates (refractory accretionary component) making up the planet as calculated by Dreibus et al. (1976). The approximate bulk compositions estimated by Dreibus et al. (1976) for Earth (E), Moon (M), and 4 Vesta (4V) are also plotted. References for calculated bulk planetary compositions in are numbered in both diagrams: 1= Morgan and Anders (1980), 2 = Morgan and Anders (1979), 3 = Anders (1977), 4 = Taylor (1982), 5 = Ringwood (1991), 6 = Taylor and McLennan (1985), Crust = Taylor and McLennan (1985), and KREEP = Warren (1988).

Although there are substantial errors in estimating the bulk compositions of planets and biases introduced in the calculation when it is based on cosmochemical models, the data calculated above do show some interesting systematics (Fig. 16). The bulk compositions of the terrestrial planets are depleted in B (volatile) and Li relative to Be (refractory) compared to CI chondrite. Depletions in other moderately volatile elements occur to varying degrees in the terrestrial planets and meteorites. Like B/Be, the ratios K/U and Mn/Fe (BVSP 1981), Rb/Sr and U/Pb isotopic systematics, and noble gases inform us of a nebula-wide loss of volatiles, particularly in the inner solar system. This process appears to have occurred fairly early in the evolution of the solar system. Rb-Sr and Pb-Pb ages of meteorites (which reflect the time of the depletion of volatile Pb and Rb relative to refractory U and Sr) suggest an episode of volatile depletion close to the time of CAI formation. The cause of this volatile loss has been attributed to intense solar activity involving strong stellar winds and flares.

Planetary Be abundances have been used to estimate the proportions of primitive solar system components that were accreted during the formation of the planets and asteroids. In their approach to estimating the bulk compositions of the Earth, the Moon and the parent body for the HED meteorites (EPB), Wänke et al. (1973, 1975) and Dreibus et al (1976) used certain observed element correlations of refractory to non-refractory elements to calculate the primary proportion of high temperature condensates (HTC, refractory component) to Mg-silicates (less refractory component). For example, Dreibus et al. (1976) suggested that the ratio of Li to refractory elements (Be or Zr) is a direct measure of the ratio of the Mg-silicates to the high temperature condensates (HTC) during planetary accretion. They used the following relationship to calculate that ratio

$$(Li/Be)_{planet} = [Li_{mg} \cdot (1-x)] / [Be_{HTC} \cdot x]$$

where (from Dreibus et al. 1976):

x = refractory portion; $(1-x)$ = non-refractory portion.

Li_{mg} = Li content (2.6 ppm) of the Mg-silicate (i.e., non-refractory portion).

Be_{HTC} = Be content (0.88 ppm) of the HTC (i.e., refractory portion).

$(Li/Be)_{planet}$ = ratio of Li to Be in a planetary body calculated from basalts.

In retrospect, some aspects of the logic behind this approach appear to be flawed. They approximated the Be_{HTC} by assuming the Be abundance of CAIs to be enriched 22 times greater than their approximated CI abundance of 40 ppm. Based on the study of Spivack et al. (1989), this enrichment value is approximately correct, but their CI abundance for Be is higher than those determined by Anders and Grevesse (1989) and Wasson and Kallemeyn (1988). They also approximated Li_{Mg} by normalizing the Li abundance in CI chondrites to the Mg content of what they assumed to be primary solar system material preserved in the lunar highlands. Although the assumption that the lunar highland contains remnants of primary solar system material is based on refuted lunar models, the Li_{Mg} value falls within the upper range of the Li content of Mg silicates in chondrules (0.2 to 3 ppm) (Hanon et al. 1999). Another assumption made by Dreibus et al. (1976) is that both elements are highly incompatible and therefore the basalts have a similar Li/Be as the mantle. Partitioning studies show that this is not entirely correct (Ryan and Langmuir 1987, 1988, 1993; Brenan et al. 1998). These studies imply that partial melting of a planetary mantle and fractional crystallization of a primary magma will decrease the Li/Be ratio of the basalt relative to its mantle source.

Based on the available Li and Be data, Dreibus et al. (1976) calculated the abundances of high temperature condensates during the planetary accretion of the Earth, Moon, and the eucrite parent body. The proportions of the HTC were calculated to be 40% for the Moon, 22% for the Earth and 13% for the eucrite parent body.

The Li/Be for the Moon based on the mare basalts is approximately 3 to 5 (Dreibus et al. 1976). Additional data collected by Shearer et al. (1994) on more primitive mare basaltic glasses indicate that the Li/Be of less fractionated lunar basalts range from 10 to 34. If interpreted in the same manner as Dreibus et al. (1976), the existence of a high Li/Be mantle source indicates that a less refractory component exists in the lunar mantle and that the bulk Moon is less refractory than previous studies suggest. Compared to other bulk lunar compositions, the estimates calculated by Dreibus et al. (1976) from Li/Be are slightly higher in SiO_2, CaO, and Al_2O_3 and lower in MgO and Mg′ [= Mg/(Mg + Fe^{2+})] compared to bulk lunar models proposed by O'Neill (1991), Ringwood (1979), and Jones and Delano (1989). Minor and trace elements such as Ti, Cr, Mn, Sr, Th, U, Zr, and Ni are very similar to those values proposed by O'Neill (1991).

Applying this approach to other planetary bodies, the plot of estimated planetary Li/Be was contoured with the percentage of the refractory component making up the planet (Fig. 16B). Most of the estimates for Li/Be of planetary bodies plot at very low percentages of refractory components compared to the Dreibus et al. (1976) estimates. At the very least, this suggests that the Be and Li values for the refractory and non-refractory components used by Dreibus et al. (1976) are incorrect and that the Li/Be ratio can be used only as a qualitative index for the composition of accretionary components. Also, the calculations assume the same accretionary model is valid for all the terrestrial planetary bodies. In the case of the Moon, prevalent models postulate that the Moon accreted from planetary material produced from the collision of the Earth and a Mars-size body. Both of these sources had already experienced extensive accretion and differentiation.

BEHAVIOR OF Be DURING PLANETARY MAGMATISM. EXAMPLE: MOON

Introduction

Beryllium and the other light lithophile elements (LLE) exhibit some unique geochemical behaviors that have provided invaluable information for interpreting terrestrial magmatic processes in mid-ocean ridge and subduction zone settings (Tera et al. 1986; Ryan and Langmuir 1987, 1988, 1993; Morris et al. 1990; Gill et al. 1993; Leeman et al. 1994; Brenan et al. 1998). The LLE behave as moderately (Li) to highly incompatible (Be, B) elements in basaltic systems (Ryan and Langmuir 1987, 1988, 1993). In addition, Li and B are highly soluble in hydrothermal solutions, whereas Be is insoluble in most hydrothermal solutions (Shearer et al. 1986; Ryan and Langmuir 1987, 1988, 1993). In contrast to the terrestrial basalts, lunar basalts provide an opportunity to observe the behavior of the LLE in a truly anhydrous magmatic environment. In addition, the behavior of Be in mare basalts provides information about the character of the lunar mantle and the petrogenetic relationship among mare basalts.

Mare basalts are exposed over 17% of the lunar surface and are thought to make up less than 1% of the volume of the lunar crust (Head 1974, 1976; Head and Wilson 1992). They primarily fill multi-ringed basins and irregular depressions on the Earth-facing hemisphere of the Moon. On the lunar far side, limited patches occur in younger craters and basins. The thickness of these flood basalts ranges from 0.5 to 1.3 km in irregular basins and up to 4.5 km in the central portions of younger basins (Bratt et al. 1985). The general style of volcanic activity was the eruption of large volumes of magma from relatively deep sources (not shallow crustal reservoirs) with very high effusion rates (Head and Wilson 1992, 1997). Evidence of lunar fire-fountaining is preserved in the form of spherical glass beads and dark mantling deposits (Head 1974, 1976; Delano 1986; Head and Wilson 1992, 1997). The radiometric ages for the mare basalts range from 3.8 to 3.16 Ga (i.e., Nyquist and Shih 1992). Basaltic volcanism prior to basin formation has been documented to be as old as 4.2 Ga through dating of clasts in breccias (Taylor et al. 1983). Photogeologic observations of basaltic flows provide additional support for basaltic volcanism prior to 3.9 Ga and flows as young as 0.9 Ga (Schultz and Spudis 1983).

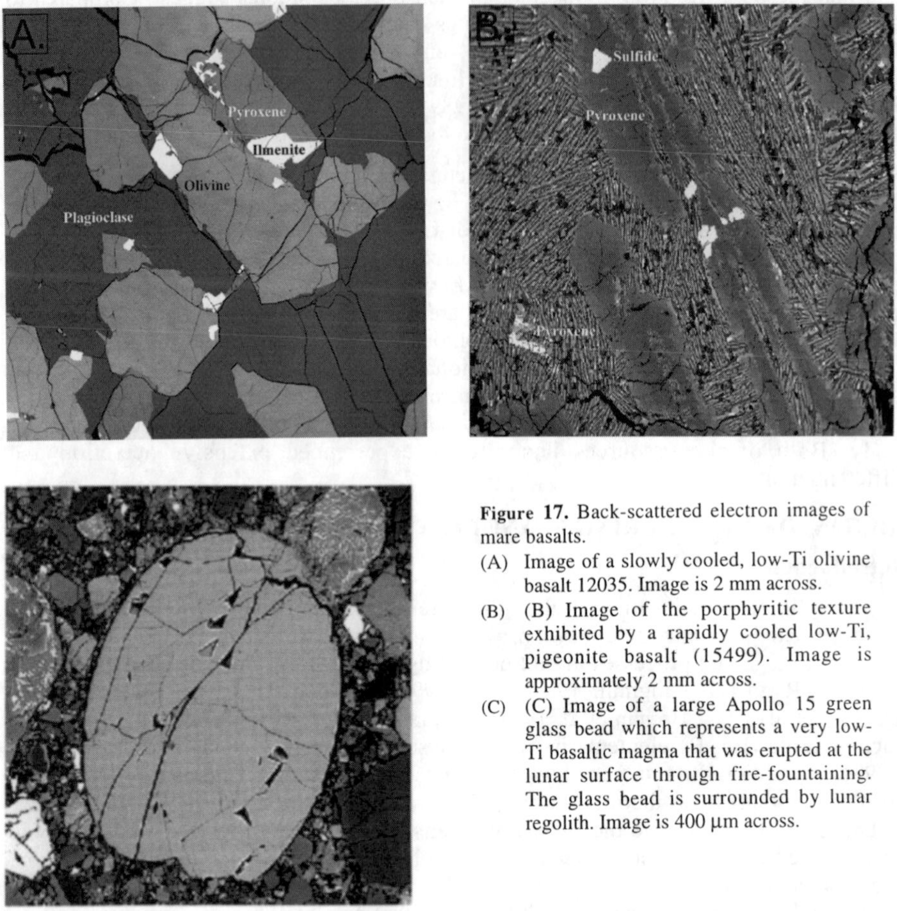

Figure 17. Back-scattered electron images of mare basalts.
(A) Image of a slowly cooled, low-Ti olivine basalt 12035. Image is 2 mm across.
(B) (B) Image of the porphyritic texture exhibited by a rapidly cooled low-Ti, pigeonite basalt (15499). Image is approximately 2 mm across.
(C) (C) Image of a large Apollo 15 green glass bead which represents a very low-Ti basaltic magma that was erupted at the lunar surface through fire-fountaining. The glass bead is surrounded by lunar regolith. Image is 400 μm across.

Mare basaltic magmatism is represented by both crystalline basalts and spherical basaltic glass beads that occur in the lunar regolith. These spherical glass beads have been referred to as pyroclastic glasses, volcanic glasses, pristine glasses, and picritic glasses. Examples of low-Ti to very low-Ti mare basalts are shown as back-scattered electron images in Figure 17. Characteristics of lunar basalts have been reviewed by Papike et al. (1976, 1998), BVSP (1981), Neal and Taylor (1992), Taylor et al. (1991), and Shearer and Papike (1993, 1999). Within the context of terrestrial basalt classification, the normative mineral assemblages of the mare basalts range from hypersthene + quartz to hypersthene + olivine. Lunar basalts that are critically silica undersaturated have not been identified. Relative to terrestrial basalts, lunar basalts exhibit (1) a spectacular range in TiO_2 content, from 0.20 to 17 wt % (Fig. 18); (2) reduced valance states for Fe, Ti, and Cr reflecting low magmatic oxygen fugacities; (3) depletions of alkali, siderophile, and volatile elements; (4) non-chondritic element ratios with ubiquitous negative

Eu anomalies, and (5) absence of water and hydrous minerals. These characteristics have been interpreted as indicating that the mare basalts were produced by melting of a highly heterogeneous, anhydrous lunar mantle that had experienced an early- stage (~4.5 Ga) of differentiation. The generation of a lunar magma ocean (LMO) that was responsible for producing an early anorthositic crust and a differentiated mantle consisting of LMO mafic cumulates is an elegant explanation of early lunar differentiation. The extent of early lunar melting and mantle processing strongly depends on the melting mechanism and source of primordial heat. Estimates for the time over which the LMO crystallized range from tens (Shearer and Newsom 2000) to hundreds of millions of years. A summary of possible models for early lunar differentiation and its role in the generation of lunar magmas was presented by Shearer and Papike (1999).

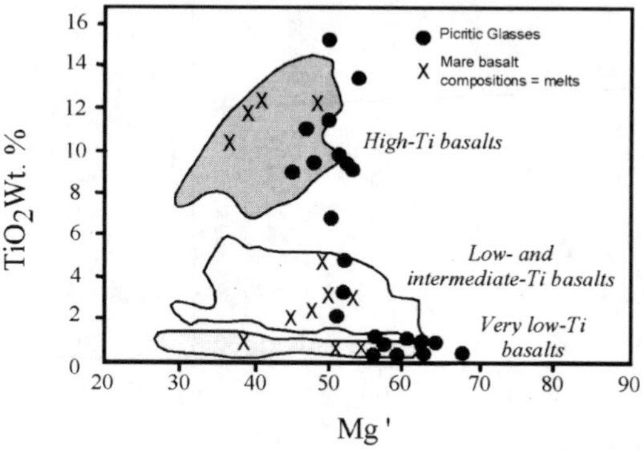

Figure 18. Plot of TiO_2 vs. Mg′ (= $Mg/(Mg+Fe^{2+})$) for mare basalts and lunar volcanic glasses (filled circles). The fields for mare basalts are from Neal and Taylor (1992) and are defined by whole-rock analyses. The mare basalt analyses do not necessarily represent liquid compositions. These analyses include rocks that have crystal accumulations and also include unrepresentative samples due to small sample sizes. Vitrophyric and fine-grained (non-cumulate) mare basalts are plotted in the figure as (×) (after Shearer et al. 1994). The glasses presented here are average values. Individual glass beads range in TiO_2 from 0.2 to 17.0%.

The compositional diversity of the lunar basalts is partially illustrated by a plot of TiO_2 wt % versus Mg′ (Fig. 18). Although the range of TiO_2 contents of the volcanic glass beads overlap with that of the crystalline mare basalts, other characteristics such as Mg′, CaO, and Al_2O_3 are consistently different. Basaltic magmas represented by the glass beads have higher Mg′ and are lower in CaO and Al_2O_3. The relatively higher Mg′ values of the volcanic glass beads make them the best approximations of primary melts that are parental to the crystalline mare basalts. However, several lines of evidence suggest that the glasses and the crystalline mare basalts thus far sampled are not related by fractional crystallization. This evidence includes (1) parallel but not common liquid lines of descent, (2) distinct chemical differences, (3) differences in Pb isotopic systematics, (4) differences in experimentally determined depths of multiple saturation, and (5) mode of eruption (e.g., Terra and Wasserburg 1976; Delano 1986; Tatsumoto et al 1987; Longhi 1987; Shearer and Papike 1993, 1999).

Be in lunar basalts

Beryllium contents of the mare basalts have been determined in only a few studies (Dreibus et al. 1976; Shearer et al. 1994). The study of Dreibus et al. (1976) emphasized bulk analyses of crystalline mare basalts. The ion microprobe study of Shearer et al.

(1994) reported the Be content of individual glass beads. In the mare basalts, Be shows considerable variation (Dreibus et al. 1976). In the glasses, Be ranges from 0.06 to 3.09 ppm. At individual landing sites, Be in the glasses is positively correlated with TiO_2 and negatively correlated with SiO_2 (Figs. 19 and 20). At similar TiO_2 content, the crystalline mare basalts are enriched in Be relative to the glasses. In the glasses, the B/Be ranges from 0.4 to 4.6 and the Li/Be values range from 14 to 34. The high TiO_2 glasses have Li/Be at the low end of that range (less than 18), whereas the very low TiO_2 glasses are at the high end of that range (generally greater than 25). Only the very low TiO_2 glasses collected at the Apollo 14 landing site have Li/Be values that overlap with the high TiO_2 glasses. Both the mare glass beads and crystalline basalts (Li/Be ≈ 3 to 5) have Li/Be lower than CI chondrite (Li/Be ≈ 60).

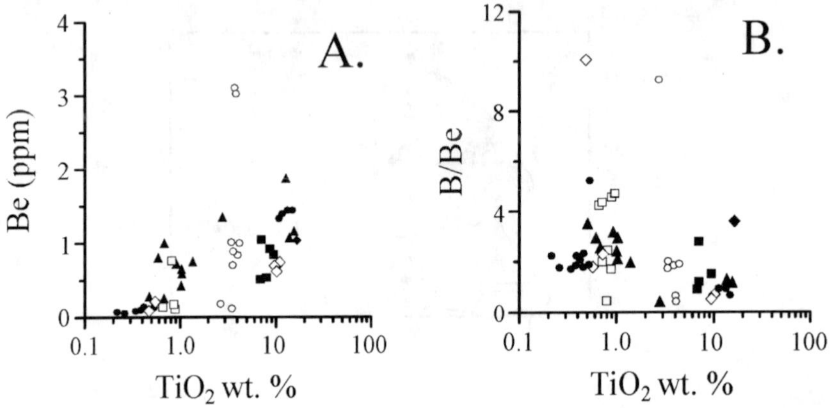

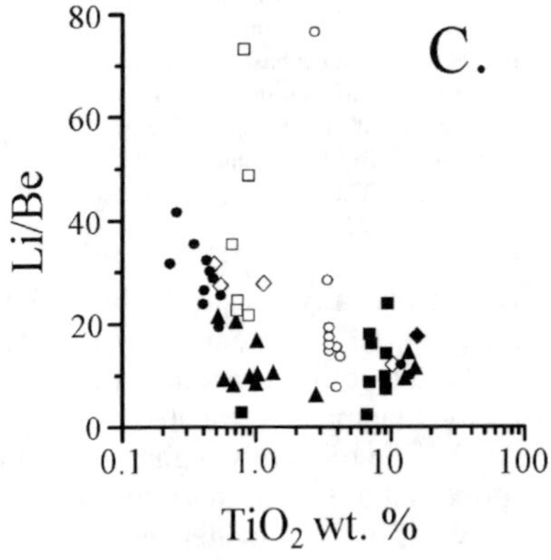

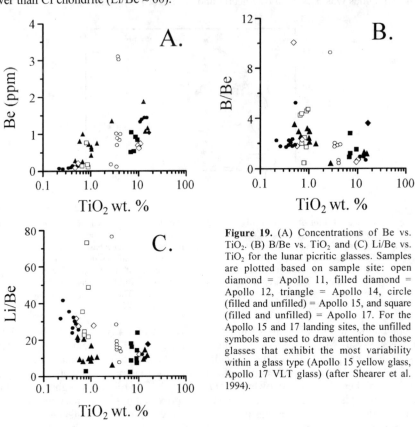

Figure 19. (A) Concentrations of Be vs. TiO_2. (B) B/Be vs. TiO_2 and (C) Li/Be vs. TiO_2 for the lunar picritic glasses. Samples are plotted based on sample site: open diamond = Apollo 11, filled diamond = Apollo 12, triangle = Apollo 14, circle (filled and unfilled) = Apollo 15, and square (filled and unfilled) = Apollo 17. For the Apollo 15 and 17 landing sites, the unfilled symbols are used to draw attention to those glasses that exhibit the most variability within a glass type (Apollo 15 yellow glass, Apollo 17 VLT glass) (after Shearer et al. 1994).

Comparison between lunar and terrestrial basalts

In Figure 21, LLE abundances for the lunar basaltic glasses and terrestrial basalts are normalized to chondrite and plotted in the sequence Be-Li-B. This sequence represents increasing volatility during condensation processes and increasing solubility in many hydrothermal solutions. Compared to the MORB, the lunar basalts show a similar slope for the LLE ($Be_N > Li_N > B_N$), but a much larger range for Be and B. In the lunar glass beads, Be generally increases with decreasing SiO_2 (Fig. 20). This contrasts with observations made for terrestrial MORB (Ryan and Langmuir 1987, 1988, 1993; Brenan et al. 1998). This is a result of different processes reflected in the MORB and lunar data sets. The dispersion of Be-Li-SiO_2 in MORB samples primarily reflects fractional crystallization processes, whereas the dispersion in the lunar glasses reflects both source composition and partial melting processes instrumental in generating a compositionally diverse suite of near primary magmas.

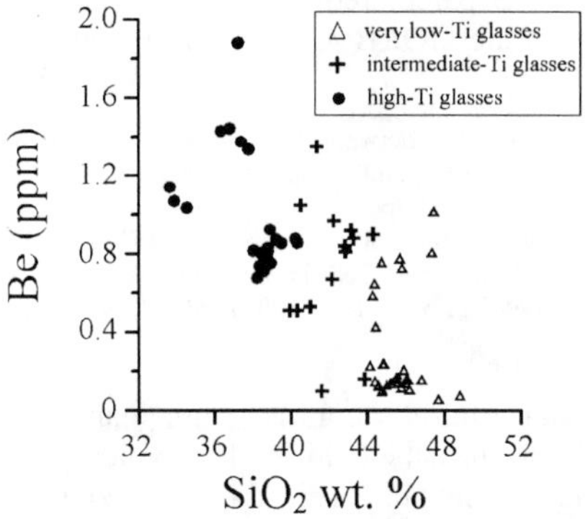

Figure 20. Be vs. SiO_2 content in lunar volcanic glasses (after Shearer et al. 1994).

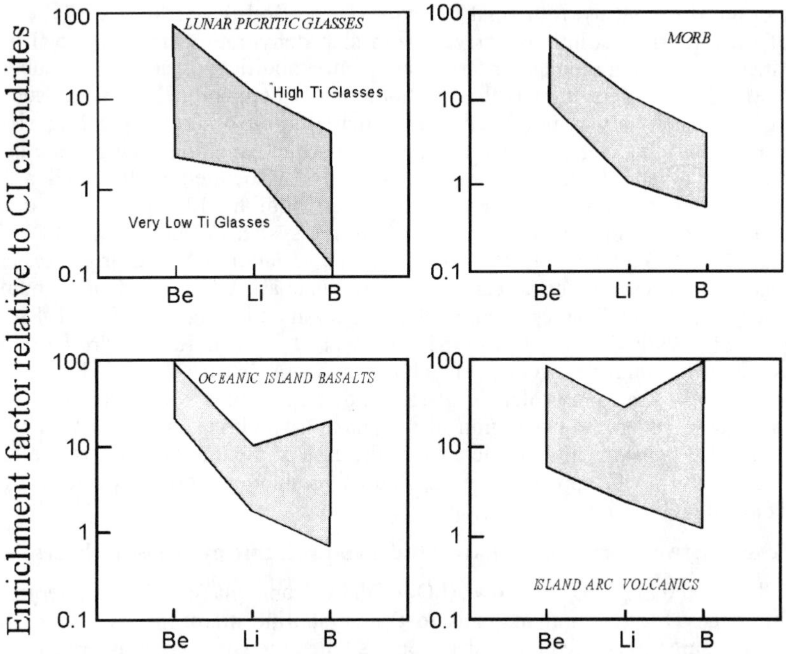

Figure 21. Chondrite-normalized LLE abundances in of lunar volcanic glasses, MORB, oceanic island basalts, and island arc volcanics. The terrestrial LLE data are from Ryan and Langmuir (1987, 1988, 1993) (after Shearer et al. 1994).

This highly inhomogeneous nature of the lunar mantle with regards to the LLE is illustrated by back-calculating mantle characteristics from the basaltic glass data. This calculation is based on the following assumptions: 10% partial melting of an olivine dominant mantle, olivine + orthopyroxene residuum, minimal fractional crystallization following melting, and appropriate D values from Ryan and Langmuir (1987, 1988, 1993) and Brenan et al. (1998). The results indicate that the range in Be concentrations in the lunar mantle is < 0.01 ppm (early cumulates) to 10 ppm (late cumulates). This wide range in the anhydrous lunar mantle contrasts with the much narrower range estimated for the terrestrial upper mantle source for MORB calculated by Ryan and Langmuir (1987, 1988, 1993). Their calculated Be content of the MORB source was 0.05 ppm.

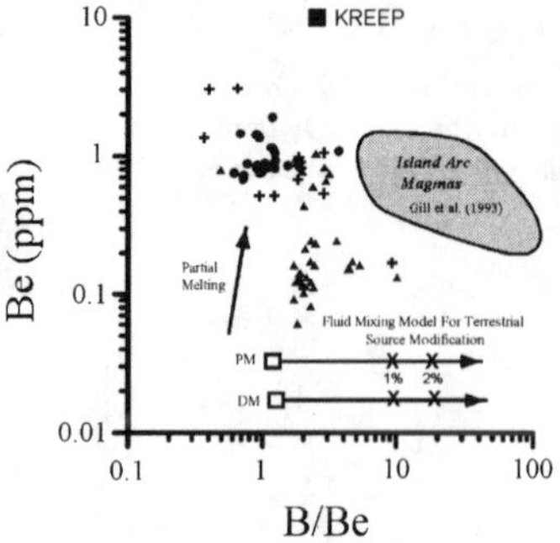

Figure 22. Comparison between Be abundances in the lunar picritic glasses (plotted points) and island arc volcanics (Gill et al. 1993). Also superimposed are terrestrial mantle mixing lines between terrestrial mantle components (PM, DM) and hydrothermal solutions released through the dehydration of a subducting plate (Leeman et al. 1994) (after Shearer et al. 1994).

The mare basalts contrast with many of the oceanic island basalts and island arc volcanics that exhibit enrichments in B (Fig. 21). This contrasting LLE behavior in island arc volcanics such as decoupling of elements with similar magmatic behavior but different hydrothermal behavior (Be:B, Be:Li), has been attributed to the role of hydrothermal solutions derived from slab dehydration. Leeman et al. (1994) illustrated fluid mixing models for terrestrial mantle modification adjacent to subduction zones (Fig. 22). In these models the "hydrothermally compatible" elements, such as B and Li, are selectively remobilized during dehydration of the subducting slab and incorporated into the overlying mantle wedge. Subsequent melting results in the increase of B/Be, B/Yb, and Li/Be ratios in the magmas associated with island arcs. As demonstrated in the lunar volcanic glass beads, similar although not as dramatic enrichments are also observed in anhydrous basaltic systems of the Moon. B/Be (0.4 to 10) (Fig. 22) and B/Yb (0.08 to 1.0) show as much variation in the lunar glasses as they do in magmas associated with volcanic arcs (Leeman et al. 1994). Based on the numerous studies of island arc volcanics (Tera et al 1986; Ryan and Langmuir 1987, 1988, 1993; Morris et al. 1990; Gill et al. 1993; Leeman et al. 1994), such hydrothermal-type mechanisms for mantle wedge enrichments seem highly likely. However, as demonstrated by the lunar volcanic glasses, other mechanisms in anhydrous basaltic systems may result in the generation of magmas in which changes in B/Yb and B/Be mimic the LLE behavior in volcanic arcs. In the case of the lunar basalts, crystallization processes in the LMO appear to be capable of producing a broad spectrum of LLE distinct mantle reservoirs.

Differences between crystalline mare basalts and volcanic pyroclastic glasses

Although the high Mg′ and low Al_2O_3 of the volcanic glasses may be interpreted as indicating they are parental magmas to the crystalline mare basalts, as previously mentioned, major-trace element and isotopic studies by Tera and Wasserburg (1976), Tatsumoto et al. (1987), Longhi (1987, 1992) and Shearer and Papike (1993) strongly suggest that this is not the case. The LLE data also illustrate this point. Lithium and Be data show that the mare basalts (Dreibus et al. 1976) and volcanic glasses are chemically offset from one another (Fig. 23). At the same Li value, the crystalline mare basalts are enriched in Be (Li/Be ≈ 3 to 5) relative to the volcanic glasses (Li/Be ≈ 10 to 34). Differences in $D^{(olivine/melt)}$ for Li and Be (Ryan and Langmuir 1987, 1988) indicate that fractional crystallization and partial melting processes involving olivine will modify the Li/Be ratio. Calculated fractional crystallization trajectories for 0.1 to 90% (latter highly unlikely) run approximately parallel to the distribution of mare basalts and volcanic glasses, indicating that they cannot be related by fractional crystallization. Assuming a batch partial melting process, the Li/Be value will change abruptly at less than 5% partial melting with olivine in the residuum (Fig. 23). Unfortunately, any such model relating the mare basalts to the glasses in this manner would require the mare basalts to be produced by incredibly small degrees of partial melting (<1%). The likelihood of consistently extracting large volumes of melt at these low degrees of partial melting is extremely low. In addition, most trace element and isotopic studies indicate the mare basalts were produced by higher degrees of partial melting than the volcanic glasses (Hughes et al. 1988, 1989; Delano 1986; Shearer and Papike 1993).

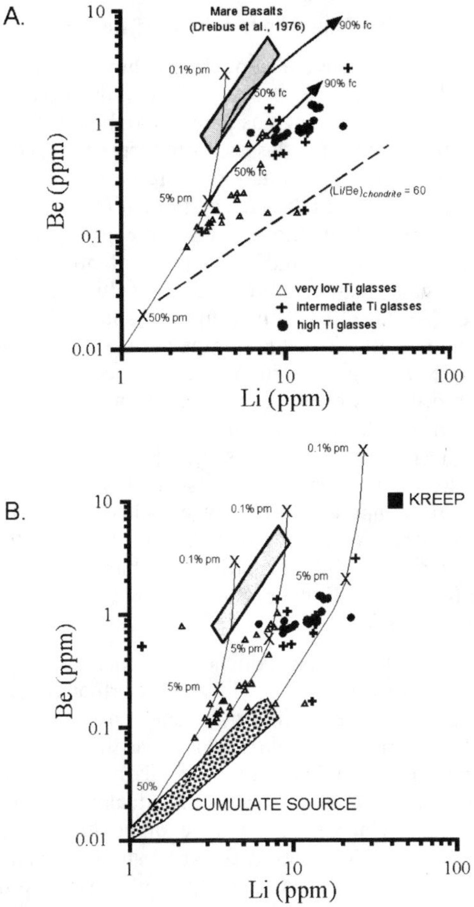

Figure 23. Beryllium vs. Li for the crystalline mare basalts and volcanic glass beads.

(A) Superimposed on this plot are partial melting and fractional crystallization trajectories. The picritic glasses are not related to the crystalline mare basalts by fractional crystallization of olivine ± pyroxene. It appears that different degrees of partial melting may be a possible petrogenetic link between the glasses and crystalline basalts. However, producing the mare basalts through smaller degrees of melting does not agree with the relative abundances of volcanic glasses to mare basalts (mare basalts >>> volcanic glasses) or other trace element ratios (Shearer and Papike 1993, 1999).

(B)Relatively small degrees of partial melting of a compositionally diverse lunar magma ocean cumulate source (LMO) will produce many of the Be and LI systematics of lunar basalts. The field of LMO cumulates was estimated assuming 3 times CI chondritic abundances of Li and Be and lunar magma ocean crystallization models of Hughes et al. (1988, 1989) and Snyder et al. (1992) (after Shearer et al. 1994).

As illustrated in Figure 23, partial melting of different lithologies of the LMO cumulate pile may produce a wide spectrum of picritic magmas that are equivalent to the volcanic glasses *or* are parental to the crystalline mare basalts. As argued by Shearer and Papike (1993), Shearer et al. (1994), Ringwood (1992), and Misawa et al. (1993), the magmas represented by the volcanic glass beads were derived from a low μ (= U/Pb), relatively volatile-enriched LMO cumulate source in the lunar mantle. LMO cumulates are required as sources for these magmas because they have LMO chemical signatures, such as negative Eu anomalies and a wide range in Ti and incompatible concentrations (Shearer and Papike 1993). Shearer and Papike (1993) and Shearer et al. (1994) have suggested that these "less depleted" lunar mantle sources for the basaltic glasses are either lunar magma ocean (LMO) cumulates that experienced less processing during post-LMO periods of magmatism or a "mixture" of LMO cumulates and a more fertile mantle. Further evidence for this "less depleted" lunar mantle component is the enrichment of the more volatile LLE (i.e., Li/Be, Li/B, B/Be) compared to the mare basalts. This "mixture" of components may be a result of mantle overturning (Ringwood and Kesson 1976; Shearer and Papike 1993, 1999; Spera 1992; Hess and Parmentier 1995) or polybaric melting (Longhi 1992).

Models for lunar basaltic magmatism as implied from the behavior of Be

Shearer and Papike (1993, 1999) summarized several models for the origin of the basaltic magmas represented by the pyroclastic glass beads and their relationship to the crystalline mare basalts. These models imply: (1) The magmas parental to the crystalline mare basalts are similar in composition to the pyroclastic glasses. Both were produced by melting of LMO cumulates. Yet, the glass beads and crystalline mare basalts thus far sampled are not related to each other by shallow fractional crystallization or partial melting processes. (2) The magmas represented by the glasses were produced by relatively small degrees of partial melting leaving behind a residuum of olivine and orthopyroxene. (3) The glasses represent magmas that were produced by melting that was initiated deep within the lunar mantle. (4) The deep lunar mantle source for the pyroclastic glass beads is highly inhomogeneous. This inhomogeneity was partially derived from both large- and small-scale mixing of a lunar magma ocean (LMO) cumulate pile through gravitational instability. This resulted in the transport of evolved, late-stage LMO crystallization products into the deep lunar mantle. Alternatively, some shallow mixing may have occurred through assimilation of late-stage LMO cumulates by magmas produced in the deep lunar mantle. (5) The magmas represented by the glasses were derived from cumulate mantle sources that were slightly enriched in volatile elements relative to the cumulate mantle source for the crystalline mare basalts. How does Be and the other LLE behave during LMO crystallization-basalts source differentiation and is there any indication from the LLE that cumulate source mixing is an important process during lunar basalt petrogenesis?

In a dynamically simple LMO, a LMO crystallization sequence advocated for many bulk composition models (Taylor 1982; Snyder et al. 1992) is: olivine → orthopyroxene ± olivine → clinopyroxene + orthopyroxene ± olivine → clinopyroxene + plagioclase + orthopyroxene ± olivine → ilmenite + clinopyroxene + plagioclase + orthopyroxene. Incompatible trace elements excluded from the crystal structures of olivine, pyroxene, plagioclase, and ilmenite were concentrated in the very last stages of residual melt. These residual liquids are referred to as KREEP (an acronym for an evolved lunar crust component that is enriched in potassium (K), rare earth elements (REE) and phosphorus (P) and other incompatible elements). Ideally, trace element signatures of lunar basaltic magmas should reflect the location in the cumulate pile stratigraphy that the magmas were derived. For example, very low-Ti basaltic magmas, which have low incompatible element abundances, should have been derived by melting of early and deep LMO cumulates, whereas high-Ti basaltic magmas, which have fairly high incompatible element abundances, should have been derived by melting of late and shallow cumulates (Taylor 1982). One problem with this simple scenario is that multiple-saturation depths that were experimentally determined for the various pyroclastic glasses have been interpreted as indicating that both the very low-Ti and high-Ti basalts were produced in the deep lunar mantle (>400 km) (Delano 1986; Longhi 1992; Shearer and Papike 1993, 1999; Papike et al. 1998).

Within the context of this very simple LMO crystallization model, the LLE should be increasingly enriched in later LMO cumulates, highly enriched in KREEP (Be = 10 ppm, Li = 40 ppm, B = 25 ppm) (Warren 1988) and Li should be enriched in the residual LMO melt to lesser a degree than B or Be (Fig. 24). The latter is strictly a result of the contrasting behavior of Li compared to B and Be in olivine. Li is moderately incompatible, whereas B and Be are highly incompatible (Ryan and Langmuir 1987, 1988, 1993; Brenan et al 1998). This will result in the modification of the Li/Be and Li/B signatures of the mantle cumulate sources. Therefore, the abundance of the LLE and their ratios reflects characteristics of the lunar mantle sources. Based on the pyroclastic glass data, the early crystallization cumulate component of the LMO (i.e., olivine and olivine + orthopyroxene cumulates) as reflected by many of the very low-Ti glasses have higher Li/Be (>30) and Li/B (>25) than late stage products of LMO crystallization (i.e., KREEP with Li/Be ≈ 1.6; Warren 1988).

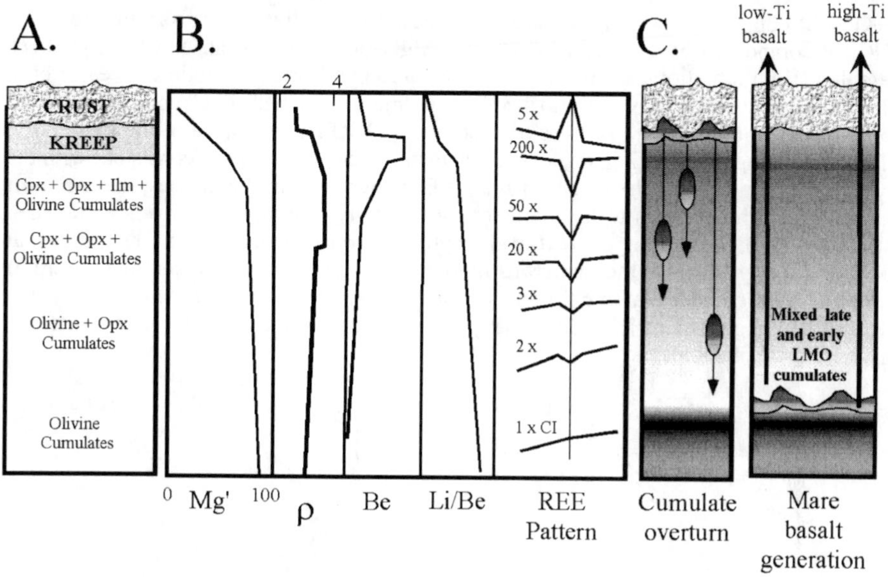

Figure 24. The behavior of Be, and Li within the context of lunar magma ocean crystallization and mare basalts petrogenesis models. (A) Simplified sequence of cumulate mineral assemblages produced during the crystallization of a lunar magma ocean. (B) Variation in properties (Mg′, density (ρ), Be, Li/Be, and REE pattern) of the lunar magma ocean cumulate pile based on the simple crystallization sequence in A. In the cumulate pile Be and Li should increase upward and Li/Be should decrease. (C) Due to density contrasts in the LMO cumulate pile, gravitational destabilization of the cumulate pile may result in the transport of the late, denser high-Ti cumulates (also high in Be, Li, B, and heat producing elements) into the deep lunar mantle. Decay of the heat producing elements will result in melting of the heterogeneous lunar mantle to produce a wide range of mare basalts (after Shearer et al. 1994).

An interesting contrast to this behavior is the variation in Li/B in calculated mantle cumulate sources for the glass beads. In the glasses from a given landing site, the Li/B remains nearly constant over the complete range in Ti. This implies unusual behavior of B during LMO cumulate crystallization, overturning or partial melting.

As shown in Figure 25, both the very low-Ti (Fig. 25A) and high-Ti (Fig. 25B) Apollo 14 glasses plot on a mixing line between KREEP and glasses low in incompatible elements (i.e., Apollo 15 Green Glasses and Apollo 11 Orange). This observation for the LLE is similar to the observations made by Shearer et al (1991) and Shearer and Papike (1993) for other incompatible lithophile elements (i.e., Sr, Ba, Zr, REE). Based on these mixing curves, the Apollo 14 glasses have between 3% and 10% KREEP component assuming 0% in the Apollo 15 Green C glasses. The incorporation of the KREEP signature in these glasses may be attributed to either (1) shallow assimilation and fractional crystallization (AFC) processes analogous to those proposed by Neal and Taylor (1992) for mare basalts, or (2) hybridization of mafic cumulates following catastrophic LMO cumulate overturning (Ringwood and Kesson 1976; Hughes et al. 1988, 1989; Spera 1992; Shearer and Papike 1993; and Hess and Parmentier 1993). Based on the modeling of Shearer et al (1991) and Shearer and Papike (1993) (see Fig. 12 in Shearer and Papike 1993), the KREEP signature in these glasses most likely reflects the recycling of KREEP and/or other late stage LMO cumulates into the deep lunar mantle (Shearer and Papike 1993). AFC models require either 2% to 25% assimilation of a KREEP component with no resulting fractional crystallization or assimilation of a KREEP component by a series of extremely MgO-rich magmas that have not yet been sampled. Both situations seem unlikely. In source hybridization models, Shearer et al (1991) calculated that only 0.3% to 1.5% intercumulus "KREEP-like" component needs to be incorporated into the sources to produce the KREEP signature in the pyroclastic glasses. Although the LLE data and trace element modeling are more consistent with the hybridization model for the origin of the KREEP signature in the glasses, the extent of KREEP enrichment tends to be characteristic of many lithologies at a single site (i.e., Apollo 14 site). Within the context of the hybridization models, it could be argues that the KREEP signature for the numerous lithologies of a site may represent the nature of the lunar mantle under that site for an extended period of magmatism. Alternatively, the evolved LLE signature of KREEP in many of these magmas may be a localized, shallow, mantle-crustal phenomena.

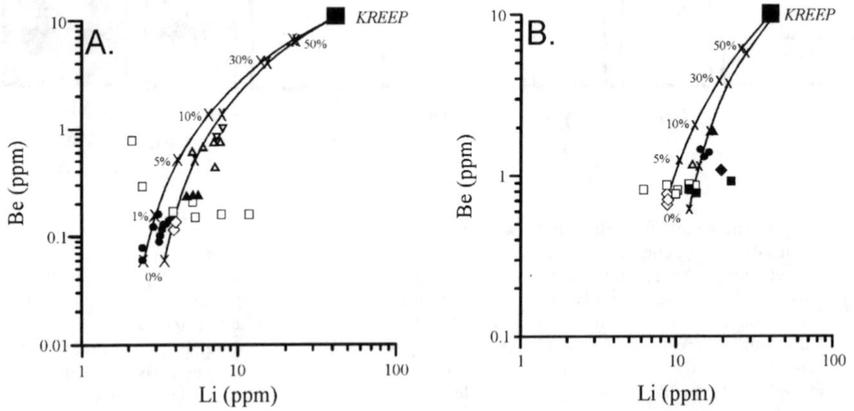

Figure 25. Plot of Be vs. Li for the (A) low-Ti glasses and (B) high-Ti glasses illustrating possible mixing between low Be-Li glasses and a KREEP component (after Shearer et al. 1994). Glasses are indexed based on sampling site: open diamond = Apollo 11, filled diamond = Apollo 12, triangle = Apollo 14 (In A, filled = Green B glass, open = Green A glass, pointing down = VLT glass. In B, filled = Orange glass, open = Black glass), circle = Apollo 15, and square = Apollo 17 (In B, open = Orange glass, filled = Black glass).

BE IN MARTIAN METEORITES: EVIDENCE FOR MARTIAN WATER AND SOIL

Introduction

Although numerous spacecraft have orbited and landed on Mars, none of these missions were designed for sample return. However, samples of the Martian crust have been studied. In the late 1970s and early 1980s differentiated meteorites referred to as the SNC (shergottites, nakhlites, and chassignites) group were recognized as our first samples of the Martian crust (McSween et al. 1979; Walker et al. 1979; Wasson and Wetherill 1979; Bogard and Johnson 1983; Becker and Pepin 1984). This conclusion was based on the young crystallization ages (<1.3 Ga) of the SNCs available at the time, and the match of several isotopic and geochemical fingerprints between measurements made by the Viking missions and the SNC (Bogard and Johnson 1983; Becker and Pepin 1984; Swindle et al. 1986). A more detailed discussion of these arguments is presented by McSween and Treiman (1998).

The suite of Martian samples consists of 1over 26 meteorites. The first SNC sample that was collected was the Chassigny meteorite. It was seen to fall in Haute-Marne, France on October 3, 1815 (Graham et al. 1985). A total of six Martian meteorites were recovered between 1977 and 1994 in Antarctica. In the late 1990s and early 2000, additional Martian samples were collected from California and the deserts of northern Africa and the Middle East. Sizes of the Martian samples range from a known mass of ~18 kg for Zagami to the 12-g sample of QUE94201. All of the Martian samples are igneous in origin and include basalts, lherzolites, clinopyroxenites-wehrlite, a dunite and an orthopyroxenite. Several of these assemblages have superimposed upon their magmatic lithologies subtle to obvious secondary mineralization that is Martian in origin. These meteorites also exhibit varying degrees of shock and terrestrial alteration.

Only recently has Be been measured in Martian meteorites and used to better understand processes on Mars. Lentz et al. (2000) and McSween et al. (2000) measured Be, B, and Li in pyroxene in four Martian meteorites using an ion microprobe. Shearer et al. (2000) measured Be and other trace elements in shock-produced melt in ETTA79001 and Newsom et al. (2001) reported preliminary Be values for alteration phases in Lafayette.

Water on Mars

Two of the important unknowns concerning the evolution of Mars are its volatile inventory and outgassing history. Not only are these two variables important for our understanding of the mantle, crustal, and atmospheric evolution of Mars, but they are also important for evaluating the possible existence of life during its evolution. Previous attempts to understand the global volatile inventory of the planet have used estimates of atmospheric gas abundances, surface morphologies, and products of Martian magmatism (volume, characteristics of eruptions, water content of magmas as preserved in Martian meteorites) (McSween and Harvey 1993). The study by McSween and Harvey (1993) and further interpretation of their observations (Watson et al. 1994; Popp et al 1995) suggest that either the Martian magmas represented by the SNC meteorites were water poor (130 to 350 ppm), or that the record of the water content of the magmas was poorly preserved. Studies of MORB suggest that the latter is a viable possibility. Sisson and Grove (1993) and Sisson and Layne (1993) illustrated that although the water contents of the MORBs were poorly recorded in bulk basalt samples, melt inclusions in the basalts indicated a higher water content.

In their preliminary ion probe study, Lentz et al (2000) and McSween et al. (2000) measured the LLE in pyroxene from Martian meteorites Nakhla, Lafayette, Zagani and

Shergotty. Beryllium in the pyroxene from these four meteorites varies from less than 5 ppb to ~50 ppb and B/Be ranges from 109 to 757. Beryllium content increases from core to rim of pyroxene crystals, whereas B/Be decreases from core to rim (Fig. 26). In addition, the zoning of Be, B, and Li in the pyroxene was substantially different in Shergotty and Zagami (crystallized in fairly shallow crustal environment) compared to Nakhla and Lafayette (crystallized in deeper crustal regimes). In Nakhla and Lafayette, these three elements behave as incompatible elements increasing from core to rim. On the other hand, pyroxenes in Shergotty and Zagami have B and Li at the pyroxene rims decreasing with increasing Be (Fig. 26).

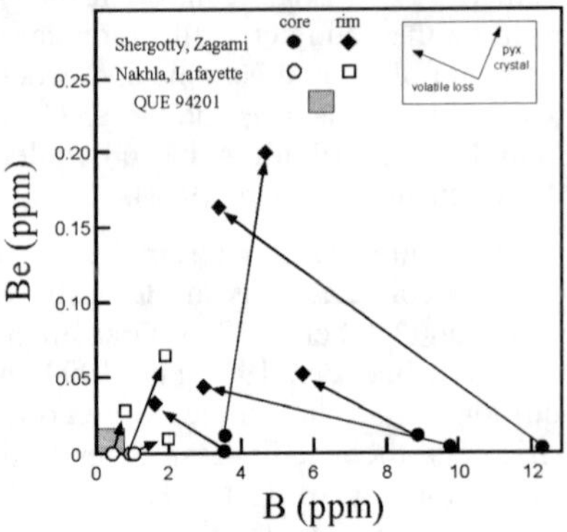

Figure 26. Be vs. B for high Ca clinopyroxene cores and rims from Nakhla, Lafayette, Zagami, and Shergotty Selected data from Lentz et al. (2000, 2001) plotted to illustrate differences in zoning charateristics. Also included is unpublished data from Shearer illustrating the range of B and Be in low- and high-Ca pyroxene from QUE 94201.

Based on the observation that $D^{pyroxene}$ for B and Be are both incompatible, they concluded that the B/Be ratio in pyroxene should be similar to the ratio in the magma. This value (109-757) is significantly different than typical MORB, which have B/Be of approximately 1 to 5 (Ryan and Langmuir 1987, 1988, 1993). They also interpreted the contrasting behavior in the Shergotty and Zagami pyroxene as reflecting volatile loss during magma ascent. In their model, they suggested that the pyroxenes in Nakhla and Lafayette and the Mg-rich pyroxene cores in Shergotty grew at depth in equilibrium with a pressurized magma. Upon either eruption or emplacement into lower pressure environment, the basaltic magma represented by Shergotty experienced exsolution of a volatile phase which depleted the magma in Li and B while the iron-rich rims of the pyroxene continued to crystallize. This is the result of the contrasting behaviors of Be (insoluble), B (highly soluble in certain fluids above 200°C), and Li (highly soluble in certain fluids above 350°C) in hydrothermal solutions (Ryan and Langmuir 1987, 1993).

Martian soil?

In several of the Martian meteorites, the igneous lithologies have been invaded by pockets and veins of shock produced melt. In Martian meteorite ETTA79001, the melt consists of brown vesicular glass containing relict igneous crystals and skeletal olivine and pyroxene that crystallized from the shock melt. It has been calculated that ETTA79001 experienced equilibrium shock pressures of ~34 GPa (Stöffler et al. 1986). Gooding and Muenow (1986) reported that this shock melt lithology also contained calcium carbonate and sulfate presumably of Martian origin. The composition of the shock melt is similar to the bulk composition of the host lithology, but is somewhat enriched in the plagioclase component (McSween and Jarosewich 1983). The melt pockets have high concentrations of noble gases with elemental and isotopic compositions similar to the Martian atmosphere (Bogard and Johnson 1983; Becker and Pepin 1984). This noble gas signature led Bogard et al. (1986) to suggest that this shocked glass may contain a Martian soil component. A detailed microprobe study by Rao et at. (1999) revealed that in addition to a plagioclase component, this impact melt glass was also substantially enriched in sulfur compared to the bulk igneous lithology. Based on mixing models, Rao et al. (1999) concluded that these enrichments could be attributed to a 6% Martian soil component.

Understanding the mineralogy and chemical characteristics of the Martian soil is critical to our interpretation of the orbital mapping of Mars, the search for life and for future pre-manned mission experiments. Yet, a Martian soil sample is not in our meteorite sample suite and it has been analyzed only on the surface of Mars during the Viking and Pathfinder missions. Therefore, the potential presence of a Martian soil component in a shock melt is important. As a further investigation of this possible soil component in the impact melt from ETTA79001, Shearer et al. (2000) analyzed these melts for Be and additional trace elements (Li, B. Ba, Th, and REE) using ion microprobe techniques. This impact melt is depleted in LREE (La 1 to 3 × CI) relative to the HREE (Yb = 5 to 7 × CI). The Eu anomaly is positive with Eu_N/Sm_N ranging from 1.7 to 1.1. Glasses with minor Eu anomalies tend to have higher HREE abundances and REE patterns similar to the bulk igneous lithology. The Be contents of the impact melt range from 47 to 73 ppb. This is similar to the Be concentrations found in alteration phases in Martian meteorite Lafayette (Newsom et al. 2001). The Be content of the plagioclase component is higher than that of the impact melt and is ~130 ppb. The plagioclase component is also slightly higher in Li. The Be content in the glass does not correlate with the size of the Eu anomaly. Based on these results, Shearer et al. (2000) were unable to distinguish a soil component that had substantially different Be, Li, and REE signatures than the bulk lithology of ETTA79001 or its plagioclase component. They concluded that if a soil component was present, it was similar in composition to a ETTA79001 lithology that was altered at the Martian surface.

USEFULNESS OF BE IN FUTURE PLANETARY STUDIES

In the previous text, I attempted to summarize the very broad use of Be and the other LLE in studying planetary materials and processes. There are numerous problems in planetary science that could benefit from additional LLE data in general and Be data in particular. Some areas that come to mind are early solar system processes, volatile elements in the lunar and Martian mantles, and surficial geochemical processes on Mars.

McKeegan et al. (2000) reported ^{10}B isotopic anomalies in a CAI from Allende. It is important to further document how wide spread ^{10}Be was in the early solar system, if any correlations exist between ^{10}Be and other CAI mineralogical or isotopic properties and if similar anomalies exist in chondrules. These studies may shed additional light upon early high temperature condensation and melting events relevant to CAIs and chondrules, identify and fingerprint different environments in the early solar nebula, and p:rovide additional insight into the early life of the sun.

As was mentioned earlier, the character and content of volatiles in both the lunar and Martian mantles are hotly debated. Fire-fountaining occurred on the Moon and is represented by the pyroclastic glass beads, yet the glasses are volatile poor. A detailed study of volatile elements such as S and the LLE in these glasses would be extremely helpful in identifying lunar volatiles and understanding their behavior during fire-fountaining. Studies by Lentz et al (2000) and McSween et al. (2000) suggest a possible means of unraveling the volatile record preserved in Martian basalts. Instrumental in interpreting these LLE data is a better understanding of the partitioning behavior of Be, Li, and B in melts that approximate Martian basalt compositions and conditions of crystallization.

Finally, our understanding of the Martian soil is in its infancy. Yet, to understand its character and evolution is critical to interpreting remotely sensed data collected by

Martian orbital missions and reconstructing the evolution of the Martian surface. Determining the abundance of mobile/non-mobile elements such as the LLE in soil samples returned from Mars during the second decade of the 21st century will provide information about aqueous processes in the Martian crust and processes that formed the soil. This may allow us to eventual understand the evolution of the Martian surface environment and aid us in the exploration for Martian life. Until then, studying the behavior and distribution of Be and the other LREE in shock melts and Martian alteration products will be a good first step.

ACKNOWLEDGMENTS

This chapter benefited greatly from reviews by Steve Simon, Kevin Righter, and Ed Grew. I also thank Rhian Jones for access to samples from the Institute of Meteoritics collection, and Graham Layne and Adrian Brearly for discussions of the behavior of the LLE in chondritic meteorites. The author was supported for this work by NASA grant #NAG5-4253 and the Institute of Meteoritics.

REFERENCES

Anders E (1977) Chemical compositions of the Moon, Earth, and eucrite parent body. Phil Trans Roy Soc A285:23-40

Anders E, Grevesse N (1989) Abundances of the elements: Meteoritic and solar. Geochim Cosmochim Acta 53:197-214

Atkinson R d'E, Houtermans FG (1929) Zur Frage der Aufbaumöglichkeit der Elemente in Sternen. Z Phys 54:656-665

Audouze J, Truran JW (1975) P-process nucleosynthesis in postshock supernova envelope environments. Astrophy J 202:204-213

Basaltic Volcanism Study Project (1981) Basaltic Volcanism on the Terrestrial Planets. Pergamon, New York

Becker RH, Pepin RO (1984) The case for a Martian origin of the shergottites: Nitrogen and noble gases in EETA 79001. Earth Planet Sci Lett 69:225-242

Beckett JR, Spivack AJ, Hutchon ID, Wasserburg GJ, Stolper EM (1990) Crystal chemical effects on the partitioning of trace elements between mineral and melt: An experimental study of melilite with application to refractory inclusions from carbonaceous chondrites. Geochim Cosmochim Acta 54:1755-1774

Bernas R, Gradeztajn E, Reeves H, Schatzman E (1967) On the nucleosynthesis of lithium, beryllium, and boron. Ann Phys 44:426-478

Bischoff A, Keil K (1983) Ca-Al-rich chondrules and inclusions in ordinary chondrites. Nature 303: 588-592

Bischoff A, Keil K (1984) Al-rich objects in ordinary chondrites: Related origin of carbonaceous and ordinary chondrites and their constituents. Geochim Cosmochim Acta 48:693-709

Bischoff A, Rubin AE, Keil K, Stöffler D (1983) Lithification of gas-rich chondrite regolith breccias by grain boundary and localized shock melting. Earth Planet Sci Lett 66:1-10

Bogard DD, Johnson P (1983) Martian gases in an Antarctic meteorite. Science 221:651-654

Bogard DD, Hörtz F.P, Johnson P (1986) Shock-implanted noble gases: An experimental study with implications for the origin of Martian gases in shergottite meteorites. J Geophys Res 91:E99-E114

Boss AP (1996) A concise guide to chondrule formation models. *In* Chondrules and the Protoplanetary Disk. Hewins RH, Jones RH, Scott ERD (eds) Cambridge Univ Press, Cambridge, p 257-263

Boynton WV (1989) Cosmochemistry of the rare earth elements: Condensation and evaporation processes. Rev Mineral 21:1-24

Bratt SR, Solomon SC, Head JW, Thurber CH (1985) The deep structure of lunar basins: Implications for basin formation and modification. J Geophys Res 90:3049-3064

Brearley AJ, Jones RH (1998) Chondritic meteorites. Rev Mineral 36:3-1 – 3-398

Brearley AJ, Layne GD (1996) Light lithophile element (Li, Be, B) abundances in microchondrules in CH chondrites: Insights into volatile behavior during chondrule formation. Lunar Planet Sci XXVI: 167-168

Brenan JM, Neroda E, Lundstrom CC, Shaw HF, Ryerson FJ, Phinney DL (1998) Behavior of boron, beryllium, and lithium during melting and crystallization: Constraints from mineral-melt partitioning experiments. Geochim Cosmochim Acta 62:2129-2141

Burbridge EM, Burbridge GR, Fowler WA, Hoyle F (1957) Synthesis of the elements in stars. Rev Modern Phys 29:547-650

BVSP (Basaltic Volcanism Study Project) (1981) Basaltic Volcanism on the Terrestrial Planets. Pergamon, New York

Cameron AGW (1973) Abundance of elements in the solar system. Space Sci Rev 15:121-146

Cameron AGW (1995) Accounting for light-element abundances. Nature 373:286

Cameron AGW, Colgate SA, Grossman L (1973) Cosmic abundances of boron. Nature 243:204-207

Canup RM, Agnor C (1998) Accretion of terrestrial planets and the Earth-Moon system. *In* Origin of the Earth and Moon, LPI Contrib 957, Lunar and Planetary Institute, Houston, p 4-6

Canup RM, Agnor C (2000) Accretion of terrestrial planets and the Earth-Moon system. *In* Origin of the Earth and Moon. RM Canup, K Righter (eds), p 113-129

Cassé M, Lehoucq R, Vanglonl-Flam E (1995) Production and evolution of light elements in active star-forming regions. Nature 373:318-319

Chaussidon M, Robert F (1995) Nucleosynthesis of ^{11}B-rich boron in the pre-solar cloud recorded in meteoritic chondrules. Nature 374:337-339

Clayton DD (1988) Stellar nucleosynthesis and chemical evolution of the solar neighborhood. *In* Kerridge JF, Matthews MS (eds) Meteorites and the Early Solar System, p 1021-1062

Cox PA (1989) The Elements. Their Origin, Abundance and Distribution. Oxford Science Publications, New York, 207 p

Delano JW (1986) Pristine lunar glasses: Criteria, data and implications. Proc Lunar Planet Sci Conf 16:D201-D213

Deloule E, Robert F (1996) Origin of water in the solar system: Ion probe determination of the D/H ratios in chondrules. Meteoritics Planet Sci 31:A36

Deloule E, Doukhan J-C, Robert F (1997) An interstellar isotopic signature recorded in altered pyroxene chondrules. 28th Lunar Planet Sci Conf, p 291-292

Dodd RT, Walter LS (1972) Chemical constraints on the origin of chondrules in ordinary chondrites. *In* On the Origin of the Solar System. E Reeves (ed) Paris, CNRC, p 293-300

Dreibus G, Spettel B, Wänke H (1976) Lithium as a correlated element its condensation behavior, and its use to estimate the bulk composition of the moon and the eucrite parent body. Proc 7th Lunar Sci Conf, p 3383-3396

Ebel DS, Grossman L (2000) Condensation in dust-enriched systems. Geochim Cosmochim Acta 64: 339-366

Eugster O, Tera F, Burnett DS, Wasserburg GJ (1970) Isotopic composition of Gd and neutron-capture effects in some meteorites. J Geophys Res 75:2753-2768

Fegley B Jr, Palme H (1985) Evidence for oxidizing conditions in the solar nebula from Mo and W depletions in refractory inclusions in carbonaceous chondrites. Earth Planet Sci Lett 72:311-326

Field BD, Olive KA, Schramm DN (1995) Implications of a high population II B/Be ratio. Astrophys J 439:854-859

Fowler WA (1984) The quest for the origin of the elements. Science 226:922-935

Fowler WA, Burbridge GR, Burbridge EM (1955) Steller Evolution and the synthesis of the elements. Astrophys J 122:271

Fowler WA, Greenstein JL, Hoyle F (1962) Nucleosynthesis during the early history of the solar system. Geophys J Roy Astron Soc 6:148-220

Gamow G (1928) Zur Quantentheorie des Atomkernes. Z Phys 51:204-212

Gamow G (1938) Nuclear energy sources and stellar evolution. Phys Rev 53:595-604

Ganapathy R, Anders E (1974) Bulk composition of the moon and earth, estimated from meteorites. Geochim Cosmochim Acta, Suppl 5:1181-1206

Geiger H, Nuttall JM (1912) The ranges of the α particles from U. Phil Mag 23:439-445

Gill JB, Morris JD, Johnson RW (1993) Timescale for producing the geochemical signature of island arc magmas; U-Th-Po and Be-B systematics in recent Papua, New Guinea lavas. Geochim Cosmochim Acta 57:4269-4284

Goldschmidt VM (1926) Probleme und Methoden der Geochemie. Gerlands Beitr Geophys 15:38-50

Goldschmidt VM (1954) Geochemistry. Alex Muir (ed) Claredon Press, Oxford

Gooding JL, Muenow DW (1986) Martian volatiles in shergottite EETA 79001: New evidence from oxidized sulfur and sulfur-rich aluminosilicates. Geochim Cosmochim Acta 50:1049-1059

Graham AL, Bevan AWR, Hutchinson R (1985) Catalogue of Meteorites, 4th Edn. British Museum, London, and Univ of Arizona Press, Tucson

Greenberg RJ, Wacker J, Chapman CR, Hartmann WK (1978) Planetesimals to planets: Numerical simulationsof collisional evolution. Icarus 35:1-26

Grossman JN, Brearley AJ (1997) Contrasting styles of alteration in radial pyroxene and low-iron-oxide porphyritic chondrules in Seemarkona. Meteoritics Planet Sci 32:A53

Grossman JN, Wasson JT (1983) The composition of chondrules in unequilibrated chondrites: An evaluation of models for the formation of chondrules and their precursor materials. *In* King EA (ed) Chondrules and Their Origin. Houston, Lunar and Planetary Institute, p 88-121

Grossman JN, Rubins AE, Nagahara H, King EA (1988a) Properties of chondrules. *In* Kerridge JF, Matthews MS (eds) Meteorites and the Early Solar System. Tucson, Univ Arizona Press, p 243-253

Grossman JN, Rubin AE, MacPherson GJ (1988b) ALH 85085: A unique volatile-poor carbonaceous chondrite with implications for nebular fractionation processes. Earth Planet Sci Lett 91:33-54

Grossman L (1972) Condensation in the primitive solar nebula. Geochim Cosmochim Acta 49:2433-2444

Grossman L (1975) Petrography and mineral chemistry of Ca-rich inclusions in the Allende meteorite. Geochim Cosmochim Acta 39:433-453

Grossman L, Ganapathy R (1976) Trace elements in the Allende meteorite-II. Fine-grained, Ca-rich inclusions. Geochim Cosmochim Acta 40:967-977

Grossman L, Larimer JW (1974) Early chemical history of the solar system. Rev Geophys Space Sci 12: 71-101

Gurney RW, Condon EU (1928) Wave mechanics and radioactive disintegration. Nature 122:439

Hanon P, Chaussidon M, Robert F (1999) Distribution of lithium, beryllium, and boron in meteoritic chondrules. Meteoritics Planet Sci 34:247-258

Harkins WD (1917) the evolution of the elements and the stability of complex atoms. J Am Chem Soc 39:856-879

Hayakawa S (1968) Origin of the light elements in the solar system. Prog Theoret Phys, Suppl 42:156-169

Head JW (1974) Lunar dark mantle deposits: Possible clues to the distribution of early mare deposits. Proc Lunar Sci Conf 5:207-222

Head JW (1976) Lunar volcanism in space and time. Rev Geophys Space Phys 14:265-300

Head JW, Wilson L (1992) Lunar mare volcanism: Stratigraphy, eruption conditions, and the evolution of secondary crusts. Geochim Cosmochim Acta 56:2155-2175

Head JW, Wilson L (1997) Lunar mare basalt volcanism: Early stages of secondary crustal formation and implications for petrogenetic evolution and magma emplacement processes (abstr). *In* Lunar Planet Sci XXVIII, p 545-546, Lunar and Planetary Institute, Houston

Hess PC, Parmentier EM (1995) A model for the thermal and chemical evolution of the Moon's interior: Implications for the onset of mare volcanism. Earth Planet Sci Lett 134:501-514

Hewins R (1988) Experimental studies of chondrules. *In* Kerridge JF, Matthews MS, (eds) Meteorites and the Early Solar System, p 660-679. Tucson, University of Arizona

Hughes SS, Delano JW, Schmitt RA (1988) Apollo 15 yellow-brown volcanic glass: Chemistry and petrogenetic relations to green volcanic glass and olivine-normative mare basalts. Geochim Cosmochim Acta 52:2379-2391

Hughes SS, Delano JW, Schmitt RA (1989) Petrogenetic modeling of 74220 high-Ti orange volcanic glasses and the Apollo 11 and 17 high-Ti mare basalts. Proc 19th Lunar Planet Sci Conf, p 175-188

Jacobs WW, Bodansky D, Chamberlin D, Oberg DL (1974) Production of Li and B in proton and alpha-particle reactions on ^{14}N at low energies. Phys Rev 9C:2134-2143

Jones JH, Delano JW (1989) A three component model for the bulk composition of the Moon. Geochim Cosmochim Acta 53:513-527

Jones RH (1990) Petrology and mineralogy of type II, FeO-rich, chondrules in Semarkona (LL3.0): Origin by closed-system fractional crystallization, with evidence for supercooling. Geochim Cosmochim Acta 54:1785-1802

Jones RH, Lee T, Connolly HC, Love SG, Shang H (2000) Formation of chondrules and CAIs: Theory vs observation. Protostars and Planets IV:927-962. Tucson, University of Arizona Press

Jones RH, Shearer CK, Schilk AJ (2001) Trace element distribution in an Al-rich chondrule from the Mokoia CV 3 chondrite. Lunar Planet Sci XXXIII: Abstr #1338

Krot AN, Scott ERD, Zolensky ME (1995) Mineralogical and chemical modification of components in CV3 chondrites: Nebular or asteroidal processing? Meteoritics Planet Sci 30:748-775

Lattimer JM, Schramm DN, Grossman L (1978) Condensation in supernova ejecta and isotopic anomalies in meteorites. Astrophys J 219:230-249

Lauretta DS, Lodders K (1997) The cosmochemical behavior of beryllium and boron. Earth Planet Sci Lett 146:315-327

Leeman WP, Carr MJ, Morris JD (1994) Boron geochemistry of the central American volcanic arc: Constraints on the genesis of subduction-related magmas. Geochim Cosmochim Acta 58:149-168

Lentz RCF, Ryan JG, Riciputi LR, McSween HY Jr (2000) Water in the Martian mantle: Clues from light lithophile elements in Martian meteorites. Lunar Planet Sci XXXI: Abstr #1672

Lissauer JJ, Stewart GR (1993) Formation of the planetesimals in the solar nebula. *In* EH Levy, JI Lunine (eds) Protostars and Planets III:1061-1088. Tucson, University of Arizona Press

Lofgren GE, Russell WJ (1986) Dynamic crystallization of chondrule melts of porphyritic and radial pyroxene composition. Geochim Cosmochim Acta 50:1715-1726

Longhi J (1987) On the connection between mare basalts and picritic volcanic glasses. Proc 17th Lunar Planet Sci Conf, p E349-E360

Longhi J (1992) Experimental petrology and petrogenesis of mare volcanics. Geochim Cosmochim Acta 56:2235-2251

MacPherson GJ, Davis AM (1994) Refractory inclusions in the prototypical CM chondrite, Mighei. Geochim Cosmochim Acta 58:5599-5626

MacPherson GJ, Grossman L (1981) A once molten, coarse-grained, Ca-rich inclusion in Allende. Earth Planet Sci Lett 52:16-24

MacPherson GJ, Huss GR (2001) Extinct ^{10}Be in CAIs from Vigarano, Leoville, and Axtell. Lunar Planet Sci XXXI: Abstr #1882

MacPherson GJ, Grossman L, Hashimoto A, Bar-Matthews M, Tanaka T (1984) Petrographic studies of refractory inclusions from the Murchison meteorite. Proc 15th Lunar Planet Sci Conf, p C299-C312

MacPherson, GJ, Wark DA, Armstrong JT (1988) Primitive material surviving in chondrites: Refractory inclusions. *In* Kerridge JF, Matthews MS (eds) Meteorites and the Early Solar System. Tucson, University of Arizona Press, p 746-807

Mason B, Martin PM (1974) Minor and trace element partitioning between melilite and pyroxene from the Allende meteorite. Earth Planet Sci Lett 22:141-144

Mason B, Taylor SR (1982) Inclusions in the Allende meteorite. Smithsonian Contrib Earth Sci 25:1-30

McKeegan KD, Chaussidon M, Robert F (2000) Incorporation of short-lived 10Be in a calcium-aluminum-rich inclusion from the Allende meteorite. Science 289:1334-1337

McKeegan KD, Chaussidon M, Krot AN, Robert F, Goswami JN, Hutcheon ID (2001) Extinct radionuclide abundances in Ca, Al-rich inclusions from the CV chondrites Allende and Efremovka: A search for synchronicity. Lunar Planet Sci XXXI: Abstr #2175

McSween HY Jr, Harvey RP (1993) Outgassed water on Mars: Constrains from melt inclusions in SNC meteorites. Science 259:1890-1892

McSween HY Jr, Jarosewich E (1983) Petrogenesis of the elephant Moraine A79001 meteorite: Multiple magma pulses on the shergottite parent body. Geochim Cosmochim Acta 47:1501-1513

McSween HY Jr, Treiman AH (1998) Martian meteorites. Rev Mineral 36:6-1 – 6-53

McSween HY Jr. Stolper EM, Taylor LA, Muntean RA, O'Kelley GD, Eldridge JS, Biswas S, Ngo HT, Lipschutz ME (1979) Petrogenetic relationship between Allan Hills 77005 and other achondrites. Earth Planet Sci Lett 45:275-284

McSween HY Jr. Lentz RCF, Grove TL, Dann JC (2000) Magmatic water in Shergotty, inferred from light-lithophile-element patterns and crystallization experiments. Meteoritics Planet Sci 35:A107

Meeker GP, Wasserburg GJ, Armstrong JT (1983) Replacement textures in CAI and implications for planetary metamorphism. Geochim Cosmochim Acta 47:707-721

Meneguzzi M, Audouze J, Reeves H (1971) The production of the elements Li, Be, B by galactic cosmic rays in space and its relation with stellar observations. Astron Astrophys 15:337-359

Misawa K, Nakamura N (1988) Highly fractionated rare-earth elements in ferromagnesian chondrules from Felix (CO3) meteorite. Nature 334:47-50

Misawa K, Tatsumoto M, Dalrymple GB, Yana K (1993) An extremely low U/Pb source in the moon: U-Th-Pb, Sm-Nd, Rb-Sr, V and ^{40}Ar/^{39}Ar isotopic systematics and the age of lunar meteorite Asuka 881757. Geochim Cosmochim Acta 57:4687-4702

Mitler HE (1972) Cosmic-ray production of deuterium, He^3, Li, Be, and B in the galaxy. Astrophys Space Science 17:186-218

Morgan JW, Anders E (1979) Chemical composition of Mars. Geochim Cosmochim Acta 43:1601-1610

Morgan JW, Anders E (1980) Chemical composition of Earth, Venus, and Mercury. Proc Nat'l Acad Sci 77:6973-6977

Morris J, Leeman WP, Tera F (1990) The Subducted Components in Island Arc Lavas: Constraints from Be isotopes and B-Be systematics. Nature 344:31-36

Murthy VR, Schmitt RA (1963) Isotope abundances of rare-earth elements in meteorites. J Geophys Res 68:911-917

Neal CR, Taylor LA (1992) Petrogenesis of mare basalts: A record of lunar volcanism. Geochim Cosmochim Acta 56:2177-2211

Newsom HE (1995) Composition of the solar system, planets, meteorites, and major terrestrial reservoirs. *In* Ahrens TJ (ed) Global Earth Physics, A Handbook of Physical Constants, p 159-189

Newsom HE, Shearer CK, Treiman AH (2001) Mobile elements determined by SIMS analysis in hydrous alteration materials in the Lafayette Martian Meteorite. Lunar Planet Sci XXXII: Abstr #1396

Noonan AF (1975) The Clovis (no. 1) New Mexico, meteorite and Ca, Al and Ti-rich inclusions in ordinary chondrites. Meteoritics 10:51-59

Noonan AF, Nelen JA (1976) A petrographic and mineral chemistry study of the Weston, Connecticut, chondrite, Meteoritics 11:111-130
Nyquist LE, Shih C-Y (1992) The isotopic record of lunar volcanism. Geochim Cosmochim Acta 56: 2213-2234
Oddo G (1914) Radioactivity and atoms. Gazz Chim Ital 44:200-218
O'Dell CR, Beckwith SVW (1997) Young stars and their surroundings. Science 276:1355-1359
O'Neill H StC (1991) The original of the Moon and the early history of the Earth—A chemical model; Part 1. The Moon. Geochim Cosmochim Acta 55:1135-1157
Papike JJ, Hodges FN, Bence AE, Cameron M, Rhodes JM (1976) Mare basalts: Crystal Chemistry, mineralogy, and petrology. Rev Geophys Space Phys 14:475-540
Papike JJ, Ryder G, Shearer CK (1998) Lunar samples. Rev Mineral 36:5-1 to 5-234
Popp RK, Virgo D, Yoder HS Jr, Hoering TC, Phillips MW (1995) An experimental study of phase equilibria and Fe oxy-component in kaersutitic amphibole: Implications for the f_{H2} and α_{H20} in the upper mantle. Am Mineral 80:534-548
Quandt U, Herr W (1974) Beryllium abundance of meteorites determined by "non-destructive" photon activation. Earth Planet Sci Lett 24:53-58
Rao MN, Borg LE, McKay DS, Wentworth SJ (1999) Martian soil component in impact glasses in a Martian meteorite. Geophys Res Lett 26:3265-3268
Ramaty R, Kozlovsky B, Lingenfelter RE, Reeves H (1997) Light elements and cosmic rays in the early galaxy. Astrophys J 488:430-748
Ramaty R, Kozlovsky B, Lingenfelter RE, Reeves H (1998) Cosmic rays, nuclear gamma rays and the origin of the light elements. Phys Today 51:30-35
Reeves H (1994) On the origin of the light elements ($Z < 6$). Rev Modern Phys 66:193-216
Reeves H, Fowler WA, Hoyle F (1970) Galactic cosmic ray origin of Li, Be, and B in stars. Nature 226:727-729
Reeves H, Audouze J, Fowler WA, Schramm DN (1973) On the origin of light elements. Astrophys J 179:909-930
Ringwood AE (1979) Origin of the Earth and Moon. Springer-Verlag, Heidelberg
Ringwood AE (1991) Phase transformations and their bearing on the constitution and dynamics of the mantle. Geochim Cosmochim Acta 55:2083-2110
Ringwood AE (1992) Volatile and Siderophile Element Geochemistry of the Moon: A Reappraisal. Earth Planet Sci Lett 111:537-555
Ringwood AE, Kesson SE (1976) A dynamic model for mare basalt petrogenesis. Proc 7th Lunar Sci Conf, p 1697-1722
Robert F, Hanon P, Chaussidon M (1999) A summary of the Li-Be-B isotope compositions and elemental ratios in chondrites. Implications for light element nucleosynthesis. Lunar Planet Sci XXX: Abstr #1188
Rubin AE, Wasson JT (1987) Chondrules, matrix and coarse-grained chondrule rims in the Allende meteorite: Origin, interrelationships and possible precursor components. Geochim Cosmochim Acta 51:1923-1937
Russell HN (1929) On the composition of the Sun's atmosphere. Astrophys J 70:11-89
Ryan JG, Langmuir CH (1987) The systematics of lithium abundances in young volcanic rocks. Geochim Cosmochim Acta 51:1727-1741
Ryan JG, Langmuir CH (1988) Beryllium systematics in young volcanic Rocks: Implications for ^{10}Be. Geochim Cosmochim Acta 52:237-244
Ryan JG, Langmuir CH (1993) The systematics of boron abundances in young volcanic rocks. Geochim Cosmochim Acta 57:1489-1498
Schultz PH, Spudis PD (1983) The beginning and end of mare volcanism on the Moon. Lunar Planet Sci Lett XIV, p 676-677
Shearer CK, Newsom HE (2000) W-Hf isotope abundances and the early origin and evolution of the Earth-Moon system. Geochim Cosmochim Acta 64:3599-3613
Shearer CK, Papike JJ (1993) Basaltic Magmatism on the Moon: A Perspective from volcanic, picritic, glass beads. Geochim Cosmochim Acta 57:4785-4812
Shearer CK, Papike JJ (1999) Magmatic evolution of the Moon. Am Mineral 84:1469-1494
Shearer CK, Papike JJ, Simon SB, Laul JC (1986) Pegmatite-wallrock interactions, Black Hills, South Dakota: Interaction between pegmatite-derived fluids and quartz-mica schist wallrock. Am Mineral 71:518-539
Shearer CK, Papike JJ, Galbreath KC, Shimizu N (1991) Exploring the lunar mantle with secondary ion mass spectrometry: A comparison of lunar picritic glass beads from the Apollo 14 and Apollo 17 sites. Earth Planet Sci Lett 102:134-147

Shearer CK, Layne GD, Papike JJ (1994) The systematics of light lithophile elements in lunar picritic glasses: Implications for basaltic magmatism on the Moon and the origin of the Moon. Geochim Cosmochim Acta 58:5349-5362

Shearer CK, Papike JJ, Rietmeijer FJM (1998) The planetary sample suite and environments or origin. Rev Mineral 36:1-01 – 1-28

Shearer CK, Papike JJ, Borg L, Rao MN, Schwandt C (2000) Trace element characteristics of lithology C in Martian meteorite EETA 79001. Implications for the composition of Martian soils. Abstr Astrobiol Sci Conf, p 248

Sill CW, Willis CP (1962) The beryllium content of some meteorites. Geochim Cosmochim Acta 26: 1209-1214

Sisson TW, Grove TL (1993) Experimental investigations of the role of H_2O in calc-alkaline differentiation and subduction zone magmatism. Contrib Mineral Petrol 113:143-166

Sisson TW, Layne GD (1993) H_2O in basalt and basaltic andesite glass inclusions from four subduction related volcanoes. Earth Planet Sci Lett 117:619-635

Snyder GA, Taylor LA, Neal CR (1992) A chemical model for generating the sources of mare basalts: Combined equilibrium and fractional crystallization of the lunar magmasphere. Geochim Cosmochim Acta 56:3809-3823

Spera FJ (1992) Lunar magma transport phenomena. Geochim Cosmochim Acta 56:2253-2266

Spivack AJ, Beckett JR, Hutchon ID, Wasserburg GJ, Stolper EM (1988) The partitioning of trace elements between melilite and liquid: An experimental study with applications to type B CAI. Chem Geol 70:155

Spivack AJ, Gnaser H, Beckett JR, Measures CI, Hutchon ID, Wasserburg GJ (1989) The abundance and distribution of Be in Allende inclusions. Lunar Planet Sci XVIII:938-939

Stewart GR, Canup RM (1998) Can an early-formed Moon avoid siderophile contamination by subsequent impacts? *In* Origin of the Earth and Moon, LPI Contrib 957, Lunar and Planetary Institute, Houston, Texas, p 4

Stöffler D, Ostertag R, Jammes C, Pfannschmidt G, Sen Gupta PR, Simon SB, Papike JJ, Beauchamp RH (1986) Shock metamorphism and petrography of the Shergotty achondrite. Geochim Cosmochim Acta 50:889-913

Stolper E (1982) Crystallization sequence of Ca-Al-rich inclusions from Allende. The effect of cooling rate and maximum temperature. Geochim Cosmochim Acta 46:2159-2180

Sugiura N (2001) Boron isotopic compositions in chondrules: Anorthite-rich chondrules in the Yamato 82094 (CO3) chondrite. Lunar Planet Sci XXXI, abstract #1277

Swindle TD, Caffee MW, Hohenberg CM (1983) Radiometric ages of chondrules. *In* King EA (ed) Chondrules and Their Origin. Lunar and Planetary Institute, Houston, p 246-216

Swindle TD, Caffee MW, Hohenberg CM (1986) Xenon and other noble gases in shergottites. Geochim Cosmochim Acta 50:1001-1015

Tatsumoto M, Premo WR, Unruh DM (1987) Origin of lead from green glass of Apollo 15426: A search for primitive lunar lead. Proc 17th Lunar Planet Sci Conf *in* J Geophys Res 92:E361-E371

Taylor GJ, Warren PH, Ryder G, Pieters C, Lofgren G (1991) Lunar Rocks. *In* Heiken GH, Vaniman DT, French BV (eds) Lunar Sourcebook: A User's Guide to the Moon, p 183-284, Cambridge University Press, Cambridge

Taylor LA, Shervais JW, Hunter RH, Shih CY, Bansal BM, Wooden J, Nyquist LE, Laul JC (1983) Pre 4.2 AE mare basalt volcanism in the lunar highlands. Earth Planet Sci Lett 66:33-47

Taylor SR (1982) Planetary Science: A Lunar Perspective. Lunar and Planetary Institute, Houston

Taylor SR (1992) Solar System Evolution: A new perspective. Cambridge Univ Press, Cambridge, 307 p

Taylor SR, McLennan SM (1985) The continental crust: Its composition and evolution. Blackwell Sci Publ, Oxford, 330 p

Tera F, Wasserburg GJ (1976) Lunar ball games and other sports (abstract). *In* Lunar Sci VII:858-860, Lunar Science Institute, Houston

Tera F, Brown L, Morris J, Sacks IS (1986) Sediment incorporation in island arc magmas: Inferences from ^{10}Be. Geochim Cosmochim Acta 50:535-550

Trimble V (1975) The origin and abundance of the chemical elements. Rev Modern Phys 27:877-976

Wagoner RV, Fowler, WA, Hoyle F (1967) On the synthesis of elements at very high temperature. Astrophy J 148:3-17

Walker D, Stolper EM, Hays JF (1979) Basaltic volcanism: The importance of planet size. Proc Lunar Planet Sci Conf 10:1995-2015

Wänke H, Baddenhausen H, Dreibus G, Jagoutz E, Kruse H, Palme H, Spettel B, Teschke F (1973) Multielement analysis of Apollo 15, 16, and 17 samples and the bulk composition of the Moon. Proc 4th Lunar Sci Conf, p 1461-1481

Wänke H, Palme H, Baddenhausen H, Dreibus G, Jagoutz E, Kruse H, Palme C, Spettel B, Teschke F, Thacker R (1975) New data on the chemistry of lunar samples: Primary matter in the lunar highlands and the bulk composition of the Moon. Proc. 6th Lunar Sci Conf, p 1313-1340

Wänke H, Dreibus G (1988) Chemical composition and accretion history of terrestrial planets. Phil Trans Roy Soc A325:545-557

Wark DA (1987) Plagioclase-rich inclusions in carbonaceous chondrite meteorites: Liquid condensates? Geochim Cosmochim Acta 51:221-242

Warren PH (1988) KREEP: Major element diversity, trace element uniformity (almost). Workshop on Moon in Transition: Apollo 14, KREEP, and Evolved Lunar Rocks, p 106-110

Wasson JT, Kallemeyn GW (1988) Composition of chondrites. Phil Trans Roy Soc Lond A325:535-544

Wasson JT, Wetherill GW (1979) Dynamical, chemical, and Isotopic evidence regarding the formation locations of asteroids and meteorites. *In* Gehrels T (ed) Asteroids, p 926-974. University Arizona Press, Tucson

Watson LL, Hutcheon ID, Epstein S, Stolper EM (1994) Water on Mars: Clues from deuterium/hydrogen and water contents of hydrous phases in SNC meteorites. Science 265:86-90

Webber WR (1997) New experimental data and what it tells us about the source and acceleration of cosmic rays. Space Sci Rev 81:107-142

Weidenschilling SJ, Cuzzi JN (1993) Formation of the planetesimals in the solar nebula. *In* Levy EH, Lunine JI (eds) Protostars and Planets III:1031-1060. Tucson, University of Arizona Press

Weller MR, Furst M, Tombrello TA, Burnett DS (1978) Boron concentrations in carbonaceous chondrites. Geochim Cosmochim Acta 42:999-1009

Wetherill GW (1988) Accumulation of Mercury from planetesimals. *In* Vilas F, Chapman CR, Mathews MS (eds) Mercury. Tucson, University of Arizona Press, p 670-691

Wetherill GW (1992) An alternative model for the formation of the asteroid belt. Icarus 100:307-325

Wetherill GW, Stewart GR (1993) Formation of planetary embryos: Effects of fragmentation, low relative velocity, and independent variation of eccentricity and inclination. Icarus 106:190-209

Wood JA (1988) Chondritic meteorites and the solar nebula. Ann Rev Earth Planet Sci 16:53-72

Woolum DS (1988) Solar-system abundances and processes of nucleosynthesis. *In* Kerridge JF, Matthews MS (eds) Meteorites and the Early Solar System, p 995-1020

Woosley SE, Weaver TA (1995) The evolution and explosion of massive stars. II. Explosive hydrodynamics and nucleosynthesis. Astrophys J 101:181-235

3 Trace-Element Systematics of Beryllium in Terrestrial Materials

Jeffrey G. Ryan

Department of Geology, University of South Florida
4202 East Fowler Avenue
Tampa, Florida 33620

ryan@chuma.cas.usf.edu

INTRODUCTION

Although often studied as key constituent in minerals, the low mass element (Z = 4; M = 9.0012182 amu) beryllium is most commonly encountered as a very low abundance trace metal in rocks and fluids. In terms of atomic structure and valence state, Be is affiliated with alkaline earth elements; but geologically and chemically beryllium shows greater affinities to the major cations Al, Si, and Mn, and the lithophile trace elements Nd and Zr. Low beryllium abundances prevented the precise measurements of Be contents in rocks and fluids until relatively recently. Intense geochemical interest in the cosmogenic radioisotope ^{10}Be has led to advances in the measurement of stable beryllium. High quality Be data now exist for a variety of natural materials found at or near the Earth's surface. We have a good general understanding of the distribution of beryllium in the solid earth, and in surficial systems. New data for both mineral/melt and mineral/fluid partitioning of beryllium offer the possibility of a detailed assessment of the geochemical behavior of Be in the Earth's major geologic environments.

HISTORY OF BERYLLIUM ANALYSIS AND PAST STUDIES

Goldschmidt and Peters (1932), using a flame-emission method, first measured trace-levels of beryllium in common rocks and minerals. They examined a variety of igneous rocks and their associated minerals, sedimentary rocks, and meteorites. While they were able to recognize differences in Be contents among materials (Be is higher in granites and alkaline igneous rocks than in mafic and ultramafic rocks; is higher still in pegmatitic rocks and minerals, and is not detectable in meteorites), they were unable to make precise measurements on most samples due to the poor sensitivity of their method.

Sandell (1952) obtained somewhat more precise Be data on igneous rocks using a fluorimetric method. He confirmed the Be abundance variations reported by Goldschmidt and Peters (1932), and inferred a bulk crustal Be content of 2 μg/g (ppm). Beus (1954, 1962, 1966) examined beryllium contents in Russian igneous and hydrothermal rocks, and noted that Be tends to be sequestered in micas and feldspars. Merrill et al. (1960) reported seawater Be contents of ~0.4 pg/g, river water contents of 0.1 ng/g, and marine sediment concentrations between 2 and 3 ppm. They calculated a Mean Oceanic Residence Time (MORT) value for Be of 150-570 years based on their results, and noted that Be readily precipitates in neutral and alkaline solutions.

Beus (1966) provides an overview of the abundance variations of Be in major terrestrial rock types as it was known circa 1960, including discussions of the various types of economic Be deposits. A more accessible summary of Be data for terrestrial rocks and meteorites through the late 1960s and for lunar rocks was compiled by Hörmann (1978). Grigoriev (1982, 1986) updates Beus's review of the Soviet literature on Be variations in rocks, also with an emphasis on economic Be resources. Beus and Grigoriev note the close affinities between Be and F in many geologic settings, as well as the fact that Be enrichments may occur during the formation of skarns, or as a byproduct of coal formation. Most

1529-6466/00/0050-0003$05.00

Be in vein and contact metamorphic assemblages was found to be sequestered in the major rock-forming minerals, as opposed to Be-rich mineral phases, and was little affected by weathering processes (Grigoriev 1982, 1986).

None of these earlier works placed useful constraints on the Be contents of mantle or basaltic rocks. Studies of Be in geologic materials at contents <1 ppm were impractical, as the sensitivities of the available analysis methods (flame photometry, atomic absorption, fluorimetry) were too poor to measure Be in such rocks without first conducting a dramatic pre-concentration step. As Be is monoisotopic - only ^{9}Be is stable, though the radionuclides ^{10}Be ($t_{1/2} \approx 1.5$ Ma) and ^{7}Be ($t_{1/2} = 55$ days) are present in very small quantities at or near the Earth's surface—Be concentrations cannot be determined via traditional high-sensitivity methods, such as isotope dilution mass spectrometry. The only successful study of low-level Be abundances before the 1980's was on meteorites (Quandt and Herr 1974). This work involved a very specialized use of gamma ray (photon) activation, pushed to the extremes in terms of sensitivity and sample size (15-20 g), to make measurements at the 10-100 ng/g Be level.

In the early 1980's, developments in the measurement of ^{10}Be in igneous rocks and sediments (Tera et al 1986; see also Morris 1991; Morris et al., this volume) served to spur interest in the measurement of Be abundances in mafic rocks, and in fluids. Plasma emission spectrometric methods (either inductively coupled (ICP) or direct current-generated (DCP) plasma sources) produce dramatically higher excitation energies suitable to low-level Be analysis. Beryllium detection limits via plasma emission spectrometry are in the 0.1 ng/g range, making measurements of rocks with <0.1 ppm Be possible at high accuracy and precision. Most hydrous fluid samples still require pre-concentration before Be analysis, but this process is now much less arduous.

Modern quadrupole ICP-sourced mass spectrometers are theoretically capable of measuring Be to even lower levels than plasma emission methods. Mass interferences are minimal at very low masses (i.e., M = 9 for Be) using ICP-MS, and while these instruments are somewhat less sensitive at the lower end of the mass range than at M > 100 amu, solution concentrations of 0.01 ng/g Be may be measured routinely. As ICP-MS approaches are both newer and more costly on a per-sample basis than plasma emission techniques, the majority of existing Be data on rocks has been collected via ICP or DCP emission spectrometry. However, ICP-MS Be data are now available as a part of commercial multi-element analysis packages.

Ryan and Langmuir (1988; see also Ryan 1989), using a DCP-based analytical method, measured Be abundances in mafic volcanic rocks, including mid-ocean ridge basalts (MORBs), some intraplate lavas, and mafic lavas from a selection of volcanic arcs. Serendipitous sample selection allowed Ryan and Langmuir (1988) to calculate ^{10}Be/Be ratios ($\approx$ ^{10}Be/^{9}Be) for one arc volcano (Bogoslov island, Aleutians) using published ^{10}Be data from Tera et al. (1986). This "isotopic" ratio for Be in the lavas permitted a quantification of the Be input from the slab (~14% of Bogoslov Be must be slab-derived), and the modeling of several slab addition processes: bulk addition, addition of a slab melt, and addition of a slab fluid. This work and the detailed assessment of ^{10}Be/Be ratios in oceanic and arc volcanic lavas by Morris and Tera (1989) which followed, established the ^{10}Be/Be ratios of recently erupted arc lavas as a measure of subducting slab chemical inputs to convergent margin magma sources. A further use of Be abundance data, in determining the B/Be ratios of arc lavas (Morris et al. 1990), made knowing the Be concentrations of rocks and fluids a matter of high priority to many igneous petrologists and geochemists. Trace-level Be data for igneous rocks are thus becoming a routine implement in the geochemical "toolbox," especially in studies of magmatism at convergent plate margins.

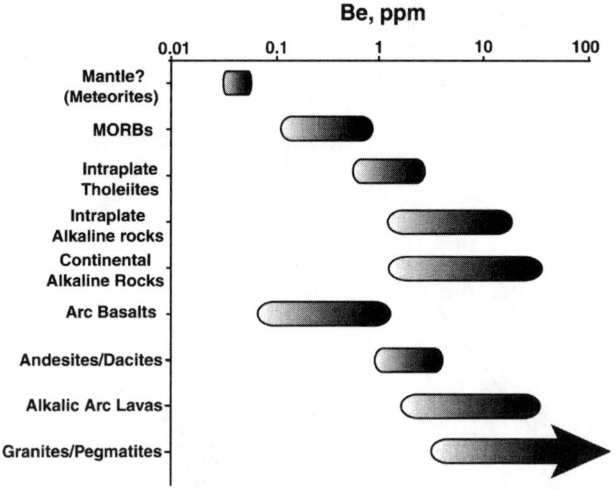

Figure 1. Ranges of Be concentration in meteorites and igneous rocks. Data sources: Quandt and Herr (1974); Ryan and Langmuir (1988) Ryan (1989); Morris and Tera (1989); Morris et al. (1990); Gill et al. (1993); Edwards et al. (1993, 1994); Ryan et al. (1995, 1996a); Dostal et al. (1996); Hochstaedter et al. (1996).

BERYLLIUM IN MAJOR GEOLOGICAL RESERVOIRS

Beryllium in volcanic rocks—analytical studies

Mid-ocean ridge basalts (MORBs). Beryllium abundances in MORBs range from ~0.15 ppm in the most primitive MORB glasses (FAMOUS area and Tamayo Fracture Zone magnesian MORBs) to as high as 2.5 ppm Be in "Fe-Ti" basalts from the Juan de Fuca Ridge and the East Pacific Rise (Fig. 1). Beryllium behaves as a strongly incompatible trace element during both melting and crystallization processes in MORB suites. As noted in Figure 2, Be varies similarly in MORBs to Zr and Nd. Ryan and Langmuir (1988) inferred that a Be/Nd ratio of ~0.05 was characteristic of the MORB source region, and Be/Zr ratios in MORB ≈ 0.0047. Inverse modeling results from primitive and fractionated MORB suites by Ryan and Langmuir suggest that the bulk solid/melt distribution coefficient of beryllium [D_{Be}] ≈ 0.02-0.04 during partial melting of the mantle, and ≈ 0.06 during fractional crystallization of basaltic magmas

Intraplate and alkaline basalts. Beryllium levels in intraplate basalts are almost a factor of five higher than those observed in MORBs (Figs. 1 and 2). Intraplate tholeiites (from Hawaii and Reunion) range from 0.5-1.0 ppm Be. Alkaline intraplate basalts range from 1-10 ppm Be. Higher Be contents in these lavas correlate closely with higher contents of the light rare earth elements, and with elevated concentrations of high field strength elements, such as Zr. In Figures 3a and 3b when Be is plotted vs. Nd or Zr, intraplate basalts show greater scatter, and follow a slightly different trajectory than MORBs overall. Based on limited data, Ryan and Langmuir (1988) inferred that the Be/Nd ratios of intraplate lavas were similar to those of MORB. The larger data set plotted in Figure 3 shows that overall, intraplate lavas may have slightly lower Be/Nd than MORBs, at ~0.045. Beryllium still appears to behave similarly to Nd in intraplate rocks, though the data for individual suites is too limited to allow a rigorous examination. Be/Zr ratios in intraplate rocks are also different: with differentiation samples approach a Be/Zr ratio of 0.006, though the most evolved intraplate rocks trend toward higher ratios (possibly related to the crystallization of Zr-bearing minerals), and others scatter to lower values. Beryllium and Zr appear to behave similarly in the less evolved intraplate rocks; though again, the data for some islands show scatter. The small differences in Be/Nd and Be/Zr between MORBs and

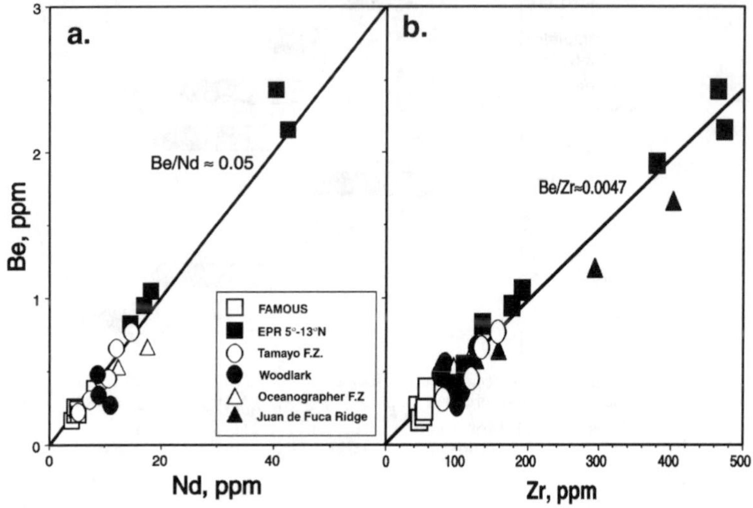

Figure 2. Beryllium covariation diagrams for mid-ocean ridge basalt (MORB) glasses. Beryllium data sources are Ryan and Langmuir (1988) and Ryan (1989); Nd and Zr data are from Bryan and Moore (1977), Bryan et al. (1979), Langmuir et al. (1977, 1986), Bender et al. (1984), Kappel (1985), Johnson et al. (1987) and Shirey et al. (1987, and unpublished). EPR: East Pacific Rise. F.Z.: Fracture Zone. (a) Be vs. Nd. Line represents Be/Nd = 0.05. (b) Be vs. Zr. Line represents Be/Zr = 0.0047.

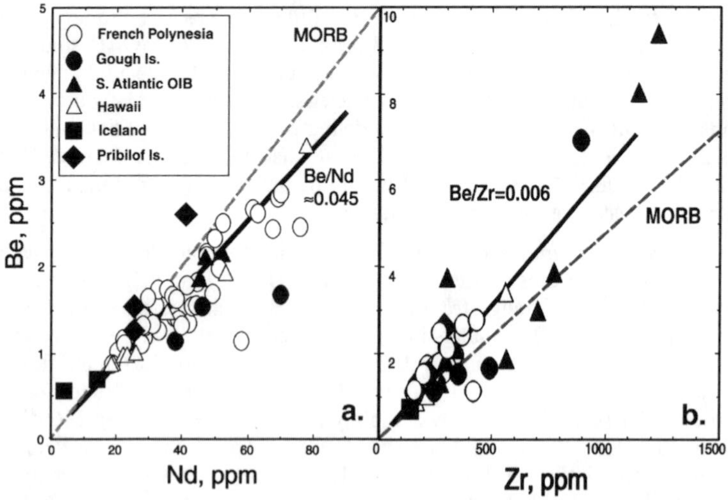

Figure 3. Beryllium covariation diagrams for intraplate and alkaline volcanic rocks. Dashed gray lines represent the MORB data arrays from Figure 2. OIBs = intraplate (ocean island) basalts. (a), (b) Be vs. Nd and Be vs. Zr, respectively for ocean island volcanic rocks. Sources for French Polynesia data are Dostal et al (1996) and Dupuy et al. (1989, 1993); other data are from Ryan and Langmuir (1988), Ryan et al. (1996a) and references therein.

intraplate rocks may relate to (a) small differences in the relative abundances of Be, Nd and Zr in MORB and intraplate source regions, (b) small differences in the solid/melt partitioning of the three elements in intraplate settings relative to MORB, or c) both effects, to some degree. However, as the differences between Be behavior in MORBs and intraplate rocks are so small, neither the abundance levels nor the distribution behaviors of these elements during melting of MORB or intraplate mantle sources seems likely to produce strong distinctions.

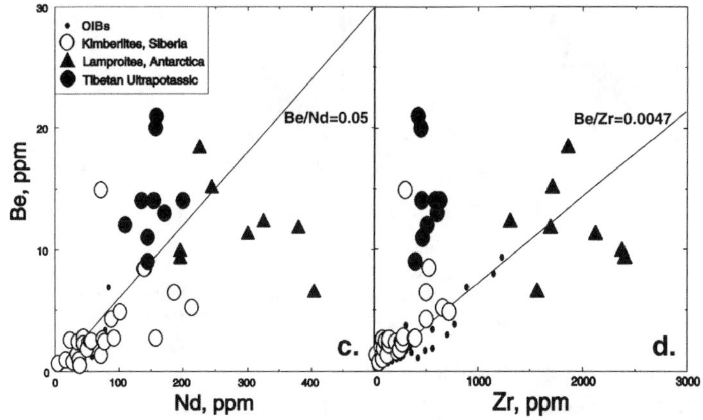

Figure 3, continued. Beryllium covariation diagrams. (c), (d) Be vs. Nd and Be vs Zr, respectively, for evolved alkaline lavas. Data sources are Norrell (1998), Miyamoto et al (2000), and Miller et al. (1999).

Evolved alkaline lavas. in evolved alkaline volcanic systems, partitioning differences between Be and other incompatible lithophile elements with progressive differentiation become more apparent. Beryllium concentrations in continental alkaline suites range from 1-45 ppm, with much of the variation at lower Be contents related to admixture of crustal rocks with lower Be contents (Norrell 1998; Mitchell 1986; Mitchell and Bergmann 1991; Miller et al. 1999; Miyamoto et al. 2000). Datasets for kimberlitic rocks from eastern Siberia, lamproites from Antarctica, and ultra-potassic lavas from Tibet (Norrell 1998; Miyamoto et al 2000; Miller et al. 1999; Fig. 3c,d) suggest some behavioral similarity between Be, Nd and Zr at concentration levels lower than about 5 ppm Be, and Be/Nd and Be/Zr ratios similar to those of intraplate basalts. At higher Be levels, divergence occurs: ultra-potassic rocks and kimberlites move to higher Be/Nd and (especially) Be/Zr ratios, while lamproites scatter toward lower Be/Zr and Be/Nd. The probable origin for this divergence of Be, Nd, and Zr abundances in alkaline lavas lies in changing crystallizing assemblages: in evolved alkaline suites, the crystallization of minor phases (such as zircon, apatite, and monazite: Miyamoto et al. 2000; Miller et al. 1999) can have a dramatic effect on whole-rock abundances of Zr and Nd, but will have little effect on Be, which is sequestered in plagioclase, amphiboles and micas (see below). The divergence in the partitioning of these elements should result in scattered correlations for whole-rock samples once minor phases stabilize, which in each of these suites seems to occur at about the 5 ppm Be abundance level.

Volcanic arc lavas. An extensive database now exists for Be abundances in volcanic arc lavas. Beryllium contents in arc volcanic rocks range widely, from <0.1 ppm Be in chemically depleted basalts from Torishima in the Izu arc, to over 10 ppm Be in alkalic arc lavas from the Mexican Volcanic Belt and the Sunda arc (Fig. 1). Figure 4 compares variations in Be to those of Nd and Zr in basaltic to andesitic lavas from arc suites.

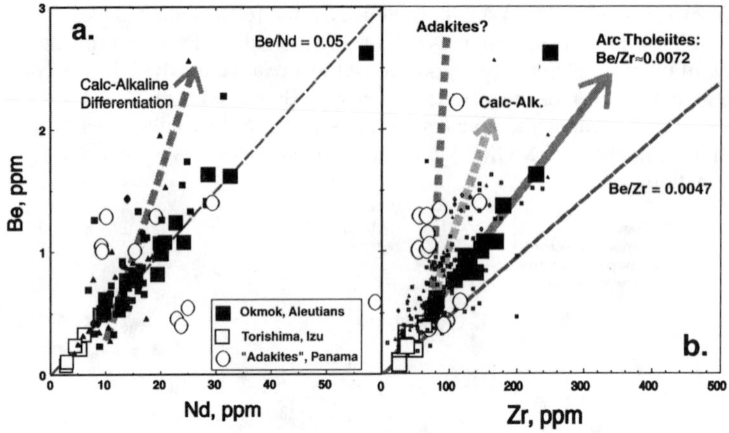

Figure 4. Beryllium covariation diagrams for arc volcanic rocks. Data sources are Ryan and Langmuir (1988); Ryan (1989) and references therein; Miller et al. (1992); Norrell et al. (1995); Ryan et al. (1995); and Tomascak et al. (2000). The thinner dashed gray lines represent the MORB data arrays from Figure 2. (a) Be vs. Nd. (b) Be vs. Zr.

Several different patterns are evident in these data:

1. A scattered "calc-alkaline lava" array, which includes most of the data.
2. A "tholeiitic" trend, seen primarily in relation to the recent basalt/basaltic andesite eruptives of Okmok volcano in the Aleutians, and of Torishima volcano in the Izu arc. Both of these suites follow a tholeiitic trend of magmatic differentiation on classical discrimination plots, such as an AFM diagram (Miller et al.1992).
3. A highly scattered "adakite" array (see Defant and Drummond 1990), defined here by Plio-Pleistocene dacitic lavas from the Panamanian segment of the Central American Arc (Defant et al 1991). These data show much less coherence than the other arc trends, and are distinguishable primarily by their lack of a consistent pattern.

Be/Nd ratios in tholeiitic trend arc lavas are similar to those in MORBs at ~0.05. Lavas in the calc-alkaline array have variably higher Be/Nd. Be/Zr ratios are higher than in MORBs for both tholeiitic and calc-alkaline array lavas.

A complicating factor in examining covariation diagrams such as Figure 4 for arc lavas is the possible influence of minor phases, such as apatite, monazite, or zircon, which may be introduced into melts via crustal assimilation (an important process in volcanic arc petrogenesis), and could serve to modify the the rare earth elements (REE) and the high field strength elements (HFSE) contents of erupting lavas relative to Be, thus changing Be/Nd and Be/Zr ratios. Watson and coworkers (Watson and Harrison 1984; Rapp and Watson 1986; Ryerson and Watson 1987) have documented that melts of basaltic to andesitic compositions are generally under-saturated in zircon, rutile, and other minor phases, so crystallization of these phases in all but the most evolved (i.e., rhyolite) arc lavas is unlikely. Addition of such phases from the overlying crust via assimilation processes would also involve the addition of Be from assimilated feldspars and micas, so the effect of assimilation would be equivalent to mixing reservoirs with mantle-derived and crust-derived Be/Nd and Be/Zr, respectively. One would expect assimilatory inputs to be most pronounced in the more siliceous lavas (i.e., lavas with higher Be, Zr, and Nd contents), so a change in the trajectory of the data arrays in Figure 4 should be very evident at higher concentrations if the effects of assimilation are important. For tholeiitic arc lavas, a change in trend may be indicated for the highest Be sample (a rhyolite from Okmok volcano), but all of the other samples from Okmok fall along a single, tight array. The "calc-alkaline" array lavas as a group form a very poorly defined trend, but also show no clear changes in trend with evolution. Based on the lack of clear changes related to minor phase/assimilation effects in all but the most siliceous lavas plotted in Figure 4, it is safe to assume that for Be, Zr, and Nd, the effects of open system processes are secondary, and that the data arrays (and implicit Be/Nd or Be/Zr ratios) are generally representative of melting and crystallization processes in sub-arc magma chambers and mantle source regions.

From the trends of the arrays in Figure 4, one can infer that $D_{Be} = D_{Nd}$ for arc tholeiites, as was the case with MORBs, but $D_{Be} < D_{Nd}$ for calc-alkaline arc rocks (see Hanson 1989). Be/Zr ratios are higher in all arc rocks than in MORBs. The linear array of the Okmok and Torishima data suggests that $D_{Be} = D_{Zr}$ during tholeiitic differentiation in these suites. Generally, $D_{Be} < D_{Zr}$ in calc-alkaline lavas.

Beryllium concentrations in highly siliceous arc rocks (i.e., arc-related granites and rhyolites) begin at 2 ppm Be and may range to values ≥10 ppm, depending on the type of granite, the nature of the primary magma, and the nature of the crustal assimilants (Beus 1966; Grigoriev 1986; Tepper and Ryan 1995; London and Evensen, this volume). Beryllium variations in granites, pegmatites, and other high-Si rocks are examined in detail by London and Evensen (this volume), and will not be considered further here.

Experimental studies

Experimental examination of Be partitioning in relevant mineral/melt systems is a relatively recent development, and as such, partition coefficient data are limited. Ryan (1989) conducted one-atmosphere equilibration experiments on natural MORB starting compositions, examining Be partitioning between olivine, plagioclase, clinopyroxene and melt, whereas Hart and Dunn (1993) reported data for $D_{Be}^{CPX/melt}$ at higher pressures. More recently, Brenan et al. (1998a) have produced a set of high-quality D_{Be} values for olivine/melt, clinopyroxene/melt, and amphibole/melt systems, under 1-atmosphere conditions for olivine and pyroxene, and at 10-15 kbar for olivine, clinopyroxene, and amphibole. Critical to the successful determinations of D_{Be} has been the advent of SIMS analysis via the ion microprobe, which has allowed the *in situ* analysis of Be in experimental run products at natural abundance levels.

Table 1 outlines the existing mineral/melt partitioning data for Be. Compared to Nd or Zr, which show similar mineral/melt systematics in natural systems, the mineral-melt affinities of Be are quite distinctive. Figure 5 is a set of composite "Onuma" diagrams (after Onuma et al. 1968), in which the mineral/melt partition coefficients of major and trace elements for specific minerals are plotted against the six-fold-coordinated ionic radii of the elements. Elements of similar valence are connected by curves, which serves to highlight the relative sizes of the major cation sites in each mineral, and permit assessment of the relative roles of ionic radius and charge in the partitioning of trace elements between the mineral and melt. The small ionic radius of beryllium (Be^{2+} = 0.45 Å in six-fold coordination, 0.27 Å in four-fold coordination (Shannon 1976)) is very similar to the ionic radius of Si^{4+}, and thus presents the possibility that Be may substitute into tetrahedral cation sites in silicate minerals. Brenan et al (1998a) successfully modeled Be substitution into olivine based on a Be-Mg exchange reaction, and its substitution into pyroxene based on the paired substitution: $Be^{2+} + Ca^{2+}$ for $^{[6]}Al^{3+} + Na^{+}$. This modeling implicitly presumes Be is substituting into octahedral sites in these minerals. However, Be occurs exclusively at tetrahedral sites in minerals (Hawthorne and Huminicki, this volume). On size-charge arguments, it seems likely that Be is substituting into tetrahedral sites in most mafic minerals, even as a trace constituent. Where the tetrahedral site is also charge-specific, as it is with olivine (and to some extent, pyroxene), Be substitutes poorly, resulting in low values of D_{Be}. In other minerals, such as plagioclase and amphibole, paired substitutions involving both tetrahedrally coordinating ions and ions of higher coordination are possible, and charge balance constraints are functionally relaxed. It is thus possible that the larger D_{Be} values for plagioclase/melt and amphibole/melt systems are due to Be substituting into these minerals for $^{[4]}Al^{3+}$, either paired with Ca^{2+} for Na^{+} (plagioclase) or with Si^{4+} for $^{[4]}Al^{3+}$ (amphibole).

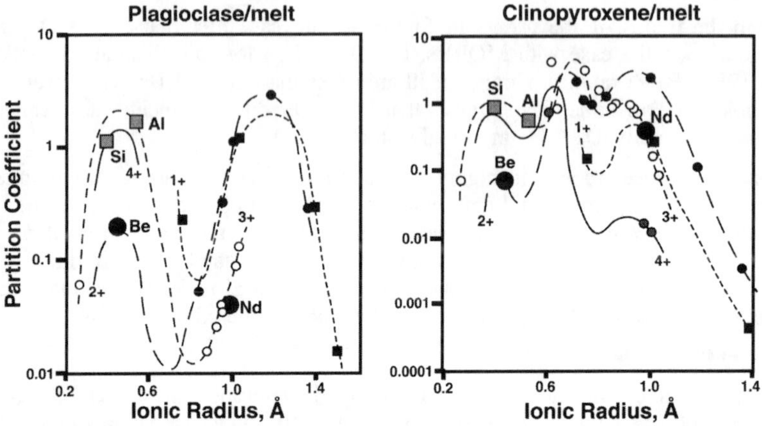

Figure 5. Schematic "Onuma" diagrams in which partition coefficient are plotted vs. ionic radius for different elements in plagioclase/melt and clino-pyroxene/melt systems. Diagrams are modified from those in Matsui et al. (1977). Data sources are: Matsui et al. (1977), Ryan (1989) and Brenan et al. (1998a). Black squares are 1+ cations; black circles 2+ cations; white circles 3+ cations; gray circles 4+ cations. Symbols for Be^{2+} and Nd^{3+} (black circles) and for Si^{4+} and Al^{3+} (gray squares) are larger for reading convenience. The small ionic radius of Be^{2+}, between that of Si^{4+} and Al^{3+}, suggests that it substitutes into the tetrahedral sites in minerals. Nd^{3+} and other large-ion lithophile trace elements typically substitute into sites of higher coordination.

Table 1. Mineral/melt partition coefficient data for beryllium.

Ol/L	*Pl/L*	*Cpx/L*	*Amph/L*	*Opx/L*	*Source (system)*
~ 0 ~	0.21-0.33	0.03-0.13			Ryan 1989 (MORBs: 1 bar)
					Taura et al 1998 (high P ol/l)
0.0015-0.0029		0.011-0.025	0.21-0.26		Brenan et al. 1998a (basalt-andesite)
	≤ 0.35	≤ 0.3			Monaghan et al. 1988 (basalts: min/matrix)
	≤ 0.8	≤ 1.0	≤ 1.5		Monaghan et al. 1988 (dacite: min/matrix)
	≤ 0.77			≤ 0.2	Monaghan et al. 1988 (andesite: min/matrix)

Note:

Ol – olivine, Pl – plagioclase. Cpx – clinopyroxene, Amph – amphibole, Opx – orthopyroxene, L – liquid

Implications of beryllium partitioning for Be-systematics in volcanic rocks

The markedly different mineral/melt partitioning behavior of Be, as compared to the light rare earth elements and other large-ion lithophiles, is to a great extent expressed in its abundance patterns in volcanic rocks from different tectonic settings. The Be/Nd ratios of the most primitive lavas from ocean ridges, ocean islands, and volcanic arc settings are generally similar. However, with differentiation Be/Nd ratios show distinctive variations. When plagioclase is not an important, early-crystallizing phase, $D_{Be} < D_{Nd}$, and Be/Nd ratios increase with crystallization. Thus, in calc-alkaline arc suites, where dissolved H_2O in the melts suppresses plagioclase crystallization, we see increases in Be/Nd as a function of progressive differentiation. In the tholeiitic Okmok arc suite, and in MORBs, where plagioclase crystallizes early, Be/Nd ratios remain relatively constant with evolution. In intraplate lavas, Be/Nd ratios, though apparently constant, are slightly lower than in MORBs. This difference may speak to mineralogical differences between MORB and intraplate source regions: the presence of garnet as the aluminous phase in intraplate mantle sources and/or the presence of a jadeite component in deep mantle clinopyroxenes may result in a different initial Be/Nd during intraplate melting, though when these melts enter shallow crustal magma chambers, the two elements appear to vary similarly. The relative variations of Be and Zr in Figures 2-4 probably reflect both changes in residual mineralogy, and changes in source abundances: higher Be/Zr ratios in evolved intraplate and alkaline lavas most likely relate to changes in the crystallizing mineral assemblages, in particular the stabilization of Zr-bearing phases. However, the uniformly higher Be/Zr of even the most primitive arc lavas require higher initial Be/Zr ratios in arc sources.

The adakite data in Figure 4 provide the most dramatic evidence for differences in the partitioning behavior of Be versus ostensibly similar trace species. Adakites form a near-vertical array on a plot of Be vs. Zr, indicating the presence of minerals during adakite evolution that strongly retain Zr. By contrast, adakite data scatters on the Be vs. Nd plot. Defant and Drummond (1990) and Drummond et al. (1996) suggest that adakites are partial melts of subducted oceanic crust, and call upon melt residues rich in garnet and amphibole to explain the anomalous trace element characteristics of these rocks. Given that adakites are high-silica rocks ($SiO_2 \approx 60$ wt %), inferred solid/melt distribution coefficients for Be and Nd approach 1.0, as the structures of the melt and the residual solids become more similar (Lesher 1986; Monaghan et al. 1988). Thus, variable Be/Nd in adakites may reflect inherent variability in source materials. The near-constant values of Zr in the adakite samples suggests some sort of buffering effect, i.e., possibly the precipitation of zircon, or another phase that sequesters Zr, but not Be or Nd.

Overall, the systematics of Be in volcanic rocks indicate that Be behaves much like the light rare-earth elements during melting and crystallization processes, though in detail its uniquely small ionic radius and charge/mass ratio lead to distinctive variations during the crystallization of H_2O-rich parent magmas. Be/Nd ratios in primitive basaltic arc lavas are not markedly different from those of primitive MORBs or intraplate lavas. Given that $^{10}Be/^{9}Be$ systematics in arcs indicate that 5-15% of the Be in arc lavas originates from the slab (Tera et al. 1986; Ryan and Langmuir 1988; Morris and Tera 1989), similarities between Be/Nd ratios in primitive arc lavas relative to MORBs or intraplate lavas can be interpreted as indicating that a similar percentage of Nd and, by extension, of the other light rare earth elements (LREE) in arc sources may also be slab-derived. Thus, the chemical mechanisms for moving Be from the slab into the mantle beneath arcs must also be moving the LREE.

The higher Be/Zr ratios of arc lavas relative to MORBs, by extension, may partially reflect slab additions of Be. The Be/Zr ratio of the Okmok arc suite, at ~0.0075, is 50% higher than the MORB ratio of ~0.0047. An ~15% addition of slab-derived Be can account for part of this increase, but some manner of Zr depletion in the sub-arc mantle also appears to be necessary. The relative depletion of Zr and other high field strength elements is a widely recognized characteristic of arc magmas (Pearce and Peate 1995; and references therein), though the origins of these depletions continue to be controversial (i.e., relative insolubility of HFSE versus retention in residual phases; see Pearce and Peate 1995). Beryllium systematics arc lavas point to some slab-to-mantle mobility, and as such may help constrain the manner by which chemical transport from the slab into arc source regions occurs (see below).

Sediments and sedimentary rocks

The data base for beryllium concentrations in unmetamorphosed and unmodified sedmimentary rocks is rather limited, as most of the literature focuses on Be mineralization in sedimentary strata (see Salisbury 1964; Ottoway et al. 1994). Beryllium abundances in sedimentary rocks range widely with composition: sandstones, siltstones, shales and other clastic sedimentary rocks generally fall between 1 and 2 ppm Be, the Be content varying with clay content. Limestones, evaporates, and other chemical sediments often show very low Be (Grigoriev 1982, 1986; Gao et al. 1991, 1992; Koeberl et al. 1996; Graham et al. 1998). Bauxite deposits can have Be contents as high as 7 ppm, indicating that like Al, Be is concentrated in the insoluble residuum of laterites (Bronevoy et al. 1985; Fransceschelli et al. 1998). Other sedimentary deposits that may contain elevated Be are rich in marine-derived Mn-oxyhydroxide mineral phases (Fransceschelli et al. 1996). Koeberl et al (1996) report Be data for a variety of unshocked and shocked sedimentary lithologies from the Manson impact site in South Dakota. In these rocks Be contents are consistently higher (2-4 ppm) in shales and shale clasts than in sandstone or limestone clasts or country rocks ($\leq$1 ppm Be.)

Marine sediments from the major oceans show relatively uniform Be contents, except where alkaline volcanogenic materials are being shed into the oceans: here, Be contents can be elevated. Marine sediments in general possess Be contents in the 1.5-2 ppm range, though samples with large percentages of Mn-oxyhydroxides can be more Be-enriched (Tanaka et al. 1982; Dean and Parduhn 1979; Sharma and Somayajulu 1982; Zheng et al. 1994). Atlantic Ocean bottom sediments collected offshore from the Azores may contain up to 4 ppm Be, reflecting inputs of Be-rich volcanic detritus from the Azores (Ryan 1989). Be is generally sequestered in the silicate fractions of marine sediments, specifically clay phases. Mn-oxyhydroxide phases (as in manganese nodules) strongly concentrate Be: Be contents of Mn nodules range from 2-15 ppm, a factor of 10^7 enrichment over seawater, and a factor of 5-10 times higher than clays (Sharma and Somayajulu 1982; Sharma et al. 1987; Ku et al. 1982; Krishnaswami et al. 1982). However, as a proportion of the total sedimentary column, Mn-rich phases are a relatively minor constituent, and the Be budget of marine sediments is dominated by the clays.

Metamorphic rocks and minerals, and mineral/fluid partitioning

Beryllium concentrations in metamorphic rocks are, in general, strongly correlated to protolith. The overall range of Be in metamorphic rocks is ~0.5 to 10 ppm, with most common metamorphic rocks in the 1-3 ppm range (Shaw et al. 1967; Sighinolfi 1973; Lebedev and Nagaytsev 1982; Sheraton et al. 1984; Bushlyakov and Grigoriev 1989; Gao et al. 1991, 1992; Bea et al. 1994; Henry et al. 1996; Bea and Montero 1999; Baba et al. 2000; Grew et al. 2000; Grew, this volume). Granitic, migmatitic and pegmatitic protoliths often preserve the considerably higher Be concentrations encountered in these rock types (Bea et al. 1994, Baba et al. 2000; Grew et al. 2000), as do metamorphosed bauxite and Mn-oxy-hydroxide deposits (Fransceschelli et al. 1996, 1998).

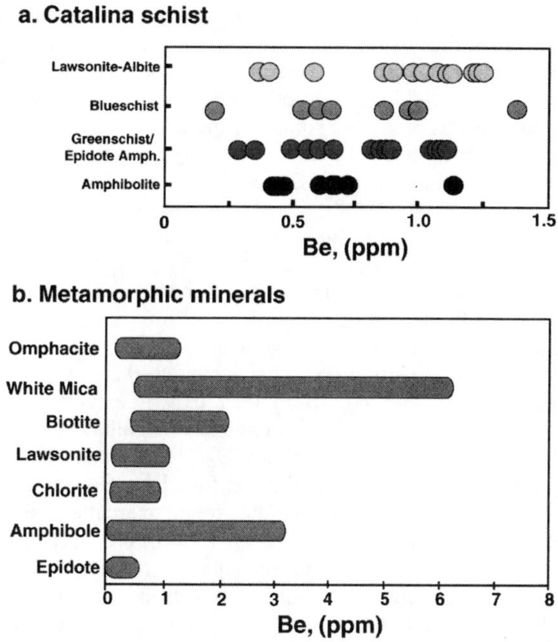

Figure 6. (a) Beryllium concentrations in metasedimentary whole-rock samples from the Catalina Schist "subduction complex" prograde metamorphic suite. Data for Be contents are from Bebout et al. (1993). (b) Be concentration ranges in common metamorphic minerals. Data from Domanik et al. (1993).

Subduction complex associations. Much recent work on Be variations in metamorphic rock assemblages has focused on "subduction complex" massifs, toward understanding the mobilization of Be from slab sediments and altered basalts during subduction processes (Domanik et al. 1993; Bebout et al. 1993, 1999). Variably metamorphosed sedimentary rocks of the Catalina Schist complex show a limited range in beryllium contents, all clustering around 0.75 ppm Be (Fig. 6a). Such variability as exists appears to be dependent on protolith (i.e., silicic metasediments contain less Be than pelites). The range in concentration does not vary significantly with grade (Fig. 6a). Mafic Catalina Schist lithologies show comparable, if variable, Be contents (0.2-2 ppm). Vein minerals show lower overall Be abundances (0.1-0.6 ppm). In pegmatitic leucosomes, Be is notably higher (0.6-4 ppm).

In subduction-complex rocks, Be appears to be housed dominantly in white micas, much like boron (Domanik et al. 1993; Domanik and Holloway 1996; Bebout et al. 1999; Fig. 6b). However, unlike B, significant Be also resides in feldspars and amphiboles, and sometimes in lawsonite and high-pressure pyroxenes. Beryllium is thus more uniformly distributed in metamorphic rocks than boron or other lithophile trace elements. Given that Be contents of metamorphic whole rocks show little change with increasing grade, the implication is that if Be is liberated during the prograde breakdown of metamorphic minerals, it is readily taken up by other phases.

Experimental studies of Be solid/fluid partitioning. Data for subduction-complex metamorphic rocks point to only a modest degree of solubility for Be in H_2O-dominated fluids; i.e., while Be contents may be elevated in vein minerals, it is substantially less mobile than other alkaline elements or boron. Experimental studies confirm this inference: Tatsumi and Isoyama (1988), in studies of serpentine hydration, found beryllium to be one of the least soluble species they studied (at <5% mobilized in liberated fluids). You et al. (1996) conducted low temperature (25-350°C) hydrothermal experiments on sediments and found that while Be contents in the produced solutions increased from 2 pg/g to 130 pg/g as temperatures increased, the amount of Be mobilized in hydrous fluids was negligible relative to the content of the sediment (2.2 ppm Be, 5-6 orders of magnitude greater than the associated fluids). Brenan et al (1998b) have conducted the only explicit mineral/fluid equilibrium partitioning study of Be, examining the mineral/fluid affinities for clinopyroxene and garnet, and then extending their results (using mineral/mineral partitioning constraints for slab materials) to phases such as amphibole, lawsonite, and mica. Their measured mineral/fluid partition coefficients for beryllium in clinopyroxene ($D_{Be}^{cpx/fluid}$) are quite high (1.8, range 0.77-4.5), while $D_{Be}^{grt/fluid}$ values are low (≈ 0.0024). Their calculated mineral/fluid partition coefficients are ~0.26 (amphibole/fluid), ~0.64 (lawsonite/fluid) and ~1.2 (mica/fluid). Both their experimentally determined and their calculated $D_{Be}^{mineral/fluid}$ values were more than an order of magnitude larger than those for Li, and over two orders of magnitude larger than any $D_{B}^{mineral/fluid}$ value. The marked difference in the solid/fluid partitioning of Be and boron was found by Brenan et al. (1998b) to be consistent with the marked elevations in B/Be ratios found in arc magmas, and with declines in B/Be ratios observed across arcs (see Ryan et al. 1996b); basically slabs will lose boron preferentially to Be during devolatilization, such that boron is more-or-less completely removed from slabs during subduction, and a significant amount of Be is retained. Thus, during subduction-related metamorphic devolatilization processes, Be appears to behave as a fluid-immobile element, at least relative to boron, lithium and other alkaline elements.

High Be rocks such as granitic pegmatites are beyond the scope of this contribution: however, the mechanisms by which such enrichments can occur are of interest. With increasing system SiO_2, Be partition coefficients approach 1.0, as do those of many trace elements, because differences between the structure of siliceous liquids and minerals in terms of elemental partitioning are minimal (see Lesher 1986). Monaghan et al. (1988) measured feldspar- and amphibole-matrix coefficients for an evolved arc lava, and found values between 0.6 and 1.2 for both phases, considerably higher (for plagioclase) than the $D_{Be}^{Plag/melt}$ values of Ryan (1989)(Table 1). In high-pressure, hydrous studies of sediment melting, Johnson and Plank (2000) found that the bulk $D_{Be}^{solid/melt}$ ranged between 0.2 and 0.7.

Be can, however, be mobilized in fluid phases rich in Cl and especially F. Greisen deposits with elevated Be, Sn and W abundances are reported globally (Barton 1987). In all cases these deposits are associated with F rich, metasomatizing fluids derived from granitoid rocks (i.e., Shawe and Bernold 1966; Sillitoe and Bonham 1984; Voncken et al. 1986; Webster et al 1987). In North America, Be mineralization is commonly associated with granitic intrusion into carbonate-rich sedimentary strata (Salisbury 1964; Barton 1987). Late-stage fluids and pegmatites can interact with country rocks to produce a variety of metasomatic deposits. In all such cases, Be enrichments are associated with enrichments in other metals, including Sn, Mo, and W.

Cratonic, migmatitic, metasomatic and other metamorphic rocks. The Be contents of cratonic rocks can range from <1 ppm to concentrations high enough to stabilize beryllium mineral phases (Shaw et al 1967; Sighinolfi 1973; Gao et al. 1991, 1992, 1998; Miyamoto et al. 2000; Baba et al. 2000; London and Evensen, this volume; Grew, this volume). Granulitic meta-igneous rocks from Brazil, Antarctica, and Minnesota (USA) range from 0.3 to 2.1 ppm Be (Sighinolfi 1973; Sheraton et al. 1984; Ryan 1989). In the Minnesota suite, Be shows a strong, positive correlation with SiO_2 content; though in the larger Brazilian and Antarctic suites, no clear pattern relative to SiO_2 is evident (see Sighinolfi 1973; Fig. 7). Lebedev and Nagaytsev (1982) report some extreme Be abundances (up to 41 ppm Be) from mineralized zones of the Ladoga metamorphic complex, Russia, though most of their reported Be contents lie between 1 and 4 ppm. Henry et al. (1996) compared vein and wallrock Be abundances for a transect across the western Alps. They found that wallrock Be contents ranged between 0.1 and 4 ppm generally, varying inversely with the percentage of carbonate in the samples. Associated vein Be contents were much lower, such that vein/wallrock abundance ratios for Be were consistently <<0.5. Bea et al (1994) conducted mineral/leucosome partitioning studies on migmatitic metamorphic rocks of the Peña Negra Complex in central Spain. They observed that Be abundances were relatively similar (at 1-4 ppm Be) in feldspars and garnets; biotites varied between 1 and 10 ppm Be, and cordierite more strongly concentrated Be (7-16 ppm). Leucosome Be contents ranged from 0.3 to 2 ppm, mesosomes from 1 to 5 ppm Be, and melanosomes from 3.5 to 8 ppm Be, indicating that Be is to some degree sequestered in the mafic residuum during migmatite development.

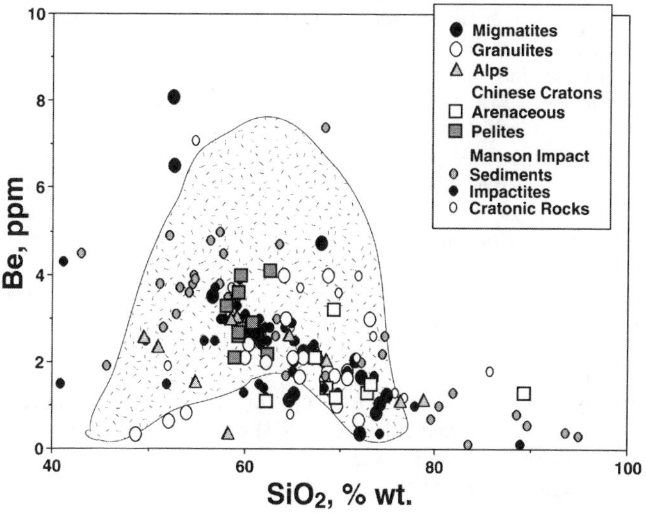

Figure 7. Plot of Be vs. SiO_2 for various cratonic metamorphic rock suites. Data from Sheraton et al. (1984); Ryan (1989); Gao et al. (1991, 1992); Bea et al. (1994); Henry et al (1996); and Koeberl et al. (1996). Field represents range of granulites and cratonic metamorphic rocks examined by Sighinolfi (1973).

Highly shocked and partially melted metamorphic rocks from the Manson impact site in South Dakota (USA) preserve an interesting variation in Be concentrations (Koeberl et al. 1996). Fragmental and fallback breccias range from <1 to 5 ppm Be, to a degree varying inversely with SiO_2 content (i.e., samples with <2 ppm Be have high SiO_2, indicating a higher percentage of sandstone clasts). Impact melt breccias and suevites (high P-T breccias consisting of clasts of variably melted target rocks in a fine-grained matrix of rock fragments, spherules, and glasses) from the Manson site are, with only a few exceptions, very uniform in their beryllium abundances, ranging from 1.5 to 4 ppm Be, with an average close to 2.5 ppm Be. Gneissic and granitic clasts from Manson, presumably representing cratonic material beneath the thick sedimentary cover of this site, scatter to much higher Be contents, i.e., from <1 ppm to 9.5 ppm Be. The melt-bearing breccias from impact sites probably reflect the averaged composition of the country rocks from the site, while the various types of lithic breccias should show compositional variations due to differing proportions of cratonic and sedimentary clasts. The samples examined by Koeberl et al. (1996) were obtained from drill cores through the Manson impact structure, which is buried beneath younger strata. Beryllium concentrations are relatively uniform in all the melt-bearing Manson samples, but are variable on a <1 m scale in those cores sampling lithic and fallback breccias. From the perspective of Be systematics, what the Manson suite indicates is that even under conditions of extreme pressures and temperatures Be contents in metamorphic rocks and anatectic melts reflect protolith composition.

Be concentration in the continental crust. Estimation of the Be abundance of the continental crust is ostensibly straightforward relative to other alkaline trace elements, given the limited degree to which Be may be redistributed via metasomatic processes. Studies of granulite facies rocks indicate that Be concentrations may be similar in the lower and upper crust, provided the overall composition of the lower and upper crust are similar. Shaw et al (1967) estimated that the exposed Precambrian Canadian Shield had an average Be concentration of 1.3 ppm. The crustal averages of Gao et al. (1998), based on the cratons of central China, suggest a lower Be content in the lower crust than in the overall crust (0.5-2.0 ppm Be vs. 1-3 ppm Be), consistent with the somewhat more mafic lower crust composition that they infer. Taylor and McLennan (1995) and Wedepohl (1995) both suggest a significant difference in the Be abundances of the lower and upper continental crust, again based largely on their overall different compositions for these units (i.e., the lower crust is more mafic in character than the upper crust). Both suggest an upper crustal Be abundance of 3-3.1 ppm, but propose very different lower crustal Be concentrations (1 ppm Be for Taylor and McLennan (1995); 1.7 ppm Be for Wedepohl (1995)). Their resultant overall averages for Be in the continental crust therefore differ significantly (1.5 vs. 2.4 ppm Be, respectively). The differences in published Be crustal estimates, however, lie in the presumed overall compositions the various crustal components, and not in disagreements over the Be abundances of crustal rocks.

Oceans and the hydrosphere

Beryllium occurs as a very low abundance species in the marine environment, typically at less than pg/g levels. Measures and Edmond (1983) examined Be abundances in seawater and found concentrations between 0.2-0.3 pg/g (20-30 pM). Broecker and Peng (1982) report a value of ~0.6 pg/g, and Quinby-Hunt and Turekian (1983) suggest a mean of 0.2 pg/g Be. Beryllium contents increase with increasing water depth, following a "nutrient" like pattern (Fig. 8), suggesting that some near-surface agent sequesters Be from the water column in the oceans (Raisbeck et al. 1979; Quinby-Hunt and Turekian 1983). Sharma et al (1987) found strong correlations between the Be isotopes and Al in sediment trap samples from the MANOP project, which suggested that Be was being scavenged by aluminosilicate phases. Anderson et al. (1990) explicitly studied the effects of particulate scavenging on Be (specifically ^{10}Be) in the oceans, and found that ^{10}Be removal from the water column and deposition in sediments near the margins of ocean basins was an order of magnitude greater than in the open ocean. While the sources of ^{10}Be and ^{9}Be fluxes into the oceans differ in detail, it is probable that boundary scavenging effects impact stable Be to a comparable degree. Marine pore waters are higher in Be than seawater, reaching values of ~2 pg/g (250 pM) (Bourles et al. 1989)(Table 2). Beryllium in porewaters increases with sediment depth, implying some mobilization of Be out of the sediments into the waters. Beryllium shows similar systematics to Mn in porewaters, perhaps reflecting the equilibration of these waters with authigenic Mn oxyhydroxide phases in which Be is compatible.

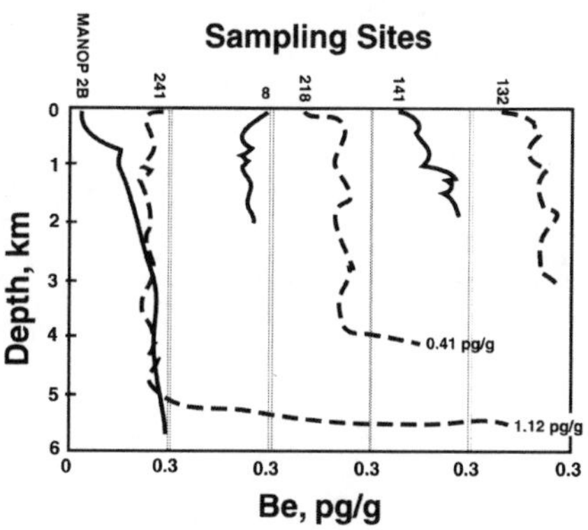

Figure 8. Beryllium concentration profiles in seawater, from Measures and Edmond (1983). MANOP B site is a Pacific station; all other sites are in the Atlantic. Two sites (241 and 218) exceed 0.3 pg/g Be at depth; the maximum Be concentrations noted for each are indicated on their respective depth patterns.

River waters are variable in Be, but are on average higher than the oceans. Acidic rivers may contain as high as 45 pg/g (5 nM) Be. Alkaline river systems with high suspended loads may be two orders of magnitude lower in Be content, on the order of 0.5 pg/g (50 pM: Measures and Edmond 1983; Brown et al. 1992). Beryllium contents in river waters appear to be a function of the bedrock types drained, the mean sediment loads, and the mean pH of the waters: as solution pH changes from 5 to >7, Be speciation in solution shifts from high levels of soluble Be^{2+}, to essentially 100% speciation as the more readily adsorbable $BeOH^+$. When river waters reach the oceans, significant amounts of dissolved Be may sorb onto particles as a function of increasing pH in estuaries, resulting in inverse correlations between Be content and salinity (Measures and Edmond 1983; Brown et al. 1992). Measures and Edmond (1983) estimated a global average river concentration of 1.2 pg/g (150 pM) and a river flux of 4.5×10^6 mol Be/yr.

Aside from rivers, significant inputs of Be to the oceans include continentally derived aerosols and marine hydrothermal exchanges. Brown et al. (1992) compared the Be fluxes associated with aeolian deposition in the Mediterranean, and Be outputs of the Amazon and Ganges/Brahmaputra river systems. The Mediterranean shows Be contents that are nearly a factor of three higher than average seawater values (i.e., 0.8-0.9 pg/g Be), and essentially all of this difference appears to be due to solubilized Be from aerosols off the Sahara desert. They estimated that approximately 10% of the Be in Saharan aerosols (i.e. 0.2 ppm) was soluble. If this level of Be solubility applies to the global aerosol flux to the oceans, then the continental Be influx is 8-27 $\times$ 10^6 mol Be/yr, or 2-5 times the amount of Be provided to the oceans by rivers.

The first results reported for Be in marine hydrothermal solutions were by Measures and Edmond (1983), but Von Damm et al. (1985a,b) have produced the most comprehensive Be data set on both sedimented (Guayamas Basin) and non-sedimented (East Pacific Rise) ocean ridge systems. On the East Pacific Rise, Be contents in high temperature vent fluids (i.e, those fluids in which $[Mg^{2+}] = 0$) vary from 0.1 to 0.4 ng/g (10 to 40 nM), three orders of magnitude higher than seawater, and 100 times as high as the most acidic river system. Beryllium shows a scattered negative correlation with dissolved Mg in vent fluids, indicating that Be is being liberated from basalts during high-temperature

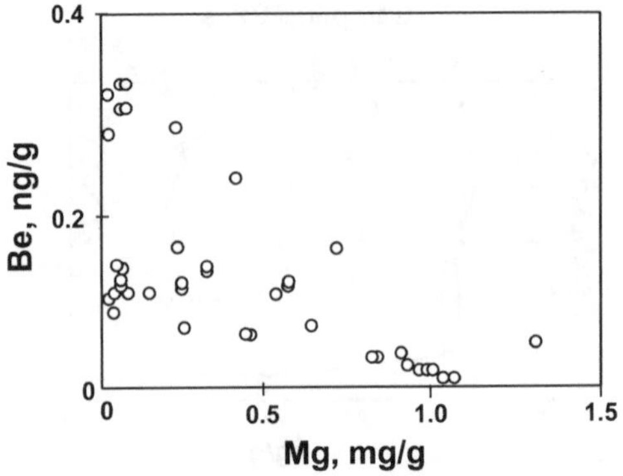

Figure 9. Beryllium in hydrothermal pore waters from the East Pacific Rise. Data from Von Damm et al. (1985a).

hydrothermal alteration (Fig. 9). The calculated "extraction efficiency" for Be (i.e., the water/rock ratio multiplied by the net additions of Be to hydrothermal solution/initial rock concentration) is ~1.4×10^{-4}, lower than for all other elements except Al (Von Damm et al. 1985a). Vent fluids from the highly sedimented Guaymas Basin reach much higher Be contents, up to ~0.9 ng/g (90 nM) Be, probably due to exchanges between the fluids and the higher Be detrital sediments that blanket the vents (Von Damm et al. 1985b). The calculated hydrothermal flux of Be into the oceans (based on oceanic vent data from Von Damm et al. 1985a and Edmond et al. 1979) lies between 1.4 and 5.3×10^6 mol/yr, comparable to the estimated river flux of Measures and Edmond (1983). This flux, like most other fluxes of soluble Be in the oceans, appears to be very efficiently scavenged from the water column and into local sediments.

The mean oceanic residence time (MORT) of Be, based on the various flux estimates, is widely believed to be very short—less than the mixing time of the oceans. Measures and Edmond (1983) estimated the oceanic residence time of Be to be 3600 years, while Anderson et al. (1990) suggests a variable Be residence time: ~100 years near coastlines, where high particle fluxes rapidly pull Be out of the water column; and 500-3000 years in the open ocean. Heterogeneities in Be abundances both with depth in the oceans, and from ocean basin to ocean basin, indicate inefficient mixing of the oceans with respect to Be (and thereby a MORT value less than that of oceanic mixing). Estimated ^{10}Be residence times are equally short (100-4000 years; Morris 1991 and Morris et al. this volume). As such, it is possible that the Be isotopes will preserve input-related heterogeneities, though such data as we have (i.e., Mn-nodules and sediments) suggest relatively constant ratios within the (at ±20%, relatively poor) precision of ^{10}Be/Be determinations.

DISCUSSION

Beryllium abundances in the mantle

To date, no reliable measurements of Be concentrations have been made on mantle-derived rocks, as probable mantle Be levels are still below the practical detection limits of all available analytical methods. Most estimates of mantle Be have been based on the relatively limited data available from silicate meteorites (Quandt and Herr 1974; Anders and Ebihara 1982; Anders and Grevasse 1989). The average Be abundance of ordinary chondrites has been reported to be ~74 ng/g (Quandt and Herr 1974) and ~25 ng/g (Anders and Grevasse 1989). Primitive mantle estimates for Be (i.e., McDonough and Sun 1995; Taylor and McLennan 1995) have typically been made by multiplying the chondritic value by a standard factor related to volatile loss and core separation (typically, 2× to 2.5× chondrite values). Recent reported values for the primitive mantle have ranged between 60 and 70 ng/g Be (McDonough and Sun 1995; Taylor and McLennan 1995). These values are consistent with the Be abundance inferred from the uniform Be/Nd ratios of primitive basalts (~66 ng/g Be), assuming a chondritic REE abundance pattern.

An alternative approach to estimating the Be content of mantle reservoirs is to utilize inverse modeling methods on suites of primitive, cogenetic basaltic lavas (e.g., Minster and Allègre 1978), and also to infer source Be abundances based on Nd abundances, as the two elements vary similarly. Examinations of primitive MORB suites from the FAMOUS area and the Tamayo Fracture Zone suggest a Be content in MORB mantle source regions to be 20-30 ng/g (Ryan and Langmuir 1988; Ryan 1989). Primitive intraplate tholeiites from Kilauea and Reunion islands are four to five times higher in Be than comparable MORBs, and as such suggest intraplate mantle sources contain ~80-100 ng/g Be. Modeling-based estimates of Be source contents necessarily have large errors, but in combination with chondrite-based inferences, they suggest that Be variability in the mantle is coherent with that of other incompatible lithophile elements.

Insights from beryllium behavior into subduction zone chemical fluxes

The systematics of ^{10}Be/Be ratios in arc lavas, especially as related to B/Be ratios, indicate the existence of measurable subduction-zone fluxes of Be off the down-going plate (specifically, subducting sediments) into the mantle wedge. However, Be concentrations are not elevated in arc volcanic rocks relative to the LREE, and beryllium is relatively immobile during metamorphism, and in H_2O-rich fluids, even under high temperature hydrothermal conditions. In what manner, then, is beryllium transported from slab to mantle wedge beneath arcs?

Figure 10 is a series of "cross-arc" variation plots examining Be/Nd and Be/Zr ratios in andesitic lavas from volcanic cross-chains across the Kuriles, and in basaltic lavas from Okmok and Bogoslov volcanoes across the Aleutian arc. With two exceptions (relatively evolved andesites from Chirinkotan and Bogoslov), Be/Nd ratios in these lavas fall within the field for MORB and intraplate mantle sources, and overall no changes in Be/Nd are evident with increasing depth. In light of the ^{10}Be/Be systematics of arc lavas, uniform Be/Nd ratios across arcs necessitates that Be and the LREE be transported off the slab into the mantle wedge to a similar (small) degree.

By contrast, Be/Zr ratios show a distinct increase across the Kurile arc, and may also increase across the Aleutians. Increases in Be/Zr can be generated during calc-alkaline differentiation through the stabilization of zircon in evolved lavas, which "buffers" magmatic Zr contents; or by the accumulation of plagioclase, which will increase Be concentrations, but not Zr. Data from Okmok volcano point to plagioclase accumulation as the source of Be/Zr variation, as these rocks are far too poor in SiO_2 for zircon to be stable (Miller et al. 1992, Watson and Harrison 1984). Several of the Kurile volcanic centers show a spread of Be/Zr ratios, which may also be due to variable amounts of accumulated feldspar. If one "looks through" this and other effects of differentiation by examining the lowest Be/Zr lavas, a significant (up to a factor of three) increase in Be/Zr across the Kurile arc is evident. If Be and Zr behave similarly during arc melting, as appears to be the case (see Fig. 4), then this increase cannot be attributed purely to mineral/melt equilibria, and may thus reflect changes in the chemical characteristics of the mantle source regions across these arcs. These changes lead to increases in the Be content of source regions above deeper slabs relative to increases in the contents of high field strength elements (HFSE), a pattern that is essentially the reverse of that observed in "fluid mobile" elements such as B, Cs, and Sb (Ryan et al. 1996b; Leeman 1996).

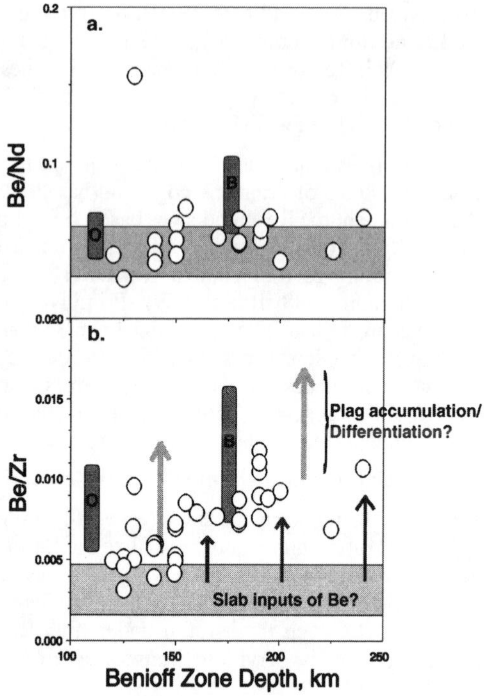

Figure 10. Cross-arc variation diagrams of Be/Nd and Be/Zr ratios, based on data from the Kurile arc (open circles: Ryan et al. 1995), and from Okmok and Bogoslov volcanoes in the Aleutians (Ryan and Langmuir 1988; Ryan 1989; Miller et al. 1992).

Increasing Be/Zr ratios and relatively constant Be/Nd ratios along arc volcanic cross-chains suggests progressively greater slab-derived inputs of beryllium and the light rare earth elements (LREE) to the mantle wedge as the slab travels deeper, and slab temperatures increase. The mechanism of Be transfer from slab to mantle thus must increase in efficiency with increasing pressure and temperature. The processes of slab fluid release, at least as documented by "fluid mobile" elements, are inconsistent with Be enrichment patterns across arcs. However, fluid phases high in fluorine and other halogens can clearly mobilize Be, along with a distinctive menu of other metals (Sn, Mo, and W: Salisbury 1964; Barton 1987). Noll et al. (1996) examined Sn and Mo variations across the Kuriles and other arcs, and proposed a hydrothermal "ore" fluid as a probable transfer medium for soluble species, based largely on the systematics of the soluble chalcophile elements As, Pb, and Sb. If the mechanism for Be mobilization in arcs is a halogen-rich ore fluid, then one should presumably see comparable systematics in Sn, Mo, or W variations.

In Figure 11, Sn/Ti and Mo/Zr ratios are examined versus depth-to-slab for the Kuriles cross-arc data set. Neither ratio shows any regular changes across this arc. While the behaviors of Sn and Mo are not well constrained during melting and crystallization, and the mantle values of these ratios are not well constrained, the lack of any clear cross-arc increases or decreases in these ratios suggests that neither Sn or Mo are being enriched in arc sources relative to the HFSE. Beryllium is behaving differently in arc settings than elements with which it is often mobilized in ore systems, which probably means that a halogen-rich fluid phase is not a likely means for Be transport from the slab.

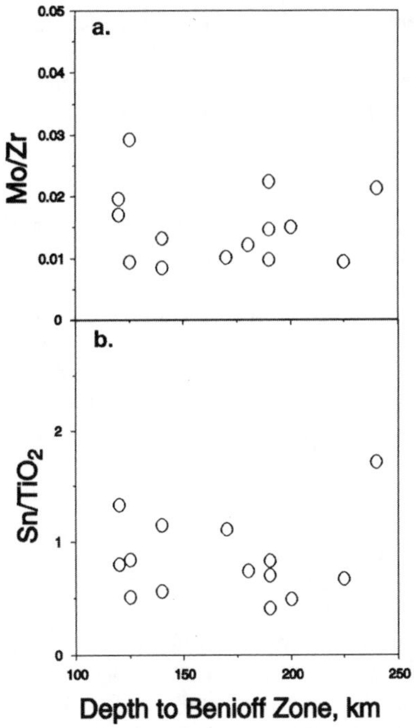

Figure 11. Sn/Ti and Mo/Zr ratios vs. depth to slab in the Kurile volcanic arc. Sn and Mo data from Noll et al. (1996); Ti and Zr data: sources as in Figure 10. The denominator elements were chosen from among the "high field-strength" elements (HFSE: Ti, Y, Zr, Ta, Nb) to minimize the effects of mineral-melt fractionation on the ratio values.

If beryllium is not transported in hydrous (± halogen) fluids, then it is possible that melt or melt-like phases (i.e., pegmatitic fluids) may be responsible. The solubility of Be in Si-rich melts and fluids has been documented—Bebout et al. (1993, 1999) found B/Be ratios in pegmatitic leucosomes from the Catalina schist to be consistently <4, due in large part to relatively high (2-4 ppm) Be contents in these rocks. The high concentration of Be in pegmatites generally (see London and Evensen, this volume), and its higher inferred distribution coefficients in siliceous melts point to the potential for Be inputs to arc sources via melt phases. Plank and Langmuir (1993) and Johnson and Plank (2000) have suggested that much of the distinctive chemical signature in arc lavas may be due to inputs to the mantle wedge of high-Si melts from subducted sediments. Schiano et al. (1994) and Kepezhinskas et al (1995) have suggested that small amounts of high-Si, high-Na melts of subducted ocean crust may be part of the slab input "package" in the Philippine and Kamchatka arcs. The observation in Figure 10 that the relative enrichment of Be in arc lavas shows a modest increase with slab depth is consistent with a model for inputs of slab-derived melts or Si-rich fluids: temperatures on the slab increase with depth, and hydrous slab phases, such as phengitic white mica, can persist to great depths to provide fluid "fluxes" for melt generation from sediments or crustal rocks (Domanik and Holloway 1996). The sediment-melting experiments of Johnson and Plank (2000) show that at temperatures above the sediment solidus between 2 and 3 GPa, as much as 15% of sedimentary Be may be sequestered in such melts. The mobile phases that liberate trace metals from slabs at depth must obviously be complex and variable mixtures, given the range of elements that are mobilized, and the changes with depth in what is transported. Ascribing melt-like characteristics to the slab fluid phase beneath arcs seems to be necessary to explain the slab-to-mantle wedge transport of Be and the rare earth elements, modest though that degree of transport is.

The global beryllium inventory

Table 2 provides estimates of the beryllium inventories of major terrestrial reservoirs, as well as estimates for major Be fluxes. The most distinctive characteristic of the terrestrial Be "cycle" is that essentially all H_2O-based reservoirs and fluid-mediated fluxes are minor by comparison to the solid-Earth reservoirs and fluxes of the mantle and crust. The transport of Be off slabs via arc magmatism will result in only relatively small additions of Be to the crust, as compared to boron, potassium, and other large-ion lithophile elements. The majority of subducted oceanic beryllium (largely Be in ocean crustal rocks + a subducted sediment contribution amounting to ~15% of the total subducted inventory) is returned to the mantle in processed slabs. Based on the factor-of-three difference between the beryllium contents of primitive MORBs and intraplate tholeiites, the Be abundances of deep mantle sources appear to be higher than that of the mantle source of MORBs. This difference in Be content does not appear to be an artifact of changes in residual mantle mineralogy, given the mineral-melt partitioning characteristics of Be; but it is comparable to the differences in the LREE signatures between MORBs and intraplate rocks. The origins of these differences in the mantle may relate to the time-integrated effects of subduction: inputs of relatively high Be ocean crustal materials on down-going slabs, during the formation of continents and thereafter, may have led to a buildup of Be in the deeper mantle. Direct, high-precision measurements of Be in mantle-derived rocks are needed to assess such distinctions, but such determinations still present a substantial analytical challenge.

CONCLUSIONS

The geochemical behavior of beryllium as a trace species in the Earth is relatively straightforward. To a great degree Be distribution patterns in rocks and fluids follow those of the light rare-earth elements and (to a lesser extent) the HFSE. Only with the advent of Be isotope ratio determinations have we been able to document the relatively modest open-system behavior of this element during subduction, but even subduction does not lead to extreme fractionations of beryllium between the crust and mantle. The limited ways in which Be can be strongly concentrated make it a useful prospecting tool for rare-metal pegmatites and vein deposits. The highly predictable, yet unique way in which Be varies in igneous systems makes it a useful tracer for studying the evolution of magmas.

ACKNOWLEDGMENTS

Thanks first and foremost to Ed Grew for his infinite patience with the slow progress on the construction and revision of this manuscript, and to Bill Leeman and Denis Shaw for valuable critiques in review. New (old!) data presented for Okmok volcano were collected during my DTM postdoctoral fellowship, and new adakite and kimberlite data presented herein was funded by NSF-EAR 9205804 and 9304133 to Ryan et al. and EAR 9405197 to Kepezhinskas and coworkers – however, the actual adakite and kimberlite analyses were all performed by Suzy Norrell as part of an REU Supplement Award project and her MS thesis research, respectively. Thanks also to Sean Solomon, Rick Carlson, Fouad Tera, and Steve Shirey of DTM, where I completed the initial draft of this chapter during my sabbatical semester.

REFERENCES

Anders E, Ebihara M (1982) Solar system abundances of the elements. Geochim Cosmochim Acta 46:2362-2380

Anders E, Grevasse N (1989) Abundances of the elements: meteoric and solar. Geochim Cosmochim Acta 53:197-214

Anderson RF, Lao Y, Broecker WS, Trumbore SE, Hofmann HJ, Wolfi W (1990) Boundary scavenging in the Pacific Ocean: a comparison of ^{10}Be and ^{231}Pa. Earth Planet Sci Lett 96:287-304

Baba S, Grew ES, Shearer CK, Sheraton JW (2000) Surinamite: a high-temperature metamorphic beryllosilicate from Lewisian sapphirine-bearing kyanite–orthopyroxene–quartz–potassium-feldspar gneiss at South Harris, NW Scotland. Am Mineral 85:1474-1484

Barton MD (1987) Lithophile-element mineralization associated with Late Cretaceous two-mica granites in the Great Basin. Geology 15:337-340

Bea F, Montero P (1999) Behavior of accessory phases and redistribution of Zr, REE, Y, Th, and U during metamorphism and partial melting of metapelites in the lower crust: An example from the Kinzigite Formation of Ivrea-Verbano, NW Italy. Geochim Cosmochim Acta 63:1133-1153

Bea F, Pereira MD, Stroh A (1994) Mineral/leucosome trace-element partitioning in a peraluminous migmatite (a laser ablation-ICP-MS study). Chem Geol 117:291-312

Bebout GE, Ryan JG, Leeman WP (1993) B-Be systematics in subduction-related metamorphic rocks: characterization of the subducted component. Geochim Cosmochim Acta 57:2227-2237

Bebout GE, Ryan JG, Leeman WP, Bebout AE (1999) Fractionation of trace elements by subduction zone metamorphism: significance for models of crust-mantle mixing. Earth Planet Sci Lett 177:69-83

Bender JF, Langmuir CH, Hanson GN (1984) Petrogenesis of basalt glasses from the Tamayo region, East Pacific Rise. J Petrol 25:213-254

Beus AA (1954) Geochemistry of beryllium. Geochem Int'l 5:511-531

Beus AA (1962) Beryllium: Evaluation of Deposits during Prospecting and Exploratory Work. W.H. Freeman and Co., London, 161 p

Beus AA (1966) Geochemistry of Beryllium and Genetic Types of Beryllium Deposits. W.H. Freeman and Co., London, 401 p

Bourles DL, Klinkhammer G, Campbell AC, Measures CI, Brown ET, Edmond JM (1989) Beryllium in marine pore waters: geochemical and geochronological implications. Nature 341:731-733

Brenan JM, Neroda E, Lindstrom CC, Shaw HF, Ryerson FJ, Phinney DL (1998a) Behavior of boron, beryllium and lithium during melting and crystallization: constraints from mineral-melt partitioning experiments. Geochim Cosmochim Acta 62:2129-2141

Brenan JM, Ryerson FJ, Shaw HF (1998b) The role of aqueous fluids in the slab-to-mantle transfer of boron, beryllium and lithium during subduction: experiments and models. Geochim Cosmochim Acta 62:3337-3347

Broecker WS, Peng T-H (1982) Tracers in the Sea. Eldigio Press, Palisades, New York, 690 p

Bronevoy VA, Zibermints AV, Tenyakov VA (1985) Average chemical compositions of bauxites and its change over time. Geochem Int'l 22:77-82

Brown ET, Measures CI, Edmond JM, Bourles DL, Raisbeck GM, Yiou F (1992) Continental inputs of beryllium to the oceans. Earth Planet Sci Lett 114:101-111

Bryan WB, Moore JG (1977) Compositional variations of young basalts in the Mid-Atlantic Ridge rift valley, near lat. 36° 49'N. Geol Soc Am Bull 88:556-570

Bryan WB, Thompson G, Michael PJ (1979) Compositional variation in a steady state zoned magma chamber: Mid-Atlantic Ridge at 36°50'N. Tectonophys 55:63-85

Bushlyakov IN, Grigoriev NA (1989) Beryllium in Ural metamorphites. Geochem Int'l 26:57-61

Dean WE, Parduhn NL (1979) Inorganic geochemistry of sediments and rocks recovered from the southern Angola Basin and Adjacent Walvis Ridge, Sites 530 and 532, Deep Sea Drilling Project Leg 75. *In* Hay WW, Sibuet J-C et al. (eds) Init Repts DSDP 75:923-958

Defant MJ, Drummond M (1990) Derivation of some modern arc magmas by melting of young subducted lithosphere. Nature 347:662-665

Defant MJ, Clark LF, Stewart RH, Drummond MS, de Boer JZ, Maury RC, Bellon H, Jackson TE, Restrepo JF (1991) Andesite and dacite genesis via contrasting processes; the geology and geochemistry of El Valle Volcano, Panama. Contrib Mineral Petrol 106: 309-324

Domanik K, Holloway JR (1996) The stability and composition of phengitic muscovite and associated phases from 5.5 to 11 GPa; implications for deeply subducted sediments. Geochim Cosmochim Acta 60:4133-4150

Domanik KJ Hervig RL and Peacock SM (1993) Beryllium and boron in subduction zone minerals: An ion microprobe study. Geochim Cosmochim Acta 57:4997-5010

Dostal J, Dupuy C, Dudoignon P (1996) Distribution of boron, lithium and beryllium in ocean island basalts from French Polynesia: implications for the B/Be and Li/Be ratios as tracers of subducted components. Mineral Mag 60:563-580

Drummond MS, Defant MJ, Kepezhinskas PK (1996) Petrogenesis of slab-derived trondhjemite-tonalite-dacite/adakite magmas. *In* Brown M, Candela PA, Peck DL, Stephens WE, Walker RJ, Zen E-an (eds) The third Hutton symposium on the Origin of granites and related rocks. Geol Soc Am Spec Paper 315:205-215

Dupuy C, Barsczus HG, Dostal J, Vidal P, Liotard J-M (1989) Subducted and recycled lithosphere as the mantle source of ocean island basalts from southern Polynesia, central Pacific. Chem Geol 77:1-18

Dupuy C, Vidal P. Maury RC, Guille G (1993) Basalts from Mururoa, Fangataufa and Gambier islands (French Polynesia): geochemical dependence on the age of the lithosphere. Earth Planet Sci Lett 117:89-100

Edwards CMH, Morris JD, Thirwall MF (1993) Separating mantle from slab signatures in arc lavas using B/Be and radiogenic isotope systematics. Nature 362:530-533

Edwards CMH, Menzies MA, Thirlwall MF, Morris JD, Leeman WP, Harmon RS (1994) The transition to potassic alkaline volcanism in island arcs: The Ringitt-Beser Complex, east Java, Indonesia. J Petrol 35:1557-1595

Fransceschelli M, Puxeddu M, Carcangiu G, Gattiglio M, Pannuti F (1996) Breccia-hosted manganese-rich minerals of Alpi Apuane, Italy: a marine, redox-generated deposit. Lithos 37:309-333

Fransceschelli M, Puxeddu M, Memmi I (1998) Li, B rich Rhaetian metabauxite, Tuscany, Italy: reworking of older bauxites and igneous rocks. Chem Geol 144:221-242

Gao S, Zhang B, Xie Q, Ouyang J, Wang D, Gao C (1991) Average chemical compositions of post-Archean sedimentary and volcanic rocks from the Qinling orogenic belt and its adjacent North China and Yangtze cratons. Sed Geol 92:261-282

Gao S, Zhang B, Luo T, Li Z, Xie Q, Gu X, Zhang H, Ouyang J, Wang D, Gao C (1992) Chemical composition of the continental crust in the Qinling orogenic belt and its adjacent North China and Yangtze cratons. Geochim Cosmochim Acta 56:3933-3950

Gao S, Luo T-C, Zhang B, Zhang H, Han Y, Zhao Z, Hu Y (1998) Chemical composition of the continental crust as revealed by studies in east China. Geochim Cosmochim Acta 62:1959-1975

Gill JB, Morris JD, Johnson RW (1993) Timescale for producing the geochemical signature of island arc magmas; U-Th-Po and Be-B systematics in Recent Papua New Guinea lavas. Geochim Cosmochim Acta 57:4269-4283

Goldschmidt VM, Peters Cl (1932) Zur Geochemie des Berylliums. Nachr Ges Wiss Göttingen Math Phys Kl III:360-376

Graham IJ, Dichtburn RG, Whitehead ME (1998) ^{10}Be spikes in Plio-Pleistocene cyclothems, Wanganui Basin, New Zealand: identification of the local flooding surface (LFS). Sed Geol 122:193-215

Grew ES, Yates MG, Barbier J, Shearer CK, Sheraton JW, Shiraishi K, Motoyoshi Y (2000) Granulite-facies beryllium pegmatites in the Napier Complex in Khmara and Amundsen Bays, western Enderby Land, East Antarctica. Polar Geosci 13:1-40

Grigoriev NA (1982) The beryllium mineral balance in primary aureoles and its changes during the formation of weathering crusts, diluvium, and soils. Geochem Int 21:130-136

Grigoriev NA (1986) Distribution of beryllium at the surface of the Earth. Int Geol Rev 28:27-179

Hanson GN (1989) An approach to trace element modeling using a simple igneous system as an example. Rev Mineral 21:79-97

Hart SR, Dunn T (1993) Experimental cpx/melt partitioning of 24 trace elements. Contrib Mineral Petrol 113:1-8

Henry C, Burkhard M, and Goffe B (1996) Evolution of synmetamorphic veins and their wallrocks through a Western Alps transect: no evidence for large-scale fluid flow. Stable isotope, major and trace-element evidence. Chem Geol 127:81-109

Hochstaedter AG, Ryan JG, Luhr JF, Hasenake T (1996) On B/Be ratios in the Mexican Volcanic Belt. Geochim Cosmochim Acta, 60:613-628

Hörmann PK (1978) Beryllium. *In* Wedepohl KH (ed) Handbook of Geochemistry II/1. Springer, Berlin, p 4-B-1 to 4-O-1, 1-6

Johnson M, Plank T (2000) Dehydration and melting experiments constrain the fate of subducted sediments. Geochemistry Geophysics Geosystems (http://gcubed.magnet.fsu.edu/)

Johnson RW, Jaques AL, Langmuir CH, Perfit MR, Staudigel H, Dunkley PN, Chappell BW, Taylor SR, Baekisapa M (1987) Ridge subduction and forearc volcanism: Petrology and geochemistry of rocks dredged from the western Solomon arc and Woodlark Basin. *In* Taylor B, Exon NF (eds) Marine Geology, Geophysics and Geochemistry of the Woodlark Basin-Solomon Islands. Circum-Pacific Council for Energy and Mineral Resources Earth Science Ser 7:155-226

Kappel E (1985) Evidence for volcanic episodicity and a non-steady state rift valley. PhD dissertation, Columbia University, New York

Kepezhinskas PK, Defant MJ, Drummond MS (1995) Na metasomatism in the island-arc mantle by slab melt-peridotite interaction; evidence from mantle xenoliths in the North Kamchatka Arc. J Petrol 36:1505-1527

Koeberl C, Reimold WU, Kracher A, Tråxler A, Voermaier A, Korner W (1996) Mineralogical, petrological, and geochemical studies of drill core samples from the Manson impact structure, Iowa. *In* Koerberl C, Anderson RR (eds) The Manson Impact Structure, Iowa: Anatomy of an Impact Crater. Geol Soc Am Spec Paper 302:145-219

Krishnaswami S, Mangini A, Thomas JH, Sharma P, Cochran JK, Turekian KK, Parker PD (1982) ^{10}Be and Th isotopes in manganese nodules and adjacent sediments: nodule growth histories and nuclide behavior. Earth Planet Sci Lett 59:217-234

Ku TL, Kusakabe M, Nelson DE, Southon JR, Kortelling RG, Vogel J, Nowikow I (1982) constancy of oceanic deposition of ^{10}Be as recorded in manganese crusts. Nature 299:240-242

Langmuir CH, Bender JF, Bence AE, Hanson GN, Taylor SR (1977) Petrogenesis of basalts from the FAMOUS area: Mid-Atlantic Ridge. Earth Planet Sci Lett 36:133-156

Langmuir CH, Bender JF, Batiza R (1986) Petrologic and tectonic segmentation of the East Pacific Rise, 5°30'-14°30'N. Nature 322:422-429

Lebedev VI, Nagaytsev YV (1982) Minor elements in metamorphic rocks as an ore-material source for certain deposits. Geochem Int'l 17:31-39

Leeman WP (1996) Boron and other fluid-mobile elements in volcanic arc lavas; implications for subduction processes. *In* Bebout GE, Scholl DW, Kirby SH, Platt JP (eds) Subduction Top to Bottom. Am Geophys Union Geophys Monogr 96:269-276

Lesher CE (1986) Effects of silicate liquid composition on mineral-liquid element partitioning from Soret diffusion studies. J Geophys Res, B91:6123-6141

Matsui Y, Onuma N, Nagasawa H, Higuchi H, Banno S (1977) Crystal structure control in trace element partition between crystal and magma. Bull Soc franç Minéral Cristallogr 100:315-324

McDonough, WF, Sun S-S (1995) The composition of the Earth. Chem Geol 120:223-253

Measures CI, Edmond JM (1983) The geochemical cycle of ^{9}Be: a reconnaissance. Earth Planet Sci Lett 66:101-110

Merrill JR, Lyden EFX, Honda M, Arnold JR (1960) The sedimentary geochemistry of the beryllium isotopes. Geochim Cosmochim Acta 18:108-129

Miller DM, Langmuir CH, Goldstein SJ, Franks AL (1992) The importance of parental magma composition to calc-alkaline and tholeiitic evolution: evidence from Umnak Island in the Aleutians. J Geophys Res 97:321-343

Miller C, Schuster R, Klotzi U, Frank W and Purtscheller F (1999) Post-Collisional Potassic and Ultrapotassic Magmatism in SW Tibet: Geochemical and Sr–Nd–Pb–O isotopic constraints for mantle source characteristics and petrogenesis. J Petrol 40:1399-1424

Minster JF, Allègre CJ (1978) Systematic use of trace elements in igneous processes; Part III, Inverse problem of batch partial melting in volcanic suites. Contrib Mineral Petrol 68:37-52

Mitchell RH (1986) Kimberlites: Mineralogy, Geochemistry and Petrology. New York, Plenum Press

Mitchell RH, Bergman SC (1991) Petrology of Lamproites. New York, Plenum Press, 447 p

Miyamoto T, Grew ES, Sheraton JW, Yates MG, Dunkley DJ, Carson CJ, Yoshimura Y, Motoyoshi Y (2000) Lamproite dikes in the Napier complex at Tonagh Island, Enderby Land, East Antarctica. Polar Geosci 13:41-59

Monaghan MC, Klein J, and Measures CI (1988) The origin of ^{10}Be in island-arc volcanic rocks. Earth Planet Sci Lett 89:288-298

Morris JD (1991) Applications of cosmogenic ^{10}Be to problems in the earth sciences. Ann Rev Earth Planet Sci 19:313-350

Morris JD, Tera F (1989) ^{10}Be and ^{9}Be in mineral separates and whole rocks from volcanic arcs: implications for sediment subduction. Geochim Cosmochim Acta 53:3197-3206

Morris, JD, Leeman WP, Tera F (1990) The subducted component in island arc lavas: constraints from B-Be systematics. Nature 344:31-35

Noll PD, Newsom HE, Leeman WP, Ryan JG (1996) The role of hydrothermal fluids in the production of subduction zone magmas: evidence from siderophile and chalcophile trace elements and boron. Geochim Cosmochim Acta 60:587-611

Norrell S (1998) Petrogenesis of Alkaline and Sub-Alkaline Volcanic Rocks from the Russian Far East. MS thesis, Univ South Florida, Tampa

Norrell S, Ryan JG, Defant MJ (1995) On the origins of adakites: evidence from B-Be-Li systematics in young lavas from Panama. EOS Suppl 76:S289

Onuma N, Higuchi H, Nakita H, Nagasawa H (1968) Trace element partition between two pyroxenes and the host lava. Earth Planet Sci Lett 5:47-51
Ottoway TL, Wicks FJ, Bryndzia LT, Kyser TK, Spooner ETC (1994) Formation of the Muzo hydrothermal emerald deposit in Colombia. Nature 369:552-554
Pearce JA, Peate DW (1995) Tectonic implications of the composition of volcanic arc magmas. Ann Rev Earth Planet Sci, 23:251-285
Plank T, Langmuir CH (1993) Tracing trace elements from sediment input to volcanic output at subduction zones. Nature 362:739-743
Press F, Siever R (1978) Earth, 2nd Edn. San Francisco, W.H. Freeman, 649 p
Quandt U, Herr W (1974) Beryllium abundance in meteorites determined by "non-destructive" photon activation. Earth Planet Sci Lett 24:53-58
Quinby-Hunt MS, Turekian KK (1983) Distribution of elements in sea water. EOS Trans Am Geophys Union 64:130-131
Raisbeck GM, Yiou F, Fruneau M, Loiseaux JM, Lieuvin M (1979) ^{10}Be concentration and residence time in the ocean surface layer. Earth Planet Sci Lett 43:237-240
Rapp RP, Watson EB (1986) Monazite solubility and dissolution kinetics: Implications for the thorium and light rare earth chemistry of felsic magmas. Contrib Mineral Petrol 94:304-316
Reymer A, Schubert G (1984) Phanerozoic addition rates to the continental crust and crustal growth. Tectonics 3:63-77
Ryan JG (1989) The Systematics of Lithium, Beryllium, and Boron in Young Volcanic Rocks. PhD dissertation, Columbia Univ, New York, 310 p
Ryan JG, Langmuir CH (1988) Beryllium systematics in young volcanic rocks: implications for ^{10}Be. Geochim Cosmochim Acta 52:237-244
Ryan JG, Leeman WP, Morris JD, Langmuir CH (1996a) The boron systematics of intraplate lavas: implications for crust and mantle evolution. Geochim Cosmochim Acta 60:415-422
Ryan JG, Morris J, Bebout G, Leeman WP (1996b) Describing chemical fluxes in subduction zones: insights from "depth profiling" studies of arc and forearc rocks. *In* Bebout GE, Scholl DW, Kirby SH, Platt JP (eds) Subduction Top to Bottom. Am Geophys Union Geophys Monogr 96:263-268
Ryan JG, Morris JD, Tera F, Leeman WP, Tsvetkov A (1995) Cross-arc geochemical variations in the Kurile arc as a function of slab depth. Science 270:625-627
Ryerson FJ, Watson EB (1987) Rutile saturation in magmas: implications for Ti-Nb-Ta depletion in orogenic rock series. Earth Planet Sci Lett 86:225
Salisbury CL (1964) Association of beryllium with tin deposits rich in fluorite. Econ Geol 59:920-926
Sandell EB (1952) The beryllium content of igneous rocks. Geochim Cosmochim Acta 2:211-216
Schiano P, Clocchiatti R, Shimizu N, Maury RC, Jochum KP, Hofmann AW (1994) Hydrous, silica-rich melts in the sub-arc mantle and their relationship with erupted arc lavas. Nature 377:595-600
Shannon RD (1976) Revised effective ionic radii and systematic studies of interatomic distances in halides and chalcogenides. Acta Crystallogr A32:751-767
Sharma P, Somayajulu BLK (1982) ^{10}Be dating of large manganese nodules from world oceans. Earth Planet Sci Lett 59:235-244
Sharma P, Mahanna R, Moore WS, Ku TL, Southon JR (1987) Transport of ^{10}Be and ^{9}Be in the ocean. Earth Planet Sci Lett 86:69-76
Shaw DM, Reilly GA, Muysson JR, Pattenden GE, Campbell FE (1967) An estimate of the chemical composition of the Canadian Precambrian Shield. Can J Earth Sci 4:829-853
Shawe DR, Bernold S (1966) Beryllium content of volcanic rocks. U S Geol Surv Bull Rep B1214C:C1-11
Sheraton JW, Black LP, McCullough MT (1984) Regional geochemical and isotopic characteristics of high-grade metamorphics of the Prydz Bay area: The extent of Proterozoic reworking of Archean continental crust in east Antarctica. Precamb Res 26:169-198
Shirey SB, Bender JF, Langmuir CH (1987) Three component isotopic heterogeneity near the Oceanographer transform, Mid-Atlantic Ridge. Nature 325:217-223
Sighinolfi GP (1973) Beryllium in deep-seated crustal rocks. Geochim Cosmochim Acta 37:702-706
Sillitoe RH, Bonham HF (1984) Volcanic landforms and ore deposits. Econ Geol 79:1286-1298
Tanaka S, Inoue T, Huang Z-Y (1982) ^{10}Be and ^{10}Be/^{9}Be in near Antarctica sediment cores. Geochem J 16:321-325
Tatsumi Y, Isoyama H (1988) Transportation of beryllium with H_2O as high pressures: implication for magma genesis in subduction zones. Geophys Res Lett 15:180-183
Taura H, Yurimoto H, Kurita K, Sueno S (1998) Pressure dependence on partition coefficients for trace elements between olivine and the coexisting melts. Phys Chem Minerals 25:469-484
Taylor SR, McLennan SM (1995) The geochemical evolution of the continental crust. Rev Geophys 33:241-265

Tepper JH, Ryan JG (1995) Temporal trends in geochemistry of Cascade arc granitoids: Correlation with subduction rate and fracture zone position. EOS Suppl 76:F658

Tera F, Brown L, Morris J, Sacks IS, Klein J, Middleton R (1986) Sediment incorporation in island arc magmas: Inferences from ^{10}Be. Geochim Cosmochim Acta 50:636-660

Tomascak P, Ryan JG, Defant MJ (2000) Lithium isotopes and light elements depict incremental slab contributions to the subarc mantle in Panama. Geology 28:507-510

Voncken JHL, Vriend SP, Kocken JWM, Jansen JBH (1986) Determination of beryllium and its distribution in rocks of the Sn-W granite of Regoufe, northern Portugal. Chem Geol 56:93-103

Von Damm KL, Edmond JM, Grant B, Measures CI, Walden B, Weiss RF (1985a) Chemistry of submarine hydrothermal solutions at 21°N, East Pacific Rise. Geochim Cosmochim Acta 49:2197-2220

Von Damm KL, Edmond JM, Measures CI, Grant B (1985b) Chemistry of submarine hydrothermal solutions at Guayamas Basin, Gulf of California. Geochim Cosmochim Acta 49:2221-2237

Watson EB, Harrison, TM (1984) Accessory minerals and the geochemical evolution of crustal magmatic systems: a summary and prospectus of experimental approaches. Phys Earth Planet Int 35:19-30

Webster JD, Holloway JR, Hervig RL (1987) Phase equilibria of a Be, U and F-enriched vitrophere from Spor Mountain, Utah. Geochim Cosmochim Acta 51:389-402

Wedepohl KH (1995) The composition of the continental crust. Geochim Cosmochim Acta 59:1217-1239

You C-F, Castillo PR, Gieskes JM, Chan L-H, Spivack AJ (1996) Trace element behavior in hydrothermal experiments: implications for fluid processes at shallow depths in subduction zones. Earth Planet Sci Lett 140:41-52

Zheng S, Morris J, Tera F, Klein J, Middleton R (1994) Beryllium isotopic investigation of sedimentary columns outboard of subduction zones. U S Geol Survey Circ Rept C 1107:366

4 Rates and Timing of Earth Surface Processes From *In Situ*-Produced Cosmogenic Be-10

Paul R. Bierman

Geology Department and School of Natural Resources
University of Vermont, Burlington, Vermont 05405
pbierman@zoo.uvm.edu

Marc W. Caffee

Center for Accelerator Mass Spectrometry
Lawrence Livermore National Laboratory
Livermore, California 94550
Now at: *PRIME Lab, Purdue University, West Lafayette, Indiana 47907*

P. Thompson Davis

Department of Natural Science, Bentley College
Waltham, Massachusetts 02152

Kim Marsella

Geology Department, Skidmore College
Saratoga Springs, New York 12866

Milan Pavich

National Center, U S Geological Survey
Reston, Virginia 22092

Patrick Colgan

Department of Geology, Northeastern University
Boston, Massachusetts 02115

David Mickelson

Department of Geology & Geophysics, University of Wisconsin
Madison, Wisconsin 53706

Jennifer Larsen

Geology Department, University of Vermont
Burlington, Vermont 05405

INTRODUCTION

Beryllium-10 is the longest-lived of the seven known unstable isotopes of Be; it results mostly from the interaction of cosmic radiation, primarily neutrons, with a variety of target atoms by spallation, the splitting of nuclei (Lal 1988). It can also be produced at very low levels by radio-disintegration of U and Th (Sharma and Middleton 1989). Although all radioactive Be isotopes are produced in the atmosphere via cosmic-ray reactions (Arnold 1956; Lal and Peters 1967; Morris et al., Chapter 5, this volume), in this paper, we are most interested in ^{10}Be produced in rock, which we refer to as *in situ*-produced. Such *in situ*-produced ^{10}Be is a cosmic-ray dosimeter functioning as a quantitative monitor of near-surface residence time. In some geologic situations, ^{10}Be can be used as a chronometer, allowing one to estimate the duration of surface exposure. In other settings, ^{10}Be can be used as a tracer allowing one to estimate the rate, distribution, and behavior of Earth surface processes.

1529-6466/00/0050-0004$05.00

Rates of ^{10}Be production are very low in rocks exposed at Earth's surface, $<10^1$ to $>10^2$ atoms g^{-1} y^{-1} resulting in typical *in situ*-produced ^{10}Be concentrations, termed activities, of 10^3 to 10^7 atoms g^{-1}. The development of accelerator mass spectrometry (AMS) in the 1980s allowed routine measurement of ^{10}Be at levels typical of rocks exposed near Earth's surface (Elmore and Phillips 1987). Analytical refinements in the 1990s lowered detection limits by an order of magnitude, allowing dating of exposure periods $<<10^3$ years in favorable geologic and geographic situations (simple exposure histories at high altitude and latitude).

Most *in situ* ^{10}Be is measured in quartz mineral separates where the primary spallation target is O (Lal and Arnold 1985). Quartz is used because the composition of quartz is simple, the O content is uniform (Nishiizumi et al. 1986, 1989), quartz is widely distributed on Earth's surface, large quantities of quartz can be easily separated from other minerals (Kohl and Nishiizumi 1992), and quartz can easily be cleaned of adhered, atmospherically-produced ^{10}Be by sequential acid etching (Nishiizumi et al. 1989; Brown et al. 1991; Kohl and Nishiizumi 1992). Measurements of ^{10}Be have also been made in pyroxene, olivine, and diamond (Lal et al. 1987; Nishiizumi 1990; Shepard et al. 1995; Seidl et al. 1997; Ivy-Ochs et al. 1998), although both chemical separation of Be from other cations and effective removal of atmospherically produced ^{10}Be, incorporated into rock during weathering and surface exposure, have proven troublesome in both olivine and pyroxene (Seidl et al. 1997; Ivy-Ochs et al. 1998).

Since the first large-scale campaign to measure *in situ*-produced ^{10}Be by AMS (Klein et al. 1986), application of this nuclide to problems in Earth Science has grown steadily. Measurements of ^{10}Be in rock have been used to date glacial events (e.g., Brown et al. 1991; Brook et al. 1993, 1995a,b, 1996; Gosse et al. 1995a,b; Ivy-Ochs et al. 1995; Davis et al. 1999; Bierman et al. 1999; Marsella et al. 2000; Phillips et al. 2000), alluvial fan surfaces and tectonic offset (e.g., Bierman et al. 1995b; Ritz et al. 1995; Brown et al. 1998; Zehfuss et al. 2001), bolide impacts (Nishiizumi et al. 1991b), river and marine terraces (Anderson et al. 1996; Phillips et al. 1997; Repka et al. 1997; Hancock et al. 1999; Perg et al. 2001), loess and eolian sand histories (Nishiizumi et al. 1993; Bierman and Caffee 2001), volcanic eruptions (Shepard et al. 1995), river incision (Burbank et al. 1996; Leland et al. 1998; Hancock et al. 1998; Granger et al. 1997, 2001a; Weissel and Seidl 1998), and landslides (Nichols et al. 2000).

In situ-produced ^{10}Be has been used to monitor rates of rock surface erosion (Nishiizumi et al. 1986; Bierman and Turner 1995; Lal et al. 1996; Small et al. 1997; Summerfield et al. 1999; Bierman and Caffee 2001, 2002; Granger et al. 2001b), understand the production of sediment from bedrock (Heimsath et al. 1997, 1999; Small et al. 1999), monitor the growth of deforming structures (Molnar et al. 1994; van der Woerd et al. 2000; Jackson et al. 2002), track the development of soils (Brown et al. 1994; Braucher et al. 1998a,b, 2000), disprove the survival of Cambrian land surfaces (Belton et al. 2000), and determine the effectiveness of glacial erosion (Davis et al. 1999; Colgan et al. in press).

This chapter presents, specifically for ^{10}Be produced in rock, detailed sample preparation methods as currently practiced at the University of Vermont with reference to other published approaches, a summary of isotopic analytical protocols, an overview of commonly employed interpretive models, and selected illustrative examples of applications we have made of ^{10}Be to problems in the Earth Sciences with references to and comparison with numerous other studies. A complementary review paper details the application of *in situ*-produced ^{10}Be measured in sediments (Bierman et al. 2001a) while another by Granger and Muzikar (2001) reviews the use of ^{10}Be for burial dating. Other review papers pertinent to the use of *in situ*-produced cosmogenic nuclides include early articles by Middleton and Klein (1987), Raisbeck and Yiou (1989), and Morris (1991), later works by Finkel and Suter (1993), Nishiizumi et al. (1993), Bierman (1994), Cerling and Craig (1994), Kurz and Brook (1994), Lal (1998), Zreda and Phillips (1998), Gosse and Phillips (2001), and a review of cosmogenic approaches limited to glacial studies provided by Fabel and Harbor (1999).

METHODS

The sample collection and analysis methods described below are designed to characterize accurately and precisely the ^{10}Be activity in quartz extracted from a rock sample and to allow interpretation of the data. Separation methods useful for olivine are presented by Seidl et al. (1997) and briefly in Nishiizumi (1990); for pyroxene preparation, consult Ivy-Ochs et al. (1998); for diamond, brief methods are provided by Lal et al. (1987).

Figure 1. Sample collection from a glacially rounded outcrop along trend with the type-Duval moraines on Baffin Island, above the hamlet of Pangnirtung. Quartz vein retains glacial polish and striations. Samples collected from vein (KM-95-40) and from rock surface (KM-95-39) give similar ages for ^{10}Be (22, 25 ky) and ^{26}Al (25, 26 ky) using Nishiizumi et al. (1989) production rates, respectively (Marsella et al. 2000).

Sample collection is typically accomplished with a hammer and chisel, with sample sites located using a Global Positioning System (GPS; Fig. 1). Sample thickness is noted in the field so that normalization for surface activity can be done during data reduction. Different applications demand different sampling approaches. For dating fans and moraines, top surfaces of the largest and best-preserved boulders are typically selected (e.g., Phillips et al. 1990). For determining rates of bedrock erosion, large open expanses of rock are usually sampled (e.g., Bierman and Caffee 2001). A detailed and useful treatise on the intricacies of sample collection is provided by Gosse and Phillips (2001).

Sample preparation

Sample preparation involves a series of chemical and physical processes designed to separate quartz from rock (Fig. 2) and extract Al and Be from the quartz (Figs. 3 and 4). The process described below is designed to separate reliably ^{10}Be and ^{26}Al from quartz

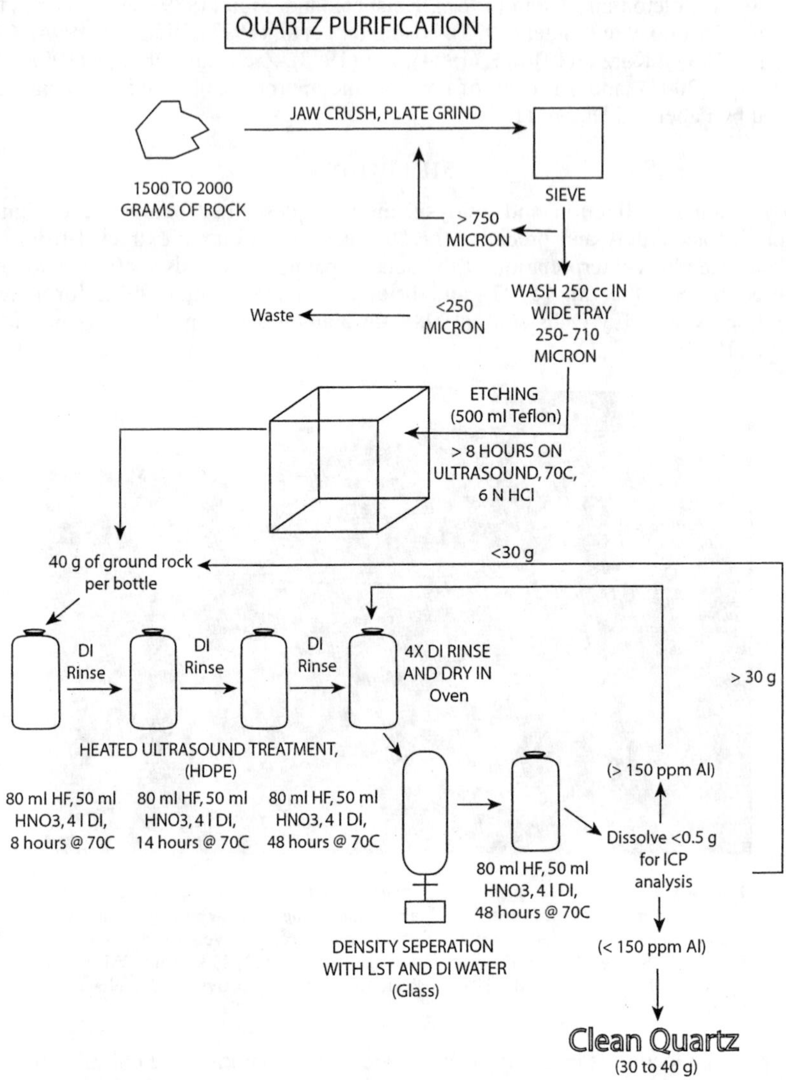

Figure 2. Diagrammatic representation of quartz extraction and purification process as currently performed at the University of Vermont.

(Figs. 3 and 4) and is the one currently followed at the University of Vermont; we have processed over 1500 samples using it with a target failure rate (beam current too low to measure) of less than 1%. Our process is adapted from those developed at the University of Pennsylvania (J. Klein, pers. comm.), Lawrence Livermore National Laboratory (LLNL; R. Finkel, pers. comm.), the Australian National University and University of Washington (J. Stone, pers. comm.), and Kohl and Nishiizumi (1992). While differing in details, it is similar overall to that used currently in most laboratories.

Only a few detailed chemical procedures have been published. Gosse and Phillips (2001) provide a schematic overview of processing procedures. An extensive description

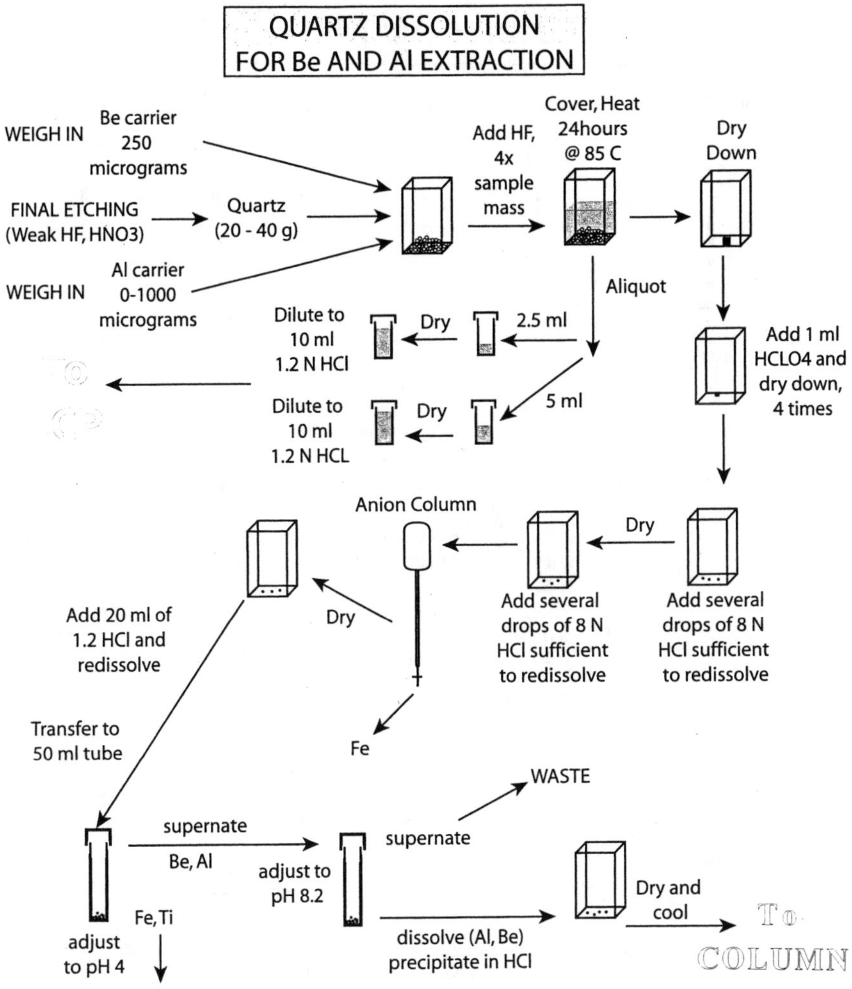

Figure 3. Diagrammatic representation of quartz dissolution and Be and Al purification process as currently performed at the University of Vermont.

of the procedure followed at ETH (Swiss Federal Institute of Technology, Zürich) is provided by Ivy-Ochs (1996). Descriptions and models of chemical processes useful for removing Be from silicate rocks and quartz using HF dissolution are provided in varying levels of detail by Arnold (1956), Nishiizumi et al. (1984, 1989), Tera et al. (1986), Brown et al. (1991), Kohl and Nishiizumi (1992), and Ochs and Ivy-Ochs (1997). Stone (1998) provides an alternative and more rapid method, relying on fluxes rather than HF; however, the flux method does not recover Al and thus does not allow for measurement of ^{26}Al, a useful and easily obtained compliment to ^{10}Be in most quartz samples.

Quartz isolation. Depending on AMS performance, chemical yield, and sample activity, reliable analyses for most terrestrial samples require 5 to >40 g of purified quartz, ideally containing <100 ppm of ^{27}Al, the stable isotope, and <500 ppm total

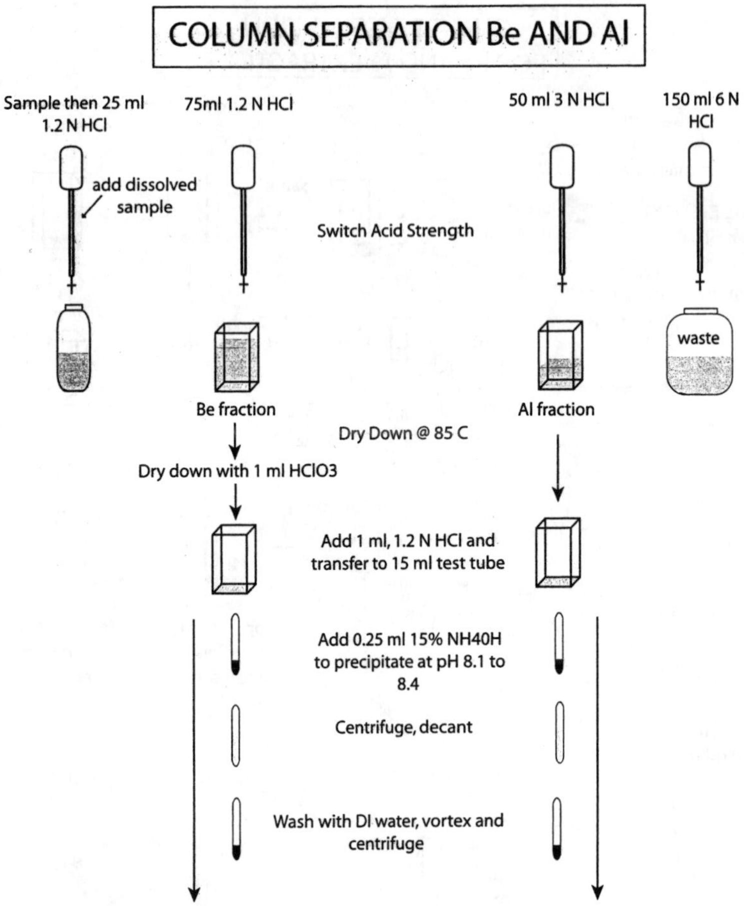

Figure 4. Diagrammatic representation of column separation process and hydroxide preparation for Al and Be as currently performed at the University of Vermont.

impurities. Samples containing higher levels of ^{27}Al will have lower $^{26}Al/^{27}Al$ ratios and thus be more difficult to measure by AMS requiring longer counting times and often resulting in lower precision (e.g., Shepard et al. 1995; Seidl et al. 1997). Samples containing higher levels of other cations, particularly Ti and Fe, are more difficult to process and usually result in lowered Be yields and thus lower beam currents and poorer counting statistics.

To achieve a high level of quartz purity, the rock is ground, sieved, and selective acid etching (Kohl and Nishiizumi 1992) is used to dissolve other minerals leaving the resistant quartz as a residue (Fig. 2). Some labs employ magnetic separation to remove mafic minerals and pyrophosphoric acid to assist in the removal of feldspars (Gosse and Phillips 2001). For most rocks, we retain the 500 to 800 μm size, although fine-grained rocks may require the retention of smaller grain sizes in order to isolate monomineralic quartz. Smaller grain sizes are more difficult to purify with density separation and more likely to react violently when dissolved in HF.

Figure 5. Mineral separation using density contrast (University of Vermont laboratory). In this feldspar-rich sample from Baffin Island, Canada, quartz has sunk and feldspars are floating.

After crushing, grinding, and sieving, several hundred grams of sample material are immersed in 6 N HCl, and etched for 6 to 24 hours in a heated ultrasonic bath in order to remove Fe and Al grain coatings (Kohl and Nishiizumi 1992). Rinsed and dried, 50 g aliquots of sample are added to water-washed 4 L plastic, wide-mouth bottles along with 4 L of 1% HF and 1% HNO_3. Depending on the quartz content of the original material, one to as many as 15 bottles may be processed to purify 40 g of quartz. The mixture is ultrasonically stirred with heat (75°C) for 8 hours, the solution washed out with deionized water, and the bottle refilled; treatment in the heated ultrasonic bath continues for another 14 hours. Once again the sample is washed, fresh acid is added, and treatment continued for an additional 24 hours. At this stage, most samples have been reduced to quartz and mafic minerals. A density separation, in a water-soluble lithium tungstate solution or any heavy liquid of density ~2.7 g cm^{-3}, removes minerals denser than quartz (Fig. 5); for extremely feldspar-rich samples, the density separation may be carried out before dilute acid etching to increase quartz yield. After density separation, the sample is again sonicated in 4 L of heated 1% HF/HNO_3 solution for an additional 48 hours.

Various analytic techniques can be used to determine the purity of each quartz mineral separate. A small (< 0.5 g) aliquot of quartz is dissolved in HF, dried, and diluted in HCl. We analyze the resulting solution using Inductively Couple Plasma Spectrometry Optical Emission (ICP-OE); others use ICP mass spectrometry (ICP-MS; Small et al. 1999) or Atomic Absorption (AA; Kohl and Nishiizumi 1992). Samples that contain high levels of Al or other cations are returned for an additional 48-hour etching after which Al levels are usually reduced. Greater than 80% of samples are cleaned to <150 ppm Al by four etches lasting a total of 96 hours; only quartzites and some sediment samples retain higher levels of Al despite extensive etching. If the quartz is sufficiently pure, the sample is etched in 1% HF/HNO_3 for another 6 hours in an acid-cleaned Teflon beaker, washed, and dried.

The sequential leaching process can effectively remove from quartz and probably olivine the ^{10}Be produced in the atmosphere and adhered to rock surfaces (Nishiizumi et al. 1989; Brown et al. 1991; Kohl and Nishiizumi 1992; Seidl et al. 1997). Experiments using a similar approach with pyroxene have not been as effective (Ivy-Ochs et al. 1998).

Separation of Be and Al. AMS analysis requires the preparation of purified Al_2O_3 and BeO targets. When processing samples, we work in batches of eight; each batch contains 6 or 7 samples and 1 or 2 full process blanks. Each batch moves through all steps of chemical processing and isotopic analysis together.

Purified quartz (usually 40 g) is covered with deionized water and dissolved in 48% HF (4 times the sample mass) in the presence of 250 μg of Be carrier and sufficient Al carrier to ensure that the total sample contains >3000 μg Al. While one can currently measure targets with less Al, beam currents are lower and such small targets may not last long enough to generate high precision analyses. We use commercially available ICP standards (SPEX brand) as our carriers. The Be carrier is required because most quartz mineral separates contain no Be, an assumption we verify by measuring Be in all samples we dissolve. After analyzing over 1500 samples of quartz, we have found less than half a dozen that contain significant amounts of Be. These quartz samples were extracted from Namibian and Himalayan granites and contain 2 to 37 ppm Be.

There is a complex trade off between the amount of carrier added and the precision, accuracy, and difficulty of making isotopic measurements. Adding additional carrier will result in higher beam currents and targets that allow extended analysis times before all the material is sputtered away for analysis by the Cs source (see extended discussion in Gosse and Phillips 2001); however, additional carrier lowers isotopic ratios and thus usually makes blank corrections more substantial. For most terrestrial samples, 250 mg of Be carrier gives reasonable beam currents (for example, 10 to 25 μA at LLNL) and allows three or four, five-minute analyses of each target. For Al, 3000 to 5000 μg samples typically generate 1.5 to 3 μA of beam current at LLNL and last for three to four, ten to fifteen-minute analyses.

Samples, contained in covered Teflon beakers and bathed in HF, are warmed slowly (over 24 hours) to just below 100 °C and then, without boiling, are allowed to dissolve for 24 hours (Fig. 6). When fully dissolved, two different-sized aliquots (2.5 and 5 ml) of warm HF solution are drawn from the beakers and dried off. These aliquots are brought

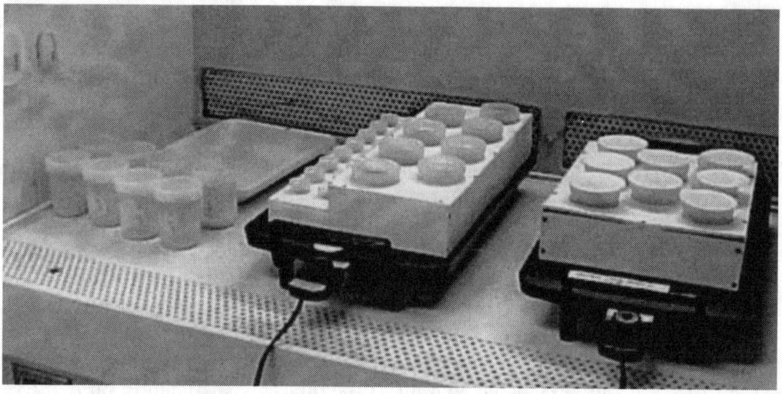

Figure 6. Three sets of samples in different stages of chemical processing. Samples in a fully exhausting laminar flow hood (University of Vermont laboratory). At left are seven samples and a blank waiting for HF. Samples are contained in Teflon 240 ml beakers with screw-on lids into which a small hole has been drilled to allow aliquots of samples for ICP analysis to be removed. In center, a batch of 8 samples are drying off after aliquots have been removed. On the right, another batch of samples is ready for anion columns. White Teflon beaker holder prevents spills and increases dry down speed by keeping sides of beakers warm. Teflon-coated, aluminum hotplates made by Presto.

up in 1.2 N HCl, heated with tight covers until they reflux aggressively, and used to quantify total Al and Be yield for each sample as well as Al and Be yield for the process blank(s). Solution chemistry is determined by ICP-OE with samples run in groups of four; each group is accompanied by an acid blank and two bracketing standards run as unknowns. We normalize the results using linear regression of the known, and the observed values of the standards and blank run as unknowns. Average, normalized Be yields for samples typically range from 97 to 102%. Al yields on the processing blanks typically range from 98 to 101%.

The remaining solution, 100 to 160 ml, is evaporated to dryness and fumed four times with $HClO_4$ at a hotplate temperature of 250°C to convert fluorides to perchlorates (Yokoyama et al. 1999). Following this, the sample is dried, then re-diluted with 8N HCl and then passed through an anion column (AG 1X8 resin 100-200 mesh) in 8N HCl to remove Fe. In order to remove Ti and any Fe that remains, the sample is precipitated at pH 4.0; the floc, which contains Ti and Fe, is discarded after centrifuging, and the remaining supernatant, containing Be and Al, is precipitated at pH 8.5. This second floc, which now contains the Be and Al, is brought up in 1.2 N HCl, dried down, re-dissolved, and placed on the cation column (AG 50WX8 resin 100-200 mesh, hydrogen form). Be is removed from the column in 1.2 N HCl and Al is removed using 3 N HCl (Fig. 7). Column chemistry can also be done using weak HF (Tera et al. 1986).

The resulting solutions are dried off, the Be fraction is fumed again with 1 ml of $HClO_4$, a step we and others (Tera et al. 1986) find necessary to reduce boron levels, and both the Al and Be are brought up in several ml of 1.2 N HCl. The pH is adjusted to 8.5 and Be and Al precipitate as gels. The gels are centrifuged, washed, transferred to acid-washed quartz crucibles, and dried. The crucibles are heated for one minute after they begin to incandesce over an open flame to convert the hydroxides to oxides. The resulting oxides of Be and Al are mixed with Nb and Ag powder, respectively, at a ratio of about 1 part oxide per volume to 2 parts metal. Mixing with metal helps to cool the oxides when

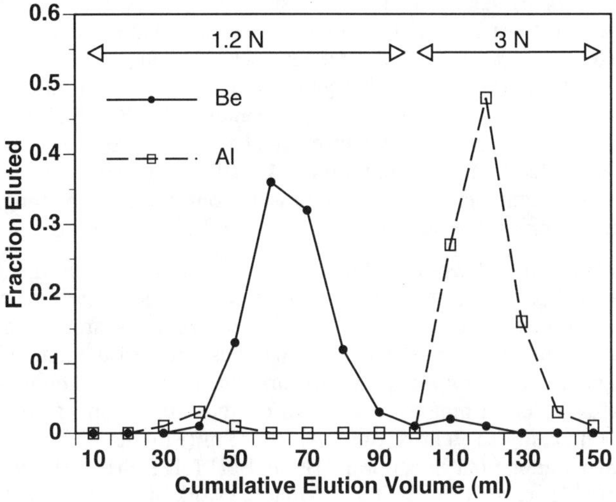

Figure 7. Example elution curve of Al and Be through cation column as chemistry is currently performed at the University of Vermont. Curve indicates concentration of these cations in acid passing through the column. Acid normality changed after 100 ml from 1.2 N HCl to 3 N HCl. Be and Al are almost fully resolved.

they are being sputtered in the source and results in higher beam currents. The oxide-metal mixtures are transferred along with steel and aluminum cathodes to a glove box where the samples are ground and packed with tampers into labeled targets.

Isotopic analysis of ^{10}Be

Beryllium-10 has a half-life of 1.5 $\times 10^6$ y (Yiou and Raisbeck 1972) and decays via β^- emission. The mode of decay, in conjunction with the long half-life, makes decay counting difficult, although not impossible. Accordingly, the widespread application of ^{10}Be to the Earth surface studies awaited the advent of AMS (Elmore and Phillips 1987). Since publication of this seminal paper, continually improving technology has allowed routine measurement of ^{10}Be produced *in situ* in surface rocks and soils (Finkel and Suter 1993; Tuniz et al. 1998).

History of measurement. The measurement of long-lived ^{10}Be has occurred in two epochs: one dominated by heroic decay-counting techniques, and a latter dominated by AMS. The limitations imposed on the application of ^{10}Be in the geosciences prior to the advent of AMS is perhaps best illustrated by an example. McCorkle et al. (1967) measured ^{10}Be by determining the beta activity in 200-year-old Greenland ice. The ^{10}Be activity was counted in a detector having a background of 0.042 count/min. This low background was achieved with the aid of a ring of anti-coincidence counters and a 15-cm thickness of iron shielding. The measurement required 1.2 $\times 10^6$ liters of water melted from the ice. The ^{10}Be activity of the glacial ice was 18.4 (+8.4, -4.8) 10^{-6} disintegrations per minute per liter of water (dpm l^{-1}). This activity corresponds to ~10^4 ^{10}Be atoms l^{-1}, suggesting that this particular measurement required ~10^{10} atoms of ^{10}Be. In contrast, such a measurement using AMS would today take four to five orders of magnitude fewer atoms and be far more precise. This example clearly illustrates that, for radionuclides having half-lives similar to ^{10}Be, counting atoms is far more efficient than counting decays.

The pioneering AMS measurements of ^{10}Be utilized a cyclotron (Raisbeck et al. 1978) and resulted in substantial increases in sensitivity. Raisbeck et al. demonstrated that ^{10}Be/^{9}Be ranging from 10^{-8} to 10^{-10} could be easily measured achieving a detection sensitivity of ~10^9 atoms. This early advance represented an improvement of three orders of magnitude—more would come. Subsequent work demonstrated the utility of using tandem (two-stage) accelerators for the measurement of ^{10}Be (Nelson et al. 1979; Turekian et al. 1979; Kilius et al. 1980; Kutschera et al. 1980). These early measurements further reduced the detectable ^{10}Be/^{9}Be ratio to <1 $\times 10^{-11}$. Turekian et al. (1979) reported a ^{10}Be count rate of 1 count per 100 sec in contrast to contemporary counting rates, which are 10 to 1000 times higher, the result of today's high efficiency ion sources.

Current AMS state-of-the-art. Berllyium-10 is measured in numerous AMS facilities throughout the world. Both high-voltage (>5MV) and low-voltage (2-3 MV) tandem accelerators are routinely used. AMS measurements are based on tandem accelerators that require injection of negative ions; thus, the measurement of any nuclide using AMS hinges critically on the ability to form such a negative ion efficiently. The entire AMS endeavor is, in many ways, based on the foundation of "ion sorcery" as defined by the University of Pennsylvania AMS group (Middleton 1990). While some early measurements attempted injection of Be as BeH (cf. Litherland 1980), all recent measurements of ^{10}Be inject Be as BeO, following the work of Turekian et al. (1979).

The measurement of ^{10}Be/^{9}Be requires the injection of both nuclides. Most laboratories employ rapid sequential injection of ^{10}Be and ^{9}Be. Rapid switching between isotopes is accomplished by changing the ion energies at the injection magnet using an electrostatic acceleration gap. This technique allows the measurement of both isotopes at least once per second and, because electrostatic fields switch much faster than magnetic fields, it is possible to switch at several hertz and still spend >95% of the time on the radioisotope measurement cycle. Because the output of the ion source can vary over time scales of seconds, the measurement of precise isotopic ratios hinges critically on the frequent measurement of the stable isotope.

The measurement of ^{10}Be by large accelerators has the advantage of yield; the peak in $^{10}Be^{3+}$ abundance (the preferred charge state for analysis) occurs at an accelerating voltage of about 8 MV (Tuniz et al. 1998). The higher terminal voltage also facilitates the separation of ^{10}Be from ^{10}B in a dE/dx (energy loss over distance) detector. An additional advantage in using $^{10}Be^{3+}$ is the option of placing a thin stripper foil at the object point of a magnetic analyzer. This foil strips all the $^{10}Be^{3+}$ to $^{10}Be^{4+}$ while stripping most of the $^{10}B^{3+}$ to $^{10}B^{5+}$; magnetic analysis separates the $^{10}Be^{4+}$ from the $^{10}B^{5+}$, further suppressing this interfering, isobaric contaminant.

Many laboratories have employed smaller accelerators to measure ^{10}Be. The terminal voltage attainable in these instruments ranges from 2 to 3 MV. Measurements at these energies typically utilize $^{10}Be^{2+}$. The reduction of the isobar ^{10}B is achieved by post-acceleration stripping of the $^{10}Be^{2+}$ to $^{10}Be^{3+}$. Although the ^{10}B is similarly stripped, it loses more energy than the ^{10}Be and is deflected by a magnetic analyzer placed after the stripper foil. While this technique is not as efficient in transporting ^{10}Be atoms to the detector as is measurement at higher energies, for high activity samples the precision of both approaches is comparable. An advantage of using the lower energy tandems is that less interfering ^{10}B is produced via nuclear reactions in the detector window.

Detection of Be and discrimination against B. The accurate detection of ^{10}Be encompasses two distinct goals: identification of ^{10}Be and rejection of ^{10}B. Both are accomplished using a gas ionization detector. An ion traveling through detector gas (of varying composition depending on the laboratory, usually P10, 90% argon and 10% methane) produces electron-ion pairs. The electrons and ions are separated by an electric field and the charge is collected on discrete anodes. The unequivocal identification of ^{10}Be requires measurement of dE/dx (energy loss over distance in the detector) and total E (energy of the incoming particle), which can be accomplished with a detector having two anodes. The ^{10}Be is stopped under the second anode so the sum of these two signals is the total energy. Alternatively, a detector having one stage of dE/dx and a solid state detector to measure total E is sufficient.

The rejection or reduction of ^{10}B is accomplished with a thick detector window in which the ^{10}B is stopped. The simplest scheme for this is to use Havar, a synthetic metal alloy foil. However, because such foils have non-uniform thickness, it is desirable to use several layers that together have the proper thickness, thereby lessening the non-uniformity in total thickness. A potential problem with this technique is that ^{7}Be ions are produced by the interaction of ^{10}B with hydrogen, $^{10}B(p,\alpha)^{7}Be$, in the Havar window. The interference from this reaction is most significant with higher energies. For those systems running ^{10}Be at 8 MV, this reaction occurs prodigiously; however the ^{7}Be and ^{10}Be are easily resolved for most samples. Some laboratories use an Ar cell for a detector window, substantially reducing this reaction. AMS systems based on tandems running at 2-3 MV are only minimally impacted by this reaction.

Backgrounds and sensitivities. AMS, in comparison to counting, has enabled a 6 to 7 orders of magnitude reduction in the isotopic ratio ($^{10}Be/^{9}Be$) that can be routinely measured. The techniques available to suppress interference from B, in conjunction with increased beam currents from better ion sources, have resulted in substantial increases in ^{10}Be detection sensitivity over the last decade. Ion sources are now capable of producing in excess of 25 μA of analyzed ^{9}Be. Presently, the best typical $^{10}Be/^{9}Be$ that can be measured in beryl blanks is 5×10^{-16}. These circumstances lead to ^{10}Be detection limits of approximately 5×10^4 atoms. It is likely that further improvements in sensitivity will occur; however, most of the ongoing measurements use far more atoms and have ratios far above the detection limit. The challenge for the next few years is to utilize fully this new sensitivity.

Standards and precision of measurement. All AMS measurements that form the basis of exposure age and erosion rate calculations are referenced to standards measured repeatedly over the course of a run in order to correct for drift. These standards include those prepared by AMS laboratories and by outside agencies (Middleton et al. 1993), typically from dilution of irradiated materials characterized by decay counting. Primary standards are usually of high enough activity that Poisson-determined counting statistics allow precision better than 1%. If unknowns are of sufficiently high activity, and enough counts can be obtained, internal errors (counting statistic-based limits) are often below 2% for ^{10}Be. External errors (multiple analyses of a single target) usually reproduce to a similar precision. In one case (Bierman and Caffee 2002), we re-measured ^{10}Be in 26 high-activity granite samples from Australia, preparing new quartz and targets 5 years later. The original data set had an average precision of 3.9%; the newly measured data set was, on average, slightly more precise (3.0%). The average difference between the new and old measurements was 2.8%, fully consistent within measurement uncertainty. The resulting r^2 value when comparing the two data sets was 0.995 implying excellent long-term reproducibility both in sample processing and in AMS measurements at LLNL.

Data reduction. AMS data are returned as ratios of $^{10}Be/^9Be$. Using the mass of sample quartz and the carrier addition, one calculates the activity of ^{10}Be in units of atoms (gram quartz)$^{-1}$. This activity is normalized for sample altitude, latitude, thickness, and exposure geometry so that samples collected at different locations can be compared. By convention, activities are normalized to rock surface values at sea level and high latitude. Nishiizumi et al. (1989), Dunne et al. (1999), and Gosse and Phillips (2001) provide methods for geometric correction. Lal (1991) and Dunai (2000) provide differing methods for altitude and latitude normalization that have proven controversial (Lifton 2000; Desilets et al. 2001; Dunai 2001). Lal (1991) is most often used.

Because process blanks, similar in size to the carrier added to each sample, are run with each batch of samples, the blank ratio is subtracted directly from the measured ratios to account for ^{10}Be in the carrier and that gained by the samples during processing. Many labs, including ours, use commercially available Be carrier with contains detectable amounts of ^{10}Be (Middleton et al. 1984). For example, the University of Vermont lab's long-term (since 1996) blank ratio ($^{10}Be/^9Be$) is $2.4\pm0.8\times10^{-14}$ (n = 172) measured in SPEX 1000 ppm carrier. It appears that an improvement in the separation of ^{10}Be from ^{10}B, made in 2000 by increasing the energy of accelerated particles at the Livermore facility, lowered our average blank ratio of $^{10}Be/^9Be$ by 30% from 2.7 ± 0.8 to $1.8\pm0.5\times10^{-14}$ (Fig. 8). Blank ratios are rarely published; however, ours appear similar to those determined on blanks at both LLNL and PRIME lab.

At present, beam currents at LLNL allow the ^{10}Be blank to be determined within 10% or better, thus reducing the impact of blank subtraction on the uncertainty of low-activity unknowns. In batches of extremely low-activity samples, we run two blanks and six samples in order to define better the blank subtraction and reduce the resultant uncertainty (Fig. 9); some other laboratories use carrier containing less ^{10}Be (e.g., Brown et al. 1991). Usually this carrier is made from beryl obtained from deep mines or from older stocks of Be which appear to have lower $^{10}Be/^9Be$ ratios (E. Brown, pers. comm.). Apparently, several decades ago, Be reagents were manufactured from different feed stocks.

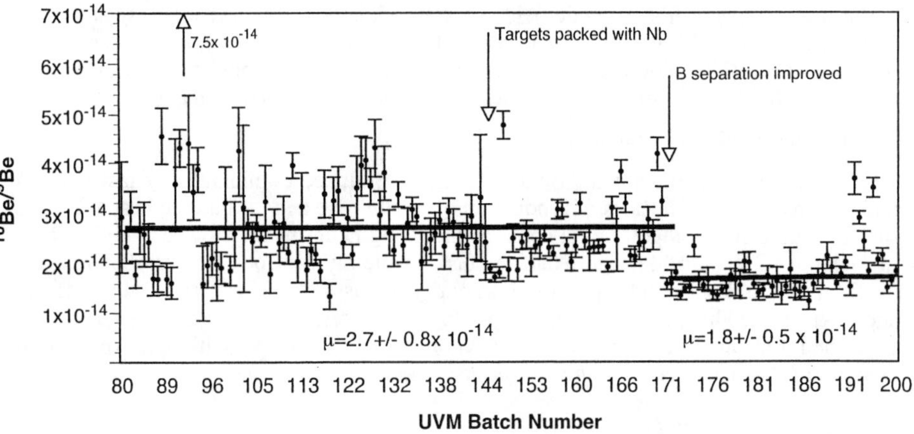

Figure 8. Be-carrier process blanks for the University of Vermont laboratory since September 1997, n = 172, measured at Lawrence Livermore National Laboratory. One blank that exceeds the range of the graph is shown by upward pointed arrow with blank value to right. Labels to right of downward vertical arrows indicate changes in target preparation and AMS configuration. Bold horizontal lines indicate average blank for period before and after boron separation improved. Uncertainties are 1σ counting statistics, including a correction for ^{10}B isobaric interference. Symbol μ is the average with one standard deviation.

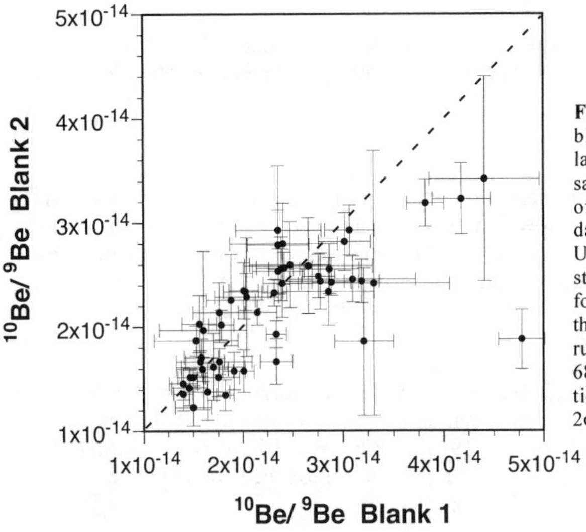

Figure 9. Scatter plot of process blanks (University of Vermont laboratory) run together in the same batch of samples; 53 pairs of blanks are included. Thick, dashed line is 1:1 correlation. Uncertainties are 1σ counting statistics, including a correction for ^{10}B isobaric interference. Of the 53 two-blank batches we have run, 41 pairs of blanks overlap at 68% confidence (77% of population) and 50 pairs overlap at 2σ (94% of population).

INTERPRETING NUCLIDE DATA

Nuclide activities, calculated from isotope ratios measured by AMS and corollary laboratory data, are interpreted using a variety of analytic and numerical models (Lal 1991; Gosse and Phillips 2001). These models are based on assumptions about exposure and irradiation history, many of which are unverifiable and thus introduce difficult-to-quantify uncertainty into the model results (Bierman and Gillespie 1991; Gillespie and Bierman 1995). Many assumptions are geologic (such as sample site history, including lack of burial, erosion, and inheritance of nuclides from a prior period of irradiation). Other uncertainties relate to radiation dosing over time and space, and include variation in the strength of Earth's magnetic field (which modulates cosmic ray intensity at Earth's surface; Guyodo and Valet 1996; Masarik et al. 2001), changes in the primary cosmic ray flux over time (Nishiizumi et al. 1980; Lal 1988), and the model used to normalize nuclide production over differing altitudes and latitudes (Lal 1991; Dunai 2000).

Depth – production relationship

The cosmogenic nuclides ^{10}Be and ^{26}Al are produced dominantly by fast neutron interactions (Lal 1988). Rates of production of *in situ* ^{10}Be and ^{26}Al are highest at Earth's surface and decrease rapidly with depth as incoming neutrons are slowed and absorbed. In rock (ρ = 2.7 g cm^{-3}), roughly half the fast cosmic ray neutrons are absorbed above a depth of 45 cm. For nuclides produced primarily by fast neutron spallation, such as ^{10}Be and ^{26}Al, the production rate (P_x) at depth (x) in a material of given density (ρ) can be described reasonably, over the first several meters of rock, with an exponential expression (Lal 1988) considering the surface production rate (P_0):

$$P_x = P_0\, e^{-(x\rho/\Lambda)} \tag{1}$$

The characteristic attenuation length (Λ) for fast neutrons is between 150 and 170 g cm^{-2} (see compiled table in Gosse and Phillips 2001) and has been shown to be similar at different locations on Earth's surface (Kurz 1986; Lal 1991; Brown et al. 1992; Sarda et al. 1993).

Model exposure ages

The model exposure age for a surface sample is calculated from the measured nuclide activity (N) and the decay constant (λ) of the nuclide if it is unstable (Lal 1988),

$$N = \frac{P}{\lambda}\left(1 - e^{-\lambda t}\right). \tag{2}$$

Such a model assumes that the nuclide production rate (P) is representative of the time-frame (t), latitude, and altitude over which the exposure occurred and that no erosion has occurred since initial exposure. In reality, magnetic field intensity fluctuations alter nuclide production rates in a complex pattern over time as a function of altitude, latitude, and duration of exposure (Clark et al. 1995; Gosse and Phillips 2001).

Model erosion rates

The activity of ^{10}Be in a surface sample can also be interpreted as a steady state erosion rate (ε). Such an approach has been considered in some detail (Lal and Peters 1967; Nishiizumi et al. 1986, 1991a; Lal 1987, 1988, 1991; Bierman and Turner 1995; Bierman and Caffee 2001). For the case of the spallation-produced nuclides, such as ^{10}Be, the following model (Lal 1991) is applicable:

$$N = \frac{P}{\varepsilon\rho\Lambda^{-1} + \lambda} \tag{3}$$

Implicit in the derivation and application of this equation is the assumption of steady or high-frequency, periodic erosion of slabs much thinner than the penetration depth of the cosmic radiation, as well as a constant rate of nuclide production. For rocks that disintegrate grain by grain, such assumptions are reasonable. However, for rock surfaces that lose mass in thick slabs, measured nuclide activities, and the model erosion rates derived from them, may over- or underestimate long-term erosion rates, depending on when the sampled surface last lost mass by episodic erosion (Lal 1991; Small et al. 1997).

Erosion after exposure

In many cases, samples are exposed instantaneously and then weather slowly; for example, lava flows erupt, weather, and erode (Shepard et al. 1995). Assuming a constant rate of nuclide production and steady erosion, Nishiizumi et al. (1991a) present the following model for such surface samples which are approaching, but have yet to reach, a steady state of nuclide activity:

$$N = \frac{P}{(\rho\varepsilon\Lambda^{-1} + \lambda)}(1 - e^{(\rho\varepsilon\Lambda^{-1} + \lambda)t}) \tag{4}$$

If the activity of nuclides having different half-lives has been measured in the same sample, it should, in principle, be possible to solve both for erosion rate (ε) and for the exposure duration (t). Considering the current uncertainties in production rates (~10%) and the precision with which nuclide activity can be measured in most samples (2-4%), solutions of this equation are relatively imprecise and of limited use in most geologic situations (Nishiizumi et al. 1991a; Gillespie and Bierman 1995).

Muons

The equations presented above (Eqns. 1-4) were derived considering the behavior of fast neutrons, which dominate the production of ^{10}Be and ^{26}Al near the Earth's surface at all latitudes and altitudes. Muons (small, charged particles that are much more weakly attenuated than fast neutrons) also produce ^{10}Be and ^{26}Al on and below Earth's surface (Lal 1987; Nishiizumi et al. 1989; Brown et al. 1995a); however, recent studies suggest that muons produce only 3% or less of total ^{10}Be nuclide activity in surface samples exposed at sea level and high latitude (Brown et al. 1995a; Braucher et al. 1998b; Granger and Smith 2000; Stone 2000). At higher altitudes, the muon contribution to total nuclide production in surface samples is even a smaller percentage.

In contrast to samples exposed at or near the surface, at depths below several meters of rock, muon production dominates, albeit at very low levels. Such production, parameterized most recently by Heisinger et al. (1997), Granger and Smith (2000), and Granger and Muzikar (2001), can be used to interpret measured nuclide activities in shielded samples as illustrated in a section to follow. However, for surface samples, muon contributions are usually ignored and the analytic equations reproduced above are typically applied. Failing to consider muons in modeling data from surface samples will result in the most significant errors for samples collected from low-elevation areas that are eroding rapidly. In samples collected from slowly eroding sites or samples collected from high elevation, production induced by spallation near the surface overwhelms that induced by muons at depth. However, error can be introduced for all types of samples if muon-induced production is incorrectly parameterized as part of the production rate calibration (Nishiizumi et al. 1989; Stone 2000).

Utilizing the $^{26}Al/^{10}Be$ ratio

Because the half-lives of ^{10}Be and ^{26}Al differ by a factor of about two (~1500 kyr and ~700 kyr, respectively), the ratio $^{26}Al/^{10}Be$ in long-lived samples will decrease over time from the production value of ~6 (Klein et al. 1986). There are two end-member cases leading to such ratio changes. If a sample has been exposed and production ceases abruptly (a once-exposed sample has been buried and is shielded from cosmic rays for $>10^5$ y), nuclide inventories are unsupported, ^{26}Al will decay more rapidly that ^{10}Be, and $^{26}Al/^{10}Be$ will drop below the production ratio (Bierman et al. 1999; Granger et al. 2001a). If the sample site is stable or only very slowly eroding, then nuclides produced early in the irradiation history or at depth are lost by radio-decay as the sample ages or approaches the surface and the $^{26}Al/^{10}Be$ ratio drops (Klein et al. 1986; Nishiizumi et al. 1991a). In either case, exposure and burial histories need to be on the order of hundreds of thousands of years before $^{26}Al/^{10}Be$ ratio changes can be detected reliably, considering current measurement uncertainty.

The two-nuclide plot (Fig. 10) is useful for visualizing such ratio changes over time (Klein et al. 1986; Lal 1991; Nishiizumi et al. 1991a; Bierman et al. 1999; Granger et al. 2001a). The plot has two bounding curves derived from the analytic solutions for the constant-exposure, no-erosion case (Eqn. 2) and for the steady-erosion case (Eqn. 3). Samples with simple exposure histories plot only within these bounds, whereas samples plotting below the lower bounding curve defined by the steady-erosion case either must have had complex histories (burial during or after exposure) or the isotopic measurements were affected by laboratory errors, the most common of which is incomplete recovery of stable Al leading to an underestimate of ^{26}Al activity (Fig. 10A). Samples plot above the constant exposure curve only as a result of laboratory errors, errors in altitude/latitude normalization protocols, or incorrect estimation of sample irradiation altitude.

For samples that plot below the bounding curves and for which analytical work has been done correctly, one can make calculations that constrain sample history (Fig. 10D). Although such solutions are non-unique, they are nevertheless useful for setting limits on sample history. For example, the shortest possible near-surface history for a sample plotting below the lower bounding curve includes an initial period of exposure (during which the sample's nuclide activity increases), followed by burial and complete shielding from cosmic rays (during which the nuclide activity and $^{26}Al/^{10}Be$ ratio diminish). Using iterative solutions, one can work back from measured nuclide activities to define a minimum burial time and either an initial exposure duration (Bierman et al. 1999) or an initial erosion rate (Granger et al. 2001a).

The location of the bounding curves is dictated by assumed nuclide production rates. For example, using the currently accepted, high-latitude, sea-level integrated production estimate of 5.2 atoms g^{-1} y^{-1} for ^{10}Be (Bierman et al. 1996; Stone et al. 1998; Stone 2000; Gosse and Stone 2001) rather than the previously accepted production rate estimates of Nishiizumi et al. (1989), shifts the bounding curves to the left. Similarly, if a sample site were buried during its dosing history by shallow soil cover, production rates would effectively be reduced proportional to cover thickness and density (Eqn. 1) and the bounding curves would also shift to the left (see Fig. 3C in Bierman et al. 1999). Sample position on the two-nuclide diagram, and thus interpretation of sample history, are also affected by the choice of elevation/latitude normalization (Lal 1991; Dunai 2000).

Nuclide production rates

Any exposure age or erosion rate based on measured ^{10}Be activity requires an estimate of the rate at which the nuclide is produced at the sampling site (see Gosse and Phillips 2001). Production rate estimates for ^{10}Be and other nuclides have been made in three different ways (Nishiizumi 1996):

Figure 10 (opposite page). Explanation of two-nuclide diagram for ^{26}Al and ^{10}Be using production rates of Nishiizumi et al. (1989). Left column figures are two-nuclide plots. Right column figures are evolution of nuclide activity and $^{26}Al/^{10}Be$ ratio over time. Dashed lines in (C) and (D) represent nuclide activity for the no erosion, constant exposure case (B). SL, 60° indicates that nuclide data have been normalized to sea level at high latitude. (A) Regions of the diagram are defined by the analytical equations for constant-exposure, no-erosion case that provides the upper bound and the steady-erosion case which provides the lower bound. (B) The constant-exposure case results in nuclide activities that increase over time as $^{26}Al/^{10}Be$ ratio decreases. (C) Samples eroding steadily

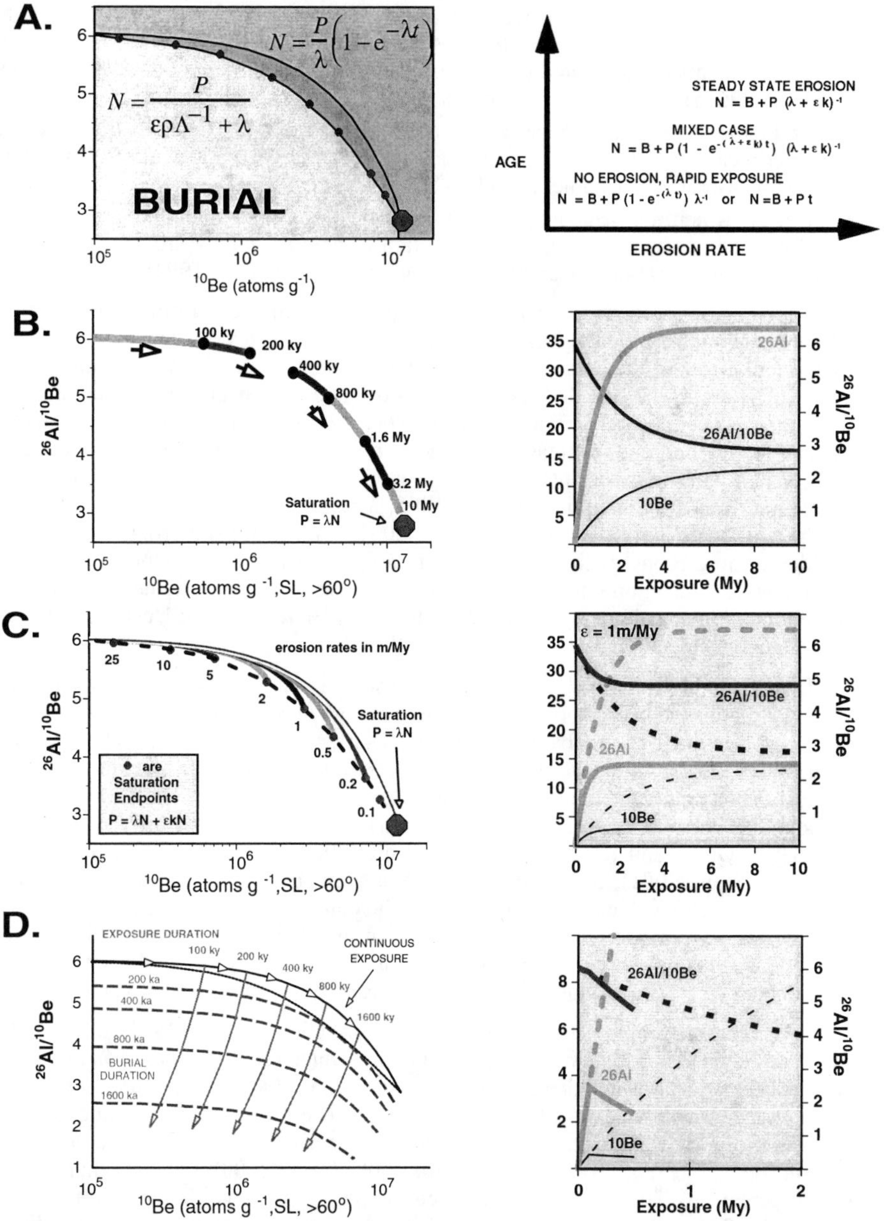

begin on a trajectory that is similar to those that are exposed and not eroding, but fall off that trajectory to an endpoint (filled circle) where nuclide production is equal to loss by decay and shedding of mass from the eroding surface. (D) Samples that plot below the curves defined in (B) and (C) have a history of burial either after or during exposure. The shortest total history consistent with a sample showing nuclide evidence for burial includes an initial exposure period followed by a period of burial. Dashed lines are burial isochrons. Lines with arrowheads are time/burial trajectories (from different initial exposure/erosion conditions) that describe the evolution of a sample's ^{10}Be activity and ^{26}Al/^{10}Be ratio as burial time increases (see Bierman et al. 1999 and Granger and Muzikar 2001).

1. Long-term, integrated estimates are made by measuring nuclide activity in rock, the exposure age of which is typically constrained by other dating methods including ^{14}C measurements of associated organic material (e.g., Nishiizumi et al. 1989; Bierman et al. 1996; Stone et al. 1998; Kubik et al. 1998).
2. Production rates over the short term are measured directly (such as the water bed experiments of Nishiizumi (1996) and Brown et al. (1996)) or calculated from short-term measurements of other nuclides (Yokoyama et al. 1977).
3. Models of neutron behavior in the atmosphere along with quantitative descriptions of nuclide production in rock as a function of neutron energy are used to estimate production rates (Masarik and Reedy 1994, 1995; Reedy et al. 1994).

Each of the methods described above provides different information and has different uncertainties. Geologically determined production rates are subject to systematic and random errors in the assumed calibration age (Clark et al. 1995), uncertainty in the exposure history of the sampled rocks in terms of burial or erosion, and the potential for inheritance of nuclides from prior periods of exposure (Bierman et al. 1998; Colgan et al. in press). Production rates estimated over short time frames reflect time-specific magnetic field intensity and must be interpreted with models before they can be generalized to longer time frames (Clapp and Bierman 1995; Masarik et al. 2001); such modeling is fraught with uncertainty (Goose and Phillips 2001). Model-derived production rates are considered more robust if they can be verified by empirical measurements. No matter how a nuclide production rate has been measured, there remain uncertainties in scaling as a function of time, latitude, and altitude (Clapp and Bierman 1995; Desilets and Zreda 2000; Dunai 2000; Lifton 2000; Gosse and Phillips 2001).

The most widely cited ^{10}Be production rate study (Nishiizumi et al. 1989) is founded on glacially polished bedrock outcrops exposed at high elevation (~3000 m) and relies on calibration ages that are probably too young (Clark et al. 1995). Using more recently acquired radiocarbon ages suggests that the age of the calibration sites was underestimated, and thus, the production rate derived from these sites is an overestimate by about 20%. Now estimated at several locations and over different time frames (Larsen et al. 1995; Gosse and Klein 1996; Stone et al. 1998; Klein et al. 2000; Stone 2000; Gosse and Stone 2001), integrated and instantaneous production rates of ^{10}Be, when normalized to sea level and high latitude as is the convention, appear to have converged on ~5.2 atoms (g quartz)$^{-1}$ y^{-1}. There remains an uncertainty of perhaps 10%. Even at such low annual production rates, with careful laboratory work, AMS sensitivity is sufficient to detect ^{10}Be produced in quartz over less than a millennium at sea level.

Most workers employ the altitude/latitude corrections of Lal (1991) to normalize data to sea-level exposure at high latitude, although Dunai (2000) proposed an alternative scaling scheme about which there has been considerable debate (Desilets and Zreda 2000; Lifton 2000; Desilets et al. 2001; Dunai 2001). The contribution of muons to the production of ^{10}Be remains uncertain, but much recent work (e.g., Brown et al. 1995a; Braucher et al. 1998b; Granger and Smith 2000; Stone 2000) suggests that at most several percent of ^{10}Be produced in surface samples at sea level is the result of muon interactions. The wide choice of approaches for dealing with muons, as well as altitude, latitude, and geomagnetic field strength scaling, allows for significantly different production rate parameterizations to be derived from the same data set.

New analyses presented in this paper are interpreted using a sea-level, high-latitude production rate for ^{10}Be of 5.17 atoms (g quartz)$^{-1}$ y^{-1} (Bierman et al. 1996; Stone 2000) scaled with Lal's (1991) polynomials for neutrons only, assuming that for samples exposed at Earth's surface, the muon contribution to the measured activity is inconsequential. For deeply shielded samples, we explicitly consider muon-induced production using the approximation of Granger and Smith (2000). We caution the reader that most (but not all) of the papers we cite used the original sea-level, high-latitude production rate estimate of 6.03 atoms (g quartz)$^{-1}$ y^{-1} made by Nishiizumi et al. (1989). Many of these papers also used the scaling of Lal (1991, his Table 1), which includes a substantial contribution from muons.

ILLUSTRATIVE CASE STUDIES

To demonstrate the power of *in situ*-produced ^{10}Be for characterizing the behavior of Earth's surface over time and space, we present a series of examples, each of which illustrates a different application or method of interpreting nuclide data. We have selected examples that illustrate the dynamic range of problems in surficial geology that can be approached using measurements of ^{10}Be produced in rock including studies of mid-Holocene surfaces and of surfaces little changed since the Tertiary. For each example, we present cosmogenic data gathered by our selves or by our co-workers. Some of these data have been published previously, in which case we reference the original publication. Data published here for the first time are presented in tables.

The five examples we present include: the use of nuclides for dating landforms including moraines and alluvial fans, constraining the magnitude and distribution of glacial bedrock erosion, understanding clast histories, estimating bedrock erosion rates, and understanding fluvial dynamics, specifically the incision of rivers into rock. For each example, we cite and/or discuss a variety of relevant or related studies that also use *in situ*-produced ^{10}Be. The measurement of ^{10}Be and ^{26}Al in fluvial, hillslope, and cave sediments is another geologically important use of these nuclides. Because Bierman et al. (2001a) and Granger and Muzikar (2001) have recently reviewed the measurement and application of ^{10}Be in quartz-bearing sediments; such use is not reviewed here in any detail.

Dating landforms

In situ-produced cosmogenic nuclides have allowed numerical dating of landforms (including moraines and alluvial fans) that for many years were dated only by relative means because they lacked organic carbon for ^{14}C analysis or well-bleached materials for luminescence dating. Indeed, the refereed literature related to *in situ*-produced ^{10}Be is dominated by glacial dating studies (we count more than 30; there would be many more if ^{3}He and ^{36}Cl were included). Other frequently dated landforms include alluvial fans and fluvial terraces. The paucity of quartz in volcanic rocks and the difficulty of extracting ^{10}Be from olivine have limited this nuclide's use in eruption dating to one study (Shepard et al. 1995).

Most glacial dating using ^{10}Be has been directed at boulders exposed on glacial moraines; only a few studies have examined glacially eroded bedrock. The early work of Phillips et al. (1990) using ^{36}Cl in the Sierra Nevada of California, USA, quickly lead to a series of studies using ^{10}Be to date boulders including other Sierra Nevada moraines (Nishiizumi et al. 1993). Many of the early studies were done in Antarctica and demonstrated the extreme antiquity (millions of years) of glacial deposits there (Brown et al. 1991; Brook et al. 1993, 1995a,b; Ivy-Ochs et al. 1995, 1997; Summerfield et al. 1999). Corollary studies used ^{10}Be in exposed Antarctic bedrock (Brown et al. 1992) and offer some constraint on rock erosion rates in this polar desert (dm Ma^{-1}), the upper limit of ice coverage, and thus the stability of regional ice sheets (Nishiizumi et al. 1991a). Recent reanalysis of production rate data by Stone (2000) suggests that these Antarctic surfaces and boulders may not be quite as long-lived as originally inferred from the very high measured ^{10}Be and ^{26}Al activities.

Studies of glacial moraines and boulders using ^{10}Be have continued in less extreme environments. Dating of moraines in western North America (Gosse et al. 1995a,b; Chadwick et al. 1997; Phillips et al. 1997; Licciardi et al. 2001) identified glacial deposits from both the latest glacial maximum and the preceding advance. Dating using ^{10}Be has begun on European glacial sequences (Ivy-Ochs et al. 1996; Brook et al. 1996; Stone et al. 1998; Tschudi et al. 2000; Stroeven at al. 2002) and in the Himalayas (Phillips et al. 2000; Owen et al. 2002). There has been a significant amount of ^{10}Be data collected in the Canadian arctic, particularly on Baffin Island (Marsella 1998; Steig et al. 1998; Davis et al. 1999; Kaplan 1999; Bierman et al. 1999; Marsella et al. 2000; Wolfe et al. 2001; Kaplan et al. 2001; Bierman et al. 2001b; Miller et al. 2002). Only limited ^{10}Be data have been published from the southern hemisphere (Ivy-Ochs et al. 1999; Barrows et al. 2001, 2002), eastern North America (Clark et al. 1995; Bierman et al. 2000; Balco et al. 2001), and central North America (Gosse et al. 1997; Bierman et al. 1999).

Alluvial and debris fans have become another favorite target for study using *in situ*-produced ^{10}Be. Most of these fan studies have been linked to tectonics with the dating of offset fan surfaces allowing authors to estimate fault slip rates. Asian fans have been dated (Ritz et al. 1995; Brown et al. 1998), as have those in southwestern North America (Nishiizumi et al. 1993; Bierman et al. 1995b; Zehfuss et al. 2001). There are two studies from South America which report exceptionally long exposure ages for some Argentinean and Chilean samples (Siame et al. 1997; Nishiizumi et al. 1998). In a related and unique study, Jackson et al. (2002) use exposure ages of residual silica-cemented boulders to estimate rates of deformation on an eroding anticline in New Zealand. A similar approach for estimating tectonic rates has been taken with offset fluvial terraces (e.g., van der Woerd et al. 1998, 2000); see discussion in a following section.

Some glacial dating studies have attempted to correlate increasingly precise cosmogenic ages (Gosse et al. 1995a; Ivy-Ochs et al. 1999), either with climate proxy data generated by other means or with nuclide measurements made on samples collected from differing altitudes and latitudes. All such studies are hampered by the uncertainty in nuclide production rates over space and time (Clark et al. 1995; Stone 2000; Dunai 2000) and by the variance in nuclide activities, and thus calculated sample ages, on single landforms (Gosse et al. 1995b). Such variance may be caused by inheritance that differs from boulder to boulder or outcrop to outcrop (Colgan et al., in press), by time-transgressive deposition on a single landform (Gosse et al. 1995b), or by the effects of boulder-specific rates of surface weathering and mass loss (Bierman and Gillespie 1991; Hallet and Putkonen 1994; Zimmerman et al. 1994).

Using cosmogenic ^{10}Be and ^{26}Al to identify reliably short-lived climate events such as the Younger Dryas climate oscillation (Gosse et al. 1995a), or to date accurately very old landforms, such as the >200 ka Mono Basin moraines (Phillips et al. 1990), will remain difficult. The accuracy of cosmogenic age estimates on young landforms is diminished by AMS measurement difficulties in low-activity samples, significant blank corrections, and time-varying Holocene nuclide production rates as well as the potential for small amounts of inherited nuclides to lever significantly, in percentage terms, model exposure ages. For older landforms (>50 to 100 ka?), such problems are usually of little consequence but the dynamic nature of Earth's surface means that exposure histories are increasingly difficult to quantify as rocks, and the landforms on which they sit, erode (Bierman and Gillespie 1991; Hallet and Putkonen 1994; Zimmerman et al. 1994).

In all fan and moraine studies, inheritance of nuclides from periods of exposure prior to emplacement may inflate model ages (Briner and Swanson 1998; Colgan et al. in press) whereas erosion of sampled surfaces will reduce apparent ages; both effects are difficult to detect and to compensate for accurately. The effect on modeled ages of both landform and boulder lowering can be significant (Hallet and Putkonen 1994). There are situations where, despite the uncertainties described above, cosmogenic nuclide analysis can disprove hypotheses conclusively. Such situations typically involve a choice between two significantly different ages for a deposit. The type Duval and associated moraines on Baffin Island, in the eastern Canadian arctic, provide just such an optimal situation. Below we describe applications of ^{10}Be to this geologic setting.

Figure 11. Baffin Island, Canada. Mt. Duval with type-Duval moraines at top of ridge in foreground to right. Smaller recessional moraines on slope below. Deeply incised drainage supplies material to alluvial fan shown in Figure 12.

Baffin Island geologic setting. On southeastern Baffin Island, along Pangnirtung and adjacent fiords, are discontinuous moraine segments that can be traced along the upper fiord walls for many kilometers (Fig. 11 and Dyke 1977). These Duval moraines were thought to mark the greatest extent of late Pleistocene ice (Dyke 1979; Dyke et al. 1982). Above the moraines are heavily weathered bedrock terrains that were believed either to have been glaciated much earlier in the Pleistocene or were never ice-covered, remaining as ice-free refugia where populations of plants and some animals survived glaciation. Near the hamlet of Pangnirtung, the moraines are deeply incised by a stream and the material removed from the moraines has been deposited in snouts and levees on bouldery debris fans on the valley bottom (Fig. 12). Below the Duval moraines, but above the fan deposits, are a series of smaller, recessional moraines.

Figure 12. Baffin Island, Canada. Bouldery fan, near Pangnirtung hamlet (in background), where three samples for cosmogenic nuclide analysis were collected (Table 1). Kim Marsella (165 cm. tall) for scale.

Until the application of ^{10}Be and ^{26}Al (Marsella 1998; Marsella et al. 2000), none of the moraines or fans had ever been directly dated, but a glacio-marine delta that was fed by meltwater from the type-Duval moraines was dated to >60 ka by the amino acid racemization technique applied to a shell fragment (Dyke 1977, 1979). This >60 ka age was correlated to other sites on Baffin Island where radiocarbon dating and amino acid analysis of shells suggested that glacial ice was most advanced prior to the last glacial maximum (~21 ka, LGM) of the Laurentide ice sheet (Loken 1966; Miller and Dyke

1974; Miller et al. 1977). Expansion of Laurentide ice was believed to have progressively starved Baffin Island of moisture causing the glaciers to shrink back from their maximum positions after 60 ka. The existing paradigm thus clearly predicted that the age of the Duval moraines should be ≥60 ka and that the recessional moraines and alluvial fan deposits should be younger.

Baffin Island study methods. In order to understand the glacial and postglacial history of the area around Pangnirtung Fiord, we collected numerous samples from boulders and bedrock on the highlands above the Duval moraines (Bierman et al. 1999, 2001b), boulders and polished bedrock on and adjacent to the type-Duval moraines and Duval moraine equivalents (Marsella et al. 2000), boulders and bedrock on and near the associated glacio-marine delta (Marsella et al. 2000), boulders on the recessional moraines (Marsella et al. 2000), and boulders on the alluvial fan (Table 1). We extracted and purified quartz and measured ^{10}Be and ^{26}Al activity in all the samples (Marsella 1998). In this polar desert, erosion of rock surfaces was likely inconsequential and thus had little effect on measured ages. Boulders were only lightly weathered. Some bedrock outcrops still retained striations.

Results #1—LGM glaciers were larger than previously believed. Cosmogenic model ages for boulders on, and polished bedrock associated with, the Duval moraines indicate unambiguously that the moraines are much younger than 60 ka (Fig. 13 and Marsella et al. 2000). Boulders on the Duval moraines have two age modes, one centered at about 11 ka, the other at about 22 ka (Fig. 13A) using ^{10}Be production rates of Nishiizumi et al. (1989). The recessional moraines below the Duval moraines (R1 and R2) and the boulders on the glacio-marine delta fed by Duval moraine melt-water have cosmogenic model ages indistinguishable from the younger Duval mode (Fig. 13 A-D). These data, despite uncertainties in nuclide production rates and sample exposure histories, show conclusively that ice filled Pangnirtung Fiord as recently as 11,000 years ago. Either the >60 ka shell amino acid date is flawed or more probably the shell fragment was transported and redeposited by ice during the last glacial maximum.

Results #2—Alluvial fans are probably neoglacial age. Cosmogenic nuclide analysis demonstrates that the alluvial fan deposits that we sampled are much younger than expected, Neoglacial (the past several thousand years) rather than early Holocene (nearly 10,000 years ago; Table 1). All three samples have both ^{10}Be and ^{26}Al model ages that are statistically indistinguishable. The average age for the deposit, considering all six ^{26}Al and ^{10}Be data points and the ^{10}Be production rates of Bierman et al. (1996), is 5.1±0.6 ka; the average age would be about 800 years younger using the production rates of Nishiizumi et al. (1989). It appears that major deposition occurred on the fan when Neoglaciation, a mid to late Holocene cooling, began. This age is indistinguishable from the estimate made by Davis et al. (1999) for the onset of Neoglaciation in the Pangnirtung Pass area (5.4±2.0 ka), also based on cosmogenic nuclide data but recalculated using ^{10}Be production rates of Bierman et al. (1996). The alluvial fan data agree with earlier estimates for the onset of Neoglaciation on Baffin island (Andrews and Barnett 1979; Davis 1980; Miller 1973; Davis and Jacobson 1985), lichenometric ages of debris flows (Davis and Fabel 1988), radiocarbon ages of debris avalanches (Davis and Kihl 1985), and pollen data (Davis 1980; Short et al. 1985).

Model fan boulder ages calculated from ^{26}Al activities are, on average, slightly older and significantly more precise than ages calculated from ^{10}Be activities. These differences reflect uncertainties propagated for significant blank and boron corrections made to the ^{10}Be measurements. Nevertheless, these fan boulder data demonstrate that one can make meaningful measurements of samples exposed for only a few thousand years at sea level and high latitude. Dramatic improvements in Be beam currents since

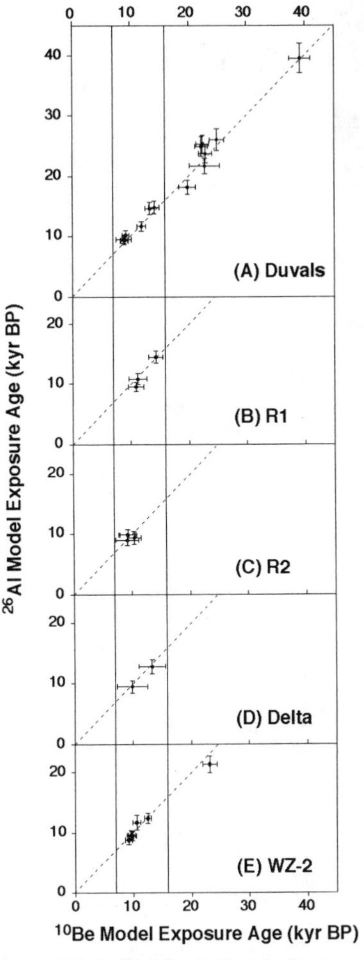

Figure 13. Baffin Island, Canada. Scatter plots of ^{26}Al and ^{10}Be model exposure ages (ka BP, production rates of Nishiizumi et al. 1989) for (A) type-Duval and Duval-equivalent moraines, including two bedrock samples, (B) first recessional moraine below type-Duval moraine, R1, (C) second recessional moraine below Duval moraine, R2, (D) 99-m raised glacio-marine delta fed by meltwater from type-Duval moraine position, and (E) WZ-2 (above and outside Duval moraines). Error bars represent a production rate uncertainty of 3% (for elevation) propagated along with analytical uncertainties. Bracketed area represents latest deposition on Duval moraines, recessional moraines, and 99-m delta. Dashed lines represent 1:1 ratio of ^{26}Al and ^{10}Be model ages. See Marsella et al. (2000) for further explanation. [Used by permission of the editor of *Geomorphology*, from Bierman et al. (2001b), Fig. 2, p. 260.]

Table 1. Cosmogenic nuclide data, Pangnirtung alluvial fan, Baffin Island, Canada

Sample[1]	^{10}Be atoms g^{-1} quartz	^{26}Al atoms g^{-1} quartz	$^{26}Al/^{10}Be$	^{10}Be model age[2] ky	^{26}Al model age[2] ky
KM-95-118	21000 ± 10000	161000 ± 20000	7.80 ± 3.70	4.0 ± 1.9	5.3 ± 0.8
KM-95-119	24000 ± 9000	168000 ± 18000	7.10 ± 2.70	4.6 ± 1.7	5.4 ± 0.8
KM-95-120	28000 ± 10000	167000 ± 26000	6.00 ± 2.20	5.6 ± 2.0	5.5 ± 1.0
			average	4.7 ± 0.8	5.4 ± 0.1

Notes [1] Latitude 66° N; elevation 16 m

[2] Model ages calculated assuming production rates of Bierman et al. (1996) and altitude/latitude correction factors of Lal (1991) for neutrons only; 10% 1σ production rate uncertainty fully propagated along with measurement uncertainty and stable Al (4%) and stable Be (2%)

Figure 14. Weathered gneissic tor (weathered, rounded, rock outcrop) on upper elevation surface above Pangnirtung hamlet, Baffin Island. C. Massey (180 cm tall) for scale.

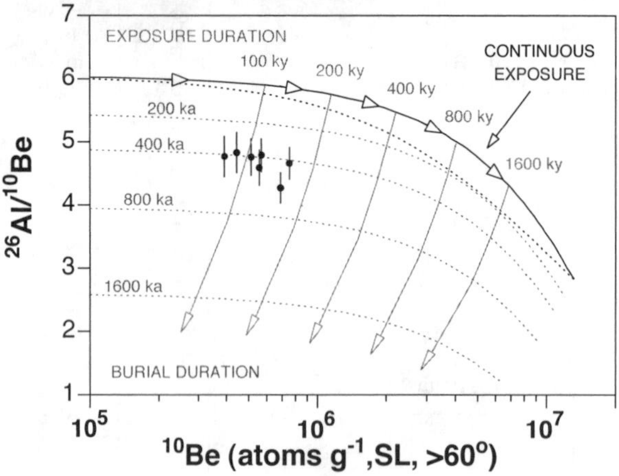

Figure 15. Two-nuclide diagram to explain burial history inferred from nuclide activities. Error bars are 1σ total analytic uncertainty. Black symbols represent Baffin Island, upper elevation bedrock samples. Surface exposure followed by full shielding represents minimum total sample history. Samples initially exposed at the surface track along the constant exposure line for various exposure times. After burial, sample trajectory is toward lower left, crossing burial isochrons. These samples plot such that they are consistent with 75 to 175 ka of exposure followed by ~400 ka of burial. Figure, modified from Bierman et al. (1999), uses the production rates of Nishiizumi et al. (1989).

these samples were measured in 1996 have improved sensitivity by nearly a factor of 10 and dramatically reduced the uncertainty in blank corrections (Fig. 8).

Results #3—Ice overrode uppermost surface without erosion. Upper elevation bedrock surfaces not only appear more weathered (Fig. 14), but also have far greater activities of ^{26}Al and ^{10}Be than fiord-bottom bedrock surfaces, a trend reported elsewhere in far northern latitudes (Gosse et al. 1993; Brook et al. 1996; Stone et al. 1998; Steig et al. 1998). The ^{26}Al/^{10}Be ratio in all Baffin Island upper elevation bedrock samples is significantly below 6, implying, when analyzed on a two-isotope plot, a complex exposure history (Fig. 15) interpreted as intermittent or continuous burial by ice or snow with little erosion (Bierman et al. 1999, 2001b). A variety of models suggest that upper elevation surfaces are extremely stable, with effective near-surface exposure times near and above half a million years (Bierman et al. 1999, 2001b). Erosion rates for these upper

Figure 16. Sample sites on the glaciated highlands above Pangnirtung Fiord, Baffin Island. (A) Erratic, subangular boulder (sample KM-95-111, average ^{26}Al and ^{10}Be exposure age 9.4 ka using production rates of Nishiizumi et al. 1989) sitting on weathered bedrock surface after transport by glacier. Sample site on weathered bedrock tor (KM-95-110, total history > 700 ky) labeled in background. (B) Different view of same samples emphasizing difference in rounding of edges between the tor and the erratic. [Used by permission of the editor of *Geomorphology*, from Bierman et al. (2001b), Fig. 1, p. 259.]

elevation bedrock samples are at most a few meters per million years based on the very high measured nuclide activities.

In contrast, an erratic boulder sitting directly on the weathered and highly dosed bedrock has concordant ^{10}Be and ^{26}Al exposure ages of 9.2±0.5 and 9.5±0.9 ka, respectively (using ^{10}Be production rates of Nishiizumi et al. (1989); Fig. 16). The young age and simple exposure history of the erratic mandate that ice was present on the uplands far beyond the Duval limits in the latest Pleistocene or earliest Holocene. Similarly young ages for samples just outside the Duval moraines (WZ-2, Fig. 13E) support this conclusion that ice cover was widespread until much later in the glacial cycle support this conclusion that ice cover was widespread until much later in the glacial cycle than previously believed. The combination of young, latest glacial erratics overlying old weathered rock mandates that cold-based, non-erosive ice covered at least some highlands on Baffin Island during the latest Pleistocene. Analysis of ^{10}Be from Fennoscandia shows similarly preserved ancient landscapes which must have survived beneath young, non-erosive ice sheets, presumably because such ice was frozen to the bed (Stroeven et al. 2002).

Constraining the magnitude of bedrock erosion by glaciers, including the Laurentide Ice Sheet, North America

Ice sheets repeatedly covered large areas of the northern hemisphere continents during the Pleistocene (Flint 1971). The ice removed much of the saprolite that once covered the glaciated areas; however, isolated pockets of weathered rock were left behind, suggesting that the efficacy of glacial erosion varied spatially in New England, USA, arctic Canada, and elsewhere (Goldthwait and Kruger 1938; Davis et al. 1993; Kleman 1994; Klemen and Stroeven 1997). Understanding the magnitude of glacial erosion is important for deciphering landscape history, as well as the interactions between mountain building, erosion, and geochemical cycles (e.g., Raymo and Ruddiman 1988) as related to glaciation. Furthermore, most cosmogenic dating studies assume that the inheritance of nuclides from prior periods of exposure is negligible; testing this assumption is important to ensure the veracity of calculated model ages.

Previously published data germane to understanding nuclide inheritance include samples from boulders riding on Antarctic ice giving an effective ^{3}He age of 9000 y BP (Brook and Kurz 1993), boulders buried >10 m in a Sierra Nevada, California, USA moraine having effective ^{36}Cl ages of only a few thousand years (Bierman 1993; Bierman et al. 1995a), and the extensive and significant inheritance detected in outcropping Puget Lowland, Washington, USA, bedrock (Briner and Swanson 1998). Nishiizumi et al. (1989) also found significant inheritance in bedrock samples collected at June Lake, Sierra Nevada, USA, in or near the ablation zone of a former alpine glacier.

Study methods for glacial erosion and nuclide inheritance. To evaluate the importance of nuclide inheritance, once can consider the analyses of quartz-bearing samples from 2 different sites covered by ice during the last advance of the Laurentide Ice Sheet ~21,500 years ago in Wisconsin and New Jersey, USA, and one site in Baffin Island, Canada covered by valley glaciers since about 4500 years ago. In Wisconsin, we collected 22 glacially polished bedrock samples from five sites, 10 to 100 km up flow lines from the paleo-ice margin (Bierman et al. 1998; Colgan et al. in press). We sampled three different types of quartz-bearing bedrock (quartzite, meta-rhyolite, and granite) along flow lines at different distances from the ice margin. In New Jersey, we collected 16 samples of bedrock and boulders at several different sites near and on the Laurentide terminal moraine (Larsen 1995). On Baffin Island, we collected 9 samples from an area deglaciated in the early Holocene, reglaciated during the Neoglacial about 4500 y BP, and deglaciated again during the last 20 years (Davis et al. 1999). Samples included polished gneissic bedrock outcrops, as well as cobbles and morainal boulders of the same lithology.

In Wisconsin and in New Jersey, we compared cosmogenic model ages with the accepted, although somewhat uncertain, radiocarbon-based chronologies. Because the Baffin Island site had been deglaciated so recently, any cosmogenic nuclides in the samples would have been produced during prior periods of irradiation and thus inherited. Using the Wisconsin and Baffin data, we calculated minimum limiting rates of glacial erosion using assumptions about the duration of ice cover, the nuclide activity prior to the most recent glacial advance, and the following equation:

$$x = -\frac{\Lambda}{\rho}\ln\left(\frac{N_m - N_e}{N_0}\right) \tag{5}$$

We define x as the depth of rock eroded (cm), Λ as the neutron absorption coefficient (165 g cm^{-2}), ρ as rock density (2.7 g cm^{-3}), N_m as measured nuclide activity (atoms g^{-1}), N_0 as the pre-erosion nuclide activity (atoms g^{-1}), and N_e as the expected nuclide activity for the period of exposure since deglaciation (atoms g^{-1}).

Results for Wisconsin. We found significant inheritance of nuclides in three of the five Wisconsin outcrops we sampled. Nuclide activity was up to 8 times higher than expected from radiocarbon dating, which indicates the outcrops were overrun by ice during the last glacial maximum about 21,500 y BP (Bierman et al. 1998; Colgan et al. in press). Such inheritance indicates that glacial erosion removed little rock from these outcrops. Minimum limiting glacial erosion rates range from 0.01 to 0.25 mm y^{-1} for these extremely strong rocks.

Rock properties, sample location on outcrops, and outcrop proximity to the former ice margin control the magnitude of inheritance of cosmogenic nuclides from prior periods of exposure. Four of five samples from meta-rhyolite outcrops with widely spaced joints contain inherited nuclides; two samples carry the equivalent of >150,000 years of surface exposure (10^5 to 10^6 atoms g^{-1} ^{10}Be and ^{26}Al). Samples highest in the landscape, or in plucked areas, carry less inheritance than those from the lee sides of landforms or lower in the landscape. The most important variable seems to be proximity to the former ice margin. Samples collected from three outcrops less than 30 km from the former ice margin showed significant inheritance and contained several times the expected nuclide activity. In contrast, samples collected from two other outcrops farther than 50 km behind the former ice margin, contain $\leq 10^5$ atoms g^{-1} ^{10}Be, consistent with late Pleistocene exposure and little, if any, nuclide inheritance.

Previous work has shown that the area glaciated by the Green Bay Lobe in Wisconsin can be divided into two major landform zones; one of net erosion and one of net deposition (Colgan and Mickelson 1997; Mickelson et al. 1983). The two sites with the least inheritance are located in the zone of erosion near the axis of the former lobe and greater than 50 km from the former margin. The three sites with significant inheritance are located in a zone of deposition near the margin. Geomorphic evidence and ice sheet modeling studies show that ice in Wisconsin advanced over permafrost as it reached its maximum extent (Mickelson et al. 1983; Cutler et al. 2000). A zone of overridden permafrost up to 100 km behind the ice margin would have initially protected bedrock from erosion. During deglaciation, this frozen zone would have begun to thaw from its up-glacier side first. This would have allowed sliding and bed erosion at sites farther from the former ice margin for a greater period of time than sites near the ice margin that would have remained frozen. Our nuclide date are consistent with this model describing spatial variation in glacial erosion caused by a large ice lobe.

These data, together with simple modeling of nuclide production by deeply penetrating muons (Fig. 17), suggest that meters of rock must be removed to reduce inheritance to negligible levels (<1000 years). The Wisconsin results, along with those of Bierman et al. (1999) from Baffin Island and Stroeven et al. (2002) from Fennoscandia, indicate that cosmogenic dating of exposed bedrock surfaces near former ice margins or in highland areas where ice was frozen to the bed is likely to be uncertain, and in some cases impossible, because nuclides are inherited from prior periods of cosmic-ray exposure. Simple age models that assume no inheritance of nuclides have the potential to

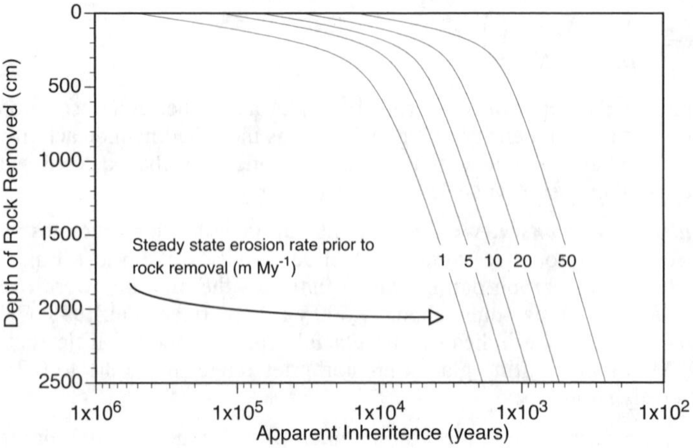

Figure 17. Apparent inheritance (^{10}Be years) resulting from instantaneous removal (glacial erosion) of rock (ρ = 2.7) from a landscape that was in erosional steady state. Each line represents a different steady-state erosion rate (m Myr^{-1}). Increase in line slope near surface reflects change from predominately neutron to predominately muon production at a depth of about 3 m. Calculations made for ^{10}Be assuming parameterization of Granger and Smith (2000). Model is strictly correct only for first glaciation following steady state erosion without ice cover.

yield erroneous and inflated cosmogenic model exposure ages and production rate estimates if such estimates are made on samples containing inherited nuclides. The nuclide data mandate large differences in erosion rates at the base of the Laurentide ice sheet, dependent on both underlying bedrock lithology and basal thermal regime.

Results for New Jersey. In contrast to the Wisconsin results, measurements of ^{10}Be and ^{26}Al in hard quartzite and lightly weathered gneiss samples from the former Laurentide Ice Sheet margin in New Jersey show no evidence of significant inheritance of nuclides from prior exposure periods (Larsen 1995; Larsen and Bierman 1995; Larsen et al. 1995; Bierman et al. 1996); rather, nuclide activities were less than expected considering radiocarbon age constraints on the timing of glacial retreat and the best nuclide production rate estimates available at the time (Nishiizumi et al. 1989).

In New Jersey, there was no significant difference between the average nuclide activity in bedrock and boulders; nor was there any significant difference between average nuclide activity in gneiss and quartzite samples (Fig. 18) as might have been expected if there were a lithologic or rock strength control on inheritance. Perhaps the proximity of the New Jersey margin to the maritime influence of the Atlantic Ocean kept the area warmer and ice/rock interface unfrozen during glaciation, allowing a sliding bed and more efficient erosion for a longer time. Whatever the cause, the New Jersey data suggest that near the ice margin in some glaciated terrains, cosmogenic nuclides can provide accurate age determinations while in others, inheritance prevents any meaningful age assignments.

Results for Baffin Island. The Baffin Island experiment also suggests that warm-based ice erodes rock efficiently (Davis et al. 1999). Samples collected from bedrock and boulders at Crater Lake and its adjacent moraines within Pangnirtung Pass have very low nuclide activities indicating that any inheritance at this site, and by analogy other fiord-bottom sites, is inconsequential (Fig. 19). Of the 18 analyses (paired ^{10}Be and ^{26}Al) that

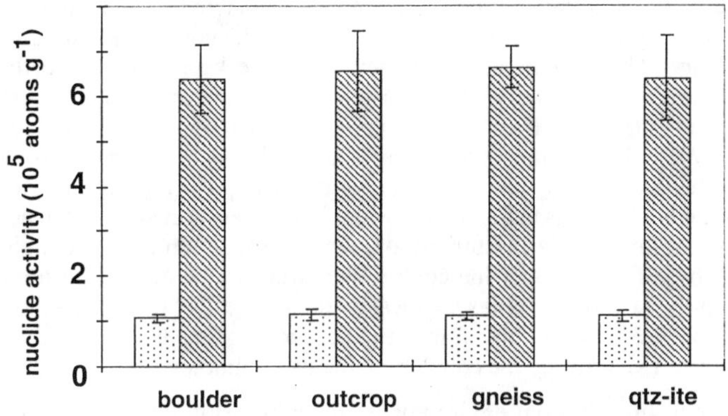

Figure 18. Data from Laurentide ice sheet terminal zone, New Jersey. Pair-wise comparison of average ^{10}Be (dots) and ^{26}Al (hatchure) abundance in gneiss and quartzite (qtz-ite) and boulders and outcrops shows no significant differences. Error bars indicate 1 standard deviation of average.

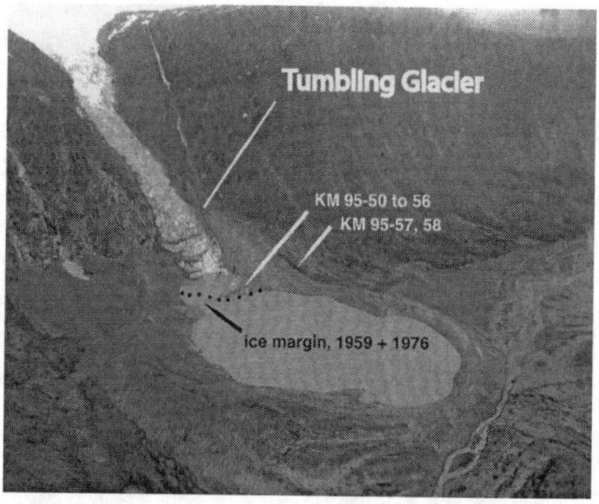

Figure 19. Baffin Island, Canada. Oblique air photograph (1994) from helicopter looking northwest at snout of Tumbling Glacier, which has retreated about 100 m since 1976, leaving exposed a promontory of roches moutonnees and overlying glacial boulders. Ice margin from 1959 and 1976 shown by dotted line. Cosmogenic nuclide sample sites are numbered and indicated with arrow. Modified from Davis et al. (1999).

we made on nine samples, only seven give finite results 2σ over the uncertainty generated by blank subtraction and counting statistics. Because Baffin Island samples have low stable Al (<140 ug g^{-1}) and because our ^{26}Al blanks are lower than our blanks for ^{10}Be, ^{26}Al is a more sensitive chronometer than ^{10}Be. At this near sea-level, high-latitude site, we have successfully measured ^{26}Al model ages and set ^{26}Al age limits less than 1000 years. It is possible that some of the ^{26}Al we measured is generated by radioactive decay rather than cosmogenically (Sharma and Middleton 1989).

Three samples of bedrock, striated and polished by the Tumbling Glacier and recently re-exposed, contain very low levels of ^{10}Be and ^{26}Al. The ages and age limits cited below have been recalculated from data in Davis et al. (1999) using the production rates of Bierman et al. (1996). ^{26}Al data suggest inheritance equivalent to less than 900 years (exposure age and age limits: <0.6, <0.7, 0.9±0.3 ka); we could not reliably detect ^{10}Be in these samples (exposure age limits: <2.1, <1.9, <2.7 ka). Samples from three boulders have similarly low nuclide activity. We were unable to detect confidently ^{10}Be in any of the boulder samples (exposure age limits: <3.6, <2.9, <3.5 ka). ^{26}Al data better constrain the cosmic ray exposure history of the boulders to the equivalent of less than a thousand years (exposure ages and age limit: 1.0±0.5, 0.6±0.3, and <0.9 ka).

Two cobbles (0.25 and 0.11 m maximum diameter, respectively) have differing histories; one carries significant inheritance. We were unable to detect either ^{10}Be (<1300 y) or ^{26}Al (<1000 y) in the larger cobble, implying that the cobble had no significant prior exposure history. The smaller cobble appears to have been exposed to the sea-level equivalent of several thousand years of cosmic-ray bombardment at the surface. It has concordant and finite ^{10}Be and ^{26}Al exposure ages of 3.5±1.0 and 3.7±1.3 ka, respectively. We collected the cobble near sea level and have used sea-level nuclide production rates to express inheritance in terms of exposure duration. However, the cobbles, and the boulders discussed above, were likely transported from the accumulation zone of the Tumbling Glacier, ~2000 m asl, where production rates are higher than the sampling site by a factor > 4. If the cobble were transported after exposure at a higher elevation, then the nuclide activities we measured could reflect less than a thousand years of surface exposure. Both cobbles and all the surrounding bedrock are all gneissic, preventing us from identifying the cobbles source using lithology.

In this arctic fiord-bottom environment, glacial erosion has removed bedrock that was dosed significantly by cosmic radiation during ice-free periods, suggesting that inheritance does not interfere with accurate dating of bedrock outcrops; glacial erosion has "reset" the cosmogenic clock to near zero. On the uplands surrounding the fiord, where the ice was probably cold-based, erosion was ineffective and inheritance is significant (Bierman et al. 1999).

Understanding the history of clasts exposed at and near Earth's surface

Beryllium-10 has been measured in clasts and boulders exposed at and near Earth's surface by non-glacial processes. Most applications have sought the ages of depositional landforms, predominantly fluvial terraces in western North America (e.g., Repka et al. 1997; Phillips et al. 1997; Chadwick et al. 1997; Hancock et al. 1999) and in Asia (Molnar et al. 1994; van der Woerd et al. 1998, 2000); others have considered beach ridges (Nishiizumi et al. 1993; Trull et al. 1995; Gosse et al. 1998; Matmon et al. in press). To perform this type of dating, one can consider explicitly the nuclide activity that the sediment contained when it was deposited (Anderson et al. 1996; Hancock et al. 1999). Other studies have calculated model ages assuming that such inheritance was negligible but recognizing the potential for it to occur (Molnar et al. 1994; Phillips et al. 1997; Chadwick et al. 1997; van der Woerd et al. 1998, 2000). Such an approach can also be taken using fine-grained sediment; consider Perg et al.'s (2001) dating of a flight of marine terraces near Santa Cruz, California.

Clasts have also been used to define burial ages $\geq 10^5$ y (Granger et al. 1997, 2001a; Granger and Smith 2000; Granger and Muzikar 2001). This technique relies on exposure and then rapid burial beyond the depth of significant cosmic ray exposure, typically in a cave system or thick gravel deposit. Because of the contrasting half lives of ^{10}Be and ^{26}Al, one can back calculate both burial time and initial nuclide activity, which can be interpreted as a rate of source rock erosion (Granger et al. 1997).

In some cases, clasts were analyzed one by one (Molnar et al. 1994; Phillips et al. 1997; Repka et al. 1997; van der Woerd et al. 1998, 2000). In other cases, workers combined numerous clasts to generate an amalgamated sample in order to reduce the number of analyses required (Nishiizumi et al. 1993; Repka et al. 1997; Hancock et al. 1999; Granger et al. 2001a). Such an approach is economical but masks the clast-to-clast variance in nuclide activity (Repka et al. 1997). Furthermore, if individual clasts have differing ^{26}Al/^{10}Be ratios, the ^{26}Al/^{10}Be ratio of the amalgamated sample cannot be understood quantitatively because physical mixing in a linear process and any calculation involving decay correction is non linear (Bierman and Steig 1996).

Piedmonts fringe many of the world's mountain ranges. In arid and semiarid regions, these piedmonts are particularly well developed. On some, bedrock crops out or is just below the surface (Denny 1967). Such piedmonts, termed pediments, have been studied using cosmogenic nuclide analysis of outcropping rock surfaces (Bierman 1993; Cockburn et al. 1999, 2000). Other piedmonts have thicker covers of sediment, alluvial, and debris flow deposits. These bajadas, or surfaces where fans have coalesced into a broad apron, have also been studied cosmogenically with measurements made on samples collected from boulders and cobbles (e.g., Bierman et al. 1995b; Siame et al. 1997; Brown et al. 1998; Zehfuss et al. 2001). Some piedmonts, particularly those developed below granitic source basins, are fine-grained and homogeneous. On two of these piedmonts, Nichols et al. (2002) measured ^{10}Be in the coarse sand fraction to track the movement of sediment down the surface; they determined that the average down-piedmont sediment speed was 20 to 100 cm y^{-1}. In southern Africa and Australia, many piedmonts expose quartz clasts at the surface. Clast-by-clast analyses of samples collected from two of these long-lived desert surfaces illustrate the power of ^{10}Be for understanding surface histories

Geologic setting for the Namib surfaces and the Brachina pediments. The Flinders Range National Park in southern Australia includes spectacular piedmonts bordering the Heysen Range near Brachina. These piedmonts are several kilometers wide and dissected by the modern drainage system leaving remnants at different elevations. The piedmont surfaces appear to be eroding, thereby exposing bright red soils on which sit rounded, white and gray quartzite clasts of varying sizes, although many can be held in the hand (Fig. 20).

Figure 20. Dissected piedmont at Brachina, Australia, from which samples designated 965 were collected, is the sloping accordant surface in the middle ground.

Namibia is dominated by the Great Escarpment, a several- to 10-km wide zone where elevation changes ~1000 m. The escarpment separates the hyper-arid coastal plain, which receives < 100 mm y^{-1} of precipitation, from the semi-arid highlands which receive several hundred mm of precipitation yearly. Below and within the escarpment zone are extensive, low relief gravel surfaces covered with polished, ventifacted, and competent quartz and chert clasts (Ward 1987). These largely unvegetated surfaces also appear to be eroding but are underlain by thick calcium-carbonate cemented soils in many places (Fig. 21). The gravels are thought to be Miocene in age and to reflect a more moisture-effective climate regime at that time (Ward 1987). Their isolation above regional drainages has been attributed to broad uplift in the Tertiary and/or Quaternary (Partridge and Maud 1987; Ward 1987).

Study methods for clast histories. We and colleagues collected quartz-bearing clasts from piedmont surfaces in southern Australia and west-central Namibia. We analyzed 12 quartz clasts, collected from 4 different surfaces in the Namib desert during December 1998 (Bierman and Caffee 2001). Six quartzite clast samples from three different Brachina surfaces were collected by J.A. Bourne (University of Adelaide, Department of Geology) in April 1993. All clasts were exposed on the surface when they were collected. Namib clasts were crushed in their entirety. The Australian clasts were cut in half before

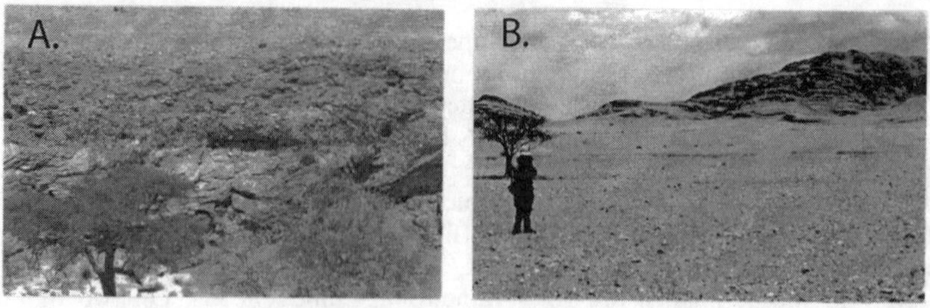

Figure 21. Namibian piedmont surface (sample site NAM 13). (A) View in cross-section showing carbonate-cemented gravel overlying beveled bedrock surface. (B) Surface from which clasts were collected.

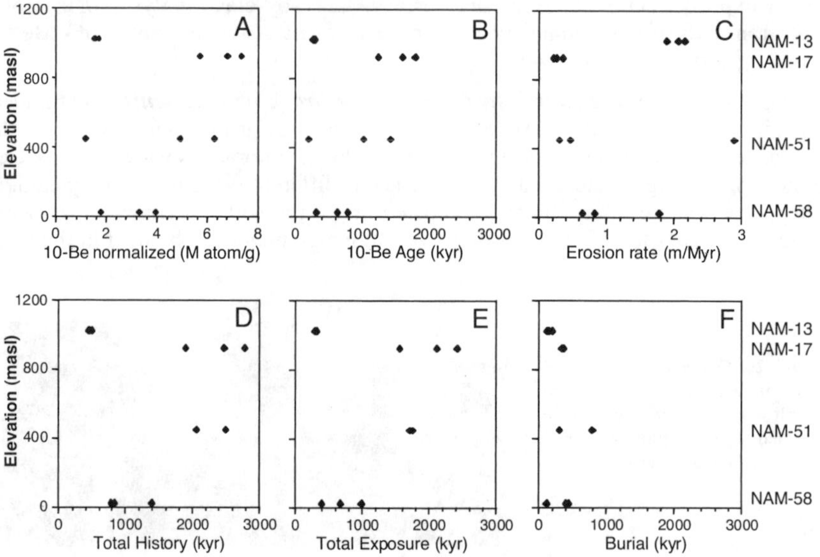

Figure 22. Namibian clast data as a function of elevation. Nuclide activity and surface age/stability appear to increase with elevation except for site NAM-13, which is actively eroding as nearby gullies cut into the escarpment zone landscape. (A) Normalized ^{10}Be activity (sea-level, high latitude). (B) ^{10}Be model exposure ages for clasts assuming no inheritance from prior exposure and no clast erosion. (C) Maximum limiting erosion rate for surface allowed by ^{10}Be clast data assuming that clasts, and the geomorphic surface upon which they sit, lower together. (D) Minimum total clast history (burial and exposure) determined from ^{26}Al/^{10}Be ratio data (Bierman et al. 1999). (E) Total minimum surface exposure history determined from ^{26}Al/^{10}Be ratio data (Bierman et al. 1999). (F) Minimum total burial history determined from ^{26}Al/^{10}Be ratio (Bierman et al. 1999). All calculations made using Nishiizumi et al. 1989 production rates. [Used by permission of the editor of *American Journal of Science*, from Bierman and Caffee (2001), Fig. 13, p. 351.]

being shipped to us; the halves that we received were crushed and analyzed.

Results for Namibian clasts. Many Namibian clasts have very high activities of ^{10}Be and ^{26}Al (Fig. 22). In fact, Namibian clasts have the highest activities we have ever measured; single-nuclide model ^{10}Be exposure ages approach two million years (Bierman and Caffee 2001 using production rates of Nishiizumi et al. 1989). We made calculations assuming that the clasts were dosed *in situ* on these nearly flat surfaces, which are now completely disconnected from the modern drainage system. If these clasts reflect the stability of the geomorphic surfaces on which they lie, erosion rates are on the order of only decimeters per million years. Such low rates have been found previously only in Antarctica and the Atacama Desert (Nishiizumi et al. 1991a, 1998). In general, cosmic-ray dosing of clasts increases as one moves away from the coast toward the escarpment; this pattern suggests that inland surfaces are either older or more stable. Samples collected within the escarpment zone (NAM-13) have the lowest nuclide activity and the least clast-to-clast variance. Both of these characteristics reflect the more rapidly eroding nature of the sampling site within the high-relief escarpment zone.

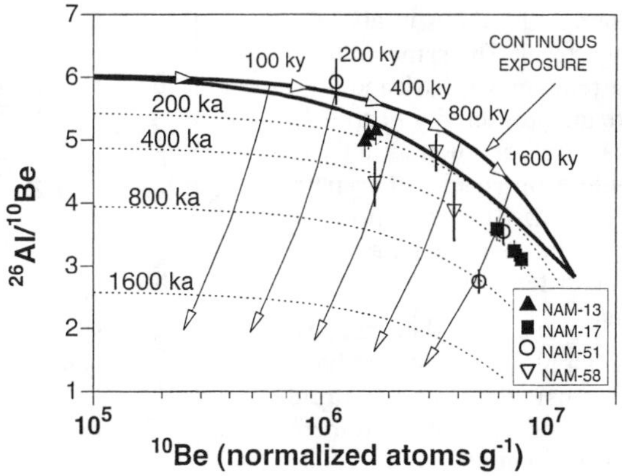

Figure 23. Two-nuclide plot for 12 clast samples collected from four surfaces in Namibia, including exposure and burial isochrons, shows that clasts have a variety of histories. Seven clasts have histories that include significant burial after or during exposure to cosmic rays because they plot more than 1 σ below the envelope consistent with steady erosion and continuous surface exposure using production rates of Nishiizumi et al. (1989). Minimum initial surface exposure times indicated in ka. Minimum burial times indicated in ka. Only samples from surface NAM-13, collected within the escarpment zone and shown by upright triangles, are consistent with steady erosion without complex exposure and burial history. For additional explanation see Bierman and Caffee (2001). [Used by permission of the editor of *American Journal of Science*, from Bierman and Caffee (2001), Fig. 14, p. 352.]

Considering both ^{26}Al and ^{10}Be activity suggests that most clasts we sampled have complex exposure histories including at least one episode of burial (Fig. 23). Analysis following Bierman et al. (1999) suggests that many clasts have extremely long near-surface residence times (Bierman and Caffee 2001), some approaching 3 million years. Total clast histories and total exposure times are higher for inland clasts; burial times are not correlated to landscape position (Fig. 22). This might be expected because burial is likely caused by burrowing animals or small, shifting sand dunes that we commonly observed. However, the relative importance of burial depth and burial time cannot be readily distinguished isotopically. The integrated shielding history as recorded by the nuclides can be the same for different near-surface histories.

Results for Brachina piedmonts. Australian piedmonts at Brachina have quartz clasts that contain significant amounts of ^{10}Be (0.9 to 1.8×10^{6} atoms g^{-1}) and ^{26}Al (5.0 to

9.4×10^6 atoms g^{-1}), the equivalent of 180 to 360 kyr of surface exposure at the elevation and latitude where they are currently exposed, considering the production rate estimates of Bierman et al. (1996). The two samples from the main piedmont surface (965-85 and 965-86) have very similar nuclide activities. Nuclide activity is lower but similar in the two samples (965-89 and 965-90) collected from the lowest piedmont remnant sampled, which is also the farthest from the range front. The samples from the highest piedmont remnant (965-81 and 965-82) have intermediate nuclide activity (Table 2). Each sample pair has similar nuclide activity and is distinguishable from each other pair, suggesting that the 3 distinct geomorphic surfaces, from which the sample pairs were collected, are behaving differently.

Together, the ^{10}Be and ^{26}Al data suggest that all the Brachina samples we analyzed have nuclide activities consistent with simple exposure histories. Considering measurement uncertainties, none of the samples plot outside the bounds set by the constant-exposure and steady-erosion models. It is clear from the nuclide data that clasts on the Australian Brachina piedmonts have very different, much shorter, and apparently much simpler, near-surface histories than those collected from the Namibian plains (Fig. 24).

Given the measurement and current production rate uncertainties, it is not possible to decide, using isotopic evidence alone, whether a steady erosion or rapid exposure model is a more reasonable interpretation of the Brachina data; however, field observations, primarily the dissected nature of the piedmonts and the exposure of weathered and reddened soils, suggest that the surfaces are eroding, slowly, but continually, exposing clasts. If we accept a model of steady erosion, in which clasts are exposed by erosion and then transported off the surfaces, then the nuclide data suggest that the main piedmont surface is eroding about 1.6 m Myr^{-1}. The other remnants are eroding somewhat more quickly, 2 and 3.0 m Myr^{-1}.

These rates of surface degradation are consistent with others measured in Australia and are quite low for unconsolidated materials (Stone et al. 1994; Bierman and Turner 1995;

Table 2. Cosmogenic nuclide data, Brachina Piedmont clasts, South Australia

Sample[1]	Elevation	^{10}Be	^{26}Al	$^{26}Al/^{10}Be$	^{10}Be model age[2]	^{26}Al model age[2]	^{10}Be model ε[2]	^{26}Al model ε[2]
	meters asl	atoms g^{-1} quartz	atoms g^{-1} quartz		ky	ky	(m My^{-1})	(m My^{-1})
965-81	320	1520000 ± 35000	8160000 ± 360000	5.37 ± 0.27	292 ± 32	280 ± 35	2.0 ± 0.3	1.9 ± 0.3
965-82	320	1310000 ± 41000	7450000 ± 360000	5.67 ± 0.33	250 ± 28	252 ± 32	2.3 ± 0.3	2.1 ± 0.4
965-85	290	1710000 ± 53000	9250000 ± 560000	5.38 ± 0.37	342 ± 39	333 ± 46	1.7 ± 0.2	1.5 ± 0.3
965-86	290	1780000 ± 54000	9460000 ± 490000	5.31 ± 0.32	356 ± 40	343 ± 46	1.6 ± 0.2	1.5 ± 0.3
965-89	260	940000 ± 56000	5800000 ± 300000	6.16 ± 0.49	185 ± 23	201 ± 25	3.2 ± 0.5	2.7 ± 0.4
965-90	260	no data	4960000 ± 380000	no data	no data	178 ± 24	no data	3.1 ± 0.5

Notes [1] Latitude 32°, South

[2] Model ages and erosion rates calculated as described in Table 1.

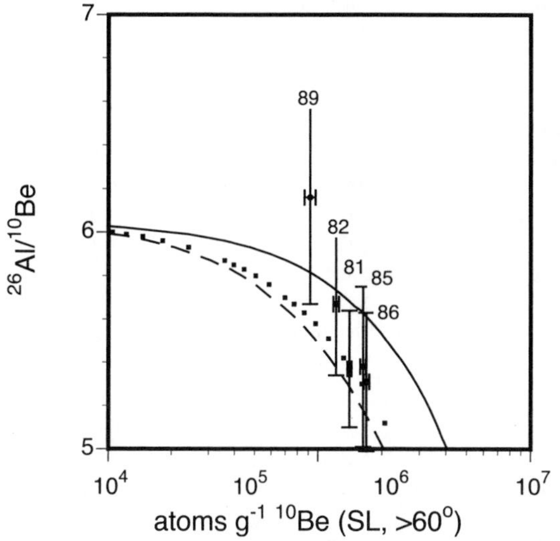

Figure 24. Two-nuclide diagram for Brachina piedmont samples (Australia) suggests that all samples have nuclide activities consistent with simple exposure history. Uncertainty is plotted at 1σ based on counting statistics and stable nuclide uncertainty, fully propagated. Sample number is plotted above and next to each error bar. Solid line and black squares are exposure and erosion cases, respectively, based on production rates of Nishiizumi et al. (1989). Dashed model curve defines erosion endpoints calculated with production rates of Bierman et al. (1996). SL, 60° indicates that nuclide data have been normalized to sea level at high latitude.

Nott and Roberts 1996; Bierman and Caffee 2002). The erosion rates are significantly higher than those implied by measurements made in Namibian clast samples consistent with the degraded appearance of the Brachina surfaces, their much higher surface slopes, and the wetter Brachina climate.

Estimating bedrock erosion rates

Quantifying the rate at which bedrock erodes is critical for understanding the tempo and means by which landscapes change, sediment is generated, and Earth's surface responds to tectonic forcing. Measuring bedrock erosion rates is challenging (Saunders and Young 1983). Different techniques, such as fission track analysis, U/He geochronology, and sediment yield determination, provide information over differing timescales and rely upon different and typically difficult-to-verify assumptions; yet, it is the integration of data gathered over varying time scales that provides the most useful long term analysis of landscape change (e.g., Nott and Roberts 1996; Cockburn et al. 2000; Matmon et al. 2001a).

In 1955, Davis and Schaeffer realized that cosmogenic nuclides could rapidly provide useful data to quantify rates of bedrock erosion. Bedrock erosion rates have been measured directly using ^{10}Be (often paired with ^{26}Al) following the pioneering laboratory work of Nishiizumi et al. (1986) and theoretical work of Lal (1986, 1987, 1988, 1991). Long-lived nuclides, such as ^{10}Be, integrate erosion rates over timescales ranging from centuries in areas where erosion is extraordinarily rapid, such as the Himalayas (Duncan et al. 2001) to millions of years where erosion is extremely slow (Nishiizumi et al. 1991a; Bierman and Turner 1995; Cockburn et al. 1999; Bierman and Caffee 2001, 2002).

Samples collected from exposed bedrock surfaces can be used to calculate rates of erosion directly. Such an approach has been taken in the arid, semi-arid, and Mediterranean environments of western North America on granitic and quartz-bearing sedimentary and volcanic rocks (Nishiizumi et al. 1986, 1993; Albrecht et al. 1993; Bierman and Turner 1995; Small et al. 1997, 1999; Heimsath et al. 1997, 1999; Clapp et al. 2001, 2002; Granger et al. 2001b) where erosion rates are typically in the range of meters to tens of meters per million years except for weakly cemented arkosic sandstones which appear to be eroding about 100 m Myr^{-1}. The few measurements made so far in eastern North America suggest similar rates of landscape change (Bierman 1993; Lal et al. 1996; Matmon et al. 2001a,b; Schroeder et al. 2001)

Measurements of ^{10}Be in samples collected from a bedrock tor, an outcropping tower of granite exposed along the escarpment characteristic of the southeastern coast of Australia, suggest sequential exposure from a lowering soil mantle at ~25 m Myr^{-1}, whereas the bedrock of the tor is eroding much more slowly, only ~8 m Myr^{-1} (Heimsath et al. 2000); similarly slow rates of erosion were inferred from ^{10}Be and ^{36}Cl measurements on tors in the arid, granitic Alabama Hills of southern California and the semi-arid Llano uplift of central Texas, USA (Bierman 1993; Bierman and Turner, 1995). Siliceous bedrock in the hyper arid southern Negev Desert of Israel appears to be eroding somewhat faster (meters to tens of meters per million years; Clapp et al. 2000). The most stable bedrock surfaces have been found in the Antarctic polar desert; ^{10}Be-based erosion rate estimates there range down to fractions of meter per million years (Nishiizumi et al. 1991a; Summerfield et al. 1999).

Bedrock erosion rates on drainage basin scales can also be calculated indirectly using the measured activity of ^{10}Be in fluvial sediments including sand and gravels (Brown et al. 1995b; Bierman and Steig 1996; Granger et al. 1996). Some of this work was reviewed in Bierman et al. (2001a), but more recent studies include an erosion rate survey based on sediments of some European rivers (Schaller et al. 2001), a mountain scale study in the Great Smoky Range of southeastern North America (Matmon et al. 2001a,b), a study of the 16,000 km^2 Rio Puerco basin (northern New Mexico, USA; Bierman et al. 2001c), and recent work in the Sierra Nevada of California documenting the influence of tectonics and the lack of influence of climate on rates of bedrock erosion (Riebe et al. 2000, 2001a,b). Analysis of ^{10}Be and ^{26}Al in buried pebbles allowed calculation of both source rock erosion rates (mostly 2-12 m Myr^{-1}) and rates of river incision (~30 m Myr^{-1}) in the southern Appalachian Mountains of eastern North America, presumably in response to climate and long-term isostatic uplift (Granger et al. 1997, 2001a).

Southern Africa and south central Australia are both stable cratons where active tectonics and glaciation have not played a role in landform development for many millions of years. Cratons of the southern hemisphere, where extensive low relief geomorphic surfaces are plentiful, offer ample opportunities to measure long-term bedrock erosion rates using cosmogenic nuclides. Overall, we suspect that many southern surfaces (excepting the Antarctic) have less complicated Neogene histories than northern hemisphere cratons, which were glaciated repeatedly during the Pleistocene, thus violating the necessary assumption of steady and uniform erosion underlying the calculation of cosmogenically-based erosion rates. Below we review findings related to ^{10}Be and bedrock erosion in Australia and southern Africa.

Geologic setting for Namibia and Australia. The shield areas on both Australia and Africa have widespread exposures of quartz-bearing crystalline rocks amenable to the analysis of ^{10}Be and ^{26}Al in quartz. These granitic and gneissic rocks commonly generate landscapes dominated by inselbergs, isolated rock outcroppings that stand above the otherwise low-relief plains (Twidale 1982). The most famous inselberg, and one of the largest, is Uluru, formerly known as Ayer's rock. In many cratonic locations, lateritic soils have developed. These soils have been the focus of several studies which demonstrated, using *in situ*-produced ^{10}Be, significant volume loss during weathering and soil formation (Brown et al. 1994; Braucher et al. 1998a,b; 2000). Similarly, a paired study of granitic inselbergs and soils utilizing ^{10}Be and ^{36}Cl has been done in humid southeastern North America (Bierman 1993; Bierman et al. 1995a; Schroeder et al. 2001).

On both the African and Australian continents, earlier workers ascribed stratigraphic ages to levels preserved on inselbergs flanks (Twidale et al. 1985) and to the plains on which the inselbergs sit (Partridge and Maud 1987) implying erosion rates so low that pristine rock surfaces could be preserved for $>10^7$ y or that landforms erode in a self-similar manner, so that landform shape is preserved even as the entire landscape erodes. Measurement of *in situ*-produced cosmogenic nuclides allows direct testing of these earlier assertions.

Inselberg surfaces have varying appearances. They range in area from 10s to 1000s of m^2 and in height from meters to 100s of meters. Some inselbergs are solid rock with few joints. Others are heavily jointed and covered with residual boulders. Most inselbergs we sampled are losing mass in a variety of ways including grain-by-grain detachment, cm-scale thin sheeting, and loss of exfoliation slabs up to several meters thick.

Figure 25. View from Turtle Rock of Mt. Wudinna, a large granitic inselberg on the Eyre Peninsula, South Australia. A sample from Turtle Rock had more ^{10}Be than any other Australian sample, consistent with erosion at only 0.35 m Ma^{-1} (Bierman and Caffee 2002).

Study methods for measuring Namibian and Australian bedrock erosion. We collected 68 samples from 9 bedrock sites and 2 quarries in Australia (Bierman and Caffee 2002). Six of the sites were inselbergs rising above the semi-arid Eyre Peninsula (Fig. 25). The remaining outcrops were inselbergs and boulders farther north. We collected four shielded samples from a quarry below one inselberg, Yarwondutta Rock, and two shielded samples from a quarry adjacent to three other sampled inselbergs. In Namibia, we collected 47 bedrock samples from granite, gneiss, quartz vein, and quartzite outcrops along several transects across the Namib desert and adjacent highlands. The goals of our sampling were to determine the range of bedrock erosion rates characterizing both field areas, to determine if there were any dependence of bedrock erosion rate on mean annual precipitation, and to use the $^{26}Al/^{10}Be$ ratio to investigate the history of bedrock outcrops in terms of now-vanished cover. A pair of shielded quarry samples allow us to check our laboratory work and estimate longer-term erosion rates based on quantifying muonic production of ^{10}Be and ^{26}Al (Granger and Smith 2000).

Results for Australian bedrock erosion. Exposed granitic bedrock, sampled at sites scattered about Australia, is eroding very slowly (Fig. 26A), in some places only 10s of centimeters per million years (Bierman and Caffee 2002) if one accepts as upper limits calculations made using the steady state model of Lal (1991). Samples from individual inselbergs show significant variability in nuclide activity. Some variability is clearly related to elevation above the adjacent plains such as at Yarwondutta Rock, and is probably best explained by later Pleistocene, perhaps climate-controlled, stripping of a few meters of colluvial or saprolite mantle to expose well-preserved granite, the weathering front (Fig. 27; Bierman and Caffee 2002). Other variability probably results from the episodic loss of granitic slabs 50 to 100 cm thick (Lal 1991). There is a positive correlation between mean annual precipitation and the rate at which the most stable

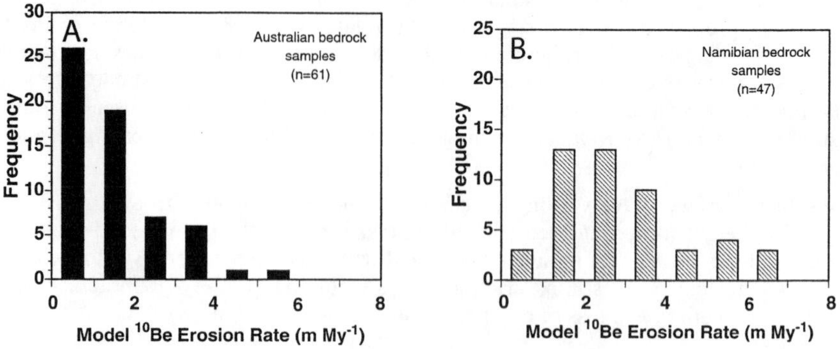

Figure 26. Summary histograms of bare-rock model erosion rates calculated from ^{10}Be activity using steady erosion model of Lal (1991) and nuclide production rates of Bierman et al. (1996). A. Australia (n = 61); overall mean erosion rate is 1.5±1.2 m Myr^{-1}. B. Namibia (n = 47); overall mean rock erosion rate is 3.0±1.6 m Myr^{-1}.

bedrock surface at any site we sampled in Australia erodes (Fig. 28).

It appears that most Australian outcrops we sampled have a simple exposure history. Most samples we analyzed plot within 1σ of the surface exposure envelope on the two-nuclide diagram (Fig. 29), particularly if one accepts the production rate estimate of Bierman et al. (1996) rather than the estimate of Nishiizumi et al. (1989). Analysis of the quarry samples from Yarwondutta, shielded by about 12 m of rock, reveals significant nuclide activity, consistent with muon-induced nuclide production at depth (Granger and Smith 2000) and long term erosion rates of 0.6 to 1.3 m Myr^{-1} (Bierman and Caffee 2002). Such erosion rates, calculated from deeply penetrating muons, are quite uncertain because parameterizations of muon intensity and nuclide production at depth remain uncertain (J. Stone and D. Granger, personal. communication). For Yarwondutta, the average long-term erosion rate (calculated by averaging the measured ^{10}Be activity induced by muons in the four shielded samples) is 0.81 (+0.20, -0.18) m Myr^{-1}, quite similar to the average of 0.61 ± 0.01 m Myr^{-1} determined from three exposed samples (in which most ^{10}Be is spallation-produced) collected at the top of Yarwondutta Rock. The similarity in erosion rates calculated from surface and shielded samples suggests that erosion rates at and near Yarwondutta Rock have been steady over millions of years.

The exceptionally high nuclide activity measured in samples collected from Australian inselbergs demands that these are long-lived landforms; yet, every sample measured has significantly less than the saturation values (P/λ) of both ^{10}Be and ^{26}Al, indicating that all sampled inselbergs are dynamic, eroding landforms, the surfaces of which lose mass over time (Bierman and Caffee 2002). Such dynamism calls into question the meaning of stratigraphic ages based on correlations of inselberg forms with inferred ages of weathering horizons (silcretes and calcretes; e.g., Twidale et al. 1985). Similar conclusions of dynamism have been reached elsewhere in Australia where ^{10}Be and fission track analyses, made on landforms alleged to be exposed continuously since the Cambrian (Stewart et al. 1986), indicate finite erosion rates of decimeters to meters every million years (Belton et al. 2000). Measurements of *in situ*-produced ^{10}Be suggest unambiguously that even in arid, tectonically quiescent cratons, surface processes act rapidly enough when considered over millions of years, to substantially resurface Earth's crust. Such a conclusion, based firmly upon numerous reproducible isotopic measure-

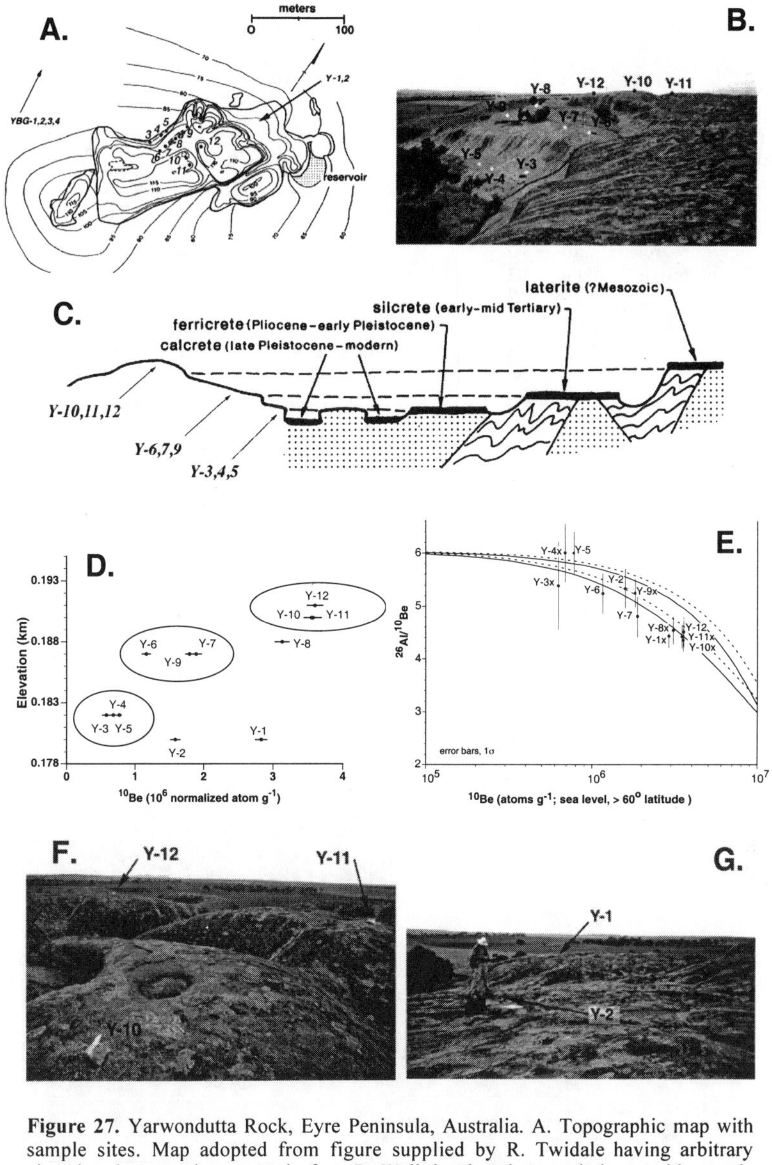

Figure 27. Yarwondutta Rock, Eyre Peninsula, Australia. A. Topographic map with sample sites. Map adopted from figure supplied by R. Twidale having arbitrary elevation datum and contours in feet. B. Well developed stepped slopes with sample sites, on platforms, marked by dots. C. Diagrammatic correlation of flared slopes with duricrusts and erosion surfaces on the Eyre Peninsula, adapted from Twidale et al. (1985). D. Comparison of ^{10}Be activity and elevation suggests sequential exposure of lower Yarwondutta platforms. Ovals surround samples from the same platform. E. Two-nuclide diagram showing most samples have nuclide activity consistent with continuous surface exposure. Dashed lines represent Nishiizumi et al. (1989) production rates; solid line is Bierman et al. (1996) rates. F. Samples Y-10, Y-11, and Y-12 collected from the deeply weathered, upper surface of Yarwondutta Rock. G. Samples sites Y-1 and Y-2 on heavily weathered, lower inselberg surface. [Used by permission of the editor of *Geological Society of America Bulletin*, from Bierman and Caffee (2002), Fig. 7, p. 794.]

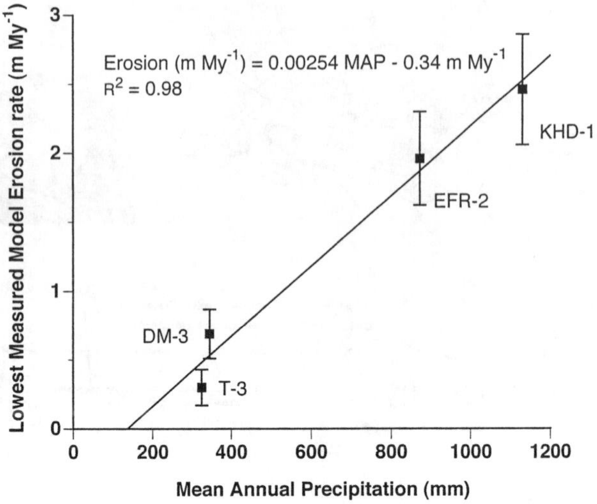

Figure 28. Mean annual precipitation and the lowest ^{10}Be model erosion rate of exposed granite surfaces are linearly related. Error bars are uncertainty of erosion rate. [Used by permission of the editor of *Geological Society of America Bulletin*, from Bierman and Caffee (2002), Fig. 12, p. 800.]

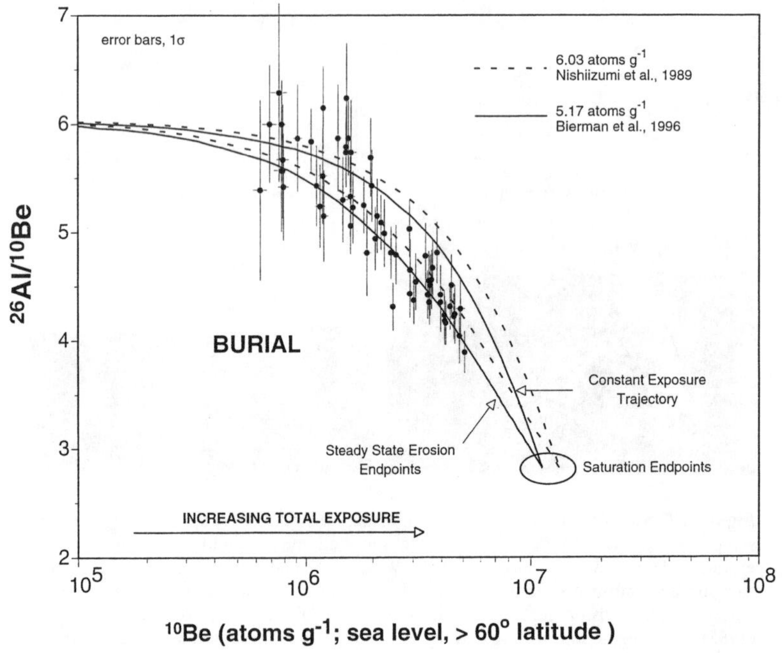

Figure 29. Two-nuclide diagram for 61 exposed Australian bedrock samples. Most are consistent with a simple exposure history. Constant exposure trajectory (upper) and line of steady state erosion endpoints (lower) plotted using Nishiizumi et al. (1989) production rates (dashed lines) and Bierman et al. (1996) production rates (solid lines). [Used by permission of the editor of *Geological Society of America Bulletin*, from Bierman and Caffee (2002), Fig. 5, p. 793.]

ments, has not always been well received (e.g., Twidale 1997a,b; Watchman and Twidale 2002).

Results for Namibian bedrock erosion. Exposed granite, gneiss, quartzite and quartz veins in hyperarid Namibia are eroding more quickly than many Australian outcrops we sampled despite receiving less rainfall (Cockburn et al. 1999; Bierman and Caffee 2001). For the 47 bedrock samples we collected, the average rate of bare rock erosion, recalculated using the Bierman et al. (1996) production rates, is ~3.0 m Myr^{-1}; the most stable Namibian rock outcrop we sampled is eroding ~1 m Myr^{-1}, nearly three times the rate (0.34 m Myr^{-1}) at which the most stable Australian outcrop (Turtle Rock, Fig. 25), is losing mass (Fig. 26B). Analysis of the $^{26}Al/^{10}Be$ ratio for all Namibian bedrock samples suggests that most outcrops have simple exposure histories, similar to the Australian case (Fig. 30).

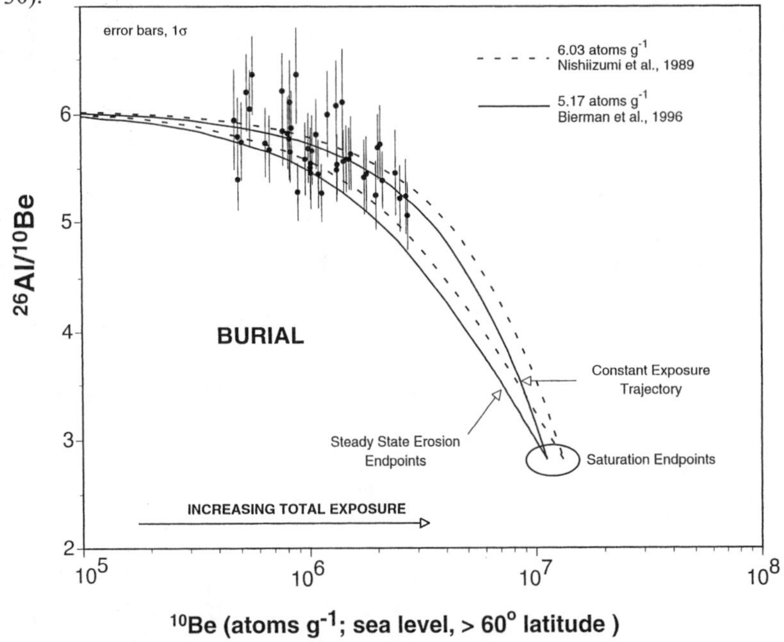

Figure 30. Two-nuclide diagram for 47 exposed bedrock samples collected in Namibia. Constant exposure trajectory (upper) and line of steady state erosion endpoints (lower) plotted using Nishiizumi et al. (1989) production rates (dashed lines) and Bierman et al. (1996) production rates (solid lines). Two of the samples plotting below the simple exposure history window were collected from the top of a quarry wall (NAMBG-1 and NAMBG-2) where a thin veneer of regolith may have been removed by quarry operations prior to our sampling; the third, NAM 50, was collected from the top of an inselberg where there is no evidence for prior burial or episodic loss of mass in thick sheets. Modified from Bierman and Caffee (2001).

There is no relationship between erosion rate and lithology excepting the massive quartz veins we sampled and the Gamsburg quartzite sampled by Cockburn et al. (2000); nor is there any relationship between model erosion rate and mean annual precipitation or distance from the coast, a proxy for the frequency of coast fogs (Fig. 31). Perhaps the lack of any relationship between these variables is due to the rapid drainage of rainwater from bare rock outcrops, effectively creating microclimates of similar aridity across the 5-fold rainfall gradient that we sampled. Alternatively, the coastal fogs and coastal salt spray so common near the coast may compensate for the lack of rainfall there.

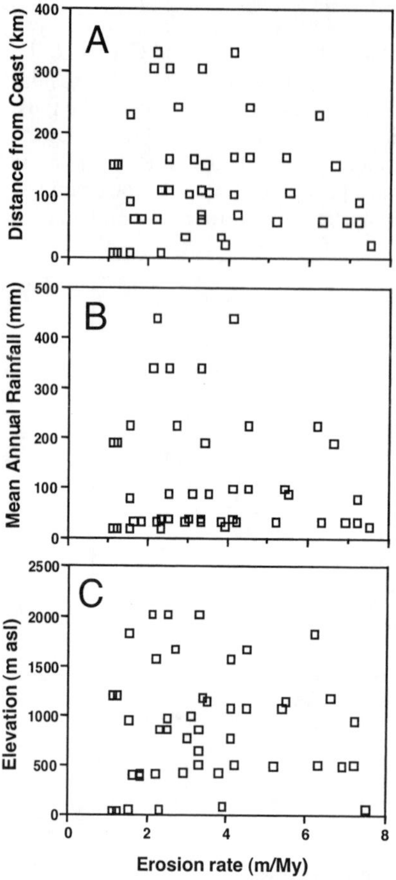

Figure 31. There is no relationship between bare rock erosion rates and gradients in three landscape-scale parameters. Regression analyses of all three graphs indicate correlation coefficients, r^2, < 0.01. (A) Sample site distance from coast, a proxy for fog frequency and salt abundance. (B) Estimated annual rainfall at sample sites (Jacobsen et al. 1995). (C) Sample site elevation above sea level. [Used by permission of the editor of the *American Journal of Science*, from Bierman and Caffee (2001), Fig. 9, p. 345.]

Table 3. Cosmogenic nuclide data, quarry samples, South Australia and Namibia

Sample[1]	Latitude (degrees S)	Sheilding Depth $(m)^2$	Elevation $(m\ asl)^3$	^{10}Be measured (atoms g^{-1} quartz)	^{26}Al measured (atoms g^{-1} quartz)	$^{26}Al/^{10}Be$	^{10}Be model ε^4 $(m\ My^{-1})$	^{26}Al model ε^4 $(m\ My^{-1})$
nambg1a	22	0.6	430	893000 ± 23000	4710000 ± 210000	5.28 ± 0.27	3.2 ± 0.4	3.3 ± 0.5
nambg1b	22	0.6	430	1130000 ± 29000	5990000 ± 270000	5.27 ± 0.27	2.4 ± 0.3	2.5 ± 0.4
nambg2a	22	13.5	430	27600 ± 1700	112000 ± 13000	4.05 ± 0.52	1.0	0.1
nambg2b	22	13.5	430	26400 ± 1500	102000 ± 11000	3.88 ± 0.49	1.4	1.3

Notes [1] more details for nambg samples in Bierman and Caffee (2001)

[2] For nambg1a,b cover is weathered granite (grus), assumed density 2.0 g/cc; for nambg2a,b cover is solid granite, assumed density = 2.7 g/cc

[3] elevation of sampled surface or of surface above quarry

[4] Model surface erosion rates calculated assuming production rates of Bierman et al. (1996) and altitude/latitude correction factors of Lal (1991) for neutrons only; deep erosion rates calculated using parameterization of Granger and Smith (2000). Latitude correction for muons follows Lal (1991) and is 0.71 at sea level. Elevation corrected using muon attenuation in air of 247 g cm^{-2} and site elevations (factor = 1.23). Resulting correction factor for the A_j and B_j coefficients is 0.873. For surface samples, we assume 10% 1s production rate uncertainty fully propagated along with measurement uncertainty and stable Al (4%) and stable Be (2%). No uncertainties provided in Granger and Smith (2000).

The two quarry samples, which were shielded by more than 13 m of rock, have low but measurable inventories of both ^{10}Be and ^{26}Al, 27,000 and 107,000 atoms g^{-1} quartz, respectively (Table 3). Using the parameterization of Granger and Smith (2000), these activities can be interpreted as long-term surface lowering rates of 0.1 to 1.4 m Myr^{-1}, on average about a factor of two lower than the rates estimated using nearby surface samples. Given the current uncertainty with which deep, muon-induced ^{10}Be and ^{26}Al production is currently parameterized, this difference may not be significant.

Data exist for Namibia which allow us to place the cosmogenic model erosion rates in a longer term perspective. Fission track data and modeling (Cockburn et al. 2000) suggest that long-term (over 36 Myr) erosion rates near the coast average several meters per million years, rising to between 8 and 15 m Myr^{-1} in the highlands. These rates are consistent with basin scale rates of denudation determined using ^{10}Be in fluvial sediment (Bierman and Caffee 2001) suggesting that landscape scale erosion rates, integrated over both the 10^5- and 10^7-y time scales, are similar. The average erosion rate of the Namibian bedrock outcrops that we sampled (as determined by cosmogenic analysis) is somewhat less than that determined both by fission tracks and by ^{10}Be analysis of fluvial sediment. This discrepancy makes sense and suggests that bedrock outcrops, for as long as they last, are more stable than the landscape as a whole, probably because they shed water efficiently. Our data imply that the relief of sampled bedrock outcrops increases until they are consumed by erosion from the sides.

River incision into rock

The rate at which rivers incise rock is poorly known. Measurement of ^{10}Be, in river-worn rock (Seidl et al. 1997; Hancock et al. 1998; Weissel and Seidl 1998), in cave deposits (Granger et al. 1997, 2001a), and in samples collected from strath terraces abandoned as rivers incise (Burbank et al. 1996; Leland et al. 1998), is one tool that can help quantify such rates. For example, Granger et al. (1997, 2001a) use cosmogenic burial dating of cave sediments to document down-cutting rates of 20 to 30 m Myr^{-1} for several rivers in the tectonically stable southern Appalachians. In contrast, Leland et al. (1998) and Burbank et al. (1996) use ^{10}Be measurements made on samples collected from strath terraces to suggest rates many times higher (1 to 12 km Myr^{-1}) along the Indus River in the tectonically active Himalayas. Hancock et al. (1998) measured ^{10}Be in samples collected directly from the Indus river bed; they calculate rates of lowering an order of magnitude less (hundreds of meters/million years) than Leland et al. (1998) and Burbank et al. (1996) perhaps due to changes in sediment loading and thus the effectiveness of erosion over time.

The channels of major rivers that cross the mid-Atlantic, piedmont-coastal plain boundary of eastern North America are commonly incised into rock. These rivers run through relatively stable passive margin settings in humid temperate environments (Pazzaglia and Gardner 1993). Many rivers (e.g., the James, Rappahannock, Potomac, and Susquehanna) that drain the central Appalachians cross the fall line in gorges bordered by flights of terraces where polished bedrock outcrops are frequent (Figs. 32 and 33). Presumably, bedrock terraces record the passage of knickpoints as cataracts migrated upstream from the piedmont-coastal plain boundary, which forms a stark lithologic and thus, erodibility contrast (Fig. 34). At the heads of these gorges are falls, the most famous of which is the *Great Falls of the Potomac*, preserved in the Chesapeake and Ohio Canal National Historical Park less than 20 km from the nation's capitol.

We know little about when and how these major landscape features formed, because, until recently, direct dating of bedrock terraces, and thus calculation of bedrock incision, mass removal, and knickpoint retreat rates was impossible. Using measurements of ^{10}Be

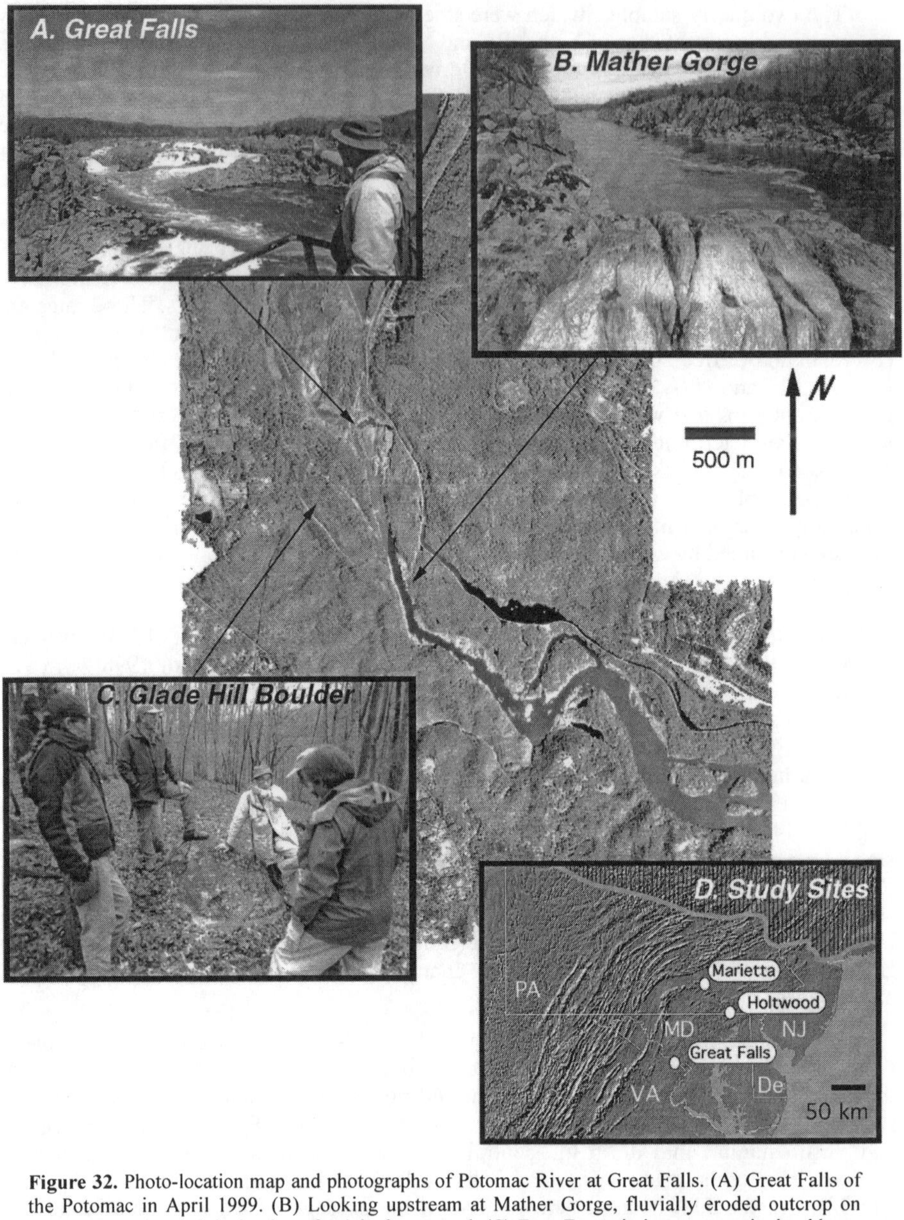

Figure 32. Photo-location map and photographs of Potomac River at Great Falls. (A) Great Falls of the Potomac in April 1999. (B) Looking upstream at Mather Gorge, fluvially eroded outcrop on strath terrace (smooth bedrock surface) in foreground. (C) E-an Zen pointing to quartzite boulder at Glade Hill similar to sample EZ-1 which gave ^{10}Be model exposure age of 253±28 ka (Bierman et al. (1996) production rates). (D) Study sites. Generalized latest Wisconsinan glacial limit shown by line and hatchures. Air photo base map of Potomac River near Great Falls from USGS digital orthophoto quadrangle 1988. Scale and orientation for base map at right center.

and ^{26}Al in quartz extracted from terrace rock surfaces and isolated boulders, one can constrain the exposure age of strath terraces and verify fundamental assumptions underlying the use of cosmogenic nuclides for understanding river incision; in particular, we sought to verify again the assumption that when terraces are first exposed they contain negligible inventories of cosmogenic nuclides (cf. Leland et al. 1998).

Great Falls geologic setting. The Potomac River has a cataract-gorge system that is well developed in schist, metagreywacke, lesser amounts of amphibolite, and granitic dikes and plugs. The river drops 20 m over Great Falls, a complex, multi-level cataract (Fig. 32), which stream power and sediment transport calculations suggest is still retreating (Zen 1997). Below the falls is Mather Gorge, which is flanked by seven flights of bedrock terraces that run parallel to the river from Great Falls to the Coastal Plain boundary (Zen 1997). Two terrace levels are above the gorge (200 ft or 61 m and 155 ft or 47 m), one defines the top of the gorge (140 ft or 43 m), and four are nested within (115, 95, 77 and 68 ft or 35, 29, 23, and 21 m).

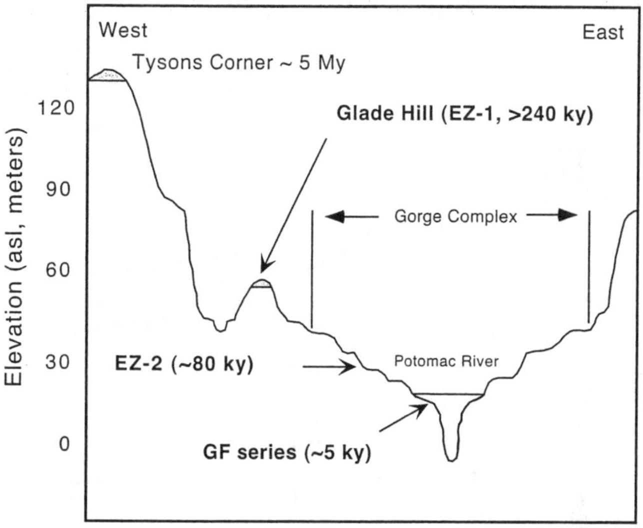

Figure 33. Schematic line of section across the Potomac River showing bedrock strath terraces, Glade Hill boulder bed, and location of samples including: EZ-1 quartzite boulder sampled from Glade Hill, EZ-2 bedrock knob, and GF series from within low-flow Potomac channel. Tysons Corner unconformity provides upper limit for age of Potomac incision. Modified from Figure 11 of Zen (1997).

The terraces are defined by accordant schist outcrops and bedrock benches that retain water polish in places and are punctuated by large, water-worn potholes. Quartz veins on outcrops show little relief (at most, a few cm) suggesting inconsequential mass loss since continuous subaerial exposure began. On the highest and oldest terrace, there are isolated stream-rounded boulders of quartzite or quartz-rich metaconglomorates, some of which are >2 m in diameter. These boulders are extremely hard, erosion-resistant, quartz-rich, and thus amenable to cosmogenic isotopic analysis, although the magnitude of nuclide activity inherited from prior periods of exposure is unknown.

The processes of river incision, in particular, the retreat of knickpoints at least several meters high, suggests that strath surfaces, newly born at the base of cataracts, should contain only low levels of cosmogenic nuclides. This assertion is further supported on the Potomac by the observation (Zen 1997) that most mass is removed at Great Falls by the quarrying of blocks several meters wide. This width is set by the most common spacing between joints and by the scale of large, rapidly drilled potholes, both

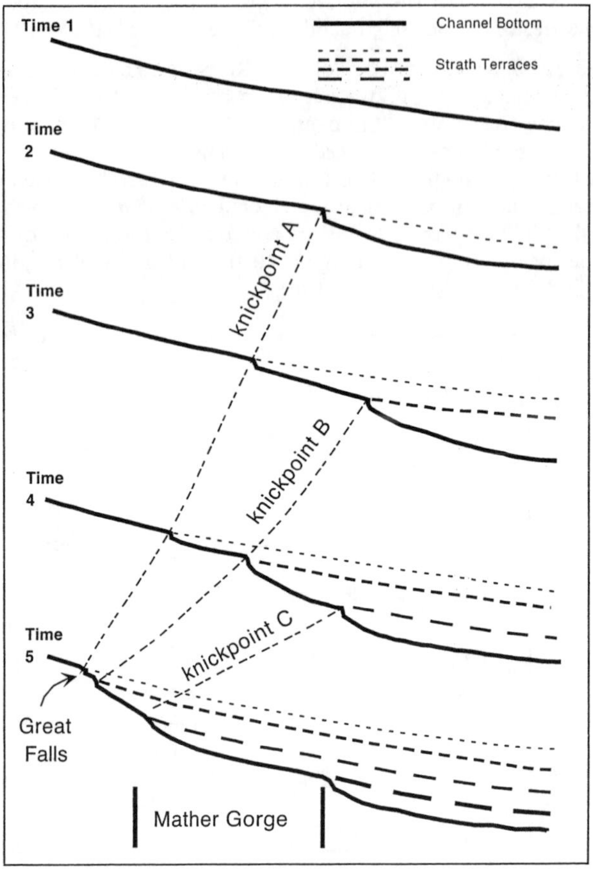

Figure 34. Model of knickpoint retreat, generation of strath terrace flights, and Great Falls by knickpoint bunching. Five sequential time steps are represented. Bottom of river channel is shown as a solid line. Strath terraces are shown as dashed lines; each strath is shown with different dash pattern. At step 1, river is at grade. At step 2, a single knickpoint progrades upstream. At step 3, a second knickpoint moves upstream leaving a second strath. By step five, the knickpoints are bunching at Great Falls. This model suggests that exposure ages on a single strath should increase downstream—a testable hypothesis. Figure from E-an Zen (1999, written communication).

of which facilitate mass removal (Zen and Prestegaard 1994). The size of the failing blocks is sufficiently large to remove most nuclides produced by irradiation prior to knickpoint retreat.

Great Falls study methods.. During 1999, we collected six samples near Great Falls on the Potomac River. In March, we collected one sample from each of two different terraces (Fig. 33). In August, during the extremely low-flow occasioned by a severe summer drought, we collected four samples within 10 meters of each other (GF-10, 11, 12, 13) from exposed bedrock platforms within Mather Gorge and the low-flow channel of the river. The accordant platforms we sampled are typically under at least 1 m of water. If the river were to incise, the samples would become exposed on a terrace surface.

We purified 10 to 40 g of quartz, reducing stable Al sufficiently to make reliable AMS measurements (65-140 μg g^{-1}) at very low nuclide activities (Table 4).

Results #1—No appreciable inheritance. Fluvially beveled bedrock along the side of the main channel, which might represent a future terrace level, contains near-background levels of cosmogenic ^{10}Be and ^{26}Al. Beryllium-10 activity in these samples is very low (20,000 to 30,000 atoms g^{-1}), indicating that rock in the active channel of the Potomac River is barely dosed by cosmic radiation (Table 4). Aluminum-26 activity is similarly low and the $^{26}Al/^{10}Be$ ratios of the samples range, with significant uncertainty, from 5.04 to 6.17. These measurements strongly support the assumption of inconsequential nuclide inheritance for older strath terrace surfaces. If the river were to incise today, exposing the accordant platform we sampled as a terrace, nuclide activities in these samples would represent, on average, only several kyr of exposure.

Results #2—Higher and older terraces. Model ages correspond to the stratigraphic sequence of terraces. Samples (EZ-1 and EZ-2) collected from two higher terraces (Fig. 33) demonstrate that nuclide activity and terrace height are positively correlated indicating longer subaerial exposure of the higher terrace surface (Table 4).

Sample EZ-2, collected from a water worn schist knob at 95 ft (29 m) above sea level (asl), has closely concordant ^{10}Be and ^{26}Al exposure ages of 79±10 and 83±11 ka using the production rate estimates of Bierman et al. (1996). Sample EZ-1, collected from the highest preserved terrace, a boulder bed at 200' (61 m) asl, contains substantially more nuclides than the other Potomac samples we collected; from the measured activity, one can calculate a minimum ^{10}Be model exposure age of 253±28 ka and a minimum ^{26}Al exposure age of 228±28 ka. The discordance of these age estimates, reflected in the $^{26}Al/^{10}Be$ ratio that we measured for this sample (5.13±0.27), suggests that it has a complex exposure history when compared to other Potomac River terrace samples (Fig. 35).

Table 4. Cosmogenic nuclide data, Great Falls area, Potomac River, USA

Sample	Location[1]	Elevation	^{10}Be	^{26}Al	$^{26}Al/^{10}Be$	^{10}Be model age[2]	^{26}Al model age[2]
		feet asl	atoms g^{-1} quartz	atoms g^{-1} quartz		ky	ky
EZ-1	Glade Hill	60	1190000 ± 33000	6040000 ± 200000	5.13 ± 0.27	253 ± 28	228 ± 28
EZ-2	Nr. Lock 12	30	370000 ± 23000	2300000 ± 110000	6.12 ± 0.59	79 ± 10	83 ± 11
GF-10	Mather Gorge	<17	21000 ± 2600	114000 ± 15000	5.45 ± 0.98	4.5 ± 0.7	4.1 ± 0.7
GF-11	Mather Gorge	<17	22500 ± 1500	139000 ± 16000	6.17 ± 0.83	4.8 ± 0.6	4.9 ± 0.8
GF-12	Mather Gorge	<17	30500 ± 1500	165000 ± 19000	5.42 ± 0.69	6.5 ± 0.7	5.9 ± 0.9
GF-13	Mather Gorge	<17	19800 ± 1400	100000 ± 14000	5.04 ± 0.79	4.2 ± 0.5	3.5 ± 0.6

Notes [1] Latitude 39°N

[2] Model ages calculated as described in Table 1.

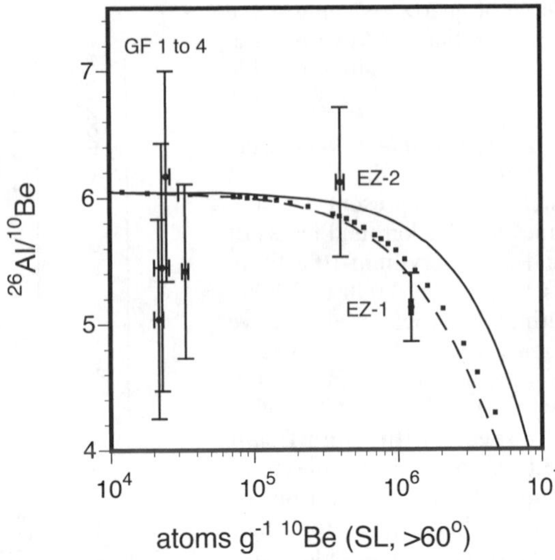

Figure 35. Log-normal, two-nuclide diagram for samples from Great Falls. GF samples have very low nuclide activities, consistent with very short near-surface residence time and no appreciable inheritance of nuclides. All four GF samples are consistent at 2σ with continuous surface exposure; three are consistent at 1σ. EZ samples have much higher nuclide activity. EZ-2 is most consistent with continuous exposure and no erosion, both isotopically and from field evidence of water smoothing. EZ-1 plots below both continuous exposure lines suggesting that the sample has been buried during part or all of its irradiation history. Samples are plotted with 1σ uncertainties. Exposure curve (solid line) and erosion endpoints (black squares) defined by production rates of Nishiizumi et al. (1989). Dashed line is erosion endpoints defined by Bierman et al. (1996) production rate estimates. SL, 60° indicates that nuclide data have been normalized to sea level at high latitude.

Interpreting the complex history of sample EZ-1. The ratio, $^{26}Al/^{10}Be$, helps constrain the near-surface history of sample EZ-1. There is no unique inverse solution to the measured nuclide activities; therefore, we discuss several end-member interpretations below, the calculations for which are based on the production rate estimate of Bierman et al. (1996). Because the sample was taken from an erratic quartzite boulder, fluvially transported to its current location on a high terrace above the Potomac River, there is the added complexity of considering where the boulder received some or all of its cosmic-ray dosing.

The simplest and shortest duration scenarios involve a period of exposure followed by deep burial as explained in Bierman et al. (1999). One possible scenario begins with the boulder accumulating initial nuclide activity upstream, before being deposited on the boulder bar, either by a period of surface exposure or from slow, steady erosion on the basin hillslopes. Because we do not know the elevation at which such dosing might have occurred, we can set only an upper limit for the duration of this initial exposure period, considering the elevation and corresponding nuclide production at the site where the boulder was sampled. Because the sample site must be equal or lower in elevation than where the dosing occurred, the calculated production rates are also lower limits.

Initial boulder exposure of ~280 kyr (cf. Fig. 10D) or erosion at >2.2 m Myr^{-1} (Granger et al. 2001a) are consistent with a two-step model in which the boulder is then buried for a period ranging from 170 kyr in the exposure case or 70 kyr in the erosion case, before being rapidly exposed and then sampled. Such burial might have been caused by the blanket of loess that currently covers the boulder bed in places; perhaps, the boulder we sampled was exposed as the loess was eroded in response to Holocene/Pleistocene climate change. These exposure-prior-to-delivery scenarios suggest that the boulder bed was deposited between 70 and 170 kyr ago. Conversely, we could assume that the boulder arrived on the bar with a negligible inventory of cosmogenic nuclides; such a scenario might imply that it was delivered to the Potomac channel by rock fall and rapidly moved downstream. Once deposited, the boulder might have been irradiated before being deeply buried by loess and recently re-exposed; if this were the case, then the bar was deposited about 450 kyr ago (280+170 ka).

While none of these cases allows us to estimate the age of the boulder bed with any certainty, they all indicate that the exposure history of the boulder was not simple and that its total near-surface residence time exceeds what one would calculate from the measurement of only a single nuclide. Analysis of additional boulders, analysis of subsurface boulders, luminescence dating of the loess, analysis of a shorter-lived *in situ*-produced cosmogenic nuclide such as ^{14}C (Lifton et al. 2001), and analysis of similar boulders perched on lower strath terraces as well as in the modern channel, would help to determine which of the above scenarios is most reasonable and provides more constraints for modeling the duration of different exposure periods.

LOOKING BACKWARD AND FORWARD

Over the last decade, analysis of ^{10}Be, along with the complementary nuclide ^{26}Al, has greatly expanded our understanding of the rate and distribution of processes active at the Earth's surface. Using ^{10}Be, we are now able to constrain the age of numerous landforms that only a decade or two ago could not be dated (Nishiizumi et al. 1991b; Bierman et al. 1995b; Gosse et al. 1995b; Anderson et al. 1996). Cosmogenic ^{10}Be has provided quantification of long-term rock erosion, sediment production, and soil generation rates around the world (Nishiizumi et al. 1986, 1991a; Bierman and Turner 1995; Small et al. 1997, 1999; Heimsath et al. 1999; Bierman and Caffee 2001, 2002). Additional applications have illustrated the source and movement of sediment (Clapp et al. 2000, 2001, 2002), constrained offset rates of faults (Siame et al. 1997; Brown et al. 1998), and illustrated the relative importance of tectonics and climate in controlling erosion rates (Riebe et al. 2000, 2001a,b). Advances in accelerator mass spectrometry and chemical preparation now allow measurement of extremely low activities of both nuclides, on the order of 10^4 atoms g^{-1}. Such sensitivity allows finite measurement of samples exposed for <1000 years at all latitudes and altitudes.

In situations where geologic evidence indicates exposure by a discreet event and minimal erosion of the surface to be sampled, ^{10}Be can be a useful chronometer if inheritance of nuclides from prior periods of irradiation can be shown to be minimal. Experiments in glaciated terrains have returned mixed results in terms of inheritance. In settings where ice was not frozen to the bed and for softer rocks, inheritance appears to be inconsequential (Larsen et al. 1995; Davis et al. 1999). In areas where ice was cold-based or the rocks were exceptionally hard, inheritance may prevent accurate dating of deglaciation (Bierman et al. 1999; Stroeven et al. 2002; Colgan et al. in press). Similarly detailed inheritance studies are being expanded into other geologic environments including fans, river terraces, and paleo beaches. Initial studies show inheritance in these settings can be significant, must be accounted for, and probably limits the applicability of

^{10}Be for dating fluvial and littoral deposits (Nishiizumi et al. 1993; Trull et al. 1995; Anderson et al. 1996; Matmon et al. in press).

In other situations, where rock surfaces are eroding steadily and losing mass in slabs thinner than the penetration depth of cosmic-ray neutrons, ^{10}Be activities can be used to monitor the rate at which rock erodes. Such analyses can lead to a better understanding of the linkages between climate, tectonics, and erosion. For example, cosmogenic nuclide measurements have shown that in most places on Earth, even those areas that are tectonically stable and arid, bare rock surfaces are eroding at rates of at least a few decimeters to meters per million years (Small et al. 1997; Cockburn et al. 1999; Belton et al. 2000; Bierman and Caffee 2001, 2002). This unequivocal indication of landscape dynamism needs to be reconciled with stratigraphic and geomorphic observations suggesting long-term stability of mega-scale landforms (Twidale 1976; Watchman and Twidale 2002).

There is much to do in the future. Nuclide production rates remain uncertain at a level that far exceeds measurement precision; in effect, we are now capable of making very precise measurements, which cannot be interpreted with similar accuracy. While dating of landforms will certainly continue, the many new and diverse applications of ^{10}Be provide an increasingly useful means by which to understand the rate and distribution of Earth surface processes. Measuring rates of soil formation (Heimsath et al. 1999), understanding tectonic and climatic controls on landscapes (Riebe et al. 2001a,b), and estimating the large scale production and movement of sediment (Clapp et al. 2000, 2001, 2002; Nichols et al. 2002) are just a few of the many ways in which the behavior of Earth systems can be understood better by measuring *in situ*-produced ^{10}Be.

ACKNOWLEDGMENTS

Sample analysis was supported by NSF Polar Programs grants to Davis and Bierman, NSF Hydrologic Sciences grants EAR-9628599 and EAR-9396261 to Bierman, NSF Geology and Paleontology grant EAR-9627798 to Mickelson, National Geographic Society grant 5858-97 to Bierman, and DOD grants DAAD-199910143 and DAAH-049610036 to Bierman. Fieldwork in Australia was supported by an ARC grant to R. Twidale. Measurements also were supported in part by DOE grant to LLNL. The U.S. ARO supported time during which Bierman wrote part of this chapter under grant DAAD-199910143. We thank E-an Zen for invaluable field assistance on the Potomac River, his idea to apply ^{10}Be to this site, and his thorough editing of this paper. We thank K. Nichols, E. Grew, J. Beer, and two anonymous reviewers for editing earlier drafts of this manuscript. We are deeply indebted to William Amidon for his tireless work identifying just about every ^{10}Be-related reference that exists. Thanks, Willy!

REFERENCES

Albrecht A, Herzog GF, Klein J, Dezfouly-Arjomandy B, Goff F (1993) Quaternary erosion and cosmic-ray-exposure history derived from ^{10}Be and ^{26}Al produced *in situ*; an example from Pajarito Plateau, Valles Caldera region. Geology 21:551-554

Anderson RS, Repka JL, Dick GS (1996) Explicit treatment of inheritance in dating depositional surfaces using *in situ* ^{10}Be and ^{26}Al. Geology 24:47-51

Andrews JT, Barnett DM (1979) Holocene (Neoglacial) moraine and proglacial lake chronology, Barnes Ice Cap, Canada. Boreas 8:341-358

Arnold JR (1956) Beryllium-10 production by cosmic rays. Science 124:584

Balco G, Stone J, Porter S, Caffee M (2001) Cosmogenic-isotope ages for last-glacial-maximum moraines of the Laurentide ice sheet, southern New England. Geol Soc Am Abstr Progr 32:A-317

Barrows TT, Stone JO, Fifield LK, Cresswell RG (2001) Late Pleistocene glaciation of the Kosciuszko Massif, Snowy Mountains, Australia. Quaternary Res 55:179-189

Barrows TT, Stone JO, Fifield LK, Cresswell RG (2002) The timing of the last glacial maximum in Australia. Quaternary Sci Rev 21:159-173

Belton DX, Brown RW, Kohn BP, Fink D (2000) The first quantitative erosion rate estimates from "the oldest persisting landforms in the world". Fission track 2000; 9th international conference on fission track dating and thermochronology Abstr Geol Soc Aust 58:19-21

Bierman PR (1993) Cosmogenic isotopes and the evolution of granitic landforms. PhD dissertation, University of Washington, St. Louis, Missouri

Bierman PR (1994) Using *in situ* cosmogenic isotopes to estimate rates of landscape evolution: A review from the geomorphic perspective. J Geophys Res 99:13885-13896

Bierman PR, Caffee MW (2001) Steady state rates of rock surface erosion and sediment production across the hyperarid Namib desert and the Namibian escarpment, southern Africa. Am J Sci 301:326-358

Bierman PR, Caffee MW (2002) Cosmogenic exposure and erosion history of ancient Australian bedrock landforms. Geol Soc Am Bull 114:787–803

Bierman PR, Gillespie A (1991) Range Fires: a significant factor in exposure-age determination and geomorphic surface evolution. Geology 19:641-644

Bierman PR, Steig E (1996) Estimating rates of denudation and sediment transport using cosmogenic isotope abundances in sediment. Earth Surf Processes Landforms 21:125-139

Bierman PR, Turner J (1995) ^{10}Be and ^{26}Al evidence for exceptionally low rates of Australian bedrock erosion and the likely existence of pre-Pleistocene landscapes. Quaternary Res 44:378-38

Bierman PR, Gillespie A, Caffee M, Elmer D (1995a) Estimating erosion rates and exposure ages with ^{36}Cl produced by neutron activation. Geochim Cosmochim Acta 59:3779-3798

Bierman PR, Gillespie A, Caffee M (1995b) Cosmogenic age-estimates for earthquake recurrence intervals and debris-flow fan deposition, Owens Valley, California. Science 270:447-450

Bierman PR, Larsen P, Clapp E, Clark D (1996) Refining estimates of 10-Be and 26-Al production rates. Radiocarbon 38:149

Bierman PR, Davis PT, Marsella K, Colgan P, Mickelson DM, Larsen P, Caffee M (1998) What do glaciers take away? What do they leave behind? Geol Soc Am Abstr Progr 30:A-299

Bierman PR, Marsella KA, Davis PT, Patterson C, Caffee M (1999) Mid-Pleistocene cosmogenic minimum-age limits for pre-Wisconsinan glacial surfaces in southwestern Minnesota and southern Baffin Island -- a multiple nuclide approach. Geomorphology 27:25-40

Bierman PR, Davis PT, Caffee MW (2000) Old surfaces on New England summits imply thin Laurentide ice. Geol Soc Am Abstr Progr 31:A-330

Bierman PR, Clapp EM, Nichols KK, Gillespie AR, Caffee M (2001a) Using cosmogenic nuclide measurements in sediments to understand background rates of erosion and sediment transport. *In* Harmon RS, Doe WM (eds) Landscape Erosion and Evolution Modelling, p 89-116

Bierman PR, Marsella KA, Davis PT, Caffee M (2001b) Reply to comment on "Mid-Pleistocene cosmogenic minimum-age limits for pre-Wisconsinan glacial surfaces in southwestern Minnesota and southern Baffin Island -- a multiple nuclide approach". Geomorphology 39:255-261

Bierman PR, Pavich M, Gellis A, Caffee M (2001c) Erosion of the Rio Puerco Basin, New Mexico; first cosmogenic analysis of sediments from the network of a drainage large basin. Geol Soc Am Abstr Progr 32:A-314

Braucher R, Bourles DL, Colin F, Brown ET, Boulange B (1998a), Brazilian laterite dynamics using *in situ*-produced ^{10}Be. Earth Planet Sci Lett 163:197-205

Braucher R, Colin F, Brown ET (1998b) African laterite dynamics using *in situ*-produced ^{10}Be. Geochim Cosmochim Acta 62:1501-1507

Braucher R, Bourles DL, Brown ET, Colin F, Muller J-P, Braun JJ, Delaune M, Edou Minko A, Lescouet C, Raisbeck GM, Yiou F (2000) Application of *in situ*-produced cosmogenic ^{10}Be and ^{26}Al to the study of lateritic soil development in tropical forest; theory and examples from Cameroon and Gabon. Chem Geol 170:95-111

Briner JP, Swanson TW (1998) Using inherited cosmogenic ^{36}Cl to constrain glacial erosion rates of the Cordilleran ice sheet. Geology 26:3-6

Brook EJ, Kurz MD (1993) Surface-exposure chronology using *in situ* cosmogenic ^{3}He in Antarctic quartz sandstone boulders. Quaternary Res 39:1-10

Brook EJ, Kurz MD, Denton GH, Ackert RPJ (1993) Chronology of Taylor Glacier advances in Arena Valley, Antarctica using *in situ* cosmogenic ^{3}He and ^{10}Be. Quaternary Res 39:11-23

Brook EJ, Kurz MD, Ackert RP, Raisbeck G, Yiou F (1995a) Cosmogenic nuclide exposure ages and glacial history of late Quaternary Ross Sea drift in McMurdo Sound, Antarctica. Earth Planet Sci Lett 131:41-56

Brook EJ, Brown ET, Kurz MD, Ackert RP, Raisbeck GM, Yiou F (1995b) Constraints on age, erosion, and uplift of Neogene glacial deposits in the Transantarctic Mountains determined from *in situ* cosmogenic ^{10}Be and ^{26}Al. Geology 23:1063-1066

Brook EJ, Nesje A, Lehman SJ, Raisbeck GM, Yiou F (1996) Cosmogenic nuclide exposure ages along a vertical transect in western Norway: Implications for the height of the Fennoscandian ice sheet. Geology 24:207-210

Brown ET, Edmond JM, Raisbeck GM, Yiou F, Kurz MD, Brook EJ (1991) Examination of surface exposure ages of Antarctic moraines using *in situ*-produced ^{10}Be and ^{26}Al. Geochim Cosmochim Acta 55:2269-2283

Brown ET, Brook EJ, Raisbeck GM, Yiou F, Kurz MD (1992) Effective attenuation of cosmic rays producing ^{10}Be and ^{26}Al in quartz: Implications for exposure dating. Geophys Res Lett 19:369-372

Brown ET, Bourles DL, Colin F, Sanfo Z, Raisbeck GM, Yiou F (1994) The development of iron crust lateritic systems in Burkina Faso, West Africa examined with *in situ*-produced cosmogenic nuclides. Earth Planet Sci Lett 124:19-33

Brown ET, Bourles DL, Colin F, Raisbeck GM, Yiou F, Desgarceaux S (1995a) Evidence for muon-induced production of ^{10}Be in near surface rocks from the Congo. Geophys Res Lett 22:703-706

Brown ET, Stallard RF, Larsen MC, Raisbeck GM, Yiou F (1995b) Denudation rates determined from the accumulation of *in situ*-produced ^{10}Be in the Luquillo Experimental Forest, Puerto Rico. Earth Planet Sci Lett 129:193-202

Brown ET, Trull TW, Jean-Baptiste P, Bourles DL, Raisbeck GM, Yiou F (1996) Direct examination of terrestrial cosmogenic nuclide production rates of ^{10}Be, ^{3}He and ^{3}H. Radiocarbon 38:151

Brown ET, Bourles DL, Burchfield C, Qidong D, Jun L, Molnar P, Raisbeck GM, Yio F (1998) Estimation of slip rates in the southern Tien Shan using cosmic ray exposure dates of abandoned alluvial fans. Geol Soc Am Bull 110:377-386

Burbank DW, Leland J, Fielding E, Anderson RS, Brozovic N, Reid MR, Duncan C (1996) Bedrock incision, rock uplift and threshold hillslopes in the northwestern Himalayas. Nature 379:505-510

Cerling TE, Craig H (1994) Geomorphology and *in situ* cosmogenic isotopes. Ann Rev Earth Planet Sci 22:273-317

Chadwick OA, Hall RD, Phillips FM (1997) Chronology of Pleistocene glacial advances in the Central Rocky Mountains. Geol Soc Am Bull 109:1443-1452

Clapp E, Bierman P (1995) First geomagnetic-based, *in situ*-produced cosmogenic isotope calibration program. Geol Soc Am Abstr Progr 27:A-59

Clapp EM, Bierman PR, Schick AP, Lekach J, Enzel Y, Caffee MW (2000) Differing rates of sediment production and sediment yield. Geology 28:995-998

Clapp E, Bierman PR, Pavich M, Caffee M (2001) Rates of sediment supply to arroyos from uplands determined using *in situ*-produced cosmogenic ^{10}Be and ^{26}Al in sediments. Quaternary Res 55:235-245

Clapp E, Bierman PR, Caffee M (2002) Using ^{10}Be and ^{26}Al to determine sediment generation rates and identify sediment source areas in an arid region drainage basin. Geomorphology 45:67-87

Clark D, Bierman PR, Larsen P (1995) Improving *in situ* cosmogenic chronometers. Quaternary Res 44:366-376

Cockburn HAP, Seidl MA, Summerfield MA (1999) Quantifying denudation rates on inselbergs in the central Namib Desert using *in situ*-produced cosmogenic ^{10}Be and ^{26}Al. Geology 27:399-402

Cockburn HAP, Brown RW, Summerfield MA, Seidl MA (2000) Quantifying passive margin denudation and landscape development using a combined fission-track thermochronology and cosmogenic isotope approach. Earth Planet Sci Lett 79:429-435

Colgan PM, Mickelson DM (1997) Genesis of streamlined landforms and flow history of the Green Bay lobe, Wisconsin, USA. Sediment Geol 111:7-25

Colgan PM, Bierman PR, Mickelson DM, Caffee MW (in press) Variation in glacial erosion near the southern margin of the Laurentide Ice Sheet, south central Wisconsin, USA: implications for cosmogenic dating of glacial terrains. Geol Soc Am Bull

Cutler PM, MacAyeal DR, Parizek BR, Colgan PM (2000) A numerical investigation of ice-lobe-permafrost interaction around the southern Laurentide ice sheet. J Glaciol 46:311-325

Davis PT (1980) Late Holocene glacial, vegetational, climatic history of the Pangnirtung and Kingnait Fiord area, Baffin Island, NWT, Canada. PhD dissertation, University of Colorado, Boulder

Davis PT, Fabel J (1988) Lichenometric dating of debris flow levees, Pangnirtung Pass, Baffin Island: more data and new interpretations. Geol Soc Am Abstr Progr 20:14-15

Davis PT, Kihl R (1985) Late Holocene record of debris avalanche deposits interbedded with lake sediments, Baffin Island, Canada. Geol Soc Am Abstr Progr 17:14

Davis PT, Thompson WB, Stone BD, Newton RM, Fowler BK (1993) Multiple glaciations and deglaciation along a transect from Boston, Massachusetts, to the White Mountains, New Hampshire *In* Cheney JT Hepburn JC (eds) Field Trip Guidebook for the Northeastern United States: 1993 Boston GSA. Geological Society of America, Boulder 2:EE1-27

Davis PT, Bierman PR, Marsella KA, Caffee MW, Southon JR (1999) Cosmogenic analysis of glacial terrains in the eastern Canadian Arctic: A test for inherited nuclides and the effectiveness of glacial erosion. Ann Glaciol 28:181-188

Davis R, Schaeffer OA (1955) Chlorine-36 in nature. Ann New York Acad Sci 62:105-122

Davis RB, Jacobson GL (1985) Late glacial and early Holocene landscapes in northern New England and adjacent areas of Canada. Quaternary Res 23:341-368

Denny CS (1967) Fans and pediments. Am J Sci 265:81-105

Desilets D, Zreda M (2000) Scaling production rates of terrestrial cosmogenic nuclides for altitude and geomagnetic effects. Geol Soc Am Abstr Progr 31:A-400

Desilets D, Zreda M, Lifton NA (2001) Comment on 'Scaling factors for production rates of *in situ*-produced cosmogenic nuclides: a critical reevaluation' by TJ Dunai. Earth Planet Sci Lett 188:283-287

Dunai TJ (2000) Scaling factors for production rates of *in situ*-produced cosmogenic nuclides: a critical reevaluation. Earth Planet Sci Lett 176:157-169

Dunai TJ (2001) Reply to comment on `Scaling factors for production rates of *in situ*-produced cosmogenic nuclides: a critical reevaluation'. Earth Planet Sci Lett 188:289-298

Duncan C, Masek J, Bierman P, Larsen J, Caffee M (2001) Extraordinarily high denudation rates suggested by ^{10}Be and ^{26}Al analysis of river sediments, Bhutan Himalayas. Geol Soc Am Abstr Progr 32:A-312

Dunne A, Elmore D, Muzikar P (1999) Scaling factors for the rates of production of cosmogenic nuclides for geometric shielding and attenuation at depth on sloped surfaces. Geomorphology 27:3-11

Dyke AS (1977) Quaternary geomorphology, glacial chronology, and climatic and sea-level history of southwestern Cumberland Peninsula, Baffin Island, Northwest Territories, Canada. PhD dissertation, University of Colorado, Boulder

Dyke AS (1979) Glacial and sea-level history of southwestern Cumberland Peninsula, Baffin Island, NWT, Canada. Arctic Alpine Res 11:179-202

Dyke AS, Andrews JT, Miller GH (1982) Quaternary geology of the Cumberland Peninsula, Baffin Island, District of Franklin. Geol Surv Canada

Elmore D, Phillips F (1987) Accelerator mass spectrometry for measurement of long-lived radioisotopes. Science 236:543-550

Fabel D, Harbor J (1999) The use of *in situ*-produced cosmogenic radionuclides in glaciology and glacial geomorphology. Ann Glaciol 28:103-110

Finkel R, Suter M (1993) AMS in the Earth Sciences: Technique and applications. Adv Analyt Geochem 1:1-114

Flint RF (1971) Glacial and Quaternary Geology. Wiley, New York

Gillespie AR, Bierman PR (1995) Precision of terrestrial exposure ages and erosion rates from analysis of *in situ*-produced cosmogenic isotopes. J Geophys Res 100:24637-24649

Goldthwait JW, Kruger FC (1938) Weathered rock in and under the drift in New Hampshire. Bull Geol Soc Am 49:1183-1198

Gosse JC, Klein J (1996) Production rate of *in situ* cosmogenic ^{10}Be in quartz at high altitude and mid latitude. Radiocarbon 38:154-155

Gosse JC, Phillips FM (2001) Terrestrial *in situ* cosmogenic nuclides: theory and application. Quaternary Sci Rev 20:1475-1560

Gosse JC, Stone JO (2001) Terrestrial cosmogenic nuclide methods passing milestones toward paleo-altimetry. EOS Trans Am Geophys Union 82:82,86,89

Gosse JC, Grant DR, Klein J (1993) Significance of altitudinal weathering zones in Atlantic Canada, inferred from *in situ*-produced cosmogenic radionuclides. Geol Soc Am Abstr Progr 25:A-394

Gosse JC, Evenson EB, Klein J, Lawn B, Middleton R (1995a) Precise cosmogenic 10-Be measurements in western North America: support for a Younger Dryas cooling event. Geology 23:877-880

Gosse JC, Klein J, Evenson EB, Lawn B, Middleton R (1995b) Beryllium-10 dating of the duration and retreat of the last Pinedale glacial sequence. Science 268:1329-1333

Gosse J, Dort W, Sorenson C, Steeples D, Grimes J, Hecht G, Klein J, Lawn B (1997) Insights on the depositional age and rate of denudation of pre-Illinoian till in Kansas from terrestrial cosmogenic nuclides. Geol Soc Am Abstr Progr 29:17

Gosse JC, Hecht G, Mehring N (1998) Comparison of radiocarbon- and *in situ* cosmogenic nuclide-derived postglacial emergence curves for Prescott Island, central Canadian Arctic. Geol Soc Am Abstr Progr 30:A-298

Granger DE, Muzikar PF (2001) Dating sediment burial with *in situ*-produced cosmogenic nuclides: theory, techniques, and limitations. Earth Planet Sci Lett 188:269-281

Granger DE, Smith AL (2000) Dating buried sediments using radioactive decay and muonogenic production of ^{26}Al and ^{10}Be. Nucl Instr Methods Phys Res 172:822-826

Granger DE, Kirchner JW, Finkel R (1996) Spatially averaged long-term erosion rates measured from *in situ*-produced cosmogenic nuclides in alluvial sediments. Journal of Geology 104:249-257

Granger DE, Kirchner JW, Finkel R (1997) Quaternary downcutting rate of the New River, Virginia, measured from differential decay of cosmogenic ^{26}Al and ^{10}Be in cave-deposited alluvium. Geology 25:107-110

Granger DE, Fabel D, Palmer AN (2001a) Pliocene-Pleistocene incision of the Green River, Kentucky, determined from radioactive decay of cosmogenic ^{26}Al and ^{10}Be in Mammoth Cave sediments. Geol Soc Am Bull 113:825–836

Granger DE, Riebe CS, Kirchner JW, Finkel RC (2001b) Modulation of erosion on steep granitic slopes by boulder armoring, as revealed by cosmogenic ^{26}Al and ^{10}Be. Earth Planet Sci Lett 186:269-281

Guyodo Y, Valet JP (1996) Relative variations in geomagnetic intensity from sedimentary records: the past 200,000 years. Earth Planet Sci Lett 143:23-36

Hallet B, Putkonen J (1994) Surface dating of dynamic landforms: young boulders on aging moraines. Science 265:937-940

Hancock GS, Anderson RS, Whipple, KX (1998) Beyond power; bedrock river incision process and form. Geophysical Monogr 107:35-60

Hancock GS, Anderson RS, Chadwick OA, Finkel RC (1999) Dating fluvial terraces with ^{10}Be and ^{26}Al profiles; application to the Wind River, Wyoming. Geomorphology 27:41-60

Heimsath AM, Dietrich WE, Nishiizumi K, Finkel RC (1997) The soil production function and landscape equilibrium. Nature 388:358-361

Heimsath AM, Dietrich WE, Nishiizumi K, Finkel RC (1999) Cosmogenic nuclides, topography, and the spatial variation of soil depth. Geomorphology 27:151-172

Heimsath AM, Chappell J, Dietrich WE, Nishiizumi K, Finkel RC (2000) Soil production on a retreating escarpment in southeastern Australia. Geology 28:787-790

Heisinger B, Niodermayer M, Hartmann F, Korschinek G, Nolte E, Morteani G, Neumaier S, Patitjean C, Kubik P, Synal A, Ivy-Ochs S (1997) *In situ* production of radionuclides at great depths. Nucl Instr Methods Phys Res 123:341-346

Ivy-Ochs S (1996) The dating of rock surfaces using *in situ*-produced ^{10}Be, ^{26}Al and ^{36}Cl, with examples from Antarctica and the Swiss Alps. PhD dissertation, Swiss Federal Institute of Technology (ETH), Zürich

Ivy-Ochs S, Schluchter C, Kubik PW, Dittrich-Hannen B, Beer J (1995) Minimum ^{10}Be exposure ages of early Pliocene for the Table Mountain plateau and the Sirius Group at Mount Fleming, Dry Valleys, Antarctica. Geology 23:1007-1010

Ivy-Ochs S, Schluchter C, Kubik PW, Synal HA, Beer J, Kerschner H (1996) The exposure age of an Egesen moraine at Julier Pass, Switzerland, measured with the cosmogenic radionuclides ^{10}Be, ^{26}Al and ^{36}Cl. Eclogae geol Helvetica 89:1049-1063

Ivy-Ochs S, Schluchter C, Prentice M, Kubik P, Beer J (1997) ^{10}Be and ^{26}Al Exposure Ages for the Sirius Group at Mount Fleming, Mount Feather and Table Mountain and the Plateau Surface at Table Mountain. *In* Ricci CA (ed) The Antarctic Region: Geological Evolution and Processes. Terra Antarctica, Siena, Italy, p 1153-1158

Ivy-Ochs S, Kubik PW, Masarik J, Wieler R, Bruno L, Schluchter C (1998) Preliminary results on the use of pyroxene for ^{10}Be surface exposure dating. Schweiz mineral petrogr Mitt 78:375-382

Ivy-Ochs S, Schluchter C, Kubik PW, Denton GH (1999) Moraine exposure dates imply synchronous Younger Dryas glacier advances in the European Alps and in the Southern Alps of New Zealand. Geografiska Annaler Ser A. Physical Geogr 81:313-323

Jackson J, Ritz JF, Siame L, Raisbeck G, Yiou F, Norris R, Youngson J, Bennett E (2002) Fault growth and landscape development rates in Otago, New Zealand, using *in situ* cosmogenic ^{10}Be. Earth Planet Sci Lett 195:185-193

Jacobsen PJ, Jacobsen KM, Seely MK (1995) Ephemeral rivers and their catchments. Desert Research Foundation of Namibia, Windhoek

Kaplan MR (1999) The last glaciation of the Cumberland Sound region, Baffin Island, Canada, based on glacial geology, cosmogenic dating, numerical modeling. PhD dissertation, University of Colorado, Boulder

Kaplan MR, Miller GH, Steig EJ (2001) Low-gradient outlet glaciers (ice streams?) drained the Laurentide ice sheet. Geology 29:343-346

Kilius LR, Beukens RP, Chang KH, Lee HW, Litherland AE, Elmore D, Ferraro R, Gove HE, Purser KH (1980) Measurement of ^{10}Be/^{9}Be ratios using an electrostatic tandem accelerator. Nucl Instr Methods Phys Res 171:355-360

Klein J, Giegengack R, Middleton R, Sharma P, Underwood JR, Weeks RA (1986) Revealing histories of exposure using *in situ*-produced ^{26}Al and ^{10}Be in Libyan desert glass. Radiocarbon 28:547-555

Klein J, Gosse J, Davis PT, Evenson EB, Sorenson CJ (2000) Younger Dryas in the Rocky Mountains and calibration of ^{10}Be/^{26}Al production rates. Geol Soc Am Abstr Progr 31:A-473

Kleman J (1994) Preservation of landforms under ice sheets and ice caps. Geomorphology 9:19-32

Klemen J, Stroeven AP (1997) Preglacial surface remnants and Quaternary glacial regimes in northwestern Sweden. Geomorphology 19:35-54

Kohl CP, Nishiizumi K (1992) Chemical isolation of quartz for measurement of *in situ*-produced cosmogenic nuclides. Geochim Cosmochim Acta 56:3583-3587

Kubik PW, Ivy-Ochs S, Masarik J, Frank M, Schluchter C (1998) ^{10}Be and ^{26}Al production rates deduced from an instantaneous event within the dendro-calibration curve, the landslide of Koefels, Oetz Valley, Austria. Earth Planet Sci Lett 161:231-241

Kurz MD (1986) *In situ* production of terrestrial cosmogenic helium and some applications to geochronology. Geochim Cosmochim Acta 50:2855-2862

Kurz MD, Brook EJ (1994) Surface exposure dating with cosmogenic nuclides. *In* Beck C (ed) Dating in exposed and surface contexts. University of New Mexico Press, Albuquerque, p 139-159

Kutschera W, Henning W, Paul M, Stephenson EJ, Yntema JL (1980) Radioisotope detection with the Argonne FN tandem accelerator. Radiocarbon 22:807-815

Lal D (1986) On the study of continental erosion rates and cycles using cosmogenic ^{10}Be and ^{26}Al and other isotopes, *In* Hurford, AJ, Jaeger, E, Ten Cate, J AM (eds) Dating young sediments CCOP/TP 16 (1986) Office of the Project Manager/Coordinator, UNDP Technical Support for Regional Offshore Prospecting in East Asia. United Nations, Economic and Social Commission for Asia and the Pacific, Bangkok, p 285-298

Lal D (1987) Cosmogenic nuclides produced *in situ* in terrestrial solids. Nucl Instr Methods Phys Res B29:238-245

Lal D (1988) *In situ*-produced cosmogenic isotopes in terrestrial rocks. Ann Rev Earth Planet Sci 16: 355-388

Lal D (1991) Cosmic ray labeling of erosion surfaces: *In situ* production rates and erosion models. Earth Planet Sci Lett 104:424-439

Lal D (1998) Cosmic ray-produced isotopes in terrestrial systems. Proc Indian Acad Sci (Earth Planet Sci) 107:241-249

Lal D, Arnold JR (1985) Tracing quartz through the environment. Proc Indian Acad Sci (Earth Planet Sci) 94:1-5

Lal D, Peters B (1967) Cosmic ray-produced radioactivity on the earth. *In* Sitte K (ed) Handbuch der Physik, Springer-Verlag, New York, p 551-612

Lal D, Nishiizumi, K, Klein, J, Middleton, R, Craig, H (1987) Cosmogenic ^{10}Be in Zaire alluvial diamonds: implications for ^{3}He contents of diamonds. Nature 328:139-141

Lal D, Pavich M, Gu ZY, Jull AJT (1996) Recent erosional history of a soil profile based on cosmogenic *in situ* radionudides ^{14}C and ^{10}Be. Geophysical Monogr 95:371-376

Larsen P (1995) *In situ* production rates of cosmogenic ^{10}Be and ^{26}Al over the past 21.5 kyr from the terminal moraine of the Laurentide ice sheet, north-central New Jersey. MS thesis, University of Vermont, Burlington

Larsen P, Bierman PR (1995) Cosmogenic ^{26}Al chronology of the late Wisconsinan glacial maximum in north-central New Jersey Geol Soc Am Abstr Progr 27:63

Larsen PL, Bierman PR, Caffee M (1995) Preliminary *in situ* production rates of cosmogenic ^{10}Be and ^{26}Al over the past 21.5 kyr from the terminal moraine of the Laurentide ice sheet, north-central New Jersey. Geol Soc Am Abstr Progr 27:A59

Leland J, Reid MR, Burbank DW, Finkel R, Caffee M (1998) Incision and differential bedrock uplift along the Indus River near Nanga Parbat, Pakistan Himalaya, from ^{10}Be and ^{26}Al exposure age dating of bedrock straths. Earth Planet Sci Lett 154:93-107

Licciardi JM, Clark P, Brook EJ, Pierce KL, Kurz MD (2001) Cosmogenic ^{3}He and ^{10}Be chronologies of the late Pinedale northern Yellowstone ice cap, Montana, US. Geology 29:1095–1098

Lifton NA (2000) A robust scaling model for *in situ* cosmogenic nuclide production rates. Geol Soc Am Abstr Progr 31:A-400

Lifton NA, Jull AJT, Quade J (2001) A new extraction technique and production rate estimate for *in situ* cosmogenic ^{14}C in quartz. Geochim Cosmochim Acta 65:1953-1969

Litherland AE (1980) Ultrasensitive mass spectrometry with accelerators. Ann Rev Nuclear Particle Sci 30:437-473

Loken OH (1966) Baffin Island refugia older than 54,000 years. Science 153:1378-3380

Marsella KA (1998) Timing and extent of glaciation in the Pangnirtung Fjord region, southern Cumberland Peninsula: determined using *in situ*-produced cosmogenic ^{10}Be and ^{26}Al. MS thesis, University of Vermont, Burlington

Marsella K, Bierman PR, Davis PT, Caffee MW (2000) Cosmogenic ^{10}Be and ^{26}Al ages for the last glacial maximum, eastern Baffin Island, Arctic Canada. Geol Soc Am Bull 112:1296-1312

Masarik J, Reedy RC (1994) Simulation of cosmogenic nuclide production in terrestrial rocks. Abstr 8th Intl Conf Geochronology, Cosmochronology, and Isotope Geology. U S Geol Surv Circular 1107:204

Masarik J, Reedy RC (1995) Terrestrial cosmogenic-nuclide production systematics calculated from numerical simulations. Earth Planet Sci Lett 136:381-396

Masarik J, Frank M, Schafer J, Wieler R (2001) Correction of *in situ* cosmogenic nuclide production rates for geomagnetic field intensity variations during the past 800,000 years. Geochim Cosmochim Acta 65:2995-3003

Matmon A, Bierman PR, Southworth S, Pavich M, Caffee M (2001a) Temporally and spatially uniform rates of erosion in the Great Smoky Mountains, Tennessee and North Carolina. Geol Soc Am Abstr Progr 33:A-315

Matmon A, Bierman PR, Southworth S, Pavich M, Caffee M, Finkel R (2001b) Rates of erosion determined from ^{10}Be analysis of sediments, Great Smoky Mountains, Tennessee and North Carolina. Eos Trans Am Geophys Union 82:455

Matmon A, Crouvi O, Enzel Y, Bierman PR, Larsen J, Porat N, Amit R, Caffee M (in press) Complex exposure histories of chert clasts in the late Pleistocene shorelines of Lake Lisan, southern Israel. Earth Surf Processes Landforms.

McCorkle R, Fireman EL, Langway CC (1967) Aluminum-26 and Beryllium-10: Greenland Ice. Science 158:1690-1692

Mickelson DM, Clayton L, Fullerton DS, Borns HW (1983) Late glacial record of the Laurentide Ice Sheet in the United States. *In* Wright HE (ed) Late Quaternary Environments of the United States. University of Minnesota Press, Minneapolis, p 3-37

Middleton R (1990) A negative ion cookbook. Dept of Physics, University of Pennsylvania, Philadelphia

Middleton R, Klein J (1987) ^{26}Al: measurement and applications. Phil Trans Royal Soc London A 323: 121-143

Middleton R, Klein J, Brown L, Tera F (1984) ^{10}Be in commercial beryllium. Nucl Instr Methods Phys Res B5:511-513

Middleton R, Brown L, Dezfouly-Arjomandy B, Klein J (1993) On ^{10}Be standards and the half-life of ^{10}Be. Nucl Instr Methods Phys Res B82:399-403

Miller GH (1973) Late Quaternary glacial and climatic history of northern Cumberland Peninsula, Baffin Island, NWT, Canada. Quaternary Res 3:561-583

Miller GH, Dyke AS (1974) Proposed extent of late Wisconsin Laurentide ice on Baffin Island. Geology 2:125-130

Miller GH, Andrews JT, Short SK (1977) The last interglacial-glacial cycle, Clyde foreland, Baffin Island, NWT Stratigraphy, biostratigraphy, and chronology. Canad J Earth Sci 14:2824-2857

Miller GH, Wolfe AP, Steig EJ, Sauer PE, Kaplan MR, Briner JP (2002) The Goldilocks dilemma; big ice, little ice, or "just-right" ice in the eastern Canadian Arctic. Quaternary Sci Rev 21:33-48

Molnar P, Brown ET, Burchfield C, Qidong F, Xiayne L, Jun L, Raisbeck G, Jianbang S, Zhangming W, Yiou F, Huichuan Y (1994) Quaternary climate change and the formation of river terraces across growing anticlines on the north Flank of the Tien Shan, China. J Geol 102:583-602

Morris JD (1991) Applications of cosmogenic ^{10}Be to problems in the earth sciences. Annual Review of Earth and Planetary Science 19:313-350

Nelson DE, Korteling R, Southon J, Nowikow I, Hammaren E, Burke DG, McKay JW (1979) Annual program report McMaster Accel Lab, Hamilton, p 93-96

Nichols KK, Bierman PR, Caffee M (2000) The Blackhawk keeps its secrets: landslide dating using *in situ* 10-Be. Geol Soc Am Abstr Progr 32:A-400

Nichols KK, Bierman PR, Hooke RL, Clapp E, Caffee M (2002) Quantifying sediment transport on desert piedmonts using ^{10}Be and ^{26}Al. Geomorphology 45:105-125

Nishiizumi K (1990) Cosmogenic 10-Be, 26-Al, and 3-He in olivine from Maui lavas. Earth Planet Sci Lett 98:263-266

Nishiizumi K (1996) Production rate of ^{10}Be and ^{26}Al on the surface of the Earth and underground. Radiocarbon 38:164-165

Nishiizumi K, Regnier S, Marti K (1980) Cosmic ray exposure ages of chondrites, pre-irradiation and constancy of cosmic ray flux in the past. Earth Planet Sci Lett 50:156-170

Nishiizumi K, Elmore D, Ma XZ, Arnold JR (1984) ^{10}Be and ^{36}Cl depth profiles in Apollo 15 drill core. Earth Planet Sci Lett 70:157-163

Nishiizumi K, Lal D, Klein J, Middleton R, Arnold JR (1986) Production of ^{10}Be and ^{26}Al by cosmic rays in terrestrial quartz *in situ* and implications for erosion rates. Nature 319:134-136

Nishiizumi K, Winterer EL, Kohl CP, Klein J, Middleton R, Lal D, Arnold JR (1989) Cosmic ray production rates of ^{10}Be and ^{26}Al in quartz from glacially polished rocks. J Geophys Res 94: 17907-17915

Nishiizumi K, Kohl CP, Arnold JR, Klein J, Fink D, Middleton R (1991a) Cosmic ray-produced ^{10}Be and ^{26}Al in Antarctic rocks: exposure and erosion history. Earth Planet Sci Lett 104:440-454

Nishiizumi K, Kohl CP, Shoemaker EM, Arnold JR, Klein J, Fink D, Middleton R (1991b) *In situ* ^{10}Be-^{26}Al exposure ages at Meteor Crater, Arizona. Geochim Cosmochim Acta 55:2699-2703

Nishiizumi K, Kohl CP, Arnold JR, Dorn R, Klein J, Fink D, Middleton R, Lal D (1993) Role of *in situ* cosmogenic nuclides ^{10}Be and ^{26}Al in the study of diverse geomorphic processes. Earth Surface Processes and Landforms 18:407-425

Nishiizumi K, Finkel R, Brimhall G, Mote T, Mueller G, Tidy E (1998) Ancient exposure ages of alluvial fan surfaces compared with incised stream beds and bedrock in the Atacama Desert of north Chile. Geol Soc Am Abstr Progr 30:A-298

Nott J, Roberts R (1996) Time and process rates over the past 100 Myr; a case for dramatically increased landscape denudation rates during the late Quaternary in northern Australia. Geology 24:883-888

Ochs M, Ivy-Ochs S (1997) The chemical behavior of Be, Al, Fe, Ca, and Mg during AMS target preparation from terrestrial silicates modeled with chemical speciation calculations. Nucl Instr Methods Phys Res B123:235-240

Owen LA, Finkel R C, Caffee MW, Gualtieri L (2002) Timing of multiple late Quaternary glaciations in the Hunza Valley, Karakoram Mountains, northern Pakistan: defined by cosmogenic radionuclide dating of moraines. Geol Soc Am Bull 114:593–604

Partridge TC, Maud RR (1987) Geomorphic evolution of Southern Africa since the Mesozoic. Trans Geol Soc South Africa 90:179-208

Pazzaglia F, Gardner T (1993) Fluvial terraces of the lower Susquehanna River. Geomorphology 8:83-113

Perg LA, Anderson RS, Finkel RC (2001) Use of a new ^{10}Be and ^{26}Al inventory method to date marine terraces, Santa Cruz, California, USA. Geology 29:879-882

Phillips FM, Zreda MG, Smith SS, Elmore D, Kubik PW, Sharma P (1990) Cosmogenic Chlorine-36 chronology for glacial deposits at Bloody Canyon, eastern Sierra Nevada. Science 248:1529-1532

Phillips FM, Zreda MG, Gosse JC, Klein J, Evenson, EB, Hall RD, Chadwick OA, Sharma P (1997) Cosmogenic ^{36}Cl and ^{10}Be ages of Quaternary glacial and fluvial deposits of the Wind River Range, Wyoming. Geol Soc Am Bull 109:1453-1463

Phillips WM, Sloan VF, Shroder JF, Sharma P, Clarke ML, Rendell HM (2000) Asynchronous glaciation at Nanga Parbat, northwestern Himalaya Mountains, Pakistan. Geology 28:431-434

Raisbeck GM, Yiou F (1989) Dating by cosmogenic isotopes ^{10}Be, ^{26}Al and ^{41}Ca. *In* Roth E, Poty B, Menager MT, Coulomb J (eds) Nuclear methods of dating. Kluwer Acad Publ, Dordrecht, Netherlands, p 353-378

Raisbeck GM, Yiou F, Fruneau M, Loiseaux JM (1978) Beryllium-10 mass spectrometry with a cyclotron. Science 202:215-217

Raymo ME, Ruddiman WF (1988) Influence of late Cenozoic mountain building on ocean geochemical cycles. Geology 16:649-653

Reedy RC, Nishiizumi, K, Lal D, Arnold JR, Englert PAJ, Klein J, Middleton R, Jull AJT, Donahue D J (1994) Simulations of terrestrial *in situ* cosmogenic-nuclide production. Nucl Instr Methods Phys Res B 92:297-300

Repka JL, Anderson RS, Finkel RC (1997) Cosmogenic dating of fluvial terraces, Fremont River, Utah. Earth Planet Sci Lett 152:59-73

Riebe CS, Kirchner JW, Granger DE, Finkel RC (2000) Erosional equilibrium and disequilibrium in the Sierra Nevada, inferred from cosmogenic ^{26}Al and ^{10}Be in alluvial sediment. Geology 28:803-806

Riebe CS, Kirchner JW, Granger DE, Finkel RC (2001a) Strong tectonic and weak climatic control of long-term chemical weathering rates. Geology 29:511–514

Riebe CS, Kirchner JW, Granger DE, Finkel RC (2001b) Minimal climatic control on erosion rates in the Sierra Nevada, California. Geology 29:447-450

Ritz JF, Brown ET, Bourles DL, Philip H, Schlupp A, Raisbeck GM, Yiou F, Enkhtuvshin B (1995) Slip rates along active faults estimated with cosmic-ray exposure dates: application to the Bogd fault, Gobi-Altai, Mongolia. Geology 23:1019-1022

Sarda P, Staudacher T, Allegre C, Lecomte A (1993) Cosmogenic neon and helium at Reunion: measurement of erosion rate. Earth Planet Sci Lett 119:405-417

Saunders I, Young A (1983) Rates of surface processes on slopes, slope retreat, and denudation. Earth Surface Processes and Landforms 8:473-501

Schaller M, von Blanckenburg F, Hovius N, Kubik PW (2001) Large-scale erosion rates from *in situ*-produced cosmogenic nuclides in European river sediments. Earth Planet Sci Lett 188:441-458

Schroeder PA, Melear ND, Bierman PR, Kashgarian M, Caffee MW (2001) Apparent gibbsite growth ages for the regolith in the Georgia Piedmont. Geochim Cosmochim Acta 65:381-386

Seidl MA, Finkel RC, Caffee MW, Hudson B, Dietrich WE (1997) Cosmogenic isotope analyses applied to river longitudinal profile evolution: problems and interpretations. Earth Surf Processes Landforms 22:195-209

Sharma P, Middleton L (1989) Radiogenic production of ^{10}Be and ^{26}Al in uranium and thorium ores: implications for studying terrestrial samples containing low levels of ^{10}Be and ^{26}Al. Geochim Cosmochim Acta 53:709-716

Shepard MK, Arvidson RE, Caffee M, Finkel R, Harris L (1995) Cosmogenic exposure ages of basalt flows; Lunar Crater volcanic field, Nevada. Geology 23:21-24

Short SK, Mode WN, Davis PT (1985) The Holocene record from Baffin Island: Modern and fossil pollen studies. *In* Andrews JT (ed) Quaternary Environments: Eastern Canadian Arctic, Baffin Bay, and West Greenland 22:608-642, Allen & Unwin

Siame LL, Bourles DL, Sebrier M, Bellier O, Castano JC, Araujo M, Perez M, Raisbeck GM, Yiou F (1997) Cosmogenic dating ranging from 20 to 700 ka of a series of alluvial fan surfaces affected by the El Tigre Fault, Argentina. Geology 25:975-978

Small EE, Anderson RS, Repka JL, Finkel R (1997) Erosion rates of alpine bedrock summit surfaces deduced from *in situ* ^{10}Be and ^{26}Al. Earth Planet Sci Lett 150:413-425

Small EE, Anderson RS, Hancock GS (1999) Estimates of the rate of regolith production using ^{10}Be and ^{26}Al from an alpine hillslope. Geomorphology 27:131-150

Steig EJ, Wolfe AP, Miller GH (1998) Wisconsinan refugia and the glacial history of eastern Baffin Island, Arctic Canada; coupled evidence from cosmogenic isotopes and lake sediments. Geology 26:835-838

Stewart AJ, Balke DH, Ollier CD (1986) Cambrian river terraces and ridgetops in central Australia, oldest persisting landforms? Science 233:758-760

Stone J (1998) A rapid fusion method for separating ^{10}Be from soils and silicates. Geochim Cosmochim Acta 62:555-561

Stone J (2000) Air pressure and cosmogenic isotope production. J Geophys Res 105:23753-23759

Stone J, Allan GL, Fifield LK, Evans JM, Chivas AR (1994) Limestone erosion measurements with cosmogenic chlorine-36 in calcite - preliminary results from Australia. Nucl Instr Methods Phys Res B92:311-316

Stone J, Ballantyne CK, Fifield LK (1998) Exposure dating and validation of periglacial weathering limits, northwest Scotland. Geology 26:587–590

Stroeven A, Fabel D, Hättestrand C, Harbor J (2002) A relict landscape in the center of Fennoscandian glaciation; cosmogenic radionuclide evidence of tors preserved through multiple glacial cycles. Geomorphology 45:145-154

Summerfield MA, Sugden DE, Denton GH, Marchant DR, Cockburn HAP, Stuart F M (1999) Cosmogenic isotope data support previous evidence of extremely low rates of denudation in the Dry Valleys region, southern Victoria Land, Antarctica. *In* Smith BJ, Whalley WB, Warke PA (eds) Uplift, erosion and stability; perspectives on long-term landscape development. Geol Soc London Spec Pub 162:255-267

Tera F, Brown L, Morris J, Sacks IS, Klein J, Middleton R (1986) Sediment incorporation in island-arc magmas: inferences from ^{10}Be. Geochim Cosmochim Acta 50:535-550

Tschudi S, Ivy–Ochs S, Schluchter C, Kubik P, Rainio H (2000) ^{10}Be dating of Younger Dryas Salpausselka I formation in Finland. Boreas 29:287-294

Trull TW, Brown ET, Marty B, Raisbeck GM, Yiou F (1995) Cosmogenic ^{10}Be and ^{3}He accumulation in Pleistocene beach terraces in Death Valley, California: implications for cosmic-ray exposure dating of young surfaces in hot climates. Chem Geo 119:191-207

Tuniz C, Bird JR, Fink D, Herzog GF (1998) Accelerator Mass Spectrometry: 'Ultrasensitive analysis for global science' CRC Press, Boca Raton, Florida

Turekian KK, Cochran JK, Krishnaswami S, Langford WA, Parker PD, Bauer KA (1979) The measurement of ^{10}Be in manganese nodules using a tandem Van De Graaff accelerator. Geophys Res Lett 6:417-420

Twidale CR (1976) On the survival of paleoforms. Am J Sci 276:77-95

Twidale CR (1982) Granite Landforms Elsevier, Amsterdam

Twidale CR (1997a) Limitations and applications of, and alternatives to, *in situ* cosmogenic nuclide dating. Assoc Am Geographers 93rd Ann Meeting Abstr:270-271

Twidale CR (1997b) Comment on "^{10}Be and ^{26}Al Evidence for Exceptionally Low Rates of Australian Bedrock Erosion and the Likely Existence of Pre-Pleistocene Landscapes" (Bierman and Turner 1995). Quaternary Res 48:381-385

Twidale CR, Campbell EM, Foale MR (1985) Uncontitchie Hill Schmucker, Adelaide

van der Woerd J, Ryerson FJ, Tapponnier P, Gaudemer Y, Finkel R, Meriaux AS, Caffee M, Zhao G, He Q (1998) Holocene left-slip rate determined by cosmogenic surface dating on the Xidatan segment of the Kunlun Fault (Qinghai, China). Geology 26:695-698

van der Woerd J, Ryerson FJ, Tapponnier P, Meriaux AS, Gaudemer Y, Meyer B, Finkel RC, Caffee MW, Gouguang Z, Zhiqin X (2000) Uniform slip-rate along the Kunlun Fault; implications for seismic behavior and large-scale tectonics. Geophy Res Lett Research Lett 27:2353-2356

Ward JD (1987) The Cenozoic succession in the Kuiseb Valley, central Namib Desert. Geol Surv Namibia Mem 9:124

Watchman A, Twidale CR (2002) Relative and 'absolute' dating of land surfaces. Earth-Sci Rev 58:1-49

Weissel JK, Seidl MA (1998) Inland propagation of erosional escarpments and river profile evolution across the southeast Australian passive continental margin. Geophysical Monogr 107:189-206

Wolfe AP, Steig E, Kaplan MR (2001) An alternative model for the geomorphic history of pre-Wisconsinan surfaces on Baffin Island: a comment on Bierman et al. (Geomorphology 25:25-39). Geomorphology 39:251-254

Yiou F, Raisbeck GM (1972) Half life of ^{10}Be. Phys Rev Lett 29:373-375

Yokoyama Y, Reyss J, Guichard F (1977) Production of radionuclides by cosmic rays at mountain altitudes. Earth Planet Sci Lett 36:44-50

Yokoyama T, Makishima A, Nakamura E (1999) Evaluation of the coprecipitation of incompatible trace elements with fluoride during silicate rock dissolution by acid digestion. Chem Geol 157:175-187

Zehfuss PH, Bierman PR, Gillespie AR, Burke RM, Caffee MW (2001) Slip rates on the Fish Springs fault, Owens Valley, California deduced from cosmogenic ^{10}Be and ^{26}Al and relative weathering of fan surfaces. Geol Soc Am Bull 113:241-255

Zen E-A (1997) The seven-story river: Geomorphology of the Potomac Gorge channel between Blockhouse Point, Maryland, and Georgetown, District of Columbia, with emphasis on the gorge complex below Great Falls. U S Geol Surv Open File Report 77

Zen E-A, Prestegaard KL (1994) Possible hydraulic significance of two kinds of potholes: examples from the paleo-Potomac River. Geology 22:47-50

Zimmerman SG, Evenson EB, Gosse JC, Erskine CP (1994) Extensive boulder erosion resulting from a range fire on the type-Pinedale moraines, Fremont Lake, Wyoming. Quaternary Res 42:255-265

Zreda M, Phillips F (1998) Quaternary dating by cosmogenic nuclide buildup in surficial materials. *In* Sowers, JM, Noller JS, Lettis WR (eds) Dating and earthquakes: Review of Quaternary geochronology and its application to paleoseismology. U S Nuclear Regulatory Comm, p 2-101–2-127

5 Cosmogenic Be-10 and the Solid Earth: Studies in Geomagnetism, Subduction Zone Processes, and Active Tectonics

Julie D. Morris

Department of Earth and Planetary Sciences, One Brookings Drive, CB 1169
Washington University, St. Louis, Missouri 63130
jmorris@levee.wustl.edu

John Gosse

Department of Geology, 120 Lindley Hall
University of Kansas, Lawrence, Kansas 66045
Now at: Dept. Earth Sciences, Dalhousie University
3006 LSC, Halifax, Nova Scotia, Canada B3H 3J5

Stefanie Brachfeld

Institute for Rock Magnetism, University of Minnesota
310 Pillsbury Drive, Minneapolis, Minneapolis 55455
Now at: Byrd Polar Research Center, Ohio State University
1090 Carmack Road, Columbus, Ohio 43210

Fouad Tera

Department of Terrestrial Magnetism
Carnegie Institution of Washington
5241 Broad Branch Road
Washington, DC 20015

INTRODUCTION

Since the mid-1980s, cosmogenic ^{10}Be, with a 1.5 Myr half-life, has proven to be an extremely useful tool for studies of the solid Earth and surface processes. Measured at very low concentrations using a particle accelerator, ^{10}Be reveals tantalizing clues to the behavior of the Earth's geodynamo, permits "geochemical imaging" of physical and magmatic processes in subduction zones, and provides ages and uplift and incision rates essential for understanding active tectonic processes. This paper emphasizes the utility of atmospheric and *in situ*-produced ^{10}Be in understanding these solid Earth processes, using igneous and sedimentary rocks, deep-sea and lacustrine sediment, and ice cores as archives.

One focus of this contribution is on studies of the geodynamo and subduction zone processes that utilize ^{10}Be produced by galactic cosmic radiation (GCR) in the atmosphere and subsequently adsorbed onto marine sediments. Following the extensive treatment of geomagnetic effects on cosmic radiation by Størmer (1955), Elsasser et al. (1956) recognized that variations in cosmogenic nuclide production rates could be used as proxies for changes in intensity of the Earth's magnetic field. A year earlier, Peters (1955) had also proposed that ^{10}Be may be favorable for recording Cenozoic marine sedimentation rates and other geophysical variations. High-resolution ^{10}Be-depth profiles in marine sediments have the potential to evaluate the suggestion of asymmetric sawtooth variation in the geomagnetic field paleointensity and to assess the speculation that Milankovitch variations in Earth's orbital parameters may influence the geodynamo. Globally coherent and systematic co-variations in paleointensity and ^{10}Be concentrations

1529-6466/00/0050-0005$10.00

in marine sediments and ice cores offer promise as a tool for correlating and dating marine sediments unsuited to other methods. The use of geomagnetic field paleointensity variations, derived magnetically and from cosmogenic nuclides, shows promise as an interhemispheric correlation tool at sub-Milankovitch time scales.

Coupled studies of ^{10}Be in sediments supplied to subduction trenches, archived in any accreted sediments, and erupting in volcanic arcs allows "geochemical imaging" of sediment accretion, underplating, subduction, and recycling at convergent margins. Flux balances for ^{10}Be on the incoming plate vs. that erupted from the volcanic arc constrain the volumes of sediment subducted to the depths of magma generation. High ^{10}Be concentrations in lavas erupting behind the volcanic front, as well as at it, map the path of slab-derived elements through the mantle. Combined U-series and ^{10}Be studies constrain time scales of magmatic processes.

Studies of late Cenozoic tectonics use ^{10}Be (and other cosmogenic nuclides) produced *in situ* through cosmic ray bombardment of rocks and minerals exposed at Earth's surface. Secondary nucleons produced in the atmosphere have sufficient energy to produce ^{10}Be through spallation and nuclear capture interactions. The concentration of ^{10}Be produced *in situ* in a rock is proportional to the time the surface has been exposed and decreases with decay during burial or with surface erosion. However, due to erosion on Earth's surface, the *in situ* ^{10}Be clock is useful only for the last few million years, unlike its atmospheric counterpart. Exposure ages for tuffs and lavas have been used to assess geologic hazards, while exposure histories for bedrock and faulted alluvial fans constrain Quaternary paleoseismicity and slip rates. ^{10}Be dating of otherwise undatable raised shorelines holds great promise for improving boundary conditions for mantle rheology and lithospheric flexure models. Bedrock incision rates and paleoaltitudes determined from cosmogenic radionuclides are central to studies of mountain evolution.

Following a brief discussion of analytical methods, this paper is divided into three major subsections, focusing on geomagnetism, subduction zone processes, and recent tectonic processes. Each section provides pertinent background, a review of published literature, and discussion of work in progress. Each also proposes future directions and any technical developments necessary. The reader is referred to Bierman et al. (this volume) for applications of ^{10}Be to studies of climate and Earth surface processes. Other articles reviewing the application of ^{10}Be to the Earth sciences include McHargue and Damon (1991), Morris (1991), Cerling and Craig (1994), Bierman (1994), Gillespie and Bierman (1995), Lal (1996) and Gosse and Phillips (2001).

BACKGROUND

This section provides an overview of ^{10}Be systematics as they relate to laboratory measurements and field applications. First is a brief discussion of laboratory methods, detection limits and analytical uncertainties. This is followed by a short section on the behavior of atmospherically produced ^{10}Be in the oceans and during sedimentation. A review of the principles for applying ^{10}Be produced *in situ*, with a focus on recent progress and aspects of the systematics that remain poorly known, concludes this section.

^{10}Be measurements

^{10}Be concentrations are determined by isotope dilution measurements using a tandem accelerator configured as a mass spectrometer (Finkel and Suter 1993). Stable ^{9}Be is added as a spike/carrier, equilibrated with the sample, and extracted for measurement as BeO. Sample preparation techniques vary widely depending on application. Leaching or total dissolution methods are typically used to extract ^{10}Be and ^{9}Be from 0.1-1 g of marine sediments. Volcanic rocks for subduction studies are prepared by total digestion of 5-g samples, followed by selective precipitation or column extraction of Be. Studies of *in situ* ^{10}Be typically use selective dissolution to isolate ~30-50 g of quartz, which is then completely digested for analysis. BeO is selectively precipitated following ion chromatography.

Measurable ^{10}Be concentrations range from 10^4 atoms g^{-1} to 10^9 atoms g^{-1}, with the lower concentrations at, or very near, blank-determined detection limits. Expressed more conventionally, these concentrations correspond to sub-femtogram to picogram levels. Spiked samples presented to the accelerator generally have $^{10}Be/^{9}Be$ ratios in the range 10^{-10} to 10^{-15}. Such low ratios are measurable by Accelerator Mass Spectrometry (AMS) because ^{10}BeO acceleration allows highly sensitive energy separation to be combined with mass separation to identify and count ^{10}Be. Recent results show that spiked samples with atom ratios $>10^{-15}$ can be measured to ±2-5% (1 σ) at the AMS facility at Lawrence Livermore National Laboratory. These uncertainties are large relative to those routinely achieved by thermal ionization mass spectrometry or multi-collector inductively coupled plasma mass spectrometry, which are used for isotope systems with larger ratios (typically $>10^{-6}$). The uncertainties in AMS ^{10}Be measurements are small, however, relative to the range observed in nature. Every study should be closely scrutinized for blank levels and sample reproducibility relative to the range reported for the samples of interest. In the following, ^{10}Be ages or dates are expressed as "ka" or "Ma", while "kyr" and "Myr" refer to the duration of an event or period.

Atmospheric ^{10}Be in marine sediments and glacial ice

^{10}Be in marine sediments, measured by beta counting, was the focus of some of the earliest published works in this field (e.g., Somyajulu 1977; Tanaka and Inouye 1979). More recently, studies of geomagnetism and subduction zone processes, discussed below, use the atmospherically produced ^{10}Be as a proxy or tracer. Atmospheric cosmogenic nuclides accumulate in natural archives such as lacustrine and marine sediment, ice sheets, and loess sequences. This paper will focus on cosmogenic nuclide time series data obtained from water-lain sediment and ice cores. Discussions regarding cosmogenic nuclides as paleoclimate tracers in marine sediment and in loess-soil sequences can be found in Aldahan et al. (1997), Chengde et al. (1992), Beer et al. (1992), Ning et al. (1994), and Gu et al. (1996).

Atmospherically produced ^{10}Be is made by spallation (high energy fission) reactions on O and N, at a production rate second only to that of ^{14}C. Monaghan et al. (1985/86) estimated the globally averaged ^{10}Be atmospheric production rate at 1.2 million atoms cm^{-2} a^{-1}. The bulk of the ^{10}Be (~75%) is produced in the troposphere (Lal and Peters 1967), where a 1-year residence time allows the latitudinal variation in production rate to be largely mixed out (Raisbeck et al. 1981a).

The concentration of a cosmogenic nuclide in sediment or ice depends in part on production and in part on the subsequent transport of the nuclide to the site of deposition. The transfer of cosmogenic isotopes to the Earth's surface is controlled by mixing in the stratosphere and troposphere, with the maximum deposition of cosmogenic isotopes predicted to occur at approximately 40° latitude (Lal and Peters 1967). ^{10}Be (and ^{36}Cl) and ^{14}C are subject to different processes after production. ^{14}C is oxidized to $^{14}CO_2$, which becomes homogenized in the atmosphere with a residence time of 6-7 years. $^{14}CO_2$ then exchanges between the atmosphere, biosphere, and ocean (Siegenthaler et al. 1980). The movement of ^{14}C through the global carbon sample dampens short period variations in the production rate.

^{10}Be does not form gaseous compounds at atmospheric conditions. ^{10}Be atoms in the atmosphere become attached to aerosols, micron- and sub-micron-sized solid and liquid particles, which are carried to the Earth's surface by wet and dry precipitation (McHargue and Damon 1991). ^{10}Be does not participate in the global carbon cycle. Since the residence time of ^{10}Be in the atmosphere is only 1-2 years (Raisbeck et al. 1981a) it is better suited to monitor short period variations in production.

The short residence time in the atmosphere means that ^{10}Be deposition at a given site is strongly dependent on the local precipitation rate. This is critically important for ice core records, in which the source of moisture, transport path, and precipitation rate at the sample site will all affect the concentration of cosmogenic nuclides in the ice. A 30,000 year record of ^{10}Be accumulation from Dome C, Antarctica, indicated that ^{10}Be concentrations in late Pleistocene ice were a factor of 2 to 3 higher than in Holocene ice (Raisbeck et al.1981b). This was attributed to lower precipitation during the late Pleistocene. Similar correlations between ^{10}Be concentration and precipitation have been observed in ice cores from Vostok and Dome C, Antarctica (Raisbeck et al.1987) and in modern firn cores from Renland, East Greenland and Droning Maud Land, Antarctica (Aldahan et al. 1998).

In addition, ^{10}Be accumulating on ice sheets may have multiple sources. When used as a tracer in ice core records, ^{10}Be is assumed to have come directly from production in the atmosphere. However, wind-borne dust particles may bring inherited ^{10}Be to the ice (McHargue and Damon 1991; Baumgartner et al. 1997a). Dust particles generated at the Earth's surface carry old ^{10}Be from atmospheric rain out or ^{10}Be produced *in situ*. Inherited ^{10}Be can potentially be several million years old when it is deposited on the ice sheet. Once in the ice, it is not known if ^{10}Be is mobile. Inherited ^{10}Be on dust may desorb into the ice, or dust particles trapped in the ice may scavenge the freshly produced atmospheric ^{10}Be. The inherited ^{10}Be contribution of terrestrial dust in polar ice cores is often neglected since the dust content itself is very small in polar ice, particularly during the Holocene (Thompson et al. 1994). However, low latitude ice cores have a very high dust content (e.g., Thompson et al. 2000a,b). Efforts to use ^{10}Be as a geophysical tracer in these records will require a means of separating fresh atmospheric ^{10}Be from inherited ^{10}Be in terrestrial dust; Baumgartner et al. (1997a) give a very thorough discussion of this matter, and address potential methods of quantifying inherited versus fresh ^{10}Be in ice.

Once introduced to the ocean ^{10}Be does not re-exchange with the atmosphere. Dissolved ^{10}Be is removed from the ocean by scavenging to particles and burial in sediments. If ^{10}Be deposition to the seafloor were globally uniform, the steady-state ^{10}Be inventory (total ^{10}Be in the sediment column, calculated for a column 1 cm^2, from the seawater-sediment interface to basement) everywhere would be $\sim 2.6\times10^{12}$ atoms cm^{-2}. However, dissolved ^{10}Be is laterally transported in the ocean by different water masses and preferentially deposited in regions of high particle flux (boundary scavenging) (Anderson et al. 1990). ^{10}Be in sediment is redistributed by bottom currents, which cause sediment focusing or winnowing. These processes can cause apparent ^{10}Be accumulation rates in excess of global production (e.g., Anderson et al. 1990; Lao et al. 1992; 1993; Frank et al. 1994, 1995, 1999, 2000; Chase et al. in press). Continental margin localities typically have higher deposition rates (1.5 to 6 times the globally averaged value), with much lower rates observed in open ocean locations (as low as 0.2 times the globally averaged value). Because boundary scavenging is important in ^{10}Be transport to the sea floor, large variations are observed in the amount of ^{10}Be supplied to different subduction trenches.

^{10}Be concentrations in very young marine sediments also vary as a function of lithology, grain size, and secular variation in production rate. The magnitude of atmospheric production rate variations over the last 800 kyr is discussed below. The effect of particle lithology on ^{10}Be concentrations in the sediment column can be dramatic. For example, Henken-Mellies et al. (1990) reported that ^{10}Be concentrations in zero age *Globigerina* oozes are 10% of ^{10}Be concentrations in associated pelagic clays. Volcaniclastic arc turbidite sediments can have 20-50% as much ^{10}Be as nearby pelagic sediments (Zheng et al. 1994, Morris et al. 2002; Valentine and Morris, submitted a). The study of ^{10}Be as a proxy for geomagnetic variation faces the challenge of isolating a geomagnetic signature from those caused by changing lithology or sedimentation rate.

The goal of obtaining high-resolution records of paleoclimatic and paleoceanographic processes has led programs such as the Ocean Drilling Program, IMAGES, and the Antarctic programs of several countries to select sites in high sedimentation rate areas. These include continental margins and inner shelf basins where the particle flux is high, and sediment drifts and contourite deposits created by bottom currents that focus sedimentation. These are precisely the settings where boundary scavenging and sediment redistribution are maximized and the accumulation history is the most complicated.

Isolation of the ^{10}Be production signal in marine sediment may be possible by using ^{230}Th to correct for sediment redistribution (Bacon and Rosholt 1982; Bacon 1984). ^{230}Th is introduced into the water column by decay of ^{234}U, which is the only source of ^{230}Th in the ocean. ^{230}Th is insoluble and highly particle reactive, being scavenged on timescales of 10 to 50 years, much shorter than its half life of 75,200 years (Anderson et al.1983a,b; Chase et al. in press). The rain rate of ^{230}Th to the seafloor is nearly independent of the particle flux, and approximately equal to its rate of production by decay of ^{234}U in the water column. In contrast, ^{10}Be is less particle-reactive. Its residence time in the ocean is 500 to 1000 years (Anderson et al. 1990), making it much more susceptible to lateral redistribution. Therefore, ^{230}Th may serve as a reference when examining the accumulation rates of other components of the sediment assemblage. Frank et al. (1994, 1995, 1997, 1999, 2000) recommend the use of the $^{10}Be/^{230}Th_{ex}$ ratio in deep-sea sediment profiles to remove the sediment redistribution effects before interpreting a ^{10}Be record in terms of paleoproductivity or geomagnetic field variations.

The sediment supply to subduction zones is integrated over the last 12 Myr, such that the net effect of secular variation in production rate is probably less than 15% (Tanaka and Inouye 1979; Gosse and Phillips 2001). The challenge for subduction zone studies is that variations in lithology or sedimentation rate can change the ^{10}Be inventory in the subducting sediment over time, leading to a departure from the steady-state conditions assumed in quantitative models.

In situ cosmogenic ^{10}Be

Previous reviews of geological applications of ^{10}Be produced *in situ* include Lal (1988, 1991, 1996, 2000a,b), Bierman (1994), Bierman and Steig (1996), Gillespie and Bierman (1995), Gosse and Phillips (2001), and Granger and Muzikar (2001). This section provides the rudimentary principles for the use of *in situ* cosmogenic nuclides, in order to acquaint the reader with basic approaches and limits to using ^{10}Be for tectonic studies. The use of *in situ* ^{10}Be in studies of other surficial processes is described by Bierman et al. (this volume). The production of *in situ* terrestrial cosmogenic nuclides (TCN) is similar in many respects to the production of extra-terrestrial cosmogenic isotopes in meteorites. Extraterrestrial cosmogenic nuclides and those produced in Earth's atmosphere (described in the preceding sections) are primarily produced from high-energy primary GCR. Interactions with atmospheric molecules produce neutrons, muons, and other particles that form a cascade of secondary radiation falling toward the Earth. It is this secondary radiation that produces on Earth's surface a variety of different nuclides and subatomic particles. The atmospheric cosmic ray flux attenuation causes the *in situ* production rate of ^{10}Be on a landform surface to be lower (by a factor of $\sim 10^{-3}$) than the integrated atmospheric production of ^{10}Be above the surface. Although a wide variety of different isotopes and subatomic particles can be produced during cosmic-ray interactions, relatively few *in situ* isotopes are useful for tectonic studies over time-scales of 10^3 to 10^7 years. Besides ^{10}Be, only five other *in situ* cosmogenic nuclides (^3He, ^{14}C, ^{21}Ne, ^{26}Al, and ^{36}Cl) have been used for Late Cenozoic landform studies because their half-lives are suitably long (or they are stable), the sample preparation and analysis is reproducible, and non-cosmogenic sources of the nuclides are negligible or can be determined.

Tectonic studies have relied on the measurement of *in situ* ^{10}Be produced at and below a landform surface. The cosmogenic isotope concentration in a rock at earth's surface, N (atoms g^{-1}), can be used to calculate the duration of exposure, T (annum, a) to cosmic radiation if the total rate of the nuclide production at the surface, P_o (atoms g^{-1} a^{-1}), from all possible interactions, is known. As discussed below, at any given time P_o will vary as a function of latitude, elevation, and depth below the rock surface. If the nuclide is stable,

$$T = \frac{N}{P_o} \tag{1}$$

at the rock surface. If the isotope is radioactive with a decay constant, λ:

$$T = -\ln\left(1 - \frac{N\lambda}{P_o}\right) \Big/ \lambda \tag{2}$$

Some studies of tectonic geology have required the measurements of ^{10}Be below the surface of the Earth (for instance, when dealing with the exposure of a bedrock fault scarp or when dating alluvial fan and terrace sediments). For subsurface exposures in rock, the production rate decreases by a factor of 1/e approximately every 56 cm (the production of ^{10}Be in shallow rock is dominated by fast neutrons, with Λ_n in rock similar to that in air, ~160±20 g cm^{-2}, and a rock density ρ = 2.7 g cm^{-3}). Currently the community tends to ignore the possibility of a small deviation from a simple exponential for fast neutron attenuation due to boundary effects at the atmosphere-rock surface. Recent work by Heisinger et al. (1997) and J. Stone (Stone 1999, 2000; Gosse and Stone 2001) has shown that only about 2.2% of ^{10}Be production (on a rock surface at sea level) is produced through muon capture. Compared to fast neutrons, the attenuation length of muons is much longer (Λ_μ = 247 g cm^{-2} in air, longer in rocks), so the relative muonic contribution of ^{10}Be production increases with depth. Burial by rock, sediment, snow, ice, water, or other materials likewise needs to be considered in using *in situ* ^{10}Be to date a surface because all shield the surface. The subsurface production rate (P_Z) due to fast neutrons and muons at depth Z (cm) is given by:

$$P_o^Z = P_n e^{\left(-Z\rho/\Lambda_n\right)} + P_\mu e^{\left(-Z\rho/\Lambda_\mu\right)} \tag{3}$$

which sums the fast neutron (n) and muon (μ) contributions. Derivations of these equations and expressions for natural exposure scenarios can be found in Bierman (1994), Cerling and Craig (1994), Gosse and Phillips (2001), and Lal (1988, 1991). Radioactive ^{10}Be (λ = 4.62×10^{-7} a^{-1}) can be produced from any exposed targets with atomic masses > 10. Quartz is the mineral of choice for most ^{10}Be exposure studies because of its crustal abundance, resistance to weathering, and simple chemical formula that makes production systematics relatively straightforward. Production rates of ^{10}Be in other minerals such as magnetite (D. Lal, pers. comm.) and olivine (Nishiizumi et al. 1990; Kong et al. 1999) are currently being determined. In quartz, ^{10}Be is produced primarily from spallation interactions between the fast (>10 MeV) secondary neutrons and Si and O nuclei. There are no significant radiogenic or nucleogenic sources of ^{10}Be, unlike the other five commonly used TCN.

Applications of ^{10}Be exposure dating require some assumptions about the surface system that is being dated. As warned by D. Lal (1996; 2000a) and others, it is important to realize that in most natural settings the sampled surface was not always at the surface of the landform during the exposure history. With rare exception (indicated by the presence of surface preservation indicators such as striae or growth fibers on a fault scarp), *in situ* ^{10}Be abundances measured in a surface or subsurface mineral must be interpreted to represent some dynamic record of exposure history that involves surface processes as well as time. Typically, we simplify the interpretation by assuming an uncomplicated exposure history for the landform we wish to date. That is, during exposure, we assume that the rock was never shielded from cosmic radiation by snow, ice, or sediment, the rock was not eroded, and the rock was continually exposed at its present position (i.e., there was no rolling, isostatic uplift, or tilting). If any of these assumptions are known to be invalid, estimates of the effects of the condition must be taken into account. Finally, it is assumed that the system is closed (both geologically and isotopically), in the sense that there are no additions or depletions of the ^{10}Be concentration other than by cosmogenic production and radioactive decay, and that there are no significant nucleogenic sources of ^{10}Be. Also implied is that any atmospheric ^{10}Be adhered to the quartz was removed during sample preparation.

As the exposure age equations above reveal, the precision and accuracy of the chronometer will also depend on the certainty in the time-integrated production rates at a given geomagnetic latitude, altitude, and rock depth. On Earth's surface, the production rate of ^{10}Be in quartz ranges between 3 and 150 atoms g^{-1} a^{-1} (corresponding respectively to production on the equator at sea level and on a mountain peak at the poles). The spatial variation of *in situ* production is due to two factors. First, the production rate increases with elevation because of the corresponding decrease in atmospheric attenuation of the cosmic ray flux. Secondly, the geomagnetic field interacts with the primary and perhaps secondary cosmic ray flux. At low geomagnetic latitudes, where the field lines are approximately parallel to the earth, the lower energy spectrum of the cosmic radiation tends to be strongly deflected. In this sense, the cosmic ray flux above the equator is said to be 'harder.' Contrarily, at the poles where the field lines tend to converge, lower energy cosmic rays can make it to the atmosphere (discussed in next section). Therefore, sea level cosmogenic production rates are highest above about 58° latitude because there is no deflection of the component of the cosmic ray flux with sufficient energy to produce the secondary cascades and ultimately *in situ* ^{10}Be.

Time-integrated production rates of the most commonly utilized TCN are believed to be known to within 15% (Gosse and Phillips 2001) and some nuclides (^{10}Be and ^{3}He) may be known within 5% (Stone 1999, 2000; Gosse and Stone 2001; Liccardi et al. 1999). An important resolution in ^{10}Be production rates was accomplished when J. Stone (1999, 2000) recognized that the disparity in published production rate estimates (4.7 to 6.1 atoms g^{-1} yr^{-1}) appears only to be an artifact of how the production rates from different sites were scaled (normalized) to high latitude sea level. Brown et al. (1995) and Heisinger et al. (1997) independently showed that muon interactions account for a smaller proportion of the total production of ^{10}Be at the Earth's surface. Using a 2.2% muon contribution to normalize the site production rates (Lal 1991; Table 1 assumed 15.6%), all of the production rates converged on one rate: 5.1 atoms ^{10}Be g^{-1} quartz a^{-1}.

Table 1. ^{10}Be flux calculations.

	Aleutian	Guatemala-Nicaragua	Costa Rica	Honshu, Japan	Mariana
DSDP/ODP SITE	183	495	1039	436	777B
Subducting ^{10}Be					
^{10}Be inventory, 10^{13} atoms cm^{-2}	0.97	1.35	1.42	1.3	0.11
^{10}Be Depth in Sediment*	>210	203	192	246	25
Convergence Rate, cm yr^{-1}	7.9	8.1	8.7	10.5	9
Subduction time, Myr	2.6	2.5	2.2	2.6	2
^{10}Be Flux to Trench, 10^{24} atoms/km-arc-Myr	7.7	10.9	12.5	13.9	0.99
^{10}Be Flux to arc, decay corrected	2.3	3.2	4.5	4.2	0.39
Erupting ^{10}Be					
Avg. basaltic ^{10}Be, million atoms g^{-1}	3	10	1	<1	0.33
Magma Prod. Rate, km^3/km-arc-Myr	28	40	40	25	30
Volcanic ^{10}Be Flux, 10^{24} atoms/km-arc-Myr	0.22	1	<0.1	<0.07	0.03
^{10}Be Flux out/Flux in	0.096	0.31	0.02	<0.017	0.077
References for sediment data	1, 2	1, 3	3, 4	1, 4	1, 5

Sediment depth above which ^{10}Be > 10^7 a cm^{-3}

Magma production rates from Reymer and Schubert (1984)

References: (1) Zheng et al. (1994); (2) George et al. (submitted); (3) Valentine and Morris (submitted a); (4) Morris et al. (2002); (5) Valentine and Morris (submitted b)

The flux of cosmic radiation over the exposure duration must be known or considered constant to use ^{10}Be for exposure dating. The integrated GCR flux to Earth is probably constant over the time scales of interest (10^4 to 10^7 a) even considering the effects of solar modulation and supernovae (Gosse and Phillips 2001). This is probably not true for the secondary cosmic radiation flux. The secondary flux is influenced by temporal variations in the geomagnetic field strength, secular variations of the position of the geomagnetic dipole axis, influences of persistent non-dipolar field attributes, and variations in atmospheric shielding due to climatic or isostatic changes. The effects of variation in geomagnetic paleointensity and dipole axis position on production rates over shorter intervals are still uncertain, and may have caused variations in integrated production rates as high as 20% if we use the relationship between geomagnetic field and cosmic ray flux estimated by Elsasser et al. (1956). Fortunately, the axis position variation influences only Holocene ages because the integrated *in situ* production rate is less variable over longer exposure periods. Despite these uncertainties, ^{10}Be ages on boulders are found to be concordant with calibrated radiocarbon ages (e.g., Gosse et al. 1995c) and other cosmogenic nuclide ages (Ivy-Ochs et al. 1998; Phillips et al. 1997). There is also reasonable consistency of ^{3}He, ^{10}Be, and ^{36}Cl production rates over different exposure durations (Liccardi et al. 1999; Phillips et al. 1996; Gosse and Phillips 2001; Gosse and Stone 2001). This consistency in production rates suggests that the effect of geomagnetic field fluctuations at the base of the atmosphere may be considerably less than predicted by the widely used approximation of the paleointensity effect on atmospheric production (Elsasser et al. 1956). Relying on transport codes to simulate cosmic ray particle interactions in the atmosphere, Masarik et al. (2001) also found that the geomagnetic field appears to have a smaller influence on production rates than Elsasser et al. suggested. With levels of precision of multiple exposure ages on a single landform approaching 3% and the persistent need for high accuracy in the ages, it is necessary to re-evaluate the influence of dipole and non-dipole variations and reconsider using the Elsasser et al. equation, which was never meant for this purpose.

The cosmogenic ^{10}Be dating method is capable of dating exposure durations ranging between 5 ka and 5 Ma. This range overlaps and bridges the gaps between the practical ranges of other radiometric dating methods commonly used in Cenozoic tectonic investigations, such as radiocarbon dating (currently calibrated to 22 ka), optically stimulated luminescence, U-series, and $^{40}Ar/^{39}Ar$. The analytical limit on the resolution of the ^{10}Be technique is about 500 years at mountain elevations; however, uncertainties in the geomagnetic field secular variation effects may become important for exposures of such short (less than 6000 year) duration. Unlike the atmospheric cosmogenic radiocarbon chronometry, which is sensitive to high frequency variations in geomagnetic paleointensity, *in situ* ^{10}Be chronometry improves with age as the geomagnetic field influences are integrated over longer exposure durations. Sample preparation and analytical techniques have improved to the point that ^{10}Be concentrations in deep (>20 m) and young (<1000-yr-old) quartz targets can be measured (Gosse and Phillips 2001). The upper limit of the dating method is controlled by geologic factors and by the achievement of radioactive secular equilibrium (> 5 half-lives). For example, with the exception of unique climatic environments such as the Antarctic dry valleys (where Pliocene surfaces have been dated (Brook et al. 1995a,b) erosion will significantly alter landform surfaces beyond a few million years. Erosion will have several effects on the ^{10}Be system:

(1) The concentration (Eqn. 3) on an eroded surface will be less than a surface on that same landform that was not eroded (and therefore the eroded surface age will need to be interpreted as a minimum bracketing age).

(2) Differential erosion will increase variability among multiple ages from the same landform.

(3) Secular equilibrium of ^{10}Be concentration due to decay will be attained more rapidly.

(4) Older surfaces will be more sensitive to the influence of erosion (more time means more rock mass removed).

Fortunately, this sensitivity of ^{10}Be (and other TCN) to erosion provides a unique means of directly determining erosion rates on landforms (see below, and Bierman et al., this volume). The chronometric technique is thus optimum for dating tectonic events in the 10^4 to 10^5 year span.

^{10}BE AND GEOMAGNETISM

Introduction

The ability of rocks and sediment to act like tape recorders of the geomagnetic field has made key contributions in the earth sciences. Magnetic recording in oceanic crust and continental rocks was a key element in the development of the theory of plate tectonics. Paleomagnetic recording enables the tracking of continental motions through time, and the geomagnetic polarity timescale (GPTS) provides a global chronological tool covering the past ~160 million years (Myr).

Observation and study of the geomagnetic field for its own sake has a long history, due to in part to its use as a navigational aid to sailors. Presently, direct measurements of the Earth's magnetic field are made at geomagnetic observatories and during aircraft and satellite surveys. These systematic measurements of the full geomagnetic field vector exist in a select few areas, covering only the past several centuries.

Efforts to extend the record of geomagnetic field observations back in time is motivated by a desire to understand the origin and evolution of the geodynamo. Observations of the geomagnetic field made at and above the Earth's surface provide a means of remotely observing the Earth's outer core where the field originates. Paleomagnetic recording in rocks and sediments provides a means of extending geomagnetic field records back through time. In this respect paleomagnetism is unique among geophysics disciplines, as it provide a means to observe past states of the Earth's interior. Geological materials have revealed a wide range of geomagnetic field behaviors, from century-scale secular variation to complete reversals of the geomagnetic dipole to long periods of field stability known as superchrons (see Merrill et al. 1996 for full discussion).

Both igneous rocks and sediments are capable of producing high fidelity records of the orientation of the geomagnetic field vector. It is more challenging to extract the intensity of the ancient field from rocks and sediment because the intensity of the signal is dependent on the properties of the recorder. Absolute paleointensity data, the intensity of the ambient field at the time of remanence acquisition, can only be obtained from volcanic rocks and archeological baked clays (pottery, bricks). These materials have been heated above the Curie temperature of the constituent magnetic minerals and cooled rapidly in the presence of the geomagnetic field, a solid-state process that imparts a thermoremanent magnetization (TRM). TRM represent an instantaneous spot reading of the geomagnetic field. Unfortunately, volcanic samples and baked clays are unequally distributed temporally and spatially. There are several sites from volcanic islands that have yielded high resolution volcanic sequences (Tric et al. 1992, 1994; Raïs et al. 1996; Brassart et al. 1997; Valet et al. 1998; Laj et al. 1997; Laj and Kissel 1999; Laj et al. 2002), but even these records are less continuous and of lower resolution than sediment sequences.

Sediments are attractive geomagnetic field recorders due to their continuity, high temporal resolution, and global availability. However, sediments can only provide relative variations in the geomagnetic field strength. The intensity of the natural remanent magnetization (NRM) recorded in sediments is strongly affected by mineralogical content and physical properties of the sediment, which are unrelated to the geomagnetic field at the time of deposition. Extracting accurate paleointensity data from sediment requires a means of removing these non-field effects.

Advances in instrumentation enabling the rapid measurement of continuous u-channel sub-samples (Tauxe et al. 1983; Nagy and Valet 1993; Weeks et al. 1993) has led to a large increase in the number of long, continuous, sedimentary records analyzed for geomagnetic paleointensity variations. The emerging database of globally distributed records has revealed the global nature of certain geomagnetic field features, as well as some unexpected behaviors. Many of the current controversies in geomagnetism center on sedimentary records of geomagnetic field behavior, mainly due to the non-field effects that enter into the recording process. Therefore, alternate tracers of geomagnetic field behavior have been sought in order to verify the geomagnetic origin of the observed records.

^{14}C, ^{36}Cl, and ^{10}Be are widely used as chronological tools in the geosciences, requiring knowledge of their production rates back through time and hence a knowledge of solar activity and geomagnetic field variations that modulate production rates. Conversely, cosmogenic nuclide production rates can be used to track solar variability and geomagnetic field variations back through time. The possibility of using radionuclide production rates as proxy indicators of these processes has stimulated measurements of ^{10}Be profiles in deep-sea sediment, lacustrine sediment, and ice cores. The results of some of these studies have been ambiguous or even contradictory, owing to the many variables that influence ^{10}Be accumulation in sediments and the magnetic recording process in sediments. However, there are many current controversies in geomagnetism whose study would benefit from the combined application of paleomagnetic and cosmogenic isotope analyses. This approach has been undertaken by several groups that focused on geomagnetic paleointensity reconstructions for the last ~100 to 200 kyr, for which there are a relatively large number of globally-distributed sediment cores and ice cores possessing their own independent chronologies (e.g., Henken-Mellies et al. 1990; Mazaud et al. 1994; Robinson et al. 1995; Guyodo and Valet 1996, 1999; Frank et al. 1997; Baumgartner et al. 1998; Frank 2000; McHargue et al. 2000; Stoner et al. 2000, 2002; Wagner et al. 2000a). It is our purpose in the following sections to address the potential of ^{10}Be to contribute to building time-series data of geomagnetic field variations, and to contribute to several current debates in geomagnetism. We will review advances on the subject, limitations, and discuss future directions for this approach.

Magnetic modulation of the primary galactic cosmic ray flux

The production rates of radionuclides such as ^{10}Be, ^{14}C, and ^{36}Cl are modulated by magnetic fields from two main sources. The primary galactic cosmic ray flux (GCR) is comprised of charged particles such as protons (in addition to lesser amounts of alpha particles and other stellar-synthesized elements) that have been accelerated to high energies (up to 10^{20} eV) by astrophysical phenomena. The GCR itself may have varied over time as a consequence of supernovae (Sonett et al. 1987). However, the flux on a timescale of $>10^6$ years appears to have been constant within ±20% (Vogt et al. 1990).

Any charged particle traveling through a magnetic field will experience a force that acts at right angles to both the magnetic field direction and the direction of motion of the particle, as occurs inside an accelerator mass spectrometer. The simplest description of this force is given by:

$$\mathbf{F} = q\mathbf{v} \times \mathbf{B} \tag{4}$$

where q is the electrical charge of the particle, **v** is the particle's velocity vector, and **B** is the geomagnetic field vector. Consequently, the paths of charged particles are altered by magnetic fields, with the amplitude of the deflection dependent on the field strength, the particle's electrical charge, the energy of the particle, and the angle between the particle's trajectory and the ambient magnetic field vector.

The GCR first encounters the solar wind as the particles travel towards the Earth. On timescales of years to hundreds of years the GCR is "filtered" by solar activity. Solar wind is a plasma with very high electrical conductivity. The very high conductivity (very low diffusivity) results in the "freezing" of magnetic field lines into the plasma, which literally carries the solar magnetic field out into space far past the orbit of Pluto. Magnetic fields carried by the solar wind interact with the energetic particles, causing deflections, diffusion, scattering and energy loss (Lal 1988). Some of the particles are swept out of the solar system. Consequently, radionuclide production rates are inversely correlated with solar activity. Carbon-14 activity ($\Delta^{14}C = {}^{14}C/{}^{12}C$) records obtained from tree rings and ^{10}Be records from polar ice cores (Stuiver 1961; Eddy 1976; Stuiver and Quay 1980; Raisbeck et al. 1981b; Beer et al. 1988; Damon and Sonnett 1991; Stuiver and Braziunas 1993; Bard et al. 1997) indicate that high production rates correspond to periods of low sunspot activity. In addition, periodicities of 11, 22, 200, 400, and ~2400 years have been observed in these records (Beer et al. 1983, 1988, 1990; Sonnett and Seuss 1984; Damon et al. 1989; Stuiver and Braziunas 1989, 1993; Suess and Linick 1990; Haubold and Beer, 1992; Steig et al. 1996, 1998; Finkel and Nishiizumi 1997; Aldahan et al.,1998; Bard 1998; Cini Castagnoli et al. 1998; Wagner et al. 2001).

The charged particles comprising the solar wind itself consist of energetic protons and heavier nuclei. These particles are typically of lower energy (1 to 100 MeV) than the GCR. Therefore, these particles only contribute to nuclear reactions at the very top of the Earth's atmosphere at high geomagnetic latitudes (>60°) where the amplitude of deflection is extremely low (Masarik and Beer 1999). Solar particles are not expected to make a significant contribution to radionuclide production, except in instances of extreme solar events that create higher than average proton fluxes (Lal 1988; Shea and Smart 1992).

The GCR next encounters the Earth's magnetosphere, in which the geomagnetic field alters the trajectories of charged particles and "shields" the Earth by preventing low-energy cosmic particles from entering and interacting with the atmosphere. The latitudinal dependence of this shielding can be visualized by considering the case of a charged particle that is vertically-incident at the geomagnetic equator, and another charged particle that is vertically incident at the geomagnetic pole. In a dipolar field, field lines are horizontal at the geomagnetic equator. The angle between the incident particle's trajectory and the geomagnetic field is 90°, and the force $F = qV\sin\theta$ is a maximum. The geomagnetic field orientation is vertical at the magnetic poles. At the magnetic pole, the angle between the incident particle's trajectory and the geomagnetic field is zero, and the force $F = qV\sin\theta$ is mathematically zero. Therefore, the shielding effect is greatest at low latitudes, even though the intensity of the geomagnetic field is only half as strong as at the poles. Consequently, the energy spectrum of particles penetrating the atmosphere varies with latitude. Again using the simplest conceptual model, strong shielding at the equator deflects the low energy particles, and the particle flux that penetrates the atmosphere at the geomagnetic equator has a higher average energy. The high geomagnetic latitudes receive a wider spectrum of energies.

The latitudinal effect of geomagnetic shielding results in a latitudinal dependence of cosmogenic nuclide production rates, with higher production rates at the poles and lower production rates at the equator (Lal and Peters 1967; O'Brien 1979; Castagnoli and Lal 1980; Lal 1988; Masarik and Beer 1999 and references therein). A first-order relationship between the geomagnetic field dipole moment (M) and the global average cosmogenic nuclide production rate (Q) was estimated by Elsasser et al. (1956) as:

$$Q/Q_O = (M/M_O)^{-1/2} \tag{5}$$

where M_O and Q_O are the present-day geomagnetic dipole moment and present-day production rate, respectively. This equation unsuccessfully predicted Q for very small values of M. However, there appeared to be a good correspondence between the available records of geomagnetic field paleointensity and ^{14}C activity in tree rings.

The geomagnetic field is not a pure dipole. Approximately 80% of the present day geomagnetic field can be explained with a dipole axis inclined 11° to the Earth's rotation axis. Approximately 20% of the field at the Earth's surface remains after the best fitting geocentric axial dipole (GAD) field is subtracted from the actual mapped field (Merrill et al.1996). This remaining part of the field is the non-dipole field, which consists of continent sized features where inclination and intensity are both higher than predicted for a GAD field (as over North America) and lower than expected for a GAD field (as over central Africa) (see Merrill et al. 1996).

Both the dipole and non-dipole parts of the geomagnetic field vary with time. However, the non-dipole components of the geomagnetic field are often ignored in cosmogenic nuclide studies for a number of reasons. First, the time-averaged field is assumed to be a geocentric axial dipole. This means that over a sufficiently long observation time, on the order of 10^4 years, the average vector orientation at any given location should be that of a geocentric axial dipole. Second, the GCR encounters the Earth's magnetosphere long before it encounters the atmosphere. Charged particles begin to experience geomagnetic deflections when they are still several Earth radii away from the surface of the Earth. Given that the intensity of the geomagnetic field falls off as $1/R^3$, it is typically assumed that the weaker non-dipole features of the geomagnetic field are attenuated and the field seen by the GCR can be represented as dipolar.

The validity of both assumptions is a matter of serious debate. High resolution archives such as sediments and ice cores have millennial-scale down to annual-scale resolution, and therefore a single sample does not span the requisite 10^4 years needed to average out the non-dipole field effects. Further, the non-dipole field features become much more significant near the Earth's surface where the secondary particles generate *in situ* radionuclides. The strategy for working with or without these assumptions and simplifications depends in part on the goal.

The "forward problem" is defined here as using geomagnetic field models to calculate cosmogenic nuclide production rates. Størmer (1955) used a geocentric axial dipole model to calculate the cosmic ray "cutoff rigidity" as a function of position on Earth. Rigidity is defined as:

$$R = p\,c\,/\,Ze \qquad (6)$$

where p is momentum, c is the velocity of light, and Ze is the charge of the particle. The magnetic rigidity, a particle's momentum per unit charge, is the quantity used to describe the particle's ability to penetrate the geomagnetic field. Depending on geomagnetic latitude and the particle's angle of incidence, there is a critical energy below which the particle cannot penetrate into the Earth's atmosphere. Simulations of cosmogenic nuclide production have used variations of Størmer's formula for cutoff rigidity to determine the latitudinal energy spectrum of GCR particles allowed into the simulation to produce cosmogenic nuclides (see Bhattacharyya and Mitra 1997; Masarik and Beer 1999 and references therein). However, Størmer's formula was shown to be invalid at low geomagnetic latitudes (see Bhattacharyya and Mitra 1997 for full discussion). Subsequent models based on a dipole field (e.g., Lal 1988; Masarik and Beer 1999; Wagner et al. 2000a) have refined the Elsasser et al., (1956) relationship. The physical models of Masarik and Beer 1999, and Wagner et al. (2000a) predict slightly higher production rates for very low field intensities, and predict a smaller modulation effect for very strong field intensities.

A more accurate way of determining cutoff rigidity is to start with a particle at the Earth's surface, for which the geomagnetic field has been mapped, and trace the trajectory as the particle spirals upward through the magnetic field. The drawback of this method is the large amount of computer time needed to calculate cutoff rigidities at enough locations on Earth to develop a global picture. Further, an enormous amount of computer time is needed to consider all possible angles of incidence. For this reason, modelers use the vertical cutoff rigidity (rigidity for a vertically-incident particle) as a representative value for a given location on Earth. Bhattacharyya and Mitra (1997) presented a method for calculating vertical cutoff rigidities using an "eccentric dipole," a spherical harmonics representation of the geomagnetic field including a quadrupole term. Isorigidity contours were calculated for four time periods. From 1835 up to 1985 the authors observed a westward drift in the location of maximum in vertical cutoff rigidity, and a decrease in the amplitude of the vertical cutoff rigidity. This suggests that effects of the non-dipole component of the geomagnetic field on cosmogenic isotope radionuclide production rates are non-negligible and need to be studied.

The inverse problem involves using a natural archive of cosmogenic nuclide production rates to extract geomagnetic field paleointensity. The inverse problem involves a single geological record from a single location on Earth, which is not sufficient to determine the global structure of the geomagnetic field. Therefore, the default assumption is that these archives are recording variations in the geomagnetic dipole moment. The similarity of geomagnetic paleointensity features with wavelengths longer than 10^3 years, collected from widely distributed sites, appears to confirm the global nature of the signal in these records (e.g., Guyodo and Valet 1996, 1999; Laj et al. 2000; Stoner et al. 2002).

Relative paleointensity recording in sediment

The potential to use sediments as geomagnetic field paleointensity recorders was convincingly demonstrated by Kent (1973). A slurry of marine sediment was repeatedly mixed and then allowed to settle in the presence of a laboratory field of varying strengths. When an invariant sediment assemblage was repeatedly re-deposited, a clear linear relation was observed between the ambient field and the intensity of remanence recorded in the sediment.

In nature, the intensity of the magnetization measured in sediments is related to the ambient field at the time of deposition, but not necessarily in a simple or linear manner. This is caused by variations in the sediment assemblage. The intensity of natural remanent magnetization (NRM) measured in sediments depends on the concentration of the magnetic material present, the composition of the magnetic material, and the grain size of the magnetic material, in addition to other influences related to the non-magnetic sediment matrix (see Tauxe 1993 for full discussion). These parameters are often influenced by environmental processes, and therefore the measured NRM does not solely reflect the ambient geomagnetic field at the time of deposition.

Relative paleointensity is obtained from sediments through "normalization," which entails dividing the intensity of the measured NRM by a parameter that tracks the concentration, composition, and grain size of those magnetic grains that carry the NRM. The most common normalization parameters in use are the anhysteretic remanent magnetization (ARM) and isothermal remanent magnetization (IRM) (see Tauxe 1993 for full discussion). The normalization parameter used in any given sedimentary record depends on the magnetic characteristics of that particular sediment. Unfortunately, the amplitude of normalized intensity features can be quite different depending on which normalization method is used (Schwartz et al. 1996; Brachfeld and Banerjee 2000). Further, normalization may not completely remove the environmental effects (e.g., Schwartz et al. 1996; Kok 1999; Lund and Schwartz 1999). Consequently, there are several debates concerning whether certain features observed in relative paleointensity records represent true field behavior or artifacts resulting from either the magnetic recording or an incomplete normalization process.

The accumulation of cosmogenic isotopes in geological archives has the potential to yield independent records of geomagnetic field behavior over time, and therefore to help to resolve controversies in geomagnetism. Records of cosmogenic isotope production rates are not influenced by the same complications involved with magnetic recording. However, as discussed earlier, isolating a geomagnetic field signal from a record of ^{10}Be accumulation in sediment or ice is a difficult task. ^{10}Be accumulation in marine sediment and ice is influenced by oceanographic, biogeochemical, and atmospheric processes that complicate, or even obscure the record of production rates. Although both proxies have their complications, a combined approach may ultimately lead to more accurate time-series data of geomagnetic field behavior.

Paleointensity as a correlation tool

The dipole component of the geomagnetic field exhibits temporal variations in its intensity on timescales of 10^2 to 10^6 years. Variations in the geomagnetic dipole are synchronously experienced everywhere on the globe. Therefore, a specific geomagnetic field paleointensity feature represents the same "instant" in time everywhere on Earth. High-resolution sedimentary records of geomagnetic field paleointensity from the Mediterranean Sea (Tric et al. 1992), the Somali Basin (Meynadier et al. 1992), Lake Baikal, Russia (Peck et al. 1996), Lac de Bouchet, France (Williams et al. 1998), the North Atlantic Ocean (Channell et al. 1997; Channell and Kleiven 2000; Laj et al. 2000), the Labrador Sea (Stoner et al. 1998, 2000), the South Atlantic Ocean (Channell et al. 2000; Stoner et al. 2002), and the Southern Ocean (Guyodo et al. 2001; Sagnotti et al. 2001) demonstrate the global coherence of features with wavelength of 10^4 to 10^5 years.

A common characteristic of these records is the presence of several intervals of very low intensity during the past 800 kyr. These are interpreted as geomagnetic excursions, which involve large but short-lived directional changes (>45° deviation of the virtual geomagnetic pole from its time-averaged position) followed by a return of the vector to its previous state. Several of the sedimentary records listed above suggest that excursions are associated with a reduction in field strength to less than 50% of its present day value.

The potential to use excursions as chronostratigraphic markers has been long been recognized. Given that excursions are associated with intervals of weak paleointensity, these features are also manifested as abrupt increases in radionuclide production rates. ^{10}Be concentration peaks observed in ice cores from Vostok and Dome C were seen as possible marker horizons that could be used to synchronize records from the Northern and Southern hemispheres, particularly in cases where low snow accumulation rates precluded the identification of annual layers (e.g., Mazaud et al. 1994). The Vostok record displays a peak in ^{10}Be concentration at 60,000 years B.P. with a duration of 1000 to 2000 years, and a second peak at 35,000 years B.P. (Fig. 1) (Yiou et al. 1985; Raisbeck et al. 1987). The 35-ka peak is also seen in the Dome C, Byrd, and Camp Century ice

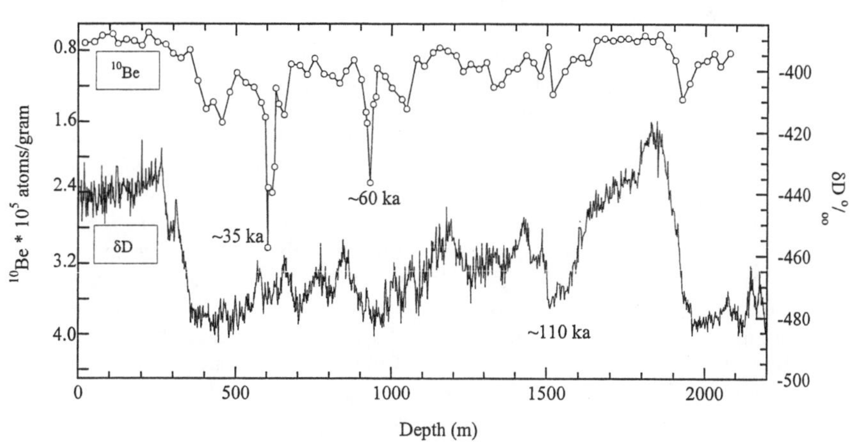

Figure 1. ^{10}Be concentration from the Vostok ice core showing 2 peaks at ~35ka and 60 ka. The dD record, a proxy of local temperature change, is shown for comparison (Raisbeck et al. 1987; Petit et al. 1999). There are no climate features that coincide with the abrupt, narrow ^{10}Be spikes. Mazaud et al. (1994) suggested a geomagnetic field origin for these features, and demonstrated the potential to use geomagnetic field paleointensity variations as a means of constraining the ages of these features.

cores (Beer et al. 1984, 1988, 1992). A peak in ^{10}Be concentration is observed in the Summit GRIP ice core at ~40 ka (Yiou et al. 1997). Two peaks in ^{36}Cl concentration at 32 ka and ~35-39 ka are seen in the Summit GRIP ice core (Baumgartner et al. 1997b, 1998; Wagner et al. 2000b). Enhanced ^{10}Be concentrations have also been observed in sedimentary records from Lake Baikal at 40 ka (Aldahan et al. 1999), the Caribbean Sea at 35-40 ka (Aldahan and Possnert 1998), the Gulf of California at 32 and 43 ka (McHargue et al. 1995), and the Mediterranean Sea at 34±3 ka (Cini Castagnoli et al. 1995).

Other hypotheses independent of the geomagnetic field have been proposed to explain these features including a long interval of low solar activity (Raisbeck et al. 1987), and cosmic ray shock way (Sonett et al. 1987), or a supernova explosion (Kocharov 1990). A climatic origin for some the features in the ice cores was ruled out by examining δ^{18}O and sulfate ion concentrations, which showed no evidence of a change in precipitation or atmospheric transport of aerosols coinciding with the ^{10}Be spikes (Beer et al. 1992). In a series of papers, Aldahan and coworkers (Aldahan et al. 1994, 1999; Aldahan and Possnert 1998) demonstrated that variations in sediment lithology and sediment grain size can also lead to enhancement of ^{10}Be content in sediments, and thus far only the ~32 ka feature and ~39 ka feature appear to have a robust, global signal.

While these individual ^{10}Be spikes provide chronostratigraphic tie points, a more powerful correlation and dating tool involves using a time series of geomagnetic paleointensity variations and cosmogenic nuclide production variations. Mazaud et al. (1994) observed that the ^{10}Be flux versus age in the Vostok ice core largely co-varied with a synthetic curve generated by assuming that ~75% of ^{10}Be production was modulated by geomagnetic field intensity. Starting with the geomagnetic paleointensity records of Meynadier et al. (1992) and Tric et al. (1992), the authors calculated the expected ^{10}Be production rate according to Lal (1988). The authors used this curve as a reference for tuning the ice paleoaccumulation rates. The resulting Vostok chronology improved the correlation between climatic signals in the Vostok ice core and in marine sediment records. Since geomagnetic modulation is minimal at the poles, this study implies that aerosols carrying ^{10}Be are transported from low latitudes to the Antarctic ice sheet (Mazaud et al. 1994).

One single record of geomagnetic field variability is not an ideal reference curve. Any single record may contain subtle flaws in its chronology or remanent magnetization. Guyodo and Valet (1996) proposed using a "stack," a weighted-average of several time-correlative records, to enhance the signal to noise ratio, average out any flaws present in the individual records, and extract the broad-scale, yet true global features of the geomagnetic field. Guyodo and Valet (1996) confirmed that a distinctive pattern of geomagnetic field paleointensity was observable in deep-sea sedimentary records from around the globe (Fig. 2). They produced a 200-kyr global stack of 19 paleointensity records, named Sint-200, which they suggested could be used as a millennial-scale correlation and dating tool. Each of the 19 constituent records in Sint-200 had its own oxygen-isotope stratigraphy. By correlating paleointensity variations in an undated core with the Sint-200 target curve, one could import the oxygen isotope stratigraphy to one's core site. For example, Stoner et al. (1998) successfully developed a geomagnetic paleo-intensity-based chronology for the Labrador Sea by correlating distinct intensity highs and lows in Labrador Sea sediment cores with Sint-200.

The constituent records of Sint-200, and its subsequent extension to the 800-kyr Sint-800 (Guyodo and Valet 1999) were all deep-sea sediment cores with relatively low sedimentation rates. A newer stack with higher temporal resolution, the 75-kyr North Atlantic Paleointensity Stack (NAPIS-75), was constructed from high sedimentation

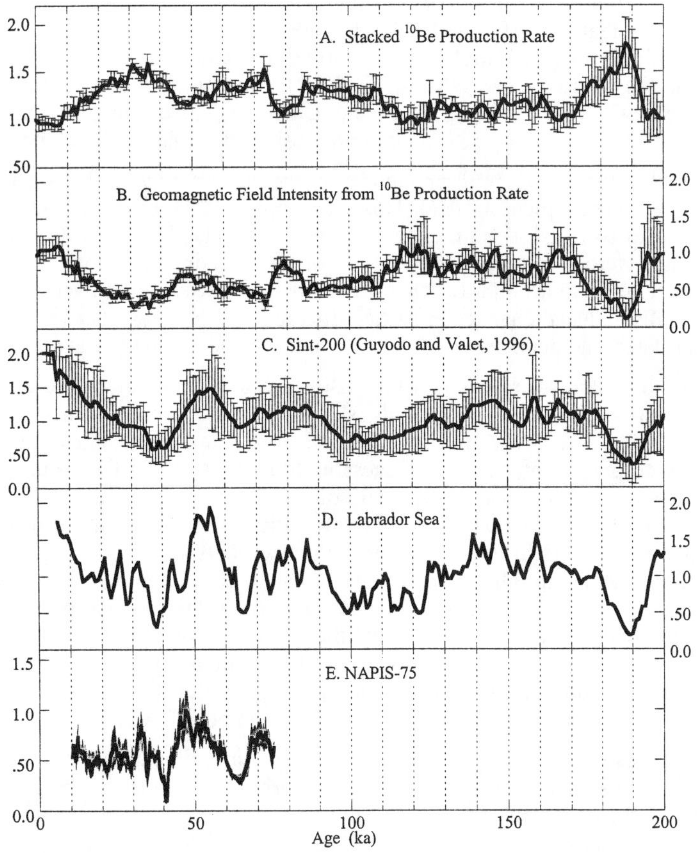

Figure 2. (a) Global stacked ^{10}Be production rate (Frank et al. 1997) (b) geomagnetic paleointensity derived from ^{10}Be production rates (Frank et al. 1997) (c) global stacked record of geomagnetic field paleointensity derived from paleomagnetic data (Guyodo and Valet 1996) (d) Labrador Sea relative paleointensity record for the past 200 kyr (Stoner et al. 1998) (e) NAPIS-75 geomagnetic paleointensity stack (Laj et al. 2000). Paleointensity is given in dimensionless, relative units. Geomagnetic field variations derived from paleomagnetic data and ^{10}Be production rates appear to be globally coherent over the past 200 kyr, which can be exploited as a correlation and dating tool. The higher resolution NAPIS-75 record suggests that the Laschamp Event at 40 to 41 ka is only 1000 years in duration.

rate cores from sediment drifts in the North Atlantic Ocean (Laj et al. 2000). NAPIS-75 is a stack of six individual high-resolution records (sedimentation rates = 20 to 30 cm/kyr) from cores recovered from the Nordic seas and the North Atlantic. The stack covers the time interval 10-75 ka, providing partial overlap with the radiocarbon timescale. NAPIS-75 was placed on the GISP2 age model by correlating the planktonic foraminifera $\delta^{18}O$ record in one of the sediment cores with the $\delta^{18}O$ record in the GISP2 ice core (Voelker et al. 1998; Kissel et al. 1999).

There are striking similarities between NAPIS-75 and the synthetic geomagnetic field record calculated from ^{36}Cl and ^{10}Be data obtained from the GRIP/GISP2 ice cores (Baumgartner et al. 1997, 1998; Finkel and Nishiizumi 1997; Yiou et al. 1997; Wagner et al. 2000a,b), which suggests a geomagnetic origin for the variations seen in NAPIS-75.

Features with a 1000 to 2000-year wavelength can be recognized in both records. The millennial scale features of NAPIS-75, coupled to the precise GISP2 time-scale, constitutes a highly efficient tool for correlating and dating cores in different oceans basins around the world (Stoner et al. 2000, 2002), particularly in the Arctic Ocean and Southern Ocean where $\delta^{18}O$ stratigraphy is not available. The extension of NAPIS-75 to NAPIS-300, a 300,000 year stack, is now in progress (C. Laj and C. Kissel, pers. comm.).

Frank et al. (1997) used similar methods to generate a 200-kyr global stack record of ^{10}Be production rates (named Sint-Be). This stack applied ^{230}Th-normalization of the ^{10}Be fluxes in order to correct for sediment redistribution effects. The ^{230}Th-normalized ^{10}Be deposition rates were then normalized to their mean value in each core and averaged in 1000-year increments to account for changes in boundary scavenging over time resulting from climatically-induced changes in particle flux and composition. The resulting record was then translated into geomagnetic field paleointensity according to Lal (1988).

Several paleointensity stacks and the ^{10}Be stack are shown in Figure 2. There are several interesting features to note. First, the peak-to-trough amplitudes of the geomagnetic records are very similar. This is encouraging and suggests that relative paleointensity normalization has been successful. Sint-200 and Sint-Be are in particularly good agreement over the past 20 kyr, one of the intervals of greatest discrepancies amongst sedimentary records using relative intensities. Second, all of the records show decreased geomagnetic field intensity over the intervals 30-42 ka, 60-75 ka, 85-110 ka, and 180-192 ka, which are all interpreted as geomagnetic excursions. Third, variations with wavelengths of ~10 kyr persist in both records. As has been noted previously, even shorter period features are correlative in the higher resolution NAPIS-75 and the GRIP/GISP2 ice cores The latter point is the basis for proposing that geomagnetic paleointensity has potential as a millennial-scale global correlation and dating tool (Guyodo and Valet 1996, 1999; Frank et al. 1997; Stoner et al. 2002).

In some cases the paleomagnetic record of deep-sea sediment may be compromised by post-depositional diagenesis or sediment disturbance. Iron-sulfur diagenesis (e.g., Karlin 1990; Leslie et al. 1990) may result in the dissolution of the magnetic grains that carry the NRM. Alternatively, magnetite authigenesis may result in the acquisition of a chemical overprint that obscures the primary NRM (e.g., Tarduno and Wilkison 1996). In rare cases, weakly magnetized sediments have been completely overprinted by a remanence acquired during the coring process (Fuller et al. 1998; Acton et al. 2002). In these cases the paleomagnetic-derived paleointensity record cannot be trusted, but paleointensity variations could be determined through ^{10}Be production rates, enabling the importation of the SPECMAP or GISP2 chronology to one's study area.

The asymmetric sawtooth

The asymmetric sawtooth pattern observed in deep sea sediments (Ninkovich et al. 1966; Kobayashi et al. 1971; Opdyke et al. 1973; Valet and Meynadier 1993; Meynadier et al. 1992; Yamazaki et al. 1995; Verosub et al. 1996) consists of an apparent rapid rise in relative paleointensity immediately after a geomagnetic field reversal, followed by a slow decay until the next reversal (Fig. 3). However, this pattern is absent in several records (e.g., Laj et al. 1996; Kok and Tauxe 1999), leading to uncertainties in its origins.

If the sawtooth represents true geomagnetic field behavior then it has strong implications for our understanding of the geodynamo (McFadden and Merrill 1997). Whether or not a relationship exists between dipole intensity and reversal frequency, or core processes in general, has long been a topic of debate in geomagnetism (see Merrill et al. 1996 for a full discussion). The sawtooth pattern displays a positive correlation between the magnitude of the post-reversal intensity recovery and the duration of the

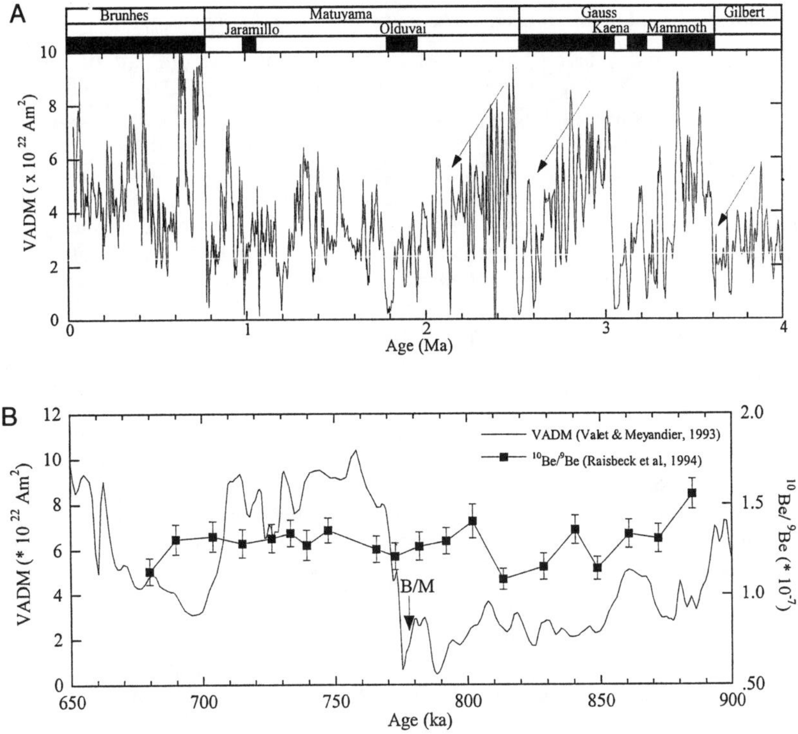

Figure 3 (A) Composite relative paleointensity variations from Ocean Drilling Program (ODP) Leg 138 Sites 848, 851, and 852, equatorial Pacific Ocean. The virtual apparent dipole moment (VADM) appears to gradually decay over time (arrows), then rebound immediately following a geomagnetic field reversal (Valet and Meynadier 1993). (B) ^{10}Be variations from Leg 138 cores 851C and 851E (Raisbeck et al. 1994) are incompatible with the paleomagnetic data. The $^{10}Be/^9Be$ ratio *decreases* leading up to the Brunhes-Matuyama reversal at ~778 ka. Further, the production-rate change at the reversal itself is substantially smaller than predicted by the paleomagnetic data. The apparent discrepancies may be due in part to lithological differences between the cores.

subsequent interval of constant polarity. This suggests that a strong field inhibits a future reversal and requires the geodynamo to have a memory of when the previous reversal occurred. From this we might suspect that geomagnetic field intensity is unusually high during superchrons. However, this suggestion is not supported by the available, and very limited, absolute paleointensity data from the Cretaceous Normal Superchron (Juárez et al. 1998).

McFadden and Merrill (1993, 1997) examined the frequency of reversals and duration of constant polarity intervals (chrons) in the Cenozoic and compared these statistics with models of inhibition in the reversal process. These studies found that a reversal at time t_1 can inhibit the next reversal at time t_2 for, at most, 50,000 years after time t_1. The sawtooth requires inhibition on timescales of 10^5 to 10^6 years. While the observational data and model results disagree in this particular instance, the sawtooth pattern presented by Valet and Meynadier (1993) represents a continuous well-dated paleointensity time series that can provide important input and "ground-truthing" for theoretical, numerical, and statistical geomagnetic field models. Therefore, it is critical that the observational data represents true geomagnetic field behavior.

Other mechanisms have been proposed to explain the sawtooth pattern, all of which invoke magnetic recording artifacts in sediments. These include intensity decay as resulting from the vector sum of oppositely aligned (normal and reversed) magnetization in the sediment (Ninkovich et al. 1966; Opdyke et al. 1973; Kobayashi et al. 1971; Mazaud 1996), "chemical lock-in" of a magnetization carried by authigenic magnetite (Tarduno and Wilkison 1996), or thermoviscous overprinting of a magnetization with a secondary component acquired in a field of opposite polarity (Kok and Tauxe 1996a,b). All of these arguments have their own complications, requiring magnetic grains to remain mobile and reorient several meters below the sediment-water interface or requiring an unrealistically narrow distribution of magnetic grain sizes (Meynadier and Valet 1996; Meynadier et al. 1998). There are presently more arguments and mechanisms for an artifact origin of the sawtooth, but debate continues over its origins (e.g., Valet and Meynadier 2001; Kok and Ynsen 2002).

^{10}Be has the potential to provide corroborating or repudiating evidence of the sawtooth pattern. Henken-Mellies et al. (1990) investigated ^{10}Be/^{9}Be ratios across the BM reversal and the Matuyama-Gauss (MG) reversal. In contrast to the predictions of the sawtooth paleointensity model, the ^{10}Be/^{9}Be ratio decreased leading up to the BM reversal and remained roughly constant prior to the MG reversal. However, the ^{10}Be/^{9}Be ratio gradually increased following the MG reversal, which is in agreement with the sawtooth pattern.

Raisbeck et al. (1994) measured the ^{10}Be/^{9}Be ratio at the same site as Valet and Meynadier (1993) spanning the interval 650 to 900 ka. They observed that the ^{10}Be/^{9}Be variations at the Brunhes-Matuyama (BM) reversal were far smaller than those expected on the basis of relative paleointensity estimates (Fig. 3). Further, the ^{10}Be/^{9}Be ratio decreased leading up to the BM reversal, in direct opposition to the paleomagnetic record.

Aldahan and Possnert (2000) constructed a 3.5-Myr record of ^{10}Be flux in deep-sea sediment from the Caribbean Sea. This is one of the longest sedimentary records of ^{10}Be flux back through time, and spans several geomagnetic polarity chrons. The site was chosen for its uniform sedimentation rate and apparent lack of climatic overprinting. This record consists of 90 samples, with the highest density of samples taken from the Brunhes chron (0 to 780 Ma). The ^{10}Be flux was elevated during geomagnetic reversals and excursions (Aldahan and Possnert 2000). However, the record does not match the pattern predicted by asymmetric sawtooth behavior of the geomagnetic field. The ^{10}Be flux appears to decrease leading up to each reversal. However, the sample density is very low, particularly prior to the Brunhes chron, making evaluation of sawtooth pattern difficult in those intervals.

Sawtooth opponents point to the absence of the pattern during the Brunhes chron, arguing that Brunhes age sediments have not yet seen an oppositely oriented field that could cause either the mechanical, chemical, or viscous realignment of the magnetization. Therefore, the Brunhes-Matuyama reversal and the Brunhes Chron may not be the ideal place to confirm or deny the sawtooth. Sub-chrons within the Matuyama Chron would be ideal intervals for investigation using ^{10}Be. It is critical that any such paired paleomagnetic and ^{10}Be study be done on the same core. Raisbeck et al. (1994) suggested that the discrepancies between the Valet and Meynadier (1993) paleomagnetic data and ^{10}Be data may have resulted from the fact that the data were gathered from two different cores with different sediment textures. The paleomagnetic data were collected from a laminated core and the ^{10}Be data gathered from a heavily bioturbated core. Assuming that the complicating influences of lithology can be minimized by careful core selection, a paired paleomagnetic and ^{10}Be study across entire sub-chrons, including but not limited to the bounding reversals, should aid in investigation the origin of the sawtooth pattern.

Milankovitch periodicities in geomagnetic paleointensity records

The cause and significance of Milankovitch periodicities in geomagnetic field timeseries data has been debated for more than 30 years. The question of whether variations in Earth's orbital parameters perturb fluid motions in the outer core (e.g., Malkus 1968; Rochester et al. 1975) appeared to be answered affirmatively when Kent and Opdyke (1977) observed a 43-kyr periodicity in a geomagnetic field paleointensity record from a Pacific Ocean sediment core. Subsequently, other records suggest the presence of 100-kyr and 41-kyr periodicities in geomagnetic paleointensity timeseries (e.g., Tauxe and Wu 1990; Tauxe 1993; Channell et al. 1998; Yamazaki 1999; Channell and Kleiven 2000; Yamazaki and Oda 2002) within different intervals within the Brunhes and Matuyama chrons. This has led to speculation that variations in earth's orbital parameters may indeed influence the geodynamo.

Tauxe (1993) tested the coherence of normalized intensity with the various normalization parameters. It was observed that periodicities in normalized intensity were also present in the normalization parameters, and the two time-series were coherent at Milankovitch frequencies. Tauxe (1993) proposed that the presence of Milankovitch periodicities in paleointensity records were due to "contamination" of the paleointensity record by climate-induced variations in the concentration and grain size of the magnetic minerals in the sediment. In this instance, normalization was not completely efficient. Subsequent studies of sediment cores from the North Atlantic (Channell et al. 1998; Channell and Kleiven 2000) and North Pacific (Yamazaki 1999) conducted the same analyses and demonstrated the presence of Milankovitch periodicities (41 kyr and 100 kyr, respectively) in the paleointensity records but not in the bulk magnetic parameters. In both cases, the authors suggested that the paleointensity records were therefore free of contamination, and represented orbital modulation of the geomagnetic field.

The paleointensity records presented by Channell et al. (1998) and Channell and Kleiven (2000), were particularly intriguing. These records were generated from Ocean Drilling Program Site 983, the Gardar Drift in the North Atlantic. This site has the benefit of high temporal resolution and superior age control from a high density of oxygen isotope measurements. The power spectra of these paleointensity records contained significant peaks at 100 kyr and 41 kyr. A similar analysis was performed on magnetic parameters such as IRM and magnetic susceptibility, which track the concentration of magnetic material. The 100-kyr cycles were present in the bulk magnetic properties, leading the authors to interpret the 100-kyr paleointensity features as a lithologic contamination. However, the 41-kyr cycles were *not* observed in the bulk magnetic parameters, leading the authors to conclude that the 41-kyr paleointensity cycles were due to true geomagnetic field behavior. Guyodo et al. (2000), re-examined these records using wavelet analysis. Wavelet analysis enables the detection of non-stationarity in a signal and identifies the specific time intervals when a given frequency is present in a record. Guyodo et al. (2000) identified the presence of periodic, though non-stationary, signals at 100 kyr and 41 kyr. Whereas global Fourier analysis failed to detect the 41-kyr signal in the bulk magnetic parameters, wavelet analysis of a magnetic grain size proxy revealed the 41-kyr signal in three discrete time intervals. These authors concluded that a lithologic overprint remained in the normalized intensity record, creating the 41-kyr variations in the paleointensity record. The authors then filtered these wavelet components from the paleointensity record, demonstrating that the secondary overprint had a minor effect on the overall paleointensity profile.

Presently, there is no known mechanism that conclusively couples earth's orbital parameters with the geodynamo. However, this particular problem and the case study from ODP Site 983 raises the issue of how to confidently identify any real geomagnetic field periodicities that might be similar or identical to any real periodicities associated with other forcing mechanisms. Guyodo and Valet (1999) investigated this question using Sint-800, the global stack record of 33 geomagnetic field paleointensity variations for the past 800 kyr. The majority of the constituent records were dated by correlating their oxygen isotope ($\delta^{18}O$) stratigraphies to those of the reference curve termed SPECMAP (Martinson et al. 1987; Bassinot et al. 1994). Stacking the 33 records together should average out any non-field effects in a given constituent record, while preserving the broad scale features of the geomagnetic field. The authors tested the validity of this assumption by re-calculating the stack 33 times, each time leaving out one of the constituent records. In each case the deviation between Sint-800 and the re-calculated stack was within the 2σ standard deviation of Sint-800, confirming the absence of significant outliers (Guyodo and Valet 1999).

Guyodo and Valet (1999) performed a spectral analysis of Sint-800 to look for periodic variations (Fig. 4). They first analyzed the entire signal, then they re-analyzed the signal in 400 kyr increments with a time step of 100 kyr (i.e., 0 to 400 kyr, 100 to 500 kyr, etc.). For comparison, the same analysis was performed with a stacked record of $\delta^{18}O$ values that was dated by orbital tuning (e.g., Imbrie et al. 1984; Martinson et al. 1987; Bassinot et al. 1994). The stacked $\delta^{18}O$ curve, named SPECMAP, is a proxy for global ice volume, which in turn is controlled by the Earth's orbital geometry (Milankovitch 1941). Orbital tuning involves synchronizing the proxy of global ice volume, typically $\delta^{18}O$ or $CaCO_3$ content in sediment, with the known history of orbital forcing, i.e., the calculated time series of summer solar irradiance at 65°N latitude, or a model of the climate response signal (e.g., Imbrie and Imbrie 1980) (see Martinson et al. 1987 for full discussion).

The spectral content of successive 400-kyr intervals of Sint-800 are very different from one another, confirming the absence of any stable periodicity (Fig. 4). In contrast, significant peaks in spectral power at the Milankovitch periodicities (23, 41, and 100 kyr) are seen in each increment of SPECMAP. Given that the SPECMAP chronology is derived from orbital tuning, it may seem circular to perform a spectral analysis, as the forcing function periodicities are guaranteed to be present. However, the purpose here is to illustrate the differences between the non-periodic Sint-800 signal and the periodic SPECMAP signal. While Sint-800 reveals large amplitude changes in geomagnetic field paleointensity, there is no evidence of a stable, dominant periodicity (Guyodo and Valet 1999).

Corroborating evidence could be obtained by performing the same analysis on a stack of ^{10}Be production. Unfortunately, neither the ^{230}Th-normalized 200-kyr Sint-Be record (Frank et al. 1997) nor the ~225-kyr Taylor Dome ^{10}Be record (Steig et al. 2000) is long enough to confidently identify the 100-kyr Milankovitch periodicity. An extension of Sint-Be would be ideal for investigation of the presence of Milankovitch cycles. The 3.5-Myr record of ^{10}Be flux in deep sea sediment from the Caribbean Sea of Aldahan and Possnert (2000) would also be well-suited to this purpose, given the site's uniform sedimentation rate and apparent lack of climatic overprinting. Another candidate is the ~400-kyr Vostok ice core (Petit et al. 1999). Both of these records would need to be enhanced with a higher sampling density in order to confidently identify Milankovitch periodicities.

Assuming that these existing records can be enhanced, or that new high resolution records will be generated, there is a good basis for expecting that geomagnetic

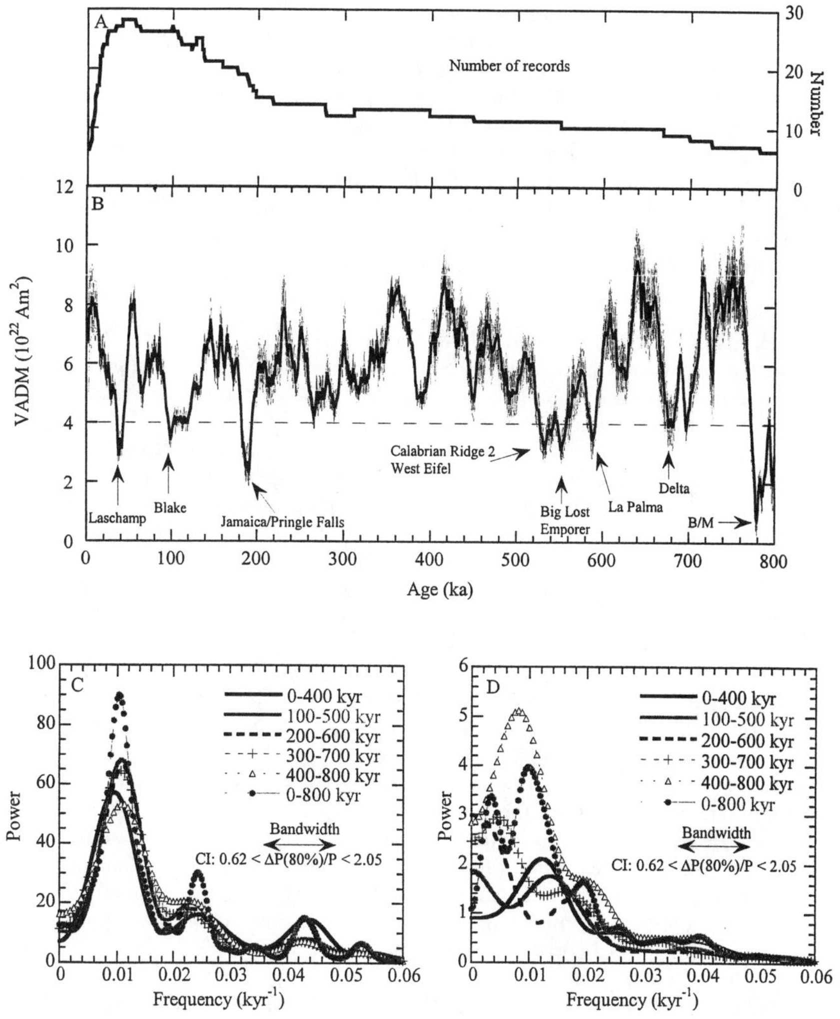

Figure 4. (a) Distribution of constituent records in Sint-800. (b) Sint-800 (Guyodo and Valet 1999) reveals 8 excursions during the Brunhes chron, each associated with a >50% reduction in the strength of the dipole moment. (c) Spectral analysis of Sint-800, and successive 400-kyr increments of Sint-800. There are no significant peaks in spectral power in the Sint-800 record. (d) In contrast, the 23-kyr, 41-kyr, and 100-kyr periodicities are present in each sub-set of the SPECMAP curve (redrawn after Guyodo and Valet 1999). Spectral analysis was performed using Analyseries software (Paillard et al., 1996).

modulation cycles can be identified in ^{10}Be production rates. Radionuclide investigations using tree rings and ice cores have revealed century and decade scale variations in the production rates of ^{14}C and ^{10}Be (e.g., Beer et al. 1988; Stuiver and Braziunas 1989; Beer et al. 1990; Stuiver and Braziunas 1993; Steig et al. 1998; Wagner et al. 2001) which have been attributed to solar modulation of the primary cosmic ray flux. A similar approach in lithologically uniform, high-deposition rate sediments could potentially resolve long-period variations in the geomagnetic field.

Any such effort must remember that observed periodicities in ^{10}Be concentration in sediments need not be due to the geomagnetic field. Henken-Mellies et al. (1990) observed a strong anti-correlation between ^{10}Be concentration and $CaCO_3$ content. Further, Kok (1999) proposed that the excellent agreement between Sint-200 and Sint-Be is due to the paleoclimate contamination present in both records rather than true geomagnetic field behavior. Henken-Mellies et al. (1990) attempted to remove climatic influences by calculating the ^{10}Be concentration on a carbonate-free basis. The presence of periodicities in both a normalized relative intensity record and a ^{230}Th-normalized ^{10}Be record from the same core, while not conclusive evidence, would further the argument in favor of periodic variations in the geomagnetic dipole moment.

Summary

Variations in geomagnetic field paleointensity derived from ^{10}Be and ^{36}Cl production rates have the potential to contribute to the resolution of controversies in geomagnetism. ^{10}Be accumulation is not affected by sediment magnetic recording processes and normalization artifacts that are at the root of these controversies. The half-life of ^{10}Be (1.5 Myr) theoretically enables its application back to ~10 Ma, which would allow a study of the pattern of radionuclide production rates over several geomagnetic polarity intervals. Long records of ^{10}Be in sediments could potentially confirm or refute the asymmetric sawtooth pattern and the apparent 100-kyr periodicity in geomagnetic paleointensity variations. Using ^{10}Be and ^{36}Cl production rates obtained from glacial ice or sediment as proxies of geomagnetic field paleointensity requires the recognition and removal of complicating influences such as changing precipitation rates and moisture sources, boundary scavenging, sediment focusing and winnowing, and climatically-driven changes in ocean circulation and water mass distribution. Sedimentary profiles of ^{10}Be can be normalized to ^{230}Th to correct for sediment redistribution effects. However, the relatively short half-life of ^{230}Th (75 kyr) means that sedimentary records cannot be corrected for sediment redistribution effects beyond ~300 ka. The combined application of paleomagnetic and ^{10}Be methods has already enhanced the reconstruction of past geomagnetic field intensity variations over the past ~200 kyr. Comparison of ice core profiles of ^{10}Be and ^{36}Cl accumulation with high-resolution sediment sequences has confirmed the geomagnetic origin of high-frequency variations in the geomagnetic dipole moment, and revealed the very abrupt nature of the Laschamp geomagnetic excursion. These types of records are contributing to the development of geomagnetic paleointensity as a millennial-scale global correlation and dating tool.

SUBDUCTION AND MAGMATISM AT CONVERGENT MARGINS

Introduction

Cosmogenic ^{10}Be, with high concentrations in young marine sediments, is an outstanding tool for tracing sediment subduction and recycling at convergent margins. Decaying with a 1.5-Myr half-life (Yiou and Raisbeck 1972), high ^{10}Be concentrations are measured only in sediments younger than 12 Ma, and do not build up in the mantle over time. High ^{10}Be concentrations in arc lavas thus require the subduction to depth of the youngest part of the sedimentary veneer, and the transport of slab-derived elements to the mantle wedge and thence to the surface. As a result, ^{10}Be studies can constrain a wide variety of physical and magmatic processes operating at convergent margins, from the trench to the back-arc.

^{10}Be on the subducting plate

This section provides an overview of the ^{10}Be transport in subduction zones. It discusses the variations in ^{10}Be supplied to subduction trenches around the world, the mineralogical hosts for ^{10}Be in the downgoing sediment column and evidence for and against ^{10}Be mobility in fluids leaving the slab at shallow levels.

The ^{10}Be cycle. Figure 5 illustrates a series of on-off switches in the subduction cycle, all of which must be in the "on" position for high ^{10}Be concentrations to be observed in arc lavas. ^{10}Be inventories in the incoming sediment column must be high enough that a small amount of young sediment mixed with large amounts of older sediment and mantle (or other diluents) can still produce a measurable ^{10}Be signal in the arc lava. Those uppermost ^{10}Be-rich sediments must largely escape frontal accretion and basal underplating as the plate subducts beneath the fore-arc. The convergence rate must be fast enough that the subducting ^{10}Be doesn't decay away during subduction (i.e., the ^{10}Be clock is still applicable). The ^{10}Be must be extracted from the downgoing slab and be transferred through the mantle and to the surface before the signal decays away (<2 Myr). The following section will discuss the global distribution of ^{10}Be in volcanic arcs in terms of these controlling factors.

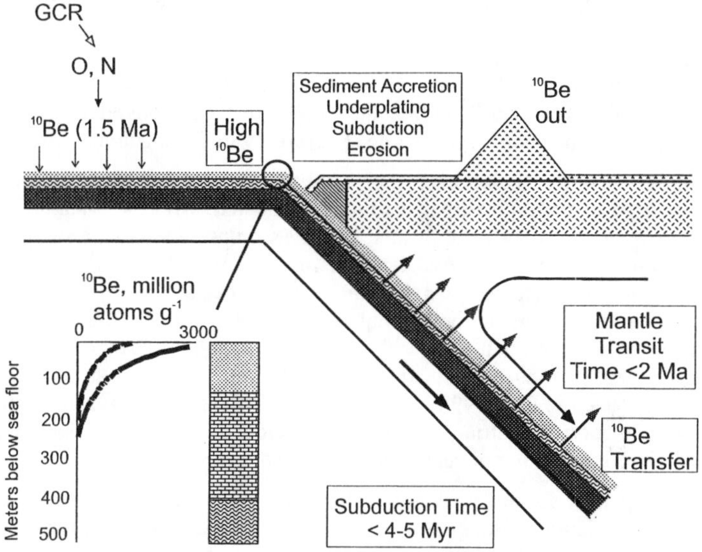

Figure 5. ^{10}Be cycle through subduction zone. ^{10}Be is created by spallation reactions on O and N in the atmosphere. It is strongly adsorbed onto settling sediment particles, and carried with the plate into the subduction zone. High ^{10}Be values in arc lavas require that (1) the incoming sediments have high ^{10}Be, (2) that most of the uppermost sediment column is not accreted but rather subducted to the depths of magma generation; (3) that the subduction time be less than 4-5 Myr, and that the ^{10}Be be extracted from the downgoing sediments and moved through the mantle within about a ^{10}Be half life. The lower left inset shows the distribution of ^{10}Be in DSDP Site 495, outboard of Guatemala. Higher values are those measured at the trench; the line displaced to lower concentrations is the calculated value for sediments subducted to a point beneath the volcanic arc and reflects the effect of ^{10}Be decay during subduction.

^{10}Be in subducting sediments. A quantitative comparison of the sediment hosted ^{10}Be flux into the trench with the ^{10}Be flux out of the volcanoes (the ^{10}Be flux balance) can be used to constrain the absolute volumes of sediments subducted to the depths of magma generation, if the necessary data are available and certain assumptions are met. A series of papers (Zheng et al 1994; Morris et al 2002; George et al., submitted; Valentine and Morris, submitted a,b) show ^{10}Be inventories and/or ^{10}Be-depth profiles for sediments entering the Mariana, Aleutian, Japan, and Middle America (Guatemala and Costa Rica) trenches, using DSDP/ODP drill core. The data, summarized in the upper part of Table 1 (p. 214), show that the total amount of ^{10}Be in the sediment column (the inventory) supplied to different trenches can vary by an order of magnitude, e.g., the incoming sediment column off the Mariana trench has a ^{10}Be inventory close to the globally averaged ^{10}Be production rate (Monaghan et al. 1985/1986), while those for Japan and Central America are ten times greater. In general, margins studied to date have values comparable to or greater than that corresponding to the globally averaged ^{10}Be production rate, implying ^{10}Be deficits in much of the open Pacific Ocean. Table 1 also shows that the sediment layer tagged by high ^{10}Be can vary in thickness from ~25 m (Marianas) to 250 m (Japan) as a function of Plio-Pleistocene sedimentation rates; preliminary data for the Tonga section suggest all ^{10}Be will be within the upper 15 m (Morris, unpubl.). Sediment lithologies for the uppermost part of the cores reported here are pelagic to hemi-pelagic, often rich in volcanic detritus or with ash layers deposited as the incoming plate approaches the arc. Considering the heterogeneity in the total amount of ^{10}Be subducted and the thickness of the ^{10}Be-tagged sediment layer, ^{10}Be measurements in the arc and also in the specific sediment column entering the associated trench are necessary for detailed interpretations of sediment dynamics or flux balance (e.g., Plank and Langmuir 1993). The ^{10}Be inventory in Table 1 was converted to a ^{10}Be flux to the trench using published convergence rates (Jarrard 1986; Zhang and Schwartz 1992).

Quantitative ^{10}Be flux balances can be used to estimate the relative proportions of sediments accreted or underplated versus those subducted, with the assumption that the margin is in an approximately steady state. The lavas erupting today are sampling a plate that began subducting 1.8- to 4-Myr ago (for margins with high ^{10}Be). The sediment composition, supply rate and sediment dynamics need to be approximately constant over this duration in order to compare the amount of ^{10}Be subducted with that erupted from the arc. In some cases (e.g., Reagan et al. 1994; Plank et al., submitted) the sediment and arc data themselves argue strongly for steady state conditions. In other cases (see Von Huene and Scholl 1991; Rea and Ruff 1996) steady state conditions may not have applied due to recent changes in tectonic setting or sedimentological conditions. In such cases, flux balances for ^{10}Be (or any other tracer) must be evaluated carefully.

The bottom part of Table 1 shows the ratio of the ^{10}Be flux out of the arc, relative to the ^{10}Be flux to the depths of magma generation. The ratio is corrected for decay during subduction, using a simplest model where the ^{10}Be subducts at the convergence rate to a point beneath the volcanic front, and magma ascent times are short relative to the ^{10}Be half life. Where a high proportion of the subducted ^{10}Be is erupted from the arc (e.g., Nicaragua) then most of the sediments must have been subducted to the depths of magma generation. Low values (e.g., Costa Rica, Honshu) require that much of the incoming young sediment be accreted or diluted by subduction erosion. These results are discussed in more detail below, but this brief discussion highlights the way in which the sediment signature recorded in the lavas allows the volcanic arc to serve as a "flow meter" for the volume and lithology of the sediments carried through the seismogenic zone to the depth of magma generation (see also Morris et al. 2002; Valentine and Morris, submitted a,b).

^{10}Be and fluid mobility. Quantitative flux balance calculations also require that ^{10}Be is not leaving the slab to be stored in areas where it cannot be sampled, measured, and included in the flux balance. This assumption may be evaluated through studies of prism sediments, subduction zone metamorphic rocks and experiments. You et al. (1994) showed lower than expected ^{10}Be concentrations below the décollement at Nankai (ODP Site 808), for which their preferred interpretation was ^{10}Be fluid mobility. Alternative interpretations are possible, given that similar offsets in the ^{10}Be-depth profile are observed at lithologic boundaries at other depths in the column, removed from areas of fluid flow. The magnitude of fluid flow along and below the décollement also has been debated (Taira and Hill 1991). By contrast, the reconnaissance-scale ^{10}Be-depth profiles from ODP Leg 170 (Sites 1040 and 1043) sampled the sediments immediately above and below the décollement. The subducting sediments (Valentine et al. 1997; Morris et al. 2002) show no deviation from the expected profile, despite strong evidence here for fluid flow along, and just below, the décollement (Kimura et al. 1997; Silver et al. 2000; Kastner et al. 2000). ^{10}Be analysis of very closely spaced samples in regions of documented fluid flow, together with ^{9}Be analysis of associated pore fluids, would help evaluate the possibility of ^{10}Be fluid mobility.

Experimental studies have been conducted on Be partitioning between sediments or igneous/metamorphic minerals, and fluids of varying compositions. At low temperatures (<150°C) and moderate to high pH (>4), Be is strongly adsorbed onto sediment surfaces, with concentrations on the particles 10^5 times that in the associated seawater (Nyffler et al 1984; You et al. 1994, 1996). Accretionary prism fluids have geochemical indicators suggesting that some fraction of the fluids may have originated at regions with temperatures as high at 110-150° (Kimura et al. 1997) and their pH is 7-8. The experimental results suggest that any Be mobility under these conditions should be an insignificant part of the total Be budget. At higher temperatures, Be becomes slightly more mobile, e.g., 10^4 times greater in the particle than the fluid. Hydrothermal fluids reacted with sediments at temperatures of about 300°C carry approximately 0.1 ppb Be (You et al. 1996). Sediment-fluid experiments at 650-700°C begin to show some fluid mobility, with Be in the sediment being 2-4 times that of the associated fluid (Johnson and Plank 1999). Serpentine dehydration experiments (Tatsumi et al. 1986) and studies of Be solubility in equilibrium with mantle minerals (Brenan et al. 1998) suggest that Be is not particularly fluid mobile under mantle conditions.

Mineralogical and geochemical studies of metamorphic assemblages exhumed from the hanging wall of paleo-subduction zones show that white mica is the primary host for Be in the sediment column (e.g., Bebout 1996; Bebout et al 1993, 1999; Grew this volume). Older studies of Be adsorption between seawater and sediment particles suggest that some fraction of adsorbed Be moves into 'unexchangeable' structural sites within the crystals (Nyffler et al. 1984). Bebout et al (1993) reported B, Be and Ba concentrations in subduction assemblage metamorphic rocks thought to have similar protoliths. Water and boron concentrations decrease systematically with increasing grade, while Be and Ba show slight variability about a constant concentration, suggesting no significant Be loss from progressively metamorphosed sediments at conditions up to approximately 40 km depth and about 700°C (see also Grew, this volume).

^{10}Be was measured in serpentinite muds recovered from the top of Conical Seamount during ODP Leg 125, to assess the possibility that ^{10}Be was extracted from the slab at relatively shallow levels (Benton 1997). The seamount is a very large (1 km high by 30-40 km across) serpentine mud volcano located approximately 20 km above the downgoing slab (Fryer et al. 1995). A large number of geochemical tracers in actively venting and pore fluids sampled there suggest that some fraction of the fluid supply derives from the subducting slab (e.g., Haggerty 1991; Haggerty and Fisher 1992; Benton 1997). Processed at Washington University under very low blank conditions (blank $^{10}Be/^{9}Be = 7\text{-}9\times10^{-16}$, < 1000 atom g^{-1}, and analyzed at University of Pennsylvania with a detection limit of 1×10^{4} atoms g^{-1}), the serpentinite muds have very low, but measurable ^{10}Be, in the range $5\text{-}7\times10^4$ atoms g^{-1} (Benton and Morris, unpubl.). ^{9}Be concentrations were below detection limits (Benton 1997). Be/Nd ratios are rather constant in many subduction zone lavas (Ryan and Langmuir 1988), but highly variable Nd concentrations in the serpentine muds (e.g., 0.03-2.6 ppm) make it difficult to predict a meaningful ^{9}Be concentration or ^{10}Be/^{9}Be ratio. The ^{10}Be in the serpentinite samples could originate from the downgoing slab, or through very minor near-surface interaction with marine sediments, not currently recognized. The serpentinite mud samples were chosen for initial analysis because of the ease of sample preparation; solid clasts of variably serpentinized harzburgite make a logical next sample set.

In summary, arc lavas sample the downgoing plate, and the chemistry of the lavas may be used to place constraints on the composition of the slab at depth (Morris et al 1990; Plank and Langmuir 1993, 1998; Armstrong 1971). All models using flux balances require the assumption or demonstration of approximately steady-state conditions during the time required for the plate to subduct from the trench to the depths of magma generation. The flux of elements to depth in the subduction zone can be uncertain if they are mobile at shallow levels in the subduction zone (e.g., B, Cs, U, Rb, As) or are contained in sediments that may be scraped off. Set against these uncertainties is the fact that drilling and seismic imaging provide constraints on sediment dynamics only in the upper 2 km and ca. 10-20 km, respectively. Any constraints on sediment transport to the deeper subduction zone must come from the chemistry of fore-arc serpentinites, arc lavas and the deep mantle. An internally consistent model of sediment transport that simultaneously satisfied the results of drilling, imaging, sediment and arc chemistry (e.g., Morris et al. 2002; Valentine and Morris, submitted a,b) gives confidence that this approach using ^{10}Be can be successfully applied.

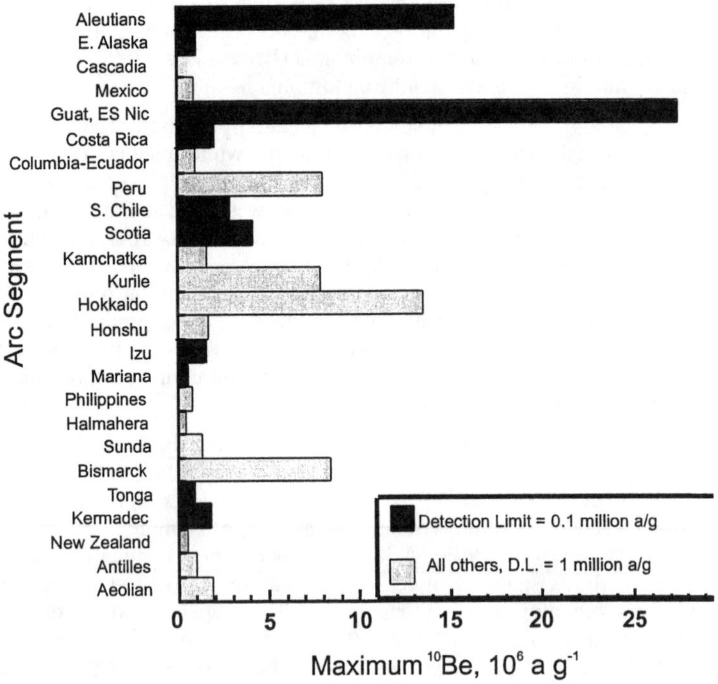

Figure 6. Bar graph showing the maximum ^{10}Be measured in volcanic arcs around the world. The data is taken from Table 2 and references therein. Almost all arcs with high ^{10}Be have a wide range of ^{10}Be values, extending from near background to the maximum value reported.

^{10}Be in volcanic arcs: A global summary

Table 2 and the bar diagram in Figure 6 show the measured concentrations of ^{10}Be in volcanic arc lavas from around the world, in comparison to the detection limits appropriate at the time of measurement. Note that detection limits have improved over time. A detection limit of about 1 million atoms g^{-1} applies to measurements made prior to 1994; more recent work has a detection limit of less than about 0.1 million atoms g^{-1} and exceptionally low-blank work can achieve detection limits of 10^4 atoms g^{-1}. In some cases, the changing detection limit makes a significant difference; note that Marianas samples measured previously were considered barren of ^{10}Be (Tera et al. 1986) while newer measurements show small but real enrichments.

The bulk of the samples shown in Figure 6 are from historic eruptions (1-300 years old). As such, they are too young to have built up *in situ* ^{10}Be through cosmic ray bombardment of the rock or through surface alteration. Mineral separate studies of ^{10}Be and ^{9}Be distribution in volcanic arc rocks (Monaghan et al. 1988; Morris and Tera 1989) showed that the two isotopes were in equilibrium in all phases measured for the 7 historic lavas analyzed. Approximately 20 % of the samples shown are prehistoric, being several hundred years to less than 50 kyr old. These samples are from localities that are geographically critical for testing geodynamic or sediment dynamic models, and for which historic eruptions do not exist. All such samples were leached with 1N HCl in an ultrasonic bath for 1 hr, and both the leachate and the leached sample analyzed. All non-historic samples reported in Figure 6 have negligible amounts of leachable ^{10}Be.

^{10}Be-barren arcs. A number of arcs (E. Alaska, Cascadia, Mexico, Honshu, Philippines, Halmahera, Sunda, New Zealand, Lesser Antilles and Aeolian) show no significant ^{10}Be enrichment. The absence of ^{10}Be in these arcs could be due to either ^{10}Be decay in transit from the trench through the mantle to the surface, or to an absence of young sediment recycling in a particular arc.

Convergence rates are low, and subduction times are long for eastern Alaska, Cascadia, and the Lesser Antilles. Because the ^{10}Be clock begins keeping time when the sediment column passes beneath the fore-arc (i.e., once ^{10}Be decay is no longer offset by its deposition), these lavas should contain no subducted ^{10}Be, which would have decayed during subduction. The lavas from these segments do not, in fact, have elevated ^{10}Be, indicating that any sediment/crust assimilation by magmas on their way to the surface did not add ^{10}Be to the lavas (see also George et al., submitted). Most other arcs in Table 2 have rapid enough convergence rates that not all incoming ^{10}Be would decay during subduction to the depths of magma generation.

The expected extent of ^{10}Be decay during transport through the mantle often can be constrained using U-series isotopes. U-series disequilibria isotope measurements show that most arcs include lavas that have excess U. U excesses are seen only in arc lavas. They are generally interpreted to mean that U but not Th was transferred from the slab to the arc mantle wedge and thence to the surface in arc lavas within the last 300 kyr, a short period relative to the ^{10}Be half-life (Gill and Williams 1990; McDermott and Hawkesworth 1991; Hawkesworth et al. 1997; Turner et al 1997; Turner and Hawkesworth 1997, 1998). Lavas from Japan, Kamchatka and Indonesia have U-excesses, indicating that magma ascent is rapid enough that ^{10}Be should not decay away in transit through the mantle.

Arcs such as Japan, Kamchatka, Halmahera, Columbia, and Indonesia show no ^{10}Be enrichment, although other geochemical tracers indicate some sediment subduction and recycling (e.g., Plank and Langmuir 1998; Kersting et al. 1996; Cousens et al. 1994; Edwards et a. 1993; Morris et al. 1983). All sediment columns yet measured outboard of

Table 2. ^{10}Be in volcanic arcs around the world.

Volcanic Arc	Min. $^{10}Be^{\ddagger}$	Max. $^{10}Be^{\ddagger}$	D.L.‡	Conv. rate†	Refs.	Volcanoes analyzed
Aleutians	0.4	15.3	1	7.5	1, 2, 4, 5, 6, 16, 18	Cold Bay, Amak, Shishaldin, Westdahl, Akutan, Makushin, Bogoslof, Okmok, Recheschnoi, Vsveidof, Atka, Kastochi, Seguam, Kanaga, Kiska
E. Alaska	0.1	0.7, 8.4*	0.1	6.4	5, 6, 18	Spurr, Redoubt, Augustine, Trident, Ukinrek, Aniakchak
Cascadia	0.1	0.5	1	3.5	2, 5	Indian Heaven, Mt. St. Helens
Mexico	0.3	0.9	1	7	1, 2	Ceburoco, Cuicuilo, El Chichón
Guatemala, El Salvador, Nicaragua	2.4	27.1	0.1-1	8.3	1 2, 5, 6, 12, 21, 22	Santa Maria, Pacaya, Santa Ana, Izalco, Boqueron, San Miguel, Conchagua, Cosiguina, Telica, Cerro Negro, Asosoca, Momotombo, Nejapa, Masaya, Mombacho, Zapatera, Concepción
Costa Rica	0.8	1.8, 8.4*	0.1	8.8	1, 2, 5, 6, 13, 21, 22	Hacha, Cerro Chopo, Orosi, Miravalles, Arenal, Platenar, Poas, Irazu,
Columbia-Ecuador	0.3	1	1	7	17	Ruiz, Purace, Galeras, Cotopaxi
Peru	0.3	8	1	8.2	2, 4	Chachani, El Misti, Ubinas
C & S. Chile	1.3	2.1	1	8.5	5, 6 , 7 ,23	San Jose, Chillan, Antuco, Lonquimay, Villarrica, Mocho, Mirador, Osorno, Calbuco, Calburgua, La Barda, Huellemole
Scotia	0.5	4.1	0.1	7.7	24	Bellingshausen, Saunders, Zavaroski, Montague, Candlemass
Kamchatka	0.5	1.6, 3.8$^{\$}$	1	8	9	Shiveluch, Klychevskoi, Bezimyanniy, Tolbachik, N. Ichinski, Ichinski, Kizmen, Kangar, Krashennikova, Karimski, Avachinksi, Gorely, Opala, Ksudach
Kurile	1	8	1	8.5	9	Alaid, Parmushir, Chirinkotan, Smt. 2.3, Onekotan, Aekarma, Lovushki, Raikoke, Sarychev, Berg, Smt 7.7, Smt 8.8, Atsonpuri, Lvinaya, Kunashir

Hokkaido	0.5	13.5	1	8.7	2, 3, 15, 20	Rishiri, Tarumai, Usu, Rausu-dake, Komaga-take, E-San, Oshim-Oshima
Honshu	0.2	2.4$^{\$}$	1	10	2, 3 , 15, 20	Osorean, Moriyoshi, Kanpu, Iwate, Akitkomatake, Kiroma, Zao, Nasu, Funagata, Fuj
Izu	0.8	1.4	0.1	10	25	Oshima, Miyake-Jima, Tori-shima
Mariana	0.1	1	1	9	2	Iwo-Jima, Uracas, Pagan, Guguan
Mariana	0.2	0.5	0.1	9	2, 14 , 26	Asuncion, Uracas, Pagan, Alamagan, Guguan
Philippines	0.8		1	9	2,	Mayon
Halmahera	0.1	0.5	1		2	Ibu, Ternate, Makian
Sunda	0.1	0.4	1	7.7	2, 10	Papandjan, Sundoro, Sumbing, Ungaran, Guntur, Galunggung, Cereme, Merbabu, Batur, Ebulobo, Keli-Mutu, Lewotobi
Bismarck	0.5	8.4	1	11	2, 6, 11	Kadovar, Manam, Karkar, Long, Langila, Garove, Makalia, Lolobau, Pago, Bamus, Ulawun, Rabaul
Tonga	0.2	2.2	0.1	<17	24	Niafoua, Tafahi, Fonualei, Late, Metis Shoal, Tofua, Ata
Kermadec	0.4	2.3	0.1	6	24	Raoul, L'Espérance, Valu Fa, Curtis, Macauley, Rumble IV, Clark
New Zealand	0.4	0.5	1	6	18	Tarawera
Antilles		<1	1	4	19	
Aeolian	0.1	1.9$^{\$}$	1	2.5	8, 25	Lipari, Alucudi, Volcano, Stromboli, Vesuvius

‡^{10}Be for arcs and detection limit (D.L.) reported in units of 10^6 atoms g^{-1}

†Convergence rates, in cm yr^{-1}, are from Zhang & Schwartz (1992) and Jarrard (1986)

* refers to a single outlier value; $ refers to values from non-historic lavas with leachable ^{10}Be

References: 1). Brown et al. (1982); 2) Tera et al. (1986); 3.) Imamura et al. (1984); 4) Monaghan et al. (1988); 5) Morris & Tera (1989); 6) Morris et al. (1990); 7) Sigmarsson et al. (1990); 8) Morris et al. (1993); 9) Tera et al. (1993); 10) Edwards et al. (1993); 11) Gill et al. (1993); 12) Reagan et al. (1994); 13) Herrstrom et al. (1995); 14) Morris (1996); 15) Shimaoka (1999); 16) Ryan (unpubl.); 17) Edwards (unpubl.); 18) George et al. (submitted); 19) Valette-Silver (unpubl.); 20) Shimaoka (in press); 21) Morris et al. (2002); 22) Valentine and Morris (submitted a) 23) Hickey-Vargas et al. (in press); 24) Morris and Tera (2000); 25) Morris (unpubl.); 26) Valentine and Morris (submitted b) subduction trenches have ^{10}Be inventories comparable to, or greater than that which is predicted by a model of globally uniform ^{10}Be production and deposition. In these margins, with moderate-fast convergence rates, these inventories would be high enough to produce ^{10}Be enrichments in arc lavas if the uppermost part of the sediment column were subducted to the depths of magma generation.

The previous discussion suggests that sediment dynamics is the likeliest explanation for an absence of ^{10}Be in these arcs. In Honshu, Table 1 shows that less than 1.7% of the subducting ^{10}Be is erupted in the arc, after correction for ^{10}Be decay during subduction. A small amount of frontal accretion (Morris et al. 2002) and large amounts of subduction erosion of the Cretaceous accretionary prism (von Huene et al. 1994) would minimize any ^{10}Be enrichment in the arc. The very low but not zero values for ^{10}Be in Costa Rican lavas have been attributed to the basal underplating of most, but not all, of the hemi-pelagic sediment section (Valentine et al. 1997; Morris et al. 2002; Valentine and Morris, submitted a,b), perhaps related to the style of graben development on the downgoing plate (Kelly and Driscoll 1998). Subduction erosion and dilution of the incoming ^{10}Be with old, barren forearc sediment could also explain the low ^{10}Be in Costa Rica (Vannuchi et al. 2001). Large accretionary prisms outboard of Indonesia, Kamchatka and Halmahera (Von Huene and Scholl 1991; Moore and Silver 1983) suggest that low values in these margins reflect frontal accretion. The absence of ^{10}Be in Columbian volcanoes still needs an explanation, as this is a non-accretionary margin with a convergence rate fast enough to get ^{10}Be down and back to the surface. If re-measured with very low detection limits, the Columbia lavas might have just a little ^{10}Be, rather than being totally barren. Except in cases where the ^{10}Be inventory in subducting sediments is low (e.g., Marianas) the difference between very low and zero ^{10}Be concentrations in the arc lavas will not generally change the foregoing interpretations. The differences could, however, allow calculated volumes of sediment accretion and erosion to be better constrained.

^{10}Be-rich arcs. Real ^{10}Be enrichments (See Table 2 and Fig. 6) are seen in a number of volcanic arcs, including the Aleutians, Middle America, Scotia, Kuriles, Izu, Marianas, Bismarck, and Tonga (see references in Table 2). All these margins have convergence rates in excess of 4-7 cm yr^{-1} and are classified as non-accretionary margins (Von Huene and Scholl 1991). U series excesses in these arcs (George et al., submitted; Reagan et al 1994; Elliott et al. 1997; Turner et al. 1997, 2000; Gill et al. 1993; Gill, pers. comm.; Elliott, pers. comm.) indicate that magma ascent times are less than 300 kyr, short with respect to the ^{10}Be half-life. The lower part of Table 1 shows the results of ^{10}Be flux modeling. The ^{10}Be flux out/flux in calculates the amount of ^{10}Be erupting from the volcanic arc relative to the amount subducted to the depths of magma generation after decay in transit has been taken into account. For the Aleutians, the Guatemala-Nicaragua segment and the Mariana arc, the calculations suggest that 10 to 30% of the subducted ^{10}Be is recycled in the arc (Zheng et al. 1994; Valentine et al. 1997; Morris et al. 2002; George et al., submitted; Valentine and Morris, submitted a,b).

The fraction of the subducted ^{10}Be flux returned to the surface in arc volcanism will be a function of the portion of the sediment column subducted to depth, and the efficiency with which Be is extracted from the subducted sediment and fed to the arc (^{10}Be recycling efficiency). The estimate of 10-30% ^{10}Be return flux can be met by 100% sediment subduction and 10-30% ^{10}Be recycling efficiency; lesser amounts of sediment subduction require greater ^{10}Be recycling efficiency. A ^{10}Be recycling efficiency of ~30% is at the high end of the range estimated by arc geochemists (e.g., Plank and Langmuir 1993; 1998). Coupled isotopic variations seen in some arcs (e.g., Reagan et al. 1994) together with the high ^{10}Be return flux are best satisfied by subduction of >95% of the incoming sediment column at the Aleutian, Mariana and middle Central America arcs.

^{10}Be enrichment in Tonga and the Marianas, where only the top 15-25 m of the incoming sediments are young enough to contain ^{10}Be, requires that virtually the entire section be subducted. Even in the absence of ^{10}Be data for the incoming sediment column, general estimates of this sort can be made about the volumes of subducting sediments.

The results in Tables 1 and 2 and Figure 6 show the highly heterogeneous nature of contemporary sediment subduction. For example, Mexico has no ^{10}Be enrichment in arc lavas, Nicaragua has the highest values yet measured in any arc, and Costa Rica has low but real enrichments of ~1 million atoms g^{-1}. These observations can be explained by >50% sediment accretion off Mexico (Tera et al 1986), complete sediment subduction beneath Guatemala, El Salvador and Nicaragua (Reagan et al. 1994) and no frontal accretion but ca 30% underplating of the incoming section beneath Costa Rica (Morris et al. 2002; Valentine and Morris, submitted a). Kamchatka has no ^{10}Be enrichment, while the adjacent Kurile arc has high ^{10}Be that requires nearly complete subduction of the incoming sediment column. These variations along strike within an arc system point to the complexity of processes that control sediment subduction and accretion, poorly understood at present. They also show that it can be difficult to arrive at a single value for the percentage of sediments (or volume of sediments) recycled within a given arc. That being the case, any global estimate of the amount of sediment subducted annually, a necessary value for models of continental growth through time, will have very large uncertainties.

^{10}Be and magmatic processes

Entering the mantle wedge only from the sedimentary veneer of the downgoing plate, and carrying a clock as its travels through the wedge, ^{10}Be provides some unique constraints on magmatic processes. This section first examines the cross-arc distribution of ^{10}Be in the Kurile, Bismarck and Aleutian arcs, with implications for hydrous mineralogy of the slab and mantle and for mantle melting processes. The following section reviews combined studies of ^{10}Be and U-series isotopes, and the time-scales for subduction modification of the mantle.

^{10}Be in cross-arc transects. Many convergent margins have volcanoes located behind the main volcanic front, derived from melting of a mantle that was modified above a deeper subduction zone. As such, these rear-arc volcanoes can provide insight into the changing composition of the subduction component between ca. 110-km depth and 200-km depth, with implied constraints on slab surface temperature, mineralogy and the medium (e.g., aqueous supercritical fluid or siliceous melt) that transfers elements from slab to mantle. Rear-arc volcanoes also provide a “last glimpse” of the slab before it heads into the deeper mantle, and give some clues as to the composition of the deeply subducted slab.

A number of studies have built upon the initial work on potassium concentration vs. depth (K-h) relationships (Dickinson 1975), examining the changing chemistry of arc lavas with increasing depth to the slab. Several recent studies (e.g., Ryan et al. 1995, 1996; Bebout et al. 1999) have examined the concentration variation of elements such as B, ^{9}Be, Rb, Cs, K, Ba, As, and Sb in cross-arc volcano suites and also in the prograde metamorphic rocks of subduction assemblages. These studies show that the absolute concentrations of elements such as B and Sb are lower in the rear-arc than at the front. This gradient is striking because the generally lower degrees of mantle melting in the rear arc would produce higher concentrations of incompatible elements in rear-arc lavas, if the mantle source composition were the same across the arc. The rear-arc lavas formed above a slab some 180-200-km deep in the Kurile arc have no distinctive subduction enrichment in elements thought to be particularly soluble in hydrous fluids (fluid mobile) such as B, Cs, Rb, and Sb. This is in contrast to lavas from the volcanic front, where high B content and high B/Be ratios are observed and are correlated with high $^{10}Be/^{9}Be$ (Morris et al. 1990). These results strongly suggest that the most fluid mobile elements are extracted from the downgoing slab beneath the fore-arc and volcanic front, with little, if any, mobile elements remaining in the slab to feed the rear-most part of the volcanic arc. This picture of progressive distillation of fluid mobile elements from the slab as it subducts is also borne out by B and Li isotopic studies (Ishikawa and Nakamura 1994; Ishikawa and Tera 1997; Nakamura and Moriguti 1998).

^{10}Be follows a different path through the subduction zone, and tells a different story. Figure 7 shows the cross-arc variations in ^{10}Be in the Kurile volcanic arc (Tera et al. 1993; Morris and Tera 2000). ^{10}Be and ^{9}Be concentrations were measured in lavas from five volcanic cross-chains (i.e., paired volcanoes above a slab of increasing depth along approximately the same perpendicular line from the trench). In three of the five cases shown, the rear-arc volcano has ^{10}Be and $^{10}Be/^{9}Be$ ratios that are comparable, to or higher than, the associated volcanic front locality. The ^{10}Be clock begins keeping time when the sediment column passes beneath the accretionary prism; higher ^{10}Be concentrations in the rear-arc are striking, because the path length from trench to volcano must be longer to the rear arc than to the volcanic front. These high rear-arc ^{10}Be concentrations thus require either a faster transport rate to the rear-arc (i.e., less ^{10}Be decay in transit) or larger sediment Be contribution to the lavas in the rear-arc. The Umnak-Bogoslof cross arc pair in the Aleutians also show $^{10}Be/^{9}Be$ ratios in the rear arc comparable to the volcanic front, and rear-arc lavas in the Bismarck arc have more ^{10}Be than predicted by assuming constant sediment contribution across the arc (Morris and Tera 1989; Gill et al. 1993). High ^{10}Be concentrations in rear-arc lavas thus appears to be fairly commonplace, rather than an anomalous feature of just one arc.

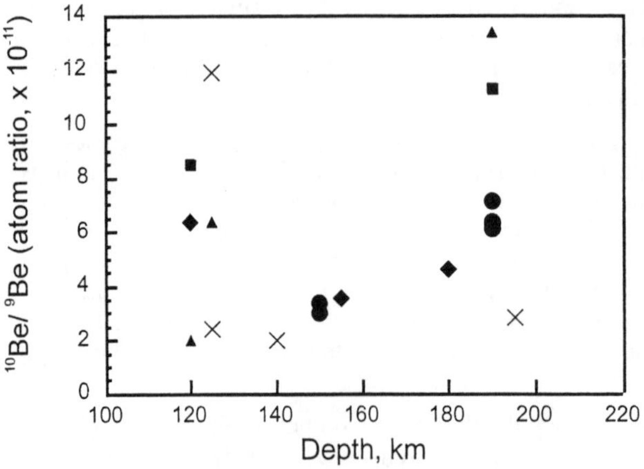

Figure 7. Plot of $^{10}Be/^{9}Be$ ratios vs. depth for Kurile cross-arc volcanic pairs. Data from Tera et al. (1993). ● = data from Paramushir and Alaid (50.5-50.8°N latitude). ◆ = Onekotan-Seamount 2.3 (49.4-49.7°N). ■ = Aekarma-Chirinkotan (48.9°-49° N). Solid triangles = Lovush-ki-Raikoke (48.1-48.4° N) and **X** = Lvinaya-Seamount 8.8 (44.6-45°N). Noteworthy is that rear-arc volcanoes Alaid, Chirinkotan and Raikoke have measured $^{10}Be/^{9}Be$ ratios and ^{10}Be concentrations that are higher than the associated volcanoes at the front, despite longer subduction paths to rear-arc localities.

It is unlikely that the presence of high ^{10}Be in the rear-arc is due to faster transport rates through the mantle. It is conceivable that ^{10}Be transport from the slab to the volcanic front could be slower, if the upward fluid/melt transport were opposing the downward convection of mantle wedge that was approximately coincident with the volcanic front. However, as noted previously, lavas from the Kurile, Bismarck, and Aleutian volcanic front are all characterized by U excesses, suggesting that the transport of slab-derived elements through the mantle to the surface was <300 kyr, a time short relative to the ^{10}Be half-life.

The alternative is that a larger proportion of the Be in the rear-arc mantle is derived from the ^{10}Be-rich sedimentary veneer of the slab. Several scenarios could satisfy this observation. The first is where a single subduction component is derived from the slab up-dip of the volcanic front and its addition to the mantle creates amphibole and phlogopite (e.g., Davies and Stevenson 1992; Tatsumi and Eggins 1997). Dehydration of mantle amphibole beneath the volcanic front and of mantle phlogopite at ca. 180 km depth would produce 2 volcanic chains, both with ^{10}Be. In the Kuriles, however, volcanoes occur over a slab with a variety of depths (e.g., 120, 130, 140, 150, 160, 180 ad 190 km above the slab), and all lavas erupting over a slab <180-km deep contain ^{10}Be. Neither the distribution of volcanoes nor the presence of ^{10}Be in volcanoes across the entire width of the arc are consistent with this model, where dehydration of just two minerals in the mantle control the slab contribution to arc lavas. In some models, contribution from two separate slab components, a fluid from the altered basaltic crust and a hydrous melt from the subducting sedimentary veneer, is invoked to explain the enrichment in elements such as Th and Ce beneath the volcanic front (e.g., Elliott et al. 1997; Johnson and Plank 1999, Class et al. 2000). In this case, it is difficult to explain the observed correlations in lavas at the volcanic front between elements such as B and U (thought to be fluid mobile) and ^{10}Be, thought to be only sparsely fluid mobile but incompatible during sediment melting. In S. Chilean lavas, strongly correlated enrichments in ^{10}Be, B, U excess and Ra excess (Sigmarsson et al. 1990, 2002) suggest that a single subduction component capable of mobilizing all four elements was involved. If sediment melting beneath the volcanic front is invoked, high ^{10}Be in rear-arc lavas requires that enough ^{10}Be and water be retained in the sediment to allow sediment melting as the slab continues to descend.

Morris and Tera (2000) considered an alternative explanation to be more likely. One possibility is that there is continuous distillation of elements out of the slab as it continues to subduct to higher pressures and temperatures. With increasing pressure and temperature, mineral solubilities in the aqueous supercritical fluid increase and the ability of the fluid to transport a wider spectrum of elements increase (e.g., Ryan et al. 1995). A more specific version of this scenario is where the subduction component beneath the volcanic front is a fluid enriched in mobile elements such as B, Cs, U and only sparsely enriched in ^{10}Be. At greater depths slab surface temperatures are hot enough to allow the sediments to melt, leading to a sediment component that is demonstrably enriched in ^{10}Be, as well as Ba, K, and Th (Ryan et al. 1995). Note that if the latter scenario is correct, it implies that the slab is too cool to allow sediment melting beneath the Kurile, Bismarck or Aleutian volcanic front, but is hot enough by ca. 135 km depth to allow sediment melting.

^{10}Be and U-Series studies. ^{10}Be data can be combined with U-series studies to better investigate the timescales and mechanisms of element transfer from the slab to the mantle. If U but not Th were transferred from the slab to the sub-arc mantle in a single event, then U-series disequilibria isotope characteristics of arc lavas may represent the time since subduction modification of the mantle. Alternative possibilities are that hydrous mantle melting could fractionate U from Th in such a way as to create the U excess, that the excess could result from dynamic melting in the mantle (Spiegelman and Elliott 1993), or that the very young ages seen for some arcs could result from crustal level processes. Because the U-series "ages" have major implications for geodynamics of the mantle wedge and melt generation/migration (e.g., Gill et al. 1993; Herrstrom et al. 1995; Turner et al. 1997, 2000; Bourdon et al. 1999), its critical to know how they originate.

Combined ^{10}Be and U-series data have been published for lavas from the S. Chile (Sigmarsson et al. 1990, 2002), Bismarck (Gill et al. 1993), Costa Rica (Herrstrom et al.

1995), Nicaragua (Reagan et al. 1994) and Aleutian (George et al. 2000) arc segments. Similar studies are underway for the Tonga (Turner et al. 1997, 2000), Scotia (Leat et al. 2000, Elliott, pers. comm.), Kurile and Kamchatka (J.Gill, pers. comm.), Philippine (Asmerom, pers. comm.) and Mariana (Elliott et al. 1997) arcs. In the case of the Costa Rica, Kamchatka and Mariana arc segments, ^{10}Be concentrations are either negligible, or too low to show meaningful variation with respect to either Th isotopic compositions or U/Th ratios. The Kurile data show no coherent variations with ^{10}Be/^{9}Be ratios (Gill, pers. comm.).

Other suites show co-variation between ^{10}Be and U-series isotopes. The Nicaragua and Bismarck arcs form trends with ca. 90- and 200-ka "ages," respectively (Reagan et al. 1994; Gill et al. 1993). ^{10}Be/^{9}Be ratios of the lavas increase systematically, by a factor of four, as the Th activity ratio and U excess increase. The Aleutian data set (George et al., submitted) define a nearly horizontal array on a U-Th disequilibrium diagram equivalent to an age of <30 ka; there is a general tendency for the highest ^{10}Be concentrations to be in lavas from the oceanic segment with the greater U excesses, but the two data sets are not highly correlated. Tonga lavas form a U-series slope of approximately 60 ka (Turner et al. 1997); there is again a general tendency for highest ^{10}Be/^{9}Be ratios to be seen in lavas with greater U excess, but the data sets are again not highly correlated. The S. Chile data also have a (^{238}U)-(^{230}Th) disequilibrium "age" of <30 ka; the magnitude of the U excess is very well correlated with (^{226}Ra)-(^{230} Th) excess, a characteristic which will decay away within 8 kyr. Both have strong positive correlations with ^{10}Be/^{9}Be ratios and ^{10}Be concentrations (Sigmarsson et al. 1990, 2002).

The results from southern Chile and Nicaragua are particularly striking. The very good correlations with ^{10}Be strongly suggest that the excess U and Ra in southern Chile are a subduction signature rather than a result of dynamic or hydrous melting or crustal contamination. The evidence for correlated transfer from slab to mantle of U, Ra and Be, but not Th, is consistent with a single aqueous fluid derived from the slab rather than a sediment melt or both a melt and a fluid. The straightforward interpretation of the data is that fluid addition and mantle melting occurred within the last 8,000 years (Sigmarsson et al. 2002). If so, several geodynamic consequences follow: (1) fluid fluxing triggered immediate melting of the mantle; (2) fast melt migration pathways were required, likely being channelized through the asthenosphere (rather than via porous flow) (Turner et al 2000) and through fractures in the lithosphere; and (3) the residence time in magma chambers was negligible. An alternative scenario to consider would be one in which the extremely young 8-ka "age" is an artifact. In this case, the processes of extracting a component from the slab, mixing it with the mantle, possible storage and subsequent melting plus magma time in transit to the surface would need to produce strong correlations between elements of different partitioning behaviors and half-lives.

For Nicaragua, ^{10}Be/^{9}Be ratios correlate better with Th isotopic composition than with U excess (Reagan et al.1994). The Nicaragua data set suggests that ^{10}Be, Ba, B, U and Th were added from the slab sometime in the last several million years, possibly in a sediment melt. Another addition of U, perhaps in a fluid, was required sometime in the last several hundred kyr (Reagan et al. 1994).

Where both U and Th may have been added from the slab, the disequilibria systematics cannot be interpreted as a simple age, but with ^{10}Be studies or other geochemical data (e.g., Elliott et al. 1997) the systematics can reveal several episodes of subduction modification of the mantle and provide general timing constraints. The general tendency for high ^{10}Be to go along with greater U excesses does suggest that the U-series characteristics derive from the subducting slab. For arcs with high ^{10}Be, there is often an order of magnitude variation in ^{10}Be/^{9}Be ratios and ^{10}Be concentrations observed for lavas from a specific arc or arc segment (Table 2 and references therein). For lavas derived by similar degrees of partial melting (e.g., the Aleutians, George et al., submitted), the ^{10}Be concentration range requires varying amounts of sediment addition to the mantle, rather than an approximately constant subduction flux to a variably depleted mantle.

The question of whether or not sediments melt beneath the volcanic arc is extensively debated. The argument hinges around detailed discussion of how elements such as Th, Ce, Be, Ba, and U behave in melts vs. fluids, but the outcome has broad implications for the thermal structure of the downgoing plate. In some arcs, Th addition is interpreted as the result of sediment melting because of its immobility in fluids (e.g., Brenan et al. 1995); U excess is thought to represent the addition of a fluid from the altered oceanic crust, based on multi-element correlations (e.g., Elliott et al. 1997). In S. Chile, the ^{10}Be from the sediment correlates with the U rather than the Th enrichment, requiring that U plus Be, without Th, be mobilized out of the slab. This could be a fluid from the sediment or one from the basaltic crust that also extracts Be from the overlying sediment column; it is unlikely to be a sediment melt. Where Th isotopic composition and ^{10}Be co-vary as in Nicaragua or the Bismarck arc, the data better fit a scenario of sediment melting. The observations of different signatures in the ^{10}Be-U series systematics and the resulting interpretations may indicate that we don't understand well the partitioning of key elements during interactions between fluid, melt, sediment, altered oceanic crust and mantle. More interesting is the possibility that there are real differences in the thermal structure of different subducting plates, such that some are hot enough beneath the arc (ca. 700°C) to permit sediment melting, while others are not.

Future directions

The new capability to measure ^{10}Be at low concentrations (ca. 0.1 million atoms g^{-1}) with very low blanks (ca. 20,000 atoms g^{-1}) opens the door to a large number of new possibilities. In cases where a tectonic, geodynamic or sediment dynamic model can be evaluated from the sediment signature of the arc lavas, re-measurement of some of the low ^{10}Be arc lavas might be useful. Variations along strike (e.g Nicaragua-Costa Rica) can be better evaluated in terms of variations in sediment dynamics, perhaps related to the abundance and depth of grabens on the downgoing plate (e.g., Kelly and Driscoll 1998). Serpentinites in the Mariana fore-arc, thought to reflect extensive fluxing of the shallow mantle by fluids from the slab, now become amenable to study. Low but real ^{10}Be enrichment in young back-arc lavas not overlying the downgoing slab could be used to map transport through the wedge and constrain timescales thereof. In places such as the Chile Ridge spreading center, trace element systematics suggest flow of subduction modified mantle around the slab to the near-trench part of the Ridge; measurable ^{10}Be in the Ridge lavas would confirm such mantle flow and provide a speedometer for flow rates. ^{10}Be enrichment in MORB or OIB lavas could be used to identify recent sediment assimilation, as opposed to ancient sediment subduction. It is probably also worth visiting the question of whether ^{10}Be could be produced in some minerals via nuclear reactions at levels that are now measurable, and could be used as a geochronometer. The latter possibility highlights the need to be cautious in interpreting very low ^{10}Be concentrations in terms of a subduction signature, given that sample alteration, cosmic ray bombardment and possibly nuclear reactions could produce small ^{10}Be enrichments in any tectonic setting.

TECTONIC APPLICATIONS OF *IN SITU* ^{10}BE

Introduction

In this section we review tectonic geology applications of cosmogenic ^{10}Be produced in rock (see Gosse and Phillips 2001, and Bierman et al., this volume, for recent reviews of other applications of *in situ* ^{10}Be to surface processes). The contributions of atmospheric ^{10}Be to understanding subduction zone tectonics have been discussed in the previous section. The *in situ* ^{10}Be method provides both chronological and denudation information—from landform to orogen spatial scales over relatively short (generally <1 Myr) timescales. In this context, *in situ* ^{10}Be complements other dating methods used in Cenozoic tectonics studies, and provides the niche-filling record of rock uplift (as interpreted from stream incision histories, for example) between the timescales of modern GPS geodetics and studies of long term (>4 Myr) orogen-scale exhumation histories using thermochronology (e.g., Ar/Ar, (U-Th)/He, and fission track analyses). Over the past 15 years, the number of tectonic geology applications of the method (Fig. 8) has expanded geometrically.

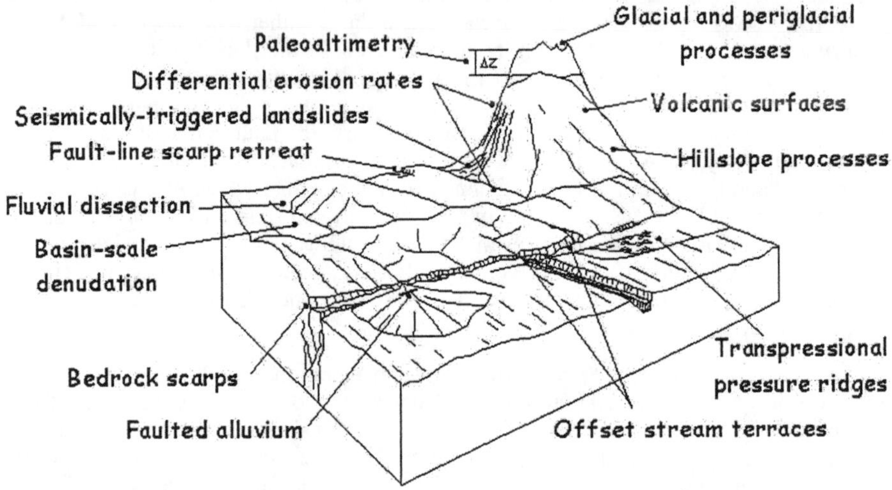

Figure 8. Published and potential applications of *in situ* cosmogenic ^{10}Be for studying tectonic geology.

For this review, we group the tectonic applications of *in situ* ^{10}Be under three general 'approaches': (i) those using ^{10}Be to bracket or directly date the age of a tectonic event such as a seismogenic landslide, fault slip, or volcanic eruption; (ii) those using ^{10}Be to constrain the incision rate of streams or denudation of entire catchments in tectonically active regions; and (iii) paleoaltimetry. Approach (i) has provided precise estimates of the timing of paleoseismic and volcanic events, their recurrence intervals, and slip rates. The chronology of offset alluvial fan surfaces can assist in showing trends in slip activity and provide timing constraints on fault slip needed for strain partitioning studies. Cosmogenic ^{10}Be exposure ages have also been used in geological hazard assessments (Gosse et al. 1996; Taylor et al. 1998) to determine the ages of lava and fault motion near Yucca Mountain, Nevada, the proposed U.S. high-level nuclear waste repository. Recognizing that incision and denudation rates calculated with Approach (ii) do not directly provide tectonic uplift rates, the approach provides meaningful data that can be used to understand the rates of orogen evolution in response to rock uplift. Ideally, incision rates are calculated by dating stream straths (erosional surface representing the bedrock floor of ancient streams, cf. Pazzaglia and Brandon 2001) so that the rate of incision into bedrock (not just the alluvial fill) above the modern stream is measured. Likewise, denudation experiments measure the mass flux through an entire drainage, and ^{10}Be in modern stream sediment has been used to calculate basin-scale denudation rates and relate those rates to rock uplift. Denudation studies provide short-term ($\sim10^4$ yr) accounts of sediment fluxes and variations in their spatial distributions, which can provide useful information for testing questions of steady state erosion and topographic equilibrium. Incision and denudation rate data can be used in geodynamic studies to evaluate rates and styles of surficial processes at orogen-scales. The use of cosmogenic nuclides in paleoaltimetry (Approach iii) has not yet been accomplished (Gosse and Stone 2001). We outline a basic strategy for this approach, and point out the steps being taken to launch the method.

Exposure chronology of tectonic events

This special volume is dedicated to applications of Be. For this reason we are not reviewing applications of other cosmogenic isotopes, which in some cases may be more useful for dating tectonic events (e.g., in cases where quartz is absent). The principles of interpreting concentrations of ^{10}Be as exposure ages have been discussed earlier in this paper. Applications of *in situ* ^{10}Be to address questions related to tectonic histories are numerous. Chronologies of alluvial fan surfaces have been used to decipher Quaternary paleoseismicity and slip rates (Bierman et al. 1995; Brown et al. 1998a; Granger et al. 1996; Nishiizumi et al. 1993; Siame et al. 1997; Van der Woerd et al. 1998, 2002). To date basaltic lava, ^{10}Be production rates of whole rock samples and olivine have been estimated using nuclear cross sections and comparisons with quartz (Gosse et al. 1996; Nishiizumi et al. 1990). The exposure history of a bedrock surface has mapped the progressive retreat of an early- to mid-Pleistocene fault line scarp to provide an estimate of the timing of the last movement along a fault with low recurrence interval where topographic evidence has been eroded (Taylor et al. 1998). Deformed alluvium has been dated to constrain rates and timing of recent transpression between topographically elusive strike slip segments (Spies et al. 2000; Spies et al. submitted) and above a blind reverse fault (Jackson et al. 2002). By dating raised beaches in the central Arctic, the method has recently been shown to yield emergence curves (Gosse et al. 1998) similar to those based on radiocarbon-dated driftwood. The potential to provide emergence histories for raised shorelines that have been otherwise undatable has global implications for improving boundary conditions for mantle rheology and lithospheric flexure models as well as ice sheet dynamics.

Volcanic surfaces. The most direct application of *in situ* cosmogenic ^{10}Be in tectonics is to interpret its concentration in a manner to provide chronological control on tectonic landforms. In this regard, ^{10}Be and other isotopes have been used to determine the age of Quaternary lava flow surfaces and cinder cones. The ^{10}Be method has primarily been restricted to quartz-bearing lithologies. However, the production systematics of ^{10}Be in olivine are similar to those in quartz. Although the mineral chemistries differ, ^{10}Be in both is produced mainly through spallogenic interactions on oxygen, plus smaller contributions from the other elements. Olivine-bearing lavas on Maui were dated (Nishiizumi et al. 1990) by summing production rates based on nuclear cross section estimates for fast neutrons on the four target elements. Recognizing the need for highly precise production rates in olivine, Kong et al. (1999) have begun measuring the production rate of ^{10}Be in basaltic olivines over a narrow Fe/Mg range. Unlike ^{36}Cl, empirically derived production rates of ^{10}Be in whole rock samples have not been determined. In an attempt to date a basalt lava flow surface of Black Cone, in southwestern Nevada, Gosse et al. (1996) with J. Klein of the University of Pennsylvania used a 'whole rock' procedure, necessary because of the absence of olivine phenocrysts. More than 35% of the original sample mass was leached before complete dissolution to remove atmospheric ^{10}Be. Diagnostic textures on the surface of the lava flow suggested that erosion may have been negligible. The lava surface exposure age was 0.88±0.22 Ma (error reflects uncertainty in production rate in addition to analytical errors; production rate calculation according to Nishiizumi et al. 1990) on lava that had been dated at

1.0±0.1 Ma with $^{40}Ar/^{39}Ar$ methods by others (Taylor et al. 1998). The ^{10}Be analysis of about 20 step-leached aliquots from whole rock samples showed that whole rock samples could lose their atmospheric ^{10}Be after 35% leaching. Unfortunately, some samples apparently retained the atmospheric signal even after 90% of the rock was removed by chemical leaching (Klein et al. 1997). In those cases the *in situ* ^{10}Be component could not be isolated and no other whole rock *in situ* ^{10}Be experiments have since been published. Far more exposure chronologies of lavas are derived from cosmogenic ^{36}Cl and ^{3}He (Cerling et al. 1999; Fenton et al. 2001; Kurz et al. 1990; Liccardi et al. 1999; Marti and Craig 1987; Phillips et al. 1996; Sheppard et al. 1995; Zreda et al. 1993) because their production rate systematics in basalts are better established.

Bedrock fault scarps. Exposed bedrock fault scarps afford the opportunity to date the most recent offset and yield information on the timing of multiple rupture events and recurrence frequency. Scarps produced from even high magnitude pure dip slip displacements rarely exceed a few meters height. In the simplest case of a single rupture event, the ^{10}Be concentration measured on the scarp face will record ^{10}Be produced in the subsurface rock prior to faulting, plus the ^{10}Be produced on the exposed scarp since faulting. The pre-faulting concentration can be determined by measuring subsurface profiles in the hanging wall or footwall. In the cases of multiple events on the same scarp, vertical profiles can be sampled along exposed scarp segments and exposure age clusters can be grouped into isochronal zones (Zreda and Noller 1998). The highest exposure zone should appear to be the most weathered and will have the oldest exposure ages. The duration of exposure in each zone provides a means to estimate the recurrence frequency and trends of pre-historic earthquake events. A scarp along the Solitario Canyon fault, a major block bounding fault in the basin and range province of Nevada, was dated by Harrington, Whitney, and Jull (unpubl.) in 1993 using cosmogenic ^{14}C in quartz. They demonstrated that the surface was a pre-Holocene scarp that had been subsequently exhumed (a fault-line scarp). However, due to difficulties in constraining the erosion rates and the component of the measured ^{14}C that was produced when the scarp surface was still underground, no additional paleoseismic information could be extracted. In a subsequent study, Zreda and Noller (1998) sampled spatially separate scarp segments that were interpreted to have been produced during recurrent displacements along the Hebgen Lake fault in Montana. After adjusting for subsurface production prior to rupture, the height vs. age distributions showed a pattern that led them to recognize at least 6 events over a 24 kyr period, with an increasing recurrence frequency.

***Fault displacement in unconsolidated sediment*.** Direct measurements of ^{10}Be on faults scarps in unconsolidated sediments (alluvial fans, terraces, moraines) has not been attempted because fault scarps are disequilibrium features that are susceptible to high rates of erosion, and are therefore not stable surfaces. Instead, it is possible to bracket the timing of single or multiple slip events by exposure dating surfaces of two or more alluvial fans, moraines, or beaches that were deposited before and after the strain event. The success of alluvium chronologies by Bierman et al. (1995), Brown et al. (1998a), Siame et al. (1997), Van der Woerd et al. (1998, 2002), Zehfuss et al. (2001), and Spies et al. (submitted) attest to the plausibility of surface clast exposure dating of offset alluvial surfaces. Slip rate analyses require highly precise chronologies and a means of distinguishing single event from multiple event histories. In alluvium, both of these quantities are difficult to obtain due to post-offset degradation of the surface.

Two instructive examples of ^{10}Be measured in clasts on the surface of alluvium to constrain slip rates in the Tibet region were published by Brown et al. (1998a) and Van der Woerd et al. (1998). The strategy employed was to determine the ages of fan and terrace surfaces (respectively) to bracket the timing of displacement. Brown et al.

calculated a fault slip rate of 2 mm a^{-1} based on measured heights of faults scarps on three late Pleistocene alluvial fans. In addition to the tectonic implications for shortening across the Tien Shan, they demonstrated that the amount of ^{10}Be inherited from exposure prior to deposition on the alluvial fan was low (corresponding to 2 kyr of exposure, based on modern wash cobbles). Similarly, Van der Woerd et al. established post-glacial slip rates on the Kunlun fault in northeastern Tibet, even from remarkably young (late Holocene) surfaces of fans and moraines (ranging from ca 1.5 to 600 ka). Using combinations of radiocarbon and *in situ* isotopes, they calculated that the slip rates over different periods throughout this duration have been remarkably similar (average 11.5±2 mm a^{-1} over the past 600 kyr). In North America, over 30 ^{26}Al and ^{10}Be exposure ages on large surface boulders in glacial debris flows on fans in Owens Valley were used to calculate a 0.24±0.04 mm a^{-1} slip rate over the past 300 kyr, consistent with rates determined previously (Zehfuss et al. 2001).

Despite these successes, an obstacle to exposure dating of unconsolidated sediment often results from not knowing the amount of ^{10}Be inherited from exposure of the clast at its source locality or during transportation, prior to its final deposition. This inherited component will result in calculated exposure ages that are older than the actual surface being dated (e.g., Trull et al. 1995). There are several ways to isolate and adjust for an inherited nuclide component. Active stream sediment (or sediment in active washes cut through alluvial fans) are too young to acquire measurable ^{10}Be concentrations since deposition, but can have a concentration that reflects ^{10}Be inherited from pre-depositional exposure in the catchment, prior to reworking of older fans, and during transport. The inherited concentration (N_{inh}) measured in the modern sediment can be subtracted from concentrations on older surfaces of fans sourced from the same catchment (Brown et al. 1998a; Spies et al., submitted). However, it is difficult to prove that the amount of inheritance in modern alluvium is the same as the inheritance in sediment deposited previously under different climate and geomophological conditions. A second means of determining the inheritance in single thick alluvial units is by the 'depth-profiling' technique (Anderson et al. 1996; Ayarbe et al. 1998; Repka et al. 1997; Hancock et al. 1999). The premise is that the production of cosmogenic nuclides should decrease exponentially with depth (Eqn. 3). In alluvium with a bulk density of 2 g cm^{-3}, the concentration at 4 m depth is less than 0.5% that at the surface. Assuming that the N_{inh} is constant throughout the aggradational unit, a high concentration at depths greater than 4 m corresponds to the amount of inherited ^{10}Be. The geometry of the concentration vs. depth profile can be modeled to precisely calculate the probable inherited component. In simple cases, a shielded sample at the base of a thick unit could be used to indicate the average inheritance of ^{10}Be in the sediment. However, more is gained from the profile method because the geometry of the concentration vs. depth profile provides information on erosion and burial history of the sediment. The necessity and advantages of using cosmogenic nuclide concentration-depth profiles are described in a number of recent studies (Anderson et al. 1996; Ayarbe et al. 1998; Repka et al. 1997; Hancock et al. 1999).

Some effort has been made to optimize the dating of alluvium. Anderson et al. (1996) proposed a means to minimize the scatter in sample concentrations from the same alluvial landform. Their sampling approach, now widely used (Gosse and Phillips 2001), is to amalgamate a minimum of 40 pebbles of equal quartz mass for surface or subsurface samples. The variance among the inherited concentrations in the amalgamated samples will be smaller than if single cobbles or boulders are sampled. Zentmire et al. (1999) showed that alluvium from once glaciated catchments may not have significant inheritance due to the rejuvenation of the surface by glacial erosion and to glacial shielding of the surface from cosmic radiation. The ^{10}Be concentrations in modern outwash clasts from the Matanuska Glacier were consistent with the chemical blank, indicating negligible inheritance from previous exposure prior to deposition (Zentmire et al. 1999). Gosse and McDonald (submitted) measured ^{10}Be concentrations of amalgamated surface samples from two alluvial fans from adjacent tributaries in Fish Lake valley. The catchment of one fan was glaciated during the last glaciation, the other fan had little or no glacial influence. The fan ages were bracketed with radiocarbon (Reheis and Sawyer 1997) and fall around the Pleistocene-Holocene transition. The ^{10}Be concentrations in samples from the fan with a large glacial influence in its catchment showed a smaller variance and mean age closer to the accepted radiocarbon chronology than the samples from the fan with little or no glacial influence. These observations, together with the lack of significant inheritance in published measurements on alluvium from glaciated catchments (e.g., Bierman et al. 1995; Brown et al. 1998b; Van der Woerd et al. 1998), show that while depth profiles may be necessary in arid regions with non-glaciated catchments, alluvial surfaces from glaciated catchments may not have large inherited concentrations.

Cosmogenic nuclides have also been used to establish the ages of faulting events and of colluvial wedges on the hanging wall of normal faults. The colluvial wedges form as a result of mass wasting off a fault scarp onto the down-dropped block. One of the best examples used ^{36}Cl (arguably ^{10}Be would have yielded similar results) to determine the age of colluvial wedges and the timing of recurrent ruptures of alluvium by the Socorro fault scarp, New Mexico (Ayarbe et al. 1998). Depth profiles of ^{36}Cl concentrations in agglomerate samples (150 pea-sized pebbles) showed rupture trends that are supported by pedological and geomorphological observations along a vertical section through the alluvium cut perpendicular to the fault plane. On the undisturbed footwall alluvium, subsurface concentrations of *in situ* ^{36}Cl define a depth profile consistent with the depositional age of the alluvium. However, the subsurface samples in the hanging wall section, including within colluvial wedges, show a complicated geometry indicative of more recent sedimentation. Ayarbe et al. (1998) argued that the hanging wall data pointed to a history of erosion of the alluvium fault scarp in a manner consistent with that predicted by simple diffusion modeling, although more data are needed to establish the fault history.

Spies et al. (submitted) have used a combination of detailed soils analyses (including profile development indices), geomorphometry, and cosmogenic nuclides to determine the rupture history of the Carrara Fault, near Yucca Mountain, in the Walker Lane, Nevada (Slemmons 1997; Stamatakos et al. 1997). Transpression along the fault system has produced a ridge of Pleistocene alluvium that has subsequently been cut by combinations of dip-slip and dextral strike slip motions on fault segments. On the southeast end of the feature, drainage density and topography express more active faulting where the ridge is cut by oblique faults that seem to displace Holocene soils. The concentrations of ^{10}Be in three amalgamated surface samples on the north end are higher (more than 1σ) than the concentrations on the southeast end of the same alluvial unit. Interpreting the concentrations as exposure ages (Fig. 9), the ages fall within the range of soil age determined by profile development indices. Spies et al. attribute the differences in concentration to an increase in erosion associated with the active faulting (roughly 18 cm more erosion would have occurred on the southeast end relative to the less active north end of the ridge.

Seismicity inferred from area affected by land-sliding. Empirical datasets have been used to correlate the area containing earthquake-triggered landslides and the magnitude of the historical seismic ground acceleration (McCalpin 1996). A logical next step is to

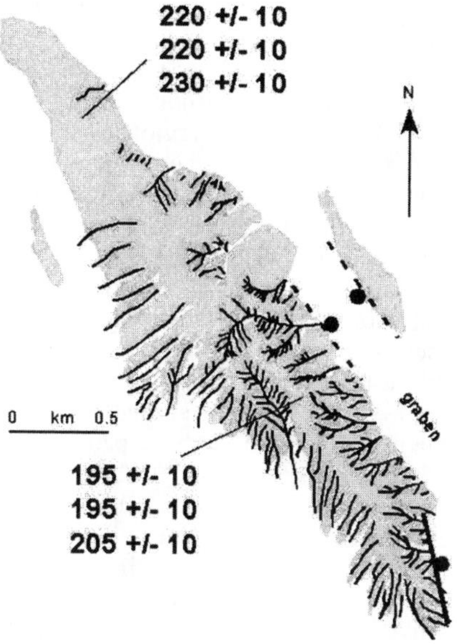

Figure 9. Cosmogenic ^{10}Be exposure ages and stream drainage pattern (to stream order 3) on a transpressional ridge in the Amargosa Valley, Nevada (Spies et al., submitted). Strike slip faults not shown because their position is weakly defined from geophysics. The ridge has an axial drainage divide that averages 16 m above the local valley floor. The ridge has been truncated by oblique slip faults (dashed and solid line) with significant normal slip component (dot represents hanging wall). Uncertainties in age only reflect random error associated with measurement of concentration. ^{10}Be concentration was adjusted for inheritance based on a measurement of modern stream sediment (Spies et al., submitted).

date pre-historic landslides and use the recent empirical datasets to reconstruct paleoseismic history. Measurements of ^{10}Be on the surfaces of boulders on an early Holocene Austrian rock slide yielded internally consistent exposure ages (Kubik et al. 1998) and demonstrated the reliability of the method to date ancient landslides. Hermanns et al. (2001) used ^{21}Ne in quartz in a similar approach, and documented the westward propagation of reverse faulting in the Argentine Puna Plateau. Large >0.3 km^3 landslides were used in all of these studies. In a Colorado Front Range study using smaller landslides, Gosse, Madole, and Klein (unpubl.) found that debris slide boulders can contain inherited concentrations of ^{10}Be that are greater than the concentration produced since the landslide. A difference between the Colorado study and the previous landslide studies is the volume and deformation of the wasted material (e.g., the Austrian and Argentine rock slides were deeper and more pervasively deformed). These results suggest that many measurements may be necessary on debris flows and similar deposits to statistically define inherited samples as outliers.

A related study in western Nevada inferred the minimum time since seismicity of a calculated magnitude occurred (Bell et al. 1998). The local moment paleo-magnitude was estimated by calculating the minimum acceleration required to topple precariously perched boulders in tectonically active areas. ^{36}Cl exposure ages on the boulders were used to estimate the amount of time the boulder may have been perched. Although assumptions regarding the relevance of the exposure ages were made, the data lend support to other paleoseismic inferences based mostly upon trench logs.

***Coastal emergence*.** Paleoshorelines and raised marine terraces dated with *in situ* ^{10}Be can provide constraints to determine (1) the rate of tectonic tilting (Marshall and Anderson 1995) and fault displacement (Perg et al. 2001); (2) the nature of emergence history (constant, exponential, or episodic; Dyke et al. 1991, 1992); differential movement of fault blocks where none was previously recognized (Sissons and Cornish 1982); and (4) physical properties controlling the rheology of the crust, upper mantle, and lower mantle (Peltier 1998).

In a study in progress, Gosse and others collected cobbles and boulders on a sequence of raised beaches and glacially-plucked bedrock surfaces on an emerging sea cliff face along a Holocene coastline on the east coast of Prince of Wales Island, Central Arctic, Canada (Gosse et al. 1998). Cosmogenic ^{10}Be measurements in boulders were used to show that the distribution of exposure age vs. elevation of shoreline can be fit with a simple exponential curve that is indistinguishable from the regional calibrated radiocarbon-derived sea level curve. The ^{10}Be concentrations were corrected to compensate for the fact that the beaches were not at their present elevation throughout their exposure history (<3%, greatest for the first beaches to be emerged due to exponential isostacy). This sensitivity to changes in atmospheric shielding is the basis for the paleoaltimetry approach discussed below. A subtraction of a small component of ^{10}Be produced subaqueously before emergence was also made, which was greatest for the most recent beaches because they spent considerable time in shallow water prior to emergence. In this study these corrections could be based on a published relative sea level curve based on driftwood, but in areas with no radiocarbon chronology, these adjustments will be made iteratively. Measurements of ^{10}Be in cobbles, pebbles, and bedrock cliff surfaces showed more scatter than in the boulders. Geological factors that may have contributed to the scatter in the dates include post-depositional beach sediment movement due to ice push or gelifluction, differential partial shielding due to snow cover, and anthropogenic displacement of quartz clasts. Sensitivity tests for the effects of surface erosion and partial shielding due to snow cover and sea water indicate the combined effects could result in a less than 3% overestimation in exposure age. These results suggest a reasonable reliability of cosmogenic nuclide exposure dating to build emergence curves from Holocene and older emergent shorelines where sufficient high quality radiocarbon datable material is absent. However, the variation about the mean age for a shoreline elevation will be prohibitively large for most tectonic applications unless a sufficient number of samples are dated. The technique will probably be most useful when it is supplemented with radiocarbon dates.

Perg et al. (2001) suggested another approach to dating terraces. In a study to determine the ages of five marine terraces along the active coast of California north of Santa Cruz, ^{10}Be was measured in surface and subsurface amalgamated samples. The exposure ages correspond with Quaternary sea-level high stands, with an implied tectonic uplift rate of 1.1 mm a^{-1} which was higher than previously published rates for the area because the exposure ages were younger than previously estimated. Their approach was used to date sediments that have been bioturbated, but could equally be applicable in areas with other mixing processes (cryoturbation, mass wasting).

Bedrock erosion, stream incision, terrace deformation, and orogen-scale denudation

In the previous section we emphasized uses of ^{10}Be to provide chronometric control on Cenozoic tectonic events. In this section we review applications that interpret the concentration of *in situ* ^{10}Be in terms of an erosion rate, not exposure duration. The recent rejuvenation of interest in landscape modeling may in part be due to the utility of cosmogenic nuclides in providing erosion rate information on scales of outcrops to orogens. In turn, this information can be used to estimate rates of rock uplift and provide surface process information for geodynamic models. Terrestrial cosmogenic nuclides produced *in situ* have really just begun to contribute to orogen-scale investigations. The techniques can provide ages of incised straths (Pazzaglia and Brandon 2001), rates of weathering and erosion on individual bedrock summits and hillslopes, and rates of basin-scale denudation. The availability of a technique that can be used to independently quantify erosion rates on different surfaces is invaluable in areas of spatially variable erosion rates, where isostatic uplift may contribute to net surface uplift of the summits. Cosmogenic ^{10}Be is well suited to such studies because (1) it is radioactive, providing information that stable cosmogenic isotopes cannot, and (2) its half-life is longer than any other commonly used *in situ* cosmogenic nuclide, so it is useful for averaging over longer time periods. An additional advantage is that a second cosmogenic nuclide, ^{26}Al, can be measured with ^{10}Be in virtually all samples containing quartz. In addition to an internal check on the chemical preparation and calculated age, the second nuclide is useful for calculating erosion rates, styles of erosion, and other features of a surface's exposure history.

Determining local bedrock erosion rates with a single isotope. The measurement of ^{10}Be alone can provide estimates of erosion rates, under certain circumstances. By assuming an erosional equilibrium scenario where erosion is controlling the concentration, and setting exposure duration (t) to infinity, erosion rate (ε) can be expressed as (Lal 1991):

$$\varepsilon = \left(\frac{P_{(0)}}{N} - \lambda\right)\mu \qquad (7)$$

where the absorption coefficient $\mu = \Lambda/\rho$. This is useful on surfaces where loss of ^{10}Be due to erosion is much higher than by decay, or on surfaces known to have been exposed for a long duration. In those cases where it is clear that the concentration has reached erosional steady state, the calculated erosion rate is the average rate. However, even if steady state has not been attained, the single ^{10}Be measurement will establish the maximum erosion rate permitted by the measured concentration. By making multiple measurements over different surfaces (peaks and depressions) it is possible to calculate the maximum Quaternary erosion rate of a larger area. This approach substantiated geomorphological estimates of the maximum possible erosion rate for Yucca Mountain and surrounding areas (0.3 to 3 mm kyr^{-1}), critical to evaluating the possibility that a high level nuclear waste repository at 200-m depth could be breached within a 100-kyr period (Gosse et al. 1996). A similar approach was used by Burbank et al. (1996) and Leland et al. (1998) to determine maximum erosion rates of strath surfaces in the Himalayas that appeared to be responding more to climate change than to tectonic forces. Notably, the $^{26}Al/^{10}Be$ ratios they measured plot within the erosional steady state island of a $^{26}Al/^{10}Be$ vs normalized ^{10}Be diagram (next section). This indicates that the surfaces probably had simple exposure histories as the authors assumed.

Determining local erosion rates with two cosmogenic nuclides. Nishiizumi, Lal, Arnold, Klein, and Middleton first used two isotopes (^{26}Al and ^{10}Be) to determine the exposure history of surfaces of unknown age and unknown erosion rate in the Antarctic Dry Valleys (Nishiizumi et al. 1991; Lal 1991). The strategy is straightforward. A pair of cosmogenic nuclides is measured with at least one being radioactive. The shorter-lived isotope will reach radioactive secular equilibrium before the longer-lived isotope. Therefore the ratio of the two isotopes will decrease over time (upper curve in Fig. 10a). In the case of two radioisotopes, the system reaches a constant ratio when both isotopes are at steady state. If the surface is eroding, the concentration of both isotopes will be proportionally lower because of the decrease in production with depth in the rock. Erosion will cause the radionuclides to reach steady state sooner than in an absence of erosion, and a family of curves describe the ratio of $^{26}Al/^{10}Be$ for a wide range of gradual and constant erosion rates. The lower curve in Figure 10a connects the ends of these erosion curves. Ignoring uncertainties in the measurement, surfaces with ratios that plot within the area between the two curves have therefore experienced erosion (although in some cases burial of the surface can also explain such ratios). The isotope ratio method can provide estimates of differential bedrock erosion rates by determining multiple local erosion rates over large areas (e.g., Small et al. 1997). The maximum non-glacial erosion rate derived from the cosmogenic $^{26}Al/^{10}Be$ measurements on bedrock surfaces of flat summit plateaus in eastern Canada, (all older than 25 ka) is 0.7 cm kyr^{-1}, but the

Quaternary long-term average erosion rate is probably less than 0.2 cm kyr^{-1} (Gosse et al. 1995c).

Despite the desire for and implications of this application of *in situ* TCN, very few studies successfully use the method for a number of reasons:

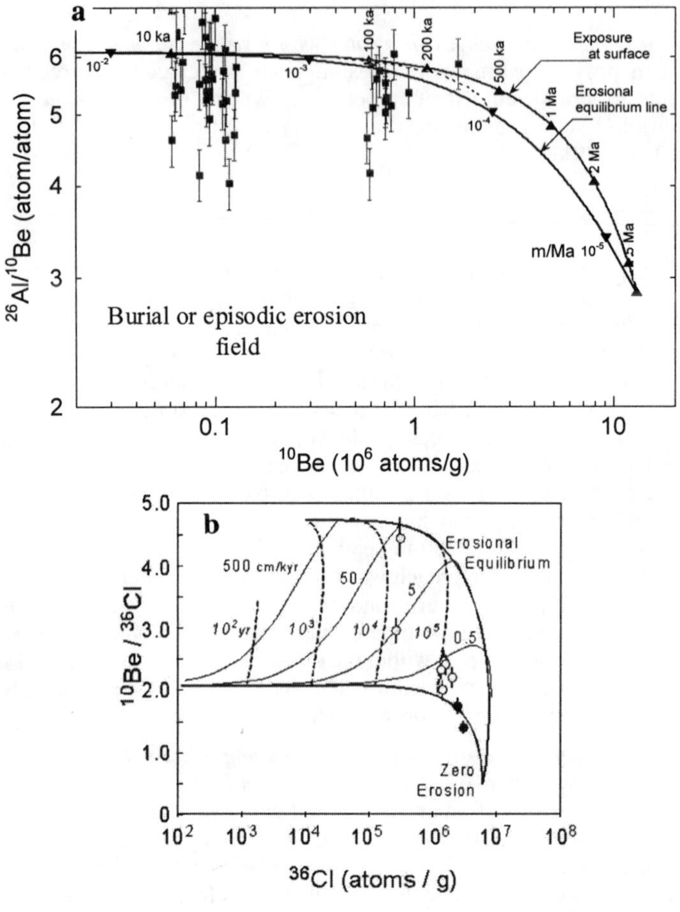

Figure 10. *In situ* isotope ratio plots. (a) $^{26}Al/^{10}Be$ vs. ^{10}Be (log concentration). The data are from granitic and granodiorite boulders and bedrock in the Wind River Range, Wyoming (Gosse et al. 1995a,b). Upper curve represents the path of a surface that is continuously exposed and not eroded. The area between the curves represents possible paths of surfaces that were eroded at a constant rate during continuous exposure. Samples that plot between the curves or below the lower curve may also have experienced an interruption in their exposure history, such as burial with sufficient shielding to significantly reduce the cosmic ray flux. Episodic erosion, such as glacial plucking, can also yield ratios that fall below the lower curve. (b) Ratio diagram, $^{10}Be/^{36}Cl$ vs. ^{36}Cl (log concentration, normalized, as in previous figures) for boulder samples from moraines in the Wind River Range, Wyoming (Phillips et al. 1997). The lower curve in Figure 10b corresponds to the upper curve in Figure 10a. The ^{36}Cl isotope is partially produced through thermal neutron capture so the width of the erosion island is greater (relative to the analytical accuracy of the ratio) than a ratio plot for two spallogenic isotopes. Position of samples indicates generally very low erosion rates for boulder surfaces.

(1) The method works well for some surfaces of long exposure duration (>10^6 a), but for younger surfaces the analytical uncertainty in the ratio measurement is prohibitively large ($1\sigma \sim 8\%$ or higher were typical until very recently).

(2) The strategy requires that the surface has been eroding at a constant rate, rather than in the episodic pulses more common in mountainous regions.

(3) The isotopic ratio is influenced by partial or complete burial and inheritance from previous exposure events.

(4) If erosion rates are greater than 1 mm kyr^{-1} or the exposure less than 1 Myr, the steady state assumption cannot be tested.

As an example of the approach, Gosse (1994) and Gosse et al. (1995a,b) measured more than 50 $^{26}Al/^{10}Be$ ratios on bedrock and boulder surfaces throughout the Wind River Range to help determine erosion rates of the Wind River Range granitic gneiss and to determine if the sampled surfaces had complicated (e.g., by burial) exposure histories (Fig. 10a). In an ideal case, the data would be sufficiently precise to define an erosion curve (such as the dashed curve), which would allow simultaneous estimates of erosion and age. The majority of the ratios do not plot within the erosion-equilibrium island, and therefore erosion and exposure age cannot be solved. This was the fate of even recent attempts to determine erosion rates of alpine bedrock summits (Small and Anderson 1998; Small et al. 1997), which prompted some of the criticism by Schaffer (1998). The point here is that even with improvements in the AMS precision, the $^{26}Al/^{10}Be$ will not provide a meaningful resolution of erosion rates on late Pleistocene surfaces.

However, Liu et al. (1994) showed how a combination of ^{36}Cl and ^{10}Be could be a useful alternative for bedrock erosion rates greater than 1 mm kyr^{-1} and surfaces less than 1 Ma (Fig. 10b). Phillips et al. (1997) used the $^{10}Be/^{36}Cl$ technique to simultaneously measure rock erosion rates and ages in the Wind River Range. Their average erosion rate on the medium grained foliated granitic gneiss was 0.6 mm kyr^{-1}. One of the powerful advantages of the cosmogenic nuclide method, not unlike U-series dating methods, is that the production ratios of different nuclides are becoming better known, allowing the flexibility of choosing the most appropriate nuclide ratio for the age and erosion rate of the surface of interest.

Denudation history. Denudation is the long-term erosion of the landscape resulting from multiple weathering and erosion processes over a large spatial scale (typically drainage basin to orogen scale). If loss of ^{10}Be from the regolith by erosion exceeds the loss due to radioactive decay (requiring >1 mm kyr^{-1} for a typical regolith), the distribution of quartz is uniform throughout the surface of the basin, and long term sediment storage is minimal, then the denudation rate $\overline{\varepsilon}$ of the basin can be expressed as (Lal 1991; Bierman 1994; Bierman and Steig 1996; Cerling and Craig 1994):

$$\overline{\varepsilon} = \frac{\overline{P}\mu}{\overline{N}} \tag{8}$$

where P is the average production rate for quartz in the basin (weighted according to the basin hypsometric curve, atoms g^{-1} a^{-1}), N is the measured (basin averaged) concentration (atoms g^{-1}), and μ is as defined previously. Investigations using this approach of estimating basin-scale denudation history have attempted to compare the results with independent estimates to evaluate the validity of the above assumptions (Brown et al. 1998b; Granger et al. 1996; Phillips et al. 1998). In the Brown et al. study, differences in the trends in two basin-scale denudation styles were found to be consistent with contrasting debris slide records. Sedimentation and erosion histories based on detailed interpretations of the soils records supported the conclusions of Granger et al. and Phillips et al. Other studies are currently underway to examine the extent to which

(1) temporary storage in valley bottoms in a catchment prior to alluvial fan deposition, and (2) episodic or low erosion rates affect the dating and denudation modeling of sediment in arid regions. Bierman and Steig (1996) proposed a model to interpret concentrations of *in situ* ^{10}Be in sediment derived from basin. Their calculations show how cosmogenic nuclides can confirm predictions of G.K. Gilbert that landscapes achieve equilibrium states in shorter time periods when erosion rates are higher (a conclusion made by Burbank et al. 1996 based on ^{10}Be measurements on strath terraces of the Indus River). Comparisons of the time-integrated denudation over the past 10^4 years calculated from ^{10}Be in stream sediments with records of erosion over different timescales (e.g., decades: Kirchner et al. 2001; or longer (Pazzaglia and Brandon 2001) have been used to infer that erosion rates may be sustained at different timescales and that topographic steady state conditions may be reached in short time periods. The ^{10}Be catchment denudation method has not yet been used to determine if a region undergoing orogeny has maintained dynamic equilibrium (topographic steady state) over periods beyond a single glacial cycle (Burbank and Anderson 2000).

Brown et al. (1998b) pointed out a useful means of determining the style of weathering and erosion in a basin, and if the style had changed over time. They measured the relative ratio of the ^{10}Be concentration in fine and coarse alluvial sediment. The ratio increases with an increasing contribution of weathering and erosion processes (e.g., debris slides) that bring coarse material to the surface. This strategy will also be useful in areas subject to periods of intense frost shattering processes, which may increase the overall erosion rate of basins, particularly for basins with a large fraction of their hypsometric curve above the zero degree isotherm. If widely applicable, it may provide a means of relating climate variability to changes in exhumation rates (part of an ongoing dilemma).

Bedrock incision rates. In a hallmark study, Burbank et al. (1996) and Leland et al. (1998) used ^{10}Be and ^{26}Al to calculate ages of strath terraces (fluvial erosional surfaces in bedrock) in the Indus River of northern Pakistan. ^{26}Al and ^{10}Be ages were sufficiently concordant to establish that erosion and burial effects were negligible after the strath was abandoned. They demonstrated, by measuring ^{10}Be in modern straths, that there was either a low probability of inheritance or a complicated exposure history that, for instance, would have resulted from the burial of the strath by enough fluvial sediment to shield the surface. They documented that bedrock incision has increased from 1-6 m kyr^{-1} to 9-12 m kyr^{-1} over a short (15 kyr) time. They could also identify an increase in incision rate downstream toward an active fault. The incision rates can be placed into context with the regional exhumation history determined by thermochronometry, helped to document the coeval maintenance of steady state hill-slopes by mass wasting into the deepening valleys, and has implications for assessing the role of climate on the evolution of mountains in tectonically active regions.

Paleoaltimetry

Paleoaltimetry is the study of the rate and nature of surface elevation change due to a combination of tectonic and isostatic uplift and erosion (i.e., surface uplift, not rock uplift). It differs from studies of the history of exhumed rocks (e.g., thermochronology, and many of the studies cited in the above section). The latter does not necessarily involve any change in the vertical position (elevation) of a topographic surface. Although geodetic analyses are providing data for the very recent history of surface uplift, only cosmogenic nuclides may provide a means of determining pre-historic vertical motion without resorting to models driven by erosion rate. In any case, paleoaltitudes must be measured at many sites over large (>1000 km^2) areas to be useful on an orogen scale (England and Molnar 1990).

The basis of paleoaltimetry studies using cosmogenic nuclides is the dependence of *in situ* production rates on atmospheric depth. Beginning with Hess's balloon missions in 1912 to document the increases in cosmic radiation with altitude, studies have shown how the secondary cosmic neutron and muonic fluxes are attenuated as they penetrate the atmosphere. The cosmic ray flux responsible for the production of ^{10}Be is predictably attenuated by interactions with atmospheric particles so that production rates at sea level are more than an order of magnitude less than rates at 4 km elevation. Consider a simple example for an uplifted surface of negligible or known erosion rate with an independently known age (or one in which the concentration of cosmogenic radionuclides have reached steady state). The difference between the calculated concentration (for the given exposure duration at its current elevation) and the actual measured concentration can be attributable to the lower time-averaged production rate, reflecting the surface's prior position at lower elevation.

The potential of cosmogenic nuclides in paleoaltimetry was introduced during the inception of the TCN exposure dating technique in 1986 when H. Craig and R.J. Poreda considered the effects of elevation changes of Hawaiian lavas during the production of ^{3}He (Craig and Poreda 1986). To estimate uplift rates, Lal (1991) described how uplift rate ($U = \Delta Z_s/\Delta t$) could be calculated from the TCN concentration, N, in the ideal case of steady state concentration, where an exponential change in production rate as a function of elevation is similar to the change due to erosion:

$$U = \frac{\bar{Z}P_{(o)}}{N_{(o)}} - \bar{Z}(\lambda + \mu\varepsilon) \tag{9}$$

where $\bar{Z}$ is the mean height (m), and $\lambda + \mu\varepsilon$ is a time term (yr^{-1}) from the effects of decay and constant erosion. The lower limit for uplift that may be constrained using *in situ* TCN paleoaltimetry is a function of the uplift rate and duration of exposure. In the ideal case above, the uplift rate must be greater than the $\bar{Z}\mu\varepsilon$. Deviations from these simple scenarios will require other models to fit the scenario.

Despite the fact that much of the theoretical and experimental foundations are complete, the direct use of cosmogenic nuclides for paleoaltimetry has never been successfully applied for several good reasons. First, finding a surface with independent exposure age and known erosion rate in an area that is useful for paleoaltimetry is no easy feat. Second, the assumption that the surface was never shielded by ice or snow cover is difficult to verify, but crucial, because the effects of small amounts of shielding obscure the slight isotope change due to uplift. Third, the reliability of the atmospheric scaling model is only recently being refined to the necessary level of reliability. Ongoing studies are directed toward understanding the latitudinal change in atmospheric attenuation and to developing a more accurate global atmospheric pressure model. These developments, along with improved precision of AMS and better understanding of production rates (including geomagnetic influences), will accommodate the resolution needed for paleoaltimetry analyses.

With the possible exception of a few ideal instances, future applications of *in situ* TCN paleoaltimetry will need to avoid the influences of unknown erosion and intermittent burial effects. A recent approach is to measure the concentration of the ^{10}Be in a laterally extensive surface (e.g., a lava or tuff) that is significantly displaced by a fault (M. Caffee, pers. comm.). Sampling of the down-dropped surface must be outside the shielding effects of any colluvial wedge. Because the surfaces are in close proximity, the influence of snow, erosion, and magnetic field effects may be assumed to be constant for both and therefore cancel out. The difference between the TCN concentrations of the two surfaces straddling the fault, with all things equal, should therefore reflect the change in altitude over time. The next logical step is to model the variability of the uplift rate over the exposure duration. It might become routine to couple the TCN paleoaltimetry with a late Cenozoic thermochronological method that has a low closure temperature ($T_c \sim$ 70°C) such as the (U-Th)/He technique for exhumation histories (House et al. 1998).

Summary

The interpretation of ^{10}Be measurements as exposure ages is non-trivial, so the tectonic community will need to be aware of existing and new strategies to use the isotope for chronology while recognizing the hazards of surface erosion and inheritance. Although measurable progress was made over the past decade in AMS analysis, sample target preparation, and our understanding of ^{10}Be production systematics (Lal 2000b), more technique development is necessary to improve precisions and spawn new applications. Three new tectonic applications of ^{10}Be are currently being developed. In settings with high surface uplift rates and with constrained or negligible erosion rates, *in situ* ^{10}Be and other long-lived or stable isotopes will be used to estimate Late Cenozoic paleoaltimetry. Cosmogenic ^{10}Be will be used to establish a late Quaternary denudation history to compare with low temperature thermochronology exhumation studies for longer durations. The concentrations can also provide information on the relative timing and direction of development of sub-orogen scale tectonic landforms such as propagations of fold and thrust belts.

CONCLUSIONS

The last decade has seen a dramatic increase in the applications of cosmogenic ^{10}Be to studies of the solid Earth. Greatly improved analytical detection limits, better integration of ^{10}Be data with other tracers in carefully designed field programs, and the development of theoretical and analytical methods for using ^{10}Be produced in situ have all had a major impact on the field.

Atmospherically produced ^{10}Be has been used to great effect in studies of Earths' geodynamo and subduction zone processes. Well-correlated variations in geomagnetic paleointensity and ^{10}Be concentrations in marine sediments and ice cores over the last 200 kyr (e.g., SINT-200, SINT-800, NAPIS-75, SINT Be, and cosmogenic nuclide profiles from the GRIP/GISP2 ice cores) show that ^{10}Be can be used as a proxy for geomagnetic field variations, if the analytical program is carefully designed. This suggests that ^{10}Be concentrations in sediment cores can be used to construct a record of geomagnetic field paleointensity variations, even where the paleomagnetic record itself may have been compromised. Further, synchronization of geomagnetic paleointensity features permits the transference of very precise ice core chronologies to lacustrine and deep-sea sediments. Future work can build on this success with studies that use ^{10}Be to test the hypothesis of asymmetric sawtooth variations in the magnetic field, and to further test the suggestion of Milankovitch forcing of the geodynamo.

At convergent margins, the concentrations of ^{10}Be erupting in arc lavas can now be compared with ^{10}Be profiles in the incoming sediment column to quantify the extent of sediment accretion, underplating and erosion beneath the fore-arc as well as sediment subduction to the depths of magma generation. Nearly all of the incoming sediment must be deeply subducted to explain high ^{10}Be in the Aleutian, Central America (except Mexico and Costa Rica), Scotia, Kurile, Izu, Mariana, Bismarck and Tonga volcanic arcs. Rear arc lavas from the Aleutian, Bismarck and Kurile arcs often have as much ^{10}Be as the associated volcano at the front, despite the longer path and greater ^{10}Be decay in transit to the rear arc. The high values in the rear arc require greater sediment Be input behind the front; one explanation is a transition from sediment dehydration beneath the front to sediment melting beneath the rear-arc. Well-correlated variations in U-series systematics and ^{10}Be for some but not all arcs suggest that the U excess in arc lavas may originate as elements leave the downgoing slab. Where U has been mobilized, but not Th, the time elapsed since element fractionation from the slab appears to be very short (~20 to 30 kyr) for some arcs.

Greatly improved detection limits of ~0.1 million atoms per gram open the door to a range of new work. For example, high ^{10}Be in fore-arc serpentinites would establish the role of the subducting slab in releasing hydrous fluids to the overlying serpentinizing mantle. ^{10}Be could help distinguish between recent and ancient sediment incorporation in contributing to the geochemical characteristics of some OIB suites, and some MORB lavas. It could also be used to track and time any recent mantle flow away from subduction zones.

The cosmogenic nuclides produced in rocks, including ^{10}Be, are now being used extensively in the study of active tectonics. A growing number of studies have successfully provided paleo-seismic constraints, dating times of repeated motions on exposed fault scarps, establishing slip rates from evolution of fault-cut alluvial surfaces, and determining the history of movement on fault-line scarps. With a useful time frame of ~5 ka to ~5 Ma, ^{10}Be provides dates and slip rates in a time window not easily datable otherwise. Combined ^{10}Be and ^{26}Al dating in principle allows erosion rates and exposure ages to be determined simultaneously, an approach suitable for long exposure histories. Where concordance between the two isotopes indicates negligible erosion, exposure ages can be used to provide uplift and incision rates.

As refinements in production rate scaling, AMS, and sampling strategies continue, studies employing *in situ* ^{10}Be measurements will have the accuracy and precision required to resolve higher frequency tectonic events. As the (U-Th)/He thermochronological method develops concurrently (e.g., House et al. 2000), there is a great potential for the pairing of these methods to provide complementary insight into late Cenozoic low temperature tectonic evolution of orogens. When coupled with detailed soils stratigraphy, geomorphological evidence, and supporting chronologies, ^{10}Be and other cosmogenic nuclides will begin to provide more insights into spatial variations of denudation rates and styles, contribute to paleoseismic databases, extend records of the timing and slip rates of Quaternary faults for examining strain partitioning among multiple faults, and to determine whether orogens have been evolving under sustained topographic equilibrium or erosional steady state conditions.

ACKNOWLEDGMENTS

All authors are deeply grateful and indebted to the accelerator crews at the University of Pennsylvania (J. Klein, H. White and R. Middleton), Lawrence Livermore National Labs (M. Caffee, R. Finkel, J. Southon) and Purdue University (D. Elmore, P. Sharma, S. Vogt). Over many years, they have provided scientific insight and inspiration, technical development and accelerator operations, and meticulous attention to sample analysis and data quality. JM acknowledges collaborators F. Tera, L. Brown, I.S. Sacks, C. Edwards, S.H. Zheng, J. Ryan, W. Leeman, R. Valentine, R. George and R. Kelly for their many contributions to ^{10}Be studies of subductology. JG thanks J. Klein, F. Pazzaglia, M. Brandon, J. Stamatakos, J. Whitney, and C. Harrington for inspiring neotectonic discussions while in the field. SB thanks Martin Frank, Bob Anderson, Zanna Chase, Yohan Guyodo, Catherine Kissel, Carlo Laj, Laure Meynadier, Joe Stoner, Lisa Tauxe, and Jean-Pierre Valet for making their datasets and reprints/preprints available. Thoughtful and constructive reviews by Devendra Lal, Jim Channell and Jeff Ryan are greatly appreciated, as is the meticulous editing provided by Ed Grew.

REFERENCES

Acton GD, Okada M, Clement BM, Lund SP, Williams T (2002) Paleomagnetic overprints in ocean sediment cores and their relationship to shear deformation caused by piston coring. J Geophys Res 107(B4):3-1 to 3-15

Aldahan A, Possnert G (1998) A high-resolution ^{10}Be profile from deep sea sediment covering the past 70 ka: Indication for globally synchronized environmental events. Quat Geochron 17:1023-1032

Aldahan A, Possnert G (2000) The ^{10}Be marine record of the last 3.5 Ma. Nucl Inst Methods Phys Res B 172:513-517

Aldahan AA, Possnert G, Gard G (1994) ^{10}Be in two sediment cores from the north Atlantic and chronological implications for the late Quaternary. N Jahrb Geol Paläont Monats H7:418-433

Aldahan AA, Ning S, Possnert G, Backman J, Boström K (1997) ^{10}Be records from sediments of the Arctic Ocean covering the past 350 ka. Marine Geol 144:147-162

Aldahan A, Possnert G, Johnsen SJ, Clausen HB, Isaksson E, Karlen W, Hansson M (1998) Sixty year ^{10}Be record from Greenland and Antarctica. Proc Indian Acad Sci 107:139-147

Aldahan A, Possnert G, Peck J, King J, Colman S (1999) Linking the ^{10}Be continental record of Lake Baikal to marine and ice archives of the last 50 ka: Implication for the global dust-aerosol input. Geophys Res Lett 26:2885-2888

Anderson RF, Bacon MP, Brewer PG (1983a) Removal of ^{230}Th and ^{231}Pa at ocean margins. Earth Planet Sci Lett 66:73-90

Anderson RF, Bacon MP. Brewer PG (1983b) Removal of ^{230}Th and ^{231}Pa from the open ocean. Earth Planet Sci Lett 62:7-23

Anderson RF, Lao Y, Broecker WS, Trumbore SE, Hofmann HJ, Wölfli W (1990) Boundary scavenging in the Pacific Ocean: a comparison of ^{10}Be and ^{231}Pa. Earth Planet Sci Lett 96:287-304

Anderson RS, Repka JL, Dick GS (1996) Explicit treatment of inheritance in dating depositional surfaces using *in situ* ^{10}Be and ^{26}Al. Geology 24:47-51

Armstrong, RL (1971) Isotopic and chemical constraints on models of magma genesis in volcanic arcs, Earth Planet. Sci. Lett 12:137-142

Ayarbe JP, Phillips FM, Harrison JBJ, Elmore D, Sharma D (1998) Application of cosmogenic nuclides to fault-scarp chronology: preliminary results from the Socorro Canyon fault. Soil, water, and earthquakes around Socorro, New Mexico. *In* Harrison JBJ (ed) 1998 Rocky Mountain Cell 'Friends of the Pleistocene' Guide, p 39

Bacon MP (1984) Glacial to interglacial changes in carbonate and clay sedimentation in the Atlantic estimated from thorium-230 measurements. Isotope Geosci 2:97-111

Bacon MP, Rosholt JN (1982) Accumulation rates of ^{230}Th and ^{231}Pa and some transition metals on the Bermuda Rise. Geochim Cosmochim Acta 46:651-666

Bard E (1998) Geochemical and geophysical implications of the radiocarbon calibration. Geochim Cosmochim Acta 62:2025-2038

Bard E, Raisbeck GM, Yiou F, Jouzel J (1997) Solar modulation of cosmogenic nuclide production over the last millennium: comparison between ^{14}C and ^{10}Be records. Earth Planet Sci Lett 150:453-462

Bassinot FC, Labeyrie KL, Lancelot Y, Quidelleur X, Shackleton NJ, Vincent E (1994) The astronomical theory of climate and the age of the Brunhes-Matuyama magnetic reversal. Earth Planet Sci Lett 126:91-108

Baumgartner S, Beer J, Wagner G, Kubik P, Suter M, Raisbeck GM, Yiou F (1997a) ^{10}Be and dust. Nucl Instr Meth Phys Res B. 123:296-301

Baumgartner S, Beer J, Suter M, Diitrich-Hannen B, Synal H-A, Kubik PW, Hammer C, Johnsen S (1997b) Chlorine-36 fallout in the Summit Greenland Ice Core Project ice core. J Geophys Res 102: 26659-26662

Baumgartner S, Beer J, Masarik J, Wagner G, Meynadier L, Synal H-A (1998) Geomagnetic modulation of the ^{36}Cl flux in the GRIP ice core, Greenland. Science 279:1330-1332

Bebout, GE (1996) Volatile transfer and recycling at convergent margins: Mass balance and insights from high-P/T metamorphic rocks. *In* Bebout G, Scholl D, Kirby S, Platt J (eds) Subduction: Top to Bottom. Am Geophys Union Monogr 96:179-194

Bebout, GE, Ryan JG, Leeman WP (1993) B-Be systematics in subduction related metamorphic rocks: characterization of the subduction component. Geochim Cosmochim Acta 57:2227-2238

Bebout GE, Ryan JG, Leeman WP, Bebout A (1999) Fractionation of trace elements by subduction-zone metamorphism-effect of convergent margin thermal evolution. Earth Planet Sci Lett 171:63-78

Beer J, Andrée M, Oeschger H, Stauffer B (1983) Temporal ^{10}Be variations in ice. Radiocarbon 25:269-278

Beer J, Andrée M, Oeschger H, Siegenthaler U, Bonani G, Hofmann H, Morenzoni E, Nessi M, Suter M, Wölfli W, Finkel R, Langway, Jr. C (1984) The Camp Century ^{10}Be record: implications for long-term variations of the geomagnetic dipole moment. Nucl Instr Phys Meth B5:380-384

Beer J, Siengenthaler U, Bonani G, Finkel RC, Oeschger H, Suter M, Wölfi W (1988) Information on past solar activity and geomagnetism from ^{10}Be in the Camp Century ice core. Nature 331:675-679

Beer J, Blinov A, Bonani G, Finkel RC, Hofmann HJ, Lehmann B, Oeschger H, Sigg A, Schwander J, Staffelbach T, Stauffer BR, Suter M, Wölfli W (1990) Use of ^{10}Be in polar ice to trace the 11-year cycle of solar activity. Nature 347:164-166

Beer J, Bonani GS, Dittrich B, Heller F, Kubik PW, Tungsheng L, Chengde S, Suter M (1992) ^{10}Be and magnetic susceptibility in Chinese loess. Geophys Res Lett 20:57-60

Bell JW, Brune JN, Liu T, Zreda M, Yount JC (1998) Dating precariously balanced rocks in seismically active parts of California and Nevada. Geology 26:495-498

Benton L (1997) Origin and Evolution of Serpentine Seamount Fluids, Mariana and Izu-Bonin Forearcs: Implications for the recycling of subducted material. PhD dissertation, University of Tulsa, Tulsa, Oklahoma

Bhattacharyya A, Mitra B (1997) Changes in cosmic ray cut-off rigidities due to secular variations of the geomagnetic field. Ann Geophysicae 15:734-739

Bierman PR (1994) Using *in situ* produced cosmogenic isotopes to estimate rates of landscape evolution: A review from the geomorphic perspective. J Geophys Res 99:13885-13896

Bierman PR, Steig EJ (1996) Estimating rates of denudation using cosmogenic isotope abundances in sediment. Earth Surf Proc Landforms 21:125-139

Bierman PR, Gillespie AR, Caffee MW (1995) Cosmogenic ages for earthquake recurrence intervals and debris flow fan deposition, Owens Valley, California. Science 270:447-450

Bourdon B, Turner S, Allegre C (1999) Melting dynamics beneath the Tonga-Kermadec island arc inferred from ^{231}Pa-^{235}U systematics. Science 286:2491-2493

Brachfeld S, SK Banerjee (2000) A new high-resolution geomagnetic paleointensity record for the North American Holocene: A comparison of sedimentary and absolute intensity data. J Geophys Res B 105:821-834

Brassart JE, Tric E, Valet JP, Herrero-Bervera E (1997) Absolute paleointensity between 60 and 400 ka from the Kohala Mountain (Hawaii). Earth Planet Sci Lett 148:141-156

Brenan JM, Shaw HF, Ryerson FJ, Phinney DL (1995) Mineral-aqueous fluid partitioning of trace elements at 900°C and 2.0 GPa: Constraints on the trace element geochemistry of mantle and deep crustal fluids. Geochim Cosmochim Acta 59:3331-3350

Brenan JM, Ryerson FJ, Shaw H (1998) The role of aqueous fluids in the slab-to-mantle transfer of boron, beryllium and lithium during subduction: Experiments and models. Geochim Cosmochim Acta 62:3337-3347

Brook EJ, Brown ET, Kurz MD, Ackert RP, Raisbeck GM, Yiou F (1995a) Constraints on erosion and uplift rates of Pliocene glacial deposits in the Transantarctic Mountains using *in situ*-produced ^{10}Be and ^{26}Al. Geology 23:1063-1066

Brook EJ, Kurz MD, Ackert RP, Raisbeck GM, Yiou F (1995b) Cosmogenic nuclide exposure ages and glacial history of late Quaternary Ross Sea drift in McMurdo Sound, Antarctica. Earth Planet Sci Lett 131:41-56

Brown L, Klein J, Middleton R, Sacks IS, Tera F (1982) ^{10}Be in island arc volcanoes and implications for subduction. Nature 299:718-720

Brown ET, Bourles DL, Colin F, Raisbeck GM, Yiou F, and Desgarceaux S (1995) Evidence for muon-induced *in situ* production of ^{10}Be in near-surface rocks from the Congo. Geophys Res Lett 22:703-706

Brown ET, Bourles DL, Burchfiel BC, Oidong D, Jun L, Molnar P, Raisbeck GM, Yiou F (1998a) Estimation of slip rates in the southern Tien Shan using cosmic ray exposure dates of abandoned alluvial fans. Geol Soc Am Bull 110:377-386

Brown ET, Stallard RF, Larsen MC, Bourles DL, Raisbeck GM, Yiou F (1998b) Determination of predevelopment denudation rates of an agricultural watershed (Cayaguas River, Puerto Rico) using *in situ*-produced ^{10}Be in river-borne quartz. Earth Planet Sci Lett 160:723-728

Burbank DW, Anderson RS (2000) Tectonic Geomorphology: Blackwell Scientific, Oxford, UK, 270 p

Burbank DW, Leland J, Fielding E, Anderson RS, Brozovic N, Reid MR, Duncan C (1996) Bedrock incision, rock uplift and threshold hillslopes in the northwestern Himalayas. Nature 379:505-510

Castagnoli G, Lal D (1980) Solar modulation effects in terrestrial production of carbon-14. Radiocarbon 11:133

Cerling TE, Craig H (1994) Geomorphology and *in-situ* cosmogenic isotopes. Ann Rev Earth Planet Sci 22:273-31

Cerling TE, Webb RH, Poreda RJ, Rigby AD, Melis TS (1999) Cosmogenic ^{3}He ages and frequency of late Holocene debris flows from Prospect Canyon, Grand Canyon, USA. Geomorphology 27:93-111

Channell JET, Kleiven HF (2000) Geomagnetic palaeointensity and astrochronological ages for the Matuyama-Brunhes boundary and boundaries of the Jaramillo subchron: palaeomagnetic and oxygen isotope records from ODP Site 983. Phil Trans R Soc Lond A 358:1027-1047

Channell JET, Hodell DA, Lehman B (1997) Relative geomagnetic paleointensity and ^{18}O at ODP Site 983 (Gardar Drift, North Atlantic) since 350 ka. Earth Planet Sci Lett 153:103-118

Channell JET, Hodell DA, McManus J, Lehman B (1998) Orbital modulation of the Earth's magnetic field intensity. Nature 394:464-468

Channell JET, Stoner JS, Hodell DA, Charles CD (2000) Geomagnetic paleointensity for the last 100 kyr from the sub-Antarctic South Atlantic: a tool for interhemispheric correlation. Earth Planet Sci Lett 175:145-160

Chase Z, Anderson RF, Fleisher MQ, Kubik PW (in press) Scavenging of ^{230}Th, ^{231}Pa and ^{10}Be in the Southern Ocean (SW Pacific sector): The importance of particle flux and advection. Deep-Sea Res II

Chengde S, Beer J, Bonani G, Liu T, Oeschger H, Suter M, Wölfli W (1992) ^{10}Be in Chinese loess. Earth Planet Sci Lett 109:169-177

Cini Castagnoli G, Albrecht A, Beer J, Bonino G, Shen CH, Callergari E, Taricco C, Dittrcih-Hannen B, Kubik P, Suter M, Zhu GM (1995) Evidence for enhanced ^{10}Be deposition in Mediterranean sediments 35 Kyr BP. Geophys Res Lett 22:707-710

Cini Castagnoli G, Bonino G, Della Monica P, Procopio S, Taricco C (1998) On the solar origin of the 200 yr Suess wiggles: evidence from thermoluminescence in sea sediments. Il Nuovo Cimento 21 C:237-241

Class C, Miller DM, Goldstein, SL, Langmuir CH (2000) Distinguishing melt and fluid components in Umnak volcanics, Aleutian arc. Geochem Geophys Geosystems 1, paper #1999GC000010

Cousens BL, Allan JF, Gorton M P (1994) Subduction-modified pelagic sediments as the enriched component in back-arc basalts from the Japan Sea; Ocean Drilling Program sites 797-794. Contrib Mineral Petrol 117:421-434

Craig H, Poreda R (1986) Cosmogenic ^{3}He in terrestrial rocks: the summit lavas of Maui. Proc Natl Acad Sci (USA) 83:1970-1974

Damon PE, Sonett CP (1991) Solar and terrestrial components of the atmospheric C-14 variation spectrum. *In* Sonett CP, Giampapa MS, Mathews MS (eds) The Sun in Time. The University of Arizona, Tucson, p 360-388

Damon PE, Cheng S, Linick T (1989) Fine and hyperfine structure in the spectrum of secular variations of atmospheric ^{14}C. Radiocarbon 31:704-718

Davies JH, Steventson DJ (1992) Physical model of source region of subduction zone volcanics. J Geophys Res 97:2037-2070

Dickinson WR (1975) Potash-depth (K-h) relations in continental margin and intra-oceanic magmatic arcs. Geology 3:53-56

Dyke AS, Morris TF, Green DEC (1991) Postglacial tectonic and sea level history of the central Canadian Arctic. Geol Surv Can Bull 397:56

Dyke AS, Morris TF, Green DEC, England J (1992) Quaternary Geology of Prince of Wales Island, Arctic Canada. Geol Surv Can Mem 433:142

Eddy JA (1976) The Maunder minimum. Science 192:1189-1201

Edwards CMH, Morris JD, Thirlwall MF (1993) Separating mantle from slab signatures in arc lavas using B/Be and radiogenic isotope systematics. Nature 362:530-533

Elliot T, Plank T, Zindler A, White W, Bourdon B (1997) Element transport from slab to volcanic front at the Mariana arc. J Geophys Res 102:14991-15019

Elsasser WM, Ney EP, Wenkker JR (1956) Cosmic ray intensity and geomagnetism. Nature 178:1226

England P, Molnar P (1990) Surface uplift, uplift of rocks, and exhumation of rocks. Geology 18: 1173-1177

Fenton CR, Webb RH, Pearthree PA, Cerling TE, Poreda RJ (2001) Displacement rates on the Toroweap and Hurricane faults; implications for Quaternary down-cutting in the Grand Canyon, Arizona. Geology 29:1035-1038

Finkel RC, Nishiizumi N (1997) Beryllium-10 concentrations in the Greenland Ice Sheet Project 2 ice core. J Geophys Res 102:26699-26706

Finkel R, Suter M (1993) AMS in the Earth Sciences: Technique and applications. Adv Analyt Geochem 1: 1-114

Frank M (2000) Comparison of cosmogenic radionuclide production and geomagnetic field intensity over the last 200,000 years. Phil Trans R Soc Lond A 358:1089-1107

Frank M, Eckhardt J-D, Eisenhauer A, Kubik PW, Dittrich-Hannen B, Segl M, Mangini A (1994) Beryllium-10, thorium-230, and protactinium-231 in Galapagos microplate sediments: Implications of hydrothermal activity and paleoproductivity changes during the last 100,000 years. Paleoceanography 9:559-578

Frank M, Eisenhauer A, Bonn WJ, Walter P, Grobe H, Kubik PW, Dittrich-Hannen B, Mangini A (1995) Sediment redistribution versus paleoproductivity change: Weddell Sea margin sediment stratigraphy and biogenic particle flux of the last 250,000 years deduced from $^{230}Th_{ex}$, ^{10}Be and biogenic barium profiles. Earth Planet Sci Lett 136:559-573

Frank M, Schwarz B, Baumann S, Kubik PW, Suter M, Mangini A (1997) A 200 kyr record of cosmogenic radionuclide production rate and geomagnetic field intensity from ^{10}Be in globally stacked deep-sea sediments. Earth Planet Sci Lett 149:121-129

Frank M, Gersonde R, Mangini A (1999) Sediment redistribution, $^{230}Th_{ex}$ - normalization and implications for the reconstruction of particle flux and export paleoproductivity. *In* Fischer G, Wefer G (eds) Use of Proxies in Paleoceanography: Examples from the South Atlantic. Springer-Verlag, Berlin–Heidelberg, p 409-426

Frank M, Gersonde R, van der Loeff MR, Bohrmann G, Nürnberg CC, Kubik PW, Suter M, Mangini A (2000) Similar glacial and interglacial export bioproductivity in the Atlantic sector of the Southern Ocean: Multiproxy evidence and implications for glacial atmospheric CO_2. Paleoceanography 15: 642-658

Fryer P, Mottl M, Johnson L, Haggerty J, Phipps S, Maekawa H (1995) "Serpentine bodies in the forearcs of Western Pacific convergent margins: origin and associated fluids." *In* Taylor B, Natland J (eds) Active Margins and Marginal Basins of the Western Pacific. Washington, DC, Am Geophys Union, p 259-279

Fuller M, Hastedt M, Herr B (1998) Coring-induced magnetization of recovered sediment. Proc Ocean Drill Prog Sci Res 157:47-56

George R, Turner S, Nye C, Hawkesworth C (2000) Along-arc U-Th-Ra disequilibria in the Aleutians: Rapid timescales of fluid transfer. V M Goldschmidt 2000 J Conf Abstr 5:436

George R, Turner S, Hawkesworth CJ, Morris J, Nye C, Ryan JG, Zheng SH (submitted) Melting processes and fluid and sediment transport rates along the Alaska-Aleutian arc from an integrated U-Th-Ra-Be isotope study. J Geophys Res

Gill JB, Williams RN (1990) The isotope and U-series studies of subduction-related volcanic rocks. Geochim Cosmochim Acta 54:1427-1442

Gill JB, Morris JD, Johnson RW (1993) Timescale for producing the geochemical signature of island arc magmas: U-Th-Po and Be-B systematics in recent Papua New Guinea lavas. Geochim Cosmochim Acta 57:4269-4283

Gillespie AR, Bierman P (1995) Precision of terrestrial exposure ages and erosion rates estimated from analysis of cosmogenic isotopes produced *in situ*. J Geophys Res 100:24637-24649

Gosse, JC (1994) Alpine glacial history reconstruction: 1. Application of the cosmogenic ^{10}Be exposure age method to determine the glacial chronology of the Wind River Mountains, Wyoming, USA; 2. Relative dating of Quaternary deposits in the Rio Atuel Valley, Mendoza, Argentina. PhD dissertation, Lehigh University, Bethlehem, Pennsylvania

Gosse JC, McDonald E (submitted) Variation of inheritance in alluvium as a function of catchment glaciation: Implication for glacial erosion and exposure age dating. Radiocarbon

Gosse JC, Phillips FM (2001) Terrestrial cosmogenic nuclides: theory and applications. Quat Sci Rev 20:1475-1560

Gosse JC, Stone JO (2001) Terrestrial cosmogenic nuclide methods passing milestones toward palaeo-altimetry. EOS Trans Am Geophys Union 82-7:82,86,89

Gosse JC, Evenson EB, Klein J, Lawn B, Middleton R (1995a) Precise cosmogenic ^{10}Be measurements in western North America: Support for a global Younger Dryas cooling event. Geology 23:877-880

Gosse JC, Klein J, Evenson EB, Lawn B, Middleton R (1995b) Beryllium-10 dating of the duration and retreat of the last Pinedale glacial sequence. Science 268:1329-1333

Gosse JC, Grant DR, Klein J, Lawn B (1995c) Cosmogenic ^{10}Be and ^{26}Al constraints on weathering zone genesis, ice cap basal conditions, and Long Range Mountain (Newfoundland) glacial history. CANQUA-CGRG Conf Abstr, Memorial University of Newfoundland, St. Johns, Canada, p 19

Gosse JC, Harrington CD, Whitney JW (1996) Applications of *in situ* cosmogenic nuclides in the geologic site characterization of Yucca Mountain, Nevada. Materials Res Soc Symp Proc 412:799-806

Gosse JC, Hecht G, Mehring N, Klein J, Lawn B, Dyke A (1998) Comparison of radiocarbon and *in situ*-cosmogenic nuclide-derived postglacial emergence curves for Prescott Island, Central Canadian Arctic. Geol Soc Am Abstr Progr 30(7):A-298

Gosse JC, Dyke AS, Klein J (in prep.) Crustal emergence curve for central Arctic based on terrestrial cosmogenic nuclide chronology of a sea cliff and raised beaches. Geophys Res Lett

Granger DE, Muzikar PF (2001) Dating sediment burial with *in situ*-produced cosmogenic nuclides: theory, techniques, and limitations. Earth Planet Sci Lett 188:269-281

Granger DE, Kirchner JW, Finkel R (1996) Spatially averaged long-term erosion rates measured from in-situ produced cosmogenic nuclides in alluvial sediment. J Geology 104:249-257

Gu ZY, Caffee MW, Chen MY, Guo ZT, Lal D, Liu TS, Southon J (1996) Five million year ^{10}Be record in Chinese loess and red clay: climate and weathering relationships. Earth Planet Sci Lett 144:273-287

Guyodo Y, Valet J-P (1996) Relative variations in geomagnetic intensity from sedimentary records: the past 200 thousand years. Earth Planet Sci Lett 143:23-36

Guyodo Y, Valet J-P (1999) Global changes in intensity in the Earth's magnetic field during the past 800 kyr. Nature 399:249-252

Guyodo Y, Gaillot P, Channell JET (2000) Wavelet analysis of relative geomagnetic paleointensity at ODP Site 983. Earth Planet Sci Lett 184:109-183

Guyodo Y, Acton GD, Brachfeld S, Channell JET (2001) A sedimentary paleomagnetic record of the Matuyama chron from the western Antarctic margin (ODP Site 1101). Earth Planet Sci Lett 191:61-74

Haggerty JA (1991) Evidence from fluid seeps atop serpentine seamounts in the Mariana Forearc: Clues for emplacement of the seamounts and their relationship to forearc tectonics. Marine Geol 102:293-301

Haggerty JA, Fisher JB (1992) Short-chain organic acids in interstitial waters from Mariana and Bonin forearc serpentines. *In* Fryer P, Pearce JA, Stokking LB et al. (eds) Proc Ocean Drilling Program Sci Results 125:387-395

Hancock GS, Anderson RS, Chadwick OA, Finkel RC (1999) Dating fluvial terraces with ^{10}Be and ^{26}Al profiles: application to the Wind River, Wyoming. Geomorphology 27:41-60

Haubold HJ, Beer J (1992) Solar activity cycles revealed by time series analysis of argon-37, sunspot-number, and beryllium-10 records. Proc IUGG, Vienna, Solar-Terrestrial Varia Glob Chan, p 11-34

Hawkesworth CJ, Turner SP, McDermott F, Peate DW, van Calsterern P (1997) U-Th isotopes in arc magmas: implications for element transfer from the subducted crust. Science 276:551-555

Heisinger B, Niedermayer M, Hartmann JF, Korschinek G, Nolte E, Morteani G, Neumaier S, Petitjean C, Kubik P, Synal A, Ivy-Ochs S (1997) In-situ production of radionuclides at great depths. Nucl Instr Meth Physics Res B123:341-346

Henken-Meillies WU, Beer J, Heller F, Hsu KJ, Shen C (1990) ^{10}Be and ^{9}Be in South Atlantic DSDP Site 519: relation to geomagnetic reversals and to sediment composition. Earth Planet Sci Lett 98:267-276

Hermanns RL, Niedermann S, Garcia AV, Gomez JS, Strecker MR (2001) Neotectonics and catastrophic failure of mountain fronts in the southern intra-Andean Puna Plateau, Argentina. Geology 29:619-623

Herrstrom EA, Reagan MK, Morris JD (1995) Variations in lava composition associated with flow of asthenosphere beneath southern Central America. Geology 23:617-620

Hickey-Vargas R, Murong S, Lopez-Escobar L, Roa HM, Morris J, Reagan R, Ryan J (in press) Multiple subduction components in the mantle wedge: Evidence from eruptive centers in the Central SVZ, Chile. Chem Geol

House MA, Wernicke BP, Farley KA (1998) Dating topography of the Sierra Nevada, California, using apatite (U-Th)/ He ages. Nature 396:66-69

House MA, Farley KA, Stockli DF (2000) Helium chronometry of apatite and titanite using Nd-YAG laser heating. Earth Planet Sci Lett 183:365-368

Imamura M, Hashimoto Y, Yoshida K, Yamane I, Yamashita, Inoue T, Tanaka S, Nagai H, Honda M, Kobayashi K, Takaoka, Ohba Y (1984) Tandem accelerator mass spectrometry of ^{10}Be/^{9}Be with internal beam monitor methods. Nucl Inst Meth B6:211-216

Imbrie J, Imbrie JZ (1980) Modelling the climate response to orbital variations. Science 207:942-953

Imbrie J, Hays JD, Martinson DG, McIntyre A, Mix AC, Jorley JJ, Pisias NG, Prell WL, Shackleton NJ (1984) The orbital theory of Pleistocene climate: Support from a revised chronology of the marine d^{18}O record. *In* Berger AL et al. (eds) Milankovitch and Climate, Part 1. NATO ASI Ser 126. Reidel, Dordrecht, The Netherlands, p 269-305

Ishikawa T, Nakamura E (1994) Origin of the slab component inferred in arc lavas from across-arc variation of Band Pb isotopes. Nature 370:205-208

Ishikawa T, Tera F (1997) Source composition and distribution of the fluid in the Kurile mantle wedges: Constraints from across-arc variations of B/Nb and B isotopes. Earth Planet Sci Lett 152:123-138

Ivy-Ochs S, Schluchter C, Kubik PW, Synal H-A, Beer J, Kerschner H (1998) The exposure age of Egesen moraine at Julier Pass, Switzerland, measured with the cosmogenic radionuclides ^{10}Be, ^{26}Al, and ^{36}Cl. Eclogae Geologica Helvetica 89:1049-1063

Jackson J, Ritz JF, Siame L, et al. (2002) Fault growth and landscape development rates in Otago, New Zealand, using *in situ* cosmogenic Be-10. Earth Planet Sci Lett 195:85-193

Jarrard RD (1986) Relations among subduction parameters. Rev Geophys 24:217-284

Johnson M, Plank T (1999) Dehydration and melting experiments constrain the fate of subductd sediments. Geochem Geophy Geosystems (G3)1, #14

Juárez MT, Tauxe L, Gee JS, Pick T (1998) The intensity of the Earth's magnetic field over the past 160 million years. Nature 394:878-881

Karlin R (1990) Magnetic mineral diagenesis in marine sediments from the Oregon continental margin. J Geophys Res 95:4405-4419

Kastner M, Morris J, Chan LH, Saether, O, Luckge, A (2000) Three distinct fluid systems at the Costa Rica Subduction Zone: Chemistry, hydrology, and fluxes. V M Goldschmidt 2000 J Conf Abstr 5:572

Keller EA, Gurrola L, Tierney TE (1999) Geomorphic criteria to determine direction of lateral propagation of reverse faulting and folding. Geology 27:515-518

Kelly RK, Driscoll NW (1998) Structural controls on Be-10 occurrences in arc lavas. EOS Trans Am Geophys Union 79:45

Kent DV (1973) Post depositional remanent magnetization in deep-sea sediment. Nature 246:32-34

Kent DV, Opdyke ND (1977) Paleomagnetic field intensity variation recorded in a Brunhes epoch deep-sea sediment core. Nature 266:156-159

Kersting AB, Arculus RJ, Gust D (1996) Lithospheric contributions to arc magmatism; isotope variations along strike in volcanoes of Honshu, Japan. Science 272:1464-1468

Kimura G, Silver EA, Blum P, et al. (1997) Proceeding of the Ocean Drilling Program. *In* Initial Reports 170. College Station, Texas, p 458

Kirchner JW, Finkel RC, Riebe CS, Granger DE, Clayton JL, Megahan WF, (2001) Episodic mountain erosion inferred from sediment yields over 10-year and 10,000-year timescales. Geology 29: 591-594.

Kissel C, Laj C, Labeyrie L, Dokken T, Voelker A, Blamart D (1999) Rapid climatic variations during marine isotopic stage 3: magnetic analysis of sediments from nordic seas and North Atlantic. Earth Planet Sci Lett 171:489-502

Klein J, Lawn B, Gosse J, Harrington C (1997) Can terrestrial cosmogenic Be-10 be measured in whole rock samples to decipher surface exposure histories. Geol Soc Am Abstr Progr 29:A-346

Kobayashi K, Kitazawa K, Kanaya T, Sakai T (1971) Magnetic and micropaleontological study of deep sea sediments from the western equatorial Pacific. Deep-Sea Res 18:1045-1062

Kocharov GE (1990) Investigation of astrophysical and geophysical problems by AMS: Successes achieved and prospects. Nucl Instr Meth B52:583-587

Kok YS (1999) Climatic influence in NRM and ^{10}Be-derived geomagnetic paleointensity data. Earth Planet Sci Lett 166:105-119

Kok YS, Tauxe L (1996a) Saw-toothed pattern of relative paleointensity records and cumulative viscous remanence. Earth Planet Sci Lett 137: 95-99

Kok YS, Tauxe L (1996b) Saw-toothed pattern of sedimentary paleointensity records explained by cumulative viscous remanence. Earth Planet Sci Lett 144:E9-E14

Kok YS, Tauxe L (1999) A relative geomagnetic paleointensity stack from Ontong-Java plateau sediments for the Matuyama. J Geophys Res 104:25401-25413

Kok YS, Ynsen I (2002) Reply to comment by J-P Valet and L Meynadier on "A relative geomagnetic paleointensity stack from Ontong-Java plateau sediments for the Matuyama." J Geophys Res 107(B3):3-1-2

Kong P, Nishiizumi K, Finkel RC, Caffee MW (1999) *In situ*-produced cosmogenic ^{10}Be and ^{26}Al in olivine. EOS Trans Am Geophys Union 80:F1166

Kubik PW, Ivy-Ochs S, Masarik J, Frank M, Schlüchter C (1998) ^{10}Be and ^{26}Al production rates deduced from an instantaneous event within the dendro-calibration curve, the landslide of Köfels, Ötz Valley, Austria. Earth Planet Sci Lett 161:231-241

Kurz MD, Colodner D, Trull TW, Moore RB, O'Brien K (1990) Cosmic ray exposure dating with *in situ*-produced cosmogenic ^{3}He: results from young Hawaiian lava flows. Earth Planet Sci Lett 97:177-189

Laj C, Kissel C (1999) Geomagnetic field intensity at Hawaii for the last 420 kyr from the Hawaii Scientific Drilling Project core, Big Island, Hawaii. J Geophys Res 104:15317-15338

Laj C, Kissel K, Garnier F (1996) Relative geomagnetic field intensity and reversals for the last 1.8 My from a central equatorial Pacific core. Geophys Res Lett 23:3393-3396

Laj C, Raïs A, Surmont J, Gillot PY, Guillou H, Kissel C, Zanella E (1997) Changes of the geomagnetic field vector obtained from lava sequences on the island of Volcano (Aeolian Islands, Sicily). Phys Earth Planet Inter 99:161-177

Laj C, Kissel C, Mazaud A, Channell JET, Beer J (2000) North Atlantic palaeointensity stack since 75 ka (NAPIS-75) and the duration of the Laschamp event. Phil Trans R Soc Lond A 358:1009-1025

Laj C, Kissel C, Scao V, Beer J, Thomas DM, Guillou H, Muscheler R, Wagner G (2002) Geomagnetic intensity and inclination variations at Hawaii for the past 98 kyr from core SOH-4 (Big Island): a new study and a comparison with existing contemporary data. Phys Earth Planet Inter 129:205-243

Lal D (1988) *In situ*-produced cosmogenic isotopes in terrestrial rocks. Ann Rev Earth Planet Sci 16: 355-388

Lal D (1991) Cosmic ray labeling of erosion surfaces: *in situ* nuclide production rates and erosion rates. Earth Planet Sci Lett 104:424-439

Lal D (1996) On cosmic-ray exposure ages of terrestrial rocks: a suggestion. Radiocarbon 37:889-898

Lal D (2000a) Cosmogenic ^{10}Be: A critical view on its widespread dominion in geosciences. Proc Indian Acad Sci (Earth Planet Sci) 109:181-186

Lal D (2000b) Cosmogenic nuclide production rate systematics in terrestrial materials: present knowledge, needs, and future actions for improvement. Nucl Inst Methods Phys Research B 172:772-781

Lal D, Peters B (1967) Cosmic ray produced radioactivity on the Earth. *In* Encyclopedia of Physics, vol 46(2). Springer-Verlag, New York, p 581-588

Lao Y, Anderson RF, Broecker WS, Trumbore SE, Hofmann HJ, Wolfli W (1992) Transport and burial rates of particulate fluxes of ^{10}Be and ^{231}Pa in the Pacific ocean during the Holocene period. Earth Planet Sci Lett 113:173-189

Lao Y, Anderson RF, Broecker WS, Hofmann HJ, Wolfli W (1993) Particulate fluxes of ^{230}Th, ^{231}Pa, and ^{10}Be in the northeastern Pacific ocean. Geochim Cosmochim Acta 57:205-217

Leat PT, Livermore RA, Millar, IL (2000) Magma supply in back-arc spreading centre segment E2, East Scotia Ridge. J Petrology 41:845-866

Leland J, Reid MR, Burbank DW, Finkel R, Caffee M (1998) Incision and differential bedrock uplift along the Indus River near Nanga Parbat, Pakistan Himalaya, from ^{10}Be and ^{26}Al exposure age dating of bedrock straths. Earth Planet Sci Lett 154:3-107

Leslie BW, Hammond DE, Berelson WE, Lund SP (1990) Diagenesis in anoxic sediments from the California continental borderland and its influence on iron, sulfur, and magnetite behavior. J Geophys Res 95:4453-4470

Liccardi JM, Kurz MD, Clark PU, Brook EJ (1999) Calibration of cosmogenic 3He production rates from Holocene lava flows in Oregon, USA, and effects of the Earth's magnetic field. Earth Planet Sci Lett 172:261-271

Liu B, Phillips FM, Fabryka-Martin JT, Fowler MM, Stone WD (1994) Cosmogenic 36Cl accumulation in unstable landforms I: Effects of the thermal neutron distribution. Water Resources Res 30:3115-3125

Lund SP, Schwartz M (1999) Environmental factors affecting geomagnetic field palaeointensity estimates from sediments. *In* Maher BA, Thompson T (eds) Quaternary Climates, Environments and Magnetism. Cambridge University Press, Cambridge, UK, p 323-311

Malkus WVR (1968) Precession as the cause of geomagnetism. Science 160:250-264

Marshall JS, Anderson RS (1995) Quaternary uplift and seismic cycle deformation, Península de Nicoya, Costa Rica. Geol Soc Am Bull 107:463-473

Marti K, Craig H (1987) Cosmic-ray produced neon and helium in the summit lavas of Maui. Nature 325:335-337

Martinson DG, Pisias NG, Hays JD, Imbrie J, Moore TC, Shackleton NJ (1987) Age dating and the orbital theory of the ice ages: Development of a high-resolution 0 to 300,000-year chronostratigraphy. Quat Res 27:1-29

Masarik J, Beer J (1999) Simulation of particle fluxes and cosmogenic nuclide production in the Earth's atmosphere. J Geophys Res 104:12099-12112

Masarik J, Frank M, Schaffer JM, Wieler R (2001) Correction of *in situ* cosmogenic nuclide production rates for geomagnetic field intensity variations during the past 800,000 years. Geochim Cosmochim Acta 65:2995-3003

Mazaud A (1996) "Sawtooth" variation in magnetic intensity profiles and delayed remanence acquisition in deep sea cores. Earth Planet Sci Lett 139:379-386

Mazaud A, Laj C, Bender M (1994) A geomagnetic chronology for Antarctic ice accumulation. Geophys Res Lett 21:337-340

McCalpin JP (1996) Paleoseismology. Academic Press, San Diego, 588 p

McDermott F, Hawkesworth C J (1991) Th, Pb and Sr isotope variations in young island arc volcanics and ocean sediments. Earth Planet Sci Lett 104:1-15

McFadden PL, Merrill RT (1993) Inhibition and geomagnetic field reversals. J Geophys Res 98:6189-6199

McFadden PL, Merrill RT (1997) Sawtooth paleointensity and reversals of the geomagnetic field. Phys Earth Planet Inter 103:247-252

McHargue LR, Damon PE (1991) The global beryllium-10 cycle. Rev Geophys 29:141-158

McHargue LR, Damon PE, Douglas J (1995) Enhanced cosmic-ray production of ^{10}Be coincident with the Mono Lake and Laschamp geomagnetic excursions. Geophys Res Lett 22:659-662

McHargue LR, Donahue D, Damon PE, Sonett CP, Biddulph D, Burr G (2000) Geomagnetic modulation of the late Pleistocene cosmic-ray flux as determined by ^{10}Be from Blake Outer Ridge marine sediment. Nucl Instr Methods Phys Res B 172:555-561

Merrill RT, McElhinny MW, McFadden PL (1996) The magnetic field of the Earth. In: Paleomagnetism, the core, and the deep mantle. Academic Press, San Diego, p 531

Meynadier L, Valet J-P (1996) Post-depositional realignment of magnetic grains and asymmetrical saw-tooth patterns of magnetization intensity. Earth Planet Sci Lett 140:123-132

Meynadier L, Valet J-P, Weeks R, Shackleton NJ, Hagee VL (1992) Relative geomagnetic paleointensity of the field during the past 140 ka. Earth Planet Sci Lett 114:39-57

Meynadier L, Valet J-P, Bassinot FC, Shackelton NJ, Guyodo Y (1994) Asymmetrical saw-tooth pattern of the geomagnetic field intensity from equatorial sediments in the Pacific and Indian Oceans. Earth Planet Sci Lett 126:109-127

Meynadier L, Valet J-P, Guyodo Y, Richter C (1998) Saw-toothed variations of relative paleointensity and cumulative viscous remanence: testing the records and the model. J Geophys Res 103:7095-7105

Milankovtich M (1941) Kanon der Erdbestrahlung und seine Andwendung auf das Eiszeitenproblem. Belgrade: Königlich Serbische Akademie. Published in English as Canon of Insolation and the Ice-Age Problem (1969). Translation by Israel Program for Scientific Translations, U S Department of Commerce, and U S National Science Foundation, Washington, DC

Monaghan MC, Krishnaswami S, Turekian KK (1985/86) The global-average production rate of ^{10}Be. Earth Planet Sci Lett 76:279-287

Monaghan MC, Klein J, Measures C (1988) The origin of ^{10}Be in island-arc volcanic rocks. Earth Planet Sci Lett 89:288-298

Moore GF, Silver EA (1983) Collision processes in the northern Molucca Sea. *In* Hayes D (ed) The Tectonic and Geologic Evolution of Southeast Asian Seas and Islands: Part 2. Am Geophys Union Monogr 27:360-372

Morris J (1991) Applications of cosmogenic ^{10}Be to problems in the Earth sciences. Ann Rev Earth Planet Sci 19:313-350

Morris JD (1996). The subducted component in Mariana Island arc lavas: Constraints from Pb and Be isotopes and light element systematics. Izu-Bonin-Mariana Workshop, August 1996, Japan. RJ Stern, M Arima, convenors

Morris J, Tera F (1989) ^{10}Be and ^{9}Be in mineral separates and whole rocks from volcanic arcs: Implications for sediment subduction. Geochim Cosmochim Acta 53:3197-3206

Morris JD, Tera F (2000) Beryllium isotope systematics of volcanic arc cross-chains. V M Goldschmidt 2000 J Conf Abstr 5:720

Morris JD, Jezek PA, Hart SR, Gill J (1983) The Halmahera island arc, Molucca Sea collision zone, Indonesia: A geochemical survey. *In* Hayes DE (ed) The Tectonics and Geologic Evolution of Southeast Asian Seas and Islands. Am Geophys Union Monogr 27:373-387

Morris JD, Leeman WP, Tera F (1990) The subducted component in island arc lavas: constraints from Be isotopes and B-Be systematics. Nature 344:31-36

Morris J, Ryan J, Leeman WP (1993) Be isotope and B-Be investigations of the historic eruptions of Mt. Vesuvius. J. Volcanol Geotherm Res 58:345-358

Morris J, Valentine R, Harrison T (2002) ^{10}Be imaging of sediment accretion, subduction and erosion, NE Japan and Costa Rica. Geology 30:59-62

Nagy E, Valet J -P (1993) New advances for paleomagnetic studies of sediment cores using U-channels. Geophys Res Lett 20:671-674

Nakamura E, Moriguti T (1998) Across-arc variation of Li isotopes in lavas and implications for crust/mantle recycling at subduction zones. Earth Planet Sci Lett 163:167-174

Ning S, Aldahan AA, Haiping Y, Possnert G, Königsson L-K (1994) ^{10}Be in continental sediments from North China: Probing into the last 5.4 MA. Quat Geochron 13:127-136

Ninkovich D, Opdyke ND, Heezeh BC, Foster JH (1966) Paleomagnetic stratigraphy, rates of deposition and tephrachronology in North Pacific deep-sea sediments. Earth Planet Sci Lett 1:476-492

Nishiizumi K, Klein J, Middleton R, Craig H (1990) Cosmogenic ^{10}Be, ^{26}Al, and ^{3}He in olivine from Maui lavas. Earth Planet Sci Lett 98:263-266

Nishiizumi, KC, Kohl P, Arnold JR, Klein J, Fink D, Middleton R (1991) Cosmic ray produced ^{10}Be and ^{26}Al in Antarctic rocks: exposure and erosion history: Earth Planet Sci Lett 104:440-454

Nishiizumi K, Kohl CP, Arnold JR, Dorn R, Klein J, Fink D, Middleton R, Lal D (1993) Role of *in situ* cosmogenic nuclides ^{10}Be and ^{26}Al in the study of diverse geomorphic processes. Earth Surf Proc Land 18:407-425

Nyffler UP, Li Y-H, Santschi PH (1984) A kinetic approach to describe trace-element distribution between particles and solution in natural aquatic systems. Geochim Cosmochim Acta 48:1513-1522

O'Brien K (1979) Secular variations in the production of cosmogenic isotopes in the Earth's atmosphere. J Geophys Res 84:423-431

Opdyke ND, Kent DV, Lowrie W (1973) Details of magnetic polarity transitions recorded in a high deposition rate deep-sea core. Earth Planet Sci Lett 20:315-324

Paillard DL, Labeyrie L, Yiou P (1996) Macintosh program performs time-series analysis. EOS Trans Am Geophys Union 77:379

Pazzaglia F, Brandon M (2001) A fluvial record of long-term steady-state uplift and erosion across the Cascadia Forearc High, Western Washington State. Am J Sci 301:385-431

Peck JA, King JW, Colman SM, Kravchinsky VA (1996) An 84-kyr record from the sediments of Lake Baikal, Siberia. J Geophys Res 101(B5):11365-11385

Peltier WR (1998) Postglacial variations in the level of the sea: implications for climate dynamics and solid-Earth geophysics. Rev Geophys 36:603-689

Perg, LA, Anderson RS, Finkel RC (2001) Young ages of the Santa Cruz marine terraces determined using ^{10}Be and ^{26}Al. Geology 29(10):879-882

Peters B (1955) Radioactive beryllium in the atmosphere and on the earth. Proc Indian Acad Sci 41:67-71

Petit JR (and 18 others) (1999) Climate and atmospheric history of the past 420,000 years from the Vostok ice core, Antarctica. Nature 399:420-436

Phillips FM, Zreda MG, Elmore D, Sharma P (1996) A reevaluation of cosmogenic ^{36}Cl production rates in terrestrial rocks. Geophys Res Lett 23:949-952

Phillips FM, Zreda MG, Gosse JC, Klein J, Evenson EB, Hall RD, Chadwick OA, Sharma P (1997) Cosmogenic ^{36}Cl and ^{10}Be ages of Quaternary glacial and fluvial deposits of the Wind River Range, Wyoming. Geol Soc Am Bull 109:1453-1463

Phillips WM, McDonald EV, Reneau SL, Poths J (1998) Dating soils and alluvium with cosmogenic ^{21}Ne depth profiles: case studies from the Pajarito Plateau, New Mexico, USA. Earth Planet Sci Lett 160:209-223

Plank T, Langmuir CH (1993) Tracing trace elements from sediment input to volcanic output at subduction zones. Nature 362:739-743

Plank T, Langmuir CH (1998) The chemical composition of subducting sediment and its consequences for the crust and mantle. Chem Geol 145: 325-394

Plank T, Balzer V, Carr M (submitted) Nicaraguan volcanoes record paleoceanographic changes accompanying closure of the Panama Gateway. Science

Raïs A, Laj C, Surmont J, Gillot PY, Guillou H (1996) Geomagnetic field intensity between 70,000 and 130,000 years B.P. from a volcanic sequence on La Réunion, Indian Ocean. Earth Planet Sci Lett 140:173-189

Raisbeck GM, Yiou F, Fruneau M, Loiseaux JM, Lieuvin M, Ravel JC (1981a) Cosmogenic $^{10}Be/^{7}Be$ as a probe of atmospheric transport process. Geophys Res Lett 8:1015-1018

Raisbeck GM, Yiou F, Fruneau M, Loiseaux JM, Lieuvin M, Ravel JC, Lorius C (1981b) Cosmogenic ^{10}Be concentrations in Antarctic ice during the past 30,000 years. Nature 292:825-826

Raisbeck GM, Yiou F, Zhou SZ (1994) Paleointensity Puzzle. Nature 371:207-208

Raisbeck GM, Yiou F, Bourles D (1985) Evidence for an increase in cosmogenic ^{10}Be during a geomagnetic reversal. Nature 315:315-317

Raisbeck GM, Yiou F, Bourles D, Lorius C, Jouzel J, Barkov NI (1987) Evidence for two intervals of enhanced ^{10}Be deposition in Antarctic ice during the last glacial period. Nature 326:273-277

Rea D, Ruff LJ (1996) Composition and mass flux of sediment entering the world's subduction zones: Impolications for global sediment budgets, great earthquakes and volcanism. Earth Planet Sci Lett 140:1-12

Reagan MK, Morris JD, Herrstrom EA, Murrell MT (1994) Uranium series and beryllium isotope evidence for an extended history of subduction modification of the mantle beneath Nicaragua. Geochim Cosmochim Acta 58:4199-4212

Reheis MC, Sawyer TL (1997) Late Cenozoic history and slip rates of the Fish Lake Valley, Emigrant Peak, and Deep Springs fault zones, Nevada and California. Geol Soc Am Bull 109:280-299

Repka JL, Anderson RS, Finkel RC (1997) Cosmogenic dating of fluvial terraces, Fremont River, Utah. Earth Planet Sci Lett 152:59-73

Reymer A, Schubert G (1984) Phanerozoic addition rates to the continental crust and crustal growth. Tectonics 3:63-77

Robinson C, Raisbeck GM, Yiou F, Lehman B, Laj C (1995) The relationship between ^{10}Be and geomagnetic field strength records in central North Atlantic sediments during the last 80 ka. Earth Planet Sci Lett 136:551-557

Rochester MG, Jacobs JA, Smytie DE, Chong KF (1975) Can precession power the geomagnetic dynamo? Geophys J R Astron Soc 43:661-678

Ryan JG, Langmuir CH (1988) Beryllium sysematics in young volcanic rocks: implications for ^{10}Be. Geochim Cosmochim Acta 52:237-244

Ryan JG, Morris J, Tera F, Leeman WP, Tsvetkov A (1995) Cross-arc geochemical variations in the Kurile arc as a function of slab depth. Science 270:625-627

Ryan J, Morris J, Bebout G, Leeman W (1996) Describing chemical fluxes in subduction zones: Insights from "depth-profiling" studies of arc and forearc rocks. *In* Bebout G, Scholl D, Kirby S, Platt J (eds) Subduction Top to Bottom. Am Geophys Union Monogr 96:263-268

Sagnotti L, Macrí P, Camerlenghi A, Rebessco M (2001) Antarctic environmental magnetism of Antarctic Late Pleistocene sediments and interhemispheric correlation of climatic events. Earth Planet Sci Lett 192:65-80

Schaffer J (1998) Comment to Small, E. E. and Anderson, R.S.: Pleistocene relief production in Laramide Mountain Ranges, western U.S. Geology 26:121-123

Schwartz M, Lund SP, Johnson TC (1996) Environmental factors as complicating influences in the recovery of quantitative geomagnetic field paleointensity estimates from sediments. Geophys Res Lett 23:2693-2696

Shea MA, Smart DF (1992) Recent and historical solar proton events. Radiocarbon 34:255-262

Sheppard MK, Arvidson RE, Caffee M, Finkel R, Harris L (1995) Cosmogenic exposure ages of basalt flows: Lunar Crater volcanic field, Nevada. Geology 23:21-24

Shimaoka AK (1999) Be Isotopic Ratios in Island-Arc Volcanic Rocks from the North-East Japan: Implications for Incorporation of Oceanic Sediments into Island-Arc Magma. PhD dissertation, The University of Tokyo

Shimaoka AK, Imamura M, Kaneoka I (in press) Investigation of acid leaching conditions for obtaining the primary ^{10}Be signatures in volcanic materials. Chem Geol

Siame LL, Bourlés DL, Sébrier M, Bellier O, Castano JC, Araujo M, Perez M, Raisbeck GM, Yiou F (1997) Cosmogenic dating ranging from 20 to 700 ka of a series of alluvial fan surfaces affected by the El Tigre fault, Argentina. Geology 25:975-978

Siegenthaler U, Heimann M, Oeschger H (1980) ^{14}C variations caused by changes in the global carbon cycle. Radiocarbon 22:177-191

Sigmarsson O, Condomines M, Morris JD, Harmon RS (1990) Uranium and ^{10}Be enrichments by fluids in Andean arc magmas. Nature 346:163-165

Sigmarsson O, Chmeleff J, Morris J, Lopez-Escobar L (2002) Rapid magma transfer from slab derived ^{226}Ra-^{230}Th disequilibria in lavas from southern Chile. Earth Planet Sci Lett 196:189-196

Silver EA, Kastner M, Fisher A, Morris J, McIntosh K, Saffer D (2000) Fluid flow paths in the Middle America Trench and Costa Rica margin. Geology 28:679-682

Sissons JB, Cornish R (1982) Differential glacio-isostatic uplift of crustal blocks at Glen Roy, Scotland. Quat Res 18:268-288

Slemmons DB (1997) Carrara Fault, in southern Nevada from paleoseismic, geologic and geophysical evidence: implications to the earthquake hazards and tectonics near Yucca Mountain, Nevada. EOS Trans Am Geophys Union 78:F453

Small E, Anderson RS (1998) Pleistocene relief production in Laramide mountain ranges, western United States. Geology 26:123-126

Small EE, Anderson RS, Repka JL, Finkel R (1997) Erosion rates of alpine bedrock summit surfaces deduced from *in situ* ^{10}Be and ^{26}Al. Earth Planet Sci Lett 150:413-425

Somayajulu BLK (1977) Analysis of causes for the beryllium-10 variations in deep sea sediments. Geochim Cosmochim Acta 41:909-913

Sonett CP, Seuss HE (1984) Correlation of bristlecone pine ring widths with atmospheric ^{14}C variations: A sun-climate relation. Nature 307:141-143

Sonett CP, Morfill GE, Jokipii RR (1987) Interstellar shock waves and ^{10}Be from ice cores. Nature 330:458-460

Spiegelman M, Elliott T, (1993) Consequences of melt transport for uranium series disequilibrium in young lavas. Earth Planet Sci Lett 118:1-20

Spies CS, Whitney JW, Gosse J, Slemmons DB, Caffee M (2000) Terrestrial cosmogenic nuclide dating of deformed alluvium along the Carrara Fault in the northern Amargosa Desert, Nye County, Nevada. Geol Soc Am Abstr Progr 32(7):A-166

Spies CS, Whitney JW, Gosse J, Slemmons DB, Caffee M (submitted) Tectonic geomorphology and evolution of an active Walker Lane transpressional ridge feature. Geomorphology

Stamatakos J, Connor CB, Hill BE, Lane Magsino S, Ferrill DA (1997) The Carrara Fault in southwestern Nevada revealed from detailed gravity and magnetic results: implications for seismicity, volcanism, and tectonics near Yucca Mountain, Nevada. EOS Trans Am Geophys Union 78:F453

Steig EJ, Polissar PJ, Stuiver M, Grootes PM, Finkel, RC (1996) Large amplitude solar modulation cycles of ^{10}Be in Antarctica: implications for atmospheric mixing processes and interpretation of the ice core record. Geophys Res Lett 23:523-526

Steig EJ, Morse DL, Waddington ED, Polissar, PJ (1998) Using the sunspot cycles to date ice cores. Geophys Res Lett 25:163-166

Steig EJ, Morse DL, Waddington ED, Stuiver M, Grootes PM, Mayewski PA, Twickler MS, Whtilow SI (2000) Wisconsin and Holocene climate history from an ice core at Taylor Dome, western Ross embayment, Antarctica. Geografiska Annaler 82A:213-235

Stone JO (1999) A consistent Be-10 production rate in quartz—muons and altitude scaling. AMS-8 Proc Abstr Vol, Vienna, Austria

Stone JO (2000) Air pressure and cosmogenic isotope production. J Geophys Res B105 10:23753-23759

Stoner JS, Channell JET, Hillaire-Marcel C (1998) A 200 ka geomagnetic chronostratigraphy for the Labrador Sea: Indirect correlation of the sediment record to SPECMAP. Earth Planet Sci Lett 159: 165-181

Stoner JS, Channell JET, Hillaire-Marcel C, Kissel C (2000) Geomagnetic paleointensity and environmental record from Labrador Sea core MD95-2024: global marine sediment and ice core chronostratigraphy for the last 110 kyr. Earth Planet Sci Lett 183:161-177

Stoner JS, Laj C, Channell JET, Kissel C (2002) South Atlantic and North Atlantic geomagnetic paleointensity stacks (0-80 ka): implications for inter-hemispheric correlation. Quat Sci Rev 21: 1141-1151

Størmer C (1955) On the trajectories of electric particles in the field of a magnetic dipole with applications to the theory of cosmic radiation. Astrophysica Norvegica 1:115-167

Stuiver M (1961) Variations in radiocarbon concentration and sunspot activity. J Geophys Res 66:273-276

Stuiver M, Braziunas TF (1989) Atmospheric ^{14}C and century-scale solar oscillations. Nature 338:405-408

Stuiver M, Braziunas TF (1993) Sun, ocean, climate and atmospheric $^{14}CO_2$, an evaluation of causal and spectral relationships. The Holocene 3:289-305

Stuiver M, Quay P (1980) Changes in atmospheric carbon-14 attributed to a variable sun. Science 207: 11-19

Suess HE, Linick TW (1990) The ^{14}C record in bristlecone pine wood of the past 8000 years based on the dendrochronology of the late C.W. Ferguson. Phil Trans R Soc London A 330:403-412

Taira A, Hill I (1991) Proc. Ocean Drilling Program, Initial Reports 131. College Station, Texas

Tanaka S, Inoue T (1979) ^{10}Be dating of North Pacific sediment cores up to 2.5 million years B.P. Earth Planet Sci Lett 49:34-38

Tarduno JA, Wilkison SL (1996) Non-steady state magnetic mineral reduction, chemical lock-in and delayed remanence acquisition in pelagic sediments. Earth Planet Sci Lett 144:315-326

Tatsumi Y, Eggins S (1997) Subduction Zone Magmatism. Blackwell Science, Oxford, UK, 211 p

Tatsumi Y, Hamilton DL, Nesbitt RW (1986) Chemical characteristics of fluid phase released from a subducted lithosphere and origin of arc magmas: evidence from high-pressure experiments and natural rocks. J Volcanol Geothermal Res 29:293-309

Tauxe L (1993) Sedimentary records of relative paleointensity and the geomagnetic field: theory and practice. Rev Geophys 31:319-354

Tauxe L, Wu G (1990) Normalized remanence in sediments of the western equatorial Pacific: Relative paleointensity of the geomagnetic field? J Geophys Res 95:12337-12350

Tauxe L, LaBrecque JL, Dodson R, Fuller M (1983) U-channels—a new technique for paleomagnetic analysis of hydraulic piston cores. EOS Trans Am Geophys Union 64:219

Taylor EM, plus 12 others (1998) Quaternary Geology of the Yucca Mountain Area, Southern Nevada. *In* 'Friends of the Pleistocene' Pacific Cell, 1998 Ann Mtg, p 223

Tera F, Brown L, Morris J, Sacks IS, Klein J, Middleton R (1986) Sediment incorporation in island-arc magmas: inferences from ^{10}Be. Geochim Cosmochim Acta 50 :535-550

Tera F, Morris J D, Ryan J, Leeman WP, Tsvetkov A (1993) Significance of $^{10}Be/^{9}Be$-B correlation in lavas of the Kurile-Kamchatka Arc. EOS Trans Am Geophys Union 74:674

Thompson LG, Peel DA, Mosley-Thompson E, Mulvaney R, Dai J, Lin PN, Davis ME, Raymond CF (1994) Climate since AD 1510 on Dyer Plateau, Antarctic Peninsula: Evidence for recent climate change. Ann Glaciol 20:420-426

Thompson LG, Mosley-Thompson E, Henderson KA (2000a) Ice core palaeoclimate records in tropical South America since the Last Glacial Maximum. J Quat Sci 15:377-394

Thompson LG, Yao T, Mosley-Thompson E, Davis ME, Henderson KA, Lin PN (2000b) A high resolution millennial record of the South Asian monsoon from Himalayan ice cores. Science 289:1916-1919

Tric E, Valet J-P, Tucholka P, Paterne M, Labeyrie L, Guichard F, Tauxe L, Fontugne M (1992) Paleointensity of the geomagnetic field during he last 80,000 years. J Geophys Res 97:9337-9351

Tric E, Valet J-P, Gillot PY, Lemeur I (1994) Absolute paleointensities between 60 and 160 kyrs B.P. from Mount Etna, Sicily. Phys Earth Planet Inter 85:113-129

Trull TW, Brown T, Marty B, Raisbeck GM, Yiou F (1995) Cosmogenic ^{10}Be and ^{3}He accumulation in Pleistocene beach terraces in Death Valley, California, USA: Implications for cosmic-ray exposure dating of young surfaces in hot climates. Chem Geol 119:191-207

Turner S, Hawkesworth C (1997) Constraints on flux rates and mantle dynamics beneath island arcs from Tonga-Kermadec lava geochemistry. Nature 389:568-573

Turner S, Hawkesworth C (1998) Using geochemistry to map mantle flow beneath the Lau Basin: Geology 26:1019-1022

Turner S, Hawkesworth C, Rogers N, Barlett J, Worthington T, Hergt J, Pearce J, Smith I (1997) ^{238}U-^{230}Th disequilibria, magma petrogenesis and flux rates beneath the depleted Tonga-Kermadec island arc. Geochim Cosmochim Acta 61:4855-4884

Turner S, Bourdon B, Hawkesworth C, Evans P (2000) ^{226}Ra-^{230}Th evidence for multiple dehydration events, rapid melt ascent and the time scales of differentiation beneath the Tonga-Kermadec island arc. Earth Planet Sci Lett 179:581-593

Valentine R, Morris J, Duncan D (1997) Sediment subduction, accretion, underplating, and arc volcanism along the margin of Costa Rica: Constraints from Ba, Zn, Ni and ^{10}Be concentrations. EOS Trans Am Geophys Union 78:673

Valentine R, Morris J (submitted a) Sediment accretion, erosion and subduction along the Costa Rica convergent margin: Constraints from geochemical imaging using ^{10}Be. J Geophys Res

Valentine R, Morris J (submitted b) ^{10}Be estimates of sediment subduction in the Izu-Mariana volcanic arc system. Geochem Geophys Geosystems

Valet J-P, Meynadier L (1993) Geomagnetic field intensity and reversals during the past four million years. Nature 366:234-238

Valet J-P, Meynadier L (2001) Comment on "A relative geomagnetic paleointensity stack from Ontong-Java plateau sediments for the Matuyama" by YS Kok and L Tauxe, J Geophys Res 106:11013-11015

Valet J-P, Herrero-Bervera E, Lockwood JP, Meynadier L, Tric E (1998) Absolute paleointensity from Hawaiian lavas younger than 35 ka. Earth Planet Sci Lett 161:19-32

Van der Woerd J, Ryerson FJ, Tapponnier P, Gaudemer Y, Finkel R, Meriaux AS, Caffee M, Guaguang Z, Qunlu H (1998) Holocene left-slip rate determined by cosmogenic surface dating on the Xidatan segment of the Kunlun fault (Qinghai, China). Geology 26:695-698

Van Der Woerd J, Tapponnier P, Ryerson FJ, et al. (2002) Uniform postglacial slip-rate along the central 600 km of the Kunlun Fault (Tibet), from Al-26, Be-10 and C-14 dating of riser offsets, and climatic origin of the regional morphology. Geophys J Intl 148:356-388

Vannuchi P, Scholl DW, Meschede M, McDougall-Reid K (2001) Tectonic erosion and consequent collapse of the Pacific margin of Costa Rica: Combined implications from ODP Leg 170, seismic offshore data and regional geology of the Nicoya Peninsula. Tectonics 20:649-688

Verosub KL, Herrero-Bervera E, Roberts AP (1996) Relative geomagnetic paleointensity across the Jaramillo subchron and the Matuyama/Brunhes boundary. Geophys Res Lett 23:467-470

Voelker A, Sarnthein M, Grootes PM, Erlenkeuser H, Laj C, Mazaud A, Nadeeau MJ, Schleicher M (1998) Correlation of marine ^{14}C ages from the Nordic Sea with GISP2 isotope record: implication for ^{14}C calibration beyond 25 ka BP. Radiocarbon 40:517-534

Vogt S, Herzog GF, Reedy RC (1990) Cosmogenic nuclides in extraterrestrial materials. Rev Geophys 28:253-275

von Huene R, Scholl DW (1991) Observations at convergent margins concerning sediment subduction, subduction erosion and the growth of continental crust. Rev Geophys 29: 279-316

von Huene R, Klaeschen D, Cropp B, Miller J (1994) Tectonic structure across the accretionary and erosional parts of the Japan Trench Margin: J Geophys Res 99:22349-22361

Wagner G, Masarik J, Beer J, Baumgartner S, Imboden D, Kubik PW, Synal H-A, Suter M (2000a) Reconstruction of the geomagnetic field between 20 and 60 kyr BP from cosmogenic radionuclides in the GRIP ice core. Nucl Instr Meth Phys Rev B 172:597-604

Wagner G, Beer J, Laj C, Kissel C, Masarik J, Muscheler R, Synal H-A (2000b) Chlorine-36 evidence for the Mono Lake event in the Summit GRIP ice core. Earth Planet Sci Lett 181:1-6

Wagner G, Beer J, Masarik R, Muscheler R, Kubik PW, Mende W, Laj C, Raisbeck GM, Yiou F (2001) Presence of the solar de Vries cycle (205 yr) during the last ice age. Geophys Res Lett 28:303-306

Weeks R, Laj C, Endignoux L, Fuller M, Roberts A, Manganne R, Blanchard E, Goree W (1993) Improvements in long-core measurement techniques: applications in palaeomagnetism and palaeoceanography. Geophys J Intl 114:651-662

Williams T, Thouveny TN, Creer KM (1998) A normalised intensity record from Lac du Bouchet: geomagnetic palaeointensity for the last 300 kyr. Earth Planet Sci Lett 156:33-46

Yamazaki T (1999) Relative paleointensity of the geomagnetic field during Brunhes Chron recorded in North Pacific deep-sea sediment cores: orbital influence? Earth Planet Sci Lett 169:23-35

Yamazaki T, Oda H (2002) Orbital influence on Earth's magnetic field: 100,000-year periodicity in inclination. Nature 295:2435-2438

Yamazaki T, Ioka N, Eguchi N (1995) Relative paleointensity of the geomagnetic field during the Brunhes chron. Earth Planet Sci Lett 136:525-540

Yiou F, Raisbeck GM (1972) Half-life of ^{10}Be. Phys Rev Lett 29:372-375

Yiou F, Raisbeck GM, Bourles D, Lorius C, Barkov NI (1985) ^{10}Be in ice at Vostok Antarctica during the last climatic cycle. Nature 316:616-617

Yiou F, Raisbeck GM, Baumgartner S, Beer J, Hammer C, Johnsen S, Jouzel J, Kubik, PW, Lestringuez J, Stiévenard M, Suter M, Yiou P (1997) Beryllium-10 in the Greenland Ice Core Project ice core at Summit, Greenland. J Geophys Res 102:26783-26794

You C-F, Morris JD, Geiskes JM, Rosenbauer R, Zheng SH, Xu X, Ku TL, Bischoff JL (1994) Mobilization of beryllium in the sedimentary column at convergent margins. Geochim Cosmochim Acta 58:4887-4897

You C-F, Castillo P, Geiskes JM, Chan LC, Spivack AF (1996) Earth Planet Sci Lett 140:41-52

Zehfuss PH, Bierman PR, Gillespie AR, Burke RM, Caffee MW (2001) Slip rates on the Fish Springs fault, Owens Valley, California, deduced from cosmogenic ^{10}Be and ^{26}Al and soil development on fan surfaces. Geol Soc Am Bull 113:241-255

Zentmire KN, Gosse JC, Baker C, McDonald E, Wells S (1999) The problem of inheritance when dating alluvial fans and terraces with TCN: Insight from the Matanuska Glacier. Geol Soc Am Abstr Progr 31(5):A81

Zhang Z, Schwartz S (1992) Depth distribution of moment release in underthrusting earthquakes at subduction zones. J Geophys Res 97:537-544

Zheng S, Morris J, Tera F, Klein J, Middleton R (1994) Beryllium isotopic investigation of sedimentary columns outboard of subduction zones. ICOG 1994 Abstr 8:366

Zreda M, Noller JS (1998) Ages of prehistoric earthquakes revealed by cosmogenic chlorine-36 in a bedrock fault scarp at Hebgen Lake. Science 292:1097-1099

Zreda MG, Phillips FM, Kubik PW, Sharma P, Elmore D (1993) Cosmogenic ^{36}Cl dating of a young basaltic eruption complex, Lathrop Wells, Nevada. Geology 21:57-60

6 Environmental Chemistry of Beryllium-7

James M. Kaste

Department of Earth Sciences
Dartmouth College
Hanover, New Hampshire 03755
james.m.kaste@dartmouth.edu

Stephen A. Norton

Department of Geological Sciences
University of Maine
Orono, Maine 04469

Charles T. Hess

Department of Physics and Astronomy
University of Maine
Orono, Maine 044699

INTRODUCTION

In addition to the stable isotope ^{9}Be, Be is also formed as two cosmogenic isotopes of interest to earth scientists. Cosmogenic Be is formed primarily in the stratosphere from cosmic-ray spallation of oxygen and nitrogen, but some is produced in the troposphere and *in situ* on the surface of the earth. Production of cosmogenic Be was hypothesized by Peters (1955). Soon after, naturally occurring ^{7}Be was identified in precipitation by Arnold and Al-Salih (1955) and essentially concurrently and independently by Goel et al. (1956). ^{10}Be ($T_{1/2}$ = 1.5×10^{6} yr) was first detected by Arnold (1956) in marine sediment cores. The ^{10}Be / ^{7}Be production ratio is approximately 0.5 (Lal and Peters 1967).

After its formation in the atmosphere, cosmogenic Be adsorbs electrostatically to aerosols that may be washed out by precipitation and delivered to ecosystems. The amount of cosmogenic Be that reaches the surface of the earth is a function of production rate (cosmic-ray intensity), stratosphere-troposphere mixing, circulation and advection within the troposphere, and efficiency of removal from the troposphere (wet and dry deposition) (Feely et al. 1989). Since the discovery of cosmic-ray produced Be, numerous researchers have gathered data on its (1) production, (2) fluxes to the oceans and terrestrial ecosystems, (3) distribution and inventory in soils, snow, sediments, and vegetation, and (4) geochemical behavior. In 1955, Peters suggested that the relatively long half-life of ^{10}Be could make it useful for quantifying Tertiary sedimentation rates and other surficial processes. The focus of this chapter is the short-lived radionuclide ^{7}Be ($T_{1/2}$ = 53.12±0.07 days; Jaeger et al. 1996). Due to its short half-life, relative ease of measurement, and well-defined source term, ^{7}Be serves as a useful tool for tracing and quantifying environmental processes on the <1 year timescale (Lal et al. 1958; Young and Silker 1980; Krishnaswami et al. 1980; Russell et al. 1981; Turekian et al. 1983; Dutkiewicz and Husain 1985; Olsen et al. 1985; Wallbrink and Murray 1993; Bonniwell et al. 1999), and has applications in meteorology, soil science, sedimentology, geomorphology, hydrology, geochemistry, and nuclear physics.

ANALYSIS FOR ^{7}BE

^{7}Be decays to stable ^{7}Li by electron capture (λ = 4.766 yr^{-1}). Eighty-nine and one-half percent of the atoms decay directly to the ground state of ^{7}Li, while 10.5% decay first to the excited state of ^{7}Li, which decays to ground state ^{7}Li via gamma-ray emission at 477.6 keV (Ajzenberg-Selove 1988). ^{7}Be activities are normally determined in

1529-6466/00/0050-0006$05.00

environmental samples using gamma spectrometers that detect the 477.6 keV gamma. In earlier investigations, ^{7}Be was detected using thallium (Tl) activated sodium-iodide (NaI) detectors (Arnold and Al-Salih 1955; Goel et al. 1956; Walton and Fried 1962; Schumann and Stoeppler 1963, and Krishnaswami et al. 1980). However, the relatively low resolution of these instruments requires chemical separation of the Be before the gamma analysis because the NaI(Tl) detector cannot distinguish the ^{7}Be photopeak from other radionuclides which decay in the same energy region (^{228}Ac at 462 keV; ^{103}Ru at 497 keV). The techniques used for the purification and concentration of ^{7}Be from various samples generally utilize cation exchange columns and/or coprecipitation of Be with Fe hydroxides and other hydroxide phases. Details of these methods are described in the references mentioned above.

More recently, high-purity germanium (Ge) detectors or lithium (Li)-drifted Ge detectors have been employed to determine ^{7}Be activity. These detectors generally have high resolution (<2 keV) and high counting efficiencies. In many cases, precipitation, soil, and sediment samples can be analyzed directly for ^{7}Be. Larson and Cutshall (1981) describe a method for detecting ^{7}Be directly in sediment samples using Marinelli Beakers. They reported a detection limit of approximately 0.015 Bq g^{-1} for a 250 g sediment sample counted 200 min (1 Bq = 1 becquerel = 1 disintegration per second, the SI unit for radioactivity. 1 Bq = 27.03 pCi; no concentration equivalent such as ng kg^{-1} or pg kg^{-1} would be applicable due to the very small number of atoms present in environmental samples). When counting samples high in ^{232}Th, one should be cautious of the 478.3 keV decay energy of ^{228}Ac. Although the relative intensity of this photon is only 0.215%, high activities of ^{228}Ac found in mineral soil and sediment samples can create a disturbance in this region of the spectrum. Surface waters are generally low in ^{7}Be, so pre-concentration is necessary for efficient detection. Kostadinov et al. (1988) suggested ^{7}Be adsorption on activated charcoal to concentrate ^{7}Be from surface waters.

For over five decades, ^{7}Be has received attention from scientists investigating the possible variability in a radioactive nuclide's decay constant. Emilio Segrè (1947) first suggested that the decay constant of a nuclide that decays by electron capture should be influenced by the density of its electrons. He suggested that, for a light element such as ^{7}Be, this change could be "appreciable", and by putting ^{7}Be in various chemical compounds, a slight change in half-life could potentially be observed. A few years later, Leininger et al. (1949; 1951) reported that $[\Delta\lambda/\lambda] \times 10^{-3}$ for $\lambda(^{7}Be) - \lambda(^{7}BeF_2)_{hex} = 0.84 \pm 0.10$. Between 1950 and 1970, several researchers reported a deviation of λ for ^{7}Be in different Be compounds (Table 1). These reports received little attention from the scientific community because the differences in λ were detected using NaI detectors or ionization chambers. The analytical error of these instruments approached the magnitude of changes in λ they were trying to resolve. In 1999, using a high purity intrinsic Ge detector, Huh (1999) reported that the half-lives of ^{7}Be in $Be(OH_2)^{+2}$, $Be(OH)_2$, and BeO are 53.69, 53.42, and 54.23 d (±0.1%), respectively. Changes in the chemical form of ^{7}Be may not be the only variable affecting its decay constant. Hensley et al. (1973) found that the decay constant of ^{7}Be in BeO increased slightly under high pressure in a diamond-anvil press. Liu and Huh (2000) found similar results as Hensley et al. (1973), but they noted that the rate of increase of λ with pressure in a $Be(OH)_2$ gel decreased with increasing pressure.

PRODUCTION AND DELIVERY OF ^{7}BE TO THE EARTH'S SURFACE

Production of ^{7}Be depends on the cosmic-ray flux, which varies with latitude, altitude, and solar activity. Lal et al. (1958) calculated production rates for cosmogenic Be. Production in the troposphere increases by a factor of approximately 1 to 3 from the equator to the poles. In the stratosphere, production can increase by a factor of 4 to 5 from the equator to the poles. Production is greatest between 12 and 20 km altitude (depending on latitude), decreasing nearly exponentially by about 3 orders of magnitude to the earth's surface (Lal et al. 1958; Bhandari et al. 1970). Cosmogenic Be production varies with the 11-year solar cycle. Solar activity maximums result in increased deflection of cosmic rays from the solar system (Lal and Peters, 1967) that decreases the cosmic-ray flux to the earth, and thus decreases ^{7}Be production. Several authors have demonstrated an inverse relationship between cosmogenic Be concentrations in air and on the surface of the earth with solar activity (Fig. 1) (Beer et al. 1990; Hötzl et al. 1991; Ďurana et al. 1996). Concentration fluctuations in surface air resulting from the 11-year solar cycles are generally on the order of 15 to 25% (Koch and Mann 1996) (Fig. 1).

Table 1. Variation in the decay constant of ^{7}Be in various chemical compounds. All values are $\Delta\lambda/\lambda \times 10^{-3}$. Modified from Johlige et al. (1970).

Source pair	*Leininger et al. (1949); Segrè & Wiegand (1951)*	*Kraushaar et al. (1953)*	*Bouchez et al. (1956)*	*Johlige et al. (1970)*	*Huh (1999)*
$\lambda(BeO)-\lambda(BeF_2)_{hex}$	0.69±0.03	0.609±0.055			
$\lambda(BeO)-\lambda(BeF_2)_{am}$				1.130±0.058	
$\lambda(Be)-\lambda(BeF_2)_{hex}$	0.84±0.10	0.741±0.047			
$\lambda(Be)-\lambda(BeF_2)_{am}$			1.2±0.1		
$\lambda(BeO)-\lambda(Be)$	-0.15±0.09	-0.131±0.051			
$\lambda(BeO)-\lambda(Be(OH)_2)$					-15.177±0.1

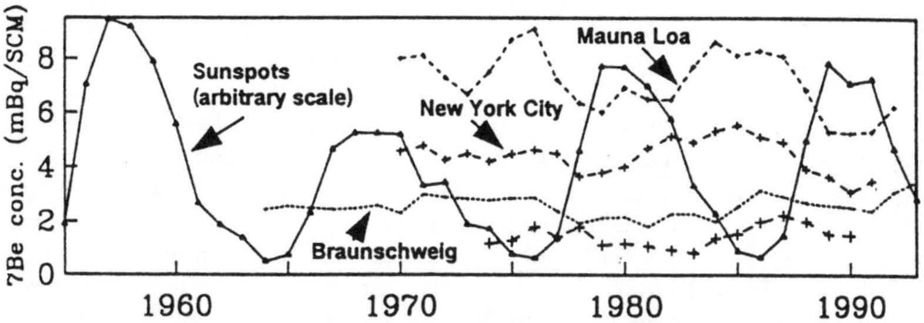

Figure 1. Annual variation of ^{7}Be in surface air plotted with solar activity, 1955-1993. ^{7}Be units are millibecquerels m^{-3} at standard temperature and pressure. Figure used by permission of the editor of *Journal of Geophysical Research*, from Koch et al. (1996), Fig. 2, p. 18,653.

Because the stratospheric residence time of aerosols (approximately 14 months [Reiter, 1975]) exceeds the half-life of ^{7}Be by more than six-fold, equilibrium activity is assumed to occur. In the troposphere, however, production of ^{7}Be is significantly lower, and the residence time of ^{7}Be is much shorter (22 to 48 days [Bleichrodt 1978; Ďurana et al. 1996] due to rapid wash-out. This results in a fairly high concentration gradient between the stratosphere and the troposphere, with tropospheric air generally containing one to two orders of magnitude less ^{7}Be (Bq m^{-3}) than the stratosphere. Rama and Honda (1961) and Bhandari and Rama (1963) reported ^{7}Be activity for air sampled at various points in the atmosphere. Stratosphere air sampled at 18-20 km ranged from approximately 0.16 to 0.58 Bq ^{7}Be m^{-3}. Troposphere air ranged from approximately 0.005 to 0.02 Bq m^{-3}. Dutkiewicz and Husain (1985) found that ^{7}Be activity in lower stratospheric air averaged approximately 0.17 Bq m^{-3}, while the upper troposphere averaged 0.02 Bq m^{-3}.

While the activity of ^7Be in the stratosphere remains fairly constant, concentrations in the troposphere and near-surface air, and thus the amount of ^7Be available to ecosystems exhibit seasonal fluctuations (Feely et al. 1989). ^7Be concentrations in near-surface air generally range from 0.001 to 0.007 Bq m^{-3} (Feely et al. 1989; Dueñas et al. 1999). Stratosphere-troposphere exchange can increase ^7Be concentrations in the troposphere and near-surface air. Husain et al. (1977) correlated high concentrations of ^7Be at Whiteface Mountain, New York, USA with stratospheric air masses as indicated by potential vorticities. Viezee and Singh (1980) correlated high concentrations of ^7Be in surface air with low-pressure troughs. Maximum mixing between the stratosphere and the troposphere generally occurs in the spring of each year at mid-latitudes, and near-surface air at the middle latitudes generally has higher concentrations of ^7Be at this time (Feely et al. 1989). Intense thunderstorms may also mix stratospheric air downward, thus increasing the amount of ^7Be available for scavenging by precipitation (Dingle 1965; Noyce et al. 1971). Several authors have suggested that vertical mixing within the troposphere also governs the concentration of ^7Be in rain and near-earth surface air (Feely et al. 1989; Baskaran 1995; Ďurana et al. 1996; Dueñas et al. 1999). Warming of the earth's surface during the spring and summer will increase convection, which would transport ^7Be from the upper troposphere to near-surface air. At latitudes >60°, seasonal variations in surface air result from the transport poleward of mid-latitude air. Arctic sites commonly have peak ^7Be air concentrations in the late winter or early spring, probably a result of the arrival of mid-latitude tropospheric air high in ^7Be (Feely et al. 1989). Sites with a pronounced seasonal variation in rainfall amount show an inverse relationship between rainfall rate and ^7Be concentration in air, which demonstrates that washout of ^7Be can have a significant impact on surface air concentration (Feely et al. 1989). Koch and Mann (1996) used a singular value decomposition (SCD) technique to analyze the spatial and temporal variability of near-surface air concentrations of ^7Be. They found that up to 47% of the monthly air concentration variability could be explained by 4 "modes" which correspond to the 11-year solar cycles, El Niño Southern Oscillation cycles (2 to 7 years), annual, and semi-annual cycles. Zanis et al. (1999) measured ^7Be at an alpine research station in Switzerland from 4 April 1996 to 1 January 1997, and compared this with meteorological parameters and composite 500 hPa geopotential height maps. They concluded that downward transport of air from the upper troposphere and wet scavenging of aerosols were main processes controlling the ^7Be concentrations measured at the site. Papastefanou (1991) found that a supernova explosion in 1987 increased the ^7Be activity in air by up to four times the normal activity found in temperate air masses. Nuclear detonations do not contribute to ^7Be production.

Arnold and Al-Salih (1955) were the first to report cosmic-ray-produced ^7Be in wet precipitation. They collected large volumes of rain and snow in Indiana and Illinois from October 1953 until April 1954. ^7Be was concentrated from the samples by adding a stable Be carrier and co-precipitating $Be(OH)_2$ along with $Fe(OH)_3$ and other hydroxides at a pH of 9. The precipitate was then purified to contain only BeO and counted using a 2.54 cm NaI (Tl) crystal scintillation spectrometer with an overall counting efficiency of 0.99%. ^7Be activities ranged from 0.16±17% to 2.40±8% Bq L^{-1}, with a volume-weighted mean of approximately 0.95 Bq L^{-1}. While the Arnold and Al-Salih (1955) paper was in press, Goel et al. (1956) independently demonstrated the presence of ^7Be in Bombay, India. During the months of July, August, and September of 1955, they documented ^7Be in monsoon rain with a volume-weighted mean activity of 0.36 Bq L^{-1}. Since then, numerous researchers have reported the activity of ^7Be in wet-precipitation, with annual volume-weighted means generally ranging from 0.75 to 2.75 Bq L^{-1}. Baskaran et al. (1993) tabulated the volume-weighted ^7Be activities in precipitation reported between 1983 and 1993. Some studies prior to 1980 did not acidify their precipitation samples during collec-

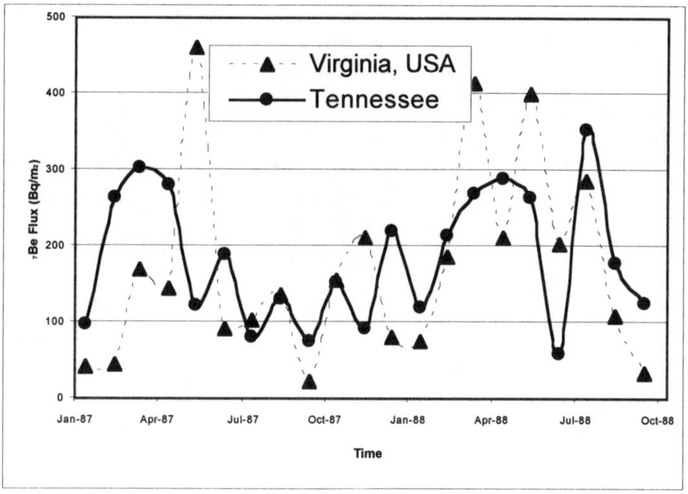

Figure 2. Seasonal variation of ^{7}Be flux at two sites. Data plotted from Olsen et al. (1985).

tion, which may have resulted in poor recovery of the ^{7}Be from the precipitation collectors due to adsorption.

As with tropospheric air, the concentration and flux of ^{7}Be in rainfall varies seasonally as well (Fig. 2). Maximum fluxes and ecosystem inventories are often recorded in the spring to early summer (Olsen et al. 1985; Brown et al. 1989; Dibb 1989; Bachuber and Bunzl 1992; Baskaran 1995). Brown et al. (1989) and Harvey and Matthews (1989) determined that ^{7}Be activities in rainfall can vary by a factor of greater than 20 between individual rainstorms. This suggests that individual storms can potentially dominate ^{7}Be deposition. Therefore, to obtain an accurate flux of ^{7}Be to the earth's surface, samples must be collected from every storm for at least 1 full year. For investigators who have sampled at least 12 months and acidified their samples in the collectors, annual atmospheric flux of ^{7}Be observed from bulk atmospheric deposition collectors has ranged from approximately 1000 to 6500 Bq m^{-2} (Turekian et al. 1983; Olsen et al. 1985; Dominik et al. 1987; Schuler et al. 1991; Wallbrink and Murray 1994). Vogler et al. (1996) reported an annual flux at Lake Constance, Germany of 2250 Bq m^{-2}. Deposition was highest during the summer and related strongly to rainfall (mm d^{-1}). Harvey and Matthews (1989) measured ^{7}Be for a year on the Hokitika, South Island, New Zealand. The annual precipitation was about 2.8 m. Wet deposition accounted for about 97% of the total deposition that was 6350 Bq m^{-2}, and the ^{7}Be flux was proportional to the precipitation volume. Dry deposition apparently contributes <10% of ^{7}Be inventories (Brown et al. 1989; Harvey and Matthews (1989); Todd and Wong (1989); Walbrink and Murray 1994; Benitez-Nelson and Buesseler 1999; Kaste 1999, Kaste et al. 1999).

^{7}Be is delivered to ecosystems primarily as Be^{+2} in slightly acidic (pH<6) rainfall. The Be^{+2} ion is extremely competitive for cation exchange sites because of its high charge density. As $^{7}Be^{+2}$ comes in contact with soils and vegetation, it is rapidly sequestered by exchange surfaces. Forest canopies may decrease the amount of ^{7}Be that reaches soils and streams, in contrast to the increases observed for most other analytes (Kaste et al. 1999). After deposition, most ^{7}Be decays in the watershed, but some may be exported in particulate or dissolved form.

Table 2. Terrestrial inventories of ^{7}Be.

Collection date	*Location*	*Approx latitude*	*Substrate analyzed (n)*	*Total ^{7}Be inventory Bq m^{-2}*	*Source*
7/82	Delaware, USA	39°N	marsh + overlying grass	207±13%	Olsen et al. (1985)
7/84	Oak Ridge, TN, USA	36°N	soil + overlying grass	673±3%	Olsen et al. (1985)
1/85	Wallops Island, VA, USA	38°N	marsh + overlying grass	673±7%	Olsen et al. (1985)
1/85	Wallops Island, VA, USA	38°N	unvegetated marsh	107±17%	Olsen et al. (1985)
9/88	Black Mt, Australia	35°S	soil + overlying grass (3)	200±10%	Wallbrink & Murray (1996)
9/88	Black Mt, Australia	35°S	alluvial bare soil (3)	130±20%	Wallbrink & Murray (1996)
5/89	Black Mt, Australia	35°S	soil + overlying grass (3)	400±10%	Wallbrink & Murray (1996)
5/89	Black Mt, Australia	35°S	alluvial bare soil (3)	155±15%	Wallbrink & Murray (1996)
5/96	Idaho, USA	44°N	soil and overlying vegetation (6)	139±16%	Bonniwell et al. (1999)
6/-8/98	Maine, USA	45°N	forested soil (8)	165±40%	Kaste (1999)
12/98	Maine, USA	45°N	bog core + overlying vegetation	554±26%	Kaste (1999)

^{7}BE DISTRIBUTION IN VEGETATION AND SOILS

^{7}Be surface inventories (activity of ^{7}Be per unit area regardless of its vertical distribution) generally range from 100 to 700 Bq m^{-2}. Table 2 summarizes some terrestrial inventories of ^{7}Be for soil and wetland profiles. Vegetation may be a very important sink for ^{7}Be, and a significant portion of the ^{7}Be surface inventory may reside in grasses and forest canopies. Olsen et al. (1985) collected 2 marsh cores in early January 1985 just one meter apart. The vegetated core site had approximately six times the inventory of ^{7}Be found in the unvegetated core site, with approximately 40% of the ^{7}Be activity in the grass. Wallbrink and Murray (1996) determined the inventory of ^{7}Be from soil+grass cores from Black Mountain, Australia from 1988 to 1989. They found ^{7}Be inventories to double in the same plot in less than one year, and attribute the increase to grass growth (50 to 90 cm increase) which would increase the efficiency of precipitation scavenging. Russell et al. (1981) determined that White Pine (*Pinus strobus* L.) needles from MA, USA contained approximately 1.9 Bq ^{7}Be g^{-1} ash, and found approximately 6 to 9 Bq ^{7}Be g^{-1} ash in twigs from the same tree. They concluded that in a climate with 1.2 meters of rain/year, the forest canopy typically retains 5 to 20% of the open field flux of atmospherically deposited radionuclides. Wallbrink and Murray (1996) calculated that a Eucalypt (*Eucalyptus*) forest canopy retained 4 to 27% of the total ^{7}Be inventory calculated from fallout measurements. Kaste et al. (1999) reported dried red spruce (*Picea rubens* Sarg.) needles containing approximately 0.15 Bq ^{7}Be g^{-1}, and estimated that over 150 Bq ^{7}Be m^{-2} could reside in the forest canopy (ca. 50% of the total surface inventory). ^{7}Be accumulates over time on vegetation, and is balanced by wash off (precipitation) and decay. Leaves apparently reach steady-state (adsorption = decay) with respect to atmospherically delivered ^{7}Be before twigs and bark (Fig. 3) (Russell et al. 1981; Norton and Perry, unpublished data 1999).

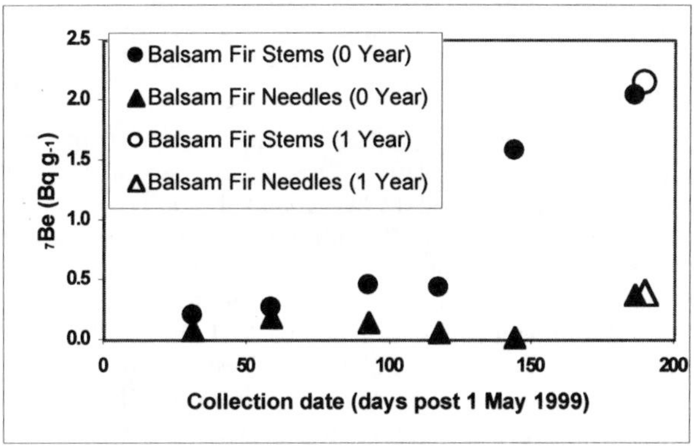

Figure 3. ^{7}Be activity on 0 year balsam fir (*Abies balsamea* L. Mill) needles and stems compared with activities on 1 year old balsam fir needles and stems. (Norton and Perry, unpublished data).

^{7}Be may be a useful tracer for studying erosional processes (Murray et al. 1992). However, such investigations require a detailed understanding of the depth distribution of ^{7}Be. Wallbrink and Murray (1996) found ^{7}Be no deeper than 2 cm in bare soil, grassland soil, and eucalyptus forest soil in Australia. Olsen et al. (1985) reported ^{7}Be was retained in the upper 2 cm of soil in Tennessee, USA and a marsh core collected from Wallops Island, Virginia, USA. However, a second marsh core collected in the summer from Wallops Island had ^{7}Be as deep as 10 cm. They suggested that the depth of the groundwater table could influence the depth distribution of ^{7}Be. In central Idaho, Bonniwell et al. (1999) reported that the entire ^{7}Be inventory was retained in the top 1.2 cm of a soil profile. Kaste (1999) found that the ^{7}Be residing in the upper 8 cm of the organic horizon at the Bear Brook Watershed in Maine, USA (64 to 300 Bq m^{-2}) plus that predicted in the canopy matched the surface inventories predicted by input measurements. Apparently the drainage structure, moisture status of the soil, and overlying vegetation govern the inventory and depth distribution of ^{7}Be.

^{7}BE IN FRESHWATERS

You et al. (1989) determined partitioning coefficients (K_d = [Bq kg^{-1}]/[Bq L^{-1}]) for ^{7}Be between river water and various substrates in the laboratory, and found that most mud, silt, and clay minerals have a K_d near 10^5 L kg^{-1} under neutral to alkaline conditions (pH > 6; see Fig. 9 in Vesely et al., this volume). They reported that the K_d decreases by four orders of magnitude as pH decreases from 6 to 2, and found no significant effect of DOC on the adsorption of ^{7}Be. Hawley et al. (1986) and You et al. (1989) both found that K_d varies inversely with the amount of suspended particles in freshwater. Li et al. (1984) suggested that higher concentrations of suspended particles increase particle aggregation, thereby decreasing the effective exchange surface area g^{-1} of suspended matter. Olsen et al. (1986) and Bonniwell et al. (1999) determined field K_d values for ^{7}Be in freshwaters to be $>10^4$. Vogler et al. (1996) reported that K_d in the water column of Lake Constance, Germany ranged from $10^{5.3}$ to 10^6.

^{7}Be mobility may be enhanced in natural waters by the formation of soluble fluoride (F) and organic acid complexes (Veselý et al. 1989; Veselý et al. this volume). If free F anions are available in the pH range 4.5 to 7, BeF^+ may be a significant fraction of the total Be species in solution. This would decrease the charge density of the total ^{7}Be in solution, and reduce its adsorption to the solid phase. Below pH 4, aluminum (Al) competition for F decreases Be-F complexes (Veselý et al. 1989). Above pH 7, Be mobility decreases as hydroxide complexes of Be become significant and Be is adsorbed by sediment and precipitating iron (Fe) and Al hydroxides. Veselý et al. (1989) suggested that Be complexes with fulvic acids may increase Be mobility in surface waters by reducing the fraction of the Be^{+2} in solution

Despite numerous publications documenting the production of ^{7}Be and the flux to the surface of the earth, there are relatively few studies documenting its mobility and export from watersheds. Dominik et al. (1987) found ^{7}Be activity in filtered streamwater from Switzerland to be $<1.1 \times 10^{-3}$ Bq L^{-1}, but found fluvial suspended matter to have an activity of up to 0.49 Bq g^{-1}, averaging 0.132 Bq g^{-1}. Bonniwell et al. (1999) reported that dissolved Be in a mountain stream in central Idaho was below detection limit ($<3 \times 10^{-4}$ Bq L^{-1}), but suspended sediment generally had 0.05 to 1 Bq g^{-1}. Kaste (1999) found that unfiltered streamwater draining an artificially acidified catchment (pH 4.7) in central Maine had up to 1 Bq L^{-1} during a large snowmelt/rain event. ^{7}Be was up to 0.75 Bq L^{-1} in unfiltered streamwater from a nearby stream with 20 mg DOC L^{-1} during high discharge. Olsen et al. (1986) reported activity of dissolved ^{7}Be in the lower Susquehanna and Raritan Rivers eastern USA ranging from approximately 0.002 to 0.003 and 0.002 to 0.014 Bq L^{-1}, respectively. ^{7}Be activities on suspended particles ranged from ~0.26 to 0.44 Bq g^{-1}. In both the Raritan and the Susquehanna, dissolved ^{7}Be activities (Bq L^{-1}) were higher when the total particle concentration (mg L^{-1}) was higher. In a ^{7}Be budget for the James River, Virginia, USA, Olsen et al. (1986) determined that less than 5% of the ^{7}Be in the estuary was from the rivers, and suggested that soils and vegetation in watersheds sequester most of the ^{7}Be delivered by rainfall. Cooper et al. (1991) studied ^{7}Be export during a snowmelt event in an arctic watershed. Even though $^{18}O/^{16}O$ ratios indicated that >85% of the discharge during the melt was from the snowpack, over 90% of the ^{7}Be inventory in the snowpack remained in the watershed after the event.

The efficiency of removal of ^{7}Be from the water column by particulates may vary among lakes. Krishnaswami et al. (1980) found that the ^{7}Be inventory in a sediment core from Lake Whitney closely matched the atmospheric flux of ^{7}Be. They suggested that if sediment focusing effects could be ruled out, the residence time of ^{7}Be in the lake was very short because of its very high reactivity with particles in the water column. However, in Lake Zurich, Switzerland, Schuler et al. (1991) found that ^{7}Be fluxes into sediment, measured by sediment traps, accounted for only 30% of the measured atmospheric flux of ^{7}Be. They proposed that the residence time of ^{7}Be in the lake was greater than its mean life ($1/\lambda = 76.6$ days). Wan et al. (1987) determined the residence time of ^{7}Be in Greifensee, Switzerland to be approximately 75 days using calculated atmospheric fluxes and measured ^{7}Be inventories in sediment. Vogler et al. (1996) observed that the concentration of total ^{7}Be in Lake Constance, Germany varied with atmospheric input and particle flux. The proportion of ^{7}Be associated with particles varied with biological productivity, and the residence time calculated with a particle-scavenging model ranged from 40 to 340 days. Steinmann et al. (1999) reported a similar residence time of 50 to 230 days for ^{7}Be in the water column at Lake Lugano in Switzerland. They found that log K_d for ^{7}Be on colloids was approximately 6, while log K_d on particulates (>1 μm) was between 4 and 5. Dominik et al. (1989) calculated the residence time of ^{7}Be in Lake Geneva to be 60 to 1100 days, using four different methods. They suggested that the most accurate method to determine residence time of ^{7}Be is to compare its atmospheric flux with ^{7}Be fluxes in sediment traps. Residence times calculated from ^{7}Be inventories in bottom sediments were generally within the range of those calculated by sediment traps (400 to 930 days), but the authors warned that small scale heterogeneity could make sediment inventory calculations unreliable.

^{7}Be inventories from lake sediment cores range from 35 to 875 Bq m^{-2}, with ^{7}Be activities in the upper few centimeters of the cores generally ranging from 0.02 to 0.55 Bq g^{-1} (Krishnaswami et al. 1980; Wan et al. 1987; Dominik et al. 1989). Krishnaswami et al. (1980) found ^{7}Be as deep as 3 cm in Lake Whitney, Connecticut, USA, while Wan et al. (1987) reported ^{7}Be no deeper than 1 cm in a lake core from Greifensee, Switzerland. Wan et al. (1987) found that ^{7}Be inventories in cores ranged by a factor of 4, and were higher where sedimentation rates were higher. Schuler et al. (1991) found a strong seasonality for ^{7}Be fluxes in sediment traps in Lake Zurich, with highest fluxes in July and August over three consecutive years. Maximum ^{7}Be atmospheric flux and maximum particle flux also occurred during this time period.

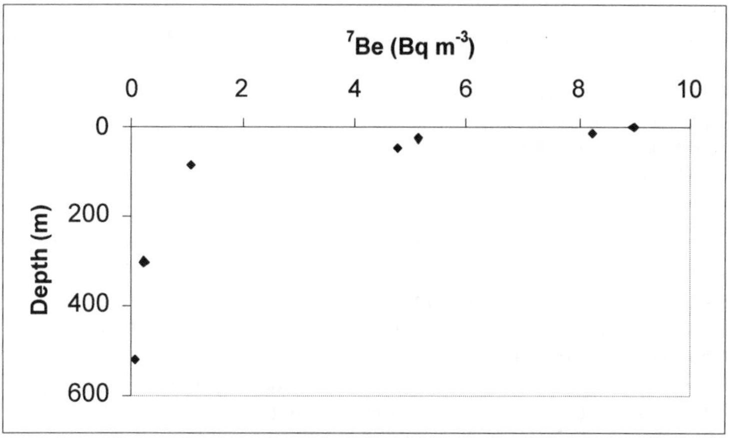

Figure 4. Depth-distribution of ^{7}Be in the Pacific Ocean. Data Plotted from Silker (1972).

^{7}BE IN THE MARINE ENVIRONMENT

Young and Silker (1980) compiled a significant amount of data on dissolved ^{7}Be in the Atlantic and Pacific Oceans. They reported that ^{7}Be activities in seawater generally ranged from 0.0017 to 0.0117 Bq L^{-1}. Total inventories ranged from approximately 80 to 600 Bq m^{-2}. They found that concentrations and inventories were highest in areas associated with high rainfall, and increased with latitude. Silker (1972) reported on the depth distribution of ^{7}Be in the North Pacific Ocean (Fig. 4), and determined that ^{7}Be distribution generally reflected the temperature profile. In cases where there was a strong thermocline, ^{7}Be activity was relatively uniform in the mixed (upper) layer, and concentrations decreased rapidly below the thermocline.

Since Young and Silker (1980), several authors have reported ^{7}Be activities in seawater (Table 3). Comparing the activities of dissolved ^{7}Be and K_d values reported for marine waters with the few data available for freshwater, it appears that ^{7}Be can be partitioned favorably into the aqueous phase in the marine environment. Considering the strong pH dependence of Be partition coefficients (You et al. 1989; Veselý et al. 1989; Veselý et al., this volume) the high pH of seawaters relative to many freshwaters should decrease ^{7}Be in the aqueous phase. Dibb and Rice (1989a) found that sorption of ^{9}Be in laboratory experiments increased with increasing salinity. The wide range of dissolved ^{7}Be and K_d values indicates that several variables are likely governing the solubility and partitioning of Be in seawater including pH, salinity, concentrations of suspended matter concentrations (especially Fe) and DOC, competing cations, and the residence time of the suspended load.

Table 3. Dissolved and particulate ^{7}Be activities in seawater.

Location [USA unless indicated] (time collected)	*Dissolved ^{7}Be(Bq L^{-1})*	*^{7}Be on suspended matter [(Bq g^{-1}), % of total ^{7}Be]*	*log K_d*	*Suspended load (mg L^{-1})*	*Source*
Sequim Bay, WA	0.0009±50%	0.26±24%, 50	5.4	3.55	Bloom & Crecelius (1983)
Sequim Bay, WA	0.0009±38%	0.28±78%, 24	6.0	1.07	Bloom & Crecelius (1983)
Gironde Estuary, France	0.0004 to .002	0.007-0.41	4.5- 5	n.a.	Martin et al. (1986)
Chesapeake Bay, MD (Spring)	0.013±3% to 0.020±2%	0.41-0.47±2%, 36	4.4-4.5	11 to 24	Olsen et al. (1986)
Chesapeake Bay, MD (Fall)	0.001±33% to 0.003±14%	0.17±11% to 0.43±4%, 20 to 54	4.8-5.3	4 to 6	Olsen et al. (1986)
Chesapeake Bay, MD (Year)	<0.0008 to 0.048±2%	<0.007-2.5, 1 to 86	2.8- 6	1 to 121	Dibb & Rice (1989b)
Sabine-Neches Estuary, TX	0.003±15% to 0.035±4%	0.08-0.55±9%, 7 to 57	3.2-4.9	8 to 92	Baskaran et al. (1997)

Bloom and Crecelius (1983) determined that ^{7}Be in seawater had a "limited affinity" for suspended matter, and the percent ^{7}Be adsorbed to particulates is directly proportional to suspended load. Using laboratory experiments with ^{7}Be and natural "detritus", they demonstrated that high-suspended loads (20 mg L^{-1}) had approximately 50% of the total ^{7}Be adsorbed to suspended matter. At approximately 1 mg L^{-1} suspended matter, only 20% of the total ^{7}Be in the water column was adsorbed. They also showed that ^{7}Be adsorbed similarly to detritus and amorphous $Fe(OH)_3$, while adsorption to the algal culture *Pavlova lutheri* (monochrysis) was significantly lower. Although the percent ^{7}Be adsorbed to suspended matter was directly proportional to suspended load concentrations, the K_d was inversely proportional to suspended load concentrations for their field data (Table 3). Martin et al. (1986) also noted that the K_d was lowest in the area with the maximum suspended load. Li et al. (1984) found that K_d for ^{7}Be and several other particle-reactive nuclides in seawater was inversely proportional to the suspended load concentrations, in agreement with Hawley et al. (1986) and You et al. (1989) for freshwater systems.

Olsen et al. (1986) found ^{7}Be activities in filtered river-estuarine and coastal waters along the northeastern USA seaboard ranging from 1.1×10^{-3} to 2×10^{-2} Bq L^{-1}, with total (unfiltered) ^{7}Be reaching 3×10^{-2} Bq L^{-1}. Most of the ^{7}Be was dissolved, typically 60 to 90%. A major factor controlling the concentration of ^{7}Be on suspended matter was the amount of time the particles remained in the water column. ^{7}Be activities on suspended particles were highest in high-energy areas, and lowest in low energy areas with high sedimentation rates. Dissolved ^{7}Be averaged 60% of total ^{7}Be in Chesapeake Bay (Dibb and Rice 1989a). However, activities of ^{7}Be on suspended matter reached 2.5 Bq g^{-1}, and partition coefficients ranged by over 3 orders of magnitude up to log K_d = 6. They concluded, as did Olsen et al. (1986), that rapid variations in the supply of particulates in estuaries and bays may not permit equilibrium to be reached between dissolved ^{7}Be and the solid phase.

Olsen et al. (1986) determined log K_d for river-estuarine and coastal waters from the northeastern USA coast to range from 3.85 to 5.3, with a median of 4.6. They found a positive correlation between K_d and the Fe concentration of the suspended matter, and suggested that the precipitation of Fe may be an important factor governing the scavenging of ^{7}Be. Baskaran et al. (1997) reported relatively low K_d values for ^{7}Be in the Sabine-Neches estuary, Texas, USA. They found log K_d to range from 3.18 to 4.94. There was no significant correlation between suspended load and K_d. Baskaran et al. (1997) attributed their low K_d values to the high concentrations of DOC (5 to 20 mg L^{-1}) and suggested that DOC could play a significant role in the fate and the residence times of particle-reactive nuclides such as ^{7}Be in the water column.

Inventories of ^{7}Be in near-shore marine sediments range from <30 to 6000 Bq m^{-2}, with ^{7}Be activities ranging up to ~0.65 Bq g^{-1} (Krishnaswami et al. 1980; Olsen et al. 1986; Dibb and Rice 1989b; Canuel et al. 1990, and Summerfield et al. 1999). Maximum inventories of ^{7}Be in marine sediment commonly occur in the spring, and in areas where fine particles are accumulating rapidly. Canuel et al. (1990) examined the spatial and vertical distribution of ^{7}Be in sediments of Cape Lookout Bight, NC, USA. They sampled 10 cores in September, 1987, and found ^{7}Be inventories ranging from approximately 650 to 1375 Bq m^{-2} in a <30 m^2 area. ^{7}Be penetration reached a maximum depth in the sediment profile during the summer months. ^{7}Be activity reached 0.033 Bq g^{-1} 13 cm deep in the core. The authors attributed this seasonal pattern in ^{7}Be depth distribution to the formation of methane bubble tubes during the summertime. Olsen et al. (1986) found that ^{7}Be reached a depth of 8 cm in coastal sediments of the eastern USA, but generally was retained in the upper 4 cm. Sneed (1986) determined ^{7}Be and ^{210}Pb in cores from a dredging site with a high sedimentation rate in New York City harbor. She found ^{7}Be to a depth of 4 cm in an area where the sedimentation rate was approximately 7 cm yr^{-1}.

APPLICATIONS OF ^{7}BE

A detailed understanding of the transport and residence time of aerosols in the atmosphere is desired because particles in the micron and sub-micron range probably play a significant role in climate change and trace metal cycling. ^{7}Be a very useful tracer in atmospheric transport studies because it (1) has a short half-life, (2) is produced at a relatively constant rate, (3) rapidly attaches to aerosols after formation, and (4) has a large concentration gradient between the stratosphere and troposphere (Bhandari and Rama 1963; Junge 1963; Dutkiewicz and Husain 1979,1985; Viezee and Singh 1980). In the past decade in particular, ^{7}Be measurements in the atmosphere have been used to test and validate chemical transport models in the atmosphere (Brost et al. 1991; Rehfeld and Heimann 1995; Koch et al. 1996). Dibb and Jaffrezo (1993) used ^{7}Be and $^{210}Pb_{ex}$ (decay product of ^{222}Rn, $T_{1/2}$ = 22.3 yr) to investigate short term "communication" between the atmosphere and the Greenland Ice Sheet. They found that concentrations of ^{7}Be and ^{210}Pb in fresh snow did not have the spring and fall peaks that ground level aerosol concentrations displayed. On the basis of these data, Dibb and Jaffrezo (1993) suggested that caution must be used when inferring atmospheric chemistry from fresh snow.

The potential for using ^{7}Be in atmospheric transport studies is particularly great when it is coupled with a nuclide with a contrasting source function and/or a different half-life. The initial production ratio of $^{10}Be/^{7}Be$ is approximately 0.5 (Lal and Peters 1967). As air moves away from the region of maximum production, the $^{10}Be/^{7}Be$ ratio on aerosols in the air mass will increase, because ^{7}Be has a much shorter radioactive half-life ($T_{1/2}$ = 53 days) than ^{10}Be ($T_{1/2}$ = 1.5 m.y.). Thus the $^{10}Be/^{7}Be$ ratio in the atmosphere can be used as a "clock" to determine air mass age, and has been used to indicate the intrusion of stratospheric air into the troposphere (Raisback et al.1981; Dibb et al. 1994; Rehfeld and Heimann 1995; Koch and Rind 1998). Because the residence time of aerosols in the troposphere is fairly short (ca.1 month [Durana et al. 1996; Bleichrodt 1978]) compared to those in the stratosphere (about 1 year), stratospheric air masses characteristically have a relatively high $^{10}Be/^{7}Be$ ratio. Dibb et al. (1994) found that the $^{10}Be/^{7}Be$ ratio was nearly constant at 2.2 throughout the year in northwestern Canada. They concluded that a significant portion of the aerosols in the arctic troposphere must be derived from the stratosphere, likely a result of vertical mixing between the two air masses rather than stratospheric injections. Koch and Rind (1998) used a General Circulation Model (GISS GCM) to suggest that the spring maximum in the $^{10}Be/^{7}Be$ ratio observed by Dibb et al. (1994) in the lower stratosphere may have resulted from the transport of equatorial air to higher latitude during the previous winter.

The contrasting source terms of ^{7}Be and ^{210}Pb make the $^{7}Be/^{210}Pb$ ratio a useful tool for investigating air mass sources and vertical transport within the troposphere (Dibb et al. 1992; Baskaran et al. 1993; Rehfeld and Heimann 1995; Koch et al. 1996; Graustein and Turekian 1996, and Benitez-Nelson and Buesseler 1999). While ^{7}Be is produced primarily in the upper atmosphere, ^{210}Pb is derived from the decay of ^{222}Rn gas ($T_{1/2}$=3.8 days) that is emitted primarily from the surface of the continents. Therefore, ^{210}Pb concentrations in air should decrease with altitude and distance from land. However, ^{7}Be concentrations in air will be uniform over land and sea, always increasing with altitude. High $^{7}Be/^{210}Pb$ ratios may be an indication of stratosphere-troposphere exchange, or an air mass with components originating from the open ocean. Benitez-Nellson and Buesseler (1999) found that the average $^{7}Be/^{210}Pb$ ratio in rainfall from Woods Hole, MA and Portsmouth, NH (USA) was 12.6 and 20.1, respectively. These data suggest that Woods Hole received air from a more continental source than Portsmouth during the sampling period. The efficiency and rate of vertical transport (convection) within the troposphere will also affect the $^{7}Be/^{210}Pb$ depositional ratio. Increased vertical transport should increase the amount of ^{7}Be delivered to the ground, and increase the rate at which ^{222}Rn moves upwards. That is, ^{7}Be is enriched in the lower troposphere while ^{210}Pb is diluted. Baskaran (1995) found that the $^{7}Be/^{210}Pb$ ratio (measured in depositional fluxes) was highest in the summertime (14.4 to 23.9) and lowest in the winter (12.8 to 13.7). He attributed the increased $^{7}Be/^{210}Pb$ ratios during the warmer months to decreased stability of the troposphere (increased convection).

^{7}Be has useful applications in sedimentological studies. Olsen et al. (1989) used the ratio of short-lived ^{7}Be to longer-lived excess ^{210}Pb ($^{210}Pb_{ex}$) to quantify the amount of re-suspended sedimentary material in coastal waters. ^{7}Be is very low on bottom sediments relative to its activity on suspended particulates, while $^{210}Pb_{ex}$ is nearly the same on bottom sediments and suspended particulates. Using $^{7}Be/^{210}Pb_{ex}$, they determined that re-suspended bottom sediments may account for more than 80% of the suspended matter in the Savannah Estuary. Canuel et al. (1990) concluded that in coastal sediments unaffected by bioturbation, ^{7}Be is a useful tracer of sediment accumulation. They found that sediment cores taken before and after storm events can be used to quantify the impacts of individual storms by identifying “new” pulses of sediment. Fitzgerald et al. (2001) used this technique to determine monthly sediment deposition and re-suspension rates in PCB-rich sediments in the Fox River system in Wisconsin, USA. They found that short-term accumulation rates determined by ^{7}Be were up to 130 times higher that those determined with ^{137}Cs, indicating the dynamic nature of sediment transport in this impounded system. Sommerfield et al. (1999) used ^{7}Be as a tracer of fine-grained river sediment on the northern California, USA continental margin. They found that ^{7}Be was present in shelf and slope deposits only after periods of extremely high river discharge, and determined that storms and floods play a significant role in sedimentary processes in modern environments in their study area. ^{7}Be has been used to calculate sediment accumulation rates in shallow marine environments (Sneed 1986). Canuel et al. (1990) and Dibb and Rice (1989b) found that short-term sediment accumulation rates determined with ^{7}Be were comparable to longer-term sediment accumulation rates determined from ^{210}Pb. Dibb and Rice (1989b) suggested that the general agreement between the ^{7}Be and

^{210}Pb derived sedimentation rates could indicate that the processes operating on the Chesapeake Bay sediments have remained relatively constant for the past 100 years. Steinmann et al. (1999) used ^{7}Be to calculate the coagulation rate of colloids in the epilimnion of Lake Lugano. They measured the sedimentary flux of ^{7}Be, as well as concentrations of dissolved, colloidal, and particulate ^{7}Be in the epilimnion. Based on the 2.2 to 16.3 day residence time of ^{7}Be in the colloidal fraction, they reported coagulation rates of 0.06 to 0.46 d^{-1}.

^{234}Th ($T_{1/2}$= 24 days; derived from decay of ^{238}U) can be used in conjunction with ^{7}Be to investigate sedimentation processes in areas with a salinity gradient. Both nuclides are particle-reactive, but the source of ^{234}Th is ^{238}U, which is positively correlated with salinity. Feng et al. (1999a, 1999b) utilized the $^{234}Th_{ex}/^{7}Be$ activity ratios on suspended and bottoms sediments as a tracer of particle sources and transport in the Hudson River estuary, where Th_{ex} is the total Th activity of the sediment minus that supported by ^{238}U. By using the activity ratio of these two nuclides, the apparent effects of grain size, composition, and concentration on adsorption are minimized. They analyzed sediment during different flow regimes and tidal cycles to determine the relative importance of local sediment re-suspension and advection of sediments through the estuary. They found that both radionuclides were removed from the water column on time scales from <1 to 13 days. They also noted that the $^{234}Th_{ex}/^{7}Be$ tended to be equal or greater than that predicted by local equilibrium (*in situ* production of $^{234}Th_{ex}$, atmospheric inputs of ^{7}Be, scavenging, particles settling and re-suspension) during low flow, which can be explained by transport of particles up-estuary by estuarine circulation. During high flows they observed lower activity ratios, indicating the importance of re-suspension. Feng et al. (1999b) reported that during the course of tidal cycles, the $^{234}Th_{ex}/^{7}Be$ values were comparable in suspended particles from surface and bottom water with different salinities, which indicated that particles are mixed fairly well in the water column.

The strong affinity of ^{7}Be for small particles and its steady production make ^{7}Be a unique tool for hydrologists and geomorphologists. Walling et al. (1999) used ^{7}Be measurements to document event-based soil redistribution. They found that the spatial patterns of ^{7}Be and ^{137}Cs (fission product, $T_{1/2}$ = 30.2 yr) total inventories were similar. By comparing the inventories of these nuclides to a reference inventory (taken in a stable area where neither erosion or aggradation is occurring) before and after a storm, they could identify areas of soil loss and accumulation on short (a few days) and medium (a few decades) time scales. Bonniwell et al. (1999) used ^{7}Be and other fallout nuclides to trace sediment and determine erosion rates in a small watershed in Idaho. They used ^{7}Be as an indicator of "new" sediment during snowmelts, by assuming that "old" sediment that gets re-suspended in the water column would have a negligible concentration of ^{7}Be. Bonniwell et al. (1999) calculated erosion rates in the watershed by comparing the total export of radionuclides with the measured inventories in the soils.

The different depth distributions of ^{7}Be, $^{210}Pb_{ex}$ and ^{137}Cs (fission product, $T_{1/2}$ = 30.2 yr) can be used to fingerprint the soil depth from which suspended sediment was originally removed (Burch et al. 1988; Walling and Woodward 1992; Wallbrink et al. 1999) (Fig. 5). Because ^{7}Be is often concentrated in the uppermost soil horizons, Murray et al. 1992 suggested using ^{7}Be as a tracer of topsoil movement. For example, suspended sediment with a high activity of ^{7}Be (and/or $^{210}Pb_{ex}$) could indicate an initial source area at the top of the soil profile, while sediment with a lower activity of ^{7}Be (and/or $^{210}Pb_{ex}$) and high activity of ^{137}Cs could have been derived from deeper in the profile. Wallbrink and Murray (1993) extended the use of these fallout nuclides to identify the type erosion process (sheet flow, rill or gully erosion) removing soil from the landscape (Fig. 6).

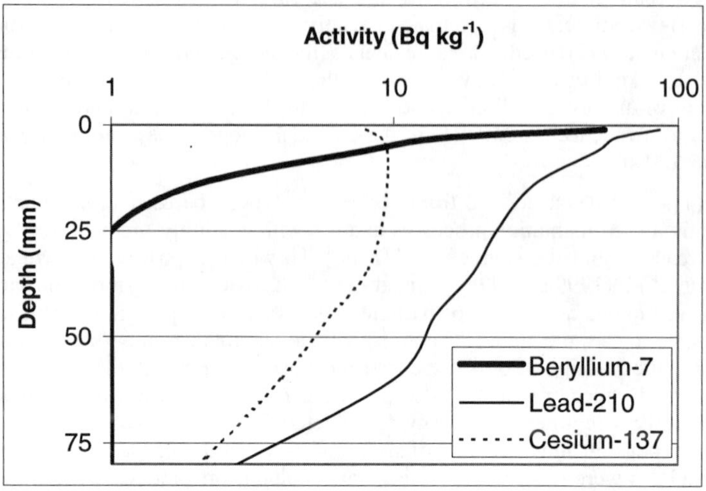

Figure 5. Depth-distribution of fallout radionuclides applicable for erosion studies. Generalized from Wallbrink et al. (1999).

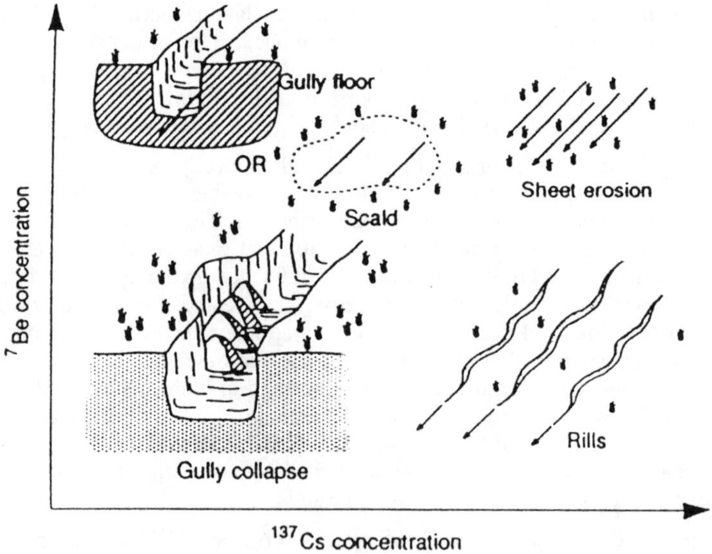

Figure 6. Generalized ^{7}Be and ^{137}Cs labeling of sediment (after Burch et al. 1988). This concept is based on the theory that sediment retains the radionuclide signature originally obtained when it was soil. Used by permission of the editor of *Hydrological Processes*, from Wallbrink and Murray (1993), Fig. 1c, p. 298.

SUMMARY AND CONCLUDING REMARKS

^{7}Be is a short-lived naturally occurring radionuclide with a unique source term and a relatively constant production rate. After ^{7}Be is formed in the atmosphere, it rapidly adsorbs to aerosols. Because the residence time of aerosols in the stratosphere is greater than the half-life ^{7}Be by more than a factor of six, steady-state (production = decay) equilibrium maintains nearly constant concentrations above the tropopause. Tropospheric concentrations of ^{7}Be are more dynamic and are generally 1 to 2 orders of magnitude lower than in the stratosphere due to wash out and decay. Near surface air exhibits fluctuations on decadal, semi decadal (4 to 7 years), annual, and semi-annual timescales. The short-term variations of ^{7}Be concentration in near surface air are a result of variation in production, wash out by precipitation, vertical mixing between and within the stratosphere and troposphere, and horizontal air mass movement/mixing within the troposphere. ^{7}Be terrestrial inventories and ecosystem fluxes (measured by bulk atmospheric deposition collectors) vary seasonally due to variation in tropospheric ^{7}Be and precipitation amount. ^{7}Be is deposited on ecosystems primarily by wet deposition; dry deposition is apparently <10% of the total flux.

^{7}Be deposited from rainfall rapidly adsorbs to vegetation, soils, and sediment, and in many cases will remain adsorbed to the solid phase until decay. In the soil profile, ^{7}Be is generally retain in the upper layers. Most of the movement of ^{7}Be within and from watersheds appears to be in particulate form. The residence time of ^{7}Be in the water column of lakes is uncertain, reports range from a few weeks to a few months. The partitioning of ^{7}Be in the marine environment is also poorly understood, with reported K_d values ranging over three orders of magnitude. In estuaries, a significant portion of the ^{7}Be may be in dissolved form, especially in the presence of DOC. Sediment inventories of ^{7}Be in both lakes and shallow ocean water exhibit seasonal variations.

^{7}Be studies have provided scientists with much insight into atmospheric, geochemical, erosional, sedimentological, and nuclear processes. Its decay energy is fairly easy and inexpensive to measure. Therefore some of the more recently developed applications of ^{7}Be, particularly in atmospheric, erosion, and sedimentation studies, have significant potential for evaluation of environmental problems. However, most of the applications of ^{7}Be require a very detailed understanding of the distribution and geochemistry of ^{7}Be, which is still somewhat incomplete in terrestrial ecosystems. In particular, more information is needed on the depth distribution of ^{7}Be in soils with respect to different drainage conditions, bulk density, and amount of macropore flow channels. Also, the partitioning of ^{7}Be and its residence time in lakes and the ocean is poorly understood. The effect of competing ligands and DOC on the adsorption behavior needs further documentation, as is evident in the wide range of K_ds reported (three orders of magnitude) in estuaries and tidal bays. With further studies on the geochemical behavior (see Veselý et al., this volume) and distribution of ^{7}Be under various conditions, it will be possible to take full advantage of this nuclide in earth science investigations.

ACKNOWLEDGMENTS

Our research was supported by the University of Maine and by the National Science Foundation (Grant EAR-9725705). In addition, this review paper was greatly improved by the comments of Peter Wallbrink, Curtis Olsen, and Jack Dibb.

REFERENCES

Ajzenberg-Selove F (1988) Energy levels of the light nuclei A = 5-10. Nuclear Physics A 490:1-225

Arnold JR (1956) Beryllium-10 produced by cosmic rays. Science 124:584-585

Arnold JR, Al-Salih H (1955) Beryllium-7 produced by cosmic rays. Science 121:451-453

Bachhuber H, Bunzl K (1992) Background levels of atmospheric deposition to ground and temporal variation of ^{129}I, ^{127}I, ^{137}Cs, and ^{7}Be in a rural area of Germany. J Environ Radioact 16:77-89

Baskaran M (1995) A search for the seasonal variability on the depositional fluxes of ^{7}Be and ^{210}Pb. J Geophys Res 100:2833-2840

Baskaran M, Coleman CH, Santschi PH (1993) Atmospheric depositional fluxes of ^{7}Be and ^{210}Pb at Galveston and College Station, Texas. J Geophys Res 98:20555-20571

Baskaran M, Ravichandran M, Bianchi TS (1997) Cycling of ^{7}Be and ^{210}Pb in a high DOC, shallow, turbid estuary of southeast Texas. Estuarine Coastal Shelf Sci 45:165-176

Beer J, Blinov A, Bonani G, Finkel RC, Hofmann HJ, Lehmann B, Oeschger H, Sigg A, Schwander J, Staffelbach T, Stauffer B, Suter M, Wolfli W (1990) Use of ^{10}Be in polar ice to trace the 11-year cycle of solar activity. Nature 347:164-166

Benitez-Nelson C, Buesseler KO (1999) Phosphorous-32, phosphorous-37, beryllium-7, and lead-210: Atmospheric fluxes and utility in tracing stratosphere/troposphere exchange. J Geophys Res 104:11745-11754

Bhandari N, Rama (1963) Atmospheric circulation from observations of sodium-22 and other short-lived natural radioactivities. J Geophys Res 68:1959-1966

Bhandari N, Lal, D, Rama (1970) Vertical structure of the troposphere as revealed by radioactive tracer studies. J Geophys Res 75:2974-2980

Bleichrodt JF (1978) Mean tropospheric residence time of cosmic-ray-produced beryllium-7 at north temperate latitudes. J Geophys Res 83:3058-3062

Bloom N, Crecelius EA (1983) Solubility behavior of atmospheric ^{7}Be in the marine environment. Marine Chem 12:323-331

Bonniwell EC, Matisoff G, Whiting PJ (1999) Determining the times and distances of particle transit in a mountain stream using fallout radionuclides. Geomorph 27:75-92

Bouchez PR, Tobailem J, Robert J, Muxart R, Mellet R, Daudel P, Daudel R (1956) Nouvelle determination de la difference des periodes de ^{7}Be metallique et de BeF_2. J Phys Radium 17:363-365

Brost RA, Feichter J, Heimann M (1991) Three-dimensional simulation of ^{7}Be in a global climate model. J Geophys Res 96:22423-22445

Brown L, Stensland GJ, Klein J, Middleton R (1989) Atmospheric deposition of ^{7}Be and ^{10}Be. Geochim Cosmochem Acta 53:135-142

Burch GJ, Barnes CJ, Moore ID, Barling RD, Mackenzie DJ, Olley JM (1988) Detection and prediction of sediment sources in catchments: Use of ^{7}Be and ^{137}Cs. *In* Hydrology and Water Resources Symposium. Aust Inst Eng National Conf Publ 88:146-151

Canuel EA, Martens CS, Benninger LK (1990) Seasonal variations in the ^{7}Be activity in the sediments of Cape Lookout Bight, North Carolina. Geochim Cosmochem Acta 54:237-245

Cooper LW, Olsen CR, Solomon DK, Larsen IL, Cook RB, Grebmeier JM (1991) Stable isotopes of oxygen and natural and fallout radionuclides used for tracing runoff during snowmelt in an arctic watershed. Water Resources Res 27:2171-2179

Dibb JE (1989) Atmospheric deposition of beryllium-7 in the Chesapeake Bay Region. J Geophys Res 94:2261-2265

Dibb JE, Jaffrezo, JL (1993) Beryllium-7 and lead-210 in aerosol and snow in the Dye 3 Gas, Aerosol, and Snow Sampling Program. Atmos Environ 27A:2751-2760

Dibb JE, Rice DL (1989a) The geochemistry of beryllium-7 in Chesapeake Bay. Estuarine, Coastal, Shelf Sci 28:379-394

Dibb JE, Rice DL (1989b) Temporal and spatial distribution of beryllium-7 in the sediments of Chesapeake Bay. Estuarine Coastal Shelf Sci 28:395-406

Dibb JE, Talbot RW, Gregory GL (1992) Beryllium-7 and lead-210 in the western arctic atmosphere: Observations from three recent aircraft-based sampling programs. J Geophys Res 97:16709-16715

Dibb JE, Meeker DL, Finkel RC, Southon JR, Caffee MW (1994) Estimation of stratospheric input to the arctic troposphere: ^{7}Be and ^{10}Be in aerosols at Alberta, Canada. J Geophys Res 99:12855-12864

Dingle AN (1965) Stratospheric tapping by intense convective storms: Implications for public health in the United States. Science 148:227-229

Dominik J, Burrus D, Vernet J-P (1987) Transport of the environmental radionuclides in an alpine watershed. Earth Planet Sci Lett 84:165-180

Dominik J, Schuler C, Santschi PH (1989) Residence times of ^{234}Th and ^{7}Be in Lake Geneva. Earth Planet Sci Lett 93:345-358

Dueñas C, Fernández MC, Liger E, Carretero J (1999) Gross alpha, gross beta activities and ^{7}Be concentrations in surface air: Analysis of their variations and prediction model. Atmos Environ 33:3705-3715

Ďurana L, Chudy M, Masarik J (1996) Investigation of ^{7}Be in the Bratislava atmosphere. J Radioanal Nucl Chem 207:345-356

Dutkiewicz VA, Husain L (1979) Determination of stratospheric and tropospheric components of ^{7}Be in surface air. Geophys Res Lett 6:171-174

Dutkiewicz VA, Husain L (1985) Stratospheric and tropospheric components of ^{7}Be in surface air. J Geophys Res 90:5783-5788

Feely HW, Larsen RJ, Sanderson CG (1989) Factors that cause seasonal variations in beryllium-7 concentrations in surface air. J Environ Radioact 9:223-249

Feng H, Cochran JK, Hirschberg DJ (1999a) ^{234}Th and ^{7}Be as tracers for the transport and dynamics of suspended particles in a partially mixed estuary. Geochem Cosmochem Acta 63:2487-2505

Feng H, Cochran JK, Hirschberg DJ (1999b) ^{234}Th and ^{7}Be as tracers for the sources of particles to the turbidity maximum of the hudson river estuary. Estuarine Coastal Shelf Sci 49:629-645

Fitzgerald SA, Val Klump J, Swarzenski PW, Mackenzie RA, Richards KD (2001) Beryllium-7 as a tracer of short-term sediment deposition and resuspension in the Fox River, Wisconsin. Environ Sci Technol 35:300-305

Goel PS, Jha S, Lal D, Radhakrishna PR (1956) Cosmic ray produced beryllium isotopes in rain water. Nuclear Phys 1:196-201

Graustein WC, Turekian KK (1996) ^{7}Be and ^{210}Pb indicate an upper troposphere source for elevated ozone in the summertime subtropical free troposphere of the eastern North Atlantic. Geophys Res Lett 23: 539-542

Harvey MJ, Matthews KM (1989) ^{7}Be deposition in a high-rainfall area of New Zealand. J Atmos Chem 8:299-306

Hawley N, Robbins JA, Eadie BJ (1986) The partitioning of ^{7}Be in fresh water. Geochim Cosmochim Acta 50:1127-1131

Hensley WK, Bassett WA, Huizenga JR (1973) Pressure dependence of the radioactive decay constant of beryllium-7. Science 181:1164-1165

Hötzl H, Rosner G, Winkler R (1991) Correlation of ^{7}Be concentrations in surface air and precipitation with the solar cycle. Naturwissenschaften 78:215-217

Huh C-A (1999) Dependence of the decay rate of ^{7}Be on chemical forms. Earth Planet Sci Lett 171: 325-328

Husain L, Coffey PE, Meyers RE, Cederwall RT (1977) Ozone transport from stratosphere to troposphere. Geophys Res Lett 4:363-365

Jaeger M, Wilmes S, Kolle V, Staudt G, Mohr P (1996) Precision measurements of the half-life of ^{7}Be. Phys Rev 54:423-424

Johlige HW, Aumann DC, Born H-J (1970) Determination of the relative electron density at the Be nucleus in different chemical combinations, measured as changes in the electron-capture half-life of ^{7}Be. Phys Rev 2:1616-1621

Junge CE (1963) Studies of global exchange processes in the atmosphere by natural and artificial tracers. J Geophys Res 68:3849-3856

Kaste JM (1999) Dynamics of cosmogenic and bedrock-derived beryllium nuclides in forested ecosystems in Maine, U.S.A. Unpublished MSc thesis, University of Maine

Kaste JM, Norton SA, Fernandez IJ, Hess CT (1999) Delivery of cosmogenic beryllium-7 to forested ecosystems in Maine, USA. Geol Soc Am Abstr 31:A305

Koch DM, Mann, ME (1996) Spatial and temporal variability of ^{7}Be surface concentrations. Tellus 48B:387-396

Koch DM, Rind D (1998) Beryllium-10/ beryllium-7 as a tracer of stratospheric transport. J Geophys Res 103:3907-3917

Koch DM, Jacob DJ, Graustein WC (1996) Vertical transport of aerosols as indicated by ^{7}Be and ^{210}Pb in a chemical tracer model. J Geophy Res 101:18651-18666

Kostadinov KN, Yanev YL, Mavrodiev VM (1988) A method for the preconcentration of cosmogenic beryllium-7 in natural waters. J Radioan Nuclear Chem Articles 121:509-513

Kraushaar JJ, Wilson ED, Bainbridge KT (1953) Comparison of the values of the disintegration constant of ^{7}Be in Be, BeO, and BeF_2. Phys Rev 90:610-614

Krishnaswami S, Benninger LK, Aller RC, Damm KL (1980) Atmospherically-derived radionuclides as tracers of sediment mixing and accumulation in near-shore marine and lake sediments: Evidence from ^{7}Be, ^{210}Pb, and ^{239,240}Pu. Earth Planet Sci Lett 47:307-318

Lal D, Peters B (1967) Cosmic ray produced radioactivity on earth. *In* Handbuch der Physik 46:551-612, Sitte K (ed) Springer-Verlag, New York

Lal D, Malhotra PK, Peters B (1958) On the production of radioisotopes in the atmosphere by cosmic radiation and their application to meteorology. J Atmos Terrest Phys 12:306-328

Larsen IL, Cutshall NH (1981) Direct determination of ^{7}Be in sediments. Earth Planet Sci Lett 54:379-384

Leininger RF, Segrè E, Wiegand C (1949) Experiments on the effect of atomic electrons on the decay constant of ^{7}Be. Phys Rev 76:897-898

Leininger RF, Segrè E, Wiegand C (1951) Erratum: Experiments on the effect of atomic electrons on the decay constant of ^{7}Be. Phys Rev 81:280

Li Y-H, Burkhardt L, Buchholtz M, O'Hara P, Santschi P (1984) Partition of radiotracers between suspended particles and seawater. Geochim Cosmochem Acta 48:2011-2019

Liu LG, Huh CA (2000) Effect of pressure on the decay rate of ^{7}Be. Earth Planetary Sci Lett 180:163-167.

Martin JM, Mouchel JM, Thomas AJ (1986) Time concepts in hydrodynamic systems with an application to ^{7}Be in the Gironde Estuary. Marine Chem 18:369-392

Murray AS, Olley JM, Wallbrink PJ (1992) Natural radionuclide behavior in the fluvial environment. Radiat Prot Dosim 45:285-288

Noyce JR, Chen, TS, Moore DT, Beck JN, Kuroda PK (1971) Temporal distributions of radioactivity and ^{87}Sr/^{86}Sr ratios during rainstorms. J Geophys Res 76:646-656

Olsen C, Larsen IL, Lowry PD, Cutshall NH, Todd JF, Wong GTF, Casey WH (1985) Atmospheric fluxes and marsh-soil inventories of ^{7}Be and ^{210}Pb. J Geophys Res 90:10487-10495

Olsen CR, Larsen IL, Lowry PD, Cutshall NH (1986) Geochemistry and deposition of ^{7}Be in river-estuarine and coastal waters. J Geophys Res 91:896-908

Olsen CR, Thein M, Larsen IL, Lowry PD, Mulholland PJ, Cutshall NH, Byrd JT, Windom, HL (1989) Plutonium, lead-210, and carbon isotopes in the Savannah Estuary: Riverborne versus marine sources. Environ Sci Technol 23:1475-1481

Papastefanou C (1991) High-level ^{7}Be concentrations in air after the SN 1987: A supernova outburst in the large magellanic cloud on 23 February 1987. Nucl Geophys 5:51-52

Peters B (1955) Radioactive beryllium in the atmosphere and on the earth. Proc Indian Acad Sci 41:67-71

Raisbeck GM, Yiou F, Fruneau M, Loiseux JM, Lieuvin M, Ravel JC (1981) Cosmogenic ^{10}Be/^{7}Be as a probe of atmospheric transport processes. Geophys Res Lett 8:1015-1018

Rama, Honda M (1961) Natural radioactivity in the atmosphere. J Geophys Res 66:3227-3231

Rehfeld S, Heimann H (1995) Three dimensional atmospheric transport simulation of the radioactive tracers ^{210}Pb, ^{7}Be, ^{10}Be, and ^{90}Sr. J Geophys Res 100:26141-26161

Reiter ER (1975), Stratospheric-tropospheric exchange processes. Rev Geophys Space Phys 13:459-474

Russell IJ, Choquette CE, Fang S-L, Dundulis WP, Pao AA, Pszenny AAP (1981) Forest vegetation as a sink for atmospheric particulates: Quantitative studies in rain and dry deposition. J Geophys Res 86:5247-5363

Schuler C, Wieland E, Santschi PH, Sturm M, Lueck A, Bollhalder S, Beer J, Bonani G, Hofmann HJ, Suter M, Wolfli W (1991) A multitracer study of radionuclides in Lake Zürich, Switzerland 1. Comparison of atmospheric and sedimentary fluxes of ^{7}Be, ^{10}Be, ^{210}Pb, ^{210}Po, and ^{137}Cs. J Geophys Res 96:17051-17065

Schumann G, Stoeppler M (1963) Beryllium 7 in the atmosphere. J Geophys Res 68:3827-3830

Segrè E (1947) Possibility of altering the decay rate of a radioactive substance. Phys Rev 71:274-275

Segrè E, Wiegand CE (1951) Erratum: Experiments on the effect of atomic electrons on the decay constant of ^{7}Be. Phys Rev 81:284

Silker WB (1972) Horizontal and vertical distributions of radionuclides in the North Pacific Ocean. J Geophys Res 77:1061-1070

Sneed SB (1986) Sediment chronologies in New York Harbor Borrow Pits. Unpublished MSc thesis, State University of New York, Stony Brook

Sommerfield CK, Nittrouer CA, Alexander CR (1999) ^{7}Be a tracer of flood sedimentation on the northern California continental margin. Cont Shelf Res 19:355-361

Steinmann P, Billen T, Loizeau JL, Dominik J (1999) Beryllium-7 as a tracer to study mechanisms and rates of metal scavenging from lake surface waters. Geochem Cosmochem Acta 63:1621-1633

Todd JF, Wong TF (1989) Atmospheric depositional characteristics of beryllium 7 and lead 210 along the southeastern Virginia coast. J Geophys Res 94:11106-1116

Turekian KK, Benninger LK, Dion EP (1983) ^{7}Be and ^{210}Pb total deposition fluxes at New Haven, Connecticut, and at Bermuda. J Geophys Res 88:5411-5415

Veselý J, Benes P, Sevcik K (1989) Occurrence and speciation of beryllium in acidified freshwaters. Water Resources Res 23:711-717

Viezee W, Singh HB (1980) The distribution of ^{7}Be in the troposphere: Implications on stratospheric/tropospheric air exchange. Geophys Res Lett 7:805-808

Vogler S, Jung M, Mangini (1996) Scavenging of ^{234}Th and ^{7}Be in Lake Constance. Limnol Oceanogr 41:1384-1393

Wallbrink PJ, Murray AS (1993) Use of fallout radionuclides as indicators of erosion processes. Hydro Process 7:297-304

Wallbrink PJ, Murray AS (1994) Fallout of ^{7}Be in south Eastern Australia. J Environ Radioact 25:213-228

Wallbrink PJ, Murray AS (1996) Distribution and variability of ^{7}Be in soils under different surface cover conditions and its potential for describing soil redistribution processes. Water Resources Res 32: 467-476

Wallbrink PJ, Murray AS, Olley JM (1999) Relating suspended sediment to its original soil depth using fallout radionuclides. Soil Sci Soc Am J 63:369-378

Walling DE, Woodward JC (1992) Use of radiometric fingerprints to derive information on suspended sediment sources. *In* Erosion and Sediment Transport Monitoring Programmes in River Basins. Int'l Assoc Hydrological Sci Publ 210:153-164

Walling DE, He Q, Blake W (1999) Use of ^{7}Be and ^{137}Cs measurements to document short- and medium-term rates of water-induced erosion on agricultural land. Water Resources Res 35:3865-3874

Walton A, Fried RE (1962) The deposition of beryllium-7 and phosphorous-32 in precipitation at north temperate latitudes. J Geophys Res 67:5335-5340

Wan G, Santschi PH, Sturm M, Farrenkothen K, Lueck A, Werth E, Schuler C (1987) Natural (^{210}Pb, ^{7}Be) and fallout (^{137}Cs, ^{239,240}Pu, ^{90}Sr) radionuclides as geochemical tracers of sedimentation in Greifensee, Switzerland. Chem. Geology 63:181-196

You CF, Lee T, Li YH (1989) The partition of Be between soil and water. Chem. Geology 77:105-118

Young J, Silker W (1980) Aerosol depositional velocities on the Pacific and Atlantic Oceans calculated from ^{7}Be measurements. Earth Planet Sci Lett 50:92-104

Zanis P, Schuepbach E, Gaggeler HW, Hübener S, Tobler L (1999) Factors controlling beryllium-7 at Jungfraujoch in Switzerland. Tellus 51B:789-805

7 Environmental Chemistry of Beryllium

J. Veselý[1], S.A. Norton[2], P. Skřivan[3], V. Majer[1], P. Krám[1], T. Navrátil[3], J.M. Kaste[4]

[1] Czech Geological Survey, Klárov 3, Prague 1, 118 21, C.R.

[2] Dept. Geological Sciences, University of Maine, Orono, Maine 04469

[3] Geological Institute, AS C. R., Rozvojová 135, Prague 6, C.R.

[4] Dept. Earth Sciences, Dartmouth College, Hanover, New Hampshire 03755

INTRODUCTION

Beryllium is a rare but widely distributed element. As a consequence of rapid developments in the capabilities to determine low concentrations of Be in inorganic solids, tissue, air, and water, it is now possible to characterize the distribution of Be in various environmental compartments, to describe the fluxes among these compartments, and to determine the controls on these fluxes. Increased human activity, including land disturbance, metallurgical processes, the burning of fossil fuels (with associated air pollution, including Be, and related terrestrial acidification), has generally increased the flow of Be through ecosystems. This fact was recognized several decades ago (e.g., Kubizňáková 1983). The distribution and redistribution of Be nuclides are important to understand because of Be toxicity to fauna (including humans) and possibly to flora. The radioisotopes (^{7}Be and ^{10}Be) are used to understand environmental and geologic processes. Because of the very different half lives (^{7}Be = 53 days; ^{10}Be = 1.5 million years), Be nuclides have been used to study earth-surface processes as short as diurnal variation (EL-Hussein et al. 2001) to exposure age-dating (Braucher et al. 2000) and erosional processes (Small et al. 1999). Consequently, we review the literature from the perspective of the interaction of geologic materials, aqueous solutions, and biota, at or near Earth´s surface. This chapter is organized parallel to the hydrologic pathways from the atmosphere, through ecosystems, to estuaries. We focus on the environmental chemistry of the stable nuclide ^{9}Be, whereas Kaste et al. (this volume) focus on the cosmogenic isotope ^{7}Be and Morris et al. and Bierman et al. focus on ^{10}Be. Effects of point-source pollution on the immediate environment, such as in the vicinity of smelters, are not discussed except where the effects may be transmitted regionally by the atmosphere or surface water.

BERYLLIUM IN THE ATMOSPHERE

The atmospheric flux of ^{9}Be (henceforth Be) is derived primarily from industrial emissions (particularly from coal combustion, Moore 1991), eolian dust, and metallurgical processes in the space and weapons industry (Zorn et al. 1988). Oil and natural gas have significantly lower concentrations of Be than coal. Most coals contain 0 to 100 mg Be kg^{-1} (Wedepohl 1969) although Fishbein (1984) reported coals with Be as high as 1000 mg Be kg^{-1}. The accumulation of Be in coal results from the affinity of Be for organic matter before and during accumulation, as well as metasomatism of the coal precursors by groundwater.

The mean concentration of Be in coal ash ranges from 0.7 mg kg^{-1} to 2000 mg kg^{-1} in coals from the North Bohemian Basin (Bouška 1981). Emission factors for coal combustion range from 1 to 5.8% of the total Be, substantially lower than for Cd or Pb, 50 to 90% of which are in flue gases (Bezačínský et al. 1984; Šebestová et al. 1996). Emissions factors for Be are substantially lower for coal that has higher ash concentration (Dubanský et al. 1990). Be may be mobilized from ash and slag, especially by acidic

1529-6466/00/0050-0007$05.00

water. For example, 2.6 mg Be L^{-1} occurs in pond waste-water at pH 2.5 in the Czech Republic (Kubizňáková 1987). Acidic water releases Be best from fine-grained ash, whereas the coarse-grained fraction is strongly attacked by alkaline solutions (Kubizňáková 1987). Concentrations up to 29 mg Be kg^{-1} occur in industrial particles (e.g., from iron production, Seth and Pandey 1985).

Kubizňáková (1987) estimated annual emissions of Be in Europe to be 30 tons yr^{-1}; Pacyna (1990) suggested 50 tons yr^{-1}. For 1980 in the U.S.A., coal consumption was approximately 500×10^6 tons (Husar 1986). With an assumed average of 10 mg Be kg^{-1} (Phillips, 1973) and an emission factor of 5%, Be emissions in the United States would have been approximately 250 tons yr^{-1}. Bezačínský et al. (1984) found that atmospheric deposition of Be during 1980 was three times higher per unit area in the former Czechoslovakia than in the United States. Yamagata (1981) reported that background concentrations of Be in Japan's ambient air ranged from 0.04 to 0.261 ng m^{-3}. Thorat et al. (2001) reported mean Be in air at the perimeter of a Be processing plant to be about 0.5 ng m^{-3}, also well below recommended levels for industrial and ambient air quality. Thorat et al. attributed the values to mobilized dust, based on Be/Fe ratios in the suspended particulate material. However, air in industrial environments may have much higher Be values. For example, Nakamura et al. (1983) found up to 6.6 μg m^{-3}, with a mean of 0.21 μg m^{-3} inside a Cu-Be casting factory.

Paleolimnological studies of remote lakes in upper New York State indicate that atmospheric deposition of Be started increasing in eastern North America about 1910 (Heath et al. 1988), twenty years later than in central Europe (see below). Such changes in atmospheric burden may explain the higher Be concentrations in the North Atlantic versus the Pacific Ocean. On the other hand, major input of Be to the surface waters of the South Atlantic may be from the partial dissolution of eolian dusts, with the proportional dissolution being about seven times that of Al (Measures et al. 1996).

BERYLLIUM IN PRECIPITATION

Beryllium concentration in precipitation reflects regional atmospheric emissions from coal combustion in power plants (Bezačínský 1980; Kubizňáková 1987). Solid atmospheric particles of soil dust and anthropogenically emitted ^{9}Be, as well as cosmogenic ^{7}Be and ^{10}Be, are transported to the Earth's surface mostly via wet precipitation (Baskaran 1995; Gaffney et al. 1994). The residence time of Be in the atmosphere can be inferred from ^{7}Be radioisotope data (Kaste et al., this volume). ^{7}Beryllium forms primarily in the stratosphere (residence time of ^{10}Be, attached to particles, is about 14 months) and has an estimated residence time in the lower troposphere of 10.3 to 48 days (Baeza et al. 1996; Ďurana et al. 1996). The value for the stable lithospheric isotope ^{9}Be should be shorter, because it is injected into the troposphere (with the coal fly ash and terrigenous dust) on coarser particles. The residence time of ^{9}Be may be about nine days, the same as atmospheric water. The depositional flux of Be is related to the distance from and strength of emission sources.

Volume-weighted concentrations of Be in bulk precipitation range from 0.007 μg L^{-1} in Wales to 0.22 μg Be L^{-1} in the most polluted area of the Czech Republic during 1985 to 1989 (Table 1). Concentrations of Be are much higher in cloud moisture (Neal et al. 1992). They reported a mean value of 0.36 μg L^{-1} for mist in an upland catchment in Wales. Atmospheric fluxes of Be in regions not under the influence of point sources of pollution range from $<10 \times 10^{-6}$ g m^{-2} yr^{-1} to approximately 50×10^{-6} g m^{-2} yr^{-1}. Values reflect the importance of the wash-out effect of the wet precipitation, the amount of precipitation, and the impact of local and regional anthropogenic emission point sources of ^{9}Be into the atmosphere. Skřivan et al. (2000) documented a steady decrease of the

Table 1. Concentration of Be in atmospheric deposition.

Precipitation	*Locality*	*Concentration Range*, µg L^{-1} *Mean*, µg L^{-1}	*Deposition* 10^{-6}g/m^2/yr	*Remarks*	*Reference*
Cloud moisture	Wales - Plynlimon	0.0 to 0.50 0.36	3		Neal et al. 1992
	Wales - Plynlimon	0.0005 to 1.5 0.039	1 to 6		Wilkinson et al. 1997
Wet deposition	Norway - northern		7.7 to 10	1989-90	Berg et al. 1994
	Norway - Nordmoen		26	1989-90	Berg et al. 1994
	Norway - Birkenes		91	1989-90	Berg et al. 1994
	Bohemia - Lysina	0.04	45	1991-92	Krám et al. 1998
	Bohemia - Lesní Potok	0.0065	40 to < 10	1995-99	Skřivan et al. 2000, 2001
	Bohemia - 4 stations	0.02 to 0.22		1988	Dvořáková & Moldan 1989
	Bohemia - 14 stations		42 to 324	1986-87	Škoda unpub
	Maine - Bear Brook	<0.01		1998	Kaste 1999
	Wales - Plynlimon	0.0001 to 0.2 0.017	50	1992-95	Wilkinson et al. 1997
	Wales - Plynlimon	0.00 to 0.060 0.007			Neal et al. 1992
	Wales - Plynlimon	0.00 to 0.24 0.01			Neal et al. 1997
Throughfall	Wales - Plynlimon	0.0 to 0.13 0.011	~20		Neal et al. 1992
	Bohemia - Lysina	0.06	42	1991-92	Krám et al. 1998
	Bohemia - Lysina	0.043	25	1991-98	Krám & Hruška, unpub
	Bohemia - Lesní Potok	Beech 0.017 Spruce 0.032	4 to 16 beech, 8 to 13 spruce	1997-99	Skřivan et al. 2000, 2001
	Bohemia	0.03	30	1990-91	Veselý et al. 1998

annual Be flux in bulk precipitation from 40×10^{-6} g m^{-2} yr^{-1} for 1995 to less than 10×10^{-6} g m^{-2} yr^{-1} in 1998-1999 at the Lesní Potok watershed (Central Bohemia, Czech Republic). The decrease is attributed to reduced emissions, particularly fly ash, from the

Bohemian coal-burning power plants.

BERYLLIUM IN THROUGHFALL

Throughfall is net precipitation that interacts with a forest canopy. Interactions of the atmosphere and wet deposition with the canopy alter the chemistry of water reaching the forest floor (the organic-rich layer typically overlying mineral soil) and generally reduce the volume. Modification is caused by (1) evaporation/sublimation of water/snow/ice from foliar surfaces, (2) wash-off of solids, liquids, and gases deposited from the atmosphere in either dry or wet form, (3) leaching or ion exchange of leaf metabolite compounds, (4) adsorption of compounds, and (5) active uptake of nutrients from precipitation. The concentration of most chemical substances in water below the canopy (throughfall) exceeds that in the bulk precipitation. The increase depends on the intensity of interaction between the aqueous phase and foliage, and on the leaf area index of the forest canopy (Robson et al. 1994; Kostelník et al. 1989). Kaste et al. (this volume) demonstrate strong adsorption of ^{7}Be by foliage. Skřivan et al. (2000) indicate that significant Be is taken up from soil solutions and incorporated into foliage, providing a possible source of Be to throughfall leaching. However, throughfall fluxes are not substantially different from bulk precipitation.

The Be concentration in throughfall and total flux of Be to the forest floor are poorly documented. Neal et al. (1992) reported the volume-weighted mean concentration of Be in throughfall and stemflow was 1.5 and 3 times higher than in rain, respectively, in mid-Wales. Beryllium in mist was five times higher than in rain. Total Be deposition on the moorland and forested area was about 20×10^{-6} g m^{-2} yr^{-1}. Fluxes of Be in throughfall in beech (*Fagus sylvatica* L.) and Norway spruce (*Picea abies*) forest in the Lesní Potok watershed (Central Bohemia) (Table 1) were comparable. However, Skřivan et al. (2000) documented for 1999 that the flux of Be to the forest floor was dominated by litterfall (averaging 177×10^{-6} g m^{-2} yr^{-1} and 136×10^{-6} g m^{-2} yr^{-1}, for beech and Norway spruce, respectively). These net fluxes are too high, compared to the Be fluxes in throughfall and in precipitation, to be attributed solely to dry deposition of Be to foliage and woody tissue. Total deposition of Be (wet plus dry) may be up to two times higher than is reported due to Be in particulate matter that is not normally included in analytical determinations (Rose 1990). Veselý (1990) showed that the concentration of dissolved Be in precipitation substantially increased after chemical digestion of precipitation samples in a microwave oven. These results suggest that particulate matter with low solubility contributed to the total Be flux in precipitation.

BERYLLIUM IN FLORA AND FAUNA

Some forms of aqueous Be are toxic. Ezawa et al. (1999) described the inhibitory effect of Be on the activity of alkaline phosphatase. They examined arbuscules of the mycorrhizal fungus *Glomus etunicatum,* which affect sugar metabolism. Mathur et al. (1994) observed that Be was associated with significant depletion of alkaline phosphatase, elevation of acid phosphatase activities in liver, and increased concentration of Be in blood and other soft organs. A series of experiments in the 1970s (e.g., Slonim 1973; Slonim and Slonim 1973) demonstrated that acute toxicity of Be to freshwater organisms is strongly decreased by increased water hardness (Ca + Mg), presumably because of the formation of non-toxic complexes between Be and HCO_3^- or other ligands. Extrapolation of their findings to soft water suggests that toxicity may occur at concentrations of Be > 5 μg L^{-1}. Jagoe et al. (1993) demonstrated that aqueous Be at 10 μg L^{-1} creates the same symptoms on fish gill surfaces as 50 μg Al L^{-1}. Formation of complexes with SO_4 and CO_3 apparently eliminates toxicity of Be concentrations as high as 10 mg L^{-1}. Beryllium toxicity in soft water is comparable with that of Cd, but in water with high concentrations of Ca and Mg it is one to two orders of magnitude less toxic (Moore 1991).

Plants normally contain <1 mg Be kg^{-1} ashed weight (Meehan and Smythe 1967). Maňkovská (1995) found Be in wide concentration (0.0001 to 0.464 mg Be kg^{-1} dry weight (dw), mean 0.030) in 2-year old spruce needles from Slovakia. Only a slight enrichment of Be in the phytomass of bulk feeds (0.005 to 0.28 mg kg^{-1}) and cereal grains (0.005 to 0.019 mg kg^{-1} dw) occurred in the severely polluted north-Bohemian region. The (Be in soil)/(Be in plant) concentration ratio was 0.004 to 0.01 for feeds crops and about 0.0015 for cereal grain, generally lower than transfer factors obtained in the laboratory, where Be was added to soils in soluble forms (Němeček et al. 1994). The Be concentrations in rice and vegetables affected by wastewater were 13 and 1.8 to 6.2 times those of the controls (Long et al. 1989). The Be concentrations in potato and barley samples, collected in Germany, were 0.8 to 7 µg kg^{-1} dw with an average concentration of 2 to 3 µg Be kg^{-1} (Grote et al. 1996). Sajwan et al. (1996) assessed the effects of soil clay concentration on uptake of Be by soybeans. The most toxic effects of soluble Be occurred in soils low in clay; plant biomass was reduced by as much as 90%, as Be concentration in the tissues reached 226 mg kg^{-1} dw. Liming of soils (increasing the pH) resulted in lowered concentrations of Be in plant tissue.

Certain aquatic macrophytes and other aquatic plants accumulate Be in polluted environments. Sarosiek and Kosiba (1993) reported Be ranging from <1 to nearly 100 mg kg^{-1} dw for a variety of submerged, floating, and emergent plants. The Be may be bound in tissue, adsorbed from the water column, or adhered to particulate material. Kaste (1999) and Norton et al. (2000) found considerable exchangeable (labile) Be (up to 4×10^{-2} cmoles(+) kg^{-1} dw) and Al (ca. 10 cmoles(+) kg^{-1} dw) in aquatic vegetation (especially *Sphagnum*) of an acidic Be-rich stream. The *Sphagnum* also provided considerable buffering capacity for Ca, Mg, and pH. Leaves from poplar (*Populus nigra*) may be useful as a bioindicator of Be pollution (Dittmann et al. 1984). Mean Be concentrations in tissue of typical tree species of the Czech Republic, beech (*Fagus sylvatica* L.) and Norway spruce (*Picea abies* L. Karst), were determined for areas underlain by base-poor (Ca, K, Mg, and Na) granitic rocks, and base-rich calcareous rocks. The granitic rocks had a relatively high concentration of Be (13 mg kg^{-1}), mostly in plagioclase (Skřivan et al. 2000). The concentration of Be increased from 3.8 mg kg^{-1} near the top of the soil to 7.4 mg kg^{-1} at 1 m depth. The calcareous limestone bedrock had 0.38 mg Be kg^{-1}. The upper 0.65 m of soil had 3.7 to 5.5 mg Be kg^{-1}. The concentrations of Be in stem wood, bark, beech leaves, and spruce needles at the two sites are given in Table 2. The stem wood and bark were sampled in the Fall. Beech tree foliage was collected monthly from one tree throughout one year from May till December. The spruce needles were sampled monthly throughout one hydrological year. Samples were

Table 2. Be in plant tissue from the Czech Republic (µg kg^{-1} dry wt) (from Skřivan et al. 2000).

Tissue/species	*Spruce*	*Beech*	*Spruce*	*Beech*
Stem wood	13.8	20.5	3.0	3.2
Bark	78.3	68.2	10.7	18.7
Leaves or needles	308	231	No data	No data
Bedrock	Granitic	Granitic	Calcarous	Calcareous

dried in a flow box at room temperature, acid digested (HNO_3) and analyzed by ICP-MS. (Skřivan and Samek, unpublished). The large differences between the Be concentration in tree components at the two localities are probably caused by different availability of Be because of different soil pH.

BERYLLIUM IN SOIL

The average upper continental crust abundance of Be is 3.1 mg kg^{-1}, about twice that of the lower continental crust (Wedepohl 1995). Granitic rocks, shales, and sandstone and limestones typically have between 2.5 and 5, 2 and 5, and <0.2 to 4 mg Be kg^{-1}, respectively (Wedepohl 1969; Grigor´yev 1984). Clay sediments generally contain more Be than sandy sediment. Enrichment of Be occurs in manganese nodules of the Pacific Ocean (Merrill et al. 1960) and in limonitic concretions of near-shore sediments (Hirst 1962). Average Be concentration in metamorphic pelitic rocks is about 3.5 mg kg^{-1}, ranging from 0.5 to 8 mg Be kg^{-1}. The concentration of Be in granites may exceed 10 mg kg^{-1} (Macháček et al. 1966; London and Evensen, this volume). In some granitic pegmatites in Maine and New Hampshire, U.S.A., Be concentrations may locally exceed 10,000 mg kg^{-1} (Barton and Goldsmith 1968; Burr 1931). In these Be-rich terranes, the Be occurs predominantly in the mineral beryl ($Be_3Al_2Si_6O_{18}$); however, six other beryllium minerals are known to occur in the pegmatites (Burr 1931). Many of the Be-rich pegmatite occurrences in the northeastern U.S.A. are associated with the granitic bodies (Barton and Goldsmith, 1968). Some minerals with non-essential Be have high Be concentrations, for example, zinnwaldite and fluorite: 50 to 80 mg kg^{-1}; topaz: 50 mg kg^{-1} (Dubanský et al., 1990). Shilin and Tsareva (1957) reported a Be concentration of 860 mg kg^{-1} for lepidolite (see also Grew, this volume).

Total Be in soil ranges from 0.1 to 300 mg kg^{-1}, with the world average near 6 mg kg^{-1} (Drury et al. 1978). Shacklette (1984) reported <1 mg Be kg^{-1} to 15 mg Be kg^{-1} in soils from about 1,500 localities throughout the U.S.A. Anderson et al. (1990) observed Be in the range 0.3 to 30.5 mg kg^{-1} in Piedmont and Coastal Plain soils (southeastern U.S.A.), with a tendency to increase with soil depth. Chen et al. (1999) reported on 417 acidic soils (pH_{H2O} = 5.04±0.97) from Florida that ranged from 0.01 to 5.92 mg Be kg^{-1}. Soils in the Lost River Valley, Alaska contain from <1 to 300 mg Be kg^{-1}, and average about 60 mg kg^{-1} (Sainsbury et al. 1986). Total and acid-soluble concentrations (2M HNO_3) of Be were investigated in agricultural soils (A_p horizon) of the north-Bohemian region which has experienced high atmospheric pollution (Podlešáková and Němeček 1994). There the average total Be concentration ranged from 2.1 to 5.2 mg kg^{-1}. Acid soluble Be ranged from 0.53 to 0.98 mg kg^{-1} (22 to 27% of total Be). The degree of soil pollution reflected the isolines of fly ash deposition. Dudka and Markert (1992) reported a geometric mean of 0.71 mg Be kg^{-1} for 130 soil samples distributed across Poland. Concentrations in the range 0.46 to 4.7 mg Be kg^{-1} occurred at the Nevada Nuclear Weapons Test Site, U.S.A., where Be may have been used in past operations (Patton 1992).

Data regarding the distribution and mobility of Be in soils are limited. Inorganic and organic ligand anions may increase, decrease or have no effect on the sorption of metals (including Be) in soils (Naidu et al. 1998). Important controls are pH, organic matter, and hydrous oxides (particularly of Al, Fe, and Mn). Anderson et al. (1990) determined exchangeable Be (5 g of soil and 25 mL of 1-M NH_4Cl) to be from <0.9 to 20 μg kg^{-1}, comprising 0.0003 to 0.0083% of CEC in southeastern U.S.A. soils. Exchangeable Be was weakly negatively correlated with soil depth, whereas total Be increased with depth. Exchangeable Be and Al were unrelated. Kaste (1999) and Pellerin et al. (1999) found exchangeable Be forming 0.003 to 0.35% and 0.001 to 0.01% of CEC in spodic soils of

Halfmile (granite bedrock) and Bear Brook (quartzite and metapelitic bedrock) watersheds in Maine. Exchangeable Be in the forest floor increased nearly 300% on a transect of 17 m from well-drained upland soil to poorly drained riparian soil. Exchangeable Al followed a similar trend, but Ca was the opposite, decreasing towards the riparian zone. The increased Al and Be concentrations occur where the soils have a higher pH because of mixing of emerging groundwater and degassing of CO_2 in the riparian zones. Concentrations of exchangeable Be were markedly different in the podzolic brown earths and peaty gley at Lysina watershed, in the Czech Republic (Krám and Kaste, unpublished). Podzolic brown earths had lower concentrations of exchangeable Be in the forest floor (O horizons; 36 $\mu g\ kg^{-1}$ versus 163 $\mu g\ kg^{-1}$) and in the mineral soil (E+B horizons; 27 $\mu g\ kg^{-1}$ versus 77 $\mu g\ kg^{-1}$) than in the peaty gley in the riparian zone of the stream. In the podzolic brown earths of the upper 40 cm of the mineral soil (E and B horizons), exchangeable Be was 5.9 mg m^{-2} versus 0.5 mg m^{-2} in the forest floor, which was 7 cm thick. Zajíc (unpublished) compared total and extractable Be in two soil profiles. Soils over Be-rich granite (15 mg Be kg^{-1}) had 3.8 to 7.3 mg Be kg^{-1}, 3 to 17% of which was extractable (0.5 g of soil with 100 mL 0.1 M HNO_3). The soil developed on limestone (0.38 mg Be kg^{-1}) had 0.38 to 5 mg Be kg^{-1}, 19 to 25% of which was extractable.

Óvári et al. (2001) determined the speciation of Be in soil material (0-20 cm) from Hungarian sites polluted by atmospheric deposition. They found that 15 to 50% of the total Be could be extracted by sequential extraction of exchangeable, acid soluble, easily reducible, moderately reducible, and oxidizable fractions. Total Be ranged from 0.17 to 0.60 mg kg^{-1}.

Beryllium is released from the Al-rich layers in the soil during acidification (Neal et al. 1992). Profiles of concentrations of Be in two acid-sensitive soils of forested regions receiving different loading of acid deposition suggested the depletion of Be in humic and upper mineral soils in more acidified areas (Veselý 1987; Gooddy et al. 1995; Veselý and Majer 1996). Kaste (1999) found that Be mobilization from an artificially acidified catchment in Maine, U.S.A. was twice that of the adjacent reference watershed.

CHEMICAL WEATHERING

Beryllium occurs in trace amounts in silicate minerals such as feldspar, biotite, and muscovite, most commonly substituting for Si and Al in tetrahedral positions. Chemical weathering of Be-bearing silicate minerals involves a set of incongruent reactions followed by the interactions of their hydrolysis products. Laboratory studies of Be release rates from granites from Maine and Bohemia (Fig. 1) and their individual minerals (0.1-0.2 mm fraction) at constant pH (4.3 and 3.8) indicate that release of Be increases as follows:

quartz < (microcline + plagioclase) < bulk granite < plagioclase < biotite

Release rates increase with increasing Be concentration in the mineral and with decreasing pH. Rate of total Be weathering ranged from 0.07 to 3.2×10^{-6} g m^{-2} yr^{-1}. The amount of Be leached from biotite and feldspars was comparable, in spite of substantially lower amount of biotite (5 to 10%), compared with the more abundant feldspar (about 60% of the granite) (Majer et al., unpublished). Plagioclase contained up to 20 mg Be kg^{-1}, exceeding microcline (up to 4 mg kg^{-1}). The fluoride (F^-) concentration of biotite may be crucial to the high release rate of Be during weathering (Minařík et al. 1997), possibly because of complexation of Be with F^- in soil leachate. In older soils where biotite has been preferentially weathered, release rates of Be by primary weathering should decrease as other factors determine the export from catchments.

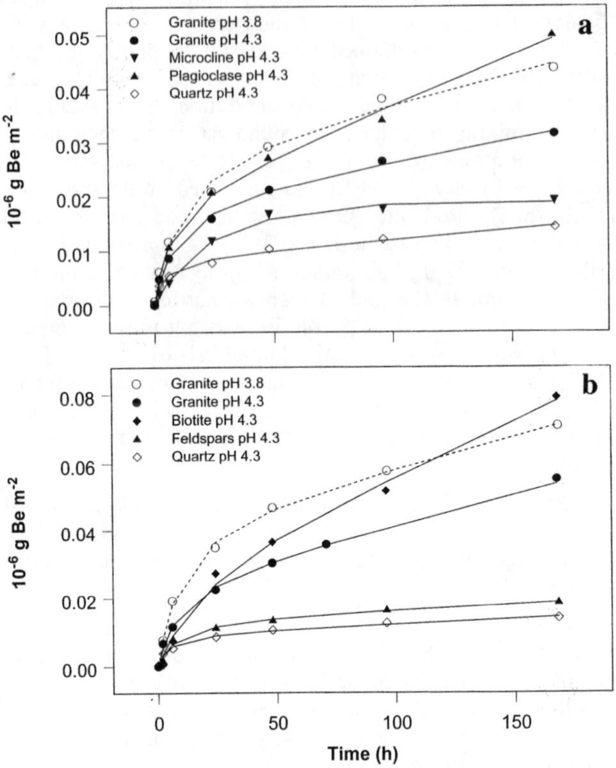

Figure 1. Experimental leaching of Be from whole rock and mineral separates from (a) Lucerne Granite (2), Maine and (b) monzogranite, Lesní Potok, Czech Republic. (0.1-0.2 mm fraction, constant pH = 4.3 or 3.8; Majer, unpubl.). Total release (10^{-6} g Be m^{-2}) versus time (hours).

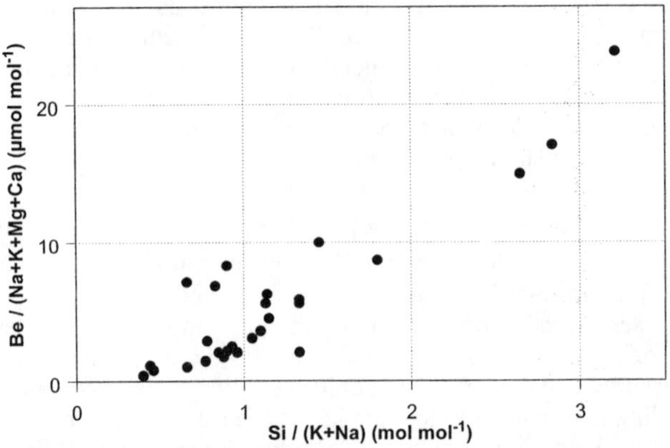

Figure 2. Dissolved Be/(Na+K+Mg+Ca) vs. Si /(K+Na) in tropical rivers with pH ≥ 6.7 flowing through sedimentary plains. Beryllium is normalized to compensate for dilution, or concentration through evapotranspiration. The trend indicates that in rivers where there has been significant formation of secondary minerals, i.e., Si/(K+ Na) is low, there is less normalized, dissolved Be (modified from Brown et al. 1992).

Brown et al. (1992) demonstrated the influence of the extent of soil development on concentrations of Be in rivers draining sedimentary basins in South America (Fig. 2). The ratio Si/(K+Na) in the water may be used as an indicator of the extent of clay mineral formation in floodplain regions. In Figure 2, Be is normalized to total cations (Na+K+Mg+Ca) to account for dilution by rain or concentration by evapotranspiration. The lower ratio occurred in rivers where there was significant formation of secondary minerals, i.e., where the ratio Si/(K+Na) was low. This relationship suggested that adsorption of dissolved Be onto secondary minerals reduced the concentrations of dissolved Be. In rivers with abundant sediment, the concentration of dissolved ^{10}Be is far lower than in the incoming rainwater, indicating that a substantial proportion of ^{10}Be is retained within the soils of the basin or is adsorbed onto suspended particulate material. However, in more acidic drainages, the concentration of ^{9}Be was generally higher and ^{10}Be in the rivers was apparently in steady state with rain input to the watersheds (Brown et al. 1992). Kaste et al. (this volume) found that even in acidic watersheds, ^{7}Be is strongly retained by exchange with soils.

BERYLLIUM IN SOIL WATER AND GROUNDWATER

Beryllium released primarily from Na-feldspar and biotite during chemical weathering is sorbed in soils or leached to soil water, groundwater, and surface water. The highest Be concentrations occur in soil water, and in shallow and deep groundwaters in non-carbonate terrain (Edmunds et al. 1992; Edmunds and Trafford 1993). They reported concentrations from samples in shallow drilled wells in England at generally less than 0.04 μg L^{-1}, but found up to 135 μg Be L^{-1} in one weathering zone (Fig. 3). The peak for Be corresponded to that of Mg, perhaps because of weathering of biotite (see *Chemical Weathering* section). High concentrations of Al in most samples suppressed the formation of Be-F complexes. Beryllium reached maximum concentrations somewhat deeper and at slightly higher pH than Al. Profiles of water chemistry in unsaturated zones

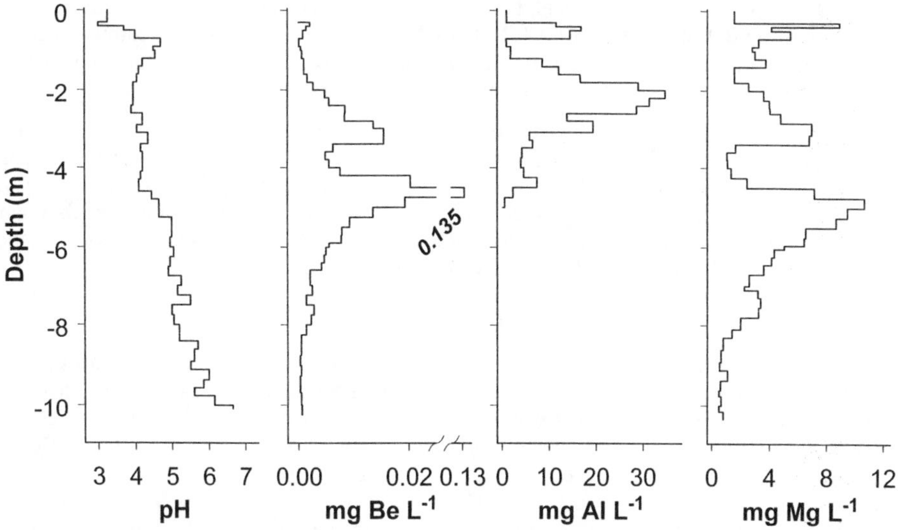

Figure 3. Soil water profile for pH plus Be, Al and Mg, Enville Forest, southern Britain, Triassic Sandstone. In the unsaturated zone, pH gradually increases downward. Significantly high Be anomalies occurred at 3 m and 5 m, deeper than Al maxima. The behavior of Be closely follows that of Mg (modified from Edmunds et al. 1992).

of the Lower Greensand (Surrey) and Triassic sandstone (West Midlands, U.K.) had >10 μg Be L^{-1} within the top 10 m (Edmunds and Trafford 1993). Be was <0.005 μg L^{-1} in carbonate aquifers. In contrast, Be in groundwater from 145 drilled bedrock wells in eastern and western Norway had a median value of 0.04 μg L^{-1}, whereas the F median concentration was 330 μg L^{-1} (Reimann et al. 1996). Banks et al. (1998b) reported that all but 1 of 72 ground water wells from Quaternary moraines and stratified sand and gravel had Be less than 0.5 μg L^{-1}. Banks et al. (1998a) reported Be from 1604 bedrock wells in Norway to range from less than detection limit (0.5 μg L^{-1}) to 11 μg L^{-1}. The Be concentrations had a minimum in the 40-60th percentile for pH (7.99-8.14) (Frengstad et al. 2001). The Be was below detection in most wells. Frei et al. (2000) reported that Be concentrations in springs were strongly inversely correlated with pH and pH related to bedrock lithology. Granites yielded spring water with Be up to 11 μg L^{-1}; basalts yielded water with <0.05 μg Be L^{-1}.

High concentrations of Be generally occur in deep groundwater where concentrations of F are high, e.g., from a spring in Karlovy Vary, Czech Republic, 77.5 μg Be L^{-1} and 6.6 mg F L^{-1} (Macháček et al. 1966). Kubizňáková (1983) reported mine water with Be as high as 100 to160 μg L^{-1}. Also, stability of Be complexes with HCO_3^- and F^- increases with temperature (Stunzas and Govorov 1981), enhancing mobility of Be in groundwaters.

Gooddy et al. (1995) documented that Be is leached to acidic soil water from acidic sandy soil (humus iron podzol) in the U. K. There, high concentrations of Cl (up to 400 mg L^{-1}) suggest a long residence time for the water providing ample opportunity for release of Be from primary weathering. At Lysina watershed, Czech Republic, soil solutions are extremely acidic. Volume-weighted mean pH was 3.4 in organic (O) horizon soil solutions, 3.2 in the uppermost mineral soil (E horizon), and 4.4 in the C horizon. Concentrations of Be in soil solution markedly increased with depth. Volume-weighted concentrations of Be in soil solutions were 0.21 μg L^{-1} in O horizon leachate, 0.41 μg L^{-1} in the E horizon, and 1.1 μg L^{-1} in the C horizon. The arithmetic mean of shallow ground water was 3.3 μg L^{-1} at pH 5.2 (Krám et al. 1998). Streamwater at the same watershed had a volume-weighted average of about 1 μg L^{-1} suggesting that the runoff was a combination of groundwater and soil water.

Acidifying watersheds may release Be at rates higher than is provided by primary chemical weathering, leading to decreased exchangeable Be in the soils and sediments (Veselý, 1994; Veselý and Majer 1996). Areas of concern for high Be occur where the water table is shallow, groundwater is acidic, and complexing ligands are abundant. Conversely, deacidification may reduce the export of Be (and base cations) to values below those provided by weathering (Galloway et al. 1983; Norton et al. 2000).

BERYLLIUM IN LAKES AND STREAMS

The concentration of Be ranges over three orders of magnitude in natural fresh water systems, i.e., from a few ng L^{-1} to a few μg L^{-1}. Concentrations of Be typically decrease in the order: soil water > groundwater > low-order stream water > river water > estuary water > ocean. Typical headwater circum-neutral stream water has 20 to 30 ng Be L^{-1}. Dissolved Be in rivers is typically around 10 ng L^{-1} (Table 3), whereas estuaries have about 1 ng L^{-1}, and the oceans have 0.05 to 0.3 ng Be L^{-1}. Almost 90% of dissolved riverine ^{9}Be is removed during estuarine mixing. In contrast, ^{10}Be does not decrease so dramatically from estuaries to the ocean (Kusakabe et al. 1991). This seaward decrease of ^{9}Be in surface waters implies that Be is strongly retained by solid particles in natural environments under neutral conditions (You et al. 1989; Veselý et al. 1989; Brown et al. 1992). Von Blanckenburg and Igel (1999) used the concentration of ^{9}Be (derived from the continents) and ^{10}Be (derived largely from direct precipitation) to draw inferences about lateral mixing and advection within ocean basins.

Table 3. Concentration of Be in rivers.

Locality	*Concentration, µg L⁻¹ range or median*	*Remarks*	*Reference*
Eastern United Kingdom	0.01 to 0.09	Dissolved	Neal & Robson 2000
	0.00 to 0.03	Particulate	Neal & Robson 2000
Czech Republic	<0.02 to 0.32 <0.02	Dissolved	Veselý et al. 2001
	<0.02 to 0.49 <0.02	Particulate	Veselý et al. 2001
North America	0.007 ± 0.007	Dissolved	Kusakabe et al. 1991
South America			
Amazon River + tributaries	0.014 ± 0.004	Dissolved	Measures & Edmund 1983
Amazon River tributaries	<0.04 to 0.23 <0.07	Acid soluble	Konhauser et al. 1994
Orinoco River + tributaries	0.021 ± 0.014	Dissolved	Brown et al. 1992
Pearl River, China	0.010	Dissolved	Kusakabe et al. 1991
Wales	0.02 to 0.54 0.01 to 0.58	Low flow High flow	Neal et al. 1998

In headwater streams in upland acidified areas of mid-Wales, Neal et al. (1992) found Be in the range <10 to 250 ng L^{-1}. Rissberger (1993) and Heath et al. (1988) reported Be ranging up to 2 µg L^{-1} in Maine. However, the concentrations of dissolved Be in unpolluted water were only 0.25 to 0.63 ng L^{-1} in Japan (Shimizu et al. 1999). Beryllium concentration in a stream may vary widely with discharge. Skřivan et al. (1994) reported 2 to 24 µg L^{-1} in the acidified stream at Lesní Potok, Czech Republic, draining granite with high F concentrations (mean 0.49 mg L^{-1}) and with highly variable pH (4.6 to 5.9). Navrátil et al. (2000) found a strong inverse relationship between annual volume-weighted means for Be and pH in the stream draining the Lesní Potok catchment. Total concentrations of Be sharply decreased with increasing pH, but Be^0 (uncharged aqueous species) concentration increased above pH = 6. Many other authors have reported inverse relationship between Be in water and pH (Measures and Edmond 1983; Veselý et al. 1989; Rissberger 1993; Veselý 1994; Neal et al. 1997; Kaste 1999). The elevated Be concentrations (and perhaps Al) associated with very low pH may be short-lived as soils and sediments become depleted of exchangeable Be. The regional freshwater survey (13,000 water samples) in the Czech Republic showed a clear inverse relationship between the concentration of Be and pH (Veselý and Majer 1996) (Fig. 4). Beryllium concentrations increased approximately by 100 times from pH 7 to 4. The median concentration of Be was about 0.03 µg L^{-1}, but the median concentration at pH 4.0 was about 1 µg L^{-1} Be. The difference between the upper and lower quartiles in Be concentrations increased with decreasing pH. The suggestion of a decrease in Be concentration at pH < 4.5 was attributed to depletion of the soils of their exchangeable Be because of anthropogenic acidification (Veselý, 1994). Kaste (1999) found that Be concentrations were increased approximately 100% in the stream water of the artificially acidified watershed at Bear Brook, Maine, U.S.A. (Norton et al. 1999). Veselý et al.

(1989) found that the relationship between Be concentration and pH for Czech surface waters is not straightforward. There was substantial scatter at lower pH, indicating that other factors, including soil acidification, influence concentrations of dissolved Be.

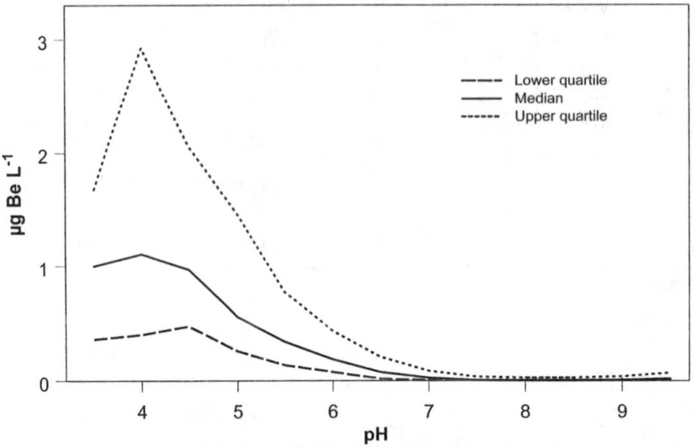

Figure 4. Relationship between the median and, upper and lower quartile concentrations of Be and pH in 13,000 Czech Republic fresh waters (Majer, unpubl.).

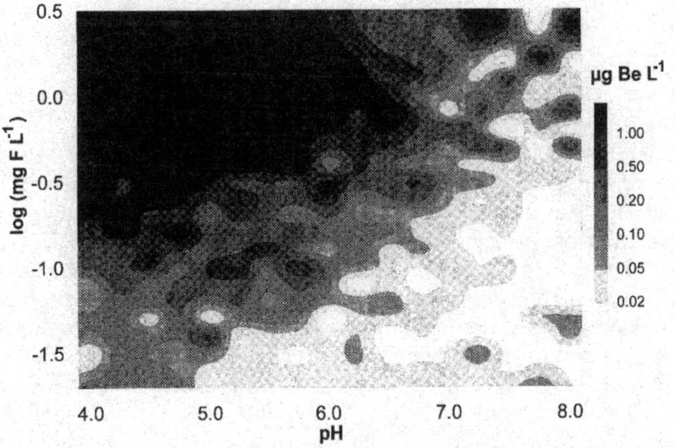

Figure 5. Beryllium in streams, contoured on pH and log F^- concentration, for 13,000 Czech Republic streams (Majer and Veselý, unpubl.).

Mobility of Be appears to be enhanced by formation of soluble F^- and organic complexes, and is suppressed by adsorption on solids at higher pH, where transport as Be-F complexes diminishes (Fig. 5). The concentration of Be in unfiltered water strongly correlates with F^- at pH < 6.0-6.5 and inversely with the pH. However, at Lysina, Be concentrations did not correlate with either pH or Al concentrations in strongly acidic water with high dissolved organic carbon (DOC) (Krám et al. 1998). There, concentrations of the Be in stream water were positively correlated with F^- and SO_4^{2-} (Fig. 6). The source of stream water Be at Lysina seems to be a mixture of acidic soil water with relatively low Be and less acidic ground water high in Be. It is likely that labile soil Be is dominated by soil cation exchange sites. Inflow of water with high concentrations of SO_4^{2-} to the mineral soil enhances the mobility of soil Be, resulting in high Be concentrations in stream water. Kaste (1999) and Norton et al. (2000) found parallel behavior between dissolved Al and Be in a low DOC stream in Maine, U.S.A. during natural and experimental acidification (Fig. 7), caused by increasing discharge and acid (HCl) addition to the stream, respectively. The investigations at Bohemia, Wales (Neal et al. 1992), and Maine indicate that the relationships between discharge and Al and Be in streams differ from watershed to watershed. Different chemistry along flow paths may explain differences in behavior among watersheds.

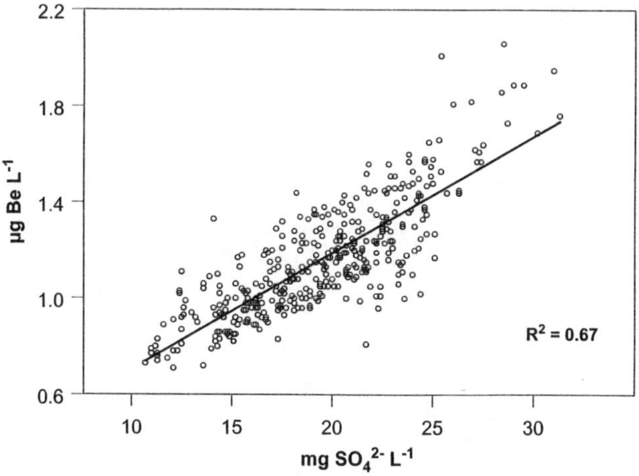

Figure 6. Relationships between concentrations of Be and SO_4 at Lysina, Czech Republic, April 1991 to February 1999 (Krám and Hruška, unpubl.).

Data regarding concentrations of Be in lake water are scarce. Skjelkvåle et al. (1996) reported a median Be (<10 ng L^{-1}) for 473 Norwegian lakes. Higher concentrations occurred in southern and eastern Norway, corresponding to the area impacted by acidic deposition. Tao et al. (1988) found Be from 0.9 to 65.7 ng L^{-1} in five lakes in Japan. Concentrations were inversely related to pH. Veselý et al. (1998) monitored Be in six acidified lakes and their tributaries in the Czech Republic from 1984 to 1995. Median concentrations in each lake were 60 to 460 ng L^{-1}. Over the 11 year record, Be decreased in the lakes by 22 to 37%, and in tributaries by 27 to 68%. The strong decrease of Be was concurrent with a sharp decline in acidic deposition after 1986, the result of reduction of emissions directed by European countries. These reductions produced decreased mobilization of Be from soil and bedrock.

Beryllium in river water was first measured by Merrill et al. (1960). Concentrations of dissolved Be reflect regional geology (Brown et al. 1992). For all studies, higher concentrations of Be generally occurred in rivers draining basins dominated by granite and where pH was lower, whereas lower concentrations occurred in higher pH rivers draining carbonate and mafic terrain. Konhauser et al. (1994) suggest that Be export from Amazon River tributaries may slightly exceed that of atmospheric input but it is not known if the export was particulate or dissolved Be. Some rivers that receive industrial discharge are enriched in Be. The highest concentrations of dissolved Be in such rivers occur during baseflow when dilution of point sources is lower (Neal and Robson 2000). Long et al. (1989) reported substantial reduction of Be in a river polluted by industrial wastewater discharge 500 to 1000 m below the industrial discharge.

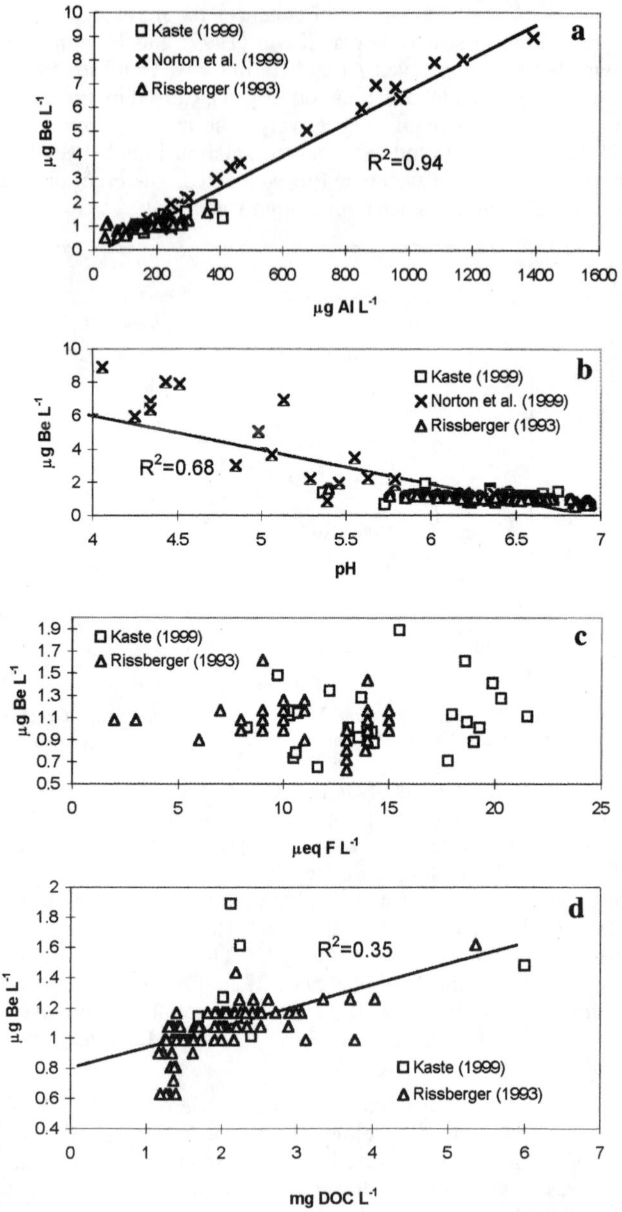

Figure 7. Relationships between Be and (a) Al, (b) pH, (c) F, and (d) Dissolved organic carbon (DOC) under natural conditions (from Kaste 1999 and Rissberger 1993) and during artificial acidification (Norton et al. 2000).

Determination of dissolved Be and suspended particulate Be in rivers with a wide range of chemical composition indicates that its geochemistry is primarily controlled by four factors. (1) Abundance of Be in the rocks and soil of the watershed (Measures and Edmond 1983), (2) the extent of Be adsorption onto particle surfaces (Veselý et al. 1989;

Brown et al. 1992), (3) discharge (Neal et al. 1997), and (4) water chemistry. Partitioning between solid and aqueous phases in soil-, ground-, and surface water systems is affected by the abundance of organic matter and secondary minerals because Be is strongly adsorbed. Beryllium distribution in acidic waters which have had little interaction with fluvial sediments is dominated by the composition of the rocks and soil of the drainage basin, whereas Be in neutral to alkaline rivers in contact with sediment is inversely related to the degree of mineral alteration (Brown et al. 1992).

Beryllium concentrations in the western North Atlantic Ocean surface water (between 33° and 42°N) are about five times those in the surface Pacific (Measures and Edmond 1983; Ku et al. 1990). There is a systematic increase in the ratio $^{10}Be/^{9}Be$ [(atom/(atom)] $\times 10^7$ along the advective flow lines of the surface North Atlantic (0.4) $\rightarrow$ deep North Atlantic (0.6) $\rightarrow$ circumpolar (1.0) $\rightarrow$ deep Pacific (1.2) $\rightarrow$ surface Pacific (1.3). This distribution pattern and contrasting North Atlantic-Pacific chemistry may be explained by a strong fluvial and continental/anthropogenic dust input of ^{9}Be to the North Atlantic (analogous to Pb) and an ocean-wide more or less uniform fluvial and atmospheric input of cosmogenically produced ^{10}Be (Ku et al. 1990).

SPECIATION OF BERYLLIUM IN WATER

Speciation of Be in water is important because toxicity varies according to aqueous speciation and mobility is largely controlled by the formation of various Be-complexes and adsorption-desorption phenomena. Equilibrium constants for various inorganic complexes have been compiled by many authors (e.g., Smith and Martell 1976; Högfeldt 1982; Rai et al. 1984). Consistent equilibrium constants may be incorporated into various chemical equilibrium models such as MINEQL+ (Schecher and McAvoy 1998). The quantitatively most important species in soft waters are Be-F and Be-OH complexes. Equilibrium constants in the Be-F-H_2O system were redetermined by Anttila et al. (1991):

$$Be^{2+} + F^- = BeF^+ \qquad \log K = 5.21$$
$$Be^{2+} + 2F^- = BeF_2^0(aq) \qquad \log K = 9.57$$

The binding constants for the ternary complexes

$$Be_3(OH)_3^{3+} + F^- = Be_3(OH)_3F^{2+} \qquad \log K = 4.48$$
$$Be_3(OH)_3F^{2+} + F^- = Be_3(OH)_3F_2^{+} \qquad \log K = 3.51$$

are about one tenth of the corresponding stepwise constants for the formation of BeF^+ and $BeF_2^0{}_{(aq)}$. High concentrations of Al in waters (e.g., shallow groundwaters, Edmunds and Trafford 1993) greatly suppress the formation of BeF^+ complex ions due to formation of stronger Al-F complexes and depletion of free F^- ions. Formation of Be-F complexes is most favored between pH 4.2 and 5.0 (Veselý et al. 1989). In waters with high total F^-, BeF^+ and $BeF_2^0{}_{(aq)}$ may dominate up to pH ~ 6.0 (Fig. 5). At pH < 4, Be occurs dominantly as free uncomplexed ions (Be^{2+}). At progressively higher pH, the Be becomes progressively hydroxylated, analagous to Al. Coordination chemistry of Be with organic ligands is reviewed by Schmidbaur (2001). He emphasizes that Be interacts with carboxylic and hydrocarboxylic acids and polyols. Binding to fulvic acids, among the strongest naturally occurring organic ligands, occurs above pH 6 (Fig. 8). Conditional stability constants for the complexation of Be with fulvic acids (log K = 5.9 to 6.5) at pH 6 to 7 (Esteves da Silva et al. 1996) are high, even larger than for the Cu(II) system. Log K for the Be-organic complex decreased with decreasing pH at pH < 6, and is slightly lower at pH 7 than at pH 6. The common trend is probably a result both of the acid-base properties of the ligand structures and of the Be hydrolysis. Some of the binding sites are protonated and unavailable for the complexation reactions. Beryllium strongly associates with soil fulvic acid, forming soluble complexes in soil water, but they decrease in relative abundance with lower pH. The relative significance of this association when compared to adsorption by inorganic (e.g., FeO(OH)) or organic (humic matter) solid phases in soils has not been estimated because no information on adsorption is available (Esteves da Silva et al. 1996). However, Be quantitatively co-precipitates with $Fe(OH)_3$ at pH 8 to 13 (Novikov et al. 1977) and Be in unfiltered waters with pH 7.8 to 8.8 correlates with Fe concentration (Veselý et al. 1989).

Veselý et al. (1989) studied the speciation of total Be (particulate, colloidal, and dissolved) in fresh waters. Aqueous speciation of Be is most directly affected by forming soluble complexes with F^- (Fig. 5) and dissolved organic matter. Rissberger (1993) and Kaste (1999) found that Be concentrations in a headwater stream in Maine related directly to discharge, the concentrations of dissolved organic carbon, and total F, and inversely to pH. Be export in the stream was partly as Be adsorbed on suspended particulate matter, dominantly $Al(OH)_3$. The percentage of total Be exported in this form decreased with decreasing pH. The adsorption of Be onto $Al(OH)_3$ occurs during degassing of CO_2 from Al-saturated groundwater emerging to surface water. The degassing raises pH and precipitates $Al(OH)_3$, which adsorbs the Be in the emerging groundwater (Norton and Henriksen 1983; Norton et al. 1992, 2000).

Beryllium is transported mostly in soluble form in acidic waters. As water is neutralized, dissolved Be becomes bound to the suspended particular matter, particularly organic matter, $Al(OH)_3$, and FeO(OH), in acid-soluble forms. In circum-neutral fresh waters (e.g., rivers) most Be migrates as suspended (non-labile) alumino-silicate minerals. Analyzing ashed suspended solids from 169 principal rivers and lakes in Japan, Teraoka and Kobayashi (1980) found Be distribution divided across Japan. Beryllium is markedly lower in the eastern half. Concentrations of Be were positively correlated with Al and Ti. Also, total exchangeable Be (dissolved + hydroxylamine-leachable) particulate phase was primarily dependent on the concentration of suspended particles. The residual Be constitutes a large proportion of the particulate Be but a relatively small fraction of the particulate ^{10}Be (Brown et al. 1992).

PARTITION (DISTRIBUTION) COEFFICIENTS OF BERYLLIUM BETWEEN WATER AND PARTICULATE MATTER

Adsorption-desorption processes strongly affect Be mobility in soil and leaching to surface or groundwater. Increasing soil pH leads to a rapid increase in net negative surface charge (Naidu et al. 1998) causing enhanced affinity for metal ions, i.e., an increase in cation exchange capacity. Changes in ambient solution chemistry can markedly influence the concentrations and nature of the Be species in soil water and fresh water and, consequently, their transport through the soil and into surface water.

Beryllium is strongly bound to solids under neutral conditions in natural systems. In waters of pH about 7.5, practically all Be occurs in/on suspended matter. The partitioning of Be between solid and water phases is characterized by its partition (distribution) coefficient K_d^{Be}, where $K_d^{Be} = C_S/C_L$, C_S = the concentration of Be on solids (mg kg^{-1}), and C_L = the concentration of Be in water (mg L^{-1}). The K_d^{Be} (L kg^{-1}) increases by 10^4 from pH 2 to 6. Thus, there is a 10^4 greater affinity of Be for solids at pH 6 than at 2. The partition coefficient of Be between solid and water is a function of pH, charge density of the solid surface, mineralogy, equilibration time, concentration of solid in the mixture, and composition and ionic strength of the solution. Log K_d^{Be} in rivers ranges from 4.20 to 5.30 (Olsen et al. 1986) and 4.08 to 5.97 (Veselý et al. 2001). Log K_d^{Be} is relatively high in lakes, 5.30 to 5.95 (Vogler et al. 1996). Hawley et al. (1986) determined log K_d^{Be} to be 6.3 to 4.48 for particle concentrations of 0.6 to 2 mg L^{-1}. You et al. (1989) experimentally

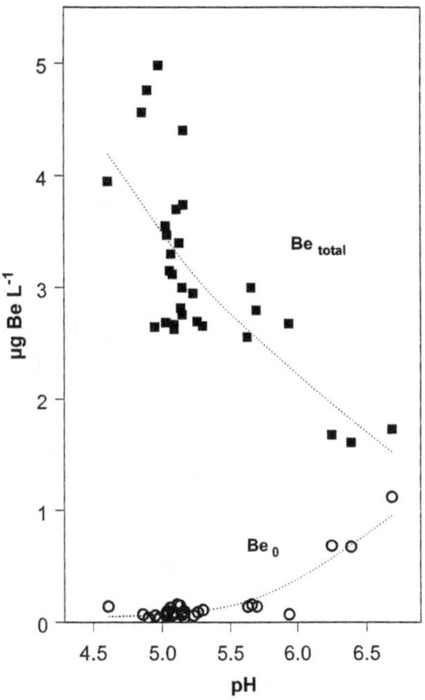

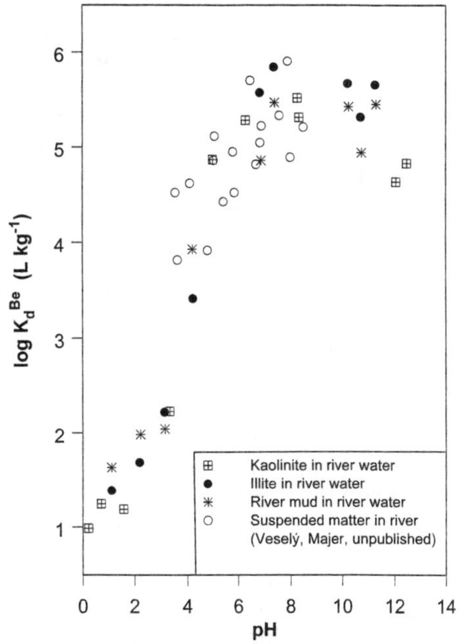

Figure 8 (left). Dependence of the concentration of Be_0 (uncharged aqueous Be) on pH at Lesní Potok, Czech Republic (Navrátil, unpubl).

Figure 9 (right). Relationship between the water-sediment partition coefficient (K_d^{Be}) and pH. K_d^{Bc} (L kg^{-1}) = ratio of solid (particulate) and dissolved fraction of Be. Data from You et al. (1989) and Majer (unpubl.) for Czech rivers.

determined log K_d^{Be} to be 4.64 to 6.08 for different solids suspended at concentrations of 200 mg L^{-1}. Aldahan et al. (1999) found that Be was up to 30 times more strongly sorbed by biotite than by albite, and not sorbed at all by quartz. They also reported that K_d^{Be} increased up to pH = 6; at higher pH, Be was controlled by the solubility of $Be(OH)_2$. Decreased K_d^{Be} with increasing concentrations of solids is commonly observed, but there is considerable scatter. Coarse-grained lake sediment with less clay had a lower K_d^{Be} (Hawley et al. 1986). K_d^{Be} decreases log linearly with declining pH below approximately 6 (You et al. 1989) with the steepest decrease ("adsorption edge") around pH 5 (Fig. 9). Equilibrium between the dissolved and particulate phases requires three or more weeks. You et al. (1989) reported that K_d^{Be} for river mud increased with temperature (log K_d^{Be} at 60°C was 6.26 compared to 5.49 at 25°C). K_d^{Be} derived from ^{7}Be in water column samples from Lake Michigan taken over a period of a few hours at the same depth and location varied by as much as a factor of 30 (Hawley et al.1986) and similar scatter of K_d^{Be} value has been observed in other studies. Olsen et al. (1986) found that the amount of Fe on particles correlates with the K_d^{Be} value. Uptake of Be by hydrous oxides may be enhanced by the presence of certain adsorbed organic ligands on their surface (Davis and Leckie 1978).

Many studies of partitioning have omitted the effect of filter-passing colloids (particles >1 kD (kiloDalton) = molecular weight > 1000, and diameter < 0.4 μm). Although colloids > 10 kD make up <15% of the total mass of suspended solids, the ^{7}Be associated with colloids and particles >1 μm was comparable in Lake Lugano (Switzerland/Italy). Partition coefficients of Be were higher for lacustrine colloids (log

K_d^{Be} ~ 6) than for particles >1 μm (log K_d^{Be} = 4 to 5) (Steinmann et al. 1999). The residence time of Be in the colloidal phase in water is typically higher than residence time in the particulate phase but one tenth the residence time for the truly dissolved Be. The rate-limiting step for scavenging was the adsorption of dissolved Be onto colloids. The coagulation rate of colloids <10 kD may be the limiting step in Be sedimentation from lake surface water. Coagulation rates ranged between 0.06 to 0.46 d^{-1}, corresponding to a residence time for ^{7}Be in the colloidal fraction of 2.2 to 16.3 days. The highest coagulation rates occurred after an algal bloom in spring (Steinmann et al. 1999).

BERYLLIUM IN SEDIMENT

^{7}Be and ^{10}Be, cosmogenic radionuclides with half lives of 53 days and 1.5 million years, respectively (see Kaste et al., this volume), are used to study sedimentation processes. ^{7}Be was used for determination of (1) rates of metal scavenging from lake surface waters (Steinmann et al. 1999), (2) sedimentation rates, and (3) recovery of the most recent sediments in lake and marine sediment cores (Moor et al. 1996; Sneed, 1986). The ^{10}Be flux to sediment of Lake Baikal (Russia) increases toward the last-glacial stage coinciding with an increase of the sediment accumulation rate. During the last-glacial stage, arid conditions in the lake's watershed caused a marked increase in flux of clay particles with low ^{10}Be concentration to the lake, resulting in an increased flux but lowered concentrations of ^{10}Be in the sediments (Horiuchi et al. 1999). Industrialization and urbanization around San Francisco Bay, California, as well as mining and agriculture in the upstream watersheds of the Sacramento and San Joaquin Rivers, have profoundly increased the sediment accumulation rate of Be throughout the estuary. The transient nature of increased ^{10}Be input (subsurface maximum) suggests that deforestation and agricultural development caused basin-wide erosion of surface soils enriched in ^{10}Be, probably before the turn of the century (van Geen et al. 1999). Silicate minerals from the Union Lake sediments (New Jersey) have a surface concentration of about 5×10^5 atoms ^{10}Be cm^{-2}. Both ^{10}Be and ^{9}Be were greatly enriched in organic matter (Lundberg et al. 1983). Bloom and Crecelius (1983) reported a much stronger affinity of ^{7}Be for inorganic particles (presumably clay minerals) than for organic matter, in marine conditions. Kaste (1999) found that the log K_d^{Be} of ^{7}Be between stream water and stream sediment or *Sphagnum* was less than for ^{9}Be but had no explanation, other than possibly an artifact of collection methods.

The origin of labile Be in sediment is basic to understanding biogeochemical cycling of Be and anthropogenic inputs of Be. Total Be in sediment of the Elbe River, Czech Republic ranged between 1.9 and 8.3 mg kg^{-1}, increasing at sites below seepage from power-plant ash dumps (Borovec 1996). Canney et al. (1987) reported stream sediment Be concentrations in northern New Hampshire ranging from 2 to 20 mg kg^{-1}, a granitic Be-rich province. Typically, ≥ 50% of Be in marine (Bourlès et al. 1992) and river (Veselý 1995) sediments is in insoluble detrital particles. Acid soluble Be (0.06M HNO_3) in sediment of a small lake in the Czech Republic was 1.5 to 8.9 mg kg^{-1} (Skřivan et al. 1996). Beryllium is bound in the organic fine-grained fraction of sediment (Lundberg et al. 1983; Veselý 1995; Borovec 1996; Skřivan et al. 1996), and to Fe-(hydr)oxides (Borovec 1996) and $Al(OH)_3$ (Kaste 1999).

Acid soluble Be in sediment from large circum-neutral man-made Czech lakes correlates with organic carbon. The highest slopes (acid soluble Be/$C_{org.}$) and concentrations of Be (up to 48 mg kg^{-1}) occur in sediments from two chain lakes in western Bohemia (Fig. 10). The ratio of acid soluble Be/C_{org} is six times higher for these two lakes because of higher Be air pollution and higher background Be concentrations from the Be-rich granite and lignite bedrock, and Be-rich groundwater (Veselý 1995).

The organic matter may serve as the carrier for a Be-rich phase of unknown composition, perhaps Fe hydroxides. Linkage to FeO(OH) is more probable in weakly alkaline water and in oxic sediment. The strength of binding by sediment organic matter is greatest at pH 6.0 to 6.5, and decreases at lower pH. At Čertovo Lake, Bohemia, which is strongly acidified (pH ~ 4.4) by air pollution, an increase in concentration and flux of organic matter in sediments dilutes the Be concentration, as does $CaCO_3$ in sediment of weakly alkaline lakes (Southon et al. 1987; Bourlès et al. 1989). At Plešné Lake, Czech Republic, concentrations of Be in sediment cores with continuous high organic matter (64 to 74% during the last 4500 years) correlate more closely and inversely with MgO concentration (Fig. 11) until about 1900 A.D. Thereafter, anthropogenic pollution dominates the Be flux.

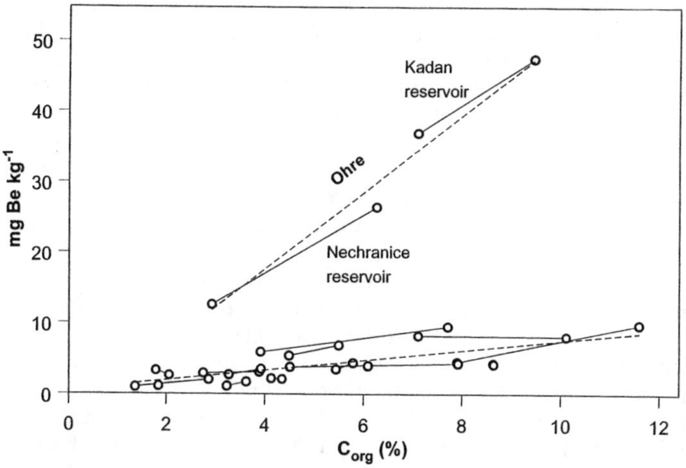

Figure 10. Relationship between concentrations of Be and organic carbon (C_{org}) in lake sediment. Values for pairs of samples from the same lake are connected. Two linear regressions (dashed lines) are shown, one for two reservoirs in the Ohře River and the other for other reservoirs from the Czech Republic. The higher slope implies higher available Be in the Ohře River region (Veselý 1995).

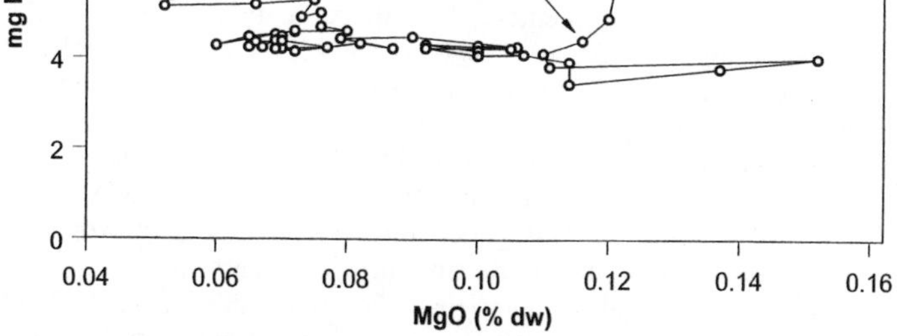

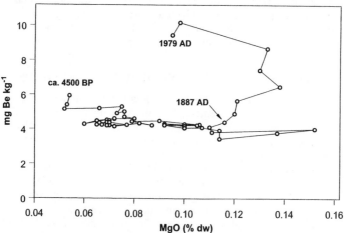

Figure 11. Relationship between concentrations of Be and MgO concentrations in sediment from Plešné Lake, Czech Republic since 4500 years B.P. (Veselý, unpubl).

Diagenetic processes affecting Be in sediment are poorly known. Isotopic exchange of Be between authigenic and detrital phases of the sediments appears negligible (Wang et al. 1996). Beryllium released to pore water in marine sediment is associated with dissolved Si. Profiles of depth versus dissolved concentrations of Be and Si are probably linked to the solubility of amorphous silica (Bourlès et al. 1989). The general decrease of the pore-water Be/Mn ratio with increasing depth indicates that Be solubilized during diagenesis is partitioned onto sediments more efficiently than Mn. Nevertheless, the flux of Be out of marine sediments at oxic and suboxic sites may be large enough to support the near-bottom Be increases observed in the water column (Bourlès et al. 1989).

Temporal changes in concentration and flux of Be in rivers and to lake sediment may provide information about intensity and patterns of pollution emission through time. For example, acid deposition and Be emissions decreased in the Czech Republic from 1986 to 1995. Concurrently, concentrations of acid soluble Be in sediment of the Elbe River (Labe) near the Czech-German border decreased by 0.25 mg kg^{-1} yr^{-1}, or by about 65% (Veselý, unpublished). Kemp et al. (1978) found little difference in the concentration of Be between surface sediment and older sediment (at ca. 20 cm depth) for a series of cores from Lakes Superior and Huron, U.S.A. In sediment of strongly acidified Plešné Lake, Czech Republic, Be concentrations correlate inversely with MgO concentration before 1880. Since 1900 the relationship was changed by atmospheric anthropogenic inputs of acids and Be (Fig. 11). Normalization of the Be sediment flux, to the concentration of MgO, yielded a calculated anthropogenic Be-flux of 0.25 mg Be m^{-2} yr^{-1}. These fluxes exceed those of modern atmospheric deposition (Table 3). Accumulation rates of Be in dated sediment cores from Lake Sagamore in the Adirondack Mountains, New York State and Lake Popradské in the High Tatras, Slovakia indicate accelerated sediment Be accumulation, caused by coal combustion, starting about 1910 A.D. in the northeastern U.S.A. and before 1900 A.D. in central Europe (Fig. 12). Net fluxes of sediment Be peaked in the second half of the 20th century. At Sagamore Lake, the maximum anthropogenic component was approximately 0.09 mg Be m^{-2} yr^{-1}. For Plešné, Sagamore, and Popradské Lakes, it is not known how much of the increased Be accumulation in sediment is caused by increased atmospheric deposition of Be, and how much is due to increased mobilization of Be from upland watersheds. The enhanced mobilization is caused by soil acidification, with subsequent deposition (adsorption onto sedimenting particles) in the higher pH regime of downstream lakes. Downward particulate fluxes of ^{10}Be at 130 m depth in Lake Zurich, Switzerland were 20% higher than fluxes at 50 m, probably due to lateral input (Schuler et al. 1991). Such focusing may enhance the accumulation rate of Be in deeper parts of lakes, which are the typical sediment coring sites.

MASS BALANCE OF BERYLLIUM IN WATERSHEDS

Export of Be from major catchments in humid environments in circum-neutral surface water, is not well documented. The sparse data suggest that the export of dissolved Be, averaging <10 ng L^{-1}, relative to Al, is approximately in the same proportion as Be:Al in bedrock. The export is also higher than the atmospheric deposition flux (e.g., Konhauser et al. 1994) except for severely polluted areas with low Be in the bedrock. Both Al and Be are preferentially removed as particulate matter in surface water, or retained in soils. In contrast, major base cations (Ca, Mg, K, and Na) are selectively removed in solution. This geochemical separation is narrowed in acidic or acidifying environments. Output of Be in surface waters from small acidified or acidifying watersheds is typically higher than the atmospheric flux. Neal et al. (1992) documented export of 42 to 149 $\times 10^{-6}$ g m^{-2} yr^{-1} in Wales (total atmospheric bulk deposition was 20 $\times 10^{-6}$ g m^{-2} yr^{-1}). About 52%, 18%, and 14% of the Be flux for

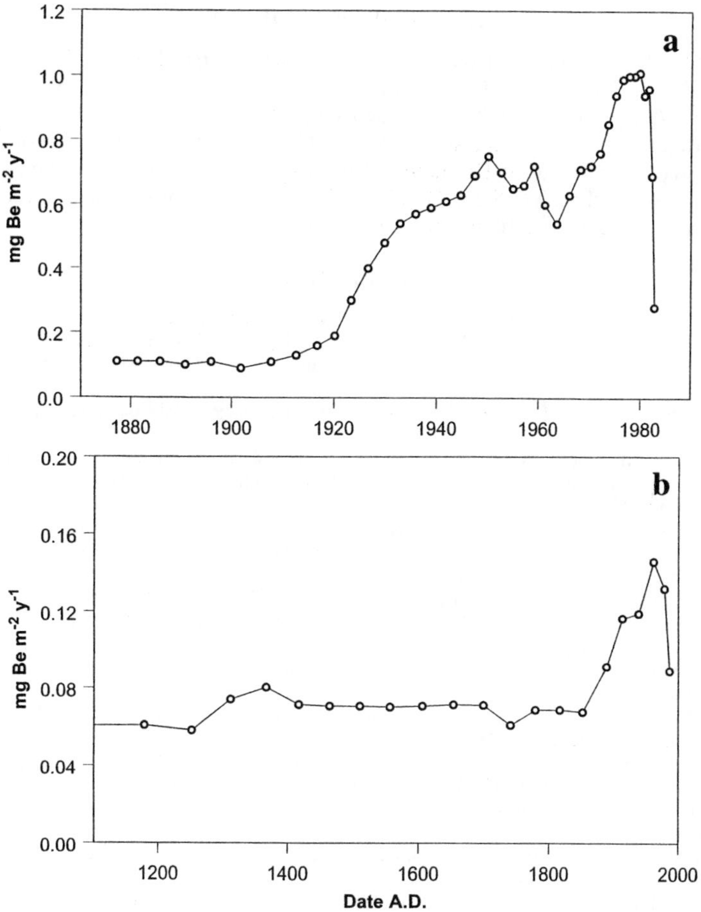

Figure 12. Be accumulation rates in lake sediment from (a) Sagamore Lake, Adirondack Mountains, New York, USA (Norton, unpubl) and (b) Popradské Lake, the High Tatras, Slovakia (Veselý, unpubl.).

drainage from the moorland, forested, and deforested areas, respectively, was from rainfall input. Veselý et al. (1989) estimated atmospheric input of Be at <20% of output for forested watersheds in Bohemia. More recently the percentage has decreased to 3 to 7%. For example, in the mid-1990s wet deposition at Lysina was 45×10^{-6} g Be m^{-2} yr^{-1} and drainage outflow was 586×10^{-6}g m^{-2} yr^{-1} (Krám et al. 1998). The annual export of Be from the acidified Lesní Potok watershed equaled $1{,}300 \times 10^{-6}$ g m^{-2} yr^{-1} (Skřivan et al. 1996). The average export of Be in the Czech Republic was estimated as 900×10^{-6} g m^{-2} yr^{-1} from very acidified mountainous terrain (pH 4.0 to 4.2), and about 120×10^{-6} g m^{-2} yr^{-1} dissolved Be at a stream pH of 6.0 to 6.2 (Veselý and Majer, unpublished). There may also be significant export as particulate matter especially at high discharge. These fluxes are substantially higher than that determined for the Elbe River, Czech Republic (6×10^{-6} g Be m^{-2} yr^{-1}; Moldan 1991), reflecting the strong influence of bedrock chemistry, surface water acidity, and recent acidification on Be mobilization. Only in acidified watersheds or severely polluted situations has human influence substantially altered the biogeochemistry of Be.

ACKNOWLEDGMENTS

The Czech Republic Ministry of Education (Grant ME147, 1998), Czech Geological Survey, Charles University, the University of Maine, and the U. S. National Science Foundation (Grant EAR-9725705) provided support for the collaborative research among Veselý, Majer, Skřivan, Krám, and Navrátil (Czech Republic), and Norton and Kaste (U.S.A.). We are grateful to Dr. Jakub Hruška for permission to use unpublished data from Lysina Watershed. We thank Drs. Jana Kubizňáková, Colin Neal, and Edward Grew for their thorough and constructive review of our manuscript.

REFERENCES

Aldahan A, Haiping Y, Possnert G (1999) Distribution of beryllium between solution and minerals (biotite and albite) under atmospheric conditions and variable pH. Chem Geol 156:209-229

Anderson MA, Bertsch PM, Miller WP (1990) Beryllium in selected southeastern soils. J Environ Qual 19:347-348

Anttila R, Grenthe I, Glaser J, Bruno J, Lagerman B (1991) Ternary Metal Complexes. 1. The Be(II)-F(⁻)-H_2O System. Acta Chem Scand 45:523-525

Baeza A, Delrio LM, Jimenez A, Miro C, Paniagua JM, Rufo M (1996) Analysis of the temporal evolution of atmospheric Be-7 as a vector of the behavior of other radionuclides in the atmosphere. J Radioanal Nucl Chem 207:331-344

Banks D, Frengstad B, Midtgård AK, Krog, JR, Strand T (1998a) The chemistry of Norwegian groundwaters: I. The distribution of radon, major and minor elements in 1604 crystalline bedrock groundwaters. Sci Total Environ 222:71-91

Banks D, Midtgård AK, Frengstad B, Krog, JR, Strand T (1998b) The chemistry of Norwegian groundwaters: II The chemistry of 72 groundwaters from Quaternary sedimentary aquifers. Sci Total Environ 222, 93-105

Barton WR, Goldsmith CE (1968) New England beryllium investigations. U S Bur Mines Rept Invest 7070:51-122

Baskaran M (1995) A search for the seasonal variability on the depositional fluxes of Be-7 and Pb-210. J Geophys Res 100(D2):2833-2840

Berg T, Røyset O, Steinnes E (1994) Trace elements in atmospheric precipitation at Norwegian background stations (1989-1990) measured by ICP-MS. Atmosph Environ 8:3519-3536

Beus AA, Sobolev BP (1963) Geochemistry of beryllium in high temperature postmagmatic mineralisation. Geochemistry (USSR) 8:316-323 (English version)

Bezačinský M (1980) Behaviour of some trace elements in combustion of Czechoslovak brown coals. Ochrana Ovzduší 12:3-7 (in Czech)

Bezačinský M, Pilátová B, Jiřele V, Bencko V (1984) To the problem of trace elements and hydrocarbons emissions from combustion of coal. J Hyg Epidemiol Microbiol Immunol 28:129-138

Bloom N, Crecelius EA (1983) Solubility behavior of atmospheric ^{7}Be in the marine environment. Mar Chem 12:323–331

Borovec Z (1996) Distribution of toxic metals in stream sediments. Acta Univ Carol Geol 38:91-103

Bourlès DL, Raisbeck GM, Yiou F (1989) ^{10}Be and ^{9}Be in marine sediments and their potential for dating. Geochim Cosmochim Acta 53:443-452

Bourlès DL, Brown ET, Raisbeck GM, Yiou F, Gieskes JM (1992) Beryllium isotope geochemistry of hydrothermally altered sediments. Earth Planet Sci Lett 109:47-56

Bouška V (1981) Geochemistry of Coal. Academia Press, Prague

Braucher R, Bourlès DL, Brown ET, Colin F, Muller J-P, Braun J-J, Delaune M, Minko AE, Lescouet C, Raisbeck GM, Yiou F (2000) Application of *in situ*-produced cosmogenic ^{10}Be and ^{26}Al to the study of lateritic soil development in tropical forest: Theory and examples from Cameroon and Gabon. Chem Geol 170:95-111

Brown ET, Edmond JM, Raisbeck GM, Bourlès DL, Yiou F, Measures CI (1992) Beryllium isotope geochemistry in tropical river basins. Geochim Cosmochim Acta 56:1607-1624

Burr FF (1931) Beryllium in Maine. Rocks and Minerals 6:8-9

Canney FC, Howd FH, Domenico JA, Nakagawa, HM (1987) Geochemical Survey Maps of the Wildernesses and Roadless Areas in the White Mountain National Forest, Coos, Grafton, and Carroll Counties, New Hampshire. U S Geol Surv Misc Field Studies Map MF-1594C

Chen M, Ma LQ, Harris WG (1999) Baseline concentrations of 15 trace elements in Florida surface soils. J Environ Qual 28:1173-1181

Davis JA, Leckie JO (1978) Effect of adsorbed complexing ligands on trace metal uptake by hydrous oxides. Environ Sci Techn 12:1309-1315

Dittmann J, Höffel I, Müller P, Neunhoeffer O (1984) Use of poplar leaves for the monitoring of environmental beryllium. Naturwiss 71:378-379

Drury JS, Shriner CR, Lewis EB, Towill LE, Hammons AS (1978) Reviews of the Environmental Effects of Pollutants: VI Beryllium. (unpubl work)

Dubanský A, Němec J, Jahoda K (1990) Geochemical role of beryllium in coals of the Northern Bohemian Lignite District. Acta Montana 82:5-50 (in Czech)

Dudka S, Markert B (1992) Baseline concentrations of As, Ba, Be, Li, Nb, Sr, and V in surface soils of Poland. Sci Total Environ 122:279-290

Ďurana L, Chudý M, Masarik J (1996) Investigation of Be-7 in the Bratislava atmosphere. J Radioanal Nucl Chem 207:345-356

Dvořáková M, Moldan B (1989) Atmospheric deposition to small watersheds monitored by the Czech Geological Survey in period 1985 to 1988. Ochrana Ovzduší 6:158-160 (in Czech)

Edmunds WM, Trafford JM (1993) Beryllium in river baseflow, shallow groundwaters and major aquifers of the U.K. Appl Geochem 219:223-233

Edmunds WM, Kinniburgh DG, Moss PD (1992) Trace metals in interstitial waters from sandstones: Acidic inputs to shallow groundwaters. Environ Pollut 77:129-141

EL-Hussein A, Mohamemed A, EL-Hady MA, Ahmed AA, Ali AE, Barakat A (2001) Diurnal and seasonal variation of short-lived radon progeny concentration and atmospheric temporal variations of ^{210}Pb and ^{7}Be in Egypt. Atmos Environ 35:4305-4313

Esteves da Silva JCG, Machado AASC, Ramos MA, Arce F, Rey F (1996) Quantitative study of Be(II) complexation by soil fulvic acids by molecular fluorescence spectroscopy. Environ Sci Techn 30:3155–3160

Ezawa T, Kuwahara S, Sakamoto K, Yoshida T, Saito M (1999) Specific inhibitor and substrate- specificity of alkaline-phosphatase expressed in the symbiotic phase of the arbuscular mycorrhizal fungus, *Glomus etunicatum*. Mycologia 91:636-641

Fishbein L (1984) Overview of analysis of carcinogenic and/or mutagenic metals in biological and environmental samples. Intl J Environ Analyt Chem 12:113-170

Frei M, Bielert U, Heinrichs H (2000) Effects of pH, alkalinity and bedrock chemistry on metal concentrations of springs in an acidified catchment (Ecker Dam, Harz Mountains, FRG). Chem Geol 170:221-242

Frengstad B, Banks D, Siewers U (2001) The chemistry of Norwegian groundwaters: IV. The pH-dependence of element concentrations in crystalline bedrock groundwaters. Sci Total Environ 277: 101-117

Gaffney JS, Orlandini KA, Marley NA, Popp CJ. (1994) Measurements of Be-7 and Pb-210 in rain, snow, and hail. J Appl Meteorol 33:869-873

Galloway JN, Norton SA, Church MR (1983) Fresh water acidification from atmospheric deposition of H_2SO_4 – A conceptual model. Environ Sci Tech 17:541A-545A

Gooddy DC, Shand P, Kinniburgh DG, van Riemsdijk WH (1995) Field-based partition coefficients for trace elements in soil solutions. Eur J Soil Sci 46:265-285

Grigor′yev NA (1984) Distribution of Beryllium on the Earth's Surface. Nauka, Moscow. 117 p (in Russian)

Grote M, Hofele J, Sietz M (1996) The uptake of beryllium by agricultural plants. Distribution and element speciation. Bioforum 19:556-567 (in German)

Hawley N, Robbins JA, Eadie BJ (1986) The partitioning of ^{7}Be in freshwater. Geochim Cosmochim Acta 50:1127–1131

Heath RC, Miller LM, Perry, CM, Norton SA (1988) Be in surface water: Sources, sinks, mobilization, and potential toxicity. *In* Trace Metals in Lakes (abstr). SETAC Intl Symp, Burlington, Ontario

Hirst DM (1962) The geochemistry of modern sediments from the Gulf of Paria—II. The location and distribution of trace elements. Geochim Cosmochim Acta 26:1147-1187

Högfeldt E (1982) Stability Constants of Metal-Ion Complexes. Part A: Inorganic Ligands. Pergamon Press, Oxford

Horiuchi K, Minoura K, Kobayashi K, Nakamura T, Hatori S, Matsuzaki H, Kawai T (1999) Last-glacial to post-glacial ^{10}Be fluctuations in a sediment core from the Academician Ridge, Lake Baikal. Geophys Res Lett 26:1047-1050

Husar RB (1986) Emissions of sulfur dioxide and nitrogen oxides and trends for eastern North America. *In* Acid Deposition—Long-Term Trends. National Academy Press, Washington, DC, p 48-92

Jagoe C, Haines TA, Matey V (1993) Beryllium ion effects on fish in acid waters analogous to aluminum toxicity. Aq Toxic 24:241-256

Kaste JM (1999) Dynamics of cosmogenic beryllium-7 and bedrock-derived beryllium-9 in forested ecosystems in Maine, U.S.A. Unpubl MSc thesis, University of Maine, Orono, ME, 96 p

Kemp ALW, Williams JDH, Thomas RL, Gregory ML (1978) Impact of man's activities on chemical composition of the sediments of the Lake Superior and Huron. Water Air Soil Pollut 10:381-402

Konhauser KO, Fyfe WS, Kronberg BI (1994) Multi-element chemistry of some Amazonian waters and soils. Chem Geology 111:155-175

Kostelnik KM, Lynch JA, Grimm JW, Corbett ES (1989) Sample size requirements for estimation of throughfall chemistry beneath a mixed hardwood forest. J Environ Qual 18:274-280

Krám P, Hruška J, Driscoll CT (1998) Beryllium chemistry in the Lysina catchment, Czech Republic. Water Air Soil Pollut 105:409-415

Ku TL, Kusakabe M, Measures CI, Southon JR, Cusimano G, Vogel JS, Nelson DE, Nakaya S (1990) Beryllium isotope distribution in the western North Atlantic: A comparison to the Pacific. Deep-Sea Res 37(5A):795-808

Kubizňáková J (1983) The occurrence, determination and screening of beryllium in various environmental compartments. PhD dissertation, Charles University, Praha (in Czech)

Kubizňáková J (1987) Beryllium pollution from slag and ashes from thermal power station. Water Air Soil Pollut 34:363-367

Kusakabe M, Ku TL, Southon JR, Shao L, Vogel JS, Nelson DE, Nakaya S, Cusimano G (1991) Be isotopes in rivers/estuaries and their oceanic budgets. Earth Planet Sci Lett 102:265-276

Long S, Lu Z, Hu Z, Xiao G, Tang L (1989) Study on beryllium pollution in China and the background levels of beryllium in the surface waters. Huanjing Kexue 10:83-85 (in Chinese)

Lundberg L, Ticich T, Herzog GF, Hughes T, Ashley G, Moniot RK, Tuniz C, Kruse T, Savin W (1983) ^{10}Be and Be in the Maurice River-Union Lake system of southern New Jersey. J Geophys Res 88 (C7):4498-4504

Macháček V, Šulcek Z, Václ J (1966) Geochemistry of beryllium in the Sokolov Basin. Sbor Geol Věd. TG 7:33-39 (in Czech)

Maňkovská B (1995) Mapping of forest environment load by selected elements through the leaf analysis. Ekologia 14:205-213

Mathur S, Flora SJS, Mathus R, Kannan GM, Dasgupta S (1994) Beryllium-induced biochemical alterations and their prevention following coadministration of meso-2,3-dimercaptosuccinic acid or 2,3 dimercaptopropane sulfonate in rats. J Appl Toxicol 14:263-267

Measures CI, Edmond JM (1983) The geochemical cycle of ^{9}Be: A reconnaissance. Earth Planet Sci Lett 66:101-110

Measures CI , Ku TL, Luo S, Southon JR, Xu X, Kusakabe M (1996) The distribution of ^{10}Be and ^{9}Be in the South Atlantic. Deep-Sea Res 43:987-1009

Meehan WR, Smythe LE (1967) Occurrence of beryllium as a trace element in environment materials. Environ Sci Tech 1:839–844

Merrill JR, Lyden EFX, Honda M, Arnold JR (1960) The sedimentary geochemistry of the beryllium isotopes. Geochim Cosmochim Acta 18:108-129

Minařík L, Burian M, Novák JK (1997) Experimental acid leaching of metals from biotite of the Říčany monzogranite. Bull Czech Geol Survey 72:239-244

Moldan B (1991) Atmospheric Deposition: A Biogeochemical Process. Rozpravy ČSAV, MPV Vol 101. Academic Press, Praha

Moor HC, Schaller T, Sturm M (1996) Recent changes in stable lead isotope ratios in sediments of Lake Zug, Switzerland. Environ Sci Techn 30:2928-2933

Moore JW (1991) Inorganic contaminants of surface water–Research and monitoring priorities (Chapter Beryllium) in Springer Series of Environmental Management, Springer Verlag, New York, p 50-56

Naidu R, Summer ME, Harter RD (1998) Sorption of heavy metals in strongly weathered soils: An overview. Environ Geochem Health 20:5-9

Nakamura I, Maroyama H, Nishida N, Sakaguchi T, Kagami M, Tada O (1983) Atmospheric beryllium concentrations in a copper-beryllium casting factory. Sci Mar Ika Daigaku Zasshi 11:6-13 (in Japanese)

Navrátil T, Skřivan P, Fottová D (2000) Human- and climate-induced changes in the surface stream activity affecting the element cycling. GeoLines 11:45-47

Neal C, Robson AJ (2000) A summary of river water quality data collected within the Land-Ocean Interaction Study: Core data for eastern UK rivers draining to the North Sea. Sci Total Environ 251/252:585-665

Neal C, Jeffery HA, Conway T, Ryland GP, Smith CJ, Neal M, Norton SA (1992) Beryllium concentrations in rainfall, stemflow, throughfall, mist and stream waters for an upland acidified area of mid-Wales. J Hydrol 136:33-49

Neal C, Wilkinson J, Neal M, Harrow M, Wickham H, Hill L and Morfitt C (1997) The hydrochemistry of the headwaters of the River Severn, Plynlimon. Hydrol Earth Syst Sci 1:583-617

Neal C, Reynolds B, Wilkinson J, Hill T, Neal M, Hill S, Harrow M (1998) The impacts of conifer harvesting on runoff water quality: A regional survey for Wales. Hydrol Earth Syst Sci 2:323-344

Němeček J, Podlešáková E, Pastuszková M (1994) Contamination of feed crops and cereal grains in the north-Bohemian emissions impact region. Rostliná Výroba 40:555-565 (in Czech)

Norton SA, Henrikson A (1983) The importance of CO_2 in evaluation of effects of acidic deposition. Vatten 39:346-354

Norton SA, Brownlee JC, Kahl JS (1992) Artificial acidification of a non-acidic and an acidic headwater stream in Maine, U.S.A. Environ Pollut 77:123-128

Norton S, Kahl J, Fernandez I, Haines T, Rustad L, Nodvin S, Scofield J, Strickland T, Erickson H, Wigington P, Lee J (1999) The Bear Brook Watershed, Maine (BBWM), U.S.A. Environ Monit Assess 55:7-51

Norton SA, Wagai R, Navrátil T, Kaste JM, Rissberger FA (2000) Response of a first-order stream in Maine to short-term in-stream acidification. Hydrol Earth Syst Sci 4:383-391

Novikov AI, Kononenko VA, Egorova LA (1977) Coprecipitation of beryllium with iron hydroxide. Radiokhimiya 19:160-165 (in Russian)

Olsen CR, Larsen IL, Lowry PD, Cutshall NH (1986) Geochemistry and deposition of ^{7}Be in river-estuarine and coastal waters. J Geophys Res 91(C1):896-908

Óvári M, Csukás M, Záray G (2001) Speciation of beryllium, nickel, and vanadium in soil samples from Csepel Island, Hungary. Fres J Anal Chem 370:768-775

Pacyna JM (1990) Estimation of the atmospheric emissions of trace elements from anthropogenic sources in Europe. Atmos Environ 18:41-50

Patton SE (1992) Beryllium in soils of the Nevada Test Site: A preliminary assessment. Unpubl work

Pellerin B, Kaste J, Fernandez IJ, Norton SA, Kahl JS (1999) Soil cation distribution in the near-stream zone of New England forested watersheds (abs). Soil Soc Am Ann Mtg, Argonne, Illinois, Abstr 295

Phillips MA (1973) Investigations into levels of both airborne beryllium and beryllium in coal at the Hayden Power Plant near Hayden, Colorado. Environ Letters 5:183-188

Podlešáková E, Němeček J (1994) Contamination of soils in the North-Bohemian Region by hazardous elements. Rostl Výr 40:123-130 (in Czech)

Rai D, Zachara J, Schwab A, Schmidt R, Girvin D, Rogers J (1984) Chemical attentuation rates, coefficients, and constants in leachate migration. 1. A critical review. Report EA-3356 to Electric Power Res Inst, Palo Alto, California

Reimann C, Hall GEM, Siewers U, Bjorvatn K, Morland G, Skarphagen H, Strand T (1996) Radon, fluoride and 62 elements as determined by ICP-MS in 145 Norwegian hard rock groundwater samples. Sci Total Environ 192:1-19

Rissberger FA (1993) Beryllium transport as a result of episodic acidification of a small watershed in Hancock County, Maine. MSc thesis, University of Maine, Orono

Robson AJ, Neal C, Ryland GP, Harrow M (1994) Spatial variations in throughfall chemistry at the small plot scale. J Hydrol 158:107-122

Rose NL (1990) A method for the extraction of carbonaceous particles from lake sediment. J Paleolimnol 3:45-53

Sainsbury CL, Hamilton JC, Huffmann C (1986) Geochemical Cycle of Selected Trace Elements in the Tin-tungsten District, Western Seward Peninsula, Alaska, US. *In* U S Geol Surv Bull 1242-F. U S Govt Printing Office, Washington, DC

Sajwan KS, Ornes WH, Youngblood TV (1996) Beryllium phytotoxicity in soybeans. Water Air Soil Pollut 86:117-124

Sarosiek J, Kosiba P (1993). The effects of water and hydrosol chemistry on the accumulation of beryllium and germanium in selected species of macrophytes. Water Air Soil Pollut 69:405-411

Schecher WD, McAvoy DC (1998) MINEQL^{+}: A chemical equilibrium program for personal computers. Environmental Research Software, Hallowell, Maine

Schuler C, Wieland E, Santschi PH, Sturm M, Lueck A, Bollhalder S, Beer J, Bonani G, Hofmann HJ, Suter M, Wolfli W (1991) A multitracer study of radionuclides in Lake Zurich, Switzerland 1. Comparison of atmospheric and sedimentary fluxes of ^{7}Be, ^{10}Be, ^{210}Pb, ^{210}Po and ^{137}Cs. J Geophys Res 96 C:17051-17065

Schmidbaur H (2001) Recent contributions to the aqueous coordination chemistry of beryllium. Coor Chem Rev 215:223-242

Šebestová E, Machovic V, Pavlíková H, Lelák J, Minařík L (1996) Environmental impact of brown coal mining in Sokolovo basin with especially trace metal mobility. J Environ Sci Health A31:2453-2463

Seth PC, Pandey GS (1985) Beryllium as pollutant in steel plant particulate fallout. Pollut Res 4:17-20

Shacklette HT (1984) Element Concentrations in Soils and other Surficial Materials of the Conterminous United States. *In* US Geol Surv Prof Paper 574-D, I-105. U S Govt Printing Office, Washington DC

Shilin LL, Tsareva LP (1957) The abundance of beryllium in rocks and minerals of pegmatites from the Lovozero and Khibina Tundra. Geochemistry (USSR) 2:383-392 (English version)

Shimizu T, Ohya K, Kawaguchi H, Shijo Y (1999) Determination of ultratrace beryllium in natural water by electrothermal atomic absorption spectrometry after preconcentration with one-drop solvent. Bull Chem Soc Japan 72:249-252

Skjelkvåle BL, Henriksen A, Vadset M, Røyset O (1996) Trace elements in Norwegian lakes. NIVA Report 3457-96. Norway (in Norwegian)

Skřivan P, Minařík L, Burian M, Vach M (1994) Cycling of beryllium in the environment under anthropogenic impact. Scientia Agric Bohem 25:65-75

Skřivan P, Šťastný M, Kotková P, Burian M (1996) Partition of beryllium and several other trace elements in surface waters. Scientia Agric Bohem 27:131-145

Skřivan P, Navrátil T, Burian M (2000) Ten years of monitoring the atmospheric inputs at the Černokostelecko region, Central Bohemia. Scientia Agric Bohem 31:139-154

Skřivan P, Minařík L, Burian M, Martínek J, Žigová A, Dobešová I, Kvídová O, Navrátil T, Fottová D (2001) Biogeochemistry of beryllium in an experimental forested landscape of the "Lesní potok " watershed in Central Bohemia, C R GeoLines 12:41-62.

Slonim ARJ (1973) Acute toxicity of beryllium sulfate to the common guppy. Wat Pollut Cont Fed 45:2110-2122

Slonim CB, Slonim AR (1973) Effect of water hardness on the tolerance of the guppy to beryllium sulfate. Bull Environ Contam Toxic 10:295-301

Small EE, Anderson RS, Hancock GS (1999) Estimate of the rate of regolith production using ^{10}Be and ^{26}Al from an alpine hillslope. Geomorph 27:131-150.

Smith RM, Martell AE (1976) Critical Stability Constants. 4. Inorganic Complexes. Plenum Press, New York

Sneed SB (1986) Sediment chronologies in New York Harbor borrow pits. Unpubl MS thesis. State University of New York, Stony Brook

Southon JR, Ku TL, Nelson DE, Reyss JL, Duplessy JC, Vogel JS (1987) ^{10}Be in a deep-sea core: Implications regarding ^{10}Be production changes over the past 420 ka. Earth Planet Sci Lett 85:356-364

Steinmann P, Billen T, Loizeau J-L, Diminik, J (1999) Beryllium-7 as a tracer to study mechanisms and rates of metal scavenging from lake surface waters. Geochim Cosmochim Acta 63:1621-1633

Stunzas AA, Govorov IN (1981) Complex carbonate, fluor-carbonate compounds of Be and their role in the Be migration in natural waters. Geokhimiya 4:517-524 (in Russian)

Tao H, Miyazaki A, Bansho K (1988) Determination of trace beryllium in natural waters by gas chromatography-helium microwave Induced Plasma Emission Spectrometry. Analyt Sci 4:299-302

Teraoka H, Kobayashi J (1980) Concentrations of 21 metals in the suspended solids collected from the principal 166 rivers and 3 lakes in Japan. Geochem J 14:203-226

Thorat DD, Mahadevan TN, Ghosh DK, Narayan S (2001) Beryllium concentrations in ambient air and its source identification. Environ Mon Assess 69:49-61

van Geen A, Valette-Silver NJ, Luoma SN, Fuller CC, Baskaran M, Tera F, Klein J (1999) Constraints on the sedimentation history of San Francisco Bay from ^{14}C and ^{10}Be. Mar Chem 64:29-38

Veselý J (1987) Influence of emissions on the chemical composition of forest soils. Lesnictví 33:385-398 (in Czech)

Veselý J (1990) Stability of the pH and the contents of ammonium and nitrate in precipitation samples. Atmosph Environ 24A:3085-3089

Veselý J (1994) Effects of acidification on trace metal transport in fresh waters. *In* Steinberg CEW, Wright RF (eds) Acidification of Freshwater Ecosystems: Implications for the Future. Wiley, New York, p 141-151

Veselý J (1995) Drainage sediments in environmental and explorative geochemistry. Bull Czech Geol Survey 70:1-8

Veselý J, Majer V (1996) The effect of pH and atmospheric deposition on concentrations of trace elements in acidified freshwaters: A statistical approach. Water Air Soil Pollut 88:227-246

Veselý J, Beneš P, Ševčík K (1989) Occurrence and speciation of beryllium in acidified freshwaters. Water Res 23:711-717

Veselý J, Hruška J, Norton SA (1998) Trends in water chemistry of acidified Bohemian lakes from 1984 to 1995: II. Trace elements and aluminum. Water Air Soil Pollut 108:425-443

Veselý J, Majer V, Kučera J, Havránek V (2001) Solid-water partitioning of elements in Czech freshwater. Appl Geochem 16:437-450

Vogler S, Jung M, Mangini A (1996) Scavenging of ^{234}Th and ^{7}Be in Lake Constance. Limnol Oceanogr 41:1384–1393

Von Blanckenburg F, Igel H (1999) Lateral mixing and advection of reactive isotope tracers in ocean basins: Observations and mechanisms. Earth Planet Sci Lett 169:113-128

Wang L, Ku, TL, Luo S, Southon JR, Kusakabe M (1996) ^{26}Al-^{10}Be systematics in deep-sea sediments. Geochim Cosmochim Acta 60:109-119

Wedepohl KH (1969) Handbook of Geochemistry. II-1. Springer-Verlag, Berlin

Wedepohl KH (1995) The composition of the continental crust. Geochim Cosmochim Acta 59:1217-1232

Wilkinson J, Reynolds B, Neal C, Hill S, Neal M and Harrow M (1997) Major, minor and trace element composition of cloudwater and rainwater at Plynlimon. Hydrol Earth Syst Sci 1(3):557-569

Yamagata T (1981) Background level of beryllium in the ambient air. Tachikawa Tandai Kiyo 14:103-106 (in Japanese)

You C-F, Lee T, Li Y-H (1989) The partition of Be between soil and water. Chem Geol 77:105-118

Zorn HR, Stiefel T, Beurs J, Schlegelmilch R (1988) Beryllium. *In* Seiler HG, Sigel H, Sigel A (eds) Handbook on Toxicity of Inorganic Compounds. Marcel Dekker, New York, p 105-114

8 Beryllium Analyses by Secondary Ion Mass Spectrometry

Richard L. Hervig

Center for Solid State Science
Arizona State University
Tempe, Arizona 85287
richard.hervig@asu.edu

INTRODUCTION

Secondary ion mass spectrometry (SIMS, or ion microprobe) represents an extremely sensitive technique for the microanalysis of beryllium. Positive ions of beryllium are readily formed and its analysis by SIMS is not overly complicated. Matrix effects appear to be small (<20%). In this chapter I will describe SIMS instrumentation, the problems facing Be analysis, solutions to these problems, calibrations for Be and limits of detection.

INSTRUMENTATION

As of the writing of this chapter (mid-2000), SIMS instruments broadly useful to geochemists are commercially available from two companies. Australian Scientific Instruments (ASI, at www.anutech.com.au/asi/shrimpii.htm) sells the SHRIMP (Super High Resolution Ion Micro Probe). The SHRIMP was designed to allow high transmission of secondary ions at high mass resolution and permits U-Pb ages on zircons to be determined. Beryllium analyses are possible on the SHRIMP, but none have been obtained to date (Vickie Bennett and Ian Williams, Australian National University, pers. comm.). Cameca Instruments Inc. (www.cameca.fr) markets an instrument (the IMS 3f-6f series) that focuses on the needs of the semiconductor industry, and is also the most common SIMS instrument in earth science laboratories. A second Cameca instrument (IMS 1270) is designed for applications demanding high transmission at high mass resolving power (as in the SHRIMP). A tutorial on SIMS can be found on the world-wide web (www.cea.com), but a brief description follows.

In SIMS, a high-energy primary ion beam is directed at a sample. The impacts of these ions with the target result in the ejection of atoms from the top two or three monolayers of the target surface (Dumke et al. 1983). Some of the ejected atoms are ionized during these collisions, and these "secondary ions" can then be accelerated into a mass spectrometer for isotope selection and analysis. Below I describe primary ion beams and the path followed by secondary ions in the Cameca instruments.

Primary ion beams

This description follows the diagram of the Cameca SIMS shown in Figure 1. Primary ions are generated in a duoplasmatron (for Ar^+, O^-, O_2^+) or a cesium source (Cs, other alkali metals). A primary beam mass filter (P.B.M.F.) allows both ion sources to be mounted on the instrument. The choice of the primary beam depends on the elements being ionized. Compared to using a noble gas primary beam (like Ar^+), Cs^+ greatly increases the negative ion signal, enhancing sensitivity to electronegative elements like oxygen and sulfur. Again comparing to Ar^+, when an oxygen beam is used, there is a large increase in the positive ion signal, enhancing the sensitivity for electropositive elements like lithium, strontium, rare earth elements, and other lithophile elements.

1529-6466/00/0050-0008$05.00

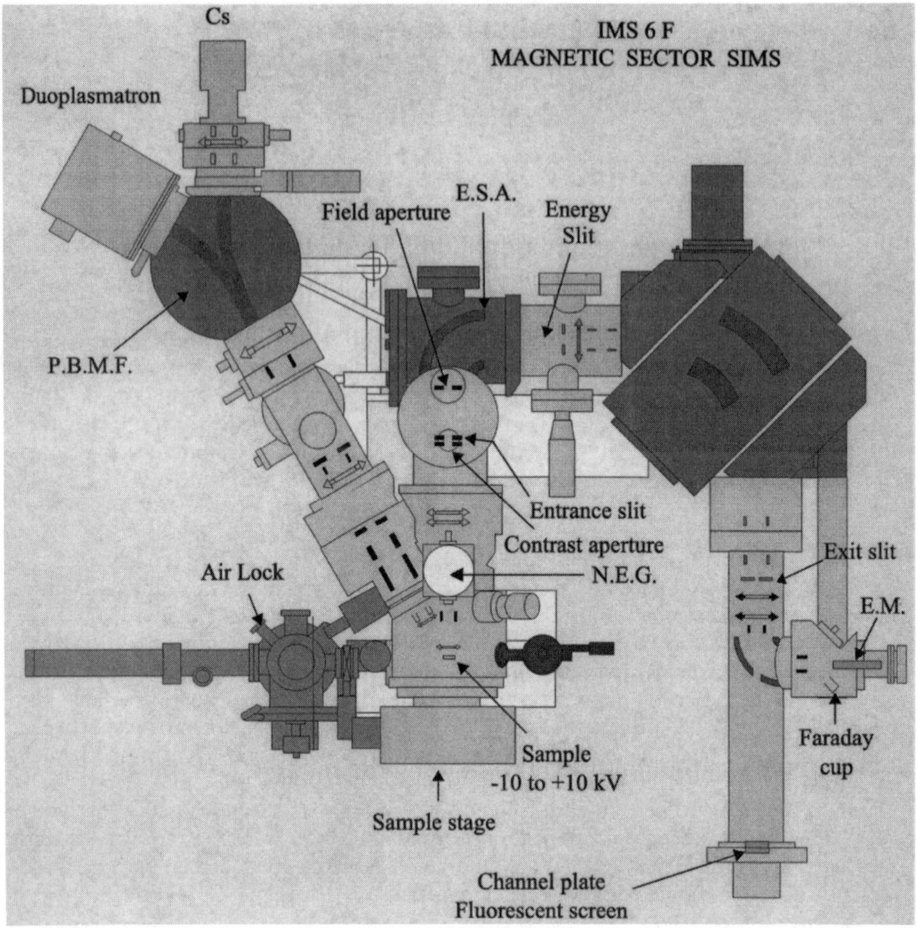

Figure 1. Schematic of the Cameca IMS 6f SIMS (perspective from the top). See text for detailed description. Reproduced by permission of Cameca S.A.

Lateral resolution of the analysis depends on the diameter of the primary ion beam, which can be focused to <1 μm in some of the newest instruments, but at the loss of primary current. Currents used for typical analyses range from 0.2 to 10 nA with beam diameters between ~3 and 40 μm. The choice of a primary beam is also important because of the problem of sample charging. Most minerals are insulators, and the addition of charge to the mineral during bombardment by the primary beam can result in the sample charging enough to deflect the secondary ions away from the mass spectrometer (Werner and Morgan 1976). Geochemists often use a negative oxygen beam because the resulting negative charge build-up can be alleviated by ejection of secondary electrons that are generated during primary ion impact. If a positive primary beam is directed at the sample, positive charge build-up will be increased by ejection of secondary electrons. In this case, an auxiliary electron flood gun to provide negative charge to the analysis point would be required (e.g., the normal incidence electron gun, N.E.G., in Fig. 1).

Secondary ions

A high voltage (typically ±4500 V, but variable from ~±500 to ±10,000 V in the newest model) is placed on the sample to accelerate positive or negative secondary ions along the sample normal into the mass spectrometer. The ions are focused by a series of lenses in the transfer optics section past the N.E.G. This section acts like the objective lens in a light microscope, collecting ions from a maximum field of 250 µm adjustable down to 35-µm diameter. A contrast aperture can be used to restrict the signal intensity and limit the collection of ions deviating from the ion optical axis. Next to the contrast aperture is the entrance slit of the mass spectrometer. Closing the slit results in increased mass resolving power (defined as $M/\Delta M$). A field aperture then limits the area on the sample from which ions are collected as well as selecting the angular acceptance of the mass spectrometer. After passing through the field aperture, the energy of the ions is selected by the electrostatic analyzer (E.S.A.) and energy slit assembly. The energy-filtered ion beam then enters the magnet for mass selection and passes through the exit slit (the width of which also determines the mass resolution). The secondary ion beam can be allowed to strike an ion to electron converter (a channel plate assembly that converts ions to electrons with a gain of >1000). The electrons are then accelerated into a fluorescent screen which displays the ions in the same spatial orientation as when they were ejected from the sample. Because of the isotopic specificity of the mass spectrometer, this image allows the SIMS to be used as a "chemical microscope." Alternatively, the ion beam can be deflected into the counting system using a second electrostatic analyzer between the exit slit and the imaging assembly. The counting system consists of an electron multiplier (e.m.) for detection of secondary ion signals with intensities of <0.1 to ~10^6 counts/s and a faraday cup electrometer for measurement of signal intensities greater than 10^6.

ANALYSES FOR BERYLLIUM

The aluminum problem

Early studies of sputtered ions showed that the mass spectrum contained not just elemental ions, but complex molecular and multiply charged species as well. These species may have the same nominal mass/charge as the elemental ion of interest, requiring some method to discriminate them. In the case of beryllium, with one stable isotope at mass 9, the only significant interference is from $^{27}Al^{3+}$. The mass of triply-charged aluminum is ~0.018 amu lighter than ^{9}Be, requiring a mass resolving power ($M/\Delta M$) of ~500. This interference is shown on Figure 2 where three mass spectra are superimposed. The line with short dashes shows the secondary ion signal obtained with the instrument set up for high transmission (largest contrast aperture and the entrance and exit slits in the mass spectrometer fully open). The transfer optics were configured to collect ions from a circular area 60 µm in diameter (150-µm image field coupled with a 750-µm field aperture). Note the shoulder at light masses representing the triply-charged Al ion. On this sample (NIST SRM 610, a glass containing 450 ppm Be and ~2.1 wt % Al_2O_3), the aluminum interference is small, and Grew et al. (1998) have shown that Al^{3+} does not significantly affect the analysis when Be contents greater than ~30 ppm. However, for an aluminous mineral containing 1 ppm Be the interference would overlap seriously with the Be peak. The line with long dashes on Figure 2 shows the result of partly closing the entrance and exit slits in the mass spectrometer (increasing the mass resolving power) to clearly resolve Be from Al^{3+}. Upon good separation, the Be signal is decreased by approximately a factor of 2. The solid line on Figure 2 shows the signal resulting from opening completely the entrance and exit slits but using a different transfer optics configuration to allow ions from a 20-µm diameter circular area (35-µm image

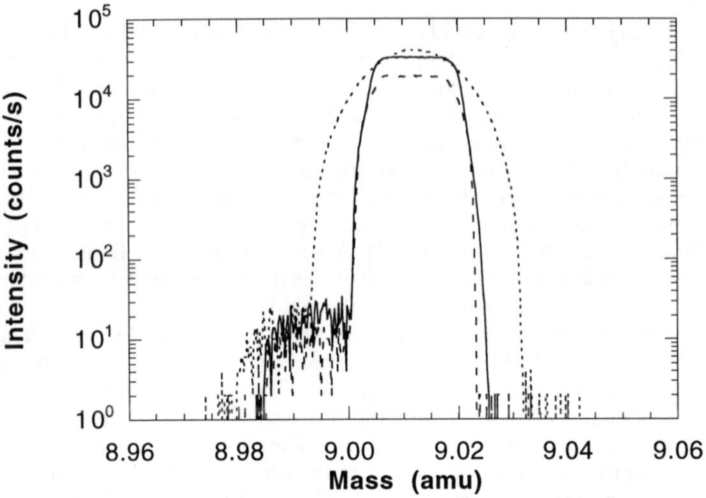

Figure 2. Mass spectrum near mass/charge = 9 amu. Peaks from $^{27}Al^{3+}$ (shoulder) and $^{9}Be^{+}$ (large peaks) are shown in NIST 610 glass. The scan indicated by the fine dashed line represents operation of the SIMS at maximum transmission using the 150-μm transfer optic setting. The scan shown by long dashes was obtained with the entrance slit partially closed to resolve the $^{27}Al^{3+}$ interference, and the solid line shows the result of opening the entrance slit but applying the 35-μm transfer optics to give high transmission plus sufficient resolution to separate the triply-charged aluminum ion. All scans obtained at the same primary current.

field and 750-μm field aperture) into the mass spectrometer. The conditions for the latter scan result in high mass resolution because the 35-μm transfer optics focus the secondary ion beam to a very small point on the entrance slit. Using approaches such as those generating the latter two scans, analyses can be obtained with good separation of the interference and high transmission.

Energy distribution of beryllium ions

Secondary ions with a range of energies are ejected from samples during primary ion bombardment. Collecting an energy spectrum of a particular species is achieved by holding the electrostatic analyzer voltages and energy slit position constant while varying the accelerating voltage applied to the sample. Energy spectra for Be and Si in phenakite (Be_2SiO_4) are shown on Figure 3. To obtain this figure, the energy slit was closed to allow ions with a range of ~2 eV kinetic energy into the magnetic sector. Note the small signal at negative energies. It represents ions that lost some energy through scattering and may possibly be joined by a small amount of sputtered neutral atoms that were ionized above the sample in the gas phase, and did not experience the full accelerating voltage. The positive kinetic energy region of Figure 3 represents ions that are ejected with energies in addition to that provided by the sample accelerating potential. These excess energies, or "initial kinetic energies" are achieved by the collisions occurring during primary ion impact. Note that the intensity of Be ions decreases with increasing initial kinetic energy (by a factor of ~20 at 50 eV and a factor of ~100 at 100 eV). Note also that Be and Si show sub-parallel behavior throughout the energy range studied. Secondary ions with high initial kinetic energies can be useful in quantitative SIMS because cluster ions with the same nominal mass/charge as elemental ions are not likely to survive the energetic collisions that provide high energies to the ions. As a result, the mass spectrum

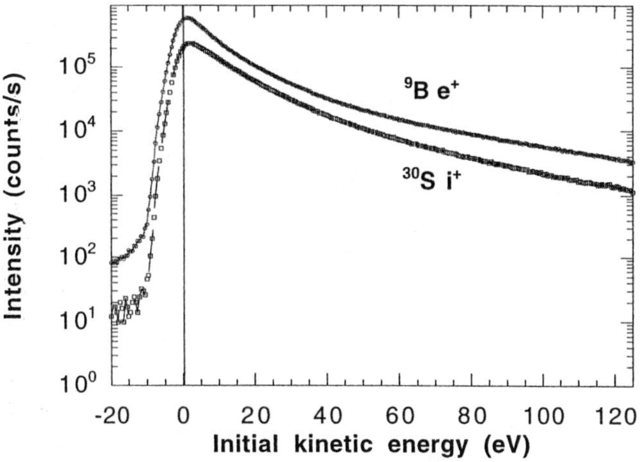

Figure 3. Initial kinetic energy spectrum for $^{9}Be^{+}$ and $^{30}Si^{+}$ secondary ions sputtered from phenakite (Be_2SiO_4). Energy slit closed to ~2 eV.

of ions with initial kinetic energies in excess of ~50-75 eV is mostly (but not completely, Zinner and Crozaz 1986) free of molecular interferences. This approach to eliminating molecular ions is known as "energy filtering." In the case of Be, there are no cluster ions to remove, and because the Al^{3+} ion has an energy spectrum much like Si or Be, it is not removed by selecting high energy ions. However, if the phase of interest is being analyzed for several additional elements, it is probably best (and most convenient) to obtain all the secondary ion intensities at the same conditions. Thus high energy ions may be used for the analysis of Be because heavy elements benefit from using energy filtering (as long as the Al^{3+} ion is removed by simultaneous operation at high mass resolution).

Quantification of the beryllium signal

Working curves. The most common approach to obtaining quantitative analyses of geological phases using SIMS is to develop a working curve. Several samples with known quantities of an element of interest (obtained through bulk analyses of natural or synthetic materials) are studied for this element. A graph showing the change in elemental concentration with the ion signal of that element is generated and the slope of the curve (if linear) provides a calibration factor. The working curve approach is important because it can help to determine whether changing the chemical make-up of minerals changes the calibration factor (matrix effects). Positive y-intercepts on working curves warn the analyst about potential backgrounds (either from a signal for that element coming from the SIMS or from an unidentified and unresolved interference). Negative y-intercepts most likely indicate that the bulk analyses are in error. In general, the raw count rate for the isotope is not used on these curves. The strong extraction field between the sample and the first lens in magnetic sector SIMS make the secondary ion intensities very sensitive to variations in the sample-extraction lens distance that result from minor changes in sample topography. The typical solution is to normalize the signal of the analyzed constituent to another signal coming from the sample matrix, as sample topography may affect most elements in the same manner. The reference element usually selected is silicon, but Ca, Al, and O have also been used. An example of a working curve is presented on Figure 4 using data from Ottolini et al. (1993). The reference data for NIST SRM 610 and 612 were updated from the nominal values used by Ottolini et al.

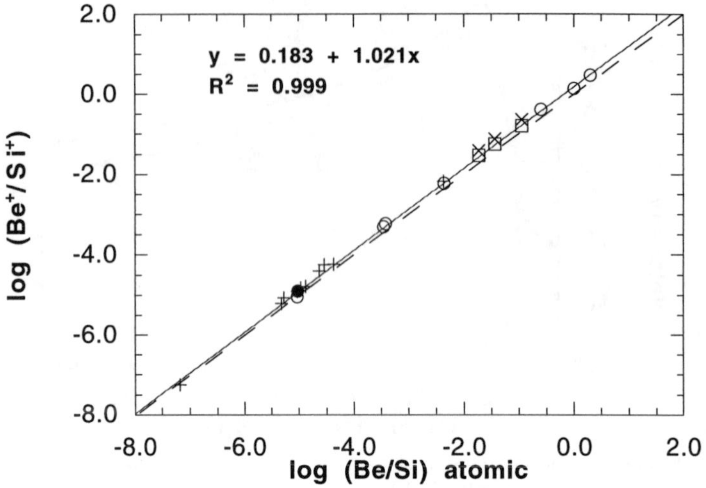

Figure 4. Log of the Be^+/Si^+ secondary ion ratio (corrected for Si isotope abundance) vs. log of the Be/Si atomic ratio for several minerals and glasses. ○ represent data modified from Ottolini et al. (1993) after recalculating for atomic ratios and using new measurements of standards (Pearce et al. 1997). ● is a basaltic glass from Shearer et al. (1994), × from Evensen et al. (1999), □ from Evensen (unpublished analyses from Arizona State University), and + from L. Riciputi (Oak Ridge National Laboratory, pers. comm., 2000). The dashed line shows a 1:1 relation; the solid line is a least-squares linear regression through the data from Ottolini et al. (1993).

(1993) on the basis of the results from Pearce et al. (1997). The standards were also presented as atomic ratios of Be/Si instead of weight ratios. This figure contains beryllium minerals and Be-bearing glasses. The samples range from high silica to low silica and alkali-rich to alkali-poor. Ottolini et al. (1993) used energy filtering, selecting ions with 100±25 eV excess kinetic energy.

Matrix effects. Geochemists using SIMS are aware that the chemical make-up of the sample can influence the yield of ions for a particular element. These matrix effects can make the working curve shown on Figure 4 inappropriate to apply to unknown samples with different chemistry than the standards used. There is a good correlation on Figure 4 over a range in Be contents from 1 ppm to >150000 ppm. This is obtained despite significant changes in major element chemistry, and is strong evidence for small matrix effects in this particular case (<20%). Two examples support this: (1) Ottolini et al. (1993) tested their calibration by applying it to hornblende and the Be mineral genthelvite, obtaining agreement with bulk analyses to better than 10%, and (2) SIMS analyses for Be in hellandite (Oberti et al. 1999) are supported by structural refinement and crystal chemical constraints, despite the lack of a similar standard and the complex character of the hellandite matrix (REE up to ~21 wt %, actinides up to ~11 wt %).

Variations among SIMS. Another question that arises is if the calibration obtained by Ottolini et al. (1993) can be applied on other SIMS instruments. This can be tested by examining the Be calibration curve derived by Shearer et al. (1994). This calibration was obtained in a different laboratory (same model SIMS) at conditions quite similar to Ottolini et al. (1993) with the exception that ions with 70±25 eV initial kinetic energy were analyzed. Shearer et al. (1994) used only basalts with very low Be. One sample in their study (JDF2) contains 50.1 wt % SiO_2 and 0.72 ppm Be, with a $Be^+/^{30}Si^+$ ratio of

0.0004. After correcting for Si isotopic abundance and normalizing to atomic Be/Si, this point was placed on Figure 4, and the agreement with the least-squares regression on the data of Ottolini et al. (1993) is good. As another example, Evensen et al. (1999) used the same SIMS as Shearer et al. (1994) to study synthetic Be-rich rhyolitic glasses (0.5 to >3 wt % BeO). The analysis conditions were modified from Shearer et al. (1994) in that ions with 100 eV initial kinetic energy were detected (energy range not given). These same samples were studied by Evensen (unpublished data) at Arizona State University (detecting ions with 75±20 eV initial kinetic energy). One other data set was kindly provided by Dr. L. Riciputi (Oak Ridge National Laboratory). These analyses were obtained on a Cameca 4f instrument using 100±20 eV ions at a mass resolving power sufficient to resolve the $^{27}Al^{3+}$ interference. Replotting the latter three data sets on Figure 4 shows the general agreement between different SIMS labs.

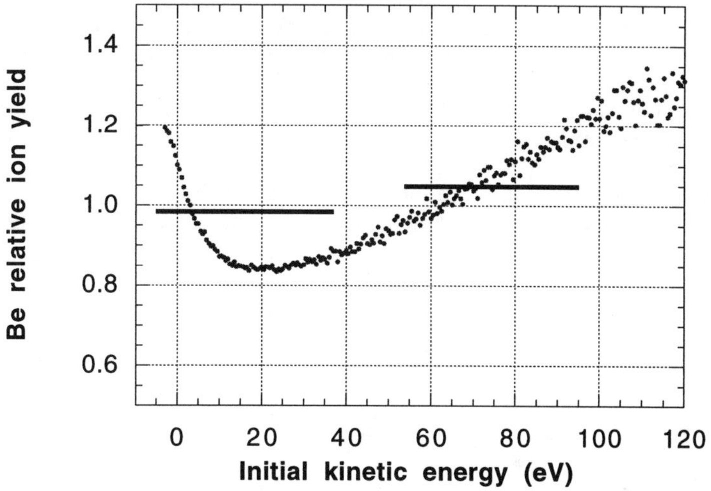

Figure 5. Relative ion yield for Be in phenakite (normalized to Si) vs. initial kinetic energy. Data from Figure 3. The horizontal bars represent the relative ion yield when the secondary ion signals are integrated over ~0-40 eV and 55-95 eV initial kinetic energy. See text for discussion.

Effect of secondary ion energy. The effect of initial kinetic energy on the relative ion yield is shown on Figure 5. The relative ion yield is determined by normalizing the intensity of Be^+ to that for Si^+ (from Fig. 3—after correcting for Si isotopic abundance), and then normalizing this ratio to the atomic Be/Si ratio in the phenakite (= 2). The values close to unity on Figure 5 indicate that Be and Si ions are formed with nearly the same efficiency regardless of the energy of secondary ions selected, up to energies around ~90 eV. The shape of this curve matches very well with those shown by Ottolini et al. (1993) for amphibole, NIST 610 glass, and Fe-rich mica. Note that the data on Figure 3 were obtained with the energy window closed down to allow ions with a range of ~2 eV of kinetic energy into the mass spectrometer. To apply the relative ion yields for Be on Figure 5 however, it is necessary to first integrate the ion signals over the same energy range as those used to obtain ion intensities on unknown phases. Two examples of such an integration are shown on Figure 5. For analyses using the maximum intensities (~0-40 eV) the relative ion yield determined on phenakite is 0.98. For analyses using energy filtering (~75±25 eV) the relative ion yield is 1.05. These two values can be compared with an earlier study by Hinton (1990) on NIST 610 glass, who measured relative ion yields for Be of 0.95 (for 0±19 eV energy ions) and 1.27 (for 77±19 eV ions).

The average relative ion yield determined on the wide range of samples studied by Ottolini et al. (1993) is 1.4±0.2. The Be-rich synthetic rhyolites studied by Evensen et al. (1999) and Evensen (unpublished data from ASU) give relative ion yields of 2.0 and 1.4, respectively, and the samples studied by Riciputi (Fig. 4) show a mean relative ion yield of 1.4±0.3). The similarity of relative secondary ion yields for Be measured on different instruments, selecting somewhat different initial kinetic energies is very good, as was noted for boron analyses by SIMS (Hervig 1996). Variations in ion yields may be caused by the selection of different secondary ion energies, the small matrix effects, and variations in electron multiplier responses with age (Zinner et al. 1985).

Ion implantation. An alternative approach to the quantification of the Be secondary ion signal is ion implantation (Leta and Morrison 1980). This method is commonly used in the semiconductor industry and has been found to be very reliable in characterizing Si and GaAs wafers for dopant concentrations and impurity levels. In ion implantation, an ion beam of the species of interest is generated and accelerated to high energy (40-200 keV). After accurately measuring the current, the beam is directed at the target for a set period of time. The current density is sufficiently low that little if any material is removed from the target, but rather the incoming ion beam is implanted into the target in a roughly gaussian distribution with depth. The average depth depends on the ion energy and mass, and the atomic density of the target. The sample is then analyzed by SIMS in depth-profile mode, and the integrated ion intensity of the implanted species is normalized to the known number of atoms introduced into the target. An example for Be is described below.

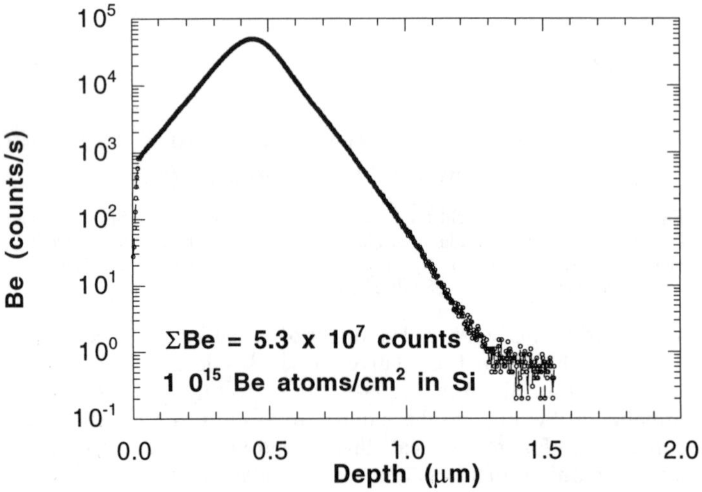

Figure 6. Depth profile of a Be-implanted Si wafer. Depth determined using surface profilometry. See text for discussion.

A silicon wafer was implanted with 1×10^{15} atoms/cm^2 of ^{9}Be at 40 keV (Kroko Inc., Tustin, California). The sample was analyzed using a $^{16}O^-$ beam rastered over an area 250×250 μm^2. Positive secondary ions from a circular area 60 μm in diameter in the center of the crater were accepted into the mass spectrometer. Secondary ions with 0-40 eV initial kinetic energy were detected. After the analysis, the depth of the crater was determined by surface profilometry (Dektak III), and the resulting graph of depth vs. Be ion intensity is shown on Figure 6. The integrated counts for Be^+ are 5.24×10^7, the total depth is 1.54×10^{-4} cm, and the integrated counts for Si (corrected for isotopic abundance) were 8.54×10^7. We can calculate a "relative sensitivity factor" (RSF) in the following way:

$$RSF_{Be} = (1 \times 10^{16} \text{ Be atoms/cm}^2 * \text{Si counts})/(\text{Be counts} * \text{crater depth, cm})$$
$$= 6.75 \times 10^{22} \text{ (atoms/cm}^3)$$

This factor can then be multiplied by the Be^+/Si^+ ratio obtained on other Si wafers at the same analytical conditions to determine Be concentrations (in atoms/cm^3). In this case, however, the application of this calibration can be expanded. The depth profile was obtained using a 12.5 kV beam of $^{16}O^-$ primary ions striking at 28° to the sample normal (Reed and Baker 1983). For impact angles less than about 30°, oxygen bombardment of Si will produce nearly SiO_2 stoichiometry at the surface (Morgan et al. 1981), making this implant potentially useful for the analysis of trace Be in quartz. Dividing the above RSF by the atomic density of SiO_2 (7.05×10^{22} atoms/cm^3) gives a relative ion yield of 0.93. Because this yield is quite close to the value of 0.98 observed (for the same energy range) on phenakite (Fig. 5) and the value of 0.95 determined by Hinton (1990) on NIST 610 (also using low energy secondary ions), it is suggested that these implants can make high quality standards for analysis of geological samples.

Negative secondary ions

Many elements give very low yields for positive secondary ions (e.g., C, O, S) and it is advantageous to use negative secondary ions for their analysis. In the case of beryllium, negative ions might be attractive because elements will not form as multiply charged negative species (i.e., $^{27}Al^{3-}$ is not stable). However, nearly all group II elements do not make negative ions, i.e., Mg, Ca, Sr, and Ba cannot be detected as elemental negative ions. Negative Be can be detected, but when the Be-implanted Si wafer was analyzed in this manner, the integrated signal for Be^- was ~300 counts. This is approximately 10^5 times less than for positive ions. Clearly, Be^- is not suitable for the analysis of most geological phases. If negative ions must be used, however, it is possible to increase sensitivity by using a molecular ion. For example, Wilson et al. (1989) showed that (on Si wafers) by monitoring the $^{28}Si^9Be^-$ signal at mass 37, the sensitivity could be improved by a factor of 4000 compared to using elemental Be^-. Of course the possibility of interferences with (for example) $^{37}Cl^-$ and $^{18}O^{19}F^-$ in geological samples would need to be tested.

LIMITS OF DETECTION

We have measured the maximum transmission of beryllium through the Cameca IMS 3f SIMS at ASU by determining the useful yield, defined as the number of ions of a species detected per atom of that species consumed. On the Si wafer, we obtained a useful yield of 2.7%, i.e., 27 Be ions detected for every 1000 Be atoms of that species removed from that sample. This value is similar to those determined on NIST 610 glass (1.2% Hervig and Shuttleworth 1998). Hervig and Shuttleworth (1998) also measured the useful yield for Be in NIST 610 glass by laser ablation inductively coupled sector field mass spectrometry, obtaining $\sim 10^{-6}$. Because we normally reduce the transmission to eliminate the $^{27}Al^{3+}$ interference, the useful yield by SIMS is near 1%. The useful yield by SIMS is further reduced if energy filtering is used- the 100× decrease in Be count rate on Figure 2 that results from allowing only ions with 100 eV initial kinetic energy into the mass spectrometer drops the useful yield down to 10^{-4}.

The useful yield is important because it allows the analyst to determine how much material must be consumed in order to detect a statistically significant signal. We can approximate the atomic density of most geological materials as $\sim 10^{23}$ atoms/cc, or 10^{11}

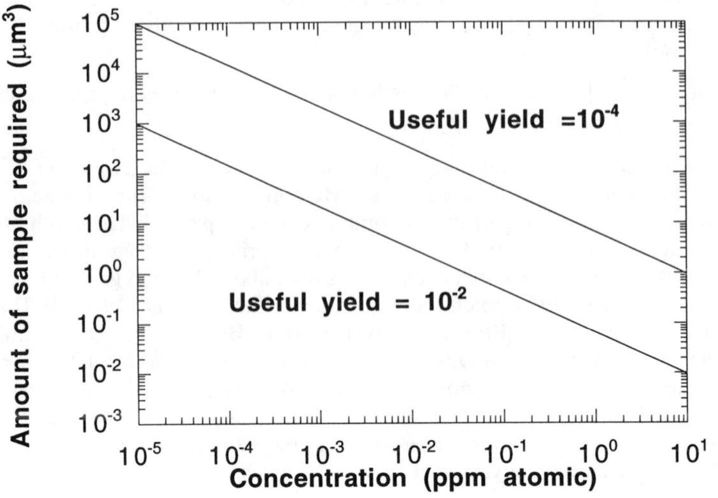

Figure 7. Amount of sample that must be sputtered (in μm^3) to give 100 total counts of Be^+ as a function of atomic concentration in the sample. Lower line assumes a useful yield of 1% and an atomic density of 10^{23} atoms/cm^3. If 100 eV initial kinetic energy ions were detected, the useful yield may decrease to a value as low as 10^{-4} (upper line; see Fig. 3) and thus 100 times more sample would be required.

atoms/μm^3. A concentration of 1 ppm Be (atomic) is thus 10^5 atoms/μm^3. Assuming a useful yield of 1%, we would obtain 1000 counts of Be per μm^3 of material consumed. If the sample contained only 1 ppb Be, we would get 1 count of Be per μm^3. So the detection limit depends on how much sample is available for consumption. An example is shown on Figure 7. Following Williams (1985) we calculate the amount of material that must be sputtered to obtain 100 counts of Be (a 10% error based on counting statistics) as a function of atomic Be content. To get a feeling as to how much material is implied, consider a 20-µm-diameter primary ion beam generating a cylindrical crater. This type of analysis will consume 100 μm^3 of sample with every ~1/3 µm increment in the depth of the crater. When the crater reaches 3 µm in depth, ~1000 μm^3 of sample will have been consumed. If the crater becomes too deep, it may penetrate into a different phase, and it is likely that if the depth of the crater approaches its width it will be difficult for ions to escape. However, for the vast majority of geochemical problems, the high useful yield suggests that a detection limit less than 1 ppb atomic (~0.2 ppb by weight) is achievable. If analyses are obtained with a 100-V offset on the sample voltage, the useful yield decreases to 10^{-4}, and limits of detection correspondingly increase (Fig. 7). However, at such low concentrations, surface contamination and memory effects will be critically important. Surface contamination results from polishing a sample containing Be-rich phases, so that Be-rich material is "impregnated" into other Be-poor phases (e.g., Shaw et al. 1988). Memory effects result when Be atoms sputtered from Be-rich materials coat the cover plate on the extraction lens in front of the sample. During subsequent analyses of Be-poor phases, sputtered neutral species may strike the cover plate and knock off Be atoms which may land in the sample crater and be analyzed. This can be an extremely important source of contamination (Wilson et al. 1989) but can be alleviated by using a high primary current on, for example, a grain of quartz (or other suitable Be-poor phase) to coat the cover plate with other material. The problem can also be solved by removing the plate from the instrument and cleaning it (Wilson et al. 1989).

APPLICATIONS

Beryllium concentrations

There have been several applications of the SIMS technique to the determination of Be in natural phases over the past 20+ years. One common application is to glassy samples, including quenched melts trapped in phenocrysts and as matrix glass (Kovalenko et al. 1988; Webster and Duffield 1991, 1994; Webster et al. 1993, 1995, 1996, 1997; Shearer et al. 1994; Kovalenko et al. 1995). In these applications, Be can reveal crystal fractionation trends. Another use is in the analysis of experiments. Here, the high sensitivity and spatial resolution provided by SIMS may be required to analyze fine-grained run products. As examples, London et al. (1988) and Webster et al. (1989) measured Be partition coefficients between silicate melt and hydrous vapor (finding a strong preference for the melt). Evensen et al. (1999) determined the solubility of beryl in silicate melt and Evensen and London (1999) measured partition coefficients for Be between feldspar (strongly incompatible), and biotite (moderately incompatible). Because SIMS allows individual minerals to be analyzed for Be contents, researchers can study the major hosts of this element, and couple this data with petrologic textures to understand the evolution of Be in rock-forming processes. For example, Domanik et al. (1993) determined that phengitic muscovite and jadeitic pyroxene were the most important phases carrying Be in subducted slabs. Grew et al. (2000) used whole-rock and microanalyses of phases (such as sapphirine, khmaralite, musgravite, and cordierite) in beryllium pegmatites to help constrain their origin by anatexis and subsequent partial modification by metamorphic reactions. In addition, SIMS analyses of kornerupine has revealed its significance in the Be budget of some metamorphic rocks (Grew et al. 1990, 1995). In many cases, SIMS has provided the important (if negative) information that the minerals studied may *not* contain significant Be. Examples of this include calcite (Mason 1987), alkali feldspar from syenite (Mason et al. 1985), Fe-cordierite (sekaninaite) (Černý et al. 1997), as well as several minerals studied by Domanik et al. (1993). Even when Be contents are very low, however, they can still be of use in modelling planetary melting events (Shearer et al. 1994) and in testing models of irradiation of extraterrestrial samples (Phinney et al. 1979).

Analyses of beryllium isotope ratios

The low abundance of ^{10}Be restricts the use of SIMS to measure Be isotope ratios directly. $^{10}Be/^{9}Be$ ratios in young arc volcanic rocks, for example, are typically 2-40 $\times 10^{-11}$ (Morris et al. 1990). The largest commonly occurring ratios in geologic materials are measured on oceanic sediments at 5×10^{-8} (Morris et al. 1990). If the useful yield described above were appropriate, obtaining a signal for ^{10}Be on these samples (where total Be contents are typically only a few ppm) would require consuming thousands of μm^3 of sample. More importantly, separating $^{10}Be^+$ from $^{10}B^+$ would require operation at a mass resolving power of ~17,000 (compared to about 500 for resolving $^{27}Al^{3+}$ from $^{9}Be^+$). This would reduce the useful yield by several orders of magnitude! While direct measurements of naturally occurring isotope ratios is not possible, it is possible to measure ratios on pre-concentrated materials. In fact, SIMS was used to characterize the National Institute of Standards and Technology standard reference material (SRM) for Be isotopes, SRM 4325. This ^{10}Be-concentrated sample was analyzed using an ARL IMMA SIMS (no longer available) following procedures similar to those described by Christie et al. (1981). In a later study, Belshaw et al. (1995) described in detail a technique for analyzing Be isotope ratios after chemical separation of Be from geologic samples. Briefly, ~100 ng of Be was loaded onto a Ta filament. The filament was placed in a VG Isolab 120 SIMS (Belshaw et al. 1994) and sputtered by a primary ion beam of O^+. Minor boron contamination was detected by measuring ^{11}B, and the ^{10}B

was correspondingly subtracted from the signal at mass 10 assuming a $^{11}B/^{10}B$ ratio of 4.04 (after correction for instrumental fractionation). The precision of measurements on samples with $^{10}Be/^{9}Be$ of 10^{-7} is ~4% (1σ).

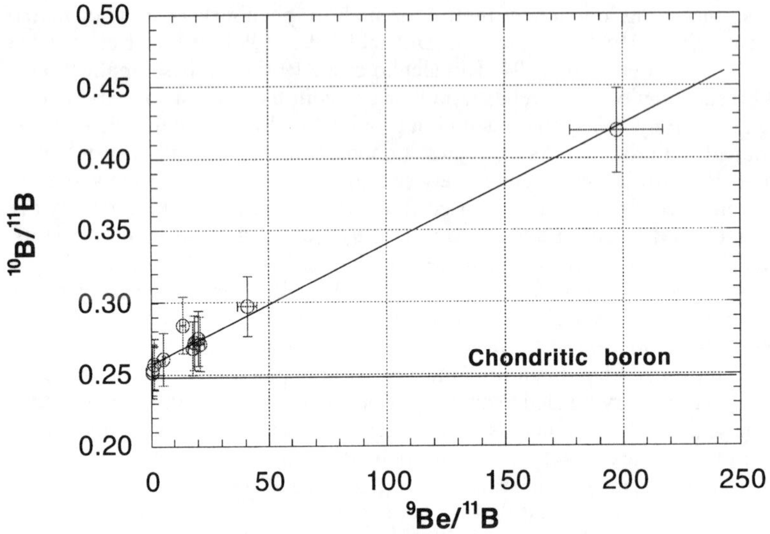

Figure 8. Boron isotope ratio vs. $^{9}Be/^{11}B$ ratio in minerals from a Ca-, Al-rich inclusion from the Allende meteorite (Redrawn from data presented in McKeegan et al. 2000). The correlated excess of ^{10}B with Be content indicate the *in situ* decay of ^{10}Be trapped in the minerals during formation early in the history of the solar system.

McKeegan et al. (2000) has shown that it is possible to detect the in-situ decay of ^{10}Be to ^{10}B by comparing $^{9}Be/^{11}B$ with $^{10}B/^{11}B$ ratios (Fig. 8). Correlated excesses in ^{10}B compared to chondritic boron isotope ratios were observed, showing that the short-lived isotope ^{10}Be was present in minerals from some calcium-aluminum-rich inclusions (CAI) with 0.05-3 ppm Be and 0.1-1 ppm B. McKeegan et al. (2000) used the high-sensitivity Cameca 1270 SIMS to make these measurements; such analyses would not be possible on the smaller SIMS models. The results are important because they provide a constraint on the time of formation of these primitive materials in the early solar system.

Ion imaging of beryllium

The distribution of elements over a sample can be imaged using SIMS to give a chemical map of the surface (e.g., Massare et al. 1982; Guan and Crozaz 2000). These images can be obtained in two ways: (1) direct ion microscopy, where the image on the fluorescent screen assembly (Fig. 1) is captured (on film or digital media) and (2) scanning ion imaging, where the secondary ion signal on the electron multiplier is synchronized with the position of the primary beam as it is swept over the sample. The high sensitivity for Be suggests that such chemical maps can be obtained on natural samples, but I am not aware of any published examples.

CONCLUSIONS

Secondary ion mass spectrometry is a very sensitive analytical technique for the microanalysis of Be. The ion ratio of Be^+/Si^+ (normalized to 100% isotopic abundance) is

~1.2±0.2 times the Be/Si atomic ratio over a wide range of secondary ion energies and for all silicate mineral and glass compositions studied to date. In the absence of surface contamination and memory effects, Be can be determined at levels less than 1 part-per-billion while consuming only a millionth of a mm^3 of sample.

ACKNOWLEDGMENTS

The author thanks Darby Dyar for supplying the phenakite sample and Philip Kyle for the NIST SRM glass. Thanks to J. Evensen and L. Riciputi for allowing Be calibrations to be shared. Reviews by E.S. Grew, L. Riciputi, and L. Ottolini were greatly appreciated. The National Science Foundation has generously supported research in SIMS at the ASU laboratory (NSF EAR 9305201 and EAR 9615982).

REFERENCES

Belshaw NS, O'Nions RK, Martel DJ, Burton KW (1994) High resolution SIMS analysis of common lead. Chem Geol 112:57-70

Belshaw NS, O'Nions RK, von Blanckenburg F (1995) A SIMS method for $^{10}Be/^{9}Be$ ratio measurement in environmental materials. Int'l J Mass Spec Ion Proc 142:55-67

Černý P, Chapman R, Schreyer W, Ottolini L, Bottazzi P, McCammon CA (1997) Lithium in sekaninaite from the type locality, Dalni Bory, Czech Republic. Can Mineral 35:167-173

Christie WH, Eby RE, Warmack RJ, Landau L (1981) Determination of boron and lithium in nuclear materials by secondary ion mass spectrometry. Analyt Chem 53:13-17

Domanik KJ, Hervig RL, Peacock SM (1993) Beryllium and boron in subduction zone minerals: An ion microprobe study. Geochim Cosmochim Acta 57:4997-5010

Dumke MF, Tombrello TA, Weller RA, Housley RM, Cirlin EH (1983) Sputtering of the gallium-indium eutectic alloy in the liquid phase. Surf Science 124:407-412

Evensen JM, London D (1999) Beryllium reservoirs and sources for granitic melts: The significance of cordierite. Geol Soc Am Abstr Prog 31(7):A-305

Evensen JM, London D, Wendlandt RF (1999) Solubility and stability of beryl in granitic melts. Am Mineral 84:733-745

Grew ES, Chernosky JV, Werding G, Abraham K, Marquez N, Hinthorne JR (1990) Chemistry of kornerupine and associated minerals, a wet chemical, ion microprobe, and x-ray study emphasizing Li, Be, B, and F contents. J Petrol 31:1025-1070

Grew ES, Hiroi Y, Motoyoshi Y, Kondo Y, Jayatileke SJM, Marquez N (1995) Iron-rich kornerupine in sheared pegmatite from the Wanni Complex, at Homagama, Sri Lanka. Eur J Mineral 7:623-636

Grew ES, Yates MG, Barbier J, Shearer CK, Sheraton JW, Shiraishi K, Motoyoshi Y (2000) Granulite-facies beryllium pegmatites in the Napier Complex in Khmara and Amundsen Bays, Western Enderby Land, East Antarctica. Polar Geosci 13:1-40

Grew ES, Yates MG, Huijsmans JP, McGee JJ, Shearer CK, Wiedenbeck M, Rouse R (1998) Werdingite, a borosilicate new to granitic pegmatites. Can Mineral 36:399-414

Guan Y, Crozaz G (2000) Light rare earth element-enrichments in ureillites: A detailed ion microprobe study. Meteoritics Planet Sci 35:131-144

Hervig RL (1996) Analyses of geological materials for boron by secondary ion mass spectrometry. Rev Mineral 33:789-803

Hervig RL, Shuttleworth S (1998) Useful yields in microbeam mass spectrometry. Geol Soc Am Abstr Prog 30(7):A80

Hinton RW (1990) Ion microprobe trace-element analysis of silicates: Measurement of multi-element glasses. Chem Geol 83:11-25

Kovalenko VI, Hervig RL, Sheridan MF (1988) Ion microprobe analyses of trace elements in anorthoclase, hedenbergite, aenigmatite, quartz, apatite and glass in pantellerite: Evidence for high water contents in pantellerite melts. Am Mineral 73:1038-1045

Kovalenko VI, Tsaryeva GM, Goreglyad AV, Yarmolyuk VV, Troitsky VA, Hervig RL, Farmer GL, 90, 530-547. (1995) The peralkaline granites-related Khaldzan-Buregtey rare element (Zr, Nb, REE) deposit, Western Mongolia. Econ Geol 90:530-547

Leta DP, Morrison GH (1980) Ion implantation for in-situ quantitative ion microprobe analysis. Analyt Chem 52:277-280

London D, Hervig RL, Morgan GB, VI (1988) Melt-vapor solubility and elemental partitioning in peraluminous granite-pegmatite systems: Experimental results with Macusani glass at 200 MPa. Contrib Mineral Petrol 99:360-373

Lowenstern JB, Bacon CR, Calk LC, Hervig RL, Aines RD (1994) Major-element, trace-element, and volatile concentrations in silicate melt inclusions from the tuff of Pine Grove, Wah Wah Mountains, Utah. U S Geol Surv Open-File Rpt 94-242, 20 p

Mason R (1987) Ion microprobe analysis of trace elements in calcite with an application to the cathodoluminescence zonation of limestone cements from the Lower Carboniferous of South Wales, UK. Chem Geol 64:209-224

Mason RA, Parsons I, Long JVP (1985) Trace and minor element chemistry of alkali feldspars in the Klokken layered syenite series. J Petrol 26:952-970

Massare D, Havette A, Slodzian G (1982) Analyse quantitative de feldspaths synthetiques par images ioniques. J Microsc Spectrosc Electron 7:477-486

McKeegan KD, Chaussidon M, Robert F (2000) Evidence for the *in situ* decay of ^{10}Be in an Allende CAI and implications for short-lived radioactivity in the early solar system. Lunar Planet Sci Conf Abstr 31, Abstr #1999

Morgan AE, de Grefte HAM, Warmoltz N, Werner HW (1981) The influence of bombardment conditions upon the sputtering and secondary ion yields of silicon. Appl Surf Sci 7:372-392

Morris JD, Leeman WP, Tera F (1990) The subducted component in island arc lavas: Constraints from Be isotopes and B-Be systematics. Nature 344:31-36

Oberti R, Ottolini L, Camara F, Della Ventura GC (1999) Crystal structure of non-metamict Th-rich hellandite-(Ce) from Latium (Italy) and crystal chemistry of the hellandite-group minerals. Am Mineral 84:913-921

Ottolini L, Bottazzi P, Vannucci R (1993) Quantification of lithium, beryllium, and boron in silicates by secondary ion mass spectrometry using conventional energy filtering. Analyt Chem 65:1960-1968

Pearce NJG, Perkins WT, Westgate JA, Gorton MP, Jackson SE, Neal CR, Chenery SP (1997) A compilation of new and published major and trace element data for NIST SRM 610 and SRM 612 glass reference materials. Geostand Newslett 21:115-144

Phinney D, Whitehead B, Anderson D (1979) Li, Be, and B in minerals of a refractory-rich Allende inclusion. Proc Lunar Planet Sci Conf 10:885-905

Reed DA, Baker JE (1983) Use of ISS and doubly charged secondary ions to monitor surface composition during SIMS analyses. Nucl Inst Meth Phys Res 218:324-326

Shaw DM, Higgins MD, Truscott MG, Middleton TA (1988) Boron contamination in polished thin sections of meteorites: Implications for other trace-element studies by alpha-track image or ion microprobe. Am Mineral 73:894-900

Shearer CK, Layne GD, Papike JJ (1994) The systematics of light lithophile elements (Li, Be, and B) in lunar picritic glasses: Implications for basaltic magmatism on the Moon and the origin of the Moon. Geochim Cosmochim Acta 58:5349-5362

Webster JD, Duffield WA (1991) Volatiles and lithophile elements in Taylor Creek Rhyolite: Constraints from glass inclusion analysis. Am Mineral 76:1628-1645

Webster JD, Duffield WA (1994) Extreme halogen abundances in tin-rich magma of the Taylor Creek rhyolite, New Mexico. Econ Geol 89:840-850

Webster JD, Holloway JR, Hervig RL (1989) Partitioning of lithophile trace elements between H_2O + CO_2 fluids and topaz rhyolite melt. Econ Geol 84:116-134

Webster JD, Taylor RP, Bean C (1993) Pre-eruptive melt composition and constraints on degassing of a water-rich pantellerite magma, Fantale volcano, Ethiopia. Contrib Mineral Petrol 114:53-62

Webster JD, Congdon RD, Lyons PC (1995) Determining pre-eruptive compositions of late Paleozoic magma from kaolinitized volcanic ashes: Analysis of glass inclusions in quartz microphenocrysts from tonsteins. Geochim Cosmochim Acta 59:711-720

Webster JD, Burt DM, Aguillon RA (1996) Volatile and lithophile trace-element geochemistry of Mexican tin rhyolite magmas deduced from melt inclusions. Geochim Cosmochim Acta 60:3267-3283

Webster JD, Thomas R, Rhede D, Forster H-J, Seltmann R (1997) Melt inclusions in quartz from an evolved peraluminous pegmatite: Geochemical evidence for strong tin enrichment in fluorine-rich and phosphorus-rich residual liquids. Geochim Cosmochim Acta 61:2589-2604

Werner HW, Morgan AE (1976) Charging of insulators by ion bombardment and its minimization for secondary ion mass spectrometry (SIMS) measurements. J Appl Phys 47:1232-1242

Williams P (1985) Limits of quantitative microanalysis using secondary ion mass spectrometry. Scan Elect Microsc 1985/II:553-561

Wilson RG, Stevie FA, Magee CW (1989) Secondary Ion Mass Spectrometry: A Practical Handbook for Depth Profiling and Bulk Impurity Analysis. John Wiley & Sons, New York, 225 p

Zinner E, Crozaz G (1986) A method for the quantitative measurement of rare earth elements in the ion microprobe. Int'l J Mass Spec Ion Proc 69:17-38

Zinner E, Fahey AJ, McKeegan KD (1985) Characterization of electron multipliers by charge distributions. Secondary Ion Mass Spectrometry (SIMS V), p 170-172, Springer-Verlag, Berlin

9 The Crystal Chemistry of Beryllium

Frank C. Hawthorne and Danielle M.C. Huminicki

Department of Geological Sciences
University of Manitoba,
Winnipeg, Manitoba, Canada R3T 2N2

frank_hawthorne@umanitoba.ca

INTRODUCTION

Beryllium is not a very abundant element in the Earth, but, being an incompatible element in common rock-forming silicate minerals, it is susceptible to concentration *via* fractionation in geochemical processes. Moreover, its properties are such that Be does not tend to show extensive solid-solution with other elements, and hence usually forms minerals in which it is a discrete and essential constituent. Beryllium (atomic number 4) has the ground-state electronic structure [He]$2s^2$ and is the first of the group IIA elements of the periodic table (Be, Mg, Ca, Sr, Ba, Ra). The first (899 kJ/mol) and second (1757 kJ/mol) ionization enthalpies are sufficiently high that the total energy required to produce the Be^{2+} ion is greater than the compensating energy of the resulting ionic solid, even when the latter involves extremely electronegative elements. Hence bond formation involves covalent (rather than ionic) mechanisms. However, Be has only two electrons to contribute to covalent bonding involving four orbitals (s, p_x, p_y, p_z), resulting in Be being a strong Lewis acid (i.e., a strong electron-pair acceptor) with a high affinity for oxygen. The cation radius of Be^{2+} is 0.27 Å, (Shannon 1976) and Be is known only in tetrahedral coordination in minerals, although BeO_3 groups are known in synthetic compounds (e.g., $Ca_{12}Be_{18}O_{29}$, Y_2BeO_4 and $SrBe_3O_4$; Harris and Yakel 1966, 1967, 1969). The BeO_4 and SiO_4 groups have a marked tendency to polymerize in the solid state. Although $^{[4]}Be$ (r = 0.27 Å) and $^{[4]}Si$ (r = 0.26 Å) are very similar in size, solid solution between Be and Si is inhibited by the difference in formal charge of the two species (Be^{2+} vs. Si^{4+}).

CHEMICAL BONDING

We adopt a pragmatic approach to chemical bonding, using bond-valence theory (Brown 1981) and its developments (Hawthorne 1985, 1994, 1997) to consider structure topology and hierarchical classification of structure. We will use molecular-orbital theory to consider aspects of structural energetics, stereochemistry and spectroscopy of beryllium minerals. These approaches are not incompatible, as bond-valence theory can be considered as a simple form of molecular-orbital theory (Burdett and Hawthorne 1993; Hawthorne 1994).

STEREOCHEMISTRY OF $Be\varphi_4$ POLYHEDRA IN MINERALS

The variation of Be–φ (φ: O^{2-}, OH^-) distances and φ–Be–φ angles is of great interest for several reasons:

(1) mean bond-length and empirical cation and anion radii play a very important role in systematizing chemical and physical properties of crystals;

(2) variations in individual bond-lengths give insight into the stereochemical behavior of structures, particularly with regard to the factors affecting structure stability;

(3) there is a range of stereochemical variation beyond which a specific oxyanion or cation-coordination polyhedron is not stable; it is obviously of use to know this range, both for assessing the stability of hypothetical structures (calculated by DLS [Distance Least-Squares] refinement, Dempsey and Strens 1976; Baur 1977) and for assessing the accuracy of experimentally determined structures.

1529-6466/00/0050-0009$10.00

Here, we examine the variation in Be–φ distances in minerals and review previous work on polyhedral distortions in $Be\varphi_4$ tetrahedra. Data for 89 $Be\varphi_4$ tetrahedra were taken from 58 refined crystal structures with $R \leq 6.5\%$ and standard deviations of ≤0.005 Å on Be–φ bond-lengths; structural references are given in Appendix A.

Variation in <Be–φ> distances

The variation in <Be–φ> distances (< > denotes a mean value; in this case, of Be in tetrahedral coordination) is shown in Figure 1a. The grand <Be–φ> distance (i.e., the mean value of the <Be–φ> distances) is 1.633 Å, the minimum and maximum <Be–φ> distances are 1.598 and 1.661 Å, respectively (the 4 or more smallest and 2 or 3 largest values in Fig. 1a are considered unreliable), and the range of variation is 0.063 Å. Shannon (1976) lists the radius of $^{[4]}$Be as 0.27 Å; assuming a mean anion-coordination number of 3.25 and taking the appropriate O/OH ratio, the sum of the constituent radii is 0.27 + 1.360 = 1.63 Å, in accord with the grand <Be–φ> distance of 1.633 Å. Brown and Shannon (1973) showed that variation in <M–O> distance correlates with bond-length distortion Δ (= $\Sigma[l(\text{o}) - l(\text{m})]/l(\text{m})$; $l(\text{o})$ = observed bond-length, $l(\text{m})$ = mean bond-length) when the bond-valence curve of the constituent species shows a strong curvature, and when the range of distortion is large. There is no significant correlation between <Be–φ> and Δ; this is in accord with the bond-valence curve for Be–O given by Brown (1981).

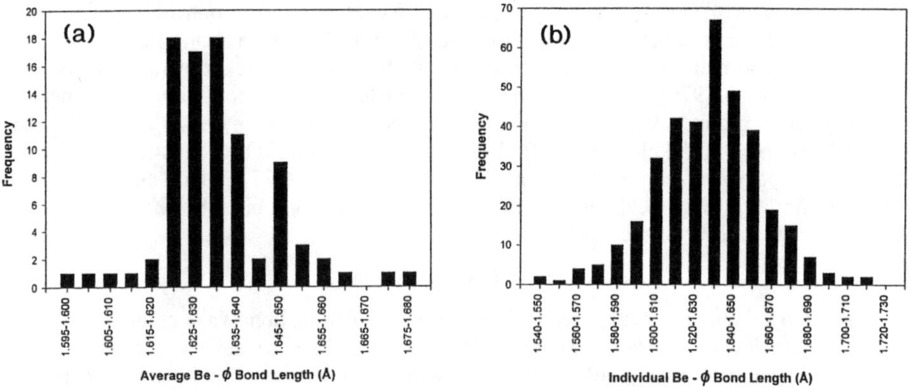

Figure 1. (a) Variation in average Be–O distance in minerals containing $Be\varphi_4$ tetrahedra; (b) variation in individual Be–O distance in minerals containing $Be\varphi_4$ tetrahedra.

Variation in Be–φ distances

The variation in individual Be–φ distances is shown in Figure 1b; the grand mean Be–φ distance is 1.633 Å. The minimum and maximum observed Be–φ distances are 1.545 and 1.714 Å, respectively, and the range of variation is 0.169 Å; the distribution is a skewed Gaussian. According to the bond-valence curve for Be (the universal curve for first-row elements) from Brown (1981), the range of variation in Be–φ bond-valence is 0.63–0.41 vu (valence units).

General polyhedral distortion in Be-bearing minerals

There is no general correlation between <Be–φ> and bond-length distortion. However, as pointed out by Griffen and Ribbe (1979), there are two ways in which polyhedra may distort (i.e., depart from their holosymmetric geometry): (1) the central cation may displace from its central position [bond-length distortion]; (2) the anions may displace from their ideal positions [edge-length distortion]; Griffen and Ribbe (1979) designate these two descriptions as BLDP (Bond-Length Distortion Parameter) and ELDP (Edge-Length Distortion

Parameter), respectively. Figure 2 shows the variation in both these parameters for the second-, third- and fourth-period (non-transition) elements in tetrahedral coordination. Some very general features of interest (Griffen and Ribbe 1979) are apparent from Figure 2:

(1) A BLDP value of zero only occurs for an ELDP value of zero; presuming that ELDP is a measure also of the O–T–O angle variation, this is in accord with the idea that variation in orbital hybridization (associated with variation in O–T–O angles) must accompany variation in bond-length.

(2) Large values of BLDP are associated with small values of ELDP, and vice versa. The variation in mean ELDP correlates very strongly with the grand mean tetrahedral-edge length for each period (Fig. 3).

Griffen and Ribbe (1979) suggested that the smaller the tetrahedrally coordinated cation, the more the tetrahedron of anions resists edge-length distortion because the anions are in contact, whereas the intrinsic size of the interstice is larger than the cation which can easily vary its cation-oxygen distances by 'rattling' within the tetrahedron.

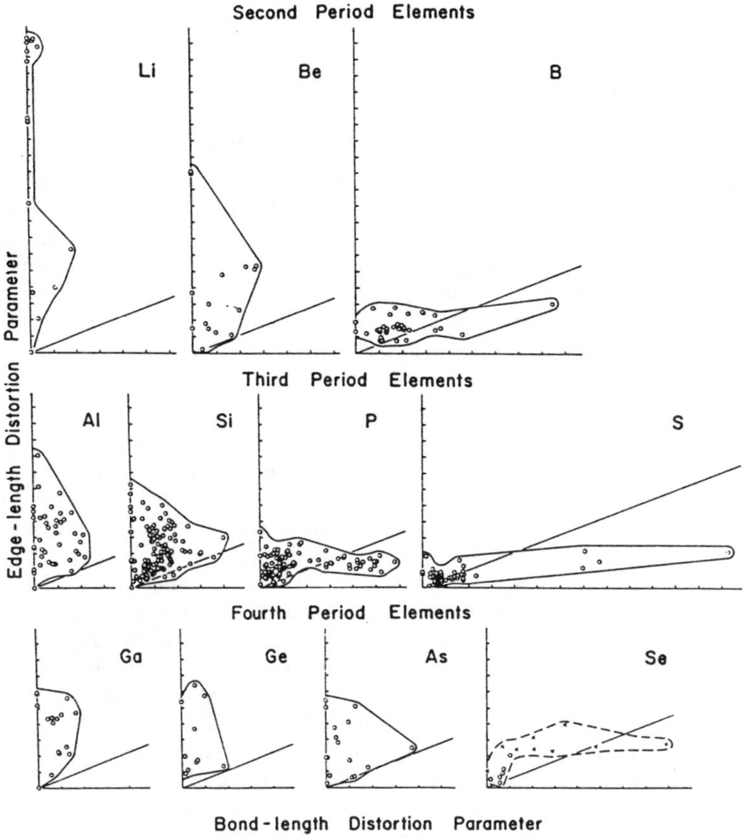

Figure 2. Variation in BLDP (Bond-Length Distortion Parameter) and ELDP (Edge-Length Distortion Parameter) for second-, third- and fourth-period non-transition elements in tetrahedral coordination by oxygen; after Griffen and Ribbe (1979).

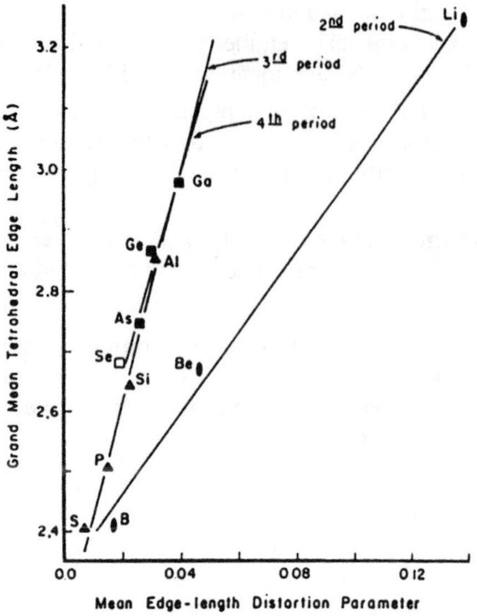

Figure 3. Variation in grand mean tetrahedral edge-length with mean ELDP for the second-, third- and fourth-period elements of the periodic table; after Griffen and Ribbe (1979).

MOLECULAR-ORBITAL STUDIES OF Beφ_4 POLYHEDRA

Molecular-orbital (MO) calculations have been used by theoretical chemists for many years, primarily to predict geometries, energetics and stabilities of molecules. MO methods are based on quantum mechanics, and range from empirical and semi-empirical methods, which include an experimentally determined component, to *ab-initio* methods, which include no experimentally determined parameters (see Tossell and Vaughan (1992) for an excellent summary of these methods). MO methods have been applied to the study of small molecules with considerable success, and available computational sophistication and power have permitted their application to mineralogically relevant problems for the past two decades (Gibbs 1982).

Beryllium-oxygen structures have been the topic of a small number of MO calculations, and are ideal in this regard due to the relatively small number of electrons involved in a BeO4 polyhedron. To date, most MO calculations for these structures have been done using molecular clusters which are designed to be an approximation of local conditions within a crystal structure. These clusters are only an approximation of the *local* environment in a structure, and many long-range effects in a periodic structure are ignored by such calculations. The first MO calculations for Be structures involved *ab-initio* MO methods applied to small clusters that modeled both individual Bej 4 polyhedra and larger clusters, allowing the study of such structural aspects as Be-O-Si bond-angles. The most rigorous approach is an *ab-initio* periodic Hartree-Fock MO calculation that includes the entire crystal structure (with appropriate boundary conditions). These calculations are computationally very demanding, and have been restricted to a small number of minerals; crystal *ab-initio* MO calculations have been done for BeO, the synthetic analogue of bromellite (Lichanot et al. 1992, Lichanot and Rerat 1993). MO calculations have provided the energetics and geometries of clusters, have given insight into the Be-O chemical bond, the molecular-orbital structures of BeO4 polyhedra, and have contributed significantly to the interpretation of various spectra of Be minerals.

Prediction of equilibrium geometries

MO calculations have often been used as an aid in understanding the geometry of Beφ_4

polyhedra, and MO arguments have frequently been used in explaining observed stereochemical variations in minerals.

Beφ_4 groups. Schlenker et al. (1978) examined the geometry of bond-bending and bond-stretching in BeO_4^{6-} clusters using extended-Hückel calculations. Molecular-orbital theory predicts that the bond-overlap population for a $^{[4]}$T–O bond should be related to <O–T–O>$_3$, the average of the three O–T–O angles involving the bond. Figure 4a shows the variation in bond-overlap population with the average of the O–T–O angles common to the bond for the BeO^{6-}_4 groups, calculated for a series of idealized polyhedra with C_{2v} and C_{3v} point symmetries. This suggests that the Be–O bond-lengths should be inversely related to <O–$^{[N]}$Be–O>$_{N-1}$, and Figure 4b shows that this is the case. There is more scatter in the trend of Figure 4b than is present in the results of the calculations in Figure 4a; however, the latter data are derived from molecules, whereas the observed stereochemical data come from crystal structures in which the polyhedra are in various strained configurations.

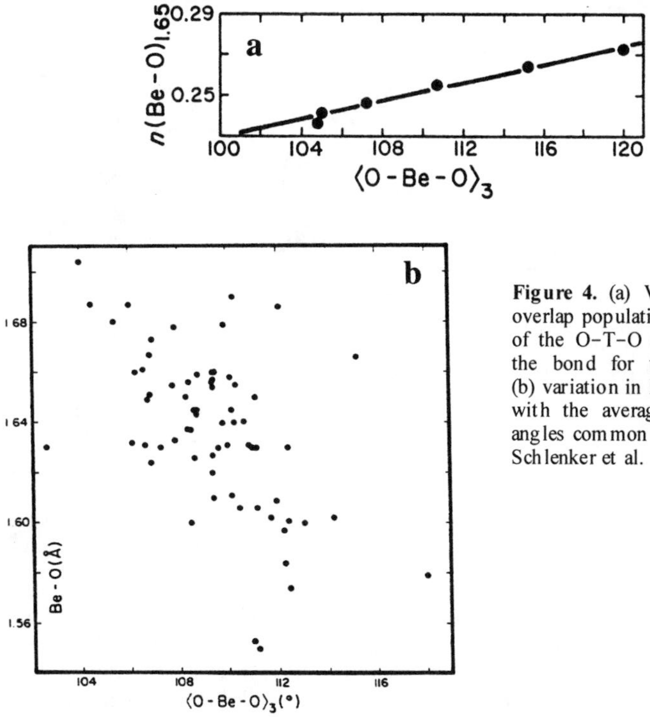

Figure 4. (a) Variation in bond-overlap population with the average of the O–T–O angles common to the bond for the $Be\varphi_4^{6-}$ group; (b) variation in Be–φ bond-lengths with the average of the φ–T–φ angles common to the bond; after Schlenker et al. (1978).

Gupta et al. (1981) reported SCF–HF (Self-Consistent Field Hartree-Fock) calculations on the $Be(OH)_4^{2-}$ tetrahedron, and observed that it is important to have a low cluster-charge in order to reproduce experimental bond-lengths. A minimal basis-set, STO–3G (Slater-Type Orbitals expanded by three Gaussians; Tossell and Vaughan (1992) give a readable summary of MO methods), reproduces experimental distances quite closely for $Be(OH)O_4^-$ polyhedra: Be–φ (exp) = 1.64, Be–φ (calc) = 1.63 Å.

The Be–O–Si linkage. In many minerals, the Beφ_4 tetrahedron polymerizes with SiO_4 tetrahedra, and hence the Be–O–Si linkage is of great interest. Tossell and Gibbs (1978) showed that Be–O–Si linkages produce a range of Be–O–Si bond-angles from ~114–138° with a maximum frequency at ~129° (see also Ganguli 1979). Downs and Gibbs (1981) reported HF calculations done using a STO–3G basis set on $BeSi(OH)_7$ and H_6BeSiO_7, both of which contain the Be–O–Si linkage. In $BeSi(OH)_7$, the bridging oxygen is formally overbonded (with an incident Pauling bond-strength sum of 2.5 vu), whereas in H_6BeSiO_7, the bridging oxygen is formally underbonded (with an incident Pauling-bond-strength sum of 1.5 vu). The optimized geometries for the two clusters (Fig. 5) show long Be–O and Si–O bonds for the overbonded bridging anion (Fig. 5a) and short Be–O and Si–O bonds for the underbonded bridging anion (Fig. 5b), as expected. The actual values of the bond-lengths are not truly representative of the local arrangement as incident bond-valence sums calculated from the values of Brown (1981) give values at the bridging anion of 1.40 (ignoring H) and 1.66 vu, respectively. Obviously the O–H bond-length used (as fixed in the optimization) is too short for the actual local arrangement in Figure 5a, and the bridging anion in Figure 5b requires another coordinating cation. Nevertheless, the Be–O–Si angles of 129° and 131° are close to the grand mean observed value reported by Tossell and Gibbs (1978).

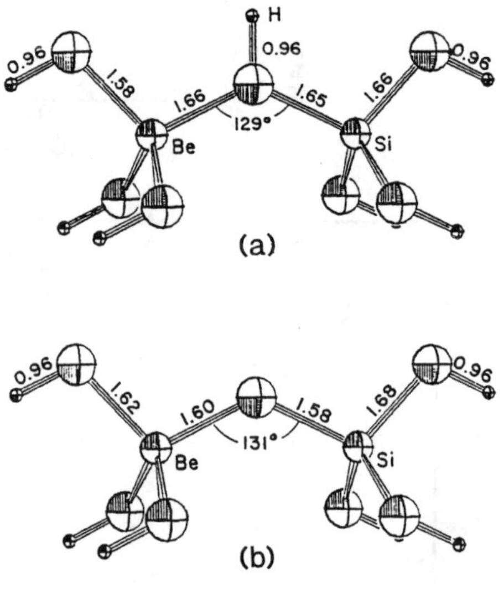

Figure 5. Optimized geometries for (a) $BeSi(OH)_7$ and (b) H_6BeSiO_7, calculated using a minimal STO-3G basis set. Large spheres = O, small spheres = H, with O–Be–O = O–Si–O = 109.47° (fixed) and O–H = 0.960 Å (fixed). After Downs and Gibbs (1981).

Downs and Gibbs (1981) also examined the energetics of bond bending for these two clusters (Fig. 6a) and showed that the Boltzmann distribution curve for the energies of $BeSi(OH)_7$ (i.e., $\exp[-\Delta E(\angle Be–O–Si)/kT]$, where T = 300 K) are in accord with the distribution of Be–O–Si angles in beryllosilicates (Fig. 6b). Similar results were reported by Geisinger et al. (1985).

Lichanot et al. (1992) and Lichanot and Rerat (1993) examined the structure and properties of bromellite using the *ab-initio* method in the Hartree-Fock approximation with an all-electron basis set and periodic boundary conditions to model the complete crystal structure. The crystallographic parameters are in fairly close agreement (Table 2, below); note that the experimental values have been extrapolated to 0 K using the thermal-expansion coefficient 2.66×10^{-5} K^{-1} for *a* and *c*, and 1.5×10^{-6} K^{-1} for *u* (Hazen and Finger 1986). The calculated elastic constants are compared with experimental values in Table 1. The experimental values of Sirota et al. (1992) were adjusted to 0 K using the model of Garber and Granato (1975a,b). The agreement is quite close for the diagonal components of the elastic tensor, but the discrepancies for the off-diagonal elements are large: C_{12} 24%, C_{13}

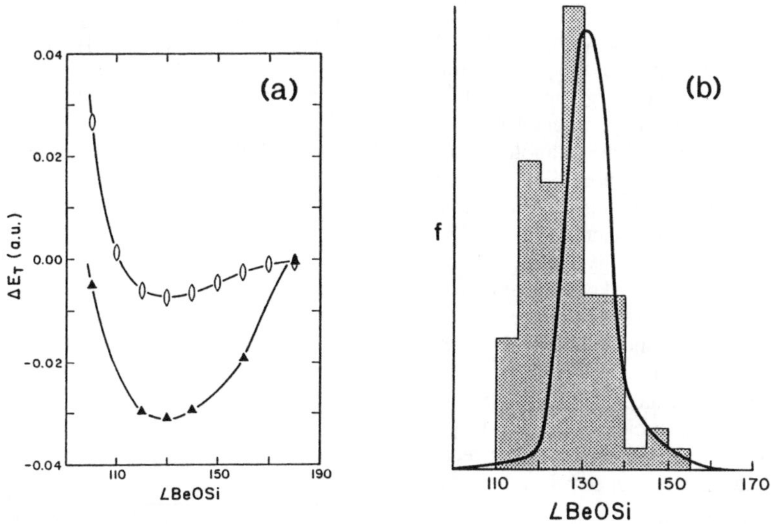

Figure 6. (a) Potential-energy curves, ΔE_T = E(180°) – (∠Be–O–Si), as a function of Be–O–Si angle in $BeSi(OH)_7$ (triangles) and H_6BeSiO_7 (lozenges) with bond lengths fixed at the optimized values; (b) histogram of Be–O–Si angles in beryllosilicates with the Boltzmann distribution curve for energies of $BeSi(OH)_7$. After Downs and Gibbs (1981).

48%, although we note that some of the uncorrected experimental values are much closer to the calculated values than the 'corrected' experimental values. This range in accuracy of the calculated values is not surprising; there are usually problems calculating off-diagonal tensor components from theoretical models, a problem that may relate to the difficulty of dealing with non-central forces.

Deformation electron-density maps

A deformation electron-density map is obtained by subtracting the electron density associated with spherically symmetrical free atoms from the total electron-density of the cluster. In such maps, a positive electron-density indicates that the electron density in this region is increased due to bonding effects, whereas a negative electron-density occurs where the electron density has been depleted by bonding effects. Such maps show the location of chemical bonds and lone-pair electrons, and can be obtained from experiment, from first-principles calculations, or from a combination of experiment and calculation. Gupta et al. (1981) used the HF MO descriptions of the electronic wavefunctions to calculate deformation-electron-density maps for $Be(OH)O_4^{2-}$ clusters. Despite ambiguities in the definition of reference state, there is a depletion of charge at the central Be atom, charge transfer to oxygen, concentration of charge density on the oxygen lone-pair region, and an increase in charge in the T–O and O–H bond regions. Downs and Gibbs (1987) obtained experimental deformation-density maps for phenakite, and showed that there are accumulations of deformation density associated with Be–O and Si–O bonds, and that short bonds show greater accumulations of deformation density than long bonds.

Interpretation of spectroscopic data

MO methods provide the molecular-orbital structure of atom clusters, and hence are of considerable use in the interpretation of many types of spectra for minerals and glasses.

Orbital energies. Many forms of spectroscopy involve the absorption and/or emission of energy as a result of transitions of the experimental system from one energetic state to another. The assignment and interpretation of such absorptions or emissions requires quantitative (or semi-quantitative) understanding of the energy levels of the system. In general, a knowledge of the molecular-orbital structure of the system provides this understanding, either directly when the transitions involve electrons, or indirectly *via* calculation of a physical parameter (e.g., the EFG: Electric-Field Gradient) that affects other (e.g., nuclear) spectral transitions.

Table 1. Physical properties of bromellite calculated with the 'crystal' Hartree-Fock method (1), (2).

	(1)	*(2)*	*Experimental*
a (Å)	2.680	2.697	2.675
c	4.336	4.361	4.344
c/a	1.618	1.617	1.624
u	0.3798	0.3791	0.3765
*K	251	244	210
C_{11}	–	526	518/461/470
C_{12}	–	110	144/127/168
C_{13}	–	92	176/88/119
C_{33}	–	556	516/492/494
C_{44}	–	148	152/148/153
C_{66}	–	–	274/167/152

(1) Lichanot et al. (1992); (2) Lichanot and Rerat (1993)
*K = bulk modulus in GPa; C_{ij} = elastic constants, three sets of experimental values are shown: Sirota et al. (1992), Cline et al. (1967) and Bentle (1966).

Vaughan and Tossell (1973) calculated the orbital energies for BeO_4^{6-} using the CNDO/2 (Complete Neglect of Differential Overlap) method (Tossell and Vaughan 1992), and obtained results that are in accord with the OK_α X-ray emission spectrum of BeO (Fig. 7); note that peak B in Figure 7 is thought to be a reflectivity spike from the analyzing crystal used in the measurement of the spectrum (Liefeld et al. 1970).

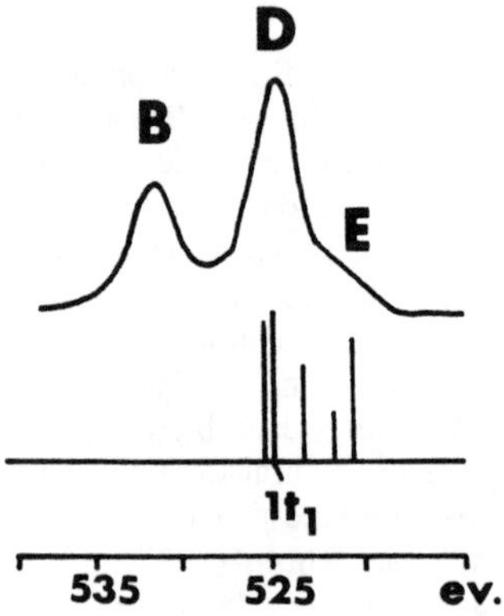

Figure 7. The OK_α spectrum of BeO (after Koster 1971); the positions and relative intensities of the predicted X-ray emission peaks are shown below; after Vaughan and Tossell (1973).

BERYLLIUM MINERALS AND THE IONIC MODEL

In the *ionic model*, a crystal structure is considered as an array of spheres that bear the formal charges of the ions constituting the crystal. Originally, the energy of the crystal was then calculated as the sum of the 'electrostatic' interactions over the whole crystal plus the sum of the short-range repulsive interactions (usually represented as an exponential or inverse-power function) (Pauling 1929; O'Keeffe 1981). This model is *empirical* in that it requires experimental observations (i.e., interatomic distances), in addition to universal constants, as input to the calculations. Moreover, it shows only semi-quantitative agreement with even the most ionic materials.

Developments of the ionic approach in the last 30 years have greatly increased its rigor and power. The key development was the proof of a universal functional relation between the potential energy (total ground-state energy) and the electron density of an array of atoms. This provided the impetus to develop *ab-initio* models that need no experimental input apart from the universal constants. A model widely used in mineralogy is the *Modified Electron Gas* (MEG) model of Gordon and Kim (1972). This model defines the ionic-limit electron density at any point in a crystal as the sum of the electron densities from the component free-ions at their sites in the structure; these latter electron densities result from quantum-mechanical calculations for the individual free-ions. For such models, the free oxide-ion (O^{2-}) wave-function is not stable; however, it can be stabilized by placing it in a potential well produced by a shell of surrounding positive charge (Watson 1958). Such a shell can be fixed, the *rigid-ion* model of Cohen and Gordon (1976), or can include spherical-charge relaxation, e.g., the *Potential Induced Breathing* (PIB) model of Boyer et al. (1985).

Table 2. Calculated (MEG method) and experimental properties for BeO (from Tossell 1980).

[cn]		*+2*	*+1*	*Experimental*
[4]	Be–O (Å)	1.7	1.84	1.64
	ΔH (kcal mol^{-1})	-212	-89	-142
	B (megbars)	-	2.58	2.2
[6]	Be–O (Å)	1.79	1.99	-
	ΔH (kcal mol^{-1})	-232	-73	-

Table 3. Physical properties* of bromellite calculated with the PIB model (after Cohen et al. 1987).

Property	*PIB*	*Experimental*
Volume ($Å^3$)	29.24	27.6
c/a	1.58	1.62
z (Be)	0.385	0.378
K (GPa)	186	210
K′	3.72	5.1
Murnaghan fit range	-10 to -47	
Max. error in P	0.08	
*E*2 (cm^{-1})	333 [0.31]	338 [0.03]
*A*1 (TO)	555 [1.54]	682 [1.57]
*E*1 (TO)	637 [1.59]	723 [1.52]
*E*2	645 [1.68]	682 [1.73]
*A*1 (LO)	1347 [0.78]	1095 [0.92]
*E*1 (LO)	1383 [0.84]	1097 -
C_{11} (Gpa)	371	518/461/470
C_{12}	108	144/127/168
C_{13}	90	176/88/119
C_{33}	361	516/492/494
C_{44}	132	152/148/153
C_{66}	131	274/167/152

* K = bulk modulus; K′ = derivative of K; *E* and *A* = Raman frequencies (Jephcoat et al. 1986; Arguello et al. 1969), values in [ ] are associated Grüneisen parameters; C_{ij} = elastic constants, three sets of experimental values are shown: Sirota et al. (1992), Cline et al. (1967) and Bentle (1966).

Tossell (1980) examined several mineralogically important oxides, including BeO, from the viewpoint of the *Modified Electron Gas* (MEG) model. He used shells of +2 and +1 charge, respectively, to stabilize the free-oxide (O^{2-}) wave-function in his calculations, and in addition, a free unstabilized wave-function. The results for BeO are summarized in Table 2. The experimental heat of formation lies between the values calculated for the +2 and +1 stabilized-ion models, somewhat closer to the latter results. This result is in accord with the fact that the +1 stabilized-ion model predicts the correct coordination-number ([4]) for BeO (whereas the +2 stabilized-ion model predicts [6] to be more stable than [4]). The calculated bulk modulus for BeO is somewhat larger than the experimental value. The measure of agreement for BeO is less than that obtained for the alkali halides but greater than that obtained for ZnO, SiO_2 and TiO_2, suggesting that the deviation between calculated and observed properties is a measure of the degree of covalency in the metal-oxygen bonds.

Cohen et al. (1987) examined several oxide minerals, including bromellite, using the *Potential Induced Breathing* (PIB) model; the results for bromellite are given in Table 3. For the zero-pressure structure and equation of state, the agreement with experiment is reasonably good and much improved over rigid-ion models. BeO is calculated to be stable in the wurtzite arrangement (i.e., bromellite) at zero pressure, in agreement with observation. The agreement for the calculated and observed elastic constants is reasonable (Table 3), as is the agreement between the calculated and observed Raman frequencies and Gruneisen parameters.

HIERARCHICAL ORGANIZATION OF CRYSTAL STRUCTURES

Ideally, the physical, chemical and paragenetic characteristics of a mineral should arise as natural consequences of its crystal structure and the interaction of that structure with the environment in which it occurs. Hence an adequate structural hierarchy of minerals should provide an epistemological basis for the interpretation of the role of minerals in Earth processes. We have not yet reached this stage for any major class of minerals, but significant advances have been made. For example, Bragg (1930) classified the major rock-forming silicate minerals according to the geometry of polymerization of $(Si,Al)O_4$ tetrahedra, and this scheme was extended by Zoltai (1960) and Liebau (1985); it is notable that the scheme parallels Bowen's reaction series (Bowen 1928) for silicate minerals in igneous rocks. Much additional insight can be derived from such structural hierarchies, particularly with regard to controls on bond topology (Hawthorne 1983a, 1994) and mineral paragenesis (Moore 1965, 1973; Hawthorne 1984; Hawthorne et al. 1987).

Hawthorne (1983a) proposed that structures be ordered or classified according to the polymerization of those cation coordination polyhedra with higher bond-valences. Higher bond-valence polyhedra polymerize to form *homo-* or *heteropolyhedral clusters* that constitute the *fundamental building block (FBB)* of the structure. The *FBB* is repeated, often polymerized, by translational symmetry operators to form the *structural unit*, a complex (usually anionic) polyhedral array (not necessarily connected) the excess charge of which is balanced by the presence of *interstitial* species (usually large low-valence cations) (Hawthorne 1985). The possible modes of cluster polymerization are obviously (1) unconnected polyhedra; (2) finite clusters; (3) infinite chains; (4) infinite sheets; (5) infinite frameworks.

POLYMERIZATION OF $Be\varphi_4$ AND OTHER $T\varphi_4$ TETRAHEDRA

Bond valence is a measure of the strength of a chemical bond, and, in a coordination polyhedron, can be approximated by the formal valence divided by the coordination number. Thus, in a BeO_4 group, the mean bond-valence is 2/4 = 0.5 vu. The valence-sum rule (Brown 1981) states that the sum of the bond valences incident at an atom is equal to the magnitude of the formal valence of that atom. Thus any oxygen atom linked to the central Be cation receives only ~0.50 vu from the Be^{3+} cation, and hence must receive ~1.50 vu from other coordinating cations. In most oxysalt structures, the coordination number of oxygen is most commonly [3] or [4]. This being the case, the *average* bond-valence incident at the oxygen atom bonded to one Be cation is ~0.50 vu for the other three cation-oxygen bonds. There are three general ways in which this bond-valence requirement may be satisfied:

(1) the oxygen atom is bonded to three additional Be atoms to produce [4]-coordination by $^{[4]}Be$, with an incident bond-valence at the oxygen atom of $4 \times 0.50 = 2.00$ vu; this is the case in bromellite: BeO.

(2) the oxygen atom is bonded to an additional Be atom and one H atom to produce an ideal incident bond-valence arrangement of $2 \times 0.50 + 1.00 = 2.00$ vu. This idealized situation is somewhat perturbed by the presence of hydrogen bonds, which provide an additional ~0.20 vu to the anion, with a concomitant lessening of the strength of the O(donor)–H bond: $2 \times 0.50 + 0.20 + 0.80 = 2.00$ vu. This is the case in behoite and clinobehoite, $Be(OH)_2$.

(3) the oxygen atom bonds to a [4]-coordinated high-valence cation (S^{6+}, As^{5+}, P, Si, Al) to produce a tetrahedral polymerization; this is the most common structural arrangement in the Be minerals. This mechanism can involve

    (a) solely a tetrahedral framework, e.g., phenakite: Be_2SiO_4;

    (b) a tetrahedral framework with interstitial cations, e.g., anhydrous beryl: $Be_3Al_2Si_6O_{18} = Al_3[Be_3Si_6O_{18}]$;

    (c) discontinuous polymerization of tetrahedra linked by interstitial cations, e.g., gugiaite: $Ca_2[BeSi_2O_7]$;

    (d) a combination of (b) and (c) with (2), e.g., väyrynenite: $Mn[Be(PO_4)OH]$.

Thus most Be minerals consist of $Be\varphi_4$ tetrahedra polymerizing with other $T\varphi_4$ tetrahedra, and hence we will organize their structures on this basis.

A STRUCTURAL HIERARCHY FOR BERYLLIUM MINERALS

In accord with the above discussion, Be minerals are classified into five distinct groups according to the polymerization of $T\varphi_4$ tetrahedra (T = Be, Zn, B, Al, Si, As, P, S) in the crystal structure:

(1) unconnected tetrahedra (Table 4);
(2) finite clusters of tetrahedra (Table 4);
(3) infinite chains of tetrahedra (Table 5, below);
(4) infinite sheets of tetrahedra (Table 6, below);
(5) infinite frameworks of tetrahedra (Table 7, below).

Within each class, structures are arranged in terms of increasing bond-valence within the constituent tetrahedra. Detailed chemical and crystallographic information, together with references, are given in Appendix A.

The structure diagrams presented here generally have the following shading pattern:

- BeO_4 tetrahedra are shaded with small crosses;
- SiO_4 tetrahedra are shaded with a random-dot pattern;
- PO_4 tetrahedra are shaded with broken lines;
- $M^{2+}O_4$ octahedra are shaded with orthogonal trellis;
- $M^{3+}O_6$ octahedra are shaded with lines.

Table 4. Beryllium minerals based on isolated Beφ_4 groups and finite Tφ_4 clusters.

Mineral	*Interstitials*	*Figure*
	Isolated tetrahedra	
Chrysoberyl	$^{[6]}$Al	–
Magnesiotaaffeite-6*N*′3*S* [1]	$^{[6]}$Al	8a,b,c
Magnesiotaaffeite-2*N*′2*S* [2]	$^{[6]}$Al	8a,b,c
	Finite cluster	
Gainesite	Zr, Na, K	9a,b

[1]previously known as musgravite-18*R*. [2]previously known as taaffeite-8*H*.

Isolated *T*φ_4 groups

Chrysoberyl, $Al_2[BeO_4]$, is isostructural with forsterite, $Mg_2[SiO_4]$, a member of the olivine group (structure not shown). The Be atom occupies the tetrahedron analogous to the SiO_4 group in forsterite, and Al occupies the octahedrally coordinated M1 and M2 sites. The overall structure is a close-packed array of O anions with Al and Be occupying the interstices. The bond valences are uniform and fairly strong, $^{[4]}$Be–O = $^{[6]}$Al–O ≅ 0.50 vu, and hence the structure has no cleavage and is hard (Mohs hardness = 8.5).

Magnesiotaaffeite-6*N*′3*S* (formerly musgravite-18*R*), $MgAl_5[BeMgAlO_{12}]$, the isostructural **ferrotaaffeite-6*N*′3*S*** (formerly pehrmanite-18*R*), $Fe^{2+}Al_5[BeFe^{2+}AlO_{12}]$, and **magnesiotaaffeite-2*N*′2*S*** (formerly taaffeite-8*H*), $MgAl_7[BeMg_2AlO_{16}]$, are all close-packed structures that are regarded as members of a *polysomatic series* (Armbruster 2001) or *polytypoids*. The structure of magnesiotaaffeite-6*N*′3*S* consists of an 18-layer repeat of O atoms in the sequence (BACACBACBCBACBABAC; *hhcccc*). There are three types of polyhedral layers in the structure: (1) an interrupted sheet of edge-sharing octahedra (Fig. 8a, shown with BeO_4 tetrahedra from an adjacent layer) that is topologically the same as the octahedral sheet in dioctahedral sheet-silicates (e.g., kaolinite); this is called the O sheet; (2) a layer of one octahedron and two tetrahedra; each MgO_6 octahedra links to three adjacent BeO_4 tetrahedra by corner-sharing to form a continuous polyhedral sheet (Fig. 8b); the additional tetrahedra within this layer, denoted T_1' by Nuber and Schmetzer (1983), occur in the interstices of the T_1' sheet (as shown in Fig. 8a) but are actually attached to the O sheet; (3) a layer of one AlO_6 octahedron and two MgO_4 tetrahedra, the T_2 layer, that topologically resembles the T_1' layer. These layers stack along the *c*-axis (Fig. 8c) in the sequence (O–T_1'–O–T_1'–O–T_2–O–T_1'–O–T_1'–O–T_2–O–T_1'–O–T_1'–O–T_2). The structure of mag-nesiotaaffeite-2*N*′2*S* consists of an 8-layer repeat of O-atoms in the sequence (BCABCBAC; *hccc*). As with magnesiotaaffeite-6*N*′3*S*, the structure consists of O, T_1' and T_2 layers, again stacked along the *c*-axis, but now in the sequence (O–T_2–O–T_1'–O–T_2–O–T_1').

Armbruster (2001) has shown that minerals of the högbomite, nigerite and taaffeite groups form polysomatic series involving spinel (*S*) and nolanite (*N*) or modified nolanite (*N*′) modules. For the Be minerals, the root name *taaffeite* is used, together with a prefix that denotes the composition of the spinel module: $MgAl_2O_4$ = *magnesio*; $FeAl_2O_4$ = *ferro*. The various polysomes are characterized by a hyphenated suffix that indicates the numbers of nolanite (*N*), modified nolanite (*N*′) and spinel (*S*) modules.

As with chrysoberyl, the bond valences are uniform and fairly strong, and it is notable that both chrysoberyl and magnesiotaaffeite-6*N*′3*S* can form rare and expensive gemstones.

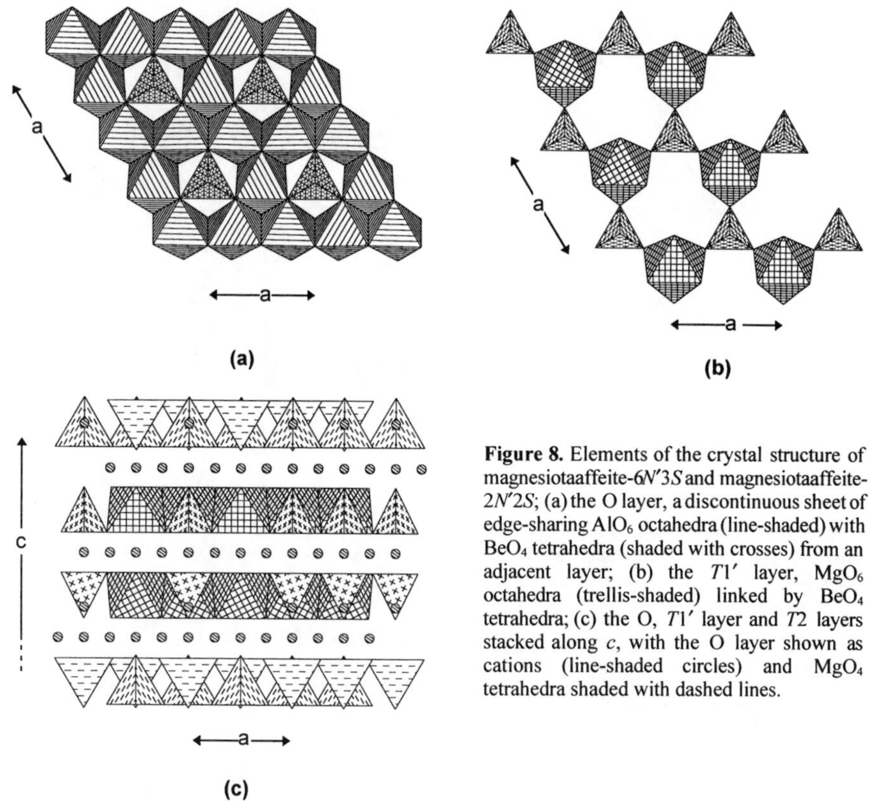

Figure 8. Elements of the crystal structure of magnesiotaaffeite-6*N*′3*S* and magnesiotaaffeite-2*N*′2*S*; (a) the O layer, a discontinuous sheet of edge-sharing AlO_6 octahedra (line-shaded) with BeO_4 tetrahedra (shaded with crosses) from an adjacent layer; (b) the *T*1′ layer, MgO_6 octahedra (trellis-shaded) linked by BeO_4 tetrahedra; (c) the O, *T*1′ layer and *T*2 layers stacked along *c*, with the O layer shown as cations (line-shaded circles) and MgO_4 tetrahedra shaded with dashed lines.

Finite clusters of $T\varphi_4$ groups

Gainesite, $Na_2Zr_2[Be(PO_4)_4](H_2O)_{1.5}$, is the only finite-cluster structure known so far among the Be minerals. A BeO_4 tetrahedron links to four PO_4 tetrahedra to form the pentameric cluster $[BeP_4O_{16}]$ (Fig. 9a). These clusters are linked into a continuous framework through ZrO_6 octahedra (Fig. 9b) topologically identical to the $[Si_5O_{16}]$ cluster in *zunyite*, $Al_{13}O_4[Si_5O_{16}](OH)_{18}Cl$. Note that the Be and P sites in the gainesite structure are only half-occupied, and in the tetrahedral-octahedral framework, tetrahedral clusters alternate with cavities occupied by interstitial Na atoms.

Infinite chains of $T\varphi_4$ tetrahedra

The minerals in this class can be divided into two broad groups based on the (bond valence) linkage involved in the infinite chains: (1) structures with Be–Si (and Si–Si) linkages; (2) structures with Be–P (or Be–As) linkages. Minerals of this class are listed in Table 5. It is interesting to note that those structures in the first group have no or only minor H-bonding, whereas the minerals of the second group have extensive H-bonding; the reason(s) for this are not yet apparent.

Chains involving Be–Si linkages.

"**Makarochkinite**," $Ca(Fe^{2+}{}_4Fe^{3+}Ti^{4+}O_2[(BeAlSi_4)O_{18}]$, contains decorated chains of BeO_4, SiO_4 and $(Si,Al)O_4$ tetrahedra; the basic chain resembles a pyroxenoid-like TO_3 chain consisting of alternating dimers of $(Be_{0.5}Si_{0.5})O_4$ tetrahedra and $(Si_{0.75}Al_{0.25})O_4$ tetrahedra, and

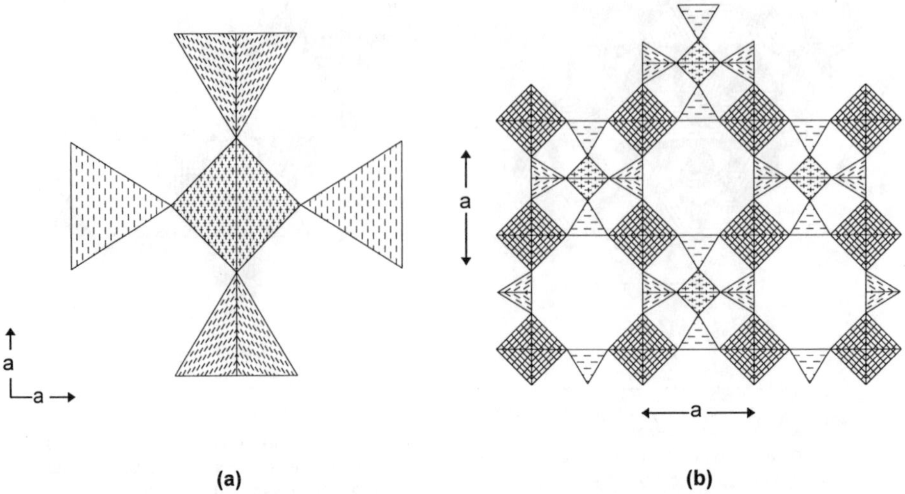

Figure 9. The crystal structure of gainesite. (a) The $[BeP_4O_{16}]$ pentamer that is the finite tetrahedron cluster in gainesite. (b) The tetrahedral-octahedral framework in gainesite; ZrO_6 octahedra are line-shaded.

Table 5. Beryllium minerals based on chains of $T\varphi_4$ tetrahedra.

Mineral	*Linkage*	*Interstitials*	*Figure*
"Makarochkinite"	Be–Si–Si–Al	$^{[6]}$[Ti,Mg,Fe]	10a,b
Khmaralite	Be–Si–Si–Al	$^{[6]}$[Mg,Fe^{2+},Al]	10c,d
Sverigeite	Be–Be–Si–Si	$^{[6]}$Sn + $^{[6]}Mn^{2+}$	11a,b,c
Joesmithite	Be–Si–Si	$^{[6]}$[Mg,Fe^{2+},Al] + $^{[8]}$Ca + Pb	11d
Surinamite	Be–Si–Si	$^{[6]}$[Mg,Al] + O	12a,b
Euclase	Be–Si	$^{[6]}$Al + H-bonding	13a,b
Bearsite	Be–As	H-bonding	14a,b
Moraesite	Be–P	H-bonding	14c,d
Fransoletite	Be–P	H-bonding	15a,b
Parafransoletite	Be–P	Ca + H-bonding	15a,b
Väyrynenite	Be–P	$^{[6]}Mn^{2+}$+ H-bonding	13c,d
Roscherite	Be–P	$^{[6]}$Al + Ca + H-bonding	15c,d

it is decorated with side-groups of SiO_4 tetrahedra attached to each of the $(Be_{0.5}Si_{0.5})O_4$ tetrahedra. These chains extend along the *a*-axis (Fig. 10a), forming layers of tetrahedra orthogonal to [011] (Fig. 10b). The layers of tetrahedra are linked by ribbons of edge-sharing $Fe^{2+}O_6$ and $Fe^{3+}O_6$ octahedra (with minor substitution of Ti^{4+} and Mg), and between these ribbons in the layer of octahedra are interstitial sites containing [7]-coordinated Ca (with minor substitution of Na).

"Makarochkinite" has not been approved as a new mineral by the Commission on New Minerals and Mineral Names of the International Mineralogical Association. However, it is a Be-bearing structure of the aenigmatite type, and it is included here on this basis. There are two additional Be-bearing minerals with the aenigmatite structure, **høgtuvaite** and **welshite**, but the crystal structures of these minerals have not yet been refined. Grauch et al. (1994) state that "makarochkinite" is identical to høgtuvaite; this issue will be discussed later.

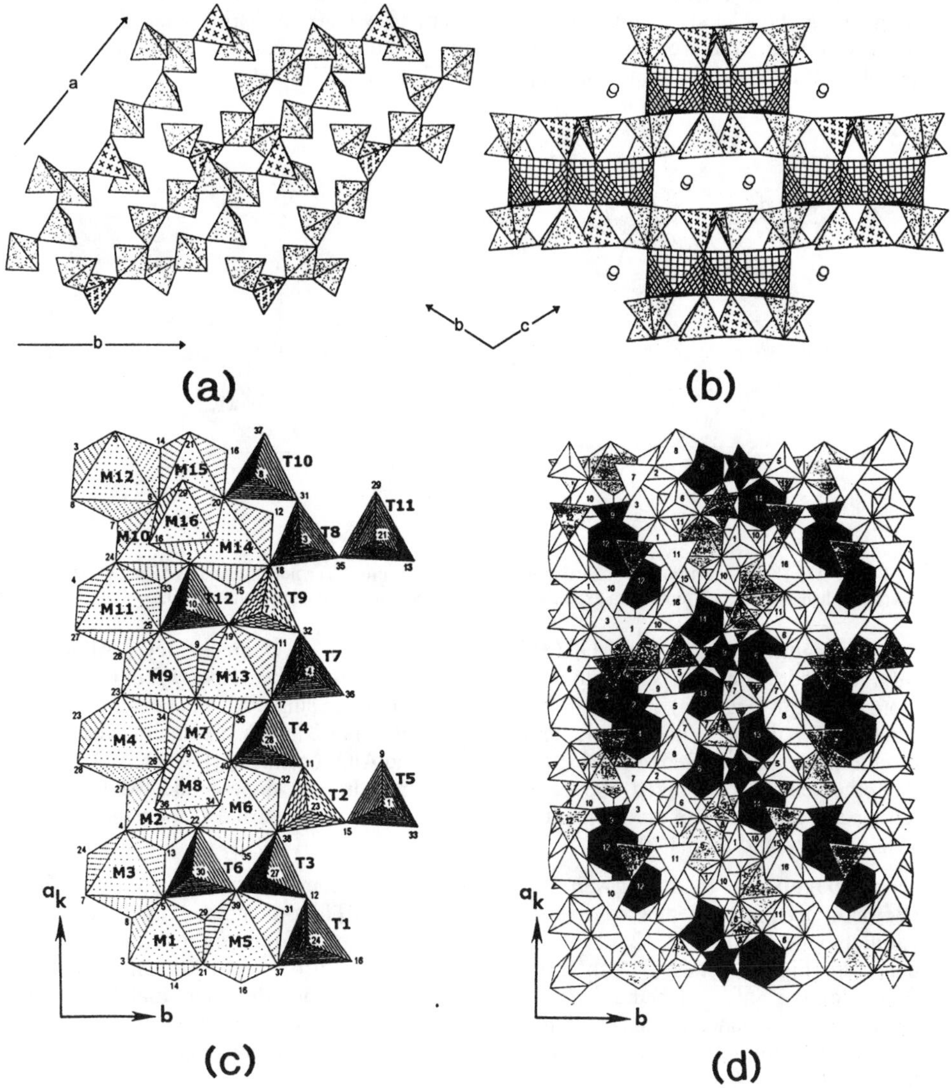

Figure 10. The crystal structures of "makarochkinite" and khmaralite; (a) "makarochkinite" projected onto (001); note the chains of BeO_4 and SiO_4 tetrahedra extending parallel to the *a*-axis; (b) "makarochkinite" projected onto (100); circles in (b) represent interstitial Ca atoms; (c) khmaralite projected onto (001); Be is concentrated in the *T*2 and *T*9 tetrahedra; (d) khmaralite projected onto (001); black, grey and white tetrahedra are enriched in Be, Si and Al, respectively. Used by permission of the Mineralogical Society of America, from Barbier et al. (1999), *American Mineralogist*, Vol. 84, Figs. 4 and 5, pp. 1656-57.

Khmaralite, $Mg_7Al_9O_4[Al_6Be_{1.5}Si_{4.5}O_{36}]$, is a close-packed Be-bearing silicate mineral that is closely related to sapphirine. It was originally noted by Grew (1981) as a beryllian sapphirine, and later reported as a new mineral by Barbier et al. (1999). The structure consists of open-branched chains of tetrahedra (Fig. 10c) that are occupied by Be, Al and Si. Some *T* sites are extensively disordered, whereas others are fairly ordered, and Be is dominant at the *T*2 and *T*9 sites. These chains combine with edge-sharing octahedra to form a dense-packed structure (Fig. 10d). The superstructure in khmaralite corresponds to a doubling of the *a* axis in monoclinic sapphirine-2*M* (in the $P2_1/c$ setting).

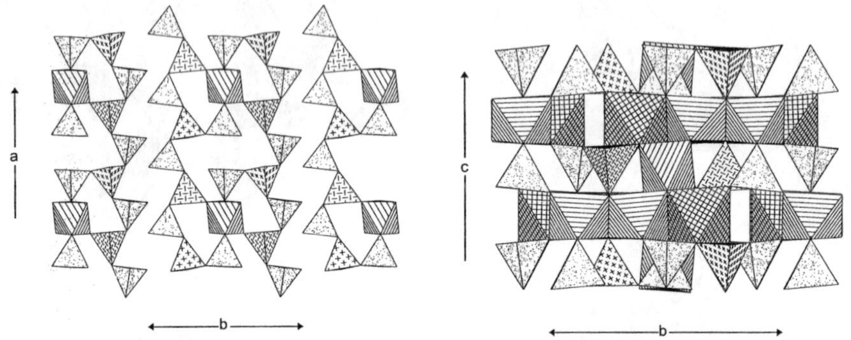

Figure 11. The crystal structure of surinamite; (a) projected onto (001); note the chains of BeO_4, AlO_4 (dash-shaded) and SiO_4 tetrahedra extending in the *a*-direction; (b) projected onto (100).

Surinamite, $(Mg,Fe^{2+})_3Al_3O[BeAlSi_3O_{15}]$, consists of $T\varphi_3$ chains of BeO_4, AlO_4 and SiO_4 tetrahedra extending parallel to *a* and decorated by SiO_4 tetrahedra every four tetrahedra along its length (Fig. 11a). Adjacent chains are linked through single AlO_6 octahedra to form ribbons that define a dominantly tetrahedral layer. These layers are interleaved with discontinuous layers of edge-sharing $(Mg,Fe^{2+})O_6$ and AlO_6 octahedra (Fig. 11b). As with the other chains considered above, the BeO_4 tetrahedron is three-connected to other tetrahedra. The structure of surinamite is related to that of sapphirine (Moore and Araki 1983; Christy and Putnis 1988).

Sverigeite, Na $(Mn^{2+},Mg)_2Sn^{4+}[Be_2Si_3O_{12}(OH)]$, has complex chains of corner-sharing $[Be_2SiO_8(OH)]$ three-membered rings and $[Be_2Si_2O_{11}(OH)]$ four-membered rings extending parallel to the *a*-axis (Fig. 12a). These chains are cross-linked in the *b*-direction by SnO_6 octahedra (Fig. 12a) and by $[(Mn^{2+},Mg)_2O_{10}]$ dimers (Fig. 12b). Viewed down the *b*-axis (Fig. 12c), the SnO_6 octahedra form a body-centered array, and the interstitial Na atoms occur between octahedra adjacent in the *c*-direction. The coordination around the Na atom is somewhat unusual with four short bonds (2.43 Å) in a square-planar arrangement and four long bonds (3.07 Å).

Joesmithite, ideally $Pb^{2+}Ca_2(Mg_3Fe^{3+}{}_2)[(Si_6Be_2)O_{22}](OH)_2$, is a novel amphibole with Be ordered at the *T*2 tetrahedron of the normal *C*2/*m* amphibole structure (Hawthorne 1983b). Double $[T_8O_{22}]$ chains of tetrahedra extend along the *c*-axis (Fig. 12d), linking ribbons of edge-sharing $(Mg,Fe^{2+})O_6$ and $(Mg,Fe^{3+})O_6$ octahedra that also extend along *c*. Occupying the [8]-coordinated M4 sites is Ca, and Pb^{2+} occupies the *A* cavity between the back-to-back chains of tetrahedra. The Pb^{2+} atom is displaced toward the Be site, bonding strongly to the chain-bridging anions involving the BeO_4 tetrahedron and also producing a local stereo-

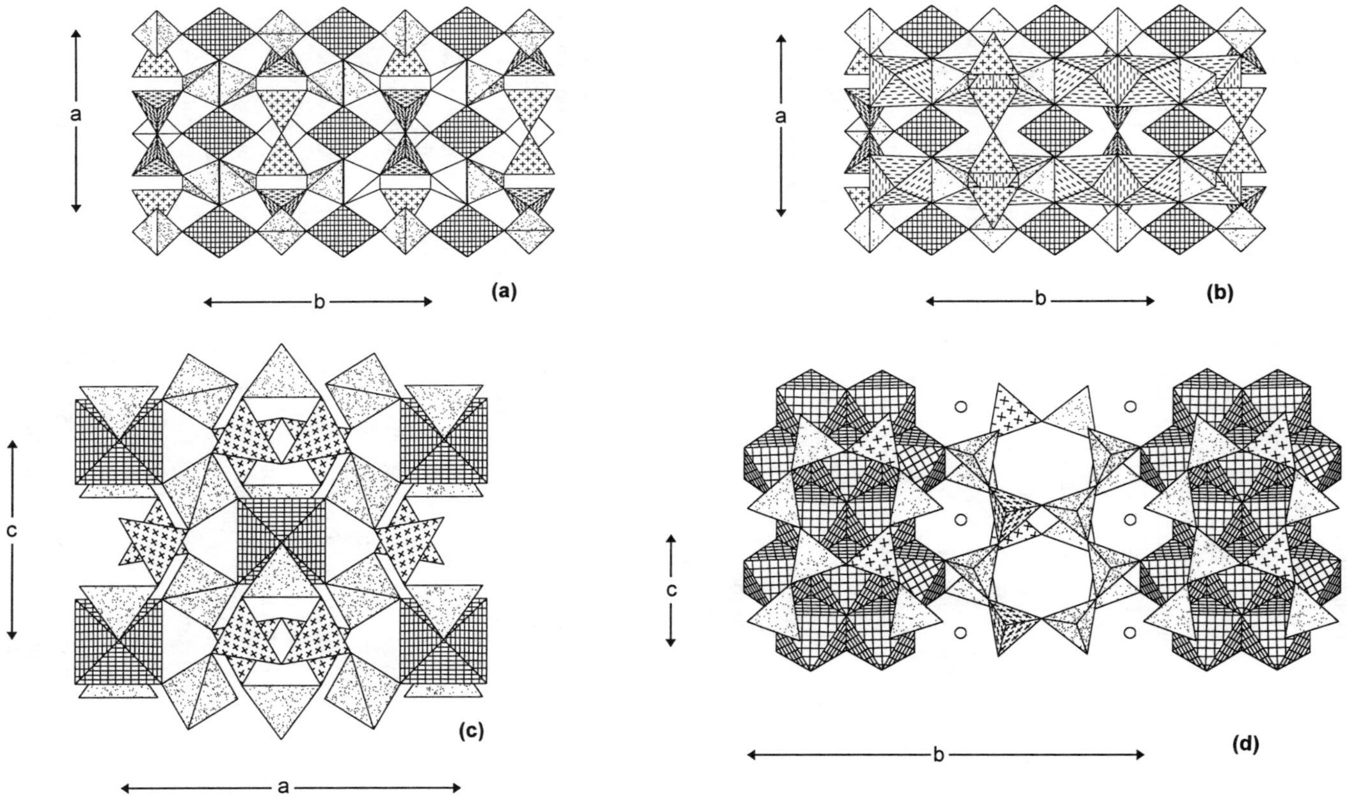

Figure 12. The crystal structures of sverigeite and joesmithite; (a) sverigeite projected onto (001); note the complex chains of corner-sharing [$Be_2SiO_8(OH)$] three-membered rings and [$Be_2Si_2O_4(OH)$] four-membered rings extending along the *a*-direction; the SnO_6 octahedron is trellis-shaded; note that the SiO_4 tetrahedra do *not* share edges; the [$(Mn^{2+},Mg)_2O_{10}$] dimers are omitted for clarity; (b) sverigeite projected onto (001); note the dimers of MnO_6 octahedra (dash-line shaded); (c) sverigeite projected down *b*; (d) joesmithite projected onto (100); unshaded circles are Ca.

chemistry that satisfies the spatial requirements for stereoactive lone-pair behavior for Pb^{2+} (Hawthorne 1983b; Moore et al. 1993).

Euclase, Al[$BeSiO_4(OH)$], with minor substitution of F for (OH), contains chains of BeO_4 and SiO_4 tetrahedra extending in the *a*-direction (Fig. 13a): BeO_4 tetrahedra link by corner-sharing to form a (fully rotated) pyroxene-like TO_3 chain that is decorated on both sides to form a ribbon in which the BeO_4 tetrahedra are three-connected to SiO_4 tetrahedra and the SiO_4 tetrahedra are two-connected to BeO_4 tetrahedra. In addition, the anion bridging the BeO_4 tetrahedra also belongs to an SiO_4 group. The ribbon-like chains of tetrahedra are arranged in modulated sheets parallel to {001} and linked by edge-sharing pyroxene-like chains of AlO_6 octahedra (Fig. 13b).

Chains involving Be–P and Be–As linkages.

Bearsite, [$Be_2(AsO_4)(OH)$](H_2O)$_4$, is made up of chains of BeO_4 and AsO_4 tetrahedra. Two BeO_4 and one AsO_4 tetrahedra alternate along the length of a simple TO_3 chain; two of these chains meld *via* sharing corners such that the BeO_4 tetrahedra are three-connected and the AsO_4 tetrahedron is four-connected. The resultant ribbons extend in the *a*-direction (Fig. 14a). The ribbons form a face-centered array when viewed down [100] (Fig. 14b), and the resultant tunnels between the ribbons are filled with interstitial H_2O groups that are held in the structure purely by H-bonding.

Moraesite, [$Be_2(PO_4)(OH)$](H_2O)$_4$, has a structure extremely similar to that of bearsite, as might be expected by the similarity of the chemical formulae, unit-cell dimensions and space groups ($C2/c$ vs. $P2_1/a$; see Appendix A). [$Be_2(PO_4)(OH)$] ribbons extend along the *c*-direction (Fig. 14c), and are topologically identical to the [$Be_2(AsO_4)(OH)$] chains in bearsite (Fig. 14a). The ribbons form a face-centered array in moraesite (Fig. 14d) similar to that in bearsite (Fig. 14b). The difference between the two structures seems to arise from the slight difference in placement of the H_2O groups within the channels (compare Figs. 14b and 14d).

Fransoletite and **parafransoletite** are dimorphs which have the composition $Ca_3[Be_2(PO_4)_2(PO_3\{OH\})_2](H_2O)_4$. The principal motif in each structure is a complex chain of tetrahedra consisting of four-membered rings of alternating BeO_4 and PO_4 tetrahedra that linked through common BeO_4 tetrahedra; these chains extend in the *a*-direction (Fig. 15a). Viewed end-on (Fig. 15b), the chains form a square array and are linked by [6]- and [7]-coordinated Ca atoms that form sheets parallel to {001}; further interchain linkage occurs through H-bonding involving H_2O groups. The fransoletite and parafransoletite structures differ only in the relative placement of the octahedrally coordinated Ca atom and the disposition of adjacent chains along their length (Kampf 1992).

Väyrynenite, Mn^{2+}[$Be(PO_4)(OH)$], contains chains of BeO_4 and PO_4 tetrahedra extending in the *a*-direction (Fig. 13c): BeO_4 tetrahedra link by corner-sharing to form a pyroxenoid-like TO_3 chain that is decorated on both sides by PO_4 tetrahedra to form a ribbon in which the BeO_4 tetrahedra are four-connected and the PO_4 tetrahedra are two-connected. These ribbon-like chains are linked by edge-sharing pyroxene-like chains of $Mn^{2+}O_6$ octahedra that also extend parallel to the *a*-axis. The resulting structural arrangement consists of modulated sheets of tetrahedra and octahedra (Fig. 13d) that are similar (Mrose and Appleman 1962) to the analogous sheets in euclase (Fig. 13a).

Roscherite and **zanazziite** are composed of very convoluted chains of $Be\varphi_4$ and PO_4 tetrahedra extending in the [101] direction (Fig. 15c; note that in this view, the two chains appear to join at a mirror plane parallel to their length; however, the plane in question is a glide plane and the two chains do not join at this plane, they are displaced in the *c*-direction). The chain consists of four-membered rings of alternating $Be\varphi_4$ and PO_4 tetrahedra linked through PO_4 tetrahedra that are not members of these rings (Fig. 15c). These chains are linked

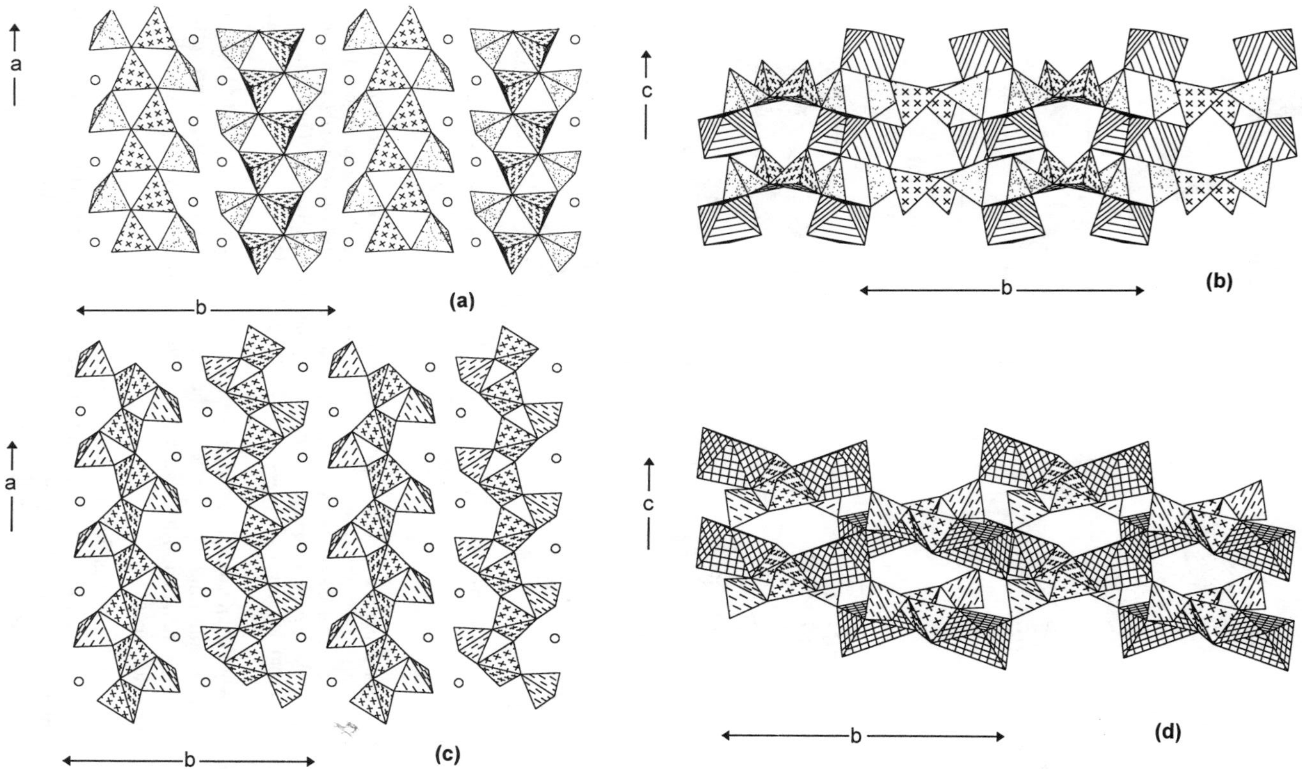

Figure 13. The crystal structures of euclase and väyrynenite; (a) euclase projected onto (001); note the chains of BeO_4 and SiO_4 tetrahedra extending in the *a*-direction; Al atoms are shown as circles; (b) euclase projected onto (100); AlO_6 octahedra are line-shaded; (c) väyrynenite projected onto (001); chains of BeO_4 and PO_4 tetrahedra extend in the *a*-direction; circles are Al atoms; (d) väyrynenite projected onto (100); note the similarity of euclase and väyrynenite in this view (although the chains of tetrahedra are topologically distinct).

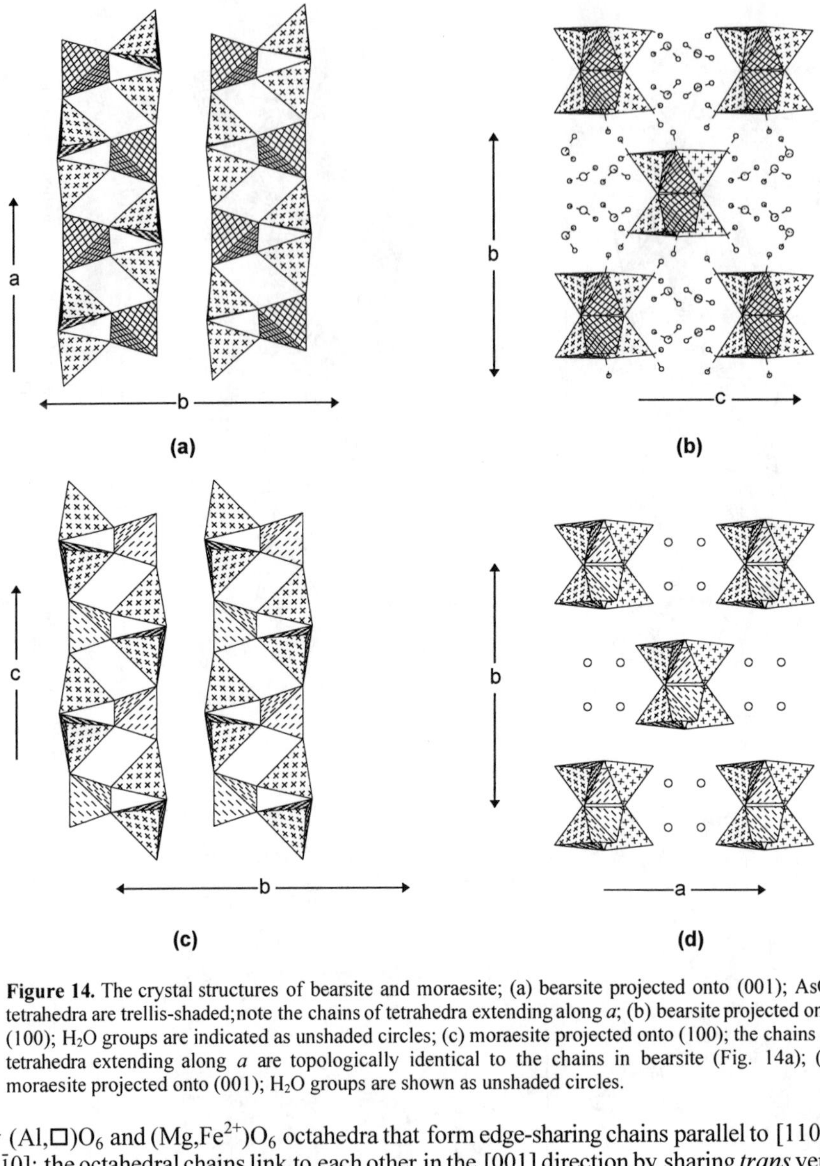

Figure 14. The crystal structures of bearsite and moraesite; (a) bearsite projected onto (001); AsO_4 tetrahedra are trellis-shaded; note the chains of tetrahedra extending along *a*; (b) bearsite projected onto (100); H_2O groups are indicated as unshaded circles; (c) moraesite projected onto (100); the chains of tetrahedra extending along *a* are topologically identical to the chains in bearsite (Fig. 14a); (d) moraesite projected onto (001); H_2O groups are shown as unshaded circles.

by (Al,□)O_6 and (Mg,Fe^{2+})O_6 octahedra that form edge-sharing chains parallel to [110] and [1$\bar{1}$0]; the octahedral chains link to each other in the [001] direction by sharing *trans* vertices (Fig. 15d). The resultant octahedral-tetrahedral framework is strengthened by [7]-coordinated Ca occupying the interstices.

The structure and composition of these minerals is not completely understood. Roscherite (Slavík 1914) is the Mn^{2+}-dominant species and zanazziite (Leavens et al. 1990) is the Mg-dominant species. Lindberg (1958) also reports an Fe^{2+}-dominant species from the Sapucaia pegmatite, Minas Gerais, that is currently unnamed. The situation is complicated by the fact that the original crystal-structure determination of roscherite (Fanfani et al. 1975) was done on a crystal of what was later determined to be zanazziite with the ideal end-

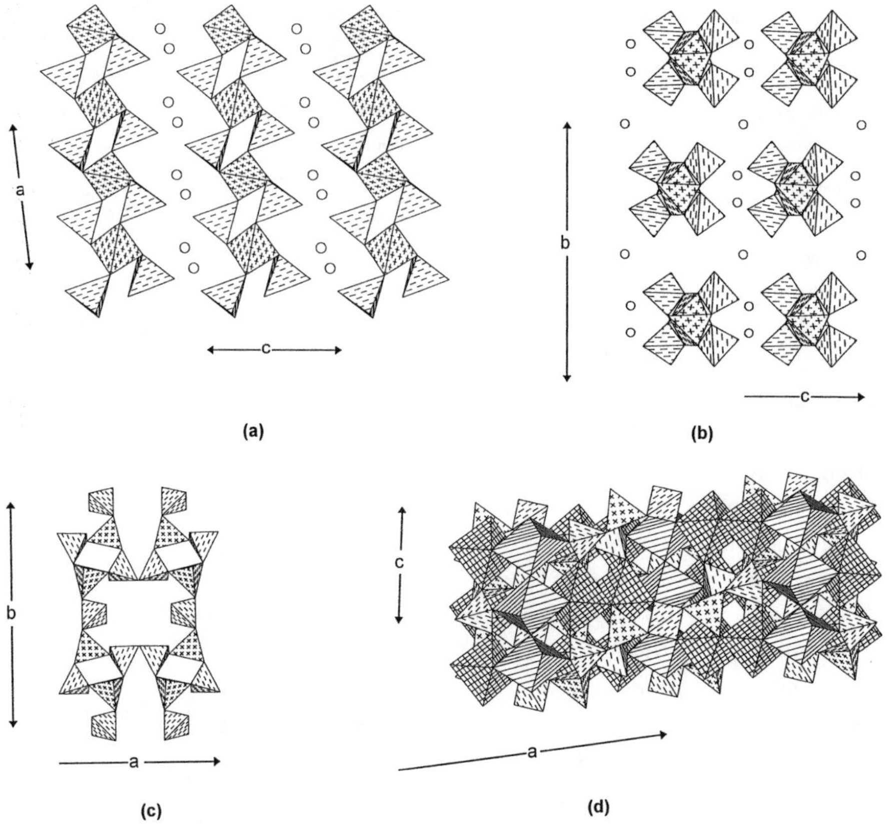

Figure 15. The crystal structure of fransoletite and roscherite; (a) fransoletite projected onto (010); circles are Mn^{2+} atoms; chains of BeO_4 and PO_4 tetrahedra extend along *a*; (b) fransoletite projected onto (100), a view in which the chains are seen end-on; (c) the structural unit in roscherite projected onto (001); chains of BeO_4 and PO_4 tetrahedra extend along *b*; note that the PO_4 tetrahedra in the center of the figure do *not* share a common anion, but are separated in the *c*-direction; (d) roscherite projected onto (010); note that the trivalent octahedra (trellis-shaded) are only two-thirds occupied (by Al); Ca atoms are omitted for clarity.

member formula $Ca_2Mg_4(Al_{0.67}\square_{0.33})_2[Be_4(PO_4)_6(OH)_6](H_2O)_4$. Fanfani et al. (1977) report a triclinic structure for roscherite that is Mn^{2+} dominant, i.e., roscherite with the ideal end-member formula $Ca_2Mn^{2+}{}_4(Fe^{3+}{}_{0.67}\square_{0.33})(\square)[Be_4(PO_4)_6(OH)_4(H_2O)_2](H_2O)_4$. Note that the trivalent-cation content ($Al_{1.33}$ vs. $Fe^{3+}{}_{0.67}$) and type are different in the two species, and electroneutrality is maintained by replacement of OH by H_2O: $Fe^{3+} + \square$ (vacancy) + 3 H_2O $\rightarrow Al^{3+}{}_2 + 3$ OH. Whether the monoclinic $\rightarrow$triclinic transition is caused by the $Mn^{2+} \rightarrow Mg$ replacement or by the reaction noted above is not yet known.

Infinite sheets of $T\varphi_4$ tetrahedra

The minerals in this class can be divided into three broad groups (Table 6) based on the bond-valences of the linkages involved in the infinite sheets: (1) structures with Be−Be linkages; (2) structures with Be−B linkages; (3) structures with Be−Si (and Si−Si) linkages; (4) structures with Be−As and Be−P linkages.

Table 6. Beryllium minerals based on sheets of $T\varphi_4$ tetrahedra.

Mineral	*Linkage*	*Interstitials*	*Figure*
Clinobehoite	Be–Be	H-bonding	16a,b
Berborite-1T	Be–B	H-bonding	17a
Berborite-2T	Be–B	H-bonding	17b
Berborite-2H	Be–B	H-bonding	17c
Gugiaite	Be–Si–Si	$^{[8]}$Ca	–
Meliphanite	Be–Si–Si	$^{[8]}$Ca, Na	–
Leucophanite	Be–Si–Si	$^{[8]}$Ca	18a,b
Jeffreyite	Be–Si–Si	$^{[8]}$Ca, Na	–
Gadolinite-(Ce)	Be–Si–Si	Ce, Fe^{2+}	–
Gadolinite-(Y)	Be–Si–Si	Y, Fe^{2+}	–
"Hingganite-(Y)"	Be–Si–Si	Y	18c,d
Hingganite-(Yb)	Be–Si–Si	Yb	18c,d
"Calcybeborosilite"	Be–Si–Si	Ca,Y	–
Minasgeraisite-(Y)	Be–Si–Si	Ca, Y	–
Semenovite	Be–Si–Si	Ca, Y Fe^{2+}, Na + H-bonding	19a,b
Aminoffite	Be–Si–Si	$^{[8]}$Ca + H-bonding	20a,b
Harstigite	Be–Si–Si	$^{[6]}$Mn + H-bonding	20c,d
Samfowlerite	Be–Si–Si–Zn	Ca, $^{[6]}$Mn + H-bonding	21a,b
Bityite	Be–Si–Si–Al	Ca, Al + H-bonding	–
Sorensenite	Be–Si–Si	Na, $^{[6]}$Sn + H-bonding	21c,d
Bergslagite	Be–As	$^{[8]}$Ca	22a,b
Herderite	Be–P	$^{[8]}$Ca + H-bonding	22c,d
Hydroxylherderite	Be–P	$^{[8]}$Ca + H-bonding	22c,d
Uralolite	Be–P	Ca	23a,b
Ehrleite	Be–P	$^{[6]}$Zn + H-bonding	23c,d
Asbecasite	Be–As, Be–Si, Si–Si	$^{[6]}$[Ti,Sn]	24a,b,c,d

Sheets involving Be–Be linkages.

Clinobehoite, [$Be(OH)_2$], consists of $Be(OH)_4$ tetrahedra linked into sheets by sharing vertices. $Be\varphi_4$ tetrahedra share corners to form [$Be\varphi_3$] chains that resemble almost fully rotated pyroxene chains that extend in the *b*-direction (Fig. 16a) and are cross-linked into a thick slab by other corner-linked $Be\varphi_4$ tetrahedra. When the sheets are seen edge-on (Fig. 16b), stacked along the *a*-direction, it can be seen that the cross-linking $Be\varphi_4$ tetrahedra share vertices and form undulating chains along *b*. The sheets are held together solely by H-bonding.

Sheets involving Be–B linkages.

Berborite, [$Be_2\{BO_3(OH)\}$](H_2O), has three polytypes, designated 1*T*, 2*T* and 2*H* by Giuseppetti et al. (1990). All three structures consist of sheets of corner-sharing BeO_4

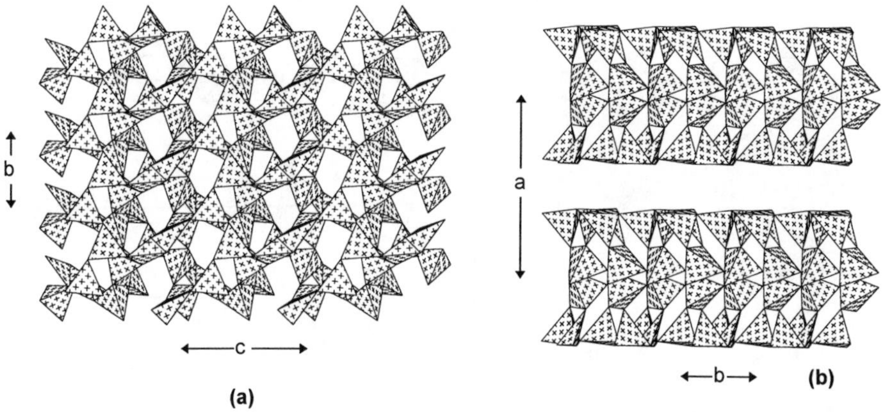

Figure 16. The crystal structure of clinobehoite: (a) projected onto (100); chains of $Be(OH)_4$ tetrahedra (fully rotated pyroxene-like conformation) are cross linked with $Be(OH)_4$ tetrahedra to form sheets parallel to (100); (b) projected onto (001), showing two sheets stacked along *a*; cf. Figures 25a and 25b for behoite.

tetrahedra and BO_3 triangles, and the sheets are identical in all three polymorphs; the structures differ in the relative positioning of adjacent layers in each polytype. The unit cell has a dimer of $Be\varphi_4$ tetrahedra surrounded by BO_3 triangles (Fig. 17a), such that the structure in the {001} plane is a 6_3 net of corner-connected $Be\varphi_4$ tetrahedra with each hexagonal hole containing a BO_3 triangle such that the polyhedra define a 3^6 net with adjacent BeO_4 tetrahedra pointing in opposite directions along *c*; this layer is the first layer in all the polytypes. In berborite-1*T*, the layers shown in Figure 17a are linked by a rather disordered arrangement of H-bonding involving the OH and H_2O groups. In berborite-2*T*, the structure consists of alternations of the layers shown in Figures 17a and 17b, again linked by H-bonding. In berborite-2*H*, the structure consists of alternations of the layers shown in Figures 17a and 17c. It is to be emphasized that the layers shown in Figure 17 are identical; their differing appearance results from different orientation and placement with regard to the unit cell. This general building principle for berborite polytypes indicates that other polytypes are (at least geometrically) possible.

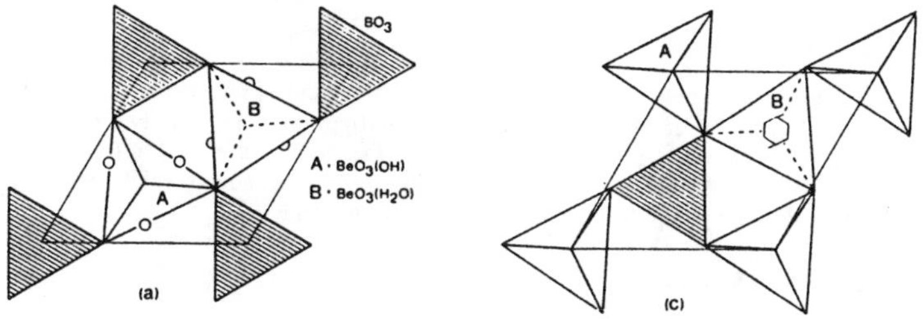

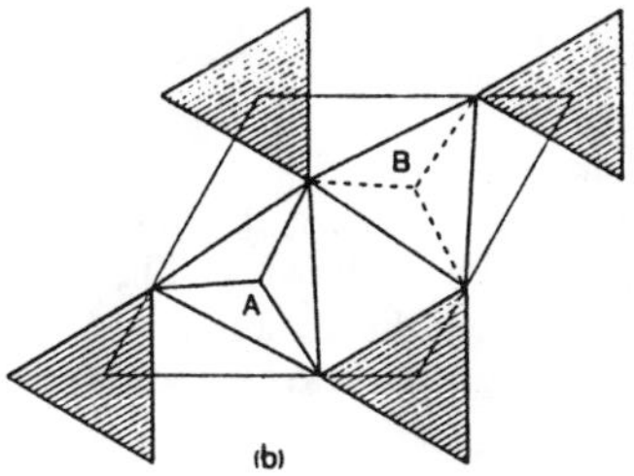

Figure 17. The crystal structures of the berborite polytypes projected onto (001); (a) the basic sheet in all three polytypes; this is the single layer in berborite-1*T*; (b) the second sheet in berborite-2*T*; (c) the second sheet in berborite-2*H*; all sheets are linked by H-bonding.

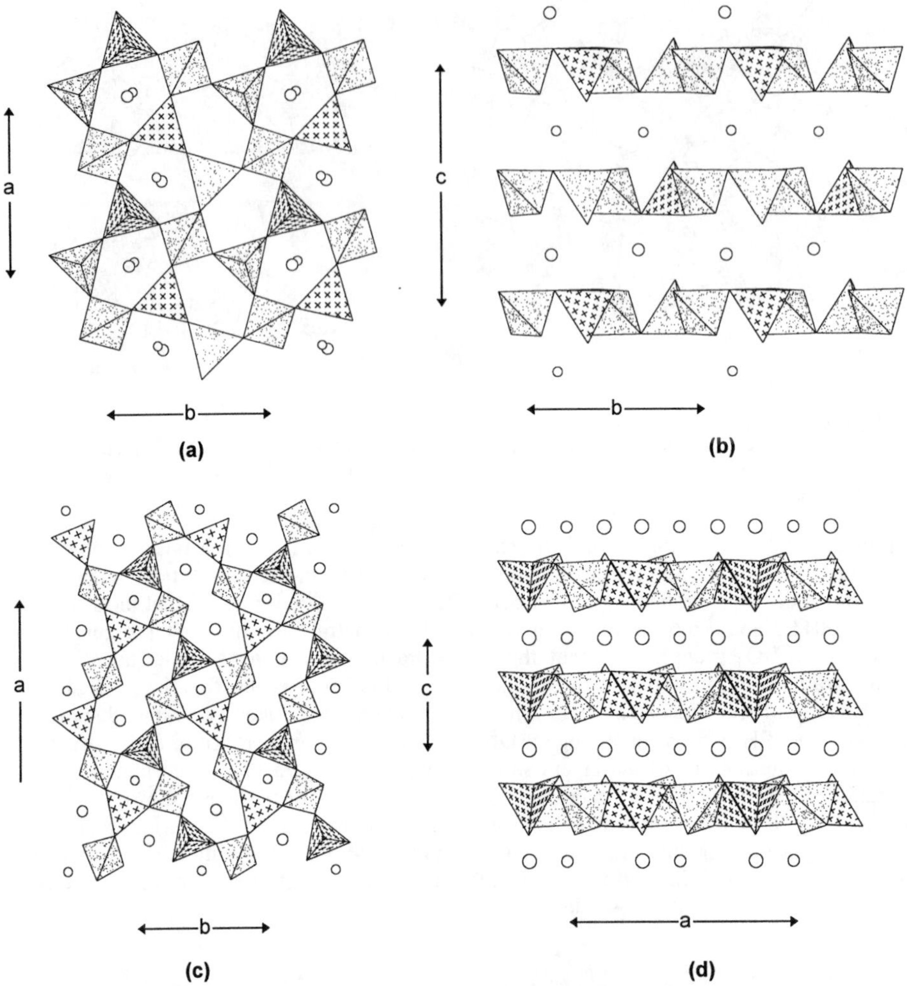

Figure 18. The crystal structures of leucophanite and hingganite-(Yb); (a) leucophanite projected onto (001); BeO_4 and SiO_4 tetrahedra occupy the vertices of a $(5^{3}_{2}5^{4}_{1})$ two-dimensional net; (b) leucophanite projected onto (100); large circles are Na, small circles are Ca; (c) hingganite-(Yb) projected onto (001); large circles are Y, small circles are □ (vacancies) which are filled in other minerals of the gadolinite group; (d) hingganite-(Yb) projected onto (010).

Sheets involving Be–Si linkages.

Gugiaite, $Ca_2[BeSi_2O_7]$,
meliphanite, $(Ca,Na)_2[Be(Si,Al)_2O_6(O,OH,F)]$,
leucophanite, $(Ca,Na)_2[Be(Si,Al)_2O_6(O,F)]$, and
jeffreyite, $(Ca,Na)_2[(Be,Al)Si_2(O,OH)_7]$,

all have structures topologically related (identical sheets) to that of *melilite*, $(Ca,Na)_2(Al,Mg)(Si,Al)_2O_7$. The structure of leucophanite is shown in Figures 18a and 18b. BeO_4 and SiO_4 tetrahedra occur at the vertices of a two-dimensional net (Fig. 18a) in which half of the SiO_4 tetrahedra are four-connected and the rest of the tetrahedra are three-connected. These sheets are stacked in the *c*-direction and are linked by [8]-coordinated interstitial cations (Fig. 18b). In leucophanite, the interstitial Ca and Na atoms are ordered into different sheets (Fig. 18b) as a result of the orthorhombic symmetry of this particular mineral. In gugiaite, Ca occupies all layers between the sheets, and gugiaite is tetragonal. In meliphanite, Ca and Na are disordered, and meliphanite also has tetragonal symmetry. Although the structure of jeffreyite is not known, it must consist of a derivative of the gugiaite arrangement with ordered replacement of some Be by Al. Rare-earth elements may substitute for Ca in leucophanite, and Cannillo et al. (1992) showed that REE-bearing leucophanite is actually triclinic; thus the structure described by Cannillo et al. (1992) is actually that of an unnamed new mineral.

Gadolinite-(Ce), $Ce_2Fe^{2+}[Be_2Si_2O_{10}]$,
gadolinite-(Y), $Y_2Fe^{2+}[Be_2Si_2O_{10}]$,
hingganite-(Yb), $Yb_2\square[Be_2Si_2O_8(OH)_2]$,
"**hingganite-(Y)**," $Y_2\square[Be_2Si_2O_8(OH)_2]$,
"**calcybeborosilite**," $CaY\square[BeBSi_2O_8(OH)_2]$, and
minasgeraisite-(Y), $CaY_2[Be_2Si_2O_{10}]$

all belong to the *gadolinite group.* The basic unit of the sheet is a four-membered ring of two BeO_4 and two SiO_4 tetrahedra in the sequence [Be–Si–Be–Si]. These rings link by sharing tetrahedral vertices to generate eight-membered rings of alternating BeO_4 and SiO_4 tetrahedra (Fig. 18c). These sheets stack in the *c*-direction (Fig. 18d) and are linked by [8]-coordinated rare-earth cations and [6]-coordinated Fe^{2+}. The different minerals have the same (topological) structural unit and differ in the interstitial species linking the sheets together. "Hingganite-(Y)" is referred to in the literature (Ximen and Peng 1985; Peterson et al. 1994) but is not yet an accredited mineral species. The crystal structure (Ximen and Peng 1985) shows it to be isostructural with hingganite-(Yb).

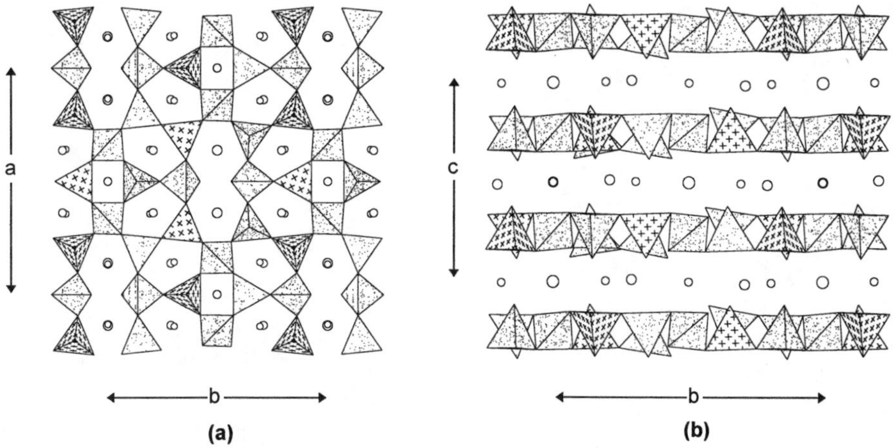

Figure 19. The crystal structure of semenovite; (a) projected onto (001); $Be\varphi_4$ and $Si\varphi_4$ tetrahedra occur at the vertices of a two-dimensional net; large circles are RE, small circles are (Ca,Na); (b) projected onto (100); Fe atoms are omitted for clarity.

Semenovite, $(RE)_2Fe^{2+}Na_{0-2}(Ca,Na)_8[Be_6Si_{14}O_{40}(OH)_4F_4]$, has ordered $Be\varphi_4$ and $Si\varphi_4$ tetrahedra occurring at the vertices of a two-dimensional net (Fig. 19a). In the structure refinement of Mazzi et al. (1979), the Be and Si occupancies are partly disordered (either 80:20 or 20:80); however, in view of the pervasive twinning present, it is possible that this partial disorder is not correct (Mazzi et al. 1979). For simplicity, we have assumed complete Be–Si order here. There are two distinct Beφ 4 tetrahedra, both of which are three-connected within the sheet, and five distinct Siφ 4 tetrahedra, one of which is four-connected and four of which are three-connected. The Beφ 4 tetrahedra involve one F and one OH group, respectively, and there is also one acid-silicate group, $SiO_3(OH)$. There is one four-membered ring consisting of one Beφ 4 and three SiO_4 tetrahedra that links to its translational equivalent along *a* through an eight-membered ring with the sequence (Si–Si–Si–Si–Si–Be–Si–Be). These chains meld in the *b*-direction by sharing vertices with adjacent chains that are displaced *a*/2 in the *a*-direction, giving rise to intermediate chains of corner-sharing five-membered rings with the sequences (Si–Si–Si–Si–Be) and (Si–Be–Si–Si–Be). These sheets stack in the *c*-direction and are linked by layers of interstitial cations (Fig. 19b), [6]-coordinated Fe^{2+}, [8]-coordinated RE, [8]-coordinated Ca and [8]-coordinated Na.

The $[Be_6Si_{14}O_{40}F_4]$ sheet in semenovite (Fig. 19a) is topologically the same as the $[Be_4Si_6O_{22}(OH)_2]$ sheet in harstigite (Fig. 20c). This equivalence is not immediately apparent from the chemical formula of the sheets, but may be seen if we write both

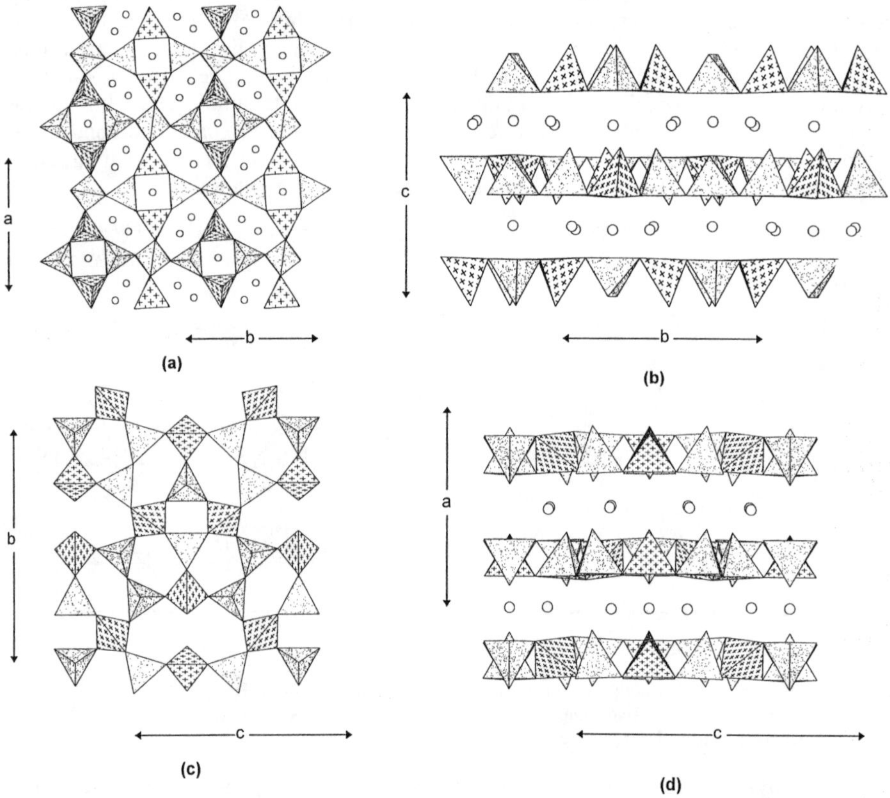

Figure 20. The crystal structures of aminoffite and harstigite; (a) aminoffite projected onto (001); $Be\varphi_4$ and SiO_4 tetrahedra occur at the vertices of a two-dimensional net; unshaded circles are Ca; (b) aminoffite projected onto (100); (c) the structural unit of harstigite projected onto (100); $Be\varphi_4$ and SiO_4 tetrahedra occur at the vertices of a net; (d) harstigite projected onto (010); circles are Ca and Mn atoms.

sheets as $[T_{20}O_{40}(OH)_4\varphi_4]$ where T = Be + Si and φ = O, F. The cell dimensions of both minerals are very close (Appendix A). However, comparison of Figures 20c and 19a shows that the pattern of ordering of Be and Si differs between the two structures. This is in accord with the different chemical compositions and space groups: $[Be_6Si_{14}O_{40}(OH)_4F_4]$, *Pmnn*, and $[Be_4Si_6O_{22}(OH)_2]$, *Pnam*.

Aminoffite, $Ca_2[Be_2Si_3O_{10}(OH)_2]$, consists of $Be\varphi_4$ and SiO_4 tetrahedra arranged at the vertices of a two-dimensional net; the resultant sheet is linked in the *c*-direction by [7]-coordinated interstitial Ca. Four-membered rings of alternating $Be\varphi_4$ and SiO_4 tetrahedra are linked by additional SiO_4 tetrahedra (Fig. 20a). The tetrahedra of the ring are three-connected, the linking SiO_4 tetrahedra are four-connected, and the tetrahedra in adjacent four-membered rings point in opposite directions (i.e., along $+c$ or $-c$). Sheets are stacked along *c* (Fig. 20b) where they are linked by layers of interstitial Ca.

Harstigite, $Ca_6Mn^{2+}[Be_4Si_6O_{22}(OH)_2]$, contains four-membered rings of alternating $Be\varphi_4$ and SiO_4 tetrahedra, as does aminoffite, but their linkage is far more complicated in the latter mineral. In harstigite, the tetrahedra occur at the vertices of a two-dimensional net (Fig. 20c). There are two distinct $Be\varphi_4$ tetrahedra, one of which is four-connected and the other of which is three-connected. The SiO_4 tetrahedra of the four-membered ring are three-connected to $Be\varphi_4$ tetrahedra; the other two SiO_4 groups of the five-membered rings form a pyro-group $[Si_2O_7]$ which links to $Be\varphi_4$ tetrahedra (Fig. 20c). The sheets are stacked along the *a*-direction (Fig. 20d) and linked by layers of [7]- and [8]-coordinated Ca and [6]-coordinated Mn^{2+} atoms.

Samfowlerite, $Ca_{14}Mn^{2+}_3[(Be_7Zn)Zn_2Si_{14}O_{52}(OH)_6]$, contains $Be\varphi_4$, SiO_4 and $Zn\varphi_4$ (broken-line shaded in Fig. 21a) tetrahedra arranged at the vertices of a two-dimensional net (Fig. 21a). Eight-membered rings of tetrahedra (Zn–Si–Be–Si–Zn–Si–Be–Si) link through four-membered (Be–Si–Be–Si) and five-membered rings to form the net; there are two types of five-membered rings: (Zn–Si–Si–Zn–Si) and (Be–Si–Be–Si–Si). These sheets (Fig. 21b) are linked by layers of [6]-coordinated Mn^{2+} and [8]-coordinated Ca atoms. Site-occupancy refinement showed some Be $\leftrightarrow$ Zn solid-solution in one of the tetrahedra. Rouse et al. (1994) raise the possibility that samfowlerite consists of domains of ordered arrangements of Zn and Be, but they have no evidence for this.

Bityite, $CaLiAl_2[BeAlSi_2O_{10}](OH)_2$, is a brittle mica, with Ca at the interlayer site. In the tetrahedral $[T_4O_{10}]$ sheet, there is almost complete ordering of (Al,Si) and Be, and the symmetry of the overall atomic arrangement is reduced from *C*2/*c* to *Cc*. There is also complete ordering between Al and (Li,□) over the octahedral sites. It seems probable that coupling of these two ordering patterns is induced by local bond-valence requirements.

Sorensenite, $Na_4Sn^{4+}[Be_2Si_6O_{18}](H_2O)_2$, contains BeO_4 and SiO_4 tetrahedra linked into a sheet in which the BeO_4 tetrahedra are four-connected and the SiO_4 tetrahedra are three-connected. Pairs of BeO_4 tetrahedra share an edge to form a $[Be_2O_6]$ group (Fig. 21c) that links to six SiO_4 tetrahedra, forming a $[Be_2Si_6O_{22}]$ cluster that also occurs in the structures of eudidymite (Fig. 34a, below) and epididymite (Fig. 34d, below). These clusters link in the *b* and *c* crystallographic directions by sharing tetrahedral corners. In the (100) plane (Fig. 21c), the clusters occur at the vertices of a centered plane orthogonal lattice with large interstices that contain octahedrally coordinated Sn^{4+} cations and [8]-coordinated Na that link the sheets in the [100] direction (Fig. 21d).

Sheets involving Be–P and Be–As linkages.

Bergslagite, $Ca[Be(AsO_4)(OH)]$,
hydroxylherderite, $Ca[Be(PO_4)(OH)]$ and

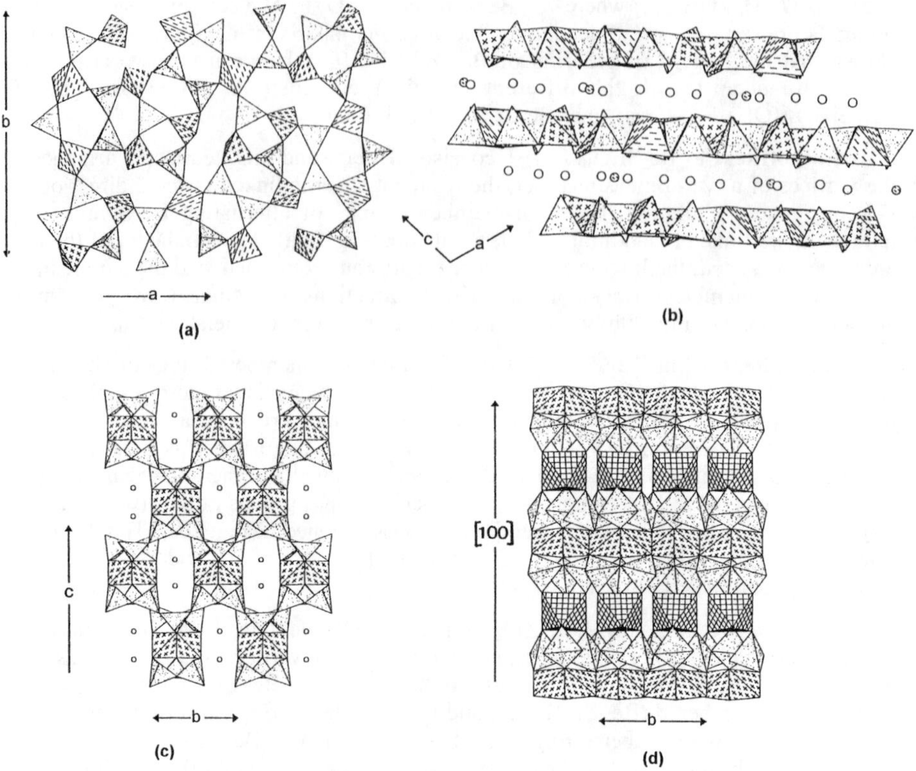

Figure 21. The crystal structures of samfowlerite and sorensenite; (a) the structural unit of samfowlerite projected onto (001); $Be\varphi_4$ and SiO_4 tetrahedra occur at the vertices of a two-dimensional net; (b) the structural unit of samfowlerite projected onto (010), with interstitial Ca (unshaded circles) and Mn (shaded circles); (c) sorensenite projected onto (100); edge-sharing $[Be_2\varphi_6]$ dimers link to triplets of SiO_4 tetrahedra to form clusters that share corners to form thick sheets; Sn are shown as small circles; (d) sorensenite projected onto (001); thick sheets are cross-linked by [6]-coordinated Sn^{4+} and Na (omitted).

herderite, $Ca[Be(PO_4)F]$,

are isostructural, although the structures of the first two were reported in different orientations: $P2_1/c$ and $P2_1/a$, respectively. The sheet unit consists of $Be\varphi_4$ and $T^{5+}O_4$ tetrahedra at the vertices of a two-dimensional net (Figs. 22a,c). Four-membered rings of alternating $Be\varphi_4$ and $T^{5+}O_4$ tetrahedra link directly by sharing vertices between $Be\varphi_4$ and $T^{5+}O_4$ tetrahedra; thus the sheet can be considered to be constructed from chains of four-membered rings that extend in the [011] and $[01\bar{1}]$ directions in bergslagite (Fig. 22a) and in the [110] and $[1\bar{1}0]$ directions in herderite (Fig. 22c). These sheets stack in the *a*-direction in bergslagite (Fig. 22b) and the *c*-direction in herderite (Fig. 22d) and are linked by layers of [8]-coordinated Ca atoms. Note that the structure reported by Lager and Gibbs (1974) seems to have been done on hydroxylherderite rather than herderite.

Uralolite, $Ca_2[Be_4P_3O_{12}(OH)_3](H_2O)_5$, contains $Be\varphi_4$ and PO_4 tetrahedra linked into a sheet (Fig. 23a). Eight-membered rings of tetrahedra (P–Be–P–Be–P–Be–P–Be) link through common PO_4 groups to form chains that extend along [101]. These chains link in the (010) plane via sharing of tetrahedral vertices, forming three-membered (Be–Be–Be and Be–Be–P) and four-membered (Be–Be–Be–P) rings. Interstitial [7]-coordinated Ca atoms lie within the eight-membered rings (in projection). The layers stack along the *b*-direction (Fig. 23b) and are linked by Ca atoms (circles) and H-bonding; in this view, the three- and four-membered rings are easily seen.

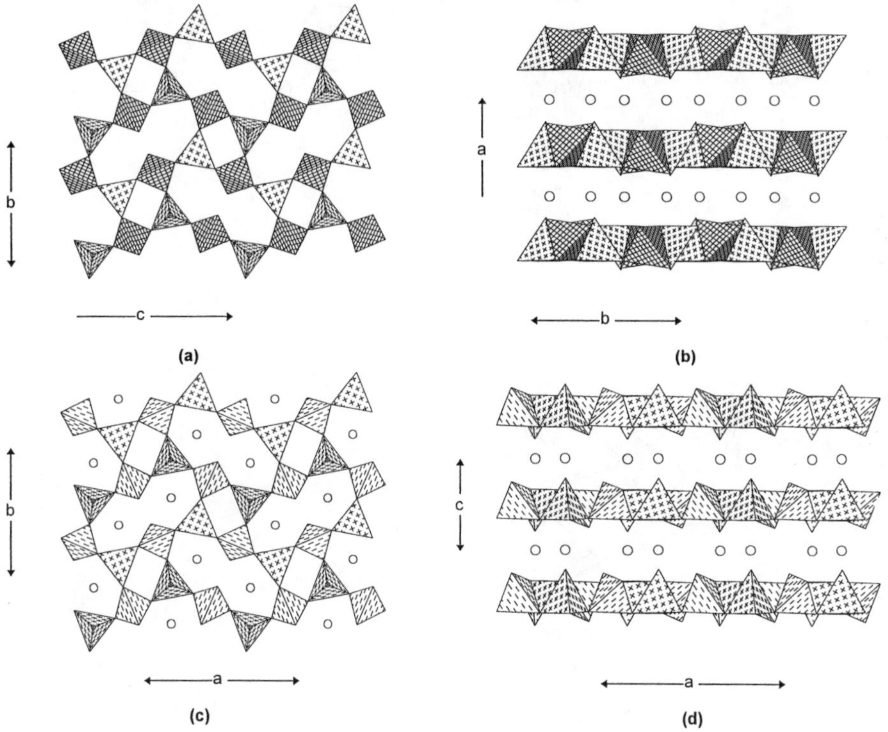

Figure 22. The crystal structures of bergslagite and herderite; (a) bergslagite projected onto (100); BeO_4 and AsO_4 (trellis-shaded) tetrahedra occur at the vertices of a net; (b) bergslagite projected onto (001); circles are Ca atoms; (c) herderite projected onto (100); BeO_4 and PO_4 tetrahedra lie at the vertices of a net; (c) herderite projected onto (010); circles are Ca atoms.

Ehrleite, $Ca_2[BeZn(PO_4)_2(PO_3\{OH\})](H_2O)_4$, has a very complicated sheet of tetrahedra, both from topological and chemical viewpoints. There is one distinct BeO_4 tetrahedron and this links to four $P\varphi_4$ groups (Fig. 23a); similarly, there is one ZnO_4 tetrahedron and this links to four $P\varphi_4$ groups. However, the $P\varphi_4$ groups link only to three or two other tetrahedra. Four-membered rings of alternating BeO_4 and PO_4 tetrahedra link through common BeO_4 tetrahedra to form chains in the *a*-direction (Fig. 23a). These chains are linked in the *c*-direction by four-membered rings of alternating ZnO_4 and PO_4 tetrahedra to form additional four-membered rings (Zn–P–Be–P). The result is an open sheet, parallel to (010), with buckled twelve-membered rings (Fig. 23a) into which project the H atoms of the acid-phosphate groups. These sheets stack along the *b*-direction (Fig. 23b) and are linked together by [7]-coordinated and [8]-coordinated interstitial Ca atoms.

Asbecasite, $Ca_3Ti^{4+}[Be_2Si_2As^{3+}{}_6O_{20}]$ contains BeO_4 and SiO_4 tetrahedra and $As^{3+}O_3$ triangular pyramids linked into thick sheets (or slabs) parallel to (001). The slab of tetrahedra and triangular pyramids consists of two sheets that meld in the *c*-direction; one

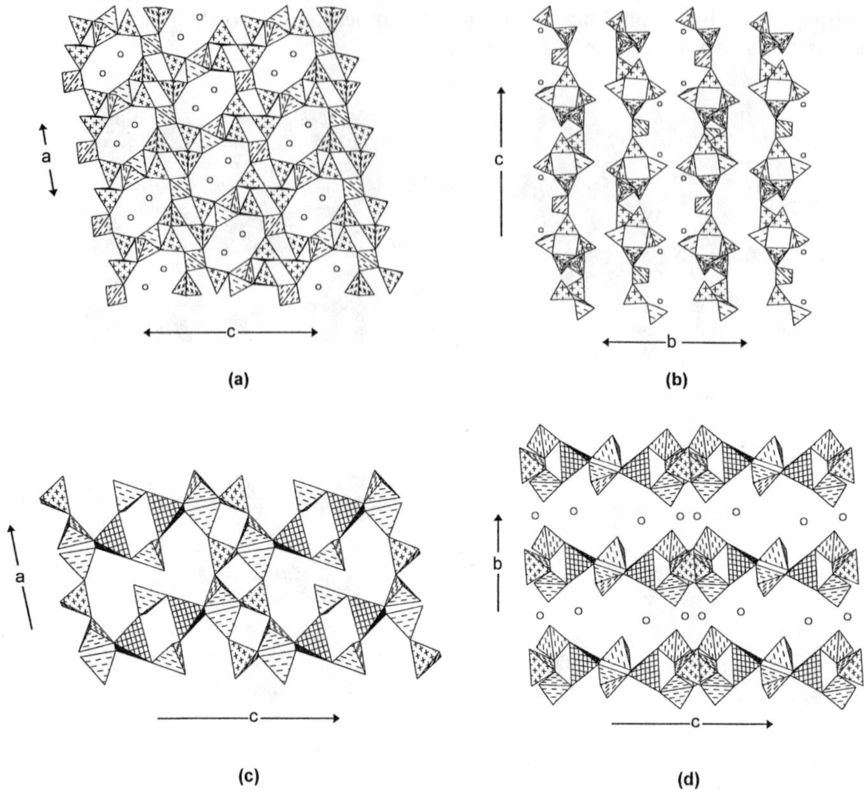

Figure 23. The crystal structures of uralolite and ehrleite; (a) uralolite projected onto (010); circles are Ca atoms; (b) uralolite projected onto (100); (c) the structural unit in ehrleite projected onto (010); BeO_4, PO_4 and ZnO_4 (trellis-shaded) tetrahedra share corners to form a sheet; (d) ehrleite projected onto (100); circles are Ca atoms.

sheet is illustrated in Figure 24a. Each tetrahedron is linked to three triangular pyramids in the plane of the sheet, and each triangular pyramid links to two tetrahedra; the result is a rather irregular-looking sheet. However, Figure 24b shows the same sheet with the $Ti^{4+}O_6$ octahedra inserted; the result is a simple 6^3 net of tetrahedra/triangles/octahedra. Hence the asbecasite sheet is simply a 6^3 net of corner-sharing tetrahedra/triangles with1/6 of the vertices omitted. Two of these layers fuse *via* linkage of BeO_4 and SiO_4 tetrahedra (Fig. 24d) to produce a thick slab that is the structural unit in asbecasite. These slabs stack in the *c*-direction and are interleaved with sheets of edge- and corner-sharing $(Ti,Sn)O_6$ octahedra and CaO_8 polyhedra (Fig. 24c).

Infinite frameworks of $T\varphi_4$ tetrahedra

The minerals of this class can be divided into seven broad groups (Table 7) based on the linkages involved in the framework: (1) structures with Be–Be linkages; (2) structures with Be–B linkages; (3) structures with Be–Be/Li–Si linkages; (4) structures with Be–Si linkages; (5) structures with Be-Si–Si–Al linkages; (6) structures with Be–Si–Si linkages; (7) structures with Be–P linkages. As with the other classes, structures with weaker *T*–*T* linkages *usually* have fewer interstitial cations

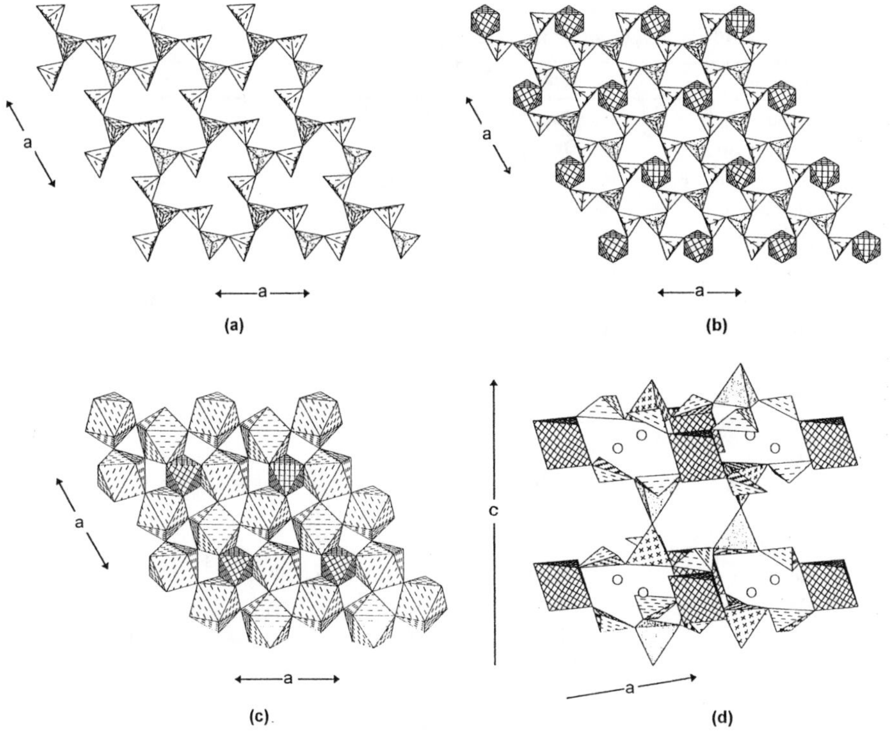

Figure 24. The crystal structure of asbecasite; (a) the sheet of BeO_4 and SiO_4 tetrahedra and $As^{3+}O_3$ triangular pyramids projected down [001]; (b) the same sheet as in (a) together with $Ti^{4+}O_6$ octahedra, showing that each polyhedron occurs at the vertices of a 6^3 net; (c) the sheet of $Ti^{4+}O_6$ octahedra and CaO_8 polyhedra projected down [001]; (d) perspective view of the way in which the sheets of tetrahedra and triangles meld to form slabs that link in the [001] direction through $Ti^{4+}O_6$ octahedra and interstitial Ca (unshaded circles); dashed triangular pyramids: $As^{3+}O_3$ groups; trellis-shaded octahedra: $Ti^{4+}O_6$ groups; broken-line-shaded polyhedra: CaO_8 groups.

(a result of the electroneutrality principle), although individual structures occur that do not follow this trend (e.g., swedenborgite).

Frameworks involving Be–Be linkages.

Behoite, [Be(OH)$_2$], consists of a simple framework of Beφ_4 tetrahedra that is closely related to the arrangement of tetrahedra in β-cristobalite. Viewed in the *a*-direction (Fig. 25a), the structure consists of BeO_4 tetrahedra at the vertices of a 6^3 net, forming six-membered rings of tetrahedra. Note that, although some tetrahedra appear to share edges, this is not the case; careful inspection of Figure 25a shows that the apparent common edge of the two tetrahedra actually inclines in opposing directions in each tetrahedron. Viewed in the *b*-direction (Fig. 25b), the tetrahedra occur at the vertices of a 4^4 net. The resultant framework is strengthened by H-bonding between the OH anions of adjacent tetrahedra in the structure.

Bromellite, [BeO] is chemically the simplest of the beryllium minerals. Viewed down [001] (Fig. 25c), BeO_4 tetrahedra occur at the vertices of a 3^6 net, forming three-

Table 7. Beryllium minerals based on frameworks of $T\varphi_4$tetrahedra.

Mineral	*Linkage*	*Interstitials*	*Figure*
Behoite	Be–Be	H-bonding	25a,b
Bromellite	Be–Be	–	25c,d
Swedenborgite	Be–Be	$^{[6]}Sb^{5+}$, Na	26a,b
Hambergite	Be–Be–B	H-bonding	27a,b
Rhodizite	Be–B	$^{[6]}Al$, K + H-bonding	27c,d
Liberite	Be–Si–Li	Li	–
Phenakite	Be–Be–Si	–	28a,b
Hsianghualite	Be–Si	Ca, Li	29a
Trimerite	Be–Si	$^{[6]}Mn$, Ca	29b,c
Danalite	Be–Si	$^{[6]}Fe$	29d,e
Genthelvite	Be–Si	$^{[6]}Zn$	29d,e
Helvite	Be–Si	$^{[6]}Mn$	29d,e
Tugtupite	Be–Si–Si–Al	Na	30a
Bavenite	Be–Si–Si–Al	Ca + H-bonding	30b,c
Roggianite	Be–Si–Si–Al	Ca + H-bonding	31a,b
Lovdarite	Be–Si–Si	Na, K + H-bonding	32a,b,c
Bertrandite	Be–Si–Si	H-bonding	33a,b
Chkalovite	Be–Si–Si	Na	33c,d
Eudidymite	Be–Si–Si	Na + H-bonding	34a,b
Epididymite	Be–Si–Si	Na + H-bonding	34c,d
Leifite	Be–Si–Si	Na + H-bonding	35a,b
Milarite	Be–Si–Si	Ca, Na, K + H-bonding	35c,d
Odintsovite	Be–Si–Si	Ca, Ti, Na, K	35e,f
Barylite	Be–Si–Si	Ba	36a,b
Chiavennite	Be–Si–Si	$^{[6]}Mn$, Ca + H-bonding	36c,d
Beryl	Be–Si–Si	$^{[6]}Al$	36e,f
Bazzite	Be–Si–Si	$^{[6]}Sc$	36e,f
Stoppaniite	Be–Si–Si	$^{[6]}Fe^{3+}$	36e,f
Beryllonite	Be–P	Na	37a,b
Tiptopite	Be–P	K, Na, Li	38a,b
Pahasapaite	Be–P	Ca, Li	38c
Weinebeneite	Be–P	Ca + H-bonding	39a,b
Hurlbutite	Be–P	Ca	39c,d
Babephite	Be–P	Ba	40a,b

membered rings of tetrahedra. With only Be and O in the structure, the bond-valence requirements of the anion have to be satisfied solely by Be. As Be–O ≅ 0.50 vu, this requires that each O anion be linked to four Be atoms. Hence bromellite differs from the most of the tetrahedral frameworks in that each tetrahedral vertex has to link to four tetrahedra, rather than two tetrahedra as is usually the case in these framework structures. Viewed along [100] (Fig. 25d), we see a similar view: tetrahedra at the vertices of a 3^6 net.

Swedenborgite, $NaSb^{5+}[Be_4O_7]$, is an extremely elegant structure, and much can be said about its architecture solely based on the chemical composition. The principal high-

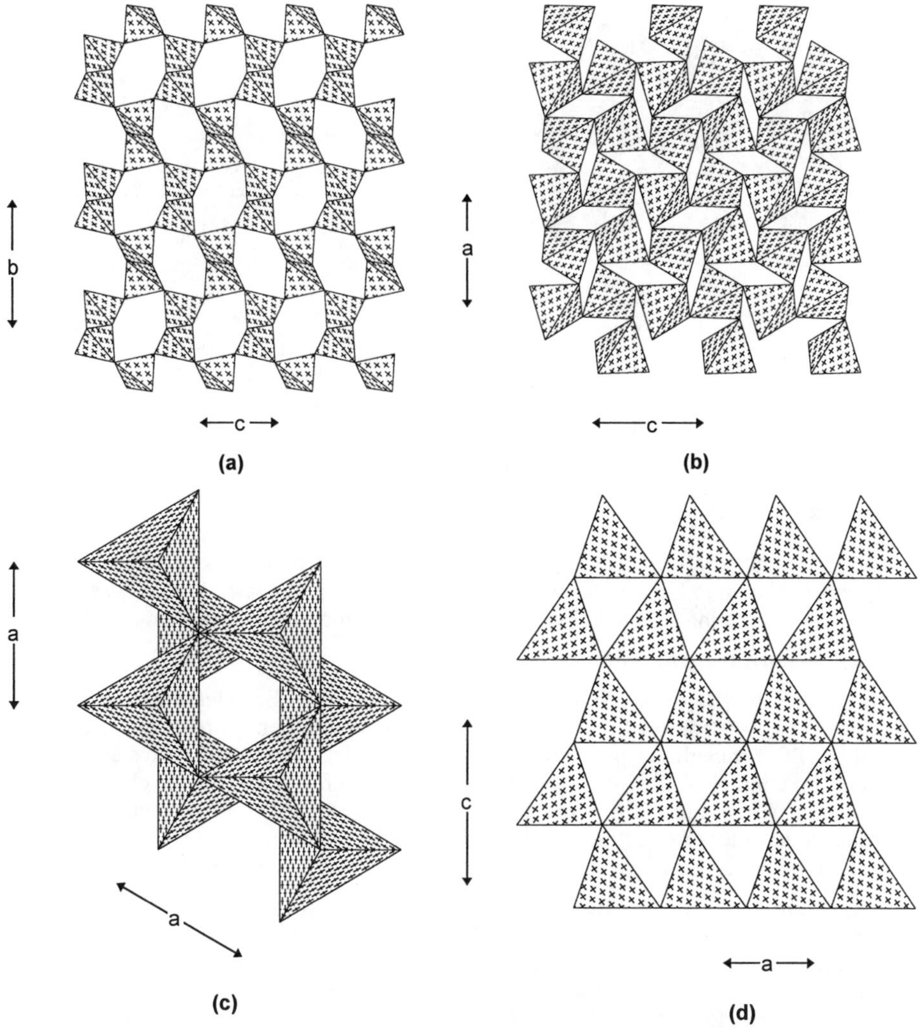

Figure 25. The crystal structures of behoite and bromellite; (a) behoite projected onto (100), a framework of corner-sharing $Be(OH)_4$ tetrahedra; the tetrahedra do *not* share an edge (they overlap in projection but are skewed in three dimensions); tetrahedra lie at the vertices of a 6^3 net; (b) behoite projected onto (010), with $Be(OH)_4$ tetrahedra lying at the vertices of a 4^4 net; (c) bromellite projected onto (001); BeO_4 tetrahedra lie at the vertices of a 3^6 net; (d) bromellite projected onto (010); BeO_4 tetrahedra lie at the vertices of a 3^6 net.

valence cation, Sb^{5+}, is usually [6]-coordinated. Presuming that this is the case in swedenborgite, this leaves an O atom whose bond-valence requirements must be satisfied by bonding to Be and Na only. The simplest way for this to occur is for one O anion to bond to four Be atoms, i.e., as in a fragment of the structure of bromellite (Fig. 25c). The resulting structure then requires linking of equal numbers of $[Be_4O_{13}]$ and (SbO_6) groups. An obvious way to arrange these motifs (because of their shape) is at the vertices of a plane hexagonal net, in which each motif is surrounded by three of the other motif.

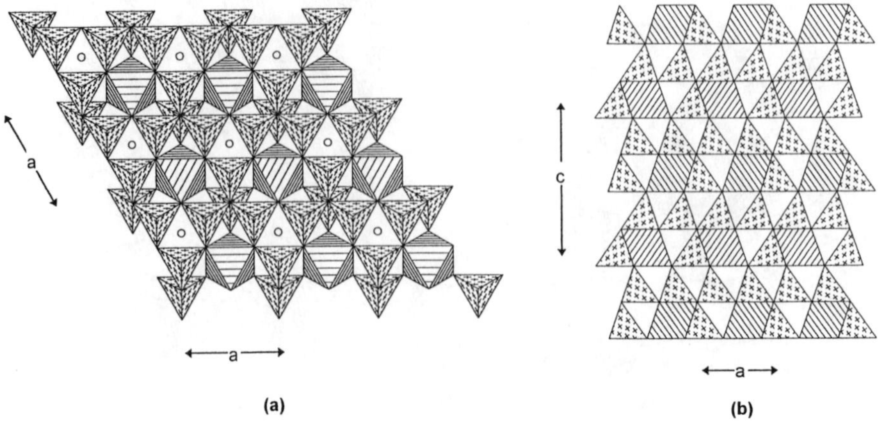

Figure 26. The crystal structure of swedenborgite; (a) projected onto (001); a framework of $[Be_4O_{13}]$ clusters, linked by sharing corners, with interstitial $^{[6]}Sb$; circles are Na atoms; (b) projected onto (010).

Linking the tetrahedral-octahedral vertices produces the stoichiometry $Sb[Be_4O_7]$ and the arrangement shown in Figure 26a, with Na atoms occupying the interstices. These layers link directly by sharing tetrahedral vertices (Fig. 26b) to form the overall framework.

Frameworks involving Be–B linkages.

Hambergite, $[Be_2(BO_3)(OH)]$, consists of a framework of $Be\varphi_4$ tetrahedra and $Be\varphi_3$ triangles. The Be:B ratio of 2:1 requires that the structure consist of $[Be_2O_7]$ dimers linked by (BO_3) triangles. This linkage can be envisaged as convoluted chains extending in the *b*-direction (Fig. 27a, in which only alternate chains are shown for clarity). The overall framework (Fig. 27b) is quite densely packed as the O atoms need to be [3]-coordinated in order to satisfy their bond-valence requirements: $O-^{[4]}Be \times 2 + O-^{[3]}B = 0.50 \times 2 + 1.00 = 2.00$ vu.

Rhodizite, $KAl_4[Be_4(B_{11}Be)O_{28}]$, and **londonite**, $CsAl_4[Be_4(B_{11}Be)O_{28}]$, also have framework-vertex coordination-numbers somewhat higher than 2.0, as is apparent from the framework formula: $[T_{16}O_{28}]$ or $[TO_{1.75}]$. In projecting down *c* (Fig. 27c), the tetrahedra form an interrupted checkerboard pattern with K and Cs occupying the vacant squares. Viewed along [011] (Fig. 27d), the tetrahedra occupy vertices of a 3^6 net in projection, with octahedrally coordinated Al occupying the interstices. The AlO_6 octahedra occur as tetramers of the form $[Al_4O_{16}]$ in which each octahedron shares an edge with the other three octahedra to form a compact cluster with a tetrahedral cavity at the center.

Frameworks involving Be–Be/Li–Si linkages.

The structure of **liberite**, $Li_2[Be(SiO_4)]$, has been solved, but there is an error in the parameters reported, and the structure should be re-examined.

Phenakite, $[Be_2(SiO_4)]$, consists of a framework of BeO_4 and SiO_4 tetrahedra with a very interesting ordering pattern of chemically different tetrahedra over the vertices of a highly connected three-dimensional net. Both the chemical formula and the local bond-valence requirements indicate that some of the anions must be more than two-connected. Viewed down the *c*-direction (Fig. 28a), the structure consists of six-membered rings of tetrahedra connected in the (001) plane by four-membered rings of tetrahedra. One-third

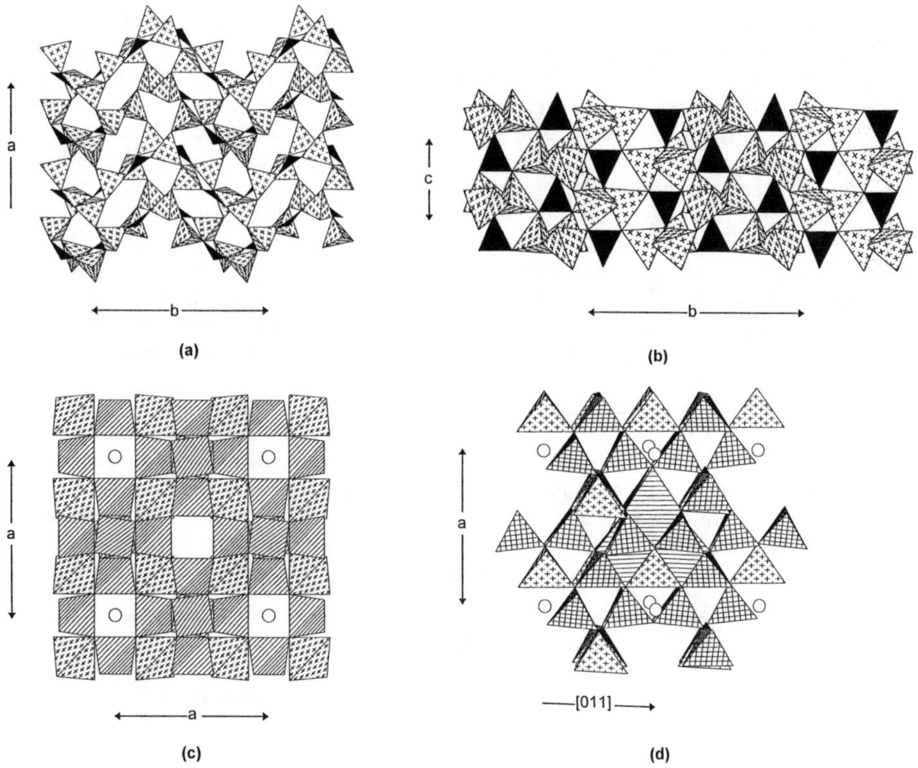

Figure 27. The crystal structures of hambergite and rhodizite; (a) hambergite projected onto (001); black triangles are $B\varphi_3$ groups; (b) hambergite projected onto (100); (c) rhodizite projected onto (001); circles are K atoms; (d) rhodizite projected onto [011]; BO_4 tetrahedra are trellis-shaded.

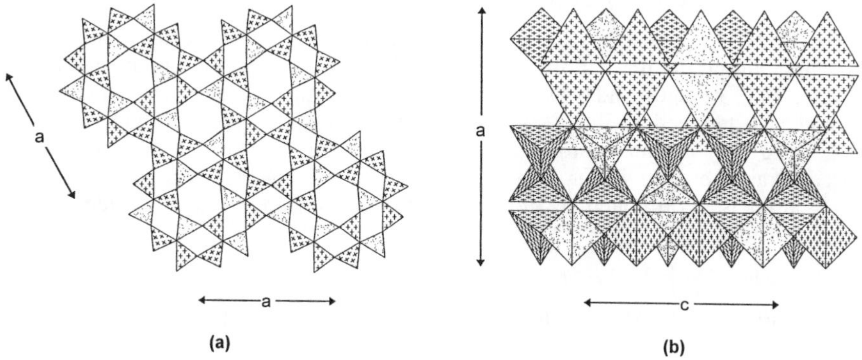

Figure 28. The crystal structure of phenakite; (a) projected onto (001); (b) projected onto (010).

of the six-membered rings consist solely of BeO_4 tetrahedra and two thirds consists of alternating BeO_4 and SiO_4 tetrahedra, giving a Be:Si ratio of 2:1. The connecting four-membered rings are of two different types: when they connect $[Be_6O_{18}]$ and $[Be_3Si_3O_{18}]$ rings, they consist of three BeO_4 tetrahedra and one SiO_4 tetrahedron; when they connect two $[Be_3Si_3O_{18}]$ rings, they consist of two BeO_4 tetrahedra and two SiO_4 tetrahedra. However, another way to look at the structure derives from considering the mineral stoichiometry and the bond-valence requirements at the anions. The simplest way in which the bond-valence requirements can be satisfied at the anions is for them to bond to two Be and one Si atoms: $2 \times 0.50 + 1.00 = 2.00$ vu. If the resulting $[Be_2SiO_{10}]^{12-}$ clusters are placed at the vertices of a plane hexagonal net such that the tetrahedra connect with each other, the structure shown in Figure 28a results. The nets of Figure 28a are stacked along the *c*-direction such that chains of tetrahedra are formed (Fig. 28b) with the sequence Be–Be–Si. The resulting structure has continuous hexagonal channels parallel to the *c*-direction.

Frameworks involving Be–Si linkages.

Hsianghualite, $Ca_3Li_2[Be_3Si_3O_{12}]F_2$, forms a framework of alternating BeO_4 and SiO_4 tetrahedra (Fig. 29a). Prominent four-membered rings of BeO_4 and SiO_4 tetrahedra are cross-linked to other four-membered rings by tetrahedra that are constituents of other four-membered rings. The result is a four-connected framework (Fig. 29a) that is very similar to the aluminosilicate framework in analcite (Mazzi and Galli 1978).

Trimerite, $CaMn^{2+}_2[Be(SiO_4)]_3$, consists of an ordered framework of four-connected BeO_4 and SiO_4 tetrahedra. Viewed in the *b*-direction, the tetrahedra occur at the vertices of a 6^3 net (Fig. 29b) and point up (*u*) or down (*d*) the *b* direction in the following sequence: [*uddudu*]. These sheets stack in the *b*-direction (Fig. 29c), linking through corner-sharing of tetrahedra from adjacent sheets at y ~0 and 1/2. Nine-coordinated Ca and [6]-coordinated Mn^{2+} occupy the interstices in the framework. Of particular interest is the octahedrally coordinated Mn^{2+} which occurs in face-sharing dimers. Trimerite is topologically isostructural with beryllonite, $Na[Be(PO_4)]$. In the latter structure, there are three independent Na sites as compared to one Ca site and two Mn sites in trimerite. In beryllonite, NaO_6 octahedra share faces, similar to the MnO_6 octahedra in trimerite.

Danalite, $Fe^{2+}_8S_2[Be_6Si_6O_{24}]$,
genthelvite, $Zn_8S_2[Be_6Si_6O_{24}]$ and
helvite, $Mn^{2+}_8S_2[Be_6Si_6O_{24}]$, are isostructural minerals of the *helvite* group with a framework similar to that of the minerals of the sodalite group. Four-membered alternating rings of BeO_4 and SiO_4 tetrahedra link to six-membered rings of alternating BeO_4 and SiO_4 tetrahedra (Fig. 29d). There are two interstitial species in danalite, Fe^{2+} and S^{2-}. The Fe^{2+} is four-coordinated by three O atoms and one S atom to form a distorted tetrahedron (which can also be described as an elongated triangular pyramid). The S^{2-} atom is coordinated by four Fe^{2+} atoms, and hence four distorted FeO_2S tetrahedra share a single vertex (S) (Fig. 29e).

Frameworks involving Be–Si–Si–Al linkages.

Tugtupite, $Na_8[Be_2Al_2Si_8O_{24}]Cl_2$, is isostructural with sodalite, $Na_8[Al_6Si_6O_{24}]Cl_2$. Four-membered rings of BeO_4, SiO_4 (× 2) and AlO_4 tetrahedra link to six-membered rings of tetrahedra (Fig. 30a). Within the four-membered rings, the linkage [Be–Si–Al–Si] does not involve Si–Si linkages. The situation in the six-membered rings is different, as there are two Si–Si linkages involved [Be–Si–Si–Al–Si–Si]. The framework charge is neutralized by [5]-coordinated Na which bonds to the interstitial Cl atom.

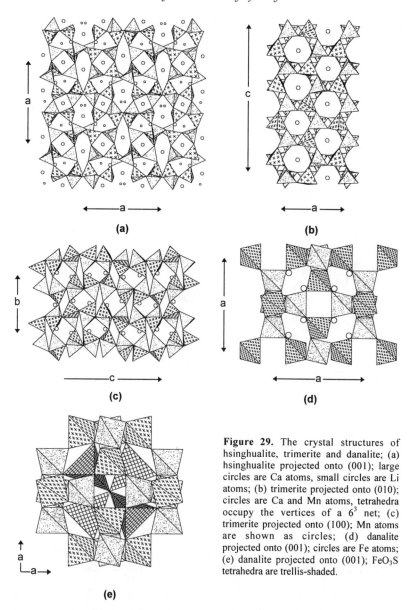

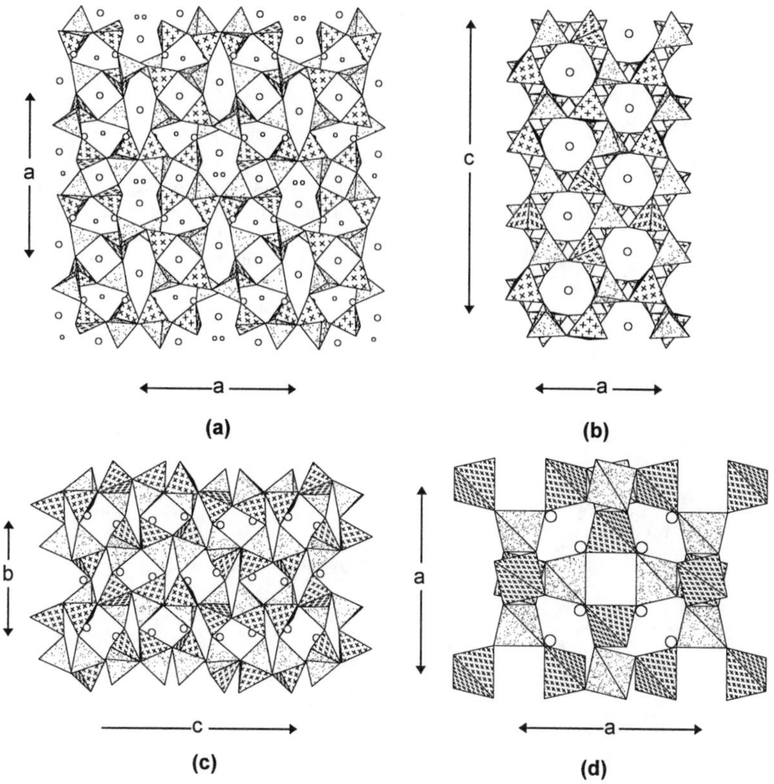

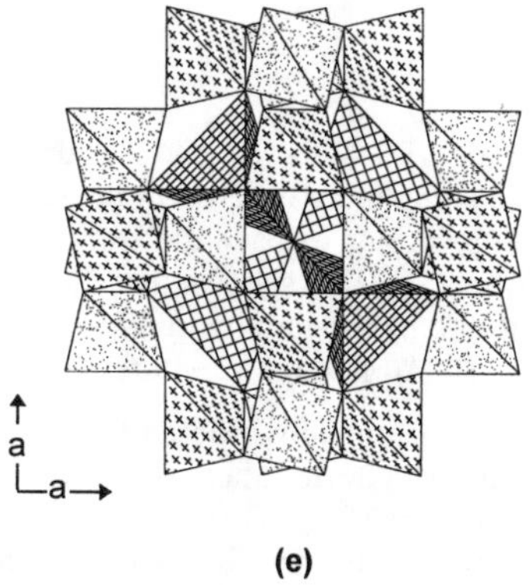

Figure 29. The crystal structures of hsinghualite, trimerite and danalite; (a) hsinghualite projected onto (001); large circles are Ca atoms, small circles are Li atoms; (b) trimerite projected onto (010); circles are Ca and Mn atoms, tetrahedra occupy the vertices of a 6^3 net; (c) trimerite projected onto (100); Mn atoms are shown as circles; (d) danalite projected onto (001); circles are Fe atoms; (e) danalite projected onto (001); FeO_3S tetrahedra are trellis-shaded.

Bavenite, $Ca_4[Be_2Al_2Si_9O_{26}(OH)_2]$, contains four-membered rings of SiO_4 and AlO_4 tetrahedra [Al–Si–Al–Si] linked through six-membered rings of SiO_4 tetrahedra to form chains that extend along the *a*-direction. Adjacent chains are linked through linear BeO_4–SiO_4–BeO_4 groups, forming two types of six-membered rings between the chains: [Be–Si–Si–Al–Si–Si] and [Be–Si–Si–Be–Si–Si]. These sheets stack in the *b*-direction, as shown in Figure 30c, with AlO_4 tetrahedra as the linking elements between the sheets. This arrangement results in large cavities that contain the [7]-coordinated interstitial

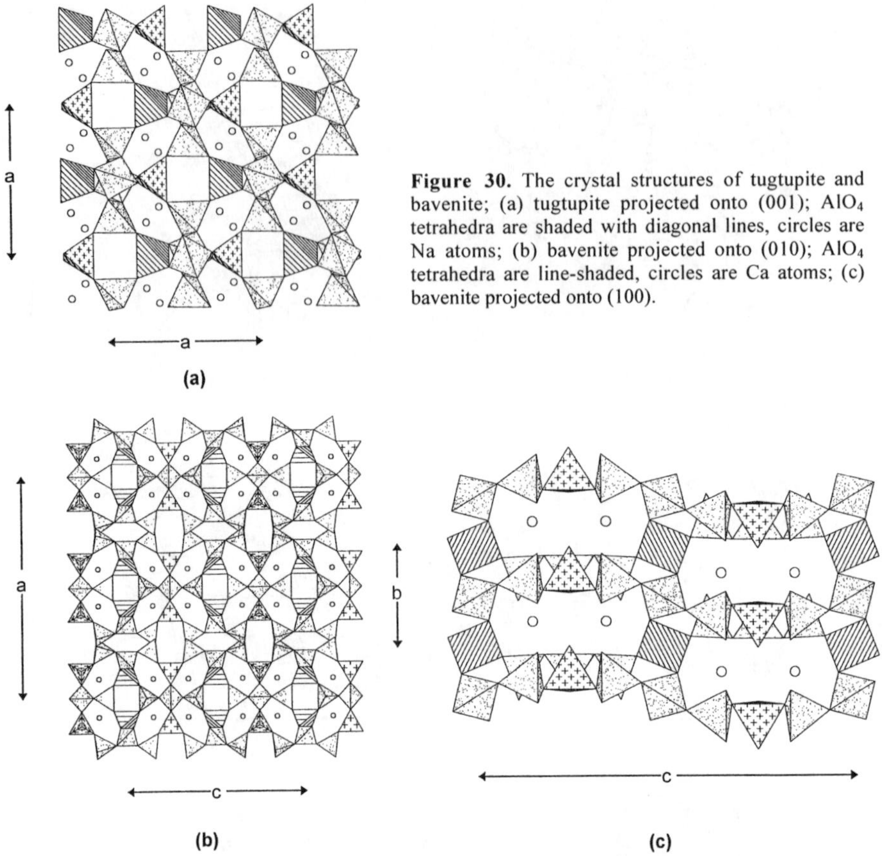

Figure 30. The crystal structures of tugtupite and bavenite; (a) tugtupite projected onto (001); AlO_4 tetrahedra are shaded with diagonal lines, circles are Na atoms; (b) bavenite projected onto (010); AlO_4 tetrahedra are line-shaded, circles are Ca atoms; (c) bavenite projected onto (100).

Ca atoms. In the structure examined by Cannillo et al. (1966), the Be and Al contents of the unit cell are close to integral, and Be and Al are completely ordered in the structure. However, Be ↔ Al substitution is indicated by compositions of bavenite from different localities, and Cannillo et al. (1966) and Kharitonov et al. (1971) suggest that the possible mechanism involves solid solution of Be and Al at one of the tetrahedrally coordinated sites, coupled to local substitution of additional H [as (OH)] into the structure.

Roggianite, $Ca_2[Be(OH)_2Al_2Si_4O_{13}](H_2O)_{2.34}$, is a completely ordered framework in which there seems to be no Be ↔ Al solid-solution. The tetrahedral framework is extremely unusual. It consists of a trellis of tetrahedra (Fig. 31a) with large inter-framework interstices that are filled with H_2O groups. Figure 31b shows that the framework consists of layers of SiO_4 and BeO_4 tetrahedra, parallel to (001), that are linked in the *c*-direction by AlO_4 tetrahedra. Within each layer are four-membered rings of SiO_4 tetrahedra and three-membered rings of BeO_4 and SiO_4 (× 2) tetrahedra (Fig. 31a). The four-membered rings stack on top of each other along the *c*-direction, linked by AlO_4 tetrahedra, and form columns that resemble the four-membered analogue of the linked six-membered rings in the structure of beryl. The smaller interstices in the framework contain [6]-coordinated Ca atoms that neutralize the framework charge.

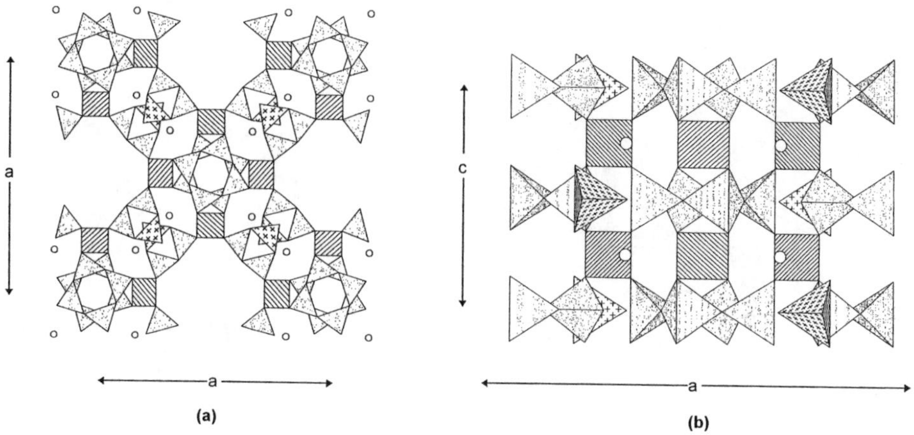

Figure 31. The crystal structure of roggianite; (a) projected onto (001); AlO_4 tetrahedra are line-shaded, circles are Ca atoms; the large interstices are filled with H_2O groups (not shown); (b) projected onto (100).

Frameworks involving Be–Si–Si linkages.

Lovdarite, $K_2Na_6[Be_4Si_{14}O_{36}](H_2O)_9$, has BeO_4 and SiO_4 tetrahedra arranged at the vertices of a 4.8^2 net when viewed in the *a*-direction (Fig. 32a). The four-membered rings consist of three SiO_4 and one BeO_4 tetrahedra, and join through Be–Si linkages along *b* and Si–Si linkages along *c*. These sheets are linked in the *a*-direction through additional SiO_4 tetrahedra (Fig. 32b) involving both Si–Si and Si–Be linkages; this linkage produces three-membered rings in the (010) plane (Fig. 32b) involving one BeO_4 and two SiO_4 tetrahedra. Adjacent sheets are shifted $(b + c)/2$ relative to each other (Fig. 32c) in order to promote intersheet linkage through the additional SiO_4 tetrahedron. The charge on the framework is neutralized by [5]- and [7]-coordinated Na and [9]-coordinated K atoms. Lovdarite shows prominent domain structures (Merlino 1990) that can be interpreted in terms of OD theory (Durovic 1997) and the occurrence of intergrowths of different polytypes.

Bertrandite, $[Be_4Si_2O_7(OH)_2]$, has the general formula $[T_6\varphi_9] = [T\varphi_{1.5}]$, indicating that the anion coordination by framework cations must be greater than 2; moreover, the H-atom must link to a Be–Be bridging anion if the anion bond-valence requirements are to be satisfied: O–Be (× 2) + O–H ≅ 0.5 × 2 + 1.0 = 2.0 vu. One anion is satisfied *via* an Si–Si linkage, and the remaining anions must be satisfied in the following way: O–Be (× 2) + O–Si ≅ 0.5 × 2 + 1.0 vu; thus four of the anions of the tetrahedral framework must be [3]-coordinated. The requirement for [3]-coordination of so many anions is satisfied by having the tetrahedra occupy the vertices of a 3^6 net (Fig. 33b). The ordering of chemical species over this net is such that the BeO_4 tetrahedra occupy the vertices of a 6^3 net and the SiO_4 tetrahedra occupy the interstices of this net. These sheets stack in the *a*-direction through Si–Si and Be–Be linkages (Fig. 33a), forming a corrugated 4.6^2 net in which the H atom links to the anion bridging the BeO_4 tetrahedra in the *a*-direction.

Chkalovite, $Na_2[BeSi_2O_6]$, must involve Si–Si linkages as well as Be–Si linkages. The tetrahedra occupy the vertices of a 4^4 net when viewed in the *b*-direction (Fig. 33c).

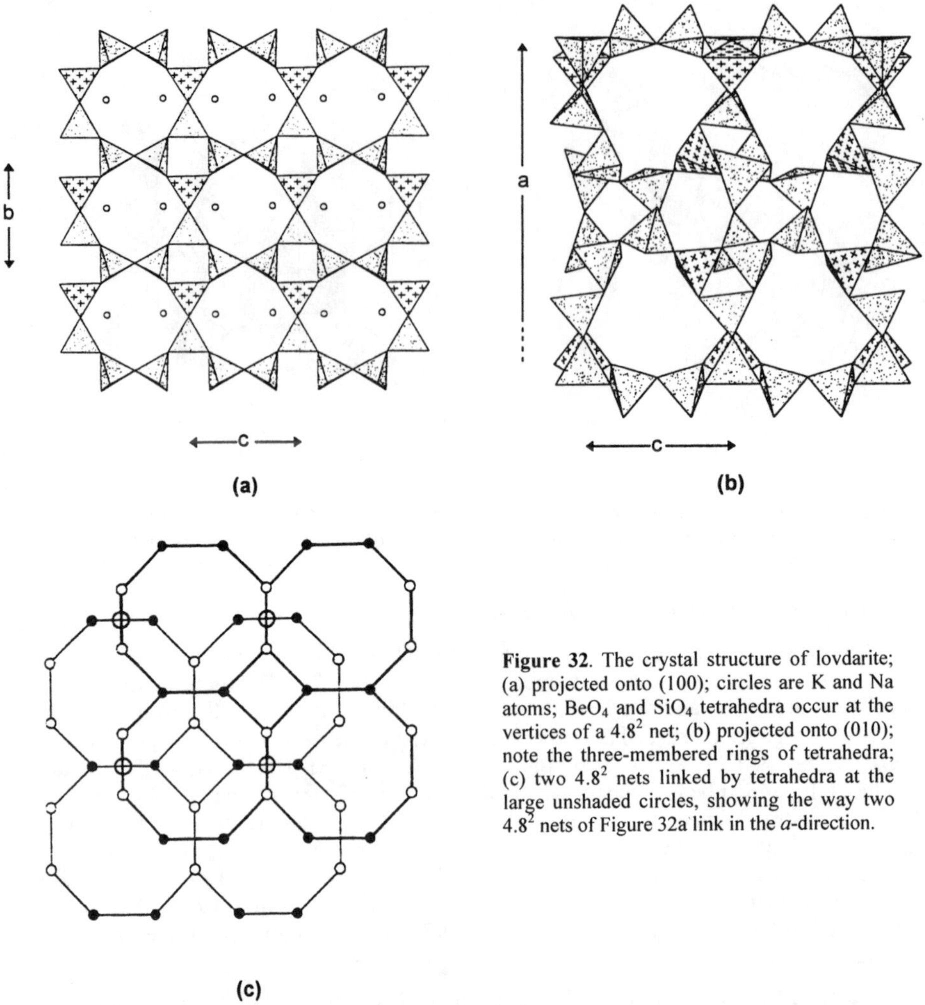

Figure 32. The crystal structure of lovdarite; (a) projected onto (100); circles are K and Na atoms; BeO_4 and SiO_4 tetrahedra occur at the vertices of a 4.8^2 net; (b) projected onto (010); note the three-membered rings of tetrahedra; (c) two 4.8^2 nets linked by tetrahedra at the large unshaded circles, showing the way two 4.8^2 nets of Figure 32a link in the *a*-direction.

There are two distinct types of four-membered ring in this net, one involving an [Si–Be–Si–Be] linkage and the other involving a [Be–Si–Si–Si] linkage. These very corrugated sheets stack in the *b*-direction, forming an 8^3 net (Fig. 33d); note that in this view, the structure also appears as $[SiO_3]$ chains extending along [021] and linked by $[Be_{0.5}\square_{0.5}O_3]$ chains. The framework charge is balanced by the presence of interstitial Na.

Eudidymite, $Na_2[Be_2Si_6O_{15}](H_2O)$, and **epididymite**, $Na_2[Be_2Si_6O_{15}](H_2O)$, are dimorphic. They both contain the $[Be_2Si_6O_{22}]$ cluster that also occurs in sorensenite (Fig. 21c). In eudidymite, these clusters link by sharing vertices of the SiO_4 tetrahedra of adjacent clusters (Fig. 34a) with non-tetrahedral cations in the interstices of the resulting network. Viewed down [010] (Fig. 34b), the structure appears as sheets of linked SiO_4 tetrahedra, joined in the [100] direction by $[Be_2O_6]$ groups. The structure of epididymite is very similar (cf. Figs. 34a and 34d). The silicate sheets contain four-membered rings (Fig. 34c) that join to form sheets which link through $[Be_2O_6]$ groups (Fig. 34e) to form a framework.

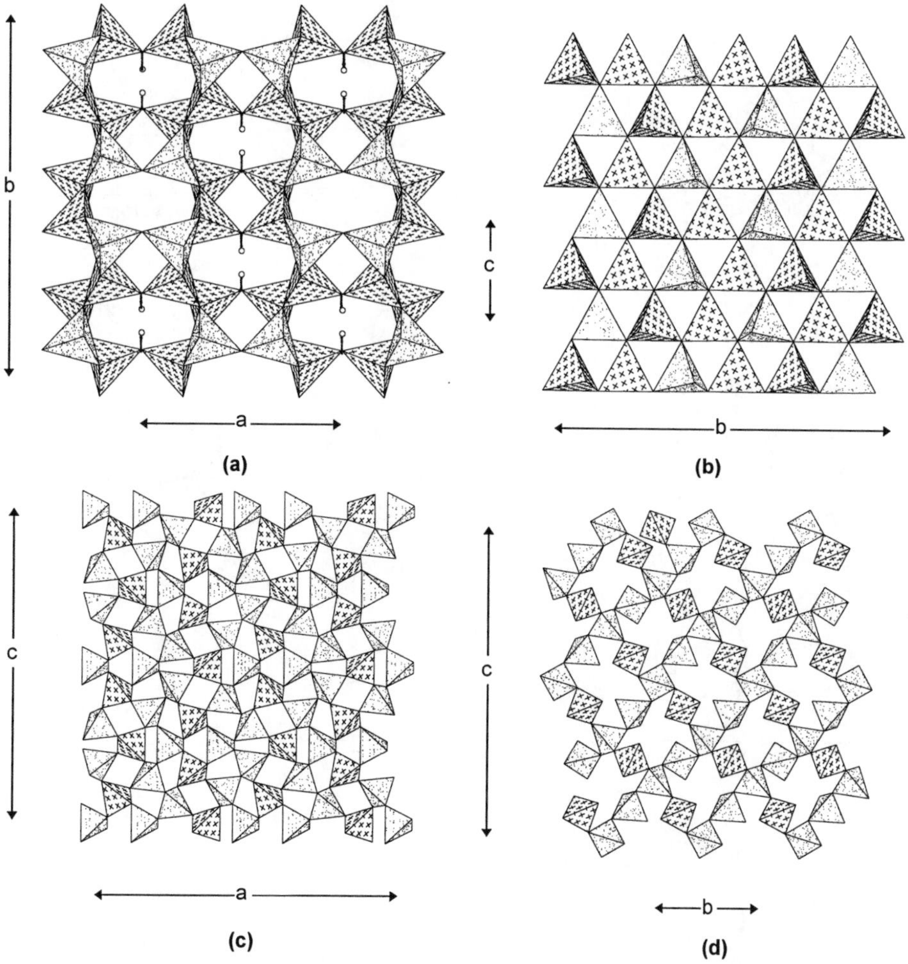

Figure 33. The crystal structures of bertrandite and chkalovite; (a) bertrandite projected onto (001); small circles are H-atoms; (b) bertrandite projected onto (100); tetrahedra occur at the vertices of a 3^6 net in this projection; (c) chkalovite projected onto (010); tetrahedra occur at the vertices of a 4^4 net; (d) chkalovite projected onto (100); tetrahedra occur at the vertices of an 8^4 net; Na atoms are omitted for clarity.

Leifite, $Na_6[Be_2Al_2Si_{16}O_{39}F_2](H_2O)_{1.6}$, has a partly ordered framework structure. There is one distinct $Be\varphi_4$ tetrahedron, but, although Al is ordered at one specific site, this site has mixed occupancy by both Al and Si, whereas the rest of the tetrahedra are SiO_4 only. Projected down the *c*-direction, the tetrahedra occur at the vertices of a very exotic two-dimensional net (Fig. 35a). Six-membered rings of $(Si,Al)O_4$ tetrahedra are linked by $[Si_4O_{11}]$ clusters consisting of an $[Si_2O_7]$ group with two additional tetrahedra sharing corners with both tetrahedra of the $[Si_2O_7]$ group. The six-membered rings occur at the vertices of a plane hexagonal net, and adjacent triplets of hexagonal rings are linked by $[Si_4O_{11}]$ groups (Fig. 35a). BeO_4 tetrahedra occur in large clover-leaf interstices

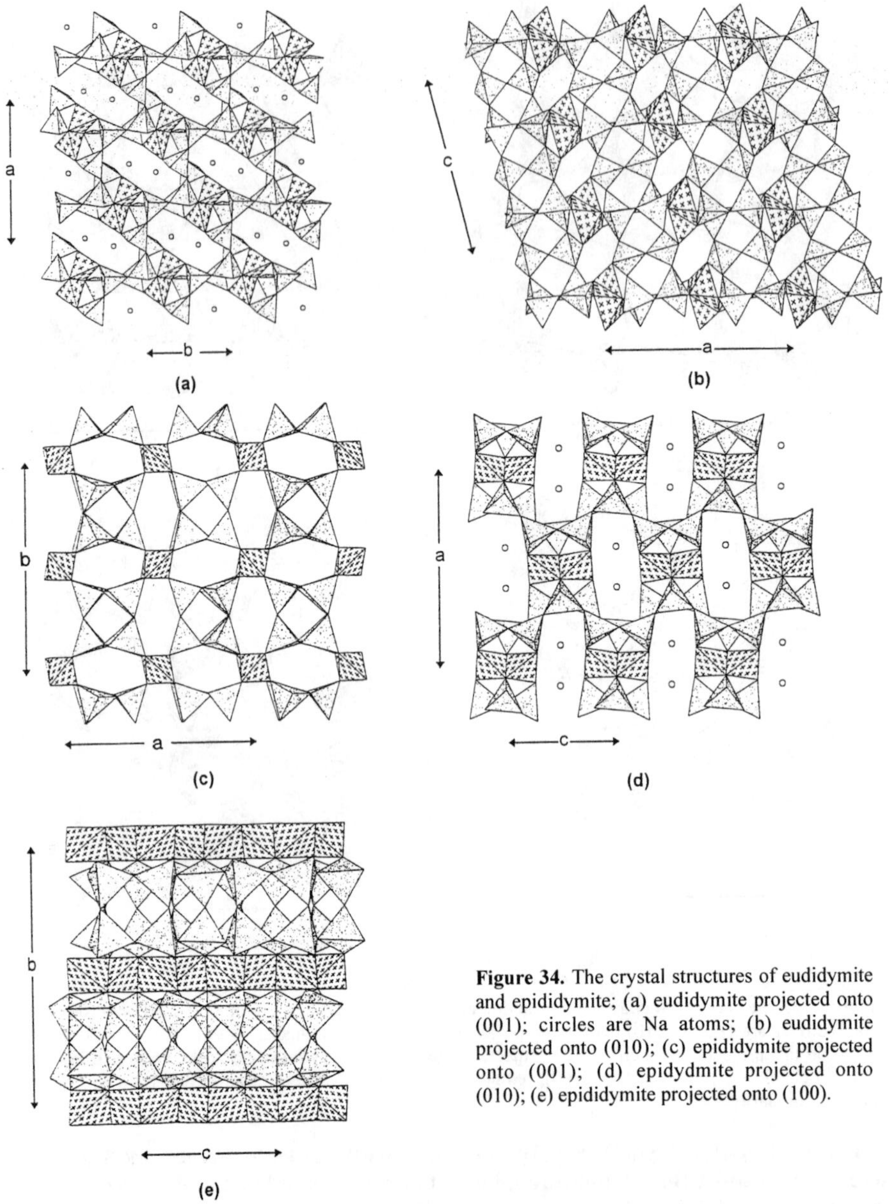

Figure 34. The crystal structures of eudidymite and epididymite; (a) eudidymite projected onto (001); circles are Na atoms; (b) eudidymite projected onto (010); (c) epididymite projected onto (001); (d) epidydmite projected onto (010); (e) epididymite projected onto (100).

of the aluminosilicate net, linking three different $[Si_4O_{11}]$ groups together. These nets stack along the *c*-direction (Fig. 35b), forming prominent $[SiO_3]$ chains. [7]-coordinated Na occurs in the interstices and H_2O groups occur down the channels formed by superposition of the six-membered rings of SiO_4 in the *c*-direction.

Milarite, $KCa_2[Be_2AlSi_{12}O_{30}](H_2O)_x$, is the type structure for a large group of minerals (Hawthorne et al. 1991). Six-membered rings of SiO_4 tetrahedra are arranged at the vertices of a hexagonal plane net (Fig. 35c) and are linked by BeO_4 tetrahedra (similar

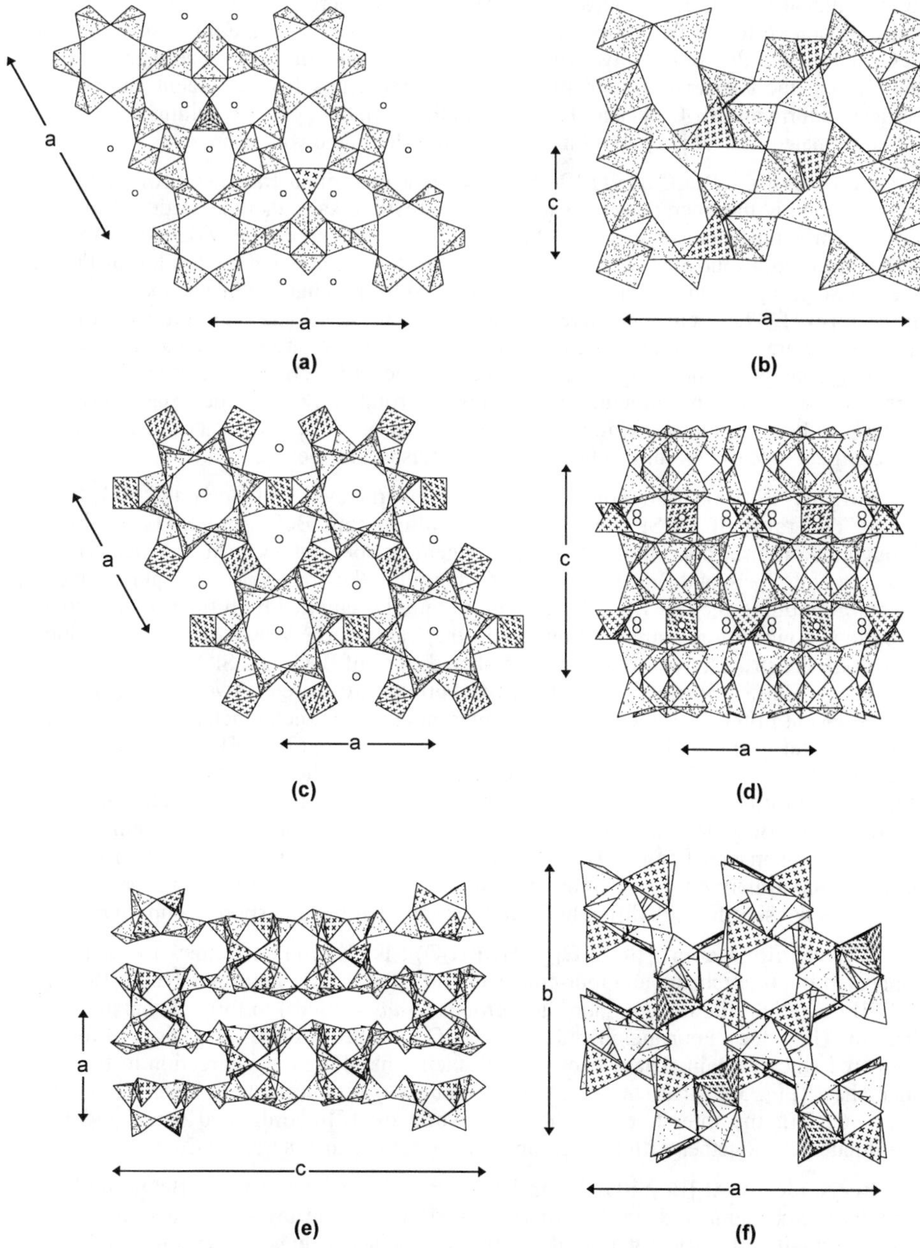

Figure 35. The crystal structures of leifite, milarite and odintsovite; (a) leifite projected onto (001); Na atoms are shown as circles; (b) leifite projected onto (010); (c) milarite projected onto (001); circles are Ca atoms; note the similarity to the structure of beryl (Fig. 36e); (d) milarite projected onto (010); note the $[Si_{12}O_{30}]$ cages; (e) the structural unit in odintsovite projected onto (010); (f) the structural unit in odintsovite projected onto (001); interstitial atoms are omitted for clarity.

to the linkage by $[Si_4O_{11}]$ groups in leifite). Note that there are pairs of six-membered rings at each vertex of the lattice, one rotated by 30° relative to the other such that the BeO_4 tetrahedra also link the two rings in the *c*-direction (Fig. 35c). These sheets then stack along the *c*-direction such that six-membered rings from adjacent sheets link vertices to form an $[Si_{12}O_{30}]$ cage (Fig. 35d). In the interstices of the resulting framework, [6]-coordinated Ca and [12]-coordinated K balance the framework charge.

Odintsovite, $K_2Na_4Ca_3Ti_2O_2[Be_4Si_{12}O_{36}]$, is a very open framework of $Be\varphi_4$ and SiO_4 tetrahedra, the interstices of which are filled with alkali, alkaline earth cations and Ti to form a fairly dense structure. Pairs of three-membered rings (Be–Si–Si) share corners to form a four-membered ring (Be–Si–Be–Si) as shown at the top left of Figure 35e. These groups meld to similar groups along *a* to form rather irregular-looking clusters of the form H. These clusters share corners to link in the *a*- and *c*-directions, forming cavities that are ~21 Å long in the *c*-direction; these cavities are arranged at the vertices of a centered plane net (Fig. 35e). Viewed in the *c*-direction (Fig. 35f), the three-membered rings are prominent. These rings are rotated 180° as they stack down *c*, forming rather irregular-looking square clusters that occur at the vertices of a centered plane lattice; note the prominent hexagonal channels in this view (Fig. 35f).

Barylite, $Ba[Be_2Si_2O_7]$, consists of an ordered framework of BeO_4 and SiO_4 tetrahedra. The ratio of Be:Si (i.e., 1:1) allows alternating BeO_4 and SiO_4 tetrahedra throughout a framework, but this type of arrangement does not occur in barylite. Instead, there are both Si–Si and Be–Be linkages; furthermore, the bond-valence requirements of the anion bridging the Be–Be linkage can also only be satisfied by further linkage to an Si cation, giving three-connected anions in the framework. Viewed in the *c*-direction (Fig. 36a), the structure consists of six-membered rings of BeO_4 and SiO_4 tetrahedra with the linkage [Be–Si–Si–Be–Si–Si]; the $[Si_2O_7]$ groups are shared between six-membered rings adjacent in the ±*a*-directions, thus maintaining ideal stoichiometry. Six-membered rings adjacent in the *b*-direction link to form four-membered rings [Be–Si–Be–Si] and three-membered rings [Be–Be–Si]. These very corrugated sheets stack in the *c*-direction (Fig. 36b); in this view, the three-membered rings and the three-connected anions are easily visible. Barylite was originally refined in the space group *Pnma* (Cannillo et al. 1969). Robinson and Fang (1977) showed barylite to be non-centrosymmetric *via* a positive second-harmonic generation test, but the crystal structure refined in space group $Pn2_1a$ was not significantly different from the structure refined in space group *Pnma*.

Chiavennite, $CaMn^{2+}[Be_2Si_5O_{13}(OH)_2](H_2O)_2$, is an ordered framework of three-connected BeO_4 tetrahedra and four-connected SiO_4 tetrahedra. Four-membered [Be–Si–Be–Si] rings link through common SiO_4 groups (*trans* vertices) to form chains in the *a*-direction (Fig. 36c). Four-membered rings of SiO_4 tetrahedra share *cis* vertices to form a staggered silicate chain. These two types of chain link along the *b*-direction to form an open sheet (Fig. 36c) that stacks in the *c*-direction to produce a very open framework (Fig. 36d). In the interstices of the framework are [6]-coordinated Mn^{2+} and [8]-coordinated Ca, together with H_2O groups that coordinate the interstitial Ca.

Beryl, ideally $Al_2[Be_3Si_6O_{18}]$, consists of an ordered framework of BeO_4 and SiO_4 tetrahedra. Six-membered rings of SiO_4 tetrahedra stack on top of one another in the *c*-direction with a relative rotation of 30° between adjacent rings. These rings are linked in the *c*- and *a*-directions by BeO_4 tetrahedra (Fig. 36e), forming a sheet that is similar in projection to that of milarite (Fig. 30c); these composite sheets stack in the *c*-direction (Fig. 36f). Interstitial [6]-coordinated Al provides further linkage both parallel and perpendicular to the *c*-axis. The resulting structure (Fig. 36f) consists of layers of SiO_4 tetrahedra interleaved with layers of BeO_4 tetrahedra and [6]-coordinated Al (shown as hollow circles in Figs. 36e,f).

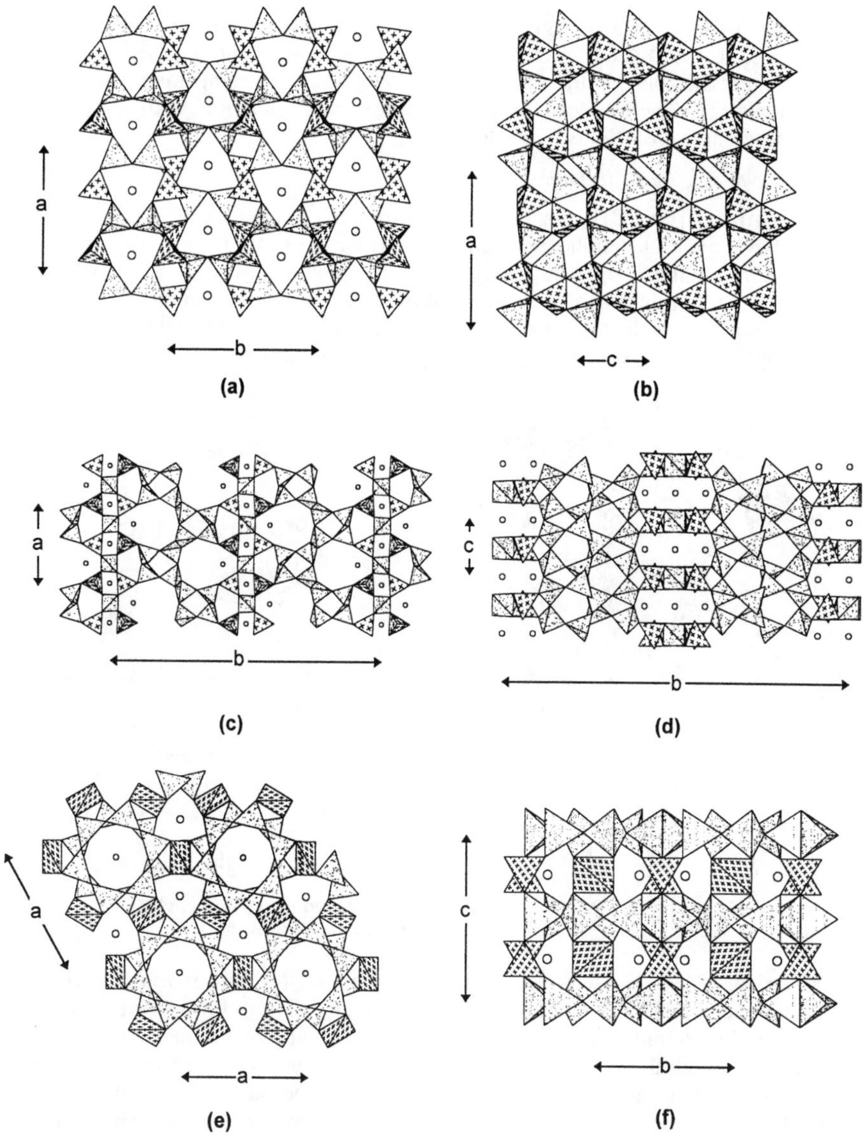

Figure 36. The crystal structures of barylite, chiavennite and beryl; (a) barylite projected onto (001); circles are Ba atoms; tetrahedra form six-membered rings in the (001) plane; (b) barylite projected onto (010); note the three-connected vertices; (c) chiavennite projected onto (001); circles are Ca and Mn atoms; (d) chiavennite projected onto (100); (e) beryl projected onto (001); large circles are Al atoms, small circles are Na, Cs and H_2O groups; note the similarity in this projection with the structure of milarite (Fig. 35c); (f) beryl projected onto (100).

There is extensive incorporation of additional constituents into the beryl structure. In particular, Na, Cs and H_2O are common substituents into the hexagonal channels through the structure (Fig. 36e), and the excess charge is compensated by substitution of Li for Be and (Mg,Fe^{2+}) for Al (Hawthorne and Černý 1977; Aurisicchio et al. 1988). Bazzite (Armbruster et al. 1995) is the Sc analogue of beryl, and stoppaniite (Ferraris et al. 1998; Della Ventura et al. 2000) is the Fe^{3+} analogue. Note that the stoppaniite described by Ferraris et al. (1998) lies close to the boundary of stoppaniite with the (as yet, hypothetical) composition $NaFe^{2+}Fe^{3+}[Be_3Si_6O_{18}]$.

Structures involving Be–P linkages.

Beryllonite, $Na[Be(PO_4)]$, consists of a well-ordered framework of alternating four-connected BeO_4 and PO_4 tetrahedra arranged at the vertices of a 6^3 net, with BeO_4 and PO_4 tetrahedra pointing in opposing directions along the *b*-axis (Fig. 37b), topologically identical to the tridymite framework. These sheets stack along the *b*-direction and share tetrahedron corners to form four-membered and eight-membered rings (Fig. 37a). The resultant framework has large channels containing [6]- and [9]-coordinated interstitial Na.

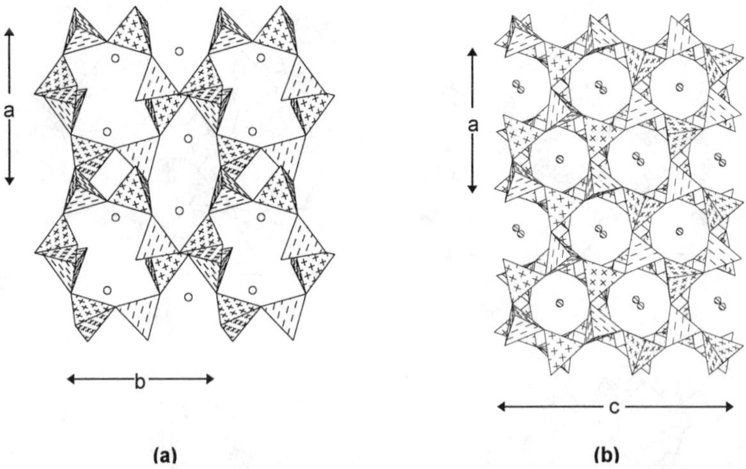

Figure 37. The crystal structure of beryllonite; (a) projected onto (001); circles are Na atoms; (b) projected onto (010); tetrahedra are arranged at the vertices of a 6^3 net.

Tiptopite, $K_2(Li_{2.9}Na_{1.7}Ca_{0.7}\square_{0.7})[Be_6(PO_4)_6](OH)_2(H_2O)_4$, is isotypic with the minerals of the cancrinite group: $Ca_2Na_6[Al_6(SiO_4)_6(CO_3)_2](H_2O)_2$ for the silicate species. The BeO_4 and PO_4 tetrahedra are arranged at the vertices of a two-dimensional net (Fig. 38a) such that all tetrahedra are three-connected when viewed down [001]. Prominent twelve-membered rings are arranged at the vertices of a 3^6 net such that they two-connect four-membered rings and three-connect through six-membered rings. These sheets link in the *c*-direction such that all tetrahedra are four-connected and, projected down the *b*-direction, form a two-dimensional net of four- and six-membered rings (Fig. 38b). The latter can be considered as a 6^3 net in which every third row of hexagons have a linear defect corresponding to an *a*-glide operation along *c*, i.e., double chains of hexagons extending in the *c*-direction and interleaved by single ladders of edge-sharing squares. Details of the rather complex relations between the interstitial species are discussed by Peacor et al. (1987).

Pahasapaite, $Ca_8Li_8[Be_{24}P_{24}O_{96}](H_2O)_{38}$, has an ordered array of BeO_4 and PO_4 tetrahedra arranged in a zeolite-rho framework, topologically similar to the minerals of the faujasite group and related to the synthetic aluminophosphate zeolite-like frameworks. Viewed along any crystallographic axis, the structure consists of prominent eight-membered rings of alternating BeO_4 and PO_4 tetrahedra (Fig. 38c) in an I-centered (F-centered in projection) array; they are connected along the axial directions by linear triplets of four-membered rings, and to nearest-neighbor eight-membered rings through six-membered rings. All tetrahedra are four-connected; BeO_4 tetrahedra link only to PO_4 tetrahedra, and vice versa. The structure has large cages (Rouse et al. 1989) and prominent intersecting channels (Fig. 38c) that contain interstitial Li, [7]-coordinated Ca and strongly disordered H_2O groups.

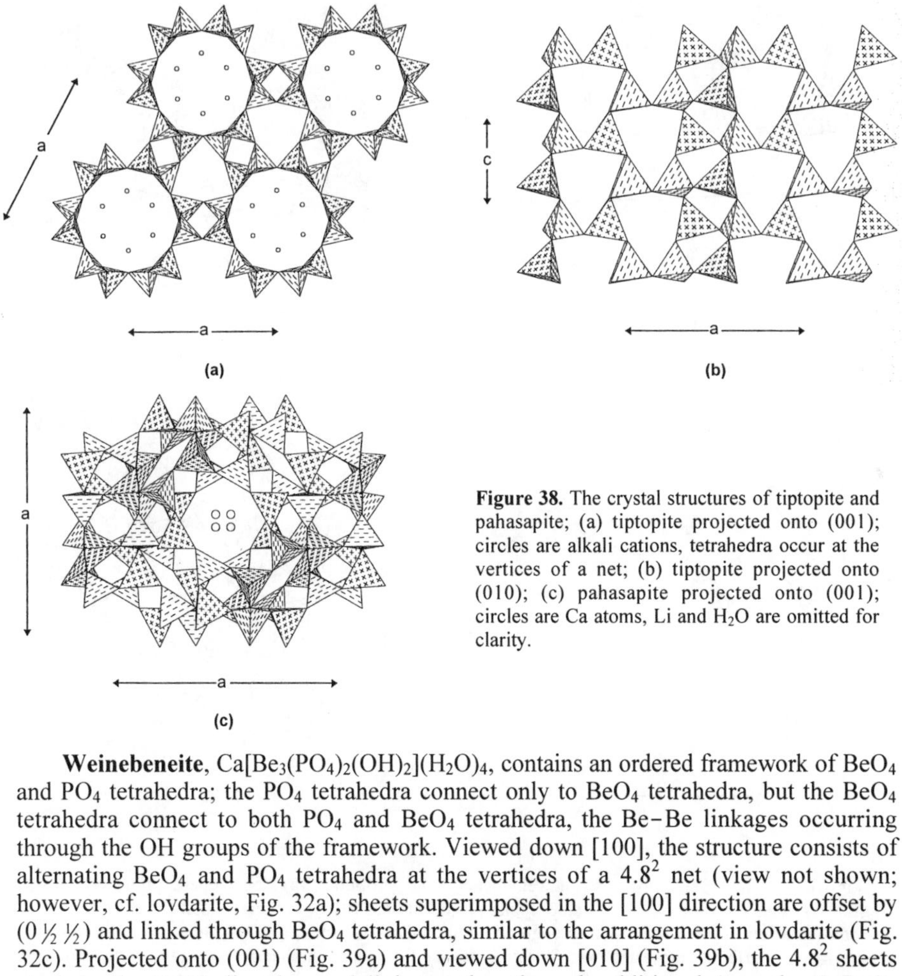

Figure 38. The crystal structures of tiptopite and pahasapite; (a) tiptopite projected onto (001); circles are alkali cations, tetrahedra occur at the vertices of a net; (b) tiptopite projected onto (010); (c) pahasapite projected onto (001); circles are Ca atoms, Li and H_2O are omitted for clarity.

Weinebeneite, $Ca[Be_3(PO_4)_2(OH)_2](H_2O)_4$, contains an ordered framework of BeO_4 and PO_4 tetrahedra; the PO_4 tetrahedra connect only to BeO_4 tetrahedra, but the BeO_4 tetrahedra connect to both PO_4 and BeO_4 tetrahedra, the Be–Be linkages occurring through the OH groups of the framework. Viewed down [100], the structure consists of alternating BeO_4 and PO_4 tetrahedra at the vertices of a 4.8^2 net (view not shown; however, cf. lovdarite, Fig. 32a); sheets superimposed in the [100] direction are offset by (0 ½ ½) and linked through BeO_4 tetrahedra, similar to the arrangement in lovdarite (Fig. 32c). Projected onto (001) (Fig. 39a) and viewed down [010] (Fig. 39b), the 4.8^2 sheets stack in the [100] direction and link together through additional (non-sheet) BeO_4 tetrahedra. Interstitial [7]-coordinated Ca is situated to one side of the large channels thus formed, with channel H_2O also bonded to the Ca.

Hurlbutite, $Ca[Be_2(PO_4)_2]$, consists of an ordered array of BeO_4 and PO_4 tetrahedra in which all tetrahedra are four-connected and there is alternation of BeO_4 and PO_4 tetrahedra in the structure. Viewed down [001] (Fig. 39c), the tetrahedra are arranged at the vertices of a 4.8^2 net with [7]-coordinated Ca occupying the interstices; these sheets link along the [001] direction by vertex-sharing (Fig. 39d). The structure is very similar to that of danburite, $Ca[B_2(SiO_4)_2]$, but has a different ordering scheme (and space group).

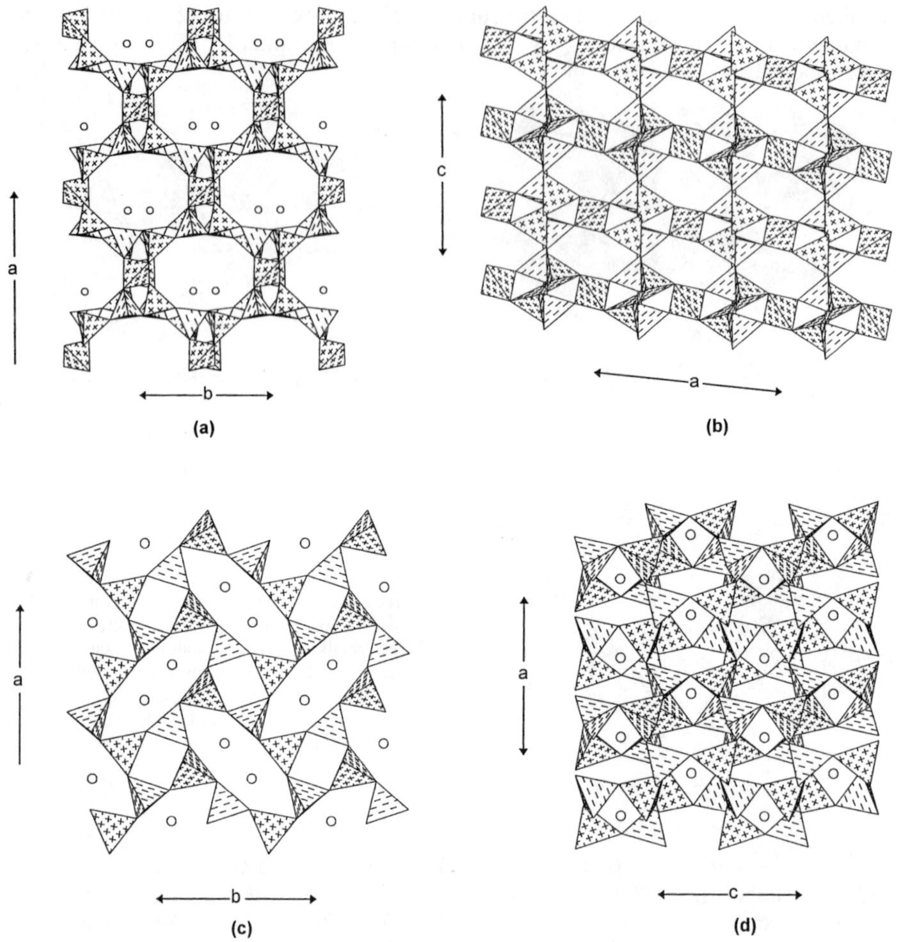

Figure 39. The crystal structures of weinebeneite and hurlbutite; (a) weinebeneite projected onto (001); circles are Ca atoms; (b) weinebeneite projected onto (010); in both (a) and (b), 4.8^2 nets of tetrahedra link in the *a*-direction through an additional BeO_4 group, (H_2O) groups are omitted for clarity; (c) hurlbutite projected onto (001); circles are Ca atoms; tetrahedra occupy the vertices of a net; (d) hurlbutite projected onto (010).

Babefphite, $Ba[Be(PO_4)F]$, is a rather unusual mineral; it is an ordered framework of PO_4 and BeO_3F tetrahedra. Projected down the *c*-direction, tetrahedra are arranged at the vertices of a 6^3 net (Fig. 40a) with the tetrahedra pointing (uuuddd). Projected down the *a*-direction, again the tetrahedra occur at the vertices of a 6^3 net (Fig. 40b) but the tetrahedra point (uuuuuu). Both the $Be\varphi_4$ and the PO_4 tetrahedra are three-connected, and the F anions are the non-*T*-bridging species in the $Be\varphi_4$ tetrahedra. The interstices of the framework are occupied by [9]-coordinated Ba.

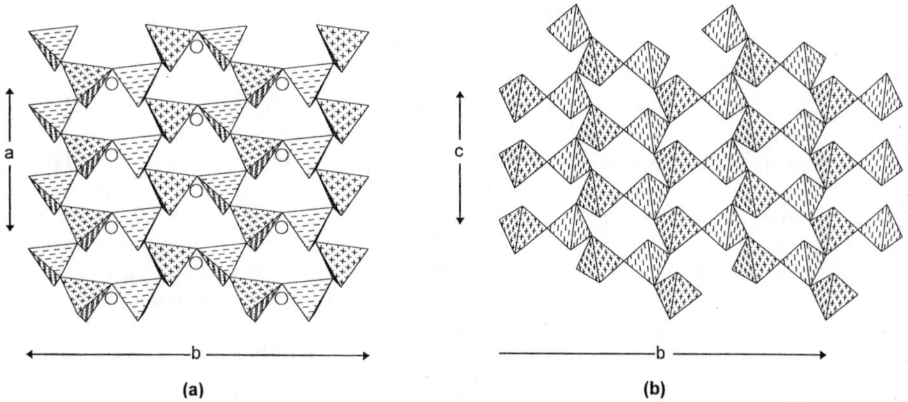

Figure 40. The crystal structure of babefphite; (a) projected onto (001); circles are Ba atoms; tetrahedra occur at the vertices of a 6^3 net; (b) projected onto (100); tetrahedra occur at the vertices of a 6^3 net.

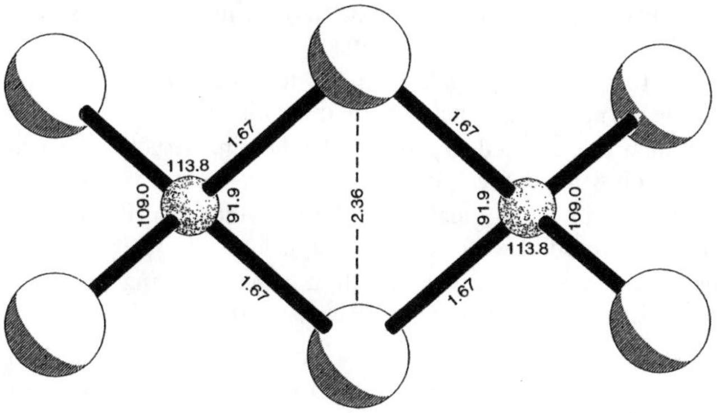

Figure 41. The $[Be_2O_6]$ group of edge-sharing BeO_4 tetrahedra:
Be, random-dot-shaded circles
O, highlighted circles.
Interatomic distances are the mean values from the structures of sorensenite, eudidymite and epididymite.

THE $[Be_2O_6]$ GROUP

An unusual stereochemical feature of a few of the Be minerals is the presence of the $[Be_2O_6]$ group (Fig. 41), a pair of edge-sharing tetrahedra. Of the oxyanion minerals, only the Be-bearing commonly show this feature; the sulfates, phosphates, silicates, aluminosilicates, etc., do not show edge-sharing tetrahedra involving their principal cations, although synthetic materials with edge-sharing LiO_4 tetrahedra are known. Pauling's third rule (Pauling 1929) predicts that high-valence low- coordination-number cations should not share coordination-polyhedra edges or faces because of the ensuing destabilization caused by strong cation-cation interactions; thus we normally do not expect tetrahedral oxyanions to share edges or faces. However, low-valence high-coordination-number cations frequently share coordination-polyhedra edges and faces, and hence the issue here is the position of the boundary between these two types of cations. Of interest in this regard is the corundum structure: This has octahedrally coordinated Al, and the octahedra share both edges and faces. Although this arrangement of face-sharing AlO_6 octahedra is unusual, it does occur in other structures besides corundum. In corundum, the mean bond-valences are 0.50 vu, similar to the mean bond-valence in BeO_4 tetrahedra. The Al–Al approach is ~2.2 Å without distortion, whereas the Be–Be approach is ~1.9 Å without distortion; however, the formal charges are higher for Al compared with Be. Thus it seems that $^{[6]}$Al and $^{[4]}$Be are where Pauling's third rule begins to lose its applicability.

The [Be_2O_6] group occurs in the structures of sorensenite (Figs. 21c,d, Table 6), eudidymite (Figs. 34a,b; Table 7) and epididymite (Figs. 34c,d,e; Table 7). Figure 41 shows the average geometry of the [Be_2O_6] group from these structures. It is immediately apparent that the two tetrahedra are significantly relaxed (i.e., distorted) relative to the holosymmetric arrangement. The bonds to the anions defining the shared edge are much longer than the mean bond-length (1.629 Å), the shared edge itself is much shorter than the mean edge-length (3.07 Å), and the angle subtended by the shared edge at the cation is much less than the holosymmetric value of 109.47°. All of these relaxations serve to increase the Be–Be distance (2.33 Å) relative to the separation of 1.88 Å for a holosymmetric arrangement, presumably stabilizing this type of linkage.

SOLID SOLUTION OF BERYLLIUM WITH OTHER CATIONS IN MINERALS: CRYSTAL CHEMISTRY

Beryllium may form distinct crystal structures of minerals in five different ways:

(1) formation of a structure that is unrelated to any non-Be-bearing structure, and in which Be occupies completely one or more distinct sites;

(2) formation of a structure that is topologically similar to other non-Be-bearing structures, and in which Be occupies completely one or more distinct sites;

(3) formation of a structure that is unrelated to any non-Be-bearing structure, but in which Be shows solid solution with other cations;

(4) formation of a structure that is topologically similar to other non-Be-bearing structures, but in which Be shows solid solution with other cations;

(5) formation of a structure that is topologically similar to a non-Be-bearing structure, but in which Be occupies a site not occupied in the Be-absent structure.

Most of the minerals of Appendix A fall into category (1). A few minerals fall into category (2); here, Be completely replaces another cation at one (or more) of a set of sites, lowering the symmetry of the atomic arrangement. An excellent example of this is the amphibole joesmithite (Fig. 11d) in which Be replaces Si completely at two of the four $T(2)$ sites of the [T_8O_{22}] chain, thereby breaking the mirror symmetry of the tetrahedral chain to produce a symmetrically (but not topologically) distinct site. Categories (1) and (2) are the most frequent for Be minerals. However, categories (3) and (4) involve solid solution and give rise to more common and flexible structures, and the details of the types of solid solution are of interest.

Beryl

The prototype mineral of the beryl group is beryl itself: Al[$Be_3Si_6O_{18}$]. Beryl may incorporate significant Li and Na (+ Cs, Rb) into its structure [i.e., is of category (4) above]. Belov (1958) proposed direct substitution of Be by Li: Li ↔ Be, and this was supported by the results of a partial refinement of a Cs-Li-enriched beryl by Bakakin et al. (1969). Conversely, Beus (1960) proposed a coupled substitution of the form $^{[6]}$Li + $^{[4]}$Al ↔ $^{[6]}$Al + $^{[4]}$Be which was supported by a preliminary refinement of a Cs-rich beryl by Evans and Mrose (1966). Hawthorne and Černý (1977) refined the crystal structure of a (Cs,Li)-rich beryl and showed definitively that Li substitutes directly for Be at the *Be* (tetrahedrally coordinated) site, i.e., ^{Be}Be. Figure 42a shows the variation in <*Be*–O> distance as a function of Be content; as $^{[4]}$Be (r = 0.27 Å) is smaller than $^{[4]}$Li (r = 0.59 Å, Shannon 1976), the <*Be*–O> distance increases with decreasing Be occupancy (and increasing Li occupancy). Moreover, the slope of the relation in Figure 42a is in accord with that expected for a hard-sphere model.

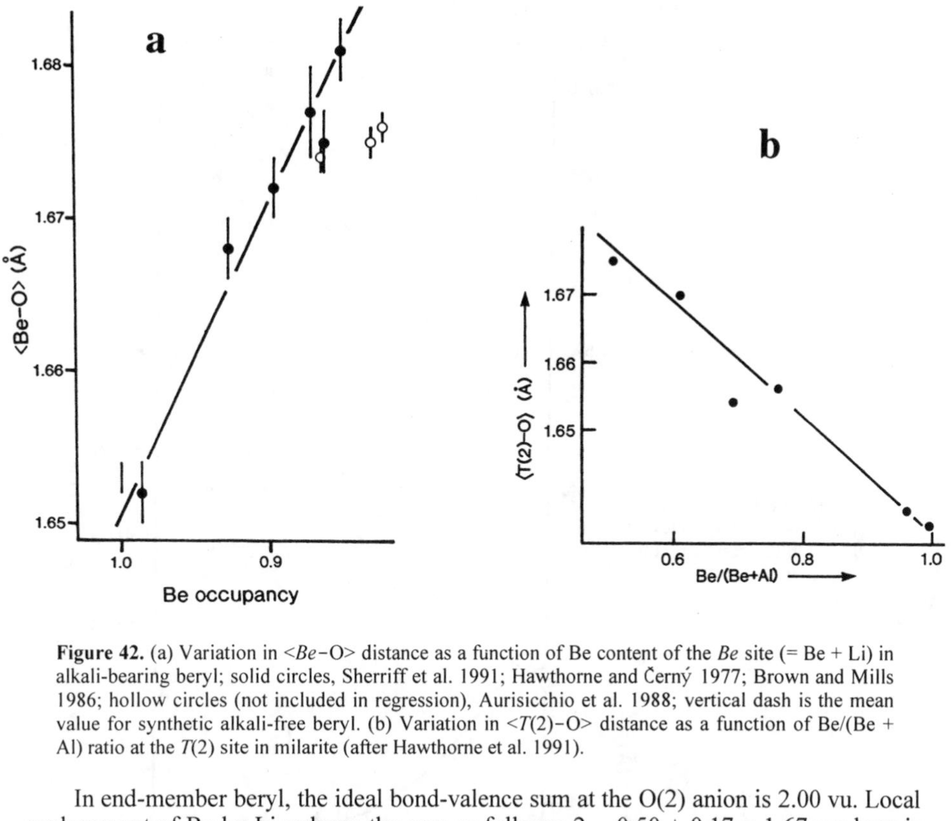

Figure 42. (a) Variation in <*Be*–O> distance as a function of Be content of the *Be* site (= Be + Li) in alkali-bearing beryl; solid circles, Sherriff et al. 1991; Hawthorne and Černý 1977; Brown and Mills 1986; hollow circles (not included in regression), Aurisicchio et al. 1988; vertical dash is the mean value for synthetic alkali-free beryl. (b) Variation in <*T*(2)–O> distance as a function of Be/(Be + Al) ratio at the *T*(2) site in milarite (after Hawthorne et al. 1991).

In end-member beryl, the ideal bond-valence sum at the O(2) anion is 2.00 vu. Local replacement of Be by Li reduces the sum as follows: 2 − 0.50 + 0.17 = 1.67 vu; how is this discrepancy compensated? To some extent, this can be done as proposed by Hawthorne and Černý (1977): The structure adjusts by shortening the Si–O(2) bond and lengthening both Si–O(1) and Si–O(1)a bonds, the resultant deficiency at O(1) being compensated by bonding from the channel alkali atoms that are part of the ^{Be}Li + ^{C}(Na,Cs) ↔ ^{Be}Be + ^{C}□ mechanism of substitution. Figure 43 shows this to be the case for continuous substitution; Si–O(1) and Si–O(1)a gradually lengthen with decreasing Be, and Si–O(2) decreases. The additional contribution of bond valence from Li (at the *Be* site) to O(2) results in satisfaction of the local short-range bond-valence requirements. However, the replacement of Be by Li leads to a net charge deficiency that is balanced by substitution of alkali cations, specifically Na and Cs, into the channel of the beryl structure. The resultant mechanism of incorporation of Li and (Na,Cs) into the beryl structure may be written as ^{Be}Li + ^{C}(Na,Cs) → ^{Be}Be + ^{C}□, where *C* denotes the channel sites.

Milarite

The prototype mineral of the milarite group is milarite: $KCa_2[AlBe_2Si_{12}O_{30}](H_2O)$, but many minerals of this group [e.g., brannockite: $KSn_2[Li_3Si_{12}O_{30}]$ and poudrettite: $KNa_2[B_3Si_{12}O_{30}]$ do not contain Be, and milarite shows extensive solid solution, with Al substituting for Be at the *T*(2) site [i.e., is of category (4) above]. Hawthorne et al. (1991)

showed that the <*T*(2)−O> distance decreases with increasing Be/(Be+Al) ratio (Fig. 42b); the slope of the relation in Figure 42b is in accord with a hard-sphere model for solid solution between Be ($^{[4]}$r = 0.27 Å) and Al ($^{[4]}$r = 0.39 Å).

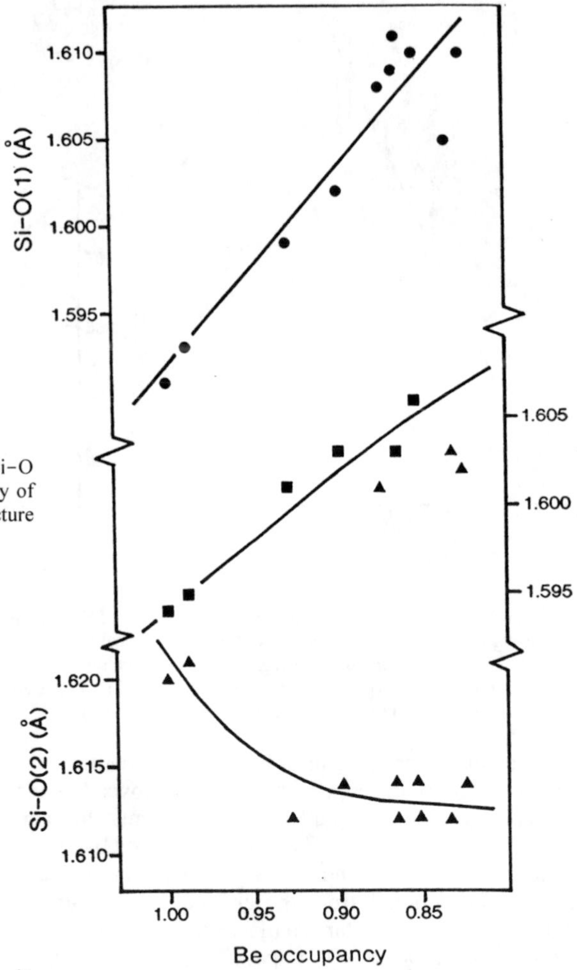

Figure 43. Variation in Si−O distances with Be occupancy of the *Be* site in the beryl structure (after Sherriff et al. 1991).

Beryllium-bearing cordierite

Cordierite has an ideal end-member formula that is usually written as $Mg_2Al_4Si_5O_{18}$. However, cordierite has a structure that is topologically identical to that of beryl (Fig. 36e), and hence it is more informative if we write their formulae as $Al_2[Be_3Si_6O_{18}]$ and $Mg_2[Al_4Si_5O_{18}]$, with the structural unit in square brackets representing the tetrahedral framework. Beryl has $[Si_6O_{18}]$ rings arranged parallel to {001} and linked into columns by BeO_4 tetrahedra, whereas cordierite has analogous $[Al_2Si_4O_{18}]$ rings linked by (AlO_4) and (SiO_4) tetrahedra. In cordierite, there is prominent long-range ordering of Al and Si in the $[T_6O_{18}]$ rings, causing cordierite to be orthorhombic (rather than hexagonal, as is beryl) (Gibbs 1966; Cohen et al. 1977).

Beryllium has been reported in cordierite by many authors. Černý and Povondra (1966) reported up to 1.94 wt % Be in cordierite, and proposed that Be was incorporated into the structure of cordierite via the substitution (Na,K) + Be ↔ □ + Al. This mechanism has been since confirmed by many authors (e.g., Povondra and Langer 1971; Schreyer et al. 1979; Armbruster and Irouschek 1983). Beryllium could potentially replace Al in the six-membered rings or in the four-membered rings. By analogy with the Na + Li → □ + Be substitution in beryl, Be should replace Al in the four-membered rings, linking the $[Al_2Si_4O_{18}]$ rings together; Armbruster (1986) has shown this to be the case.

Hölscher and Schreyer (1989) synthesized a hexagonal cordierite-like phase with the composition $Mg_2[(Al_2Be)Si_6O_{18}]$, i.e., Be is incorporated into the cordierite (indialite) structure via the substitution Be + Si ↔ Al + Al. However, as noted by Hölscher and Schreyer (1989), this substitution has not been identified as yet in cordierite: There is always sufficient Na present to accommodate the substitution (Na,K) + Be ↔ □ + Al of Černý and Povondra (1966).

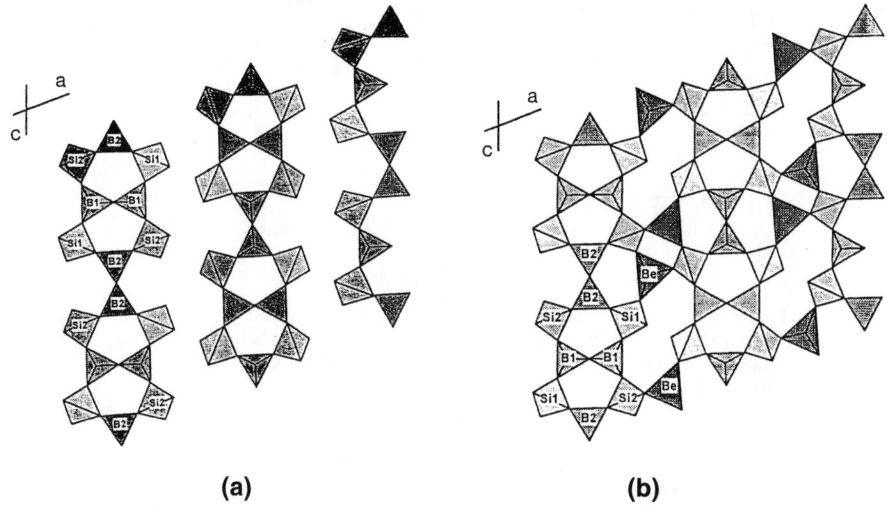

Figure 44. (a) The borosilicate chains in the structure of hellandite; (b) the *Be* sites in Be-bearing hellandite linking the borosilicate chains to form a sheet (after Oberti et al. 1999).

Beryllium-bearing hellandite

Hellandite is a borosilicate mineral with the end-member formula $Ca_5YAl[B_4Si_4O_{20}](OH)$ (Hawthorne et al. 1996). Oberti et al. (1999) refined the crystal structure of a Be-bearing hellandite and showed that Be occupies a site that is vacant in Be-absent hellandite [i.e, this Be-bearing hellandite is of category (5) above]. The borosilicate structural unit of hellandite is shown in Figure 44a; it consists of pentagonal borosilicate rings that link to form complex chains extending along the *c*-axis. In Be-bearing hellandite (Fig. 44b), Be (partly) occupies tetrahedrally coordinated sites that link the borosilicate chains into a berylloborosilicate sheet. In the crystal examined by Oberti et al. (1999), the *Be* site is also partly occupied by Li. Both the refined site-scattering values and the results of SIMS analysis suggest that occupancy of the *Be* site by Be (and Li) is locally associated with F, forming a BeO_3F group.

Oberti et al. (2002) have reported variable Be (and Li) in a suite of hellandites with Be varying between 0.02 and 1.18 apfu. The Be and Li occur at the new site (denoted *T*) of Oberti et al. (1999), and the resultant tetrahedron shows a regular variation in mean bond-length as a function of the scattering at the *T* site (Fig. 45). It is apparent from Figure 45 that there is continuous variation of Be and Li at the *T* site in beryllian hellandite. Oberti et al. (2002) identify several distinct end-member compositions with the hellandite structure (Table 8). Two of these are potential Be minerals in which the *T* site is filled with Be. Moreover, one of the compositions listed by Oberti et al. (2002) has Be > 1.0 apfu and falls within the composition field of end-member (7).

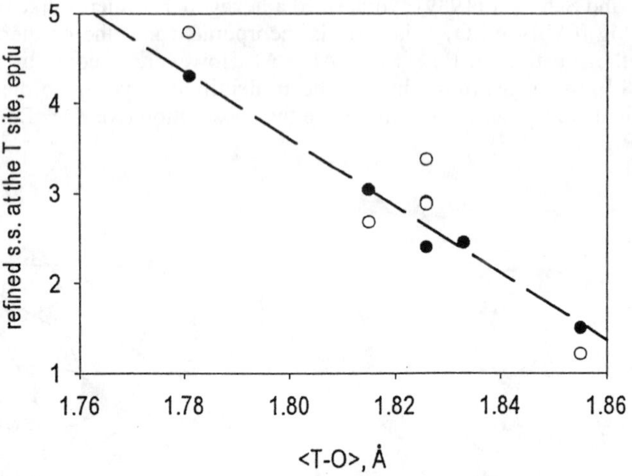

Figure 45. Variation in the site scattering at the *T* site as a function of <*T*–O> in Be-bearing hellandite structures; solid circles are SREF data, hollow circles are SIMS data. Used by permission of the Mineralogical Society of America, from Oberti et al. (2002), *American Mineralogist*, Vol. 87.

Table 8. Possible end-member compositions for hellandite-group minerals*.

	*X***	*Y***	*Z*	*T*		*W*
(1)	Ca_3Y	Y_2	Al	$□_2$	$[B_4Si_4O_{22}]$	$(OH)_2$
(2)	Ca_4	Y_2	Ti	$□_2$	$[B_4Si_4O_{22}]$	$(OH)_2$
(3)	Ca_4	ThY	Al	$□_2$	$[B_4Si_4O_{22}]$	$(OH)_2$
(4)	Ca_4	Y_2	Ti	Li_2	$[B_4Si_4O_{22}]$	O^{2-}_2
(5)	Ca_3Y	Y_2	Al	Li_2	$[B_4Si_4O_{22}]$	O^{2-}_2
(6)	Ca_4	ThY	Al	Li_2	$[B_4Si_4O_{22}]$	O^{2-}_2
(7)	Ca_4	CaY	Al	Be_2	$[B_4Si_4O_{22}]$	O^{2-}_2
(8)	Ca_4	Ca_2	Ti	Be_2	$[B_4Si_4O_{22}]$	O^{2-}_2

* from Oberti et al. (2002)
** $Y = (Y^{3+} + REE^{3+})$; $Al = (Al^{3+} + Mn^{3+} + Fe^{3+})$;
$OH = (OH^- + F^-)$

Rhodizite

Rhodizite is a berylloborate structure with the ideal end-member composition $KAl_4[Be_4(B_{11}Be)O_{28}]$. The structure is a framework (Figs. 27c,d) of BeO_4 and BO_4 tetrahedra arranged in a checker-board pattern. There is one *Be* site with a rank of 4 and a <*Be*–O> distance of 1.604 Å, and there is one *B* site with a rank of 12 and a <*B*–O>

distance of 1.492 Å. Pring et al. (1986) proposed a model with the *Be* site filled with Be and the *B* site occupied by 11.35 B + 0.50 Be, although this was not a truly least-squares solution. The <Be–O> and <B–O> distances are in reasonable accord with this model. However the observed <Be–O> distance of 1.604 Å is much less than the grand <Be–O> distance of 1.633 Å shown in Figure 1a, raising the possibility of a small amount of B being incorporated at this site [thereby accounting for the small <*Be*–O> distance]. The observed <*B*–O> distance of 1.492 Å is somewhat longer than the grand <B–O> distance of 1.476 Å reported by Hawthorne et al. (1996). This is in accord with the incorporation of some Be at this site. Thus rhodizite could possibly be slightly less ordered than suggested by Pring et al. (1986), although there is definite substitution of both Be and B at the same site in rhodizite.

Hyalotekite

Hyalotekite (Moore et al. 1982) is a complex framework of BO_4 and SiO_4 tetrahedra. One of the tetrahedral sites with *T*–O distances fairly typical of Si–O [*Si*(1)] showed lower than expected scattering, and Moore et al. (1982) assigned Be to this site. Thus Be and Si occur at the *Si*(1) site. Another possible arrangement can be suggested. The <*Si*(1)–O> distance is 1.597 Å, which is very short for a <Si–O> distance; the other <Si–O> distances in hyalotekite are 1.613, 1.610 and 1.613 Å. Now Be ($^{[4]}r$ = 0.27 Å) is marginally larger than Si ($^{[4]}r$ = 0.26 Å), and substitution of Be cannot account for the small <*Si*(1)–O> distance. However, if Be were to substitute for B at the *B* site and the displaced B were to substitute for Si at the *Si*(1) site, then the <*Si*(1)–O> distance would be shorter than the distance characteristic of <Si–O>, and the <B–O> distance should be longer than the distance characteristic of <B–O> (= 1.476 Å, Hawthorne et al. 1996). Moreover, Be substitutes for B and B substitutes for Si, as is observed in many other structures, rather than Be substituting for Si, which tends not to occur. Christy et al. (1998) re-refined the data of Moore et al. (1982) and came to this same conclusion.

Let us examine the possible end-member compositions of hyalotekite from the viewpoint of the above discussion of the site occupancies of Be, B and Si. Christy et al. (1998) write the general formula of hyalotekite as

$$(Ba,Pb,K)_4\,(Ca,Y)_2\,Si_8\,(Be,B)_2\,(Si,B)_2\,O_{28}\,F$$

Their chemical data show that Ba is often dominant over Pb^{2+}, and Ca is dominant over Y, and hence let us write a simplified formula as

$$Ba_4Ca_2\,Si_8\,(Be,B)_2\,(Si,B)_2\,O_{28}\,F$$

This is not an end-member formula as there is disorder at more than one site in the structure. This formula can be resolved into two simpler components

(1) $Ba_4\,Ca_2\,Si_8\,(BeB)\,(Si)_2\,O_{28}\,F$

(2) $Ba_4\,Ca_2\,Si_8\,(B_2)\,(SiB)\,O_{28}\,F$

These are true end-members in that neither can be resolved into more simple compositions that retain the hyalotekite structure and remain neutral.

Christy et al. (1998) provide chemical data for a range of compositions of hyalotekite. Figure 46a examines this data from the perspective of the above two end-members. The data extend between the two end-member compositions, in accord with the above discussion. Moreover, the displacement of the data from the ideal line correlates closely with the amount of K substituting for Ca, indicating the presence of a significant additional substitution. Let us next examine the character of this substitution. There are three possible ways to incorporate K into the hyalotekite structure if we ignore replacement of Ca by Y: (1) Ba + Be ↔ K + B; (2) 2Ba + Be ↔ K + Si; (3) Ba + B ↔ K + Si. The directions of these three substitutions relative to the axes of Figure 46a are

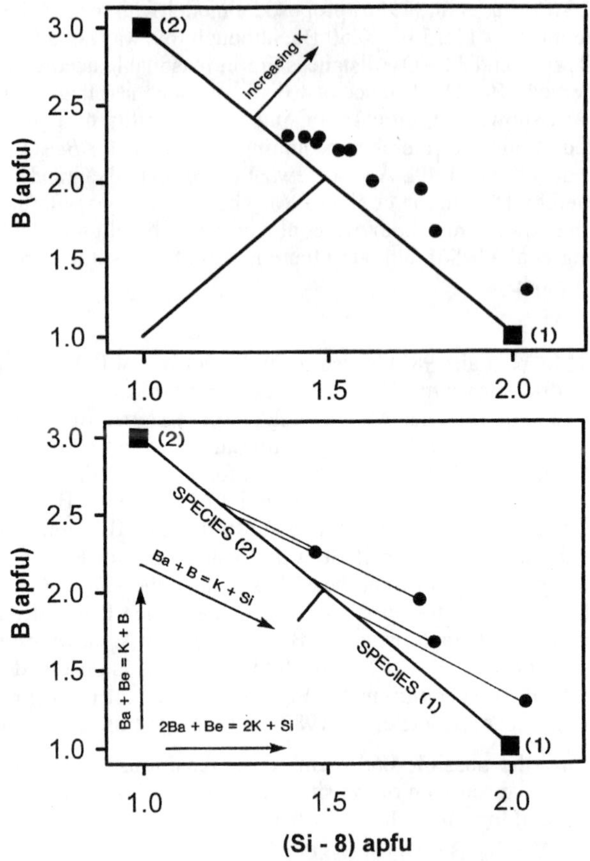

Figure 46. Variation in the chemical composition of hyalotekite; (a) B vs. [8−Si]; the compositions (1) $Ba_4Ca_2Si_8(BeB)(Si_2)O_{28}F$ and (2) $Ba_4Ca_2Si_8(B_2)(SiB)O_{28}F$ are marked by filled squares and are joined by a line representing the substitution B + B ↔ Be + Si; data from Christy et al. (1998) are shown as filled circles; (b) B vs. [8−Si]; the arrows represent the directions of the substitutions involving K, the lines join selected datapoints to points on the (1)−(2) line with the same Be content; note that these lines are parallel to the substitution Ba + B ↔ K + Si.

shown in Figure 46b. Now, if the data were to lie exactly along the line between compositions (1) and (2) in Figure 46a, the Be content should vary along this line from 1.0 apfu at composition (1) to 0.0 apfu at composition (2). Let us join the data points to the corresponding measured Be contents along the line between compositions (1) and (2) (Fig. 46b). For clarity, not all data are shown in Figure 46b. However, what is apparent is that the lines joining the observed data points with their Be content along the (1)−(2) join are all sub-parallel to the substitution direction defined by Ba + B ↔ K + Si, indicating that this is the substitution whereby K is incorporated into the hyalotekite structure.

Another important feature of Figure 46b is that is shows that *both* species (1) and (2) are represented in the compositional data of Christy et al. (1998). We may summarize the situation for hyalotekite as follows: There are two distinct end-members (1) and (2) (see above), and that compositions fall into both compositional fields. Thus there are two distinct ‘hyalotekite minerals,’ one of which [composition (1)] contains essential Be, and the other of which [composition (2)] does not contain essential Be. In this regard, kapitsaite-(Y) is Be-free and isostructural with hyalotekite. Sokolova et al. (2000) wrote the formula as

$(Ba_{3.68}K_{0.12}Pb^{2+}_{0.20})\ (Y_{1.00}Ca_{0.66}REE_{0.34})\ Si_8\ B_2\ (B_{1.7}Si_{0.3})\ O_{28}\ F$.

Thus the end-member composition may be written as

$Ba_4\ (YCa)\ Si_8\ (B_2)\ (B_2)\ O_{28}\ F$

following the scheme used above for hyalotekite, and is related to the Be-free end-member of the hyalotekite series by the substitution $Y + B \leftrightarrow Ca + Si$.

"Makarochkinite"

"Makarochkinite" is a non-accredited Be-bearing aenigmatite-like phase. The structure (Fig. 10) consists of decorated chains of BeO_4, SiO_4 and $(Si,Al)O_4$ tetrahedra extending along the *a*-axis (Fig. 10a) and forming layers of tetrahedra orthogonal to [011] (Fig. 10b). Two of the tetrahedra show disorder of Si and Be: *T*1 and *T*4. The $\langle T–O \rangle$ distances and the isotropic-displacement factors seem in accord with this assignment. However, it must be noted that multiple occupancy of sites by Be and Si is quite unusual.

It is of interest to examine the compositional relations between these minerals, as Grauch et al. (1994) state that "makarochkinite" is the same as høgtuvaite. Details of the structure of aenigmatite are given by Cannillo et al. (1971). It is triclinic, space group $P\bar{1}$, with six [4]-coordinated sites, designated *T*(1) to *T*(6), seven [6]-coordinated sites designated *M*(1) to *M*(7), and two [8]-coordinated sites designated *Na*(1) and *Na*(2). Two of the *M* sites, *M*(1) and *M*(2), occur at centers of symmetry, and thus one can write the resulting structural formula as $Na_2M_6T_6O_{22}$. In aenigmatite (Cannillo et al. 1971), Ti is strongly ordered at the *M*(7) site and Fe^{2+} occurs at the other M*(1)* to M(6) sites to give the end-member formula $Na_2(Fe^{2+}_5Ti)Si_6O_{20}$. We may ignore homovalent substitutions in the present case, and focus on "makarochkinite" and høgtuvaite, the chemical compositions and formulae of which are shown in Table 9. Following aenigmatite, we assume that Ti is ordered at the *M*(7) site. In "makarochkinite," Ti > 0.50 apfu and hence the *M*(7) site is dominated by Ti, i.e., Ti is an essential constituent of "makarochkinite." In høgtuvaite, Ti < 0.50 apfu and hence does not predominate at the *M*(7) site, i.e., Ti is *not* an essential constituent of høgtuvaite. Are "makarochkinite" and høgtuvaite distinct? Yes, according to the data currently available, as Ti is an essential constituent of "makarochkinite" and is *not* an essential constituent of høgtuvaite. The problem is most easily resolved by writing the end-member formula of each of these species, following what we have discussed above. The *T* content of each are very close (Table 9), but fall on either side of the boundary between two distinct arrangements: (Si_4BeAl) in "makarochkinite" and (Si_5Be) in høgtuvaite. The *M* content of each may be written as follows: $(Fe^{2+}_4Fe^{3+}Ti)$ for "makarochkinite" (where $Fe^{2+} \equiv Fe^{2+} + Mg + Mn$) and $(Fe^{2+}_4Fe^{3+}_2)$ for høgtuvaite (where $Fe^{2+} \equiv Fe^{2+} + Mg + Mn$). Thus we may write the end-member formulae as follows:

"makarochkinite"	$Ca_2\ (Fe^{2+}_4Fe^{3+}Ti)$	(Si_4BeAl)	O_{20}
høgtuvaite	$Ca_2\ (Fe^{2+}_4Fe^{3+}_2)$	(Si_5Be)	O_{20}

Hence these are distinct species, and the statement of Grauch et al. (1994) that the "informally proposed new mineral" (*viz.* "makarochkinite") "is the same as høgtuvaite" is not correct according to the published data, although we do note that the formula for "makarochkinite" is unsatisfactory from the excess amount of (Ca + Na + K) (Table 8).

Very recently, Barbier et al. (2001) report a refined crystal structure of "makarochkinite". Although full details of the structure are not yet available, they state that Ti is dominant at *M*(7) and Be is disordered over two *T* sites, but do not report site populations

Table 9. Chemical composition (wt %)* and unit formula (apfu) for "makarochkinite" (M) and høgtuvaite (H).

	M		*H*
SiO_2	30.09		31.60
Al_2O_3	3.55		2.64
BeO	2.32		2.65
TiO_2	6.02		2.77
Fe_2O_3	11.12		19.03
FeO	26.91		28.06
MnO	1.26		0.27
MgO	2.74		0.42
CaO	13.38		10.44
Na_2O	1.35		1.52
K_2O	0.30	SnO_2	0.53
	99.04		99.93
Si	4.39		4.60
Al	0.61		0.45
Be	0.81		0.93
ΣT	5.81		5.98
Fe^{3+}	1.22		2.09
Fe^{2+}	3.28		3.42
Mn	0.16		0.03
Mg	0.60		0.09
Ti	0.66		0.30
ΣM	5.92		**5.96
Ca	2.09		1.63
Na	0.38		0.43
K	0.06		-
Σ	2.53		2.06

* Data from Yakubovich et al. (1990) and Grauch et al. (1994)** including 0.03 Sn.

for *M*(1)–*M*(6). Moreover, the Al content is slightly less than that reported by Yakubovich et al. (1990); however, the Ca (1.77) and Na (0.20 apfu) contents of the [7]-coordinated sites are much more satisfactory than the previously reported values. Thus the "makarochkinite" reported by Barbier et al. (2001) is almost exactly half-way between the compositions

(1) Ca_2 $(Fe^{2+}{}_4Fe^{3+}Ti)$ (Si_4BeAl) O_{20}

(2) Ca_2 $(Fe^{2+}{}_5Ti)$ (Si_5Be) O_{20}

Composition (1) corresponds to that above that we derived for the "makarochkinite" of Yakubovich et al. (1990) and composition (2) is new. Both compositions (1) and (2) are *bona-fide* end-member compositions of the aenigmatite structure, provided they show ordered cation distribution at the *M*(1)–*M*(6) and/or *T*(1)–*T*(6) sites.

Table 10 summarizes the end-member compositions for aenigmatite and some Be-bearing compositions. Moore (1978) gives the end-member composition as welshite (1) (Table 10). This seems the most reasonable end-member, although the chemical composition leads to an empirical formula with only 5.5 tetrahedrally coordinated cations pfu. Grew et al. (2001) give a new empirical composition for welshite:

$Ca_2(Mg_{3.8}Mn^{2+}{}_{0.6}Fe^{2+}{}_{0.1}Sb^{5+}{}_{1.5})O_2[Si_{2.8}Be_{1.7}Fe^{3+}{}_{0.65}Al_{0.7}As_{0.17}O_{18}]$

Combining Mg, Mn^{2+} and Fe^{2+} and expressing the result as Mg, combining Si and As and expressing the result as Si, and combining Al and Fe^{3+} and expressing the result as Al, we get the following:

$Ca_2\ (Mg_{4.5}Sb^{5+}{}_{1.5})O_2\ [Si_{3.0}Be_{1.7}Al_{1.3}O_{18}]$

This formula has an excess negative charge of 0.2. Now there are two possibilities for the octahedrally coordinated cations in an end member: $(Mg_4Sb^{5+}{}_2)$ and (Mg_5Sb^{5+}). These require the charge at the tetrahedrally coordinated sites to be 18^+ and 21^+, respectively. For a charge of 18^+, the possible tetrahedrally coordinated cations are $[Al_6]$, $[SiBeAl_4]$, $[Si_2Be_2Al_2]$ and $[Si_3Be_3]$; comparison with the above empirical formula indicates that the first two are inappropriate. For a charge of 21^+, the possible tetrahedrally coordinated cations are $[Si_3Al_3]$ and $[Si_4BeAl]$. Thus we end up with four possible end-members:

(1) $Ca_2\ (Mg_4Sb_2)\ O_2\ [Si_2Be_2Al_2O_{18}]$

(2) $Ca_2\ (Mg_4Sb_2)\ O_2\ [Si_3Be_3O_{18}]$

(3) $Ca_2\ (Mg_5Sb)\ O_2\ [Si_3Al_3O_{18}]$

(4) $Ca_2\ (Mg_5Sb)\ O_2\ [Si_4BeAlO_{18}]$

Table 10. Beryllium minerals of the aenigmatite group: comparison with aenigmatite.

Aenigmatite	Na_2	$Fe^{2+}{}_4$	Fe^{2+}	Ti^{4+}	O_2	$[Si_6O_{18}]$
"Makarochkinite"(1)	Ca_2	$Fe^{2+}{}_4$	Fe^{3+}	Ti^{4+}	O_2	$[BeAlSi_4O_{18}]$
"Makarochkinite"(2)	Ca_2	$Fe^{2+}{}_5$		Ti^{4+}	O_2	$[BeSi_5O_{18}]$
Høgtuvaite	Ca_2	$Fe^{2+}{}_4$	Fe^{3+}	Fe^{3+}	O_2	$[BeSi_5O_{18}]$
Welshite (1)	Ca_2	Mg_4	Fe^{3+}	Sb^{5+}	O_2	$[Be_2Si_4O_{18}]$
Welshite (2)	Ca_2	Mg_4		$Sb^{5+}{}_2$	O_2	$[Be_3Si_3O_{18}]$

Taking compositions exactly intermediate between (2) and (3), and (1) and (4), gives the same composition:

$$Ca_2 \quad (Mg_{4.5}Sb_{1.5}) \quad O_2 \quad [Si_3Be_{1.5}Al_{1.5}O_{18}]$$

This is extremely close to the simplified composition for welshite given above. However, combining end-members (2) and (3) can produce different values of Be and Al, whereas combining end-members (1) and (4) must produce equal amounts of Be and Al. Hence combining end-members (2) and (3) in the proportions 0.57 and 0.43, respectively, produces the composition

$$Ca_2 \quad (Mg_{4.43}Sb_{1.57})\ O_2 \quad [Si_{3.00}Be_{1.70}Al_{1.30}O_{18}]$$

which is extremely close to the composition of welshite given by Grew et al. (2001). Hence the most appropriate end-member for this sample of welshite is (2), and the sample has extensive solid-solution toward end-member (3).

Sapphirine-related structures

Khmaralite, $Mg_7Al_9O_4[Al_6Be_{1.5}Si_{4.5}O_{36}]$, and surinamite, $Mg_3Al_3O[AlBeSi_3O_{15}]$, are structurally related to sapphirine, $Mg_7Al_9[Al_9Si_3O_{40}]$, where the formulae given here are the Mg–Al end-members. All three structures consist of the basic sapphirine arrangement (Moore 1969) in which different composition and ordering schemes produce different superstructures. Sapphirine itself shows extensive polytypism and disorder in the form of planar and line defects (Christy and Putnis 1988). Moreover, it can incorporate variable amounts of Be (Grew 1981; Christy 1988; Grew et al. 2000). Grew (1981) showed that sapphirine can contain variable Be (0.5–1.0 wt % BeO), and proposed the substitution Be + Si ↔ Al + Al. Grew et al. (2000) provided additional data, extended the known range of this substitution (1.1–2.5 wt % BeO), and showed that there is a continuous correlation between Be and Si from sapphirine through to khmaralite (Fig. 47) following the substitution proposed by Grew (1981). In both surinamite and khmaralite, Be occupies the three-connected tetrahedra in the branched chains, but details of the stereochemical variations with varying Be/Si contents are not available over the range of compositions represented in Figure 47.

Be-bearing micas

Details of the structure of bityite, ideally $Ca(LiAl_2)(BaAlSi_2)O_{10}(OH)_2$, are not available, but Lin and Guggenheim (1983) refined the structure of a (Li,Be)-rich brittle mica intermediate in composition between bityite and margarite, ideally $Ca(Al_2)(Al_2Si_2)$-$O_{10}(OH)_2$. The *T*-site populations given by Lin and Guggenheim are as follows: $T(1)$ = 0.24 Be + 0.71 Al + 0.05 Si, $T(11)$ = 0.04 Be + 0.15 Al + 0.81 Si, $T(2)$ = 0.12 Al + 0.88 Si, $T(22)$ = 0.30 Be + 0.70 Al. We may adjust these to bring them into accord with the chemical composition of this mica to give $T(1) = 0.32$ Be + 0.68 Al, $T(11) = 1.00$ Si, $T(2) = 1.00$ Si, $T(22) = 0.32$ Be + 0.68 Al. It is very apparent here that Be substitutes for Al and does not substitute for Si.

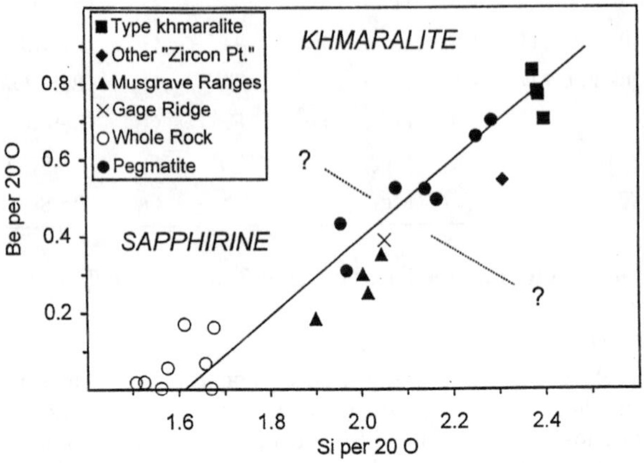

Figure 47. Variation in Be as a function of Si in sapphirine and khamaralite; the line designates the substitution Be + Si ↔ Al + Al; after Grew et al. (2000).

General observations on solid-solution relations involving beryllium

There are two issues involved in solid-solution relations involving Be: (1) at what specific sites does Be occur in the crystal structure, and (2) what cation(s) is Be replacing at its constituent sites. Huminicki and Hawthorne (2001b) have examined the first issue for Be minerals whose structural units are sheets. They represented the sheets as two-dimensional nets and looked at the local topology of the net vertices that represent Be sites in the structures. Although none of the minerals examined display solid solution between Be and other tetrahedrally coordinated cations, their work addresses the issue of the effect of network topology on the occupancy of tetrahedrally coordinated sites by Be. Huminicki and Hawthorne (2001b) note that Be has a strong preference for three-connected sites in sheet minerals.

Table 11 summarizes the substitutions involving Be for minerals in which there is solid solution involving Be. The majority of substitutions involve replacement of a cation differing from Be by one formal charge (i.e., Li^+, B^{3+} and Al^{3+}). With the exception of beryllian hellandite (for which there is a paucity of data), these solid solutions are extensive and fairly common. The effective radii of these cations in tetrahedral coordination are as follows: Li: 0.59 Å, Be: 0.27 Å, B: 0.11 Å. The differences between these radii are greatly in excess of the oft-quoted 15%, but no serious crystal chemist has considered this figure appropriate for decades. The key generalization is that *solid solutions tend to involve cations that differ in formal charge by one valence unit, and that can adopt the same coordination number.* It is also notable that all of these substitutions occur at 4-connected tetrahedra; this point is seemingly contradictory to the observation by Huminicki and Hawthorne (2001b) that Be prefers tetrahedrally coordinated sites that are 3-connected in sheet structures. A more general examination of this issue seems justified.

There are two substitutions that are not in accord with our general rule of substituents involving cations with a difference in formal charge of one valence unit. In beryllian hellandite, we have Be and Li substituting for a □, so the charge difference is

2^+. In "makarochkinite," Be substitutes for Si, again a charge difference of 2^+. Inspection of Table 11 shows that these substitutions in which there is a charge difference of 2^+ occur at tetrahedra with a connectivity of 3, whereas substitutions in which there is a charge difference of 1^+ occur at tetrahedra with a connectivity of 4. In khmaralite, Barbier et al. (1999) report Be at five tetrahedrally coordinated sites. However, the site occupancies for Be at two sites (T4 and T8) are low ($\leq$0.05) and hence cannot be considered as significant. Thus Be occurs at three sites, together with Al and Si, and hence it is not apparent whether Be is substituting for Al or Si. In surinamite, refinement of the structure indicates a site occupancy of 0.97 Be + 0.03 Si; the occupancy of this site by Si is probably not statistically significant.

Thus there seems to be some regularities in the occurrence of Be as a constituent of solid solutions (as summarized in Table 11), but the number of examples of this (as distinct from Al ↔ Si substitution, for example) is too small to be certain that these regularities will hold for other, as yet uncharacterized, solid solutions involving Be.

Table 11. Minerals with solid solutions involving Be, the principal substituents, and the connectivity of the site at which the solid solution occurs.

Mineral	*Substituents*	*Connectivity*
Beryl	Be ↔ Li	4
Milarite	Be ↔ Al	4
Beryllian cordierite	Be ↔ Al	4
(Li,Be)-mica	Be ↔ Al	3
Beryllian hellandite	Be ↔ G	3
Rhodizite	Be ↔ B	4
Hyalotekite	Be ↔ B	4
"Makarochkinite"	Be ↔ Si	3
Khmaralite	Be ↔ (Al,Si)	3

ACKNOWLEDGMENTS

We thank Sergey Krivovichev, Vanessa Gale and Editor Ed Grew for detailed reviews of this chapter. This work was supported by Natural Sciences and Engineering Research Council of Canada grants to FCH.

APPENDIX A

&

APPENDIX B

on the following four pages.

Appendix A. End-member formulae and crystallographic data for Be minerals of known structure.

Name	End-member formula	a (Å)	b (Å)	c (Å)	β (°)	Sp. Gr.	Ref.
aminoffite	$Ca_3[Be_2Si_3O_{10}(OH)_2]$	9.864(2)	*a*	9.930(2)	–	$P4_2/n$	(1)
asbecasite	$Ca_3Ti^{4+}[As_6Be_2S_2O_{20}]$	8.318(1)	*a*	15.264(2)	–	$P\bar{3}c1$	(2)
babefphite[1]	$Ba[BePO_4F]$	6.889(3)	16.814(7)	6.902(3)	90.3(1)	$F1$	(3)
barylite	$Ba[Be_2Si_2O_7]$	9.820(10)	11.670(10)	4.690(10)	–	$Pnma$	(4)
bavenite	$Ca_4[Be_2Al_2Si_9O_{26}(OH)_2]$	23.190(20)	5.005(9)	19.390(20)	–	$Cmcm$	(5)
bazzite	$Sc_2[Be_3Si_6O_{18}]$	9.510	*a*	9.110	–	$P6/mcc$	(6)
bearsite	$[Be_2AsO_4(OH)](H_2O)_4$	7.235(1)	12.686(2)	8.655(1)	98.4(0)	$P2_1/a$	(7)
behoite	$[Be(OH)_2]$	4.530(2)	4.621(2)	7.038(2)	–	$P2_12_12_1$	(8)
berborite-2*H*	$[Be_2BO_3(OH)](H_2O)$	4.433(2)	*a*	10.638(5)	–	$P6_3$	(9)
berborite-1*T*	$[Be_2BO_3(OH)](H_2O)$	4.434(1)	*a*	5.334(2)	–	$P3$	(9)
berborite-2*T*	$[Be_2BO_3(OH)](H_2O)$	4.431(1)	*a*	10.663(3)	–	$P3/c1$	(9)
bergslagite	$Ca[BeAsO_4(OH)]$	4.882(1)	7.809(1)	10.127(1)	90.2(0)	$P2_1/c$	(10)
bertrandite	$[Be_4Si_2O_7(OH)_2]$	8.716(3)	15.255(3)	4.565(1)		$Cmc2_1$	(11)
beryl	$Al_2[Be_3Si_6O_{18}]$	9.278(2)	*a*	9.195(2)		$P6/mcc$	(12)
beryllonite	$Na[BePO_4]$	8.178(3)	7.818(2)	14.114(6)	90.0(0)	$P2_1/n$	(13)
bityite	$CaLiAl_2[BeAlSi_2O_{10}](OH)_2$	5.058(1)	8.763(3)	19.111(7)	95.39(2)	Cc	(14)
bromellite	$[BeO]$	2.718(1)	*a*	4.408(1)	–	$P6_3mc$	(15)
"calcybeborosilite"	$CaY\square[BeBSi_2O_8(OH)_2]$	9.846(4)	7.600(2)	4.766(2)	90.1(0)	$P2_1/a$	(16)
chiavennite	$CaMn^{2+}[Be_2Si_5O_{13}(OH)_2](H_2O)_2$	8.729(5)	31.326(11)	4.903(2)	–	$Pnab$	(17)
chkalovite	$Na_2[BeSi_2O_6]$	21.129(5)	6.881(2)	21.188(5)	–	$Fdd2$	(18)
chrysoberyl	$Al_2[BeO_4]$	9.402(1)	5.475(0)	4.426(0)	–	$Pnma$	(19)
clinobehoite	$[Be(OH)_2]$	11.020(8)	4.746(6)	8.646(9)	98.9(1)	$P2_1$	(20)
danalite	$Fe^{2+}_8[Be_6Si_6O_{24}]S_2$	8.218(0)	*a*	*a*	–	$P\bar{4}3n$	(21)
ehrleite[2]	$Ca_2[ZnBe(PO_4)_2(PO_3OH)](H_2O)_4$	7.130(4)	7.430(4)	12.479(9)	102.1(0)	$P\bar{1}$	(22)
epididymite	$Na_2[Be_2Si_6O_{15}](H_2O)$	12.74(1)	13.63(1)	7.33(1)	–	$Pnma$	(23)
euclase	$Al[BeSiO_4OH]$	4.746(1)	14.189(11)	4.599(1)	100.2(0)	$P2_1/a$	(24)
eudidymite	$Na_2[Be_2Si_6O_{15}](H_2O)$	12.630(10)	7.380(10)	14.020(10)	103.7(1)	$C2/c$	(25)

Appendix A. cont.

Name	End-member formula	*a* (Å)	*b* (Å)	*c* (Å)	β (°)	Sp. Gr.	Ref.
fransoletite	$Ca_3[Be_2(PO_4)_2(PO_3\{OH\})_2](H_2O)_4$	7.348(1)	15.052(3)	7.068(1)	96.52(1)	$P2_1/a$	(26)
gadolinite-(Ce)	$Ce_2Fe^{2+}[Be_2Si_2O_{10}]$	4.82(2)	7.58(2)	10.01(3)	90.5(3)	$P2_1/c$	(27)
gadolinite-(*Y*)	$Y_2Fe^{2+}[Be_2Si_2O_{10}]$	4.747(1)	7.544(1)	9.931(1)	90.5(0)	$P2_1/c$	(28)
gainesite	$NaKZr_2[BeP_4O_{16}](H_2O)_{1.5}$	6.567(3)	*a*	17.119(5)	–	$I4_1/amd$	(29)
genthelvite	$Zn_8[Be_6Si_6O_{24}]S_2$	8.109(0)	*a*	8.109(0)	–	$P\bar{4}3n$	(21)
gugiaite	$Ca_2[BeSi_2O_7]$	7.419(1)	*a*	4.988(1)	–	$P\bar{4}2_1m$	(30)
hambergite	$[Be_2BO_3(OH)]$	9.754(2)	12.231(2)	4.434(1)	–	*Pbca*	(31)
harstigite	$Ca_6Mn^{2+}[Be_4Si_6O_{22}(OH)_2]$	9.793(2)	13.636(3)	13.830(3)	–	*Pnam*	(32)
helvite	$Mn^{2+}{}_8[Be_6Si_6O_{24}]S_2$	8.291(1)	*a*	*a*	–	$P\bar{4}3n$	(21)
herderite	$Ca[BePO_4F]$	–	–	–	–	–	(–)
"hingganite-(Y)"	$Y[BeSiO_4(OH)]$	9.930(6)	7.676(7)	4.768(3)	90.28(–)	$P2_1/a$	(33)
hingganite-(Yb)	$Yb[BeSiO_4(OH)]$	9.888(5)	7.607(3)	4.740(2)	90.4(0)	$P2_1/a$	(34)
høgtuvaite[3]	$Ca_2(Fe^{2+}{}_4Fe^{3+}{}_2)O_2[BeSi_5O_{18}]$	10.317(1)	10.724(1)	8.855(1)	92.21(1)	$P\bar{1}$	(35)
hsianghualite	$Ca_3Li_2[Be_3Si_3O_{12}]F_2$	12.864(2)	*a*	*a*	–	$I2_13$	(36)
hurlbutite	$Ca[Be_2P_2O_8]$	8.306(1)	8.790(1)	7.804(1)	89.5(0)	$P2_1/a$	(37)
hyalotekite[4]	$(Pb,Ba)_4Ca_2[(B,Be)_2(Si,B)_2Si_8O_{28}]F$	11.310(2)	10.955(2)	10.317(3)	90.02	$I\bar{1}$	(38)
hydroxylherderite	$Ca[BePO_4OH]$	9.789(2)	7.661(1)	4.804(1)	90.0(0)	$P2_1/a$	(39)
joesmithite	$Pb^{2+}Ca_2(Mg_3Fe^{3+}{}_2)[Be_2Si_6O_{22}](OH)_2$	9.915(2)	17.951(4)	5.243(1)	105.9(0)	$P2/a$	(40)
khmaralite	$Mg_7Al_9O_4[Al_6Be_{1.5}Si_{4.5}O_{36}]$	19.800(1)	14.371(1)	11.254(1)	125.53(1)	$P2_1/c$	(41)
leifite	$Na_6[Be_2Al_2Si_{16}O_{39}(OH)_2](H_2O)_{1.5}$	14.352(3)	*a*	4.852(3)	–	$P\bar{3}m1$	(42)
leucophanite	$CaNa[BeSi_2O_6F]$	7.401(2)	7.412(2)	9.990(2)	–	$P2_12_12_1$	(43)
liberite	$Li_2[BeSiO_4]$	4.680(20)	4.950(30)	6.130(20)	90.3(1)	*Pn*	(44)
londonite	$CsAl_4[Be_4(B_{11}Be)O_{28}]$	7.321(1)	*a*	*a*	–	$P\bar{4}3m$	(45)
lovdarite	$K_2Na_6[Be_4Si_{14}O_{36}](H_2O)_9$	39.756(1)	6.931(0)	7.153(0)	–	*Pma*2	(46)
"makarochkinite"[5]	$Ca_2(Fe^{2+}{}_4Fe^{3+}Ti^{4+})O_2[BeAlSi_4O_{18}]$	10.352(5)	10.744(3)	8.864(4)	96.2(0)	$P\bar{1}$	(47)
mccrillisite	$NaCs[BeZr_2(PO_4)_4](H_2O)_{1-2}$	6.573(2)	*a*	17.28(2)	–	$I4_1/amd$	(48)
meliphanite	$CaNa[BeSi_2O_6F]$	10.516(2)	*a*	9.887(2)	–	$I\bar{4}$	(49)

Appendix A. cont.

Name	End-member formula	*a* (Å)	*b* (Å)	*c* (Å)	β (°)	Sp. Gr.	Ref.
milarite	$KCa_2[Be_2AlSi_{12}O_{30}](H_2O)$	10.340(1)	*a*	13.758(2)	-	*P6/mcc*	(50)
minasgeraisite	$CaY_2[Be_2Si_2O_{10}]$	4.702(1)	7.562(1)	9.833(2)	90.46(6)	$P2_1/c$	(51)
moraesite	$[Be_2PO_4(OH)](H_2O)_4$	8.553(6)	12.319(6)	7.155(8)	97.9(1)	*C2/c*	(52)
magnesiotaaffeite-6*N*'3*S* *	$MgAl_5[BeMgAlO_{12}]$	5.682(1)	*a*	41.150(10)	-	$R\bar{3}m$	(53)
odintsovite	$K_2Na_4Ca_3Ti^{4+}{}_2O_2[Be_4Si_{12}O_{36}]$	14.243(3)	13.045(4)	33.484(6)	-	*Fddd*	(54)
pahasapaite	$Ca_8Li_8[Be_{24}P_{24}O_{96}](H_2O)_{38}$	13.783(1)	*a*	*a*	-	*I23*	(55)
parafransoletite[6]	$Ca_3[Be_2(PO_4)_2(PO_3\{OH\})_2](H_2O)_4$	7.327(1)	7.696(1)	7.061(1)	96.82(1)	$P\bar{1}$	(26)
ferrotaaffeite-6*N*'3*S* *	$Fe^{2+}Al_5[BeFe^{2+}AlO_{12}]$	5.70	*a*	41.16	-	$R\bar{3}m$	(56)
phenakite	$[Be_2SiO_4]$	12.485(3)	*a*	8.264(3)	-	$R\bar{3}$	(57)
rhodizite	$KAl_4[Be_4(B_{11}Be)O_{28}]$	7.319(1)	*a*	7.319(1)	-	$P\bar{4}3m$	(58)
roggianite	$Ca_2[Be(OH)_2Al_2Si_4O_{13}](H_2O)_{2.34}$	18.330(20)	*a*	9.160(10)	-	*I4/mcm*	(59)
roscherite	$Ca_2Mn^{2+}{}_5[Be_4P_6O_{24}(OH)_4](H_2O)_6$	15.88(4)	11.90(3)	6.66(3)	94.7(3)	*C2/c*	(60)
samfowlerite	$Ca_{14}Mn^{2+}{}_3[(Be_7Zn)Zn_2Si_{14}O_{52}(OH)_6]$	9.068(2)	17.992(2)	14.586(2)	104.9(0)	$P2_1/c$	(61)
selwynite	$NaK[BeZr_2(PO_4)_4](H_2O)_2$	6.570(3)	*a*	17.142(6)	-	$I4_1/amd$	(62)
semenovite	$(RE)_2Fe^{2+}Na_{0-2}(Ca,Na)_8[Be_6Si_{14}O_{40}(OH)_4F_4]$	13.879(5)	13.835(5)	9.942(6)	-	*Pmnn*	(63)
sorensenite	$Na_4Sn^{4+}[Be_2Si_6O_{18}](H_2O)_2$	20.698(17)	7.442(5)	12.037(11)	117.3(1)	*C2/c*	(64)
stoppaniite	$Fe^{3+}{}_2[Be_3Si_6O_{18}](H_2O)_2$	9.397(1)	*a*	9.202(2)	-	*P6/mcc*	(65)
surinamite	$Mg_3Al_3O[AlBeSi_3O_{15}]$	9.916(1)	11.384(1)	9.631(1)	109.3(0)	*P2/n*	(66)
sverigeite	$NaMn^{2+}{}_2Sn^{4+}[Be_2Si_3O_{12}(OH)]$	10.815(8)	13.273(8)	6.818(6)	-	*Imma*	(67)
swedenborgite	$NaSb^{5+}[Be_4O_7]$	5.470	*a*	8.920	-	$P6_3mc$	(68)
magnesiotaaffeite-2*N*'2*S* *	$MgAl_7[BeMg_2AlO_{16}]$	5.687(1)	*a*	18.337(3)	-	$P6_3mc$	(53)
tiptopite	$K_2Li_3Na_3[Be_6P_6O_{24}(OH)_2](H_2O)_4$	11.655(5)	*a*	4.692(2)	-	$P6_3$	(69)
trimerite	$CaMn^{2+}{}_2[BeSiO_4]_3$	8.098	7.613	14.065	90.0(0)	$P2_1/n$	(70)
tugtupite	$Na_8[BeAlSi_4O_{12}]_2Cl_2$	8.769	*a*	8.976	-	$I\bar{4}$	(71)
uralolite	$Ca_2[Be_4P_3O_{12}(OH)_3](H_2O)_5$	6.550(1)	16.005(3)	15.969(4)	101.6(0)	$P2_1/n$	(72)
väyrynenite	$Mn^{2+}[BePO_4(OH)]$	5.411(5)	14.490(20)	4.730(50)	102.8(1)	$P2_1/a$	(73)

Appendix A. cont.

Name	End-member formula	*a* (Å)	*b* (Å)	*c* (Å)	β (°)	Sp. Gr.	Ref.
weinebeneite	$Ca[Be_3P_2O_8(OH)_2](H_2O)_4$	11.897(2)	9.707(1)	9.633(1)	95.8(0)	*Cc*	(74)
welshite[7]	$Ca_2(Mg_4Sb^{5+}{}_2)O_2[Be_3Si_3O_{18}]$	10.381(1)	10.766(2)	8.881(1)	96.33(1)	$P\bar{1}$	(75)
zanazziite[8]	$Ca_2Mg_5[Be_4P_6O_{24}(OH)_4](H_2O)_6$	15.921(5)	11.965(4)	6.741(1)	94.3(1)	$C\bar{1}$	(76)

[1] α = 90.0, γ = 90.0°; [2] α = 94.3(0), γ = 82.7(0)°; [3] α = 105.77(1), γ = 124.77(1)°; [4] α = 90.43(2), γ = 90.16°;
[5] α = 105.7(0), γ = 124.9(0)°; [6] α = 94.90(1), γ = 101.87(1)°, [7] α = 105.92(1), γ = 124.97(1), [8] α = 91.1(1), γ = 90.0(1)°

* New nomenclature: magnesiotaaffeite-6*N*′3*S* ≡ musgravite-18*R*; ferrotaaffeite-6*N*′3*S* ≡ pehrmanite-18*R*; magnesiotaaffeite-2*N*′2*S* ≡ taaffeite-8*H*

References: (1) Coda et al. (1967), Huminicki and Hawthorne (2001b); (2) Sacerdoti et al. (1993); (3) Simonov et al. (1980); (4) Robinson and Fang (1977); (5) Cannillo et al. (1966), Kharitonov et al. (1971); (6) Armbruster et al. (1995); (7) Harrison et al. (1993); (8) Seitz et al. (1950), Stahl et al. (1998); (9) Giuseppetti et al. (1990); (10) Hansen et al. (1984); (11) Giuseppetti et al. (1992); (12) Artioli et al. (1995); (13) Giuseppetti and Tadini (1973); (14) Lin and Guggenheim (1983); (15) Hazen and Finger (1986); (16) Rastsvetaeva et al. (1996); (17) Tazzoli et al. (1995); (18) Simonov et al. (1975); (19) Pilati et al. (1993); (20) Nadezhina et al. (1989); (21) Hassan and Grundy (1985); (22) Hawthorne and Grice (1987); (23) Robinson and Fang (1970); (24) Hazen et al. (1986); (25) Fang et al. (1972); (26) Kampf (1992); (27) Segalstad and Larsen (1978); (28) Demartin et al. (1993); (29) Moore et al. (1983); (30) Kimata and Ohashi (1982); (31) Burns et al. (1995); (32) Hesse and Stuempel (1986); (33) Ximen and Peng (1985); (34) Yakubovich et al. (1983); (35) Grauch et al. (1994); (36) Rastsvetaevaet al. (1991); (37) Bakakin et al. (1974); (38) Christy et al. (1998); (39) Lager and Gibbs (1974); (40) Moore et al. (1993); (41) Barbier et al. (1999); (42) Coda et al. (1974); (43) Cannillo et al. (1967), Grice and Hawthorne (1989); (44) Chang (1966); (45) Simmons et al. (2001); (46) Merlino (1990); (47) Yakubovich et al. (1990); (48) Foord et al. (1994); (49) Dal Negro et al. (1967); (50) Hawthorne et al. (1991); (51) Foord et al. (1986); (52) Merlino and Pasero (1992); (53) Nuber and Schmetzer (1983); (54) Rastsvetsaeva et al. (1995); (55) Corbin et al. (1991); (56) Burke and Lustenhouwer (1981); (57) Tsirel'son et al. (1986); (58) Taxer and Buerger (1967), Pring et al. (1986); (59) Giuseppetti et al. (1991); (60) Fanfani et al. (1975, 1977); (61) Rouse et al. (1994); (62) Birch et al. (1995); (63) Mazzi et al. (1979); (64) Metcalf and Gronbaek (1977); (65) Ferraris et al. (1998); (66) Moore and Araki (1983); (67) Rouse et al. (1989); (68) Pauling et al. (1935), Huminicki and Hawthone (2001a); (69) Peacor et al. (1987); (70) Klaska and Jarchow (1977); (71) Hassan and Grundy (1991); (72) Mereiter et al. (1994); (73) Mrose and Appleman (1962), Huminicki and Hawthorne (2000); (74) Walter (1992); (75) Moore (1978), Grew et al. (2001); (76) Fanfani et al. (1977), Leavens et al. (1990).

Appendix B. End-member formulae and crystallographic data for Be minerals of unknown structure.

Name	*Formula*	*a*, Å	*b*, Å	*c*, Å	β °	*Sp. gr.*
Faheyite[1]	$(Mn,Mg,Na)[Be_2(PO_4)_4](H_2O)_6$	9.42(2)	9.42(2)	15.98(3)	-	$P6_422$
Jeffreyite[2]	$Ca_2[BeSi_2O_7]$	14.90(1)	14.90(1)	40.41(8)	-	$C222_1$
Tvedalite[3]	$(Ca,Mn)_4Be_3Si_6O_{17}(OH)_4(H_2O)_3$	8.724(6)	23.14(1)	4.923(4	-	*C****
Wawayandaite[4]	$Ca_{12}Mn_4B_2Be_{18}Si_{12}O_{46}(OH,Cl)_{30}$	15.59(2)	4.87(1)	18.69(4)	101.8(2)	*P2/c*

References: [1]Lindberg (1964); [2]Grice and Robinson (1984); [3]Larsen et al. (1992); [4]Dunn et al. (1990)

REFERENCES

Arguello CA, Rousseau DL, Porto SPS (1969) First-order Raman effect in wurtzite-type crystals. Phys Rev 181:1351-1363

Armbruster T (1986) Role of Na in the structure of low-cordierite: A single-crystal X-ray study. Am Mineral 71:746-757

Armbruster T (2002) Revised nomenclature of högbomite, nigerite and taaffeite minerals. Eur J Mineral (submitted)

Armbruster T, Irouschek A (1983) Cordierites from the Lepontine Alps: Na + Be → Al substitution, gas content, cell parameters, and optics. Contrib Mineral Petrol 82:389-396

Armbruster T, Libowitsky E, Diamond L, Auernhammer M, Bauerhansl P, Hoffmann C, Irran E, Kurka A, Rosenstingl H (1995) Crystal chemistry and optics of bazzite from Furkabasistunnel (Switzerland). Mineral Petrol 52:113-126

Artioli G, Rinaldi R, Wilson CC, Zanazzi PF (1995) Single-crystal pulsed neutron diffraction of a highly hydrous beryl. Acta Crystallogr B51:733-737

Aurisicchio C, Fioravanti G, Grubessi O, Zanazzi PF (1988) Reappraisal of the crystal chemistry of beryl. Am Mineral 73:826-837

Bakakin VV, Rylov GM, Belov NV (1969) Crystal structure of a lithium-bearing beryl. Dokl Acad Nauk SSSR 188:659-662 (in Russian)

Bakakin VV, Rylov GM, Alekseev VI (1974) Refinement of the crystal structure of hurlbutite $CaBe_2P_2O_8$. Kristallografiya 19:1283-1285 (in Russian)

Barbier J, Grew ES, Moore PB, Shu S-C (1999) Khmaralite, a new beryllium-bearing mineral related to sapphirine: A superstructure resulting from partial ordering of Be, Al, and Si on tetrahedral sites. Am Mineral 84:1650-1660

Barbier J, Grew ES, Yates MG (2001) Beryllium minerals related to aenigmatite. Geol Assoc Canada, Mineral Assoc Canada, Joint Ann Meet, Abstr 26:7

Baur WH (1977) Computer simulation of crystal structures. Phys Chem Minerals 2:3-20

Belov NV (1958) Essays on structural mineralogy IX. Mineral Sbornik Geol Soc Lvov 12:15-42 (in Russian)

Bentle GG (1966) Elastic constants of single-crystal BeO at room temperature. J Am Ceram Soc 49: 125-128

Beus AA (1960) Geochemistry of Beryllium and Genetic Types of Beryllium Deposits. Publ House Acad Sci, Moscow

Birch WD, Pring A, Foord EE (1995) Selwynite, $NaK(Be,Al)Zr_2(PO_4)_4 \cdot 2H_2O$, a new gainesite-like mineral from Wycheproof, Victoria, Australia. Can Mineral 33:55-58

Bowen NL (1928) The Evolution of Igneous Rocks. Princeton University Press, Princeton, New Jersey

Boyer LL, Mehl MJ, Feldman JL, Hardy JR, Flocken JW, Fong CY (1985) Beyond the rigid ion approximation with spherically symmetric ions. Phys Rev Lett 54:1940-1943

Bragg WL (1930) The structure of silicates. Z Kristallogr 74:237-305

Brown GE Jr, Mills BA (1986) High-temperature structure and crystal chemistry of hydrous alkali-rich beryl from the Harding pegmatite, Taos County, New Mexico. Am Mineral 71:547-556

Brown ID (1981) The bond-valence method: An empirical approach to chemical structure and bonding. *In* O'Keeffe M, Navrotsky A (eds) Structure and Bonding in Crystals. Academic Press, New York, 2:1-30

Brown ID, Shannon RD (1973) Empirical bond strength–bond length curves for oxides. Acta Crystallogr A29:266-282

Burdett JK, Hawthorne FC (1993) An orbital approach to the theory of bond valence. Am Mineral 78: 884-892

Burke EAJ, Lustenhouwer WJ (1981) Pehrmanite, a new beryllium mineral from Rosendal pegmatite, Kemiö Island, southwestern Finland. Can Mineral 19:311-314

Burns PC, Novak M, Hawthorne FC (1995) Fluorine-hydroxyl variation in hambergite: A crystal structure study. Can Mineral 22:1205-1213

Cannillo E, Coda A, Fagnani G (1966) The crystal structure of bavenite. Acta Crystallogr 20:301-309

Cannillo E, Giuseppetti G, Tazzoli V (1967) The crystal structure of leucophanite. Acta Crystallogr 23: 255-259

Cannillo E, Dal Negro A, Rossi G (1969) On the crystal structure of barylite. Rend Soc Ital Mineral Petrol 26:2-12

Cannillo E, Mazzi F, Fang JH, Robinson PD, Ohya Y (1971) The crystal structure of aenigmatite. Am Mineral 56:427-446

Cannillo E, Giuseppetti G, Mazzi F, Tazzoli V (1992) The crystal structure of a rare earth bearing leucophanite: $(Ca,RE)CaNa_2Be_2Si_4O_{12}(F,O)_2$. Z Kristallogr 202:71-79

Černý P, Povondra P (1966) Beryllian cordierite from Vežná: (Na,K) + Be → Al. N Jahrb Mineral Monatsh 1966:36-44

Chang HC (1966) Structural analysis of liberite. Acta Geol Sinica 46:76-86

Christy AG (1988) A new 2*c* superstructure in beryllian sapphirine from Casey Bay, Enderby Land, Antarctica. Am Mineral 73:1134-1137

Christy AG, Putnis A (1988) Planar and line defects in the sapphirine polytypes. Phys Chem Minerals 15:548-558

Christy AG, Grew ES, Mayo SC, Yates MG, Belakovskiy DI (1998) Hyalotekite, $(Ba,Pb,K)_4(Ca,Y)_2Si_8(B,Be)_2(Si,B)_2O_{28}F$, a tectosilicate related to scapolite: New structure refinement, phase transitions and a short-range ordered 3*b* superstructure. Mineral Mag 62:77-92

Cline CF, Dunegan HL, Henderson GW (1967) Elastic constants of hexagonal BeO, ZnS and CdSc. J Appl Phys 38:1944-1948

Coda A, Rossi G, Ungaretti L, Carobbi SG (1967) The crystal structure of aminoffite. Accademia Nazionale dei Lincei, Rend Classe Sci Fis, Mat Nat 43:225-232

Coda A, Ungaretti L, Guista AD (1974) The crystal structure of leifite, $Na_6Si_{16}Al_2(BeOH)_2O_{39}(H_2O)_{1.5}$. Acta Crystallogr B30:396-401

Cohen AJ, Gordon RG (1976) Modified electron-gas study of the stability, elastic properties and high-pressure and high-pressure behavior of MgO and CaO crystals. Phys Rev B14:4593-4605

Cohen JP, Ross FK, Gibbs GV (1977) An X-ray and neutron diffraction study of hydrous low cordierite. Am Mineral 62:67-78

Cohen RE, Boyer LL, Mehl MJ (1987) Theoretical studies of charge relaxation effects on the statics and dynamics of oxides. Phys Chem Minerals 14:294-302.

Corbin DR, Abrams L, Jones GA, Harlow RL, Dunn PJ (1991) Flexibility of the zeolite RHO framework: Effect of dehydration on the crystal structure of the beryllophosphate mineral, pahasapaite. Zeolites 11:364-367

Dal Negro A, Rossi G, Ungaretti L (1967) The crystal structure of meliphanite. Acta Crystallogr 23:260-264

Della Ventura G, Rossi P, Parodi GC, Mottana A, Raudsepp M, Prencipe M (2000) Stoppaniite, $(Fe,Al,Mg)_4(Be_6Si_{12}O_{36})\cdot(H_2O)_2(Na,\square)$, a new mineral of the beryl group from Latium (Italy). Eur J Mineral 12:121-127

Demartin F, Pilati T, Diella V, Gentile P, Gramaccioli CM (1993) A crystal-chemical investigation of alpine gadolinite. Can Mineral 30:127-136

Dempsey MJ, Strens RGJ (1976) Modelling crystal structures. *In* Strens RGJ (ed) Physics and Chemistry of Rocks and Minerals. Wiley, New York.

Downs JW, Gibbs GV (1981) The role of the BeOSi bond in the structures of beryllosilicate minerals. Am Mineral 66:819-826

Downs JW, Gibbs GV (1987) An exploratory examination of the electron density and electrostatic potential of phenakite. Am Mineral 72:769-777

Dunn PJ, Peacor DR, Grice JD, Wicks FJ, Chi PH (1990) Wawayandaite, a new calcium manganese beryllium boron silicate from Franklin, New Jersey. Am Mineral 75:405-408

Durovic S (1997) Fundamentals of the OD theory. Eur Mineral Union Notes in Mineralogy 1:3-28

Evans HT Jr, Mrose ME (1966) Crystal chemical studies of cesium beryl. Abstracts for 1966, Geol Soc Am Spec Paper 101:63

Fanfani L, Nunzi A, Zanazzi PF, Zanzari AR (1975) The crystal structure of roscherite. Tschermaks mineral petrol Mitt 22:266-277

Fanfani L, Zanazzi PF, Zanzari AR (1977) The crystal structure of a triclinic roscherite. Tschermaks mineral petrogr Mitt 24:169-178

Fang JH, Robinson PD, Ohya Y (1972) Redetermination of the crystal structure of eudidymite and its dimorphic relationship to epididymite. Am Mineral 57:1345-1354

Ferraris G, Principe M, Rossi P (1998) Stoppaniite, a new member of the beryl group: Crystal structure and crystal-chemical implications. Eur J Mineral 10:491-496

Foord EE, Gaines RV, Crock JG, Simmons WB Jr, Barbosa CP (1986) Minasgeraisite, a new member of the gadolinite group from Minas Gerais, Brazil. Am Mineral 71:603-607

Foord EE, Brownfield ME, Lichte FE, Davis AM, Sutley SJ (1994) McCrillsite, $NaCs(Be,Li)Zr_2[PO_4]_4\cdot 1-2\ H_2O$, a new mineral species from Mount Mica, Oxford County, Maine, and new data for gainesite. Can Mineral 32:839-842

Ganguli D (1979) Variabilities in interatomic distances and angles involving BeO_4 tetrahedra. Acta Crystallogr B35:1013-1015

Garber JA, Granato AV (1975a) Theory of the temperature dependence of second-order elastic constants in cubic materials. Phys Rev B 11:3990-3997

Garber JA, Granato AV (1975b) Fourth-order elastic constants and the temperature dependence of second-order elastic constants in cubic materials. Phys Rev B 11:3998-4007

Geisinger KL, Gibbs GV, Navrotsky A (1985) A molecular orbital study of bond length and angle variations in framework silicates. Phys Chem Minerals 11:266-283
Gibbs GV (1966) The polymorphism of cordierite: I. The crystal structure of low cordierite. Am Mineral 51:1068-1087
Gibbs GV (1982) Molecules as models for bonding in solids. Am Mineral 67:421-450
Gibbs GV, Breck DW, Meagher EP (1968) Structural refinement of hydrous and anhydrous synthetic beryl, $Al_2(Be_3Si_6)O_{18}$ and emerald $Al_{1.9}Cr_{0.1}(Be_3Si_6)O_{18}$. Lithos 1:275-285
Giuseppetti G, Tadini C (1973) Refinement of the crystal structure of beryllonite, $NaBePO_4$. Tschermaks mineral petrogr Mitt 20:1-12
Giuseppetti G, Mazzi F, Tadini C, Larsen AO, Asheim A, Raade G (1990) Berborite polytypes. N Jahrb Mineral Abh 162:101-116
Giuseppetti G, Mazzi F, Tadini C, Galli E (1991) The revised crystal structure of roggianite: $Ca_2[Be(OH)_2Al_2Si_4O_{13}]\cdot 2.5H_2O$. N Jahrb Mineral Monatsh 1991:307-314
Giuseppetti G, Tadini C, Mattioli V (1992) Bertrandite, $Be_4Si_2O_7(OH)_2$, from Val Vigezzo (NO) Italy: The X-ray structural refinement. N Jahrb Mineral Monatsh 1992:13-19
Gordon RG, Kim YS (1972) Theory for the forces between closed-shell atoms and molecules. J Chem Phys 56:3122-3133
Grauch RI, Lindahl I, Evans HT Jr, Burt DM, Fitzpatrick JJ, Foord EE, Graff P-R, Hysingjord J (1994) Høgtuvaite, a new beryllian member of the aenigmatite group from Norway, with new X-ray data on aenigmatite. Can Mineral 32:439-448
Grew ES (1981) Surinamite, taaffeite, and beryllian sapphirine from pegmatites in granulite-facies rocks of Casey Bay, Enderby Land, Antarctica. Am Mineral 66:1022-1033
Grew ES, Yates MG, Barbier J, Shearer CK, Sheraton JW, Shirashi K, Motoyoshi Y (2000) Granulite-facies beryllium pegmatites in the Napier complex in Khmara and Amundsen Bays, Western Enderby Land, East Antarctica. Polar Geosci 13:1-40
Grew ES, Hålenius U, Kritikos M, Shearer CK (2001) New data on welshite, e.g., $Ca_2Mg_{3.8}Mn^{2+}{}_{0.6}Fe^{2+}{}_{0.1}Sb^{5+}{}_{1.5}O_2[Si_{2.8}Be_{1.75}Fe^{3+}{}_{0.65}Al_{0.7}As_{0.17}O_{18}]$, an aenigmatite-group mineral. Mineral Mag 65:665-674
Grice JD, Hawthorne FC (1989) Refinement of the crystal structure of leucophanite. Can Mineral 27: 193-197
Grice JD, Robinson GW (1984) Jeffreyite, $(Ca,Na)_2(Be,Al)Si_2(O,OH)_7$, a new mineral species and its relation to the melilite group. Can Mineral 22:443-446
Griffen DT, Ribbe PH (1979) Distortions in the tetrahedral oxyanions of crystalline substances. N Jahrb Mineral Abh 137:54-73
Gupta A, Swanson DK, Tossell JA, Gibbs GV (1981) Calculation of bond distances, one-electron energies and electron density distributions in first-row tetrahedral hydroxy and oxyanions. Am Mineral 66: 601-609
Hansen S, Faelth L, Johnson O (1984) Bergslagite, a mineral with tetrahedral berylloarsenate sheet anions. Z Kristallogr 166:73-80
Harris LA, Yakel HL (1966) The crystal structure of calcium beryllate $Ca_{12}Be_{17}O_{29}$. Acta Crystallogr 20:296-301
Harris LA, Yakel HL (1967) The crystal structure of Y_2BeO_4. Acta Crystallogr 22:354-360
Harris LA, Yakel HL (1969) The crystal structure of $SrBe_3O_4$. Acta Crystallogr B25:1647-1651
Harrison WTA, Nenoff TM, Gier TE, Stucky GD (1993) Tetrahedral-atom 3-ring groupings in 1-dimensional inorganic chains: $Be_2AsO_4OH\cdot 4H_2O$ and $Na_2ZnPO_4OH\cdot 7H_2O$. Inorg Chem 32: 2437-2441
Hassan I, Grundy HD (1985) The crystal structure of helvite group minerals, $(Mn,Fe,Zn)_8(Be_6Si_6O_{24})S_2$. Am Mineral 70:186-192
Hassan I, Grundy HD (1991) The crystal structure and thermal expansion of tugtupite, $Na_8(Al_2Be_2Si_8O_{24})Cl_2$. Can Mineral 29:385-390
Hawthorne FC (1983a) Enumeration of polyhedral clusters. Acta Crystallogr A39:724-736
Hawthorne FC (1983b) The crystal chemistry of the amphiboles. Can Mineral 21:173-480
Hawthorne FC (1984) The crystal structure of stenonite and the classification of the aluminofluoride minerals. Can Mineral 22:245-251
Hawthorne FC (1985) Towards a structural classification of minerals: The ${}^{vi}M^{iv}T_2O_n$ minerals. Am Mineral 70:455-473
Hawthorne FC (1994) Structural aspects of oxide and oxysalt crystals. Acta Crystallogr B50:481-510
Hawthorne FC (1997) Short-range order in amphiboles: A bond-valence approach. Can Mineral 35: 201-216
Hawthorne FC, Černý P (1977) The alkali-metal positions in Cs-Li beryl. Can Mineral 15:414-421
Hawthorne FC, Grice JD (1987) The crystal structure of ehrleite, a tetrahedral sheet structure. Can Mineral 25:767-774

Hawthorne FC, Groat LA, Raudsepp M, Ercit TS (1987) Kieserite, a titanite-group mineral. N Jahrb Mineral Abh 157:121-132

Hawthorne FC, Kimata M, Černý P, Ball N, Rossman GR, Grice JD (1991) The crystal chemistry of the milarite-group minerals. Am Mineral 76:1836-1856

Hawthorne FC, Burns PC, Grice JD (1996) The crystal chemistry of boron. Rev Mineral 33:41-115.

Hazen RM, Finger LW (1986) High-pressure and high-temperature crystal chemistry of beryllium oxide. J Appl Phys 59:3728-3733

Hazen RM, Au AY, Finger LW (1986) High-pressure crystal chemistry of beryl ($Be_3Al_2Si_6O_{18}$) and euclase ($BeAlSiO_4OH$). Am Mineral 71:977-984

Hesse K-F, Stuempel G (1986) Crystal structure of harstigite, $MnCa_6Be_4(SiO_4)_2(Si_2O_7)_2(OH)_2$. Z Kristallogr 177:143-148

Hölscher A, Schreyer W (1989) A new synthetic hexagonal BeMg-cordierite, $Mg_2[Al_2BeSi_6O_{18}]$, and its relationship to Mg-cordierite. Eur J Mineral 1:21-37

Huminicki DMC, Hawthorne FC (2000) Refinement of the crystal structure of väyrynenite. Can Mineral 38:1425-1432

Huminicki DMC, Hawthorne FC (2001a) Refinement of the crystal structure of swedenborgite. Can Mineral 39:153-158

Huminicki DMC, Hawthorne FC (2001b) The crystal structure of aminoffite. Can Mineral (accepted)

Jephcoat AP, Hemley RJ, Hazen RM (1986) Bromellite: Raman spectroscopy to 40 Gpa. (abstr) EOS Trans Am Geophys Union 67:361

Kampf AR (1992) Beryllophosphate chains in the structures of fransoletite, parafransoletite, and ehrleite and some general comments on beryllophosphate linkages. Am Mineral 77:848-856

Kharitonov YuA, Kuz'min EA, Ilyukhin VV, Belov NV (1971) The crystal structure of bavenite. Zh Struktur Khimii 12:87-93 (in Russian)

Kimata M, Ohashi H (1982) The crystal structure of synthetic gugiaite, $Ca_2BeSi_2O_7$. N Jahrb Mineral Abh 143:210-222

Klaska KH, Jarchow O (1977) Die Bestimmung der Kristallstruktur von Trimerit $CaMn_2(BeSiO_4)_3$ und das Trimeritgesetz der Verzwillingung. Z Kristallogr 145:46-65

Koster AS (1971) Influence of the chemical bond on the K emission spectrum of oxygen and fluorine. J Phys Chem Solids 32:2685-2692

Lager GA, Gibbs GV (1974) A refinement of the crystal structure of herderite, $CaBePO_4OH$. Am Mineral 59:919-925

Larsen AO, Åsheim A, Raade G, Taftø J (1992) Tvedalite, $(Ca,Mn)_4Be_3Si_6O_{17}(OH)_4 \cdot 3H_2O$, a new mineral from syenite pegmatite in the Oslo Region, Norway. Am Mineral 77:438-443

Leavens PB, White JS, Nelen JA (1990) Zanazziite, a new mineral from Minas Gerais, Brazil. Mineral Rec 21:413-417

Lichanot A, Rerat M (1993) Elastic properties in BeO. An ab initio Hartree-Fock calculation. Chem Phys Lett 211:249-254

Lichanot A, Chaillet M, Larrieu C, Dovesi R, Pisani C (1992) Ab initio Hartree-Fock study of solid beryllium oxide: Structure and electronic properties. Chem Phys 164:383-394

Liebau F (1985) Structural Chemistry of Silicates. Springer-Verlag, Berlin

Liefeld RJ, Hanzely S, Kirby TB, Mott D (1970) X-ray spectrometric properties of potassium acid pthalate crystals. Adv X-ray Anal 13:373-381

Lin J-C, Guggenheim S (1983) The crystal structure of a Li,Be-rich brittle mica: A dioctahedral-trioctahedral intermediate. Am Mineral 68:130-142

Lindberg ML (1958) The beryllium content of roscherite from the Sapucaia pegmatite mine, Minas Gerais, Brazil, and from other localities. Am Mineral 43:824-838

Lindberg ML (1964) Crystallography of faheyite, Sapucaia pegmatite mine, Minas Gerais, Brazil. Am Mineral 49:395-398

Mazzi F, Galli E (1978) Is each analcime different? Am Mineral 63:448-460

Mazzi F, Ungaretti L, Dal Negro A, Petersen OV, Rönsbo JG (1979) The crystal structure of semenovite. Am Mineral 64:202-210

Mereiter K, Niedermayr G, Walter F (1994) Uralolite, $Ca_2Be_4(PO_4)_3(OH) \cdot 3.5(H_2O)$: New data and crystal structure. Eur J Mineral 6:887-896

Merlino S (1990) Lovdarite, $K_4Na_{12}(Be_8Si_{28}O_{72}) \cdot 18(H_2O)$, a zeolite-like mineral: Structural features and OD character. Eur J Mineral 2:809-817

Merlino S, Pasero M (1992) Crystal chemistry of beryllophosphates: The crystal structure of moraesite, $Be_2(PO_4)(OH) \cdot 4H_2O$. Z Kristallogr 201:253-262

Metcalf Johansen J, Gronbaek Hazell R (1976) Crystal structure of sorensenite, $Na_4SnBe_2(Si_3O_9)_2(H_2O)_2$. Acta Crystallogr B32:2553-2556

Moore PB (1965) A structural classification of Fe–Mn-orthophosphate hydrates. Am Mineral 50: 2052-2062

Moore PB (1969) The crystal structure of sapphirine. Am Mineral 54:31-49

Moore PB (1973) Pegmatite phosphates. Descriptive mineralogy and crystal chemistry. Mineral Rec 4: 103-130

Moore, PB (1978) Welshite, $Ca_2Mg_4Fe^{3+}Sb^{5+}O_2[Si_4Be_2O_{18}]$, a new member of the aenigmatite group. Mineral Mag 42:129-132

Moore PB, Araki T (1983) Surinamite, ca. $Mg_3Al_4Si_3BeO_{16}$: Its crystal structure and relation to sapphirine, ca. $Mg_{2.8}Al_{7.2}Si_{1.2}O_{16}$. Am Mineral 68:8804-8810

Moore PB, Araki T, Ghose S (1982) Hyalotekite, a complex lead borosilicate: Its crystal structure and the lone-pair effect of Pb(II). Am Mineral 67:1012-1020

Moore PB, Araki T, Steele IM, Swihart GH (1983) Gainesite, sodium zirconium beryllophosphate: A new mineral and its crystal structure. Am Mineral 68:1022-1028

Moore PB, Davis AM, Van Derveer DG, Sen Gupta PK (1993) Joesmithite, a plumbous amphibole revisited and comments on bond valences. Mineral Petrol 48:97-113

Mrose ME, Appleman DE (1962) The crystal structures and crystal chemistry of väyrynenite, $(Mn,Fe)Be(PO_4)(OH)$, and euclase, $AlBe(SiO_4)(OH)$. Z Kristallogr 117:16-36

Nadezhina TN, Pushcharovskii DY, Rastsvetaeva RK, Voloshin AV, Burshtein IF (1989) Crystal structure of a new natural form of $Be(OH)_2$. Dokl Akad Nauk SSSR 305:95-98 (in Russian)

Nuber B, Schmetzer K (1983) Crystal structures of ternary Be-Mg-Al oxides: Taaffeite, $BeMg_3Al_8O_{16}$, and musgravite, $BeMg_2Al_6O_{12}$. N Jahrb Mineral Monatsh 1983:393-402

Oberti R, Ottolini L, Camara F, Della Ventura G (1999) Crystal structure of non-metamict Th-rich hellandite-(Ce) from Latium (Italy) and crystal chemistry of the hellandite-group minerals. Am Mineral 84:913-921

Oberti, R, Della Ventura G, Ottolini L, Bonazzi P, Hawthorne FC (2002) Re-definition of hellandite based on recent single-crystal structure refinements, electron- and ion-microprobe analyses, and FTIR spectroscopy: A new unit-formula and nomenclature. Am Mineral (in press)

O'Keeffe M (1981) Some aspects of the ionic model of crystals. *In* O'Keeffe M, Navrotsky A (eds) Structure and Bonding in Crystals. Academic Press, New York, 1:299-322

Pauling LS (1929) The principles determining the structure of complex ionic crystals. J Am Chem Soc 51:1010-1026

Pauling L, Klug HP, Winchell AN (1935) The crystal structure of swedenborgite, $NaBe_4SbO_7$. Am Mineral 20:492-501

Peacor DR, Rouse RC, Ahn J-H (1987) Crystal structure of tiptopite, a framework beryllophosphate isotypic with basic cancrinite. Am Mineral 72:816-820

Petersen OV, Ronsbo JG, Leonardsen ES (1994) Hingganite-(Y) from Zomba-Malosa complex, Malawi. N Jahrb Mineral Monatsh 1994:185-192

Pilati T, Demartin F, Cariati F, Bruni S, Gramaccioli CM (1993) Atomic thermal parameters and thermodynamic functions for chrysoberyl ($BeAl_2O_4$) from vibrational spectra and transfer of empirical force fields. Acta Crystallogr B49:216-222

Povondra P, Langer K (1971) Synthesis and some properties of sodium-beryllium bearing cordierite, $Na_xMg_2(Al_{4-x}Be_xSi_5O_{18})$. N Jahrb Mineral Abh 116:1-19

Pring A, Din VK, Jefferson DA, Thomas JM (1986) The crystal chemistry of rhodizite: A re-examination. Mineral Mag 50:163-172

Rastsvetaeva RK, Rekhlova O Yu, Andrianov VI, Malinovskii Yu A (1991) Crystal structure of hsianghualite. Dokl Akad Nauk SSSR 316:624-628 (in Russian)

Rastsvetaeva RK, Evsyunin VG, Kashaev AA (1995) Crystal structure of a new natural K,Na,Ca-titanoberyllosilicate. Kristallografiya 40:253-257 (in Russian)

Rastsvetaeva RK, Pushcharovskii D Yu, Pekov IV, Voloshin AV (1996) Crystal structure of calcybeborosilite and its place in the datolite-gadolinite isomorphous series. Kristallografiya 41:235-239 (in Russian)

Robinson PD, Fang JH (1970) The crystal structure of epididymite. Am Mineral 55:1541-1549

Robinson PD, Fang JH (1977) Barylite, $BaBe_2Si_2O_7$: Its space group and crystal structure. Am Mineral 62: 167-169

Rouse RC, Peacor DR, Metz GW (1989) Sverigeite, a structure containing planar NaO_4 groups and chains of 3- and 4-membered beryllosilicate rings. Am Mineral 74:1343-1350

Rouse RC, Peacor DR, Dunn PJ, Su S-C, Chi PH, Yeates H (1994) Samfowlerite, a new CaMnZn beryllosilicate mineral from Franklin, New Jersey: Its characterization and crystal structure. Can Mineral 32:43-53

Sacerdoti M, Parodi GC, Mottana A, Maras A, Della Ventura G (1993) Asbecasite: Crystal structure refinement and crystal chemistry. Mineral Mag 57:315-322

Schlenker JL, Griffen DT, Phillips MW, Gibbs GV (1978) A population analysis for Be and B oxyanions. Contrib Mineral Petrol 65:347-350

Schreyer W, Gordillo CR, Werding G (1979) A new sodian-beryllian cordierite from Soto, Argentina, and the relationship between distortion index, Be content, and state of hydration. Contrib Mineral Petrol 70:421-428

Segalstad TV, Larsen AO (1978) Gadolinite-(Ce) from Skien, southwestern Oslo region, Norway. Am Mineral 63:188-195

Seitz A, Roesler U, Schubert K (1950) Kristallstruktur von $Be(OH)_2$-beta. Z Anorg Allgem Chem 261: 94-105

Shannon RD (1976) Revised effective ionic radii and systematic studies of interatomic distances in halides and chalcogenides. Acta Crystallogr A32:751-767

Sherriff BL, Grundy HD, Hartman JS, Hawthorne FC, Černý P (1991) The incorporation of alkalis in beryl: Multi-nuclear MAS NMR and crystal-structure study. Can Mineral 29:271-285

Simmons WB, Pezzotta F, Falster AU, Webber KL (2001) Londonite, a new mineral: The Cs-dominant analogue of rhodizite from the Antandrokomby granitic pegmatite, Madagascar. Can Mineral (accepted)

Simonov MA, Egorov-Tismenko YK, Belov NV (1975) Refined crystal structure of chkalovite $Na_2Be(Si_2O_6)$. Dokl Akad Nauk SSSR 225:1319-1322 (in Russian)

Simonov MA, Egorov-Tismenko YK, Belov NV (1980) Use of modern X-ray equipment to solve fine problems of structural mineralogy by the example of the crystal RE of structure of babefphite $BaBe(PO_4)F$. Kristallografiya 25:55-59 (in Russian)

Sirota NN, Kuz'mina AM, Orlova NS (1992) Debye-Waller factors and elastic constants for beryllium oxide at temperatures between 10 and 720 K. I. Anisotropy of ionic mean-square displacements. Crystallogr Res Techn 27:703-709

Slavík F (1914) Neue Phosphate vom Greifenstein bei Ehrenfriedersdorf. Ak Ceská, Bull–Bull intern ac sc Bohême 19:108-123

Sokolova EV, Ferraris G, Ivaldi G, Pautov LA, Khvorov PV (2000) Crystal structure of kapitsaite-(Y), a new borosilicate isotypic with hyalotekite—Crystal chemistry of the related isomorphous series. N Jahrb Mineral Monatsh 2000:74-84

Stahl R, Jung C, Lutz HD, Kockelmann W, Jacobs H (1998) Kristallstrukturen und Wasserstoffbrücken bindungen bei β-$Be(OH)_2$ und ε-$Zn(OH)_2$. Z anorg allg Chem 624:1130-1136

Taxer KJ, Buerger MJ (1967) The crystal structure of rhodizite. Z Kristallogr 125:423-436

Tazzoli V, Domeneghetti MC, Mazzi F, Cannillo E (1995) The crystal structure of chiavennite. Eur J Mineral 7:1339-1344

Tossell JA (1980) Calculation of bond distances and heats of formation for BeO, MgO, SiO_2, TiO_2, FeO and ZnO using the ionic model. Am Mineral 65:163-173

Tossell JA, Gibbs GV (1978) The use of molecular-orbital calculations on model systems for the prediction of bridging-bond-angle variations in siloxanes, silicates, silicon nitrides and silicon sulfides. Acta Crystallogr A34:463-472

Tossell JA, Vaughan DJ (1992) Theoretical Geochemistry. Oxford University Press, New York

Tsirel'son VG, Sokolova YV, Urusov VS (1986) An X-ray diffraction study of the electron-density distribution and electrostatic potential in phenakite Be_2SiO_4. Geokhimiya 8:1170-1180 (in Russian)

Vaughan DJ, Tossell JA (1973) Molecular orbital calculations on beryllium and boron oxyanions: Interpretation of X-ray emission, ESCA, and NQR spectra and of the geochemistry of beryllium and boron. Am Mineral 58:765-770

Walter F (1992) Weinebenite, $CaBe_3(PO_4)_2(OH)_2 \cdot 4H_2O$, a new mineral species: Mineral data and crystal structure. Eur J Mineral 4:1275-1283

Watson RE (1958) Analytic Hartree-Fock solutions for $O^{=}$. Phys Rev 111:1108-1110

Ximen L, Peng Z (1985) Crystal structure of hingganite. Acta Mineral Sinica 5:289-293 (in Chinese) [Am Mineral 73:441-442]

Yakubovich OV, Matvienko EN, Voloshin AV, Simonov MA (1983) The crystal structure of hingganite-(Yb), $(Y_{0.51}Ln_{0.36}Ca_{0.13})Fe_{0.065}Be(SiO_4)(OH)$. Kristallografiya 28:457-460 (in Russian)

Yakubovich OV, Malinovskii Yu A, Polyakov VO (1990) Crystal structure of makarochkinite. Kristallografiya 35:1388-1394 (in Russian)

Zoltai T (1960) Classification of silicates and other minerals with tetrahedral structures. Am Mineral 45: 960-973

10 Mineralogy of Beryllium in Granitic Pegmatites

Petr Černý

Department of Geological Sciences
University of Manitoba
Winnipeg, Manitoba, Canada R3T 2N2

p_cerny@umanitoba.ca

INTRODUCTION

Beryllium is one of the most widespread rare elements in granitic pegmatites. These rocks have been historically the sole industrial source of this metal (e.g., Norton et al. 1958), and they still contribute a significant proportion of the global output of beryllium ores. Hand-cobbed beryl constitutes a substantial proportion of beryllium ore concentrates in Africa, Asia and South America, although non-pegmatitic, rhyolite-related bertrandite ores are virtually the single source in North America (Petkof 1975).

The mineralogy of beryllium in granitic pegmatites is strongly diversified, but very "imbalanced" in terms of numbers of species per mineral class on one hand and of the paragenetic role, distribution and abundances on the other. Only a very few Be minerals form at the magmatic stage of pegmatite consolidation, with beryl absolutely dominant among them. Phosphates constitute most of the late subsolidus phases, with silicates a close second, but the number of the phosphate occurrences is very low and volumes are negligible. A few oxide, hydroxide and borate minerals complement the spectrum. So far, no other mineral classes are represented, although the occurrence of some arsenates is considered possible.

Part of the reason for the pattern above is the crystal-chemical behavior of Be, one of the classic amphoteric elements, which acts as a cation in acidic environments but participates in complex anions under alkaline conditions. Thus beryllo(alumino)silicates and beryllo-phosphates of alkali and alkaline-earth cations are widespread, in contrast to silicates or phosphates of beryllium with no other cations, or Al only. Be^{2+} is always tetrahedrally coordinated with oxygen $(BeO_4)^{6-}$, or with oxygen and hydroxyl $(BeO_3OH)^{5-}$.

The divalent charge on Be renders substitutions for other tetrahedrally coordinated oxycomplex-forming cations, such as Si, Al or P, difficult. It is possible only by charge-balancing *via* additional alkali or alkaline-earth cations in normally vacant sites, or *via* highly charged cations in adjacent polyhedra. However, cases of simple tetrahedral substitutions, such as $BeSiAl_{-2}$ in cordierite, seem to be rather exceptional. Also, the ionic properties of trace high-field-strength elements (HFSE), largely octahedrally coordinated, are not easily accommodated in the structures of most silicates and phosphates typical of granitic pegmatites. All of this limits the crystal-chemical roles of Be in granitic pegmatites, which typically are peraluminous to subaluminous in their bulk composition, relatively poor in highly charged cations, and carry only a few minerals with open structures rich in vacancies. In contrast, peralkaline environments enriched in, e.g., Ti, Zr or Nb generate a broad spectrum of HFSE- and Be-bearing phases, particularly silicates.

Many pegmatite minerals of Be have complex histories in the scientific literature, as beryllium has not always been recognized as their substantial component by the original describers. For example, the silicate milarite (Kuschel 1877) was initially characterized as an Al-bearing phase, and Be was detected half a century later (Palache 1931). The same is true for several other silicates (e.g., "pilinite" = bavenite from Strzegom, Poland) and some phosphates, such as roscherite. Application of emission spectrographic analysis was

1529-6446/00/0050-0010$05.00

particularly instrumental in unveiling the presence of Be, which was commonly included with Al in classical wet chemical analysis. Chances of unmasking additional new minerals of Be from among Al-rich species are quite poor today. However, our understanding of the crystal chemistry and substitution mechanisms of some (Al,Be)-bearing minerals is still far from satisfactory. Current progress is achieved mainly by application of SIMS to the analysis of Be and other light elements (e.g., Ottolini et al. 1993).

The following review of the mineralogy of Be in granitic pegmatites strongly reflects the idiosyncrasies of beryllium behavior and the imbalanced diversity of beryllium minerals, as well as the great disparities in the degree of our understanding of these phases.

CLASSIFICATION OF GRANITIC PEGMATITES

A brief review of the categories of granitic pegmatites is required here, as they will be referrred to through the text of this chapter. The terminology and definitions follow the classification proposed by Černý (1990, 1991a, 1992), rooted in part in the early scheme of Ginzburg et al. (1979).

Four geological classes of granitic pegmatites are distinguished: (i) abyssal pegmatites (segregations of anatectic leucosome in low- to high-pressure granulite to upper-amphibolite facies environments of ~700 to 800°C, ~4 to 9 kbar); (ii) deep-seated muscovite class (largely conformable and deformed bodies in high-pressure amphibolite-facies terranes with Barrovian kyanite-sillimanite succession, anatectic or granite-derived, characterized by ~580 to 650°C, ~5 to 8 kbar); (iii) moderate-depth rare-element class (granite-derived, quasi-conformable to crosscutting bodies in low-pressure amphibolite- to rarely upper-greenschist-facies environments of 500 to 650°C, ~2 to 4 kbar, with Abukuma-type andalusite-sillimanite succession); and (iv) shallow-level miarolitic class (vug-bearing pegmatitic facies or intrusive pegmatites in granites, and intrusive in schists, solidified at pressures as low as 1 kbar). Understandably enough, minerals of beryllium are most abundant, widespread and most diversified in the pegmatites of the rare-element class, and it is the pegmatites of this category that will be referred to almost exclusively throughout this paper.

Five types of rare-element pegmatites (in part split into subtypes) are defined by *mineral paragenesis and geochemical signature*, and in part by the conditions of solidification: (i) rare-earth type, with allanite and gadolinite subtypes (REE, Y, U, Th, Be, Nb>Ta, Ti, Zr, F); (ii) beryl type, with beryl-columbite and beryl-columbite-phosphate subtypes (Be, Nb-Ta, P, ± Sn, B, Li); (iii) complex type, with spodumene, petalite, lepidolite, amblygonite and elbaite subtypes (Li, Rb, Cs, Be, B, Sn, Ta>Nb, P, B, F); (iv) albite-spodumene type (Li, Sn, Be, Ta-Nb, B); and (v) albite type (Ta-Nb, Be, ± Li, Sn, B).

The rare-element pegmatites also are subdivided into *three major petrogenetic families*: (i) the LCT (lithium, cesium, tantalum) pegmatites generated by mildly to extremely peraluminous S- (to I-) granites, mainly in synorogenic to late-orogenic regimes; (ii) the NYF (niobium, yttrium, fluorine) pegmatites derived from subaluminous to metaluminous A- (to I-) granites, commonly but not exclusively in postorogenic to anorogenic settings; and (iii) the mixed-signature NYF-LCT pegmatites which bear the earmarks of both preceding categories and may be formed by several processes.

Considerations of beryllium mineralogy require a short note on two additional kinds of granitic pegmatites, which do not fit the principal categories of the above classification schemes. These are the Be-bearing pegmatites in high-grade granulite terrains studied by E.S. Grew, and the desilicated pegmatites introduced into the literature by A.E. Fersman.

Examination of localities of these two types requires utmost care and caution in interpretation.

Late-Archean granitic pegmatites in the granulite-facies terrain of the Napier complex, Antarctica, were extensively studied by E.S. Grew and associates (Grew 1981, 1998; Grew et al. 2000, and other references cited therein). Roughly corresponding to the abyssal class characterized above, these pegmatites differ by appreciable content of beryllium contained in sapphirine, surinamite, chrysoberyl and musgravite, and by distinct metamorphic overprints. Initially interpreted as possible derivatives of charnockitic magmatism, these pegmatites were recently relegated to products of regional-metamorphic anatexis (recruiting Be from metamorphic sapphirine and reprecipitating it as magmatic sapphirine and khmaralite), undergoing moderate-pressure cum high-temperature metamorphism (surinamite and khmaralite), and later decompression (generating beryllian cordierite) (Grew et al. 2000). The geochemical argument forwarded to support mobilization of Be from a (decidedly Be-poor) metamorphic protolith deserves re-consideration, in view of the recent experiments of Evensen et al. (1999) on beryl saturation of granitic melts. Nevertheless, the Antarctic localities seem to be the first examples of a rare-element pegmatite environment that survived high-grade metamorphism (>10 kbar) and retained the rare-element content (Be, and also B in related localities elsewhere; Grew 1998), as these and other components (Li, Rb, Cs) are usually mobilized and transported down the pressure gradient. Considering the complex history of these pegmatites after their initial magmatic consolidation, the controversial interpretations of their origin are not surprising.

Desilicated pegmatites are modified in their mineral modes and bulk chemical composition by more or less extensive reaction of their solidifying magma with basic to ultrabasic rocks, mainly serpentinites (Fersman 1929, 1960). Quartz-free plagioclase dikes with accessory corundum, rimmed by endocontact phlogopite and exocontact actinolite, anthophyllite and/or talc are typical products of this process, with emerald, chrysoberyl and/or phenakite in the mica-rich assemblages (see also Franz and Morteani, this volume). The emerald-bearing vein systems in the southern Ural Mountains of Russia (Emerald Mines; Vlasov and Kutukova 1960) were considered classic examples of desilicated granitic pegmatites, but Ginzburg et al. (1979) argued that these deposits resemble contaminated derivatives of greisen-type and quartz-vein mineralization. Also, Franz et al. (1986) and Grundmann and Morteani (1989; see also Herrmann 1992) interpreted the classic Habachtal locality and other occurrences in and outside of Austria as regional-metamorphic products. Nevertheless, a great number of other localities leave no doubt about their pegmatitic origin: Rila Mts. in the Rhodopes, Bulgaria (Arnaudov and Petrusenko 1971; Petrusenko and Arnaudov 1980), the Socoto mine at Salvador, Bahia (Cassedanne 1985) and other localities in Brazil (Pough 1936b), the Franqueira deposit in Spain (Martin-Izard et al. 1995, 1996; Franz et al. 1996), Haramosh Mts. in Pakistan (Laurs et al. 1996), and the Crabtree pegmatite of North Carolina (Tappen and Smith 1997), to name a few.

For further details and discussion of pegmatite classification, including examples of pegmatite bodies and populations of individual categories, the reader is referred to the papers quoted throughout this section. Of equal significance are the recent papers that expand on the above scheme (Novák and Povondra 1995; Wise 1999), or point out deficiencies of the current classification and problems to be solved: uncertainties in the poorly known abyssal class, a schism in the miarolitic category, the need for rational subdivision of the NYF pegmatites, and possibly for a new category additional to LCT and NYF (Buck et al. 1999; Kjellman et al. 1999; Martin 1999; Černý 2000). An independent, but in many respects closely comparable classification was offered recently by Zagorskyi et al. (1999).

SYSTEMATICS OF THE BERYLLIUM MINERALS FROM GRANITIC PEGMATITES

The following review is devoted to brief crystal-chemical characteristics and modes of occurrences of the beryllium minerals, with quotes of the principal references that provide further information on morphology, chemical composition, physical properties and diverse aspects of specific localities. The reader is referred to Hawthorne and Huminicki (this volume) for details on the crystal structures of individual minerals and on the structural systematics.

Oxides

Only a few minerals of this class are known from granitic pegmatites, but chrysoberyl is one of the relatively abundant primary phases. All oxide minerals are characteristic of relatively high-pressure environments.

Chrysoberyl, $BeAl_2O_4$, has the dense structure of olivine. Compositional variations are minimal, mainly restricted to low contents of Fe_2O_3 (≤6 wt %) and minor Cr_2O_3 (≤0.3 wt %). Chrysoberyl is relatively widespread in granitic pegmatites, although much less so than beryl. Experimental work and paragenetic relationships indicate that it may form at higher pressures and temperatures than some beryl (Franz and Morteani 1981), but the stability fields of both minerals extensively overlap. The relative rarity of chrysoberyl stems from its reaction with K-feldspar to beryl + muscovite at decreasing temperature and increasing $a(H_2O)$ (Evensen and London 1999). Besides its occurrences in run-of-the-mill, moderately differentiated, and relatively high-temperature pegmatites, chrysoberyl is favored in shear-stressed environments associated with sillimanite, and it also forms in phlogopite selvages of desilicated pegmatites and similar micaceous veins crosscutting ultrabasic rocks.

The first type of occurrence is amply documented in the literature (e.g., Black Hills of South Dakota, Moore and Ito 1973; Sierra Albarrana, Spain, Gonzales del Tánago 1991; southern Kerala state, India, Soman and Druzhinin 1987; see also Jacobson 1982). Effects of shear stress during or after consolidation of pegmatite bodies are preserved in the fabric of sillimanite-bearing assemblages at many localities of chrysoberyl (e.g., Nigeria, Jacobson and Webb 1948; "Central Asia", Rossovskii and Shostatskii 1964; Maršíkov in northern Moravia, Czech Republic, Dostál 1966, Černý et al. 1992; see the review by Heinrich and Buchi 1969). In this environment, chrysoberyl + quartz locally form at the expense of beryl.

Desilicated pegmatites with chrysoberyl, transecting peridotites or serpentinites, are documented from the Rila Mt. in the Rhodopes, Bulgaria (Arnaudov and Petrusenko 1971), the Socoto mine at Salvador, Bahia, Brazil (Cassedanne 1985), and the Franqueira deposit in Spain (Martin-Izard et al. 1995). Other deposits originally classified as desilicated pegmatites, such as the Emerald Mines in the Urals, Russia may have different, non-pegmatitic genesis (cf. "Classification of Granitic Pegmatites" above).

Taaffeite group of polytypoids comprises, using the symbols of McKie (1963), ***taaffeite****-4*H $Mg_3Al_8BeO_{16}$ (Anderson et al. 1951), ***musgravite****-9*R $(Mg,Fe^{2+},Zn)_2Al_6BeO_{12}$ (Hudson et al. 1967), and ***pehrmanite*** $(Fe^{2+},Zn,Mg)_2Al_6BeO_{12}$ (Burke and Lustenhouwer 1981). All three minerals are based on a layered structure with three types of polyhedral sheets (Fig. 8 of Hawthorne and Huminicki, this volume). In granitic pegmatites, taaffeite and musgravite are so far known only as products of metamorphism, and as such are dealt with by Grew (this volume).

Pehrmanite is known so far only from its type locality, the Rosendal pegmatite on Kemiö Island, southwestern Finland (Burke and Lustenhouwer 1981; cited as "unknown oxide" by Burke et al. 1977). However, it may be more widespread, as it is physically close to chrysoberyl. In the wall zone of its parent pegmatite, pehrmanite is associated with, i.a., muscovite, sillimanite, hercynite, nigerite and garnet.

Bromellite, BeO. Jacobson (1993) cites an unpublished report that claims the occurrences of bromellite as disseminated masses at several locations in the Mount Antero granite, Colorado. However, Jacobson expresses grave doubts about the identity of the mineral in these localities; bromellite is extremely scarce on global scale, known only from a few Al-poor, skarn-like environments and a syenite pegmatite (Larsen et al. 1987).

Hydroxides

Two very rare minerals are represented here, both of low-temperature and low-pressure origin.

Behoite (orthorhombic; Fig. 25 in Hawthorne and Huminicki, this volume) and ***clinobehoite*** (monoclinic; Fig. 16 in Hawthorne and Huminicki, this volume) are dimorphs of the same simple composition $Be(OH)_2$, and so far very rare minerals. Behoite was identified as an alteration product of gadolinite from the Rode Ranch pegmatite, Texas (Ehlmann and Mitchell 1970). In contrast, clinobehoite was discovered at Mursinsk, in the Ural Mountains of Russia, in hydrothermally altered zones of alleged desilicated pegmatites, associated with bavenite, bityite, analcime, phillipsite and albite (Voloshin et al. 1989).

Borates

Occurrences of Be-bearing borates in granitic pegmatites are restricted to three species, and one potential candidate. Paragenetically, they are typical of the boron-rich elbaite subtype. The structural and crystal-chemical aspects of these minerals also are discussed by Hawthorne et al. (1996) and by Hawthorne and Huminicki (this volume).

Hambergite. The first study of the $F(OH)_{-1}$ substitution in hambergite, $Be_2(BO_3)(OH,F)$, by Switzer et al. (1965) was recently expanded by Burns et al. (1995) and Novák et al. (1998). So far, the maximum F content does not significantly exceed 50 at.% of the (OH,F) site (Fig. 27 in Hawthorne and Huminicki, this volume). Unit-cell dimensions and optical properties show linear variations with the $F(OH)_{-1}$ substitution. In a few granitic pegmatites of the Czech Republic, hambergite was found as a primary component of the massive granitic border units, and of the inner graphic and blocky zones of K-feldspar and quartz (Novák et al. 1998). However, most of its infrequent occurrences are restricted to miarolitic cavities. Except for peralkaline pegmatites (including the type locality at Langesundsfjord, Norway; Brögger 1890; see also Semenov and Bykova 1965), the parent rocks correspond to granitic pegmatites of the elbaite (or transitional elbaite-lepidolite) subtype in Madagascar, California, Czech Republic and Pamir (Tajikistan) (e.g., Lacroix 1922; Ranorosoa 1986; Taylor et al. 1993; Novák et al. 1998; Dzhurayev et al. 1998).

Rhodizite series. Rhodizite, cubic $(K,Cs)Al_4Be_4(B,Be)_{12}O_{28}$ with minor Rb and Na (Fig.27 in Hawthorne and Huminicki, this volume), is a relatively rare mineral, identified to date from only four pegmatite districts worldwide, but rather abundant in Madagascar (Falster et al. 1999). This phase has long been misunderstood, mainly because faulty analytical data cited in early publications. Consequently, Pring et al. (1986) considered rhodizite a Cs-dominant phase, and they unofficially proposed a K-dominant analog as a new mineral, based on their work. More recently, a continuous substitution sequence from the potassic near-end-member to the Cs-dominant compositions was established for samples from Madagascar; subordinate Rb increases with the content of Cs (Fig. 1) (Falster et al. 1999). The recent IMA-approved proposal assigns the term rhodizite to the much more abundant K-based phase, and generates a new name for the mineral with dominant Cs: *londonite* $(Cs,K)Al_4Be_4(B,Be)_{12}O_{28}$ (Simmons et al. 2001). The parent pegmatites are invariably B-rich and all seem to be closely related to the elbaite subtype (Novák and Povondra 1995).

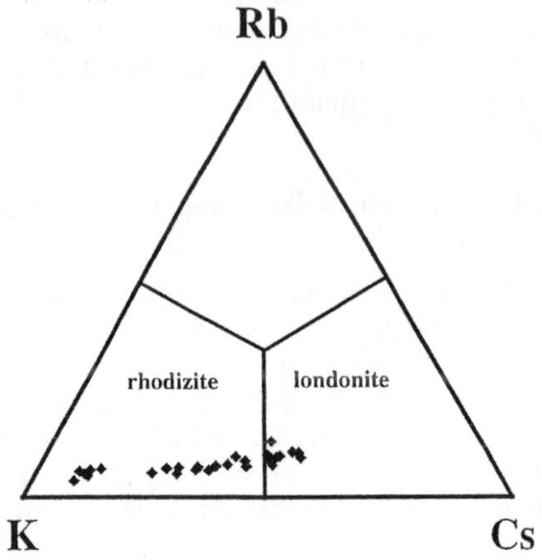

Figure 1. Minerals of the rhodizite-londonite series from Madagascar in the K-Cs-Rb diagram (from Simmons et al. 2000). Note the slight increase in Rb with increasing Cs.

Berborite. This mineral was so far found only in a skarn, associated with, e.g., hambergite, fluorite and helvite, in northwestern Russia (Nefedov 1967), and in syenite pegmatites (Giuseppetti et al. 1990). The chemical composition, $Be_2(BO_3)(OH,F)(H_2O)$, and the mineral assemblage at the type locality suggest that berborite may also be found in LCT granitic pegmatites, particularly those of the elbaite subtype (Novák and Povondra 1995).

Phosphates

The broad diversity of beryllium phosphates is not arranged here in any specific sequence, except the first half of the species involved, which belong in part to Hawthorne's and Huminicki's (this volume) phosphates with chains of Be-P linkages (A to F: roscherite to parafransoletite) and to phosphates with general Be-P linkages (G to K: hurlbutite to beryllonite). The crystal structure of some others (e.g., faheyite) is still not known. In general, all minerals considered here are beryllophosphates of a great variety of cations; moraesite is the only phase corresponding to a phosphate of beryllium (see discussion below in the introduction to silicates).

(A) Roscherite-group minerals, ***roscherite*** and ***zanazziite***, are reported to have a general composition of $Ca_2(Mn,Mg,Fe,Al)_5[Be_3(PO_4)_6(OH)_4{\cdot}6H_2O$, and are apparently dimorphous (roscherite-A and -M of Gaines et al. 1997). Probable compositional differences between the dimorphs have not yet been elucidated, and the Fe^{2+}-Mg-Mn^{2+} plus Fe^{3+}-Al systematics have not been properly defined. Additional complications include charge-balancing of variable sum of (Fe^{3+},Al) by OH substituting for H_2O (cf. Hawthorne and Huminicki, this volume, and Fig. 15 therein). So far, triclinic Mn^{2+}-Fe^{3+}-dominant roscherite is characterized (Fanfani et al. 1977), as is monoclinic Mg-Al^{3+}-dominant zanazziite (Leavens et al. 1990; Fanfani et al. 1975). The two polymorphs have not been distinguished in the literature dealing with the occurrences of roscherite-group minerals, and their chemical compositions are not usually specified. These minerals commonly form as alteration products of beryl, or as late crystals in miarolitic cavities. They are documented from Greifenstein in Saxony, from the Sapucaia pegmatite and Lavra da Ilha in Brazil, several localities in Maine and from the Foote lithium pegmatites in North

Carolina (e.g., Lindberg 1958; Cassedanne and Cassedanne 1973). Roscherite is commonly accompanied by other low-temperature secondary phosphates such as lacroixite, ježekite and childrenite.

(B) Herderite series consists of ***herderite*** $Ca[BePO_4F]$ and ***hydroxylherderite*** $Ca[BePO_4(OH)]$, with continuous solid solution between the two (Leavens et al. 1978), and has a close structural relationship to datolite (Strunz 1936, Lager and Gibbs 1974; Fig. 22 of Hawthorne and Huminicki, this volume). However, herderite with F>OH is known only from a single specimen (Dunn and Wight 1976), whereas hydroxylherderite is rather widespread. Very minor Fe, Mg or Mn may substitute for Ca, and minor Si is encountered, allegedly indicative of solid solution with ellestadite (?) (Gaines et al. 1997). Hydroxylherderite is not uncommon in granitic pegmatites of Maine as a member of albitic units (e.g. Landes 1925), and in the same paragenetic position in the Meldon "aplite", intergrown with beryllonite (von Knorring and Condliffe 1984). At Newry and in the BEP pegmatite of southeastern Manitoba, hydroxylherderite forms at the expense of beryllonite (Palache and Shannon 1928; Černá et al. 2002). However, hydroxylherderite also is widely distributed in pegmatites of Maine as a late phase in miarolitic cavities. Large vug-grown crystals are known from several pegmatites in Minas Gerais, Brazil (e.g., Dunn and Wight 1976; Dunn et al. 1979). There are minor occurrences in miarolitic cavities in a granite from the Fichtelgebirge, Germany (Strunz and Tennyson 1980). Alteration of beryl yields hydroxylherderite in the Norrö and Rånö pegmatites, Sweden (Nysten and Gustafsson 1993), in the Finnish Viitaniemi pegmatite (Volborth 1954; Lahti 1981), in Eastern Transbaikalia (Kornetova and Ginsburg 1961) and in some New England pegmatites (Yatsewitch 1935; Perham 1964). Secondary veinlets of hydroxylherderite were reported from the Forcarei Sur pegmatite field in Spain (Fuertes-Fuente and Martin-Izard 1998). Additional occurrences are listed in Burt (1975b). Understandably enough, the OH-dominant members of the series are widespread in late, low-temperature assemblages, but F-rich compositions may occur as primary phases generated in higher-temperature environments.

(C) ***Väyrynenite,*** $Mn^{2+}[BePO_4(OH,F)]$, is structurally related to the silicate euclase (Mrose and Appleman 1962; Fig. 13 of Hawthorne and Huminicki, this volume). A rare phosphate, it was discovered in the Viitaniemi pegmatite, Finland (Volborth 1954) and has been described from two related Finnish localities (Lahti 1981), from miarolitic pegmatites of Pakistan and Afghanistan (Meixner and Paar 1976), and from the Norrö dike, Sweden (Nysten and Gustafsson 1993; see also Mrose and von Knorring 1959). In the Finnish localities, väyrynenite seems to be a product of alteration of beryl, associated with amblygonite-montebrasite and commonly rimmed by chlorite. In the Pakistani pegmatites, väyrynenite crystallized in miarolitic cavities, whereas at Norrö it participates in alteration products after beryl.

(D) ***Moraesite***, $Be_2[PO_4(OH)]{\cdot}4H_2O$, is one of the numerous phosphate minerals discovered in the Sapucaia pegmatite, Minas Gerais, Brazil as a hydrothermal breakdown product of beryl and triphylite, in association with other phosphates (Lindberg et al. 1953; for crystal structure see Fig. 14 in Hawthorne and Huminicki, this volume). Also known from other Brazilian localities, such as the Humaita and Mulundu pegmatites in the Itinga-Taquaral district, where it formed with frondelite in aggregates of albite, quartz and muscovite at the expense of beryl (Cassedanne and Cassedanne 1985). A notable locality is reported from Eastern Transbaikalia (Malaya Kalenda; Kornetova 1959, Pekov 1994). Additional occurrences are located in New England at Newry and North Groton, Maine, and at Londonderry, Australia.

(E) ***Fransoletite,*** $Ca_3[Be_2(PO_4)_2(PO_3OH)_2]{\cdot}4H_2O$ is known only from its type locality, the Tip Top pegmatite in the Black Hills of South Dakota (Peacor et al. 1983). This monoclinic phase is closely associated with altered beryl, and with other secondary minerals such as tiptopite, whitlockite, montgomeryite, carbonate-apatite, hurlbutite, roscherite and englishite, but it formed very late in the crystallization sequence. The crystal structure of fransoletite is shown in Figure 15 of Hawthorne and Huminicki (this volume).

(F) ***Parafransoletite***, $Ca_3[Be_2(PO_4)_2(PO_3OH)_2]\cdot 4H_2O$, is the triclinic dimorph of fransoletite, which was identified as a new mineral from the same locality as the fransoletite quoted above (Kampf et al. 1992). Parafransoletite was found on fracture surfaces in beryl, along with roscherite, montgomeryite, robertsite, mitridatite, whitlockite and englishite; parafransoletite is again a very late phase among these secondary minerals.

(G) ***Hurlbutite***, $Ca[Be_2(PO_4)_2]$, is an orthorhombic mineral of simple composition, with negligible substitutions of other cations and minimal, if any, deviations from ideal stoichiometry (see Fig. 39 of Hawthorne and Huminicki, this volume, for crystal structure). Hurlbutite has apparently a broad crystallization span in granitic pegmatites. Mrose (1952), Staněk (1966), and Cempírek et al. (1999) provide compelling evidence that hurlbutite can participate in the tail-end of magmatic consolidation, together with blocky K-feldspar, albite, muscovite and quartz; the three localities involved here are the Smith Mine in New Hampshire, Kostelní Vydří and (probably) Otov in Czech Republic, respectively. Otherwise, it is well known as an alteration product of beryl in, e.g., the Finnish Viitaniemi pegmatite and Norrö, Sweden (Volborth 1954, Nysten and Gustafsson 1993). Additional occurrences are listed by Burt (1975b)

(H) ***Pahasapaite***, $(Ca,Li,K,Na,\square)_8Li_8[Be_{24}(PO_4)_{24}]\cdot 38H_2O$ (Fig. 38 of Hawthorne and Huminicki, this volume), is yet another beryllium phosphate defined from the Tip Top pegmatite in South Dakota (Rouse et al. 1987). Continuous loss of water between 70 and 500°C indicates the zeolitic nature of this cubic phase, which is structurally related to a synthetic Al,Si-zeolite rho. Pahasapaite is known only from its type locality, in fractured crystals of beryl, associated with tiptopite, roscherite, englishite, eosphorite-childrenite and montgomeryite.

(I) ***Tiptopite***, ideally $K_2(Li,Na,Ca,\square)_{5\text{-}6}[Be_6P_6O_{24}(OH)_4].4H_2O$, is hexagonal, structurally related to cancrinite (Grice et al. 1985, Peacor et al. 1987; Fig. 38 in Hawthorne and Huminicki, this volume). Two assemblages host tiptopite in the outer intermediate zone of the Tip Top pegmatite in South Dakota. Both consist of low-temperature secondary phosphates and both contain whitlockite, montgomeryite, englishite, hurlbutite, "roscherite" of different colors, whiteite and fairfieldite. One assemblage also carries fransoletite and robertsite-mitridatite. The other has eosphorite-childrenite and "another new mineral". It is not known from any other locality.

(J) ***Weinebenite***, $Ca[Be_3(PO_4)_2(OH)_2]\cdot 5H_2O$, is known only from its type locality, the Weinebene pegmatites in Carinthia, Austria (Walter et al. 1990). Weinebenite has a tetrahedral beryllophosphate framework, with three-membered rings of tetrahedra, and Ca plus H_2O in open cages of the structure (Walter 1992; Fig. 39 in Hawthorne and Huminicki, this volume). It occurs associated with fairfieldite, roscherite and uralolite in fractures of spodumene-rich pegmatites.

(K) ***Beryllonite***, $Na[BePO_4]$, has quasi-ditrigonal rings of Be- and P-populated tetrahedra with Na in channels (Fig. 37 in Hawthorne and Huminicki, this volume). Paragenetically, beryllonite evidently forms under a broad range of conditions. In the Viitaniemi pegmatite, beryllonite is a primary phase associated with nodules containing other Be-phosphates, or with albitic assemblages that also carry apatite, topaz, elbaite and micas (Lahti 1981). Some of its numerous occurrences in New England pegmatites also belong to these two paragenetic types (Mrose 1952; Moore 1973, 1982; King and Foord 1994). Beryllonite intergrown with herderite at Meldon, England in a late but primary assemblage of elbaite, lepidolite and quartz belongs to the second paragenetic type (von Knorring and Condliffe 1984). On the other hand, crystals in miarolitic cavities are known from New England, from Afghanistan and several pegmatites in Brazil (Cassedanne and Cassedanne 1973), and from leaching cavities as well (Mrose 1952). A fourth but apparently rare mode of occurrence is in pseudomorphs after beryl: beryllonite was observed intergrown with hurlbutite in such association at Norrö, Sweden (Nysten and Gustafsson 1993).

"Glucine", allegedly $CaBe_4(PO_4)_2(OH)_4{\cdot}0.5H_2O$, was described by Grigor'yev (1963) from a mica-fluorite greisen in Ural Mountains. Dunn and Gaines (1978) consider "glucine" to be [impure?] beryllonite, de Fourestier (1999) treats it as a poorly defined species, and Blackburn and Dennen (1997) do not list it at all. However, Anthony et al. (2000) reported it as a valid species. They also list an additional occurrence at Mt. Mica, Maine, in association with mitridatite, moraesite, siderite, tourmaline and albite, presumably in a granitic pegmatite.

Faheyite, $(Mg,Mn^{2+},Na)Fe^{3+}{}_2[Be_2(PO_4)_4]{\cdot}6H_2O$, is a late secondary mineral that follows muscovite, quartz, variscite and frondelite at its type locality, the Sapucaia pegmatite in Minas Gerais, Brazil (Lindberg and Murata 1953; Lindberg et al. 1953). It was also found in the Noumas pegmatite in Namaqualand, South Africa (Gaines et al. 1997), and with strengite in the Roosevelt pegmatite mine of the Black Hills, South Dakota.

Ehrleite, $Ca_2[ZnBe(PO_4)_2(PO_3OH)]{\cdot}4H_2O$, is a very rare triclinic phosphate, known to date only from its type locality, the Tip Top pegmatite in the Black Hills of South Dakota (Robinson et al. 1985). The formula and unit cell were re-defined by Hawthorne and Grice (1987; see Fig. 23 in Hawthorne and Huminicki, this volume, for crystal structure). It was found closely associated with an assemblage of secondary low-temperature phosphates: roscherite, hydroxyl-herderite, mitridatite, parascholzite and goyazite-crandallite.

Uralolite, $Ca_2[Be_4(PO_4)_3(OH)_3]{\cdot}5H_2O$ (Fig. 23 in Hawthorne and Huminicki, this volume), is a rare secondary phosphate, originally described as associated with beryl, "glucine" (= beryllonite; see Dunn and Gaines 1978) and moraesite from a skarn in the Ural Mountains of Russia (Grigor'yev 1964), but subsequently discovered in several granitic pegmatites. In the Dunton quarry, Maine,uralolite is the last member of an alteration sequence formed at the expense of beryllonite, after hydroxyl-herderite and in part coeval with ferroan roscherite (Dunn and Gaines 1978); also found at Taquaral and Galilea, Minas Gerais, Brazil, and with other Be-phosphates in late fractures of the Weinebene pegmatites, Austria (Mereiter et al. 1994).

Gainesite group has a general formula of $Na(K,Cs,Na,Rb)Zr_2[(Be,Li)(PO_4)_4{\cdot}1\text{-}2H_2O$, with differences among the members defined by populations of the second, larger alkali site. Besides the Na-dominant gainesite, K-dominant selwynite and Cs-dominant mccrillisite, Foord et al. (1994) suggested the possible existence of a Rb-dominant phase. A Ca-dominant mineral also could be expected, charge-balanced by vacancies or Li > Be. The tetragonal zeolite-like structure is based on BeP_4O_{16} clusters; cation sites are commonly only partially occupied (Fig. 9 of Hawthorne and Huminicki, this volume). Generally waterclear minerals, but gainesite and selwynite may display lavender to deep purple-blue color, presumably because of small amounts of Mn^{3+} or Mn^{4+}, or both, substituting for Zr.

Gainesite, $NaNa(Be,Al)Zr_2(PO_4)_4{\cdot}2H_2O$, was originally described by Moore et al. (1983) as an anhydrous mineral but recent analyses by Foord et al. (1994) established a full compliance of the formula with those of other members of the group. Gainesite was found in crevices of cleavelandite, associated with roscherite and eosphorite at Plumbago Mt. in

Maine.

Mccrillisite, $NaCsZr_2[(Be,Li)(PO_4)_4]\cdot 1\text{-}2H_2O$, is so far known only from its type locality at Mount Mica, Maine (Foord et al. 1994). This late-stage mineral is associated with about twenty other silicate, oxide, carbonate, arsenide and phosphate minerals, most of them products of hydrothermal alteration of earlier phases. Associated late phosphates include eosphorite, moraesite, Fe-rich roscherite and kosnarite. Significantly, corroded zircon is present among the altered preexisting minerals.

Selwynite, $NaKZr_2[(Be,Al)(PO_4)_4]\cdot 2H_2O$, was described by Birch et al. (1995) from small cavities in a muscovite- and schorl-bearing pegmatite at Wycheproof, Australia, closely associated with wardite and eosphorite. Also present are, i.a., rockbridgeite and kosnarite.

Babefphite. At first glance, this is an unlikely candidate for mineral assemblages of granitic pegmatites, because of its composition of $Ba[BePO_4F]$. It was discovered in Nb,W,Ti,Be-bearing fluorite deposits in Siberia. Nevertheless, enrichment of late subsolidus assemblages in Ba is known from some granitic pegmatites (e.g., London and Burt 1982; Teertstra et al. 1995); given sufficient activity of F, alteration of beryl in the presence of phosphates may lead to stabilization of this phase.

Arsenates

Arsenates of beryllium are so far not known from granitic pegmatites. However, loellingite and arsenopyrite are commonly found in granitic pegmatites of intermediate to advanced levels of geochemical evolution, and both are routinely found altered to a variety of arsenates (such as scorodite, arseniosiderite or pharmacosiderite) under hydrothermal and supergene conditions. If beryl associated with löllingite or arsenopyrite were to suffer alteration, two beryllium arsenates may be expected to form.

Bearsite, $Be_2[AsO_4(OH)]\cdot 4H_2O$, the arsenian analog of the phosphate moraesite, was found in the oxidation zone of a polymetallic deposit at Bota Burum, Kazakhstan (Kopchenova and Sidorenko 1962; Pekov 1998).

Bergslagite, $Ca[BeAsO_4(OH)]$, isostructural with the phosphate hydroxylherderite (Hawthorne and Huminicki, this volume), was found to date only at its type locality at Långban, in the Bergslagen region of Sweden (Hansen et al. 1984a,b), at two Swiss localities, and in altered rhyolite in the Spessart Mts. of Bavaria (Kolitsch 1996).

Beryllium-bearing silicates

Silicates with beryllium are currently summarily treated as beryllosilicates and aluminoberyllosilicates because of the identical fourfold coordination of Be, Si and (at least in part) Al. The tetrahedral sites are considered equal from the viewpoint of the architecture of the structures, disregarding the cations involved. If fixed differences in tetrahedral populations are encountered, they are interpreted as total ordering of the cations involved (e.g., Be and Si in phenakite Be_2SiO_4), although disordered counterparts are not known, and are probably impossible. This approach is understandable and justified from the viewpoint of *descriptive hierarchy of crystal architecture.* However, such treatment conceals some facts of crystal chemistry, significant from the genetic viewpoint.

Defining phenakite as a beryllosilicate creates a terminology problem, as this term (et similia) designates an anion, and it triggers a question—beryllosilicate of what? Chemically speaking, a designation of phenakite as a double oxide would be closer to home than beryllosilicate. However, with Be "ordered" in a specific tetrahedral site, phenakite may be best interpreted as a silicate of beryllium.

Amphoteric Be acts as an independent cation in acidic environment, but it enters complex anions in alkaline conditions. Consequently, Be generates silicates of Be ± Al (with Be clearly ordered outside the Si-populated tetrahedra) on one hand, and beryllosilicates or beryllo-aluminosilicates (of alkalis, alkaline earths or transition metals) on the other. This distinction is not always easy to make: e.g., ideal beryl corresponds to the first category but alkali enriched, Fe-, Mg- and Sc-bearing members of the beryl group fall into the second. Nevertheless, the distinction is attempted here, as it is helpful in genetic interpretation of beryllium minerals in subsequent chapters.

Silicates of beryllium

Phenakite, $Be_2[SiO_4]$ (Fig. 28 in Hawthorne and Huminicki, this volume), is present in granitic pegmatites in three strikingly different mineral associations. The most widespread type of occurrence is in miarolitic cavities, as a relatively late phase. Mount Antero in Colorado and other A-type granites with NYF-family pegmatites in Idaho are typical examples (Switzer 1939; Boggs 1986; Jacobson 1993), as are many LCT-related localities in Brazil (Cassedanne 1985). Alteration of beryl also generates phenakite, but rarely; notable occurrences were documented by Jakovleva (1961), Orlov et al. (1961), Pough (1936a), Mårtensson (1960) and Jonsson and Langhof (1997). Last but not least, phenakite is found along with greenish beryl to emerald and chrysoberyl in the phlogopite-rich endocontacts of desilicated pegmatites near Salvador, Bahia (Cassedanne 1985) and other localities in Brazil (Pough 1936a), and in the Franqueira deposit in Spain (Martin-Izard et al. 1995). Other localities initially claimed to be of this origin, such as the classic Russian emerald deposits in the Urals, were relegated to supercritical/hydrothermal and/or metamorphic processes, (cf. "Classification of Granitic Pegmatites").

Euclase, $BeAl[SiO_4](OH)$, is isostructural with the phosphate väyrynenite (Fig. 13 in Hawthorne and Huminicki 2000). Euclase is relatively widespread in alpine veins, greisens, emerald-bearing veins and other hydrothermal assemblages but rather scarce in granitic pegmatites. As in other environments, it is a low-temperature mineral formed at the expense of beryl, or in miarolitic cavities. Notable localities of the first type include the Evje-Iveland pegmatite field, southern Norway (Strand 1953) and in central Sweden (Jonsson and Langhof 1997), whereas the second style of occurrence is typical of Königshein (Hille 1990) and other localities in Bavaria (Durrfeld 1910), and some occurrences in Brazil (e.g., Graziani and Guidi 1980) and Zimbabwe (Stockmayer 1998).

Bertrandite. The structure of this orthorhombic sorosilicate, $Be_4[Si_2O_7](OH)_2$, (Ito and West 1932) was more recently refined by Hazen and Au (1986) and Downs and Ross (1987; see Fig. 33 in Hawthorne and Huminicki, this volume). The chemical composition of bertrandite is rather simple and constant, no significant substitutions have been detected. Bertrandite from non-pegmatitic deposits is attaining the status of the chief ore mineral of Be in North America and on global scale (Petkof 1975). In granitic pegmatites, bertrandite is relatively widespread as an alteration product of beryl, helvite and danalite (e.g., Černý 1968a, Lahti 1981; Novák et al. 1991; Jonsson and Langhof 1997), and less commonly as a low-temperature, cavity-lining phase with no evident Be-bearing precursors (e.g., at Mt. Antero, Colorado, Switzer 1939; Jacobson 1993).

It should be noted that Hsu (1983) synthesized an F-dominant bertrandite, which has not been so far discovered as a natural mineral; indeed, there is apparently no information available about the F content of bertrandite from granitic pegmatites or other environments.

Beryllosilicates and berylloaluminosilicates

Beryl group achieved its present status with the discovery of *stoppaniite* (Ferraris et al. 1998, Della Ventura et al. 2000), which added a third member to the previous *beryl-bazzite* series.

Beryl, ideally a silicate of Be and Al, ${}^{T(2)}Be_3{}^{O}Al_2[{}^{T(1)}Si_6O_{18}]$, is by many orders of magnitude the most abundant and widespread mineral of Be in granitic pegmatites of virtually all categories, and in most cases also the source of Be for a great variety of secondary Be-bearing minerals. The hexagonal structure of beryl has served as the prototype of ring silicates, with vertically stacked six-membered rings of Si-bearing tetrahedra *T*(1) linked laterally and vertically by Be-bearing tetrahedra *T*(2) and Al-populated octahedra *O*. Prominent hollow channels *C* run through the centers of the six-membered rings (Fig. 36 in Hawthorne and Huminicki, this volume).

Crystal chemistry. The chemical composition and structure of most beryl samples are, however, more complicated. Whereas Si in *T*(1) seems to be virtually "untouchable", extensive substitutions affect the Be- and Al- polyhedra and vacancies in the channels. Lithium and minor vacancies can substitute for Be, charge-balanced by Na, Cs and rarely Ca in the channels (Belov 1958, Bakakin et al 1969). The ionic radius of Na^+ fits the interior of the tetrahedral $[Si_6O_{18}]^{12-}$ rings, whereas Cs^+ is accomodated between the centers of these rings; the intermediate r_i of K^+ and Rb^+ excludes their entry into the channels in any significant quantities. The octahedral site accomodates Fe^{3+}, Sc, Cr, Mg, Mn, Fe^{2+} and possibly Ca, with alkalis in the channels compensating for the electrostatic charge of the divalent cations.

In addition to the alkalis, the channels also carry H_2O (Hawthorne and Černý 1977; Brown and Mills 1986; Taylor et al. 1992). Water molecules are present in two orientations, dependent on the absence or presence of the alkalis (Gibbs et al. 1968; Morosin 1972; Wood and Nassau 1968; Aurisicchio et al. 1994). Hydroxyl ions and fluorine also were identified in minor quantities as channel constituents (e.g., Wickersham and Buchanan 1967; Odikadze 1983; Aurisicchio et al. 1988). Goldman et al. (1978) and Isotani et al. (1989) provided evidence for channel-hosted Fe^{2+}. The channels also serve as a trap for migrating radiogenic He and Ar (e.g., Saito et al. 1984). Isotopic composition of H_2O in channels was examined to decipher the character of late pegmatite fluids (Taylor et al. 1992).

Thus a general formula of beryl may be written as ${}^{C}(Na,Cs)_{2X-Y+Z}{}^{C}(H_2O,He,Ar)_{\leq 2-(2X-Y+Z)+Na}{}^{T(2)}(Be_{3-X}Li_Y\square_{X-Y}){}^{O}(Al,Fe,Sc,Cr)^{3+}{}_{2-Z}{}^{O}(Fe,Mg,Mn)^{2+}{}_{Z}{}^{T(1)}[Si_6O_{18}]$, where $Y \leq 2$, $X \geq Y$, $Z \ll 2$ and $2X-Y+Z \leq 2$ (modified from Belov 1958; Bakakin et al. 1969; Hawthorne and Černý 1977; Aurisicchio et al. 1988; Sherriff et al. 1991). Hydroxyl, F, and Fe in the channels are not readily quantifiable. Because of their fundamental geochemical contrast, the tetrahedral and octahedral substitutions are in most cases divorced from each other, although not for crystal-chemical reasons (Fig. 2; see Schaller et al. 1962 for beryl with 4.80 wt % femic oxides and 7.94 wt % alkali oxides). Belov introduced the acronymic terms t-beryl and o-beryl to designate samples with different dominant substitutions; this terminology has found some usage in the literature (e.g., Aurisicchio et al. 1988).

The accomodation of Li in the structure of beryl indicated above was subject to some controversy. Beus (1966) proposed a coupled substitution $({}^{O}Li{}^{T(2)}Al)({}^{O}Al{}^{T(2)}Be)_{1-}$, which was supported by a preliminary refinement of a Cs-rich beryl by Evans and Mrose (1968). Also, Glavinaz and Lagache (1988) and Manier-Glavinaz and Lagache (1989) claimed Li to reside in the channels, although their cation-exchange experiments did not explicitly support such interpretation. However, Hawthorne and Černý (1977) showed definitely that Li substitutes for Be in the *T*(2) tetrahedra, and X-ray plus neutron diffraction studies of Artioli et al. (1993) confirmed this conclusion. Additional arguments in favor of tetrahedrally coordinated Li are summarized by Sherriff et al. (1991, who also detected minor Li in channels) and Hawthorne and Huminicki, this volume).

Unit-cell dimensions of beryl, with *a* and *c* virtually identical for near-ideal compositions, respond differenly to the tetrahedral and octahedral substitutions. An increase in *c* characterizes the Li- plus Na,Cs-enriched phases, whereas those with significant substitution for Al have the *a* parameter increased and *c* very slightly reduced (Bakakin et al. 1970).

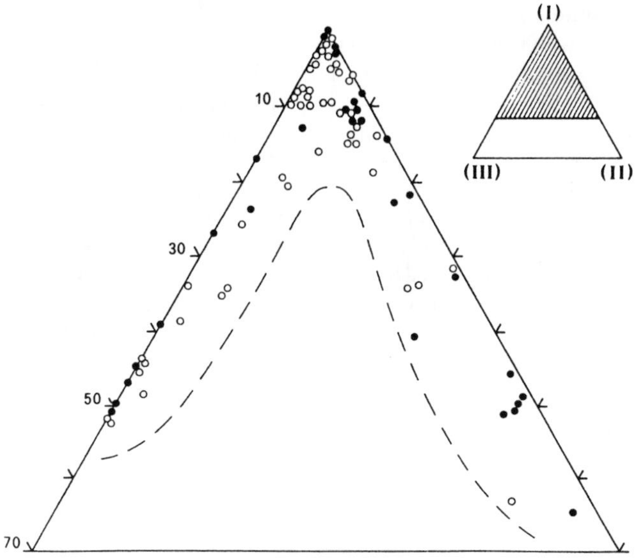

Figure 2. Separation of octahedral and tetrahedral substitutions in beryl, after Aurisicchio et al. (1988). Most samples used in this study come from granitic pegmatites. I = ideal stoichiometric beryl $Be_3Al_2Si_6O_{18}.xH_2O$, II = tetrahedrally substituted beryl with "end-member" composition of ${}^{C}(Na,Cs)^{+}Be_2LiAl_2Si_6O_{18}{\cdot}xH_2O$, III = octahedrally substituted beryl with "end-member" composition ${}^{C}(Na,Cs)^{+}Be_3(Al,Fe^{3+},Sc)(Fe,Mg,Mn)^{2+}Si_6O_{18}{\cdot}xH_2O$.

Morphology. The size of beryl crystals ranges from microscopic to tree-trunk caliber, such as the tapering 13 × 1.5 to 2 meters giant in Madagascar (De Saint Ours 1960). The habit grades from "pencil"-type crystals elongate parallel to *c*, through short prismatic and stumpy, to thick tabular and platy parallel to the basal pinakoid. This last variety is typical of alkali-rich beryl. Values of the *c/a* ratio are accordingly variable between ~20 and 0.02 (unpubl. data of the author).

Euhedral crystals commonly show faces of {10-10}, {0001}, {10-11} and {11-21}. The second-order prism {11-20} and dihexagonal prism(s) are less frequent, as are other bipyramidal and dihexagonal-bipyramidal forms, usually developed only in cavity-grown crystals. Voluminous information on the morphological crystallography of beryl, available in the literature, is well condensed in Sinkankas (1989).

Most beryl crystals in massive pegmatite assemblages are subhedral, locally almost anhedral. "Stuffed" or "shell" crystals, with interiors filled by quartz, albite, K-feldspar and/or muscovite, are typical of some pegmatite populations. Interpretations range from contemporaneous crystallization of all minerals involved (Shaub 1937) to replacement of internal parts of massive beryl crystals by the quartzofeldspathic assemblage, and to metasomatic growth of skeletal metacrystals of beryl into a solidified pegmatite matrix (Shaub 1937, Nikitin 1954, Beus 1966). The first view above is distinctly favored by recent observations.

Gentle etching to deep corrosion of beryl is typical of the cavity-grown crystals that were exposed to the action of late fluids. Coarse spongy masses are undoubtedly products of corrosion. However, delicate patterns of hillocks, pits and grooves, variable in morphology within individual and among different crystal forms, are locally difficult to assign to growth or dissolution (see, e.g., Feklitchev 1964 and Sinkankas 1989 for reviews and examples). A similar problem is posed by growth vs etching nature of tubes developed parallel to *c*, locally with trumpet-like widening at terminations both inside beryl crystals and on their basal faces (Bartoshinskyi et al. 1969, Lahti and Kinnunen 1993).

Physical properties. The refractive indices of beryl increase (and modestly also its birefringence and density) with both the alkali and octahedral substitutions. The alkali vs *n* trend is relatively smooth for most pegmatitic beryl because of the rather uniform trend in alkali substitutions (Fig. 3). However, it is disturbed by the independent variations in the degree of the octahedral substitutions (Černý and Turnock 1975, Černý and Hawthorne 1976). Optical properties occasionally betray slight deviations from hexagonal symmetry, which have not been so far confirmed by structure refinements. Biaxial optical figures are observed particularly in alkali-rich crystals (Foord and Mills 1978, Scandale et al. 1984). Stresses induced by compositional heterogeneity and consequent shifts in unit-cell dimensions may be responsible, but no systematic investigation has been conducted to date.

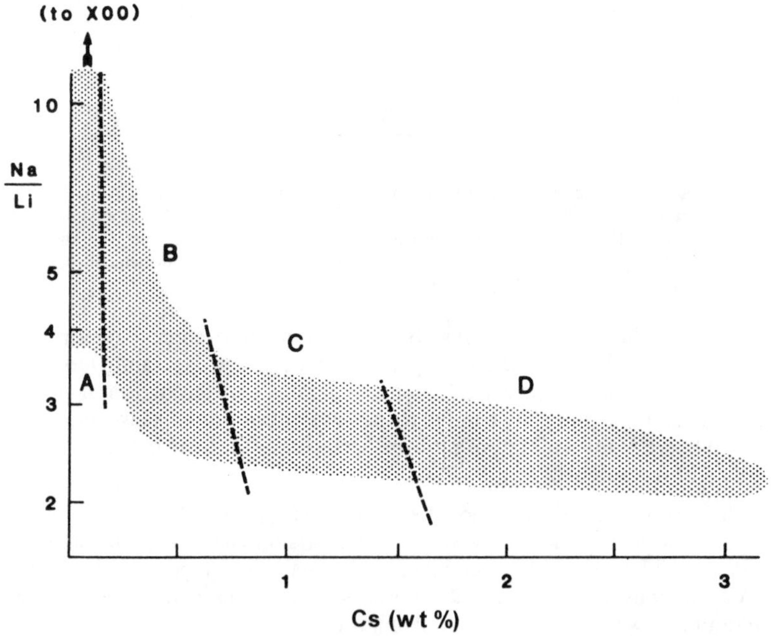

Figure 3. The Na/Li versus Cs diagram for late primary beryl from granitic pegmatites (Trueman and Černý 1982). The decrease in Na/Li parallels an increase in absolute contents of both elements. A = barren and geochemically primitive beryl-type pegmatites (including most pegmatites of the rare-earth type), B = geochemically evolved beryl-columbite and beryl-columbite-phosphate pegmatites, C = albite-spodumene and complex pegmatites, D = highly fractionated Li,Cs,Ta-rich complex pegmatites. The trend characteristic of the progressive evolution of individual pegmatite categories is also typical for sequences of beryl generations from early (outer) to late (mainly inner) zones within individual pegmatite bodies.

Pure beryl, $Be_3Al_2[Si_6O_{18}]$, is waterclear and colorless, but a great diversity of colors is encountered in natural crystals. The most common varietal names based on color are *goshenite* (colorless to white), *aquamarine* (bluish to greenish blue), *heliodor* (yellowish to orange- or greenish yellow), *morganite* or *vorobyevite* (pink), and *maxixe* (deep blue, fading in daylight) (Gaines 1976). In many cases, the color of beryl can be altered, bleached or induced by heat treatment or irradiation, as reviewed by Sinkankas (1989). Maxixe-type deep-blue color can be induced into some samples of beryl with diverse natural coloration by gamma-ray, X-ray or neutron irradiation but the nature of the color centers is different from those in natural maxixe (Nassau et al. 1976).

Chromophoric effects of minor cations, substituting mainly into the Al-populated octahedra, are discussed by, e.g., Goldman et al. (1978) and Solntsev and Bukin (1997). There is a general agreement concerning the color of emerald, generated by substitution of Cr and/or V into the octahedral site. Also, Mn^{3+} is considered responsible for the pink color of morganite. However, the interpretation of some other colors is not unanimous. For example, Wood and Nassau (1968) assign yellow color to octahedral Fe^{3+} and blue to the channel-hosted Fe^{2+}, but they do not ascribe any color effects to octahedral Fe^{2+} and tetrahedral Fe^{3+}. In contrast, Goldman et al. (1978) and Rossman (1981) explain the blue, green and yellow tones by different values of Fe^{2+}/Fe^{3+}. The deep-blue color of maxixe beryl is still enigmatic (e.g., Nassau et al. 1976).

Petrology. Beryl can form at any stage of pegmatite consolidation, from magmatic crystallization of the border and wall zones to the central core portions, during metasomatic processes and late hydrothermal events. Beryl participates in the coating of miarolitic cavities, and in endocontact assemblages of pegmatites that extensively react with host rocks of contrasting composition. Beryl is also known as a secondary alteration product of early Be-bearing phases.

Experimental work led Evensen et al. (1999) to the conclusion that temperature, activity of alumina (and silica) and addition of other components are the three factors influencing the precipitation of beryl in granitic pegmatites. Peraluminous and quartz-saturated environment is required, and the Be content required for beryl saturation of the melt decreases with decreasing temperature, but it increases with progressive fractionation and complexity of the melt. Scarce as they are, data on the bulk Be content of highly evolved pegmatites indicate that beryl saturation, achieved in simple beryl-columbite pegmatites at ~70 ppm, is maintained at even lower Be contents. This is contrary to the experimental results on the Macusani glass, which indicate that highly fractionated pegmatite melts attain beryl saturation only at substantially higher Be levels. The lower range of crystallization temperatures of the complex and volatile-enriched pegmatite melts, combined with cooling effects of host rocks, may offset the experimentally established Be values.

Consequently, an interplay of trends and/or fluctuations in temperature, alumina saturation index, degree of fractionation, volatile content and undercooling controls stabilization of beryl in essentially Be-poor pegmatite melts. This leads to dispersion of beryl crystals throughout (most of) a pegmatite body on one hand, massive precipitation at a specific stage of pegmatite consolidation on the other, and different combinations of both cases. For example, two zones of the Tanco pegmatite in Manitoba contain beryl evenly dispersed among their rock-forming phases, but aplitic albite bodies abutting against the quartz zone are rimmed by a massive "beryl fringe" (Černý and Simpson 1977, London 1986a, Thomas et al. 1988).

Paragenetic and geochemical evolution. In the majority of typical modestly fractionated granitic pegmatites, primary beryl is usually prismatic, greenish and enriched in Fe and Mg (rare-earth and beryl types), but its crystals gradually turn short prismatic to thick tabular, white, Fe- and Mg-poor but alkali-enriched in progressively more fractionated pegmatite categories (complex, albite-spodumene and albite types; Beus 1966; Černý and Turnock 1975). The same change in morphology, color and chemical attributes is commonly observed within individual moderately to highly evolved pegmatite bodies (e.g., Černý and Simpson 1977).

Early generations of beryl in border zones of pegmatite bodies may show extreme enrichment in the femic components. Blue beryl from endocontacts of an Arizona pegmatite contains 2.98 wt % Fe_2O_3, 2.24 wt % FeO, 0.29 wt % MnO, 2.16 wt % MgO, 0.10 wt % Sc_2O_3, 0.09 wt % Cr_2O_3 (but also 0.23 wt % Li_2O, 1.16 wt % Na_2O and 6.68 wt % Cs_2O, Schaller et al. 1962; see also sample #3 in Aurisicchio et al. 1988, and #20 in Bakakin et al. 1970). This type of beryl usually has some Na but negligible Li and Cs and high Na/Li. With progressive solidification of pegmatite bodies, the content of the femic components decreases, the Na and Li contents increase and the value of Na/Li decreases to ~3.5. From this stage on, later generations of beryl show only a very slow decrease in Na/Li but a remarkable enrichment in Cs (Fig. 3; e.g., Černý 1975). Considerable alkali enrichment is characteristic of beryl associated with albitic assemblages, and of late cavity-hosted crystals.

In miarolitic cavities, beryl commonly forms gem-quality crystals of a great variety of colors and complex morphology, associated with feldspars, topaz and smoky quartz in simple pegmatites (e.g., Bartoshinskyi et al. 1969, Lahti and Kinnunen 1993, Jacobson 1993) but with elbaite, spessartine and lepidolite in highly evolved dikes (e.g., Foord 1976, Foord et al. 1989). Evans and Mrose (1968) reported a Madagascar beryl of this latter category corresponding to $(Cs_{1.0}Na_{0.4})(Be_{4.3}Li_{1.7})Al_{4.1}[Si_{12}O_{36}]{\cdot}0.7H_2O$, with 11.3 wt % Cs_2O (and ~2.5 wt % Li_2O, ~1.1 wt % Na_2O estimated from the formula). Černý (1972) found 1.61 wt % Li_2O, 1.30 wt % Na_2O, 0.18 wt % Rb_2O, 0.80 wt % K_2O and 7.16 wt % Cs_2O in a late hydrothermal beryl from the Tanco pegmatite, associated with minute crystals of a later generation whose refractive indices, and presumably alkali contents, were even higher than those measured by Evans and Mrose (1968). In contrast to the alkali metals, Ca is very rarely encountered in substantial quantities, but the CaO content may be as high as 3.98 wt % (Simpson 1948).

Concentric compositional zoning is rather common in individual crystals of beryl, with classic examples of progressive alkali enrichment described by, e.g., Hurlbut and Wenden (1951), Feklitchev (1964) and Černý and Simpson (1977). This zoning usually closely follows the same trend which characterizes the compositional evolution of beryl in individual pegmatite bodies and in cogenetic populations of progressively fractionated pegmatites (Fig. 3). In other cases, the zoning proves to be largely oscillatory (e.g., Graziani et al. 1990). It becomes particularly random in cavity-grown and regenerated beryl, owing to rapid fluctuations of alkali and transition-metal activities during (repeated) ruptures and re-sealings of the pockets. Sector zoning is much less common (Scandale and Lucchesi 2000). Complex compositional heterogeneity is locally generated by deformation of primary crystals and their healing by overgrowths or recrystallization (e.g., Feklitchev 1964).

Reaction of pegmatite melts with basic to ultrabasic host rocks (amphibolites to serpentinites) generates emerald, locally associated with chrysoberyl and phenakite, in the phlogopite- or biotite-rich endocontacts ("glimmerites", and "slyudites" of Russian authors). Despite the questionable occurrences discussed in "Classification of Granitic Pegmatites" above, many localities of emerald do correspond beyond reasonable doubt to this classic concept of desilicated pegmatites: Franqueira in Spain (Martin-Izard et al. 1995, 1996), Haramosh Mts. in Pakistan (Laurs et al. 1996), Rila Planina in Bulgaria

(Arnaudov and Petrusenko 1971), Salvador in Brazil (Cassedanne 1985), and the Crabtree pegmatite of North Carolina (Tappen and Smith 1997). Pegmatitic emerald is known to contain up to 0.98 wt % Cr_2O_3, 3.37 wt % MgO, and 2.72 wt % total Fe as FeO (Aurisicchio et al. 1988, 1994, Feklitchev 1964, Bakakin et al. 1970). However, as little as 0.04 wt % Cr_2O_3 is locally sufficient to generate the emerald-green color (Petrusenko and Arnaudov 1980).

Finally, beryl is also known as a low-temperature alteration product of Be-bearing silicates. The most common case (although rarely documented) is that of beryllian cordierite, which decomposes into micaceous and chloritic assemblages with more or less microscopic beryl dispersed thoughout the pseudomorphs (Vrána 1979; Povondra et al. 1984; Jobin-Bevans and Černý 1998). Note, however, that Gordillo et al. (1985) described a similar assemblage of beryl + muscovite + chlorite + apatite as primary, coexisting with beryllian cordierite.

Fluid inclusions. Pegmatitic beryl displays the full spectrum of primary, pseudoprimary and secondary fluid inclusions, as do other pegmatite minerals such as spodumene, petalite and tourmaline. The primary inclusions range from liquid + gaseous to multiphase ones with several solids. Compared to pegmatitic quartz, which is notorious for thorough obliteration of the primaries (London 1985), the primary inclusions in beryl are more commonly preserved, possibly because beryl is more refractory or less soluble than quartz (London 1986b). They have the form of negative crystals or irregular cavities, locally aligned along growth zones,, or attached to growth tubes (Lahti and Kinnunen 1993). Primary inclusions in beryl can significantly contribute to deciphering the temperature regime and liquid + fluid composition during the evolution of the parent pegmatites (e.g., London 1986b, Ruggieri and Lattanzi 1992), but caution must be exercised in their interpretation. Preservation of a residual melt phase to 290°C at 2.9 to 2.7 kbar in beryl (and 265°C in core quartz), based on melting of multiphase inclusions, is rather questionable (Thomas et al. 1988). Secondary fluid inclusions typically occur as "veils" along more or less healed fracture planes of random orientation.

Bazzite, ideally $Be_3Sc_2[Si_6O_{18}]$, is the scandium analog of beryl, with the same aspects of crystal structure and subject to the same compositional variations as the latter. The substitution of Fe^{3+} for Sc is more extensive than that for Al in beryl (Platonov et al. 1981, Juve and Bergstøl 1990), and suggests an oxidized environment as a condition for stabilization of bazzite. Alkali metals enter the bazzite structure in proportions similar to those encountered in beryl (Juve and Bergstøl 1990). Discovered first in the NYF miarolitic pegmatites at Baveno, Italy (Artini 1915), bazzite prefers the NYF or related affiliation also in its other occurrences in granitic pegmatites: in central Kazakhstan (Chistyakova 1968), at Mount Antero, Colorado (Jacobson 1993) and in southern Norway (Juve and Bergstøl 1990).

Stoppaniite, $Be_3Fe^{3+}{}_2[Si_6O_{18}]$, is a new member of the beryl group, encountered so far only in vesicles of alkaline volcanics at Capranica in Latium, Italy (Ferraris et al. 1998, Della Ventura et al. 2000). Beryl from granitic pegmatites does display considerable Fe^{3+} in some of its occurrences, mainly in outer zones of pegmatite bodies close to contacts with mafic host rocks (e.g., Schaller et al. 1962, Aurisicchio et al. 1988). This is the only environment in which stoppaniite might possibly be found in granitic pegmatites, but chances are rather poor. Stoppaniite requires highly oxidizing conditions and, most likely, peralkaline environment. Also, transitions from bazzite into stoppaniite are not probable, because the bazzite-generating environments in granitic pegmatites are Fe-poor.

Gadolinite group has a general formula $X_2Y[Be_2Si_2O_{10}]$, with X_2Y represented by Y_2Fe^{3+} in ***gadolinite-(Y)***, Ce_2Fe^{3+} in ***gadolinite-(Ce)***, $Y_2\square$ in ***hingganite-(Y)***, $Ce_2\square$ in

hingganite-(Ce), $Yb_2\square$ in ***hingganite-(Yb)***, and CaY_2 in ***minasgeraisite***. In the hingganites, charge balance is maintained by substitution of 2 oxygens by $(OH)_2$ (see Fig. 18 in Hawthorne and Huminicki, this volume, for crystal structure). Compositional variations among the REE cations, Y and Ca are common, although the extent of substitutions is not yet completely established. Additional complication comes from the substitution of B for Be: Pezzotta et al. (1999) documented extensive solid solution of the homilite component in gadolinite, and of datolite in hingganite.

Gadolinite-group minerals (except minasgeraisite) are typical of the NYF family of granitic pegmatites, and of some transitional NYF-LCT parageneses. Three modes of occurrence are distinguished: (i) relatively early, primary crystallization in blocky feldspar-quartz zones or albitic assemblages of rare-element pegmatites, (ii) late metasomatic (?) growth along intergranular contacts of medium- to late-crystallizing phases, and (iii) late precipitation in miarolitic cavities. Many examples of the first type come from the Evje-Iveland pegmatite district of southern Norway (e.g., Bjørlykke 1935, Frigstad 1968), Llano County, Texas (Landes 1932), South Platte district in Colorado (Simmons et al. 1987), the Ytterby-type pegmatites of southern Sweden (Smeds 1990), Kangasala and Kimito in southern Finland (Vorma et al. 1966, Pehrman 1945), Kola Peninsula, Russia (Belolipetskyi and Voloshin 1996), and from the Shatford Lake pegmatite group, Manitoba (Buck et al. 1999). The intergranular coatings are demonstrated mainly by hingganite from Kola Peninsula. Gadolinite and hingganite from miarolitic cavities are best documented from the shallow-seated Golden Horn batholith of Idaho (Boggs 1986) and the Baveno and Cuasso al Monte granites in Italy (Pezzotta et al. 1999). The Kola and Baveno localities also provide the best examples of compositional evolution of the gadolinite-group minerals in the two different environments.

Minasgeraisite is known so far only from its type locality at Jaguaraçú, Minas Gerais, Brazil (Foord et al. (1986). In contrast to the mode of origin of the other members of the gadolinite group, minasgeraisite is a very late hydrothermal mineral associated with adularia, Fe-Li-bearing muscovite, and yttrian milarite (see below), in a pegmatite with the LCT signature but strongly enriched in Y.

Helvite group consists of ***helvite*** $Mn_4[Be_3Si_3O_{12}]S$, ***danalite*** $Fe_4[Be_3Si_3O_{12}]S$ and ***genthelvite*** $Zn_4[Be_3Si_3O_{12}]S$ (see Fig. 29 in Hawthorne and Huminicki, this volume, for crystal structure). Members of this compositionally exotic group of cubic minerals are occasionally found in pegmatites of different affiliations (Dunn 1976)—mainly in peralkaline and syenitic e.g., Larsen 1988), but only locally in subaluminous NYF, very rarely in peraluminous LCT, and never in any significant quantities. The compositions commonly are intermediate, particularly between danalite and genthelvite (Dunn 1976). Stabilization of these minerals is probably due to local and transient conditions of low activity of alumina, and to relatively reducing conditions which accomodate S^{2-}, generally atypical of granitic pegmatite consolidation (Burt 1980, 1988; Bilal and Fonteilles 1988). Examples from NYF-family A-granites include several localities at Rockport, Massachusetts, miarolitic pegmatites of the Sawtooth batholith, Idaho (Boggs 1986), and St. Peter's Dome in the Pikes Peak batholith of Colorado (Adams et al. 1974; Foord et al. 1984). The following localities belong to the LCT category: Nyköpingsgruvan and Stora Vika in Sweden; Pala and Rincon districts of Southern California; Salisbury, Zimbabwe; Sušice, Bohemia; and Elba (Langhof et al. 2000; Holtstam and Wingren 1991; Foord 1976; von Knorring 1959; Čech and Novák 1961; and Pezzotta 1994, respectively).

Milarite. Although discovered in alpine veins, milarite proved with time to be fairly widespread in granitic pegmatites. The general chemical formula reads A_2B_2C $[T2_3T1_{12}O_{30}](H_2O)$; the hexagonal structure is characterized by doubled six-membered tetrahedral rings, interconnected laterally and vertically by (Al,Be) tetrahedra, Ca-octahedra, and with alkali metals plus H_2O along the channels within the vertically stacked double rings (Fig. 35 of Hawthorne and Huminicki, this volume). Most compositions are loosely clustered around $(K,Na)_2Ca_4[Be_4Al_2Si_{24}O_{30}](H_2O)$. However, the most significant substitution $(^{B}Na^{T2}Be)(^{B}\square^{T2}Al)_{-1}$ leads to compositions halfway to the Al-free end-member $(Na,K)_2(K,Na)_2Ca_4[Be_6Si_{24}O_{30}](H_2O)$ on one hand, and halfway to the alkali-free end-member $Ca_4[Be_2Al_4Si_{24}O_{30}](H_2O)$ on the other (Fig. 4; Černý et al. 1980). Silicon populates doubled six-membered tetrahedral rings, whereas Be and Al are restricted to linking tetrahedra (Hawthorne et al. 1991; cf. Hawthorne and Huminicki, this volume). Several localities such as Jaguaraçú in Brazil and in Norway yielded Y,REE-enriched milarite with extensive substitution $[^{A}(Y,REE)^{T2}Be]\,(^{A}Ca^{T2}Al)_{-1}$, up to a maximum of an Al-free composition $K(Ca_{0.5}Y_{0.5})[Be_3Si_{12}]O_{30}(H_2O)$ (Černý et al. 1991; Nysten 1997).

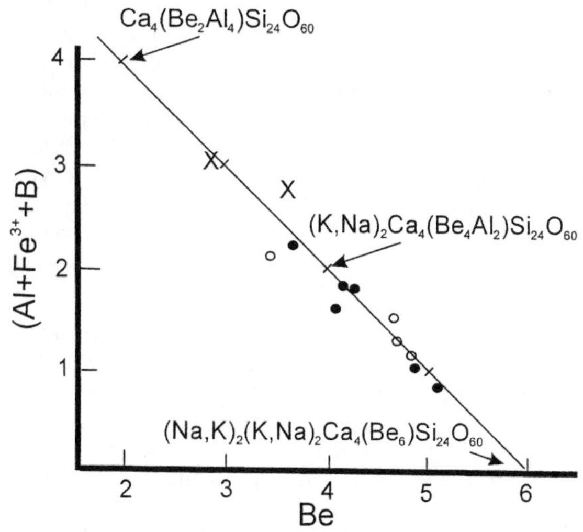

Figure 4. Beryllium versus the sum of other *T2* cations $(Al+Fe^{3+}+B)$ in Y-poor milarite. Most of the data closely follow the Be_6 - $(Al+Fe^{3+}+B)_6$ line (normalized to 60 oxygen atoms of anhydrous formula; modified from Černý et al. 1980).

Unit-cell dimensions of milarite respond to the increasing Be content of *T*(2) by reduction of *c*, and to the increase in Y in *A* by shortening of *a* (Černý et al. 1980, 1991). Hexagonal on X-ray diffraction, milarite rarely shows optical properties conformable with this symmetry; optical and chemical sector zoning indicative of lower symmetry is widespread. Strain along boundaries of sectors with different alkali and water contents, caused by dimensional misfit, is much more probable than possible differences in ordering of tetrahedrally coordinated cations (Černý et al. 1980). Janeczek's (1986) interpretation of the sector zoning as a growth phenomenon does not contradict the strain hypothesis. Optically hexagonal milarite is actually very rare. Structure refinement of such compositionally homogeneous crystals suggests positional disorder of H_2O and Ca (Černý et al. 1980; Hawthorne et al. 1991).

In LCT and NYF granitic pegmatites alike, milarite is found as an alteration product of beryl (e.g., Věžná, Czech Republic, Černý 1963, 1968a,b; Drag in Tysfjord, Norway, Raade 1966; Gruvdalen, Sweden, Nysten 1997) or beryllian cordierite (Věžná and Radkovice, Černý 1960, 1967, 1968a,b). However, at some localities milarite does not seem to be linked to a breakdown of primary Be-bearing phases: late milarite coating older minerals in miarolitic cavities is known, e.g., from Tittling in Bavaria (Tennyson 1960), Kola Peninsula, Russia (Sosedko 1960), Kent in Kazakhstan (Chistyakova et al. 1964), Strzegom, Poland (Janeczek 1986) and Högsbo, Klintberget and Utö, Sweden (Nysten 1997).

The classic Ca-based milarite is typical of the LCT pegmatites, whereas the (Y,REE)-enriched variety is encountered almost solely in the NYF and peralkaline envirinments (Černý et al. 1991, Nysten 1997, Oftedal and Saebø 1965). A remarkable exception from the latter case is the Jaguaraçú pegmatite in Brazil, which shows a typical LCT signature but extensive enrichment in Y and REE's (e.g., Foord et al. 1986).

Bityite. This trioctahedral brittle mica is, ideally, $CaLiAl_2[BeAlSi_2O_{10}](OH)_2$, but it commonly deviates toward Be- and Li-poor compositions. Solid solution with dioctahedral *margarite* $CaAl_2[Al_2Si_2O_{10}](OH)_2$ was recognized by Beus (1966): the coupled octahedral-tetrahedral substitution $(^{VI}\square^{IV}Al)\ (^{VI}Li^{IV}Be)_{-1}$ covers most of the intermediate range (cf. structure refinement of one such composition by Lin and Guggenheim 1983). However, the stoichiometry of this substitution is slanted in favor of Be in the most (Li,Be)-enriched samples (see Grew, this volume). Two modes of occurrence are dominant: late flaky aggregates replacing beryl (Huron Claim, Manitoba, Paul 1984; Eräjärvi, Finland, Lahti and Saikkonen 1985; Gruvdalen, Sweden, Nysten 1997) or coating cavities (Maharitra, Mt. Bity, Madagascar, Lacroix 1908; Tittling, Bavaria, Tennyson 1960). In contrast, coarse-flaked aggregates of intermediate bityite-margarite chemistry in contacts of "desilicated pegmatites" with ultrabasic wallrocks are less common (Ural Mts., Russia; Beus 1966; Vlasov and Kutukova 1960).

Bavenite. Ideally $Ca_4[Be_2Al_2Si_9O_{27}](OH)$, orthorhombic bavenite was long known to have variable Be, Al and H contents, leading to compositions with Be_3Al (see Fig. 30 in Hawthorne and Huminicki, this volume, for crystal structure). Substitution of Be for Al, compensated by (OH) for O, was proposed in different specific ways by Beus (1966), Berry (1963) and Canillo et al. (1966). However, current refinements of crystal structure on a wide variety of samples suggests that there is no direct $BeAl_{-1}$ substitution in a single site, but that the substitution also involves Si in the following manner: $^{[T(3)]}Be\ ^{[T(4)]}Si\ ^{[O(2)]}(OH)\ ^{[T(3)]}Si_{-1}\ ^{[T(4)]}Al_{-1}\ ^{[O(2)]}O_{-1}$ (M.A. Cooper, pers. comm. 2000). Minor Y and B were detected by emission spectrography; otherwise, no significant substitutions were observed. Similar to bertrandite, bavenite is a relatively widespread alteration product of beryl (e.g., Černý 1956, 1968a,b; Bondi et al. 1983, Paul 1984) and it is also found as a late coating on feldspars and quartz in miarolitic pegmatites (e.g., at Baveno, Tittling and Strzegom; Artini 1901, Tennyson 1960 and Janeczek 1985, respectively).

Epididymite and ***eudidymite***. These orthorhombic and monoclinic polymorphs of $Na_2[Be_2Si_6O_{15}]H_2O$ (Robinson and Fang 1970; Fang et al. 1972; Fig. 34 in Hawthorne and Huminicki, this volume) are typical components of low-temperature assemblages in peralkaline associations. However, they were also found as alteration products of beryl in a beryl-columbite-subtype pegmatite crosscutting serpentinite at Věžná, Czech Republic (Černý 1963, 1968b; Novák et al. 1991). A degree of Be substitution for Si, possibly balanced by (OH) substitution for O, is reported for eudidymite by Gaines et al. (1997).

Roggianite, $Ca_2[Be_2(OH)_2Al_2Si_4O_{13}](H_2O)_{2.5}$, has a fully ordered berylloaluminosilicate structure, and represents one of the very few Be-bearing silicate zeolites known to date (Fig. 31 in Hawthorne and Huminicki, this volume; cf. the phosphate zeolite pahasapaite above). Minor Sr, Ba, K and Fe^{3+} substitute for the major cations. It occurs as a secondary mineral in fractures of two albitic pegmatites in Val Vigezzo, Italy (Passaglia and Vezzalini 1988, Giuseppetti et al. 1991) and in "desilicated pegmatites" in the Ural Mts., Russia (former "ginzburgite", Voloshin et al. 1986).

Surinamite, $(Mg,Fe^{2+})_3Al_3[AlBeSi_3]O_{16}$, is a rare phase in nature, related to sapphirine (Moore and Araki 1983; Fig. 11 of Hawthorne and Huminicki, this volume). Beryllium is essential for stabilization of this mineral, which cannot be synthesized in Be-free systems (De Roever et al. 1981). It is a metamorphic mineral (e.g, De Roever and Vrána 1985), including its occurrences in granitic (possibly charnockitic) pegmatites located in high-pressure environments, namely in granulite-facies country rocks in Casey Bay, Antarctica (Grew 1998; Grew et al. 2000). Hölscher et al. (1986) argue that surinamite should not be expected as a primary magmatic mineral in granitic pegmatites.

Sapphirine series. Normally a Be-poor to Be-free metamorphic mineral, *sapphirine* $(Al,Mg,Fe^{2+},Fe^{3+})_8[(Al,Si)_6O_{20}]$ from pegmatites in granulite-facies terranes of Antarctica is reported to contain elevated percentage of BeO, apparently via the substitution $SiBeAl_{-2}$ (Grew 1981). Recent data confirm this substitution which generates, together with partial ordering of tetrahedrally coordinated Be, Al and Si, a series extending to *khmaralite* $(Mg,Al,Fe^{2+},Fe^{3+})_{16}[(Al, Be,Si)_{12}O_{40}]$ (Barbier et al. 1999, Grew, this volume). To date, khmaralite is known only from its type locality in East Antarctica, but its similarity to sapphirine suggests that khmaralite may be more widespread.

Høgtuvaite, $(Ca,Na)_2(Fe^{2+},Fe^{3+},Ti)_6[(Si,Be,Al)_6O_{20}]$, is a triclinic mineral in the aenigmatite group. To date, it was found only at its type locality, Høgtuva Mountain at Mo i Rana, central Norway (Grauch et al. 1994, Burt 1994). It occurs as a minor component of gneissic granite and related interior pegmatite veinlets. Granitoid pegmatites also are the parent rocks of "makarochkinite", described from the Ilmen Mts. of Russia by Polyakov et al. (1986; see also Yakubovich et al. 1990), which was not approved by IMA and is most probably identical with høgtuvaite (for crystal structure see Fig. 10 in Hawthorne and Huminicki, this volume).

Chiavennite, $CaMn^{2+}[Be_2Si_5O_{13}(OH)_2](H_2O)_2$, was so far reported only from its type locality, granitic pegmatites crosscutting ultrabasic rocks at Chiavenna in Italian Alps (Bondi et al. 1983), from syenite pegmatites of the peralkaline Oslo suite (Raade et al. 1983), and from the lithium pegmatites on the island of Utö, Sweden (Langhof et al. 2000). The mineral is a zeolite with a very open framework structure, with Mn and water-coordinated Ca in interstices (Fig 36 in Hawthorne and Huminicki, this volume). Its chemical composition is very close to the ideal formula above, with negligible Na and Al; the Utö mineral is boron-bearing. Chiavennite formed by alteration of beryl, and is closely associated with bavenite.

Liberite. To date, this mineral is known only from veins crosscutting skarns at Nanling, southern China (Chao 1964). However, the compound $Li_2[BeSiO_4]$ can reasonably be expected to occur in late assemblages of lithium-bearing LCT pegmatites.

Hsianghualite, $Ca_3Li_2[Be_3(SiO_4)_3]F_2$, associated with liberite, chrysoberyl and musgravite at the liberite locality, has not been reported from granitic pegmatites but can conceivably be expected to be found in them: high activity of Ca is reported locally in late stages of even highly fractionated pegmatite bodies. Preservation of some Ca through pegmatite crystallization may play a role (e.g., Teertstra et al. 1995, 1999), or contamination from wallrocks (Novák et al. 1999, Selway et al. 2000).

Aminoffite. So far, this calcium-rich phase $Ca_2[(Be,Al)Si_2O_7(OH)]\cdot H_2O$ (Hurlbut 1937, Coda et al. 1967, Moore 1968) has not been detected in granitic pegmatites, but it can probably form in low-temperature environments marked by increased activity of Ca (such as Huron Claim pegmatite, Manitoba). Examination of stability fields of aminoffite and the compositionally related bavenite should prove interesting.

BERYLLIUM AS A SUBSTITUENT IN SILICATES

Beryllium is widespread as a trace element in many rock-forming and accessory minerals of granitic pegmatites, and its contents rarely amount to more than a few tens of ppm. However, several exceptions do occur and should be mentioned here, either because of significantly high concentration of Be or because of the geochemical significance of the minerals as Be carriers.

Cordierite group. The close similarity of the cordierite and beryl structures suggests the possibility of Be entry into the first mineral. Cordierite, typically restricted to the peraluminous LCT family, was indeed recognized as a Be-bearing phase by Ginzburg and Stavrov (1961) and was also examined from this viewpoint by Griffitts and Cooley (1961). As much as 1.94 wt % BeO was found in cordierite from Věžná by Černý and Povondra (1965), and other discoveries of enhanced Be content in cordierite from granitic pegmatites followed later (e.g., Černý and Povondra 1967; Piyar et al. 1968; Povondra and Čech 1978; Povondra et al. 1984). Experimental work indicates that miscibility of Mg-cordierite and beryl is very limited (4 mol % beryl at 3 kbar/750°C, Povondra and Langer 1971a), but the $NaBeAl_{-1}$ substitution identified by Černý and Povondra (1965) is quite extensive and increases with pressure (up to ~69 mol % of "$NaMg_2Al_3BeSi_5O_{18}$" at 3 kbar/850°C, Povondra and Langer 1971b). The alkali-free, strictly tetrahedral substitution $BeSiAl_{-2}$ also is substantial (at least in synthetic phases). The $NaBeAl_{-1}$ substitution also is significant in cordierite from granites (e.g., Schreyer et al. 1979; Gordillo et al. 1985) and schists (Armbruster and Irouschek 1983).

Thortveitite. Beus (1966) proposed the substitution $BeZr(SiSc)_{-1}$ in thortveitite, based on the joint occurrence of Zr and Be in this mineral. However, he could not find a single composition in the literature that would confirm the expected stoichiometry of such substitution. Recent literature provides mainly the results of electron-microprobe analyses, but Be is not determined.

Hellandite. This borosilicate, $Ca_5YAl[B_4Si_4O_{20}](OH)$ (Hawthorne et al. 1996), is a rare mineral but typical of the NYF family of granitic pegmatites. Appreciable Be was established in hellandite from volcanic rocks (Oberti et al. 1999), but chances are it can also enter hellandite minerals from pegmatites (the provenance of a Be-bearing hellandite from China could not be verified). Beryllium enters a tetrahedral site that is vacant in the Be-free phase, and is charge-balanced by oxygen substituting for (OH) and, presumably, F.

The mica group. The possibility of incorporation of Be into the mica structure was emphasized by Strunz (1956), but the Be contents determined thus far were minor (except the brittle micas of the margarite-bityite series discussed earlier). Nevertheless, the micas could represent a voluminous sink for this element, and they may account for considerable quantities of dispersed Be by the sheer volume of their abundances. Černý et al. (1995) and Margison (2002) found Be preferentially concentrated in lepidolite, relative to muscovite, and they cited the crystal-chemical arguments of J.-L. Robert (pers. comm. 1994), which interpret such partitioning.

Werdingite, $(Mg,Fe)_2Al_{12}(Al,Fe)_2Si_4B_2(B,Al)_2O_{37}$, was found to contain up to 0.55 wt % BeO in granitic pegmatites residing in high-pressure environments in Madagascar and Norway, probably by the substitution $SiBe\ (Al,B)_{-2}$ (Grew et al. 1998).

Zircon. Beus (1966) noticed the common presence of subordinate Ca and Be (0.X wt %) in zircon from granitic pegmatites, confirmed by a few later data. He proposed a tripled substitution $CaBeF(ZrSiO)_{-1}$, based on the isostructural relationship between zircon and $CaBeF_4$. This argument is also strenghtened by the common presence of (OH,F) in zircon.

PARAGENETIC, PETROLOGIC AND GEOCHEMICAL RELATIONS

LTC versus NYF assemblages

Table 1, based mainly on references cited in the descriptive mineralogical section above, summarizes the main paragenetic features of beryllium minerals in granitic pegmatites.The most conspicuous difference between the two main families of granitic pegmatites is the absence of beryllium phosphates in the NYF category. This is a direct consequence of the generally low P concentrations in the NYF magmas at the putonic as

Table 1. Affiliation of minerals of beryllium with families of granitic pegmatites, and their paragenetic position

	LCT				NYF				other
	prim	ab	late	alt	prim	ab	late	alt	
chrysoberyl	O								desil.+, metam.
musgravite									metam.
pehrmanite	O								
hambergite	O		O						elb.
rhodizite series	O	O							elb.
beryllonite	O	O	O	O					
hurlbutite	O			O					
herderite series		O	O	O					
roscherite group			O	O					
väyrynenite			O	O					
weinebeneite			O	?					
faheyite			O						
ehrleite			O	?					
uralolite			O	O					
gainesite group			O	O					
moraesite				O					desil.+
fransoletite				O					
parafransoletite				O					
pahasapaite				O					
tiptopite				O					
beryl	O	O	O	O	O	O	O		desil.+
bazzite							O	O	
gadolinite group*					O	O	O		
høgtuvaite					O				
helvite group			O				O		
phenakite	O**		O	O			O	O	desil.+
euclase			O				O	O	
bertrandite			O	O			O	O	
milarite			O	O			O	O	
bityite			O				O	O	
bavenite			O	O			O	O	
roggianite			O						
epididymite			O	O					
eudidymite			O	O					
surinamite									metam.
sapphirine series									metam.+
chiavennite				O				O	desil.
minasgeraisite			O						

prim = primary magmatic phases; ab = associated with albitic units; late = in miarolitic vugs and fissures; alt = alteration of early phases; desil. = desilicated pegmatites; desil.+ = also in desilicated pegmatites; metam. = product of metamorphism; metam+ = also magmatic in anatectic pegmatites; elb. = elbaite subtype of complex LCT pegmatites. * except minasgeraisite (listed separately); **only in desilicated pegmatites. See text for behoite, clinobehoite, roggianite and minerals with subordinate Be substitution.

well as pegmatitic levels (see Pan and Černý 1989; London et al. 1990; Černý 1991b for petrogenetic and geochemical factors responsible for depletion in P in these largely A-type igneous suites).

Borates also are restricted to the LCT pegmatites, as NYF pegmatite suites commonly are very depleted in this element and rarely carry any B-bearing phases (such as tourmaline; e.g., Černý 1991a, 1992). Even in the LCT family, borates of Be are restricted to the extremely B-rich elbaite subtype (Novák and Povondra 1995), along with danburite and datolite. In other LCT categories, tourmaline is usually the only significant boron-bearing phase.

Otherwise, the differences between the LCT and NYF populations are minor. Oxide minerals of Be are so far missing in the NYF assemblages, whereas gadolinite and høgtuvaite are restricted to them. Beryl dominates the scene in both families, but bazzite is specific to the usually Sc-enriched NYF environments (cf., e.g., Novák and Černý 1998).

Among the late subsolidus minerals, the only difference between those of the LCT and NYF families is, understandably enough, the lack of phosphates in the latter. Silicate phases depend mainly on the characteristics of low-temperature fluids, which are highly variable in space and time in both of the above categories. They tend to generate the same spectrum of mineral assemblages, and beryl is the vastly predominant primary phase in both cases.

Primary versus late minerals

In all pegmatite categories of granitic pegmatites, the primary assemblages of Be-minerals are very simple, and largely represented by a single mineral. Beryl is the most widespread, chrysoberyl the distant second, and the two of them are locally associated. Beryllian cordierite can be locally significant. In contrast, the four borates and phosphates cited in Table 1 are quite rare in the role of primary phases, and they are restricted to specific categories of the LCT pegmatites: those that typically represent, or at least verge to, the elbaite and amblygonite subtypes, respectively (Novák and Povondra 1995, Černý 1991a, 1992).

Paragenetically late (supercritical to hydrothermal) minerals of Be fall into two broad categories: alteration products of early Be-bearing phases on one hand, and minerals coating miarolitic cavities and fissures on the other. In the first case, late phases are found in direct association with altered relics of their parent minerals, so that there is no reasonable doubt about their origin by decomposition of Be-bearing precursors. In the second case, the mobilization of Be (and other components) from preexisting minerals cannot be excluded, unless we can unambiguously prove their absence in the whole pegmatite—which is a daunting task under even the best circumstances. Thus, e.g., bavenite in miarolitic cavities of a pegmatite might be interpreted as precipitated from Be conserved by complexing to very low temperatures, if *all* primary beryl is fresh and unaltered, or if *no* beryl (or another primary Be-bearing phase) can be found in the whole pegmatite body. Otherwise the possibility of transport of Be (± P, Na, Ca, Li, F) from altered but remote parts of the pegmatite and deposition in late vugs or fissures remains a distinct possibility. The work of Nysten (1997) illustrates this dilemma in explicit detail.

Controls on mineral assemblages

Experimental and thermodynamic lines of evidence for general stabilities, compatibilities and reactions of beryllium minerals are presented elsewhere in this volume (Barton and Young; London and Evensen; Morteani and Franz, this volume). Thus the present notes are restricted to a few points of particular interest to pegmatite mineral assemblages and their parent environments, and the main emphasis will be focussed on empirical observations.

Primary phases. The most common primary phases—beryl and chrysoberyl—are controlled by temperature, total pressure and water pressure (Barton 1986). A general stability field of beryl is difficult to glean from the available literature, but there is no doubt about its stability under most conditions characteristic of consolidation of rare-element granitic pegmatites (~2 to 4 kbar, ~700 to 200°C). Experiental proof was provided as early as in the nineteen fifties (Wyart and Scavnicar 1957, van Valkenburg and Weir 1957). The general stability fields of beryl and chrysoberyl overlap, although chrysoberyl + quartz replace beryl + alminosilicate at high $P(H_2O)$, and at total pressures >3 kbar and low $a(H_2O)$ (Barton 1986, Barton and Young this volume). The latter conditions are, however, unrealistic in magmatic stages of granitic pegmatites. Instability of chrysoberyl + K-feldspar and their reaction to beryl + muscovite under conditions typical of pegmatite consolidation are the main reason for the relatively low abundance of chrysoberyl (London and Evensen, this volume). Metamorphism and shearing superimposed on consolidated pegmatite bodies can, however, operate under relatively dry conditions and promote breakdown of beryl into chrysoberyl-bearing assemblages (e.g., Franz and Morteani 1984, Černý et al. 1992).

Evensen et al. (1999) examined experimentally the stability of beryl in granitic melts. Only very low activities of Be are required to trigger precipitation of beryl, and they rapidly decrease with decreasing temperature and increasing peraluminosity of the melt: peraluminous magmas become saturated in beryl at realistic pegmatite-consolidating conditions, ~600°C, 2 kbar, with only about 100 ppm BeO (Fig. 2 in London and Evensen, this volume). This explains the common appearance of beryl crystals in geochemically primitive pegmatites. This low threshold also explains the multigeneration sequences of beryl, separated by barren gaps, which characterize many pegmatites with complex internal structure and history of consolidation. Fluctuations of Be activity at a mere trace levels can evidently initiate and interrupt stabilization of beryl, in conjunction with changes in bulk composition of the pegmatite melt and its temperature (London and Evensen, this volume).

Evensen and London (1999) experimentally confirmed the role of beryllian cordierite in controlling the abundance of beryl during the evolution of granitic systems. Their conclusions are also valid for individual pegmatite bodies: those with abundant beryllian cordierite carry negligible beryl, if any (Černý and Povondra 1965, 1967).

Phenakite is a relatively late mineral in the interior of rare-element (and miarolitic) pegmatites, but it is found, with or without chrysoberyl, in the outer parts of desilicated pegmatites consisting of endo- and exocontact assemblages involving phlogopite, actinolite, anthophyllite and talc (Beus 1966; Vlasov and Kutukova 1960; Martin-Izard et al. 1995). The sequence from beryl ± quartz in the feldspar-rich internal parts of desilicated pegmatites to phenakite ± chrysoberyl in the feldspar- and quartz-free endocontacts is compatible with a path from high to low silica activities at relatively high temperatures in the model of Barton (Fig. 11 in Barton 1986).

Gadolinite serves as an "index" mineral for one of the rare-element pegmatite subtypes, as it is restricted to (Y,HREE)-enriched NYF environments. The stability field of gadolinite seems to be extensive, as it is also found as an accessory mineral in granites and as crystals in miarolitic cavities and alpine veins, but experimental investigation has not been so far undertaken.

Beryllian sapphirine and khmaralite are magmatic minerals stabilized in anatectic pegmatites in Antarctica, generated from sapphirine-bearing granulite-facies lithologies. Moderate-pressure, high-temperature metamorphism of these pegmatites produced musgravite and surinamite, and subsequent decompression yielded beryllian cordierite at the expense of these minerals (Grew 1981, 1998; Grew et al. 2000). This Antarctic occurrence is so far unique, except the Be-bearing werdingite in boron-rich assemblages of anatectic pegmatites in granulite-facies rocks of Madagascar and Norway (Grew 1998; Grew et al. 1998). The overall rarity of Be-bearing minerals in high-pressure anatectic pegmatites is understandable, as Be may be retained in either cordierite or sapphirine (London and Evensen, this volume; Grew, this volume). Only a breakdown of these minerals during anatexis may release Be in quantities sufficient for beryl saturation.

Very high activity of phosphorus is required to stabilize beryllonite or hurlbutite at the magmatic stage, as beryl is commonly the only primary Be-mineral in most amblygonite-subtype pegmatites. The restriction of primary B-rich phases, rhodizite-series minerals and hambergite, to boron-rich elbaite-subtype pegmatites (or at least close to it; Novák and Povondra 1995) was already emphasized.

Late silicates. Phase relationships among low-temperature Be-bearing silicates in the BeO-Al_2O_3-SiO_2-H_2O (BASH) system were subject to experimental and thermodynamic studies of, i.a., Ganguli and Saha (1967), Seck and Okrusch (1972), Burt (1978), Franz and Morteani (1981), Hsu (1983), Hemingway et al. (1986) and Barton (1986). Systems involving F were examined by Burt (1975a) and by Hsu (1983), whereas Burt (1975b) analyzed the Al-free system of CaO-BeO-SiO_2-P_2O_5-F_2O_{-1}. Markl and Schumacher (1997) modelled breakdown of beryl and K-feldspar to bertrandite and kaolinite in greisen-generating fluids.

These and other studies indicated that beryl is stable under more or less neutral conditions. The lower thermal limit of ideal stoichiometric beryl in the presence of water was established at about 200 to 400°C, followed by breakdown to euclase-bearing assemblages. This limit is even lower for alkali-rich beryl; incorporation of volatile species and alkalis significantly enhances its stability. Euclase, however, is stabilized only at elevated alumina activities in alkali- and silica-poor environments, which explain its comparative rarity in granitic pegmatites (Barton 1986). Euclase also is considered a relatively high-density phase, stable at pressures higher than those favourable for bertrandite (Burt 1978).

Bertrandite is more likely to originate at low temperatures: it was shown to be stable below 260°C (although Hsu 1983 suggested the upper limit as high as 350°C), and to form under acidic conditions, even at high concentrations of HF. Experimental breakdown of beryl
attacked by F-bearing fluids yielded initially a quartz residuum, and betrandite + quartz in longer runs; extended reaction with beryl and K-feldspar neutralized the solution and produced bertrandite + muscovite (Beus et al. 1963; Beus and Dikov 1967).

Phenakite originates at relatively high temperatures, generally above those of euclase and bertrandite. The sequence (topaz + chrysoberyl)–(topaz + beryl)–(topaz + phenakite)–(topaz + phenakite + quartz) was shown to follow increasing acidity in the supercritical environment (Burt 1975a).

In contrast to the silicates of beryllium discussed above, Ganguli and Saha (1968) generated epididymite + analcime by severe alkaline decomposition of beryl. Eudidymite can also be presumed to precipitate under similar conditions. No attempts at synthesis of bityite, bavenite, milarite or chiavennite seem to be available in the literature, but the composition of these beryllo- and berylloaluminosilicates indicates that an alkaline, rather than acidic, environment is required to generate them in general, and at the expense of beryl in particular.

A total of 47 secondary assemblages after beryl were reviewed by Černý (1968a, p. 168-169), with some more recent references incorporated here into Table 2. This table

Table 2. Alteration products of beryl

		acidic fluids																variable conditons				alkaline fluids														
	locality #	1	2	3	4	5	6	7	8	9	10	11	12	13	14	15	16	17	18	19	20	21	22	23	24	25	26	27	28	29	30	31	32	33	34	35
silicates of Be (Al)	bertrandite	O	O	O		O	O		O	O	O	O	O	O	O		O	O	O	+	O															
	phenakite	O	+		O	O		O		+						O	O																			
	euclase	O	O	O	O																															
associated minerals	quartz		O	O			O		O				O	O		+						O														
	kaolinite																O		+																	
	muscovite			O			O		O	O	O		O		+					+	?								X							
	albite			O			O						O	O		+			O										X		+					
	K-feldspar						O		+						+	+				O	O										+					
	montmorillonite											O						+															O			
	zeolites																								O		O		X			+				
	chlorite																						O	O												
	clinozoisite																																	O		
	calcite																																		O	
beryllo- and berylloaluminosilicates	bityite																					O		O	O	+		O						O		
	bavenite																					O	O	O	O	O	O			O			O			O
	milarite																												X			O	O	O	O	
	eudidymite																															O	O			
	epididymite																				+											O	O			
	chiavennite																																			+

O = alteration product; + = mineral formed after, or at the expense of, the first alteration assemblage; X = mineral formed simultaneously with O but outside the pseudomorphs after beryl; 1 = Kolsva, Sweden (Mårtensson 1960); 2 - Sels-Vitberget, Sweden (Jonsson and Langhof 1997); 3 - Iveland, Norway (Strand 1953); 4 - Cap de Creus, Spain (Alfonso and Melgarejo 2000); 5 to 9 - southern Bohemia (Bouška and Čech 1961, Čech 1981); 10 and 11 - Huron Claim, Manitoba (Paul 1984); 12 to 15 - Eastern Transbaikalia, Russia (Timchenko 1959); 16 - Schwarzwald, Germany (Markl and Schumacher 1997); 17 - Bedford, New York (Pough 1936b); 18 - Drahonin, Czech Republic (Černy 1956); 19 - Emerald Mines, Urals, Russia (Vlasov and Kutukova 1960); 20 - Věžná, Czech Republic (Novák et al. 1991); 21 - Londonderry, Australia (Rowledge and Hayton 1950); 22 to 25 - Huron Claim, Manitoba (Paul 1984); 26 - Kola Peninsula, Russia (Matias 1959); 27 to 29 - Tittling, Germany (Tennyson 1960); 30 to 32 - Věžná,, Czech Republic (Černy 1968b); 33 - Gruvdalen, Sweden (Nysten 1997); 34 - Drag, Tysfjord, Norway (Raade 1966); 35 - Chiavenna, Italy (Bondi et al. 1983).

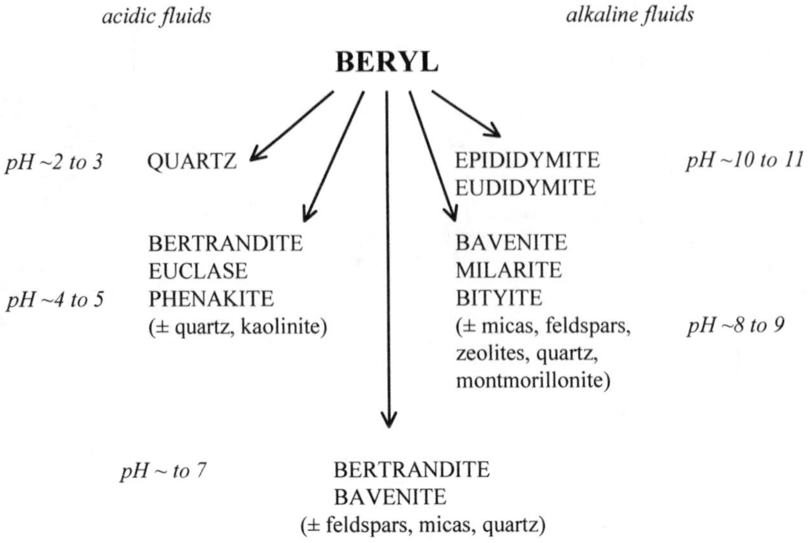

Figure 5. The most common silicate products of alteration of beryl in granitic pegmatites, generated by acidic (left side) and alkaline (right side) solutions. Modified after Černý (1970).

summarizes, on selected examples, the main features of the secondary silicate assemblages, separated into those generated by acidic decomposition of beryl from those formed by alkaline solutions, in general agreement with the results of experimental work. The Be-free phases associated with the Be-bearing minerals are well known to form under conditions established for the latter, and confirm the requirement of acidic vs alkaline environments. Figure 5 is a schematic representation of silicate products of the acidic and alkaline decomposition of beryl, in both cases approximating the sequence generated by gradually neutralized fluids. A detailed discussion of some of the secondary silicate assemblages after beryl and additional references are provided in Černý (1970).

Late phosphates. In contrast to the silicates discussed above, the secondary assemblages with phosphates of beryllium are much more diversified and they involve a much greater number of variables. Besides Be and P, additional components encountered in these phases include Li, Na, K, Rb, Cs, Ca, Mg, Fe^{2+}, Mn, Zn, Al, Fe^{3+}, Zr, F and (OH), with Ca being particularly common (66% of recognized species). Moreover, Be-phosphate occurrences are locally coupled with phases of the BASH and CBASH (CaO-BeO-Al_2O_3-SiO_2-H_2O) systems, with consequent competition for many components.

The rarity of most of the pertinent minerals, and the commonly unique occurrences of their individual assemblages severely reduce the database on natural occurrences. For example, ehrleite, fransoletite, parafransoletite, pahasapaite and tiptopite are known from only a few specimens collected at their type locality. Kampf et al. (1992) pointed out several reasons for the paucity of the beryllium phosphates, despite the fair abundance of Be, P (and Ca) in many pegmatites. The sources of mobilized P and Be (e.g., triphylite-lithiophilite, amblygonite-montebrasite and beryl) are not commonly found in close proximity, beryl is quite resistant to alteration at low temperatures and moderate pH, the behaviour of Be is very sensitive to pH, and the beryllophosphates probably form over very narrow ranges of pH.

Furthermore, exact descriptions of paragenetic sequences of secondary phosphates involving Be (and closely associated silicates, carbonates etc.) are rare in the literature, and those that are available serve mainly to demonstrate the complexity and variability of these assemblages (see, e.g., the excellent study by Campbell and Roberts 1986). Thus a simplified scheme, such as that constructed for late silicates in Table 2 and Figure 5, is impossible here, and only a few notes can be offered for some specific systems.

Interpretation of natural Be-phosphate assemblages is also complicated by a general paucity of experimental and theoretical studies. The experimental work of Harvey and Meyer (1989) was aimed at the synthesis of zeolitic framework beryllophosphates, with corresponding restrictions on the starting compositions (Be/P = 1), cation utilized (alkalis only), and P-T conditions (however, these were within the range expected for the formation of natural secondary beryllium phosphates). The framework phases developed only at pH of about 6, whereas only small amounts of unspecified anhydrous beryllophosphates with dense structures formed, and no crystalline beryllophosphates precipitated at pH > 7. Harvey and Meyer stressed the dependence of the framework type on the cation present, and the narrow range of pH conditions.

Nevertheless, the results of the above study have broader implications. To cite Kampf et al. (1992), "it is likely that solutions containing other proportions of Be and P and a variety of other cations, particularly Ca^{2+}, would yield other beryllophosphate structures over a similar range of pH and temperature. It also seems likely that the degree of condensation of Be and P tetrahedra in these compounds is dependent, in part, on pH and that less condensed structures might be expected at lower pH."

Burt (1975b) modelled the relationship of hurlbutite and herderite, and demonstrated the importance of high fluorine fugacity for the stabilization of the latter mineral, and the high activity of P_2O_5 required for both phases. He also illustrated the precipitation of herderite + topaz in the fluorapatite stability field, but was aware of the limitations imposed by the water- and alumina-free assemblages examined, as opposed to the more complex natural systems. Indeed, the common occurrence of hydroxylherderite and virtual absence of herderite in natural assemblages indicate that the fluorine fugacity is generally insufficient for production of phases with F>OH. Extending Burt's model, Charoy (1999) interpreted the breakdown of primary beryllonite in the highly evolved Beauvoir granite by increasing activity of F in hydrothermal solutions, which generated sequentially hurlbutite, hydroxylherderite (with OH/F of 1.9) and S-bearing fluorapatite in a relatively Si-poor system (absence of quartz).

CONCLUDING REMARKS

Much progress was achieved in understanding the mineralogy of beryllium in granitic pegmatites in the 1950s and 1960s, during the extensive search for deposits of this strategic element. Our understanding of the conditions of stabilization of individual minerals, and of phase reactions responding to changes in environmental parameters was further increas-ed during the past two decades. However, substantial gaps in our knowledge remain to be filled, namely in the crystal chemistry and structure of some silicates and phosphates such as the rhodizite series, aminoffite, bavenite, milarite, bertrandite, "glucine," roscherite or faheyite. Experimental and thermodynamic examination of some mineral types and assem-blages also leaves much to be desired, first of all the conditions defining the fields of the wide variety of primary and secondary phosphates. Minerals with very broad spans of crystallization in natural assemblages, such as hambergite, beryllonite, hurlbutite and hydroxylherderite are of particular interest. Several Ca-bearing silicate phases such as bavenite, milarite and roggianite also require definition of their stability fields and condi-tions of their formation at the expense of primary phases—beryl and beryllian cordierite.

ACKNOWLEDGMENTS

I am indebted to the Natural Sciences and Engineering Research Council of Canada, Canada Department of Energy, Mines and Resources, Manitoba Department of Natural Resources, Geology Office of the NWT, and the Tantalum Mining Corporation of Canada Ltd. for research support. J. Brown provided invaluable assistance with literature search, I. _erná, J. Selway and S. Mehia assisted in the final production of the manuscript. Numerous colleagues and students have contributed over the past forty years to my pegmatology studies that involved occasional examination of the mineralogy and geochemistry of beryllium. Among them, Alan Anderson, Alexei Beus, Don Burt, Scott Ercit, Joey Evensen, Dick Gaines, Ed Grew, Frank Hawthorne, Scott Jobin-Bevans, Dave London, Milan Novák, Brian Paul, Pavel Povondra, Julie Beryl Selway, Fred Simpson, Dave Trueman and Mike Wise deserve a particular thanks. Carl Francis, Ed Grew and Dave London contributed careful reviews of this chapter that much improved the content and presentation.

REFERENCES

Adams JW, Botinelly T, Sharp WN, Robinson K (1974) Murataite, a new complex oxide from El Paso County, Colorado. Am Mineral 59:172-176

Alfonso P, Melgarejo JC (2000) Boron vs. phosphorus in granitic pegmatites: The Cap de Creus case. J Czech Geol Soc 45:133-141

Anderson BW, Payne CJ, Claringbull GF (1951) Taaffeite, a new beryllium mineral, found as a cut gemstone. Mineral Mag 29:765-772

Anthony JW, Bideaux RA, Bladh KW, Nichols, MC (2000) Handbook of Mineralogy, Vol 4, Arsenates, Phosphates, Vanadates. Mineral Data Publishing, Tucson, Arizona, 680 p

Armbruster T, Irouschek A (1983) Cordierite from the Lepontine Alps: Na + Be → Al substitution, gas content, cell parameters, and optics. Contrib Mineral Petrol 82:389-396

Arnaudov V, Petrusenko S (1971) Chrysoberyl from two different pegmatite types in the Rila-Rhodope area. Bull Geol Inst Bulg Acad Sci 20:91-97 (in Bulgarian)

Artini E (1901) Di una nova specie minerale trovata nel granito di Baveno. Atti Rend Accad Lincei 10:139-145

Artini E (1915) Due minerali di Baveno contenenti terre rare: Weibyeite e bazzite. Rend Accad Lincei 24: 313-319

Artioli G, Rinaldi R, Stahl K, Zanazzi PF (1993) Structure refinement of beryl by single-crystal neutron and X-ray diffraction. Am Mineral 78:762-768

Aurisicchio C, Fioravanti G, Grubessi O, Zanazzi PF (1988) Reappraisal of the crystal chemistry of beryl. Am Mineral 73:826-837

Aurisicchio C, Grubessi O, Zecchini P (1994) Infrared spectroscopy and crystal chemistry of the beryl group. Can Mineral 32:55-68

Bakakin VV, Rylov GM, Belov NV (1969) On the crystal structure of lithium-bearing beryl. Dokl Akad Nauk SSSR 188:659-662 (in Russian)

Bakakin VV, Rylov GM, Belov NV (1970) X-ray diagnostics of isomorphous varieties of beryl. Geokhimiya 1970:1302-1311 (in Russian)

Barbier J, Grew ES, Moore PB, Su S-C (1999) Khmaralite, a new beryllium-bearing mineral related to sapphirine: A superstructure resulting from partial ordering of Be, Al and Si on tetrahedral sites. Am Mineral 84:1650-1660

Barton MD (1986) Phase equilibria and thermodynamic properties of minerals in the BeO-Al_2O_3-SiO_2-H_2O (BASH) system, with petrologic applications. Am Mineral 71:277-300

Bartoshinskyi ZV, Matkovskyi OI, Srebrodolskyi BI (1969) Accessory beryl from the chamber pegmatites of the Ukraine. Mineral Sbor Lvov Gosud Univ 23:382-397 (in Russian)

Belolipetskyi AP, Voloshin AV (1996) Yttrium and rare earth element minerals of the Kola Peninsula, Russia. *In* Jones AP, Wall F, Williams CT (eds) Rare Earth Minerals, Chemistry, Origin and Ore Deposits. Chapman & Hall, London, p 311-326

Belov NV (1958) Essays on structural mineralogy IX. Mineral Sbornik Geol Soc Lvov 12:15-42 (in Russian)

Berry LG (1963) The composition of bavenite. Am Mineral 48:1166-1168

Beus AA (1966) Geochemistry of beryllium and genetic types of its deposits. Acad Sci USSR Moscow (in Russian, 1960); Engl transl Freeman & Co, 401 p

Beus AA, Dikov YuP (1967) Geochemistry of beryllium in processes of endogenous mineralization (on the basis of hydrothermal experiments). Nedra Moscow, 160 p (in Russian)

Beus AA, Sobolev BP, Dikov YuP (1963) Contribution to the geochemistry of beryllium in high-temperature postmagmatic mineralization processes. Geochem1963:297-304 (in Russian)

Bilal E, Fonteilles M (1988) Conditions d'apparition réspectives de l'helvite, de la phénacite et du béryl dans l'environments granitiques: Example du massif de Sucuri (Brésil). C R Acad Sci Paris 307(II):273-276

Birch WD, Pring A, Foord EE (1995) Selwynite, $NaK(Be,Al)Zr_2(PO_4)_4.2H_2O$, a new gainesite-like mineral from Wycheproof, Victoria, Australia. Can Mineral 33:55-58

Björlykke H (1935) The mineral paragenesis and classification of the granite pegmatites of Iveland, Setesdal, southern Norway. Norsk Geol Tidssk 14:211-311

Blackburn WH, Dennen WH (1997) Encyclopedia of Mineral Names. Mineral Assoc Canada Spec Publ 1, 360 p

Boggs RC (1986) Miarolitic cavity and pegmatite mineralogy of Eocene granitic plutons in the northwestern U.S.A. Intl Mineral Assoc 14th General Meet, Stanford, Abstr, p 58

Bondi M, Griffin WL, Mattioli V, Mottana A (1983) Chiavennite, $CaMnBe_2Si_5O_{13}(OH)_2.2H_2O$, a new mineral from Chiavenna (Italy). Am Mineral 68:623-627

Bouška V, Čech F (1961) Pseudomorphs of orthoclase and bertrandite after beryl from a pegmatite at Rudolfov, near České Budějovice. Časopis pro Mineral Geol 6:1-6 (in Czech)

Brögger WC (1890) Die Mineralien der Syenitpegmatitgänge der südnorwegischen Augit- und Nephelinsyenite. 16. Hambergit. Z Kristallogr 16:65-67

Brown GE Jr, Mills BA (1986) High-temperature structure and crystal chemistry of hydrous alkali-rich beryl from the Harding pegmatite, Taos County, New Mexico. Am Mineral 71:547-556

Buck HM, Černý P, Hawthorne FC (1999) The Shatford Lake pegmatite group, southeastern Manitoba: NYF or not? Can Mineral 37:830-831.

Burke EAJ, Lustenhouwer WJ (1981) Pehrmanite, a new beryllium mineral from Rosendal pegmatite, Kemiö Island, southwestern Finland. Can Mineral 19:311-314

Burke EAJ, Lof P, Hazebroek HP (1977) Nigerite from the Rosendal pegmatite and aplites, Kemiö island, southwestern Finland. Bull Geol Soc Finland 49:151-157

Burns PC, Novák M, Hawthorne FC (1995) Fluorine-hydroxyl variation in hambergite: A crystal-structure study. Can Mineral 33:1205-1213

Burt DM (1975a) Natural fluorine buffers in the system $BeO-Al_2O_3-SiO_2-F_2O_{-1}$ (abstr). EOS Trans Am Geophys Union 56:467

Burt DM (1975b) Beryllium mineral stabilities in the model system $CaO-BeO-SiO_2-P_2O_5-F_2O_{-1}$ and the breakdown of beryl. Econ Geol 70:1279-1292

Burt DM (1978) Multisystem analysis of beryllium mineral stabilities.: The system $BeO-Al_2O_3-SiO_2-H_2O$. Am Mineral 63:664-676

Burt DM (1980) The stability of danalite $Fe_4Be_3(SiO_4)S$. Am Mineral 65:355-360

Burt DM (1988) Stability of genthelvite, $Zn_4(BeSiO_4)_3S$; an exercise in chalcophilicity using exchange operators. Am Mineral 73:1384-1394

Burt DM (1994) Vector representations of some mineral compositions in the aenigmatite group, with special reference to høgtuvaite. Can Mineral 32:449-457

Campbell TJ, Roberts WL (1986) Phosphate minerals from the Tip Top mine, Black Hills, South Dakota. Mineral Record 17:237-254

Canillo E, Coda A (1966) The crystal structure of bavenite. Acta Crystallogr 20:301-309

Cassedanne JP (1985) Recent discoveries of phenakite in Brazil. Mineral Rec 16:107-109

Cassedanne JP, Cassedanne JO (1973) Minerals from the Lavra da Ilha pegmatite, Brazil. Mineral Rec 4: 207-213

Cassedanne JP, Cassedanne JO (1985) La moraésite de la mine de tourmaline de Humanita, Minas Gerais, Brésil. Can Mineral 25:419-424

Čech F (1981) Pegmatites of Bohemia. *In* J Bernard et al. (eds) Mineralogy of Czechoslovakia. Academia Praha, p 100-131 (in Czech)

Čech F, Novák F (1961) Helvite from a pegmatite at Susice (Bohemia, Czechoslovakia). Acta Univ Carolinae–Geol 2:87-94

Cempírek J, Novák M, Vávra V (1999) Hurlbutite from a beryl-columbite pegmatite at Kostelní Vydří, near Telč, western Moravia. Acta Musei Moraviae Sci Geol 94:45-48

Černá I, Černý P, Selway JB, Chapman R (2002) Paragenesis and origin of secondary beryllophosphates: Beryllonite and hydroxylherderite from the BEP granitic pegmatite, southeastern Manitoba, Canada. Can Mineral (in press)

Černý P (1956) Bavenite and associated minerals from Drahonín. Časopis pro Mineral Geol 1:197-203 (in Czech)

Černý P (1960) Milarite and wellsite from Věžná. Práce brněnské základny ČSAV 32(1) #1:1-16 (in Czech)

Černý P (1963) Epididymite and milarite—alteration products of beryl from Věžná, Czechoslovakia. Mineral Mag 33:450-457

Černý P (1967) Notes on the mineralogy of some west-Moravian pegmatites. Časopis pro Mineral Geol 12: 461-464 (in Czech)

Černý P (1968a) Beryllumwandlungen in Pegmatiten—Verlauf und Produkte. N Jahrb Mineral Abh 108: 166-180

Černý P (1968b) Berylliumminerale in Pegmatiten von Věžná und ihre Umwandlungen. Ber Deutsch Ges Geol Wiss B, Mineral Lagerst 13:565-578

Černý P (1970) Review of some secondary hypogene parageneses after early pegmatite minerals. Freiberger Forschungshefte C 270 Mineralogie–Lagerstättenlehre, Probleme der Paragenese, p 47-67

Černý P (1972) The Tanco pegmatite at Bernic Lake, Manitoba. VIII. Secondary minerals from the spodumene-rich zones. Can Mineral 11:714-726

Černý P (1975) Alkali variations in pegmatitic beryls and their petrogenetic implications. N Jahrb Mineral Abh 123:198-212

Černý P (1990) Distribution, affiliation and derivation of rare-element granitic pegmatites in the Canadian Shield. Geol Rundschau 79:183-226.

Černý P (1991a) Rare-element pegmatites. Part I. Anatomy and internal evolution of pegmatite deposits. Geosci Canada 18:49-67

Černý P (1991b) Fertile granites of Precambrian rare-element pegmatite fields: Is geochemistry controlled by tectonic setting or source lithologies? Precamb Res 51:429-468

Černý P (1992) Geochemical and petrogenetic features of mineralization in rare-element granitic pegmatites in the light of current research. Appl Geochem 7:393-416

Černý P (2000) Constitution, petrology, affiliations and categories of miarolitic pegmatites. Mem Soc Ital Sci Nat Mus Civico Milano 25:5-12

Černý P, Hawthorne FC (1976) Refractive indices versus alkali contents in beryl: General limitations and applications to some pegmatitic types. Can Mineral 14:491-497

Černý P, Povondra P (1965) Beryllian cordierite from Věžná: (Na,K) + Be → Al. N Jahrb Mineral Monatsh 1966:36-44

Černý P, Povondra P (1967) Cordierite in west-Moravian desilicated pegmatites. Acta Univ Carolinae–Geol. 1967:203-221

Černý P, Simpson FM (1977) The Tanco pegmatite at Bernic Lake, Manitoba. IX. Beryl. Can Mineral 15: 489-499

Černý P, Turnock AC (1975) Beryl from the granitic pegmatites at Greer Lake, southeastern Manitoba. Can Mineral 13:55-61

Černý P, Hawthorne FC, Jarosewich E. (1980) Crystal chemistry of milarite. Can Mineral 18:41-57

Černý P, Hawthorne FC, Jambor JL, Grice JD (1991) Yttrian milarite. Can Mineral 29:533-541

Černý P, Novák M, Chapman R (1992) Effects of sillimanite-grade metamorphism and shearing on Nb-Ta oxide minerals in granitic pegmatites: Maršíkov, northern Moravia, Czechoslovakia. Can Mineral 30:699-71

Černý P, Staněk J, Novák M, Baadsgaard H, Rieder M, Ottolini L, Kavalová M, Chapman R (1995) Geochemical and structural evolution of micas in the Rožná and Dobrá Voda pegmatites, Czech Republic. Mineral Petrol 55:177-202

Chao, C-L (1964) Liberite (Li_2BeSiO_4), a new lithium-beryllium silicate mineral from the Nanling Ranges, south China. Ti Chih Hsueh Pao 44:334-342 (in Chinese). Am Mineral (1965) 50:519 (abstr)

Charoy B (1999) Beryllium speciation in evolved granitic magmas: Phosphates versus silicates. Eur J Mineral 11:135-148

Chistyakova MB (1968) Beryl and bazzite from crystal-bearing cavities of granitic pegmatites of Kazakhstan. Trudy Mineral Mus Akad Nauk SSSR 18:140-153 (in Russian)

Chistyakova MB, Osolodkina GA, Razmanova ZP (1964) Milarite from Central Kazakhstan. Dokl Akad Nauk SSSR 159:1305-1308 (in Russian)

Coda A, Rossi G, Ungaretti L (1967) The crystal structure of aminoffite. Atti Rend Accad Lincei 43:225-232

De Fourestier J (1999) Glossary of Mineral Synonyms. Mineral Assoc Canada Spec Publ 2, 496 p

Della Ventura G, Rossi P, Parodi GC, Mottana A, Raudsepp M, Prencipe M (2000) Stoppaniite, $(Fe,Al,Mg)_4(Be_6Si_{12}O_{36})^*(H_2O)_2(Na,\square)$, a new mineral of the beryl group from Latium (Italy). Eur J Mineral 12:121-128

De Roever EWF, Vrána S (1985) Surinamite in pseudomorphs after cordierite in polymetamorphic granulites from Zambia. Am Mineral 70:710-713

De Roever EWF, Lattard D, Schreyer W (1981) Surinamite: A beryllium-bearing mineral. Contrib Mineral Petrol 76:472-473

De Saint Ours J. (1960) Etude générale des pegmatites. Rap Ann Service Geol Republique Malgache 1960: 73-87

Dostál J (1966) Mineralogische und petrographische Verhältnisse von Chrysoberyll-Sillimanit-Pegmatit von Maršíkov. Acta Univ Carolinae–Geol 4:271-287

Downs JW, Ross FK (1987) Neutron-diffraction study of bertrandite. Am Mineral 72:979-983

Dunn PJ (1976) Genthelvite and the helvine group. Mineral Mag 40:627-636

Dunn PJ, Gaines RV (1978) Uralolite from the Dunton gem mine, Newry, Maine: A second occurrence. Mineral Rec 9:99-100

Dunn PJ, Wight W (1976) Green gem herderite from Brazil. J Gemmol 15:27-28

Dunn PJ, Wolfe CW, Leavens PB, Wilson WE (1979) Hydroxyl-herderite from Brazil and a guide to species nomenclature for the herderite - hydroxyl-herderite series. Mineral Rec 10:5-11

Durrfeld V (1910) Die Drusenmineralien des Waldsteingranites im Fichtelgebirge. Z Kristallogr 47:242-248

Dzhurayev ZT, Zolotarev AA, Pekov IV, Frolova LV (1998) Hambergite from pegmatite veins of the eastern Pamirs. Zapiski Vseross Mineral Obshchestva 127(4):132-139 (in Russian)

Ehlman AJ, Mitchell RS (1970) Behoite, beta-Be$(OH)_2$, from the Rode Ranch pegmatite, Llano County, Texas. Am Mineral 55:1-9

Evans HT Jr, Mrose ME (1968) Crystal chemical studies of cesium beryl. Abstracts for 1966, Geol Soc Am Spec Paper 101:63

Evensen JM, London D (1999) Beryllium budgets in granitic magmas: Consequences of early cordierite for late beryl. Can Mineral 37:821-823

Evensen JM, London D, Wendlandt RF (1999) Solubility and stability of beryl in granitic melts. Am Mineral 84:733-745

Falster AU, Pezzotta F, Simmons WB, Webber KL (1999) Rhodizite from Madagascar; new localities and new chemical data. Rocks Minerals 74:182-183 (abstr)

Fanfani L, Nunzi A, Zanazzi PF, Zanzari AR (1975) The crystal structure of roscherite. Tschermaks mineral petrogr Mitt 22:266-277

Fanfani L, Zanazzi PF, Zanzari AR (1977) The crystal structure of a triclinic roscherite. Tschermaks mineral petrogr Mitt 24:169-178

Fang JH, Robinson PD, Ohya Y (1972) Redetermination of the crystal structure of eudidymite and its dimorphic relationship to epididymite. Am Mineral 57:1345-1354

Feklitchev VG (1964) Beryl—morphology, composition and structure of crystals. IMGRE, Akad Nauk SSSR, Nauka Moscow, 124 p (in Russian)

Ferraris G, Prencipe M, Rossi P (1998) Stoppaniite, a new member of the beryl group: Crystal structure and crystal-chemical implications. Eur J Mineral 10:491-496

Fersman AE (1929) Geochemische Migration der Elemente. III. Smaragdgruben im Uralgebirge. Abh prakt Geol Bergw 1:74-116

Fersman AE (1960) Selected Works VI. Pegmatites. Akad Nauk SSSR Moscow, 747 p; reprint of the 3rd edn of 1940 (in Russian)

Foord EE (1976) Mineralogy and petrogenesis of layered pegmatite-aplite dikes in the Mesa Grande district, San Diego County, California. PhD dissertation, Stanford Univ, Stanford, California

Foord EE, Mills BA (1978) Biaxiality in 'isometric' and 'dimetric' crystals. Am Mineral 63:316-325

Foord EE, Sharp WN, Adams JW (1984) Zinc- and Y-group-bearing senaite from St. Peter's Dome, and new data on senaite from Dattas, Minas Gerais, Brazil. Mineral Mag 48:97-106

Foord EE, Gaines RV, Crock JG, Simmons WB Jr, Barbosa CP (1986) Minasgeraisite, a new member of the gadolinite group from Minas Gerais, Brazil. Am Mineral 71:603-607

Foord EE, Spaulding LBJr, Mason RA, Martin RF (1989) Mineralogy and paragenesis of the Little Three Mine pegmatites, Ramona District, San Diego County, California. Mineral Record 20:101-127

Foord EE, Brownfield ME, Lichte FE, Davis AM, Sutley SJ (1994) Mccrillisite, $NaCs(Be,Li)Zr_2(PO_4)_4 \cdot 1\text{-}2 H_2O$, a new mineral from Mount Mica, Oxford County, Maine and new data on gainesite. Can Mineral 32:839-842

Franz G, Morteani G (1981) The system $BeO\text{-}Al_2O_3\text{-}SiO_2\text{-}H_2O$: Hydrothermal investigation of the stability of beryl and euclase in the range from 1 to 6 kb and 400 to 800°C. N Jahrb Mineral Abh 140:273-299

Franz G, Morteani G (1984) The formation of chrysoberyl in metamorphosed pegmatites. J Petrol 25:27-52

Franz G, Grundmann G, Ackermand D (1986) Rock-forming beryl from a regional metamorphic terrain (Tauern Window, Austria): Parageneses and crystal chemistry. Tscherm Mineral Petrogr Mitt 35:167-192

Franz G, Gilg HA, Grundmann G, Morteani G (1996) Metasomatism at a granitic pegmatite-dunite contact in Galicia: The Franqueira occurrence of chrysoberyl (alexandrite), emerald, and phenakite: Discussion. Can Mineral 34:1329-1331

Frigstad OF (1968) En undersøkelse av cleavelanditsonerte pegmatittganger i Iveland-Evje, Nedre Setesdal. Unpubl thesis, Univ Oslo (in Norwegian)

Fuertes-Fuente M, Martin-Izard A (1998) The Forcarei Sur rare-element granitic pegmatite field and associated mineralization, Galicia, Spain. Can Mineral 36:303-325

Gaines RV (1976) Beryl—A review. Mineral Rec 7:211-223

Gaines RV, Skinner HCW, Foord EE, Mason B, Rosenzweig A (1997) Dana's New Mineralogy, 8th edn. John Wiley & Sons, New York

Ganguli D, Saha P (1967) A reconnaissance of the system $BeO-Al_2O_3-SiO_2-H_2O$. Ind Ceram Soc Trans 26: 102-110

Ganguli D, Saha P (1968) Synthetic epididymite (abst). Intl Mineral Assoc 6th Gen Meet, Prague, Abstracts, p 28

Gibbs GV, Breck DW, Meagher EP (1968) Structural refinement of hydrous and anhydrous synthetic beryl, $Al_2(Be_3Si_6)O_{18}$, and emerald, $Al_{1.9}Cr_{0.1}(Be_3Si_6)O_{18}$. Lithos 1:275-285

Ginzburg AI, Stavrov OD (1961) On the content of rare elements in cordierite. Geokhimiya 1961:183-187 (in Russian)

Ginzburg AI, Timofeev IN, Feldman LG (1979) Principles of Geology of The Granitic Pegmatites. Nedra Moscow, 296 p (in Russian)

Giuseppetti G, Mazzi F, Tadini C, Larsen AO, Åsheim A, Raade G (1990) Berborite polytypes. N Jahrb Mineral Abh 162:101-116

Giuseppetti G, Mazzi F, Tadini C, Galli E (1991) The revised crystal structure of roggianite: $Ca_2[Be(OH)_2Al_2Si_4O_{13}]$. $2,5H_2O$. N Jahrb Mineral Monatsh 1991:13-19

Glavinaz V, Lagache M (1988) Etude experimentale, en milieu hydrothermal, de l'introduction des éléments alcalins dans la structure du béryl, à 600°C et 1,5 kbar. C R Acad Sci Paris 307-II:149-154

Goldman DS, Rossman GR, Parkin KM (1978) Channel constituents of beryl. Phys Chem Minerals 3:225-235

Gonzales del Tánago J (1991) Las pegmatitas graníticas de Sierra Albarrana (Córdoba, España); mineralizaciónes de berilio. Bol Geol Minero 102:578-603

Gordillo CE, Schreyer W, Werding G, Abraham K (1985) Lithium in Na,Be cordierites from El Peñón, Sierra de Córdoba, Argentina. Contrib Mineral Petrol 90:93-101

Grauch RI, Lindahl I, Evans HT Jr, Burt DM, Fitzpatrick JJ, Foord EE, Graff P-R, Hysingjord J (1994) Høgtuvaite, a new beryllian member of the aenigmatite group from Norway, with new X-ray data on aenigmatite. Can Mineral 32:439-448

Graziani G, Guidi G (1980) Euclase from Santa do Encoberto, Minas Gerais, Brazil. Am Mineral 65:183-187

Graziani G, Lucchesi S, Scandale E (1990) General and specific growth marks in pegmatite beryls. Phys Chem Minerals 17:379-384

Grew ES (1981) Surinamite, taaffeite and beryllian sapphirine from pegmatites in granulite-facies rocks of Casey Bay, Enderby Land, Antarctica. Am Mineral 66:1022-1033

Grew ES (1998) Boron and beryllium minerals in granulite-facies pegmatites and implications of beryllium pegmatites for the origin and evolution of the Archean Napier Complex of East Antarctica. *In* Origin and Evolution of Continents. Y Motoyoshi, K Shiraishi (eds) NIPR, p 74-92 (Memoirs NIPR, Spec Issue 53)

Grew ES, Yates MG, Huijsmans JPP, McGee JJ, Shearer CK, Wiedenbeck M, Rouse RC (1998) Werdingite, a borosilicate new to granitic pegmatites. Can Mineral 36:399-414

Grew ES, Yates MG, Barbier J, Shearer CK, Sheraton JW, Shiraishi K, Motoyoshi Y (2000) Granulite-facies beryllium pegmatites in the Napier Complex in Khmara and Amundsen Bays, western Enderby Land, East Antarctica. Polar Geosci 13:1-40

Grice JD, Peacor DR, Robinson GW, van Velthuizen J, Roberts WL, Campbell TJ, Dunn PJ (1985) Tiptopite $(Li,K,Na,Ca,\square)_8[Be_6P_6O_{24}(OH)_4]$, a new mineral species from the Black Hills, South Dakota. Can Mineral 23:43-46

Griffitts WR, Cooley EF (1961) Beryllium content of cordierite. U S Geol Surv Prof Paper 424-B:259-263

Grigor'yev NA (1963) Glucine, a new beryllium mineral. Zapiski Vses Mineral Obshch 92:691-696 (in Russian)

Grigor'yev NA (1964) Uralolite, a new mineral. Zapiski Vses Mineral Obshtch 93:156-162 (in Russian)

Grundmann G, Morteani G (1989) Emerald mineralization during regional metamorphism: The Habachtal (Austria) and Leydsdorp (Transvaal, South Africa) deposits. Econ Geol 84:1835-1849

Hansen S, Fälth L, Petersen OV, Johnson O (1984a) Bergslagite, a new mineral species from Långban, Sweden. N Jahrb Mineral Monatsh 1984:257-262

Hansen S, Fälth L, Johnson O (1984b) Bergslagite, a mineral with tetrahedral berylloarsenate sheet anions. Z Kristallogr 166:73-80

Harvey G, Meier WM (1989) The synthesis of beryllophosphate zeolites. *In* Jacobs PA, van Santen, RA (eds) Zeolites: Facts, Figures, Future. Elsevier, Amsterdam

Hawthorne FC, Černý P (1977) The alkali-metal positions in Cs-Li beryl. Can Mineral 15:414-421

Hawthorne FC, Grice JD (1987) The crystal structure of ehrleite, a tetrahedral sheet structure. Can Mineral 25:767-774

Hawthorne FC, Kimata M, Černý P, Ball N, Rossman GR, Grice JD (1991) The crystal chemistry of the milarite-group minerals. Am Mineral 76:1836-1856

Hawthorne FC, Burns PC, Grice JD (1996) The crystal chemistry of boron. Rev Mineral 33:41-115

Hazen RM, Au AJ (1986) High-pressure crystal chemistry of phenakite (Be_2SiO_4) and bertrandite ($Be_4Si_2O_7(OH)_2$). Phys Chem Minerals 13:69-78
Heinrich EW, Buchi SH (1969) Beryl-chrysoberyl-sillimanite paragenesis in pegmatites. Indian Mineral 10:1-7
Hemingway BS, Barton MD, Robie RA, Haselton HTJr (1986) The heat capacities and thermodynamic functions for beryl, $Be_3Al_2Si_6O_{18}$, phenakite, Be_2SiO_4, euclase, $BeAlSiO_4(OH)$, bertrandite, $Be_4Si_2O_7(OH)_2$, and chrysoberyl, $BeAl_2O_4$. Am Mineral 71:557-568
Herrmann JR (1992) Formation of emerald under the conditions of anatexis, Capoeirana, Minas Gerais, Brazil. Geol Soc Canada - Mineral Assoc Canada Ann Meet Wolfville, Abstr 17:A48-A49
Hille K (1990) Die Königshainer Berge, eine Schatzkammer Lausitzer Pegmatitminerale. Lapis 1:37-44
Hölscher A, Schreyer W, Lattard D (1986) High-pressure, high-temperature stability of surinamite in the system $MgO\text{-}BeO\text{-}Al_2O_3\text{-}SiO_2\text{-}H_2O$. Contrib Mineral Petrol 92:113-127
Holtstam D, Wingren N (1991) Zincian helvite, a pegmatite mineral from Stora Vika, Nynashamn, Sweden. GFF 113:183-184
Hsu LC (1983) Some phase relations in the system $BeO\text{-}Al_2O_3\text{-}SiO_2\text{-}H_2O$ with comments on effects of HF. Mem Geol Soc China, p 33-46
Hudson DR, Wilson AF, Threadgold IM (1967) A new polytype of taaffeite—a rare beryllium mineral from the granulites of central Australia. Mineral Mag 36:305-310
Hurlbut CS (1937) Aminoffite, a new mineral from Långban, Sweden. GFF 59:290-292
Hurlbut CS, Wenden HE (1951) Beryl at Mt. Mica, Maine. Am Mineral 36:751-759
Isotani S, Furtado WW, Antonini R, Dias OL (1989) Line-shape and thermal kinetics analysis of the Fe^{2+} band in Brazilian green beryl. Am Mineral 74:432-438
Ito T, West J (1932) The structure of bertrandite ($H_2Be_4Si_2O_7$). Z Kristallogr 83:384-393
Jacobson MI (1982) Chrysoberyl, a U.S. review. Rocks Minerals 57:49-57
Jacobson MI (1993) Antero Aquamarines—Minerals from The Mount Antero-White Mountain Region, Chaffee County, Colorado. L.R. Ream Publishing, Coeur d'Alene, Idaho
Jacobson R, Webb JS (1948) The occurrence of nigerite, a new tin mineral in quartz-sillimanite rocks from Nigeria. Mineral Mag 28:118-128
Jakovleva ME (1961) Alterations of beryl. Trudy Mineral Mus Akad Nauk SSSR 12:145-155 (in Russian)
Janeczek J (1985) Typomorphic minerals of pegmatites from the Strzegom-Sobotka granitic massif. Geol Sudetica 20:1-68 (in Polish)
Janeczek J (1986) Chemistry, optics and crystal growth of milarite from Strzegom, Poland. Mineral Mag. 50:271-277
Jobin-Bevans S, Černý P (1998) The beryllian cordierite + beryl + spessartine assemblage, and secondary beryl in altered cordierite, Greer Lake granitic pegmatites, southeastern Manitoba. Can Mineral 36:447-462
Jonsson E, Langhof J. (1997) Late-stage beryllium silicates from the Sels-Vitberget granitic pegmatite, Kramfors, central Sweden. GFF 119:249-251
Juve G, Bergstøl S (1990) Caesian bazzite in granite pegmatite in Tørdal, Telemark, Norway. Mineral Petrol 43:131-136
Kampf AR, Dunn PJ, Foord EE (1992) Parafransoletite, a new dimorph of fransoletite from the Tip Top pegmatite, Custer, South Dakota. Am Mineral 77:843-847
King VT, Foord EE (1994) Mineralogy of Maine. Dept Conserv, Maine Geol Surv, Augusta
Kjellman J, Černý P, Smeds S-A (1999) Diversified NYF pegmatite populations of the Swedish Proterozoic: Outline of a comparative study. Can Mineral 37:832-833
Kolitsch U (1996) Bergslagit aus dem Rhyolith-Steinbruch bei Sailauf im Spessart. Mineral Welt 7:45-46
Kopchenova EV, Sidorenko GA (1962) Bearsite, an arsenic analogue of moraesite. Zapiski Vses Mineral Obshtch 91:442-446 (in Russian)
Kornetova VA (1959) A hydrous beryllium phosphate, moraesite, in eastern Transbaikalian pegmatites. Dokl Akad Nauk SSSR 129:424-427 (in Russian)
Kornetova VA, Ginsburg AI (1961) Hydroxyl-herderite from pegmatites of Eastern Transbaikalia. Trudy Mineral Mus AN SSSR 11:175-181 (in Russian)
Kuschel H (1877) Mitteilung an Prof. G. Leonhard. Milarit. N Jahrb Mineral Geol Palaeont, p 925-926
Lacroix A (1908) Les minéraux des filons de pegmatite à tourmaline lithique de Madagascar. Bull Soc fr Minéral 31:218-247
Lacroix A (1922) Minéralogie de Madagascar, Vol 1. Challamel, Paris
Lager GA, Gibbs GV (1974) A refinement of the crystal structure of herderite, $CaBePO_4OH$. Am Mineral 59:919-925
Lahti SI (1981) On the granitic pegmatites of the Eräjärvi area in Orivesi, southern Finland. Bull Geol Surv Finland 314, 82 p
Lahti SI, Kinnunen KA (1993) A new gem beryl locality: Luumaki, Finland. Gems Gemology 29:30-37
Lahti SI, Saikkonen R (1985) Bityite-2M_1 from Eräjärvi compared with related Li-Be brittle micas. Bull Geol Soc Finland 57:207-215

Landes KK (1925) The paragenesis of the granite pegmatites of central Maine. Am Mineral 10:374-411
Landes KK (1932) The Barringer Hill, Texas, pegmatite. Am Mineral 17:381-390
Langhof J, Holtstam D, Gustafsson L (2000) Chiavennite and zoned genthelvite-helvite as late stage minerals at Utö, Stockholm, Sweden. GFF 122:207-212
Larsen AO (1988) Helvite group minerals from syenite pegmatites in the Oslo region, Norway. Contributions to the mineralogy of Norway, No. 68. Norsk Geol Tidssk 68:119-124
Larsen AO, Åsheim A, Berge SA (1987) Bromellite from syenite pegmatite, southern Oslo region, Norway. Can Mineral 25:425-428
Laurs BM, Dilles JH, Snee LW (1996) Emerald mineralization and metasomatism of amphibolite, Khaltaro granitic pegmatite–hydrothermal vein system, Haramosh Mountains, northern Pakistan. Can Mineral 34:1253-1286
Leavens PB, Dunn PJ, Gaines RV (1978) Compositional and refractive index variation of the herderite - hydroxyl-herderite series. Am Mineral 63:913-917
Leavens PB, White JS, Nelen JA (1990) Zanazziite, a new mineral from Minas Gerais, Brazil. Mineral Rec 21:413-417
Lin J.-C., Guggenheim S. (1983) The crystal structure of a Li,Be-rich brittle mica: A dioctahedral-trioctahedral intermediate. Am Mineral 68:130-142
Lindberg ML (1958) The beryllium content of roscherite from the Sapucaia pegmatite mine, Minas Gerais, Brazil, and from other localities. Am Mineral 43:824-838
Lindberg ML, Murata KJ (1953) Faheyite, a new phosphate mineral from the Sapucaia pegmatite mine, Minas Gerais, Brazil. Am Mineral 38:263-270
Lindberg ML, Pecora WT, Barbosa ALdeM (1953) Moraesite, a new hydrous berullium phosphate from Minas Gerais, Brazil. Am Mineral 38:1126-1133
London D (1985) Origin and significance of inclusions in quartz: A cautionary example from the Tanco pegmatite, Manitoba. Econ Geol 80:1988-1995
London D (1986a) Magmatic-hydrothermal transition in the Tanco rare-element pegmatite: Evidence from fluid inclusions and phase-equilibrium experiments. Am Mineral 71:376-395
London D (1986b) Formation of tourmaline-rich gem pockets in miarolitic pegmatites. Am Mineral 71: 366-405
London D, Burt DM (1982) Alteration of spodumene, montebrasite and lithiophilite in pegmatites of the White Picacho district, Arizona. Am Mineral 67:97-113
London D, Černý P, Loomis PJ, Pan JC (1990) Phosphorus in alkali feldspars of rare-element granitic pegmatites. Can Mineral 28:771-786
Manier-Glavinaz V, Lagache M (1989) The removal of alkalis from beryl: Structural adjustments. Can Mineral 27:663-671
Margison SM (2002) Mineralogy and geochemistry of micas in the Tanco pegmatite, southeastern Manitoba. Unpubl MSc thesis, Univ Manitoba, Winnipeg (in preparation)
Markl G, Schumacher JC (1997) Beryl stability in local hydrothermal and chemical environments in a mineralized granite. Am Mineral 82:194-202
Mårtensson C (1960) Euklas und Bertrandit aus dem Feldspat pegmatit von Kolsva in Schweden. N Jahrb Mineral Abh 94:1248-1252
Martin RF (1999) Petrogenetic considerations: A-type granites, NYF granitic pegmatites, and beyond. Can Mineral 37:804-805
Martin-Izard A, Paniagua A, Moreiras D, Acevedo RD, Marcos-Pascual C (1995) Metasomatism at a granitic pegmatite-dunite contact in Galicia; the Franqueira occurrence of chrysoberyl (alexandrite), emerald, and phenakite. Can Mineral 33:775-792
Martin-Izard A, Paniagua A, Moreiras D, Acevedo RD, Marcos-Pascual C (1996) Metasomatism at a pegmatite-dunite contact in Galicia: The Franqueira occurrence of chrysoberyl (alexandrite), emerald, and phenakite: Reply. Can Mineral 34:1332-1336
Matias VV (1959) On a find of bavenite in the pegmatites of Kola Peninsula. Materials on Minerals of Kola Peninsula 1:130-134 (in Russian)
McKie D (1963) The högbomite polytypes. Mineral Mag 33:563-580
Meixner H, Paar W (1976) Ein Vorkommen von Väyrynenit-Kristallen aus "Pakistan." Z Kristallogr 143: 309-318
Mereiter K, Niedermayr G, Walter F (1994) Uralolite, $Ca_2Be_4(PO_4)_3(OH)_3.5H_2O$: New data and crystal structure. Eur J Mineral 6:887-896
Moore PB (1968) Relation of the manganese-calcium silicates, geigite and harstigite: A correction. Am Mineral 53:1418-1420
Moore PB (1973) Pegmatite phosphates: Descriptive mineralogy and crystal chemistry. Mineral Rec 4:103-130
Moore PB (1982) Pegmatite minerals of P(V) and B(III). Mineral Assoc Canada Short Course Hbk 8:267-291
Moore PB, Ito J (1973) Wyllieite, $Na_2Fe^{2+}{}_2Al(PO_4)_3$, a new species. Mineral Rec, p 131-136

Moore PB, Araki T (1983) Surinamite, ca. $Mg_3Al_4Si_3BeO_{16}$: Its crystal structure and relation to sapphirine, ca. $Mg_{2.8}Al_{7.2}Si_{1.2}O_{16}$. Am Mineral 68:804-810
Moore PB, Araki T, Steele IM, Swihart GH, Kampf AR (1983) Gainesite, sodium-zirconium beryllophosphate: A new mineral and its crystal structure. Am Mineral 68:1022-1028.
Morosin B (1972) Structure and thermal expansion of beryl. Acta Crystallogr B28:1899-1903
Mrose ME (1952) Hurlbutite, $CaBe_2(PO_4)_2$, a new mineral. Am Mineral 37:931-940
Mrose ME, Appleman DE (1962) The crystal structure of väyrynenite, $(Mn,Fe)Be(PO_4)(OH)$, and euclase, $AlB(SiO_4)(OH)$. Z Kristallogr 117:16-32
Mrose ME, von Knorring O (1959) The mineralogy of väyrynenite, $(Mn,Fe)Be(PO_4)(OH)$. Z Kristallogr 112:275-288
Nassau K, Prescott BE, Wood DL (1976) The deep blue Maxixe-type color center in beryl. Am Mineral 61: 100-107
Nefedov EI (1967) Berborite, a new mineral. Dokl Akad Nauk SSSR 174:189-192 (in Russian)
Nikitin VD (1954) Characteristics of processes of metasomatic mineral formation. Kristallographia, Trudy Fedorovskoi Nauchnoi Sessii 1954, Leningrad, p 4-18 (in Russian)
Norton JJ, Griffitts WR, Wilmarth VR (1958) Geology and resources of beryllium in the United States. Proc 2nd U N Intl Conf Peaceful Uses of Atomic Energy Geneva 2, Surv Raw Materials Resources, p 21-34
Novák M, Černý P (1998) Scandium in columbite-group minerals from LCT pegmatites in the Moldanubicum. Krystalinikum 24:73-89
Novák M, Povondra P (1995) Elbaite pegmatites in the Moldanubicum: A new subtype of the rare-element class. Mineral Petrol 55:159-176
Novák M, Korbel P, Odehnal, F (1991) Pseudomorphs of bertrandite and epididymite after beryl from Věžná, western Moravia, Czechoslovakia. N Jahrb Mineral Monatsh 1991:473-480
Novák M, Burns PC, Morgan GB VI (1998) Fluorine variation in hambergite from granitic pegmatites. Can Mineral 36:441-446
Novák M, Selway JB, Černý P, Hawthorne FC, Ottolini L (1999) Tourmaline of the elbaite-dravite series from an elbaite-subtype pegmatite at Bližná, southern Bohemia, Czech Republic. Eur J Mineral 11:557-568
Nysten P (1997) Paragenetic setting and crystal chemistry of milarites from Proterozoic granitic pegmatites in Sweden. N Jahrb Mineral Abh 1996:564-576
Nysten P, Gustafsson L (1993) Beryllium phosphates from the Proterozoic granitic pegmatite at Norrö, southern Stockholm archipelago, Sweden. GFF 115:159-164
Oberti O, Ottolini L, Camara F, della Ventura G (1999) Crystal structure of non-metamict Th-rich hellandite-(Ce) from Latium (Italy) and crystal chemistry of the hellandite-group minerals. Am Mineral 84:913-921
Odikadze GL (1983) On the content of rare elements and fluorine in muscovites and beryls in the pegmatites of central Sahara. Geokhimiya 1983:147-152 (in Russian)
Oftedal I, Saebø PC (1965) Contributions to the mineralogy of Norway No. 30. Minerals from nordmarkite druses. Norsk Geol Tidssk 45:171-175
Orlov, YuL, Ginsburg AI, Pinevitch GN (1961) Paragenetic relations of beryllium minerals in some pegmatites. Trudy Mineral Mus Akad Nauk SSSR 11:103-113
Ottolini L, Bottazzi P, Vannucci R (1993) Quantification of lithium, beryllium and boron in silicates by secondary ion mass spectrometry using conventional energy filtering. Analyt Chem 65:1960-1968
Palache C (1931) On the presence of beryllium in milarite. Am Mineral 16:469-470
Palache C, Shannon EV (1928) Beryllonite and other phosphates from Newry, Maine. Am Mineral 13:392-396
Pan JC, Černý P (1989) Phosphorus in feldspars of rare-element granitic pegmatites. Geol Assoc Canada – Mineral Assoc Canada Ann Meet Montreal, Progr Abstr 14:A82
Passaglia E, Vezzalini G (1988) Roggianite: Revised chemical formula and zeolitic properties. Mineral Mag 52:201-206
Paul BJ (1984) Mineralogy and petrochemistry of the Huron Claim pegmatite, southeastern Manitoba. Unpubl MSc thesis, Univ Manitoba
Peacor DR, Dunn PJ, Roberts WL, Campbell TJ, Newbury, D (1983) Fransoletite, a new calcium beryllium phosphate from the Tip Top pegmatite, Custer, South Dakota. Bull Minéral 106:499-503
Peacor DR, Rouse RC, Ahn J-H (1987) Crystal structure of tiptopite, a framework beryllophosphate isotypic with basic cancrinite. Am Mineral 72:816-820
Pehrman G (1945) Die Granitpegmatite von Kimito (S.W.-Finnland) und ihre Minerale. Acta Acad Aboensis Math et Phys 15:1-84
Pekov IV (1994) Remarkable finds of minerals of beryllium from the Kola Peninsula to Primorie. World of Stones 4:10-26
Pekov IV (1998) Minerals first discovered on the territory of the former Soviet Union. Ocean Pictures, Moscow, 369 p
Perham FC (1964) Waisanen mine operation—Summer, 1963. Rocks Minerals 39:341-347
Petkov B (1975) Beryllium. *In* Mineral Facts and Problems, 1975 edn, U S Bur Mines Bull 667:137-146

Petrusenko S, Arnaudov V (1980) Emeralds from desilicated pegmatites of Bulgaria. Proc XI Gen Meet Intl Mineral Assoc, Novosibirsk, Gem Minerals sect, Nauka Leningrad, p 74-80 (in Russian)

Pezzotta F (1994) Helvite of a M.te Capanne pluton pegmatite (Elba Island, Italy): Chemical, X-ray diffraction data and description of the occurrence. Rend Fis Accad Lincei ser 9, 5:355-362

Pezzotta F, Diella V, Guastoni A (1999) Chemical and paragenetic data on gadolinite-group minerals from Baveno and Cuasso al Monte, southern Alps, Italy. Am Mineral 84:782-789

Piyar YuK, Goroshnikov BI, Yuryev LD (1968) On beryllian cordierite. Mineral Sbor Lvov Gosud Univ 22: 86-89 (in Russian)

Platonov AN, Polshin EV, Chistyakova MB, Taran MN (1981) Isomorphism of iron in bazzite from Kazakhstan pegmatites. Geokhimiya 1981:393-398 (in Russian)

Polyakov VO, Cherepivskaya GYe, Shcherbakova YeP (1986) Makarochkinite—A new beryllosilicate. *In* New and little-studied minerals and mineral associations of the Urals. Sverdlovsk, Ural Nauch Tsentr Akad Nauk SSSR, p 108-110 (in Russian)

Pough FH (1936a) Phenakit, seine Morphologie und Paragenesis. N Jahrb Mineral Monatsh 71:291-341

Pough FH (1936b) Bertrandite and epistilbite from Bedford, N.Y. Am Mineral 21:264-265

Povondra P, Čech F (1978) Sodium-beryllium-bearing cordierite from Haddam, Connecticut, U.S.A. N Jahrb Mineral Monatsh 1978:203-209

Povondra P, Langer K (1971a) A note on the miscibility of magnesia-cordierite and beryl. Mineral Mag 38: 523-526

Povondra P, Langer K (1971b) Synthesis and some properties of sodium-beryllium-bearing cordierite, $Na_xMg_2(Al_{4-x}Be_xSi_5O_{18})$. N Jahrb Mineral Abh 116:1-19

Povondra P, Čech F, Burke EAJ (1984) Sodian-beryllian cordierite from Gammelsmorskärr, Kemiö Island, Finland and its decompostion products. N Jahrb Mineral Monatsh 1984:125-136

Pring A, Din VK, Jefferson DA, Thomas JM (1986) The crystal chemistry of rhodizite. A re-examination. Mineral Mag 50:163-172

Raade G (1966) A new Norwegian occurrence of milarite. Norsk Geol Tidssk 46:628-630 Raade G, Åmli R, Mladeck, MH, Din VK, Larsen AO, Åsheim A (1983) Chiavennite from syenite pegmatites in the Oslo Region, Norway. Am Mineral 68:628-633

Ranorosoa N (1986) Etude minéralogique et microthermométrique des pegmatites du champ de la Sahatany, Madagascar. Unpubl PhD dissertation, Univ Paul Sabatier, Toulouse

Robinson GW, Grice JD, van Velthuizen J (1985) Ehrleite, a new calcium beryllium zinc phosphate hydrate from the Tip Top pegmatite, Custer, South Dakota. Can Mineral 23:507-510

Robinson PD, Fang JH (1970) The crystal structure of epididymite. Am Mineral 74:1343-1350

Rossman GR (1981) Color in gems: The new techniques. Gems Gemology 17:60-71

Rossovskii LN, Shostatskii AN (1964) Pegmatites with chrysoberyl in one of the regions of Central Asia. Trudy Mineral Mus Akad Nauk SSSR 15:154-161 (in Russian)

Rouse RC, Peacor DR, Dunn PJ, Campbell TJ, Roberts WL, Wicks FJ, Newbury D (1987) Pahasapaite, a beryllophosphate zeolite related to synthetic zeolite rho, from the Tip Top pegmatite of South Dakota. N Jahrb Mineral Monatsh 1987:433-440

Rowledge HP, Hayton JD (1950) Two new beryllium minerals from Londonderry. J Roy Soc West Australia 33:45-52

Ruggieri G, Lattanzi P (1992) Fluid inclusion studies on Mt. Capanne pegmatites, Isola d'Elba, Tuscany, Italy. Eur J Mineral 4:1085-1096

Saito K, Alexander EC Jr, Dragon JC, Zashu S (1984) Rare gases in cyclosilicates and cogenetic minerals. J Geophys Res B 89:7891-7901

Scandale E, Lucchesi S (2000) Growth and sector zoning in a beryl crystal. Eur J Mineral 12:357-366

Scandale E, Lucchesi S, Graziani G (1984) Optical anomalies of beryl crystals. Phys Chem Minerals 11:60-66

Schaller WT, Stevens RE, Jahns RH (1962) An unusual beryl from Arizona. Am Mineral 47:672-699

Schreyer W, Gordillo CE, Werding G (1979) A new sodian-beryllian cordierite from Soto, Argentina and the relationship between distortion index, Be content, and state of hydration. Contrib Mineral Petrol 70: 421-428

Seck H, Okrusch M (1972) Phasenbeziehungen und Reaktionen im System $BeO\text{-}Al_2O_3\text{-}SiO_2\text{-}H_2O$. Fortschr Mineral 50:91-92

Selway JB, Smeds S-A,Černý P, Hawthorne FC (2000) Compositional evolution of tourmaline in the petalite-subtype Nyköpingsgruvan pegmatites, Utö Island, Sweden. GFF 124:93-102

Semenov YI, Bykova VA (1965) Beryllium borate, hambergite in alkaline pegmatite of the Baikal region. Dokl Acad Sci USSR, Earth Sci Sect 161:148-149

Shaub BM (1937) Contemporaneous crystallization of beryl and albite vs. replacement. Am Mineral 22: 1045-1051

Sherriff BL, Grundy HD, Hartman JS, Hawthorne FC, Černý P (1991) The incorporation of alkalis in beryl: Multi-nuclear MAS NMR and crystal-structure study. Can Mineral 29:271-285

Simmons WB, Lee MT, Brewster RH (1987) Geochemistry and evolution of the South Platte granite-pegmatite system, Jefferson County, Colorado. Geochim Cosmochim Acta 51:455-471

Simmons WB, Pezzotta F, Falster AU, Webber KL (2002) Londonite: The new Cs analog of rhodizite. Can Mineral 39:747-755

Simpson E (1948) Minerals of Western Australia, I. Perth.

Sinkakas J (1989) Emerald and Other Beryls. Geoscience Press, Prescott, Arizona

Smeds S-A (1990) Regional trends in mineral assemblages of Swedish Proterozoic granitic pegmatites and their geological significance. GFF 112:227-242

Solntsev VP, Bukin GV (1997) The color of natural beryls from rare-metal Mozambique pegmatites. Russ Geol Geophys 38:1661-1668

Soman K, Druzhinin AV (1987) Petrology and geochemistry of chrysoberyl pegmatites of South Kerala, India. N Jahrb Mineral Abh 157:167-183

Sosedko TA (1960) An occurrence of milarite in the Kola Peninsula. Dokl Akad Nauk SSSR 131:643-646 (in Russian)

Staněk J (1966) Scholzite and hurlbutite from the pegmatites at Otov, near Domažlice. Časopis pro Mineral Geol 11:21-26 (in Czech)

Stockmayer S (1998) Blue euclase from Zimbabwe—A review. J Gemmol 26:209-218

Strand T (1953) Euclase from Iveland, occurring as an alteration product of beryl. Norsk Geol Tidssk 31:1-5

Strunz H (1936) Datolith und Herderit. Z Kristallogr 93:146-150

Strunz H (1956) Bityit, ein Berylliumglimmer. Z Kristallogr 107:325-330

Strunz H, Tennyson C (1980) Die Kluft- und Drusenmineralien der Fichtelsgebirgsgranite. Der Aufschluss 31:419-451

Switzer G (1939) Granite pegmatites of the Mt. Antero region, Colorado. Am Mineral 24:791-809

Switzer G, Clarke RS, Sinkankas J, Worthing HW (1965) Fluorine in hambergite. Am Mineral 50:85-95

Tappen CM, Smith MS (1997) Emerald and tourmaline mineralogy of the Crabtree pegmatite, Spruce Pine district, North Carolina. Geol Soc Am Ann Meet, Progr Abstr 29:390

Taylor MC, Williams AE, McKibben MA, Kimbrough DL, Novák M (1993) Miarolitic elbaite subtype of complex rare-element pegmatites, Peninsular Ranges batholith, Southern California. Geol Soc Am Ann Meet, Progr Abstr 25:A321

Taylor RP, Fallick AE, Breaks FW (1992) Volatile evolution in Archean rare-element granitic pegmatites; evidence from the hydrogen isotopic composition of channel H_2O in beryl. Can Mineral 30:877-893

Teertstra DK, Černý P, Novák M (1995) Compositional and textural evolution of pollucite in pegmatites of the Moldanubicum. Mineral Petrol 55:37-51

Teertstra DK, Černý P, Ottolini L (1999) Stranger in paradise: Liddicoatite from the High Grade Dike pegmatite, southeastern Manitoba, Canada. Eur J Mineral 11:227-235

Tennyson C (1960) Berylliummineralien und ihre pegmatitische Paragenese in den Graniten von Tittling/Bayerischer Wald. N Jahrb Mineral Abh 94:1253-1265

Thomas AV, Bray CJ, Spooner ETC (1988) A discussion of the Jahns-Burnham proposal for the formation of zoned granitic pegmatites using solid-liquid-vapour inclusions from the Tanco pegmatite, S.E. Manitoba, Canada. Trans Roy Soc Edinburgh, Earth Sci 79:299-315

Timchenko TI (1959) Alterations of beryl in pegmatites of Eastern Transbaikalia. Trudy Mineral Muzeya Akad Nauk SSSR 9:138-145 (in Russian)

Trueman DL, Černý P (1982) Exploration for rare-element granitic pegmatites. *In* Mineral Assoc Canada Short Course Hbk 8:463-493

Van Valkenburg A, Weir CE (1957) Beryl studies $3BeO{\cdot}Al_2O_3{\cdot}6SiO_2$ (abstr). Bull Geol Soc Am 68:1808

Vlasov KA, Kutukova EI (1960) Izumrudnye Kopi [Emerald Mines]. Akad Nauk SSSR Moscow (in Russian)

Volborth A (1954) Phosphatminerale aus dem Lithiumpegmatit von Viitaniemi, Eräjärvi, Zentralfinnland. Ann Acad Sci Fennicae A III, Geol Geogr 39:5-90

Voloshin AV, Pakhomovskyi YaA, Rogachev DL, Tyusheva AL, Shishkin NM (1986) Ginzburgite—A new calcium-beryllium silicate from desilicated pegmatites. Mineral Zh 8:85-90 (in Russian)

Voloshin AV, Pakhomovskyi YaA, Rogachev DL, Nadezhina, TN, Pushcharovskyi, DY, Bakhchisaraytsev, AY (1989) Clinobehoite—A new natural modification of $Be(OH)_2$ from desilicated pegmatites. Mineral Zh 11:88-95 (in Russian)

von Knorring O (1959) Helvine from a lithium pegmatite near Salisbury, Southern Rhodesia. Mineral Mag 32:87-89

von Knorring O, Condliffe E (1984) On the occurrence of niobium-tantalum and other rare-element minerals in the Meldon aplite, Devonshire. Mineral Mag 48:443-448

Vorma A, Ojanperä P, Hoffrén V, Siivola J, Lofgren, A (1966) On the rare-earth minerals from the Pyörönmaa pegmatite in Kangasala, SW-Finland. Bull Comm Géol Finlande 222:241-274

Vrána S (1979) A secondary magnesium-bearing beryl in pseudomorphs after pegmatitic cordierite. Časopis pro Mineral Geol 24:65-69 (in Czech)

Walter F (1992) Weinebenite, $CaBe_3(PO_4)_2(OH)_2 \cdot 4H_2O$, a new mineral species: Mineral data and crystal structure. Eur J Mineral 4:1275-1283

Walter F, Postl W, Taucher J (1990) Weinebenite: Paragenesis and morphology of a new Ca-Be phosphate from a spodumene pegmatite deposit at Weinebene, Koralpe, Carinthia. Mitt Abt Mineral Landesmuseum Joanneum 58:37-43

Wickersheim KA, Buchanan RA (1967) Some remarks concerning the spectra of water and hydroxyl groups in beryl. J Chem Phys 42:1468-1469

Wise MA (1999) Characterization and classification of NYF-type pegmatites. Can Mineral 37:802-803

Wood DL, Nassau K (1968) The characterization of beryl and emerald by visible and infrared absorption spectroscopy. Am Mineral 53:777-800

Wyart J, Scavnicar S (1957) Synthèse hydrothermale du béryl. Bull Soc fr Minéral Cristallogr 80:395-400

Yakubovich OV, Malinovskii YuA, Polyakov OV (1990) Crystal structure of makarochkinite. Kristallografiya 35:1388-1394

Yatsewitch GM (1935) The crystallography of herderite from Topsham, Maine. Am Mineral 20:426-437

Zagorskyi VY, Makagon VM, Shmakin BM (1999) The systematics of granitic pegmatites. Can Mineral 37:800-802

11 Beryllium in Silicic Magmas and the Origin of Beryl-Bearing Pegmatites

David London and Joseph M. Evensen

School of Geology & Geophysics
University of Oklahoma
Norman, Oklahoma 73019

dlondon@ou.edu

INTRODUCTION

This chapter addresses the portion of the geochemical cycle of beryllium that entails its transfer through the continental crust by the generation and migration of silicic magmas, which will be referred to loosely as granitic but will include all of the textural variants of rocks derived from the granitic bulk compositions. Within the granite system, we will examine the distribution of Be between anatectic melts and residual crystals, and we will track the accumulation of Be in melt via crystal fractionation, culminating in the saturation of pegmatite-forming melts in the mineral beryl. Beryl is unquestionably the most common manifestation of Be enrichment, and the most abundant of the phases that contain essential beryllium (see Černý, this volume). To explain the presence of beryl in pegmatites, we pose four questions:

1. How much Be does it take to saturate a granitic (pegmatite-forming) melt in beryl?
2. How much Be is likely to be incorporated in silicic partial melts at their source?
3. Consequently, how much fractionation of a typical granitic magma must occur to achieve beryl saturation in pegmatitic derivatives (or is beryl saturation possible without extended magmatic fractionation)?
4. And should all granitic magmas be expected to achieve beryl saturation eventually, or are there special attributes of the source material or crystalline phases in the magma that predetermine the likelihood of forming beryl in pegmatitic differentiates?

To address these questions, we start with a review of beryl saturation in pegmatites. We then examine the likely abundance of Be derived by anatexis at the sources of granitic magmas. With the two ends of the system bracketed, we can then subject Be fractionation to conventional modeling methods to ascertain how much fractionation must take place to bring granitic melts to beryl saturation. As a review, this chapter relies as much as possible on readily available publications (monographs and journal articles, as opposed to abstracts, unpublished dissertations, and other more obscure works) that bear on these questions. We will, however, introduce new results from our laboratory that provide much of the partitioning data needed to model the fractionation of Be from anatectic source to pegmatite. We will not address liquidus effects, viscosity, and the volumetric properties of Be in granitic melts (Hess et al. 1996; Evensen et al. 1999), nor will we assess the stability or occurrence of beryllium phosphates or borates in this chapter (see Burt 1975; Charoy 1999; Černý, this volume).

BERYLLIUM CONTENTS OF PEGMATITIC MELTS AT BERYL SATURATION

Experimental evidence

Beryl (Brl) occurs principally in granitic pegmatites, but it is also found in topaz rhyolite, in hydrothermal veins associated with high-level silicic porphyries, in metamorphic rocks (gneiss), and less commonly in leucogranites (see Beus 1966 and

1529-6466/00/0050-0011$05.00

Černý, this volume). The widespread occurrence of beryl rather than other Be-silicates in granitic pegmatites stems from the stability relations of Be phases in quartz-saturated portions of the system BeO-Al_2O_3-SiO_2-H_2O (Burt 1978; Franz and Morteani 1981; Barton 1986; Cemič et al. 1986). Divariant assemblages containing beryl + quartz occur over the range of moderate pressures (~100-400 MPa) and temperatures (~400°-700°C) pertinent to the consolidation of pegmatites (Barton 1986; Černý 1991b). Figure 1 predicts three fields containing 3-phase assemblages among beryl, phenakite (Be_2SiO_4), chrysoberyl ($BeAl_2O_4$), and quartz. Beryl-bearing pegmatites are commonly peraluminous in composition, and therefore the mineral assemblage beryl + chrysoberyl (Cbr) + quartz is a possible (though uncommon) high-temperature assemblage (Burt 1978). In natural systems, however, chrysoberyl reacts with alkali feldspar plus quartz (or their components of melt) upon cooling and an increase in a_{H2O},

$$\underset{\text{chrysoberyl}}{3\,BeAl_2O_4} + \underset{\text{K-feldspar}}{2\,KAlSi_3O_8} + \underset{\text{quartz}}{6\,SiO_2} + H_2O = \underset{\text{beryl}}{Be_3Al_2Si_6O_{18}} + \underset{\text{muscovite}}{2\,KAl_3Si_3O_{10}(OH)_2} \quad (1)$$

to generate the common association of beryl and muscovite. The assemblage Cbr + Brl + Phn (phenakite) is unlikely in quartz-saturated systems such as granitic pegmatites, but it is of interest because it contains most of the important Be phases in the quaternary system cited above.

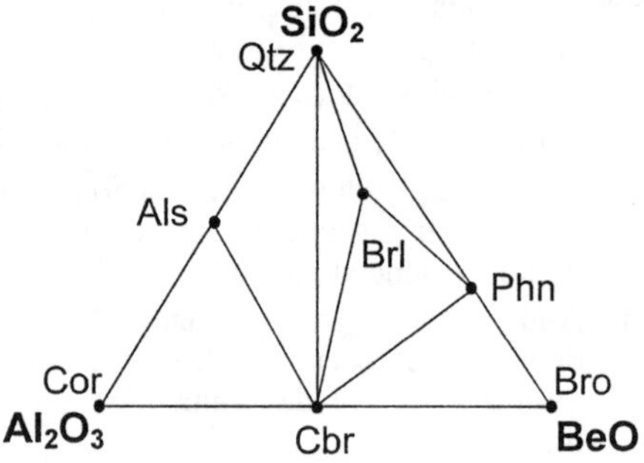

Figure 1. Proposed phase relations in the system BeO-Al_2O_3-SiO_2-H_2O at the moderate *P-T* conditions investigated in this study (685°-850°C, 200 MPa). Abbreviations are Cor (corundum), Qtz (quartz), Bro (bromellite), Als (aluminosilicate), Cbr (chrysoberyl), Brl (beryl), Phn (phenakite). The tie line Cbr-Qtz is stable at the experimental conditions investigated by Evensen et al. (1999). Bertrandite [$Be_4Si_2O_7(OH)_2$] and euclase [$BeAlSiO_4(OH)$], not shown here, become stable at lower temperatures. Mineral assemblages are from Burt (1978) and Barton (1986).

Evensen et al. (1999) determined the solubility of beryl-, chrysoberyl-, and phenakite-bearing assemblages in metaluminous to peraluminous granitic melts, and hence addressed the first question posed above: what is the Be content needed to saturate granitic melts in these phases at P-T conditions (200 MPa H_2O) that pertain to hydrous granites and pegmatites? They employed two approaches to beryl saturation, one in which forward (thermally prograde) and reverse (thermally retrograde) experiments utilized beryl as the source of Be, and replicate experiments in which a chemically equivalent assemblage containing phenakite was substituted for beryl:

$$\underset{\text{beryl}}{2\,Be_3Al_2Si_6O_{18}} = \underset{\text{phenakite}}{2\,Be_2SiO_4} + \underset{\text{alumina}}{2\,Al_2O_3} + \underset{\text{quartz}}{9\,SiO_2}, \quad (2)$$

The granitic compositions included a metaluminous mineral mixture made to the 200 MPa H_2O granite minimum (Tuttle and Bowen 1958), a similar mixture made strongly peraluminous (ASI ~ 1.3) by addition of synthetic boehmite, and the chemically complex peraluminous rhyolite obsidian from Macusani, Peru (Barnes et al. 1970; London et al. 1988).

The resultant Be mineral assemblages were the same as those predicted from the topology of Figure 1. Quartz-undersaturated experiments contained the assemblage Phn + Brl + Cbr. As predicted (by Burt 1978; Barton 1986), the assemblage Cbr-Qtz is stable, and the tie line between these two phases precludes assemblages of Brl + Als at moderate P and T. Beryl occurred with and without Cbr in quartz-saturated experiments.

The experimental study by Evensen et al. (1999) assessed Be saturation of melt not only as a function of T (at constant P_{H2O} = 200 MPa), but also as functions of the activities of other melt components. Evensen et al. (1999) recognized that beryl saturation should be determined as an activity product, i.e.,

$$\underset{\text{beryl}}{[a_{Be3Al2Si6O18}]} = \underset{\text{melt}}{[a_{BeO}]^3[a_{Al2O3}][a_{SiO2}]^6}, \quad (3)$$

in which the activity of a beryl component in melt is a function of the activities of its constituent oxides. This representation emphasizes the importance of the activities of silica and of alumina (monitored through the parameter of ASI) on the BeO content of melt at beryl saturation. It allowed Evensen et al. (1999) to consider the effects of these and other components (e.g., B, P, and F in the Macusani obsidian) on the activities of the individual beryl-forming components.

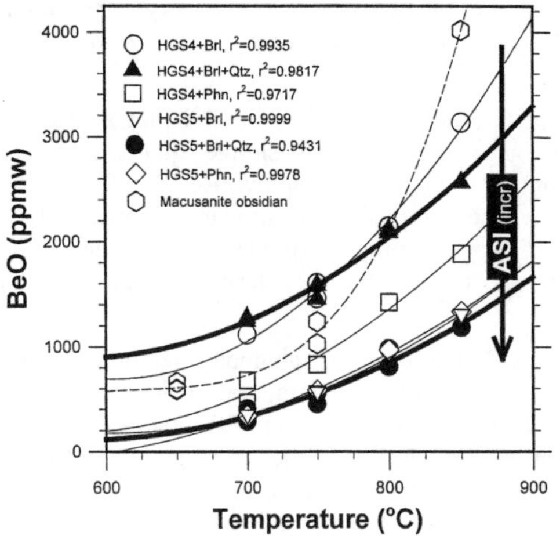

Figure 2. BeO contents of granitic melts at beryl saturation from 650-850°C and 200 MPa H_2O, from Evensen et al. (1999). The thick curves are for quartz-saturated experiments. The arrow shows the direction of increasing melt ASI from 1.00 in the simple metaluminous (HGS4) to 1.28 in the peraluminous (HGS5) melts. The experimental trend using the Macusani obsidian, with an ASI = 1.27, lies in the opposite direction of the arrow, at higher BeO contents of beryl-saturated melt, for reasons explained in the text. From Evensen et al. (1999).

Effect of silica activity. The activity of the beryl component in melt varies to the sixth power of silica activity. Consequently, variations in the activity of SiO_2 can effect major changes in the BeO content of melt at beryl saturation. This relationship is evident in the data for metaluminous granite + Brl with and without quartz saturation (Fig. 2). Without added quartz, the metaluminous haplogranite composition HGS4 + Brl, which is quartz-saturated at the solidus, becomes increasingly silica-undersaturated with increasing temperature above the solidus. When sufficient quartz was added to the mineral mixture HGS4 + Brl + Qtz to maintain quartz saturation at all the temperatures above the solidus, the resultant Be content of melt decreased, and divergence between the two saturation curves increased with increasing temperature (Fig. 2). Though the experiments illustrate an effect of silica activity on beryl saturation, the effect is negligible only because the activity of silica does not change appreciably over the range of temperature investigated. In addition, beryl almost invariably occurs with quartz in natural systems, so the activity of silica in melt or vapor is normally buffered.

Effect of alumina activity. The solubility of beryl varies to the first power of the activity of alumina in melt. In contrast to silica, however, the activity of alumina in melt, as measured by the Aluminum Saturation Index ASI (= mol $Al_2O_3/[Na_2O+K_2O+CaO]$) varies substantially from ASI values of ~1.0 to ~1.3. At the high ASI of the chrysoberyl-saturated runs, the BeO contents of peraluminous melts were displaced systematically (parallel to metaluminous results) to lower values. The lowest values were obtained in experiments with haplogranite compositions (HGS5 + Brl + Qtz) that were saturated in both chrysoberyl and quartz. Increasing the ASI of melt from 1.0 to 1.3 served to reduce the BeO content of melt at beryl saturation by over 65% at 700°C.

Effects of other components. The Macusani glass is high in silica and strongly peraluminous with an ASI = 1.27 (including Li, Rb, and Cs in the total of alkalis). In contrast to the experiments with similarly peraluminous haplogranite ± Qtz, however, the Macusani obsidian yielded the highest BeO contents of any experiments (with Cbr ± Brl). The disparity between Macusani obsidian and the peraluminous haplogranite (HGS5, with a similarly high ASI value) indicates that speciation reactions among the more numerous components of the Macusani glass (Li, B, P, F, Rb, and Cs—all at 10^4 to 10^5 ppmw levels) and any or all of the beryl-forming components lead to a net reduction of activity for beryl. Among these components, F and P are believed to react with Al to form melt species (e.g., Manning et al. 1980; Mysen and Virgo 1985; Wolf and London 1994, 1997). Hence, in reducing the activity of alumina, these components raise the BeO content of melts needed to achieve saturation in beryl or chrysoberyl. Speciation reactions involving Be, F and P may also play a role (Toth et al. 1973; Wood 1991; Charoy 1999).

Summary effects of bulk composition. Evensen et al. (1999) found that at a given temperature (Fig. 2), the solubility of beryl is lowest in simple quartz-saturated peraluminous melt compositions (~200 ppmw BeO, or ~70 ppmw Be at 700°C). The added components of the Macusani glass, whose composition is similar to many evolved Li-rich pegmatites, reduce the activities of beryl-forming components, requiring higher BeO concentrations (~900 ppmw BeO, or ~325 ppmw Be) to achieve beryl saturation in complex lithium pegmatites.

Effects of temperature. Although the activities of silica, alumina, and other added components influence the solubility of beryl, the effect of temperature is significantly greater over the range of the conditions studied by Evensen et al. (1999). For example, the Be content of melt saturated in Cbr ± Brl increases nearly an order of magnitude between 750 and 850°C. The low-temperature range of Figure 2, however, is most relevant to pegmatites. The slopes of the beryl saturation curves decrease substantially and tend to converge toward very low temperature.

COMPARISON TO BULK COMPOSITIONS OF PEGMATITES

The coarse grain size and heterogeneous zonation within common and rare-element granitic pegmatites renders estimates of bulk composition imprecise. Consequently, there are not many published attempts to derive pegmatite bulk compositions. Table 1 lists those pegmatite whole-rock estimates that were found to include Be in the totals. Methodologies for the assessments vary in detail but in most cases they utilized cored or quarried samples from which different minerals or whole-rock samples of recognizable zones were analyzed for Be, which in most cases was represented as the oxide BeO. Representative bulk compositions were obtained by assessing the Be content of a zonal assemblage and weighting its proportion to other zones. Among those sources cited in Table 1, the information presented by Stilling (1998) for the Tanco pegmatite, Manitoba, is undoubtedly the most accurate, as it is derived from data logged from 1,355 surface and underground drill cores. Interested readers should consult each source for additional details. For comparison, Table 1 also contains analyses of Be in melt inclusions found in pegmatitic quartz from a pegmatite mineralized in Sn, W, P, and F at Ehrenfriedersdorf, central Erzgebirge, Germany (Webster et al. 1997).

Table 1. Average BeO content of granitic pegmatites.

Location	Pegmatite Type (where known)	average BeO ppm	Source
Tanco	complex Li	452	Stilling (1998)
Manitoba, Canada	petalite		
Sparrow Pluton	beryl	722	Kretz et al. (1989)
Yellowknife, NWT, Canada			
Helen Beryl	beryl	254	Staatz et al. (1963)
Black Hills, SD USA			
Tin Mountain	complex Li	619	Staatz et al. (1963)
Black Hills, SD USA	spodumene		
Souchon	complex Li	560	Gallagher (1975)
Kamativi mine, Zimbabwe	lepidolite		Rijks and van der Veen (1972)
Augustus	complex Li	560	Gallagher (1975)
near Salisbury, Zimbabwe	lepidolite		Ackermann et al. (1968)
Mistress	complex Li	1200	Gallagher (1975)
near Salisbury, Zimbabwe	lepidolite		Ackermann et al. (1968)
Benson 1	complex Li	1600	Gallagher (1975)
Mtoko district, Zimbabwe	lepidolite		Ackermann et al. (1968)
Benson 4	complex Li	210	Gallagher (1975)
Mtoko district, Zimbabwe	lepidolite		Ackermann et al. (1968)
Benson 2	complex Li	100	Gallagher (1975)
Mtoko district, Zimbabwe	lepidolite		Ackermann et al. (1968)
Al Hayat	complex Li	350	Gallagher (1975)
Bikita district, Zimbabwe	spodumene		Ackermann et al. (1968)
Bikita Main	complex Li	500	Gallagher (1975)
Bikita district, Zimbabwe	petalite		Ackermann et al. (1968)
Beryl Rose		420	Gallagher (1975)
Darwin district, Zimbabwe			
unnamed pegmatite	complex Li	430	Rossovskiy and Matrosov (1974)
Gobi Field, E. Gobi, Mongol	lepidolite		
Ehrenfriedersdorf	complex Li	472	Webster et al. (1997)
Erzgebrige, Germany	Li-mica & topaz		

The compositionally simple beryl-rich pegmatites (Helen Beryl, Table 1) have BeO contents that match the experimental values expected of muscovite-bearing pegmatites at ~700°C. Similarly low BeO values are evident in some of the less-evolved Li pegmatites (lepidolite-bearing but low in Li minerals overall) of the Benson claims (Table 1). These are primarily noted for their cassiterite contents, and are not manifestly beryl-rich; thus, the low BeO values could mean that the pegmatite magmas became saturated in beryl only after extended crystal fractionation. The highly fractionated, complex Li pegmatites (Tanco, Tin Mountain, and Bikita, Table 1) possess about half of the BeO expected for such evolved melt compositions, especially considering that the pegmatites are beryl-rich. These values may be indicative of melts that crystallized at low temperatures, perhaps substantially lower than the likely liquidus temperatures near 650°C (London 1992,

1996). A few beryl pegmatites, such as the most Be-rich body associated with the Sparrow pluton in the Yellowknife field, Canada, are higher than anticipated for such chemically simple pegmatites. Overall, however, the pegmatite compositions fall within in the range expected from the experimental results of Evensen et al. (1999).

The data recorded from melt inclusions, such as in the beryl-bearing Ehrenfriedersdorf pegmatite (Webster et al. 1997), ostensibly provide the most reliable measures of trace-element abundances in magma, devoid of subsolidus hydrothermal remobilization. Melt inclusions in the Ehrenfriedersdorf pegmatite contain an average of 472 ppmw BeO (170 ppmw Be), which is comparable to other pegmatites shown in Table 3 (below). What is peculiar about these data, however, is that Be does not correlate well with other lithophile trace elements such as F, Sn, Rb, and Li (see R^2 values in Fig. 3, and further discussion below). Moreover, other element pairs that might be expected to correlate well (e.g., Li and Sn, Li and Rb) do not (Fig. 3e,f).

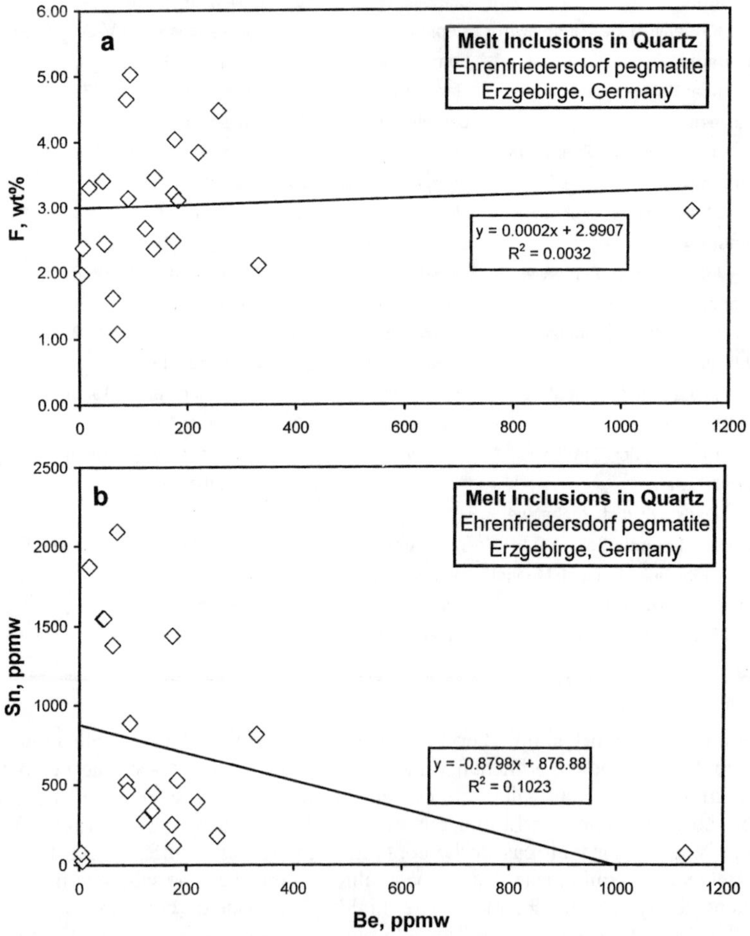

Figure 3. Plots of Be versus other elements and chemical indices of 21 melt inclusions hosted by quartz from the beryl-bearing, Sn-W Ehrenfriedersdorf pegmatite, Erzgebirge, Germany (from Webster et al. 1997). **See following pages for 3c-3f.**

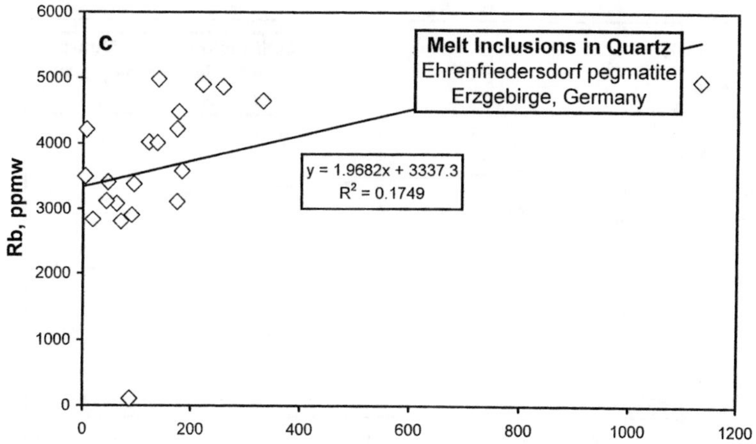
c
Melt Inclusions in Quartz
Ehrenfriedersdorf pegmatite
Erzgebirge, Germany
y = 1.9682x + 3337.3
R² = 0.1749
Rb, ppmw
0
1000
2000
3000
4000
5000
6000
0
200
400
600
800
1000
1200

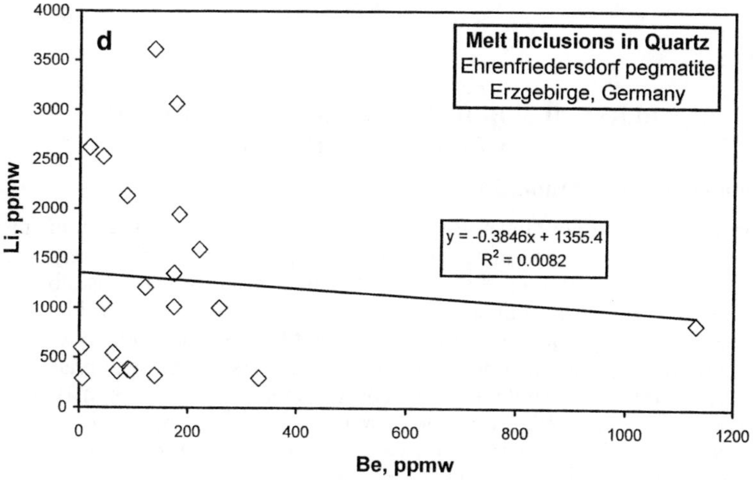
d
Melt Inclusions in Quartz
Ehrenfriedersdorf pegmatite
Erzgebirge, Germany
y = -0.3846x + 1355.4
R² = 0.0082
Li, ppmw
Be, ppmw
0
500
1000
1500
2000
2500
3000
3500
4000
0
200
400
600
800
1000
1200

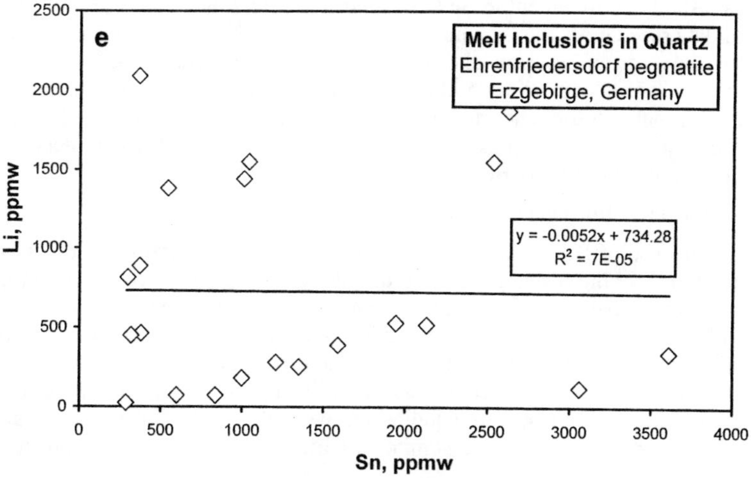
e
Melt Inclusions in Quartz
Ehrenfriedersdorf pegmatite
Erzgebirge, Germany
y = -0.0052x + 734.28
R² = 7E-05
Li, ppmw
Sn, ppmw
0
500
1000
1500
2000
2500
0
500
1000
1500
2000
2500
3000
3500
4000

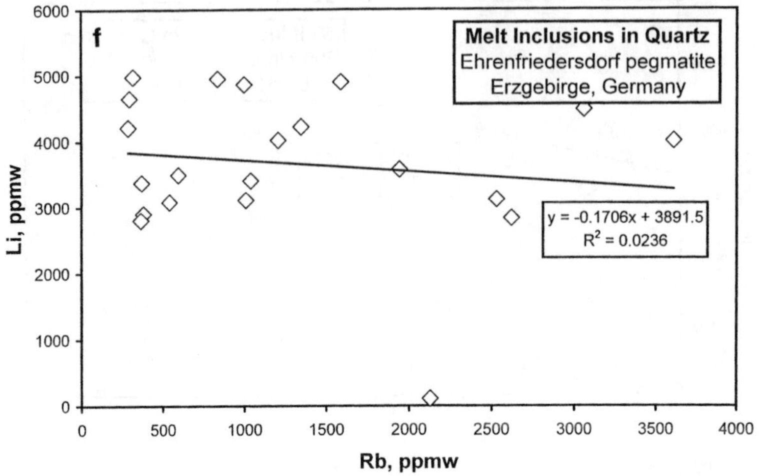

BERYLLIUM BUDGETS FOR MAGMAS AT THEIR ANATECTIC SOURCES

Beryllium contents of protoliths

Having established how much Be is required to achieve beryl saturation in granitic magmas over a suitable range of temperatures, we now examine the likely sources of magma to assess the starting Be budget. Granitic magmas can arise by extended fractionation of mafic or even ultramafic precursors (e.g., Bowen 1928), but more felsic crustal materials of igneous (I-type) or sedimentary (S-type) origin are the most productive sources because their bulk compositions are closer to those of granites. Among these potential protoliths, mantle rocks are conspicuously poor sources of Be (<1 ppmw Be), and oceanic crust, whether fresh or altered by seawater, is negligibly higher (Ryan, this volume). Psammitic sedimentary rocks (arkose, greywacke) represent fertile sources of granitic magmas, but taken as a whole, psammitic and even pelitic sediments do not manifest substantial enrichment in Be (~2-5 ppmw: Grew, this volume). From the limited sources of data (see Grew, this volume), Be does not appear to be conspicuously mobile during prograde metamorphism: the Be content of metasedimentary rocks remains essentially unchanged up to conditions of the granulite facies of metamorphism. Consequently, mafic, felsic, and micaceous metamorphic rocks arrive at the amphibolite to granulite facies of metamorphism—where anatectic reactions can commence—with a bulk Be content of ~0-5 ppmw Be.

Mineral/melt partition coefficients

The abundance of Be in partial melts will vary with the proportions of phases that participate in the melting reactions and the distribution of Be between them and the melt phase. Mineral/melt partition coefficients are applicable if the minerals and melts closely approach chemical equilibrium. Though melting intervals may be too short for mineral-melt equilibrium to be achieved by intracrystalline diffusion (Bea 1996), some minerals, such as the micas, tend to reëquilibrate rapidly by simultaneous dissolution and reprecipitation of new phases in equilibrium with the melt (Johannes et al. 1994; Icenhower and London 1995). Mineral/mineral partition coefficients are more useful for the purposes of estimating trace element abundances in systems that fail to achieve equilibrium between crystals and melt (Bea 1996). In either case, the likely Be content of an anatectic melt can be ascertained with knowledge of where Be resides in the phases of the protolith, and the proportions of phases involved in the melting reactions.

Relevant data for the distribution of Be among coexisting minerals or mineral/melt pairs are sparse. Table 2 lists reported average values, which are derived from only a few and very disparate geochemical systems and sources (i.e., experimental versus natural, mafic systems to ultrafractionated leucogranites). For phases that may be major contributors to granitic melts (e.g., quartz, feldspars, micas, and aluminous ferromagnesian phases) or restitic phases in equilibrium with those melts, the Be content of the coexisting phases decreases in the order:

Table 2. Partition coefficients for beryllium

Source Phases	D[Be] Avg (s.d.)	Reference
phenocryst/groundmass of ongonite, Mongolia		Kovalenko et al. (1977)
K-feldspar/groundmass	0.53(0.19)	
albite/groundmass	1.56(1.46)	
Li-"mica"/groundmass	6.11(2.76)	
K-feldspar/albite	0.34	
Li-"mica"/K-feldspar	11.53	
Li-"mica"/albite	3.92	
restite/leucosome Peña Negra, Spain		Bea et al. (1994b)
K-feldspar/leucosome	3.05(0.3)	
plagioclase/leucosome	3.13(0.4)	
biotite/leucosome	15.5(0.9)	
garnet/leucosome	3.23(0.25)	
cordierite/leucosome	29.1(2.1)	
experimental synthetic metapelite		Evensen and London (1999)
quartz/melt	0.24	London and Evensen (2001)
K-feldspar/melt	0.14(0.11)	Evensen (2001)
albite/melt	0.19(0.13)	
oligoclase-andesine (An_{31})/ı	1.84(0.21)	
biotite/melt	0.46(0.07)	
muscovite/melt	1.35	
cordierite/melt	201 to 6.7	
experimental basalt-andesite		Brenan et al. (1998a)
olivine/melt	0.002	
orthpyroxene/melt	0.02	
clinopyroxene/melt	0.02	mafic melt
clinopyroxene/melt	0.01	intermediate melt
amphibole/melt	0.23	

quartz monzonite		Beus (1966)
K-feldspar/plagioclase	0.25	
K-feldspar/quartz	2.86	
K-feldspar/biotite	2.00	
biotite/quartz	1.43	
biotite/plagioclase	0.13	
coarse-grained biotite granite		Beus (1966)
K-feldspar/plagioclase	2.5	
K-feldspar/quartz	25	
K-feldspar/biotite	0.5	
biotite/quartz	50	
biotite/plagioclase	5	
muscovite granite		Beus (1966)
muscovite/plagioclase	5	
muscovite/quartz	250	
plagioclase/quartz	50	
coarse-grained "albitized" granite		Beus (1966)
muscovite/feldspar	2.45	
muscovite/quartz	49	
feldspar/quartz	20	
coarse-grained "greisenized" muscovite granite		Beus (1966)
muscovite/feldspar	2	
muscovite/quartz	31	
feldspar/quartz	16	

cordierite » calcic oligoclase ≥ muscovite > biotite > quartz ≅ albite ≅ K-feldspar (4)

The position of plagioclase in this order depends on its anorthite content. The Be content varies with the An content and reaches a maximum around the oligoclase-andesine boundary (Beus 1966; Kosals et al. 1973). In our recent experimental work, we have established a partition coefficient of 1.84 for plagioclase of An_{31} (Evensen 2001), which compares with a value of 0.19 for albite (Table 2). The partitioning sequence of (4) above is essentially the same for natural versus experimentally determined values except in the relative position of quartz. Our recent experimental results (London and Evensen 2001) show Be to be more compatible in quartz than in the alkali feldspars, but this is not the case in natural minerals (Grew, this volume). We do not know if this is a vagary of our experiments, or if natural quartz tends to lose Be during retrograde recrystallization, which is prevalent in igneous and especially metamorphic rocks and manifested by ductile deformation, healed cracks, etc. The distribution of Be between quartz phenocrysts and their associated melt inclusions in volcanic ignimbrites could resolve the question if the quartz-inclusion pairs truly represent a quenched magmatic system near equilibrium, but those data do not appear to exist at the present time.

Values of $D_{Be}^{mica/fds}$ are on the order of 1 to 10, so that mica-rich protoliths (meta-pelites) generally represent the most fertile sources of Be-enriched magmas. In natural micas, $D_{Be}^{musc/biot}$ appears to be greater than 1 (Grew, this volume), which is consistent with the experimental results reported by London and Evensen (2001) in Table 2.

Partition coefficients for garnet are similar to those of the feldspars (Bea et al. 1994b).

The role of cordierite during anatexis. Cordierite stands out as an important reservoir of Be. Among the common rock-forming minerals, Be is highly compatible in cordierite, and partition coefficients for Be between cordierite and all other coexisting minerals are large. The experimentally derived values of $D_{Be}^{Crd/melt}$ cited in Table 2 (Evensen and London 1999) range from 201 to 7 between 700° and 850°C at 200 MPa H_2O. These data fit the linear regression,

$$D_{Be}^{Crd/melt} = -1.37\,T + 1145, \qquad (5)$$

with $R^2 = 0.88$ over this temperature interval. In cordierite, Be goes into a $T^{[4]}$ site for Al, with two different substitution mechanisms: $Na_{channel}BeAl_{-1}$ or $BeSiAl_{-2}$ (Hölscher and Schreyer 1989; Hawthorne and Huminicki, this volume; Grew, this volume). Experiments reveal that these two substitution mechanisms vary inversely with temperature: $Na_{channel}BeAl_{-1}$ predominates at low temperature whereas $BeSiAl_{-2}$ predominates at high temperature (Evensen and London 1999). At the temperatures of anatexis, the exchange component $BeSiAl_{-2}$ controls the compatibility of Be in cordierite and hence the distribution of Be between cordierite and melt. There is extensive miscibility between beryl and cordierite or its hexagonal high-temperature polymorph indialite, such that complete miscibility exists between these phases above 900°C at 200 MPa H_2O (Fig. 4).

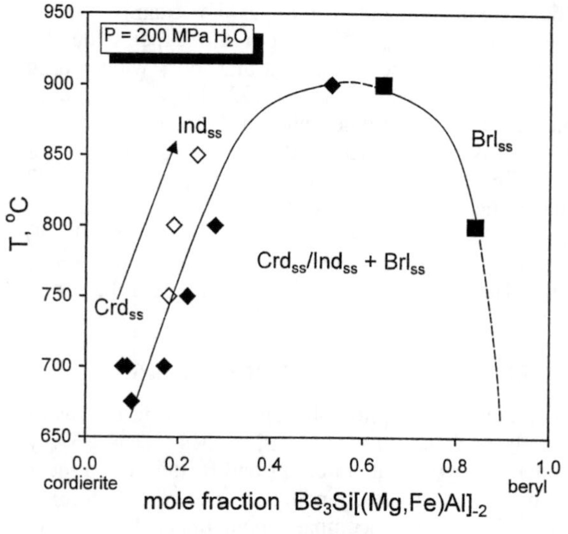

Figure 4. Experimental miscibility relations between cordierite (Crd) or indialite (Ind) [diamonds] and beryl [■] in granitic melt as a function of temperature at 200 MPa H_2O. Coexisting Crd/Ind and beryl are solid symbols; experiments reulting in only Crd/Ind are open diamonds. Uncertainties in temperature and analytical errors are smaller than the symbols used. The exchange operator which converts Crd (at $X = 0.0$) to Brl (at $X = 1.0$) accounts for ~65% to 85% of the composition of the synthetic minerals. The solvus is hand-drawn. Modified from Evenson (2001) based on the SIMS analyses of experimental run products.

Summary of partitioning data. The values of $D_{Be}^{min'l/min'l}$ summarized in Table 2 are mostly similar among the various studies. There is generally good agreement between the natural and experimental values of $D_{Be}^{min'l/min'l}$, which lends some support for the experimental $D_{Be}^{min'l/melt}$ values as well (Evensen and London 1999). The distribution of Be between minerals and groundmass or leucosome (Table 2) stand out as anomalous because they show moderately to strongly compatible behavior for Be in all rock-forming minerals except K-feldspar in the ongonites (Kovalenko et al. 1977; Bea et al. 1994b). Were these values correct, then partial melts starting with a few ppmw Be could never attain beryl saturation. We suggest that the experimental values (Evensen and London 1999; London and Evensen 2001; Evensen 2001) constitute a more reliable data set.

The Be content of aluminous anatectic melts will be controlled strongly by the participation of micas and cordierite in melting reactions. White micas are expected to melt or decompose completely at most conditions of anatexis, and the process of incomplete melting of micas followed at higher temperature by reactions such as:

$$\text{Ms} = \text{K-feldspar} + \text{Cor} + H_2O \text{ or } \text{Ms} + \text{Qtz} = \text{K-feldspar} + \text{Als} + H_2O \quad (6)$$

can effectively transfer a large fraction of the trace element content of white micas (dark micas, too) to a small fraction of partial melt (London 1995). In cordierite-bearing rocks, virtually all Be will reside in cordierite, and the Be budget of partial melts will be almost entirely controlled by the degree to which this phase participates in melting reactions. For the cordierite-bearing migmatites of Peña Negra, central Spain, Bea et al. (1994b) cite an average of 2.3 ppmw Be in the paleosome of the migmatites, and 6.0 and 1.2 ppmw Be respectively in the melanosome (restite-rich) and leucosome (melt-rich) portions of the migmatitic rocks. This yields an apparent restite(melanosome)-melt(leucosome) distribution coefficient of 5, which is close to the partition coefficient for Be between cordierite and melt at high temperature (Evensen and London 1999). Del Moro et al. (1999) also document the distribution of Be between various lithologic facies of migmatites in Italy, but the Be contents are so low throughout the different rock types (reported as 0.00-2.87 ppmw Be) that no trends are evident.

Using partition coefficients to estimate the Be contents of melts. The partition coefficients presented in Table 2 could be used to assess the Be content of melts by the combination of mineral/mineral coefficients with mineral/melt values. For example, values of $D_{Be}^{min'l/K\text{-}fds}$ are known, as are the value of $D_{Be}^{K\text{-}fds/melt}$ from the experiments by Evensen and London (1999). Thus, in relation to one of the best and most widely used indicators of chemical fractionation in granite-pegmatite systems (K-feldspar; e.g., Černý 1994), the concentration of Be in melt could be ascertained and used, for example, as a prospecting tool for Be-rich pegmatites. Taking the range of 200 to 900 ppmw BeO for beryl-saturated melts, K-feldspar in equilibrium with those melts should contain ~5 to 27 ppmw BeO. These values closely match the range of values for K-feldspar samples that have been analyzed for Be (see citations in Parsons (1994) and Roda Robles et al. (1999) for pegmatites that are not distinctly beryl-rich). This method does offer promise for exploration and evaluation of pegmatitic Be deposits specifically, but it cannot be tested with the existing database.

THE BERYLLIUM CONTENT OF SILICIC IGNEOUS ROCKS

The data base for Be in rhyolites and granites is quite robust in comparison to that of pegmatites. However, Be is not routinely analyzed in whole-rock studies (because x-ray fluorescence methods dominate in modern analytical laboratories), and Be does not appear as a keyword in citations of relevant petrologic articles. Finding these data, therefore, is hit-or-miss. The data sources gleaned for this review encompass about 400 analyses in all. The purposes of this literature survey are to establish mean Be contents of obsidians, rhyolites, and granites at various degrees of chemical fractionation, and to explore correlations among Be, other elements, and the protolith (source type) or tectonic setting of the magma source.

Obsidians and rhyolites

As a whole, obsidians represent the closest approximation to silicic melt compositions that are readily available in the geological literature. Though trapped melt inclusions may more accurately preserve the volatile components of melts, analyses of these are comparatively few.

Subalkaline silicic rhyolites. Macdonald et al. (1992) compiled a large chemical database for subalkalic silicic obsidians, which will be referred to here as the USGS (for

U S Geological Survey) data set. Their Appendix I, though heavily skewed toward North American localities, contains 152 analyses that include Be along with other trace and major elements. For the total data set, the average Be content is 4 ppmw with a standard deviation (s.d.) of 3 ppmw. In terms of tectonomagmatic groups recognized by Macdonald et al. (1992), rhyolite obsidians from continental interiors contain the highest average Be content with 6.9 ppmw (range: 1.8-32 ppmw). Note that the Be contents of melt inclusions in quartz from undifferentiated Paleozoic tonsteins (kaolinized K-rich metaluminous to mildly peraluminous rhyolites of continental origin) are similar with a mean 7.6 ppmw Be (s.d. = 4.6, N = 27: Webster et al. 1995). Next highest are obsidians from islands over "oceanic extensional zones", with an average 4.7 ppmw Be and a narrower range of 1.9-5.8 ppmw. Obsidians derived from subduction-related volcanism at continental margins and mature island arcs mostly fall well below the average for the entire data set, with means of 2.6 (1.3-5.5) and 1.6 (0.86-3.2) ppmw Be, respectively. As a whole, these trends broadly reflect the more fractionated compositions of continental rocks, and perhaps more specifically that continental crustal rocks contain a larger fraction of minerals that can more readily accommodate Be, such as the micas.

A correlation matrix for the USGS data set (Macdonald et al. 1992) reveals the most positive associations between Be and F (0.82), Sn (0.82), Rb (0.75), Pb (0.74) U (0.70), Li (0.68), and the REE (see Figs. 5e-f, 5h, and 5k). Positive correlations of Be with Sn, Rb, U, and Li are well documented in these and other geochemical discrimination plots, but these do not uniquely distinguish the sources of the magmas. Enrichment in this suite of elements is as characteristic of intracontinental rift-related rhyolites (A-types) as of collision-related metasedimentary S-types (e.g., Christiansen et al. 1986; Černý et al. 1985; Černý and Meintzer 1988; Černý 1991a; London 1995). The USGS data set can be scanned as a whole, including the data from "continental interiors" and "continental margins", to identify magmas with a large S-type component, as these are distinctly peraluminous and tend to have high K* (atomic K/[Na+K]) (White and Chappell 1983). The data set contains very few peraluminous obsidians (ASI or A/CNK $\geq$ 1.1) with the exception of the green glass pebbles from Macusani, Peru (Noble et al. 1984), which have been omitted from the plots in Figure 5. The poorly defined trend of decreasing ASI with increasing Be (Fig. 5c) may be best ascribed to Be enrichment in the mildly alkaline A-type rhyolites, which are included in the USGS data set but are discussed separately below. Although there is a weak positive correlation between K* and ASI (Fig. 5b), which might be indicative of mica-rich sources, we note that the peraluminous rhyolites with ASI > 1.1 actually lie below the trend line for the entire USGS data set. To further identify the S-type protoliths in the USGS data, the values for Be were plotted against Cs* (atomic 100*Cs/K) and Ta* (atomic Ta/[Ta+Nb]). In addition to Li (Fig. 5f), these ratios are diagnostic of the suite of mostly S-type, granite-associated ore deposits with the LCT (Li-Cs-Ta) chemical signature (Černý 1991a). Figures 5j and 5l show no correlation of Be with Cs*, or Ta* for the USGS data. In addition, the mean values of Cs* or Ta* are low, 45 and 8 respectively, in comparison, for example, to the strongly peraluminous and distinctly S-type Macusani rhyolite obsidian (e.g., Pichavant et al. 1988), for which Cs* = 4670 and Ta* = 30. We conclude that the only S-type obsidian in the USGS data set is from the Macusani rhyolite, Peru.

Figure 5. Plots of Be versus other elements and chemical indices of 151 subalkalic silicic obsidians presented in Appendix 1 of Macdonald et al. (1992). Though reported by Macdonald et al. (1992), these plots do not contain the green obsidian glass from the Macusani volcanic province of southeastern Peru (Barnes et al. 1970) because of the extreme skewness that results from including this obsidian in the plots. **Figures 5a-l are on the following four pages.**

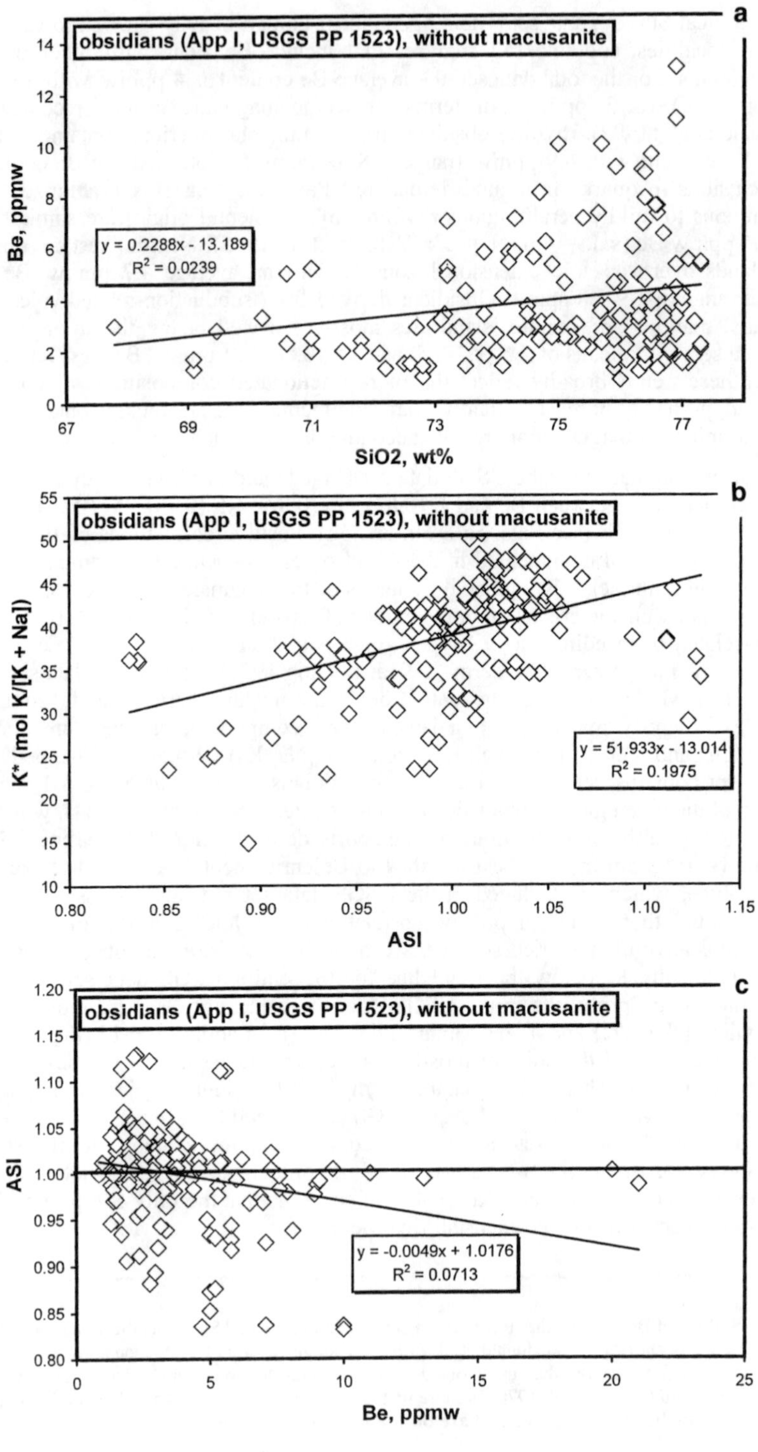
a
obsidians (App I, USGS PP 1523), without macusanite
y = 0.2288x - 13.189
R² = 0.0239
Be, ppmw
SiO2, wt%
b
obsidians (App I, USGS PP 1523), without macusanite
y = 51.933x - 13.014
R² = 0.1975
K* (mol K/[K + Na])
ASI
c
obsidians (App I, USGS PP 1523), without macusanite
y = -0.0049x + 1.0176
R² = 0.0713
ASI
Be, ppmw

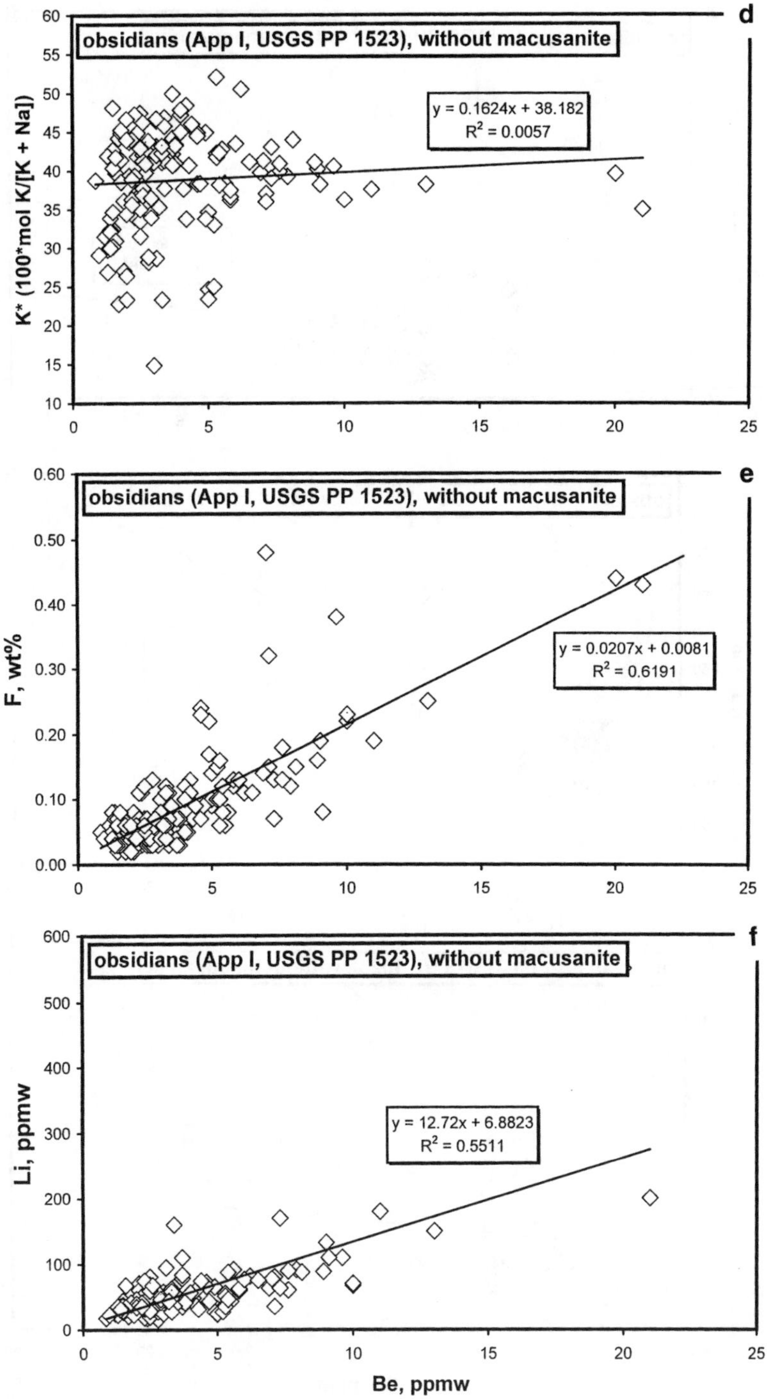
d
obsidians (App I, USGS PP 1523), without macusanite
y = 0.1624x + 38.182
R^2 = 0.0057
K* (100*mol K/[K + Na])
e
obsidians (App I, USGS PP 1523), without macusanite
y = 0.0207x + 0.0081
R^2 = 0.6191
F, wt%
f
obsidians (App I, USGS PP 1523), without macusanite
y = 12.72x + 6.8823
R^2 = 0.5511
Li, ppmw
Be, ppmw

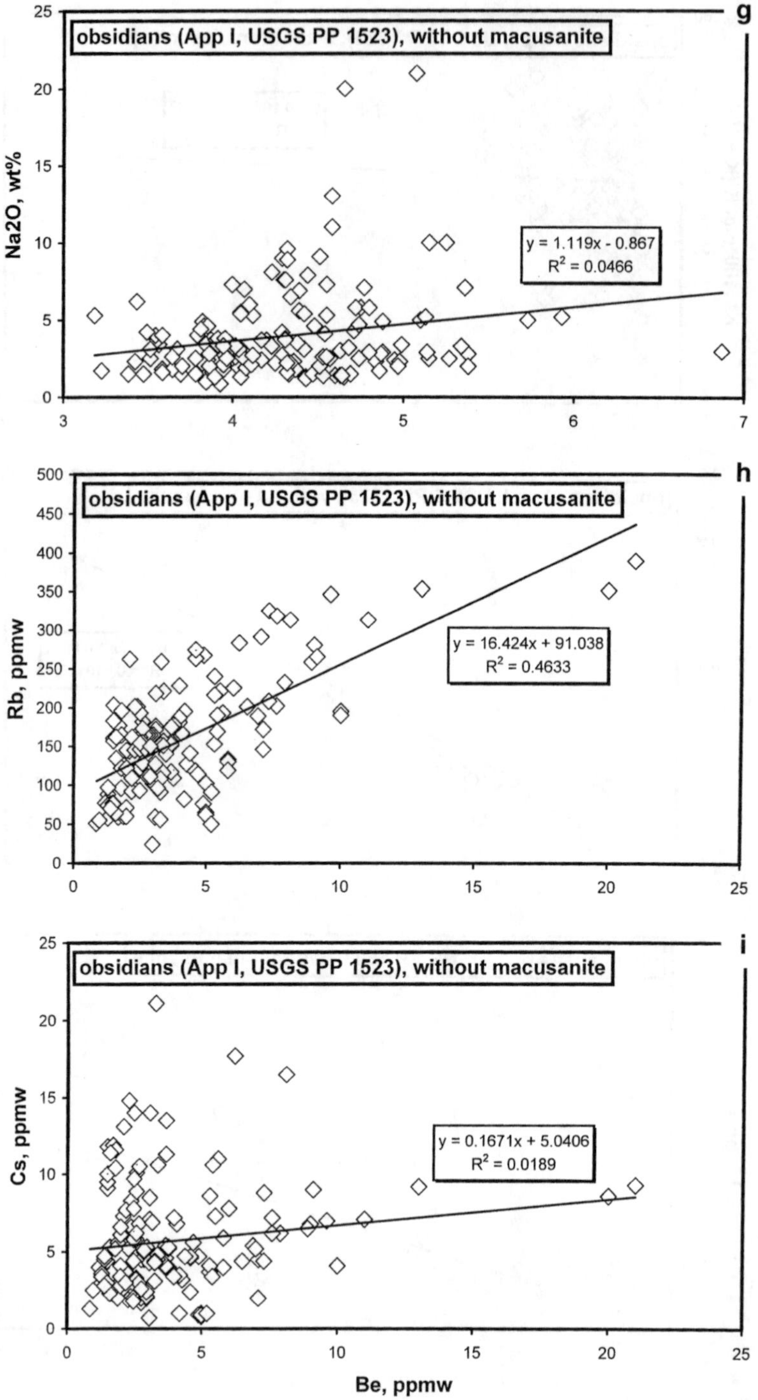
g
obsidians (App I, USGS PP 1523), without macusanite
Na2O, wt%
y = 1.119x - 0.867
R² = 0.0466
0
5
10
15
20
25
3
4
5
6
7
h
obsidians (App I, USGS PP 1523), without macusanite
Rb, ppmw
y = 16.424x + 91.038
R² = 0.4633
0
50
100
150
200
250
300
350
400
450
500
0
5
10
15
20
25
i
obsidians (App I, USGS PP 1523), without macusanite
Cs, ppmw
y = 0.1671x + 5.0406
R² = 0.0189
0
5
10
15
20
25
0
5
10
15
20
25
Be, ppmw

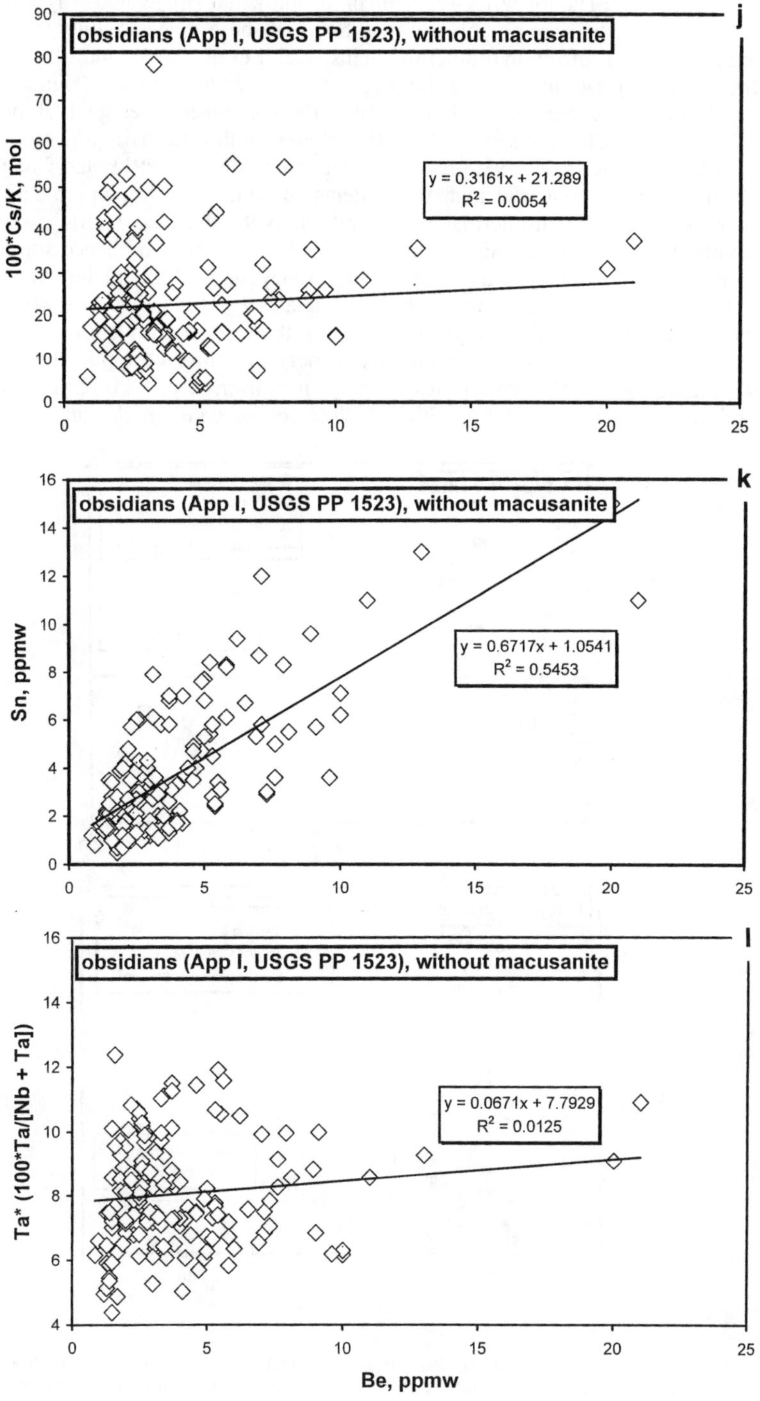
j
obsidians (App I, USGS PP 1523), without macusanite
y = 0.3161x + 21.289
R² = 0.0054
100*Cs/K, mol
k
obsidians (App I, USGS PP 1523), without macusanite
y = 0.6717x + 1.0541
R² = 0.5453
Sn, ppmw
l
obsidians (App I, USGS PP 1523), without macusanite
y = 0.0671x + 7.7929
R² = 0.0125
Ta* (100*Ta/[Nb + Ta])
Be, ppmw

A-type rhyolites. The topaz (A-type, or intracontinental rift-related) rhyolites of the western U.S. and Mexico are notably Be-rich. Some of these volcanic units in Utah have produced gem red beryl from hydrothermal veins (Keith et al. 1994), and others enriched in bertrandite comprise the largest Be deposits in North America (Lindsey 1977). Rhyolites from Utah exclusive of Honeycomb Hills contain an average 14.1 ppmw Be (s.d. = 7.5, N = 7: Christiansen et al. 1986). Honeycomb Hills is anomalous with an average 87.7 ppmw Be (s.d. = 91.3, N = 6: Congdon and Nash 1991), and Congdon and Nash (1991) proposed that the high Be contents of some samples, up to 267 ppmw, result from subsolidus hydrothermal concentration. Note that the high Be content of Honeycomb Hills is not matched by similarly high F, Li, or Sn, and hence supports the hypothesis that anomalously high values in the Honeycomb Hills are not the result of igneous fractionation. In the topaz rhyolite data set, Be shows weakly positive correlations with Rb, Cs, and Sn (Figs. 6d and 6g,h); there is no correlation of Be with Li or F (Figs. 6b,f).Melt inclusions from quartz phenocrysts in tin-rich topaz rhyolites from the Sierra Madre Occidental, central Mexico, contain an average of 18.0 ppmw Be (s.d. = 4.4, N = 35) (Webster et al. 1996), which is close to the whole-rock values for similar

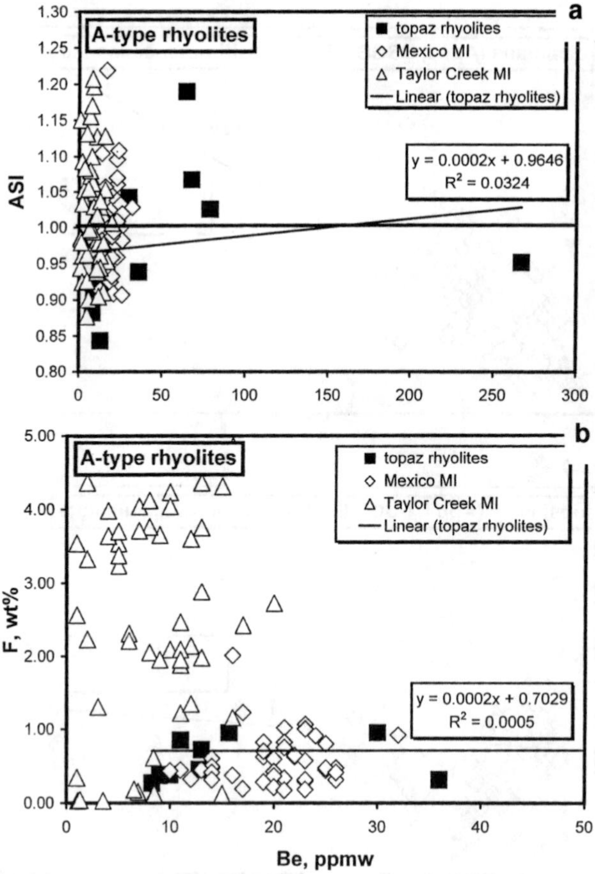

Figure 6. Plots of Be versus other elements and chemical indices of topaz rhyolites from the western U.S. (from Christiansen et al. 1986), and vitreous melt inclusions (MI) from F-rich rhyolites of New Mexico (Webster et al. 1991a) and Mexico (Webster et al. 1996).

Figure 6

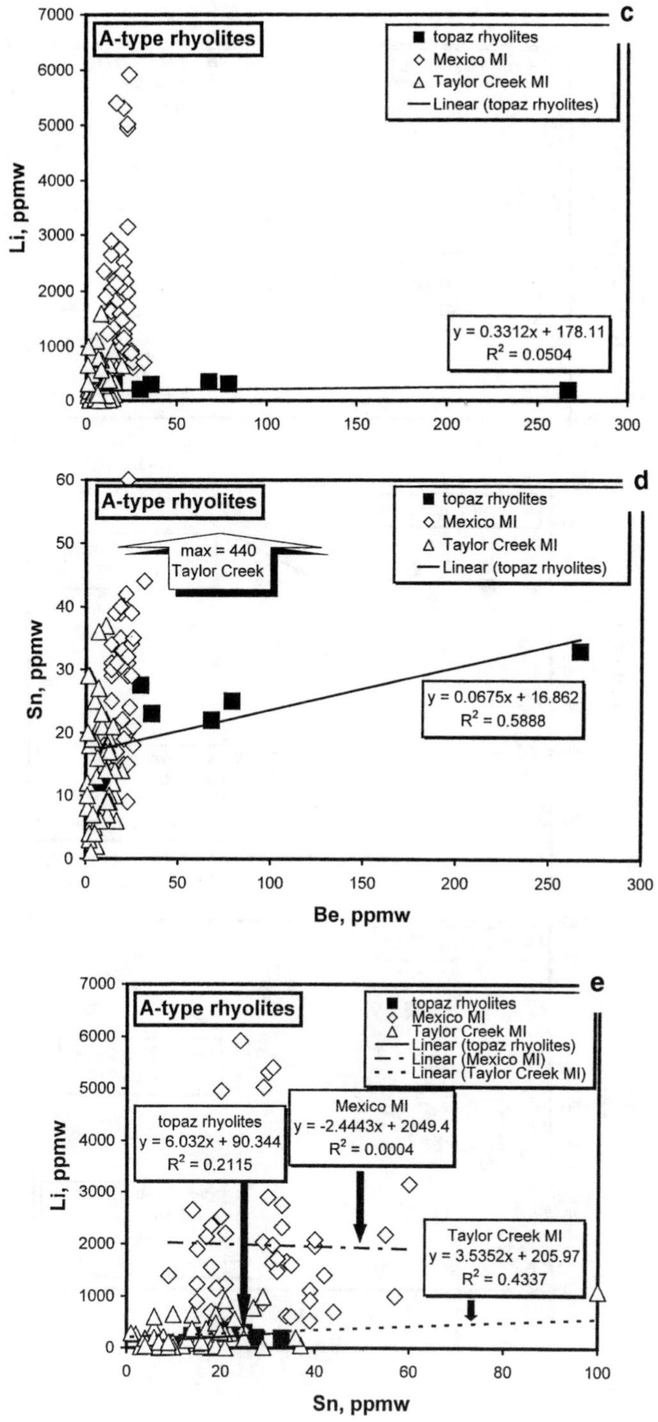
c
A-type rhyolites
topaz rhyolites
Mexico MI
Taylor Creek MI
Linear (topaz rhyolites)
y = 0.3312x + 178.11
R² = 0.0504
Li, ppmw
7000
6000
5000
4000
3000
2000
1000
0
0
50
100
150
200
250
300
d
A-type rhyolites
topaz rhyolites
Mexico MI
Taylor Creek MI
Linear (topaz rhyolites)
max = 440
Taylor Creek
y = 0.0675x + 16.862
R² = 0.5888
Sn, ppmw
60
50
40
30
20
10
0
Be, ppmw
e
A-type rhyolites
topaz rhyolites
Mexico MI
Taylor Creek MI
Linear (topaz rhyolites)
Linear (Mexico MI)
Linear (Taylor Creek MI)
topaz rhyolites
y = 6.032x + 90.344
R² = 0.2115
Mexico MI
y = -2.4443x + 2049.4
R² = 0.0004
Taylor Creek MI
y = 3.5352x + 205.97
R² = 0.4337
Li, ppmw
80
100
Sn, ppmw

Figure 6

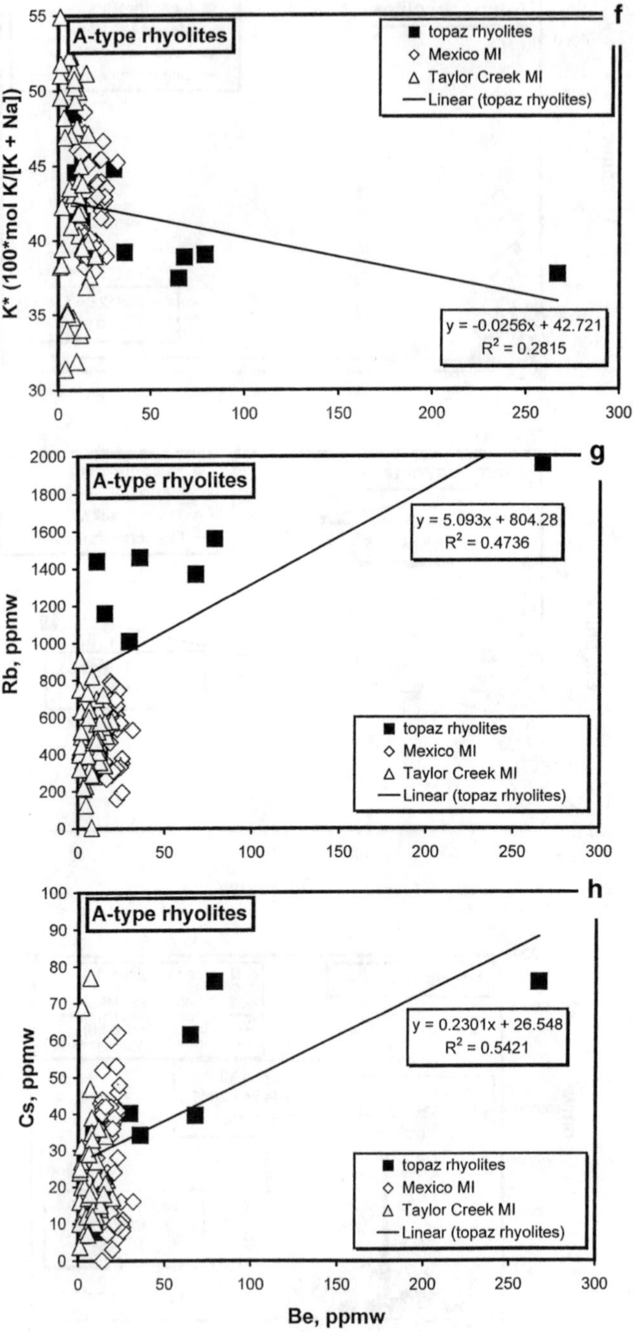
f
A-type rhyolites
topaz rhyolites
Mexico MI
Taylor Creek MI
Linear (topaz rhyolites)
y = -0.0256x + 42.721
R² = 0.2815
K* (100*mol K/[K + Na])
g
A-type rhyolites
y = 5.093x + 804.28
R² = 0.4736
Rb, ppmw
topaz rhyolites
Mexico MI
Taylor Creek MI
Linear (topaz rhyolites)
h
A-type rhyolites
y = 0.2301x + 26.548
R² = 0.5421
Cs, ppmw
topaz rhyolites
Mexico MI
Taylor Creek MI
Linear (topaz rhyolites)
Be, ppmw

rhyolites in Utah. Analyses of melt inclusions from the F- and Sn-rich Taylor Creek rhyolite, New Mexico, are grossly similar to those from the Mexican localities except for lower average Be contents of 8 ppmw (s.d. = 4.7, N = 53: Webster and Duffield 1991a,b). Figure 6 illustrates, however, that there is no correlation of Be with any of the typical fractionation indices in the melt inclusions of either data set because there is relatively little change in Be abundance while there are large variations in the other fractionation indices. The lack of correlation of Be with the other incompatible lithophile trace elements is surprising, and distinct from the trends of the USGS data set (cf. Figs. 5 and 6). While plots such as Figures 6c and 6d might give the impression that elements such as Li and Sn should be highly correlated in the melt inclusions, Figure 6e demonstrates that this is not necessarily so. Though there is a positive correlation of Li with Sn in the Taylor Creek inclusions, there is no correlation in the melt inclusions from the topaz rhyolites of Mexico. In total, the melt inclusion data appear to reflect a bulk crystal/melt distribution coefficient of 1 for Be, which is possible when, for example, calcic oligoclase represents a large component of the fractionating crystalline assemblage (discussed further below). It is not our purpose, however, to interpret the melt inclusion data in relation to the other geochemical aspects of these rhyolites. Overall, the F- and Sn-rich A-type rhyolites and their trapped melt inclusions establish a mean Be content of about 14-18 ppmw in these evolved melts, and that is the important point for this assessment.

S-type rhyolites. The highest single value in the USGS obsidian data set (32 ppmw Be) comes from the green glass nodules from the Macusani volcanic center in southeastern Peru. The less-altered tuffs and pumices from the Macusani province contain 28 (s.d. = 9, N = 4) ppmw, with a high of 37 ppmw Be (Noble et al. 1984). These rhyolites are S-type (metasedimentary protoliths) in overall chemical character with high $^{87}Sr/^{86}Sr_0$ of 0.7309, strongly peraluminous compositions, and trace-element patterns that follow from an abundance of argillic or micaceous material at the source (Pichavant et al. 1988). To the southeast in the Bolivian tin belt, the voluminous Los Frailes and Morococala volcanic fields consists of similarly peraluminous S-type ignimbrites (Ericksen et al. 1990; Morgan et al. 1998). The Los Frailes volcanics vary from dacites to rhyolites; they are, as a group, less fractionated than the Morococala volcanics, which range in composition from quartz latites to cordierite-andalusite rhyolites to muscovite-andalusite rhyolites (Morgan et al. 1998). This is evident in the trace-element systematics of the two fields. The mean Be content of the Los Frailes volcanics is 4.3 ppmw (s.d. = 1.0, N = 26: Luedke et al. 1997), whereas that of the Morococala field is 10.0 ppmw Be (s.d. = 4.6, N = 12: Luedke et al. unpublished data). Other fractionation indices, e.g., Li and Rb, similarly reveal the greater chemical fractionation of the Morococala field over that of Los Frailes (Fig. 7a). Note that although Li and Rb are well correlated in the Morococala field, this is not true of the Los Frailes volcanics. Figure 7b illustrates that Be increases with ASI, which corresponds to the general trend of increasing fractionation in both volcanic fields. The correlation of Be with ASI is more pronounced for the Los Frailes rocks than for the Morococala suite, because all of the Morococala units possess high ASI values near the magmatic saturation limit (discussed below under granites). Increasing Be is weakly to negatively correlated with K* (Fig. 7b), as is the case for the A-type rhyolites (Fig. 6f). A decrease in K* with increasing fractionation, as is seen in both S- and A-type volcanics, mirrors a similar trend observed in granitic rocks. The decrease in K* is thought to stem from homogeneous reactions in melt of anions such as F, P, and B with Na, which drives residual liquids to Na-richer compositions as more potassic phases (K-feldspar, joined by mica) increasingly dominate the crystalline mode (London 1997). The correlation of Be with Li, Rb, and Sn in the two Bolivian volcanic fields (Fig. 7c) ranges from good (with Rb in the Los Frailes volcanics) to poor to negative (with Rb in the Morococala field).

Both Bolivian volcanic fields are chemically less evolved than those of the Macusani province to the northwest (cf. Pichavant et al. 1988 with Luedke et al. 1997; Morgan et al. 1998). Given that peraluminous S-type rhyolites are, globally, comparatively few in number, the much-studied Macusani volcanics clearly should not be construed as typical of these magma types. Using the available data, the Be contents for these S-type volcanics in Bolivia and Peru span the full range seen for all obsidians in the USGS data set.

Summary of data on rhyolites. Without further assessment of the individual sources, the averages for the four groupings used by Macdonald et al. (1992) probably represent suitable starting points for modeling the fractionation of Be in most large-volume silicic magma bodies. All of the obsidians in the USGS data set possess negative Eu anomalies, which become more pronounced toward the magmas from "continental interiors". Hence, various degrees of restite separation and crystal-liquid fractionation have occurred in all of these magmas. The most fractionated rhyolites include the topaz rhyolites of the western U.S.A. and Mexico, and the Macusani rhyolite and obsidian from Peru (e.g., Christiansen et al. 1986; Pichavant et al. 1988). Their Be contents are the highest among the rhyolite data sets. We now compare these values with those from a diverse suite of plutonic rocks.

Granites

Unlike the data sets for rhyolites, which collectively give good coverage of Be in obsidians, rhyolites, and associated melt inclusions from only a few published sources, the sources of data on Be in granites are more numerous but less comprehensive—meaning that the data sets are essentially limited to individual plutons. Also, in contrast with the USGS obsidian suite, peraluminous and probably S-type granites comprise a large proportion of the data cited in Table 3. The sources of the peraluminous granites are skewed to European localities, most of which contain accessory cordierite or are likely to have originated in cordierite-bearing sources. In Figure 8, Be is plotted against the same geochemical parameters as in Figures 5-7 for most of the granite sources.

Because of the high compatibility of Be in cordierite, we can recognize in Table 3 granites that have achieved a high degree of chemical fractionation but are poor in Be. These include the granites of Spain and the Czech Republic, where at least some facies in each data set can be regarded as highly fractionated (Bea et al. 1994a; Breiter et al. 1991; Breiter and Scharbert 1995; Ramirez and Grundvig 2000). When grouped as representative of a region, the Jalama and Pedrobernardo plutons in Spain (Bea et al. 1994a; Ramirez and Grundvig 2000) and the Argemela microgranite dikes in Portugal (Charoy and Noronha 1996) display positive correlations of Be with Li, Rb, and Sn (Fig. 8a). This is actually an artifact of the two different populations, the Spanish granites on one end and the Portuguese dikes on the other. As typical representatives of the central Iberian granite system, the Jalama and Nisa-Alburquerque plutons are similar: they are peraluminous with cordierite-bearing interior facies and evolved tourmaline-bearing marginal facies (Ramirez and Menendez 1999; Ramirez and Grundvig 2000). Cordierite (whether magmatic or restitic) likely constitutes the principal mineralogical reservoir of Be, and hence variations of Be with fractionation are relatively small (i.e., note the small standard deviations for Be in Table 3). There is no correlation of Be with the typical trace elements cited above for the rhyolite data set. In the Nisa-Alburquerque batholith, for example, the correlation (R^2) of Be with Rb is 0.10 and with Sn it is 0.03. In the Pedrobernardo pluton of central Spain, Bea et al. (1994a) proposed that the uppermost of three internal facies was derived from the lower portions by a melt extraction process

Figure 7 (on the next page). Plots of Be versus other elements and chemical indices in S-type peraluminous volcanic rocks from the Los Frailes and Morococala volcanic fields of Bolivia (Luedke et al. 1997; Morgan et al. 1999).

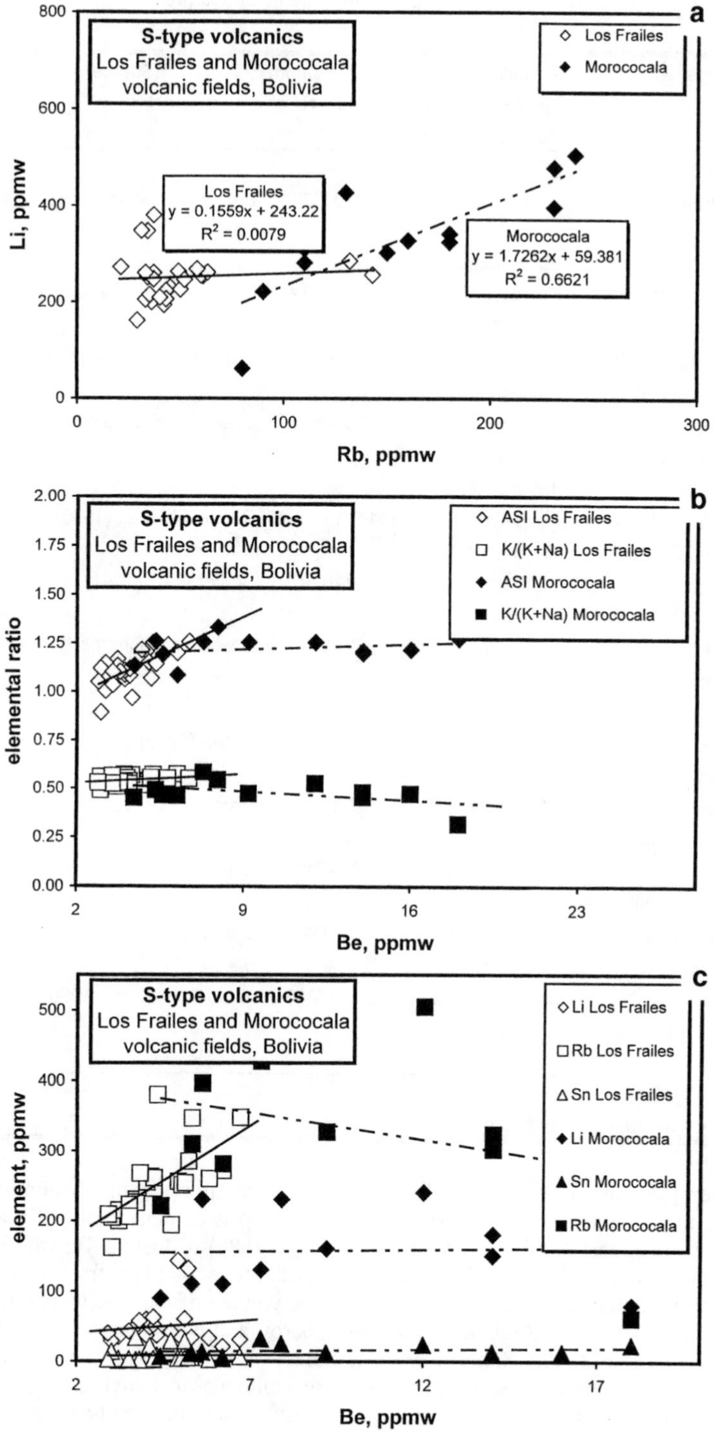
a
S-type volcanics
Los Frailes and Morococala
volcanic fields, Bolivia
Los Frailes
Morococala
Los Frailes
y = 0.1559x + 243.22
R² = 0.0079
Morococala
y = 1.7262x + 59.381
R² = 0.6621
Li, ppmw
Rb, ppmw
0
200
400
600
800
0
100
200
300
b
S-type volcanics
Los Frailes and Morococala
volcanic fields, Bolivia
ASI Los Frailes
K/(K+Na) Los Frailes
ASI Morococala
K/(K+Na) Morococala
elemental ratio
Be, ppmw
0.00
0.25
0.50
0.75
1.00
1.25
1.50
1.75
2.00
2
9
16
23
c
S-type volcanics
Los Frailes and Morococala
volcanic fields, Bolivia
Li Los Frailes
Rb Los Frailes
Sn Los Frailes
Li Morococala
Sn Morococala
Rb Morococala
element, ppmw
Be, ppmw
0
100
200
300
400
500
2
7
12
17

Table 3. Average beryllium content of granites.

Location	Granite Type Element Enrichment Other Comments	Average Be ppm (S.D.)N	Source
Pedrobernardo Spain	S-Type P Crd[1] source	4.1(0.6)11	Bea et al. (1994a)
Jalama Spain	S-type P	7.2(3.1)13	Ramirez and Grundvig (2000)
Nisa-Alburquerque Portugal and Spain	S-type Crd-bearing	5.0(3.0)9	Ramirez and Menendez (1999)
Argemela Portugal	S-type P beryl-bearing	130(99)10	Charoy and Noronha (1996)
Beauvoir Massif Cent'l, Fr.	S-type Li, Sn, Ta, P	101.6(116.0)45	Raimbault et al. (1995)
Bohemia Czech Republic	S-type? Sn, P	12(10)10	Breiter et al. (1991)
Homolka Czech Republic	S-type P Crd source	2.3(0.7)10	Breiter and Scharbert (1995)
Milevsko Czech Republic	peraluminous felsite dikes	23(19)7	Vrána (1999)
Pelhrimov Czech Republic	felsite dikes	5.5(ND)4	Vrána (1999)
Lasenice Czech Republic	felsite dikes	1.6(ND)11	Vrána (1999)
East Transbaikalia Russia	S-type (?) Li, Ta, F	28(41)11	Zaraisky et al. (1997)
Fawwarah Saudi Arabia	peraluminous Sn	6.3(2.0)18	Kamilli and Criss (1996)

[1] Crd = cordierite
ND = only average value and number of analyses cited

such as filter pressing. Hence, most fractionation indices jumped up discontinuously between the upper facies and the rocks below. This jump is evident for Li, Rb, and Sn, but there is no change in Be among any of the layers (Fig. 8b). Constant Be values speak to a buffering reaction, as a small quantity of cordierite or a large fraction of calcic oligoclase or white mica might accomplish for a large volume of magma. In contrast, the microgranitic dike swarm at Argemela, Portugal (Charoy and Noronha 1996) is strongly enriched in Be (Table 3, Fig. 8c). Argemela is notable because beryl is reported as an accessory magmatic phase in some dikes; the Be contents of individual dikes reach 385 ppmw, and we presume that these values correspond to the beryl-saturated rocks.

Table 3, continued.

Ghost Lake	S-type	3.0(1.3)14	Breaks and Moore (1992)
Ontario, Canada	Li, Sn		
	Crd and garnet		
Jarna, Dala group	I-type	1.6(0.8)22	Ahl et al. (1999)
Sweden			
Siljan, Dala group	I-type	4.15(1.62)13	Ahl et al. (1999)
Sweden	high-level		
	fractionated		
Garberg, Dala group	I-type	4.9(2.32)10	Ahl et al. (1999)
Sweden	fractionated		
Eurajoki	A-type	6(4)10	Haapala (1997)
Finland	Sn		
Silvermine	A-type	5.5(0.71)2	Lowell and Young (1999)
Missouri, USA	mafic enclaves		
Red Bluff	A-type	5.73(0.46)2	Shannon et al. (1997)
Texas, USA	mafic and felsic		
Oka	NYF	3.1(3.3)20	Gold et al. (1986)
Quebec, Canada	peralkaline		
Strange Lake	NYF	160(115)14	Salvi & Williams-Jones (1996)
Quebec-Labrador	peralkaline		
Khaldzen-Buregtey	NYF	33-92	Kovalenko et al. (1995)
western Mongolia	peralkaline		
Tuva	alkali granites	3.4(ND)30	Popolitov et al. (1967)
Siberia-Mongolia	"Phase II"		
Tuva	alkali granites	5.3(ND)15	Popolitov et al. (1967)
Siberia-Mongolia	"Phase III"		
Tuva	plagiosyenites	1.7(ND)29	Popolitov et al. (1967)
Siberia-Mongolia			
Tuva	nepheline syenites	2.9(ND)29	Popolitov et al. (1967)
Siberia-Mongolia			
Fantale	pantellerite	10.0(1.2)19	Webster et al. (1993)
Ethiopia	melt inclusions		

[1] Crd = cordierite

ND = only average value and number of analyses cited

Although the dikes at Argemela contain tourmaline as the only mafic silicate, a few studies of likely basement sources and peraluminous Hercynian plutons of Portugal report that garnet, rather than cordierite, is the prevalent aluminous mafic silicate (Godinho 1974; Reavy 1987). The dominance of garnet rather than cordierite at the source could explain the sharply higher Be contents of the Portuguese rocks compared to those in Spain, and for the occurrence of beryl-bearing pegmatites in Galicia (Fuertes-Fuente and Martin-Izard 1998), to the north of the Portuguese locality.

From the Bohemian Massif of the Czech Republic, analyses of mostly S-type granites show a large variation of Be, from anomalously low Be in the Homolka granite (average 2.3 ppmw Be, Table 3) to values as high as 60 ppmw Be in one of a suite of evolved felsite porphyry dikes (Vrána 1999) not far from Homolka. The low values for such highly fractionated plutons as the Homolka granite correlate with their cordierite-bearing source rocks (Breiter and Scharbert 1995). By implication, those granites and dikes with substantially higher Be contents had fundamentally different, or at least cordierite-absent, source materials. With the highest Be values reported in the Bohemian Massif, the Milevsko felsite porphyries (devitrified subvolcanic dikes) described by Vrána (1999), range from 8 to 60 ppmw. Beryllium, however, does not correlate with F ($R^2 = 0.15$) or with Sn ($R^2 = 0.14$).

Among the highly fractionated granites cited in Table 3, the Beauvoir granites of the Massif Central, France, stand out for their Be enrichment. Values cited for the various facies of the Beauvoir granite in the GPF drill core range from 5.7 to 494 ppmw Be (Raimbault et al. 1995). In the Beauvoir series, fractionation increases from a basal B3 unit through B2 to an upper B1 facies. These three units are interpreted as a stacked sequence of successively more fractionated magmas, and in support of this argument, the large increase of Be correlates positively, though variably, with Li and Rb from B3 to B1 (Fig. 8d). Beryl is reported as a very rare phase in pegmatitic segregations within the B1 facies (Cuney et al. 1992). A large fraction of the Be resides in micas (up to 150 ppmw Be: Monier et al. 1987; Cuney et al. 1992, Ru et al. 1992), but at least locally, the phosphates beryllonite and hydroxylherderite contribute to the high whole-rock Be values (Charoy 1999). Thus, the paucity of beryl in this Be-rich magmatic suite stems from the increased compatibility of Be in Li-micas (Grew, this volume), and from the high activity of phosphorus in the system. A large portion of the ASI values reported by Raimbault et al. (1995) lies above the likely magmatic limit, 1.4 (Fig. 8j,k). This limit is based on experimentation involving silicate liquids in equilibrium with peraluminous crystalline phases including white micas and tourmaline, where ASI values of melt reach 1.3-1.4 (e.g., Icenhower and London 1995; Wolf and London 1997; London et al. 2000, 2001).

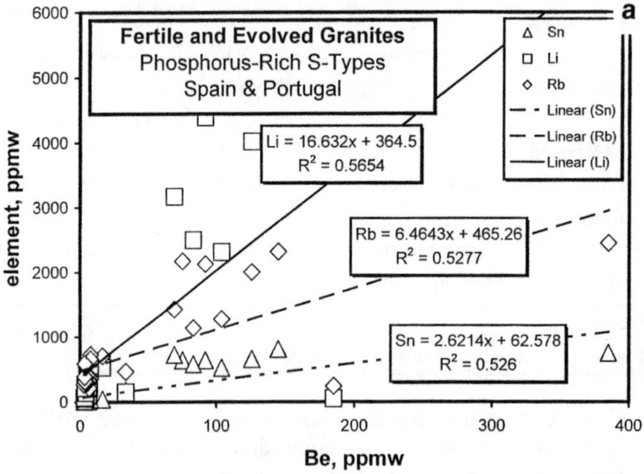

Figure 8. Plots of Be versus other elements and chemical indices in granitic rocks. Sources include the Jalama (Ramirez and Grundvig 2000) and Pedrobernardo (Bea et al. 1994a) plutons in Spain; microgranite dikes at Argemela, Portugal (Charoy and Noronha 1996); the Homolka granite (Breiter and Scharbert 1995) and various other localities in Bohemia, Czech Republic (Breiter et al. 1991); the Fawwarah granite and Silsilah Sn deposits of Saudi Arabia (Kamilli and Criss 1996); parental granites at Ghost Lake, Ontario (Kretz et al. 1989) and Lac du Bonnet, Manitoba, Canda (Černý and Meintzer 1988); rapikivi granites from Eurajoki, Finland (Haapala 1997); and peraluminous Ta-rich granites from Orlova and Etyka, Transbaikalia, Russia (Zaraisky et al. 1997). **Cont'd next 4 pages.**

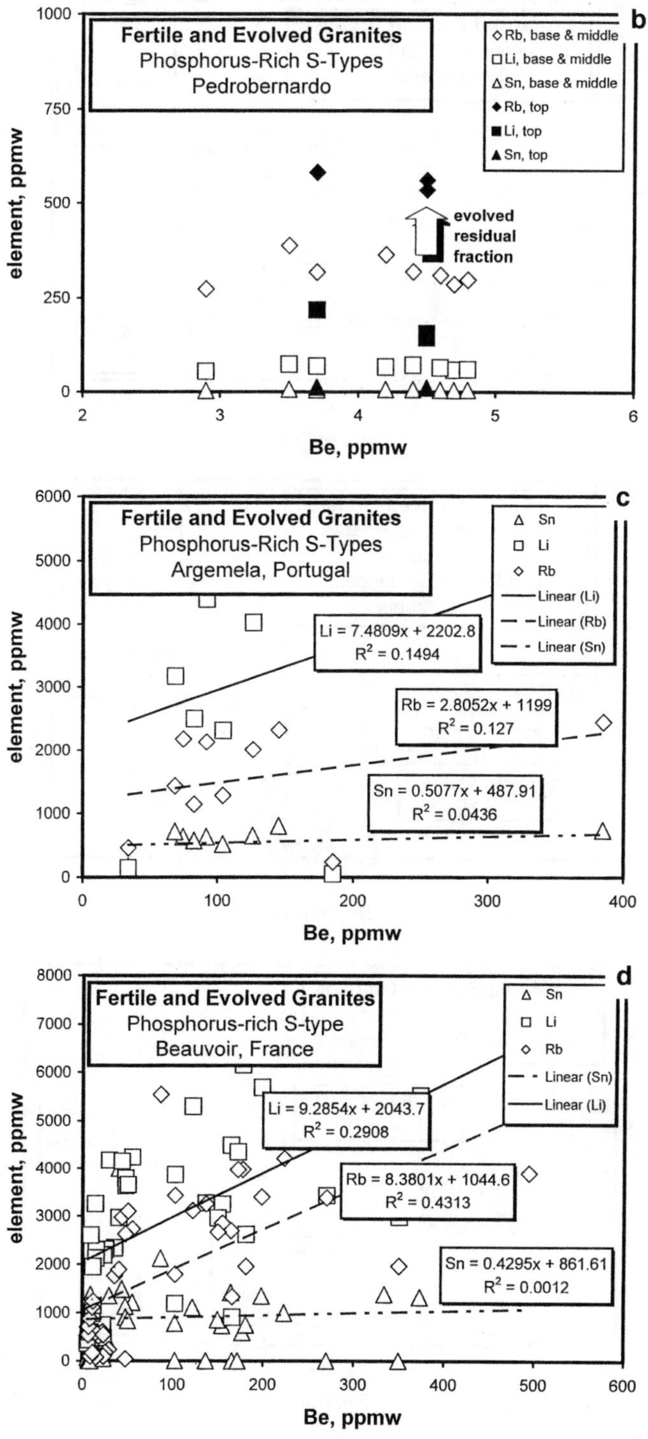
b
Fertile and Evolved Granites
Phosphorus-Rich S-Types
Pedrobernardo
Rb, base & middle
Li, base & middle
Sn, base & middle
Rb, top
Li, top
Sn, top
evolved residual fraction
element, ppmw
Be, ppmw
c
Fertile and Evolved Granites
Phosphorus-Rich S-Types
Argemela, Portugal
Sn
Li
Rb
Linear (Li)
Linear (Rb)
Linear (Sn)
Li = 7.4809x + 2202.8
R² = 0.1494
Rb = 2.8052x + 1199
R² = 0.127
Sn = 0.5077x + 487.91
R² = 0.0436
element, ppmw
Be, ppmw
d
Fertile and Evolved Granites
Phosphorus-rich S-type
Beauvoir, France
Sn
Li
Rb
Linear (Sn)
Linear (Li)
Li = 9.2854x + 2043.7
R² = 0.2908
Rb = 8.3801x + 1044.6
R² = 0.4313
Sn = 0.4295x + 861.61
R² = 0.0012
element, ppmw
Be, ppmw

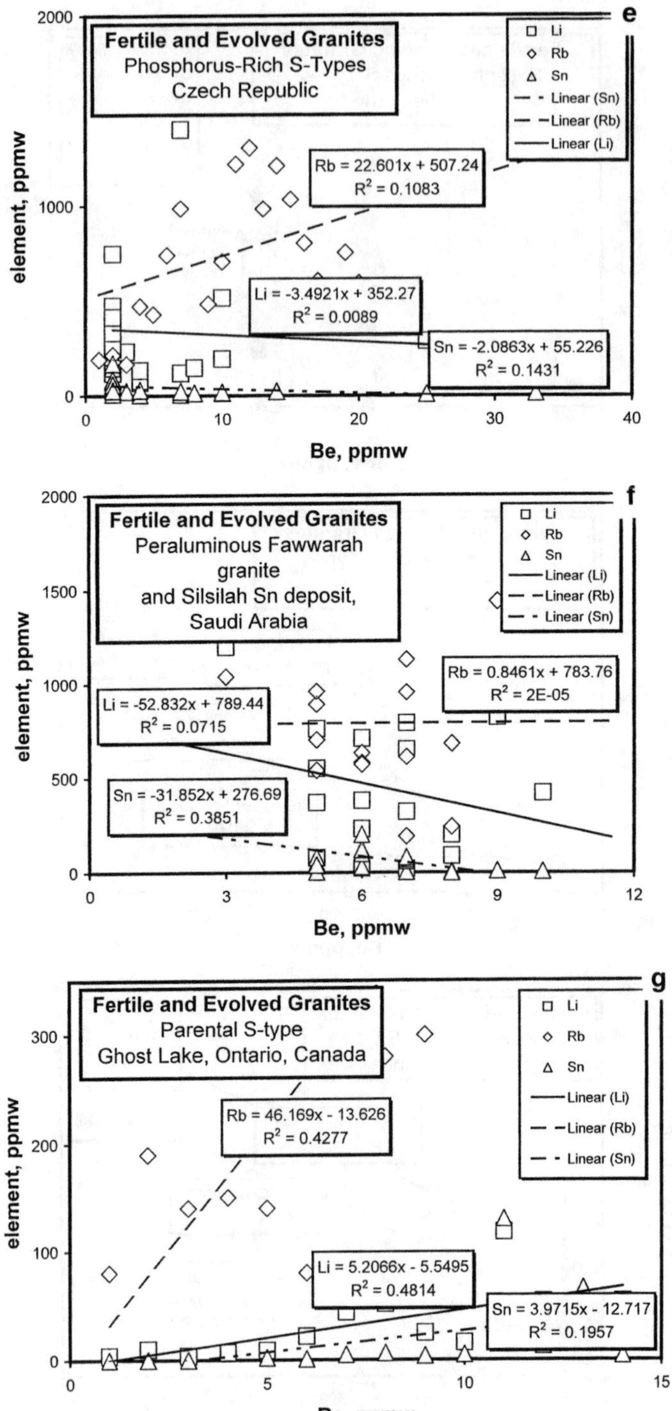
e
Fertile and Evolved Granites
Phosphorus-Rich S-Types
Czech Republic
Li
Rb
Sn
Linear (Sn)
Linear (Rb)
Linear (Li)
Rb = 22.601x + 507.24
R² = 0.1083
Li = -3.4921x + 352.27
R² = 0.0089
Sn = -2.0863x + 55.226
R² = 0.1431
element, ppmw
Be, ppmw
0
1000
2000
10
20
30
40
f
Fertile and Evolved Granites
Peraluminous Fawwarah granite
and Silsilah Sn deposit,
Saudi Arabia
Li
Rb
Sn
Linear (Li)
Linear (Rb)
Linear (Sn)
Rb = 0.8461x + 783.76
R² = 2E-05
Li = -52.832x + 789.44
R² = 0.0715
Sn = -31.852x + 276.69
R² = 0.3851
element, ppmw
Be, ppmw
0
500
1000
1500
2000
3
6
9
12
g
Fertile and Evolved Granites
Parental S-type
Ghost Lake, Ontario, Canada
Li
Rb
Sn
Linear (Li)
Linear (Rb)
Linear (Sn)
Rb = 46.169x - 13.626
R² = 0.4277
Li = 5.2066x - 5.5495
R² = 0.4814
Sn = 3.9715x - 12.717
R² = 0.1957
element, ppmw
Be, ppmw
0
100
200
300
5
10
15

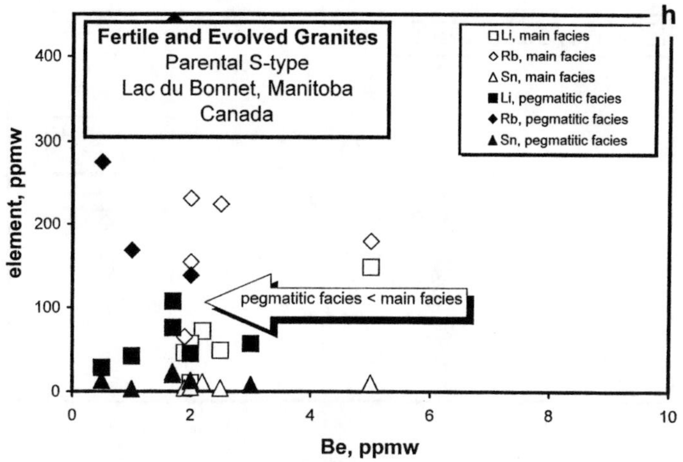
h
Fertile and Evolved Granites
Parental S-type
Lac du Bonnet, Manitoba
Canada
Li, main facies
Rb, main facies
Sn, main facies
Li, pegmatitic facies
Rb, pegmatitic facies
Sn, pegmatitic facies
pegmatitic facies < main facies
element, ppmw
Be, ppmw

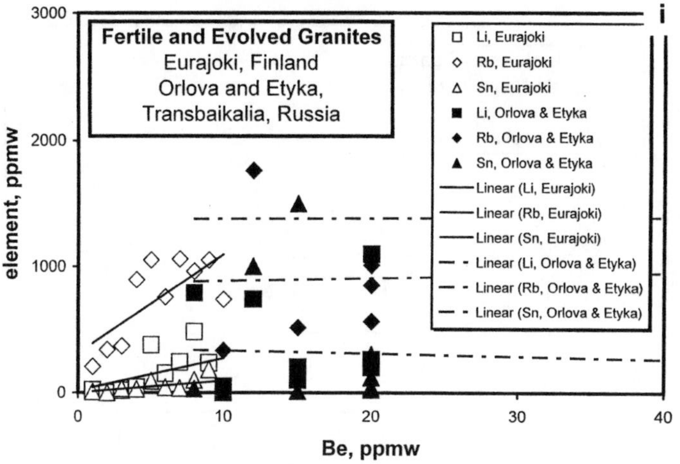
i
Fertile and Evolved Granites
Eurajoki, Finland
Orlova and Etyka,
Transbaikalia, Russia
Li, Eurajoki
Rb, Eurajoki
Sn, Eurajoki
Li, Orlova & Etyka
Rb, Orlova & Etyka
Sn, Orlova & Etyka
Linear (Li, Eurajoki)
Linear (Rb, Eurajoki)
Linear (Sn, Eurajoki)
Linear (Li, Orlova & Etyka)
Linear (Rb, Orlova & Etyka)
Linear (Sn, Orlova & Etyka)
element, ppmw
Be, ppmw

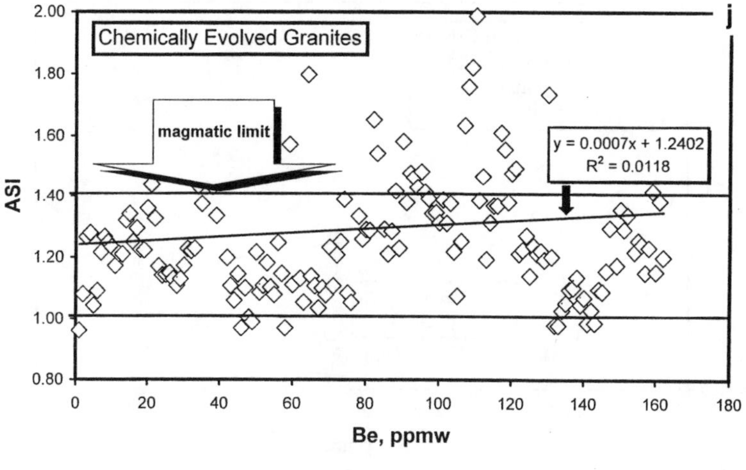
j
Chemically Evolved Granites
magmatic limit
y = 0.0007x + 1.2402
R² = 0.0118
ASI
Be, ppmw

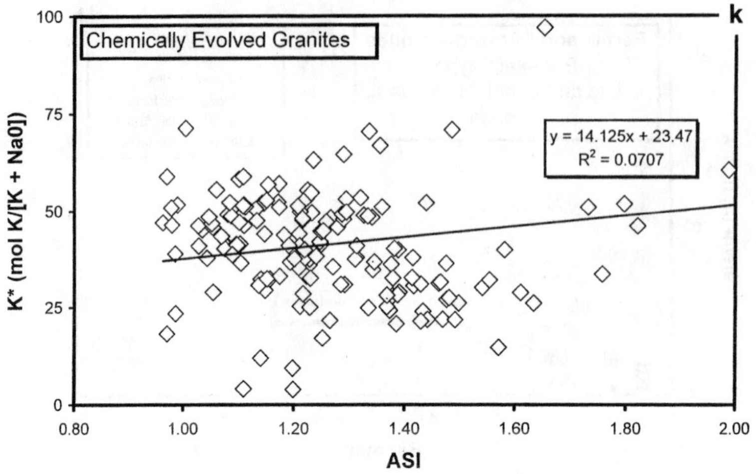
k
Chemically Evolved Granites
y = 14.125x + 23.47
R^2 = 0.0707
K* (mol K/[K + Na0])
ASI
0.80
1.00
1.20
1.40
1.60
1.80
2.00
0
25
50
75
100

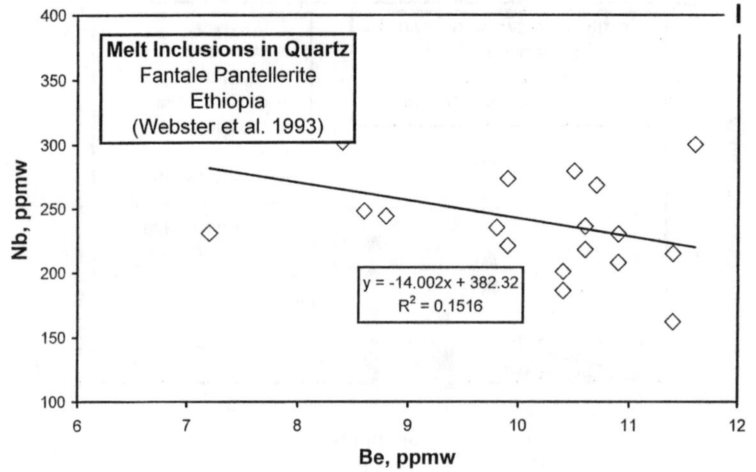
l
Melt Inclusions in Quartz
Fantale Pantellerite
Ethiopia
(Webster et al. 1993)
y = -14.002x + 382.32
R^2 = 0.1516
Nb, ppmw
Be, ppmw
100
150
200
250
300
350
400
6
7
8
9
10
11
12

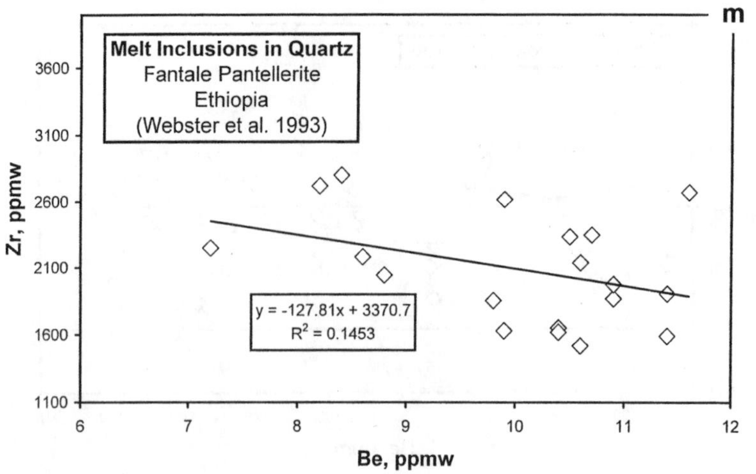
m
Melt Inclusions in Quartz
Fantale Pantellerite
Ethiopia
(Webster et al. 1993)
y = -127.81x + 3370.7
R^2 = 0.1453
Zr, ppmw
Be, ppmw
1100
1600
2100
2600
3100
3600
6
7
8
9
10
11
12

It is likely that the most aluminous and Be-rich rocks at Beauvioir contain a significant hydrothermal overprint. Raimbault and Burnol (1998), however, identified a highly fractionated rhyolitic dike at Richemont, Haute-Vienne, France, which they interpret as a quenched magmatic phase of the Beauvior granites. Like the Beauvoir granites and the Argemela dikes, this dike possesses high ASI (1.45), Be (91.6 ppmw), Li (3194 ppmw), Sn (487 ppmw), etc. Also like the Beauvior granites, the Richemont dike is enriched in P (1.18 wt % P_2O_5) and F (1.30 wt %), and hence contains abundant Li-micas and various phosphates, but not beryl.

Beryllium contents are uniformly low in the Lac du Bonnet, Greer Lake, Osis Lake, and Ghost Lake granites in Canada (Goad and Černý 1981; Černý and Meintzer 1988; Breaks and Moore 1992), all of which are linked to beryl-bearing pegmatite fields. The beryl-rich Shatford Lake pegmatite group (Buck et al. 1999) is derived from the biotite facies of the Lac du Bonnet granite (1.9 ppm Be; Goad and Černý 1981, Černý and Meintzer 1988), and the nearby but genetically independent Greer Lake (3.8 ppm) and Osis Lake (0.8 ppm) leucogranites (Goad and Černý 1981; Baadsgaard and Černý 1993) also yielded Be-rich pegmatite aureoles. It may seem logical that, as likely sources of the beryl-rich pegmatitic melts, these granites are themselves depleted in Be and other incompatible components.

I- and A-type granites. The Järna granitoids (Sweden), which are comprised of deep-level, slightly fractionated I-type monzodiorites to monzogranites, contain exceptionally low Be contents (Ahl et al. 1999). The Siljan and Garberg granite centers, which are more fractionated I-types emplaced at higher levels (Ahl et al. 1999), have relatively higher Be contents but are still near the granite and rhyolite averages of ~4-6 ppmw (Table 3), and similar to the A-type Eurajoki rapakivi granite (Fig. 8l; Haapala 1997). In the USA, two A-type granites, the Silvermine Granite (Lowell and Young 1999) in the St. Francois Mountians, Missouri, and the Red Bluff Granite suite in west Texas (Shannon et al. 1997) contain the typical and normal Be levels at 5-6 ppmw (Table 3).

Peralkaline granites and syenites. Peralkaline granites and related rocks (including syenites, pegmatites, and carbonatites) evolve toward a suite of enriched lithophile elements that includes Be along with the distinctive trilogy of Nb, Y, and F (NYF: Černý 1991a). Enrichment in Be is not diagnostic of the NYF type of ore deposits (Černý 1991a), as the following two examples illustrate (Table 3). The Oka alkaline igneous complex in southeastern Quebec, Canada (Gold et al. 1966; Treiman and Essene 1985) carries the typical NYF enrichment, but all of the various peralkaline igneous rocks (mafic to felsic) and carbonatites are low in Be (Gold et al. 1986). In contrast, the Strange Lake NYF deposit (Miller 1996; Salvi and Williams-Jones 1996) is exceeding rich in Be (Table 3), though Be does not correlate with Li (R^2 = 0.00), and it shows a weakly negative correlation with F (R^2 = 0.31). Alkali granites, both mineralized (Nb, Zr, and REE) and nonmineralized, in Siberia and Mongolia (Popolitov et al. 1967; Kovalenko et al. 1995) manifest similar contrasts in Be abundance (Table 3).

A data set of melt inclusions sampled from quartz in pantellerite from Fantale, Ethiopia, provides the same sort of controls on magmatic processes as opposed to subsolidus hydrothermal effects in a high-silica (≥70 wt % SiO_2) peralkaline magma (Webster et al. 1993). These data are included with granites (Table 3) because they reflect the compositions of melts that coexisted with phenocrysts at plutonic conditions. Abundances of Zr, Nb, Y, and Ce show strong positive correlations (Webster et al. 1993), but Be correlates negatively with these elements (Fig. 8 l,m).

Summary of data on granites. One obvious difference between the volcanic and plutonic sources is that the granites, as a group, are far more aluminous than the rhyolites (see Fig. 8j) This is mostly due to the prevalence of S-type sources in the granite database

(Table 3), but also to alkali loss via sericitic or argillic alteration that is more pervasive in plutonic rocks. The correlations of Be with the fractionation indices of Li, Rb, and Sn are poor in the granite data base as a whole, especially in comparison the USGS data set on obsidians. Regression of the data to exponential or polynomial functions does not yield much better fits. The poorer correlations follow from the more variable mineralogy of the granites, especially in the proportions and compositions of the micas, and from the likelihood of more pervasive subsolidus alteration in the granites, which may be selective by mineralogy. Nevertheless, the data for Be in granites from different ages, sources, and degrees of fractionation are grossly comparable to those of the obsidian and rhyolite data sets in these ways:

1. The correlations of Be with other incompatible elements, though poor, are generally positive. Thus, Be behaves incompatibly and increases with fractionation. One notable exception is the melt inclusion data from Fantale, Ethiopia (Fig. 8l,m), in which Be varies inversely with the other fractionation indices.
2. The Be contents of the granite data set are, on average, similar to the obsidians. Single-digit values of Be in ppmw are the norm in large plutons. Double-digit values (a few tens of ppmw Be) are typical of small, highly evolved plutons and dikes.

THE BERYLLIUM BUDGET FROM MIGMATITE TO PEGMATITE

Consider two starting melt compositions, one that corresponds to the minimum in the H_2O-saturated metaluminous haplogranite system Ab-Or-Qtz at 200 MPa, and another that reflects the 200 MPa H_2O minimum for melting of a muscovite-bearing protolith via the reaction:

$$\text{Muscovite} + \text{albite} + \text{quartz} + H_2O = \text{melt} + \text{aluminosilicate/corundum}. \qquad (7)$$

We choose the pressure of 200 MPa because the compositions of minimum-melting liquids are well constrained for Reaction (7) at that pressure (Icenhower and London 1995). For the H_2O-saturated haplogranite minimum composition at this pressure, $Ab_{38}Or_{28}Qz_{34}$, the bulk distribution coefficient for Be as derived from our experimental data in Table 2 is 0.19 (Table 4). At 200 MPa, the melt produced by Reaction (7) above has a normative composition close to $Ab_{37}Ms_{20}Or_{10}Qz_{33}$ (Icenhower and London 1995), for which the bulk distribution coefficient of Be is 0.43 (Table 4). Figure 9 illustrates that,

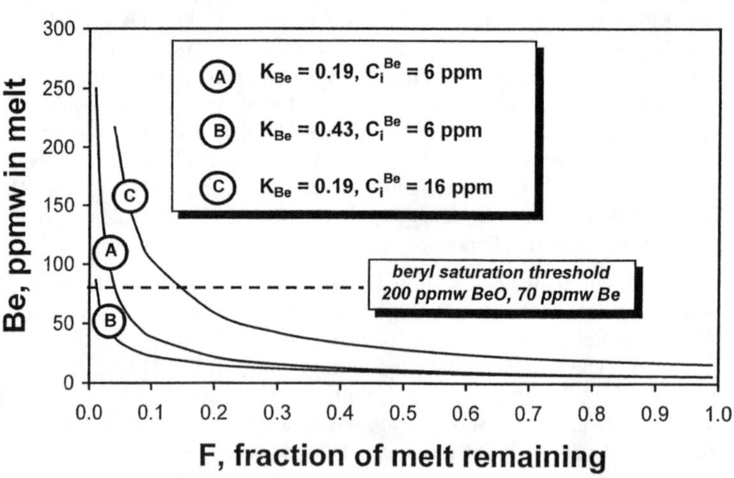

Figure 9. Rayleigh fractionation model for melts containing initially 4 or 16 ppmw BeO and bulk distribution coefficients of 0.19 (haplogranite minimum composition) and 0.43 (peraluminous haplogranite minimum). See the text for explanation.

Table 4. Bulk distribution coefficients for beryllium.

Phase	$D_{Be}^{mineral/melt}$	Normative wt fraction	Distribution coefficient $K_{Be}^{mineral/melt}$
haplogranite minimum at 200 MPa H_2O			
Qtz	0.24	0.34	0.08
Pl_{An1}	0.19	0.38	0.07
Kf	0.14	0.28	0.04
		K_{Be}	**0.19**
muscovite-saturated granite minimum at 200 MPa H_2O			
Qtz	0.24	0.33	0.08
Pl_{An1}	0.19	0.37	0.07
Kf	0.14	0.10	0.01
Ms	1.35	0.20	0.27
		K_{Be}	**0.43**
muscovite-saturated Ca-bearing granite at 200 MPa H_2O			
Qtz	0.24	0.33	0.08
Pl_{An31}	1.84	0.37	0.68
Kf	0.14	0.10	0.01
Ms	1.35	0.20	0.27
		K_{Be}	**1.04**

taking 6 ppmw Be as a starting concentration, neither magma reaches what we consider the minimum beryl saturation threshold, ~70 ppmw Be, until they are >95% crystallized. At ~80% crystallization, either magma would contain ~20 to 30 ppmw Be, a range of values that corresponds to some of the more evolved rhyolites and obsidians cited above. If that liquid were efficiently extracted, for example by filter pressing (Bea et al. 1994a), then continued fractionation of that small volume of residual melt could approach beryl saturation with ~20% melt remaining. That last fraction of residual melt, corresponding to the final few percent of the original melt volume, would correspond to the stage where pegmatite dikes are derived from an apical pluton or the cupola of a large body. There is nothing unique about the value of melt extraction at 20% of melt remaining at each stage, except that the liquids at these steps in the fractionation model correspond well with natural rocks. The extraction of ~20% or less of residual granitic melt appears to be quite feasible, especially if some shear stress is applied to the crystal-rich magma (Brown et al. 1995; Vigneresse et al. 1996).

This scenario conforms in large measure to the observed field relations: pegmatite fields commonly emanate from small, texturally and chemical evolved pegmatitic plutons (Goad and Černý 1981; Černý and Meintzer 1988; Černý 1991c), and these in turn appear to have batholithic-scale sources (Černý and Brisbin 1982; Mulja et al. 1995).

Because pegmatite fields extend above and beyond their sources, the connections between pegmatites and larger plutons or batholiths are often difficult to establish unless exposures constitute crustal cross sections. However, Kosals and Mazurov (1968) report Be values in the vertical evolution of a stacked and zoned plutonic sequence from southwest Baikalia. The Be contents, from 7 ppmw Be in the batholithic source to 21 ppmw Be in the evolved apophyses higher in the plutonic system, correspond well with the modeled values described above.

We can now answer the third and fourth questions posed at the beginning of this chapter. The plausible protoliths for granitic plutons—amphibolites, gneisses, and metapelitic schists—contain similar (order of magnitude) and uniformly low Be abundances (Grew, this volume). Despite generally low mineral-melt partition coefficients, these source rocks appear to be capable of imparting only a few ppmw Be to anatectic melts (based on the data for rhyolites and relatively undifferentiated granites presented here). This observation corroborates an important concept expressed recently by Bea (1996): that the time frames of anatexis are generally not sufficient to achieve a redistribution of trace elements between residual crystals and melt via diffusion through the residual crystalline phases. Consequently, partition coefficients are $\cong$ 1 and are determined by the proportions in which crystalline phases participate in melting. With the conclusion that most protoliths can and do impart about the same Be content to partial melts, it would appear that no one granite source rock is uniquely or specially predetermined to produce beryl-bearing pegmatites. There are, however, attributes of mineralogy at the source and during fractional crystallization of magma that do affect the accumulation of Be in derivative melts.

The composition of plagioclase. Kosals et al. (1973) noted that the Be content of plagioclase reaches a peak (in natural rocks) at ~An_{30}, and decreases in abundance in both directions away from this composition. That observation has been corroborated by the recent experimental work of Evensen and London (2001a) and Evensen (2001): as reported in Table 2, the partition coefficient for Be drops from 1.84 at An_{31} to 0.19 at An_{01}. Because of the compatibility of Be in calcic oligoclase, the Be concentration in melt cannot increase appreciably until plagioclase fractionation has removed virtually all Ca from the granitic melt.

Micas. Micas, especially white micas near muscovite in composition, represent important reservoirs of Be in crustal rocks (Grew, this volume). When white micas contribute to melting reactions such as (7) above, they can transfer a large mass fraction of Be available from the source rock into the melt. The melting of white mica also imparts a high ASI value to the anatectic melt (e.g., Icenhower and London 1995), which ultimately lowers the Be content needed to saturate derivative melt fractions in beryl. However, the crystallization of abundant white mica from granitic melts consumes Be and forestalls beryl saturation (e.g., the Li-rich granites of the Massif Central, France). Indeed, a model granitic magma that crystallizes with the assemblage of calcic oligoclase and white mica in the proportions observed in experiments would possess a distribution coefficient for Be near 1 (Table 4), and such granites would not experience any increase of Be with crystallization. The compatibility of Be in calcic oligoclase and white mica probably explains the lack of variation of Be with fractionation in the three granite facies at Pedrobernardo, Spain (Fig. 8b). The An content of plagioclase remains moderately high, with cores of $An_{20\text{-}30}$ and rims of $An_{5\text{-}10}$, throughout the crystallization of the (predominant) lower and middle zones (Bea et al. 1994a). These granites also contain appreciable muscovite (6 to 14 modal %).

The special role of cordierite. Miscibility between cordierite or indialite and beryl increases with temperature, and solid solution between indialite and beryl appears to be complete at high temperatures (Fig. 4). At the same time, the partition coefficient for Be between cordierite and granitic partial melt is extremely high (Table 2), though it decreases with increasing temperature (Evensen and London 1999). It is evident that if melting occurs in the medium- to low-pressure regime of cordierite (Mukhopadhyay and Holdaway 1994), then cordierite controls the fate of Be.

When S-type granite magmas originate from previously unmelted metapelites, then reactions similar to (7) above (but including biotite) largely determine both the volume and the Be contents of the melt, even if cordierite is present in the melting assemblage. This is because normative calculations based on experimental liquids and their crystalline products indicate that only a small fraction (<4 wt %) of cordierite is imparted to the melt up to 800°C (Evensen 2001), and hence most Be remains in residual cordierite. Melting reactions that produce cordierite (from the prograde reaction of biotite + sillimanite + quartz: Pereira and Bea 1994; Clark 1995) similarly generate granitic liquids that are low in Be and in normative Crd component.

We have already noted that restitic cordierite can be highly enriched in Be (Schreyer et al. 1979), and becomes more so as reactions (not necessarily melt-producing) consume cordierite with increasing metamorphic grade. The situation with sapphirine, another phase that concentrates Be (Grew, this volume), and the beryllosilicate surinamite (e.g., Baba et al. 2000), is less obvious. Grew et al. (2000) proposed that Be-enriched pegmatites in Enderby Land, Antarctica, formed at temperatures near 1000°C by partial melting of metapelites containing Be-bearing sapphirine, but work in progress on these rocks suggests an alternative scenario whereby the Be pegmatites formed as sapphirine was breaking down to Be-poor phases.

Cordierite versus garnet sources. The parental magmas of most beryl-bearing pegmatites have sources in the stability field of garnet ± aluminosilicate, rather than cordierite. We have noted already a possible connection between Be-poor granites in Spain and Be-rich granites in Portugal with the distinction of cordierite- or garnet-bearing source rocks. Continental terranes where cordierite-bearing S-type granites are prevalent (western Europe and eastern New South Wales, Australia) are conspicuously devoid of beryl-rich pegmatites. Conversely, in regions where peraluminous granites are marked by accessory garnet but not cordierite (North America), beryl-rich pegmatites are commonplace.

Vapor transport. So far, we have addressed only the distribution of Be between minerals and melt. It is pertinent, however, to address the partitioning and transport of beryllium by hydrothermal fluids, at least cursorily (see Ryan, this volume; Morris et al. this volume). Wood (1991) has calculated that the solubility of Be chloride complexes in equilibrium with bertrandite or phenakite is <1 ppmw in hydrothermal fluids to 300°C. Though Wood (1991) proposed that Be is far more soluble as a fluoride complex, the experimental data on melt-vapor partitioning in halide-rich systems does not reveal a strong tendency for Be to partition into the vapor phase. In the relevant experiments, London et al. (1988) examined the behavior of Be and other trace elements using the F-rich Macusani obsidian as starting material, and Webster et al. (1989), employed the F-rich Spor Mountain vitrophyre, with and without added Cl. In both experimental programs, the vapor/melt partition coefficient for Be was generally ≤1 over a range of fluid compositions, salinities, temperatures, and pressures. From the recent work of Evensen et al. (1999), we know that silicate melts are capable of dissolving hundreds to thousands of ppmw of Be. Thus, we conclude that silicate melts are more efficient agents for the transfer of Be than are aqueous fluids. This does not mean that hydrothermal fluids are unimportant in the movement of Be through the Earth's crust (Tatsumi and Isoyama 1988; Morris et al. 1990; Bebout et al. 1993; You et al. 1994; Leeman 1996;

Brenan et al. 1998a,b; Ryan, this volume), only that they are not as effective as silicate melts on an equal-volume basis.

Beryl saturation in pegmatites

In reviewing the likely conditions for beryl saturation by magmatic processes, Evensen et al. (1999) stressed the relation of beryl solubility to temperature. The Be content of melt necessary to achieve saturation in beryl decreases sharply with temperature. Recent studies of pegmatite cooling histories indicate very low temperatures of crystallization (e.g., ~400°-450°C: London 1992, 1996; Morgan and London 1999; Webber et al. 1999) for small dikes that are likely to be farthest from source (Baker 1998).

TEMPERATURE:	700°C	600°C	500°C	400°C
Be content of melt, ppmw				
Initially in melt:	**140**	**140**	**140**	**140**
In equilibrium with beryl at *T*:	**140**	**60**	**50**	**40**

Be content of minerals, ppmw, & mineral modal percent				
quartz (30 wt%)	34	14	12	10
K-feldspar (20 wt%)	27	8	7	5
plagioclase (40 wt%)	252 (An_{31})	6 (An_1)	5 (An_1)	4 (An_1)
muscovite (10 wt%)	189	81	68	54
Total Be in RFM, ppmw:	135	16	14	11
Be in beryl, % of total Be:	4%	89%	90%	92%

Pegmatite Class:	muscovite	beryl	rare-element	rare-element

NO BERYL | **BERYL SATURATED**

Figure 10. Model for the distribution of Be between rock-forming minerals and melts in a zoned pegmatite dike system, and the formation of beryl-bearing pegmatites. See the text for explanation.

Figure 10 illustrates how the distribution of Be among rock-forming minerals and the abundance of beryl might vary in a dike system that experiences the thermal gradient shown, from 700°C to 400°C. The model considers a peraluminous granitic melt of uniform composition with 140 ppmw Be (≈ 400 ppmw BeO, a representative value for the beryl class of rare-element pegmatites; Table 1) that is injected along its entire length and cools to the temperatures shown before crystallization commences. The melt is then allowed to crystallize, and partition coefficients determine the distribution of Be between rock-forming minerals. The model utilizes our experimentally measured partition coefficients presented in Table 2. The beryl saturation curve for peraluminous granitic liquid compositions (Fig. 1) is extrapolated linearly at 10 ppmw Be per 100°C from 600° to 400°C. The mineral mode (wt %) is set at 40% plagioclase, 30% quartz, 20% K-feldspar, and 10% muscovite, which is more feldspathic than the minimum melt composition in the muscovite-bearing granite system at 200 MPa H_2O (Icenhower and London 1995) but also more typical of "two-mica" granite-pegmatite systems.

Because Be is compatible in moderately calcic plagioclase (we used An_{31}, for which we have partitioning data, in this model) and in muscovite, the rock that crystallizes at 700°C is essentially beryl-undersaturated. Using the partition coefficients and the modes cited above, all but 4% of the initial Be content of the melt is accounted for in the rock-forming mineral assemblage. Upon falling to 600°C, however, the melt becomes saturated in beryl (at ~60 ppmw Be in melt), partly because of the drop in temperature and because we have adjusted the plagioclase composition to near the albite end member (An_1, again corresponding to our experimental compositions). Such an abrupt decrease in the An-content of plagioclase does have natural precedent, for example in the Pedrobernardo and Alburquerque plutons of Spain (Bea et al. 1994a; London et al. 1999). In both cases, plagioclase of the main plutonic body is calcic oligoclase (An_{10} at Alburquerque, An_{20-30} at Pedrobernardo), which falls sharply to An_1 in the marginal or upper facies of each pluton.

At 600°C, the rock-forming mineral assemblage accounts for only 16% of the total Be content, and the remainder is in beryl. The fraction of Be in beryl is largely controlled by, and varies inversely with, the abundance of white mica at this stage of the modeling. Note also in Figure 10 that the Be concentrations in mica and alkali feldspars fall sharply from the beryl-free proximal pegmatites (700°C) to the beryl-bearing distal ones (≤600°C). This pattern is in fact observed in natural occurrences (Smeds 1992). Because the mineral modes and partition coefficients are held constant, the changes in Be concentrations in the rock-forming minerals and the fraction of Be in beryl change only slightly with falling temperature according to the projection of the beryl saturation surface to 500°C and then at 400°C.

The absence of beryl in the proximal mica-rich pegmatites arises from the higher temperatures required to saturate the melt in beryl, and from the relative compatibility of Be in muscovite and in plagioclase that is typically in the range of oligoclase in composition (Kosals and Mazurov 1968; Kosals et al. 1973). The compatibility of Be in these two phases effectively suppresses Be accumulation and beryl saturation in the melt. These proximal muscovite-rich pegmatites are considered 'barren' in some exploration models (Trueman and Černý 1982). With a drop of only 100°C to 600°C, a peraluminous melt containing 140 ppmw Be and depleted in Ca would crystallize beryl in profusion. Thus, it is not surprising that the pegmatite type normally recognized just beyond the 'barren zone' of muscovite pegmatites is the beryl type (Trueman and Černý 1982). If partition coefficients do not change with temperature or composition, then the abundance and distribution of beryl in the distal pegmatites would resemble those of the beryl class. The most distal pegmatites, however, are more chemically evolved, and from the work of Evensen et al. (1999) we would expect higher concentrations of Be to saturate these melts in beryl. The net effect might be that beryl saturation occurred later in the consolidation of the pegmatites, and more concentrated and segregated into the latest internal units.

ACKNOWLEDGMENTS

Financial support for this research was provided by NSF grants EAR-9404658, EAR-9618867, EAR-9625517, INT-9603199, and EAR-990165 to D. London. We thank Luis Neves (University of Coimbra, Portugal) for some obscure but important information on the geology of Portugal, and Eric A. Fritz for assistance in transcribing the chemical data for rhyolites and granites. We also extend our thanks to Antonio Acosta and three MSA reviewers—Petr Černý, Don Dingwell, and Ed Grew—for their comments and suggestions.

REFERENCES

Ackermann KJ, Branscombe KC, Hawkes JR, Tidy AJL (1968) The geology of some beryl pegmatites in Southern Rhodesia. Trans Geol Soc South Africa 69:1-38

Ahl M, Sundblad K, Schoeberg H (1999) Geology, geochemistry, age and geotectonic evolution of the Dala granitoids, central Sweden. Precamb Res 95:147-166

Baadsgaard H, Černý P (1993) Geochronological studies in the Winnipeg River pegmatite populations, southeastern Manitoba. Mineral Assoc Can Prog Abstr 18:5

Baba S, Grew ES, Shearer CK, Sharaton JW (2000) Surinamite, a high-temperature metamorphic beryllosilicate from Lewisian sapphirine-bearing kyanite-orthopyroxene-quartz-potassium feldspar gneiss as South Harris, N.W. Scotland. Am Mineral 85:1474-1484

Baker DR (1998) The escape of pegmatite dikes from granitic plutons: Constraints from new models of viscosity and dike propagation. Can Mineral 36:255-263

Barnes VE, Edwards G, McLaughlin WA, Friedman I, Joensuu O (1970) Macusanite occurrence, age and composition, Macusani, Peru. Geol Soc Am Bull 81:1539-1546

Barton MD (1986) Phase equilibria and thermodynamic properties of minerals in the BeO-Al_2O_3-SiO_2-H_2O (BASH) system, with petrologic applications. Am Mineral 71:277-300

Bea F (1996) Controls on the trace element composition of crustal melts. Trans Royal Soc Edinburgh Earth Sci 87:33-41

Bea F, Pereira MD, Corretge LG, Fershtater GB (1994a) Differentiation of strongly perphosphrous granites: The Pedrobernardo pluton, central Spain. Geochim Cosmochim Acta 58:2609-2627

Bea F, Pereira MD, Stroh A (1994b) Mineral/leucosome trace-element partitioning in a peraluminous migmatite (a laser ablation–ICP-MS study). Chem Geol 117:291-312

Bebout GE, Ryan JG, Leeman WP (1993) B-Be systematics in subduction-related metamorphic rocks: characterization of the subducted component. Geochim Cosmochim Acta 57:2227-2237

Beus AA (1966) Geochemistry of Beryllium and Genetic Types of Beryllium Deposits. W.H. Freeman and Co., 286 p

Bowen NL (1928) The Evolution of the Igneous Rocks. Princeton University Press, Princeton, New Jersey, 332 p

Breaks FW, Moore JM Jr (1992) The Ghost Lake batholith, Superior Province of northwestern Ontario: A fertile S-type, peraluminous granite–rare-element pegmatite system. Can Mineral 30:835-875

Breiter K, Scharbert S (1995) The Homolka magmatic centre—An example of late Variscan ore bearing magmatism in the Southbohemian batholith (southern Bohemia, Northern Austria). Jahrb Geol Bund 138:9-25

Breiter K, Sokolova M, Sokol A (1991) Geochemical specialization of the tin-bearing granitoid massifs of NW Bohemia. Mineral Dep 26:298-306

Brenan JM, Neroda E, Lundstrom CC, Shaw HF, Ryerson FJ, Phinney DL (1998a) Behaviour of boron, beryllium, and lithium during melting and crystallization: Constraints from mineral-melt partitioning experiments. Geochim Cosmochim Acta 62:2129-2141

Brenan JM, Ryerson FJ, Shaw HF (1998b) The role of aqueous fluids in the slab-to-mantle transfer of boron, beryllium, and lithium during subduction: experiments and models. Geochim Cosmochim Acta 62:3337-3347

Brown M, Averkin Yu A, McLellan EI (1995) Melt segregation in migmatites. J Geophys Res 100B:15655-15679

Buck HM, Černý P, Hawthorne FC (1999) The Shatford Lake pegmatite group, southeastern Manitoba: NYF of not? (abstr) Can Mineral 37:830-831

Burt DM (1975) Beryllium mineral stabilities in the model system CaO-BeO-SiO_2-P_2O_5-F_2O_{-1} and the breakdown of beryl. Econ Geol Soc Econ Geol 70:1279-1291

Burt DM (1978) Multisystems analysis of beryllium mineral stabilities: The system BeO-Al_2O_3-SiO_2-H_2O. Am Mineral 63:664-676

Cemič L, Langer K, Franz G (1986) Experimental determination of melting relationships of beryl in the system BeO-Al_2O_3-SiO_2-H_2O between 10 and 25 kbar. Mineral Mag 50:55-61

Černý P (1991a) Fertile granites of Precambrian rare-element pegmatite fields: Is geochemistry controlled by tectonic setting or source lithologies? Precamb Res 51:429-468

Černý P (1991b) Rare-element granite pegmatites. Part I: Anatomy and internal evolution of pegmatite deposits. Geosci Can 18:49-67

Černý P (1991c) Rare-element granite pegmatites. Part II: Regional to global environments and petrogenesis. Geosci Can 18:68-81

Černý P (1994) Evolution of feldspars in granitic pegmatites. *In* Parsons I (ed) Feldspars and Their Reactions, p 501-540. NATO ASI Series. Ser C, Math Phys Sci, D. Reidel Publishing Company, Dordrecht–Boston

Černý P, Brisbin WC (1982) The Osis Lake Pegmatitic Granite, Winnipeg River District, southeastern Manitoba. *In* Cerny P (ed) Granitic pegmatites in science and industry. Mineral Assoc Can Short Course Hdbk 8:545-555

Černý P, Meintzer RE (1988) Fertile granites in the Archean and Proterozoic fields of rare-element pegmatites: Crustal environment, geochemistry and petrogenetic relationships. *In* Taylor RP, Strong DF (eds) Recent advances in the geology of granite-related mineral deposits. Can Inst Min Metal Spec Vol 39:170-207

Černý P, Meintzer RE, Anderson AJ (1985) Extreme fractionation in rare-element granitic pegmatites: Selected examples of data and mechanisms. Can Mineral 23:381-421

Charoy B (1999) Beryllium speciation in evolved granitic magmas: Phosphates versus silicates. Eur J Mineral 11:135-148

Charoy B, Noronha F (1996) Multistage growth of a rare-element volatile-rich microgranite at Argemela (Portugal). J Petrol 37:73-94

Christiansen EH, Sheridan MF, Burt DM (1986) The geology and geochemistry of Cenozoic topaz rhyolites from the Western United States. Geol Soc Am Spec Paper 205, 82 p

Clark DB (1995) Cordierite in felsic igneous rocks: A synthesis. Mineral Mag 59:311-325

Congdon RD, Nash WP (1991) Eruptive pegmatite magma: Rhyolite of the Honeycomb Hills, Utah. Am Mineral 76:1261-1278

Cuney M, Marignac C, Weisbrod A (1992) The Beauvoir topaz-lepidolite albite granite (Massif Central, France): the disseminated magmatic Sn-Li-Ta-Nb-Be mineralization. Econ Geol Bull Soc Econ Geol 87:1766-1794

Del Moro A, Martin S, Prosser G (1999) Migmatites of the Ulten Zone (NE Italy): a record of melt transfer in deep crust. J Petrol 40:1803-1826

Ericksen GE, Luedke RG, Smith RL, Koeppen RP, Urquidi F (1990) Peraluminous igneous rocks of the Bolivian tin belt. Episodes 13:3-7

Evensen JM (2001) The geochemical budget of beryllium in silicic melts, and superliquidus, subliquidus, and starting state effects on the kinetics of crystallization in hydrous haplogranite melts. PhD dissertation, Univ Oklahoma, 230 p

Evensen JM, London D (1999) Beryllium reservoirs and sources for granitic melts: The significance of cordierite. (abstr.) Geol Soc Am Abstr Prog 31:305

Evensen JM, London D, Wendlandt RF (1999) Solubility and stability of beryl in granitic melts. Am Mineral 84:733-745

Franz G, Morteani G (1981) The system $BeO-Al_2O_3-SiO_2-H_2O$: Hydrothermal investigation of the stability of beryl and euclase in the range from 1 to 6 kb and 400 to 800°C. N Jahrb Mineral Abh 140:273-299

Fuertes-Fuente M, Martin-Izard A (1998) The Forcarei Sur rare-element granitic pegmatite field and associated mineralization, Galicia, Spain. Can Mineral 36:303-325

Gallagher MJ (1975) Composition of some Rhodesian lithium-beryllium pegmatites. Trans Geol Soc S Africa 78:35-41

Goad BE, Černý P (1981) Peraluminous pegmatitic granites and their pegmatite aureoles in the Winnipeg River District, southeastern Manitoba. Can Mineral 19:177-194

Godinho MM (1974) Sobre o plutonometamorfismo da regiao de Guardao (Caramulo, Portugal). Mem e Not, Publ Mus Lab Mineral Geol Univ Coimbra, 78:37-77

Gold DP, Vallée M, Charette J-P (1966) Geology and geophysics of the Oka complex. Can Mineral Metal Bull 59, 273 p

Gold DP, Bell K, Eby GN, Vallée M (1986) Carbonatites, diatremes, and ultra-alkaline rocks in the Oka area, Quebec. Geol Soc Can Guidbk 21, 53 p

Grew ES, Yates MG, Barbier J, Shearer CK, Sheraton JW, Shiraishi K, Motoyoshi Y (2000) Granulite-facies beryllium pegmatites n the Papier Complex in Khmara and Amundsen Bays, western Enderby Land, East Antarctica. Polar Geosci 13:1-40

Haapala I (1997) Magmatic and postmagmatic processes in tin-mineralized granites: Topaz-bearing leucogranite in the Eurajoki Rapakivi Granite, Finland. J Petrol 38:1645-1659

Hess KU, Dingwell DB, Webb SL (1996) The influence of alkaline-earth oxides (BeO, MgO, CaO, SrO, BaO) on the viscosity of a haplogranitic melt: systematics of non-Arrhenian behavior. Eur J Mineral 8:371-381

Hölscher A, Schreyer W (1989) A new synthetic hexagonal BeMg-cordierite, $Mg_2[Al_2BeSi_6O_{18}]$, and its relationship to Mg-cordierite. Eur J Mineral 1:21-37

Icenhower JP, London D (1995) An experimental study of element partitioning between biotite, muscovite and coexisting peraluminous granitic melt at 200 MPa (H_2O). Am Mineral 80:1229-1251

Johannes W, Koepke J, Behrens H (1994) Partial melting reactions of plagioclases and plagioclase-bearing systems. *In* Parsons I (ed) Feldspars and Their Reactions, p 161-194. NATO ASI Se C, Math Phys Sci 421, D. Reidel Publishing Company. Dordrecht-Boston

Kamilli RJ, Criss RE (1996) Genesis of the Silsilah tin deposit, Kingdom of Saudi Arabia. Econ Geol Bull Soc Econ Geol 91:1414-1434

Keith JD, Christiansen EH, Tingey D (1994) Geological and chemical conditions of formation of red beryl, Wah Wah Mountains, Utah. *In* Blackett RE, Moore JN (eds) Cenozoic Geology and Geothermal Systems of Southwestern Utah, p 155-170. Utah Geol Assoc 23, Salt Lake City, Utah, United States

Kosals YA, Mazurov, MP (1968) Behavior of rare alkalis, boron, fluorine, and beryllyium during the emplacement of the Bitu-Dzhida granitic batholith, southwest Baykalia. Geochem Int'l 1968: 1024-1034

Kosals, YA, Nedashkovskiy, PG, Petrov, LL, Serykh, VI (1973) Beryllium distribution in granitoid plagioclase. Geochem Int'l 1973:753-767

Kovalenko, VI, Antipin, VS, Petrov, LL (1977) Distribution coefficients of beryllium in ongonites and some notes on its behavior in the rare metal lithium-fluorine granites. Geochem Int'l 14:129-141

Kovalenko VI, Tsaryena GM, Goreglyad AV, Yarmolyuk VV, Troitsky VA, Hervig RL, Farmer GL (1995) The peralkaline granite-related Khaldzan-Buregtey rare metal (Zr, Nb, REE) deposit, western Mongolia. Econ Geol Bull Soc Econ Geol 90:530-547

Kretz R, Loop J, Hartree R (1989) Petrology and Li-Be-B geochemistry of muscovite-biotite granite and associated pegmatite near Yellowknife, Canada. Contrib Mineral Petrol 102:174-190

Leeman WP (1996) Boron and other fluid-mobile elements in volcanic arc lavas: implications for subduction processes. *In* Bebout GE, Scholl DW, Kirby SH, Platt JP (eds) Subduction top to bottom, p 269-276. Geophys Monogr 96, Am Geophys Union, Washington, DC

Lindsey DA (1977) Epithermal beryllium deposits in water-laid tuff, western Utah. Econ Geol Bull Soc Econ Geol 72:219-232

London D, (1992) The application of experimental petrology to the genesis and crystallization of granitic pegmatites. Can Mineral 30:499-540

London D (1995) Geochemical features of peraluminous granites, pegmatites, and rhyolites as sources of lithophile metal deposits. *In* Thompson JFH (ed) Magmas, Fluids, and Ore Deposits, p 175-202. Mineral Assoc Can Short Course Hdbk 23:175-202

London D (1996) Granitic pegmatites. Trans Roy Soc Edinburgh, Earth Sci 87:305-319

London D (1997) Estimating abundances of volatile and other mobile components in evolved silicic melts through mineral-melt equilibria. J Petrol 38:1691-1706

London D Evensen JM (2001) The beryllium cycle from anatexis of metapelites to beryl-bearing pegmatites. (abstr) 11th Ann Goldschmidt Conf Abstr 3367, LPI Contrib 1088, Lunar Planet Inst, Houston (CD-ROM)

London D, Hervig RL, Morgan GB (1988) Vapor-undersaturated experiments with Macusani glass + H_2O at 200 MPa, and the internal differentiation of granitic pegmatites. Contrib Mineral Petrol 102: 360-373

London D, Wolf MB, Morgan GB VI, Gallego Gariddo M (1999) Experimental silicate-phosphate equilibria in peraluminous granitic magmas, with a case study of the Alburquerque batholith at Tres Arroyos, Badajoz, Spain. J Petrol 40:215-240

London D, Acosta A, Dewers T (2000) The Aluminum Saturation Index of S-Type Granites. (abstr) EOS Trans Am Geophys Union 81:1292

London D, Acosta A, Dewers TA, Morgan, GB VI (2001) Anatexis of metapelites: The ASI of S-type granites. (abstr) 11th Ann Goldschmidt Conf Abstr #3363, LPI Contrib 1088, Lunar Planet Inst, Houston (CD-ROM)

Lowell GR, Young GJ (1999) Interaction between coeval mafic and felsic melts in the St. Francois Terrane of Missouri, USA. Precamb Res 95:69-88

Luedke RG, Ericksen GE, Urquidi F, Tavera F, Smith RL, Cunningham CG (1997) Geochemistry of peraluminous volcanic rocks along the southerna margin of the Los Frailes volcanic field, central Bolivian tin belt. Bol Serv Nac Geol Mineral 23:14-36

Macdonald R, Smith RL, Thomas JE. (1992) Chemistry of the subalkalic silicic obsidians. U S Geol Surv Prof Paper 1523:214 p

Manning DAC, Hamilton DL, Henderson CMB, Dempsey MJ (1980) The probable occurrence of interstitial Al in hydrous, F-bearing and F-free aluminosilicate melts. Contrib Mineral Petrol 75: 257-262

Miller R (1996) Structural and textural evolution of the Strange Lake peralkaline rare-element (NYF) granitic pegmatite, Quebec-Labrador. Can Mineral 34:349-371

Monier G, Charoy B, Cuny M, Ohnstetter D, Robert J-L (1987) Evolution spatiale et temproelle de la composition des micas du granite albitique à topaze-lépidolite de Beauvior. Géol Fran 2-3:179-188

Morgan GB VI, London D (1999) Crystallization of the Little Three layered pegmatite-aplite dike, Ramona District, California. Contrib Mineral Petrol 136:310-330

Morgan, GB VI, London, D, Luedke, R (1998) The late-Miocene peraluminous silicic volcanics of the Morococala field, Bolivia. J Petrol 39:601-632

Morris JD, Leeman WP, Tera F (1990) The subducted component in island arc lavas: constraints from B-Be isotopes and Be systematics. Nature 344:31-36

Mukhopadhyay, B Holdaway, M J (1994) Cordierite-garnet-sillimanite-quartz equilibrium: I. New experimental calibration in the system $FeO-Al_2O_3-SiO_2-H_2O$ and certain P-T-X_{H2O} relations. Contrib Mineral Petrol 116:462-472

Mulja T, Williams-Jones AE, Wood SA, Boily M (1995) The rare-element-enriched monzogranite-pegmatite-quartz vein systems in the Preissac-Lacorne Batholith, Quebec: II, Geochemistry and petrogenesis. Can Mineral 33:817-833

Mysen BO, Virgo D (1985) Structure and properties of fluorine-bearing aluminosilicate melts: The system $Na_2O-Al_2O_3-SiO_2$-F at 1 atm. Contrib Mineral Petrol 91:205-220

Noble DC, Vogel TA, Peterson PS, Landis GP, Grant NK, Jezek PA, McKee EH (1984) Rare-element-enriched, S-type ash-flow tuffs containing phenocrysts of muscovite, andalusite, and sillimanite, southeastern Peru. Geology 12:35-39

Parsons I (ed) (1994) Feldspars and Their Reactions. NATO ASI Ser C, Math Phys Sci, 650 p, D. Reidel Publishing Co., Dordrecht-Boston

Pereira MD, Bea F (1994) Cordierite-producing reactions in the Peña Negra complex, Avila batholith, central Spain: The key role of cordierite in low-pressure anatexis. Can Mineral 32:763-780

Pichavant M, Kontak DJ, Briqueu LB, Valencia J, Clark A H (1988) The Miocene-Pliocene Macusani volcanics, SE Peru. II. Geochemistry and origin of a felsic peraluminous magma. Contrib Mineral Petrol 100:325-338

Popolitov EI, Petrov LL, Kovalenko VI (1967) Geochemistry of beryllium in the middle Paleozoic intrusives of northeastern Tuva. Geochem Int'l 4:682-689

Raimbault L, Burnol L (1998) The Richemont rhyolite dyke, Massif Central, France: A subvolcanic equivalent of rare-metal granites. Can Mineral 36:265-282

Raimbault L, Cuney M, Azencott C, Duthou JL, Joron JL (1995) Geochemical evidence for a multistage magmatic genesis of Ta-Sn-Li mineralization in the granite at Beauvoir, French Massif Central. Econ Geol Bull Soc Econ Geol 90:548-576

Ramirez JA, Grundvig, S (2000) Causes of geochemical diversity in peraluminous granitic plutons: The Jalama pluton, Central-Iberian Zone (Spain and Portugal). Lithos 50:171-190

Ramirez JA, Menendez LG (1999) A geochemical study of two peraluminious granites from south-central Iberia: The Nisa-Alburquerque and Jalama batholiths. Mineral Mag 63:85-104

Reavy RJ (1987) An investigation into the controls of granite plutonism in the Serra da Freita region, Northern Portugal. PhD dissertation, Univ St. Andrews, St. Andrews, Scotland

Rijks HRP, van-der-Veen AH (1972) The geology of the tin-bearing pegmatites in the eastern part of the Kamativi District, Rhodesia. Mineral Dep 7:383-395

Roda Robles E, Pesquera P, Velasco Rodan F, Fontan, F (1999) The granitic pegmatites of the Fregeneda area (Salamanca, Spain): Characteristics and petrogenesis. Mineral Mag 63:535-558

Rossovskiy LN Matrosov II (1974) Geochemical features of topaz-lepidolite-albite pegmatites of the East Gobi, Mongolia. Geochem Int'l 11:1323-1327

Ru CW, Fontan F, Monchoux P, Rossi P (1992) Determination of beryllium-bearing minerals within the Beauvoir Granite (Allier) by chemical analysis of the different densimetric fractions. C R Acad Sci, Sér 2, Méc Phys Chim, Sci l'Univ, Sci Terre 314:671-674

Salvi S, Williams-Jones AE (1996) The role of hydrothermal processes in concentrating high-field strength elements in the Strange Lake peralkaline complex, northeastern Canada. Geochim Cosmochim Acta 60:1917-1932

Schreyer W, Gordillo CE, Werding G (1979) A new sodium-beryllium cordierite from Soto, Argentina, and the relationship between distortion index, Be content, and state of hydration. Contrib Mineral Petrol 70:421-428

Shannon WM, Barnes CG, and Bickford ME (1997) Grenville magmatism in west Texas: Petrology and geochemistry of the Red Bluff granitic suite. J Petrol 38:1279-1305

Smeds SA (1992) Trace elements in potassium-feldspar and muscovite as a guide in the prospecting for lithium- and tin-bearing pegmatites in Sweden. J Geochem Expl 42:351-469

Staatz MB, Page LR, Norton JJ, Wilmarth VR (1963) Exploration for beryllium at the Helen Beryl, Elkhorn and Tin Mountain pegmatites, Custer County, South Dakota. U. S. Geol Surv Prof Paper 297C:129-197

Stilling A (1998) Bulk composition of the Tanco pegmatite at Bernic Lake, Manitoba, Canada. MSc thesis, Univ Manitoba, Canada, 76 p

Tatsumi Y, Isoyama H (1988) Transportation of beryllium with H_2O at high pressures: implication for magma genesis in subduction zones. Geophys Res Lett 15:180-183

Toth LM, Bates JB, Boyd GE (1973) Raman spectra of $Be_2F_7^{3-}$ and higher polymers of beryllium fluorides in the crystalline and molten state. J Phys Chem 77:216-221

Treiman AH, Essene EJ (1985)The Oka carbonatite complex, Quebec: geology and evidence for silicate-carbonate liquid immiscibility. Am Mineral 70:1101-1113

Trueman DL, Černý P (1982) Exploration for rare-element granitic pegmatites. *In* Černý P (ed) Granitic pegmatites in science and industry. Mineral Assoc Can Short Course Hdbk 8:463-493

Tuttle OF, Bowen NL (1958) Origin of granite in the light of experimental studies in the system $NaAlSi_3O_8$-$KAlSi_3O_8$-SiO_2-H_2O. Geol Soc Am Mem 74

Vigneresse JL, Barbey P, Cuney M (1996) Rheological transitions during partial melting and crystallization with applications to felsic magma segregation and transfer. J Petrol 37:1579-1600

Webber KL, Simmons WB, Falster AE, Foord EE (1999) Cooling rates and crystallization dynamics of shallow level pegmatite-aplite dikes, San Diego County, California. Am Mineral 84:708-717

Webster JD, Duffield WA (1991a) Volatiles and lithophile elements in the Taylor Creek Rhyolite: Constraints from glass inclusion analysis. Am Mineral 76:1628-1645

Webster JD, Duffield WA (1991b) Extreme halogen enrichment in tin-rich magma of the Taylor Creek Rhyolite, New Mexico. Econ Geol Bull Soc Econ Geol 89:840-850

Webster JD, Holloway JR, Hervig RL (1989) Partitioning of lithophile trace elements between H_2O and $H_2O + CO_2$ fluids and topaz rhyolite melt. Econ Geol Bull Soc Econ Geol 84:116-134

Webster JD, Taylor RP, Bean C (1993) Pre-eruptive melt composition and constraints on degassing a water-rich pantellerite magma, Fantale volcano, Ethiopia. Contrib Mineral Petrol 114:53-62

Webster JD, Congdon RD, Lyons PC (1995) Determining pre-eruptive compositions of late Paleozoic magma from kaolinized volcanic ashes: Analysis of glass inclusions in quartz microphenocrysts from tonsteins.Geochim Cosmochim Acta 59:711-720

Webster JD, Burt DM, Aguillon A (1996) Volatile and lithophile trace-element geochemistry of Mexican tin rhyolite magmas deduced from melt inclusions. Geochim Cosmochim Acta 60:3267-3283

Webster JD, Thomas R, Dieter R, Förster H-J, Seltmann R (1997) Melt inclusions in quartz from an evolved peraluminous pegmatite: Geochemical evidence for strong tin enrichment in fluorine and phosphorus-rich residual liquids. Geochim Cosmochim Acta 61:2589-2604

White AJR, Chappell BW (1983) Granitoid types and their distribution in the Lachlan Fold Belt, southeastern Australia. Geol Soc Am Mem 159:21-33

Wolf MB, London D (1994) Apatite dissolution into peraluminous haplogranitic melts: An experimental study of solubilities and mechanisms. Geochim Cosmochim Acta 58:4127-4145

Wolf MB, London D (1997) Boron in granitic magmas: Stability of tourmaline in equilibrium with biotite and cordierite. Contrib Mineral Petrol 130:12-30

Wood SA (1991) Speciation of Be and the solubility of bertrandite/phenakite mineral in hydrouthermal solutions. *In* Pagel M, Leroy J-L (eds) Source, Transport, and Deposition of Metals, p 147-150. A.A. Balkema, Rotterdam, Netherlands

You CF, Morris JD, Gieskes JM, Rosenbauer R, Zheng SH, Xu X, Ku TL, Bischoff JL (1994) Mobilization of beryllium in the sedimentary column at convergent margins. Geochim Cosmochim Acta 58:4887-4897

Vrána S (1999) Dyke swarm of highly evolved felsitic alkali-feldspar granite porphyry near Milevsko, Central Bohemian Pluton. Bull Czech Geol Surv 74:67-74

Zaraisky GP, Seltmann R, Shatov VV, Aksyuk AM, Shapovalov YuB, Chevychelov VYu (1997) Petrography and geochemistry of Li-F granites and pegmatite-aplite banded rocks from the Orlovka and Etyka tantalum deposits in Eastern Transbaikalia, Russia. *In* Papunen H (ed) Mineral Deposits: Research and Exploration. Where do They Meet? Balkema, Rotterdam, Netherlands, p 695-698

12 Beryllium in Metamorphic Environments (emphasis on aluminous compositions)

Edward S. Grew

Department of Geological Sciences
University of Maine
5790 Bryand Center
Orono, Maine 04469

esgrew@maine.edu

INTRODUCTION

Beryllium is an element thought to be largely associated with igneous and hydrothermal processes and the world's major economic Be deposits are largely igneous or hydrothermal. However, metamorphic rocks also play a major role in the Be budget of the Earth's crust. Beryllium enrichments in pegmatites and hydrothermal deposits are associated with granitic systems that many geoscientists think are derived from melting of metasedimentary rocks, with metapelites being the most fertile for Be (e.g., London and Evensen, this volume). Consequently, metamorphism plays an important role in the cycling beryllium from sediments to granitic systems.

The emphasis of my chapter is on rocks of pelitic composition or otherwise relatively rich in aluminum such as bauxite and metamorphosed peraluminous pegmatites (see also Franz and Morteani, this volume) and on the minerals making up these rocks. It concerns beryllium in metamorphic environments where beryllium was not introduced *during* metamorphism. Two major questions are addressed: (1) average and range of Be contents of metamorphic rocks, and (2) the effect of metamorphism on their beryllium content. Addressing both questions requires not only extensive data on metamorphic rocks, but also a comparable database of beryllium contents of unmetamorphosed precursor sediments. In general, beryllium is analyzed much less often than other trace elements, and as far as I am aware, there have been no systematic large-scale studies of Be in pelitic rocks or any other rock type. For example, Terry Plank (pers. comm. 2001) finds that there still is not enough high quality data to consider its behavior in marine sediments as has been done for other trace elements (Plank and Langmuir 1998).

Beryllium minerals, including both silicates and oxides, are rare in most metamorphic rocks, although a surprisingly large number of species has been found. This scarcity undoubtedly reflects the low abundance of Be in these rocks, including metapelites, metapsammites, carbonates, and metabauxites, in which Be contents rarely exceed 10 ppm. At this low concentration, Be can be accommodated in some common rock-forming minerals, notably muscovite, staurolite and cordierite, so that minerals containing essential beryllium, e.g., beryl or chrysoberyl, are relatively rare. In contrast, the situation with boron, another element characteristic of the upper continental crust and the subject of the companion *Reviews in Mineralogy*, Volume 33, is very different. Borosilicates are widespread (Henry and Dutrow 1996; Grew 1996) and B contents averaging 100 ppm are not rare in non-metasomatic metapelites (e.g., Leeman and Sisson 1996). This amount of boron generally cannot be accommodated either in quartz or in the most abundant minerals of the mica, feldspar, amphibole and pyroxene groups, so that a boron phase, in most cases a tourmaline-group mineral, is present.

1529-6466/00/0050-0012$10.00

Although well over half the known beryllium minerals (see Grew, Chapter 1, Appendix 1) have been found in metamorphic environments, I will not discuss all of them in the chapter on metamorphism, but refer the reader to the Introduction for the necessary information. I have selected only those minerals (Table 1) that are predominantly metamorphic and sufficiently studied that coverage in Appendix 1 of the Introduction is inadequate. Beryl and chrysoberyl, which are the two most widespread Be minerals, are being covered in this volume by Černý, London and Evensen, Franz and Morteani (Chapters 10, 11 and 13, respectively).

BERYLLIUM CONTENTS OF NON-PELITIC METAMORPHIC ROCKS

Carbonate sediments contain less Be than pelitic sediments, e.g., <0.2 to 4 ppm (Warner et al. 1959; Beus 1966; Hörmann 1978; Gao et al. 1991). Nonetheless, many metasomatic beryllium deposits have formed by replacement of carbonate rocks, e.g., the Yermakovskoye phenakite-bertrandite-fluorite deposit in Buratiya, Transbaikalia, Russia (Novikova et al. 1994).

Low-grade metamorphosed oolitic ironstone and ferruginous sandstone from Sardinia contain 0.24-7.29 ppm Be; the most abundant minerals are chamosite, siderite, and magnetite (Franceschelli et al. 2000).

Beryllium is not particularly enriched in many manganese deposits and Mn-rich rocks, e.g., <0.2-0.84 ppm Be, Franciscan Complex, California, (Huebner and Flohr 1990; Huebner et al. 1992); 2-5 ppm in Mn-rich metapelites ("redschist") and associated spessartine-quartzites ("coticules") from the Ardennes Mountains, Belgium (Krosse and Schreyer 1993). However, relatively elevated Be contents have been reported as the following Mn-bearing rocks, minerals, and deposits:

- 2-15 ppm Be in Mn nodules (Ryan, Chapter 3, this volume)
- 20-200 ppm Be in residual manganese ore concentrated from several U.S. manganese deposits associated with quartzite and dolomite (Warner et al. 1959)
- 0.04-0.4 wt % BeO in tilasite ($CaMgAsO_4F$), garnet and mineralized arkose from Guettara, Algeria (Agard 1965)
- 0.22-0.25 wt % BeO in rhodochrosite in a supergene deposit formed over a Be-bearing greisen (Grigor'yev 1967)
- 2-34 ppm Be in cryptomelane, the most abundant Mn oxide in volcanogenic-hydrothermal Mn deposits of the Calatrava Volcanic Field, Spain (Crespo and Lunar 1997).

Beryllium enrichment is also reported in some metamorphic manganese deposits, of which the most famous is Långban, Sweden (e.g., Magnusson 1930; Moore 1971; Sandström and Holtstam 1999). Metamorphosed manganese deposits with beryllium mineralization have been described from localities near Långban (e.g., Harstigen, Moore 1971), central French Pyrenees (Ragu 1994) and Val Ferrera, eastern Swiss Alps (Brugger and Gieré 1999; Brugger et al. 1998); Be contents range from <0.2 to 175 ppm in Fe-Mn deposits of Val Ferrara (Brugger and Gieré 2000). Prevailing opinion is that Be was introduced hydrothermally from granites into these manganese deposits during or after metamorphism (e.g., Moore 1971; Ragu 1994). However, the possibility of a pre-metamorphic origin of Be from volcanic exhalations cannot be excluded; such an origin has been proposed for As, Pb, U and other elements at Långban (Boström et al. 1979; Bollbark 1999) and for Be and W in the eastern Swiss Alps (Brugger and Gieré 1999, 2000). Krosse and Schreyer (1993) inferred a similar origin for Fe-Mn enrichment in the Mn-rich metasediments from the Ardennes, Belgium.

Table 1. Minerals containing major beryllium and mineral abbreviations used in this chapter.

Mineral (abbreviation)	*Symmetry*	*Formula plus exchange introducing Be*
AENIGMATITE GROUP AND RELATED MINERALS		
Høgtuvaite	Triclinic	$(Ca_{1.8}Na_{0.2})(Fe^{2+}_{3.55}Fe^{3+}_{2.2}Ti_{0.25})O_2[Si_{4.5}BeAl_{0.5}O_{18}]$
"Makarochkinite"	Triclinic	$(Ca_{1.76}Na_{0.19}Mn_{0.05})(Fe^{2+}_{3.66}Fe^{3+}_{1.36}Ti_{0.54}Mg_{0.25}Mn_{0.10}Nb_{0.06}Sn_{0.02}Ta_{0.01})O_2$-$[Si_{4.45}BeAl_{0.49}Fe^{3+}_{0.06}B_{0.01}O_{18}]$
Welshite	Triclinic	$Ca_2Mg_{3.8}Mn^{2+}_{0.6}Fe^{2+}_{0.1}Sb^{5+}_{1.5}O_2[Si_{2.8}Be_{1.7}Fe^{3+}_{0.65}Al_{0.7}As_{0.17}O_{18}]$
Sapphirine (Spr) – **khmaralite (Khm)**	Monoclinic	$(Al,Mg,Fe)_8O_2[(Al,Si,Fe^{3+})_6O_{18}]$ plus up to 0.78 $BeSiAl_{-2}$
Surinamite (Sur)	Monoclinic	$(Mg,Fe^{2+})_3(Al,Fe^{3+})_3O[AlBeSi_3O_{15}]$
OTHER SILICATES		
Cordierite (Mg>Fe, Crd) – sekaninaite (Fe>Mg)	Orthorhombic	$\square(Mg,Fe)_2Al_4Si_5O_{18}{\cdot}nH_2O$ plus $NaBe(\square Al)_{-1}$ and/or $BeSiAl_{-2}$
Euclase	Monoclinic	$BeAlSiO_4(OH)$
TAAFFEITE GROUP (OXIDES)		
Magnesiotaaffeite-6*N*'3*S*, formerly "**musgravite**" **(Mgr)**	Hexagonal	$(Mg,Fe,Zn)_2Al_6BeO_{12}$
Ferrotaaffeite-6*N*'3*S*, formerly "**pehrmanite**"	Hexagonal	$(Fe,Zn,Mg)_2Al_6BeO_{12}$
Magnesiotaaffeite-2*N*'2*S*, formerly "**taaffeite**" **(Tff)**	Hexagonal	$(Mg,Fe,Zn)_3Al_8BeO_{16}$

Note: Minerals in bold contain essential beryllium. Other abbreviations (based on Kretz 1983) used in this chapter are: Ab – albite, Alm – almandine, An – anorthite; And – andalusite; Ap – apatite, Bor – boralsilite, Brl –beryl, Bt – biotite, Cal – calcite, Cb – chrysoberyl, $BeAl_2O_4$; Chl – chlorite, Crn – corundum, Di – diopside, Fl – fluorite, Grt – garnet; Gdd – grandidierite, Hc – hercynite, Hem – hematite, Hög – högbomite, Ilm – ilmenite, Kfs – K-feldpsar; Ky – kyanite, Mc – microcline, Mgt – magnetite, Ms – muscovite, Mnz – monazite, Oam – orthoamphibole; Ol – olivine; Opx – orthopyroxene; Pg – paragonite, Phl – phlogopite, Pl – plagioclase, Prp – pyrope, Qtz – quartz; Rt – rutile, Sil – sillimanite; Spl – spinel, St – staurolite, Tur – tourmaline, Wrd – werdingite.

BERYLLIUM CONTENTS OF SEMI-PELITIC, PELITIC AND OTHER ALUMINOUS ROCKS

In brief, the great majority of pelitic and semipelitic sediments and their metamorphic equivalents contain 0.2 to 5 ppm Be and average near 3 ppm Be for all grades of metamorphism. Ten ppm Be represents a maximum for unmineralized pelite not associated with bauxite or metasediments unusually enriched in Be.

Unmetamorphosed rocks

Available data on pelitic sediments represent a random sampling of different lithologic types whose Be content depends on the environment of deposition. The beryllium content of pelagic clays and other marine sediments has been reported to vary within a narrow range of 0.3 to 3 ppm Be (Merrill et al 1961; Ryan and Langmuir 1988; Johnson and Plank 1999; summarized in Table 2 and Figure 1a; see also Hörmann 1978 and Ryan, this volume). In contrast, near-shore clays from Trinidad and redeposited terrestrial clays from different areas in the former USSR show a much greater range.

Carbonaceous sediments in some cases contain more beryllium than other sediments. Coal ash can have relatively high Be contents, although these vary locally, regionally and temporally, e.g., in the U.S., Late Paleozoic Appalachian coal ash averages 62 ppm Be, nearly three times the average in the Great Plains and Rocky Mountains of Mesozoic-Cenozoic age (Warner et al. 1959; Stadnichenko et al. 1961). Hörmann (1978) concluded that variations were too great to calculate a meaningful average for coal ash overall. On the other hand, carbonaceous shales are rarely enriched in Be. A comprehensive survey of U.S. black shales gave a median content of 1 ppm Be and very few contents above 10 ppm Be (Vine and Tourtelot 1970); Warner et al. (1959) reported several lignitic and carbonaceous shales and clays containing 7-14 ppm Be.

Be contents of pelites are higher than those of sands, sandstone and carbonate rocks (Hörmann 1978; Gao et al. 1991, 1998). Hörmann (1978) suggested an average of 5 ppm Be for bauxites, but a wide range of Be contents has been reported (Table 2).

Taylor and McLennan (1985 1995) noted that beryllium is one of the elements with a very small seawater-upper crust partition coefficient and relatively short residence time in seawater, and thus, its abundance in terriginous sedimentary rocks (i.e., shales and related rocks, which constitute an estimated 70% of all sedimentary rocks) should approach the average Be content of the upper continental crust. This conclusion, which is largely based on the compilations of Hörmann (1978), is supported by the trends based on new, not yet published data in Figure 1a, which shows that Be increases with Al. Wedepohl (1995) calculated 3.1 ppm Be for the upper crust using the average compositions of crustal rocks (mostly from the compilations of Hörmann 1978) multiplied by their proportion in the upper crust. Gao et al. (1998) calculated 1.95 ppm Be as the average content for the upper crust of central East China (sedimentary carbonate rock-free basis). The lower value for Be could reflect real differences between central East China and the Earth as a whole; Gao et al. (1998) noted that relatively low Al_2O_3 in East China reflects such differences.

Metamorphosed rocks

Wedepohl (1995) noted that for Be "reliable literature data on common rocks, especially on metamorphic rocks, are rather rare", and this observation still holds despite a spate of studies reporting Be data since 1995 (Table 2, Fig. 1b). Available data do not suggest that beryllium is lost from pelitic rocks during metamorphism (e.g., Lebedev and Nagaytsev 1980). The studies by Bebout et al. (1993 1999) of the Catalina Schist, California are the most systematic for low to middle-grade metapelites. Beryllium content

Table 2. Beryllium contents of rocks

Rock types and localities	*Metamorphic zone (or other information)*	*Be, ppm*		*Source*
		*Range**	*Average (n)*	
UNMETAMORPHOSED PELITIC ROCKS				
Pelagic clays in 5 sediment cores		2.0-3.0	2.6 (32)	Merrill et al. (1960)
Pelagic clay in one core		2.1-3.9	2.75(4)	
Red pelagic clay		2.16-2.53	2.32(5)	Johnson & Plank (1999)
Marine sediments		0.282-2.83	1.45 (19)	Ryan & Langmuir (1988)
Calcareous argillite, Paris basin		2.23-2.64	2.48 (3)	Henry et al. (1996)
Clays, Gulf of Paria, between Trinidad and Venezuala		1.0-8.7	6.2 (12)	Hirst (1962)
Limonitic concretions, Gulf of Paria		9.8-18	12.5(4)	
Clays, miscellaneous from former USSR		0.2-10	1.5-3(60)†	Beus (1966, Table 142)
Miscellaneous shale and clay, including carbonaceous varieties, U.S.		Mostly <4; maximum 14	<4 (181)	Warner et al. (1959)
Shale			4 (36)	Goldschmidt & Peters (1932)
Miscellaneous shale not cited above		0.5-2	(5)	Hörmann (1978)
METAMORPHOSED PELITIC ROCKS				
Geochemical reference slate			4 (1)	Flanagan (1973)
Metasediments with Mn deposit, Franciscan Complex, Calif.	Blueschist facies	Shale	2.90 (1)	Huebner et al. (1992)
		Graywacke: 0.67-1.20	0.96 (7)	
Metamorphosed shales and sandstones, Catalina Schist, California	Lawsonite-albite	0.34-1.3	0.95 (14)	Bebout et al. (1993, 1999)
	Blueschist	0.19-1.4	0.77 (8)	
	Greenschist & Epidote-amphibolite	0.3-1.1 (*4.1-5.4*)	0.73 (16+2)	
	Amphibolite	0.41-1.15	0.70 (10)	
Very weakly metamophosed carbonate-rich and siliceous black shales, emerald deposits, Colombia	Chivor and Gachala districts	3.00-4.00	3.73(7)	Giuliani et al. (1999)
	Muzo mine		3 ± 0.5	Ottaway et al. (1994)
			4.0(38)	Beus (1979)
Low-grade metapelites, western Alps		1.36-4.38	3.1(11)	Henry et al. (1996)
Black metapelite associated with oolitic ironstones, Nurra region, NW Sardinia, Italy			10.4	Franceschelli et al. (2000)

Aluminous metasediments, Quebec			1.8 (5)	Shaw et al. (1967)
Aluminous metasediments, Baffin Island			1.5 ± 1.3 (20)	Shaw et al. (1967)
Schists from "geosynclinal" regions of the former USSR		1-8	3.8 (22)	Beus (1966, Table 142)
Schists from eastern Transbaikalia, Russia		1-5	3.5 (10)	Beus (1966, Table 142)
Mica schist, different grades		0-8	~1 (7)	Hietanen (1969)
Biotite and biotite-muscovite schists with porphyroblasts of garnet, staurolite and andalusite, Ladoga complex, NW Russia		2.5-6.5	4.1 (16)	Sergeyev et al. (1967)
Meta-terrigeneous, Ladoga complex, NW Russia (by metamorphic zone)	Garnet		2.9 (5)	Lebedev & Nagaytsev (1980)
	Staurolite-andalusite		2.6 (11)	
	Sillimanite-muscovite		3.5 (7)	
	Sillimanite-K feldspar		3.0 (30)	
	Hypersthene		2.3 (28)	
Phyllite, schist, shale, greenschist- and amphibolite-facies, Urals		0.5-8	2.6 (40)	Bushlyakov & Grigor'yev (1989)
Metapelites, Kinzigite Formation, Ivrea-Verbano, Italy	"Kinzigite" = amphibolite facies	1.3-3.9	2.34 (5)	Bea & Montero (1999)
	Transitional	1.6-1.7	1.65 (2)	
	"Stronalite" = granulite facies	0.3-0.9	0.48 (5)	
Peña Negra complex, Spain	Mesosomes	1.13-4.75	2.34 (5)	Bea et al. (1994)
	Melanosomes	3.51-8.10	6.04 (3)	
	Leucosomes	0.37-2.04	1.20 (6)	
	Migmatites, undifferentiated	1.13-3.16	1.98(4)	Pereira Gómez & Rodríguez Alonso (2000)
	Restites	1.66-3.17	2.40(3)	
Whiteschist (talc-kyanite), Mulvoj, Tajikistan			2.4 (1)	Grew (unpub. data)
Amphibolite facies with relict granulite and eclogite facies, Zerendin Series, Kokchetav massif, Kazakhstan	Metasedimentary mica schists with Grt and Sil		3.4 (33)	Rozen & Serykh (1972)
	Gneisses derived from feldspathization of the above		3.2 (20)	
Amphibolite-facies inclusions in granitic intrusives, N. Victoria Land, Antarctica	Biotite-schist xenoliths	2-5	3 (9)†	Sheraton et al. (1987a)
	Biotite-rich restite		4 (2)	

Miscellaneous pelitic granulites, Prydz Bay, East Antarctica		1-6	2.4 (22)	Sheraton et al. (1984), Sheraton (unpubl. data)
Ultrahigh-*T* granulite-facies Napier complex, Enderby Land, East Antarctica	Miscellaneous pelitic granulites, S. and E. of Amundsen Bay	0-10 (*23*)	2.92 (24+1)	Sheraton (1980, 1985, unpubl. data); Sheraton et al. (1982)
	Quartzofeldspathic granulite		5 (1)	Grew (1998)
	Quartz rich-granulites, Khmara Bay and Mt. Pardoe	0.8-8.7	3.9 (24)	Grew et al. (2000, 2001b, unpubl. data)
Granulite-facies metapelite in lower crustal xenoliths in the Hannuoba basalt, Zhouba, North China craton		0.33-0.67	0.54 (3)	Liu et al. (2001)
Miscellaneous clayslate, phyllite, schist, gneiss, not cited above		0.8-10.5		Hörmann (1978)
Clastic metasediments, anchizone to Bt zone, Cinco Villas massif, Spain		<1-7	5.4 (45)	Pesquera and Velasco (1997)
Tur-rich metapelite and metapsammite, Broken Hill district, Australia		1-3	2 (3)	Slack et al. (1993)
BAUXITE AND ASSOCIATED ROCKS, UNMETAMORPHOSED AND METAMORPHOSED AT LOW-T CONDITIONS				
Bauxite, Jamaica, unmetamorphosed			180 (1)	Warner et al. (1959)
Bauxite, Dutch Guiana			4 (1)	
Metabauxite and chloritoid schists, Tuscany, Italy		0.32-6.75	4.06 (10)	Franceschelli et al. (1998)
Metabauxite and meta-aluminous argillites, western Alps, Switzerland and France		3.47-9.38	5.06 (7)	Poinssot et al. (1997)
Miscellaneous bauxite, unmetamorphosed, not cited above		0-36		Hörmann (1978)
ROCKS CONTAINING A BE-RICH MINERAL (IN SMALL CAPITALS)				
Granulite-facies quartzofeldspathic gneiss, South Harris, Scotland (SURINAMITE)			8.7, 9 (1)	Baba et al. (2000)
Quartz-rich granulites (2) and quartzofeldspathic gneiss, Napier complex, East Antarctica (SAPPHIRINE with 0.2-1.34 wt % BeO), a subset of the Napier Complex data cited above		1.4-6.9	4.4 (3)	Grew (1998); Grew et al. (2000); Sheraton (1985, unpubl. data)
Chloritoid-chlorite phyllite, Aiguille du Fruit, Vanoise (W. Alps), France (EUCLASE)		25, 61		Goffé (1980, 1982); Catel & Goffé (pers. comm. 2001)
Spinel-phlogopite rock, Mt. Painter, South Australia (MAGNESIOTAAFFEITE-2*N*'2*S*)			5.3 (1)	Teale (1980)

Note: n – number of samples. *Exceptional values (in italics) are not included in average. †Range of averages on 4 groups of samples.

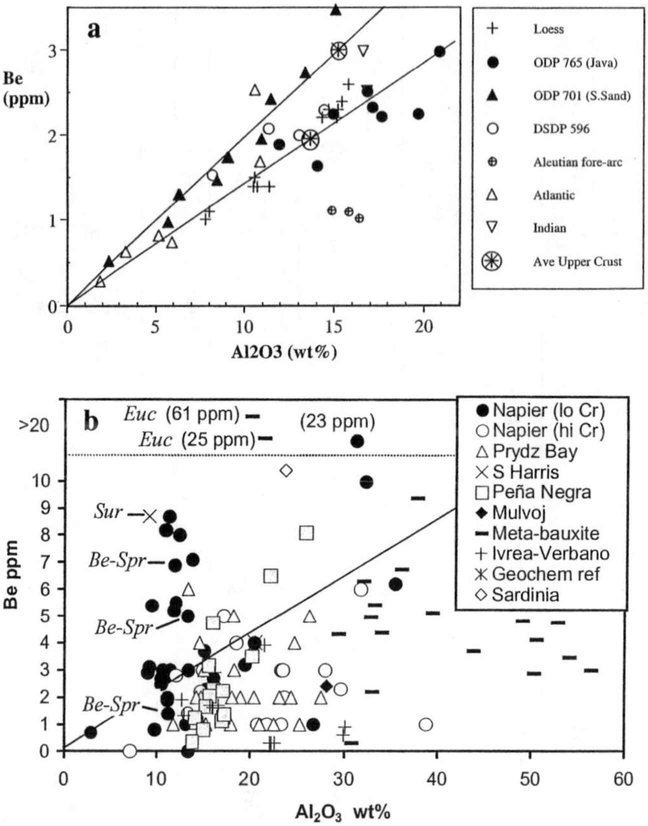

Figure 1. (a) Beryllium contents of marine and other sediments (compiled by Terry Plank; cf. Plank and Langmuir 1998). Loess is a good estimate of upper crust (Taylor et al. 1983). ODP Site 765 (Java) – Al from Plank and Ludden (1992); Be from Plank (unpublished ICP-MS data). ODP Site 701 (South Sandwich) – Al from Plank and Langmuir (1998); Be from Plank (unpubl. ICP-MS data). DSDP 596 refers to Tonga sediments from DSDP 595&6 – Al from Plank and Langmuir (1998); Be from Plank (unpublished ICP-MS data). Aleutian fore-arc sediments are dominated by arc volcaniclastics, which explains their anomalous low Be contents (data from Ryan 1989). Atlantic sediments are also from Ryan (1989), whereas the Indian Ocean sediments are from Plank (unpublished ICP-MS data). The average upper crustal values of 3.0 and 1.95 ppm Be are from Taylor and McLennan (1985) and Gao et al. (1998), respectively. (b) Whole-rock Be and Al_2O_3 contents of metapelites and related metamorphic rocks listed in Table 2, where sources of data are given. Lo Cr and Hi Cr refer to average Cr contents of Napier Complex metapelites, respectively 6 ppm and 674 ppm Cr (Sheraton 1980). Points for rocks containing euclase (Euc), surinamite (Sur) or sapphirine containing ≥0.2 wt % BeO, Be-Spr (Table 2) are labeled. The line is a reference passing through the estimated Be (3 ppm) and Al_2O_3 (15.2 wt %) contents of the upper continental crust (Taylor and McLennan 1985, 1995).

decreases with increasing grade from an average of 0.95 ± 0.31 ppm Be to 0.70 ± 0.26 (see also Ryan, this volume). However, by the authors' (Bebout et al. 1993) own admission, the overlap of Be contents from grade to grade is too great for this difference to indicate Be loss with metamorphic grade. In their comprehensive study of central East China, Gao et al. (1991 1998) reported a difference in the Be contents between unmetamorphosed, greenschist-facies and amphibolite-facies post-Archean pelites (2.69

ppm Be) and arenites (1.78 ppm Be) and metapelites and meta-arenites in amphibolite- and granulite-facies Archean complexes, respectively, 2.08 and 1.18 ppm Be. Being of different ages, the rocks are not exactly comparable, and the overall higher metamorphic grade of the Archean rocks may be only one factor behind their lower Be contents.

Co-variance of Be with Al is not evident in the metamorphic rocks overall (Fig. 1b), as it is in the unmetamorphosed sediments (Fig. 1a); nonetheless, samples from certain areas do show trends, e.g., migmatites from the Peña Negra complex, Spain, and the low-Cr metapelites from Napier Complex, Antarctica. The co-variance of Be with Al in the Peña Negra migmatites could be a relic of the original sediments, but a substantial influence of anatexis cannot be discounted.

The very weakly metamorphosed carbonaceous shales in the emerald deposits, Colombia (see Franz and Morteani, this volume) contain 3-4 ppm Be on the average, i.e., the same as shale overall, but this amount is deemed sufficient to supply Be for emerald formation (e.g., Beus 1979; Ottaway et al. 1994; Giuliani et al. 1999).

Low-grade metabauxites and associated aluminum-rich rocks on the whole contain more Be than metapelites, i.e., 5-10 ppm Be are not unusual (Table 2), whereas the 25-61 ppm Be measured in two euclase-chloritoid-chlorite rocks from the western Alps (Goffé 1982; Catel and Goffé, pers. comm. 2001) are exceptional (Fig. 1b), albeit non-metasomatic (see below under euclase).

The granulite facies is a special case. Citing older data supplemented by new results on rocks from the Brazilian shield, Sighinolfi (1973) concluded that there was no evidence for Be depletion in granulite facies rocks overall, though "acid" granulites appeared to be depleted relative to other rocks of granitic composition. Consequently, Sighinolfi (op. cit.) proposed two scenarios: (1) the Be concentrations measured in granulites represent Be concentrations of the precursors (metamorphism was isochemical) (2) Be was redistributed by dehydration reactions or anatexis during metamorphism resulting in depletions in some rocks, but enrichments in others. Overall, Be contents of granulite-facies metapelites rocks are not lower than in lower-grade rocks (Table 2, Fig. 1b), and thus are consistent with Sighinolfi's (1973) general conclusion and supportive of his first scenario. However, a closer look indicates that there are some depletions and enrichments suggesting that the second scenario might be more appropriate. For example, data of Bea and Montero (1999) for the Kinzigite Formation of Ivrea-Verbano, Italy show a decrease in Be in the granulite facies "stronalites" (0.3-0.9 ppm Be) relative to the upper-amphibolite facies "kinzigites" and transitional to granulite-facies (NB. This conclusion is based on Bea and Montero's Table 2 and not on these authors' statement in the text on p. 1137 reporting that the granulite-facies "stronalites to be significantly richer in Be"). On the other hand, the South Harris surimanite-bearing gneiss and many low-Cr Napier Complex metapelites are somewhat enriched in Be relative to most pelites. Redistribution of Be by anatexis under granulite-facies conditions is also suggested by slight enrichment in Be in two anatectic pegmatites in the Napier Complex, 11.2-11.4 ppm Be (Grew et al. 2001b and unpublished data) vs. an average of 3.9 ppm in associated granulites (Table 2).

Mica schists in the aureoles of the Be-enriched Beauvoir granite, Massif Central, France (Burnol 1974; Monier et al. 1987; Rossi et al. 1987; Piantone and Burnol 1987) deserve a brief mention. Earlier studies cited by Piantone and Burnol (1987) reported that average Be contents of metapelites in aureoles around the Beauvoir and related granites range from 3 to 6 ppm, values not much greater than for the pelites listed in Table 2. Higher Be contents were found in mica schist from the aureoles cut by the Échassières drill hole, i.e., 2 to 277 ppm; average 12 ± 32 ppm (77 samples, Piantone and Burnol

1987). Although the 277 ppm maximum represents an extraordinary enrichment of Be adjacent to the granite, Be on the whole does not vary significantly as a function of distance in the drill core from the Beauvoir granite. Multivariant statistical treatment shows that Be behaves independently of F, Li, Sn, and Rb in the aureoles.

Another example of Be enrichment are muscovite schists (10.7-71.9 ppm Be) associated with emerald deposits in Habachtal, Austria, but it is possible these are volcanogenic rocks enriched in Be prior to metamorphism (Grundmann and Morteani 1982, 1989; Franz and Morteani, this volume).

Relationship between whole-rock Be content and mineral assemblage

The observed variations in Be content with metamorphic grade can be rationalized in terms of the minerals present. White mica, which is relatively abundant in nearly all low- and medium-grade metapelites, should be able to carry the amount of Be normally present in pelites, i.e., ≤5 ppm whole-rock vs. 20-120 ppm present in some muscovite. Thus, as long as muscovite is present in reasonable abundance, no loss of Be would be expected, and none has been conclusively demonstrated in muscovite-bearing rocks. However, at temperatures above the breakdown of muscovite, retention of beryllium most likely depends on biotite or plagioclase unless a major Be sink such as cordierite or sapphirine is present. For example, cordierite and sapphirine are absent from the Ivrea-Verbano granulite-facies "stronalites"; which are richer in garnet and poorer in biotite (and to a lesser extent, poorer in plagioclase) than the lower-grade "kinzigites" (Bea and Montero 1999). Thus, loss of Be could be explained by the increase in modal garnet, a phase poor in Be, at the expense of biotite and plagioclase. In contrast, cordierite is present in all the Peña Negra (Spain) migmatites (Bea et al. 1994; Pereira Gómez and Rodríguez Alonso 2000) and in all but two of the Prydz Bay (Antarctica) metapelites (the two cordierite-free Prydz Bay rocks have only 1 ppm Be, Sheraton, unpublished data). Thus, the retention of Be in these metapelites can be attributed to the presence of cordierite. The situation with the Napier Complex is less clearly defined. Most of the Napier Complex metapelites contain cordierite or sapphirine or both, and the Be contents of these metapelites average somewhat higher than Napier Complex metapelites lacking these minerals. However, the difference is less than 1σ of the average and in the rocks lacking sapphirine and cordierite, there could be another carrier for Be that has not be elucidated to date.

In contrast to euclase-bearing rocks containing 25-61 ppm Be, other pelitic and Al-rich metamorphic rocks containing Be-rich phases are only modestly, if at all, enriched in Be (Table 2, Fig. 1b), implying that Be content is not the sole factor determining whether a discrete Be-rich phase appears. Its appearance in rocks containing only a few ppm Be could result from the very low Be concentrations accepted in the major constituents of the rock such as quartz, alkali feldspar and garnet. In rocks containing a few ppm to few tens ppm Be and abundant muscovite, a Be-rich phase would not appear because this amount of Be could be accommodated in muscovite. If a Be "sink" such as cordierite or sapphirine were present in trace amounts only with quartz, alkali feldspar and garnet, Be could reach substantial concentrations in these "sinks". One example appears to be the Napier Complex metapelites unusually poor in Cr (6 ppm Cr on the average, Sheraton 1980), V, and Ni (Fig. 1b). Because these are relatively poor in alumina and the modal amount of sapphirine correspondingly low, Be is concentrated in sapphirine. Anatexis is another process by which Be enrichments might appear in association with pelites themselves not unusually enriched in Be. Bea et al. (1994) reported substantial enrichment of Be in Peña Negra melanosomes (Table 2; Fig. 1b). Partial melting at ultrahigh temperatures (the most recent estimate at one locality is 1120 °C, Harley and Motoyoshi 2000) of Napier Complex quartz-rich granulites in Khmara Bay and Mt.

Pardoe (Table 2) has resulted in pegmatites containing beryllian sapphirine-khmaralite (Grew 1981 1998; Grew et al. 2000). Another example is the Almgjotheii, Norway, pegmatite, which contains Be-bearing sillimanite, werdingite and boralsilite. Grew et al. (1998c) suggested that B in these phases could have originated from graphite-rich layers in the metapelitic host rocks to the pegmatite, and that Be had been remobilized with B. As noted above, carbonaceous sediments can contain more Be than average metapelite, and thus, graphite-rich layers could have been a source for Be as well as B.

DESCRIPTION OF SELECTED MINERALS WITH MAJOR BERYLLIUM

Aenigmatite group and related minerals

Surinamite and sapphirine-khmaralite, two major carriers of Be in high-grade Al-rich metamorphic rocks, are related to the aenigmatite group, which includes two rare minerals found in metamorphic environments, høgtuvaite and welshite (Table 1). Aenigmatite-group minerals are triclinic chain silicates having the general formula $^{[8]}A_2{}^{[6]}B_6O_2[^{[4]}T_6O_{18}]$, where A = Ca, Na; B = Al, Cr, Fe^{3+}, Fe^{2+}, Mg, Sb, Ti; T = Al, Si, B, Be, and Fe^{3+} (e.g., Cannillo et al. 1971; Deer et al. 1978; Burt 1994; Mandarino 1999; Kunzmann 1999; Hawthorne and Huminicki this volume). Sapphirine and khmaralite differ from the aenigmatite-group minerals in that the A site is octahedral and is occupied dominantly by Mg and Fe^{2+} (e.g., Moore 1969; Barbier et al. 1999; Fig. 2).

The six-membered tetrahedral chains in aenigmatite-group minerals, sapphirine and khmaralite differ from tetrahedral chains in pyroxene by the presence of wings of single tetrahedra sharing a corner with half the chain tetrahedra. As a result, two tetrahedra

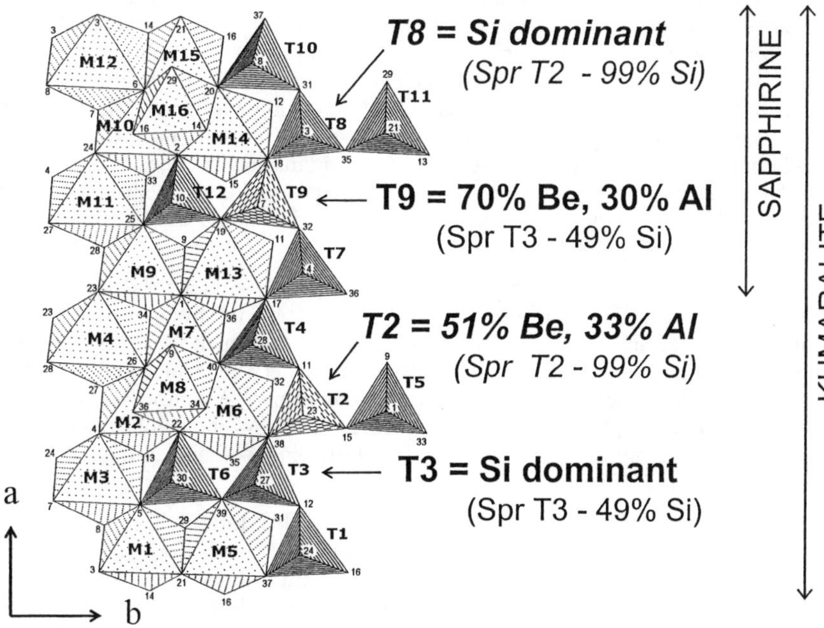

Figure 2. Crystal structure of khmaralite showing the relationship between sapphirine (Spr) and khmaralite and occupancies of the tetrahedra sharing three corners with other tetrahedra (the so called Q^3 tetrahedra). Italics emphasize the alternation of ordering schemes from one sapphirine subcell to the next, resulting in the doubling of the unit cell in khmaralite. Modified from Barbier et al. (1999).

share three corners with other tetrahedra; these are the most highly polymerized, so called Q^3 tetrahedra (Fig. 2). Two tetrahedra share two corners with other tetrahedra as in pyroxenes, and the two wing tetrahedra share only one corner with a tetrahedron. Beryllium is found to reside exclusively in the tetrahedra sharing three corners. In surinamite, only every fourth tetrahedron has a wing so that only one tetrahedron in the five-membered chain shares three corners with tetrahedra (Moore and Araki 1983). Barbier (1998) noted that tetrahedral order in surinamite is nearly complete, whereas the tetrahedral sites in sapphirine and khmaralite are never completely ordered, so that surinamite is a more favorable host for beryllium. The partial disorder in sapphirine-khmaralite could result from the availability of two competing Q^3 tetrahedra. Beryllium cannot occupy the two Q^3 tetrahedra simultaneously because the bridging oxygen between them would be seriously undersaturated (Be-O-Be bridge); this complication does not arise with the single Q^3 tetrahedron in surinamite. The need to avoid Be-O-Be bridges also limits the maximum Be content in the double-winged chain to one Be per 6 tetrahedra, i.e., 1 Be per 20 O. Be contents of høgtuvaite,"makarochkinite", synthetic beryllian sapphirine and khmaralite do not exceed 1 Be per 20 O, but welshite composition appears to violate this constraint.

Høgtuvaite and "makarochkinite". Høgtuvaite, ideally $Ca_2(Fe^{2+}{}_5Ti)(Si_5Be)O_{20}$ (Burt 1994), was first reported as a metamorphic mineral in granitic gneiss and mafic pegmatites near the eponymous Høgtuva, Mo i Rana, Nordland, Norway (Grauch et al. 1994). In the absence of a crystal structure determination, which was impossible due to pervasive polysynthetic twinning, defining høgtuvaite as a mineral distinct from rhönite was justified by the vector analysis method, whereby høgtuvaite is related to rhönite by the vector $BeSiAl_{-2}$ (Burt 1994). To date, høgtuvaite has only been reported from the Høgtuva area. However, another aenigmatite-group mineral, "makarochkinite", was described from a granite pegmatite in the Ilmeny Natural Reserve, south Urals, Russia (Polyakov et al. 1986). Grauch et al. (1994) equated it with høgtuvaite, but "makarochkinite" is untwinned and its crystal structure could be refined (Yakubovich et al. 1990; Barbier et al. 2001 and in preparation). Because no proposal for "makarochkinite" had been submitted to the IMA Commission on New Minerals and Mineral Names, it now must be demonstrated that "makarochkinite" is distinct from høgtuvaite in order for it to qualify for species status (see Hawthorne and Huminicki, this volume).

Høgtuvaite forms prismatic crystals up to 4 cm long and 6 mm across, and tends to be porphyroblastic (Grauch et al. 1994). It is nearly opaque, and its strong brown to green pleochroism was detectable only in ultra-thin sections <10-μm thick. Refractive indices had to be estimated from reflected light. The difference between calculated and measured density was attributed to inclusions and chemical heterogeneity. "Makarochkinite" forms masses 5-50 mm across. Optically, it is weakly pleochroic, shows anomalous interference colors and does not extinguish; $\gamma = 1.860$, $\alpha = 1.799$ (thickness of studied fragments not specified, Polyakov et al. 1986).

Burt (1994) proposed $Ca_2(Fe^{2+}{}_5Ti)(Si_5Be)O_{20}$ as the ideal formula for end-member høgtuvaite, whereas the analyses reported by Grauch et al. (1994) are closer to $(Ca_{1.8}Na_{0.2})(Fe^{2+}{}_{3.55}Fe^{3+}{}_{2.2}Ti_{0.25})(Si_{4.5}BeAl_{0.5})O_{20}$. Høgtuvaite varies somewhat in BeO (2.47-2.75 wt %), Na_2O, and TiO_2 contents. Høgtuvaite is enriched in a large number of trace elements, e.g., B (95 ppm), Y (1250 ppm), Nb (1110 ppm), REE (e.g., Ce 350 ppm), Th (570 ppm) and U (205 ppm).

Electron microprobe analysis of "makarochkinite" together with a crystal structure refinement gave

$$(Ca_{1.76}Na_{0.19}Mn_{0.05})(Fe^{2+}{}_{3.66}Fe^{3+}{}_{1.36}Ti_{0.54}Mg_{0.25}Mn_{0.10}Nb_{0.06}Sn_{0.02}Ta_{0.01})(Si_{4.45}BeAl_{0.49}Fe^{3+}{}_{0.06}B_{0.01})O_{20}$$

(Barbier et al. 2001 and in preparation), which is compares favorably to the wet chemical analysis reported by Polyakov et al. (1986). "Makarochkinite" contains more divalent and high-valence cations (Ti, Nb) and less Fe^{3+} and Na than høgtuvaite, but its status as a distinct species would depend on whether one M site is dominated by Ti (see Hawthorne and Huminicki, this volume). Available crystallographic data (Yakubovich et al. 1990; Barbier et al. 2001 and in preparation) do not provide convincing evidence for such dominance.

Høgtuvaite is found in the distinctive Bordvedåga mineralized quartzofeldspathic gneiss and associated pegmatites of Early Proterozoic age in the Høgtuva window, Norway (Lindahl and Grauch 1988; Grauch et al. 1994). These rocks were metamorphosed to at least lower-amphibolite-facies grade by the Caledonian event, and the pegmatites have a "metamorphic origin". Precursors to the høgtuvaite-bearing rocks could have been a stratabound beryllium deposit hosted by a highly evolved igneous suite. Associated minerals the gneisses and pegmatites include quartz, microcline, albite, biotite, epidote, magnetite, fluorite, zircon, allanite and other REE minerals, and calcic amphibole with overgrowths of sodic amphibole. The most important associated Be mineral is phenakite; gadolinite, danalite and genthelvite are minor. The gneisses are unusually enriched in trace elements such as Zr, Y, Nb, Th, U; some layers contain 0.19-0.42% Be, an enrichment that could be pre-metamorphic. In contrast to the highly aluminous rocks containing surinamite, sapphirine and khmaralite, høgtuvaite-bearing rocks have a slightly alkaline character.

"Makarochkinite" is found in a granite pegmatite; associated minerals include helvite, gadolinite and phenakite, as well as samarskite-(Y), columbite and allanite (Polyakov et al. 1986), i.e., an assemblage virtually identical to that in the Bordvedåga gneiss.

Welshite. Welshite, $Ca_2Mg_{3.8}Mn^{2+}_{0.6}Fe^{2+}_{0.1}Sb^{5+}_{1.5}O_2[Si_{2.8}Be_{1.7}Fe^{3+}_{0.65}Al_{0.7}As_{0.17}O_{18}]$, was originally described from Långban, Sweden, the only known locality, where it occurs in dolomite filling fractures in hematite ore or in calcite veinlets cutting hematite ore (Moore 1967 1971 1978; Nysten et al. 1999; Grew et al. 2001a). Associated minerals include arsenates (e.g., adelite, tilasite, manganese-hörnesite), Sb^{5+} oxides (e.g., roméite) and silicates, serpentine, and barite. It is not associated with any Be mineral other than rare swedenborgite. Moore (1971) included welshite among the Magnusson's (1930) period C or "vug minerals," for which Grew et al. (1994b), citing circumstantial evidence, inferred temperatures in the range of 500-600°C and P as low as 2 kbar.

Moore (1978) described welshite as being deep reddish-brown to reddish black; in transmitted light, brownish-orange and not discernibly pleochroic with low birefringence and high refringence (α = 1.81, γ = 1.83), whereas optical absorption spectra are consistent with weak pleochroism , absorption $\alpha \approx \gamma < \beta$ (Grew et al. 2001a).

Like høgtuvaite, welshite is polysynthetically twinned, foiling attempts to refine of its crystal structure, and thus its chemical composition has been difficult to characterize. By analogy with the aenigmatite group, Moore (1978) proposed the end-member formula $Ca_2Mg_4Fe^{3+}Sb^{5+}O_2[Si_4Be_2O_{18}]$, which is a reasonable simplification of his empirical formula, $Ca_{2.35}Mg_{3.53}Mn_{0.13}Fe^{3+}_{1.06}Sb_{1.42}As_{0.29}Al_{0.38}Si_{3.03}Be_{1.49}O_{20}$, based on an electron microprobe analysis (plus a BeO determination by an unspecified wet chemical method) totaling only 93.7 wt %. New microprobe data, in conjunction with Mössbauer spectroscopy, optical absorption spectroscopy and ion microprobe data, gave reasonable analytical totals and the formula cited above for one sample (Grew et al. 2001a). A second sample has the formula

$$Ca_2Mg_{3.8}Mn^{2+}_{0.1}Fe^{2+}_{0.1}Fe^{3+}_{0.8}Sb^{5+}_{1.2}O_2[Si_{2.8}Be_{1.8}Fe^{3+}_{0.65}Al_{0.5}As_{0.25}O_{18}],$$

which is related to the first by the substitution of approximately $0.6^{[6 \text{ and } 4]}(Fe,Al)^{3+}$ for $0.4^{[6]}(Mg,Mn,Fe)^{2+} + 0.2(^{[6]}Sb,^{[4]}As)^{5+}$, i.e., $3^{[vi,iv]}M^{3+} = 2^{[vi]}M^{2+} + {}^{[vi,iv]}M^{5+}$ (cf. Kunzmann 1999).

Khmaralite and Be-bearing sapphirine. Wilson and Hudson (1967) were the first to report significant beryllium in sapphirine, $\sim(Mg,Fe)_{3.5}(Al,Fe)_9Si_{1.5}O_{20}$: 0.65 wt % BeO in a sample from the Musgrave Ranges, Australia, and Povondra and Langer (1971b) synthesized sapphirine suspected to contain Be because of its smaller cell volume. Recognition of a "sapphirine" containing 2.5 wt % BeO from Khmara Bay (part of Casey Bay), Antarctica (Grew 1981, 1998) as the new mineral khmaralite (Barbier et al. 1999) had to wait for single-crystal X-ray technology capable of measuring the very faint superstructure reflections first seen by Christy (1988) in electron diffraction diagrams. Grew et al. (2000) reported that 0.5 Be/20 O appears to be the minimum needed for the superstructure reflections to appear.

The presence of these superstructure reflections is the critical feature distinguishing khmaralite from sapphirine. Christy et al. (2002) found no superstructure reflections in TEM images of the synthetic sapphirine with 1 Be/20 O (vs. 0.78 Be/20 O in type khmaralite, Fig. 3a). It is likely that ordering of Be into the most polymerized tetrahedral sites is as extensive in synthetic beryllian sapphirine as it is in khmaralite, but unlike khmaralite, there is no regular alternation of different ordering schemes from one chain segment to the next. The regular alternation of ordering schemes, which results in the doubling of the unit cell (Fig. 2), is undoubtedly due to annealing, either during a superimposed metamorphic event or during cooling following metamorphism.

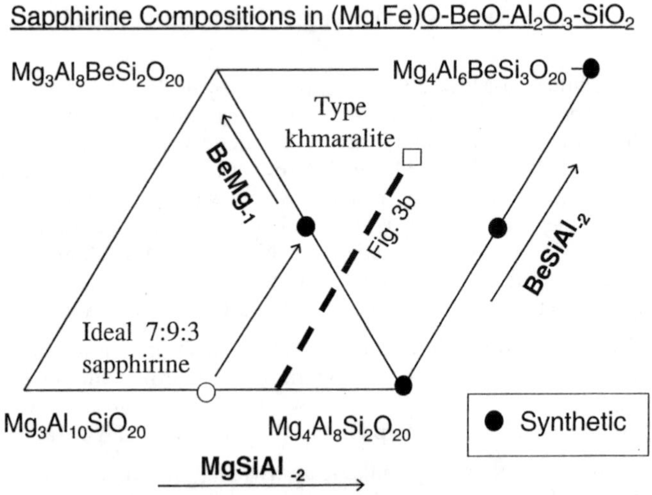

Figure 3. (a) A planar section of compositional space in the $MgO\text{-}BeO\text{-}Al_2O_3\text{-}SiO_2$ system showing the relative positions of the beryllian sapphirines synthesized by Hölscher (1987) and khmaralite with Fe^{2+} combined with Mg and Fe^{3+} combined with Al (modified from Christy et al. 2002).

Khmaralite forms masses 3-5 cm across that are deeply embayed by minerals formed from its breakdown. It is indistinguishable from iron-rich sapphirine both in hand specimen and thin section, i.e., it is nearly opaque in large pieces, blue when powdered and moderately pleochroic in shades of blue and green.

Khmaralite composition is equivalent to that of a sapphirine of composition $\sim(Mg,Fe^{2+})_{3.6}(Al,Fe^{3+})_{8.8}Si_{1.6}O_{20}$ to which $0.78BeSiAl_{-2}$ has been added (Grew et al. 2000;

Barbier et al. 2002). Ion probe (SIMS) analyses of Be-bearing sapphirine (Be > 0.1 per 20 O or about 0.3 wt % BeO) and khmaralite from several localities and different rock types suggest a continuum of compositions along a pseudobinary between ~$(Mg,Fe^{2+})_{3.6}(Al,Fe^{3+})_{8.8}Si_{1.6}O_{20}$ and ~$(Mg,Fe^{2+})_{3.6}(Al,Fe^{3+})_{7.2}Be_{0.8}Si_{2.4}O_{20}$ with no discernible break between the fields of sapphirine and khmaralite (Fig. 3b). The close approach to this pseudobinary is surprising given that sapphirine-khmaralite from both silica-saturated (Napier Complex) and silica-undersaturated assemblages (Musgrave Ranges) are included. However, several compositions of sapphirine suspected to contain about 0.35Be per 20 O (but not analyzed for Be) from silica-saturated rocks from S. Harris, Scotland, do not plot close to this pseudobinary (Fig. 3b); they are richer in the $MgSiAl_{-2}$ component (Baba et al. 2000). Compositions of synthetic beryllian sapphirine also vary in $MgSiAl_{-2}$ (Fig. 3a).

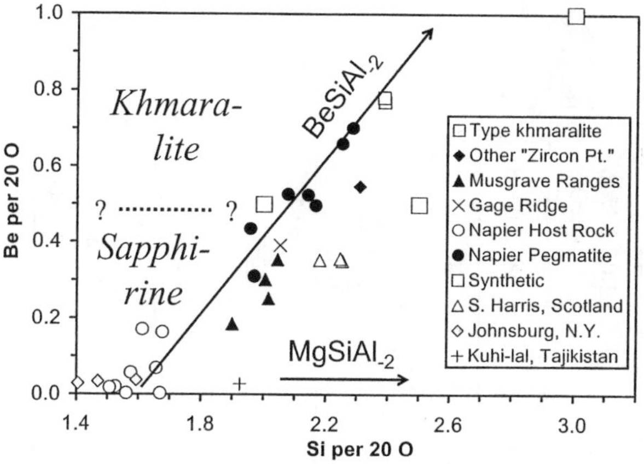

Figure 3. (b) Composition of natural and synthetic sapphirine for which ion microprobe Be measurements are available except South Harris, for which Be content was estimated, and synthetic, for which Be content was assumed to be the same as in the starting gel (extensively modified from Grew et al. 2000, their Fig. 12). The dashed line marking a tentative boundary between khmaralite and sapphirine in natural material is based on the appearance of superstructure reflections in a natural sample containing 0.52 Be/20 O (Grew et al. 2000). Type khmaralite, other "Zircon Point", Gage Ridge, Napier Host Rock and Napier Pegmatite all refer to samples from the Napier Complex, Enderby Land, East Antarctica. Sources of data are given in Table 3; also Baba et al. (2000) for South Harris, Scotland.

Sapphirine is a sink for beryllium. Its Be content can be substantial even in rocks with a modest bulk Be content if sapphirine is present in trace amounts; e.g., its Be content ranges from 60 to 1000 times whole-rock Be content in Napier Complex metamorphic rocks (Table 3). Only kornerupine-prismatine and cordierite incorporate a comparable amount of Be (see below).

Sapphirine containing more than 1Be per 20 O could not be synthesized (Hölscher 1987; Christy et al. 2002), and khmaralite in Be-rich assemblages contains even less (0.78 Be/20 O), consistent with the constraint on Be incorporation discussed above under aenigmatite-group minerals.

In both synthetic Fe-free sapphirine, and in natural Fe-rich sapphirine-khmaralite of comparable total Fe content, cell volume decreases 2.7 % from Be = 0 to Be = 1 per 20 O, i.e., the component $BeSiAl_{-2}$ has a molar volume of $-5.28\ cm^3mol^{-1}$ (Fig. 4).

Table 3. Beryllium content of sapphirine and khmaralite.

Rock type, grade		*Host rock Be, ppm (n)*	*Spr, Khm Be*(n) ppm Be*	*Source*
Napier Complex, East Antarctica.	Beryllium pegmatites, Khmara Bay	—	1.07-2.77 (6) wt % BeO	Barbier et al. (1999); Grew et al. (2000); Grew & Yates (unpubl. data); Barbier et al. (2002)
	Quartz granulite hosting pegmatite, "Christmas Point"	1.4-7.1 (5)	790-2200 (5)	Grew et al. (2000, 2001b and unpubl. data)
	Quartz granulite hosting pegmatite, Mt. Pardoe	3.7	240	Grew et al. (2000)
		—	30	
	Quartzofeldspathic granulite, Gage Ridge	5	1.34 wt % BeO	Grew (1981, 1998), Grew et al. (2000)
	Quartz-bearing granulite, Mt. Riiser-Larsen	—	~0.1 wt % BeO	Christy (1989)
"Musgravite" nodule, Musgrave Ranges, Australia (Fig. 6)		—	0.04-2.10 (8) wt % BeO	Hudson (1968); Wilson & Hudson (1967); Grew et al. (2000); Grew (unpubl. data)
Serendibite-bearing metasomatic zones, Johnsburg, NY, USA		—	330-430 (3)	Grew et al. (1991a, 1992)
Ultramagnesian metamorphic rocks, SW Pamirs, Tajikistan		—	0-300 (5)	Grew et al. (1994a, 1998b)
Kornerupine-bearing rocks, various localities		—	0-100 (14)	Grew et al. (1990, 1991b); Grew (1986)
Sapphirine-enstatite-hornblende rock, Mautia Hill, Tanzania		—	22	McKie (1963)
Sapphirine-cordierite-biotite-garnet-hypersthene rock, Val Codera, Italy		—	3	Barker (1964)

Note: n – number of samples. * 0.1 wt % BeO ≡ 360 ppm Be

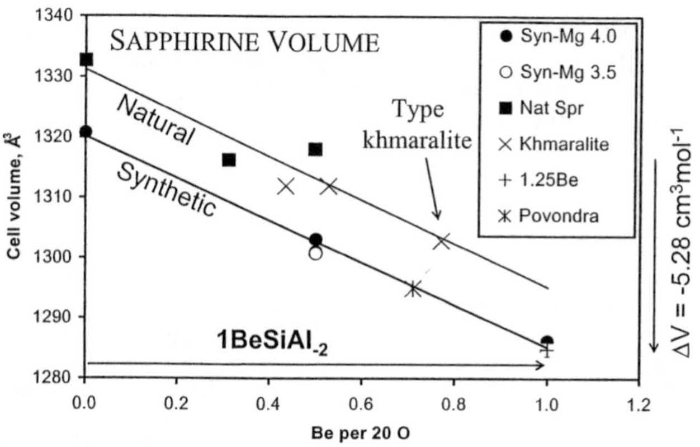

Figure 4. Effect of beryllium on sapphirine unit-cell volume for synthetic sapphirine (Hölscher 1987; Christy et al. 2002) and for natural sapphirine and khmaralite containing 8-10 wt % Fe (Sahama et al. 1974; Barbier et al. 1999; Grew et al. 2000). The lines are least-square fits to the natural data (filled squares and "X's") and the synthetics containing 4 Mg/20 O (filled circles). ΔV applies to the synthetic sapphirine (Christy et al. 2002). The Be contents of sapphirine synthesized by Povondra and Langer (1971b) and by Hölscher (1987) from a gel containing 1.25 Be/20 O were calculated by Christy et al. (2002) from cell volume and the least-square fit for the synthetics.

Khmaralite and beryllian sapphirine (≥1 wt % BeO) has been reported to date from two pegmatites metamorphosed under granulite-facies conditions in the Archean Napier Complex (Table 3). In the original description of khmaralite I interpreted it to be a metamorphic mineral that crystallized in the "Zircon Point" assemblage khmaralite + sillimanite + surinamite + magnesiotaaffeite-6*N*'3*S* ("musgravite") + garnet + biotite + rutile at $T \geq 820°C$, $P \geq 10$ kbar (Barbier et al. 1999), conditions sufficient to stabilize sillimanite + orthopyroxene (Grew 1998). However, textural relations of coarse grained khmaralite and beryllian sapphirine from "Christmas Point" suggest that these minerals crystallized from a pegmatitic magma with sillimanite in prisms up to 10 cm long and 4 cm across, quartz, alkali feldspar, biotite and wagnerite (Grew et al. 2000). The pegmatites were emplaced at the peak of metamorphism possibly at 1000-1100°C at 9-11 kbar. During subsequent metamorphism, khmaralite and beryllian sapphirine reacted with quartz to form sillimanite, surinamite and garnet, or where isolated from quartz, broke down to chrysoberyl- or magnesiotaaffeite-6*N*'3*S* -bearing assemblages. Nonetheless, sapphirine-khmaralite could have equilibrated with its breakdown products to form stable metamorphic assemblages, some of which are shown in Figure 5. An alternative scenario is that these minerals formed by reaction of sapphirine and khmaralite with the pegmatitic melt prior to crystallization of quartz (Grew et al. 2001b).

Other sapphirine relatively rich in Be (Fig. 3b) has been reported from quartz-bearing rocks in the Napier Complex, including host rocks to the beryllium pegmatites in Khmara Bay, and from quartz free rocks in the Musgrave Ranges, Australia (Table 3; Fig. 6). Napier Complex metamorphic sapphirine was stable with quartz at ultrahigh temperatures (e.g., Sheraton et al. 1987b), i.e., in the range 1000-1100 °C at pressures from 7 to 11 kbar according to Harley (1998). The Musgrave Ranges sapphirine is associated with a magnesiotaaffeite-6*N*'3*S* ("musgravite") nodule enclosed in phlogopite replacing pyroxenite (Fig. 6). Sapphirine BeO content decreases progressively outward from the nodule to the phlogopite enclosing it. Wilson and Hudson (1967) suggested that beryllium may have been introduced metasomatically with K, F, and P in the contact aureole of the nearby Ernabella ferrohypersthene adamellite; temperatures of crystallization could have been ca. 700°C, the value estimated for a similar complex about 100 km west of the sapphirine locality (Maboko et al. 1989 1991; cf. oxygen isotope temperature of 550°C reported by Wilson et al. 1970 for the sapphirine locality).

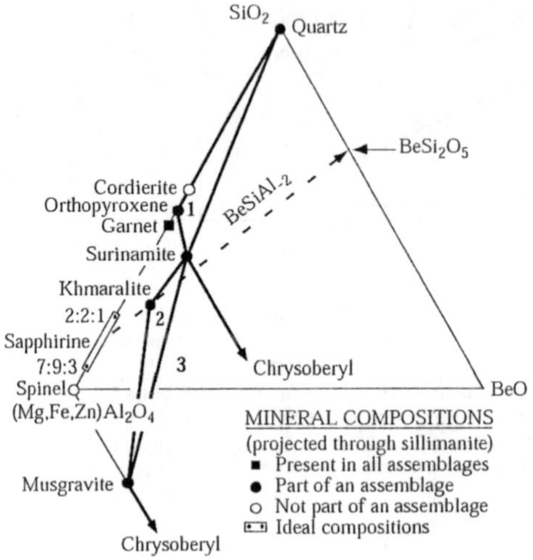

Figure 5. Possible assemblages of Be minerals (all with sillimanite) in the system $(Mg,Fe,Zn)Al_2O_4$-BeO-SiO_2 in Be-bearing assemblages, *e.g.*,pegmatites (no. 3, Grew 1981; no. 2 without garnet, Grew et al. 2000) and quartzofeldspathic gneiss (no. 1 without garnet, Baba et al. 2000). The dashed line marked $BeSiAl_{-2}$ is one of an array of lines converging on the point $BeSi_2O_5$. Copied from Barbier et al. (1999, Fig. 1).

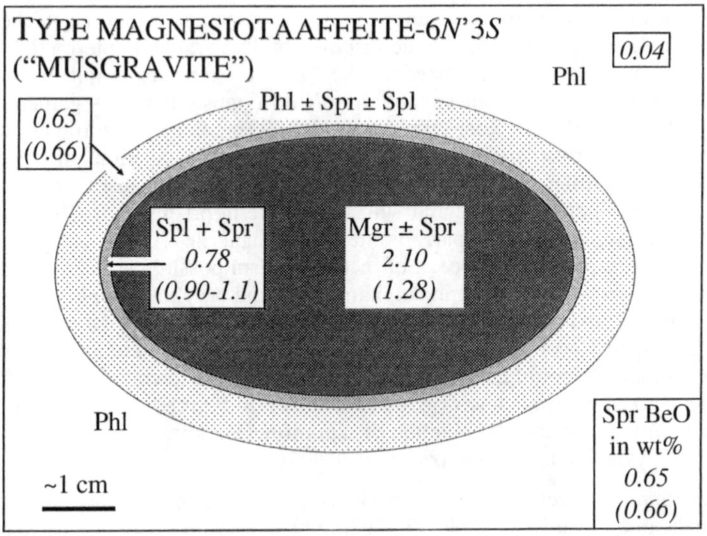

Figure 6. Schematic representation of a 10×9×5 cm nodule of magnesiotaaffeite-6*N*'3*S* (Mgr) described by Wilson and Hudson (1967), Hudson et al. (1967) and Hudson (1968) from the Musgrave Ranges, Australia. The spinel (Spl)–sapphirine (Spr) rim around magnesiotaaffeite-6*N*'3*S* is mostly 0.5-2 mm thick. Sapphirine BeO contents were measured both spectrographically (Wilson and Hudson 1967; Hudson 1968) and by ion microprobe (values in parentheses, Grew et al. 2000 on Hudson's sample 14820). The 0.04 wt % BeO value was obtained on sapphirine 20 cm from the magnesiotaaffeite-6*N*'3*S* nodule (Hudson 1968).

Two studies reported experimental syntheses of beryllian sapphirine. Povondra and Langer (1971b) reported beryllian sapphirine in run products from glasses on the join $Mg_2Al_4Si_5O_{18}$–$NaMg_2Al_3BeSi_5O_{18}$ at $T \geq 900°C$ for P up to 3 kbar. Hölscher (1987) synthesized sapphirine from gels having several starting compositions in a reconnaissance study at T from 700 to 1350°C and P from 1-13 kbar, which is being written up by Christy et al. (2002). Hölscher's (1987) syntheses lie within the stability field of sapphirine in the Be-free MgO-Al_2O_3-SiO_2-H_2O system (Seifert 1974; Ackermand et al. 1975) except for one synthesis at 750°C and 13 kbar, about 30°C below the minimum T at 13 kbar given by Ackermand et al. (1975). Determining whether Be expands the stability range of sapphirine would require systematic studies involving reversed experiments of sapphirine breakdown to surinamite and cordierite.

Surinamite. Surinamite, $(Mg,Fe^{2+})_3(Al,Fe^{3+})_3O(AlBeSi_3O_{15})$, was first described in a granulite-facies metapelite from the Bakhuis Mountains, Surinam, as a ferromagnesian aluminosilicate that closely resembled sapphirine in thin section and in crystal structure (de Roever et al. 1976; Moore 1976). Grew (1981) and de Roever et al. (1981) independently discovered that Be was a major constituent, and de Roever et al. (1981) demonstrated that Be was essential for synthesizing an Fe-bearing surinamite at 20 kbar and 800°C. Moore and Araki (1983) successfully refined the crystal structure and confirmed the stoichiometry in the formula proposed by de Roever et al. (1981), $(Mg,Fe^{2+})_3(Al,Fe^{3+})_3(AlBeSi_3O_{16})$. Hölscher et al. (1986) made a comprehensive study of surinamite stability in the MgO-Al_2O_3-BeO-SiO_2-H_2O system and showed that it formed at high temperature (≥650°C) and relatively high pressure (≥4 kbar). Barbier (1996 1998) has synthesized several Ga and Ge, but Be-free, analogues of surinamite and refined their crystal structures.

Surinamite is generally too fine-grained to be seen in hand specimen. The cm-sized surinamite-rich aggregates in beryllium pegmatites from Casey Bay in Antarctica are exceptional; these are nearly black, but show blue in powder, which distinguishes them from other dark ferromagnesian silicates other than sapphirine. In thin section surinamite resembles iron-bearing sapphirine in its platy habit, high refractive indices, low birefringence, strong dispersion, and strong color. It differs from sapphirine-khmaralite (e.g., Grew 1998, Fig. 3, where khmaralite is labeled as sapphirine) in its distinctive pleochroism, which is violet parallel to Y and ranges from greenish blue to colorless or very light greenish brown perpendicular to Y (de Roever et al. 1976; Barbier et al. 2002). The orientation of the indicatrix is $X \wedge c \approx 20°$, $Y = b$, and $Z \wedge a \approx 5°$, but the axes of the indicatrix do not correspond to those of the absorption surface in (010).

The most important variable in natural surinamite is Fe, which ranges from 5.25 to 14.17 wt % as FeO (e.g., Baba et al. 2000; Grew et al 2000). A room-temperature Mössbauer spectrum and crystal structure refinement gave $Fe^{3+}/\Sigma Fe$ = 0.31, 0.35, respectively, for two surinamite specimens from Khmara Bay for which 0.30, 0.35, respectively, were calculated by assuming 16 O and 11 cations (Barbier et al. 2002). These results indicate that calculating Fe^{3+}/Fe^{2+} ratio from stoichiometry is a servicable approximation. Surinamite $Fe^{3+}/\Sigma Fe$ increases with increasing Fe^{3+} content in associated sillimanite and reaches 0.54 in samples from "Christmas Point" (Khmara Bay), where associated oxide is ilmeno-hematite and magnetite (Grew et al. 2000).

A new crystal-structure refinement of a Khmara Bay surinamite (Barbier et al. 2002) gave exactly 1 Be per 16 O (vs. 0.945 reported by Moore and Araki 1983). Within analytical uncertainty, ion microprobe measurements are consistent with the ideal 1Be per formula in other Khmara Bay surinamite (Grew et al. 2000), but only 0.766-0.824 Be was found in South Harris surinamite. Baba et al. (2000) attributed this deficiency in Be to the coupled substitution of $^{[4]}Al$ for $^{[4]}Be$ and $^{[6]}Mg$ for $^{[6]}Al$, summing to $MgBe_{-1}$. No other surinamite has been quantitatively analyzed for Be.

Other constituents reported in surinamite include MnO (≤2 wt %), P_2O_5 (≤0.33 wt %), K_2O, CaO and ZnO (≤0.2 wt %), Na_2O (<0.1 wt %), and Cu, Pb, Sn, and V (0.01-0.05 wt %)(de Roever et al. 1976, Hålenius 1980; Grew 1981; Grew et al. 2000). Ion microprobe boron contents range from 3-12 ppm B in surinamite from quartzofeldspathic gneiss (Baba et al. 2000) to 0.09-0.15 wt % B_2O_3 in surinamite from metamorphosed pegmatite (Grew et al. 2000).

The tetrahedral sites of Khmara Bay surinamite are highly ordered, and the octahedral sites show no mixing between divalent and trivalent cations, although there is Mg-Fe^{2+} mixing on some sites and Al-Fe^{3+} mixing on others (Moore and Araki 1983; Barbier et al. 2002). The greater cation order in surinamite compared to sapphirine is reflected in the more restricted compositional variation in terms of the Tschermaks substitution, $MgSiAl_{-2}$, e.g., 2.89-3.07 Si per 16 O (Baba et al. 2000; Grew et al. 2000) vs. 1.13–1.57 Si per 16 O for Be-free sapphirine (Christy 1989).

Surinamite preferentially fractionates Mg relative to Fe^{2+} compared to associated minerals except cordierite: X(Mg) increases Grt << Spr and khmaralite < surinamite ≤ Crd in samples from Casey Bay and South Harris (Baba et al. 2000; Grew et al. 2000); some ambiguity results from the large error in estimating Fe^{3+}/Fe^{2+} ratio in surinamite from stoichiometry.

Surinamite is exclusively a metamorphic mineral reported only from metapelites and metapegmatites metamorphosed under upper-amphibolite or granulite-facies conditions (Table 4). Ramesh Kumar et al. (1995) reported surinamite rich in B and Ga from an eighth locality, the Eastern Ghats belt, India, but this identification could not be confirmed; it appears that hypersthene was misidentified as surinamite (Grew et al. 2001c). Surinamite is invariably associated with Al-rich minerals, e.g., sillimanite, kyanite, and cordierite, but never with primary muscovite. Almost all assemblages are quartz saturated, but surimanite also occurs with corundum in isolation from quartz at "Christmas Point" (Grew et al. 2000). Al_2SiO_5 + K-feldspar assemblages are characteristic of most surinamite-bearing rocks. Ferromagnesian minerals associated with surinamite typically have relatively high X(Mg), e.g., garnet (Fig. 7). In the MgO-BeO-Al_2O_3-SiO_2 model system, surinamite is equivalent to pyrope + chrysoberyl: $Mg_3Al_4BeSi_3O_{16} = Mg_3Al_2Si_3O_{12} + Al_2BeO_4$, and the assemblage Grt + Cb + Sur thus theoretically defines the limit of Fe^{2+} substitution for Mg in surinamite at a given pressure and temperature. Observed compositions for Antarctic garnet and surinamite are not entirely consistent with this relationship: several garnet-surinamite pairs in assemblages lacking chrysoberyl are more ferroan than the pair associated with chrysoberyl (Fig. 7).

The temperature estimated for surinamite formation mostly exceeds 800°C, although temperature for the upper-amphibolite-facies Chimwala, Zambia occurrence was more likely in the 600-700°C range. Estimated pressure exceeds 8 kbar. These conditions lie within the stability field Hölscher et al. (1986) determined for the Mg-end member, $T \geq 650$°C, $P \geq 4$ kbar (see also Franz and Morteani, this volume). Table 4 summarizes reactions proposed for surinamite formation from Be-bearing cordierite, generally due to an increase in pressure, and from Be-bearing sapphirine-khmaralite due a nearly isobaric temperature decrease. Theoretically, surinamite is expected to break down at high pressure to pyrope + chrysoberyl, which is denser, but Hölscher et al. (1986) encountered unidentified phases and anomalies in the XRD pattern of chrysoberyl formed from surinamite breakdown at 45 kbar.

Hölscher et al. (1986) noted that most pegmatites, if hydrous, would crystallize at too low a temperature for surinamite; in addition, their bulk X(Mg) is too low for its

Table 4. Surinamite occurrences.

Locality	*Host rock*	*Associated minerals* / *Reactions for surinamite formation*	*T, P conditions*
1. Bakhuis Mtns., Surinam	Mylonitic mesoperthite gneiss	Qtz, Kfs, Pl, Bt, Sil, Ky, Spl	Granulite facies
		Crd → Sur + Al_2SiO_5 + Qtz	
2. Strangways Ranges, central Australia	Aluminous granulite	Included in Crd. Also Qtz, Spl, Spr, Phl, Opx, Sil, opaque oxide	*T* < 900-950°C, *P* = 8-9 kbar
3. "Christmas Point", Casey Bay, Antarctica	Meta-pegmatite	Qtz, Kfs, Sil, Spr, khmaralite, wagnerite, Grt, Opx, Bt, Hem, Mgt; Spl, Mgr, Crn; 2nd Crd, And, Ky	*T* = 800-900°C, *P* = 8-9 kbar
		Khm-Spr + Qtz → Sur + Grt + Sil and *Khm-Spr → Mgr + Sil + Sur*	
4. "Zircon Point", Casey Bay, Antarctica	Meta-pegmatite	Qtz, Kfs, Pl, Sil, Ky, Khm, Mgr, dumortierite, Cb, Grt, Bt, Rt	
5. Mount Pardoe, Amundsen Bay, Antarctica	Meta-pegmatite	Enclosed in 2nd Crd with Sil, Bt. Also Qtz, Kfs, wagnerite, Grt, Opx, And, Ky	
6. Chimwala area, Zambia	Cordierite granulite	Qtz, Kfs, Pl, Bt, Ky, Grt, Mgt. Crd and a Sil-like mineral are completely replaced	Amphibolite, transitional to granulite, facies
		Crd + Kfs + H_2O → Sur + Qtz + Ky + Bt + Ab	
7. South Harris, Scotland, U.K.	Gneiss	Qtz, Kfs, Spr, Opx, Ky, Sil, 2nd Crd	*T* = 850-900°C, *P* > 12 kbar: M2 stage
		First *$BeSiAl_{-2}$ (in Spr) + 3 Sil + 3 Opx → Sur + 4 Qtz*, then *$Mg_{7.4}Al_{15.7}Be_{0.7}Si_{4.2}O_{40}$ (Be Spr) + 9.65 Qtz → 0.7 Sur + 5.3 Opx + 6.45 Ky*, which is (Crd) in Figure 8	

Note: Sources of information on surinamite localities: 1-de Roever (1973), de Roever et al. (1976). 2-Woodford and Wilson (1976), Goscombe (1992). 3, 4, 5-Grew (1981, 1998), Grew et al. (2000). 6-Vavrda and Vrána (1972), de Roever and Vrána (1985). 7- Baba (1998a,b, 1999), Baba et al. (2000). Table format modified from Baba et al. (2000).

stability. Thus, the presence of magmatic surinamite in a granite pegmatite is unlikely, and none has been reported to date. The garnet + chrysoberyl assemblage apparently takes the place of surinamite in most pegmatites, and indeed almandine-spessartine garnet and chrysoberyl are often reported from the same pegmatite (e.g., Okrusch 1971, Franz and Morteani 1984; Soman and Druzhinin 1987; Černý et al. 1992; Černý, this volume). Franz and Morteani (1984) presented evidence that many of the chrysoberyl-bearing pegmatites have been metamorphosed after emplacement. However, it is less obvious that garnet and chrysoberyl coexisted in all cases. A clear cut example of a metamorphic garnet + chrysoberyl assemblage was illustrated by Černý et al. (1992) from a muscovite pegmatite at Maršíkov, Czech Republic, which was metamorphosed at 600°C, 4-6 kbar ($P_{fluid} < P_{total}$), i.e., conditions only slightly outside the surinamite stability field reported

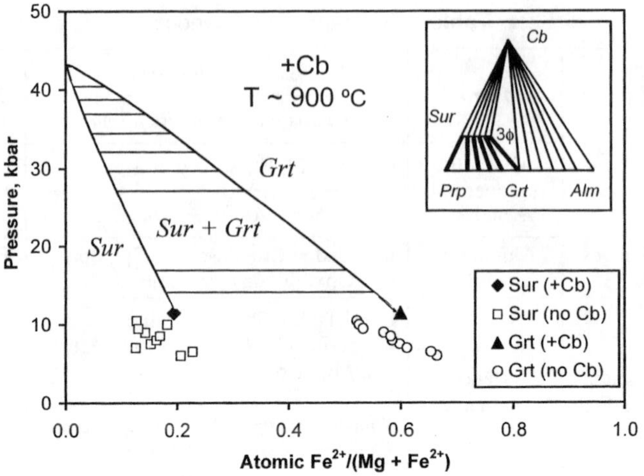

Figure 7. Schematic and hypothetical pressure-composition section for the system surinamite–Fe^{2+} analogue of surinamite modified from Hölscher et al. (1986, Fig. 13a) with compositions of a surinamite-garnet pair (+ chrysoberyl) from "Zircon Point", Khmara Bay, Antarctica (Grew 1981) and of 8 surinamite-garnet pairs (no chrysoberyl) from "Christmas Point", Khmara Bay and from Mt. Pardoe, Amundsen Bay, Antarctica (Grew 1981; Grew et al. 2000 and unpublished data). All nine pairs are presumed to have formed at the same pressure, but are plotted at slightly different pressures to show regularity of Fe-Mg distribution; scatter in surinamite data is due to uncertainties in calculating Fe^{2+} from stoichiometry. Inset shows sketch of the ternary system almandine (Alm)-pyrope (Prp)-chrysoberyl (Cb). Three-phase field Grt + Sur + Cb (3ϕ) corresponds to filled symbols and the 2-phase field Grt + Sur (heavy tielines) corresponds to unfilled symbols.

by Hölscher et al. (1986). Given the marked enrichment of Fe^{2+} and Mn in garnet, the Maršíkov assemblage is not a compositional equivalent to surinamite, and thus the Maršíkov assemblage is not a definitive test of the stability field proposed by Hölscher et al. (1986). A closer alternative to surinamite in quartz- and sillimanite-bearing pegmatites would be garnet + cordierite + chrysoberyl, but this assemblage has not yet been reported.

Most high-temperature metapelites and many peraluminous meta-pegmatites do not contain chrysoberyl. Thus, a petrogenetic grid restricted to assemblages with chrysoberyl (Hölscher et al. 1986, their Fig. 11) has limited applicability to Be-bearing assemblages in metamorphic rocks. Instead, compositional relationships can be simplified by projection through sillimanite (Fig. 5). A P-T diagram (Fig. 8) based on this projection is constructed so that surinamite is a high-pressure mineral relative to sapphirine-khmaralite and cordierite consistent with the formation of surinamite in nature (Table 4) and experiment (Hölscher et al. 1986). For example, Baba et al. (2000) reported surimanite formation from beryllian sapphirine during a pressure increase as temperature decreased between the first and second metamorphic events at South Harris, Scotland (M1 → M2, Fig. 8) and its breakdown to cordierite as pressure decreased following the second event (M2 → M3, Fig. 8). These observations are consistent with calculations involving molar volumes of surinamite, beryllian sapphirine and anhydrous beryllian cordierite (Christy and Grew in preparation). Because natural Na-Be-bearing cordierite is commonly hydrous, the breakdown of cordierite to surinamite would be a dehydration and the reactions labelled (Qtz) and (Spr) in Figure 8 could have negative slopes.

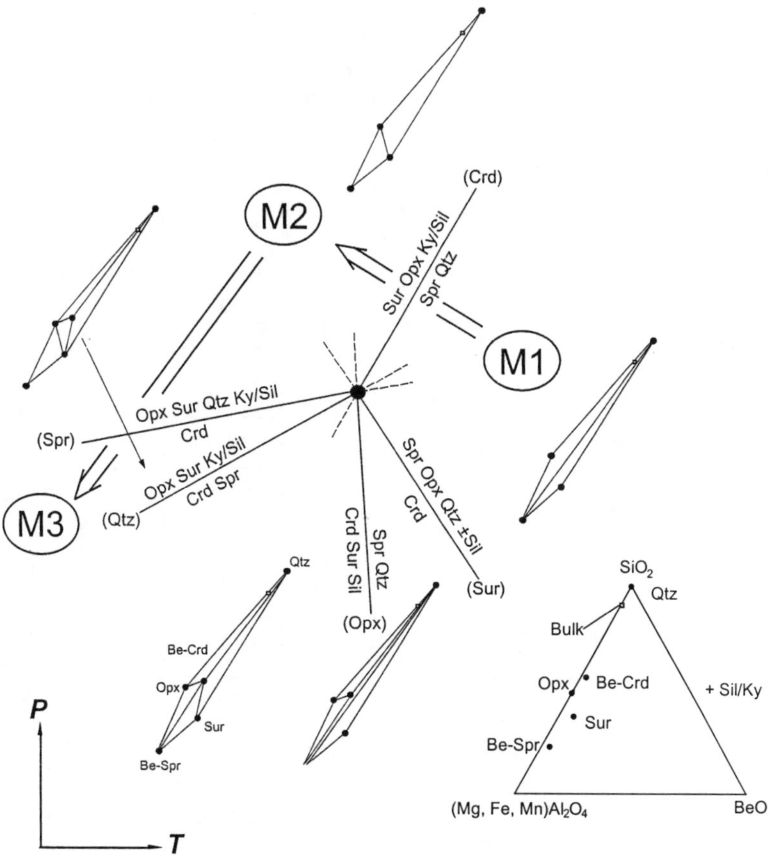

Figure 8. Pressure-temperature diagram based on a Schreinemakers net for phase relations in the system $MgAl_2O_4$-BeO-SiO_2 (cf. Fig. 5) for assemblages containing Al_2SiO_5. Reactions are labeled by the non-participating phase in parenthesis. Arrow indicates *P-T* path inferred for the surinamite-bearing rock at South Harris, Scotland (see text). Slopes of the Qtz-absent and Spr-absent reactions are shown positive as cordierite is presumed to be anhydrous. Copied from Baba et al. (2000, their Fig. 7).

Surinamite is indicative of a distinctive metamorphic environment and history. Surinamite-bearing rocks are typically polymetamorphic; in most cases, there is evidence for pre-existing cordierite or sapphirine enriched in beryllium. Mylonitic textures are characteristic of several examples, notably, South Harris and Bakhuis Mountains. In all cases, surinamite formed with an increase in pressure and/or decrease in temperature; no examples are known of surinamite forming as pressure decreased isothermally. Surinamite is not restricted to rocks enriched in Be and scarcity of Be in metamorphic systems is not the only factor controlling surinamite formation. Surinamite formation apparently depends on a mechanism for concentrating disseminated beryllium. This mechanism appears to be formation of a mineral such as cordierite or sapphirine that can scavenge Be from a large mass of rock and becomes the locus of surinamite crystallization during a succeeding metamorphic event. Conditions during this event must be such that Be is not newly dispersed. Low water activities probably restrict its mobility.

Table 5. Beryllium content of cordierite and sekaninaite (BeO ≤ 0.55 wt%, i.e., Be ≤ 2000 ppm).

Rock type		Host rock Be, ppm (n)	Crd Be ppm (n)	Source
METAMORPHIC				
Miscellaneous metasediments		—	0-140 (6)	Ginzburg & Stavrov (1961)
Miscellaneous gneiss		—	10-30 (6)	Griffitts & Cooley (1961)
Migmatite, Bavarian Forest, Germany		—	1-49 (–)	Kalt et al. (1999)
Contact metapelite T = 508 to 762°C, P = 2 kbar, Kos, Greece		—	7-30 (15)	Kalt et al. (1998)
Ladoga complex, NW Russia	Sil + Kfs zone	3.0 (30)	40 (5)	Lebedev & Nagaytsev (1980)[1]
	Hypersthene zone	2.3 (28)	18 (3)	
Rocks with kornerupine-prismatine	Ellammankovilpatti, India (granulite facies)	—	20-170 (3)	Grew et al. (1987, 1990)
	Miscellaneous (granulite facies)	—	0-30 (5)	Grew et al. (1990, 1991b)
"Kinzigite" (amphibolite-facies metapelite), Ivrea-Verbano, Italy		1.7	12.3	Bea & Montero (1999); Bea (pers comm. 2001)
Migmatite, Peña Negra Complex, Spain		0.37-8.1	1.9-16.2 (7)	Bea et al. (1994); Bea (pers comm. 2001)
		—	22	Malcherek et al. (2001)
"Cordieritite" formed by an anatectic process, El Pilón, Soto, Argentina		—	40-65 (2)	Schreyer et al. (1979)
Sapphirine-cordierite-biotite-garnet-hypersthene rock, Val Codera, Italy		—	5	Barker (1964)
Cordierite-bearing gneiss, Bamble Sector, South Norway		—	<2-257 (20)	Visser et al. (1994)
PEGMATITE, VEIN AND UNSPECIFIED				
Pegmatites, Murzinka, Urals, Russia and Turkestan Mtns., Kazakhstan		—	190-580 (2)	Ginzburg & Stavrov (1961)
Miscellaneous veins and pegmatites (unaltered only)		—	1.5-2000 (6)	Griffitts & Cooley (1961)
Pegmatite, Bjordammen ("Bjordan"), Bamble, Norway		—	430	Newton (1966)
Pegmatitic sekaninaite, Dolní Bory, Czech Republic		—	4-30 (8)	Černý et al. (1997)
		—	0 (2)	Malcherek et al. (2001)
Pegmatite, Sri Lanka		44	290	Grew et al. (1995)
Miscellaneous, possibly vein or pegmatite		—	0-270 (9)	Malcherek et al. (2001)

Note: n – number of samples. [1] Values are averages; individual analyses were not given.

It should not be forgotten that surinamite occurs very sparsely except in the pegmatites at "Christmas Point" and "Zircon Point" and could be easily overlooked because of its close resemblance to sapphirine.

Be-bearing cordierite (Mg > Fe)–sekaninaite (Fe > Mg)

The structural relationship between beryl and cordierite (indialite, the high-temperature hexagonal form of cordierite, is isostructural with beryl) and the frequent presence of cordierite in contact zones and pegmatites prompted Ginzburg and Stavrov (1961) and Griffitts and Cooley (1961) to search cordierite for beryllium. As much as 2000 ppm Be were found in pegmatitic cordierite (Table 5, see also Černý, this volume and London and Evensen, this volume). The structural relationship between cordierite and beryl also led investigators to look for evidence of solid solution between them, both in natural phases and experimentally (e.g., Newton 1966; Borchert et al. 1970; Povondra and Langer 1971a,b; Franz and Morteani, this volume). Several substitutions involving Be, Mg, Al and Si have been proposed for cordierite, notably Be + Si = 2Al (Schreyer 1964) and 3Be + Si = 2Mg + 2Al, which relates cordierite and beryl (Newton 1966). Povondra and Langer's (1971a,b) failure to confirm early evidence for extensive solid solution between beryl and cordierite is not surprising. The substitution $(Be_3Si)(Al_{-2}Mg_{-2})$ linking cordierite and beryl is equivalent to $2Al_2(MgSi)_{-1}$ (Tschermaks) + $3BeSiAl_{-2}$, and the former substitution is virtually unknown in both cordierite and beryl (e.g., Deer et al. 1986; Aurisicchio et al. 1988). Some authors (e.g., Vrána 1979) have suggested solid solution of beryl towards cordierite, but it has not been confirmed by systematic studies (e.g., Aurisicchio et al. 1988; Demina and Mikhaylov 2000; Černý, this volume; Franz and Morteani, this volume).

Černý and Povondra (1966) concluded that Na + Be → □ + Al is the most important substitution in natural cordierite, and this has been confirmed by crystal structure refinements (e.g., Armbruster 1986) and subsequent analytical studies summarized in Figure 9 in which K and 2Ca are added to Na. Povondra and Langer (1971b) showed experimentally that considerable Be could be incorporated by Na + Be → □ + Al at geologically reasonable temperatures of 700-880 °C at P_{H2O} = 1-3 kbar. Even when corrected for incorporation of Li by the substitution $(Na,K)Li(\square Mg)_{-1}$ (e.g., Gordillo et al. 1985), the sum (Na + K + 2Ca) generally exceeds Be, so this sum cannot be used as a proxy to estimate Be + Li content.

It is possible that Schreyer's (1964) proposed substitution $BeSiAl_{-2}$ could play a limited role under conditions comparable to those under which synthetic cordierite incorporates Be by this substitution, i.e., at very high temperatures in the absence of Na, K and Ca (Hölscher and Schreyer 1989). Baba et al. (2000) reported cordierite with excess Si as a corona around surinamite and estimated that it contains about 12% of the $Na(Mg,Fe)_2Al_3BeSi_5O_{18}$ end member and 12% of Hölscher and Schreyer's (1989) $Mg_2[Al_2BeSi_6O_{18}]$ end member, but had no Be analysis to substantiate this estimate.

The reduction in volume afforded by $1(NaBe)(\square Al)_{-1}$ is ~2% of the volume of the Mg end member in synthetic cordierite (Povondra and Langer 1971b) and 1.49% in natural cordierite, which is the same as the effect of Fe substitution for Mg, +1.54% for $1FeMg_{-1}$ (Armbruster and Irouschek 1983). It is thus not surprising that the volume reduction due to Be incorporation in the samples for which compositional (Fig. 9) and XRD data are available is obscured by variations in FeO, which ranges from 3.6 to 14.1 wt % (sekaninaite) in these samples. In contrast, the reduction afforded by $1BeSiAl_{-2}$ is 4.2% (Hölscher and Schreyer 1989), greater than in sapphirine, 2.7% (Fig. 4).

Most cordierite (and all sekaninaite) containing BeO > 0.1 wt % is reported from pegmatites (both unmetamorphosed and metamorphosed) and related vein deposits

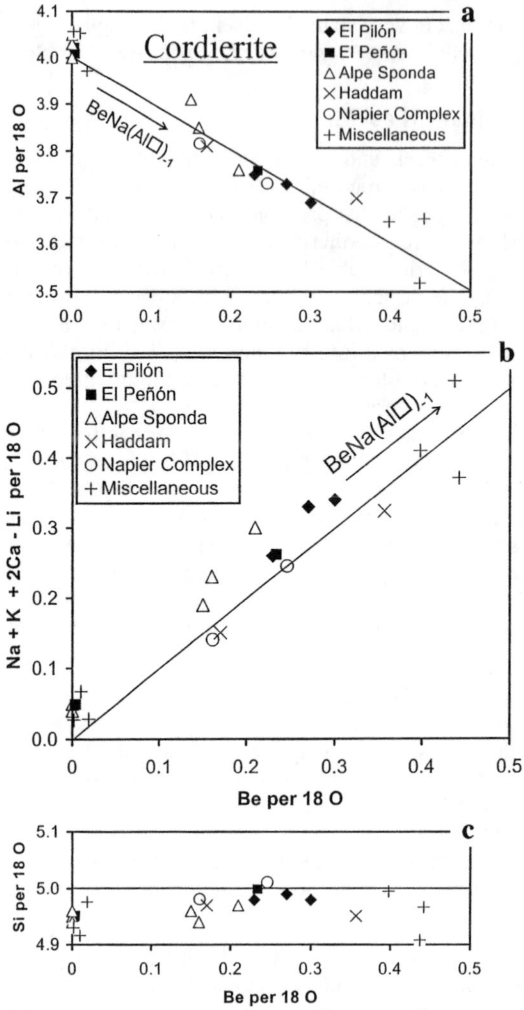

Figure 9. Plots of compositional variables vs. Be content of cordierite and sekaninaite from pegmatites and metamorphic rocks. (A) Aluminum. (B) Alkali and calcium. (C) Silicon. Sources of data: El Pilón, near Soto, Argentina – Schreyer et al. (1979); El Peñón, Sierras de Córdoba, Argentina—Gordillo et al. (1985); Alpe Sponda, Switzerland (includes data from nearby Miregn)—Armbruster and Irouschek (1983); Haddam, Connecticut – Armbruster and Irouschek (1983), Povondra and Čech (1978); Napier Complex, Antarctica—Grew et al. (2000); Miscellaneous—Povondra et al. (1984), Černý and Povondra (1966), Piyar et al. (1966), Malcherek et al. (2001).

(Tables 5, 6; Černý, this volume). At El Peñón, Sierras de Córdoba, 65 km west of Córdoba City and south of Tala Cañada (Edgardo Baldo, pers. comm. 2001), Argentina, beryllian sekaninaite is a major constituent of some nodules in pegmatites cutting a roof pendant in the Achala Batholith (Gordillo et al. 1985), while in the El Pilón complex near Soto, about 50 km to the west-northwest of El Peñón, beryllian cordierite is a porphyroblast in biotite-rich "cordieritite" a few meters from a pegmatite (Schreyer et al. 1979). The origin of the "cordieritites" (mostly Be-poor) in the Soto area, which contain subordinate biotite, quartz, fibrolite and plagioclase, is under continuing investigation by Carlos Rapela and colleagues; formation by low-pressure (garnet-absent) partial melting of metasediments in the roof zone of granite was proposed by Rapela et al. (1995 1997 1998). It has not been specified whether Be in the cordierite-sekaninaite originated from nearby pegmatites at either locality. The first and still most noteworthy metamorphic occurrence of beryllian cordierite, i.e., with no evidence for introduction of Be from an external source such as pegmatite, is kyanite-paragonite-staurolite schist from Alpe Sponda, Switzerland (Armbruster and Irouschek 1983). But there could be other

Table 6. Occurrences of beryllian cordierite and sekaninaite (BeO > 0.5 wt %)

Locality	*Host rock*	*Associated minerals*	*T, P conditions*	*BeO wt % (n)*
1. Alpe Sponda, Ticino, Switzerland	Mica schist	Pg, St, Ky, Pl, Bt, Tur	T = 600-650°C, P = 6-10 kbar	0.60-0.81 (3)
2. El Pilón, Soto, Argentina	Porphyroblast in "cordieritite"	Qtz, Bt, Pl	Upper amphibolite	0.93
3. El Peñon, Argentina (sekaninaite)	Nodules in pegmatite	Qtz, Kfs, Pl, Ms, Bt, Ap, And, Sil, Brl, Tur, uraninite, 2nd Chl	Upper amphibolite	0.87-1.16 (3)
4. Ukraine (exact locality unknown)	Vein formed by reaction of pegmatitic magma and ultramafic rock	Grt, Pl, Mc, Sil, Qtz, Spl, Mgt, Ap		1.77
5. Věžná, Czech Republic	Metasomatic zones in pegmatite	Qtz, Kfs, Pl, Rt, columbite, Brl, Mnz, Tur, Ap		1.70-1.94 (–)
6. Haddam, Connecticut, USA	Pegmatite	Qtz, Mc, Ab, Tur, Grt, Cb, columbite		0.52-1.44 (3)
7. Micanite, Colorado, USA (sekaninaite?)	Pegmatite	Qtz, Mc, Ms, Bt, Tur		0.70
8. Kemiö Island, Finland (sekaninaite)	Pegmatite	Qtz; 2nd Qtz, Ms, Pg, Chl, Brl		1.55
9. Sugamo, Japan (sekaninaite)	Pegmatite	Qtz, Pl, Kfs, Ms, Sil, Tur, Grt		~1
10. "Christmas Point", Antarctica	Metamorphosed pegmatite	Qtz, Sil, Grt, Sur, Opx	T = 600-650°C, P = 2-5 kbar	1.00
11. Mt. Pardoe, Antarctica		Qtz, Sil, Grt, Sur, Opx		0.64

Note: n – number of samples. Sources of information on cordierite composition, paragenesis and P-T estimates: 1-Irouschek-Zumthor (1983), Armbruster and Irouschek (1983), Armbruster, 1986. 2-Schreyer et al. (1979). 3-Gordillo et al. (1985). 4-Piyar et al. (1968). 5- Černý and Povondra (1966, 1967). 6-Newton (1966), Povondra and Čech (1978), Armbruster and Irouschek (1983), Heinrich (1950). 7-Newton (1966), Heinrich (1950). 8-Povondra et al. (1984). 9-Sambonsugi (1957), Miyashiro (1957), Selkregg and Bloss (1980). 10, 11-Grew et al. (2000); Harley (1985), Sandiford (1985).

examples of metamorphic beryllian cordierite but not yet analyzed, e.g., the cordierite that had broken down to surinamite (Bakhuis, Mtns., Strangways Ranges, Chimwala, Table 4). In other metamorphic rocks, cordierite is not rich in Be; nonetheless, its Be content is several times the host-rock bulk Be content (Table 5), implying that like sapphirine, cordierite is a sink for Be (see below and Evensen and London 1999, in press; London and Evensen, this volume).

Euclase

Since its discovery by Haüy (see Delamétherie 1792, 1797; Haüy 1799; Dana 1892), euclase, $AlBeSiO_4(OH)$, was long considered to be a relatively rare mineral, but now it has been found in a variety of environments, including granite pegmatites, greisens, alpine fissures and metamorphic rocks (Introduction, Appendix 1; Černý, this volume).

Composition. Most wet chemical analyses (e.g., Damour 1855; von Knorring et al. 1964) and electron microprobe analyses (e.g., Goffé 1980; Barton 1986; Hemingway et al. 1986) confirm ideal stoichiometry and the minimal presence of impurities. Nonetheless, minor amounts of Na, Ca, Fe, Ti, Ga, Ge, and Sn have been reported, e.g., Fe_2O_3 to 0.47 wt %, Fe as FeO to 0.28 wt % (respectively, Yegorov 1967; Graziani and Guidi 1980); 0.02 wt % TiO_2 (Mattson and Rossman 1987), and from 0.084 to about 0.15 wt % Ge (Yegorov 1967; Sharp 1961). Fluorine was reported in the first analysis to establish the presence of water (0.38 wt % F, Damour 1855), but few authors have sought fluorine since. Impurities such as included fluorite, which was found in euclase from Colombia (Notari et al. 2000), pose a problem when assessing the reliability of wet chemical analysis of F, e.g., Popov's (1998) report of 0.09 and 1.35 wt % in euclase from the Mariin emerald deposit (Urals, Russia). Electron microprobe analysis avoids the problem of impurities, but detection limits can be high, e.g., 0.5 wt % F in a study reporting no F detected (Hemingway et al. 1986). Hsu (1983) reported that synthetic euclase incorporated little, if any, F in the presence of HF, whereas bertrandite in this environment became F dominant.

Occurrence. Euclase has been reported from low-grade metamorphic environments. In some cases, formation of euclase is associated with hydrothermal activity and such occurrences are not strictly metamorphic. For example, Dana (1892, p. 509) mentioned an occurrence of euclase "with topaz in chloritic schist" in the mining district of Villa Rica (now Ouro Preto, possibly the type locality or near it, see Cassedanne 1989), Minas Gerais, Brazil. The origin of euclase and topaz in this area is controversial, i.e., whether they formed in a pegmatite or by metamorphism of a stratabound deposit affected by hydrothermal fluids (e.g., Olsen 1971 1972; Fleischer 1972; Keller 1983; Cassedanne 1989). According to Cassedanne (1989), who reviewed literature on the Ouro Preto topaz deposits, euclase has been found only as loose crystals in association with topaz, which is restricted to a single, endogenic horizon dominated by muscovite, sericite and quartz. In Yakutiya, Russia, euclase is found in chlorite-quartz zones that cut coarse-grained granite and are associated with its cataclasis (Yegorov 1967). Formation of euclase in these zones is attributed to low-temperature hydrothermal activity under conditions of high acidity. A third example is the development of beautifully faceted euclase crystals with quartz, rutile, albite ("pericline"), calcite, ankerite, muscovite and chlorite in fissures cutting metamorphic rocks at several localities in the Alps (e.g., Meixner 1957); these occurrences are related to postmetamorphic activity. Lastly, euclase occurs sparingly in emerald-bearing veins cutting Cretaceous sediments in Colombia (Rubiano 1990; Notari et al. 2000); these emerald deposits are now thought to be metamorphic-hydrothermal in origin (see Franz and Morteani, this volume, for a review of the Colombian emerald deposits).

Examples of metamorphic euclase include Goffé's (1980 1982) report of relatively abundant (3.6 modal % in the studied section) euclase in a Fe-rich chlorite-rich rock from Aiguille du Fruit, Vanoise, Briançonnais zone, France (W Alps) containing chloritoid and quartz. Both ferromagnesian silicates approach their respective Fe end members. The chlorite rock is associated with metabauxite and rocks containing the high-pressure/low-temperature phase magnesiocarpholite (sample Chanrossa B2, Henry et al. 1996) and resulted from metamorphism estimated at 10-11 kbar, 330-350°C by the method of Goffé and Bousquet (1997; Goffé, pers. comm. 2001). That is, euclase is also high-pressure/low-temperature phase in these rocks. This rock and a second one with the same assemblage later found on the Aiguille du Fruit contain 25-61 ppm Be (Catel and Goffé, pers. comm. 2001). Nonetheless, Goffé (1982) and Poinssot et al. (1997) inferred a sedimentary origin for this remarkable, if localized, Be enrichment; the Be-enriched rocks and associated bauxites are Al-rich materials formed by alteration of a granitic basement and subsequently washed away, transported and redeposited.

Hanson (1985) reported euclase in a quartz vein with pyrophyllite, ottrelite and davreuxite that cuts phyllites containing chloritoid, andalusite and spessartine, at Ottré, Ardennes Mountains, Belgium. The vein was emplaced during the later stages of metamorphism, which peaked near 380°C and 1-2 kbar. Beryllium could have been remobilized from the host Mn-rich metasediments, in which 2-5 ppm Be have been reported (Krosse and Schreyer 1993), and concentrated in the vein; a magmatic source of Be is unlikely because as such rocks are absent in the Ottré area.

Taaffeite group

There are three mineral species in the taaffeite group: magnesiotaaffeite-2*N*'2*S* ($Mg_3BeAl_8O_{16}$), magnesiotaaffeite-6*N*'3*S* ($Mg_2BeAl_6O_{16}$) and the Fe^{2+} analogue of the latter, ferrotaaffeite-6*N*'3*S* (Armbruster 2002). These minerals are polysomes consisting of modified nolanite (*N*') and spinel (*S*) modules, respectively $Be(Mg,Fe)Al_4O_8$ and $(Mg,Fe)Al_2O_4$ (see also Hawthorne and Huminicki, this volume). The suffix in the mineral name gives the proportions of the modified nolanite and spinel modules, whereas the chemical prefix specifies the dominant divalent cation. The taaffeite group is closely related to the högbomite and nigerite groups, which are Fe-Mg-Zn-Al oxide minerals forming polysomatic series and containing significant amounts of Ti and Sn, respectively (Armbruster 2002).

Magnesiotaaffeite-2*N*'2*S* was originally described as "taaffeite" (Anderson et al. 1951) from a cut gemstone from Sri Lanka, whereas magnesiotaaffeite-6*N*'3*S* ("musgravite") was first found by Hudson et al. (1967) in granulite-facies rocks from the Musgrave Ranges, Australia and its Fe^{2+} analogue, ferrotaaffeite-6*N*'3*S* ("pehrmanite") from a pegmatite in Finland (Burke and Lustenhouwer 1981). The compositional and structural similarity of these minerals to one another and problems in the original analysis of magnesiotaaffeite-2*N*'2*S* resulted in considerable confusion that was finally clarified by Schmetzer (1983a,b).

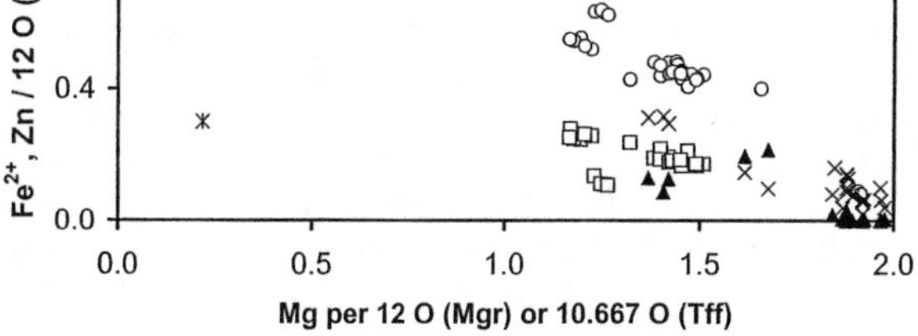

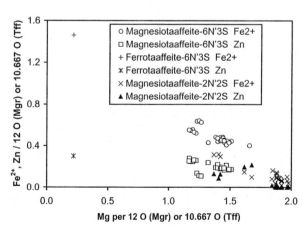

Figure 10. Ferrous iron and zinc contents of taaffeite-group minerals. Formulae of magnesiotaaffeite-2*N*'2*S* have been recalculated to a 10.667 O basis so as to be directly comparable to magnesiotaaffeite-6*N*'3*S* and ferrotaaffeite-6*N*'3*S*, i.e., with divalent cations totalling close to 2. Sources of data: Hudson et al. (1967), Teale (1980), Burke and Lustenouwer (1981), Grew (1981), Moor et al. (1981), Schmetzer (1983a), Schmetzer and Bank (1985), Chadwick et al. (1993), Rakotondrazafy (1999), Grew et al. (2000).

Composition. The three oxides are largely Fe^{2+}-Mg-Zn solid solutions (Fig. 10). Nuber and Schmetzer (1983) found no isomorphous replacement of Be and Mg, and thus Be content is fixed for each of the three minerals. Most analyses are by electron microprobe, and thus few direct determinations of ferric/ferrous ratio have been reported. In most samples, the sum of trivalent cations excluding Fe approaches the ideal 6 and 8 atoms per formula units for magnesiotaaffeite-6*N*'3*S* and magnesiotaaffeite-2*N*'2*S*, respectively, implying that ferric iron contents are low; this is consistent with wet chemical determination on the type magnesiotaaffeite-6*N*'3*S* (0.4 wt % Fe_2O_3, Hudson et al. 1967). Higher Fe_2O_3 contents, 1.5-2.0 wt %, equivalent to 0.11-0.13 Fe^{3+} per formula unit, were calculated by stoichiometry from electron microprobe analyses of magnesiotaaffeite-6*N*'3*S* associated with hematite-ilmenite, which implies a relatively oxidizing environment with high Fe^{3+}/Fe^{2+} ratios (Grew et al. 2000). Huang et al. (1988) reported wet chemical analyses of magnesiotaaffeite-2*N*'2*S* from the Hsianghualing deposit, China, in which the iron is dominantly ferric, i.e., 7.15-8.14 wt % Fe_2O_3 and 0-0.45 wt % FeO. However, these values give an excess of trivalent cations, 8.27-8.34 per formula unit and my recalculation of the analysis assuming stoichiometry gave 0.10-0.16 Fe^{3+} per formula unit, more in accord with Schmetzer's (1983a) electron microprobe analyses of magnesiotaaffeite-2*N*'2*S* from this deposit.

Maximum contents reported for other constituents include 3.03 wt % MnO, 0.64 wt % Ga_2O_3, 0.33 wt % Cr_2O_3, 0.15 wt % V_2O_3 (Schmetzer 1983a; Schmetzer et al. 2000).

The data plotted in Figure 10 suggest that Zn and Fe^{2+} increase together, e.g., the highly magnesian Sakeny magnesiotaaffeite-6*N*'3*S* contains negligible Zn (Devouard, pers. comm., 2001; Devouard et al. 2002; Rakotondrazafy, Raith, Devouard Nicollet, in preparation) as does highly magnesian magnesiotaaffeite-2*N*'2*S*. In addition, magnesiotaaffeite-6*N*'3*S* appears to incorporate more Fe^{2+} and Zn than magnesiotaaffeite-2*N*'2*S*. However, these trends could result from the geochemical environments from which the analyzed samples happen to originate rather than from crystallographic controls on the incorporation of $Fe^{2+}$and Zn. An assemblage of both magnesiotaaffeite minerals has yet to be found, and thus distribution of elements between them cannot be determined directly. When compared to spinel from separate assemblages, magnesiotaaffeite-6*N*'3*S* fractionates Mg relative to Fe^{2+} slightly more than magnesiotaaffeite-2*N*'2*S* does, i.e., $(Mg/Fe^{2+})_{Mgr}/(Mg/Fe^{2+})_{Spl}$ = 1.70-2.85 vs. $(Mg/Fe^{2+})_{Tff}/(Mg/Fe^{2+})_{Spl}$ = 1.48 (Wilson and Hudson 1967; Teale 1980; Rakotondrazafy 1999; Grew et al. 2000).

Occurrence. Both magnesiotaaffeite minerals are found in calcareous and non-calcareous silica-undersaturated environments (Table 7). Three of the four known magnesiotaaffeite-2*N*'2*S*-bearing deposits, i.e., Hsianghualing (China), Sakhir-Shulutynsky and Pitkäranta (Russia), are carbonate rocks or magnesian ("apocarbonate") skarns that have been metasomatized by hydrothermal fluids rich in F and Be, as well as Sn or Li associated with granites. Magnesiotaaffeite-2*N*'2*S* probably formed at temperatures below 400°C at these three deposits although I am not aware of temperature-pressure estimates specifically for magnesiotaaffeite-2*N*'2*S* formation at any of them. For example, temperature of beryllium mineralization at similar deposits (e.g., Yermakovskoye fluorite-phenakite-bertrandite deposit in the Transbaikal region, Russia) are estimated from fluid inclusions to have ranged from 340°C at the start to 140°C at the end at depths not exceeding 1.5 km (Bulnaev 1996; cf. 400-140°C range at $P < 1$ kbar reported by Kosals et al. 1973). Although phenakite and bertrandite are not found with magnesiotaaffeite-2*N*'2*S*, both minerals are present elsewhere in the Pitkäranta and Hsianghualing deposits (Nefedov 1967; Pekov 1994, Huang et al. 1988), and phenakite is

Table 7. Occurrences of taaffeite-group beryllium oxides.

Locality	Associated minerals	T, P	Source (*italics* – P-T data only)
MAGNESIOTAAFFEITE-6N'3S ("MUSGRAVITE")			
Musgrave Ranges, Australia (type)	Spl, Spr	~550°C (probably too low)	Hudson et al. (1967); Wilson & Hudson (1967); *Wilson et al. (1970)*
"Zircon Point", Khmara Bay, East Antarctica	Sil, Sur, Khm, Cb, Grt, Bt, Rt, Qtz	~800-900°C, ~8-9 kbar	Grew (1981, 1998); Grew et al. (2000)
"Christmas Point", Khmara Bay, East Antarctica	Sil, Sur, Spr-Khm, Ilm-Hem, Spl, Crn, Bt, wagnerite		
Dove Bugt, North-East Greenland	Cal, Norbergite, Spl†, Phl, Ap, Fl, arsenopyrite, 2nd Chl		Chadwick et al. (1993); Jensen & Stendal (1994); Jensen (1994)
Pegmatite P30, Antsofimbato, Sahatany, Madagascar	Tur, spessartine, beryl, Cb		Ranorosoa (1986)
Sakeny, ~100 km NW of Ihosy, southern Madagascar	An, Spl, Di, Crn, Spr, 2nd Ms	<~700°C, 4-5 kbar	Rakotondrazafy (1999); Rakotondrazafy & Raith (2000); Devouard et al. (2002)
FERROTAAFFEITE-6N'3S ("PEHRMANITE")			
Rosendale pegmatite, Kemiö Island, Finland	Sil, Ms, Grt, Hc, Cb, Tur, Crn, nigerite		Burke et al. (1977); Burke & Lustenhouwer (1981)
MAGNESIOTAAFFEITE-2N'2S ("TAAFFEITE")			
Hsianghualing deposit, Hunan Province, China	Fl, Cal, Li mica, Phl, nigerite, Spl, Cb, cassiterite, cancrinite		Peng & Wang (1963), Beus (1966), Vlasov (1966); Huang et al. (1988)
Sakhir-Shulutynskiy pluton*, E. Sayan, Russia	Li mica, Fl		Kozhevnikov et al. (1975a); locality from N.N. Pertsev (pers. comm. 2000)
Mount Painter, South Australia	Spl, Rt, Crn; 2nd högbomite		Teale (1980)
Pitkäranta, Karelia, Russia	Spl		Schmetzer (1983a); Pekov (1994)

Note: The great majority of reported "taaffeites" are gemstones of unknown provenance from Sri Lanka (type) and, in a few cases, Myanmar (Burma) and Tunduru, Tanzania; rare "musgravite" gemstones are also found in Sri Lanka. *For background, see Kozhevnikov et al. (1975b), Kozhevnikov and Perelyayev (1987). †Spinel identified optically in sample #339313 from the Greenland Geological Survey

reported in metasomatites in the aureole of the Sakhir-Shulutynskiy pluton (Kozhevnikov et al. 1975a). The Dove Bugt magnesiotaaffeite-6*N*'3*S* paragenesis is also a skarn, which resulted from metasomatic action of a sheet of granite, which introduced Be into underlying calcite marble (Chadwick et al. 1993; Jensen 1994; Jensen and Stendal 1994).

This metasomatic activity probably post-dates the regional upper amphibolite-facies metamorphism during which sillimanite, garnet and cordierite developed in schist and paragneiss in the Dove Bugt area (Chadwick and Friend 1991; Chadwick et al. 1990; 1993), and thus there is little information bearing on the *P-T* conditions for magnesiotaaffeite-6*N*'3*S* formation. Chadwick et al. (1993) inferred that magnesiotaaffeite-6*N*'3*S* predated introduction of F-rich fluid, but the association of magnesiotaaffeite-2*N*'2*S* and fluorite at other localities (and the association of Be and F mineralization in general) suggests that magnesiotaaffeite-6*N*'3*S* at Dove Bught is more likely coeval with fluorite and F-rich fluid activity.

In the case of the non-calcareous environments, magnesiotaaffeite minerals are found with the Al-rich minerals, chrysoberyl, spinel or sapphirine-khmaralite, mostly in high-temperature, deep-seated rocks. At the type locality magnesiotaaffeite-6*N*'3*S* constitutes a nearly monomineralic nodule 10×9×5 cm in a phlogopite replacement zone (Fig. 6). Wilson and Hudson (1967) and Hudson et al. (1967) inferred that magnesiotaaffeite-6*N*'3*S* formed from replacement of corundum during metasomatic introduction of Be (see above under *Khmaralite-sapphirine*). Parageneses in the Khmara Bay pegmatites differ in that magnesiotaaffeite-6*N*'3*S* occurs in beryllium-rich aggregates formed by reaction of sapphirine-khmaralite with quartz during a high-temperature metamorphic event following intrusion; Be enrichment is clearly premetamorphic (Grew et al. 2000; see above under *Surinamite*). Magmatic sapphirine-khmaralite reacted with quartz to form sillimanite-surinamite-garnet coronas; magnesiotaaffeite-6*N*'3*S* formed where breakdown of sapphirine-khmaralite proceeded in isolation from quartz. The single exception to silica-undersaturation in parageneses with either magnesiotaaffeite mineral is a magnesiotaaffeite-6*N*'3*S* grain enclosed in quartz in the pegmatite at "Zircon Point" (Grew 1981). At Mt. Painter, South Australia, metamorphic magnesiotaaffeite-2*N*'2*S* and corundum are completely enclosed in spinel porpyroblasts and isolated from the phlogopite matrix; a magnesiohögbomite mineral formed from magnesiotaaffeite-2*N*'2*S* during a later metamorphic event (Teale 1980). Teale (1980) reported no evidence for metasomatic activity at Mt. Painter and inferred that Be had been present in the sedimentary precursor to the magnesiotaaffeite-2*N*'2*S*-bearing rock. Ranorosoa (1986) reported from the Sahatany pegmatite district a "taafféite" with a powder X-ray diffraction pattern similar to that of magnesiotaaffeite-6*N*'3*S*. Apart from the minerals listed in Table 7, no details on the paragenesis were reported, and it is not clear whether magnesiotaaffeite-6*N*'3*S* at Sahatany is strictly pegmatitic or like ferrotaaffeite-6*N*'3*S*, a late, hydrothermal phase.

The Sakeny, Madagascar, magnesiotaaffeite-6*N*'3*S* paragenesis in "sakénite" (anorthite-rich rocks containing spinel, sapphirine or corundum, Lacroix 1939) is both Ca- and Al-rich, but calcite is absent (Rakotondrazafy 1999). Anorthite and phlogopite constitute the matrix; corundum is often isolated by coronas of spinel and/or sapphirine. Magnesiotaaffeite-6*N*'3*S* forms platelets up to 0.5×0.1 mm enclosed in a spinel corona and almost touches diopside (Rakotondrazafy 1999) or in spinel ± sapphirine + anorthite symplectite between spinel and anorthite (C. Nicollet, pers. comm. 2001; Rakotondrazafy, Raith, Devouard and Nicollet, in preparation). Both igneous (anorthosite) and sedimentary precursors have been suggested for "sakénite"; Rakotondrazafy and Raith (1997) and Rakotondrazafy (1999) cited geochemical evidence for metasomatic introduction of Si and Mg as critical in their development from sediments containing clayey and calcareous intercalations. Devouard et al. (2002) propose that Be was introduced by infiltration metasomatism related to emplacement of Be- and B-rich pegmatites associated with the Ranotsara ductile shear zone at temperatures below ~700°C and pressures of 4-5 kbar, i.e., less than the >700°C, 5.0-5.5 kbar estimated for pelitic gneiss in Ihosy, 100 km distant (Nicollet 1985, 1990).

Ferrotaaffeite-6*N*'3*S* is found in aggregates of sillimanite, garnet, muscovite, columbite-tantalite, zincian hercynite, nigerite-group minerals, chrysoberyl, corundum, tourmaline, apatite, and zincian staurolite in the wall zone of the Rosendal pegmatite, Finland (Burke et al. 1977; Burke and Lustenhouwer 1981). Ferrotaaffeite-6*N*'3*S* overgrows and replaces the nigerite-group minerals, which in turn replace hercynite. The above authors attributed formation of ferrotaaffeite-6*N*'3*S* and nigerite-group minerals to the action on hercynite of late-stage Sn- and Be-bearing hydrothermal solutions that separated from the pegmatitic melt. Although quartz and sodic plagioclase are also present in the wall zone, there is no evidence ferrotaaffeite-6*N*'3*S* coexisted with either.

The relative stabilities of the magnesiotaaffeite minerals are unknown (see also Franz and Morteani, this volume). Both minerals are found over a comparable range of *P-T* conditions in similar geochemical environments, and the difference in Be:Mg:Al ratio between them would in principle allow stable coexistence, yet the two phases have not been found together. It remains unclear which conditions favor one over the other, and experimental work sheds little light on the problem. Syntheses of both minerals have been reported in the ceramic literature (Geller et al. 1946; Reeve et al. 1969; Kawakami et al. 1986; Franz and Morteani, this volume) and gemological literature (Schmetzer et al. 1999), but at temperatures above those encountered in the crustal environments where these minerals are found, i.e., 1200 °C and higher. In addition, Hölscher (1987) listed "Taaffeit 9R", i.e., magnesiotaaffeite-6*N*'3*S*, with beryllian sapphirine and cordierite as phases synthesized at 1200°C and 1 kbar from a gel of $(Mg_{3.5}Al_{4.5})(Al_{3.5}Be_{0.5}Si_{2.0})O_{20}$ composition but did not report any details on how the magnesiotaaffeite-6*N*'3*S* was identified or on its physical properties. To date, the magnesiotaaffeite minerals are the only ternary compounds found in the MgO-BeO-Al_2O_3 system at near-liquidus temperatures. Although crystallization of a melt has yielded only magnesiotaaffeite-6*N*'3*S*, both minerals formed by exsolution from a $MgAl_2O_4$-$BeAl_2O_4$ spinel solid solution on cooling. Whether spinel can incorporate sufficient beryllium at geologically reasonable temperatures to exsolve magnesiotaaffeite minerals is another question that needs further study. The presence of magnesiotaaffeite-6*N*'3*S* as lamellae in spinel and spinel + anorthite symplectite at Sakeny, Madagascar suggests the possibility that musgravite resulted from exsolution of the Be component from an originally high-temperature beryllian spinel.

The provenance of the alluvial magnesiotaaffeite minerals that are highly prized as gemstones from Sri Lanka, Myanmar (Burma) and Tanzania has yet to be elucidated (e.g., Kampf 1991; Demartin et al. 1993; Burford 1998; Kiefert and Schmetzer 1998; Schmetzer et al. 2000).

BERYLLIUM CONTENTS OF SELECTED ROCK-FORMING MINERALS

General statement

Beryllium contents of rock-forming minerals are less often determined than the contents of other trace elements, and there have been even fewer systematic studies of the distribution of beryllium among common rock-forming minerals in metamorphic rocks, indeed in any rock type. The older studies of Be contents, which began with Goldschmidt and Peters (1932) and Goldschmidt (1958), are based on spectrographic analyses of mineral separates (e.g., Pearson and Shaw 1960; Petrova and Petrov 1965; Wenk et al. 1963; Schwander et al. 1968). The Russian-language literature on Be geochemistry is relatively large, and I have made no attempt to cover it entirely. More recent studies using mineral separates have made use of atomic absorption spectrophotometry (e.g., Lahti 1988), inductively coupled plasma emission spectroscopy (e.g., Bebout et al. 1993; Dahl et al. 1993) or direct current plasma emission spectroscopy (Kretz et al. 1989).

Investigators have also increasingly resorted to microbeam techniques of individual grains or zones in a given grain, including nuclear (e.g., particle-induced X-ray or gamma-ray emission, i.e., PIXE or PIGE, Lahlafi 1997), ion microprobe (e.g., Hervig, this volume; Domanik et al. 1993; Černý et al. 1995; Grew et al. 1998a,b,c), and laser ablation inductively coupled plasma emission mass spectroscopy, LA-ICP-MS (Bea et al. 1994; Bea, unpublished data).

The present review, while not confined to samples found in metamorphic rocks, emphasizes those minerals occurring in metapelites. The large number of studies of minerals in alkalic systems has not been covered; it should be noted that compared to metasedimentary and calc-alkalic rocks, amphiboles and pyroxenes are more enriched in Be in nepheline syenite pegmatites (reviewed by Hörmann 1978).

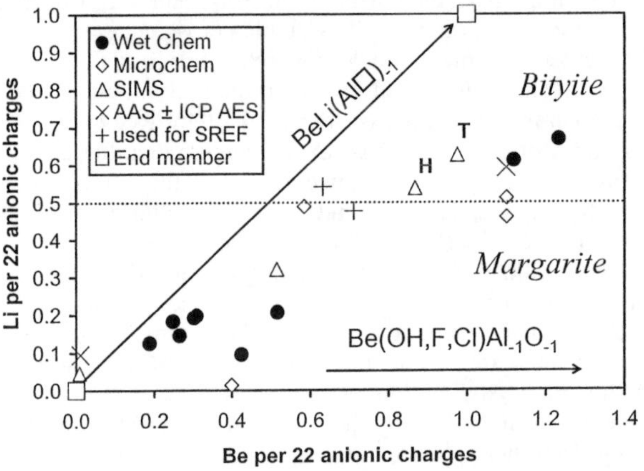

Figure 11. (a) Beryllium and lithium contents of bityite-margarite solid solutions. The substitution (BeLi)(Al□)$_{-1}$ joins the end-member compositions of margarite and bityite, respectively, $Ca\square Al_2(Al_2Si_2O_{10})(OH)_2$ and $CaLiAl_2(BeAlSi_2O_{10})(OH)_2$. Beryllium and lithium data were obtained by wet chemistry on mineral separates (Wet Chem, microchemical, atomic absorption spectrophotometry, AAS and inductively coupled plasma-atomic emission spectroscopy, ICP-AES), secondary ion mass spectroscopy (SIMS or ion microprobe), or by single crystal structural refinement (SREF). Sources of data: Arnaudov et al. (1982, wet chem), Bayrakov (1973, wet chem, one used by Sokolova et al. 1979 for SREF), Beus (1966, wet), Evensen (pers. comm. 2000, SIMS), Gallagher and Hawkes (1966, microchemical analyses), Grew et al. (1986, SIMS), Kupriyanova (1976, wet), Kutukova (1959, wet), Lahti (1988, AAS), Lahti and Saikkonen (1985, AAS and ICP-AES), Lin and Guggenheim (1983, microchem from Gallagher and Hawkes 1966, corrected for quartz; used for SREF), Rowledge and Hayton (1948, wet). Lin and Guggenheim (1983) used the same sample from the Mops pegmatite, Zimbabwe as Gallagher and Hawkes (1966). H = holotype specimen; T = specimen from near type locality in Sahatany Valley, Madagascar (J. Evensen, pers. comm.).

Mica group

Among naturally occurring micas only margarite-bityite solid solutions are known to contain a significant amount of beryllium (Fig. 11a). The Mica Subcommittee of the Commission on New Minerals and New Mineral Names (International Mineralogical Association) defined bityite as a trioctahedral brittle mica with $^{[6]}Li > {}^{[6]}\square$ and with an end-member formula $CaLiAl_2(AlBeSi_2)O_{10}(OH)_2$ (e.g., Can. Mineral. 36: 905-912 1998). This composition is linked to end-member margarite, $Ca\square Al_2(Al_2Si_2)O_{10}(OH)_2$, by the exchange vector, BeLi(Al□)$_{-1}$ (Lin and Guggenheim 1983). Most micas plotting in the bityite field have more Be and less Li than predicted by this exchange, e.g., the most beryllian bityite contains 1.23 Be and 0.67 Li per formula unit. Grigor'yev and Pal'guyeva (1980) reported "beryllium margarite", a variety in which only Be is present (Table 8). However, it is not clear whether Li was sought and not found, or simply not sought, so the possibility of a Li-free, Be-rich component in margarite remains to be demonstrated. Ginzburg (1957) suggested that Be could also be introduced with hydroxyl via $Be(OH)Al_{-1}O_{-1}$, whence his formula for margarite-bityite solid solutions, $CaLi_{x-y}Al_2[Al_{2-x}Be_xSi_2(OH)_y]O_{10-y}(OH)_2$; Grigor'yev and Pal'guyeva (1980) reported that this substitution was consistent with analytical data on "beryllium margarite". Albeit scattered, the data plotted in Figure 11b lend some credence to this suggestion if F and Cl are included with OH in computing anionic composition. Nonetheless, a crystallographic study of bityite with significant excess of Be over Li or of a Li-free "beryllium margarite" would be needed to confirm it. A similar substitution is implicit in the conclusions reached by Robert et al. (1995), who reported infrared and thermogravimetric evidence that the charge deficiency resulting from Be replacement of Si in micas synthesized in the system K_2O-MgO-BeO-SiO_2-H_2O was largely balanced by incorporation of extra OH, i.e., the most important substitution could be $Be(OH)_2Si_{-1}O_{-2}$.

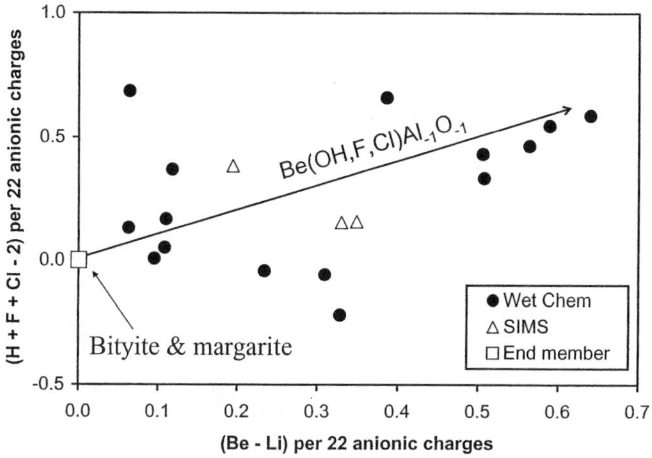

Figure 11 (b) Anion compositions of bityite-margarite solid solutions (wet chemical and SIMS data). Square marks end-member compositions.

Few investigators have sought B in bityite, but the amounts reported suggest that B could be a significant constituent, i.e., 0.48 wt % B_2O_3 in a sample from Finland (Lahti and Saikkonen 1985) and 1.45-1.51 wt % B_2O_3 in samples from the Sahatany Valley, Madagascar, including holotype material (J. Evensen, pers. comm.).

The compositions plotted in Figure 11 are for pegmatitic micas, except for the Be-poor metamorphic margarite reported from amphibolite-facies rocks in Antarctica by Grew et al. (1986) and metasomatic beryllian lithian margarite reported from the Harding Pegmatite (New Mexico) by J. Evensen (pers. comm. 2000). This bityite is an exocontact phase in a tourmalinized zone of amphibolite from 0 to about 6 cm from the pegmatite contact. The beryllium content of metamorphic margarite has overall received little attention (Table 8), although it probably contains more Be than associated paragonite (one pair reported) and muscovite.

Muscovite, paragonite and potassic lithium micas rarely contain more than 100 ppm Be even in environments rich in Be (Table 8, Fig. 12), *e.g.*, 19–68 ppm Be in muscovite from mica schists containing 10-61 ppm Be from the Habachtal emerald deposit, Austria

Table 8. Beryllium contents of micas (excluding bityite)

Rock Type		*Be, ppm (n)*	*Source*
MUSCOVITE AND PHENGITE			
Amphibolite-facies metasediment, Catalina Schist, Calif., USA		0.69 (1)	Bebout et al. (1993)
Pegmatite, Catalina Schist, Calif., USA		0.92-4.0 (3)	
Eclogite, Trescolmen, Central Alps, Switzerland		0.27-2.47 (7)	Zack (2000)
Mica schist, different grades		0-5 (3)	Hietanen (1969)
Amphibolite-facies metamorphic, Urals, Russia		2.5-4.0 (2)	Bushlyakov & Grigor'yev (1989)
Metamorphics, St + And + Sil grade, Ladoga, Russia		31 (-)[1]	Lebedev & Nagaytsev (1980)
Metagraywacke, metabasalt, etc.		0.6-6.2 (74)	Domanik et al. (1993)
Rožná pegmatite, Czech Republic (beryl)		11-18 (4)	Černý et al. (1995)
Dobrá Voda pegmatite, Czech Republic		0 (1)	
Miarolitic pegmatites, Pikes Peak, Colorado, USA		23 (1)	Foord et al. (1995)
Granite, Yellowknife, Canada		11 (1)	Kretz et al. (1989)
Pegmatite, Yellowknife, Canada		17 (1)	
Miscellaneous metamorphics, Tessin Alps, Switzerland		1-7 (55)	Schwander et al. (1968), cf. Wenk et al. (1963)
Pegmatite, Tessin Alps, Switzerland		<2-15(15)	
Schist and gneiss from Habachtal emerald deposit, Austria		19-68 (12)	Grundmann & Morteani (1989)
Staurolite zone schists, Black Hills, South Dakota, USA		2.1-17.7 (31)	Dahl et al. (1993); Dahl (pers. comm. 1999)
Sillimanite zone schists, Black Hills, South Dakota, USA	Non-metasomatic	0-14.2 (16)	
	Aureoles of pegmatite	16.9-42.5 (2)	
Yermakovskoye hydrothermal Be deposit, Transbaikalia, Russia		101	Novikova et al. (1994)
Miscellaneous granite pegmatites		12-120 (7)	Goldschmidt (1958)
Pegmatites containing beryllium minerals		20-108 (30)	Beus (1966, Table 129)
Metamorphic borian Ms in meta-pegmatite, Koralpe, Austria		49-140 (5)	Bernhardt et al. (1999); Kalt et al. (2001)
Muscovite replacing topaz and Tur in pegmatite, Orivesi, Finland		6-63 (6)	Lahti (1988)
Muscovite, pegmatite, Stoneham, Maine, USA		35 (SIMS)	Dyar et al. (2001)
		43 (ICP-AES)	

PARAGONITE		
Eclogite, Trescolmen, Central Alps, Switzerland	7.0-9.7 (3)	Zack (2000)
Mt. Bernstein, N. Victoria Land, Antarctica	30[2]	Grew et al. (1986)
MARGARITE		
Miscellaneous margarite	Between 4 and 36 (3)	Schaller et al. (1967) semiquantitative spectrographic analysis
	700-1000(2)	
Lithian, sodian margarite, pegmatite, Orivesi, Finland	255	Lahti (1988)
Mt. Bernstein, N. Victoria Land, Antarctica	200[2]	Grew et al. (1986)
"Beryllium margarite" in weathering residua from pegmatite, Urals, Russia	1.12 – 2.91 wt% BeO (2)	Grigor'yev (1980); Grigor'yev & Pal'guyeva (1980)
BIOTITE AND PHLOGOPITE		
Amphibolite-facies metamorphic, Urals, Russia	1.5-13 (6)	Bushlyakov & Grigor'yev (1989)
Metamorphics, Ladoga, Russia — with St	~2-4.5 (16)	Sergeyev et al. (1967)
Metamorphics, Ladoga, Russia — Grt to Opx zone	0.5-4.2 (5)[3]	Lebedev & Nagaytsev (1980)
Metamorphic, in amphibolite wallrock of Harding pegmatite (also bityite), New Mexico, USA	17	J. Evensen (pers. comm.. 2000)
Metasedimentary and mafic migmatites, California, USA	0.5-2.0 (5)	Domanik et al. (1993)
Pegmatite, Madagascar	~0-0.4 (2)	Grew et al. (1998c)
Whiteschist, SW Pamirs, Tajikistan	0-1 (6)	Grew et al. (1994, 1998b)
Miarolitic pegmatites, Pikes Peak, Colorado, USA	1 (1)	Foord et al. (1995)
Granite, Yellowknife, Canada	6 (1)	Kretz et al. (1989)
Granite, syenite, diorite (Dzhida complex, Trans-Baikal, Russia)	0.3-4.0 (40)	Petrova & Petrov (1965); Kostetskaya et al. (1969)
	6-10 (3)	Kosals & Mazurov (1968)
Miscellaneous plutonics, Russia, Kazakhstan	0.9-4 (5)	Petrov (1973)
Schist and gneiss from emerald deposit, Habachtal, Austria	~0-18 (38)	Grundmann & Morteani (1989)
Migmatites, Peña Negra Complex, Spain	0.11-1.43 (7)	Bea et al. (1994); Bea (pers. comm. 2001)
Metapelites, Ivrea-Verbano, Italy	0-6.14 (10)	Bea & Montero (1999); Bea (pers. comm. 2001)

Staurolite zone schists, Black Hills, South Dakota, USA		0-5.8 (31)	Dahl et al. (1993), Dahl (pers. comm., 1999)
Sillimanite zone schists, Black Hills, South Dakota, USA	Unaffected	0-5.7 (12)	
	Aureoles of pegmatite	9.1-12.3 (2)	
	Contaminated with beryl inclusions (?)	35.1-75.3 (4)	
Granite pegmatites, Norway		2 (2)	Goldschmidt (1958)
Phlogopite from fringe of beryl-bearing vein		30-600 (10)[4]	Beus (1966, Tables 132, 134)
"Lepidomelane" from nepheline syenite	Lovozero, Russia	3-30 (-)	
	Pegmatites, Låven, Norway	4-36 (-)	Goldschmidt & Peters (1932)
Pegmatite, Tessin Alps, Switzerland		<5 to 7 (5)	Wenk et al. (1963); cf. Schwander et al. (1968)
Baveno granite, Italy		<5 & 20 (2)	
Gneisses and few schists, Tessin Alps, Switzerland		Mostly <5; 8(1)	
With surinamite, pegmatite, Enderby Land, Antarctica		5	Grew (unpublished data)
LITHIUM MICAS			
Lepidolite, Rožná pegmatite, Czech Republic (beryl)		22-40 (7)	Černý et al. (1995)
Lepidolite, Dobrá Voda pegmatite, Czech Republic		0-7 (5)	
Miscellaneous, miarolitic pegmatites, Pikes Peak, Colorado, USA		5-78 (27)	Foord et al. (1995)
Lepidolite-zinnwaldite, Beauvoir granite, France		38-237(7)	Monier et al. (1987)
		<20-512 (29)	Volfinger & Robert (1994)
Lepidolite in granite pegmatite		16-126 (10)	Beus (1966, Table 129)
Zinnwaldite, Bom Futuro tin mine, Brazil		19-26 (6)	Lowell & Ahl (2000)

Note: n – number of samples. [1]Average of an unspecified number of samples. [2]Be contents reported in original sources have been corrected for revised surinamite standardization, Grew et al. (2000). [3]The ppm values given are themselves averages of 5 to 28 samples. [4]The reported Be contents are so high (average 60 ppm) that contamination is highly likely.

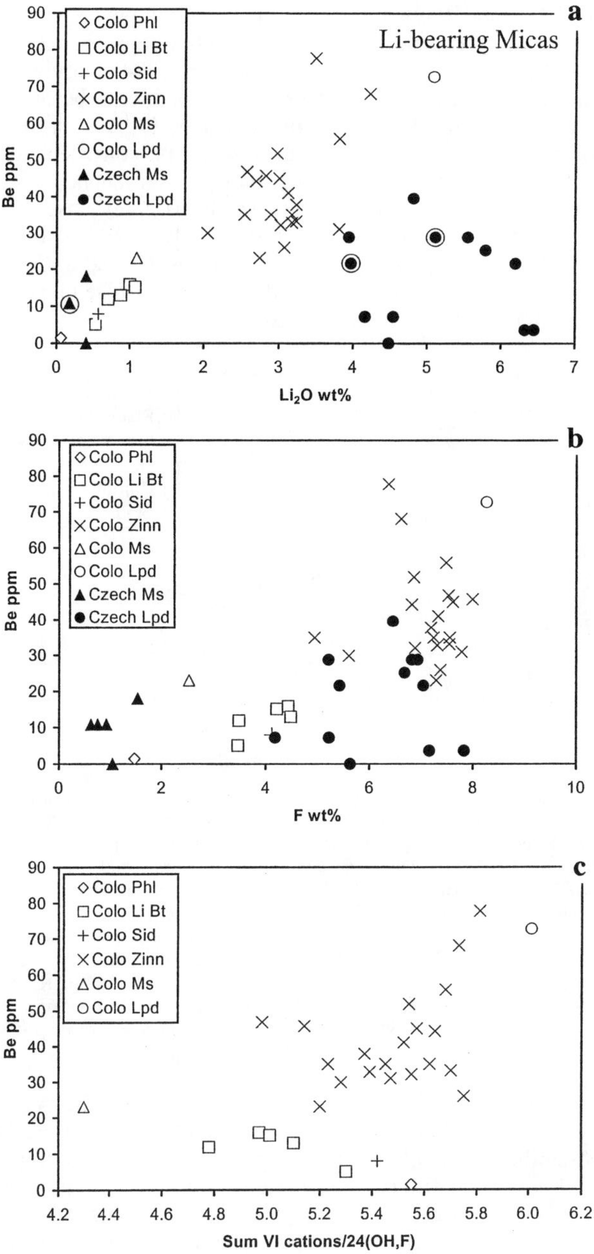

Figure 12. Beryllium, Li_2O, and F contents, as well as octahedral occupancy of Li-bearing micas from miarolitic pegmatites, Pikes Peak, Colo., USA (Foord et al. 1995) and Rožná and Dobrá Voda pegmatites, Czech Republic (Černý et al. 1995). Circled points are two coexisting muscovite-lepidolite pairs (muscovite points are superimposed). Abbreviations not in Table 1 are: Sid = siderophyllite, Zinn = zinnwaldite, Lpd = lepidolite.

(Grundmann and Morteani 1989). An exception is lepidolite with up to 512 ppm Be from the Beauvoir granite, which contains from less than 60 to more than 250 ppm Be (Monier et al. 1987; Rossi et al. 1987; Volfinger and Robert 1994). Grundman and Morteani (1989) noted that 20 ppm could be a natural limit for biotite because with increasing whole-rock Be content in the Habachtal emerald deposit (Austria), biotite Be content increases to 20 ppm and then levels out, an observation consistent with data reported elsewhere except for "lepidomelane" (Fe-rich biotite) from nepheline syenite (Table 8). In most pelitic rocks, where bulk Be rarely exceeds 10 ppm, muscovite and biotite Be contents rarely exceed 20 ppm and 10 ppm, respectively (Figs. 13, 14). Paragonite is an important host for Be in high-pressure rocks from Syros, Cyclades, Greece (Marschall et al. 2001); Zack (2000) reported five- to sixfold enrichment of Be in paragonite relative to phengite in two eclogites from the Central Alps.

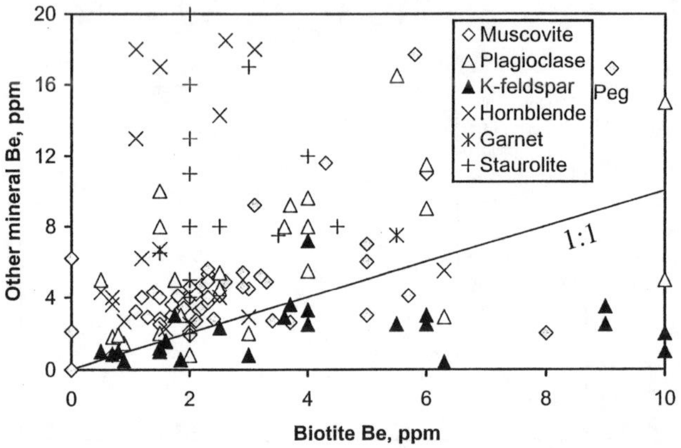

Figure 13. Mineral Be contents for plutonic and metamorphic rocks for biotite Be contents not exceeding 10 ppm. Peg refers to biotite-muscovite pair from aureole of pegmatite; a second pair, not plotted, is Bt 12.3 ppm, Ms 42.6 ppm, Dahl et al. 1993 and Dahl, pers. comm. 1999). Kosals and Mazurov (1968) reported oligoclase Be contents of 28-50 ppm, which are not plotted; neither are staurolite Be contents of 24-45 ppm (Sergeyev et al. 1967). Sources of data: Beus and Sazhina (1956; cf. Beus 1966, Table 123), Bushlaykova and Grigor'yeva (1989), Dahl et al. (1993), Dominik et al. (1993), Hügi and Röwe (1970), Kosals and Mazurov (1968), Kostetskaya et al. (1969), Kretz et al. (1989), Moxham (1965), Petrov (1973), Petrova and Petrov (1965), Sandell (1952), Schwander et al. (1968), Sergeyev et al. (1967), Wenk et al. (1963; only biotite containing measurable Be, i.e., ≥5 ppm).

Nonetheless, syntheses of K-Mg-Be mica with up to 0.5 Be per formula unit and excess hydroxyl and K-Al-Mg and K-Al-Li micas with up to 1 Be per formula unit suggests the possibility of significant Be incorporation in potassic micas (Robert et al. 1995; Lahlafi 1997). Lahlafi (1997) reported experiments during which relatively significant amounts of Be were incorporated in muscovite and biotite neoformed from mica seeds in the presence of granitic melts, including a suite of melts doped in Li, Be and F, that is, the ready availability of F and Li in the environment enhances the incorporation of Be in mica. Higher F activity in pegmatites could also contribute to incorporation of Be by faciliting mobility of Be.

Figure 12 suggests that both geochemical environment and crystallographic constraint could be important controls on Be content in Li-bearing micas. Beryllium

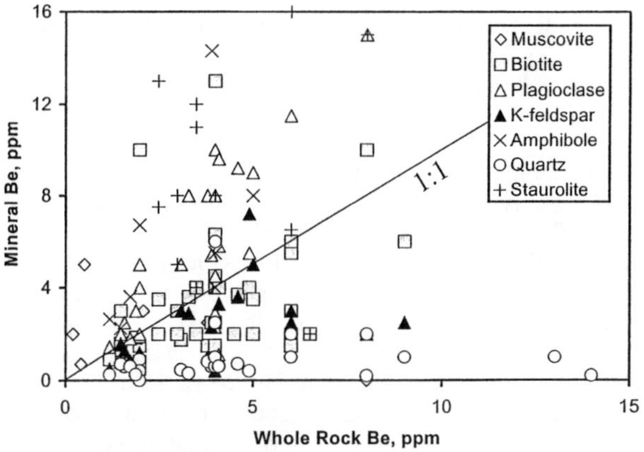

Figure 14. Mineral and whole-rock Be contents for plutonic and metamorphic rocks containing up to 16 ppm Be. Staurolite Be contents of 17-45 ppm (Sergeyev et al. 1967; Hietaten 1969) are not plotted. Sources of data: Bebout et al. (1993), Beus (1966), Beus and Sazhina (1956; cf. Beus 1966, Table 123), Bushlaykova and Grigor'yeva (1989), Hietanen (1969), Hügi and Röwe (1970), Kosals and Mazurov (1968), Kretz et al. (1989), Petrov (1973), Petrova and Petrov (1965), Sandell (1952), Sergeyev et al. (1967).

content of the micas from Pikes Peak (Colorado, USA) increases almost linearly with their Li content and less regularly with F content, both of which increased as the Pikes Peak miarolitic pegmatites evolved (Foord et al. 1995), that is, mica Be content is a function of the timing of crystallization of the mica in the Pikes Peak pegmatites. However, such an evolution is not obvious in two pegmatites from the Czech Republic (Fig. 12). Data on bulk lepidolite-zinnwaldite separates indicate simultaneous Be-Li-F enrichment in the Beauvoir granite, France, but analyses of individual lepidolite-zinnwaldite flakes in thin section showed no co-variance of Be with Li (Monier et al. 1987; Volfinger and Robert 1994).

The rather imperfect correlation of Be with F compared to the Be-Li correlation suggests another factor at work. Increasing Li content could create a crystallographic environment more conducive to Be incorporation. Because Be replaces the more highly charged Al and Si on tetrahedral sites, additional charge is needed to make up the deficit. This could be accomplished in part by filling of vacant octahedral sites by Li as in bityite. A weak correlation of Be and octahedral occupancy in the Pikes Peak, Colorado, zinnwaldite and lepidolite is consistent with this interpretation, but data on the biotite and muscovite lend no support to it (Fig. 12).

In general, receptivity to Be incorporation increases as follows: biotite < muscovite < lepidolite << margarite-bityite (Černý et al. 1995; this paper, Figs. 13 and 14). The preference for Be ↔ Al miscibility over Be ↔ Si miscibility (e.g., Barbier et al. 1999; Hawthorne and Huminicki, this volume) could explain the high receptivity in margarite-bityite, which has the most tetrahedrally coordinated Al. Refinement of tetrahedral occupancies in micas intermediate between bityite and margarite (Fig. 11) is consistent with this because two tetrahedral sites are occupied nearly exclusively by Be and Al, and these sites alternate with Si dominated sites (Sokolova et al. 1979; Lin and Guggenheim 1983). As in the case of sapphirine-khmaralite, this arrangement eliminates possible Be-O-Be bridges, minimizes the number of Be-O-Al bridges and maximizes the number of Be-O-Si bridges. In addition, predominance of Ca in the interlayer site could supply charge needed to make up the deficit resulting from Be^{2+} substituting for Al^{3+} in margarite-bityite. J.-L. Robert (pers. comm. to Černý et al. 1995) suggested that preferred incorporation of Be in lepidolite over muscovite is due to weak interaction of OH groups with tetrahedral oxygens in lepidolite.

Feldspar group

There are more Be measurements for feldspars than for any other mineral, and these data have been comprehensively reviewed by Smith (1974) and Smith and Brown (1988). In general, feldspar Be contents rarely exceed 20 ppm, even in pegmatites, and Be preferentially enters sodic plagioclase relative to calcic plagioclase and K-feldspar (see also London and Evensen, this volume). For example, in a study of plagioclase from anorthosite, "basalt", granite, and pegmatite, Steele et al. (1980; discussed in Smith 1974) found that Be content determined by ion probe increased by nearly three orders of magnitude from calcic plagioclase (~0-1 ppm Be) to sodic plagioclase (~3-30 ppm Be) with a maximum ~200 ppm in Be oligoclase from beryl-bearing pegmatite, the most Be reported in a feldspar. In addition, Be increased with albite content for anorthosite and "basalt"considered individually. While the marked trend is more likely due to structural factors, the observations do not preclude a possible role for a geochemical association of Be and Na in the host rocks of the analyzed plagioclase. More recent studies of plagioclase include Kalt et al. (2001), who reported 79-169 ppm Be (ion probe analyses) in sodian oligoclase in a metapegmatite from Koralpe, Austria in which Be-bearing muscovite (Table 8) and tourmaline (see below) are also present.

The maximum Be content in K-feldspar cited in the above compilations is 40 ppm in microcline from pegmatite. Roda Robles et al. (1999) reported 1-15 ppm Be in pegmatitic K-feldspar from Salamanca, Spain, and Foord et al. (1989) reported 6.1 ppm Be in pegmatitic microcline (ion probe) from San Diego, California, USA.

Because incorporation of Be in feldspar involves heterovalent substitutions Be $\rightarrow$ Al or Be $\rightarrow$ Si, investigators have proposed coupling with incorporation of rare earth elements for K, Na and Ca (Beus 1956, 1966) or highly charged species such as As^{5+} for Si (Smith 1974; Smith and Brown 1988).

Staurolite

Available data, albeit limited, suggest that staurolite is potentially a major carrier of Be in pelitic rocks:

- An average of 13 ppm Be for 16 staurolite samples (range: ~2 to 45 ppm Be) from mica schist containing porphyroblastic staurolite, garnet and andalusite in the Ladoga formation, Russia (Sergeyev et al. 1967). Average Be contents of the host rocks and of associated biotite are 4 and 3 ppm, respectively (Tables 2, 8). A later study of the same area reported an average of 12 ppm Be in five staurolite samples from the staurolite-andalusite zone and 4 ppm Be in one staurolite sample from sillimanite-muscovite zone (Lebedev and Nagaytsev 1980).
- From 14 to over 30 ppm Be in 22 of 27 staurolite samples from the Keivsky Block, Kola Peninsula, Russia (Shcheglova et al. 2000)
- 15-40 ppm Be in two staurolite samples from pelitic schist north of the Idaho Batholith (Hietanen 1969)
- 14-16 ppm Be in staurolite from a talc-tourmaline schist, northern Victoria Land, Antarctica and 30-40 ppm in staurolite from metapelitic schist, Alpe Sponda, Ticino, Switzerland (Armbruster and Irouschek 1983; Grew et al. 1986; Grew and Shearer, preliminary ion microprobe analyses with a surinamite standard)
- 70-140 ppm Be in Li-bearing Fe-dominant staurolite from Truchas Mountains, New

Mexico, USA; ~0 ppm Be in zincostaurolite from the Zermatt valley, Swiss western Alps; and between ~40 and ~220 ppm Be in magnesiostaurolite from the Dora-Maira massif, Italian western Alps (Holdaway et al. 1986; Chopin et al. in preparation; C. Chopin, pers. comm., ion microprobe analyses with a beryl standard). The last value could be the highest reported for staurolite, but it should be noted that the measurement is based on a scan in which the $^{9}Be^{+}$ peak was not completely resolved from the $^{27}Al^{3+}$ peak (see Hervig, this volume), and the calibration was not direct.

Hölscher (1987) attempted synthesizing a Be-bearing staurolite from gels of two compositions in the MgO-BeO-Al_2O_3-SiO_2-H_2O system at 800°C, 15 kbar and obtained staurolite, yoderite, corundum, chlorite or quartz (?), and a trace of chrysoberyl. That is, staurolite and/or yoderite must have incorporated some Be as the amount of chrysoberyl was insufficient to contain the beryllium in the starting mixture. The cell volumes of the two product staurolites (Hölscher 1987) were only 3 $Å^3$ (or 0.4%) smaller than that of synthetic Be-free magnesiostaurolite (Schreyer and Seifert 1969), i.e., 732.4(2) to 732.7(3) $Å^3$ vs. 735.6(3) $Å^3$. Assuming that the observed reduction of the staurolite cell volume is due entirely to incorporation of Be via $BeSiAl_{-2}$ and that this substitution shrinks staurolite cell volume by about the same amount as it does sapphirine cell volume (Fig. 4), the 0.4 % reduction in cell volume translates into 2000 ppm Be in Hölscher's (1987) synthetic staurolite. Thus, staurolite Be content merits further investigation with microbeam methods so the problem of inclusions can be avoided.

Almandine

Although up to 0.19 wt % BeO has been reported in grossular-andradite (e.g., Glass et al. 1944; Warner et al. 1959), available evidence suggests that garnet is not a favorable host for beryllium in metamorphic and most plutonic rocks, i.e., below detection to 5 ppm on average (Hietenan 1969; Lebedev and Nagaytsev 1980, Lyakhovich and Lyakhovich 1983; Zack 2000) and 0.12-1.81 ppm (Bea et al. 1994; Bea and Montero 1999; Bea, pers. comm. 2001). Even in granite pegmatites, no more than 20 ppm Be has been found (Beus 1966, Table 129). Compositions of the analyzed garnets were not given, but presumably they are dominantly almandine in the metamorphic and non-pegmatitic plutonic rocks, but could be spessartine-rich in the pegmatites. Beryllium contents of pyrope-almandine in 5 samples of Napier complex (Antarctica) pelitic granulites are less than 1 ppm (Grew and Shearer, unpublished ion microprobe data).

Orthopyroxene

Orthopyroxene is a potential carrier of Be in granulite-facies metapelites. In ultramafic nodules containing spinel, diopside, enstatite and olivine from the Eifel, Germany it concentrates Be to some extent: Ol (0.25-0.6 ppm Be) < Opx (0.4-1.1 ppm Be) < Di (0.6-1.2 ppm Be) ≈ Spl (1.0-1.1) (Hörmann 1966).

Pelitic orthopyroxene Be content is expected to be greater because pelitic rocks are richer in Be than ultramafics, which is borne out by preliminary analyses of orthopyroxexe in Napier Complex metapelites. Be content of this orthopyroxene ranges from 0.1 to 16 ppm, in some cases, exceeding the Be content of the rock as a whole (Grew et al. 2001b and unpublished ion microprobe data, see below).

Miscellaneous calcium silicates

Overall, Ca silicates could be somewhat more favorable hosts for Be than the most abundant rock-forming minerals in metapelites, e.g.,

- Scapolite: trace-37 ppm Be
- Clinopyroxene: trace-19 ppm Be
- Amphibole: trace-18 ppm Be; rarely 25-85 ppm

- Amphibole and clinopyroxene incorporate comparable amounts of Be, but less than scapolite

(Bushlyakov and Grigor'yev 1989; Dodge and Ross 1971; Dodge et al. 1968; Domanik et al. 1993; Grew et al. 1991a; Hepp et al. 2001; Hietanan 1971; Kostetskaya et al. 1969; Moxham 1960 1965; Petrov 1973; Petrova and Petrov 1965; Rao 1976; Shaw 1960; Shaw et al. 1963; Wenk et al. 1974). Clinopyroxene (largely omphacite and containing 1.79-4.07 ppm Be) is the main carrier of Be in eclogite from the Central Alps (Zack 2000). One of the more studied minerals with respect to Be content is vesuvianite, yet the amount that can be incorporated in this mineral remains a moot point and newer studies have failed to confirm old reports of high Be contents (Introduction, this volume).

Kyanite

In most cases, Be contents of kyanite are negligible (Table 9), e.g., spectrographic analyses of kyanite separates turned up 2.2 and 22 ppm Be in two samples out of 30 analyzed spectroscopically; Be was near ("trace") or below detection in the other 28. Pomirleanu et al. (1965) reported Be contents between 10 and 1000 ppm (plus an exceptional 10 000 ppm Be in one lot) in kyanite from mica schist in the Sebeş Mountains, southern Carpathian range, Romania and attributed the high Be contents to solutions emanating from nearby beryl-bearing pegmatites. However, contamination from beryl impurities in the kyanite samples cannot be excluded in this case.

Distribution of beryllium among common rock-forming minerals

Analytical difficulties, low concentrations and possible impurities undoubtedly contribute to the scatter in the Be data in Figures 13 and 14. The plotted data are for mineral separates. Data on plutonic rocks have been included because the data on metamorphic rocks are relatively sparse. Although some of the data are contradictory, a definite pattern emerges whether Be contents of rock-forming minerals are compared to whole-rock (Fig. 14) or to biotite (Fig. 13) Be contents (cf. London and Evensen, this volume; Ryan, Fig. 6b, this volume):

staurolite, amphibole > muscovite > plagioclase > biotite ≈ K-feldspar ≈ whole rock > quartz

Steppan et al. (2001) reported a similar sequence in amphibolite-facies metapelites from Greece and Switzerland (presumably based on ion microprobe data):

staurolite > plagioclase > muscovite > biotite > tourmaline > garnet > kyanite

This sequence is consistent with the negligible Be contents reported in most almandine-rich garnet and kyanite (see above) and schorl-dravite lacking tetrahedral boron (see below).

The data obtained using laser-ablation ICP-MS for *in situ* analyses of minerals in upper-amphibolite- and granulite-facies metamorphic rocks from Peña Negra, Spain, and the Ivrea-Verbano, Italy (Fig. 15) gave a somewhat different sequence, albeit with exceptions:

cordierite >> whole-rock ≈ sillimanite ≥ biotite ≈ garnet > plagioclase > K-feldspar

That many of the minerals contain less Be than the rock as a whole could be interpreted to indicate the presence of another carrier of Be, e.g., quartz, which was not analyzed. However, a significant role for quartz is not indicated by preliminary ion microprobe data on Napier complex granulite-facies rocks in which sapphirine takes the place of cordierite as the major carrier of Be (Grew et al. 2001b and unpublished data):

sapphirine (29-2200 ppm Be) >> orthopyroxene (0.1-16 ppm Be) ≈ sillimanite (0.1-15 ppm Be) > garnet (0.06-0.2 ppm Be) ≈ quartz (0.01-0.8 ppm Be); the position of the rock as a whole is variable (1.4-7.1 ppm Be).

Table 9. Beryllium and boron contents of Al_2SiO_5 and related phases.

Mineral, locality	B, ppm	Be, ppm (n)	Source
METAMORPHIC Al_2SiO_5 PHASES			
Sillimanite, miscellaneous	30-170	Trace-25 (3)	Pearson & Shaw (1960)
Andalusite, miscellaneous	<10-25	<0.5-2 (6)	
Kyanite, miscellaneous		<0.5-trace (11)	
Kyanite, miscellaneous, Brazil		<2-22 (17)	Herz & Dutra (1964)
Kyanite, Franciscan-Catalina, Calif., USA		0	Domanik et al. (1993)
Kyanite, Carpathian Mtns., Romania		between 10 and 1000(1)	Pomirleanu et al. (1965)
Kyanite, andalusite, Boehls Butte, Idaho, USA		0 (3)	Hietanen (1956)
Kyanite, whiteschist, Zambia and Tajikistan		0 (3)	Grew et al. (1998b)
Sillimanite, Urungwe, Zimbabwe	1400	12†	Grew et al. (1997)
Sillimanite, Port Shepstone, South Africa	1000	16	Grew et al. (1990); Grew (unpublished data)
Sillimanite, Bok se Puts, South Africa	1200	21	Grew (unpubl. data)
Sillimanite, Napier Complex, Antarctica	<1 (?)	0.1-15 (7)	Grew & Shearer (unpublished data)
Sillimanite, migmatite, Peña Negra, Spain		0.8-2.6 (2)	Bea (pers. comm.. 2001)
Sillimanite, Ivrea-Verbano, Italy		0.2	
PEGMATITIC Al_2SiO_5 PHASES, NO ASSOCIATED BE PHASES			
Sillimanite, Almgjotheii, Norway and Andrahomana, Madagascar	1100-2000	290-480(5)†	Grew et al. (1998c)
Andalusite, ditto	10-60	<0.4(2)†	
PEGMATITIC Al_2SiO_5 PHASE ASSOCIATED WITH A BE PHASE			
Sillimanite, "Christmas Point" (Napier Complex), Antarctica	6	23	Grew (unpublished data). Associated with surinamite.
Sillimanite, "Zircon Point" (Napier Complex), Antarctica	170-200	100-140(2)	
Sillimanite, Taseyevskoye muscovite deposit, Russia		20	Saltykova (1959). Possibly associated with chrysoberyl (see Beus 1960, p. 206)
BOROALUMINOSILICATES			
Boralsilite, miscellaneous		12-990(8)†	Grew et al. (1998a)
Werdingite, miscellaneous		40-2200(10)†	Grew et al. (1997, 1998c)
Grandidierite, miscellaneous		<0.4-8(3)†	

Note: n – number of samples. † Be contents reported in original sources have been corrected for revised surinamite standardization, Grew et al. (2000).

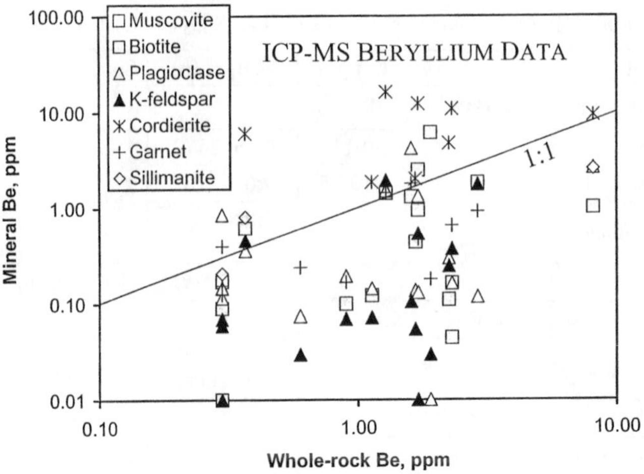

Figure 15. Mineral and whole-rock Be contents for high-grade metamorphic rocks from Peña Negra, Spain, and the Ivrea-Verbano, Italy (Bea et al. 1994; Bea and Montero 1999; F. Bea, pers. comm. 2001). Whole-rock Be contents were obtained by ICP-MS (mass spectrometry) and mineral Be contents, by laser-ablation ICP-MS. The latter are averages of analyses at several points per sample. Detection level should have been close to 0.05 ppm Be. Amounts reported as being below detection were not included in the average, unless all were, in which case, the amount was plotted as 0.01 ppm Be.

Orthopyroxene and sillimanite Be contents increase regularly with increasing sapphirine Be content, but not with whole-rock Be content, a situation consistent with approach to an equilibrium distribution of Be among the three minerals. Most likely, bulk Be content of cordierite- and sapphirine-bearing rocks is determined by the modal amount and Be content of cordierite and sapphirine, which concentrate Be so markedly relative to associated minerals.

Kornerupine-prismatine (see below) is another potential "sink" for Be in metamorphic rocks. Available data, albeit somewhat contradictory, suggest that sequence of fractionation is most likely sapphirine ≈ kornerupine-prismatine > cordierite (Grew et al. 1987, 1990, 1995, 1998b).

The special case of boron

Compared to other silicates, several borosilicates and minerals containing significant non-essential boron incorporate a relatively large amounts of Be, the converse of significant B in some beryllosilicates, e.g., bityite (see above).

Sillimanite, andalusite and related boroaluminosilicate phases. The borosilicates boralsilite, $Al_{16}B_6Si_2O_{37}$, werdingite, $(Mg,Fe)_2Al_{14}B_4Si_4O_{37}$, and grandidierite, $(Mg,Fe)Al_3BSiO_9$, together with sillimanite and andalusite, constitute a family of boroaluminosilicates based on chains of edge-sharing AlO_6 octahedra (Peacor et al. 1999). Pearson and Shaw (1960) suggested that sillimanite incorporates more Be than either andalusite or kyanite, which has been confirmed by later analyses (Table 9), including those of sillimanite and andalusite associated with borosilicates in pegmatites and metamorphic rocks (Fig. 16a). Incorporation of beryllium appears to be greater in B-bearing sillimanite, e.g., B-poor sillimanite associated with surinamite in beryllium pegmatites of the Napier Complex, Antarctica contains less beryllium than pegmatitic sillimanite richer in B, but not associated with any Be phase. However, andalusite associated with the Be- and B-bearing sillimanite from pegmatites contains negligible Be and B (Fig. 16a). Hölscher (1987) attempted synthesizing a Be-bearing sillimanite using a starting composition of $0.875Al_2SiO_5$ + $0.125BeSiAl_{-2}$ at 780°C, 6 kbar; no B was present. Cell parameters of the product sillimanite, which had formed with chrysoberyl, corundum and quartz, are identical to Be-free sillimanite, which led her to conclude that very little beryllium was incorporated in the synthetic sillimanite. In contrast Gelsdorf et al. (1958) reported that as much as 1.5 wt % BeO could be incorporated in synthetic mullite at 1500°C, which resulted in a 0.38% reduction in the cell volume, largely in the *c* direction.

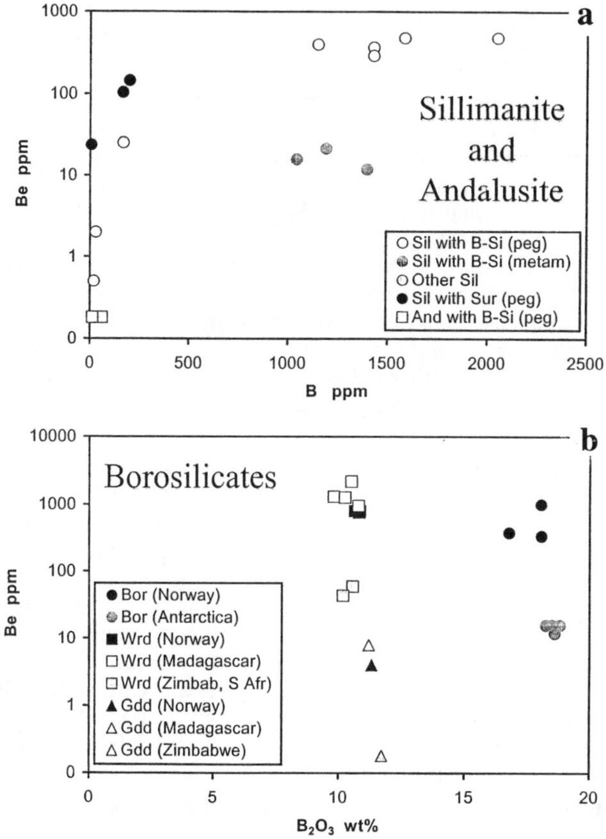

Figure 16. Beryllium (log scale) and boron contents of boroaluminosilicates. Be contents reported in original Grew et al. sources have been corrected for revised surinamite standardization, Grew et al. (2000). (A) Sillimanite (Sil) and andalusite (And) in association with borosilicate minerals (B-Si) from pegmatites and metamorphic rocks in Norway, Madagascar, Zimbabwe and South Africa (Grew et al. 1997, 1998c, unpublished data) or with surinamite (Sur) in Napier Complex beryllium pegmatites (Grew unpublished data); "other" refers to assemblages in which no mention is made of associated B or Be minerals (Pearson and Shaw 1960). (B) Boralsilite (Bor), werdingite (Wrd) and grandidierite (Gdd) from pegmatites in Norway and Madagascar and from metamorphic rocks in Zimbabwe and South Africa (Grew et al. 1997 1998a,c). The Be contents of associated werdingite and grandidierite increase as follows: Zimbabwe metamorphic rock < Norway pegmatite < Madagascar pegmatite.

Boralsilite and werdingite from pegmatites in Norway and Madagascar contain about

1000 ppm Be, over two orders of magnitude more than grandidierite from these two pegmatites (Fig. 16b), i.e., the relationship between the Be signatures of boralsilite-werdingite and grandidierite mirrors the relationship between the Be signatures of sillimanite and andalusite. Available data suggest that boralsilite and werdingite are more closely related in structure to one another and to sillimanite, e.g., werdingite-boralsilite intergrowths suggestive of exsolution (Grew et al. 1998a) and the presence of werdingite-like domains in sillimanite (Niven et al. 1991; Werding and Schreyer 1992). Grandidierite and andalusite are closer in structure to one another than either to sillimanite (Stephenson and Moore 1968; Niven et al. 1991). In summary, the crystallographic differences between boralsilite, werdingite, sillimanite, and mullite on the one hand, and andalusite and grandidierite on the other, are reflected in the amount of Be incorporated in a Be-bearing environment.

Tourmaline group. Beryllium contents of tourmaline-group minerals are generally low, including samples from pegmatites (Henry and Dutrow 1996; this paper, Table 10). The high, presumbly wet-chemical (but method not specified) Be contents reported by Otrashchenko et al. (1971) should be viewed with caution given anomalies in other aspects of the reported compositions (high SiO_2 or K_2O, low H_2O). Instead, the highest Be content (120 ppm by SIMS, A. Kalt, pers. comm.) reported in lithian excess-boron olenite containing significant ${}^{[4]}B$ from a metapegmatite at Koralpe, Austria (Kalt et al. 2001) is probably a better measure of the maximum amount of Be accepted in tourmaline-group minerals. Beryllium content tends to increase with ${}^{[4]}B$ in the Koralpe tourmalines, which range from Mg-rich schorl lacking ${}^{[4]}B$ to lithian olenite with ${}^{[4]}B/({}^{[4]}B+{}^{[4]}Al+Si) = 0.15$, a further indication that minerals containing variable amounts of tetrahedral B are more receptive to Be incorporation than the same mineral lacking tetrahedral B.

Kornerupine-prismatine series. The highest BeO contents reported in this series, $(\square,Fe,Mg)(Al,Mg,Fe)_9(Si,Al,B)_5O_{21}(OH,F)$, are 0.32 and 0.74 wt % in kornerupine from whiteschists from Tajikistan and Zambia, respectively (Grew 1996; Grew et al. 1998b; corrected for revised surinamite standardization, Grew et al. 2000). In a crystal structure refinement of the kornerupine containing 0.74 wt % BeO, which corresponds to 0.22 Be/21.5 O, Mark Cooper (pers. comm. 2001) reports that Be is found only at the third of three T sites for which he determined a preliminary occupancy of 0.17Si, 0.19Al, 0.44B, and 0.20Be. Given the absence of a Be-saturating phase in whiteschist, it is likely that kornerupine-prismatine could incorporate significantly more Be. Hölscher (1987) attempted to synthesize a Be-bearing, B-free kornerupine at 780°C, 6 kbar using a starting composition of $Mg_{3.5}Al_6Be_{0.5}Si_4O_{20}(OH)_2$, but obtained only a trace of kornerupine among the products.

Crystallographic features favoring incorporation of Be in minerals

Beryllium incorporation appears to be very site specific; substantial amounts are incorporated or certain sites, but not on others (Hawthorne and Huminicki, this volume). Because of the large difference in charge between Be^{2+} and Si^{4+}, Be would be expected to share sites with B or Al instead of Si despite the greater disparity in size (Si 0.26, Be 0.27, B 0.11, Al 0.39 Å, Shannon 1976). Nonetheless, Be-Si miscibility is observed in several minerals, particularly when Al or B is also present, e.g., khmaralite, kornerupine. Availability of abundant tetrahedral Al in margarite and tetrahedral B in olenitic tourmaline could explain the preference of Be for these minerals over micas containing less tetrahedral Al and tourmaline containing negligible tetrahedral B.

Similar constraints may be controlling the incorporation of Be in boroalumino-silicates. Boralsilite, werdingite and sillimanite contain ${}^{[4]}B$ or ${}^{[4]}Al$, which are absent in

Table 10. Beryllium content of tourmaline

Tourmaline species	Rock Type, Locality	Be, ppm (n)	Source
"Black" (schorl?)	Pegmatite, Snarum,, Norway	4	Goldschmidt (1958)
"Red" (elbaite?)	Pegmatite, San Diego, California, USA	120	
Not specified	Miscellaneous plutonic rocks	5.0(2)-17.4(9)	Lyakhovich & Lyakhovich (1983) Averages only
	Altered granites	31.2(5)	
	Pegmatite	21.8(8)	
	Quartz veins	3.2	
Schorl (?)	Quartz veins, Uzbekistan	220-360 (2)	Otroshchenko et al. (1971)
Uvite	Skarn, Talas Mtns., Kyrgyzstan	360 (1)	
Dravite	Whiteschist, Zambia and Tajikistan	0-3 (3)	Grew et al. (1998b)
Schorl-foitite	Granulite-facies pegmatite, Larsemann Hills, Antarctica & Almgjotheii, Norway	0.4-7(8)	Grew et al. (1998a)
Mg-rich schorl	Meta-pegmatite and contact-metasomatic tourmalinite, Koralpe, Austria	6-25	Kalt et al. (2001);
Lithian excess-boron olenite		40-120	A. Kalt (pers. comm..)
Dravite	Pegmatite(?), Madagascar	13	Dyar et al. (2001)
Schorl	Pegmatite, Mozambique	22	
Elbaite	Pegmatite, Brazil	19	

Note: n – number of samples.

andalusite and grandidierite (Peacor et al. 1999), and Be may be more readily accommodated at these sites rather than at $^{[4]}$Si, thereby explaining the fractionation of Be into boralsilite, werdingite and sillimanite relative to associated andalusite and grandidierite. In addition, the greater variety of polyhedra cross-linking the octahedral chains in boralsilite, werdingite and in the werdingite-like domains in B-bearing sillimanite, could play an important role by providing a flexibility that allows heterovalent substitutions involving Be, B and Al.

CONCLUSION

As noted in the Introduction to this chapter, metamorphic rocks are a critical link in the crustal Be budget, that is, getting Be from sediments to granites, where igneous processes concentrate Be (London and Evensen, this volume) into economic deposits, both pegmatitic and volcanogenic. Retention of beryllium in sediments during burial and heating depends on the minerals constituting metasedimentary rocks. Data summarized in this chapter suggest that the few ppm Be present in most terrigenous sediments could be easily accommodated in common rock-forming silicates, most importantly, muscovite, so beryllium is not expected to be lost at low to medium grades of metamorphism where muscovite is stable. Staurolite could also play a significant role in medium-grade rocks, particularly in rocks with little or no muscovite. At temperatures above which muscovite and staurolite break down, beryllium retention depends on the presence of another suitable host, such as cordierite. The breakdown of muscovite is often accompanied by anatexis, and this is a critical juncture in the fate of beryllium in buried sediments. As noted by Sighinolfi (1973), there is a possibility of both Be depletion and Be enrichment associated with anatexis (or fluid movement) in the upper amphibolite-facies and granulite-facies conditions. London and Evensen (this volume) suggest that at lower pressures, cordierite could form and would be the only common-rock forming mineral to retain Be in the restite, resulting in melts depleted at Be. Conversely, at higher pressures, where cordierite is not stable, garnet-sillimanite-biotite restites may not be able retain Be, which would be lost to the melt. At greater depths, a potential host is sapphirine, so again Be could be retained in the restite. Thus, it is possible beryllium could be retained in metasediments buried to lower crustal depths. The ultrahigh-temperature Napier Complex is an example of sapphirine-bearing lower crustal metasediments that have retained Be, indeed, Fig. 1b suggests a modest enrichment. Evidently, the Napier metapelites never experienced a melting event that depleted them in Be, i.e., these rocks most likely followed a P-T path within the stability fields of cordierite and sapphirine. However, there are anatectic Be-bearing pegmatites in the Napier. If sapphirine retains Be, how could it be released into an anatectic melt? The Strona granulites lack cordierite and sapphirine and are depleted in Be; could these have lost Be-bearing melts similar to those discussed by London and Evensen (this volume)?

The evolution outlined above is very sketchy. There are many aspects that need more detailed study. One is the Be content of the starting material. What are the average Be contents of the major sedimentary rocks? Distribution of Be among minerals and between minerals and melt is still very imperfectly known, as evident in the scatter in Figures 13-15 and in London and Evensen (this volume). Low Be concentrations and contamination has stymied the acquisition of the requisite compositional data; these studies require microbeam technologies such as LA-ICP-MS, PIGE or SIMS. More consideration needs to be given to possible Be enrichments and mobility in situations not involving obvious Be metasomatism such as greisens. For example, Figure 6 shows a decrease in sapphirine BeO content away from a nodule of magnesiotaaffeite-6*N*'3*S* ("musgravite"), implying outward migration of Be. A fourth aspect needing attention is the behavior of Be in the anatectic process, itself currently a popular field of research. Rocks undoubtedly undergo melting over a protacted extent of their metamorphic evolution at depth, resulting in many opportunities for Be loss or exchange. In order to understand how Be gets into a granitic system, one must understand how Be distributed between anatectic melts, fluid phase and the sediments undergoing metamorphism and anatexis.

ACKNOWLEDGMENTS

I thank the following individuals who provided references that were not available from Interlibrary Loan or valuable information from their unpublished work or work in advance of publication: Edgardo Baldo, Fernando Bea, Christian Chopin, Andrew Christy, Nicole Catel (analyst with Bruno Goffé), Mark Cooper, Peter Dahl, Bertrand Devouard, Joseph Evensen, François Fontan, Shan Gao, Bruno Goffé, Ulf Hålenius, Angelika Kalt, Roy Kristiansen, Christian Nicollet, Nikolay Pertsev, Terry Plank, Michael Raith, Carlos Rapela, Jean-Louis Robert, Karl Schmetzer, H. Schneider and Werner Schreyer (for copy of Hölscher 1987). I also thank Anthony Higgins for loan of #3319313 of the Greenland Geological Survey. Andrew Christy, Gerhard Franz, Giulio Morteani and Peter Nabelek are thanked for their thoughtful and detailed reviews of an earlier draft of the manuscript. The staff of the Interlibrary Loan Department of Fogler Library, University of Maine, is thanked for their willingness to search for both obscure and old articles in a variety of languages and scripts, and their patience with my continual requests. Financial support was provided from U.S. National Science Foundation grants OPP-9813569 and OPP-0087235.

REFERENCES

Ackermand D, Seifert F, Schreyer W (1975) Instability of sapphirine at high pressures. Contrib Mineral Petrol 50:79-92

Agard J (1965) Découverte de béryllium dans la minéralisation du gîte de wolfram, molybdène et cuivre d'Azegour (Haut Atlas, Maroc) et dans celle du gîte de manganèse de Guettara (Sud algérien). Comptes Rend Acad Sci Paris 261:4179-4180

Anderson BW, Payne CJ, Claringbull GF, Hey MH (1951) Taaffeite, a new beryllium mineral, found as a cut gemstone. Mineral Mag 29:765-772

Armbruster T (1986) Role of Na in the structure of low-cordierite: A single-crystal X-ray study. Am Mineral 71:746-757

Armbruster T (2002) Revised nomenclature of högbomite, nigerite, and taaffeite minerals. Eur J Mineral 14: 389-395

Armbruster T, Irouschek A (1983) Cordierites from the Lepontine Alps: Na + Be → Al substitution, gas content, cell parameters, and optics. Contrib Mineral Petrol 82:389-396

Arnaudov V, Petrusenko S, Pavlova M (1982) Beryllium-bearing margarite and fuchsite from desilicated pegmatites in the Rila Mountain. Geochemistry, Mineralogy and Petrology (Bulgarian Acad Sci) 15:33-40 (in Bulgarian)

Aurisicchio C, Fioravanti G, Grubessi O, Zanazzi PF (1988) Reappraisal of the crystal chemistry of beryl. Am Mineral 73:826-837

Baba S (1998a) Proterozoic anticlockwise *P-T* path of the Lewisian Complex of South Harris, Outer Hebrides, NW Scotland. J Metamorphic Geol 16:819-841

Baba S (1998b) Ultra-high temperature metamorphism in the Lewisian Complex, South Harris, NW Scotland. *In* Motoyoshi Y, Shiraishi K (eds) Origin and Evolution of Continents, Mem Nat Inst Polar Research Special Issue 53. Tokyo, p 93-108

Baba S. (1999) Sapphirine-bearing orthopyroxene-kyanite/sillimanite granulites from South Harris, NW Scotland: evidence for Proterozoic UHT metamorphism in the Lewisian. Contrib Mineral Petrol 136:33-47

Baba S, Grew ES, Shearer CK, Sheraton JW (2000) Surinamite: A high-temperature metamorphic beryllosilicate from Lewisian sapphirine-bearing kyanite-orthopyroxene-quartz-potassium feldspar gneiss at South Harris, N.W. Scotland. Am Mineral 85:1474-1484

Barbier J (1996) Surinamite analogs in the $MgO-Ga_2O_3-GeO_2$ and $MgO-Al_2O_3-GeO_2$ systems. Phys Chem Mineral 23:151-156

Barbier J (1998) Crystal structures of sapphirine and surinamite analogues in the $MgO-Ga_2O_3-GeO_2$ system. Eur J Mineral 10:1283-1293

Barbier J, Grew ES, Moore PB, Su S-C (1999) Khmaralite, a new beryllium-bearing mineral related to sapphirine: A superstructure resulting from partial ordering of Be, Al and Si on tetrahedral sites. Am Mineral 84:1650-1660

Barbier J, Grew ES, Yates MG, Shearer CK (2001) Beryllium minerals related to aenigmatite. Geol Assoc Canada Mineral Assoc Canada, Joint Ann Mtg Abstr Vol 26:7

Barbier J, Grew ES, Hålenius E, Hålenius U, Yates, MG (2002) The role of Fe and cation order in the crystal chemistry of surinamite, $(Mg,Fe^{2+})_3(Al,Fe^{3+})_3O[AlBeSi_3O_{15}]$: A crystal structure, Mössbauer spectroscopic, and optical spectroscopic study. Am Mineral 87:501-513

Barker F (1964) Sapphirine-bearing rock, Val Codera, Italy. Am Mineral 49:146-152

Barton MD (1986) Phase equilibria and thermodynamic properties of minerals in the $BeO-Al_2O_3-SiO_2-H_2O$ (BASH) system, with petrologic applications. Am Mineral 71:277-300

Bayrakov VV (1973) A new find of lithium-beryllium margarite in the USSR. Dopovidi Akad Nauk Ukrain RSR Ser B Geol Geofiz Khim Biol 35(6):483-486 (in Ukrainian)

Bea F, Montero P (1999) Behavior of accessory phases and redistribution of Zr, REE, Y, Th, and U during metamorphism and partial melting of metapelites in the lower crust: An example from the Kinzigite Formation of Ivrea-Verbano, NW Italy. Geochim Cosmochim Acta 63:1133-1153

Bea F, Pereira MD, Stroh A (1994) Mineral/leucosome trace-element partitioning in a peraluminous migmatite (a laser ablation-ICP-MS study). Chem Geol 117:291-312

Bebout GE, Ryan JG, Leeman WP (1993) B-Be systematics in subduction-related metamorphic rocks: Characterization of the subducted component. Geochim CosmochimActa 57:2227-2237

Bebout GE, Ryan JG, Leeman WP, Bebout AE (1999) Fractionation of trace elements by subduction-zone metamorphism—effect of convergent-margin thermal evolution. Earth Planet Sci Lett 171: 63-81

Bernhardt H-J, Brandstätter F, Ertl A, Körner W, Mikenda W, Pertlik F (1999) Untersuchungen an einem borhaltigen Muskovit-$2M_1$ von der Koralpe, Steiermark. Annal Naturhist Mus Wien 100A:1-11

Beus AA (1956) Characteristics of the isomorphous entry of beryllium into crystalline mineral structures. Geochemistry 1956:62-77

Beus AA (1966) Geochemistry of beryllium and genetic types of beryllium deposits. San Francisco and London, Freeman

Beus AA (1979) Sodium—a geochemical indicator of emerald mineralization in the Cordillera Oriental, Colombia. J Geochem Exploration 11:195-208

Beus AA, Sazhina LI (1956) The average beryllium content of acidic magmatic rocks of the USSR. Doklady Nauk SSSR 109:807-810

Bollmark B (1999) Some aspects of the origin of the deposit. *In* Holtstam D, Langhof J (eds) Långban. The Mines, their Minerals, Geology and Explorers. Swedish Museum of Natural History, Raster Förlag, Stockholm, p 43-49

Borchert W, Gugel E, Petzenhauser I (1970) Untersuchungen im System Beryll-Indialith. N Jahrb Mineral Monatsh 1970:385-388

Boström K, Rydell H, Joesuu O (1979) Långban—an exhalative sedimentary deposit? Econ Geol 74: 1002-1011

Brugger J, Gieré R (1999) As, Sb, Be and Ce enrichment in minerals from a metamorphosed Fe-Mn deposit, Val Ferrera, eastern Swiss Alps. Can Mineral 37:37-52

Brugger J, Gieré R (2000) Origin and distribution of some trace elements in metamorphosed Fe-Mn deposits, Val Ferrera, eastern Swiss Alps. Can Mineral 38:1075-1101

Brugger J, Gieré R, Grobéty B, Uspensky E (1998) Scheelite-powellite and paraniite-(Y) from the Fe-Mn deposit at Flanel, eastern Swiss Alps. Am Mineral 83:1100-1110

Bulnaev KB (1996) Origin of the fluorite-bertrandite-phenakite deposits. Geology of Ore Deposits 38: 128-136

Burford M (1998) Gemstones from Tunduru, Tanzania. Can Gemmologist 19:105-110

Burke EAJ, Lustenhouwer WJ (1981) Pehrmanite, a new beryllium mineral from Rosendal pegmatite, Kemiö Island, southwestern Finland. Can Mineral 19:311-314

Burke EAJ, Lof P, Hazebroek HP (1977) Nigerite from the Rosendal pegmatite and aplites, Kemiö Island, southwestern Finland. Bull Geol Soc Finland 4):151-157

Burnol L (1974) Géochimie du béryllium et types de concentration dans les leucogranites du Massif central français. Mém BRGM 85:1-169

Burt DM (1994) Vector representation of some mineral compositions in the aenigmatite group, with special reference to høgtuvaite. Can Mineral 32:449-457

Bushlyakov IN, Grigor'yev NA (1989) Beryllium in Ural metamorphites. Geochem Intl 26:57-61

Cannillo E, Mazzi F, Fang JH, Robinson PD, Ohya Y (1971) The crystal structure of aenigmatite. Am Mineral 56:427-446

Cassedanne JP (1989) Famous mineral localities: The Ouro Preto topaz mines. Mineral Record 20: 221-233

Černý P, Povondra P (1966) Beryllian cordierite from Věžná: (Na,K) + Be → Al. N Jahrb Mineral Monatsh 1966:36-44

Černý P, Povondra P (1967) Cordierite in west-Moravian desilicated pegmatites. Acta Univ Carolinae–Geol 3:203-221

Černý P, Novák M, Chapman R (1992) Effects of sillimanite-grade metamorphism and shearing on Nb-Ta oxide minerals in granitic pegmatites: Maršíkov, northern Moravia, Czechoslovakia. Can Mineral 30: 699-718

Černý P, Staněk J, Novák M, Baadsgaard H, Rieder M, Ottolini L, Kavalová M, Chapman R (1995) Geochemical and structural evolution of micas in the Rožná and Dobrá Voda pegmatites, Czech Republic. Mineral Petrol 55:177-201

Černý P, Chapman R, Schreyer W, Ottolini L, Bottazzi P, McCammon CA (1997) Lithium in sekaninaite from the type locality, Dolní Bory, Czech Republic. Can Mineral 35:167-173

Chadwick B, Friend CRL (1991) The high-grade gneisses in the south-west of Dove Bugt: an old gneiss complex in a deep part of the Caledonides of North-East Greenland. Rapp Grønlands Geol Undersøgelse 152:103-111

Chadwick B, Friend CRL, Higgins AK (1990) The crystalline rocks of western and southern Dove Bugt, North-East Greenland. Rapp Grønlands Geol Undersøgelse 148:127-132

Chadwick B, Friend CRL, George MC, Perkins WT (1993) A new occurrence of musgravite, a rare beryllium oxide, in Caledonides of North-East Greenland. Mineral Mag 57:121-129

Chopin C, Goffé B, Ungaretti L, Oberti R (in preparation) Magnesiostaurolite and zincostaurolite: description with a petrologic and crystal-chemical up-date. Eur J Mineral (submitted)

Christy AG (1988) A new 2*c* superstructure in beryllian sapphirine from Casey Bay, Enderby Land, Antarctica. Am Mineral 73:1134-1137

Christy AG (1989) The effect of composition, temperature and pressure on the stability of the 1*Tc* and 2*M* polytypes of sapphirine. Contrib Mineral Petrol 103:203-215

Christy AG, Grew ES (in preparation) Synthesis of beryllian sapphirine in the system MgO-BeO-Al_2O_3-SiO_2-H_2O, and comparison with naturally occurring beryllian sapphirine and khmaralite. Part 2: Composition-volume relations and a MBeAS petrogenetic grid. Am Mineral (submitted)

Christy AG, Tabira Y, Hölscher A, Grew ES, Schreyer W (2002) Synthesis of beryllian sapphirine in the system MgO-BeO-Al_2O_3-SiO_2-H_2O and comparison with naturally occurring beryllian sapphirine and khmaralite. Part 1: Experiments, TEM and XRD. Am Mineral 87:1104-1112

Crespo A, Lunar R (1997) Terrestrial hot-spring Co-rich Mn mineralization in the Pliocene–Quaternary Calatrava Region (central Spain). *In* Nicholson K, Hein JR, Bühn B, Dasgupta S (eds) Manganese Mineralization: Geochemistry and Mineralogy of Terrestrial and Marine Deposits. Geol Soc Spec Pub 119:253-264

Dahl PS, Wehn DC, Feldmann SG (1993) The systematics of trace-element partitioning between coexisting muscovite and biotite in metamorphic rocks from the Black Hills, South Dakota, USA. Geochim Cosmochim Acta 57:2487-2505

Damour A (1855) Nouvelles recherches sur la composition de l'euklase, espèce minérale. Comptes Rendus Acad Sci 40:942-944

Dana ES (1892) The System of Mineralogy, 6th Edition. Wiley, New York

Deer WA, Howie RA, Zussman J (1978) Rock-Forming Minerals, Volume 2A. Single-Chain silicates, 2nd Edition. Longman, London

Deer WA, Howie RA, Zussman J (1986) Rock-Forming Minerals, Volume 1B. Disilicates and Ring Silicates, 2nd Edition. Longman, London

Delamétherie JC (1792) De l'euclase. J Phys 41:155-156

Delamétherie JC (1797) De l'euclase. Théorie de la Terre, 2nd Edn 2:254-256

Demartin F, Pilati T, Gramaccioli CM, de Michele V (1993) The first occurrence of musgravite as a faceted gemstone. J Gemm 23:482-485

Demina TV, Mikhaylov MA (2000) Magnesium and calcium in beryl. Zapiski Vserossiyskogo Mineral Obshchestva 129:97-109 (in Russian)

de Roever EWF (1973) Preliminary note on coexisting sapphirine and quartz in a mesoperthite gneiss from Bakhuis Mountains (Suriname). Geol Mijnbouwkundige Dienst Suriname Med 22:67-70

de Roever EWF, Vrána S (1985) Surinamite in pseudomorphs after cordierite in polymetamorphic granulites from Zambia. Am Mineral 70:710-713

de Roever EWF, Kieft C, Murray E, Klein E, Drucker WH (1976) Surinamite, a new Mg-Al silicate from the Bakhuis Mountains, western Surinam. I. Description, occurrence, and conditions of formation. Am Mineral 61:193-197

de Roever EWF, Lattard D, Schreyer W (1981) Surinamite: A beryllium-bearing mineral. Contrib Mineral Petrol 76:472-473

Devouard B, Raith M, Rakotondrazafy R, El-Ghozzi M, Nicollet C (2002) Occurrence of musgravite in anorthite-corundum-spinel-sapphirine rocks ("sakenites") from South Madagascar: Evidence for a high-grade metasomatic event. Abstracts and Programme 18th Gen Meeting Intl Mineral Assoc (in press)

Dodge FCW, Ross DC (1971) Coexisting hornblendes and biotites from granitic rocks near the San Andreas fault, California. J Geol 79:158-172

Dodge FCW, Papike JJ, Mays RE (1968) Hornblendes from granitic rocks of the central Sierra Nevada batholith, California. J Petrol 9:378-410

Domanik KJ, Hervig RL, Peacock SM (1993) Beryllium and boron in subduction zone minerals: An ion microprobe study. Geochim Cosmochim Acta 57:4997-5010

Dyar MD, Wiedenbeck M, Robertson D, Cross LR, Delaney JS, Ferguson K, Francis CA, Grew ES, Guidotti CV, Hervig RL, Hughes JM, Husler J, Leeman W, McGuire AV, Rhede D, Rothe H, Paul RL, Richards I, Yates M (2001) Reference minerals for the microanalysis of light elements. Geostandards Newsletter 25:441-463

Evensen JM, London D (1999) Beryllium reservoirs and sources for granitic melts: the significance of cordierite. Geol Soc Am Abstr Progr 31:A305

Evensen JM, London D (in press) Complete cordierite-beryl solid solutions in granitic systems: Phase relations and complex crystal chemistry. Am Mineral

Flanagan FJ (1973) 1972 values for international geochemical reference samples. Geochim Cosmochim Acta 37:1189-1200

Fleischer R (1972) Origin of topaz deposits near Ouro Preto, Minas Gerais, Brazil. Econ Geol 67:119-120

Foord EE, Spaulding LB Jr, Mason RA, Martin RF (1989) Mineralogy and paragenesis of the Little Three Mine Pegmatites, Ramona District, San Diego County, California. Mineral Record 20:101-127

Foord EE, Černý P, Jackson LL, Sherman DM, Eby RK (1995) Mineralogical and geochemical evolution of micas from miarolitic pegmatites of the anorogenic Pikes Peak batholith, Colorado. Mineral Petrol 55:1-26

Franceschelli M, Puxeddu M, Memmi I (1998) Li, B-rich Rhaetian metabauxite, Tuscany, Italy: reworking of older bauxites and igneous rocks. Chem Geol 144:221-242

Franceschelli M, Puxeddu M, Carta M (2000) Mineralogy and geochemistry of Late Ordovician phosphate-bearing oolitic ironstones from NW Sardinia, Italy. Mineral Petrol 69:267-293

Franz G, Morteani G (1984) The formation of chrysoberyl in metamorphosed pegmatites. J Petrol 25:27-52

Gallagher MJ, Hawkes JR (1966) VII.–Beryllium minerals from Rhodesia and Uganda. Bull Geol Surv Great Britain 25:59-75

Gao S, Zhang B, Xie Q, Gu X, Ouyang J, Wang D, Gao C (1991) Average chemical compositions of post-Archean sedimentary and volcanic rocks from the Qinling Orogenic Belt and its adjacent North China and Yangtze Cratons. Chem Geol 92:261-282

Gao S, Luo T-C, Zhang B-R, Zhang H-F, Han Y-W, Zhao Z-D, Hu Y-K (1998) Chemical composition of the continental crust as revealed by studies in East China. Geochim Cosmochim Acta 62:1959-1975

Geller RF, Yavorsky PJ, Steierman BL, Creamer AS (1946) Studies of binary and ternary combinations of magnesia, calcia, baria, beryllia, alumina, thoria and zirconia in relation to their use as porcelains (Research Paper RP1703). J Res Nat Bur Standards 36:277-312

Gelsdorf G, Müller-Hesse H, Schweite H-E (1958) Einlagerungsversuche an synthetischem Mullit und Substitutionsversuche mit Galliumoxyd und Germaniumdioxyd Teil II. Archiv Eisenhüttenwesen 29: 513-519

Ginzburg AI (1957) Bityite – lithium-beryllium margarite. Trudy Mineral Muzyeya Akad Nauk SSSR 8:128-131 (in Russian)

Ginzburg AI, Stavrov OD (1961) Content of rare elements in cordierite. Geochemistry 1961:208-211

Giuliani G, Bourlès D, Massot J, Siame L (1999) Colombian emerald reserves inferred from leached beryllium of their host black shale. Explor Mining Geol 8(1-2):109-116

Glass JJ, Jahns RH, Stevens RE (1944) Helvite and danalite from New Mexico and the helvite group. Am Mineral 29:163-191

Goffé B (1980) Magnésiocarpholite, cookéite et euclase dans les niveaux continentaux métamorphiques de la zone briançonnaise. Données minéralogiques et nouvelles occurrences. Bull Minéral 103:297-302

Goffé B (1982) Définition du faciès à Fe-Mg carpholite–chloritoïde, un marqueur du métamorphisme de HP-BT dans les métasédiments alumineux. Thèse de doctorat d'Etat, Univ Pierre et Marie Curie, Paris

Goffé B, Bousquet R (1997) Ferrocarpholite, chloritoïde et lawsonite dans les métapélites des unités du Versoyen et du Petit St Bernard (zone valaisanne, Alpes occidentales). Schweiz mineral petrogr Mitt 77:137-147

Goldschmidt VM (1958) Geochemistry. Oxford University Press, Oxford, UK

Goldschmidt VM, Peters C (1932) Zur Geochemie des Berylliums. Nachrichten Akad Wissenschaften Göttingen, II. Math-Phys Klasse, p 360-376

Gordillo CE, Schreyer W, Werding G, Abraham K (1985) Lithium in NaBe-cordierites from El Peñón, Sierra de Córdoba, Argentina. Contrib Mineral Petrol 90:93-101

Goscombe B (1992) Silica-undersaturated sapphirine, spinel and kornerupine granulite facies rocks, NE Strangways Range, Central Australia. J Metamorphic Geol 10:181-201

Grauch RI, Lindahl I, Evans HT Jr, Burt DM, Fitzpatrick JJ, Foord EE, Graff P-R, Hysingjord J (1994) Høgtuvaite, a new beryllian member of the aenigmatite group from Norway, with new X-ray data on aenigmatite. Can Mineral 32:439-448

Graziani G, Guidi G (1980) Euclase from Santa do Encoberto, Minas Gerais, Brazil. Am Mineral 65: 183-187

Grew ES (1981) Surinamite, taaffeite and beryllian sapphirine from pegmatites in granulite-facies rocks of Casey Bay, Enderby Land, Antarctica. Am Mineral 66:1022-1033

Grew ES (1986) Petrogenesis of kornerupine at Waldheim (Sachsen), German Democratic Republic. Z Geol Wissenschaften 14:525-558

Grew ES (1996) Borosilicates (exclusive of tourmaline) and boron in rock-forming minerals in metamorphic environments. Rev Mineral 33:387-502

Grew ES (1998) Boron and beryllium minerals in granulite-facies pegmatites and implications of beryllium pegmatites for the origin and evolution of the Archean Napier Complex of East Antarctica. Mem Nat Inst Polar Res Spec Issue 53:74-92

Grew ES, Hinthorne JR, Marquez N (1986) Li, Be, B, and Sr in margarite and paragonite from Antarctica. Am Mineral 71:1129-1134

Grew ES, Abraham K, Medenbach O (1987) Ti-poor hoegbomite in kornerupine-cordierite-sillimanite rocks from Ellammankovilpatti, Tamil Nadu, India. Contrib Mineral Petrol 95:21-31

Grew ES, Chernosky JV, Werding G, Abraham K, Marquez N, Hinthorne JR (1990) Chemistry of kornerupine and associated minerals, a wet chemical, ion microprobe, and X-ray study emphasizing Li, Be, B and F contents. J Petrol 31:1025-1070

Grew ES, Yates MG, Swihart GH, Moore PB, Marquez N (1991a) The paragenesis of serendibite at Johnsburg, New York, USA: an example of boron enrichment in the granulite facies. *In* Perchuk LL (ed) Progress in Metamorphic and Magmatic Petrology. Cambridge University Press, Cambridge, UK, p 247-285

Grew ES, Yates MG, Beryozkin VI, Kitsul VI (1991b) Kornerupine in slyudites from the Usmun River Basin in the Aldan Shield. Part II. Chemistry of the minerals, mineral reactions. Soviet Geol Geophys 32:85-98

Grew ES, Yates MG, Romanenko IM, Christy AG, Swihart GH (1992) Calcian, borian sapphirine from the serendibite deposit at Johnsburg, N.Y., USA. Eur J Mineral 4:475-485

Grew ES, Pertsev NN, Yates MG, Christy AG, Marquez N, Chernosky JC (1994a) Sapphirine+forsterite and sapphirine+humite-group minerals in an ultra-magnesian lens from Kuhi-lal, SW Pamirs, Tajikistan: Are these assemblages forbidden? J Petrol 35:1275-1293

Grew ES, Yates MG, Belakovskiy DI, Rouse RC, Su S-C, Marquez N (1994b) Hyalotekite from reedmergnerite-bearing peralkaline pegmatite, Dara-i-Pioz, Tajikistan and from Mn skarn, Långban, Värmland, Sweden: a new look at an old mineral. Mineral Mag 58:285-297

Grew ES, Hiroi Y, Motoyoshi Y, Kondo Y, Jayatileke SJM, Marquez N (1995) Iron-rich kornerupine in sheared pegmatite from the Wanni Complex, at Homagama, Sri Lanka. Eur J Mineral 7:623-636

Grew ES, Yates MG, Shearer CK, Wiedenbeck M (1997) Werdingite from the Urungwe District, Zimbabwe. Mineral Mag 61:713-718

Grew ES, McGee JJ, Yates MG, Peacor DR, Rouse RC Huijsmans JPP, Shearer CK, Wiedenbeck M, Thost DE, Su S-C (1998a) Boralsilite ($Al_{16}B_6Si_2O_{37}$): A new mineral related to sillimanite from pegmatites in granulite-facies rocks. Am Mineral 83:638-651

Grew ES, Pertsev NN, Vrána S, Yates MG, Shearer CK, Wiedenbeck M (1998b) Kornerupine parageneses in whiteschists and other magnesian rocks: is kornerupine + talc a high-pressure assemblage equivalent to tourmaline + orthoamphibole? Contrib Mineral Petrol 131:22-38

Grew ES, Yates MG, Huijsmans JPP, McGee JJ, Shearer CK, Wiedenbeck M, Rouse RC (1998c) Werdingite, a borosilicate new to granitic pegmatites. Can Mineral 36:399-414.

Grew ES, Yates MG, Barbier J, Shearer CK, Sheraton JW, Shiraishi K, Motoyoshi Y (2000) Granulite-facies beryllium pegmatites in the Napier Complex in Khmara and Amundsen Bays, western Enderby Land, East Antarctica. Polar Geoscience 13:1-40

Grew ES, Hålenius U, Kritikos M, Shearer CK (2001a) New data on welshite, e.g., $Ca_2Mg_{3.8}Mn^{2+}{}_{0.6}Fe^{2+}{}_{0.1}$–$Sb^{5+}{}_{1.5}O_2[Si_{2.8}Be_{1.7}Fe^{3+}{}_{0.65}Al_{0.7}As_{0.17}O_{18}]$, an aenigmatite-group mineral. Mineral Mag 65:665-674

Grew ES, Harley SL, Sandiford M, Shearer CK, Sheraton JW (2001b) Beryllium in the Napier Complex, Antarctica: A tracer for determining the relative timing of partial melting, deformation and ultrahigh-temperature metamorphism in the lower crust. Geol Soc Am Abstr Prog 33:A-330

Grew ES, Rao AT, Raju KKVS, Yates MG (2001c) A reexamination of quartz-sillimanite-hypersthene-cordierite gneisses from the Vijayanagaram district: Does surinamite occur in the Eastern Ghats Belt? Current Science 82:1353-1358

Griffitts WR, Cooley EF (1961) Beryllium content of cordierite. U S Geol Surv Prof Paper 424-B:259

Grigor'yev NA (1967) Co-deposition of beryllium with manganese during formation of rhodochrosite under hypergene conditions. Doklady Akad Nauk SSSR 173:1411-1413 (in Russian)

Grigor'yev NA (1980) Exogenic beryllium minerals. *In* Minerals in Rocks and Ores of the Urals, Akad Nauk SSSR Ural'skiy Nauchnyy Tsentr, Sverdlovsk, p 47-51 (in Russian)

Grigor'yev NA, Pal'guyeva GV (1980) Beryllium margarite. *In* Minerals in Rocks and Ores of the Urals, Akad Nauk SSSR Ural'skiy Nauchnyy Tsentr, Sverdlovsk, p 52-56 (in Russian)

Grundmann G, Morteani G (1982) Die Geologie des Smaragdvorkommens im Habachtal (Land Salzburg, Österreich). Arch Langerst.forsch Geol B-A 2:71-107

Grundmann G, Morteani G (1989) Emerald mineralization during regional metamorphism: The Habachtal (Austria) and Leydsdorp (Transvaal, South Africa) deposits. Econ Geol 84:1835-1849

Hålenius E (1980) Zur Geochemie und Färbung von blauem Andalusit, Surinamit und Violan. Unpubl Dipl Thesis, Univ Bonn

Hanson A (1985) Découverte d'euclase dans un filon de quartz à Ottré, massif de Stavelot, Belgique. Bull Minéral 108:139-143

Harley SL (1985) Paragenetic and mineral–chemical relationships in orthoamphibole-bearing gneisses from Enderby Land, east Antarctica: a record of Proterozoic uplift. J Metamorphic Geol 3:179-200

Harley SL (1998) On the occurrence and characterization of ultrahigh-temperature crustal metamorphism. *In* Treloar PJ, O'Brien PJ (eds) What Drives Metamorphism and Metamorphic Reactions? Geol Soc (London) Spec Pub 138:81-107

Harley SL, Motoyoshi Y (2000) Al zoning in orthopyroxene in a sapphirine quartzite: evidence for >1120°C UHT metamorphism in the Napier Complex, Antarctica, and implications for the entropy of sapphirine. Contrib Mineral Petrol 138:293-307

Haüy RJ (1799) 10. Euclase (N.N.), c'est-à-dire, *facile à briser*. J Mines 5:258

Heinrich EW (1950) Cordierite in pegmatite near Micanite, Colorado. Am Mineral 35:173-184

Hemingway BS, Barton MD, Robie RA, Haselton HT Jr (1986) Heat capacities and thermodynamic functions for beryl, $Be_3Al_2Si_6O_{18}$, phenakite, Be_2SiO_4, euclase, $BeAlSiO_4(OH)$, bertrandite, $Be_4Si_2O_7(OH)_2$, and chrysoberyl, $BeAl_2O_4$. Am Mineral 71:557-568

Henry C, Burkhard M, Goffé B (1996) Evolution of synmetamorphic veins and their wallrocks through a Western Alps transect: no evidence for large-scale fluid flow. Stable isotope, major- and trace-element systematics. Chem Geol 127:81-109

Henry DJ, Dutrow BL (1996) Metamorphic tourmaline and its petrologic applications. Rev Mineral 33:503-557

Hepp S, Kalt A, Altherr R (1997). Fluids during Variscan HT metamorphism in the Schwarzwald: Li, Be and B variations in amphibole. Berichte Deutsch Mineral Gesell (Beihefte Eur J Mineral 13) 1:77 (abstr)

Herz N, Dutra CV (1964) Geochemistry of some kyanites from Brazil. Am Mineral 49:1290-1305

Hietanen A (1956) Kyanite, andalusite, and sillimanite in the schist in Boehls Butte Quadrangle, Idaho. Am Mineral 41:1-27

Hietanen A (1969) Distribution of Fe and Mg between garnet, staurolite, and biotite in aluminum-rich schist in various metamorphic zones north of the Idaho Batholith. Am J Sci 267:422-456

Hietanen A (1971) Distribution of elements in biotite-hornblende pairs and in an orthopyroxene-clinopyroxene pair from zoned plutons, northern Sierra Nevada, California. Contrib Mineral Petrol 30:161-176

Hirst DM (1962) The geochemistry of modern sediments from the Gulf of Paria—II. The location and distribution of trace elements. Geochim Cosmochim Acta 26:1147-1187

Holdaway MJ, Dutrow BL, Shore P (1986) A model for the crystal chemistry of staurolite. Am Mineral 71:1142-1159

Hölscher A (1987) Experimentelle Untersuchungen im System MgO-BeO-Al_2O_3-SiO_2-H_2O: MgAl-Surinamit und Be-Einbau in Cordierit und Sapphirin. Unpublished PhD dissertation, Ruhr-Universität Bochum

Hölscher A, Schreyer W (1989) A new synthetic hexagonal BeMg-cordierite, $Mg_2[Al_2BeSi_6O_{18}]$, and its relationship to Mg-cordierite. Eur J Mineral 1:21-37

Hölscher A, Schreyer W, Lattard D (1986) High-pressure, high-temperature stability of surinamite in the system MgO-BeO-Al_2O_3-SiO_2-H_2O. Contrib Mineral Petrol 92:113-127

Hörmann PK (1966) Die Verteilung des Berylliums in den Mafititknollen des Dreiser Weihers (Eifel). Contrib Mineral Petrol 13:374-388

Hörmann PK (1978) Beryllium. *In* Wedepohl KH (ed) Handbook of Geochemistry II/1. Springer, Berlin, p 4-B-1 to 4-O-1, 1-6

Hsu LC (1983) Some phase relationships in the system BeO-Al_2O_3-SiO_2-H_2O with comments on effects of HF. Geol Soc China Memoir 5:33-46

Huang Y, Du S, Zhou X (1988) Hsianghualing rocks, mineral deposits and minerals. Beijing Science and Technology Publication Bureau, Beijing (in Chinese with English summary)

Hudson DR (1968) Some mafic minerals from the granulites and charnockitic granites of the Musgrave Ranges, central Australia. Unpubl Ph.D. thesis, Univ Queensland

Hudson DR, Wilson AF, Threadgold IM (1967) A new polytype of taaffeite—a rare beryllium mineral from the granulites of central Australia. Mineral Mag 36:305-310

Huebner JS, Flohr MJK (1990) Microbanded manganese formations: Protoliths in the Franciscan Complex, California. U S Geol Surv Prof Paper 1502:1-72

Huebner JS, Flohr MJK, Grossman JN (1992) Chemical fluxes and origin of a manganese carbonate–oxide–silicate deposit in bedded chert. Chem Geol 100:93-118

Hügi T, Röwe D (1970) Berylliummineralien und Berylliumgehalte granitischer Gesteine der Alpen. Schweiz mineral petrogr Mitt 50:445-480

Irouschek-Zumthor A (1983) Mineralogie und Petrographie von Metapeliten der Simano Decke unter besonderer Berücksichtigung cordieritführender Gesteine zwischen Alpe Sponda und Biasca. Unpub PhD dissertation, Univ Basel

Jensen SM (1994) Lead isotope signatures of mineralised rocks in the Caledonian fold belt of North-East Greenland. Rapp Grønlands Geol Undersøgelse 162:169-176

Jensen SM, Stendal H (1994) Reconnaissance for mineral occurrences in North-East Greenland (76°-78°N). Rapp Grønlands Geol Undersøgelse 162:163-168

Johnson MC, Plank T (1999) Dehydration and melting experiments constrain the fate of subducted sediments. Geochem Geophys Geosystems 1, Paper No. 1999GC000014

Kalt A, Altherr R, Ludwig T (1998) Contact metamorphism in pelitic rocks on the Island of Kos (Greece, Eastern Aegean Sea) a test for the Na-in-cordierite thermometer. J Petrol 39:663-688

Kalt A, Berger A, Blümel P (1999) Metamorphic evolution of cordierite-bearing migmatites from the Bayerische Wald (Variscan Belt, Germany). J Petrol 40:601-627

Kalt A, Schreyer W, Ludwig T, Prowatke S, Bernhardt H-J, Ertl A (2001) Complete solid solution between magnesian schorl and lithian excess-boron olenite in a pegmatite from the Koralpe (eastern Alps, Austria). Eur J Mineral 13:1191-1205

Kampf AR (1991) Taaffeite crystals. Mineral Record 22:343-347

Kawakami S, Tabata H, Ishii E (1986) Phase relations in the pseudo-binary system $BeAl_2O_4$-$MgAl_2O_4$. Reports of the Government Industrial Research Institute, Nagoya 35:235-239 (in Japanese)

Keller PC (1983) The Capão topaz deposit, Ouro Preto, Minas Gerais, Brazil. Gems Gemology 19:12-20

Kiefert L, Schmetzer K (1998) Distinction of taaffeite and musgravite. J Gemm 26:165-167

Kosals YaA, Mazurov MP (1968) Behavior of the rare alkalis, boron, fluorine, and beryllium during the emplacement of the Bitu-Dzhida granitic batholith, southwest Baykalia. Geochem Intl 5:1024-1034

Kosals YaA, Dmitriyeva AN, Arkhipchuk RZ, Gal'chenko VI (1973) Temperature conditions and formation sequence in deposition of fluorite-phenacite(*sic*)-bertrandite ores. Intl Geol Rev 16: 1027-1036

Kostetskaya EV, Petrov LL, Petrova ZI, Mordvinova VI (1969) Distribution of beryllium and fluorine and correlation between their contents in the minerals and rocks of the Dzhida Paleozoic granitoid complex. Geochem Intl 6:72-79

Kozhevnikov OK, Perelyayev VI (1987) The Sorok-Snezhnaya Belt of plumasitic rare-metal granite and associated tin-tungsten mineralization in the eastern Sayan and the Khamar Daban. Doklady Acad Sci USSR Earth Sci Sections 296:121-123

Kozhevnikov OK, Dashkevich LM, Zakharov AA, Kashayev AA, Kukhrinkova NV, Sinkevich TP (1975a) First taafeite [sic] find in the USSR. Doklady Acad Sci USSR Earth Sci Sections 224:120-121

Kozhevnikov OK, Zakharov AA, Kukhrinkova NV (1975b) Lithium-fluorine granite in the eastern Sayan. Doklady Acad Sci USSR, Earth Sci Sections 220:197-199

Kretz, R (1983) Symbols for rock-forming minerals. Am Mineral 68:277-279

Kretz R, Loop J, Hartree R (1989) Petrology and Li-Be-B geochemistry of muscovite-biotite granite and associated pegmatite near Yellowknife, Canada. Contrib Mineral Petrol 102:174-190

Krosse S, Schreyer W (1993) Comparative geochemistry of coticules (spessartine-quartzites) and their redschist country rocks in the Ordovician of the Ardennes Mountains, Belgium. Chemie Erde 53:1-20

Kunzmann T (1999) The aenigmatite-rhönite mineral group. Eur J Mineral 11:743-756

Kupriyanova II (1976) Beryllium margarite. *In* Ginzburg AI (ed) Mineralogy of hydrothermal beryllium deposits. Nedra, Moscow, p 119-122 (in Russian)

Kutukova YeI (1959) Beryllium-bearing margarite from the middle Urals. Trudy Inst Mineral Geokhim Kristallokhim Redkikh Elementov Akad Nauk SSSR 3:79-84 (in Russian)

Lacroix A (1939) Sur un nouveau type de roches métamorphiques (sakénites) faisant partie des schistes cristallins du sud de Madagascar. Comptes Rendus Acad Sci 209:609-612

Lahlafi M (1997) Rôle des micas dans la concentration des éléments légers (Li, Be et F) dans les granites crustaux: étude expérimentale et cristallochimique. Unpublished Thèse Grade de Docteur, Université d'Orléans

Lahti SI (1988) Occurrence and mineralogy of the margarite- and muscovite-bearing pseudomorphs after topaz in the Juurakko pegmatite, Orivesi, southern Finland. Bull Geol Soc Finland 60:27-43

Lahti SI, Saikkonen R (1985) Bityite $2M_1$ from Eräjärvi compared with related Li–Be brittle micas. Bull Geol Soc Finland 57:207-215

Lebedev VI, Nagaytsev YuV (1980) Minor elements in metamorphic rocks as an ore-material source for certain deposits. Geochem Intl 17:31-39

Leeman WP, Sisson VB (1996) Geochemistry of boron and its implications for crustal and mantle processes. Rev Mineral 33:645-707

Lin J-C, Guggenheim S (1983) The crystal structure of a Li,Be-rich brittle mica: a dioctahedral-trioctahedral intermediate. Am Mineral 68:130-142

Lindahl I, Grauch RI (1988) Be-REE-U-Sn-mineralization in Precambrian granitic gneisses, Nordland County, Norway. *In* Zachrisson E (ed) Proc Seventh Quadrennial IAGOD Symp (Luleå, Sweden). E Schweizerbart'sche Verlagsbuchhandlung, Stuttgart, p 583-594

Liu Y-S, Gao S, Jin S-Y, Hu Y-H, Sun M, Zhao Z-B, Feng J-F (2001) Geochemistry of lower crustal xenoliths from Neogene Hannuoba Basalt, North China Craton: Implications for petrogenesis and lower crustal composition. Geochim Cosmochim Acta 65:2589-2604

Lowell GR, Ahl M (2000) Chemistry of dark zinnwaldite from Bom Futuro tin mine, Rondônia, Brazil. Mineral Mag 64:699-709

Lyakhovich TT, Lyakhovich VV (1983) New data on accessory-mineral compositions. Geochem Intl 20:91-108

Maboko MAH, McDoughall I, Zeitler PK (1989) Metamorphic *P–T* path of granulites in the Musgrave Ranges, Central Australia. *In* Daly JS, Cliff RA, Yardley BWD (eds) Evolution of Metamorphic Belts, Geol Soc (London) Spec Pub 43:303-307

Maboko MAH, Williams IS, Compston W (1991) Zircon U-Pb chronometry of the pressure and temperature history of granulites in the Musgrave Ranges, Central Australia. J Geol 99:675-697

Magnusson NH (1930) The iron and manganese ores of the Långban district. Sveriges Geol Undersökning Avh Ca 23:1-111 (in Swedish with extended English summary)

Malcherek T, Domeneghetti MC, Tazzoli V, Ottolini L, McCammon C, Carpenter MA (2001) Structural properties of ferromagnesian cordierites. Am Mineral 86:66-79

Mandarino JA (1999) Fleischer's Glossary of Mineral Species 1999. The Mineralogical Record, Tucson, Arizona

Marschall H, Altherr R, Ludwig T, Kalt A (2001): Li-Be-B budgets of high-pressure metamorphic rocks from Syros, Cyclades, Greece. Ber Deutsch Mineral Gesellschaft (Beih 1 Eur J Mineral) 13:117 (abstr)

Mattson SM, Rossman GR (1987) Identifying characteristics of charge transfer transitions in minerals. Phys Chem Minerals 14:94-99

McKie D (1963) Order-disorder in sapphirine. Mineral Mag 33:635-645

Meixner H (1957) Ein neues Euklasvorkommen in den Ostalpen. Tschermaks mineral petrograph Mitt 6:246-251

Merrill JR, Lyden EFX, Honda M, Arnold JR (1960) The sedimentary geochemistry of the beryllium isotopes. Geochim Cosmochim Acta 18:108-129

Miyashiro A (1957) Cordierite-indialite relations. Am J Sci 255:43-62

Monier G, Charoy B, Cuney M, Ohnenstetter D, Robert J-L (1987) Évolution spatiale et temporelle de la composition des micas du granite albitique à topaze-lépidolite de Beauvoir. Mém Géologie profonde de la France tome 1, Géologie de la France 2-3:179-188

Moor R, Oberholzer WF, Gübelin E (1981) Taprobanite, a new mineral of the taaffeite-group. Schweiz mineral petrograph Mitt 61:13-21

Moore PB (1967) Eleven new minerals from Långban, Sweden. Can Mineral 9:301 (abstr)

Moore PB (1969) The crystal structure of sapphirine. Am Mineral 54:31-49

Moore PB (1971) Mineralogy & chemistry of Långban-type deposits in Bergslagen, Sweden. Mineral Record 1(4):154-172

Moore PB (1976) Surinamite, a new Mg-Al silicate from the Bakhuis Mountains, western Surinam. II. X-ray crystallography and proposed crystal structure. Am Mineral 61:197-199

Moore PB (1978) Welshite, $Ca_2Mg_4Fe^{3+}Sb^{5+}O_2[Si_4Be_2O_{18}]$, a new member of the aenigmatite group. Mineral Mag 42:129-132

Moore PB, Araki T (1983) Surinamite, *ca.* $Mg_3Al_4Si_3BeO_{16}$: its crystal structure and relation to sapphirine, *ca.* $Mg_{2.8}Al_{7.2}Si_{1.2}O_{16}$. Am Mineral 68:804-810
Moxham RL (1960) Minor element distribution in some metamorphic pyroxenes. Can Mineral 6:522-545
Moxham RL (1965) Distribution of minor elements in coexisting hornblendes and biotites. Can Mineral 8:204-240
Nefedov YeI (1967) Berborite, a new mineral. Doklady Akad Nauk SSSR Earth Sci Sect 174:114-117
Newton RC (1966) BeO in pegmatitic cordierite. Mineral Mag 35:920-927
Nicollet C (1985) Les gneiss rubanés à cordiérite et grenat d'Ihosy: Un marqueur thermo-barométrique dans le sud de Madagascar. Precambrian Research 28:175-185
Nicollet C (1990) Crustal evolution of the granulites of Madagascar. *In* Vielzeuf D, Vidal P (eds) Granulites and Crustal Evolution. Kluwer, Dordrecht, p 291-310
Niven ML, Waters DJ, Moore JM (1991) The crystal structure of werdingite, $(Mg,Fe)_2Al_{12}(Al,Fe)_2Si_4(B,Al)_4O_{37}$, and its relationship to sillimanite, mullite, and grandidierite. Am Mineral 76:246-256
Notari F, Boillat P-Y, Grobon C (2000) L'euclase bleu-verte de Colombie. Revue Gemm 140:18-20
Novikova MI, Shpanov YeP, Kupriyanova II (1994) Petrology of the Yermakovskoye beryllium deposit, western Transbaikalia. Petrologiya 2(1):114-127 (in Russian)
Nuber B, Schmetzer K (1983) Crystal structure of ternary Be-Mg-Al oxides: taaffeite, $BeMg_3Al_8O_{16}$, and musgravite, $BeMg_2Al_6O_{12}$.N Jahrbuch Mineral Monatsh 1983:393-402
Nysten P, Holtstam D, Jonsson E (1999) The Långban minerals. *In* Holtstam D, Langhof J (eds) (1999) Långban. The Mines, Their Minerals, Geology and Explorers. Swedish Museum of Natural History, Raster Förlag, Stockholm, p 89-183
Okrusch M (1971) Zur Genese von Chrysoberyll- und Alexandrit-Lagerstätten. Eine Literaturübersicht. Z Deutsch Gemm Gesellschaft 20:114-124
Olsen DR (1971) Origin of topaz deposits near Ouro Preto, Minas Gerais, Brazil. Econ Geol 66:627-631
Olsen DR (1972) Origin of topaz deposits near Ouro Preto, Minas Gerais, Brazil—a reply. Econ Geol 67: 120-121
Otroshchenko VD, Dusmatov VD, Khorvat VA, Akramov MB, Morozov SA, Otroshchenko LA, Khalilov MKh, Kholopov NP, Vinogradov OA, Kudryavtsev AS, Kabanova LK, Sushchinskiy LS (1971) Tourmalines in Tien Shan and the Pamirs, Soviet Central Asia. Intl Geol Rev 14:1173-1181
Ottaway TL, Wicks FJ, Bryndzia LT, Kyser TK, Spooner ETC (1994) Formation of the Muzo hydrothermal emerald deposit in Colombia. Nature 369:552-554
Peacor DR, Rouse RC, Grew ES (1999) Crystal structure of boralsilite and its relation to a family of boroaluminosilicates, sillimanite and andalusite. Am Mineral 84:1152-1161
Pearson GR, Shaw DM (1960) Trace elements in kyanite, sillimanite, and andalusite. Am Mineral 45: 808-817
Pekov IV (1994) Remarkable finds of minerals of beryllium: from the Kola Peninsula to Primorie. World of Stones (Mir Kamnya) 4:10-26 (English), 3-12 (Russian)
Peng CC, Wang KJ (1963) Discovery of 8-layered closest-packing—crystal structure analysis of taaffeite. Scientia Sinica 12:276-278 (in Russian)
Pereira Gómez MD, Rodríguez Alonso MD (2000) Duality of cordierite granites related to melt-restite segregation in the Peña Negra anatectic complex, central Spain. Can Mineral 38:1329-1346
Pesquera A, Velasco F (1997) Mineralogy, geochemistry and geological significance of tourmaline-rich rocks from the Paleozoic Cinco Villas Massif (western Pyrenees, Spain). Contrib Mineral Petrol 129: 53-74
Petrov LL (1973) Behavior of beryllium in crystallization of granitoid magmas. Geochem Intl 10:627-640
Petrova ZI, Petrov LL (1965) Beryllium in the minerals of granitoids. Geochem Intl 1965:488-492
Piantone P, Burnol L (1987) Géochimie des micaschistes du sondage d'Échassières. Mémoire Géologie profonde de la France tome 1, Géologie de la France 2-3:295-309
Piyar YuK, Goroshnikov BI, Yur'yev LD (1968) Beryllian cordierite. Mineral Sbornik Lvov Gosud Univ 22:86-89 (in Russian)
Plank T, Langmuir CH (1998) The chemical composition of subducting sediment and its consequences for the crust and mantle. Chem Geol 145:325-394
Plank T, Ludden JN (1992) 8. Geochemistry of sediments in the Argo Abyssal Plain at site 765: A continental margin reference section for sediment recycling in subduction zones. *In* Gradstein FM, Ludden JN et al. (eds) Proc Ocean Drilling Program Scientific Results 123:167-189
Poinssot C, Goffé B, Toulhoat P (1997) Geochemistry of the Triassic-Jurassic Alpine continental deposits : origin and geodynamic implications. Bull Soc Géol France 168:287-300
Polyakov VO, Cherepivskaya GYe, Shcherbakova YeP (1986) Makarochkinite—a new beryllosilicate. *In* New and little-studied minerals and mineral associations of the Urals. Akad Nauk SSSR Ural'skiy Nauchnyy Tsentr, Sverdlovsk, p 108-110 (in Russian)

Pomirleanu V, Apostoloiu A, Maieru O (1965) La température de crystallisation du disthène du cristallin de Sebeş. Analele Stiint Univ "Al I Cuza" Iasi noua, Sec II, Stiin naturale B, Geol-Geograf 11:7-12

Popov MP (1998) Composition of and inclusions in euclase from the Mariin emerald deposit. *In* Mineralogy of the Urals, Materials of the 3rd Regional Meeting, 12-14 May 1998, Miass 2:72-73 (in Russian)

Povondra P, Čech F (1978) Sodium-beryllium-bearing cordierite from Haddam, Connecticut, U.S.A. N Jahrb Mineral Monatsh 1978:203-209

Povondra P, Langer K (1971a) A note on the miscibility of magnesia-cordierite and beryl. Mineral Mag 38:523-526

Povondra P, Langer K (1971b) Synthesis and some properties of sodium-beryllium-bearing cordierite $Na_xMg_2(Al_{4-x}Be_xSi_5O_{18})$. N Jahrb Mineral Abh 116:1-19

Povondra P, Čech F, Burke EAJ (1984) Sodian-beryllian cordierite from Gammelmorskärr, Kemiö Island, Finland and its decomposition products. N Jahrb Mineral Monatsh 1984:125-136

Ragu A (1994) Helvite from the French Pyrénées as evidence for granite-related hydrothermal activity. Can Mineral 32:111-120

Rakotondrazafy R (1999) Les granulites alumineuses et magnésiennes de hautes températures et roches associées du sud de Madagascar. Pétrologie, géochimie, thermobarométrie, géochronologie et inclusions fluides. Thèse Docteur ès Sciences Naturelles. Université d'Antananarivo, Madagascar

Rakotondrazafy R, Raith M (1997). Unusual anorthite-corundum-spinel-sapphirine rocks („sakenites")—products of high-grade metamorphism of argillaceous limestones. Berichte Deutsch Mineral Gesell (Beihefte Eur J Mineral 9) 1:287 (abstr)

Rakotondrazafy R, Raith M (2000) Les sakenites: anorthositites à corindon, spinelle et saphirine (N-W Ihosy, Sud de Madagascar). Une nouvelle occurrence de la musgravite. *In* 40th Anniversary of the University of Antananarivo, Abstr Vol Intl Meeting, September 18-28, 2000, p 29 [not consulted]

Ramesh-Kumar PV, Raju KKVS, Ganga Rao BS (1995) Surinamite (Be, B, Ga) from cordierite gneisses of eastern Ghat mobile belt, India. Current Science 69:763-767

Ranorosoa N (1986) Etude minéralogique et microthermométrique des pegmatites du champ de la Sahatany, Madagascar. Thèse de doctorat, Univ Paul Sabatier, Toulouse, 223 p

Rao AT (1976) Study of the apatite-magnetite veins near Kasipatnam, Visakhapatnam District, Andhra Pradesh, India. Tschermaks mineral petrogr Mitt 23:87-103

Rapela CW, Pankhurst RJ, Baldo E, Saavedra J (1995) Cordieritites in S-type granites: Restites following low pressure, high degree partial melting of metapelites. *In* Brown M, Piccoli PM (eds) The Origin of Granites and Related Rocks. Third Hutton Symp Abstr. U S Geol Surv Circular 1129:120-121

Rapela CW, Pankhurst RJ, Baldo E, Saavedra J (1997) Low-pressure anatexis during the Pampean Orogeny. VIII Congreso Geologico Chileno, 13 al 17 de Octubre 1997, Actas 3:1714-1718

Rapela CW, Pankhurst RJ, Casquet C, Baldo E, Saavedra J, Galindo C, Fanning CM (1998) The Pampean Orogeny in the southern proto-Andes: Cambrian continental collision in the Sierras de Córdoba. *In* Pankhurst RJ, Rapela CW (eds) The Proto-Andean Margin of Gondwana. Geol Soc London Spec Pub 142:181-217

Reeve KD, Buykx WJ, Ramm EJ (1969) The system $BeO-Al_2O_3-MgO$ at subsolidus temperatures. J Austral Ceram Soc 5:29-32

Robert J-L, Hardy M, Sanz J (1995) Excess protons in synthetic micas with tetrahedrally coordinated divalent cations. Eur J Mineral 7:457-461

Roda Robles E, Pesquera Perez A, Velasco Roldan F, Fontan F (1999) The granitic pegmatites of the Fregeneda area (Salamanca, Spain): characteristics and petrogenesis. Mineral Mag 63:535-558

Rossi P, Autran A, Azencott C, Burnol L, Cuney M, Johan V, Kosakevitch A, Ohnenstetter D, Monier G, Piantone P, Raimbault L, Viallefond L (1987) Logs pétrographique et géochimique du granite de Beauvoir dans le sondage «Échassières I» Minéralogie et géochimie comparées. Mémoire Géologie profonde de la France tome 1, Géologie de la France 2-3:111-135

Rowledge HP, Hayton, JD (1948) 2.–Two new beryllium minerals from Londonderry. J Royal Soc Western Australia 33:45-52

Rozen OM, Serykh VI (1972) The geochemical aspect to the question of granite formation. *In* Granitization, Granites and Pegmatites, Doklady First Intl Geochem Congress, Moscow, USSR, 20-25 July 1971, 3:95-117 (in Russian)

Rubiano M (1990) La Euclasa en la región de Chivor y su significado en el origen de las esmeraldas colombianas. Geologia Colombiana 17:239-241

Ryan JG (1989) The systematics of lithium, beryllium and boron in young volcanic rocks. Unpub PhD dissertation, Columbia University, New York

Ryan JG, Langmuir CH (1988) Beryllium systematics in young volcanic rocks: Implications for ^{10}Be. Geochim Cosmochim Acta 52:237-244

Sahama TG, Lehtinen M, Rehtijärvi P (1974) Properties of sapphirine. Ann Acad Sci Fennicae, Ser A, III Geol-Geograph 114:1-24

Saltykova VS (1959) Analyses of minerals containing rare elements carried out by the chemical laboratory of the Institute of Mineralogy, Geochemistry, and Crystallochemistry of Rare Elements of the U.S.S.R. Academy of Sciences in 1954-1957. Trudy Inst Mineral Geokhim Kristallokhim Redkikh Elementov 2:189-208 (in Russian)

Sambonsugi M (1957) Iron-rich cordierite structurally close to indialite. Proc Japan Acad 33:190-195

Sandell EB (1952) The beryllium content of igneous rocks. Geochim Cosmochim Acta 2:211-216

Sandiford M (1985) The origin of retrograde shear zones in the Napier Complex: implications for the tectonic evolution of Enderby Land, Antarctica. J Struct Geol 7:477-488

Sandström F, Holtstam D (1999) Geology of the Långban deposit. *In* Holtstam D, Langhof J (eds) Långban. The Mines, Their Minerals, Geology and Explorers. Swedish Museum of Natural History, Raster Förlag, Stockholm, p 29-41

Schaller WT, Carron MK, Fleischer M (1967) Ephesite, $Na(LiAl_2)(Al_2Si_2)O_{10}(OH)_2$, a trioctahedral member of the margarite group, and related brittle micas. Am Mineral 52:1689-1696

Schmetzer K (1983a) Crystal chemistry of natural Be–Mg–Al-oxides: taaffeite, taprobanite, musgravite. N Jahrb Mineral Abh 146:15-28

Schmetzer K (1983b) Taaffeite or taprobanite—a problem of mineralogical nomenclature. J Gemm 18: 623-634

Schmetzer K, Bank H (1985) Zincian taaffeite from Sri Lanka. J Gemm 19:494-497

Schmetzer K, Bernhardt H-J, Medenbach O (1999) Heat-treated Be–Mg–Al oxide (originally musgravite or taaffeite). J Gemm 26:353-356

Schmetzer K, Kiefert L, Bernhardt H-J (2000) Purple to purplish red chromium-bearing taaffeites. Gems Gemology 36(1):50-59

Schreyer W (1964) Synthetische und natürliche Cordierite I. Mischkristallbildung synthetischer Cordierite und ihre Gleichgewichtsbeziehungen. N Jahrb Mineral Abh 102:39-67

Schreyer W, Seifert F (1969) High-pressure phases in the system $MgO\text{-}Al_2O_3\text{-}SiO_2\text{-}H_2O$. Am J Sci, Schairer Vol 267-A:407-443

Schreyer W, Gordillo CE, Werding G (1979) A new sodian-beryllian cordierite from Soto, Argentina, and the relationship between distortion index, Be content, and state of hydration. Contrib Mineral Petrol 70: 421-428

Schwander H, Hunziger J, Stern W (1968) Zur Mineralchemie von Hellglimmern in den Tessineralpen. Schweiz mineral petrogr Mitt 48:357-390

Seifert F (1974) Stability of sapphirine: A study of the aluminous part of the system $MgO\text{-}Al_2O_3\text{-}SiO_2\text{-}H_2O$. J Geol 82:173-204

Selkregg KR, Bloss FD (1980) Cordierites: compositional controls of Δ, cell parameters and optical properties. Am Mineral 65:522-533

Sergeyev AS, Petrov VP, Galibin VA, Predovskiy AA (1967) Distribution of impurity elements in staurolite and biotite of highly aluminous schists of northern Priladozh'ye [Lake Ladoga area]. *In* Mineralogiya i Geokhimiya, Collection of Papers, no. 2. Leningrad University, Leningrad, p 136-144 (in Russian)

Shannon RD (1976) Revised effective ionic radii and systematic studies of interatomic distances in halides and chalcogenides. Acta Crystallogr A32:751-767

Sharp WN (1961) Euclase in greisen pipes and associated deposits, Park County, Colorado. Am Mineral 46:1505-1508

Shaw DM (1960) The geochemistry of scapolite. Part II. Trace elements, petrology, and general geochemistry. J Petrol 1:261-285

Shaw DM, Moxham RL, Filby RH, Lapkowsky WW (1963) The petrology and geochemistry of some Grenville skarns. Part II. Geochemistry. Can Mineral 7:578-616

Shaw DM, Reilly GA, Muysson JR, Pattenden GE, Campbell FE (1967) An estimate of the chemical composition of the Canadian Precambrian Shield. Can J Earth Sci 4:829-853

Shcheglova TP, Skublov SG, Drygova GM, Bushmin SA (2000) Peculiarities of the chemical composition of staurolite from highly aluminous rocks of the Keivsky Block. Zapiski Vserossiyskogo Mineral Obshchestva 129:71-80 (in Russian)

Sheraton JW (1980) Geochemistry of Precambrian metapelites from East Antarctica: secular and metamorphic variations. BMR J Austral Geol Geophys 5:279-288

Sheraton JW (1985) Chemical analyses of rocks from East Antarctica: Part 2. Australia Bureau of Mineral Resources, Geology and Geophysics Record 1985/12

Sheraton JW, England RN, Ellis DJ (1982) Metasomatic zoning in sapphirine-bearing granulites from Antarctica. BMR J Austral Geol Geophys 7:269-273

Sheraton JW, Black LP, McCulloch MT (1984) Regional geochemical and isotopic characteristics of high-grade metamorphics of the Prydz Bay area: the extent of Proterozoic reworking of Archaean continental crust in East Antarctica. Precambrian Res 26:169-198

Sheraton JW, Babcock RS, Black LP, Wyborn D, Plummer CC (1987a) Petrogenesis of granitic rocks of the Daniels Range, northern Victoria Land, Antarctica. Precambrian Res 37:267-286

Sheraton JW, Tingey RJ, Black LP, Offe LA Ellis DJ (1987b): Geology of an unusual Precambrian high-grade metamorphic terrane—Enderby Land and western Kemp Land, Antarctica. Austral Bur Mineral Resources Geol Geophys Bull 223:1-51

Sighinolfi GP (1973) Beryllium in deep-seated crustal rocks. Geochim Cosmochim Acta 37:702-706

Slack JF, Palmer MR, Stevens BPJ, Barnes RG (1993) Origin and significance of tourmaline rich rocks in the Broken Hill district, Australia. Econ Geol 88:505-541

Smith JV (1974) Feldspar minerals, vol 2. Chemical and Textural Properties. Berlin–New York, Springer-Verlag

Smith JV, Brown WL (1988) Feldspar minerals, vol 1 Crystal Structure, Physical, Chemical and Textural Properties, 2nd edn. Berlin–New York, Springer-Verlag

Sokolova GV, Aleksandrova VA, Drits VA, Bayrakov VV (1979) The crystal structures of two brittle lithium micas. *In* Kristallokhimiya i Strukturnaya Mineralogiya, p 55-66 (in Russian)

Soman K, Druzhinin AV (1987) Petrology and geochemistry of chrysoberyl pegmatites of South Kerala, India. N Jahrb Mineral Abh 157:167-183

Stadnichenko T, Zubovic P, Sheffey NB (1961) Beryllium content of American coals. U S Geol Surv Bull 1084-K:253-295

Steele IM, Hutcheon ID, Smith JV (1980) Ion microprobe analysis of plagioclase feldspar ($Ca_{1-x}Na_xAl_{2-x}Si_{2+x}O_8$) for major, minor, and trace elements. *In* 8th Intl Congress on X-ray Optics and Microanalysis. Pendell Publishing Co., Midland, Michigan, p 515-525

Steppan N, Kalt A, Altherr R (1997). Li, Be, B contents of metapelitic minerals—examples from Ikaria Island (Greece), Campo Tencia (Swiss Alps) and Künisches Gebirge (Germany). Berichte Deutsch Mineral Gesell (Beihefte Eur J Mineral 13) 1:179 (abstr)

Stephenson DA, Moore PB (1968) The crystal structure of grandidierite, $(Mg,Fe)Al_3BSiO_9$. Acta Crystllogr B24:1518-1522

Taylor SR, McLennan SM (1985) The continental crust: its composition and evolution. Blackwell, Oxford

Taylor SR, McLennan SM (1995) The geochemical evolution of the continental crust. Rev Geophys 33:241-265

Taylor SR, McLennan SM, McCulloch MT (1983) Geochemistry of loess, continental crustal composition and crustal model ages. Geochim Cosmochim Acta 47:1897-1905

Teale GS (1980) The occurrence of högbomite and taaffeite in a spinel-phlogopite schist from the Mount Painter Province of South Australia. Mineral Mag 43:575-577

Vavrda I, Vrána S (1972) Sillimanite, kyanite and andalusite in the granulite facies rocks of the basement complex, Chipata district. Rec Geol Surv Zambia 12:69-80

Vine JD, Tourtelot EB (1970) Geochemistry of black shale deposits—a summary report. Econ Geol 65:253-272

Visser D, Kloprogge JT, Maijer C (1994) An infrared spectroscopic (IR) and light element (Li, Be, Na) study of cordierites from the Bamble Sector, South Norway. Lithos 32:95-107

Vlasov KA, ed. (1966) Geochemistry and Mineralogy of Rare Elements and Genetic Types of Their Deposits, vol. II: Mineralogy of Rare Elements. Israel Program for Scientific Translations, Jerusalem

Volfinger M, Robert J-L (1994) Particle-induced-gamma-ray-emission spectrometry applied to the determination of light elements in individual grains of granite minerals. J Radioanal Nuclear Chem Articles 185:273-291

von Knorring O, Sahama TG, Saari E (1964) A note on euclase from Muiane mine, Alto Ligonha, Mozambique. Comptes Rendus Soc Géol Finlande 36:143-145

Vrána S (1979) A secondary magnesium-bearing beryl in pseudomorphs after pegmatitic cordierite. Časopis pro Mineral Geol 24:65-69

Warner LA, Holser WT, Wilmarth VR, Cameron EN Cameron EN (1959) The occurrence of nonpegmatite beryllium in the United States. U S Geol Surv Prof Paper 318

Wedepohl KH (1995) The composition of the continental crust. Geochim Cosmochim Acta 59:1217-1232

Wenk E, Schwander H, Hunziger J, Stern W (1963) Zur Mineralchemie von Biotit in den Tessineralpen. Schweiz mineral petrogr Mitt 43:435-463

Wenk E, Schwander H, Stern W (1974) On calcic amphiboles and amphibolites from the Lepontine Alps. Schweiz mineral petrogr Mitt 54:97-149

Werding G, Schreyer W (1992) Synthesis and stability of werdingite, a new phase in the system MgO-Al_2O_3-B_2O_3-SiO_2 (MABS), and another new phase in the ABS-system. Eur J Mineral 4:193-207

Wilson AF, Hudson DR (1967) The discovery of beryllium-bearing sapphirine in the granulites of the Musgrave Ranges (Central Australia). Chem Geol 2:209-215

Wilson AF, Green DC, Davidson LR (1970) The use of oxygen isotope geothermometry on the granulites and related intrusives, Musgrave Ranges, Central Australia. Contrib Mineral Petrol 27:166-178

Woodford PJ, Wilson AF (1976) Sapphirine, högbomite, kornerupine, and surinamite from aluminous granulites, northeastern Strangways Range, central Australia. N Jahrb Mineral Monatsh 1976:15-35
Yakubovich OV, Malinovskii YuA, Polyakov VO (1990) Crystal structure of makarochkinite. Sov Phys Crystallogr 35:818-822
Yegorov IN (1967) Euclase from quartz-chlorite zones of altered granites of Yakutiya. Doklady Akad Nauk SSSR 172:433-436 (in Russian)
Zack T (2000) Trace element mineral analysis in high pressure metamorphic rocks from Trescolmen, Central Alps. Dissertation, Universität Göttingen

13 Be-Minerals: Synthesis, Stability, and Occurrence in Metamorphic Rocks

Gerhard Franz

Technische Universität Berlin
Fachgebiet Petrologie
D 10623 Berlin, Germany

gerhard.franz@tu-berlin.de

Giulio Morteani

Technische Universität München
Lichtenbergstrasse 4
D 85747 Garching, Germany

INTRODUCTION

A major interest in beryllium minerals, which are rock-forming constituents in certain metamorphic, hydrothermal and igneous environments, comes from their use as ore materials for Be, as single-crystal material for special optical purposes, and as gemstones. Beryl, ideally $Al_2Be_3Si_6O_{18}$, bertrandite, $Be_4Si_2O_7(OH)_2$, and phenakite, Be_2SiO_4 are mined and processed as ores of Be. Beryl and chrysoberyl are used as maser and laser crystals, and bromellite BeO has outstanding properties such as exceptionally high electrical conductivity and speed-of-sound propagation in combination with great strength and chemical resistance. Be-minerals find wide use as gemstones; emerald and alexandrite are among the most highly prized. For all these reasons, a large amount of experimental work has been reported on the synthesis of these minerals.

The interest in Be-minerals for petrology comes from the fact that Be can be a minor component in several rock-forming silicates such as sapphirine, margarite, and cordierite, and can have an important influence on the phase relations of these minerals. The study of the behavior of Be and Be-minerals may give an important insight in the history of magmatic and metamorphic rocks. One of the key questions for the natural occurrences is the problem of enrichment of Be, so that either Be-minerals can be formed or that common rock-forming minerals such as cordierite contain an appreciable amount of Be.

We start with a review of experimental work about synthesis of Be-minerals in various simplified systems, their phase equilibria, and their compositional variation. Thermodynamic data and calorimetric work are reviewed by Barton and Young (this volume). In the second part we review the geological setting and mineralogical characteristics of Be-mineralizations in regional metamorphic terrains and on hydrothermal Be-occurrences. Metamorphosed Be-rich pegmatites, leucogranites and silicic volcanoclastic rocks, where the Be-enrichment is pre-metamorphic (see Černý, this volume) are reviewed first. No attempt was made to separate strictly "metamorphic" from "igneous," considering migmatization and the formation of pegmatites during this process. Descriptions of such occurrences are included here and for most of them the term "regional metamorphic," where no major chemical changes are involved, is appropriate. Then we give an overview for the typical emerald, chrysoberyl and phenakite deposits, where a combined metasomatic-metamorphic process is important. The separation between "metamorphic," "metasomatic," where material is transported by a fluid phase at scales ranging from centimeters to kilometers, and "hydrothermal," where the fluids are dominantly aqueous and temperatures relatively low, is somewhat artificial

1529-6466/00/0050-0013$05.00

for these occurrences, and gradations are possible. Material transport, which might produce the enrichment of Be to form Be-minerals, can be related to dynamic regional metamorphism, to a regional increase of the geothermal gradient or to a more localized phenomenon associated with igneous rocks. Finally some examples for more typically hydrothermal deposits are described. No attempt has been made to achieve a complete review of the large literature on hydrothermal deposits, which can be found in Barton and Young (this volume, Chapter 14).

In writing this chapter we want to encourage further more systematic experimental work on Be-minerals and their stability relations as well as more investigation of their natural occurrences. Special attention in further investigations should be given to the questions of the enrichment of Be, to the possible transport of Be with a fluid phase, and to the application of phase equilibrium diagrams for a more quantitative interpretation of the genesis of the Be-mineralizations.

SYNTHESIS AND STABILITY EXPERIMENTS

Experimental techniques

Starting materials and investigation of the run products. Synthesis experiments and stability experiments have been carried out using synthetic as well as natural phases. For Be, reagent grade BeO was used in many studies and—except for its highly toxic character—no experimental problems have been reported. It reacts easily and can be obtained as an essentially pure component. Other chemical reagents used were $Be(OH)_2$, BeF_2, $BeCl_2$, $BeCO_3$ and $Be(NO_3)_2$. Using natural Be-minerals has the advantage that Be is bound in silicate or complex oxide and therefore less toxic. We would strongly recommend not to use the highly toxic reagents such as BeF_2. Euclase, phenakite and chrysoberyl can be obtained in gem quality as essentially pure minerals, though chrysoberyl frequently contains some Fe. Beryl was often used, but commonly has minor amounts of Na, Mg and Fe and other elements, and must definitely be analyzed before using it as a starting material. Glass of pure beryl composition can be prepared from the oxides and reacts easily to beryl, which is more likely to have cation disorder, however.

Many of the reviewed studies have been carried out a rather long time ago, and identification and characterization of the run products was not complete. Optical methods and powder XRD were used for the identification, and in some cases electron microprobe analysis was performed, which of course yields only partial analyses for the Be-minerals. A complete chemical characterization of the run products including Be and H_2O by microanalysis and systematic X-ray work to obtain information about ordering states and solid solution was not made in many of the experimental studies, which leaves a great uncertainty for the interpretation of the results.

One-atmosphere techniques. For the synthesis of Be-minerals at one atmosphere simple fusion, flux-fusion and sintering are applied. Figure 1 gives an overview of some of the methods. Industrial synthetic Be-minerals are produced currently by various companies such as Gilson and Lapidaries, Lens (France), Chatham, Vacuum Ventures Inc., Allied Chemical Corp. (USA), Nakazumi Earth Crystals, Kyocera-Kyoto Ceramics Company, Suwa Seikosha Seiko (Japan), Crystals Research Company, Byron Minerals (Australia) and in Russia. Russian synthetic emeralds, which currently dominate the market with an estimated 100 kg (= 500,000 carat) annual production (Schmetzer 2001) are sold by different companies like Tairus and Crystural in Thailand, Kimberley and J.O. Crystals in the USA (Koivula and Keller 1985, Henn 1995). The synthetic emeralds are sold under different commercial names, e.g., Bijoreve, Lennix, Kiberley Created Emeralds and Empress Cultured Emerald (Sinkankas and Read 1986; Henn 1995).

The Czochralsky method, where rotating seed crystals are slowly extracted from melt of the pure composition, is described for the synthesis of chrysoberyl by Bukin (1993) and applied industrially in Japan. Bukin (1993) modified the Czochralsky method for bromellite from a saturated melt and flux (Fig. 1). The Japanese company Seiko is reported to produce alexandrite by the zone refining method (Henn 1995). Attempts to produce beryl or emerald by the flame-fusion Verneuil method failed and produced a glassy phase instead, because beryl melts incongruently (see below). Already Verneuil noted in 1911 that emeralds have not been produced unequivocally by a flame fusion method. Only Gentile et al. (1963) claimed to have produced beryl, but the results were questioned by Nassau (1976).

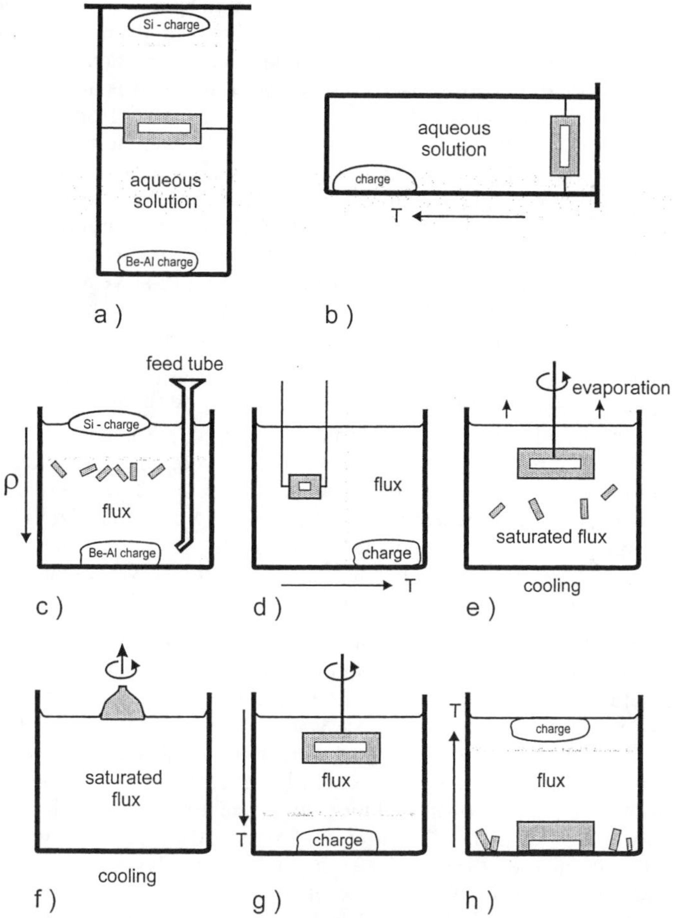

Figure 1. Schematic arrangement for the hydrothermal (a,b) and flux technique (c-h) methods for single-crystal synthesis of beryl and other Be-minerals. Spontaneous nucleation of beryl is suppressed by mechanical separation of the Si nutrient (mostly SiO_2-glass or oxides) from the Al- and Be-nutrients (a) or in a density gradient (c). When a temperature gradient is applied, the charge must be separated mechanically from the seed by a Pt-sieve. In most cases, the Be-minerals grow on a seed plate (blank rectangle in figures) mounted on Pt-wire or on a rotating rod, but spontaneous nucleation may also yield large crystals as shown in c, e and h. Modified from Nassau (1980) and Bukin (1993); for further explanation see text.

The flux-fusion method in several modifications (Fig. 1) proved successful in the production of beryl, emerald and other Be-minerals, also on the industrial scale. Beryl was synthesized from melts with lithium molybdate, potassium, lead and other vanadates, fluoride or boric acid as flux material (Linares 1967; Nassau 1976; Henn 1995). The flux technique has a long tradition and is one of the main lines in crystal growth techniques. It started with first emerald syntheses by Ebelman from 1814 to 1852 and Hautefeuille and Perrey (1888) in France, was then successfully developed between 1911 and 1934 in the project 'Igmerald' by the German company IG Farben (Espig 1960). The first 'igmeralds' appeared on the market in 1934. In the USA, Chatham grew his first emerald crystals in 1935, and in France, Gilson started his emerald synthesis work in 1950 (Gübelin 1961; Plough 1965; Henn 1995). In the flux technique, spontaneous nucleation for the growth of a single crystal is suppressed by separating the charge by a density gradient within the flux (Fig. 1c): SiO_2 is floating on the flux, Be- and Al-components are introduced by a feed tube to the bottom, and the growing beryl crystals are separated by a Pt sieve. Si, Be and Al are transported by diffusion. If a temperature gradient is applied (Fig. 1d), the charge can be, e.g., natural beryl (without separation of Si from the Be- and Al-component), and the synthetic beryl grows on a seed plate. Be-minerals can also be grown from a saturated flux on a rotating seed plate (Fig. 1e), where the saturation is achieved by evaporation and or cooling. Figure 1f shows the Czochralsky method as described by Bukin (1993) with a saturated flux, where the saturation level is kept by cooling. Growth on a rotating seed plate in a vertical temperature gradient (bottom hot; Fig. 1g) requires to separate the charge at the bottom by a Pt-sieve, or with an immersed seed plate at the top (bottom cool; Fig. 1h). Growth rates are in the order of 0.06-0.09 mm and day. Coloring agents are chromates, Ni and V compounds.

A special method is described by the Russian working groups from Novosibirsk as the 'gas transport method' (Sobolev and Novoselova 1959; Rodionov 1986). These experiments are carried out in quartz ampoules lined with pyrographite at a temperature of ~1100°C and a temperature gradient between dissolution and crystallization zones of 5-120°C. Residual pressure at room temperature is given as 10^{-1}-10^{-3} GPa. Run duration varied from 1 day to 30 days which yielded crystals of 5 × 5 × 5 mm size (Rodionov et al. 1987). Reaction of gaseous components of Be, Al, Si, O and halogens (especially F) produces beryl, chrysoberyl and phenakite. These synthetic beryls contain typically poly-phase gas-glass-crystal inclusions (Rodionov and Konovalova 1988).

Hydrothermal techniques. They have been applied for petrological purpose using cold seal pressure vessels as well as a piston cylinder apparatus. Experiments in internally heated pressure vessels at high temperature have not yet been reported. The highest pressure for hydrothermal beryl synthesis was 25 kbar (Cemic et al. 1986), for 'dry' synthesis 45 kbar (Munson 1967). Studies were carried out with pure water as well as with alkali-bearing and other solutions as a mineralizing reagent and also to study the influence of the composition of the solution on the stability (e.g., Rykl and Bauer 1969, 1971; Syromyatnikov et al. 1972). Barton (1986) used the silica buffering technique to determine the stability of the Be-phases in the system BeO-Al_2O_3-SiO_2-H_2O, other workers used the conventional technique of reversed experiments with a mixture of reactant and product phases. In beryl and emerald synthesis for commercial purpose gold lined steel autoclaves are operated at temperatures between 400 and 600°C and pressures between 0.7 and 1.5 kbar. The autoclave is filled with a neutral to alkaline solution of alkali and ammonium salts (pH 7 to 12.5), or a strongly acid medium (pH 0.2 to 4.5) using similar halide mineralizers. Beryl grows on a seed plate cut oblique to the c-axis at an angle of 15-35°, which ensures high growth rates of 0.8 mm per day. The Si-charge has to be separated by the Be-and Al-charge to prevent spontaneous nucleation of beryl (Fig. 1a). Alternatively, a temperature gradient can be applied (Fig. 1b). The nutrient for

Be, Al and Si (mostly as oxides) or natural beryl or emerald are kept at the hot end, the seed crystals at the cold end (Sinkankas and Read 1986; Henn 1995). Typical for such hydrothermally grown emerald are inclusions of phenakite ('phenakite nails' in the gemological literature) and special growth features on the seed plates. To avoid the characteristic growth features on such hydrothermally grown emeralds, optically pure emerald can be grown on seed plates cut at an angle of 45°. The first hydrothermal emeralds were produced by Lechleitener (Austria) (Gübelin 1961) and Linde (USA), later by Vacuum Ventures (USA), Byron Minerals (Australia) and Tairus (Russia). Details about the techniques of hydrothermally grown emerald were first published by Flanigen (1971). Nowadays hydrothermally grown emeralds come in considerable amounts from laboratories in Estonia, China and Russia (Eppler 1967; Bank 1976; Kane and Liddicoat 1985; Sinkankas and Read 1986; Henn and Bank 1988; Henn 1995; Schmetzer 1996, 2001).

System BeO-H_2O

The only known minerals in this system are behoite and clinobehoite, both $Be(OH)_2$, and bromellite BeO (See Introduction, this volume). Newkirk (1964) dehydrated behoite at 170°C and 1.3 kbar and 200°C and 4.1 kbar (unreversed reaction). Synthesis of single crystals of BeO with a modified Czochralsky method was described by Bukin (1993). Nothing is known about the phase relations between behoite and clinobehoite.

System BeO-Al_2O_3 (and related $R^{3+}{}_2O_3$)

Chrysoberyl, including the gem variety alexandrite. Chrysoberyl is the only known mineral in the system. Its formula is commonly written as $BeAl_2O_4$, but Al_2BeO_4 is more appropriate, because this form emphasizes the isostructural relationship to olivine (Strunz 1966). The alexandrite-effect is due to impurities of Cr^{3+}, which produces green color in daylight and red in artificial light. The generally yellowish color is attributed to Fe^{3+}.

Synthesis of chrysoberyl was reported as early as 1845 by the French chemist Ebelmen and subsequently by Deville, Caron, Hautefeulle and Perry (Nassau 1980). It melts at 1870°C at atmospheric pressure (Lang et al. 1952). Synthesis is possible from melt with the Czochralsky method and from flux (e.g., Farell et al. 1963; Sunagawa 1982; O'Donoghue 1988) or by sintering Be-carbonate and Al-hydroxide (Gjessing et al. 1943). Chrysoberyls, in which substantial amounts of trivalent Fe, Ga and Cr substitute for Al, were also synthesized with this method, whereas the same method failed to produce the analogous In-bearing compound (Gjessing et al. 1943). Sarazin and Forestier (1959) synthesized chrysoberyl with up to 70 atom % V^{3+} at 1700°C from oxide mixtures. Klyakhin (1995) reported experimental data on the solubility of chrysoberyl in 1M solutions of Na-salts of the main inorganic acids by the weight loss method, and synthesis of chrysoberyl-alexandrite, using Na_2CO_3 as a dissolution media.

Hydrothermal synthesis runs by Franz and Morteani (1981) at 4 kbar, 500°C and 750°C yielded only chrysoberyl, corundum and bromellite on the join BeO-Al_2O_3. XRD investigations gave no hint for considerable solid solution in this system. Hölscher et al. (1986) found chrysoberyl + pyrope as a breakdown product of surinamite at 850°C and 55 kbar and 1000°C and 50 kbar. However, the chrysoberyl X-ray pattern is displaced compared to ideal chrysoberyl and minor peaks disappeared. It is not clear whether chrysoberyl is stable at these high pressures or whether there exists another Al_2BeO_4 polymorph. Rykl and Bauer (1969, 1971) found chrysoberyl to be unstable in 1N and higher concentrated NaOH solutions at 450 to 550°C and 0.3 to 3 kbar. Bromellite and an unidentified phase formed instead.

No systematic investigations exist about a possible solid solution of the olivine-type Al_2BeO_4 structure with isostructural silicates, e.g., Fe_2SiO_4. In the system BeO-Al_2O_3-

SiO_2 natural (Franz and Morteani 1984; Henn 1985) as well as synthetic (Cemic et al. 1986) chrysoberyl may contain small amounts of SiO_2 (natural: up to 1.2 wt %, on the average 0.5 wt %; synthetic: 0.2 wt % at 1200°C) as indicated by electron microprobe analysis. If Fe is present in the system, it preferentially enters chrysoberyl compared to beryl, melt and phenakite at high temperatures (Cemic et al. 1986). It is conceivable that coupled Fe and Si contents in chrysoberyl are due to a solid solution with fayalite, and that in this case some of the total Fe in chrysoberyl is Fe^{2+}.

Other Be-Al oxides. Phases with the composition $Be_3Al_2O_6$ (or $3BeO{\cdot}Al_2O_3$ in the formula type of ceramic nomenclature) and $BeAl_6O_{10}$ (or $BeO{\cdot}3Al_2O_3$) were synthesized at temperatures above 1500°C (Foster and Royal 1949; Lang et al. 1952; Galakhov 1957). The compound $BeO{\cdot}3Al_2O_3$ melts at 1910°C (Lang et al. 1952). The compound $3BeO{\cdot}Al_2O_3$ melts incongruently at 1980°C into liquid and bromellite (Galakhov 1957). However, Reeve et al. (1969) could not synthesize this phase at atmospheric pressure, 1480-1600°C, and speculated that it might be either metastable or difficult to nucleate by solid state reaction. Pestryakov et al. (1997) compared the physical properties of $BeAl_6O_{10}$ grown by the Czochralsky-method with those of chrysoberyl.

System BeO-MgO-Al_2O_3

The polysomatic taaffeite-group minerals (Armbruster 2002) include magnesiotaaffeite-2*N*'2*S* (formerly "taaffeite;" end-member: $Mg_3BeAl_8O_{16}$), ferro-taaffeite-6*N*'3*S* (formerly "pehrmanite," end-member: $Fe_2BeAl_6O_{12}$) and magnesio-taaffeite-6*N*'3*S* (formerly "musgravite," end-member: $Mg_2BeAl_6O_{12}$) (see Grew, Chapter 12, this volume) are known in this system. Magnesiotaaffeite-2*N*'2*S* and magnesiotaaffeite-6*N*'3*S* on the pseudobinary chrysoberyl-spinel at 75 and 66.7 mol % spinel, respectively (Fig. 2). Geller et al. (1946) first found a ternary compound with a composition on or near the pseudobinary at $BeO{\cdot}2Al_2O_3{\cdot}MgO$ ($MgBeAl_4O_8$, or 50 mol % spinel), but Reeve et al. (1969) in a reinvestigation of the system claimed the composition to be

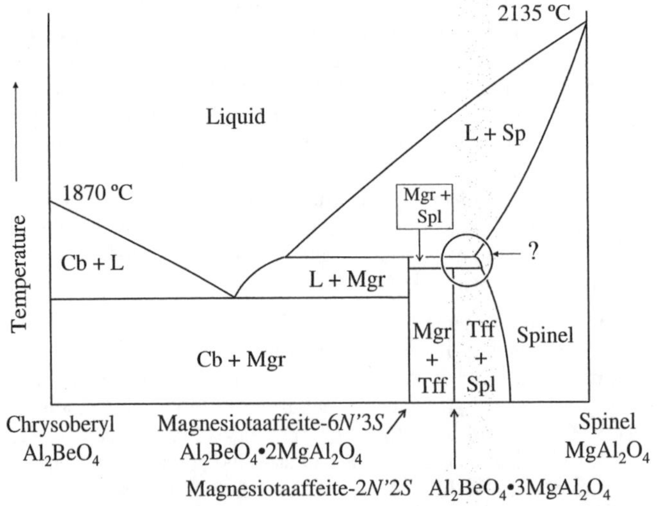

Figure 2. Schematic phase relations for the pseudobinary Al_2BeO_4 (chrysoberyl, Cb)–$MgAl_2O_4$ (spinel, Spl) at high temperature (modified from Kawakami et al. with melting points of chrysoberyl from Lang et al. 1952 and spinel from Deer et al. 1992). Temperatures for the eutectic and peritectic points could not be determined, and available data are insufficient to determine whether magnesiotaaffeite-6*N*'3*S* (Mgr) breaks down at a higher temperature than magnesiotaaffeite-2*N*'2*S* (Tff), as shown (circled area), or vice versa. Figure courtesy of E.S. Grew.

$3BeO{\cdot}8Al_2O_3{\cdot}5MgO$ ($Mg_5Be_3Al_{16}O_{32}$, or 62.5 mol % spinel). This compound is more likely one of the magnesiotaaffeite polysomes, as suggested by later experiments. The Mg end-members have also been synthesized by floating-zone technique (Kawakami et al. 1986; Teraishi 1984), by flux-pulling (Miyasaka 1987; see Schmetzer et al. 1999) and hydrothermally by Hölscher (1987). According to Kawakami et al. (1986), a spinel-phase coexists with melt at an (unspecified) very high temperature (Fig. 2). Spinel shows solid solution towards chrysoberyl and exsolves magnesiotaaffeite lamellae with decreasing temperature, whereas chrysoberyl shows negligible solid solution towards spinel. Magnesiotaaffeite-6*N*'3*S* has a eutectic relation with chrysoberyl and a peritectic one with spinel, but relationships with magnesiotaaffeite-2*N*'2*S* are not known.

System BeO-SiO_2-H_2O-(F)

Phenakite and bertrandite. Besides phenakite and bertrandite, beryllite and sphaerobertrandite belong to this system. Beryllite, for which Strunz gave the formula $Be_3SiO_4(OH)_2{\cdot}H_2O$ (see also the Introduction, this volume), is rare and nothing is known about its stability or attempts to synthesize the mineral. Lahlafi (1997) found beryllite as a run product at the synthesis of Be-micas (see below). Sphaerobertrandite (Semenov 1957) has recently been accepted as a valid mineral species (Introduction, this volume); it had been considered identical to bertrandite (Fleischer 1958; Guillemin and Permingeat 1959). The first synthesis experiments for phenakite were carried out by the French chemist Ebelmen in 1851 from the oxides and borax, and later by several other workers (see Sobolev and Novoselova 1959). Dry synthesis of phenakite from the oxides and with willemite (Zn_2SiO_4) at 1500°C and 1 atm pressure was reported by Morgan and Hummel (1949). Bukin (1967) synthesized phenakite Be_2SiO_4 and bertrandite $Be_4Si_2O_7(OH)_2$ at 270 to 500°C and 720 atm in NaCl solutions. Up to 300°C bertrandite was more abundant than phenakite, at 400 and 500°C phenakite was the dominant phase. Using NaOH-NaCl solutions, he also synthesized a fibrous, spherolitic phase that he equated with sphaerobertrandite because the X-ray powder patterns were identical, and spherical aggregates of unknown composition, his "phase A." Franz and Morteani (1981) were not able to synthesize bertrandite at 300°C and 4 kbar in pure water either from the oxides or from phenakite, though natural bertrandite crystals were used as seeds. At 500°C and 4 kbar phenakite formed in their synthesis experiments from the oxides.

Phenakite dissociates into bromellite and cristobalite at 1560°C, possibly its upper stability limit at 1 atmosphere, and bromellite and cristobalite form a eutectic at ~95 wt % SiO_2 in the system BeO-SiO_2 at a temperature of 1713°C (Morgan and Hummel 1949). At pressures of up to 45.5 kbar phenakite persists up to 2100°C (Munson 1967). Cemic et al. (1986) observed slightly non-stoichiometric synthetic phenakite with 2-3 wt % SiO_2 in excess of the stoichiometric amount and up to 2.5 wt % Al_2O_3 at 1050 to 1200°C, 10 to 20 kbar.

Sobolev and Novoselova (1959) synthesized phenakite by sintering the oxides in quartz ampoules at 650°C to 1300°C in the presence of BeF_2 and alkali fluor-beryllates. Their synthesis produced crystals up to 6-mm long in a few hours, and they concluded that Be migrates readily in the F-bearing gaseous phase within the quartz ampoule.

Dissolution experiments for phenakite in solutions of different composition were carried out by Syromyatnikov et al. (1972) at 0.5 and 1 kbar. Run time was only 5 to 6 hours, and therefore nothing can be said about the state of equilibrium in these experiments. Phenakite dissolves incongruently in pure water according to the reaction phenakite + water = bromellite + SiO_2 in aqueous solution, indicating a much lower solubility for Be than for Si; in NaCl-bearing solutions no significant increase of solubility was observed. Using Na- and K-fluoride and -carbonate increases the solubility and produces melt at 700°C. Wood (1991, 1994) showed in a study of the solubility of phenakite and bertrandite with different ligands that it was fluoride BeF_2 or mixed fluoride-carbonate complexes that are the most efficient at transporting Be. Lebedev and Klyakhin (1983) investigated the crystal habit of phenakite as a function of temperature (500-750°C) and pH of the solution. They found that phenakite changes its habit from short prismatic at pH 8 to long prismatic at pH 0.1 because pH controls the Be and Si ratio in the fluid, independently of temperature and the presence of other cations (Al, Fe, Mg, Li, Na, K).

Geologically, the most important equilibrium in this system is the upper thermal stability limit of bertrandite:

bertrandite = 2 phenakite + water (1)

which Hsu (1983) determined to be at ~350°C and 0.5 to 3.5 kbar. This result agrees with a theoretical analysis by Burt (1978), who placed bertrandite breakdown between the kaolinite-pyrophyllite-quartz (i.e., minimum temperature of ~260 to 300°C) and pyrophyllite-aluminosilicate-quartz equilibria (maximum temperature at ~400 to 450°C). Barton (1986) determined the bertrandite-phenakite equilibrium via the reactions bromellite + aqueous silica = bertrandite, and bromellite + aqueous silica = phenakite, and located the equilibrium at ~240°C, almost independent of pressure between 1 and 8 kbar (see Fig. 3). Previous studies by Ganguly and Saha (1967), who synthesized bertrandite up to 475°C at 1 kbar and transformed beryl glass into phenakite + pyrophyllite at these conditions, had assigned a temperature of 500°C for the equilibrium. Lebedev (1980) determined the upper temperature boundary of F-free bertrandite stability at 450 to 470°C (P_{H_2O} = 1 to 1.5 kbar). The partial substitution of F for OH (x_F = 0.65) raises the temperature to 540°C, but completely substituted F-bertrandite can be synthesized at 1000°C (Rodionov 1986). Hsu (1983) estimated a 20 to 35°C increase of bertrandite stability in HF-rich environments, attributed to F substitution for OH in bertrandite, which was found to be nearly complete in one case.

In summary, the P-T stability of the Reaction (1) remains uncertain, due to the well known kinetic problems in the system at temperatures below 400°C, and probably also due to the poorly defined nature of synthetic phenakite and bertrandite, their possible deviation from the stoichiometric composition, and the possible involvement of sphaerobertrandite. The spontaneous decomposition of bertrandite into phenakite at 450°C and above (Barton 1986) gives a clear upper temperature limit.

System BeO-Al_2O_3-SiO_2-H_2O-(F)

In the ternary subsystem BeO-Al_2O_3-SiO_2 beryl, ideally $Al_2[Be_3Si_6O_{18}]$, is the only known mineral. However, in the presence of H_2O, it must be considered that beryl can incorporate H_2O molecules into the channels of its structure, and therefore the stability will be strongly influenced by the water activity (or P_{H_2O}) in the system (Burt 1978). Consequently, the ternary subsystem is not considered separately. Euclase, $BeAlSiO_4(OH)$, and beryl are the only quaternary minerals (Fig. 4). Synthetic phases of unknown chemical composition, called "hybrid phases" which are possibly Be-substituted Al-silicates (? mullite-type), were found in 1 atm-experiments in the dry system by Miller and Mercer (1965). Franz and Morteani (1981) also found unidentified Be-Al-silicate phases in their hydrothermal experiments aimed to determine the stability of beryl + andalusite = chrysoberyl + quartz, at 500°C and 3 kbar, 54 days run duration (run product shown in their Fig. 9) using a mixture of natural minerals. This phase was observed at temperatures between 500 and 610°C and they proposed a stability field at 500 to 600°C and 2 to 3 kbar. Their synthesis experiments with oxide mixtures at 2 kbar, 550°C also produced unidentified phases. Barton's (1986) experiments at higher pressure did not confirm this, and it seems possible that these hybrid phases are metastable run

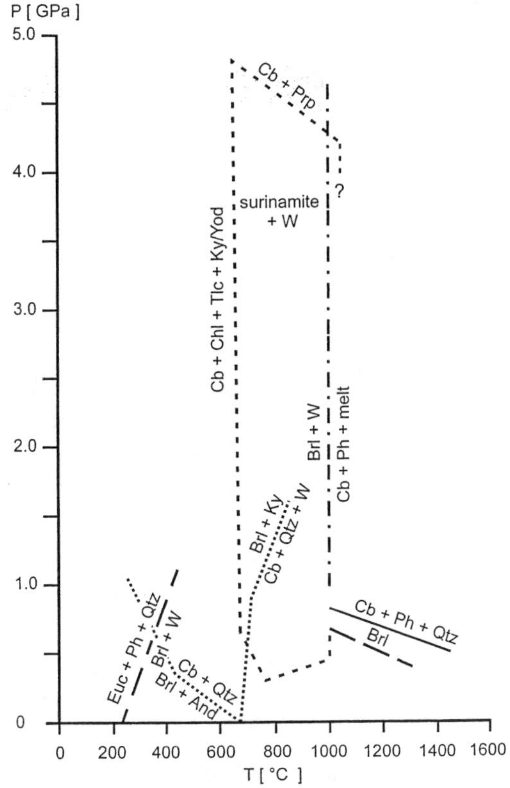

Figure 3. Stability relations of beryl and surinamite. The low temperature stability of beryl with water (dashed line, calculated) and the upper thermal stability of a natural Na-bearing beryl without water (solid line) and the calculated (dashed) line for pure beryl are taken from Barton (1986). The equilibrium beryl + aluminosilicate = chrysoberyl + quartz + water was experimentally determined at 1.5 and 1.7 GPa and thermodynamically extrapolated to lower P by Barton (1986). The dotted line, labeled Brl + Ky = Cb + Qz + W, corresponds to the calculated curve; the experimentally determined breakdown (not shown) for Na-bearing natural beryl is 20 to >40°C lower. The equilibrium beryl + aluminosilicate = chrysoberyl + quartz (dotted line, labeled Cb + Qz = Brl + And) was also calculated by Barton (1986). Kinks in the equilibrium lines correspond to the kyanite-sillimanite and andalusite-kyanite phase transitions. Synthesis experiments (– • – • –); Wilson 1965, Cemič et al. 1986) indicate beginning of melting of hydrous beryl. Surinamite (- - - - -), experimentally determined by Hölscher et al. 1986) is restricted to upper amphibolite and granulite facies temperatures of ≥650°C. However, it is stable up to very high pressure and can be expected in all rocks that have either cordierite or pyrope-almandine in the Be-free system. Abbreviations. And = andalusite, Brl = beryl, Cb = chrysoberyl, Chl = chlorite, Euc = euclase, Ky = kyanite, Ph = phenakite, Prp = pyrope, Qtz = quartz, W = aqueous phase, Yod = yoderite.

products or restricted to low pressure conditions. At atmospheric pressure, mullite can incorporate up to 1.5 mol % BeO at 1500°C, which leads to a reduced c-axis length; a higher amount of BeO leads to the formation of chrysoberyl + glass (Gelsdorf et al. 1958).

Euclase. This mineral was synthesized by Franz and Morteani (1981) and Hsu (1983), who also studied the breakdown reactions

$$20 \text{ euclase} = 3 \text{ beryl} + 7 \text{ chrysoberyl} + 2 \text{ phenakite} + 10 \text{ water} \quad (2)$$

and

$$4 \text{ euclase} + 2 \text{ quartz} = \text{beryl} + \text{chrysoberyl} + 2 \text{ water} \quad (3)$$

at low to intermediate pressure. Preliminary results for Reaction (2) were first reported by Seck and Okrusch (1972), and both reactions were redetermined by Barton (1986) at pressures up to 15 kbar. The results for Reaction (3) are in good agreement; discrepancies for Reaction (2) are in the order of approximately ±50°C at pressures of 5 kbar and below. In any case, euclase decomposes in a P-T-field where (hydrous) beryl is stable. In contrast to bertrandite, F substitution for OH in euclase in HF-rich environments was not found (Hsu 1983).

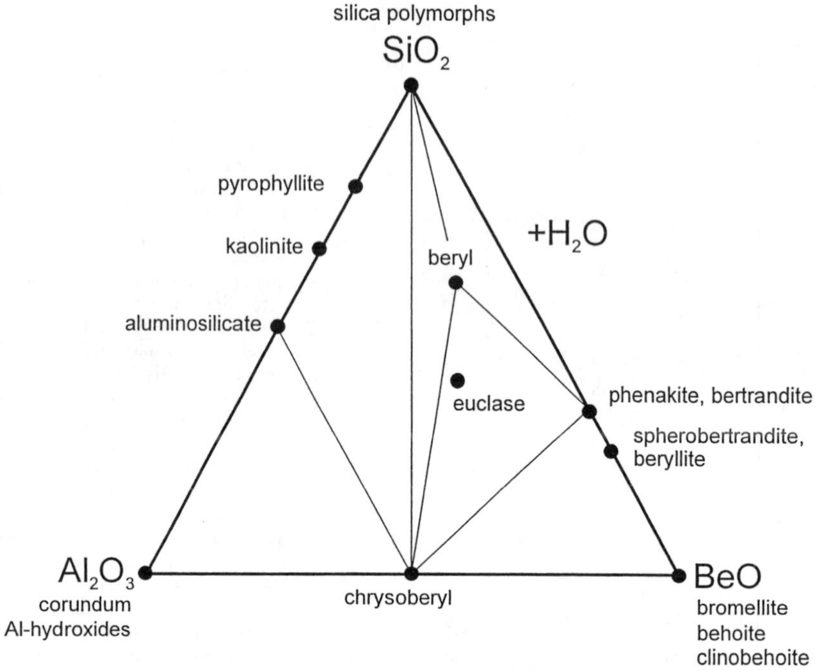

Figure 4. Minerals in the system $BeO-Al_2O_3-SiO_2-H_2O$, projected from H_2O. One set of possible stable tie lines for upper amphibolite facies conditions is shown. Phases with the composition $Be_3Al_2O_6$ (or $3BeO·Al_2O_3$; see text) and $BeAl_6O_{10}$ (or $BeO·3Al_2O_3$) have not been found in nature.

The upper-pressure limit for euclase is not known; Hölscher et al. (1986) reported that euclase persisted in experiments to pressures as high as 25 kbar (at 675°C) and nucleated spontaneously at 550°C and 20 kbar. Overall, the experimental findings are consistent with the formation of euclase in low-temperature and low- to moderate-pressure environments, e.g., in low-T and high-P rocks of the western Alps (to 10-11 kbar, see Grew, this volume).

Beryl. In the anhydrous system $BeO-Al_2O_3-SiO_2$ beryl is the only stable mineral and seems to be nearly pure $Al_2Be_3Si_6O_{18}$. Franz and Morteani (1981) speculated that the substitution 2 Al = Be + Si may be operative to a small extent. However, hydrothermally grown Cr-bearing beryl (1.5 kbar, ~600°C) is very close to the ideal composition with a range of Si between 5.97 and 6.06 apfu and 2.91 and 3.07 Be apfu (Thomas and Klyakhin 1987). At atmospheric pressure beryl decomposes according to the reaction

$$\text{beryl} = \text{chrysoberyl} + \text{phenakite} + 5\ SiO_2 \qquad (4)$$

This reaction was first investigated by Ganguly (1972) using gels and oxide mixtures with synthetic beryl seed crystals. Beryl did not react at 1300°C, the breakdown assemblage formed at 1400°C. However, Miller and Mercer (1965) reported that natural beryl (impurities not specified) begins to decompose at 1100°C with the formation of small amounts of phenakite, melt appeared first at 1460°C, and that extensive melting occurred at 1475°C. The eutectic temperature in the ternary system $BeO-Al_2O_3-SiO_2$ lies at 1515°C (Ganguly and Saha 1965).

The *dry high temperature-high pressure behavior of beryl* was determined by Barton (1986), and he located the Reaction (4) at 5 kbar and 1450°C to 7 kbar and 1150°C (Fig.

3) using natural alkali-bearing beryl in a piston cylinder apparatus, taking special care to drive off all H_2O in the starting and capsule material. The reaction has a negative slope because the relatively loosely packed structure of beryl results in negative ΔV. The results were thermodynamically fit for the pure system, and this extrapolation yielded 1000°C at 6.7 kbar and 1300°C at 4 kbar (Fig. 3; Barton 1986). With an internally consistent thermodynamic data set he calculated the breakdown at 970°C at 1 atm, much lower than the previous studies of Ganguly (1972) and Ganguly and Saha (1965) indicated. A preliminary study of the dry high temperature-high pressure behavior of the variety emerald (Cr bearing beryl) was reported by Wilson (1965). He found melting at 1410°C and 1 atm and about 2000°C and 15 kbar. However, these experiments were designed for crystal growth and not for determining equilibria, and no special care was taken to ensure a completely dry synthesis. In another preliminary study Munson (1967) described the unreversed breakdown of natural (alkali and iron bearing) beryl at 1340°C and 15 kbar and 930°C and 45 kbar, in the latter experiment with the formation of phenakite, chrysoberyl and coesite. These results are in agreement with the data by Barton (1986) that phenakite, chrysoberyl and quartz is the stable high pressure association (Fig. 3). Takubo et al. (1971) synthesized beryl in a piston cylinder apparatus from gel mixtures. Their kinetically oriented study showed that beryl crystallized within 5 min at 10 kbar, 1200°C, and within 20 min at 20 kbar, 1100°C. Melt having the composition of beryl yields a glass showing phase separation at 1800°C (Riebling and Duke 1967) with spinodal decomposition.

The *high temperature-high pressure behavior of beryl at hydrous conditions*, according to Reaction (4), was investigated Franz and Morteani (1981) and Barton (1986). At 1 kbar and 700 to 800°C, 2 kbar and 750°C, 4 kbar and 700°C and 6 kbar and 440 to 500°C, as well as at $P_{H_2O} = P_{total}$ between 10 and 20 kbar and 725 and 825°C they found that beryl is the stable phase. It is clear that water has a large effect on the position of the Reaction (4), modified to (4′)

$$\text{hydrous beryl} = \text{chrysoberyl} + \text{phenakite} + 5\ SiO_2 + \text{vapor} \qquad (4')$$

where the ΔV of solids and fluid and the modification of SiO_2 are the essential factors. An alternative reaction for the breakdown of beryl is

$$2\ \text{beryl} = 3\ \text{phenakite} + 2\ \text{sillimanite} + 7\ \text{quartz} \qquad (5)$$

(Franz and Morteani 1981), but experiments at 1 to 2 kbar and 700 to 800°C showed that beryl is the stable phase. Cemič et al. (1986) investigated the upper thermal stability at $P_{H_2O} = P_{total}$ between 10 and 25 kbar pressure. They found beryl stable up to 1000°C, beryl + chrysoberyl + phenakite + melt between 1000 and ~1100°C, and at higher temperatures only chrysoberyl + phenakite + melt (Fig. 3). They proposed that water will be distributed between melt and beryl thus creating a divariant field for beryl + chrysoberyl + phenakite + melt. In combination with Ganguly's (1972) and Ganguly and Saha's (1965) studies about the subsolidus breakdown of anhydrous beryl (Reaction 4) and the melting of chrysoberyl + phenakite + cristobalite = liquid they argued that below 10 kbar the Reaction (4′) must lie between 1000 and 1100°C.

Pankrath and Langer (2002) determined experimentally the water content in beryl as a function of P_{H_2O} up to 14.5 kbar and 400 to 900°C. The hydration behavior of beryl according to Reaction (6),

$$Al_2Be_3Si_6O_{18} + n\ H_2O = Al_2Be_3Si_6O_{18}\cdot nH_2O, \qquad (6)$$

with $0 < n < 1$ is very similar to that of cordierite. Hydration of beryl expands the structure in *c*-direction by 0.07 % and the cell volume by 0.1 % for n = 0.96. Raising water contents were found to enhance the refractive indices of beryl from $n_o = 1.5585$ to 1.5820 and n_e from 1.5617 to 1.5872, Δn from 0.0032 to 0.0052. At 14.5 kbar, the maximum amount of nearly 1 H_2O pfu was incorporated, at intermediate conditions of 5 kbar and 600°C, 0.65 H_2O pfu was incorporated. At a constant pressure of 5 kbar, the wt % H_2O content decreases from 2.4 at 400°C to 1.5 at 1000°C. Barton (1986) discussed in detail the consequences of hydration of beryl on its stability relations, based on (1) an ideal solution model between end-member hydrous and anhydrous beryl, (2) a zeolitic water model and (3) a non-energetic, volume-only interaction model. He pointed out that the influence of hydration is more important in beryl breakdown reactions than in the breakdown reactions of cordierite. ΔV of the solids is much more pronounced in reactions with cordierite, where Al changes its coordination from 4 to 6, whereas for Reactions (4) and (4′) with beryl Al does not change its coordination. Solntsev et al. (1978) found that the presence of Cs^+ substantially reduces the amount of H_2O in the beryl channels. Ammonium (together with H_2O) can also be an important constituent in beryl synthesized at 1 kbar and 600°C in the presence of NH_4Cl (Mashkovtsev and Solntsev 2002). In commercially available synthetic Biron and Regency beryls they interpreted from IR-spectroscopy that HCl (together with H_2O) is also present.

Other high pressure synthesis experiments were reported by Kodaira et al. (1982) at 10 kbar, 500 to 750°C, using oxide mixtures and pure water as well as NaOH-bearing solutions of different concentrations. Beryl forms, but with increasing NaOH concentration and with increasing temperature, phenakite forms besides beryl, and with a 1 N solution at 700°C, phenakite is the dominant phase.

Dissolution experiments for beryl in solutions of different composition were carried out by Syromyatnikov et al. (1972) at 0.5 and one kbar. Run time was only 5 to 6 hours, and therefore nothing can be said about the state of equilibrium in these experiments. Beryl dissolves incongruently in pure water according to the reaction beryl + water = chrysoberyl + phenakite + SiO_2 in aqueous solution, indicating a much lower solubility for Al and Be than for Si; in HCl-bearing solutions no phenakite formed. In 0.5 M NaCl and KCl solutions the solubility of beryl increases and other non-identified phases formed, at 800°C beryl melts. Using NaF and KF reduces the melting temperature to 700°C, with Na_2CO_3 and K_2CO_3 different breakdown products were observed. Thomas and Lebedev (1982) studied the kinetics of the hydrothermal crystallization of beryl (600°C at 1 kbar) using oxide mixtures with 50 wt % solution of Na- and Li-fluoride to explain the observation that in acid chloridic solutions chrysoberyl and phenakite form, but in fluoride solutions beryl is the dominant run product. They observed that beryl does not form directly, but via bertrandite-topaz. Betrandite disappears after 2 days, and after ~12 days the maximum yield of beryl (together with phenakite, topaz, cryolite, quartz) is reached.

The *lower temperature stability of beryl* was studied by Lebedev (1980) at temperatures below 500°C and pressures below 1.5 kbar. He concluded that the lower temperature boundary of beryl under hydrothermal conditions is near 300°C according to the reaction

3 bertrandite + 4 pyrophyllite + 2 quartz = 4 beryl + 7 H_2O.

The equilibrium temperature will be higher if F substitutes for OH in bertrandite.

Hydrothermal synthesis of beryl and emerald with trace impurities and beryl analogues. Wyart and Ščavničar (1957) first showed that beryl can easily be synthesized from a stoichiometric mixture of its compounds (Be carbonate as the Be source) at 400 to 600°C and 0.4 to 1.5 kbar. The green variety of beryl, generally Cr-bearing, but with contributions from V and Fe to the color, has been synthesized hydrothermally in several industrial laboratories (see above). Seed plates are generally used, and Cl-bearing solutions (e.g., alkali or ammonium chloride) were used as a mineralizing agent. Unpublished data (Franz and Morteani) on synthesis runs at hydrothermal conditions with pure water (0.5 to 2 kbar and 500 to 800°C) also showed that Cr-bearing beryl can easily be synthesized from oxide mixtures. However, due to the sluggish reaction behavior of the system it was not possible to determine the amount of solid solution for Al-Cr substitution as a function of P and T. Thomas and Klyakhin (1987) found that the substitution of Cr for Al in hydrothermally grown beryl (with up to 1.1 wt % Cr_2O_3 [= 0.08 Cr apfu]; 1.5 kbar, ~600°C, from oxide mixtures; run products analyzed by wet chemical methods) is accompanied by a complex substitution of Si for Al, with charge compensation by Be, or by Be plus Na or Be plus Li. However, the maximum of Si excess is only 0.06 apfu. Much of the information on emerald synthesis is for gemological purposes such as the distinction between natural and synthetic emerald (e.g., on the basis of electron microprobe analysis from the presence of Na, Mg, or Cl, or on inclusions and growth features; Hänni 1982; Schmetzer 1996; Schmetzer et al. 1998) and the reader is referred to this literature.

The incorporation of unspecified amounts of V^{3+}, Mn^{3+}, Ni^{3+} and Co^{3+} was tested in hydrothermal synthesis experiments from the oxides and $BeCO_3$, together with minerals quartz, topaz and pyrophyllite at ~1.5 kbar and 600°C, in the presence of B-bearing solutions (Emel'yanova et al. 1965). Single crystals of 1 to 1.5 mm formed spontaneously or as overgrowths on seed crystals in the same order of magnitude. Based on the absorption spectra, the authors argue for isomorphous substitution of unspecified trace amounts. Taylor (1967) also reported the synthesis of V emerald.

The Sc-dominant analogue of beryl, bazzite, ideally $Sc_2Be_3Si_6O_{18}$, with various amounts of trivalent Fe, Cr, Mn, V and Ga substituting for Sc^{3+} has been synthesized hydrothermally from gels at 450 to 750°C, 2 kbar, and also by sintering at 1020°C (Frondel and Ito 1968). It was possible to synthesize end-member bazzite and bazzite in which up to 0.5 cations pfu of Sc are replaced by Fe^{3+}, 0.67 by Cr^{3+} and 0.34 by Ga^{3+} cations, but no other end-member could be synthesized (cf. stoppianiite, the Fe^{3+}-dominant analogue of beryl, with an octahedral composition $Fe^{3+}_{2.70}Al_{0.42}Sc_{0.02}R^{2+}_{0.70}$; Della Ventura et al. 2000). Synthesis of the In-analogue of beryl was reported by Ito (1968) at 420 to 600°C, 2 to 3 kbar, but the run products also contained several unidentified phases.

Incorporation of Li into beryl at 600°C and 1 kbar (Klyakhin and Il'in 1991) is possible by the substitution

$$^{IV}Be + {}^{ch}\square = {}^{IV}Li + {}^{ch}Na + 2\,{}^{ch}H_2O.$$

The H_2O content of the Li-beryls is positively correlated with Li, but not with other cations K, Rb and Cs (Lebedev et al. 1988). Alkalis (R^+) can also be incorporated into the channels by the substitutions

$$^{IV}Si = {}^{IV}Al + {}^{ch}R^+$$

$$^{VI}Al^{3+} = {}^{VI}Fe^{2+}, Ni^{2+}, Mg^{2+}, Co^{2+}, Mn^{2+} + {}^{ch}R^+$$

(Lebedev et al. 1988; Klyakhin and Il'in 1991).

Phase equilibria among Be-minerals. Not much information is available from experimental studies. An exception is Reaction (7),

$$\text{beryl} + 2\ \text{aluminosilicate} = 3\ \text{chrysoberyl} + 8\ \text{quartz} \qquad (7)$$

because in natural rocks, chrysoberyl frequently occurs together with quartz in a stable assemblage. It was first studied by Seck and Okrusch (1972) at 2 kbar, then by Franz and Morteani (1981) between 2 and 6 kbar, and by Barton (1986) at 15 to 17 kbar (Fig. 3).

Chrysoberyl + quartz is the high temperature-low pressure assemblage, but Barton (1986) showed that the reaction is extremely dependent on the hydration state of beryl, which strongly influences the ΔV of the reaction. At low pressures of 2 to 3 kbar, Franz and Morteani (1981) found an unidentified Be-Al-silicate ("hybrid phase") in their run products which was not found by Barton (1986) at high pressures, and it is possibly metastable (see above). If, however, such a phase exists, the intersection of the univariant curves may lie at 4 to 5 kbar and 400 to 450°C. It must be mentioned that the reaction is very sluggish: In most runs reported by Franz and Morteani (1981) the reaction progress is small, even at a run duration of 30 to 70 days. Only at temperatures of 600 and 650°C at 4 kbar, almost complete reaction to chrysoberyl and quartz was found. Hölscher et al. (1986) also found that it was difficult to bracket the reactions in the system with BeO and Al_2O_3, even at temperatures as high as 950°C. Their synthesis runs for surinamite yielded chrysoberyl + quartz at 7 and 9 kbar and 600°C, which is compatible with Franz and Morteani's (1981) and Barton's (1986) data.

System BeO-MgO-Al_2O_3-SiO_2-H_2O

Surinamite, khmaralite, Be-sapphirine and Be-Mg-cordierite are quarternary and quinary phases in this system (see Grew this volume; Hawthorne and Huminicki this volume). Hölscher and Schreyer (1989) synthesized and investigated the stability range of surinamite, a chain silicate with the formula $(Mg_3Al_3)O[AlBeSi_3O_{15}]$. Electron microprobe analyses of one run product yielded near-ideal stoichiometry in terms of Mg, Al and Si, while X-ray data suggest close approach to the near-perfect tetrahedral order observed in natural material. The absence of changes in d-values and intensities in X-ray diffractograms of run products synthesized over a wide range of pressure and temperature suggest negligible change in composition. Surinamite is restricted to pressures above 4 kbar and is stable up to ~45 kbar, where it decomposes into chrysoberyl + pyrope (Fig. 3). It is also a high temperature mineral and forms from chrysoberyl + chlorite + talc + yoderite at 650-700°C. At pressures below 4 kbar, chrysoberyl, cordierite, sapphirine and enstatite constitute the alternative paragenesis to surinamite.

Povondra and Langer (1971b) found sapphirine in their attempt to synthesize Be-cordierite and suggested that it contains significant amounts of Be. Hölscher (1987) showed that sapphirine with more than one Be per 20 oxygen could not be synthesized at T from 700 to 1350°C and P from 1 to 13 kbar, and khmaralite in Be-rich assemblages contains only 0.77 Be per 20 oxygen. The influence of Be substitution on the stability range of both minerals remains to be determined. Christy et al. (2000) gave a first structural characterization of synthetic Be-Mg-sapphirine with approximately one Be pfu: it primarily consisted of the 1A polytype (with lamellae of 2M sapphirine) in which the dominant substitution vector was $(BeSi)(AlAl)_{-1}$.

System Na_2O-BeO-MgO-Al_2O_3-SiO_2-H_2O

The problem of solid solution between isostructural beryl and indialite and cordierite is illustrated in Figures 5 and 6 and in Table 1. The end-member Be-Mg-cordierite was synthesized by Hölscher and Schreyer (1989), whereas Demina and Mikhailov (1993, 2000) and Mikhailov and Demina (1998) synthesized complex solid solutions and called the hypothetical end-members 'Be-indialite' and 'Newton component'. The complexity arises from the fact that the substitution on the tetrahedral sites involves three cations with different charges 2^+, 3^+ and 4^+ in addition to the vacancy position in the channel, where the large Na with 1^+ charge can be situated. It also seems conceivable that at high temperatures above 1000°C, tetrahedral vacancies exist. Figure 5 illustrates how the different end-members are related by the exchange vectors $BeMg_{-1}$, $NaBe(\square Al)_{-1}$ and $Al_2(BeSi)_{-1}$.

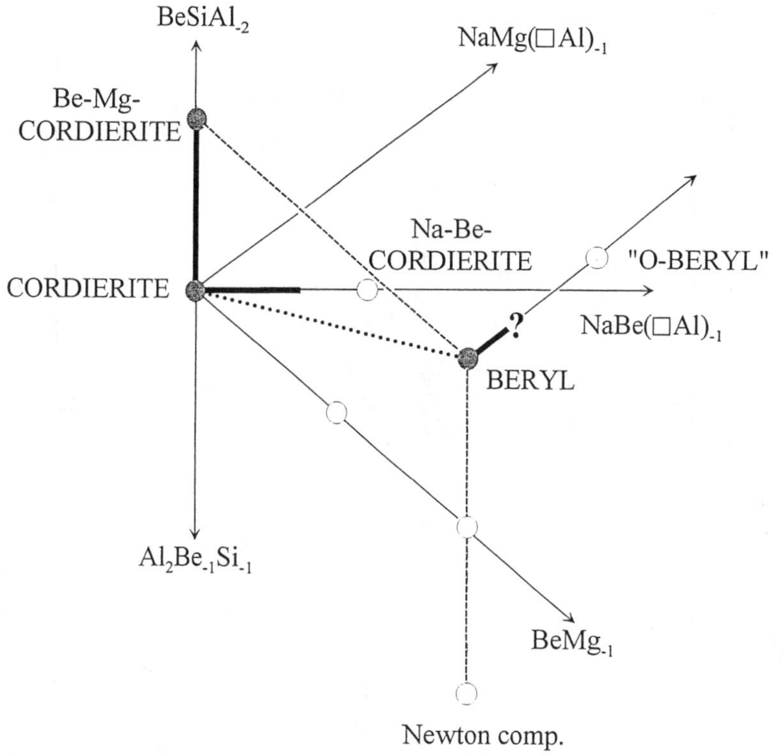

Figure 5. Substitution vectors and compositional relationships among known synthetic and natural phases (capitals and filled symbols) and hypothetical compounds related to beryl and cordierite in the system Na_2O-BeO-MgO-Al_2O_3-SiO_2. The extent of substitutions determined experimentally is schematically indicated by heavy lines. Dotted lines mark joins where less than 5% solid solution has been found. For complete solid solution between cordierite and beryl in granitic melt, see London and Evenson (this volume). Figure courtesy of E.S. Grew.

Figure 6 shows the compositional effect of the substitutions listed in Table 1 in ternary projections. The projection BeO-Al_2O_3-SiO_2 would allow to distinguish between the substitutions which include $MgBe_{-1}$ and $(\square Si)_1Be_{-2}$, from substitutions $NaBe(\square Al)_{-1}$, $NaMg(\square Al)_{-1}$, $NaMg_2Al(\square Be_{-2}Si_{-1})$ and from $Al_2(BeSi)_{-1}$. In the projection MgO-Al_2O_3-SiO_2 analytical uncertainty would probably blur any significant trend in both beryl and cordierite, except for $MgBe_{-1}$. The conversion from analytically determined wt % contents into cations per formula unit results in a very large error for the light element Be, especially if other light elements such as Li can be present, and to solve the problem which substitutions are involved requires high accuracy in the analyses. It also requires a microbeam technique, because it is known from natural crystals (see below) that zoning is an important and frequently observed phenomenon. Unfortunately there are no such data published from natural and only a few data of synthetic beryls to be plotted in these diagrams, which would allow one to derive crystal chemical conclusions.

The complexities of the substitution mechanisms are probably the reason why the first reports of considerable solid solution between cordierite and beryl (Newton 1966; Borchert et al. 1970) were erroneous and represented metastable conditions. Povondra and Langer (1971a) redetermined the extent of solid solution and found only ~4 mol %

Table 1. End-members with cordierite–indialite–beryl structures in the system Na_2O-BeO-MgO-Al_2O_3-SiO_2

End-member	Formula	additive component substitution	${}^{VI}Mg_2[Al_4Si_5O_{18}]$ exchange vector	additive component substitution	${}^{VI}Al_2[Be_3Si_6O_{18}]$ exchange vector
Orthorhombic					
cordierite	${}^{VI}Mg_2[Al_4Si_5O_{18}]$			$2\,{}^{VI}Al + 3\,{}^{IV}Be + {}^{IV}Si = 2\,{}^{VI}Mg + 4\,{}^{IV}Al$	$Be_{-3}Si_{-1}Mg_2Al_2$
Na-Be-cordierite	${}^{ch}Na\,{}^{VI}Mg_2[BeAl_3Si_5O_{18}]$	${}^{ch}\square + {}^{IV}Al = {}^{ch}Na + {}^{IV}Be$	$NaBeAl_{-1}$		
Hexagonal					
indialite	${}^{VI}Mg_2[Al_4Si_5O_{18}]$			$2\,{}^{VI}Al + 3\,{}^{IV}Be + {}^{IV}Si = 2\,{}^{VI}Mg + 4\,{}^{IV}Al$	$Be_{-3}Si_{-1}Mg_2Al_2$
Be-Mg-cordierite[1]	${}^{VI}Mg_2[BeAl_2Si_6O_{18}]$	$2\,{}^{IV}Al = {}^{IV}Be + {}^{IV}Si$	$BeSiAl_{-2}$	$2\,{}^{VI}Mg + 2\,{}^{IV}Al = 2\,{}^{VI}Al + 2\,{}^{IV}Be$	$MgBe_{-1}$
beryl	${}^{VI}Al_2[Be_3Si_6O_{18}]$	$2{}^{VI}Mg + 4{}^{IV}Al = 2{}^{VI}Al + 3{}^{IV}Be + {}^{IV}Si$	$Be_3SiMg_{-2}Al_{-2}$		
o-beryl[2]	${}^{ch}Na\,{}^{VI}MgAl[Be_3Si_6O_{18}]$			${}^{ch}\square + {}^{VI}Al = {}^{ch}Na + {}^{VI}Mg$	$NaMgAl_{-1}$
vacancy-beryl	${}^{VI}Al_2[\square BeSi_7O_{18}]$			$2\,{}^{IV}Be = {}^{IV}Si + {}^{IV}\square$	$SiBe_{-2}$
unnamed	${}^{VI}Mg\,{}^{VI}Al[BeAl_3Si_5O_{18}]$	${}^{VI}Mg + {}^{IV}Al = {}^{VI}Al + {}^{IV}Be$	$BeMg_{-1}$	${}^{VI}Al + 2\,{}^{IV}Be + {}^{IV}Si = {}^{VI}Mg + 3\,{}^{IV}Al$	$MgAl_2Be_{-2}Si_{-1}$
Newton component[3]	${}^{VI}Al_2[BeAl_4Si_4O_{18}]$	$2{}^{VI}Mg + {}^{IV}Si = 2\,{}^{VI}Al + {}^{IV}Be$	$Al_2BeMg_{-2}Si_{-1}$	$2\,{}^{IV}Be + 2\,{}^{IV}Si = 4\,{}^{IV}Al$	$Al_2Be_{-1}Si_{-1}$

Notes: **Bold** indicates an independent phase, ch = located in the channel, □ = vacancy
1 = Be-indialite; 2 = (Na-Mg-beryl); 3 defined by Demina and Mikhailov (1993).

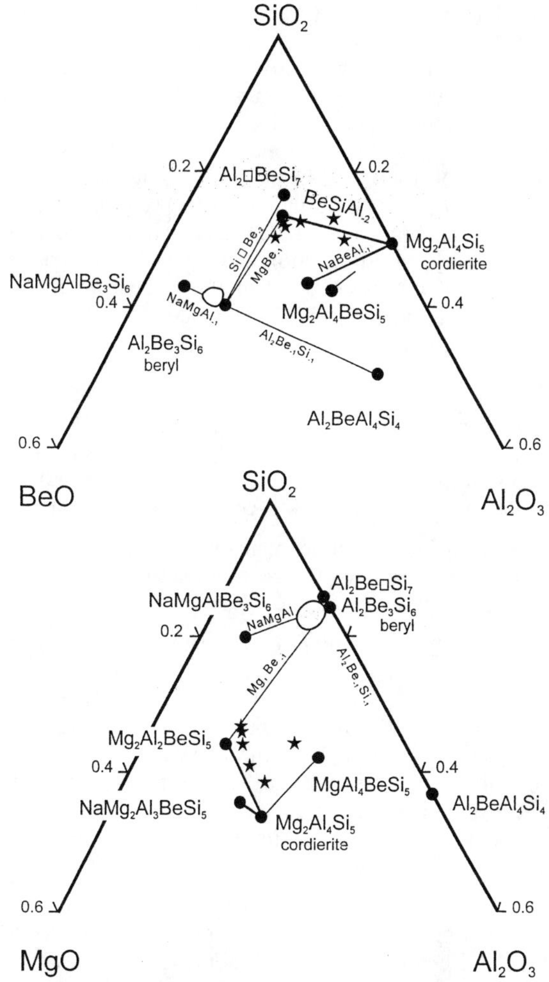

Figure 6. Projection of various beryl and cordierite end-member compositions in the system MgO-Al_2O_3-SiO_2 (a) and BeO-Al_2O_3-SiO_2 (b) due to substitutions Be-Al-Si, including Mg, Na and vacancies. Cation contents in the formulae are given for 18 oxygens. Heavy lines indicate complete solid solution according to the experimental data by Hölscher and Schreyer (1989) and Povondra and Langer (1971a). Asterisks are compositions of synthetic phases (Mikhailov and Demina 1998) and the enlarged field around beryl indicates the compositional variation in synthetic beryl according to Demina and Mikhailov (2000).

beryl in cordierite at conditions of 1 atm and 1300°C and 3 kbar and 750°C. In contrast, the solid solution between Be-Mg-cordierite and indialite (hexagonal cordierite) appears to be complete at 1 atm and 1350-1450°C (Hölscher and Schreyer 1989). Be-Mg-cordierite is hexagonal, whereas cordierite changes its structure at ~1450°C from hexagonal to orthorhombic, and therefore towards lower temperature (e.g., 1 atm at 850°C) and higher pressure (e.g., 5 kbar P_{H_2O}) a large miscibility gap opens. Beginning of melting (glass + cristobalite for Be rich assemblages, glass + mullite for cordierite rich assemblages identified) of Be-Mg-cordierite starts at 1350°C and 1 atm (Hölscher and Schreyer 1989), which is the same range of temperature as for beryl.

Demina and Mikhailov (1993, 2000) and Mikhailov and Demina (1998) conducted synthesis experiments at atmospheric pressure in the temperature range 950°C to 1400°C, from melts, fluxes and solid reactions, partly producing zoned crystals by varying the temperature during the synthesis runs. Run products were analyzed by electron microprobe and partly by spectrometry for Be. Their kinetically oriented study (Demina and Mikhailov 1993) indicated that Be-indialite (equivalent to Be-Mg-cordierite of Hölscher and Schreyer 1989) forms via enstatite, forsterite and cristobalite, and that after ~200 hours at ≥1340°C, 80-90 wt % Be-indialite was produced. The compositions given are plotted in Figure 6 and indicate a rather large compositional field for the solid solutions. In contrast, beryl solid solution with Mg-compounds is rather restricted (Demina and Mikhailov 2000). They found that all Mg-bearing beryls have some deficit in Be (≥2.64 pfu), but Si is near to the ideal value of 6.0. Up to 0.3 Mg pfu can be incorporated by the substitution

$$^{IV}Al + {}^{VI}Mg = {}^{IV}Be + {}^{VI}Al,$$

which includes in some cases also up to 0.25 Cr pfu and 0.28 V pfu. Ca, added as $CaCl_2$ to the run charge, can not be incorporated in beryl. Preliminary results of hydrothermal synthesis experiments from oxide mixtures (Franz 1982) also indicate that at P_{H_2O} = 2 kbar, 750°C solid solution in beryl with a Mg-compound is less than 5 mol %.

The relatively high MgO- and FeO-contents known in natural beryl as well as BeO-contents in natural cordierite are mostly accompanied by an alkali ion (predominantly Na) as a charge balance:

in beryl: $^{ch}Na + {}^{VI}(Mg, Fe) = {}^{ch}vacancy + {}^{VI}Al$

in cordierite: $^{ch}Na + {}^{IV}Be = {}^{ch}vacancy + {}^{IV}Al$

The first substitution leads to a theoretical end-member $Na^{VI}(MgAl)[Be_3Si_6O_{18}]$, the so-called octahedral beryl substitution in natural material ("o-beryl;" see Černý, this volume). Hydrothermal experiments by Franz and Morteani (unpublished data) at 2 kbar, 750°C, starting from gel mixtures with a composition on the join $Al_2Be_3Si_6O_{18}$-$NaMgAlBe_3Si_6O_{18}$, yielded beryl (+ phenakite + quartz) at a starting composition of up to ≈ 0.2 Na pfu, beryl + a cordierite like phase (+ phenakite + quartz) up to ~ 0.5 Na pfu, but no beryl at higher concentrations of Na.

The second substitution leads to a theoretical end-member, $Na^{VI}Mg_2[Al_3BeSi_5O_{18}]$, which may be called Na-Be-cordierite (Fig. 5). Solid solution between cordierite and Na-Be-cordierite was investigated by Povondra and Langer (1971b). At one atm, they found up to 50 mol % of Na-Be-cordierite in cordierite at temperatures of 1000 to 1300°C. The end-member Na-Be-cordierite could not be synthesized, cordierite solid solution + nepheline + an unidentified phase formed instead. Solid solution increases with increasing pressure from 50 mol % at 1 kbar P_{H_2O} to 60% at 3 kbar P_{H_2O}, with increasing temperature from 40 to 50 mol %, 600 to 1000°C at 1 kbar. At 3 kbar, no significant increase with temperature was observed, above 900°C solid solutions of cordierite with Na-Be-cordierite decompose into spinel, sapphirine and melt. In contrast to the simplified systems described above, where a miscibility gap between beryl and cordierite exists, London and Evenson (this volume) found indications for complete solid solution of beryl and cordierite coexisting with granitic melt at T > 900°C and 2 kbar. In this complex system they found a strongly asymmetric solvus between 675 and 900°C, thus confirming the earlier studies that beryl is generally more restricted in chemical composition than cordierite.

System Na_2O-K_2O-BeO-Al_2O_3-SiO_2-H_2O-(F)

Beus et al. (1963) showed experimentally at 490 to 540°C in Morey's type autoclaves (unconstrained, probably low pressure) that alkali-feldspar reacts with solutions of Na-K-fluoroberyllate and forms beryl and albite, thus demonstrating that beryl formation is possible in a post-magmatic stage during the albitization of alkali-feldspar. Solutions, which were acid at the beginning of the experiment, became basic (pH 8 to 12) at the end of the experiment.

Franz and Morteani (1984) conducted a few hydrothermal experiments between 1.5 kbar and 5 kbar, 630 to 850°C, to determine the melting reaction

alkali-feldspar + beryl + water = phenakite + chrysoberyl + melt.

They concluded that the reaction lies in the P-T range between the reactions K-feldspar + albite + quartz + water = melt and K-feldspar + quartz + water = melt.

System Li_2O-K_2O-BeO-MgO-Al_2O_3-SiO_2-H_2O-F

Synthetic Be micas were investigated by Robert et al (1995) and Lahlafi (1997), with the emphasis on the solid solution between Li micas and hypothetical Be micas. They synthesized hydrothermally at 500°C, 1 kbar, three to four weeks run duration, starting from gels. The end-member compositions

$KMg_3[Si_{3.5}Be_{0.5}O_{10}(OH)_2]$, $KMg_2Al[Si_3BeO_{10}(OH)_2]$ and $KAl_2Li[S_{i3}BeO_{10}(OH)_2]$ yielded single-phase run products as indicated by optical, XRD and IR investigations. Run products are very fine grained, however, and not suitable for microbeam chemical investigations.

Bulk compositions of $NaMg_2Al[Si_3BeO_{10}(OH)_2]$ and $NaAl_2Li[Si_3BeO_{10}(OH)_2]$ (Na-micas) did not yield single phase run products, but a mica-phase together with either beryllite and magnesiotaaffeite-6*N*'3*S*, and beryllite, eucryptite and Li-Be-silicate, respectively. Solid solution between $KMg_3[Si_{3.5}Be_{0.5}O_{10}(OH)_2]$ and $KMg_{2.5}[Si_4O_{10}(OH)_2]$ is complete and accompanied by additional protons, up to $(OH)_{2.5}$ for the Be-end-member, as determined by thermogravimetry. The most likely proton acceptors are the strongly underbonded apical oxygens of $^{IV}Be^{2+}$.

Solid solution between $KMg_3[Si_{3.5}Be_{0.5}O_{10}(OH)_2]$ and $KMg_2Li[Si_4O_{10}F_2]$ (tainiolite) is also complete, but the lattice constants do not change continuously. A discontinuity at 60 to 70 mol % of the K-Li-F-end-member is explained by local order-disorder phenomena. The K-Li-OH-end-member could not be synthesized, but bulk compositions with up to ~50 mol % yielded single phase run products. A similar result was obtained for the solid solution with polylithionite $KLi_2Al[Si_4O_{10}F_2]$; up to 80 mol % of the F-end-member can be incorporated, the (OH)-end-member can not be synthesized and solid solutions with the Be-mica yielded only minor amounts of mica.

Solid solution of F-dominant zinnwaldite $KAlMgLi[Si_3AlO_{10}F_2]$ with KAl_2Li-$[Si_3BeO_{10}F_2]$ yielded single-phase run products at bulk compositions with up to 90 mol % of the Li-F-end-member, and solid solution between F-dominant zinnwaldite $KAlMgLi[Si_3AlO_{10}F_2]$ and $KMg_2Al[Si_3BeO_{10}F_2]$ seems to be complete.

System BeO-ZnO-SiO_2

This system received attention because of the structural similarity of phenakite with willemite and their use as phosphors (Hund 1970; Fonda 1941). Hahn and Eysel (1970) concluded that despite of the structural similarities, phenakite does not take up any appreciable amount of Zn_2SiO_4, but that willemite incorporates up to 16 mol % Be_2SiO_4 at 1200°C and 22 mol % Be_2SiO_4 at 1340°C. The asymmetric solvus is due to the fact that a smaller cation can more readily substitute for the larger than the converse. Sharma and Ganguli (1973) and Chatterjee and Ganguli (1975) reinvestigated the system and found up to 30 mol % Be_2SiO_4 in willemite at 1340°C, where the eutectic temperature of the system is reached.

System BeO-MnO-FeO-ZnO-SiO_2-S_2

Furzhenko and Klyakhin (1981) and Furzhenko (1982, 1989) investigated synthesis conditions, crystal habit and the solid solution between helvite $Mn_4Be_3(SiO_4)_3S$, danalite $Fe^{2+}{}_4Be_3(SiO_4)_3S$ and genthelvite $Zn_4Be_3(SiO_4)_3S$ in dependence of oxygen fugacity, NaOH and HCl in the solution. Danalite and genthelvite can only be synthesized in solutions of halogen acids and NH_4F. In F-containing solutions, danalite formation is accompanied by fayalite formation and a Be-Fe-F-silicate. Genthelvite can be synthesized in basic solutions, in halogen acids and in alkali halogen salt solutions. Willemite and a Zn silicate form as well under these conditions. At 600°C and ~0.8 to 1 kbar pressure (with a temperature gradient of 50°C), an oxide mixture of the composition 1:1:1 of the three end-members and a starting fluid:solid ratio of 1:1 yielded ~40 mol % of both the Mn- and Zn-end-members and ~20 mol % of the Fe-end-member in the presence of a 1 molar HCl, and 40 % of the Mn-end-member, 60 % of the Zn-end-member at a composition of 1 molar HCl, 2 molar NaOH, as indicated by investigation of the run products by electron microprobe. Danalite could be synthesized at temperatures above 500°C, genthelvite above 400°C, and helvite above 300°C. Danalite crystallizes only at oxygen fugacities below the hematite-magnetite buffer, helvite below the Mn_3O_4-Mn_2O_3 buffer.

Systems BeF_2—fluorides of alkalis and alkaline earths

Hahn (1953) investigated the systems of BeF_2 with K, Na, Li and Ca fluorides as structurally analogous systems for silicates, e.g., BeF_2 for SiO_2 or Li_2BeF_4 for Mg_2SiO_4, based on the assumption that the cation and anion charge ratios and ionic ratios are similar (see also Bell 1972; Ross 1964 for Be-halides and their crystal chemistry). These compounds have melting temperatures on the order of 400 to 600°C. According to his experiments at atmospheric pressure, Li_2BeF_4 and $NaLiBeF_4$ have a phenakite-willemite structure, whereas Na_2BeF_4 is isotypic with olivine; K_2BeF_4 is isotypic with β-KSO_4. $Na_2LiBe_2F_7$, with melilite and $CaBeF_4$, is isotypic with zircon.

OCCURRENCE IN METAMORPHIC ROCKS

Metamorphosed pegmatites, granites and silicic volcanoclastic rocks

The assemblage chrysoberyl + quartz is a characteristic type of Be-mineralization in some pegmatites, described as "desilicated" pegmatites due to metasomatic exchange with SiO_2-poor country rocks or assimilation of Al_2O_3-rich country rocks (Beus 1966; Heinrich and Buchi 1969; Okrusch 1971). Franz and Morteani (1984) investigated some of these pegmatites (Kolsva, Sweden; Maršíkov, Czech Republic; Haddam, Connecticut) and emphasized the importance of a regional metamorphic overprint for the formation of chrysoberyl. However, Reaction (7), which involves an Al-silicate, is not directly applicable because Al-silicate is generally not a primary phase coexisting with beryl in the pegmatites. Therefore Franz and Morteani (1984) postulated, based on the reaction textures, the breakdown of beryl with alkali-feldspar in a hydrous fluid according to Reaction (9):

$$\begin{aligned} &1\ \text{beryl} + 4\ \text{(K,Na)-feldspar} + 4\ H^+ = \\ &\quad 3\ \text{chrysoberyl} + 14\ \text{quartz} + 4\ (K^+,Na^+) + 2\ H_2O \end{aligned} \tag{9}$$

or in combination with white mica according to Reaction (10):

$$\begin{aligned} &1\ \text{beryl} + 4\ \text{albite} + 2\ \text{muscovite} + 6\ H^+ = \\ &\quad 3\ \text{chrysoberyl} + 21\ \text{quartz} + 4\ Na^+ + 2\ K^+ + 5\ H_2O \end{aligned} \tag{10}$$

A reaction of beryl + alkali-feldspar to chrysoberyl + quartz + melt also seemed possible, because the P-T conditions of the metamorphism of these pegmatites were in the high grade amphibolite facies, but experimental investigations (Franz and Morteani 1984) showed that chrysoberyl + phenakite are the liquidus phases, produced by the melting of alkali-feldspar + beryl, not chrysoberyl + quartz. A retrograde reaction analogous to Reaction (10) can be observed in the rocks, with the formation of secondary beryl in pseudomorphs of white mica after chrysoberyl. For the Maršíkov pegmatites in Czech Republic, Černý et al. (1992) essentially confirmed the concept of metamorphic overprint on pegmatitic protoliths. They derived P-T conditions of approximately 600±50°C and 3±1 kbar for the metamorphism. The metamorphism occurred in two stages, which could be differentiated by different stress conditions, and which produced large chrysoberyl without preferred orientation in stage I and recrystallized small chrysoberyl, well aligned with fibrous sillimanite in stage II.

Similar observations led Gonzalez del Tanago et al. (1984) to conclude also that metamorphism was the essential chrysoberyl-forming process in granitic pegmatites in the polymetamorphic area of the Sierra Albarrana, Spain. The authors describe replacement textures of primary beryl by chrysoberyl + quartz. Primary aluminosilicate is absent, but in other pegmatites from the same area, the transformation of andalusite into sillimanite was described, and in combination with other mineral equilibria the authors concluded that an increase in P and T to ~550°C and 4 kbar after the cooling of the pegmatite produced chrysoberyl. They also observed the formation of secondary beryl, which replaces chrysoberyl + quartz, and ascribed it to retrograde metamorphism. In the Eastern Rhodope Mountains in Bulgaria, Arnaudov and Petrusenko (1971) also described the replacement of beryl by chrysoberyl and muscovite in a small (4.5 m × 0.3 m) pegmatite lens in biotite-gneiss and kyanite-bearing mica schist.

In contrast, Soman and Nair (1985) and Soman et al. (1986) described chrysoberyl as a primary phase crystallizing with quartz and sillimanite from an Al-saturated pegmatitic melt in an upper amphibolite facies, transitional to granulite facies area in the Western Ghats supracrustal sequence at Trivandrum (India). The pegmatites occur together with migmatites, and from fluid inclusion data the authors argue for P-T conditions of crystallization at 2.5 kbar and 600°C. However, it seems also possible that the pegmatites formed during regional metamorphism at higher P-T conditions. Retrograde formation of white mica around chrysoberyl is similar to that found at the localities described above.

Metamorphism of khmaralite- and sapphirine-bearing pegmatites under granulite-facies conditions in Antarctica has resulted in the breakdown of these minerals to magnesiotaaffeite-6*N*'3*S*, surinamite and chrysoberyl (Grew 1981, this volume; Grew et al. 2000). Rupasinghe and Dissanayake (1985) attributed the occurrence of beryl and chrysoberyl in the high-grade metamorphic terrain of Sri Lanka to a desilicification process at the contact between basic charnockites and aluminous metasedimentary country rocks.

Franz et al. (1986) investigated a Hercynian leucogranite in the Eastern Alps in Austria (Schrammacher locality, Tauern window), which is closely associated with a Mo-porphyry deposit and which was metamorphosed at Alpine amphibolite facies conditions. Beryl with inclusions of relict phenakite occurs in cm-sized nodules of white mica, garnet and chlorite. Such nodules are generally explained in the leucogranites of the Tauern window as pseudomorphs after cordierite, and in this case it probably was beryllian cordierite. The macroscopically and microscopically blue beryls have up to almost 4 wt % FeO_{tot}, very strong zoning from core to rim with the rim enriched in Na + Fe^{2+}, and an intermediate to very high Fe/(Fe+Mg) ratio. The interpretation that the nodules are pseudomorphs after beryllian cordierite is supported by observations of such a replacement of beryllian cordierite by biotite, muscovite and beryl (Jobin-Bevans and Černý 1998) in granitic pegmatites from Manitoba, Canada. The beryl-forming reaction from phenakite is schematically

$$3 \text{ phenakite} + 2 \text{ muscovite} + 9 \text{ quartz} = 2 \text{ beryl} + 2 \text{ K-feldspar} \qquad (11)$$

The incipient formation of phenakite from the possible cordierite pseudomorphs is not documented in the rocks from the Tauern area; it could have been the reverse of Reaction (11), with beryl as a solid solution component of the cordierite. A geologically similar occurrence of a Hercynian granite in the Swiss Alps was described by Abrecht and Hänni (1979). They observed beryl as a reaction product of phenakite in quartz veins due to the Alpine metamorphism.

An example for the hydrothermal formation of phenakite (as well as bertrandite in a granite), which produces the left hand side assemblage of Reaction (11), is given by Markl and Schumacher (1997). During cooling after crystallization of a melt, beryl can interact with hydrothermal solutions, e.g., in a pegmatite associated with leucogranite in the Schwarzwald, Germany (Markl and Schumacher 1997). Previously precipitated pegmatitic beryl was corroded and reprecipitated as blue beryl (aquamarine) at relatively high temperatures of 550°C. At lower temperature, beryl and alkali-feldspar react to form either phenakite + muscovite + quartz or kaolinite + bertrandite. These assemblages may form the starting assemblages for metamorphism of granitic rocks (see above). Bilal and Fonteilles (1988) argued for similar relations between beryl and phenakite in solutions with K^+ and H^+, where beryl decomposes in alkaline solutions into muscovite and phenakite. In Na^+-rich solutions, beryl decomposes into albite plus phenakite. These observations underline the importance of phenakite as a breakdown product of beryl, and thus for metamorphic assemblages phenakite (or its hydrated precursor bertrandite) will be the major Be-mineral in the protoliths, as it was observed in several occurrences of metamorphic beryl.

The localities of metamorphosed silicic volcaniclastic rocks described in the literature include the Habachtal emerald deposit and the Felbertal scheelite mine in the Eastern Alps, Austria, and Ianapera, Madagascar (Cheilletz et al. 2001). Details for the Habachtal deposit are given below together with the other emerald deposits. The Be- (and W-) mineralizations from the Felbertal scheelite mine (Franz et al. 1986) are associated with a porphyritic pre-Alpine metagranite and a suite of different volcanoclastic rocks. The pre-metamorphic enrichment of Be might be due to hydrothermal processes in connection with the igneous activity, and because fluorite is a frequent accessory mineral, transport of Be-F species was likely. The rocks now contain beryl with phenakite as a relict mineral, similar to the relationship between these minerals as in the Hercynian metagranites described above. Phenakite is always surrounded by beryl, never in contact with quartz. Beryl is interpreted as the product of the Alpine metamorphism, and as no other Al-minerals are present in the immediate vicinity of the beryl, its formation must be due to a metasomatic process during metamorphism with transport of Al. This observation underlines the importance of element transport in hydrous fluids during the metamorphic process for beryl formation. The importance of fluid-transport is documented, too, by the fact that metamorphic beryl is mostly zoned in respect to the Na contents, but no Na mineral for an exchange reaction is available in the immediate vicinity. Such metasomatic processes play a dominant role for the genesis of emerald.

Possible pre-metamorphic equivalents of the above mentioned Felbertal and Habachtal localities, Austrian Alps, are hydrothermal non-pegmatitic deposits such as the bertrandite mineralizations of Topaz-Spor Mountain (Utah, USA), Aguachile Mountain (Mexico) and Apache Warm Springs (New Mexico, USA). They formed by low temperature fluid-rock-interaction of fluids enriched in silica, alumina, alkalis with carbonate rocks and/or porous silica-rich volcaniclastic rocks (Levinson 1962; Hillard 1969; see also Barton and Young, this volume, Chapter 14). The solutions are derived from Be and fluorine rich alkaline magmas found beneath the bertrandite mineralization. At Spor Mountain evidence of a Be- (and F-) rich source beneath the deposit is given by the great volume of a topaz rhyolite that is found in the Thomas Range. The presence of fluorite in all those deposits suggests that Be was concentrated and transported by Be-F complexes (Levinson 1962). The importance of fluorine-complexes in the transport and precipitation of Be-minerals is well documented by the experiments of Sobolev and Novoselova (1959) and Govorov and Stunzhas (1963).

Metamorphic emerald-chrysoberyl-phenakite deposits

Emerald, the green gem variety of beryl, and alexandrite, the gem variety of chrysoberyl, are besides diamond, ruby and sapphire the most sought after and highly valued gemstones. The green coloring of beryl is due to Cr (with absorption at 630 nm in the visible spectrum), V and Fe (e.g., Taylor 1967). Henn (1995) argued that only beryl colored predominantly by Cr should be called "emerald," that colored by V "vanadium beryl," and that colored by Fe "green beryl." The scarcity of gem quality emerald and vanadium beryl is due to the fact that the geochemical behavior of Be as a major component and Cr or V as the coloring trace element component is very different. It requires a special geological setting that they can occur together. In contrast to beryl crystals from granitic pegmatites, which formed in a siliceous melt and which can reach up to several meters in length, beryls from the majority of emerald and alexandrite deposits show the typical appearance of metamorphic porphyroblasts with a crystal size of generally less than 2 cm in length. These "schist-type deposits," or "micaite-deposits" in the Russian literature, were formed in the context of metamorphism. The only deposit which does not belong to this type is the (small and economically unimportant) Norwegian deposit Eidsvoll. It is described as a pegmatite intrusion (related to a Permian syenite-granite suite) into Cambro-Ordovician marine alumina-rich and organic rich sediments, which are the source of predominantly V and minor Cr for the coloring of beryl (review by Schwarz 1990a). Accordingly, these beryls have up to 1.5 wt % (average 1.3) V_2O_3, and only an average of 0.2 wt % Cr_2O_3 (Schwarz 1990a). Emeralds from different deposits have a distinct trace element fingerprint, which can potentially be used to delimit the origin of a gem stone (Calligaro et al. 2000), a useful tool in archeometry.

In the schist type deposits emerald and chrysoberyl are found typically alone or in association at the contact between Cr-rich ultramafic rocks and Be rich K-Na-Al quartzofeldspathic rocks such as (meta)pegmatites, metasediments and metavolcanics in greenschist to amphibolite facies terrains. Phenakite, which is also used as a gem stone, is often found in the same association, so that the schist-type deposits can be called "emerald-chrysoberyl-phenakite-association." Examples of schist type deposits are:

- ***Africa:*** Gravelotte, South Africa (le Grange 1929; Grundmann and Morteani 1987; Nwe and Morteani 1993), Kitwe and Lake Manyara, Tanzania (Thurm 1972; Bank and Gübelin 1976), Novello Claims, Chikwanda and Sandawana, Zimbabwe (Gübelin 1960; Martin 1962; Kanis et al. 1991; Zwaan et al. 1997), Miku, Kitwe and Lafubu, Zambia (Bank 1974; Graziani et al. 1983; Sliwa and Nguluwe 1984), Umm Kabu, Sikait and Zabara, southern Egypt (Grundmann and Morteani 1993a,b), Andilana, Mananjur, and Ianapera, Madagascar (Hänni and Klein 1982; Schwarz and Henn 1992; Thomas 1993; Delbos and Rantoanina 1995; Ranorosoa 1996; Cheilletz et al. 2001) Nigeria (Kanis and Harding 1990; Henn and Bank 1991); Gilé, Mozambique (Thomas 1994).

- ***Asia:*** Panjshir Valley, Afghanistan (Bowersox et al. 1991), Swat Valley, Pakistan (Gübelin 1982; Hussain et al. 1993; Arif et al. 1996; Laurs et al. 1996), Udaipur, India (Webster 1950; Gübelin 1951; Menon et al. 1994).
- ***Australia:*** Calvert White Quartz Hill, Dowerin, Emmaville, Melville, Menzies, Pilgangoora, Poona, Torrington, Warda Warra, Wodgina (Simpson 1932, Garstone 1981; Schwarz 1990a; Grundmann and Morteani 1996, 1998; Downes and Bevan, in press)
- ***Europe:*** Mursinka and Tokowaja, Urals, Russia (Fersman 1929; Ustinov and Chizhik 1994; Laskovenkov and Zhernakov 1995), Habachtal, Austria (Leitmeier 1937; Gübelin 1956; Grundmann and Morteani 1982; Schäfer et al. 1992), Urdini Lake, Bulgaria (Petrusenko et al. 1964/5; Arnaudov and Petrusenko 1971; Petrusenko et al 1971), Franqueira, Spain (Martin-Izard et al. 1995; Franz et al. 1996)
- ***North America:*** Hiddenite and Spruce Pine, North Carolina, USA (Crowningshield 1970; O'Donoghue 1973; Tacker 1999); Crown Showing, SE Yukon, Canada (Groat et al. 2000)
- ***South America:*** Belmont, Bom Jesus da Meiras, Capoeirana, Carnaíba, Itaberai, Itabira, Nova Era, Pirenópolis, Salininha, Santa Terezinha, Santana dos Ferros, Socotó, Tauá, Brazil (Bodenlos 1954; Giuliani et al. 1990; Schwarz 1990b,c; Souza 1992; Cheilletz et al. 1993; Ribeiro-Althoff et al. 1997; Giuliani et al. 1997a,b; Preinfalk et al. in press)

  [For additional references, see Sinkankas (1981), Grundmann (2001), and Barton and Young (this volume, Chapter 14).]

The mafic to ultramafic rocks are always metamorphosed. They are mostly serpentinized peridotites, and, more rarely, komatiites, amphibolites or rocks of greenstone belts. Grade of metamorphism is mostly upper greenschist to amphibolite facies, the occurrence at Dowerin, Australia has been recently described by Downes and Bevan (in press) from an Archean granulite facies terrain. The felsic rocks are mostly granitic pegmatites or aplites, for the locality Habachtal, Austria, Grundmann and Morteani (1982) assume Be-rich metasediments and metavolcanics as the quartzofeldspathic rocks, and for the locality at Poona, Australia greisenized portions of ultrabasic rocks (Grundmann and Morteani 1996). The contact between the ultramafic and the quartzofeldspathic rocks is either tectonic or intrusive. For most localities an intrusive contact is described, but a detailed investigation of structural relationships between deformation features in the metamorphic rocks, beryl growth and intrusion is not given in many cases. The pegmatites are often described as strongly tectonized and the original K-feldspar is transformed into albite, which indicates the strong influence of deformation and metasomatism during the emerald-forming process.

At the contact between the chemically contrasting rocks multiple metasomatic reaction zones known in metamorphic petrology as "blackwall zones" (Phillips and Hess 1936; Brady 1977) are observed. These zones of chlorite-talc-tremolite-biotite are often monomineralic and can be several meters thick. Distribution of Be-minerals in the schists and chemical balancing shows that Be is involved in the metasomatic process together with the major elements Si, Na, K, Ca and Mg (Grundmann and Morteani 1982). The primary host of Be in the protolith is beryl or phenakite. Phengitic white mica is also an important carrier, e.g. up to ~40 ppm is concentrated in white mica of the emerald deposit of the Habachtal (Austria). A third carrier at Habachtal is beryllian margarite. Generally, the largest amount of beryl precipitation occurs at the contact between Be-rich felsic rocks and biotite schists which form the innermost part of the blackwall zone (Fig. 7). The beryl crystals from schist type deposits show the typical features of metamorphic porphyroblasts, and zoning and inclusion trails indicate in many cases a multistage

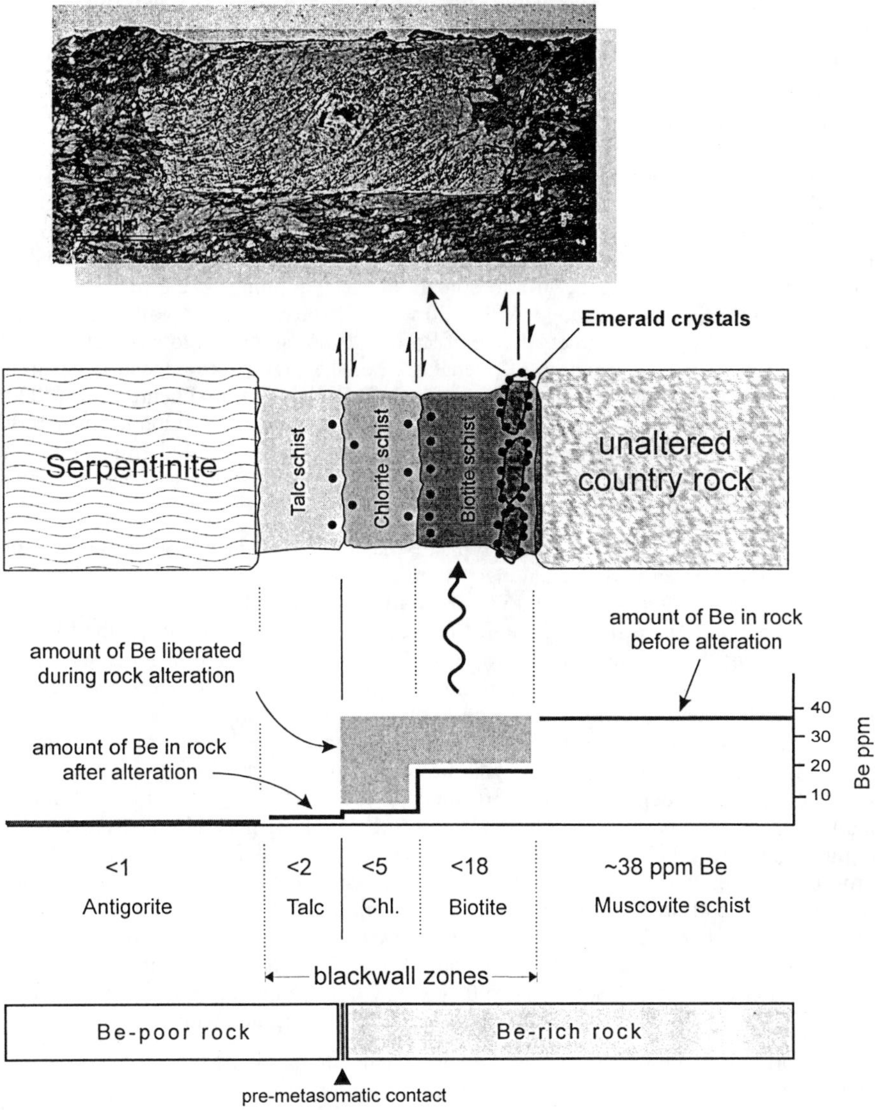

Figure 7. Schematic sketch of the evolution of a schist-type emerald mineralization in its characteristic setting, a blackwall zoning sequence between quartzofeldspathic rocks and serpentinites. The top illustration is a photomicrograph from a thin section of a typical metamorphic emerald porphyroblast showing multistage, in part syntectonic, growth and rotation, with an inclusion-rich core, an inner rim with sigmoidal distribution of the inclusions and a clear outer rim. The other illustrations show development from the original pre-metasomatic contact to the final stage and Be contents of the serpentinite, country rock and metasomatic zones. The total width of the talc zone is highly variable (between <1 m and up to 50 m) due to the tectonic processes; the other blackwall zones rarely exceed 1 m (according to Grundmann and Morteani 1993a,b; Grundmann et al 1993).

growth of emerald (Fig. 7). The major growth of beryl is syn- to post-deformational in respect to the main foliation of the schist.

Phenakite is often the precursor mineral of beryl, where it is found as corroded relics within emerald porphyroblasts. Based on fluid inclusion data, Nwe and Morteani (1993) interpreted the formation of phenakite as an effect of alkali metasomatism which also albitized the pegmatites (example from the Cobra pit, Gravelotte, South Africa). The fluid evolution in this deposit seems to be more or less continuous. Low salinity, H_2O + CH_4 fluids dominate in the first stage and evolve to more saline (>26 % NaCl equiv.) and low CH_4 fluids. A concomitant increase of the fO_2 is observed and referred to the interaction with the host rocks (Nwe and Morteani 1993). The development of phenakite rather than of an Al-bearing beryllium silicate at the beginning of the process of Be-mineral precipitation is explained with a low Al activity, and not with the thermal instability of beryl. The experimental data of Barton (1986) show that low Al activity and high alkalinity favor the formation of phenakite. For the Franqueira deposit in Spain at the contact between pegmatites and ultramafic-mafic rocks, Fuertes-Fuente et al. (2000) described the reaction chrysoberyl + phenakite + SiO_2 in aqueous solution to form beryl (emerald), based on textural replacement criteria. From their detailed fluid inclusion study they argue for a formation temperature below 400°C at decreasing pressure, and a significant contribution in the fluid from the metamorphic country rocks.

Beryl from the schist type deposits typically has high MgO and Na_2O contents of up to 3 and 2 wt % respectively (partial electron microprobe analysis, Franz 1982; Franz et al. 1986; Schwarz 1990a,b,c). FeO_{tot} is mostly lower than MgO and in the case of the locality Habachtal (Austria) the Mg and (Mg+Fe) ratio of 0.9 to 0.8 reflects the Mg and (Mg+Fe) ratio of the ultramafic rocks. Mg and (Mg+Fe) ratio in beryl is generally higher than in coexisting tremolite, chlorite, biotite and phengite (Franz et al. 1986). Element correlations between Fe-Mg-Na-Al show that the substitutions $^{ch}NaMgAl_{-1}$ and $^{ch}NaFe^{2+}Al_{-1}$ dominate, but because Na is often less than Mg + Fe, a certain amount of total Fe is assumed to be Fe^{3+}. Analysis of beryl from a large variety of the above-mentioned emerald deposits, which formed at lower to upper amphibolite facies conditions, suggests that the substitution $^{ch}Na(Mg,Fe^{2+})Al_{-1}$ is generally restricted to 0.5 Na pfu and 0.5 $(Mg+Fe^{2+})$. Lattice constant *a* increases from ~9.2 Å for pure beryl to 9.3 Å for beryl with 0.5 $Na(Mg,Fe^{2+})Al[Be_3Si_6O_{18}]$, whereas *c* remains constant (Franz 1982). Other substitutions seem to play only a minor role. The content of Cr_2O_3 is generally below 1 wt % and many crystals show chemical zoning, where Cr may be concentrated either in the core or in the rim; 1.5 wt % is reported as the highest Cr-content (Franz et al. 1986; Schwarz 1990a,b,c). Li_2O contents of up to 0.3 wt % were reported (Franz 1982), and they are generally negatively correlated with MgO.

The formation of emerald, chrysoberyl and phenakite in "schist type deposits" is still a matter of debate. Leitmeier (1937), Sinkankas (1981), Bank and Gübelin (1976), Lawrence et al. (1989), Ustinov and Chizhik (1994) and Martin-Izard et al. (1995), and many other authors interpreted the formation of these deposits according to the classical theory of exometasomatism first developed by Fersman (1929) for the emerald deposit of Tokovaya, Urals (Russia). The exometasomatism is the result of the intrusion and consequent reaction of Be-rich pegmatitic siliceous melts and hydrous fluids with Cr-rich ultramafic rocks. The pegmatitic melts and fluids may originate from granites (e.g., Garstone 1981) or from partial melting of crustal rocks (e.g., Bank and Gübelin 1976).

This model was essentially modified by Grundmann and Morteani (1989), Nwe and Grundmann (1990) and Nwe and Morteani (1993) from a study of the Habachtal and Gravelotte deposits, where they emphasized the importance of the blackwall-zoning and of regional metamorphism. The two models were discussed in detail by Franz et al.

(1996). Blackwall-zoning, which is generally the effect of regional metamorphism and often observed in ultramafic rocks in a geologic setting completely unrelated to igneous activity, is a common feature for all schist-type deposits. According to the model of Grundmann and Morteani (1989), a regional tectonometamorphic event is essential for the formation of emerald, chrysoberyl and phenakite at the contact between the Cr-rich ultramafic rocks and Be bearing K-Na-Al silicate rocks. Ribeira-Althoff et al. (1997) also described the importance of a shear zone system for the origin of the Santa Terezinha emerald deposit in Brasilia. The first step in the formation of Be-mineralization is the tectonic or intrusive juxtaposition of the Be-rich felsic rocks with ultramafic Cr- and Mg-rich rocks. The next step involves the development of multiple metasomatic reaction zones ("blackwall zones") between the chemically contrasting rocks, which is the common site of beryl, chrysoberyl and phenakite precipitation. Cheilletz et al. (1993) and Ribeiro-Althoff et al. (1997) showed by ^{40}Ar and ^{39}Ar laser spot analyses of phlogopite inclusions in emerald that the emerald precipitation and the formation of the blackwall zones were coeval. This process is responsible for the transport of Be from the primary (mostly pegmatitic) source to the ultramafic schist. An intrusion of felsic igneous rocks in the dimension of a small pegmatite body cannot supply enough energy and fluids to produce important reaction zones (Franz et al. 1996), which largely exceed the dimensions of the pegmatite. The multistage process is documented by the structural relationship between the emerald porphyroblasts and the host rocks.

Metamorphic-hydrothermal Be-mineralization

The Colombian emerald deposits are described as sedimentary-hydrothermal (Cheilletz and Giuliani 1996), because the mineralizing fluid circulation is restricted to tectonic zones and does not affect the whole sedimentary cover, and the fluids have a sedimentary origin as hydrothermal brines from evaporites. They were formed during the Andean orogeny at Eocene-Oligocene and Late Cretaceous times (Cheilletz et al. 1994; 1997). However, the conditions of emerald formation in the Coscuez deposit at about 1 kbar and 290 to 360°C (Cheilletz et al. 1994) are clearly within the regime of metamorphism. In the Eocene-Oligocene Muzo-Coscuez district they are genetically related to a compressive tectonic setting where hydrothermal breccias were formed during thrusting, folding and faulting (Laumonier et al. 1996; Branquet et al. 1999a). In the Late Cretaceous Chivor-Macanal district (Cheilletz et al. 1997) they were formed during a thin-skinned extensional tectonic event (Branquet et al. 1999b). Given the temperature of formation (which is near the lower thermal stability limit of beryl according to Lebedev 1980), the occurrence of chloritoid in a zone several hundreds of meters wide around the deposit at Muzo (Branquet 1999) and the strong deformation of the rocks, from the hand specimen (Cheilletz and Giuliani 1996) to the regional scale (Branquet et al. 1999b), these deposits are clearly metamorphic-hydrothermal, although sedimentary features are very well preserved, e.g., in the rocks illustrated in Cheilletz and Giuliani (1996). Even gastropod shells perfectly replaced by microcrystalline emerald have been reported (Schwarz and Giuliani 2001).

The mineralization occurs in a breccia-type rock (called "cenicero" in the local miner's terminology) in centimeter-wide veins and pockets, filled with calcite, dolomite, albite and pyrite (fluorite, barite and parisite are found as accessory minerals). The polygenetic breccia is composed of fragmented black shales and albitites (metasomatized rocks) cemented by albite and pyrite. The fluid composition, as determined from fluid inclusions and Raman-probe, is H_2O-$NaCl$-$CaCl_2$-CO_2-N_2-rich and similar for both emerald-bearing districts, the Yacopi-Muzo-Coscuez-Penas Blancas district, west of the Eastern Cordillera of Colombia, and the Gachalà-Chivor-Macanal district on the eastern side of the Cordillera (Cheilletz and Giuliani 1996). Koslowski et al. (1988), Ottaway et al. (1994) and Cheilletz et al. (1994) concluded that the source of Be in the deposits, which have as much as 1 wt % Be, is black shale containing 3-5 ppm Be, one of the sediments hosting the deposits. Be was extracted from the country rocks by alkaline (pH between 6.8 and 11.8) brines rich in Na, Ca and Mg that introduced albite and calcite (Cheilletz and Giuliani 1996). However, Ottaway et al. (1994) argued for an acid environment. Buffering of the fluids by the host rocks is indicated by $\delta^{18}O$ data (Giuliani et al. 1997a). Sulfate reduction by organic matter of originally evaporitic deposits provided the source for sulfide H_2S and for $(HCO_3)^-$ and the $\delta^{13}C$ and $\delta^{18}O$ values are typical for metamorphic or highly evolved sedimentary formation waters (Giuliani et al. 1995, 2000).

The precipitation of beryllium as emerald occurred in consequence of a decrease in alkalinity and pressure (Cheilletz and Giuliani 1996) and of mixing of two fluids which were derived from the black shales and from evaporites (Banks et al. 2000). Transport of Be by the hydrothermal fluid might have occurred as $(OH)^-$ complexes (Ottaway et al. 1994), but Banks et al. (2000) argued that the experimental data by Wood (1991, 1994) were in favor of fluoride or mixed carbonato-fluoride complexes, supported by the coprecipitation of fluorite and carbonate. Cheilletz and Giuliani (1996) speculated that Be was transported as carbonatoberyllates with the general formula $[Be_4(CO_3)_6(OH)_3Cl]^{8-}$. This problem of Be-species in the fluid is clearly related to the P-T-X conditions: Thermodynamic calculation of the composition of a volcanic gas from Kudryavy volcano (Kurile Islands; Churakov et al. 2000) have shown that in a predominantly hydrous fluid phase with small amounts of CO_2, SO_2, H_2S, HCl and HF, the Be-species in the fluid is BeO_2H_2 at 1 bar above 600°C, below 600°C it is BeF_2.

The spectacular color of the Colombian emeralds, as compared to that of the schist type ones, is due to a red fluorescence produced by Cr that enhances the luminosity of the green color (Nassau 1980; Ottaway et al. 1994). This fluorescence is suppressed by the high iron content in the emeralds of the schist type deposits. The very low Fe content of the Colombian emeralds is due to the removal of Fe by the coeval crystallization of pyrite.

Recently Sabot et al. (2000, 2001) reported an occurrence of emerald from Panjshir, Afghanistan, which has some similarity with the genesis of the Colombian emeralds. As with Colombian stones, the Afghan emeralds have a very low Fe content and their color is mainly due to the presence of Cr and V. The deposits are linked with hydrothermal fluids that derived their salinity from leaching of a metaevaporite sequence, an inference based on $\delta^{37}Cl$ of -0.32 ‰ in fluid inclusions and $\delta^{11}B$ of -3.3 ‰ in associated tourmaline. $\delta^{34}S$ values of +12 to +15 ‰ in associated pyrite are in the same range as for pyrite in the Colombian emeralds and suggest that sulfur originated by reduction of sulfate minerals. The deposits are associated with a strong Na-K-B metasomatism which produced a white rock called "Kamar Saphet" in the local language, comparable to the "Cenicero" of Colombia. However, the metamorphic grade is higher; the Panjshir deposits are situated in medium-grade metamorphic rocks. The process was probably triggered by the tectonism that preceded uplift of the Himalayan orogen.

Another hydrothermal occurrence of (blue) beryl in dolomitic sedimentary rocks from a hematite mine (Lassur, Ariège, France) was described by Fontan and Fransolet (1982), and supports the observation that Be can be mobilized from sedimentary rocks and concentrated in hydrothermal veins, and that the temperature for such a process can be rather low.

Ragu (1994) described the occurrence of helvite $Mn_4Be_2(SiO_4)_3S_2$ in hydrothermal veins in stratabound low grade metamorphic Mn-jasper deposits from the French

Pyrenees. He argued for a hydrothermal formation after metamorphism, where Be was introduced from near-by granites and reacted with the Mn-rich layers. However, Brugger and Gieré (2000) presented arguments for a similar occurrence in the Swiss Alps, that Be was remobilized from the Mn volcano-sedimentary layers during a late stage of the metamorphic event. This is another example for the complex interplay between hydrothermal and metamorphic processes as described above for the schist-type emerald deposits.

Summary and open questions

Experimental work coupled with mineralogical-petrological observations on natural assemblages has been a very powerful tool for gaining insight into the genetic processes of mineral formation. For the Be-minerals, an understanding comparable to that achieved for the more common rock-forming minerals is not yet possible, although a considerable amount of experimental work is already available. Straightforward application of experimental data to natural occurrence of Be-minerals is not yet possible; there is still a large gap between "experiment" and "nature." The main reason is that the natural reactions take place in an aqueous solution for which it is very difficult to specify the composition. Associated with this problem is the transport mechanism of Be in the context of metamorphic-metasomatic reactions, for example, in schist type emerald-chrysoberyl-phenakite deposits. Fluorine may play a key role in this context. A thorough understanding of the aqueous geochemistry of Be in complex systems is needed at elevated pressure and temperature.

We give here some of the unresolved experimental problems. In several studies unidentified phases were observed, especially in Al-rich bulk compositions, and it is likely that more such phases exist. Some of them seem to be metastable, but in any case a more systematic approach to investigate these systems in more detail and to expand it to other systems is necessary. This may also be of help to identify new mineral species. However, the experimentalist must be aware of sluggish reactions, even at high temperatures. Of the systems for which we did not find much experimental work, those with Ca and Fe are the most important. Moreover, there is insufficient information from experiments on the substitution OH-F-Cl in minerals such as euclase and bertrandite; this information is particularly critical because Be-mineralizations are often found in geological environments where a strong enrichment of halogens can be expected. Another hydoxyl phase of interest is the Li-Be brittle mica bityite; F was found in most analyses of this mica when it was sought (see references cited by Grew, this volume; cf. Robert et al. 1993, who reported that in contrast to other Li-micas, bityite is F-free).

Much remains to be done in order to properly understand Be-Al-Si substitution and order-disorder phenomena in Be-minerals, particularly in cases where some of the sites are not fully occupied. It is suspected that especially at very high temperature disorder on the tetrahedral sites becomes important despite the problem of local charge balance. Order-disorder has a large effect on the stability relations and may thus explain the problems encountered with the high temperature behavior of beryl. At temperatures more relevant for the majority of natural assemblages a picture is emerging, such that Be substitutes more readily for Al, B or Li than for Si. Local charge imbalance associated with Be = Si is probably the reason for the relative scarcity of this substitution (Hawthorne and Huminicki, this volume). Thus, disorder is far more likely between Al and Be than between Be and Si, as appears to be the case in beryl, cordierite and khmaralite and sapphirine. Phosphorus, another element in tetrahedral coordination, has been reported in natural surinamite (Grew et al. 2000) and beryl (3.6 wt % P_2O_5, Feklichev and Razina 1964) as well as in synthetic beryl (up to 1.65 wt % P_2O_5 at 620°C and 3 kbar in granitic melts, Lahlafi 1997), and it seems worthy to investigate the complex relations between tetrahedrally coordinated cations of different charge.

Solid solution between compounds containing beryllium and related Be-free compounds is another aspect needing detailed study. Experimental data reported to date indicate substantial solid solution between cordierite and two beryllian compounds very closely related to beryl and suggest the possibility of extensive solid solution in other systems at high temperatures, i.e., in spinel on the spinel-chrysoberyl join, in olivine on the fayalite-chrysoberyl join, and in willemite on the willemite-phenakite join. The possibility of solid solutions at low-temperatures is suggested by the margarite-bityite series and by the synthetic K-Li-Be micas. Investigations of synthetic solid solutions should include a quantitative determination of Be in run products by methods such as SIMS (see London and Evenson, this volume). Run products in the reviewed studies were not analyzed by microanalytical methods for Be, and the conclusions reached remain therefore somewhat speculative.

For beryl, the most prominent Be-mineral, the stability field needs to be determined in more detail. Despite its framework structure with large channels it seems to be stable up to extremely high pressure, where it can be synthesized easily, but nothing is known about its upper pressure limit of stability. Moreover, the high-temperature stability and melting relationships remain unclear. This is partly due to the fact that beryl incorporates up to ~3 wt % H_2O into the channels of its structure, and this water has a dramatic effect on its stability. The effect of other channel constituents such as CO_2 and alkalis, most importantly Na, is still unknown. In general, limits determined in the model BeO-Al_2O_3-SiO_2-H_2O system may have little relevance to metamorphic rocks, migmatites and acid igneous rocks, where reactions terminal to beryl undoubtedly involve muscovite, alkali feldspar, ferromagnesian silicates and granitic melt. The lack of experimental investigations for systems more complex than BeO-Al_2O_3-SiO_2-H_2O is a large obstacle for the explanation of natural occurrences by experimental data.

Beryl is not the only Be-mineral needing more study; the same applies to other Be-minerals. Bertrandite (sphaerobertrandite has a higher Be and Si ratio than bertrandite, Fig. 3) is the hydrated low-temperature equivalents of phenakite, but available experimental data on bertrandite breakdown to phenakite are contradictory. This breakdown reaction could have important applications because of the increasing importance of bertrandite as an economic source of Be. Euclase, the simplest Be-Al-silicate, is stable at lower temperature compared to beryl, but there is a large overlap in their stability fields up to 600°C. For all low temperature equilibria, experimental studies will encounter the problem of sluggish reactions and metastable formation of phases. Therefore thermodynamic calculations based on internally consistent data sets are the more promising approach (see Barton and Young, this volume). Phenakite and chrysoberyl are stable to both high temperatures and pressures; nothing is known about possible structural modifications at high pressure. Though this is not relevant for explanation of natural assemblages, it might be important for crystal chemical aspects, such as the transition from an olivine-type structure into a spinel type structure in chrysoberyl.

A sound genetic concept for metamorphic (hydrothermal-metasomatic) Be-mineralizations should start with an investigation of the large scale geological context in order to answer the questions about the possible geodynamic setting at the time of Be-mineralization and about the possible types of heat and fluid sources for mass transfer. This has been overlooked in many studies, which were dominated by the mineralogical aspect. To distinguish between metamorphic, hydrothermal, and metasomatic processes or to establish their relative importance, investigators need to search for indications of mass transfer, such as metasomatic reaction zones and for deformation, at all scales from the outcrop to the thin section. The investigator must be aware that deformation may have little effect on coarse grained pegmatites in a schistose matrix, where strain is heterogeneously distributed, and therefore the influence of metamorphism may easily be overlooked. Mineralogical investigations should concentrate on the crystallization sequence of Be-minerals, and on different growth stages distinguishable by zoning, deformation, replacement textures, inclusion patterns etc. In many studies this is not documented well enough. Electron microprobe analyses are still a good first step for information about crystal chemistry, but complete chemical characterization requires other microanalytical methods to obtain Be, Li, and B contents. Even relatively small amounts of these light elements in terms of wt % can play significant roles in the crystal chemistry. More chemical investigations about concentrations of Be in common rock-forming minerals are needed, as well as whole rock Be contents as indications for mass transfer on different scales. To hypothesize about the mode of Be-enrichment to form Be-minerals, stable isotope investigations to determine sources of the fluid has shown to be important. Fluid inclusion data are also useful, although a careful evaluation is important because of the polyphase processes during the mineralization.

In view of the widespread occurrence of metamorphic parageneses with Be-minerals, their petrologic use as indicators for mass transport phenomena, their being a source for Be, and last, but perhaps not least, the aesthetic aspect of their gem varieties, it seems highly worthwhile to investigate these parageneses in more detail.

ACKNOWLEDGMENTS

We thank Ch. Preinfalk for the help in the preparation of the figures and editing of the manuscript. Critical readings of a first draft and the help in literature research by M.D. Barton, M. Burianek, S. Young, E.S. Grew, F. Hawthorne, S. Herting-Agthe, M. Taran and V. Thomas are gratefully acknowledged; S. Churakov was of great help with the translation of Russian literature. Special thanks go to E.S. Grew for his careful editing and many suggestions to improve style and content of the manuscript.

REFERENCES

Abrecht J, Hänni H (1979) Eine Beryll-Phenakit (Be_2SiO_4)-Paragenese aus dem Rotondo-Granit. Schweiz mineral petrograph Mitt 59:1-4

Arif M, Fallick AE, Moon CJ (1996) The genesis of emeralds and their host rocks from Swat, northwest Pakistan: A stable isotope investigation. Mineralium Deposita 31:255-268.

Armbruster T (2002) Revised nomenclature of högbomite, nigerite, and taaffeite minerals. Eur J Mineral 14 (in press)

Arnaudov V, Petrusenko S (1971) Chrysoberyl from two different types of pegmatites from the Rila-Rhodope area. Izvestija na geologiceskija institut. Seiji geochimija, mineralogika i petrografiija 20: 91-97 (in Bulgarian)

Bank H (1974) The emerald ocurrence of Miku, Zambia. J Gemmol 14:8-15

Bank H (1976) Mit synthetischem Smaragd überzogene natürliche farblose Berylle (nach Lechleitner). Z Deutsch Gemmol Ges 25:107-108

Bank H, Gübelin E (1976) Das Smaragd-Alexandritvorkommen von Lake Manyara, Tansania. Z Deutsch Gemmol Ges 25:130-147

Banks DA, Giuliani G, Yardley BWD, Cheilletz A (2000) Emerald mineralisation in Colombia: Fluid chemistry and the role of brine mixing. Mineralium Deposita 35:699-713

Barton MD (1986) Phase equilibria and thermodynamic properties of minerals in the $BeO-Al_2O_3-SiO_2-H_2O$ (BASH) system, with petrologic applications. Am Mineral 71:277-300

Bell NA (1972) Beryllium halides and pseudohalides. Adv Inorg Chem Radiochem 14:255-332

Beus AA (1966) Geochemistry of beryllium and genetic types of beryllium deposits. San Francisco, Freeman, 161 p

Beus AA, Sobolev BP, Dikov YuP (1963) Geochemistry of beryllium in high-temperature postmagmatic mineralization. Geochemistry 1963:316-323

Bilal E, Fonteilles M (1988) Conditions d'apparition du béryl dans l'environement granitique: Exemple du massif de Sucurri (Brésil). C R Acad Sci Paris Série II 307:273-276

Bodenlos AJ (1954) Magnesite deposits in the Serra das Aguas, Brumado, Bahia, Brazil. U S Geol Surv Bull 975-C:87-170

Borchert W, Gugel E, Petzenhauser I (1970) Untersuchungen im System Beryll-Indialith. N Jahrb Mineral Mh 1970:385-388

Bowersox G, Snee LW, Foord EE, Seal RR (1991) Emeralds of the Panjshir valley, Afghanistan. Gems Gemology 27:26-39

Brady JB (1977) Metasomatic zones in metamorphic rocks. Geochim Cosmochim Acta 41:113-125

Branquet Y (1999) Etude structurale et métallogénique des gisements d'émeraude de Colombie: Contribution à l'histoire tectono-sédimentaire de la Cordillère Orientale de Colombie. Unpublished PhD dissertation, INPL Vandoeuvre-les-Nancy, France, 295 p

Branquet Y, Cheilletz A, Giuliani G, Laumonier B, Blanco O (1999a) Fluidized hydrothermal breccia in dilatant faults during thrusting: The Colombian emerald deposits. *In* McCaffrey KJW, Lonergan L, Wilkinson JJ (eds) Fractures, fluid flow and mineralization. Geol Soc London, Spec Publ 155:183-195

Branquet Y, Laumonier B, Cheilletz A, Giuliani G (1999b) Emeralds in the Eastern Cordillera of Colombia: Two tectonic settings for one mineralization. Geology 27:597-600

Brugger J, Gieré R (2000) Origin and distribution of some trace elements in metamorphosed Fe-Mn deposits. Val Ferrera, eastern Swiss Alps. Can Mineral 38:1075-1102

Bukin GV (1967) Crystallization conditions of phenakite-bertrandite-quartz association (experimental data). Dokl Akad Nauk SSSR 176:664-667 (in Russian)

Bukin GV (1993) Growth of crystals of beryllium oxides and silicates using fluxes. Growth of Crystals 19:95-121, Consultants Bureau, New York

Burt DM (1978) Multisystem analysis of beryllium mineral stabilities: The system $BeO-Al_2O_3-SiO_2-H_2O$. Am Mineral 63:664-676

Calligaro T, Dran J-C, Poirot J-P, Querré G, Salomon J, Zwaan JC (2000) PIXE/PIGE characterization of emeralds using an external micro-beam. Nucl Instr Meth Phys Res B 161-163:768-774

Cemič L, Langer K, Franz G (1986) Experimental determination of melting relationships of beryl in the system $BeO-Al_2O_3-SiO_2-H_2O$ between 10 and 25 kbar. Mineral Mag 50:55-61

Černý P, Novák M, Chapman R (1992) Effects of sillimanite-grade metamorphism and shearing on Nb-Ta oxide minerals in granitic pegmatites: Marsíkov, Northern Moravia, Czechoslovakia. Can Mineral 30:699-718

Chatterjee M, Ganguli D (1975) Phase relationships in the system $BeO-ZnO-SiO_2$. N Jahrb Mineral Mh 1975:518-526

Cheilletz A, Giuliani G (1996) The genesis of the Colombian emeralds: A restatement. Mineralium Deposita 31:359-364

Cheilletz A, Féraud G, Giuliani G, Ruffet G (1993) Emerald dating through $^{40}Ar/^{39}Ar$ step-heating and laser spot analysis of syngenetic phlogopite. Earth Planet Sci Lett 120:473-485

Cheilletz A, Feraud G, Giuliani G, Rodriguez C T (1994) Time-pressure-temperature constraints on the formation of Colombian emeralds: An $^{40}Ar/^{39}Ar$ laser-probe and fluid inclusion study. Econ Geol 89:362-380

Cheilletz A, Giuliani G, Zimmermann JL, Ribeiro-Althoff AM (1995) Ages, geochemical signatures and origin of Brazilian and Colombian emerald deposits: A magmatic versus sedimentary model. *In* Pasava J, Zak K (eds) Proc Third Biennial SGA Meeting. Balkema, Rotterdam, p 569-572

Cheilletz A, Giuliani G, Branquet Y, Laumonier B, Sanchez MAJ, Féraud G, Arhan T (1997) Datation K-Ar et 40Ar/39Ar à 65±3 Ma des gisements d'émeraude du district de Chivor Macanal: Argument en faveur d'une déformation précoce dans la Cordillère orientale de Colombie. C R Acad Sci Paris 324, Série Iia, p 369-377

Cheilletz A, Sabot B, Marchand P, De Donalo P, Taylor B, Archibald D, Barres O, Andrianjaffy J (2001) Emerald deposits in Madagascar: Two different types for one mineralizing event. Abstract EUG Strasbourg, Abstr Vol, p 547

Christy AG, Hölscher A, Grew ES, Schreyer W (2000) Synthetic beryllian sapphirine: Structural characterization. Geol Soc Am Abstr Progr 32:A-54

Churakov SV, Tkachenko SI, Korzhinskiy MA, Bocharnikov RE, Shmulovich KI (2000) Evolution of composition of high-temperature fumarolic gases from Kudryavy Volcano, Iturup, Kuril Islands: The thermodynamic modeling. Geochem Intl 38:436-451

Crowningshield DR (1970) North Carolina emerald. Gems Gemology 8:250-253

Deer WA, Howie RA, Zussman J (1992) An Introduction to the Rock-Forming Minerals, 2nd ed. Longman, London, 696 p

Delbos L, Rantoanina M (1995) Les gisements d´émeraude et de saphir de Madagascar. Chronique Recherche Minière, BRGM 525:54-58.

Della Ventura G, Rossi P, Parodi GC, Mottana A, Raudsepp M, Prencipe M (2000) Stoppaniite, $(Fe,Al,Mg)_4(Be_6Si_{12}O_{36})\cdot(H_2O)_2(Na,\square)$, a new mineral of the beryl group from Latium (Italy). Eur J Mineral 12:121-127

Demina TW, Mikhailov MA (1993) Formation of beryllium indialite in solid phase reactions. Mineral Zh 15:61-70 (in Russian)

Demina TW, Mikhailov MA (2000) Magnesium and calcium in beryl. Zapiski Vseross Mineral Obshchestva 129:97-109 (in Russian, English abstr)

Downes PJ, Bevan AWR (in press) Chrysoberyl and associated mineralisation in metasomatised Archean rocks at Dowerin, Western Australia. Mineral Mag

Emel'yanova EN, Grum-Grzhimailo SV, Boksha ON, Varina TM (1965) Artificial beryl containing V, Mn, Co, and Ni. Sov Phys-Crystallogr 10:46-49

Eppler WF (1967) Der synthetische Smaragd von Linde. Z Deut Ges Edelksteinkunde 61:58-66

Espig H (1960) Die Synthese des Smaragds. Chem Techn 12:327-331

Farell EF, Fang JH, Newnham RE (1963) Refinement of the chrysoberyl structure. Am Mineral 48:804-810

Feklichev VG, Razina LS (1964) On phosphorous beryl. Mineraly SSSR 15:247-250 (in Russian)

Fersman AE (1929) Geochemische Migration der Elemente, III, Smaragdgruben im Uralgebirge. Abh Prakt Geologie Bergwirtschaftslehre 18:74-116

Flanigen EM (1971) Hydrothermal process for growing crystals having the structure of beryl in an alkaline halide medium. United States Patent No. 3,567,643, issued March 2, 1971

Fleischer M (1958) New mineral names. Gelbertrandite, spherobertrandite. Am Mineral 43:1219-1220

Fonda GR (1941) Constitution of zinc beryllium silicate phosphors. J Phys Chem 45:282-288

Fontan, F, Fransolet, A-M (1982) Le béryl bleu riche en Mg, Fe et Na de la mine de Lassur, Ariège, France. Bull Minéral 105:615-620

Foster WR, Royal HF (1949) An intermediate compound in the system $BeO-Al_2O_3$ (corundum). J Am Ceramic Soc 32:26-34

Franz G (1982) Kristallchemie von Beryll, Varietät Smaragd. Fortschr Mineral 60:76-78

Franz G, Morteani G (1981) The system $BeO-Al_2O_3-SiO_2-H_2O$: Hydrothermal investigation of the stability of beryl and euclase in the range from 1 to 6 kb and 400 to 800°C. N Jahrb Mineral Abh 140:273-299

Franz G, Morteani G (1984) The formation of chrysoberyl in metamorphosed pegmatites. J Petrol 25:27-52

Franz G, Grundmann G, Ackermand D (1986) Rock-forming beryl from a regional metamorphic terrain (Tauern Window, Austria): Paragenesis and Crystal Chemistry. Tschermaks mineral petrograph Mitt 35:167-192

Franz G, Gilg HA, Grundmann G, Morteani G (1996) Metasomatism of a granitic pegmatite-dunite contact in Galicia: The Franqueira occurrence of chrysoberyl (alexandrite), emerald and phenakite: Discussion. Can Mineral 34:1329-1331

Frondel C, Ito, J (1968) Synthesis of the scandium analogue of beryl. Am Mineral 53:943-953

Fuertes-Fuente M, Martin-Izard A, Boiron MC, Mangas Viñuela J (2000) P-T path and fluid evolution of the Franqueira granitic pegmatite, central Galicia, northwestern Spain. Can Mineral 38:1163-1176

Furzhenko DA (1982) Dependence of helvite composition from amount of alkalies and acids in mineral-forming solutions. *In* Fisiko-chimicheskiye issledovaniya mineraloobrazuyuschikh system, Akad Nauk SSSR, Sibirskoye otdelniye, Inst Geologii i Geophysiki 1982:104-107 (in Russian)

Furzhenko DA (1989) Conditions of synthesis of helvite-group minerals. Trudy Instituta Geologii i Geofiziki SO AN SSSR 701:1-77 (in Russian)

Furzhenko DA, Klyakhin VA (1981) Synthesis of minerals of the helvite group. *In* Melnik YuP (ed) Vsesoyz Sovesch po experim i tekh mineral i petrog 10:119-126 (in Russian)

Galakhov FY (1957) Studies of the aluminous region of ternary aluminosilicate systems. 2. The system $BeO-Al_2O_3-SiO_2$. Izvest Akad Nauk SSSR Otdel Khim Nauk 1957:1062-1067

Ganguly D (1972) Crystallization of beryl from solid-solid reactions under atmospheric pressure. N Jahrb Mineral Mh 1972:193-199

Ganguly D, Saha P (1965) Preliminary investigations in the high-silica region of the system $BeO-Al_2O_3-SiO_2$. Trans Indian Ceram Soc 24:134-146

Ganguly D, Saha P (1967) A reconnaissance study of the system $BeO-Al_2O_3-SiO_2-H_2O$. Trans Indian Ceram Soc 26:102-110

Garstone JD (1981) The geological setting and origin of emerald deposits at Menzies, Western Australia. J Royal Soc Western Australia 64:53-64

Geller RF, Yavorsky PJ, Steierman BL, Creamer AS (1946) Studies of binary and ternary combinations of magnesia, calcia, baria, beryllia, alumina, thoria and zirconia in relation to their use as porcelains. J Res Nat Bur Standards (Res Paper RP1703) 36:277-312

Gelsdorf G, Müller-Hesse H, Schwiete H-E (1958) Einlagerungsversuche an synthetischem Mullit und Substitutionsversuche mit Galliumoxyd und Germaniumdioxyd. Teil II. Archiv Eisenhüttenwesen 29:513-519

Gentile AL, Cripe DM, Ander FH (1963) The flame fusion synthesis of emerald. Am Mineral 48:940-944

Giuliani G, D'El Rey SLJ, Couto PA (1990) Origin of emerald deposits of Brazil. Mineralium Deposita 25:57-64

Giuliani G, Sheppard MF, Cheilletz A, Rodriguez C (1992) Contribution de l'étude des phases fluides et de la géochimie isotopique $^{18}O/^{16}O$, $^{13}C/^{12}C$ à la genèse des gisements d'émeraude de la Cordillère Orientale de la Colombie. C R Acad Sci Paris Sér II, 314:269-274

Giuliani G, Cheilletz A, Arboleda C, Carrillo V, Rueda F, Baker JH (1995) An evaporitic origin of the parent brines of Colombian emeralds: Fluid inclusion and sulphur isotope evidence. Eur J Mineral 7:151-165

Giuliani G, France-Lanord C, Zimmermann JL, Cheilletz A, Arboleda C, Charoy B, Coget P, Fontan F, Giard D (1997a) Fluid composition, δD of channel H_2O, and $\delta^{18}O$ of lattice oxygen in beryls; genetic implications for Brazilian, Colombian, and Afghanistan emerald deposits. Intl Geol Rev 39:400-424

Giuliani G, Cheilletz A, Zimmermann JL, Ribeiro-Althoff AM, France-Lanord C, Féraud G (1997b) Les gisements d'émeraude du Brésil: genèse et typologie. Chronique Recherche Minière, BRGM 526: 17-60

Giuliani G, France-Lanord C, Cheilletz A, Coget P, Branquet Y, Laumonier B (2000) Sulfate reduction by organic matter in Colombian emerald deposits: Chemical and stable isotope (C, O, H) evidence. Econ Geol 95:1129-1153

Gjessing L, Larsson T, Major H (1943) Isomorphous substitute for Al^{3+} in the compound Al_2BeO_4. Norsk Geol Tidssk 22:92-99

Gonzalez del Tanago J, Martinez M, Peinado M (1984) El crisoberilo de las pegmatitas graniticas de la Sierra Albarrana. Condiciones geneticas y evolucion. J Congreso Espanol de Geología II:131-145

Govorov IN, Stunzhas AA (1963) Mode of transport of beryllium in alkali metsaomatism. Geokhimiya 1963:383-390

Graziani G, Gübelin E, Lucchesi S (1983) The genesis of an emerald from the Kitwe district, Zambia. N Jahrb Mineral Mh 1983:175-186

Grew ES (1981) Surinamite, taaffeite, and beryllian sapphirine from pegmatites in granulite-facies rocks of Casey Bay, Enderby Land, Antarctica. Am Mineral 66:1022-1033

Grew ES, Yates MG, Barbier J, Shearer CK, Sheraton JW, Shiraishi K, Motoyoshi Y (2000) Granulite-facies beryllium pegmatites in the Napier Complex in Khmara and Amundsen Bays, western Enderby Land, East Antarctica. Polar Geoscience 13:1-40

Groat LA, Ercit TS, Marshall DD, Gault RA, Wise MA, Wengzynowski W, Eaton WD (2000) Canadian emeralds: The Crown Showing, Southeastern Yukon. Newsletter, Mineral Assoc Can 63:1,12-13

Grundmann G (2001) Die Smaragdvorkommen der Welt. *In* D Schwarz, R Hochleitner (eds) extraLapis Smaragd. Der kostbarste Beryll, der teuerste Edelstein 21:26-37

Grundmann G, Morteani G (1982) Die Geologie des Smaragdvorkommens im Habachtal (Land Salzburg, Österreich). Archiv Lagerstättenforschung Geol BA Wien 2:71-107

Grundmann G, Morteani G (1987) Multistage emerald formation in regional metamorphism: Case studies from Gravelotte, South Africa and Habachtal, Austria. Terra Cognita 7:292

Grundmann G, Morteani G (1989) Emerald mineralization during regional metamorphism: The Habachtal (Austria) and Leydsdorp (Transvaal, South Africa) deposits. Econ Geol 84:1835-1849

Grundmann G, Morteani G (1993a) Die Smaragdminen der Cleopatra: Zabara, Sikait und Umm Kabo in Ägypten. Lapis 18:27-39

Grundmann G, Morteani G (1993b) Emerald formation during regional metamorphism: The Zabara, Sikeit and Umm Kabo deposits (Eastern Desert, Egypt). *In* Thorweihe U, Schandelmeier H (eds) Geoscientific Research in Northeast Africa. Balkema, Rotterdam, p 495-498

Grundmann G, Morteani G (1996) Ein neues Vorkommen von Smaragd, Alexandrit, Rubin und Saphir in einem Topas-führenden Phlogopitfels von Poona, Cue-District, West-Australien. Mitt Österr Mineral Ges 141:343-356

Grundmann G, Morteani G (1998) Alexandrite, emerald, ruby, sapphire and topaz in a biotite-phlogopite fels from Poona, Cue District, Western Australia. Australian Gemmologist 20:159-167

Grundmann G, Morteani G, Seemann R, Koller F (1993) Smaragdlagerstätte Leckbachscharte, Habachtal und Kupferlagerstätte Hochfeld, Untersulzbachtal (Habachformation, Tauernfenster). Beiheft Eur J Mineral 5:137-188

Gübelin E (1951) Some additional data on Indian emeralds. Gems Gemology 7:13-22

Gübelin E (1956) Emerald from Habachtal. J Gemm 5:342-361

Gübelin E (1960) Notes on Sandawana emeralds. Gemmologist 29:8-16

Gübelin E (1961) Beryll mit synthetischem Smaragdüberzug. Z Deutsch Ges Edelsteinkunde 37:6-12

Gübelin E (1982) Die Edelsteinvorkommen Pakistans; II, Die Smaragdvorkommen im Swat-Tal. Lapis 7:19-26

Guillemin C, Permingeat P (1959) Revue des espèces minérales nouvelles. Bull Soc fr Minéral Cristallogr 82:91

Hahn T (1953) Modellbeziehung zwischen Silikaten und Fluoberyllaten. N Jahrb Mineral Abh 86:1-65

Hahn T, Eysel W (1970) Solid solubility in the system Zn_2SiO_4-$ZnGeO_4$-Be_2SiO_4-Be_2GeO_4. N Jahrb Mineral Mh 1970:263-275

Hänni HA (1982) A contribution to the separability of natural and synthetic emeralds. J Gemmology 18:138-144

Hänni HA, Klein HH (1982) Ein Smaragdvorkommen in Madagaskar. Z Deutsch Gemmol Ges 31:71-77

Hautefeuille P, Perrey A (1888) Sur la reproduction de la phénacite et de l'émeraude. C R Acad Sci Paris 106:1801-1803

Heinrich EWM, Buchi S (1969) Beryl-chrysoberyl-sillimanite paragenesis in pegmatites. Indian Mineral 10:1-7

Henn U (1985) Vergleichende chemische und optische Untersuchungen an Chrysoberyllen verschiedener Lagerstätten. Ph D dissertation, Johannes Gutenberg Universität, Mainz, Germany

Henn U (1995) Edelsteinkundliches Praktikum. Z Deutsch Gemmol Ges 44:1-114

Henn U, Bank H (1988) Hydrothermally grown synthetic emeralds from USSR. Can Gemologist 9:6672

Henn U, Bank H (1991) Aussergewöhnliche Smaragde aus Nigeria. Z Deutsch Gemm Ges 40:181-187.

Hillard PD (1969) Geology and beryllium mineralization near Apache Warm Springs. Socorro County, New Mexico. New Mexico Bur Mines Mineral Resources Circ 103:1-16

Hölscher A (1987) Experimentelle Untersuchungen im System MgO BeO-Al_2O_3-SiO_2-H_2O: MgAl-Surinamite und Be-Einbau in Cordierit und Sapphirin. Unpublished PhD dissertation, Ruhr-Universität, Bochum

Hölscher A, Schreyer W (1989) A new synthetic hexagonal BeMg-cordierite, $Mg_2Al_2BeSi_6O_{18}$, and its relationship to Mg-cordierite. Eur J Mineral 1:21-37

Hölscher A, Schreyer W, Lattard D (1986) High-pressure, high-temperature stability of surinamite in the system MgO-BeO-Al_2O_3-SiO_2-H_2O. Contrib Mineral Petrol 92:113-127

Hsu LC (1983) Some phase relationships in the system BeO-Al_2O_3-SiO_2-H_2O with comments on the effect of HF. Geol Soc China Memoir 5:33-46

Hund F (1970) Phenakit-Mischphasen mit Zn_2SiO_4 als Wirt. Z anorg allgem Chem 374:191-200

Hussain SS, Chaudhury MN, Dawood H (1993) Emerald mineralisation of Barang, Bajaur Agency, Pakistan. J Gemmol 23:402-408

Ito J (1968) Synthetic indium silicate and indium hydrogarnet. Am Mineral 53:1663-1673

Jobin-Bevans S, Černý P (1998) The beryllian cordierite + beryl + spessartine assemblage and secondary beryl in altered cordierite, Greer Lake granitic pegmatite, Southeastern Manitoba. Can Mineral 36: 447-462

Kane RE, Liddicoat RT (1985) The Biron hydrothermal synthetic emerald. Gems Gemology 21:156-170

Kanis J, Harding RR (1990) Gemstone prospects in Central Nigeria. J Gemmol 22:195-202

Kanis J, Arps CES, Zwaan PC (1991) "Machingwe"; a new emerald deposit in Zimbabwe. J Gemmol 22:264-272

Kawakami S, Tabata H, Ishi E (1986) Phase relations in the pseudo-binary system $BeAl_2O_4$-$MgAl_2O_4$. Reports of the Government Industrial Research Institute, Nagoya, 35:235-239 (in Japanese)

Klyakhin VA (1995) Synthesis of chrysoberyl in hydrothermal solutions. Trudy RAN Sibirskogo Otdelniya Ob'yedinennogo instituta geologii geofiziki i mineralogii N832:24-26 (in Russian)

Klyakhin VA, Il'in AG (1991) Experimental study of simultaneous substitution of Mg, Na and Li in hydrothermal beryl. Vsesoyuznoye soveshchaniye po experimental'noy mineralogii Miass 24-26 Sentyabriya 1991 Tezisi dokladov. Chernogolovka, p 51 (abstract, in Russian)

Kodaira K, Iwase Y, Tsunashima A, Matsushita T (1982) High pressure hydrothermal synthesis of beryl crystals. J Crystal Growth 60:172-174

Koivula JI, Keller PC (1985) Russian flux-grown synthetic emeralds. Gems Gemology 21:79-85

Koslowski A, Metz P, Jaramillo HAE (1988) Emeralds from Somondoco, Colombia: Chemical composition, fluid inclusions and origin. N Jahrb Mineral Abh 159:23-49

Lahlafi M (1997) Rôle des micas dans la concentrations des élément légers (Li, Be et F) dans les granites crustaux: étude expérimentale et cristallochimique. PhD dissertation, Université d'Orléans, France, 362 p

Lang SM, Fillmore CL, Maxwell LH (1952) The system beryllia-alumina-titania: Phase relations and general physical properties of three-component porcelains. J Res Nat Bur Standards 48:298-312

Laskovenkov AF, Zhernakov VI (1995) An update on the Ural emerald mines. Gems Gemology 31: 106-113

Laumonier B, Branquet Y, Lopès B, Cheilletz A, Giuliani G, Rueda F (1996) Mise en évidence d'une tectonique compressive Éocène-Oligocène dans l'Ouest de la Cordillère orientale de Colombie, d'après la structure en duplex des gisements d'émeraude de Muzo et de Coscuez. C R Acad Sci Paris 323 Série IIa:705-712

Laurs BM, Dilles, JH, Snee LW (1996) Emerald mineralisation and metasomatism of amphibolite, Khaltaro granitic pegmatite hydrothermal vein system, Haramosh Mountains, northern Pakistan. Can Mineral 34:1253-1286

Lawrence RD, Kazmi AH, Snee LW (1989) Geological setting of the emerald deposits. *In* Kazmi AH, Snee LW (eds) Emeralds of Pakistan; Geology, gemmology and genesis. Van Nostrand Reinhold Co, New York, p 13-38

Lebedev AS (1980) The lower temperature boundary of beryl crystallization at hydrothermal conditions. *In* Godovikov AA (ed) Vyrashivanie kristallov berilliyevykh mineralov i issledovaniye ikh svoystv (Growth of the crystals of Be-containing minerals and investigation of their properties. Akad Nauk SSSR, Sibirskoye Otdeleniye, Institut Geologii i Geophysiki), p 11-21 (in Russian)

Lebedev AS, Klyakhin VA (1983) Influence of crystallization conditions on habit of phenakite crystals. *In* Novoye idei v geneticheskoy mineralogii (New ideas in genetic mineralogy), Nauka, Leningrad, p 72-77 (in Russian)

Lebedev AS, Klyakhin VA, Solntsev VP (1988) Crystal chemical peculiarities of hydrothermal beryl. *In* Materialy po geneticheskoi i experimental'noi mineralogii. Rost i svoistva kristallov. Nauka, Novosibirsk, p 190 (abstract, in Russian)

le Grange JM (1929) The Barbara beryls: A study of an occurrence of emeralds in the North-Eastern Transvaal, with some observations on metallogenetic zoning in the Murchison Range. Trans Geol Soc South Africa 32:1-25

Leitmeier H (1937) Das Smaragdvorkommen im Habachtal in Salzburg und seine Mineralien. Z Krist Mineral Petrogr, Abt B, Mineral petrograph Mitt 49:245-368

Levinson AA (1962) Beryllium-fluorine mineralization at Aguachile Mountain, Coahuila, Mexico. Am Mineral 47:67-74

Linares RC (1967) Growth of beryl from molten salt solutions. Am Mineral 52:1554-1559

Markl G, Schuhmacher J (1997) Beryl stability in local hydrothermal and chemical environments in a mineralized granite. Am Mineral 82:194-202

Martin HJ (1962) Some observations on Southern Rhodesian emeralds and chrysoberyl. Chamber Mines J 4:34-38

Martin-Izard A, Paniagua A, Moreiras D, Acevedo RD, Marcos-Pascual C (1995) Metasomatism at a granitic pegmatite-dunite contact in Galicia: The Franqueira occurrence of chrysoberyl (alexandrite), emerald, and phenakite. Can Mineral 33:775-792

Mashkovtsev RI, Solntsev VP (2002) Channel constituents in synthetic beryl: Ammonium. Phys Chem Minerals 29:65-71

Menon RD, Santosh, M, Yoshida, M (1994) Gemstone mineralization in Southern Kerala, India. J Geol Soc India 44:241-252

Mikhailov MA, Demina TV (1998) Evolution of the compositions and unit cell metrics within rows of the beryllium indialite, cordierite and beryl solid solutions. Zapiski Vserossiyskogo Mineral Obshchestva 127:22-37 (in Russian with English abstract)

Miller RP, Mercer RA (1965) The high temperature behaviour of beryl melts and glasses. Mineral Mag 35:250-276

Miyasaka H (1987) Production of taaffeite single crystals. Japanese Patent Application, open-laid No. 62-187200, August 15 (in Japanese)

Morgan RA, Hummel FA (1949) Reactions of BeO and SiO_2; synthesis and decomposition of phenakite. J Am Ceram Soc 32:250-255

Munson RA (1967) High-temperature behavior of beryl and beryl melts at high pressure. J Am Ceram Soc 50:669-670

Nassau K (1976) Synthetic emerald: The confusing history and the current techniques. Lapidary J 30:196-202, 468-472, 488-492

Nassau K (1980) Gems made by man. Radnor, Pennsylvania, Chilton Book Company, 530 p

Newkirk HW (1964) The system beryllium oxide-water at moderate temperatures and pressures. Inorganic Chem 3:1041-1043

Newton RC (1966) BeO in pegmatitic cordierite. Mineral Mag 35:920-927

Nwe YY, Grundmann G (1990) Evolution of metamorphic fluids in shear zones: The record from emeralds of Habachtal, Tauern Window, Austria. Lithos 25:281-304

Nwe YY, Morteani G (1993) Fluid evolution in the H_2O-CH_4-CO_2-NaCl system during emerald mineralization at Gravelotte, Murchison Greenstone Belt, Northeast Transvaal, South Africa. Geochim Cosmochim Acta 57:89-103

O'Donoghue M (1973) Emerald from North Carolina. J Gem 14:339-253

O'Donoghue M (1988) Gemstones. Chapman and Hall Ltd, London, 372 p

Okrusch D (1971) Zur Genese von Chrysoberyll- und Alexandritlagerstätten. Eine Literaturübersicht. Z Deutsch Gemmol Ges 20:114-124

Ottaway TL, Wicks FJ, Bryndzia LT, Kyser TK, Spooner, ETC (1994) Formation of the hydrothermal emerald deposits in Colombia. Nature 369:552-554

Pankrath R, Langer K (2002) Molecular water in beryl, $^{VI}Al_2[Be_3Si_6O_{18}]\cdot nH_2O$, as a function of pressure and temperature: An experimental study. Am Mineral 87:238-244

Pestryakov EV, Petrov VV, Zubrinov II, Semenov VI, Trunov VI, Kirpichnikov AV, Alimpiev AI (1997) Physical properties of $BeAl_6O_{10}$ single crystals. J Appl Physics 82:3661-3666

Petrusenko S, Arnaudov V, Kostov I (1964/5) Emerald pegmatite from Urdini Lakes, Rila Mountain. Annuaire de L'Université de Sofia, Faculté de Géologie et Géographie 59:247-268 (Bulgarian with English summary)

Petrusenko S Arnaudov V Kostov I (1971) Comparative study of beryls in Bulgaria. Bulgarska Akademia na Naukite, Geologicheski Institut, Izvestiya, Seriya Geokhimia, Mineralogiya i Petrografiya 20:45-68 (Bulgarian with English summary)

Phillips AH, Hess HH (1936) Metamorphic differentiation at the contacts between serpentinite and siliceous country rocks. Am Mineral 21:333-362

Plough FH (1965) A new synthetic emerald. J Gemology 9:426-433

Povondra P, Langer K (1971a) A note on the miscibility of magnesia-cordierite and beryl. Mineral Mag 38:523-526

Povondra P, Langer K (1971b) Synthesis and some properties of sodium-beryllium-bearing cordierite, $Na_xMg_2(Al_{4-x}Be_xSi_5O_{18})$. N Jahrb Mineral Abh 116:1-19

Preinfalk C, Kostitsyn Y, Morteani G (in press) The emerald mineralisation of Capoeirana and Belmont and the pegmatites of the Nova Era-Itabira-Ferros pegmatite district (Minas Gerais, Brazil): Rb-Sr dating and geochemistry. J South Am Earth Sci

Ragu A (1994) Helvite from the French Pyrénées as evidence for granite-related hydrothermal activity. Can Mineral 32:111-120

Ranorosoa N (1996) Notes sur le gisement d'émeraude de Mananjary (Madagascar). Rev Gemmologie 127:4-5

Reeve KD, Buykx WJ, Ramm EJ (1969) The system BeO-Al_2O_3-MgO at subsolidus temperatures. J Australian Ceramic Soc 5:29-32

Ribeiro-Althoff AM, Cheilletz A, Giuliani G, Féraud G, Barbosa Camacho G, Zimmermann JL (1997) $^{40}Ar/^{39}Ar$ and K-Ar- geochronological evidence for two periods (≈2 Ga and 650 to 500 Ma) of emerald formation in brazil. Intl Geol Rev 39:924-937

Riebling EF, Duke DA (1967) BeO-Al_2O_3-SiO_2 system: Structural relationships of crystalline, glassy and molten beryl. J Mater Sci 2:33-39

Robert J-L, Beny J-M, Della Ventura G, Hardy M (1993) Fluorine in micas: Crystal-chemical control of the OH-F distribution between trioctahedral and dioctahedral sites. Eur J Mineral 5:7-18

Robert J-L, Hardy M, Sanz J (1995) Excess protons in synthetic micas with tetrahedrally coordinated divalent cations. Eur J Mineral 7:457-461

Rodionov AYa (1986) Synthesis of Be-minerals in the system BeO-Al_2O_3-SiO_2-halogenide by the method of gas transport reactions. *In* Morfologiya i fazoviye ravnovesiya mineralov. Materialy 13 kongressa Mezhdunarodnoy Mineralogicheskoy Associatsii (IMA) Varna, 19-25 Sentyabriya Sofija, Bulgaria, p 271-277 (in Russian)

Rodionov AYa, Solntsev VP, Veis NS (1987) Crystallization and properties of colored beryls synthesized by the method of gas transport reaction. Trudy Instituta Geologii i Geofiziki SO AN SSSR 679:41-53 (in Russian)

Rodionov AYa, Konovalova TI (1988) Synthesis of beryl by the gas-transport-reaction method using a vapor-liquid-melt machanism. *In* Materialy po geneticheskoy i experimental'noy mineralogii. Rost i svoistva kristallov. Nauka, Novosibirsk, p 191 (abstract, in Russian)

Ross M (1964) Crystal chemistry of beryllium U S Geol Survey Prof Paper 468:1-30

Rupasinghe MS, Dissanayake CB (1985) Charnockites and the genesis of gem minerals. Chem Geol 53: 1- 16

Rykl D, Bauer J (1969) Hydrothermal synthesis of chrysoberyl. II. System Al_2O_3-BeO-H_2O-Na_2O at 450-600°C and 700 bar. Sbornik Vysoke Skolny Chemicko-Technologicke y Praze, Mineralogie 11:75-95 (in Czech)

Rykl D Bauer J (1971) Hydrothermal synthesis of chrysoberyl. III. System Al_2O_3-BeO-H_2O-Na_2O at 450-550°C and 300-3000 bar. Sbornik Vysoke Skolny Chemicko-Technologicke y Praze, Mineralogie 13:111-127 (in Czech)

Sabot B, Cheilletz A, de Donato P, Banks DA, Levresse G, Barrès O (2000) Afghan emeralds face Colombian cousins. Chronique de la Recherche minière, N° 541/2000:111-114

Sabot B, Cheilletz A, De Denato P, Banks D, Levresse G, Barrès O (2001) The Panjshir-Afghanistan Emerald deposits: New field and geochemical evidence for Colombian style mineralization. Abstract EUG Strasbourg, Abstr Vol, p 548

Sarazin G, Forestier H (1959) Étude de la substitution d'ions Al^{3+} dans le chrysobéryl. C R Acad Sci Paris 248:2208-2210

Schäfer W, Henn U, Schwarz D, (1992) Smaragde aus der Kesselklamm, Untersulzbachtal, Österreich; Vorkommen und Eigenschaften. Der Aufschluss 43:231-240

Schmetzer K (1988) Characterization of Russian hydrothermally grown synthetic emerald. J Gemology 21:145-164

Schmetzer K (1996) Growth method and growth-related properties of a new type of Russian hydrothermal synthetic emerald. Gems Gemology 32:40-43

Schmetzer K (2001) Smaragde aus dem Labor. *In* D Schwarz, R Hochleitner (eds) extraLapis Smaragd. Der kostbarste Beryll, der teuerste Edelstein 21:82-90

Schmetzer K, Kiefert L, Bernhardt H-J, Zhang Beili (1998) Im Hydrothermalverfahren gezüchtete synthetische Smaragde aus China. Gemm Z 9:93-90

Schmetzer K, Bernhardt H-J, Medenbach O (1999) Heat-treated Be-Mg-Al oxide (originally musgravite or taaffeite). J Gemmology 26:353-356

Schwarz D (1990a) Die chemischen Eigenschaften der Smaragde II. Australien und Norwegen. Z Deutsch Gemmol Ges 40:39-66

Schwarz D (1990b) Die chemischen Eigenschaften der Smaragde I, Brasilien. Z Deutsch Gemmol Ges 40: 233-272

Schwarz D (1990c) Die brasilianische Smaragde und ihre Vorkommen: Santa Terezinha de Goias/Go. Z Deutsch Gemm Ges 39:13-44

Schwarz D, Henn U (1992) Emeralds from Madagaskar. J Gemm 23:140-149

Schwarz D, Giuliani G (2001) Südamerika (1): Kolumbien. *In* D Schwarz, R Hochleitner (eds) extraLapis Smaragd. Der kostbarste Beryll, der teuerste Edelstein 21:38-45

Seck H, Okrusch M (1972) Die Phasenbeziehungen und Reaktionen im System BeO-Al_2O_3-SiO_2-H_2O. Fortschr Mineral 50:91-92

Semenov YeI (1957) New hydrous silicates of beryllium-gelbertrandite and spherobertrandite. Trudy Inst Mineralogii Geokhimii Kristallokhimii Redkikh Elementov 1:64-69 (in Russian)

Sharma KK, Ganguli D (1973) Crystallization of phenakite structures with Zn-Be and Mg-Be substitutions. Ceramurgia 3:155-158

Simpson ES (1932) Contributions to the mineralogy of Western Australia Series VII. J Royal Soc Western Australia 18:61-65

Sinkankas J (1981) Emeralds and other beryls. Geoscience Press, Prescott, Arizona, 665 p

Sinkankas J, Read PG (1986) Beryl. Buttherworth Gem Books, London, 225 p

Sliwa AS, Nguluwe CA (1984) Geological setting of Zambian emerald deposits. Precambrian Res 25: 213-228

Sobolev BP, Novoselova AV (1959) The role of fluorine compounds in the transportation of beryllium and the formation of phenakite. Geochemistry 1959:21-32

Solntsev VP, Harchenko EI, Bukin GV, Klyakhin VA, Lebedev AS, Lohova GG, Ripinen OI (1978) Study of micro-isomorphic substitutions in natural and synthetic beryls. *In* Issledovanija po experimental'noj mineralogii [Study on experimental mineralogy], Godovikov AA (ed) Institut geologii i geofiziki, SO AN SSSR, Novosibirsk, 1978:39-54 (in Russian)

Soman K, Nair NGK (1985) Genesis of chrysoberyl in the pegmatites of southern Kerala, Indian Mineral Mag 49:733-738

Soman K, Nair NGK, Druzhinin AV (1986) Chrysoberyl pegmatites of South Kerala and their metallogenic implications. J Geol Soc India 27:411-418

Souza JL (1992) A jazida de esmeraldas de Itabira, Minas Gerais. *In* Schobbenhaus C, Teixeira QE, Silva CCE (eds) Gems and ornamental rocks, Departamento Nacional da Producao Mineral, Brazil, Companhia de Pesquisa de Recursos Minerais, p 223-343

Strunz H (1966) Mineralogische Tabellen. Akademische Verlagsgesellschaft, Leipzig, 560 p

Sunagawa I (1982) Gem materials, natural and artificial. Current Topics Mater Sci 10:353-497

Syromyatnikov FV, Makarova AP, Kupriyanova II (1972) Experimental studies of beryl and phenacite in aqueous solutions. Intl Geol Rev 14:837-839

Tacker R (1999) Preliminary observations of the emerald deposits of Hiddenite, North Carolina, USA. Geol Soc Am Ann Meeting Abstr Progr 31:A-306

Takubo H, Kume S, Koizumi M (1971) Crystal growth of some silicate minerals under high pressure. Mineral Soc Japan, Spec Pap 1 (IMA, 7th Gen Meet, 1970, Papers Proc), p 47-51

Taylor AM (1967) Synthetic vanadium emerald. J Gemm 10:211-217

Teraishi K (1984) Synthesis of artificial taaffeite single crystals. Japanese Patent Application, open-laid No. 59-141484, August 14 (in Japanese)

Thomas A (1993) The emerald mines of Madagaskar. South African Gemmol 7:3-11

Thomas AE (1994) Moçambique emerald. South African Gemmol 8:10-11

Thomas VG, Klyakhin VA (1987) The specific feature of chromium incorporation into beryl at hydrothermal conditions. *In* Sobolev NV (ed) Mineraloobrazovaniye v endogennykh protsessakh (The mineral formation in endogenetic processes) Novosibirsk, Nauka, p 60-66 (in Russian)

Thomas VG, Lebedev AS (1982) Kinetics of hydrothermal crystallization of beryl from the oxides. *In* Fisiko-khimicheskiye issledovaniya mineraloobrazuyuschikh system, p 98-104. Novosibirsk, Akad Nauk SSSR Sibirskoye Otdeleniye, Institut Geologii i Geophysiki, Novosibirsk, p 98-104 (in Russian)

Thurm RE (1972) The lake Manyara emeralds of Tanzania. J Gemmology 13:98-99

Ustinov VI, Chizhik OYe (1994) Sequential nature of the formation of emerald and alexandrite in micaite-type deposits. Geochem Intl 31:115-118

Webster R (1950) Some notes on Indian emeralds. Gems Gemology 6:344-345

Webster R, Anderson BW (1983) Gems: Their sources, descriptions and identification. Archon Books, Hamden, Connecticut, 1006 p

Wilson W (1965) Synthesis of beryl under high pressure and temperature. J Appl Physics 36:268-270

Wood SA (1991) Speciation of Be and solubility of bertrandite/phenakite minerals in hydrothermal solutions. *In* Pagel M, Leroy JL (eds) Source, transport and deposition of metals. Balkema, Rotterdam, p 147-150

Wood SA (1994) Theoretical prediction of speciation and solubility of beryllium in hydrothermal solution to 300°C at saturated vapour pressure: Application to bertrandite/phenakite deposits. Ore Geol Rev 7:249-278

Wyart J, Ščavničar S (1957) Synthèse hydrothermale du béryl. Bull Soc franç Minéral Cristallogr 80: 395-396

Zwaan JC, Kanis J, Petsch J (1997) Update on emeralds from the Sandawana mines, Zimbabwe. Gems Gemology 33:80-101

Non-pegmatitic Deposits of Beryllium: Mineralogy, Geology, Phase Equilibria and Origin

Mark D. Barton and Steven Young

Center for Mineral Resources
Department of Geosciences
University of Arizona
Tucson, Arizona 85721

barton@geo.arizona.edu

INTRODUCTION

Non-pegmatitic occurrences of Be minerals constitute a diverse set of geologic environments of considerable mineralogical and petrological interest; they currently provide the majority of the world's Be ore and emeralds and they contain the greatest resource of these commodities. Of the approximately 100 Be minerals known (see Chapter 1 by Grew; Appendix A), most occur in hydrothermal deposits or non-pegmatitic igneous rocks, where their distribution varies systematically with the setting and origin (Table 1, Fig. 1).

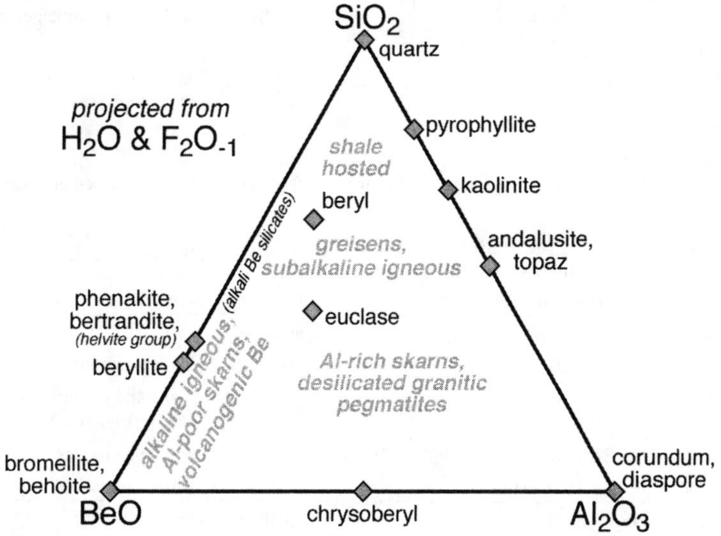

Figure 1. Chemography of the principal solid phases in the BeO-Al_2O_3-SiO_2-H_2O(-F_2O_{-1}) "BASH" system with the projected positions of helvite group and alkali Be silicates. Also shown are generalized fields for some of the major types natural of occurrences (cf. Table 1, Fig. 4; see text for discussion).

Beryllium minerals are best known from geologic systems associated with felsic magmatism. They also occur in a variety of settings that lack evident igneous associations. Environments range from the surface to the deep crust and host rocks range from feldspathic to carbonate to ultramafic in composition. Genetically related igneous rocks are felsic and share low calcium and high F contents, but are diverse in composition, setting and origin. Compositions range from strongly peraluminous to

1529-6466/00/0050-0014$10.00

Table 1. Beryllium minerals in non-pegmatitic deposits: formulas, use and occurrence.[1]

Mineral	*Formula*	*Use*	*Family*[2]	*Occurrence*
Behoite (Bht) [3]	$Be(OH)_2$	ore?	BASH	alkalic pegmatite, skarns, rhyolites
Bertrandite (Brt)	$Be_4Si_2O_7(OH)_2$	ore	BASH	granitic and alkalic pegmatites, greisens, skarns, veins, rhyolites
Beryl (Brl)	$Be_3Al_2Si_6O_{18}$	ore, gem	BASH	granitic pegmatites, greisens, skarns, veins, rhyolites
Beryl v. Emerald	$Be_3(Al,Cr,V)_2Si_6O_{18}$	gem	BASH	granitic pegmatites & metamorphosed equivalents, veins, metamorphic rocks
Beryllite (Byl)	$Be_3SiO_4(OH)_2{\cdot}H_2O$		BASH	alkalic pegmatite
Bromellite (Brm)	BeO		BASH	skarns, desilicated pegmatites
Chrysoberyl (Ch)	$BeAl_2O_4$	ore, gem	BASH	granitic pegmatites, skarns
Clinobehoite (Cbe)	$Be(OH)_2$		BASH	desilicated pegmatites
Euclase (Euc)	$BeAlSiO_4(OH)$	gem	BASH	granitic pegmatites, greisens, skarns, veins
Phenakite (Ph)	Be_2SiO_4	ore, gem	BASH	alkaline & granitic pegmatites, skarns, greisens, veins
Bazzite (Bz)	$Be_3(Sc,Al)_2Si_6O_{18}$		BASH+	alkalic and granitic pegmatites, veins
Magnesiotaaffeite-2*N*'2*S* (Taf) ("Taaffeite", "Taprobanite")	$BeMg_3Al_8O_{16}$	gem	BASH+	Mg-Al schists (metamorphosed pegmatite?), skarns
Magnesiotaaffeite-6*N*'3*S* (Mgr) ("Musgravite")	$BeMg_2Al_6O_{12}$	gem	BASH+	metamorphosed pegmatites
Stoppaniite (Spp)	$(Na,\square)(Fe^{3+},Al,Mg)_2$-$Be_3Si_6O_{18}{\cdot}H_2O$		BASH+	alkaline volcanic
Surinamite (Sur)	$Mg_3Al_4(BeSi_3O_{16})$		BASH+	metamorphosed pegmatites
Aminoffite (Am)	$Ca_3Be_2Si_3O_{10}(OH)_2$		alkaline	skarns
Barylite (Bar)	$BaBe_2Si_2O_7$	ore	alkaline	alkalic pegmatite; skarns; greisens
Bavenite (Bav)	$Ca_4Be_2Al_2Si_9O_{26}(OH)_2$		alkaline	alkalic and granitic pegmatites, veins, skarns, greisens
Chkalovite (Chk)	$Na_2BeSi_2O_6$		alkaline	alkalic pegmatites
Epididymite (Epd)	$Na_2Be_2Si_6O_{15}{\cdot}H_2O$	ore?	alkaline	alkalic pegmatites, skarns
Eudidymite (Eud)	$Na_2Be_2Si_6O_{15}{\cdot}H_2O$		alkaline	alkalic pegmatites
Gadolinite[4]**-(Y), –(Ce)** (Gad)	$Be_2Fe(Y,REE)_2Si_2O_{10}$	ore	alkaline	alkaline pegmatites and granites, veins, greisens
Gugiaite (Gug)	$Ca_2BeSi_2O_7$		alkaline	skarns
Hingganite[4]-(Y) (Hin)	$Be_2(\square,Fe)(Y,REE)_2Si_2O_8$-$(OH,O)_2$		alkaline	alkaline pegmatites
Hsianghualite (Hsh)	$Ca_3Li_2Be_3(SiO_4)_3F_2$		alkaline	skarns
Hyalotekite (Htk)	$(Ba,Pb,K)_4(Ca,Y)_2Si_8$-$(B,Be)_2(Si,B)_2O_{28}F$		alkaline	Fe-Mn "skarns"; alkaline pegmatites
Joesmithite (Jo)	$PbCa_2(Mg,Fe^{2+},Fe^{3+})_5[Si_6Be_2O_{22}](OH)_2$		alkaline	Fe-Mn "skarns"

Leifite (Lf)	$(Na,\square)(H_2O,\square)Na_6Be_2[Al,Si,Zn)_3Si_{15}O_{39}F_2]$		alkaline	alkaline pegmatites
Leucophanite (Lph)	$CaNaBeSi_2O_6F$	ore?	alkaline	alkaline pegmatites; skarns
Lovdarite (Lv)	$K_2Na_6(Be_4Si_{14}O_{36})\cdot 9H_2O$		alkaline	alkaline pegmatites
Meliphanite (Mph)	$Ca_4(Na,Ca)_4Be_4AlSi_7O_{24}(F,O)_4$		alkaline	alkaline pegmatites; skarns
Milarite (Mil)	$K(\square,H_2O,Na)_2(Ca,Y,REE)_2(Be,Al)_3Si_{12}O_{30}$		alkaline	alkaline & granitic pegmatites; skarns; veins
Odintsovite (Od)	$K_2(Na,Ca,Sr)_4(Na,Li)Ca_2(Ti,Fe^{3+},Nb)_2O_2[Be_4Si_{12}O_{36}]$		alkaline	alkaline veins
Roggianite (Rg)	$Ca_2[Be(OH)_2Al_2Si_4O_{13}]\bullet <2.5H_2O$		alkaline	veins, pegmatites
Samfowlerite (Sf)	$Ca_{14}Mn_3Zn_2(Zn,Be)_2Be_6(SiO_4)_6(Si_2O_7)_4(OH,F)_6$		alkaline	Fe-Mn "skarns"
Semenovite-(Ce) (Sem)	$(Ce,La,REE,Y)_2Na_{0\text{-}2}(Ca,Na)_8(Fe,Mn)(Si,Be)_{20}(O,OH,F)_{48}$		alkaline	alkaline pegmatites
Sorensenite (Ss)	$Na_4SnBe_2Si_6O_{18}\cdot 2H_2O$		alkaline	alkaline veins
Sverigeite (Sv)	$Na(Mn,Mg)_2Sn[Be_2Si_3O_{12}(OH)]$		alkaline	Fe-Mn "skarns"
Trimerite (Trm)	$(Mn_2Ca)[BeSiO_4]_3$		alkaline	Fe-Mn "skarns"
Tugtupite (Ttp)	$Na_4BeAlSi_4O_{12}Cl$		alkaline	alkalic pegmatites & veins
Wawayandaite (Ww)	$Ca_{12}Mn_4B_2Be_{18}Si_{12}O_{46}(OH,Cl)_{30}$		alkaline	Fe-Mn "skarns"
Welshite (Wsh)	$Ca_2Mg_{3.8}Mn^{2+}_{0.6}Fe^{2+}_{0.1}Sb^{5+}_{1.5}O_2[Si_{2.8}Be_{1.7}Fe^{3+}_{0.65}Al_{0.7}As_{0.17}O_{18}]$		alkaline	Fe-Mn "skarns"
Danalite (Dn)	$Fe_4Be_3Si_3O_{12}S$	ore	helvite	skarns, granitic pegmatites
Genthelvite (Gnt)	$Zn_4Be_3Si_3O_{12}S$		helvite	alkaline pegmatites, carbonatite
Helvite (Hlv)	$Mn_4Be_3Si_3O_{12}S$		helvite	veins, skarns, greisens, alkaline and granitic pegmatites
Babefphite (Bf)	$BaBe(PO_4)F$		non-silicate	placer (alkaline igneous?)
Bearsite (Bs)	$Be_2(AsO_4)(OH)\cdot 4H_2O$		non-silicate	polymetallic porphyry
Bergslagite (Bsg)	$CaBeAsO_4(OH)$		non-silicate	Fe-Mn "skarns"
Beryllonite	$NaBePO_4$		non-silicate	granite and granitic pegmatites
"Glucine" (Gl)	$CaBe_4(PO_4)_2(OH)_4\cdot 0.5H_2O$		non-silicate	weathering
Hambergite (Hmb)	$Be_2(OH,F)BO_3$		non-silicate	alkaline and granitic pegmatites
Herderite (Hrd)	$CaBePO_4(F,OH)$		non-silicate	greisens, granitic pegmatites
Hurlbutite (Hrb)	$CaBe_2(PO_4)_2$		non-silicate	granitic pegmatites, veins
Moraesite (Mr)	$Be_2(PO_4)(OH)\cdot 4H_2O$		non-silicate	granitic pegmatites, weathering
Swedenborgite (Sw)	$NaBe_4SbO_7$		non-silicate	Fe-Mn "skarns"
Uralolite (Ur)	$Ca_2Be_4(PO_4)_3(OH)_3\cdot 5H_2O$		non-silicate	greisen, granitic pegmatites
Berborite (Bb)	$Be_2BO_3(OH,F)\cdot H_2O$		non-silicate	pegmatite, alkaline igneous(?), skarn

[1] More common minerals in **bold**. Compiled from Mandarino (1999), Strunz and Nickel (2001) and Appendix 1 of Chapter 1 of this volume.

[2] Minerals are grouped into the four families by common chemical characteristics: (1) predominantly $BeO\text{-}Al_2O_3\text{-}SiO_2\text{-}H_2O$ the "BASH" group, including a subgroup "BASH+" for minerals also containing Mg, Fe, Sc and Na, (2) Na-Ca-K silicates—"alkaline" group, (3) $M_4Be_3Si_3O_{12}S$ helvite group, and (4) complex non-silicates (phosphates, borates, arsenates, etc.).

[3] Abbreviations for Be- and other minerals are taken from Kretz (1983) or constructed to be consistent with that paper. They are used in most figures, Table 2 and Appendix A. In alphabetical order (by abbreviation) these are: Act (actinolite), Ad (K-feldspar var. adularia), Aeg (aegerine), Agt (aegerine-augite), Ab (albite),

Table 1 footnotes, continued.

[3] Abbreviations, continued: Am (amphibole), Amz (K-feldspar var. amazonite), Anc (analcime), And (andalusite), Ap (apatite), Ath (anthophyllite), Bt (biotite), Cal (calcite), Carb (carbonates), Chl (chlorite), Chr (chromite), Col (columbite), Cpx (Ca-clinopyroxene), Crn (corundum), Cst (cassiterite), Cyl (cryolite), Dsp (diaspore), Drv (dravite), Ep (epidote), Eud (eudialyte), Fa (fayalite), Fl (fluorite), fo (forsterite)Fs (feldspar), Ghn (gahnite), Grt (garnet), Hbl (hornblende), Hdd (spodumene var. hiddenite), Hem (hematite), Kfs (K-feldspar), Kln (kaolinite), Ky (kyanite), Mag (magnetite), Mc (microcline), Mnz (monazite), Mo (molybdenite), Ms (muscovite), Ne (nepheline), Ntr (natrolite), Ofs (oligoclase), Pas (parisite), Phl (phlogopite), Pll (polylithionite), Pl (plagioclase), Prl (pyrophyllite), Px (pyroxene), Py (pyrite), Qtz (quartz), Rbk (riebeckite), Sch (scheelite), Sid (siderite), Sdl (sodalite), Sid (siderophyllite), Tlc (talc), Toz (topaz), Tr (tremolite), Ttn (titanite), Tur (tourmaline), Ves (vesuvianite [idocrase]), W (water), Wlf (wolframite), Znw (zinnwaldite), Zrn (zircon).

[4] Most investigators have not distinguished gadolinite-(Y) and gadolinite-(Ce), so gadolinite-group minerals are simply referred to in the text as “gadolinite”. Similarly, hingganite-group minerals are simply referred to in the text as “hingganite.”

peralkaline and can be silica undersaturated. Beryllium minerals also occur in metamorphic and basinal environments and are redistributed by surface processes. Table 2 summarizes the types and significance of major groups of occurrences by their lithologic associations. Figure 2 shows the global distribution of some important examples and regional belts. For most types, at least one example has been described in some detail and can be used to help evaluate general patterns; however, even in these only rarely has Be been the principal economic interest.

Few papers cover this spectrum of deposits. The classic synthesis studies are from the Soviet literature (e.g., Beus 1966; Vlasov 1968; Zabolotnaya 1977; Ginzburg et al. 1979; Grigor'yev 1986) with few extensive summaries in the western literature (e.g., Warner et al. 1959; Mulligan 1968; Sinkankas 1981). The golden age of investigation was in the 1950s and 1960s, driven by exploration interest in the U.S. and the (then) Soviet Union, with most papers published between about 1960 and 1985. Much quality work was done by Soviet scientists, a moderate amount of which is available in English translation. Unfortunately many of the detailed studies are in limited-distribution monographs and reports that are difficult to access. Many compendia of papers dealing with aspects of rare metal systems have been published that contain related papers (Evans 1982; Hutchison 1988; Taylor et al. 1988; Moeller et al. 1989; Stein et al. 1990; Seltmann et al. 1994; Pollard 1995b; Kremenetsky et al. 2000b and earlier volumes). Continuing work on Be-bearing magmatic systems, particularly pegmatites, is reviewed by _ern_ (this volume) and London and Evensen (this volume).

This chapter reviews the principal types of non-pegmatitic Be occurrences—magmatic, hydrothermal, metamorphic and surface-related—covering aspects of their mineralogy, stability, geologic framework, genesis and global distribution. Although there is a continuum between pegmatitic and non-pegmatitic occurrences, granitic pegmatites are only briefly mentioned here. In spite of the considerable study that the non-pegmatitic occurrences have received as possible sources of Be as a commodity or of Be minerals as gems or specimens, there remains a great deal to be learned about the characteristics and origins of these systems.

Economic sources of beryllium and beryllium minerals

Beryllium ore. Prior to about 1970, the main source of Be was hand-picked pegmatitic beryl typically from small, labor-intensive operations. New uses for Be in nuclear and other high-tech applications motivated extensive exploration campaigns for Be and other rare metals from the 1940s through the early 1960s. These efforts resulted in the discovery in the Soviet Union, the United States, and Canada of many significant occurrences of non-pegmatitic Be mineralization. The Spor Mountain, Utah Be deposits, the world's most important source of Be (Cunningham 2000), were discovered during intensive regional Be exploration in 1959 and began producing in 1969. This exploration was aided by the recognition of the association of Be with chemically evolved felsic igneous rocks, the occurrence with F-rich rocks, and the development of neutron-sourced gamma ray spectrometers ("berylometers", Brownell 1959), which enabled rapid semi-quantitative assay in the field of the Be content of rocks (e.g., Meeves 1966).

Global production of Be in 2000 was 226 tonnes (t) of metal equivalent of which about 75% (180 t) was produced in the U.S. from the Spor Mountain operation of Brush Wellman Corporation (Cunningham 2000). In 1998, Brush Wellman reported reserves for the Spor Mountain district of 7 million tonnes (Mt) at 0.26% Be (0.72% BeO) or about 18,300 t of contained metal. Global production was down from 289 t in 1998 and represents less than half of world capacity. Consumption in 1998 (390 t) was substantially larger and was supported by sales of ore from U.S. government stockpiles. A total value of $140 million was based on quoted prices for Be-Cu master alloy, the main product.

Presently there is little economic incentive for Be exploration, because the Spor Mountain district alone contains roughly 50 years of resource at current consumption rates and large, sub-economic resources have been identified in a number of other areas (Fig. 3, see Appendix A). Solodov (1977) gave general estimates for types of Be deposit as a function of age, setting, and type. His estimates totaled >100,000 t of contained Be metal of which half is in non-pegmatitic deposits with grades ≥0.05% Be. Many times this amount likely exist in the numerous unevaluated occurrences that resemble the better known deposits (data compiled in Appendix A indicate >200,000 t of contained Be).

Gems. Non-pegmatitic deposits are also major sources of gems, notably emerald, aquamarine, red beryl and alexandrite (chrysoberyl). Desilicated granitic pegmatites and veins in ultramafic and mafic rocks provide emerald, chrysoberyl, and some phenakite (Beus 1966; Sinkankas 1981). Shear-zone and vein-type emerald deposits are also important, especially the black shale-hosted deposits of Colombia (Snee and Kazmi 1989; Cheilletz 1998). Most aquamarine occurrences are pegmatitic, however some gem material comes from miarolitic cavities, greisens and veins, and a considerable fraction is reworked by surficial processes into placer deposits. Many of the hard rock occurrences also produce sought-after specimens of other Be minerals such as phenakite and bertrandite (Sinkankas 1981; Jacobson 1993a). In 1999 U.S. production of beryl gemstones totaled approximately $3 million and U.S. consumption of cut emeralds (~1/3 world total) amounted to about 5 million carats (1,000 kg) worth approximately $180 million (Olson 2000). Global resource estimates for Be gemstones do not exist.

Although economic deposits of Be and Be gems are limited to Spor Mountain, granitic pegmatites, and a large handful of gem producing districts, the varied occurrence of and popular and scientific interest in Be minerals merit a more general treatment.

TYPES OF DEPOSITS

We group Be deposits by geologic setting (Table 2) specifically emphasizing differences in (1) associated sources (magmas or other materials) and (2) depositional environment (magmatic or metasomatic, and the host). Figure 4 illustrates the general geologic environments for the major groups of occurrences. Beryllium deposits naturally divide into igneous-related and non-magmatic types. They divide further by the nature of the associated magma and the host rock. As explained below, host rock and magma compositions exert strong controls on Be mineralogy as a function of their acidity-

Table 2. Main types of beryllium occurrences: significance, principal beryllium minerals, and examples.[1]

association		igneous		metasomatic		
type	variety	magmatic	pegmatitic	aluminosilicate (greisen, vein)	carbonate (skarn, replacement)	mafic/ultramafic (blackwall, vein)
Igneous connection direct						
granite	metaluminous to peraluminous	• common? / Be resource? • Li micas, beryl • Beauvoir, France; Sheeprock, USA	• abundant / gems & major source of Be • beryl • Minas Gerais, Brazil; Bernic Lake, Canada	• abundant / gems±Be resource(?) • beryl, phenakite • Sherlova Gora, Russia; Mt. Antero, USA; Aqshatau, Kazakhstan	• abundant / Be resource • phenakite, chrysoberyl, bertrandite, helvite gr. • Lost River, USA; Mt. Wheeler, USA; Mt. Bischoff, Australia	• common / emerald • beryl, chrysoberyl, phenakite • Reft River, Russia; Khaltaro, Pakistan; Carnaiba, Brazil
rhyolite	metaluminous to peraluminous	• common / — • Be in glass or micas • Macusani, Peru; topaz rhyolites, western USA	• —	• rare / gem red beryl • beryl • Wah Wah Mtns, USA; Black Range, USA; (cf. Spor Mtn. USA)	• uncommon / principal source of Be • bertrandite • Spor Mtn, USA; Aguachile, Mexico	• —
granite	peralkaline	• rare? / — • ? • Khaldzan-Burgtey, Mongolia	• uncommon / Be resource • gadolinite, phenakite • Strange Lake, Canada	• rare / Be resource • phenakite, helvite, gadolinite • Verknee Espee, Kazakhstan	• rare / Be resource • phenakite, bertrandite, leucophanite • Ermakovskoe, Russia	• —
rhyolite	peralkaline	• rare? / Be resource? • ? • Brockman, Australia	• —	• —	• rare / Be resource • bertrandite • Sierra Blanca USA	• —
syenites	peralkaline	• uncommon / Be resource? • chkalovite, epididymite • Ilímaussaq, Greenland; Wind Mtn, USA	• common / — • epididymite, eudidymite, chkalovite • Lovozero, Russia; Ilímaussaq, Greenland	• rare / Be resource • barylite, eudidymite • Seal Lake, Canada; Thor Lake, Canada	• —	• —

Igneous connection indirect or absent ("non-magmatic")						
metamorphic	shear / vein	• —	• —	• uncommon / specimens • phenakite, milarite, bavenite • Swiss & Italian Alps	• uncommon? / emerald • beryl • Mingora, Pakistan; Brumado, Brazil	• uncommon? / emerald source • beryl • Habachtal, Austria;
basin	vein	• —	• —	• uncommon / premier emerald source • beryl (euclase) • Muzo & Chivor, Colombia	• —	• —
miscellaneous	hydrothermal	• common / —; Be minerals and enrichments are present in a variety of Mn-rich hydrothermal systems and Fe-Mn oxide-rich deposits including hot-spring systems (Butte, USA, Silverton, USA; Långban, Sweden; Golconda, USA)				
surface	placer	• common / source of gems; most placer Be minerals are believed to be derived from pegmatites or metamorphosed pegmatites, but they can come from many sources of resistant Be minerals (Sri Lanka; Madagascar; Minas Gerais, Brazil)				
surface	supergene	• rare(?)/—; little documented, BASH minerals survive; helvite group & Na-Ca silicates mostly weather; local supergene transport				

[1] For each type of occurrence, the first bullet summarizes the number of occurrences and the economic importance, if any; the second bullet list the main Be-bearing phases, and the third bullet names one or more prominent localities. A dash indicates absence.

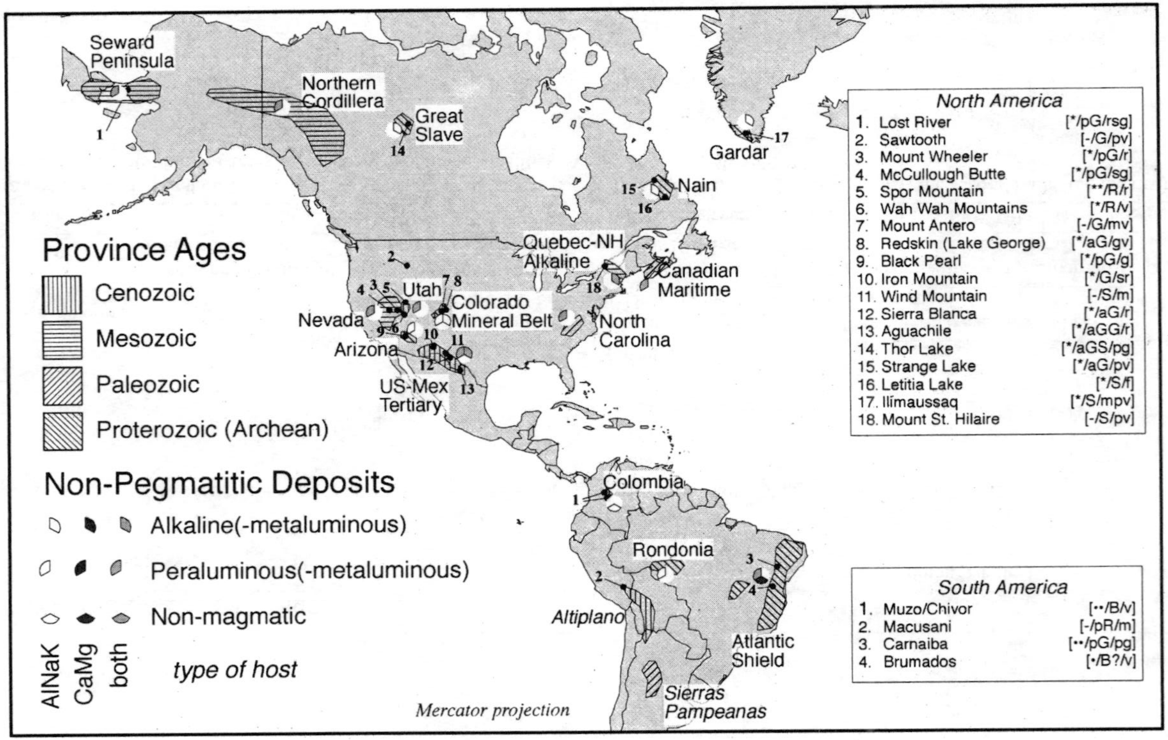

Figure 2. Map showing the location of non-pegmatitic Be occurrences and the Be-bearing belts that are mentioned in this paper. Deposit-type assignments can be uncertain or generalized depending on available information and the complexity of the region. The symbolism for the key deposits is: significance (0-2 stars) / petrologic association **(figure and caption continued next page >>)**

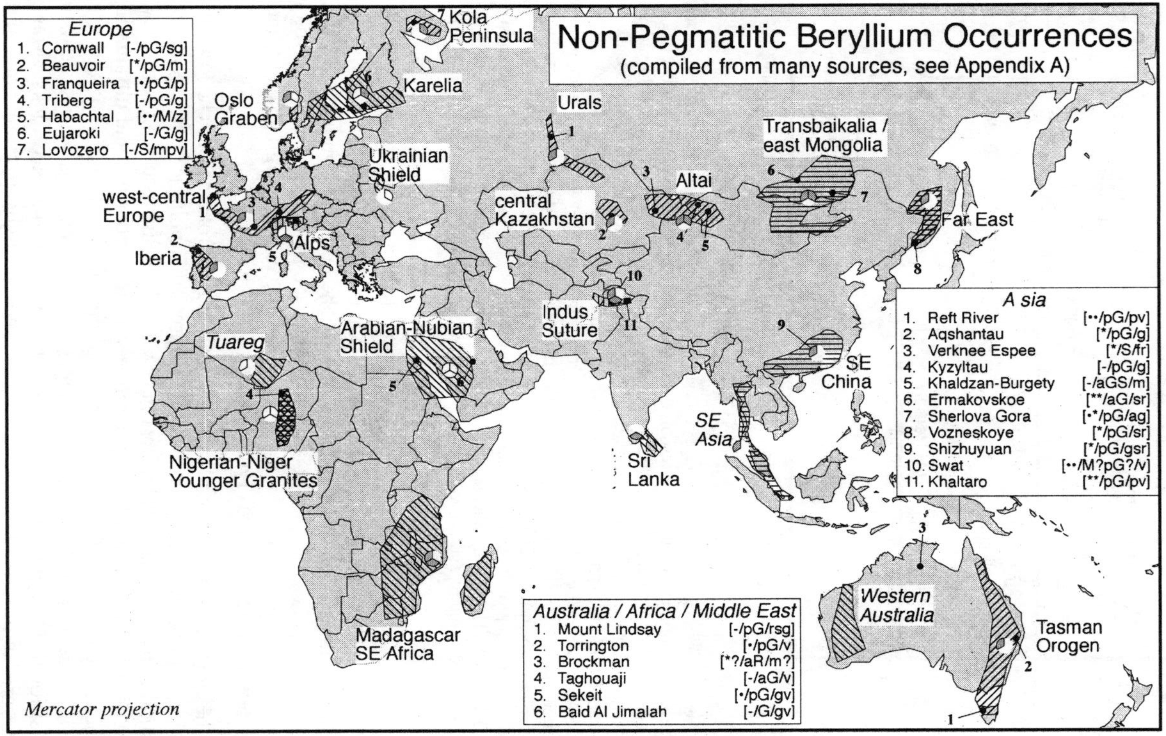

(G = granitoid, S = syenite, R = rhyolite, B = basinal; M = metamorphic; for igneous rocks: p = strongly peraluminous m = principally metaluminous; a = peralkaline) / deposit style (f = fenites, g = greisen, m = magmatic, p = pegmatitic, r = replacement, s = skarn, v = veins, z = reaction zones). See text for discussion. Compiled from multiple sources summarized in Appendix A.

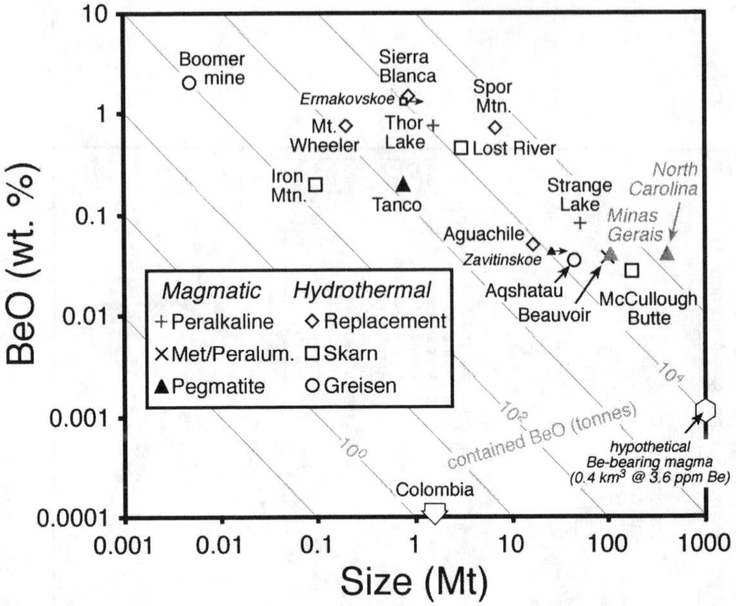

Figure 3. BeO concentrations and tonnage for some better documented Be-bearing mineral deposits. These are a mixture of published resource estimates and geologic inventories reflecting the sparse data available for Be occurrences. A considerable fraction in some systems likely resides as isomorphic substitutions in micas or other silicates (e.g., Beauvoir, McCullough Butte). Data and sources are summarized in Appendix A except for the pegmatite deposits (in black; Tanco: Sinclair 1996, Zavintoskoe: Kremenetsky et al. 2000a) or districts (in gray; Minas Gerais: Sinclair 1996; North Carolina tin belt: Griffitts 1954). Most Russian deposits lack tonnages, but grade and minimum sizes are given by Kremenetsky et al. (2000a). The two highest grade systems with the highest rank (size) are plotted at their minimum reported sizes (Zavitinskoe and Ermakovskoe, which are italicized). The point labeled "hypothetical Be-bearing magma" illustrates the small amount of magma required to make a world-class Be deposit compared to 100 km^3 or more for most other metals.

basicity and their degree of silica saturation. Emerald deposits are commonly treated as a group unto themselves (Sinkankas 1981; Snee and Kazmi 1989; Cheilletz 1998); here, we also treat them separately, but group them by origin. The text and Figure 4 are organized around this geological classification in order to emphasize mineralogical and petrological similarities, whereas Appendix A and Figure 2 are organized geographically and can serve as an index to the text via the "types" columns.

Within the igneous-related group, there is a continuum from Be-enriched magmas to complex behavior in pegmatites (London and Evensen, this volume) to the wide variety of hydrothermal deposits considered in this paper. The latter include skarns, replacement bodies, greisens and veins which form in aluminosilicate, carbonate, and ultramafic host rocks (cf. Shcherba 1970). Most non-pegmatitic accumulations form in the upper crust, typically in the upper 5 km. Mineral assemblages and compositions vary systematically with compositional variations of host rocks and related igneous rocks. Magmatic compositions are uniformly felsic but range from strongly peraluminous through metaluminous to peralkaline. Most source rocks are quartz-rich with the important exception of silica-undersaturated syenitic suites (Fig. 5A). Apart from sharing highly felsic compositions, igneous-related systems are chemically diverse (Fig. 5B). Likewise, tectonic settings are quite varied although moderately thick continental crust and late- or post-orogenic timing are common themes. It is the shared low CaO and elevated F

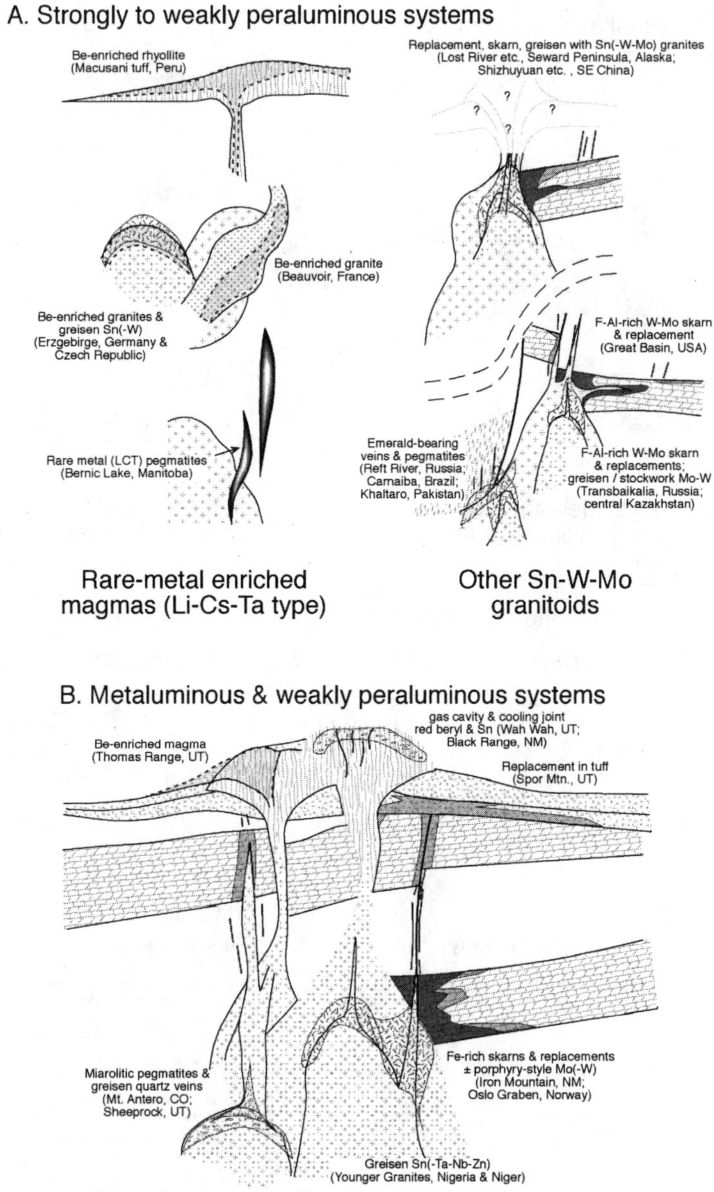

Figure 4. Sketches illustrating the main types of Be deposits. (A) Deposits associated with strongly peraluminous magmatism. The distinction between the Li-Cs-Ta enriched group and the others is gradational, see text for details. (B) Deposits associated with metaluminous to weakly peraluminous magmas. These rarely have strongly peraluminous and peralkaline phases.

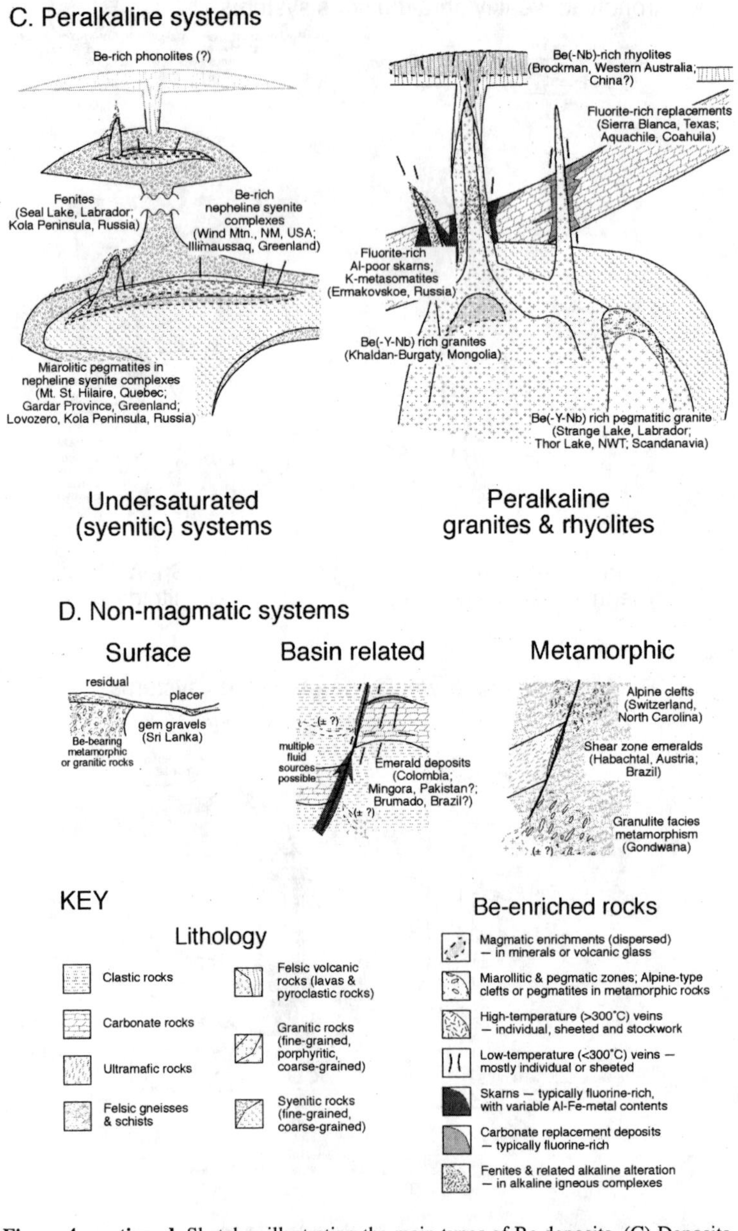

Figure 4, continued. Sketches illustrating the main types of Be deposits. (C) Deposits associated with peralkaline magma types. These are further divided by silica saturation into undersaturated (nepheline syenites) and oversaturated (granites and quartz syenites). (D) Non-magmatic systems of diverse origins. Examples are listed in Tables 2 and Appendix A; locations are shown in Figure 2. See text for further description and discussion.

contents, and not magma sources or other intensive variables such as oxidation state or water content, that probably favor magmatic and post-magmatic Be enrichment (Fig. 5B,C).

Traditionally, the magmas associated with rare metals (e.g., Li, Be, Nb, Ta, REE, W) have been divided into three broad groups by their associated enriched elements (e.g., Tischendorf 1977; Kovalenko 1978; Pollard 1989):

- normal (biotite ± muscovite) granites with or without W(-Mo-F-Bi-Sn) mineralization
- Li-F rare-metal enriched granites typically with Sn-Ta(-Nb-Cs) enrichments
- peralkaline granites with associated Nb-Ta-Zr-F concentrations.

This classification does not explicitly distinguish differences in alumina saturation ($a_{Al_2O_3}$)or silica saturation (a_{SiO_2}). Given that these variables strongly influence Be and alteration mineral stability, the three traditional groups reflect neither distinct Be mineral assemblages nor distinct types of hydrothermal alteration. In light of this, we divide igneous rocks (volcanic and intrusive, including pegmatitic) and associated Be occurrences into four groups that emphasize differences in magmatic $a_{Al_2O_3}$ and a_{SiO_2} (e.g., Shand 1927; Carmichael et al. 1974; cf. Fig. 12, below):

- strongly to weakly peraluminous suites that range from chemically non-specialized with W-Mo mineralization to Li-F-Sn so-called "specialized" granites—these have BeO-Al_2O_3-SiO_2-H_2O ("BASH") family minerals; muscovitic hydrothermal alteration is characteristic,
- metaluminous to weakly peraluminous suites with variable Nb, Ta, F, Sn, Mo and Li enrichments—these have phenakite, bertrandite, and helvite group minerals; Li-Fe micaceous hydrothermal alteration is characteristic,
- peralkaline to metaluminous quartz-saturated suites typically with Nb-Y-F enrichments—these have phenakite, bertrandite, and Ca-Na-Be silicates; feldspathic hydrothermal alteration is characteristic, and
- silica-undersaturated, generally peralkaline suites with high Nb-REE-Y—these have Ca-Na-Be silicates and helvite group minerals; feldspathic hydrothermal alteration is characteristic.

There can be a wide-range of element enrichments (geochemical specialization) within each group. Not surprisingly, this division has parallels with Černý's classification of common and rare-metal pegmatites (Černý 1991a and Chapter 10, this volume). An advantage of using this four-part classification is that it systematizes and makes predictable the principal differences in Be mineral parageneses and alteration mineralogy. Thus it is possible, in principle, to place a deposit into one of these groups based on the mineral parageneses present. These compositional variations also broadly correlate with tectonic setting and with time as is discussed in the concluding section of this paper. In contrast, more traditional approaches that focus on depositional environment (e.g., skarn, vein, replacement, greisen etc.) do not by themselves distinguish fluid sources or broader environments.

Beryllium minerals also occur in a handful of metamorphic, sedimentary and surficial environments (Table 2). At best, these have tenuous connections to felsic magmatism. Some types, such as the Colombian emerald deposits, have distinctive basin-related hydrothermal origins, whereas others, such as some of the "shear-zone" emerald deposits likely form by local redistribution of materials during metamorphism (Grundmann and Morteani 1989). Placer accumulations are best known where coarse, Be minerals are sourced from high-grade metamorphic terrains (Rupasinghe et al. 1984;

Figure 5 (opposite page). Plots summarizing whole rock chemical data for selected igneous suites associated with non-pegmatitic Be deposits. Major element data are from sources cited in Appendix A and the text. (A) Total alkalis vs. silica showing fields for rock suites grouped by alumina saturation (same as in B). Compositional ranges for alkaline and subalkaline global volcanic rocks shown for comparison (Wilson 1989). (B) Al_2O_3 and CaO contents normalized to (Na_2O+K_2O+CaO) for Be-associated igneous suites highlighting the wide range of alkalinities and aluminum saturation index (ASI = molar Al_2O_3/(Na_2O+K_2O+CaO) but low overall CaO. This projection shows the location of the boundaries for strongly peraluminous, weakly peraluminous, metaluminous, and peralkaline compositions while highlighting the relative CaO contents. (C) and (D) Beryllium, F and Li concentrations in glasses (Macusani, Spor Mtn., Topaz Mtn., Khaldzan-Buregtey), other volcanic rocks and intrusive rocks (data from Coats et al. 1962; Tauson et al. 1978; Christiansen et al. 1984, 1988; Černý and Meintzer 1988; Pichavant et al. 1988a; Trueman et al. 1988; Kovalenko et al. 1995b; Raimbault et al. 1995). Also shown on the right-hand side of (C) is beryl solubility at 650°C in granitic melt for ASI values of 1.0 and 1.3 (Evensen et al. 1999). Note the contrasting trends for magmatic evolution—strongly peraluminous systems evolve to Li-Cs-Ta-enriched compositions ("LCT"), whereas most other systems show more subdued rare alkali enrichment (cf. the Nb-Y-F = "NYF" mixed types of Černý 1991a).

Dissanayake and Rupasinghe 1995). These commonly provide outstanding gem material (Sinkankas 1981).

Emerald deposits deserve special comment because of their economic importance and popular appeal. They form with granitic pegmatites and magmatic-hydrothermal veins of many types, by local metamorphic redistribution of materials, and in basin-related and metamorphic-derived hydrothermal systems. Like other Be deposits, no single factor controls emerald formation save for the requirement of Cr (± V) from local host rocks to generate their deep green color.

BERYLLIUM MINERAL COMPOSTIONS

Most of the Be minerals listed in Table 1 exhibit little natural compositional variability (e.g., Chapter 1 by Grew, this volume; Chapter 10 by Černý, this volume). In non-pegmatitic occurrences, the main exceptions are the beryl group (beryl, stoppaniite and bazzite, plus structurally related milarite) and the helvite group (helvite, danalite, and genthelvite), plus minerals including the taaffeite group, the gadolinite group and meliphanite-leucophanite. Given the variably F-rich nature of Be occurrences, substitution of F for OH may be more common than appreciated even though evidence for this substitution mainly restricted to herderite, euclase and bertrandite (Beus 1966; Hsu 1983; Lebedev and Ragozina 1984; see Chapters 10 and 13, this volume by Černý and Franz and Morteani, respectively). A few other minerals such as chrysoberyl have minor, though petrologically and gemologically interesting variations in cation contents. Examination of compositional patterns in the beryl and helvite groups both documents systematic differences with environment and yields insight into differences in the conditions of formation.

Beryl group—$(\square,Na,Cs,H_2O)(Be,Li)_3(Al,Sc,Fe^{+3},Cr,Fe^{+2},Mg)_2[Si_6O_{18}]$

Composition. The compositions of beryl and related minerals have long been known to vary with geologic environment (Fig. 6A; Staatz et al. 1965; Beus 1966). The principal chemical substitutions in the beryl structure, ${}^{C}\square{}^{T(2)}Be_3{}^{O}Al_2[{}^{T(1)}Si_6O_{18}]$, can be represented as:

$$ {}^{C}\square{}^{O}Al^{+3} = {}^{C}(Na,K){}^{O}(Mg,Fe^{+2},Mn^{+2}) \quad (1)$$

$$ {}^{C}\square{}^{T(2)}Be^{+2} = {}^{C}(Na,Cs,Rb){}^{T(2)}Li \quad (2)$$

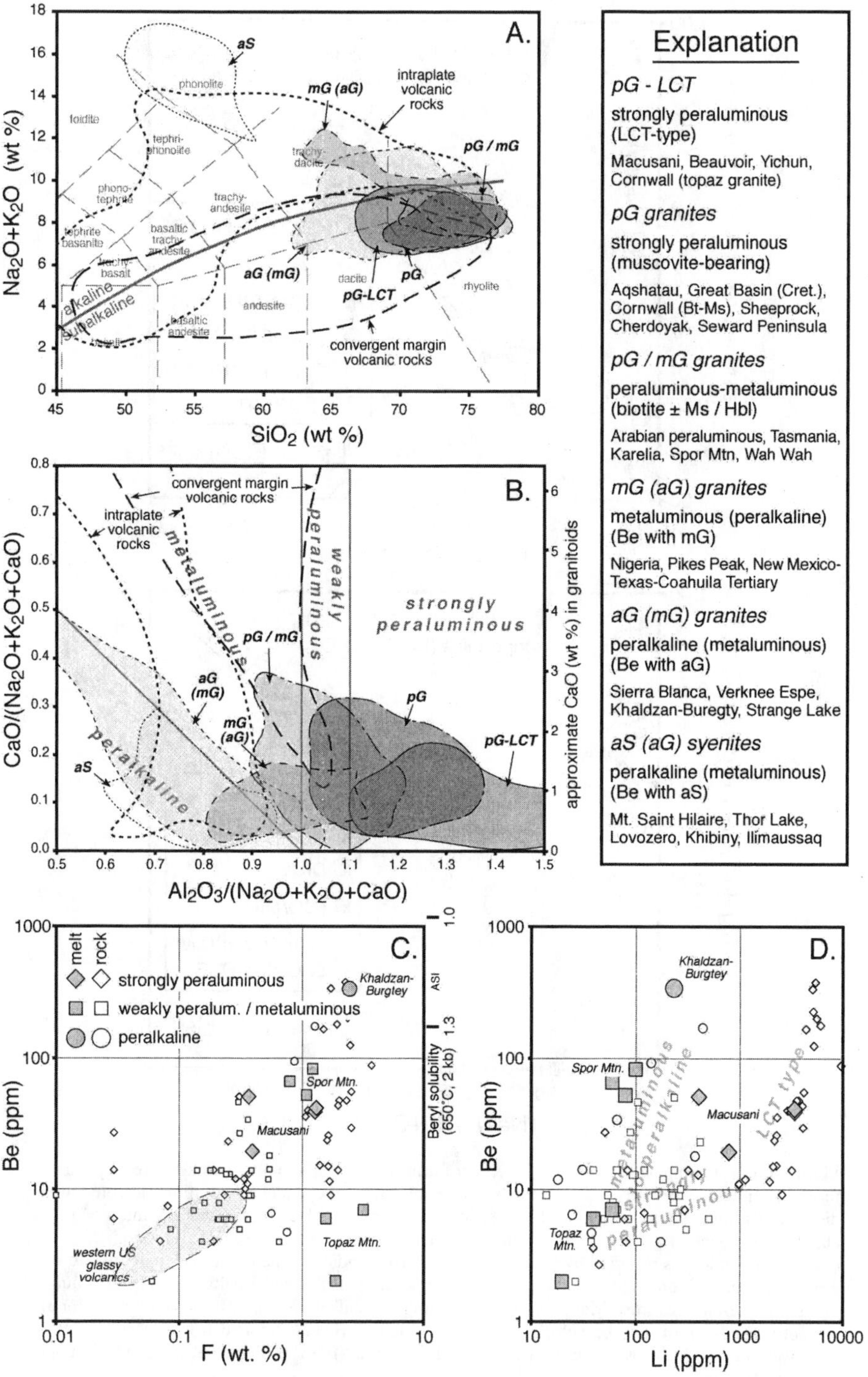
A.
aS
phonolite
foidite
tephri-phonolite
phono-tephrite
tephrite basanite
trachy-basalt
basaltic trachy-andesite
trachy-andesite
trachy-dacite
mG (aG)
intraplate volcanic rocks
pG / mG
aG (mG)
dacite
pG
pG-LCT
rhyolite
andesite
basaltic andesite
basalt
alkaline
subalkaline
convergent margin volcanic rocks
Na2O+K2O (wt %)
SiO2 (wt %)
B.
convergent margin volcanic rocks
intraplate volcanic rocks
metaluminous
weakly peraluminous
strongly peraluminous
pG / mG
aG (mG)
mG (aG)
pG
pG-LCT
aS
peralkaline
CaO/(Na2O+K2O+CaO)
Al2O3/(Na2O+K2O+CaO)
approximate CaO (wt %) in granitoids
Explanation
pG - LCT
strongly peraluminous (LCT-type)
Macusani, Beauvoir, Yichun, Cornwall (topaz granite)
pG granites
strongly peraluminous (muscovite-bearing)
Aqshatau, Great Basin (Cret.), Cornwall (Bt-Ms), Sheeprock, Cherdoyak, Seward Peninsula
pG / mG granites
peraluminous-metaluminous (biotite ± Ms / Hbl)
Arabian peraluminous, Tasmania, Karelia, Spor Mtn, Wah Wah
mG (aG) granites
metaluminous (peralkaline) (Be with mG)
Nigeria, Pikes Peak, New Mexico-Texas-Coahuila Tertiary
aG (mG) granites
peralkaline (metaluminous) (Be with aG)
Sierra Blanca, Verknee Espe, Khaldzan-Buregty, Strange Lake
aS (aG) syenites
peralkaline (metaluminous) (Be with aS)
Mt. Saint Hilaire, Thor Lake, Lovozero, Khibiny, Ilímaussaq
C.
melt rock
strongly peraluminous
weakly peralum. / metaluminous
peralkaline
Khaldzan-Burgtey
Spor Mtn.
Macusani
Topaz Mtn.
western US glassy volcanics
ASI 1.0 1.3
Beryl solubility (650°C, 2 kb)
Be (ppm)
F (wt. %)
D.
Khaldzan-Burgtey
Spor Mtn.
Macusani
Topaz Mtn.
metaluminous to peralkaline
strongly peraluminous
LCT type
Be (ppm)
Li (ppm)

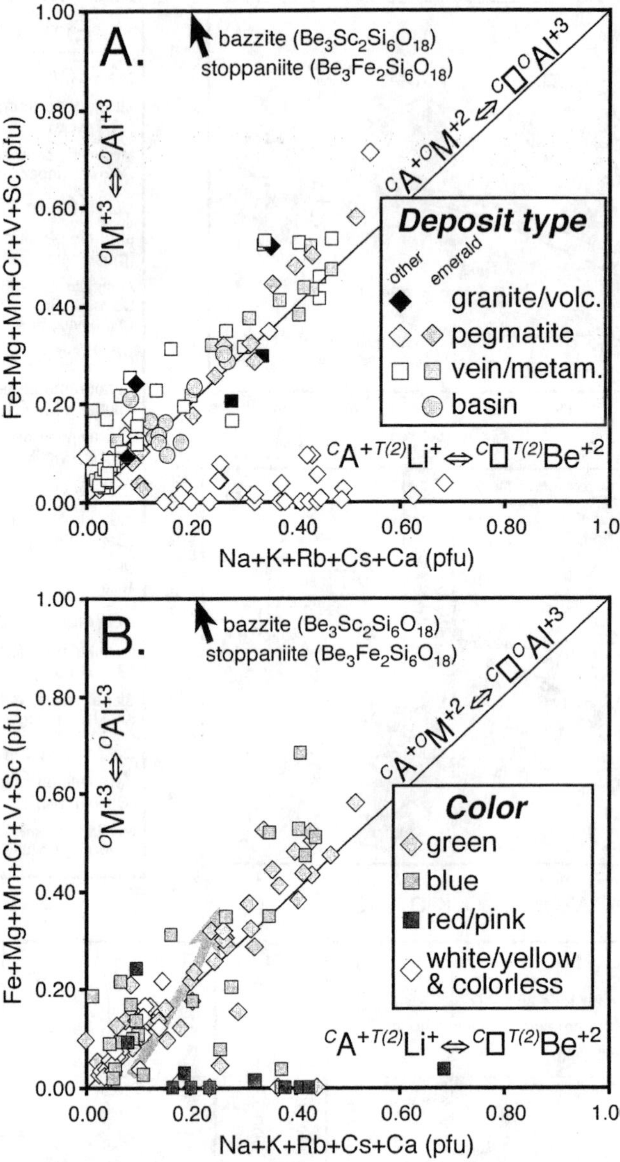

Figure 6. Beryl compositions plotted in terms of transition metal and alkali contents (except Li) per formula unit (6 Si). Broadly, this corresponds to octahedral and channel substitutions as noted on the diagram (following Aurisicchio et al. 1988; see Hawthorne and Huminicki, this volume). (A) Data classified by general geologic environment. Compare Figure 7. See text for discussion. (B) Data classified by color (as reported by the authors). The arrow indicates the trend from pale blue to dark green color in the Somondoco, Colombia (Kozlowski et al. 1988) and Khaltaro, Pakistan (Laurs et al. 1996) emerald localities. Many analyses including most alkali beryls have no reported color and are not plotted—most may be colorless or weakly colored. (Data compiled from Deer et al. 1978; Aurisicchio et al. 1988; Kozlowski et al. 1988; Laurs et al. 1996; Calligaro et al. 2000; S. Young and M.D. Barton, unpubl. analyses).

$$^{O}Al^{+3} = {}^{O}(Fe^{+3},Sc^{+3},Cr^{+3},V^{+3}) \quad (3)$$

$$^{C}\square = {}^{C}(H_2O, CO_2, Ar) \quad (4)$$

(Aurisicchio et al. 1988, Černý, this volume, Hawthorne and Huminicki, this volume). The first two coupled substitutions lead, respectively, to "octahedral" (Exchange 1) and "tetrahedral" (Exchange 2) beryls. Both are probably limited to no more than about 0.5 per formula unit (pfu) because they lead to underbonding on one of the oxygens in the beryl structure (Aurisicchio et al. 1988; cf. Fig. 6A). In contrast, exchange between Al^{+3} and other trivalent cations in the octahedral site (Exchange 3) can go to completion, as evidenced by the end-member minerals bazzite (Sc^{+3}) and stoppaniite (Fe^{+3}). Other substitutions are permissible. Li can exchange with Na and Cs in the alkali site as demonstrated by experiment (Manier-Glavinaz et al. 1989b); however, its importance in nature is unclear given that atomic Li rarely exceeds the other alkalis less divalent cations (i.e., the amount required for type 2 exchange).

Non-pegmatitic beryls range from end-member beryl to large octahedral substitutions by both Exchanges 1 and 3. In these beryls, tetrahedral substitution is minor (Fig. 6A). In contrast, pegmatitic beryls—except for pegmatite-related emeralds—range from nearly pure compositions with at most limited type 1 exchange (<0.2 pfu) to quite high values of type 2 exchange (~0.5 pfu; Fig. 6). The most extensive type (1) substitution occurs in metamorphic-hosted beryls—both emeralds and non-emeralds (e.g., Franz et al. 1986). The more extensive type (3) substitutions, up to bazzite and stoppaniite, occur in metaluminous granites and syenites as well as in some metamorphic rocks. Channel volatile contents (dominantly H_2O) can be virtually nil, as in volcanic red beryl (Shigley and Foord 1984), but fall mostly between ~0.6 and 2.8% in both pegmatitic and non-pegmatitic types. Other components typically reflect host rock compositions: V in sedimentary rocks, Mn in chemically evolved pegmatites and volcanic rocks, Cr and Mg in ultramafic, mafic and some sedimentary rocks, and Sc and in Fe^{+3} in A-type (mildly alkaline, oxidized) granites. Where chrysoberyl forms in the same settings, for example in desilicated pegmatites as the gem variety alexandrite, it accommodates Fe^{+3} and Cr to about the same degree as beryl.

Milarite, $(K,\square,H_2O,Na)_2(Ca,Y,REE)_2[(Be,Al)_3Si_{12}O_{30}]$, resembles beryl in having a structure of double six-membered rings interconnected by Be tetrahedra and Ca octahedra (Hawthorne and Huminicki, this volume). As in beryl, alkalis and water can substitute in channels which in milarite are defined by stacking of the double rings. Milarite occurs in skarns, alpine veins and various alkaline-related metasomatic rocks (Appendix A) as well as in various types of pegmatites (Černý, Chapter 10, this volume). Compositional variations of milarite are sparsely documented, but the (Y, REE)-rich varieties appear to be more common in alkaline settings (cf. Černý).

Color and composition. Not surprisingly, transition-metal-rich, octahedrally substituted beryls typically have more intense colors, mostly blues or greens, although red is characteristic of volcanic-hosted beryl (Fig. 6B). Pegmatitic beryls can be intensely colored (e.g., aquamarine and emerald), however, most tetrahedrally substituted beryls, if not colorless, tend to be pale in color, typically pink, less commonly yellow, green or blue.

In emerald, the intense green color reflects substitution of Cr^{+3} for Al^{+3} and a paucity of Fe (Fig. 7A) regardless of setting, whereas the rare alkali content does reflect their environment of origin (Fig. 7B). The latter is true in spite of the fact that all emeralds are dominated by the octahedral substitution (Fig. 6A). Given these patterns and the great interest in emeralds in the gem trade, it is obvious why chemical fingerprinting of emerald provenance has been pursed with some vigor and success (e.g., Dereppe et al.

2000). Emerald is properly restricted to beryl where Cr exceeds other coloring agents by weight (Kazmi and Snee 1989a). The analogous substitution of V^{+3} in beryl also creates an intense green coloration that is often termed emerald. Even the deep red Mn-rich volcanic-hosted beryl from Utah has been marketed, controversially, as "red emerald" (Spendlove 1992).

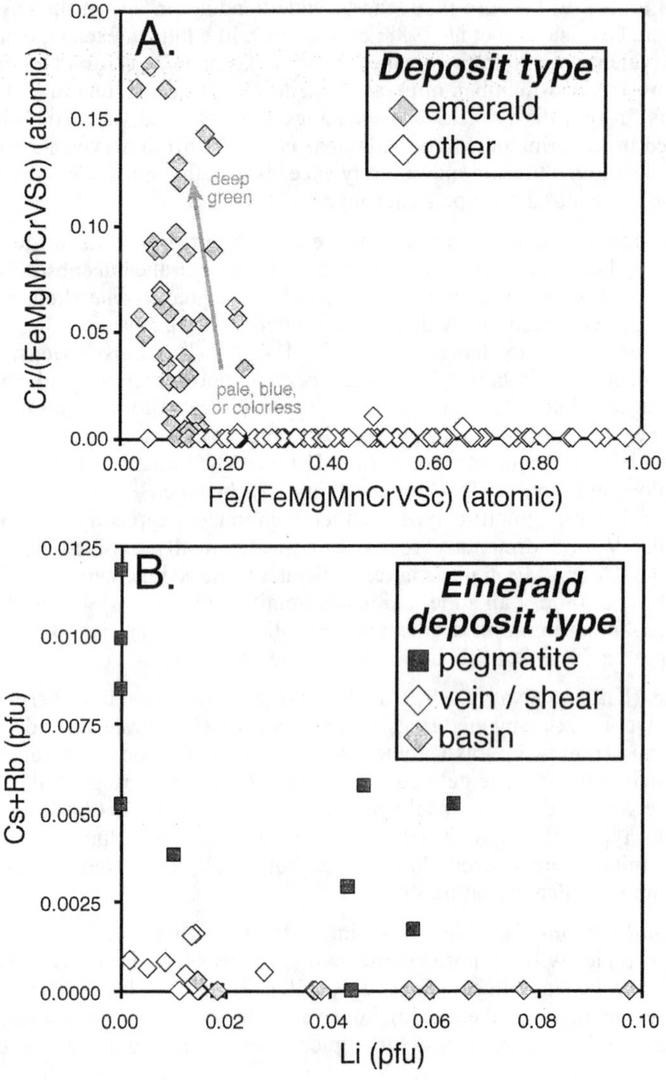

Figure 7. Emerald and other beryl compositions from the literature (see Figure 6 for sources). (A) Plot illustrating the elevated Cr contents and low Fe to Mg (etc.) ratios of emeralds compared to other types of beryls. This illustrates the main difference with other environments. Cr is not reported in many of the other analyses; it may have either been below detection or not sought. As in Figure 6, the arrow shows the trend from pale blue to dark green colored beryls at Somondoco, Colombia (basin-related) and Khaltaro, Pakistan (pegmatite). (B) Plot of rare alkalis in emeralds from various settings illustrating variations analogous to those seen in other beryls. See text for discussion.

Petrologic controls on beryl composition. A simple analysis of the common substitutions in terms of alumina activity ($a_{Al_2O_3}$) and the availability of other cations helps rationalize their correlation with geologic environment. In the simplest case, illustrated by equation 5, the type 3 substitution of trivalent Cr, Fe, V and Sc for Al will be promoted by the relative abundance of these elements in certain rocks or by decreasing $a_{Al_2O_3}$. Alumina

$$^{O}Al^{+3} + 0.5\ M_2O_3 = {}^{O}(M^{+3}) + 0.5\ Al_2O_3 \quad (5)$$

activity will be low in aluminum-deficient assemblages (e.g., many ultramafic and carbonate rocks) and in alkaline igneous rocks. Reaction (6) shows that alkalinity and alumina activity inversely correlate in feldspar-bearing rocks:

$$NaAlSi_3O_{8,plagioclase} = 0.5Na_2O + 0.5Al_2O_3 + 3\ SiO_2 \quad (6)$$

Similarly, any combination of decreasing $a_{Al_2O_3}$, increasing alkalinity, or increasing availability of (Mg, Fe, Mn)O will promote type 1 (octahedral) substitution:

$$^{C}\square^{O}Al^{+3} + 0.5\ A_2O + MO = {}^{C}A^{+1O}M^{+2} + 0.5\ Al_2O_3 \quad (7a)$$

$$^{C}\square^{O}Al^{+3} + NaAlSi_3O_8 + MO = {}^{C}Na^{+1O}M^{+2} + Al_2O_3 + 3\ SiO_2 \quad (7b)$$

Thus, as observed, beryl group minerals forming in metaluminous igneous rocks and in ultramafic or carbonate host rocks should generally have higher octahedral substitutions than beryls from peraluminous varieties. For example, emerald and green vanadian beryls are most common in rocks lacking muscovite (e.g., Kazmi and Snee 1989b). Ferric-iron-rich aquamarines, the Fe^{+3} end member stoppaniite, and the Sc^{+3} end member bazzite are most typical of metaluminous rocks—biotite granites or, in the case of stoppaniite, syenite (Ferraris et al. 1998; Della Ventura et al. 2000). Conversely, in some circumstances Fe contents may be suppressed either by intrinsically low Fe relative to other octahedral cations (as in ultramafic rocks) or by sequestration in other phases (e.g., pyrite in the Colombian emerald deposits, Ottaway et al. 1994).

The tetrahedral (type 2) substitution is common in Li-Cs-Ta pegmatites, but apparently is rare elsewhere. It logically follows Reaction (8) where availability of Li or Cs is the key.

$$^{C}\square^{T(2)}Be^{+2} + 0.5\ Li_2O + 0.5\ A_2O = {}^{C}A^{T(2)}Li + BeO \quad (8)$$

Increasing overall alkalinity (reaction 6) is not likely to be a factor given that Li-Cs-Ta pegmatites are strongly peraluminous (_ern_ 1991a), but it could contribute to tetrahedral substitution in some mildly alkaline greisen-type systems. Unfortunately very few complete beryl analyses are available for the latter. One might expect octahedral substitutions to accompany the tetrahedral except for the fact that highly evolved pegmatites with high Li and Cs have very low contents of Mg and Fe and only modest Mn. This may contribute to the separation of the field for tetrahedrally substituted beryls from the other occurrences in Figure 6A.

Helvite group—$(Mn,Fe,Zn)_4[BeSiO_4]_3S)$

Composition. Helvite-group minerals are present in minor quantities in Be-bearing skarns, alkaline igneous settings, and some hydrothermal veins. Changes in Mn-Fe-Zn ratios spanning all three end-members account for most of compositional variation in the helvite group (Fig. 8). Rarely, Al substitutes for Zn; Finch (1990) proposed that the mechanism is $2\ Al^{+3} + \square = 3\ Zn^{+2}$ based on compositional variations in hydrothermal genthelvite from the syenitic Motzfeldt intrusion, Greenland which contains to ~10 wt % Al_2O_3. Other elements might be present, for example Na given the structural similarity with tugtupite ($Na_4[BeSiO_4]_3Cl$), or Cd where genthelvite coexists with greenockite (Nechaev and Buchinskaya 1993).

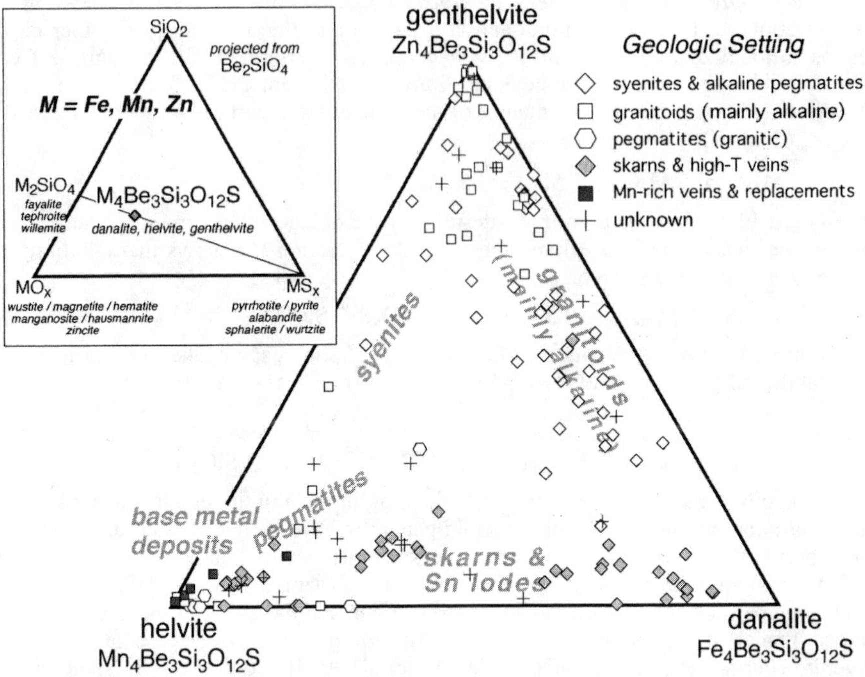

Figure 8. Analyzed helvite group minerals plotted in terms of the end member compositions and distinguished by geologic environment (sources of data include: Vlasov 1966b; Dunn 1976; Kwak and Jackson 1986; Larsen 1988; Perez et al. 1990; Ragu 1994a). The inset shows the chemographic relationship of helvite group minerals to silica, phenakite, and Mn-Fe-Zn oxides, sulfides and silicates.

As illustrated in Figure 8, helvite-group compositions differ systemati-cally between genetic environments. Zinc-rich compositions (genthelvite) with or without Al typically occur in pegmatites, miarolitic cavities or veins associated with metaluminous to peralkaline granites and syenites (Burt 1988; Larsen 1988; Perez et al. 1990). Peraluminous granitic pegmatites and occurrences in base-metal-sulfide veins and replacements are typically Mn-dominated, whereas variable Fe:Mn varies from near end-member danalite to helvite in skarns and Sn lodes (greisens), with danalite being dominant common in the more reduced systems (Burt 1980; Kwak and Jackson 1986).

Petrologic controls on helvite-group compositions. The unusual composition of the helvite group—combining Be_2SiO_4, a metal sulfide, and a metal orthosilicate (Fig. 8 inset)—means that these minerals are sensitive to redox and sulfidation states as well as to the activity of phenakite (Burt 1980, 1988). Conditions favorable for formation of the various end-members differ based on the relative stability of the related sulfides and silicates as illustrated in Figure 9. For each of the three, maximum stability occurs along the boundaries where their respective orthosilicates and monosulfides coexist along with phenakite. Departure from the ideal conditions by oxidation, reduction, gain or loss of sulfur, or reducing the activity of phenakite will all be unfavorable. Hence, low $a_{Al_2O_3}$ ("alkaline" conditions) favor helvite group minerals because beryl replaces phenakite and lowers $a_{Be_2SiO_4}$ with increasing $a_{Al_2O_3}$ (see next section). Danalite preferentially occurs in reduced and low sulfidation state environments; helvite dominates in more sulfidized,

Mn-rich settings where pyrite and sphalerite sequester Fe and Zn; and genthelvite is restricted to relatively oxidized but low sulfur settings characteristic of many (per)alkaline rocks where Fe and Mn mainly enter oxides and other silicates (cf. Burt 1980, 1988).

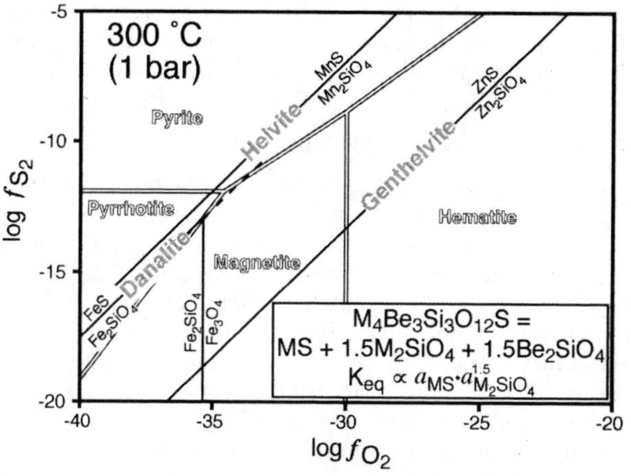

Figure 9. Helvite group mineral stability a function of oxidation and sulfidation state relative to some other zinc, iron and manganese minerals. End members should have maximum stabilities on the orthosilicate- monosulfide boundaries (inset; also see Fig. 8 inset). Note the that maximum stability for danalite would project along the dashed line were it not for magnetite formation. Calculated using thermodynamic data from Barton and Skinner (1979) and Robie et al. (1978).

Other minerals

Gadolinite group minerals, $(Y,REE)_2(Fe,\square)[Be_2Si_2O_8](O,OH)_2$, leucophanite, $CaNaBeSi_2O_6F$, and meliphanite, $Ca_4(Na,Ca)_4Be_4AlSi_7O_{24}(F,O)_4$, occur mainly in alkaline or metaluminous pegmatites or miarolitic cavities but are also found in a handful of alkaline-rock related hydrothermal deposits (Table 1, Appendix A). Little is published about gadolinite-group compositions in non-pegmatitic occurrences. Based on the study of Pezzotta et al. (1999) who studied a range of granite-related occurrences in the southern Alps, considerable variation in Y / LREE / HREE would be expected as well as variable B contents. Leucophanite and meliphanite solid solutions are reported from alkaline metasomatites (Ganzeeva et al. 1973; Novikova 1984) presumably reflecting differences in Ca/Na.

BERYLLIUM MINERAL STABILITIES

Available data on beryllium mineral stabilities, derived from experiment, theory and natural assemblages, provides a valuable framework for classification and understanding of natural occurrences. Published studies on Be mineral stabilities are summarized in Appendix B and have been reviewed extensively elsewhere (Barton 1986; Burt 1988; Wood 1992; Franz and Morteani, London and Evensen, Chapters 13 and 11, respectively, this volume). Most of this work has focused on the BeO-Al_2O_3-SiO_2-H_2O (BASH) system and coexisting melts and aqueous fluids. Here we briefly review mineral equilibria and solubilities of particular relevance to non-pegmatitic deposits and focused on BASH minerals.

The figures presented here were calculated using the internally consistent thermodynamic model for BASH phases and topaz from Barton (1982b, 1986), which were adapted to the SUPCRT database (Johnson et al., 1992) by adjusting for differences in the enthalpy of formation of Al_2O_3 between the databases, and refitting the heat capacities to the Meier-Kelly function. Presently, there is a need to reevaluate the thermodynamic data for BASH minerals by including results published since 1985 (Appendix B) in a rigorous fit. In addition, one could also build a thermodynamic model for other phases, such as the helvite group and the Na-Be silicates, by combining available experimental data with constraints from natural assemblages.

Pressure-temperature-activity relationships

P-T. Other than for the BASH system there are essentially no reversed equilibrium data for the pressure-temperature stability fields of Be minerals (Appendix B). In the BASH system, the salient characteristics of pressure-temperature phase relationships (Fig. 10) are (1) that the hydrous minerals (excepting beryl) are stable only at temperatures below 500°C and (2) that the assemblages are not distinctly pressure sensitive. Bertrandite persists only up to about 300°C. The lower limit of beryl stability is between 200 and 350°C depending on coexisting minerals (Fig. 10 inset). In quartz-bearing assemblages, chrysoberyl is restricted to near-magmatic and higher temperatures, although the position of the reaction chrysoberyl+quartz = beryl+aluminum silicate is sensitive to beryl composition and its position remains controversial. See Barton (1986) and Franz and Morteani (this volume) for further discussion of these relationships.

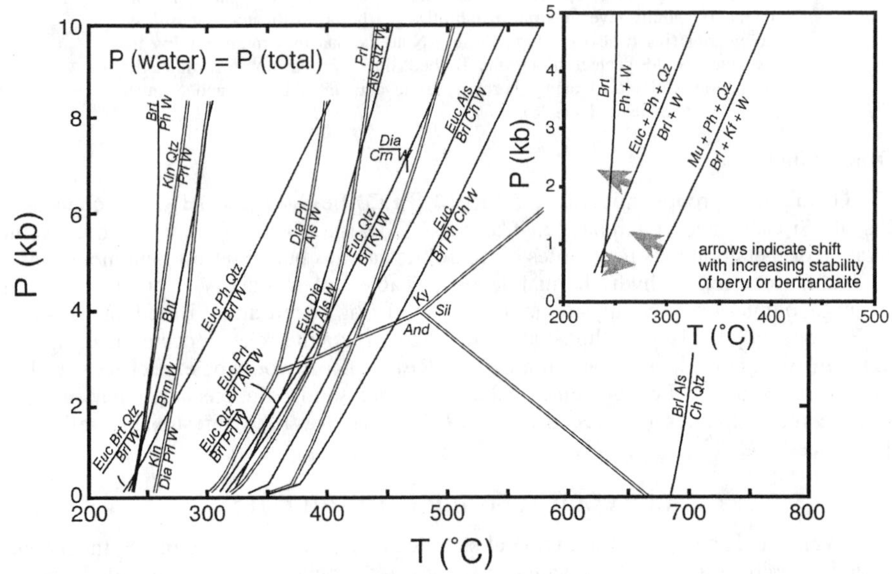

Figure 10. Pressure-temperature projection of phase relationships in the $BeO-Al_2O_3-SiO_2-H_2O$ (BASH) system. Redrawn from Barton (1986). Limiting reactions for bertrandite and beryl both can depend on solid solution effects, F for OH in bertrandite, and multiple components in beryl (inset).

T-activity. In contrast to the limited insight available from the P-T relationships, activity diagrams are of considerably greater utility in understanding the occurrence of Be minerals because of the metasomatic origin of most non-pegmatitic Be deposits (Figs. 9, 11-14). The most useful independent variables are: (1) temperature, which varies markedly in time and space in most Be-bearing geologic systems, and (2) the activities of the major components, notably alumina and silica. Silica and alumina are key because they frame the thermodynamic conditions defined by many rock-forming minerals and, in addition, can be related to alkalinity of melts and fluids through reactions (6) and (9). Reaction (9) relates fluid acidity to alkalinity in the presence of plagioclase when $a_{Al_2O_3}$ and a_{SiO_2} are defined.

$$H^+ + NaAlSi_3O_{8,plagioclase} = Na^+ + 0.5\ H_2O + 0.5\ Al_2O_3 + 3\ SiO_2 \qquad (9)$$

Figure 11 plots BASH mineral assemblages in terms of each a_{SiO_2} and $a_{Al_2O_3}$ as functions of temperature. At high T, beryl, phenakite, and chrysoberyl (T > 600°C) are stable at high silica activities (Fig. 11A,C. With decreasing silica activity beryl is replaced by chrysoberyl+phenakite and phenakite is ultimately replaced by bromellite. This is the characteristic sequence found in desilicated pegmatites. A similar progression occurs at lower temperatures except that chrysoberyl is strongly quartz undersaturated and first euclase and then bertrandite become key phases. Skarns and carbonate hosted replacement bodies typically exhibit zoning that reflects these varying degrees of silica saturation and paths from high- to low-temperature across Figure 11A. At ≤1 kbar solutions can become strongly undersaturated with respect to quartz, whereas at higher pressures they may stay closer to quartz saturation (Fig. 11A inset). These contrasting paths rationalize differences observed in carbonate-hosted hydrothermal systems.

Another useful contrast comes from consideration of $a_{Al_2O_3}$, a variable which highlights differences between Al-rich and Al-poor assemblages (Fig. 11B). The saturation surface for the Al-only phases, corundum (T > 360°C) and diaspore (T < 60°C), neither of which is stable with quartz, bounds the top of the diagram. Quartz coexists with andalusite at high temperature, but then pyrophyllite followed by kaolinite formed with decreasing temperature. Chrysoberyl and euclase are the characteristic minerals at high $a_{Al_2O_3}$, whereas beryl occupies an intermediate field (Fig. 11B). In contrast, phenakite and bertrandite are stable only at distinctly lower $a_{Al_2O_3}$ conditions until bertrandite and kaolinite become stable together at about 225°C. A key boundary is that between K-feldspar and muscovite which separates strongly peraluminous assemblages from others. Considering this reaction, it becomes clear why in most quartz-bearing rocks, beryl is the dominant silicate down to relatively low temperatures barring conditions of unusual acidity (as in some greisens) or basicity (as in peralkaline rocks). On cooling in the presence of muscovite and K-feldspar, only below T ≈ 300°C does beryl give way to phenakite+quartz (arrow in Fig. 11B). Solid solution will expand the beryl field to still lower temperatures (Fig. 10 inset).

Odintsova (1993)derived an analogous topology as a function of a_{BeO} and temperature. She subsequently use it to interpret the paragenesis of ultramafic-hosted emerald deposits in the Ural Mountains (Odintsova 1996). Because BeO is rarely more than a minor component, most assemblages will only have a single saturating Be phase, thus relationships among Be-bearing mineral assemblages are more readily applied when cast in terms of other components.

Activity-activity. Projecting the variables from Figure 11 into $a_{Al_2O_3}$ - a_{SiO_2} space (Fig. 12) provides a particularly revealing look at Be mineral assemblages because reactions among rock-forming minerals separate major rock types on the same diagrams. In Figure 12, quartz-saturated rocks (granitoids, rhyolites, etc.) lie along the top of the diagrams passing downward into undersaturated rocks. The latter are split by key reactions such as $Mg_2SiO_4 + SiO_2 = Mg_2Si_2O_6$. Saturation with muscovite and andalusite occurs along the right boundary, defining strongly peraluminous rocks, whereas peralkaline assemblages (and rocks) are located near the acmite-bearing reaction that passes diagonally across the left half of the diagram.

Figure 11 (opposite page). (A) Beryllium mineral stability as a function of temperature and concentration of aqueous silica at 1000 bars. Shading outlines the upper a_{SiO_2} limit of chrysoberyl, the lower a_{SiO_2} limit of beryl, and the upper thermal stability of euclase. The inset shows aqueous silica concentrations at two pressures and 3 alternative fluid paths. Path (a) represents cooling with little decompression and would remain quartz saturated; path (b) represents decompression, whereas path (c) represent isobaric cooling but quartz undersaturation because of the low-P retrograde solubility of quartz. See text for additional discussion. (B) Beryllium mineral stability as a function of temperature and the activity of alumina (corundum) at 1000 bars. Shading outlines stability limits for beryl, the upper $a_{Al_2O_3}$ limit for phenakite/bertrandite, and the upper thermal stability of euclase. Note the arrow and label for the lower limit of stability for beryl in the presence of K-feldspar (cf. inset in Fig. 10). (C) BASH mineral compatibilities at 225, 350 and 500°C projected from H_2O onto the BeO-Al_2O_3-SiO_2 plane (cf. Fig. 1). Mineral abbreviations from Table 1. (A) and (B) are modified from Barton (1986).

By examination of the superimposed Be mineral stability boundaries at 600°C (Fig 12A), it is clear why beryl is typical of strongly peraluminous granitoids and rocks, why phenakite (± helvite group) is common in metaluminous and peralkaline rocks, and why alkali Be silicates occur in peralkaline silica-undersaturated rocks. Chrysoberyl has a large stability field but only for unusual rocks that must be Al-rich and Si-poor (e.g., desilicated pegmatites). With decreasing temperature, the fields for euclase and, especially, the Al-free Be silicates expand at the expense of the beryl and chrysoberyl fields. Topologically-correct phase boundaries for beryllite, epididymite and chkalovite are shown in the upper left based on occurrences in peralkaline syenites. The breadth of the phenakite/bertrandite fields is consistent with widespread occurrence of these minerals in low-temperature deposits, particularly carbonate replacements. The right hand side matches assemblages found in strongly peraluminous igneous-hosted greisens (top) and in silica-undersaturated greisens developed in carbonate rocks (right; the meaning of greisen is discussed below).

Activity relationships in terms of other components are germane to a number of occurrences, particularly HF, CaO, MgO and P_2O_5. Increasing the activity of acid fluoride species leads to topaz replacing other Al-bearing silicates and fluorite replacing other Ca-bearing minerals—these are typical minerals of greisens (Burt 1975, 1981). Phenakite and bertrandite replace beryl and euclase with increasing HF as well as with increasing alkalinity (e.g., K^+/H^+, see Fig. 13, cf. Fig. 11B) consistent with their widespread occurrence in greisens of various flavors. Fluorine has a similar role in Ca-bearing rocks, where fluorite formation sequesters Ca and leads to more acid (Al-dominated) mineral assemblages. This was considered by Burt (1975) who used natural assemblages to derive topologies for activity diagrams involving P_2O_5, CaO and F_2O_{-1} and analyze the relationships between beryl, phenakite and various Be phosphates.

Beryllium mineral parageneses in the Ca-Mg silicate assemblages of ultramafic and carbonate hosted deposits can also be usefully visualized by recasting phase relationships in terms of the activities of CaO, MgO and SiO_2. For example, Figure 14 illustrates possible phase relationships and zoning paths in desilicated pegmatites or quartz-feldspar veins at 500°C and 3 kbar. Starting with a granitic/vein assemblage on the high-silica side, paths can go upward (as in a dolomitic limestone) into the actinolite (or clinopyroxene) field and yield zoning from beryl to chrysoberyl to phenakite or downward into phenakite and ultimately bromellite. Under these particular conditions, beryl is near its stability limit (Fig. 14 inset) and small differences in solid solution can have significant differences in the position of phase boundaries and thus paths.

The meaning of greisen. Many Be-bearing rocks are referred to as greisen, which refers to a broad spectrum of Al-bearing metasomatic rocks that are typically F-rich and

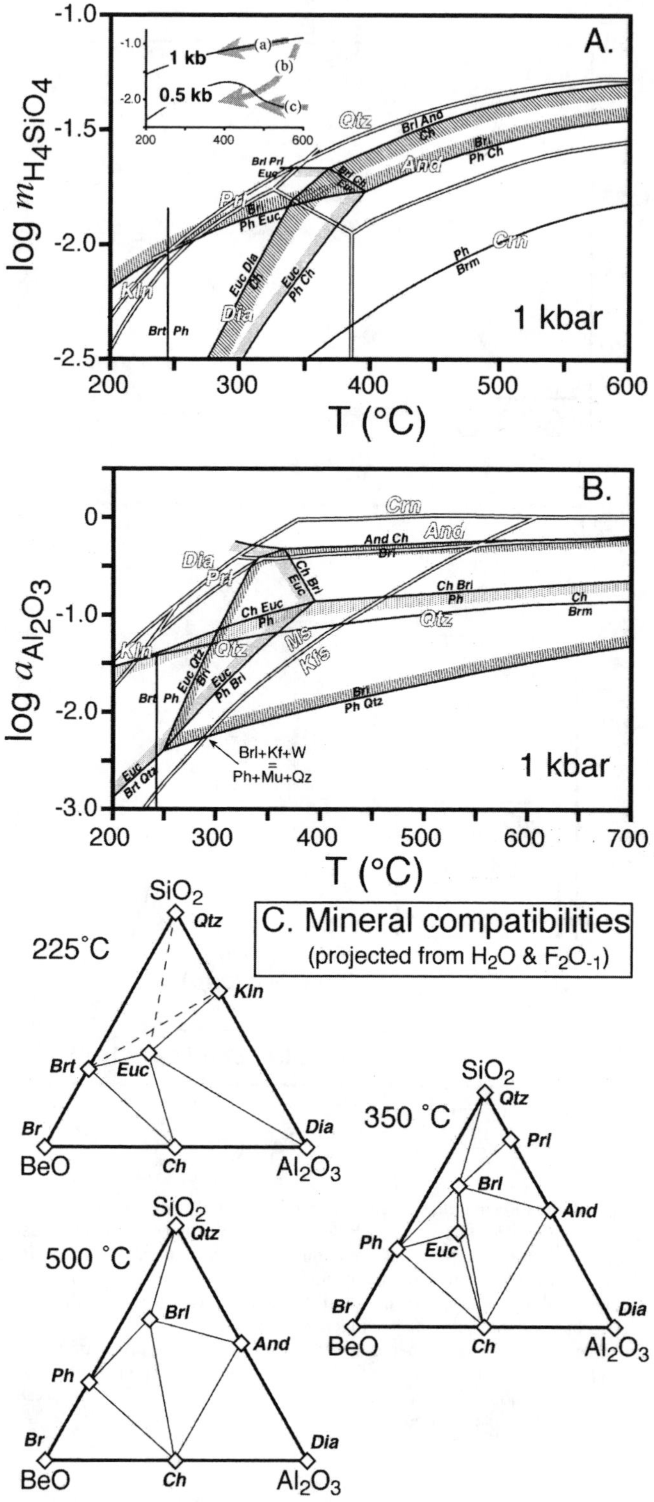
A.
1 kbar
log $m_{H_4SiO_4}$
T (°C)
1 kb
0.5 kb
(a)
(b)
(c)
Qtz
Brl And Ch
Brl Ph Ch
And
Brl Prl Euc
Brl Ch Euc
Prl
Brl Ph Euc
Kln
Euc Dia Ch
Euc Ph Ch
Dia
Brt Ph
Ph Brm
Crn
B.
1 kbar
log $a_{Al_2O_3}$
T (°C)
Crn
And
And Ch Brl
Dia
Prl
Ch Brl Euc
Ch Brl Ph
Ch Brm
Ch Euc Ph
Qtz
Ms
Kfs
Kln
Euc Qtz Brl
Euc Ph Brl
Brt Ph
Brl Ph Qtz
Euc Brt Qtz
Brl+Kf+W = Ph+Mu+Qz
C. Mineral compatibilities
(projected from H_2O & F_2O_{-1})
225°C
350 °C
500 °C
SiO_2
Qtz
Kln
Prl
And
Brl
Brt
Euc
Ph
Br
Dia
Ch
BeO
Al_2O_3

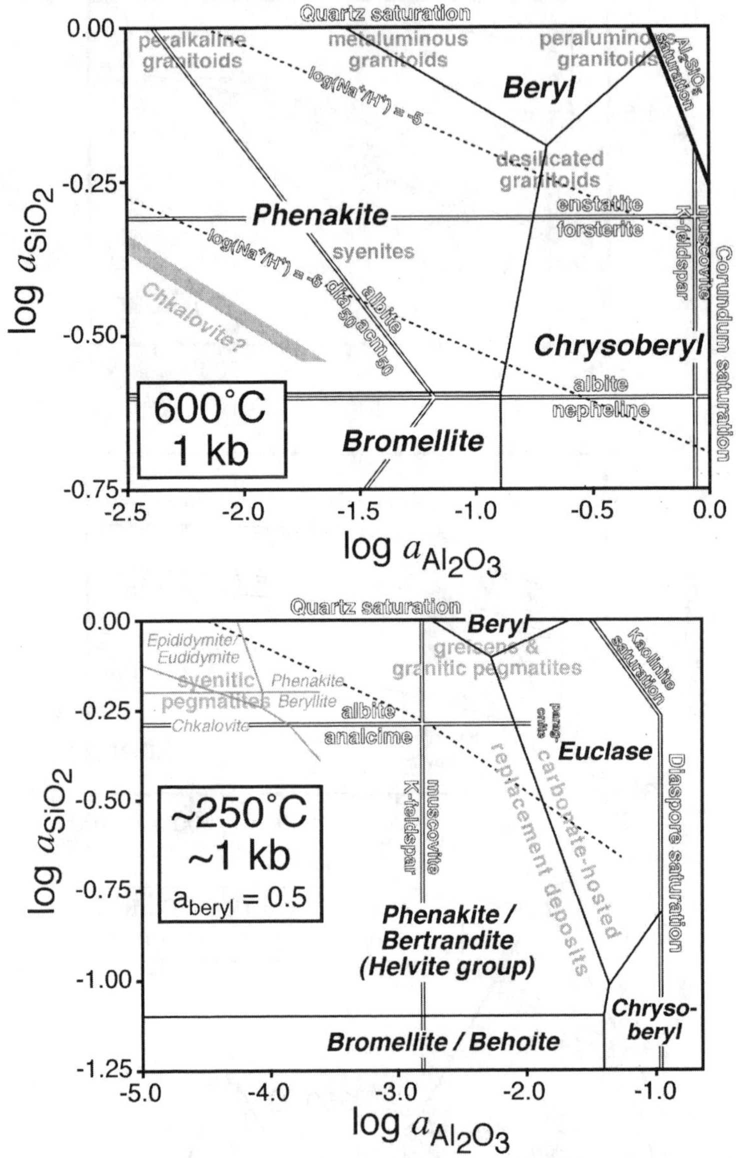

Figure 12. Beryllium mineral stabilities as a function of silica and alumina activities. The diagrams illustrate the preeminent control that these rock-defined variables have on mineral assemblages in Be-bearing hydrothermal systems. (A) Phase relationships of Be minerals as a function $a_{Al_2O_3}$ and a_{SiO_2} at 600°C and 1 kbar related to mineralogy in felsic igneous rocks. The field for chkalovite stability is speculative although topologically plausible and is consistent with the recent work by Markl (2001). (B) Phase relationships of Be minerals as a function $a_{Al_2O_3}$ and a_{SiO_2} at 250°C and 1 kbar related to mineralogy in felsic igneous rocks and some major groups of Be deposits. The activity of beryl = 0.5. The speculative fields for beryllite, chkalovite, epididymite / eudidymite are consistent with their chemography and mineral associations reported from alkaline syenitic pegmatites (stable with albite and analcime).

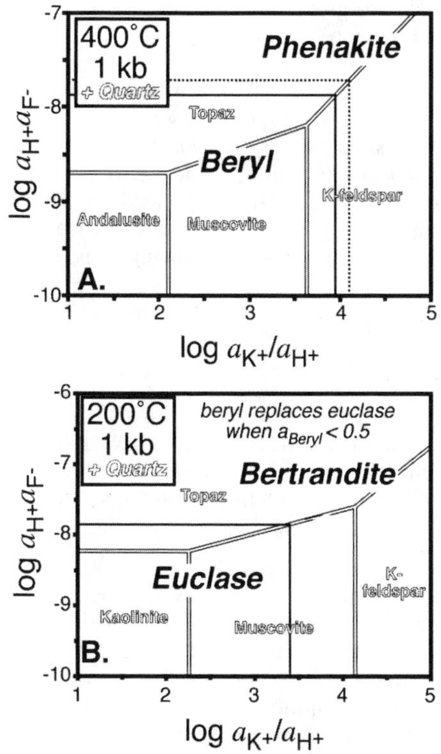

Figure 13. Calculated Be mineral stability as a function of H·F and K/H at 1000 bars and temperature of 400°C (A) and 200°C (B). The P-T dependence of limiting reactions is shown in the inset in Figure 10. Similar topologies are discussed by Kupriyanova et al. (1982) and Burt (1981).

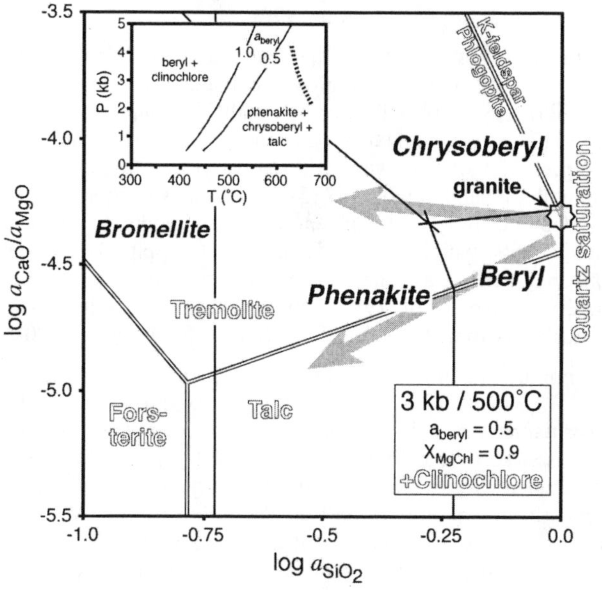

Figure 14. One topology for phase relationships of Be minerals in ultramafic-hosted deposits at 500°C and 3 kbar. The arrows represent two alternative evolutionary that are paths discussed in the text. Chlorite is present throughout and activities of beryl and clinochlore are reduced. The inset shows a schematic water-saturated granite solidus and the calculated position of the dehydration reaction for beryl+chlorite = phenakite+chrysoberyl+talc, demonstrating that on the activity diagram the field of beryl will expand significantly at lower temperatures, helping to account for the scarcity of phenakite and chrysoberyl in ultramafic-hosted occurrences.

commonly contain one or more newly formed mica group minerals (Shcherba 1970; Burt 1981; Kotlyar et al. 1995). Greisen is most common in feldspathic host rocks, but it is described in many protoliths including carbonate and ultramafic rocks ("apocarbonate" and "apoultramafic" greisens, respectively, see Shcherba 1970). This traditional, broad definition lacks the mineralogical specificity to be petrologically useful. In this paper, rather than restrict the long-ingrained usage, we simply focus on the mineral assemblages and note their implications for intensive variable such as $a_{Al_2O_3}$ or acidity. For example, whereas a beryl-bearing topaz-quartz-muscovite greisen is intrinsically acid (Fig. 13; e.g., from Aqshatau, Kazakhstan), a phenakite-bearing polylithionite greisen (e.g., from Thor Lake, Canada) is intrinsically alkaline compared to assemblages containing spodumene, as demonstrated by Reaction (10):

$$2\ LiAlSi_2O_6 + 3\ SiO_2 + K^+ + Na^+ + 2\ H_2O = KLi_2Al[Si_4O_{10}](OH)_2 + 2\ H^+ + NaAlSi_3O_8 \quad (10)$$

Solubility relationships

Another requirement in understanding Be occurrences is the behavior of Be in fluids—aqueous solutions and silicate melts. Although few experimental data exist (Appendix B; London and Evensen, this volume), the principal results merit comment here because they yield useful insight into the processes and patterns in non-pegmatitic deposits.

Aqueous fluids. BeO is only sparingly soluble in pure water, however Be compounds with F^-, CO_3^{-2}, Cl^- and SO_4^{-2} are all significantly soluble (or decompose) in water at room temperature. These potential ligands plus OH^- have received some attention from experimentalists, although not necessarily in experiments designed to yield thermodynamic data (Appendix B). The nearly ubiquitous association of F-bearing minerals with Be deposits has led many investigators to postulate that complexing by F^- is important (Beus 1966). A few others have advocated other complexes, particularly for those deposits where F is apparently absent and other potential ligands such as CO_3^{-2} or SO_4^{-2} are abundant (e.g., Griffitts 1965; Reyf and Ishkov 1999).

In his review and synthesis of the existing experimental data, Wood (1992) concluded that only F^-, F^--$CO_3^=$ and F^--OH^- complexes can generate aqueous Be concentrations >1 ppm in equilibrium with phenakite or bertrandite at temperatures up to 300°C and at plausible pH conditions. According to Wood's analysis, fluoride complexes (BeF^+, BeF_2°, BeF_3^-, BeF_4^{-2}) predominate at lower pH (2-5) whereas a mixed F^--$CO_3^=$ complexes (e.g., $BeCO_3F^-$) may dominate at higher pH (5-7), particularly where [F^-] and [$CO_3^=$] both exceed about 0.01 molal. Beryllium concentrations exceeding 1 ppm seem necessary to make many Be deposits, which commonly have >1000 ppm Be. In some settings, lower concentrations may suffice, as for instance in the case of the Colombian emerald deposits where Renders and Anderson (1987) believe that OH- complexes were sufficient to move all the Be necessary to make the emeralds (but cf. Banks et al. 2000).

In spite of their obvious importance to understanding many hydrothermal deposits, aqueous Be concentrations at T > 300°C are virtually unexplored except for a very few studies. As is the case at lower temperatures, F^- is implicated as though not proven to be the key complexing agent. Beus et al. (1963) found significant Be concentrations in F-bearing solutions that had reacted with beryl, alkali feldspar and quartz at 490-540°C. This is consistent with evidence from experiments on fluids equilibrated with Macusani rhyolite at 650°C and 2 kbar (London et al. 1988). Macusani rhyolite melt (39 ppm Be, 1.3% F) furnishes only 6 ppm Be and 0.35% F to coexisting aqueous fluid (London et al. 1988). Given these results and the fact that beryl solubility in Macusani melts is near 500 ppm Be (Evensen et al. 1999), one can speculate that a plausible maximum Be concentration in a magmatically-derived aqueous fluid would be on the order of 100 ppm. Such concentrations resemble those calculated by Wood (1992) at lower temperatures for phenakite- and bertrandite-bearing assemblages. They are more than adequate to make a major Be deposit.

Silicate melts. Beryllium solubility in felsic melts and its partitioning with coexisting minerals and aqueous fluids has been extensively studied by David London and coworkers (London et al. 1988; Evensen et al. 1999; Evensen and London 2002, London and Evensen, this volume). Others have focused on distribution of Be among silicate minerals in igneous rocks (e.g., Kovalenko et al. 1977; Bea et al. 1994).

Melting of beryllium phases in the end-member systems (Appendix B) has limited geologic relevance, whereas the principal controls on Be solubility in felsic magmas are a_{SiO_2}, $a_{Al_2O_3}$, and, more rarely other components (Evensen et al. 1999):

$$Be_3Al_2Si_6O_{18} = 3\ BeO_{melt} + Al_2O_{3,melt} + 6\ SiO_{2,melt} \qquad (11a)$$

$$Be_2SiO_{4,phenaite} = 2\ BeO_{melt} + SiO_{2,melt} \qquad (11b)$$

$$2\ NaBePO_{4,beryllonite} + Al_2O_{3,melt} + 6\ SiO_{2,melt} = 2\ NaAlSi_3O_{8,plag} + P_2O_{5,melt} \qquad (11c)$$

The first two reactions were investigated by Evensen and London (1999). They showed that Be mineral solubility is a strong function of temperature, increasing by factors of 2-10 from 650°C to 850°C, and that beryl is the saturating phase (±chrysoberyl) in metaluminous and peraluminous melts (cf. Fig. 12). Their results in compositionally simple haplogranite melts demonstrated that Be solubility decreases with the increasing $a_{Al_2O_3}$ consistent with Reaction (11a). Complexing of Be by other elements is implied by increased beryl solubility in the Li-B-P-F-rich, but nonetheless strongly peraluminous (andalusite- and sillimanite-bearing, Pichavant et al. 1988b) Macusani rhyolite.

Evensen and London's experimental results are roughly consistent with what one would expect from Reaction (11a) and the 1 to 1.5 log unit difference in $a_{Al_2O_3}$ between strongly peraluminous granites (e.g., Al_2SiO_5-saturated) and metaluminous granites (at the phenakite-beryl boundary) shown in Figure 12. Using Reaction (11a), predicted Be contents of beryl-saturated melt should increase by approximately 0.5 log units (a factor of 3) from the Al_2SiO_5 limit to phenakite-saturated conditions. This is compatible with the experimentally observed 3-8 times increase in Be solubility over a simila range of ASI. The differences likely reflect more complex speciation (and thus activity-composition relationships) than this simple analysis allows. Applying the same reasoning to the phenakite-stable field in Figure 12 and using Equation (11b), one predicts that Be contents of phenakite-saturated peralkaline granites would be the same as in metaluminous granites (barring changes in Be melt speciation). Only with decreasing a_{SiO_2}, as in undersaturated syenites, would solubilities be substantially higher, perhaps by as much as a factor of two. In melts with exceptionally high P_2O_5 activities beryllonite and possibly other Be-bearing phosphates could substitute for beryl (chrysoberyl or phenakite) as the liquidus phase (Reaction 11c, Charoy 1999).

This analysis underscores the conclusion of Evensen and London (1999) that Be mineral saturation in peraluminous melts is plausible for geologically reasonable Be contents and, furthermore that discrete magmatic Be minerals would not be expected in peralkaline and undersaturated systems except, perhaps, in very late pegmatites.

MAGMATIC BERYLLIUM ENRICHMENTS

Magmatic beryllium enrichments are apparently common, and of interest in their own right, but are they important to make Be deposits? This is uncertain. Enrichment in other elements, notably F for aqueous complexing of Be, may be more much important for igneous-related hydrothermal systems. Here we review aspects of magmatic enrichments and the compositions of igneous rocks associated with non-pegmatitic deposits.

In felsic magmatic systems, Be concentrations can exceed typical crustal compositions of 2-6 ppm by a factor of 10 or more (Beus 1966; Hörmann 1978). Magmatic Be concentration takes place in intrusive and volcanic rocks which range from strongly peraluminous to peralkaline in composition (as in pegmatites, cf. Černý 1991a, this volume; London and Evensen, this volume). Figures 5C and 5D also illustrate the range of Be contents in magmatic systems and the positive correlation between Be and F contents found in volcanic and hypabyssal rocks (Coats et al. 1962; Shawe and Bernold 1966; Kovalenko et al. 1977; Macdonald et al. 1992). The correlation with F is not seen in many deeper rock suites as in pegmatites, for example, where F may be fugitive (e.g., Černý and Meintzer 1988; London 1997). Post-eruption loss of F also likely accounts for some of the variability in volcanic rock suites.

In peraluminous rocks, magmatic Be contents appear to be limited to a few hundred ppm Be (e.g., Kovalenko and Yarmolyuk 1995; Raimbault et al. 1995) but they may exceed 1,000 ppm in some alkaline rocks (Meeves 1966; Richardson and Birkett 1996). This follows the known pattern of increasing Be solubility with increasing alkalinity of the melt (Evensen et al. 1999, cf. Fig. 5C). In metaluminous and peraluminous systems, Be enrichment commonly accompanies enrichment in Li, Cs, Ta whereas in peralkaline systems, Be enrichment sporadically accompanies enrichments in Zr, Nb, REE and others (Fig. 5D; Tischendorf 1977; Kovalenko and Yarmolyuk 1995; Pollard 1995a). The highest concentrations in most igneous environments are in late-stage pegmatites and post-magmatic hydrothermal alteration. Many systems exhibit a continuum between magmatic and hydrothermal features with Be-bearing igneous rocks having clearly post-magmatic veins and cavities with hydrothermal Be minerals. It is commonly difficult to distinguish magmatic from post-magmatic enrichment. Beryllium- and F-enriched rhyolites (topaz rhyolites, ongonites, etc.) are widespread, typically in the same regions and commonly in the same districts as hydrothermal Be deposits (Shawe 1966; Kovalenko and Yarmolyuk 1995).

Strongly peraluminous to metaluminous systems

Peraluminous magmas may or may not show strong enrichment in Li with the Be enrichment. Some follow enrichment like that in Li-Cs-Ta pegmatites ("LCT" type of Černý 1991a and this volume). Examples include a number of the highly evolved Hercynian (Variscan) granitoids of Europe (e.g., Raimbault and Burnol 1998; Charoy 1999), the Macusani rhyolite, Peru (Pichavant et al. 1988a), and the Honeycomb Hills, Utah (Congdon and Nash 1991). In contrast, many strongly peraluminous granites do not exhibit this extreme enrichment in rare elements (e.g., Transbaikalia, the western U.S.; Shaw and Guilbert 1990). Nonetheless they have high F contents and late magmatic (miarolitic to pegmatitic) beryl transitional into Be-bearing hydrothermal assemblages. They may evolve along a different path (cf. London 1992). In weakly peraluminous to metaluminous granitoids and volcanic rocks Be enrichments are not accompanied by dramatic (percent level) contents of Li, but they do have elevated values (Fig. 5).

In most peraluminous to metaluminous igneous rocks, Be is dispersed as a trace element in the rock-forming minerals, most commonly the micas and sodic plagioclase (e.g., London and Evensen, this volume; Kovalenko et al. 1977). Accessory magmatic beryl is described in some granites, aplites and miarolitic zones (e.g., Sheeprock Mountains, Utah, Christiansen et al. 1988; Rogers and Christiansen 1989; Argemela, Spain, Charoy 1999; Mt. Antero, Colorado, Jacobson 1993b). Beryllonite is apparently the principal discrete Be mineral in the Beauvoir granite, France, where only modest amounts of Be (ca. 100 ppm) occur in lepidolite (Charoy 1999).

With the possible exception of the Beauvoir granite (Fig. 15; Cuney et al. 1992; Raimbault et al. 1995), non-pegmatitic peraluminous magmatically enriched rocks lack sufficient Be to be considered Be resources (cf. North Carolina Sn-Ta belt, Griffitts 1954, Fig. 3). At Beauvoir, a composite stock of fine-grained Li-rich leucogranite contains 20-300 ppm Be (>100 ppm in the most evolved unit). It post-dates more voluminous muscovite-biotite granites which have associated greisen-style W and Sn mineralization. Similar patterns are common elsewhere in European Hercynian igneous centers (e.g., Cornwall, England; Manning and Hill 1990).

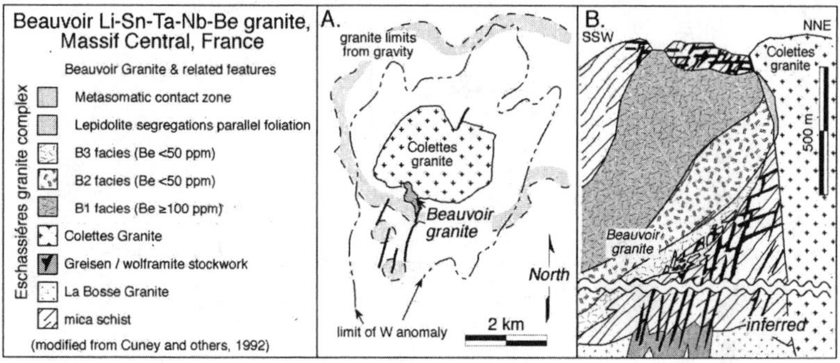

Figure 15. Geology of the Beauvoir Li-F-Sn-Ta-Be granite, a strongly peraluminous system with magmatic rare-metal enrichments. (A) The Beauvoir rare metal granite is a late, volumetrically minor phase of the Echassières leucogranite complex; Sn and W mineralization are associated with earlier phases. (B) Cross section through the Beauvoir granite showing three main phases and cross-cutting relationships with earlier granites and W mineralization. Beryllium is concentrated in lepidolite and beryllonite in B1, the final intrusive unit (Charoy 1999). Figures modified from Cuney et al. 1992.

Peralkaline-metaluminous systems

Peralkaline to metaluminous magmatic systems can have substantial Be enrichments in rocks ranging from riebeckite-aegirine granites to undersaturated syenites and their volcanic equivalents (Richardson and Birkett 1996; Sørensen 1997; Fig. 2, Appendix A). Like the magmatically enriched peraluminous suites, these rocks are typically enriched in F as well as Be but contain a different set of trace elements characterized by Y, Nb, REE with more moderate enrichment in Li (Table 2, Fig. 5; Černý 1991b; Sørensen 1992; Kovalenko et al. 1995a). Associated pegmatitic and hydrothermal deposits are common.

In alkaline granites and quartz syenites Be enrichments can be in the 100s of ppm (Fig. 5; e.g., Khaldzan-Buregtey, Mongolia, Appendix A) and have associated Be-rich alkaline pegmatites. Large deposits with pegmatitic character at Strange Lake and Thor Lake in Canada (Fig. 2, Appendix A) formed during the terminal stages of the development of rare-element-rich alkaline centers. Both have complex internal structures and prominent hydrothermal overprints and the importance of magmatic versus hydrothermal processes concentration is contentious. At Thor Lake (Fig. 16) phenakite, bertrandite, gadolinite and helvite occur in late quartz-fluorite-polylithionite "greisen" zones in a composite feldspar-dominated "pegmatite" (Trueman et al. 1988). The Be mineralization postdates Ta-Nb-Zr mineralization; both are associated with syenite breccias in syenites and peralkaline granites of the Blachford Lake complex. The deposit post-dates the youngest intrusion, a syenite, and is emplaced in somewhat older alkali granite of the same complex. At Strange Lake, gadolinite, leifite and milarite form in lenticular zones associated with the latest stages of a Zr-Nb-Y-enriched riebeckite granite complex. A hydrothermal overprint is clear, although it is debated whether the enrichments are fundamentally magmatic (Miller 1996) or hydrothermal (Salvi and Williams-Jones 1996).

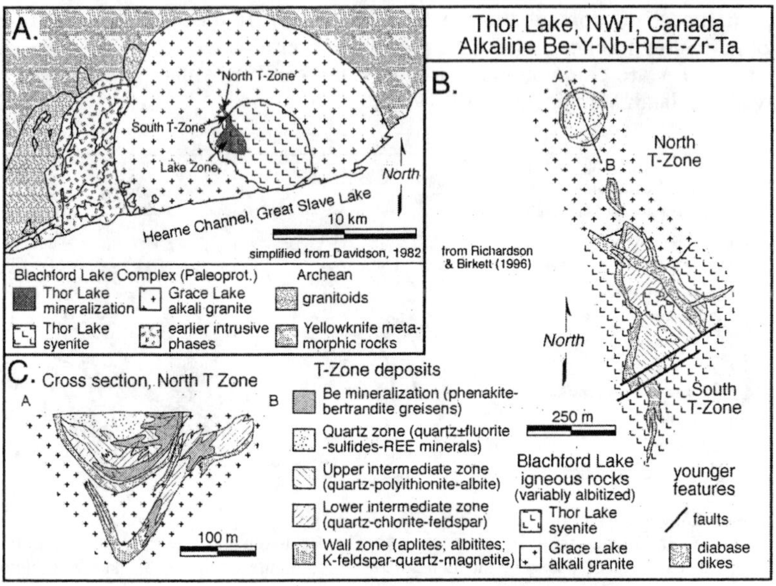

Figure 16. Geology of the Thor Lake area and rare metal deposits, Northwest Territories, Canada. (A) The Thor Lake deposits are associated with the Thor Lake syenite, the youngest member of the alkaline Early Proterozoic (2.1 Ga) Blachford Lake Complex (Davidson 1982). (B) Be mineralization occurs in the T-zone deposits near the NW margin of the Thor Lake syenite. Phenakite(-bertrandite-gadolinite)-rich hydrothermal quartz-fluorite-polylithionite pegmatitic "greisens" are superimposed on a complex set of albite, microcline, and magnetite-rich rocks (Trueman et al. 1988).

The volcanic equivalent of magmatic Be-enriched alkaline granites may be represented by the F-Nb-Zr-Ta-Y-REE-rich trachytic rocks of the Brockman deposit, Western Australia (Ramsden et al. 1993; Taylor et al. 1995a). At Brockman, the hydrothermally altered "Niobium Tuff" averages several hundred ppm Be, which is present (redistributed into?) in quartz-carbonate-bertrandite veins that are restricted to this rare-element enriched stratum. Magmatic concentrations of Be up to 180 ppm occur in the hypabyssal cryolite-bearing, Nb- peralkaline to peraluminous rhyolites of the Sierra Blanca district Texas (Price et al. 1990), which have associated Be-F replacement deposits (see below). Although some rocks from both of these areas are chemically peraluminous, their geological associations, trace-element patterns and associated minerals clearly link them to the peralkaline family.

Beryllium enrichments are also common in the late magmatic phases in undersaturated rocks including examples from the Kola Peninsula, Greenland, and the southwestern United States (Appendix A; Sørensen 1997). Lujavrites (eudialyte-acmite nepheline syenites) from Ilímaussaq, Greenland average 60 ppm Be, while contents up to 1000 ppm have been reported from pegmatitic nepheline syenite at Wind Mountain, New Mexico (Meeves 1966; Steenfelt 1991; Sørensen 1992; Markl 2001). Large Be inventories have been reported (Fig. 3; Appendix A), but none of the undersaturated alkaline systems appear to host plausible resources due to low grades and dispersion of Be in the rock-forming silicates. Large Be enrichments in phonolites have apparently not been recognized, although the elevated Be seen in shallow intrusive systems like Ilímaussaq and Wind Mountain make eruption of such magmas plausible (Fig. 4). They would be silica-undersaturated, peralkaline analogs of the Macusani rhyolites.

Post-magmatic Be enrichments are widespread in syenitic pegmatites and hydrothermal veins in these locations and others, notably the Oslo province, the Kola Peninsula and Mt. Saint-Hilaire (Appendix A; Beus 1966; Vlasov et al. 1966; Engell et al. 1971; Horváth and Gault 1990; Men'shikov et al. 1999; see below). These undersaturated, typically feldspathoid- or zeolite-bearing rocks contain a distinctive suite of Be minerals, notably the Al-poor, Na-Ca-Be silicates (e.g., epidydimite, leifite, leucophanite, chkalovite) plus others such as gadolinite, phenakite, bertrandite, genthelvite and bromellite.

Magmatic vs. metasomatic albite-rich granitoids. Albite-rich granitoids (and some albite-rich syenites) can have either magmatic or metasomatic origins. Both types commonly have Be enrichments but they can be difficult to distinguish from one another. Magmatic varieties have F-enrichments and carry considerable concentrations of rare metals such as Ta, the specific suite corresponding to the overall genetic family (Kovalenko and Yarmolyuk 1995; Pollard 1995a). Such granitoids are extreme differentiates of F-rich magmas (Manning 1982). Fluorine-rich metasomatic albitization is also common in granitic systems and can carry contain broadly similar element enrichments (Charoy and Pollard 1989; Laurs et al. 1996; Haapala 1997). Distinguishing between the two requires textural or geochemical observations (sharp versus gradational geologic contacts; petrographic evidence for replacement, dissolution of earlier minerals such as quartz; high versus low variance assemblages, uniformity of phase proportions). Beryl concentrates in both settings (e.g., Beus 1966; Charoy 1999).

HYDROTHERMAL OCCURRENCES ASSOCIATED WITH FELSIC MAGMATISM

Hydrothermal Be deposits generated by felsic magmas are numerous and diverse (Table 2, Figs. 2, 4). Depositional environments, particularly the composition of the host rocks, exert the most prominent control on the styles of mineralization regardless of magmatic compositions. Igneous compositions strongly influence mineralogy, element enrichments and zoning. Magmatic Be enrichment can be important in some cases, but overall is apparently subordinate to other factors. We group systems by igneous compositions and foremost, by the degree of alumina saturation because this is predictive of mineral associations (Fig. 12) and correlates broadly with other intensive variables and geologic setting. Boundaries between groups can be arbitrary as there is clearly a continuum among these groups and many igneous centers possess a range of compositions.

Vein, greisen and volcanic (fumarolic) deposits occur in felsic igneous and siliciclastic sedimentary host rocks. These deposits commonly have abundant F. Muscovite-rich alteration, quartz veins and variable amounts of W, Mo, Bi, and Sn typify the beryl-dominated mineralization that forms in strongly peraluminous systems. Li(-Fe) micas and alkali-feldspar alteration become characteristic with decreasing $a_{Al_2O_3}$, as metal assemblages gain Zr-REE-Nb and lose W. Fenites and quartz-absent hydrothermal veins form in silica-undersaturated systems. At low temperatures (<300°C) bertrandite-bearing quartz veins can form commonly with K-feldspar±carbonate±sericite±fluorite, in some cases with Mn-rich, base-metal sulfide-rich associations.

Skarn, greisen and replacement deposits form in carbonate-bearing host rocks where they generally contain abundant fluorite. They comprise the economically most important deposits including the fluorite-rich replacement deposits in the carbonate-lithic-rich tuff of Spor Mountain. All silica-saturated magma types can produce garnet, pyroxene and vesuvianite-rich Be-bearing skarns where mineral ratios and compositions tend to reflect the redox state of the related granites (cf. Einaudi et al. 1981). Aluminum-rich metasomatism—which produces muscovite, other micas and diaspore (cf. "apocarbonate greisen" of Shcherba 1970)—characterizes the BASH mineral-bearing deposits that form near peraluminous granites. Less aluminous magmas generate skarns and K-feldspar-bearing or Al-poor fluorite-rich replacement deposits. Typical minerals include phenakite, bertrandite, and the helvite group with less common bavenite, leucophanite, gadolinite, milarite and others. A distinctive texture found in many carbonate-hosted systems is rhythmically banded replacement containing alternating light and dark layers with combinations of fluorite, Be minerals (helvite-danalite is typical) and other minerals including silicates and magnetite ("ribbon rock" Jahns 1944b; "wrigglite" Kwak 1987; see photos in Fig. 18C and Fig. 22, below).

Mafic and ultramafic host rocks are relatively uncommon, but they can be important in that they host most emerald deposits, which form where beryl-bearing pegmatites or veins gain Cr and lose silica during original emplacement or subsequent metamorphism. Most such systems are peraluminous. Biotite-producing metasomatism is ubiquitous. This group is treated separately below.

Peraluminous magma-related systems

Hydrothermal Be mineralization occurs with many strongly peraluminous muscovite- or cordierite-bearing granites as well as with weakly peraluminous biotite granites ("pG" in Appendix A). This suite contains some of the more important non-pegmatitic Be deposits, including large sub-economic resources in the Seward Peninsula, Alaska, eastern Nevada and central Kazakhstan. It is also notable for emerald and aquamarine deposits associated with ultramafic and greisen host rocks, respectively (e.g., Reft River, Ural Mtns; Sherlova Gora, Transbaikalia). The salient characteristics of the peraluminous group are aluminum-rich hydrothermal alteration and predominance of BASH minerals.

The peraluminous family can be cast into two groups (cf. Fig. 4A): (1) specialized strongly peraluminous granites, commonly with exceptionally high Li-Cs-Ta (LCT) and other lithophile elements, locally with associated greisen Sn mineralization, and (2a) less specialized but strongly peraluminous granites with or without W-Mo(-Sn) mineralization, or (2b) weakly peraluminous Sn-W(-Mo) systems with elevated rare metal contents. The last group commonly has late muscovite-bearing leucogranites. Although this group can be considered to form a continuum with metaluminous systems, it generally has highly aluminous alteration assemblages in various rock types that are lacking in the latter. Most of hydrothermal systems formed at <5 km depth, but some, particularly those associated with strongly peraluminous muscovite-biotite granites, formed in the 5-10 km range. Geochemical data and geological associations point to a metasedimentary, perhaps dominantly pelitic, source for the magmas of group (1), a mixed crustal source for the magmas of group (2), and a hybrid crust and mantle source for the magmas of group (3) (cf. Černý 1991b; Newberry 1998).

It is unusual to find hydrothermal Be mineralization associated with the more evolved Li-Cs-Ta-type magmas even though many have substantial magmatic Be contents (Fig. 5; e.g., Macusani, Beauvoir, Richemont). For example, most specialized granites of the European Hercynian lack hydrothermal Be occurrences (Stussi 1989; Manning and Hill 1990). In the Cornubian Sn-W district Be minerals occur in greisens, veins and skarns that are related to the biotite(-muscovite) granites but do not form with the highly specialized topaz granites (Appendix A; Jackson et al. 1989; Manning and Hill 1990). Where Be minerals are mentioned with Li-Cs-Ta-type magmas they form in early assemblages, for example in the Erzgebirge where beryl and herderite are minor constituents of proximal quartz-topaz greisens at Ehrenfriedersdorf and elsewhere (Baumann 1994). Similar relationships are apparent in the Geiju district, Yunnan, China where Be-W mineralization occurs at the apex of greisenized rare-metal granites and zones outward into Sn and sulfide mineralization (Kwak 1987). The lack of hydrothermal accumulations in many of these systems may reflect their low solidus temperatures, their low water contents and limited ability to exsolve water, and the relatively low partition coefficients for Be into coexisting aqueous fluid (London et al. 1988; Raimbault and Burnol 1998). In contrast to the Li-Cs-Ta-group, the less compositionally extreme peraluminous magmas are associated with many occurrences.

Feldspathic host rocks: These rocks host three common styles of fracture-controlled Be mineralization: beryl in quartz-K-feldspar(-mica) veins, beryl in albitized rocks, beryl and other Be minerals in muscovite-topaz-fluorite-dominated greisens. The latter are by far the most important. Many areas contain all three styles in a progression from early, proximal and typically deeper K-feldspar-stable assemblages through albitization to late, commonly distal greisen associations. Occurrences are widespread, notable examples found with biotite±muscovite granites occur in China, central Asia, the North American cordillera, and western Europe (Appendix A).

Small, coarse-grained quartz-K–feldspar(-muscovite-biotite) veins containing accessory beryl, molybdenite and wolframite occur with some peraluminous pegmatites and W-Mo(-Sn) affiliated granitoids (e.g., in the Canadian cordillera and maritime provinces, Mulligan 1968). These veins typically lack fluorite and paragenetically later Be-rich veins are rare. Geological context indicates that they formed at considerable depth; they could represent root zones of other deposit types. One variant on this theme is illustrated by the large Verknee Qairaqty and Koktenkol stockwork W(-Mo) deposits in Kazakhstan where minor beryl occurs only in early, 300-400°C quartz-K–feldspar-molybdenite-scheelite veins (Mazurov 1996; Russkikh and Shatov 1996). Beryllium is distributed throughout the paragenesis at Dajishan, Jiangxi, China where quartz-feldspar-beryl veins change with time and distance into helvite-bearing fluorite-muscovite-quartz veins with wolframite, scheelite and molybenite (Raimbault and Bilal 1993). Another variant may be represented by the relatively F-poor Sn-W deposits of SE Asia where Be mainly occurs as beryl in pegmatitic bodies (Suwimonprecha et al. 1995; Linnen 1998). Feldspar-dominated veins are of little economic interest and consequently are thinly documented. Conversely, alkali feldspars are present in many F-rich greisen and albitic assemblages associated with major Be deposits.

Albitized rocks are widespread in Be-rich peraluminous systems where they grade into mica-dominated greisen assemblages. Typically, albite+muscovite±fluorite±chlorite replace igneous feldspars and micas; modal quartz also commonly decreases. These form pipes, veins, vein envelopes (commonly around mica-rich greisen veins), and pervasive zones particularly near the tops of intrusions. Accessory beryl with albitic assemblages is reported in many systems (Beus 1966; Dyachkov and Mairorova 1996). A well described example at Triberg, Germany (Markl and Schumacher 1996, 1997) formed from biotite-muscovite leucogranites that have late beryl-bearing miarolitic pegmatites. Hydrothermal beryl-albite-muscovite-fluorite alteration ultimately grades into beryl(-bertrandite-phenakite)-bearing quartz-muscovite-topaz greisen veins. Like many other beryl-rich two-mica systems, the mineralizing fluids contained <10 wt % NaCl equivalent ($NaCl_{eq}$) and Sn and W were only weakly concentrated. Rogers (1990) describes similar relationships in muscovite-bearing biotite granite in the Sheeprock Mountains, Utah.

Quartz-rich greisens with abundant accessory muscovite, topaz, fluorite and siderophyllite contain most Be minerals (beryl > phenakite, bertrandite, euclase) found in peraluminous-related deposits. Beryl can be either in the vein fill with quartz and other minerals or it can concentrate at the outer margins of the greisen envelopes against feldspar-stable assemblages (generally albite; Beus 1966). Overall Be distribution varies in greisens; it is typically distal or late within the intrusions and may or may not extend into surrounding veins or skarns. Hematite-bearing alteration and helvite group minerals are rare with the peraluminous group in contrast to greisens associated with fundamentally metaluminous biotite granites.

Many examples from around the world illustrate these patterns (Appendix A). In the Great Basin, reduced (low Fe^{+3}/Fe^{+2}) Cretaceous muscovite-biotite granites and their clastic host rocks contain minor beryl in muscovite-fluorite-quartz-pyrite±wolframite greisen veins. Paragenetically earlier quartz-K–feldspar veins and albitization typically lack beryl. This is compatible with the paucity of Be in the correlative pre-greisen stages of associated skarns. These systems contain large quantities of F, Be and Zn, with minor Mo (proximal), W and Sn (distal). Associated fluids had moderate CO_2 contents and salinities (5-10 wt % $NaCl_{eq}$) and were of magmatic derivation. In the northern Cordillera, sparse beryl occurs in proximal muscovite- or topaz-bearing alteration in W-Mo (e.g., Logtung, Yukon) and Sn-W (e.g., Lost River, Alaska) systems. Similar relationships hold along the western margin of the Pacific in southeastern China (e.g., Wangfengshan, Guangdong) and eastern Australia (e.g., Mole Granite, New South Wales)

Many Be-bearing W-Mo(-Sn) greisens occur in central and eastern Kazakhstan where they are associated with mainly Late Paleozoic biotite±muscovite leucogranites (Appendix A; Burshtein 1996; Serykh 1996; Ermolov 2000). Beryllium occurs in several modes: as beryl in muscovite-topaz-quartz greisens (e.g., Aqshatau; Fig. 17), as bertrandite±helvite (after beryl) in late fluorite-rhodochrosite-sulfide veins (e.g., East Qonyrat), and as chrysoberyl and other minerals in F-rich skarns (e.g., Qatpar).

Among peraluminous-related deposits greisen deposits are the more common than other types and the only variety from which Be has been produced. The largest reported resource is the Aqshatau district in Kazakhstan (Fig. 17; Appendix A; Beskin et al. 1996) where beryl has been produced from W(-Mo-Bi) greisen veins. Beryl formed in the distal parts of the quartz-topaz-muscovite-wolframite greisen veins where BeO contents can exceed 0.1% (Fig. 17B). These veins exhibit zoning centered on a multi-phase muscovite-biotite leucogranite complex that appears to be the fluid source. Extensive study demonstrates that saline (>30 wt % $NaCl_{eq}$) magmatic fluids account for most of the mineralization with fluid pressures fluctuating near lithostatic values. These fluids were followed by an influx of dilute, meteoric waters under hydrostatic conditions that formed late quartz-sulfide-carbonate assemblages.

The noted aquamarine locality at Sherlova Gora, Siberia (Sinkankas 1981) has extensive beryl-bearing quartz veins (with BeO $\geq$ 0.02%) with topaz, siderophyllite, fluorite and muscovite greisens (Beus 1966). These occur in the outer portion of a variably miarolitic and porphyritic biotite±muscovite granite pluton. Within the pluton, beryl is late and tends to be distal in the greisen veins. Beryllium as well as alkalis are removed from intensely greisenized rocks. In the system as whole, beryl and minor Mo-W±Sn mineralization form proximally whereas Sn-polymetallic mineralization extends well away from the intrusions (Troshin and Segalevich 1977). This district, although emplaced at shallower levels than those in the Great Basin, is also characterized by both muscovite-bearing granites and relatively low Sn-W-Mo contents.

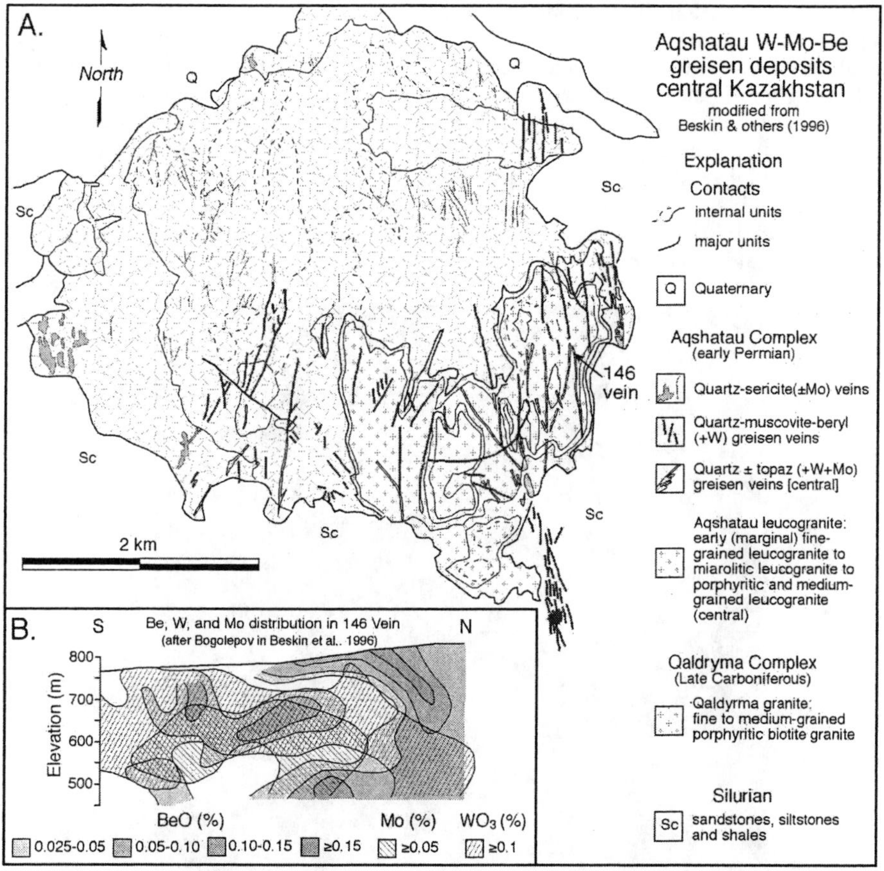

Figure 17. Geology of the Aqshatau (Akchatau) greisen W-Mo-Bi-Be district, central Kazakhstan. (A) Geology map showing the distribution of granitoids and hydrothermal features in the central part of the district. Older, Carboniferous granodiorites and granites of the Qaldyrma complex are cut by cupolas of the ore-related Permian biotite-bearing leucogranites of the Aqshatau complex. Note lateral zoning of W-Mo-Be mineralization away from the latter. (B) Zoning of WO_3, Mo and BeO (values in %) along the 146 vein (see A for location). [A,B redrawn from Beskin et al. 1996.]

Although greisen-type alteration hosts most Be minerals in peraluminous granitoids, Be minerals are commonly sparse when compared to the amounts present in adjacent carbonate hosted-mineralization. With only a handful of exceptions worldwide (Aqshatau, Sherlova Gora), the peraluminous granite-hosted deposits have not been a significant source of either Be metal or gems. This is well illustrated by Phanerozoic Be-rich magmatic-hydrothermal systems around the circum-Pacific (e.g., Cretaceous Cordillera, Tasman system, SE China; see Appendix A).

Carbonate host rocks. In carbonate rocks, Be deposits related to strongly peraluminous granitoids are characterized by exceptionally high F and Al contents and elevated contents of many other elements including Li, Sn, and W. Beryllium occurs both in skarns and in superimposed or distal apocarbonate greisen or replacement deposits. In skarns, Be is reported to isomorphically substitute in vesuvianite and other silicates (e.g., Beus, 1966), whereas in the greisen or replacement deposits Be clearly forms discrete phases. Fluorite, F-rich silicates, micas (muscovite, Li-micas, phlogopite), topaz, albite, K-feldspar and quartz can all be abundant. Typical Be minerals include chrysoberyl, phenakite, beryl, bertrandite, euclase and bertrandite. Rarely present are bavenite and Mg-bearing aluminates of the taaffeite group (Table 1). Quartz is typically sparse in carbonate-hosted greisen-style alteration. Iron and base metal contents vary considerably. Pressure (depth of emplacement) appears to be an important factor in quartz and sulfide abundances: higher salinity fluids, more extensive metal transport and silica-under-saturated assemblages occur at <1-1.5 kbar, whereas more siliceous assemblages (due to higher silica mobility) and less concentrated metals (linked to lower fluid salinities) are more common at higher pressures (cf. Fig. 11A).

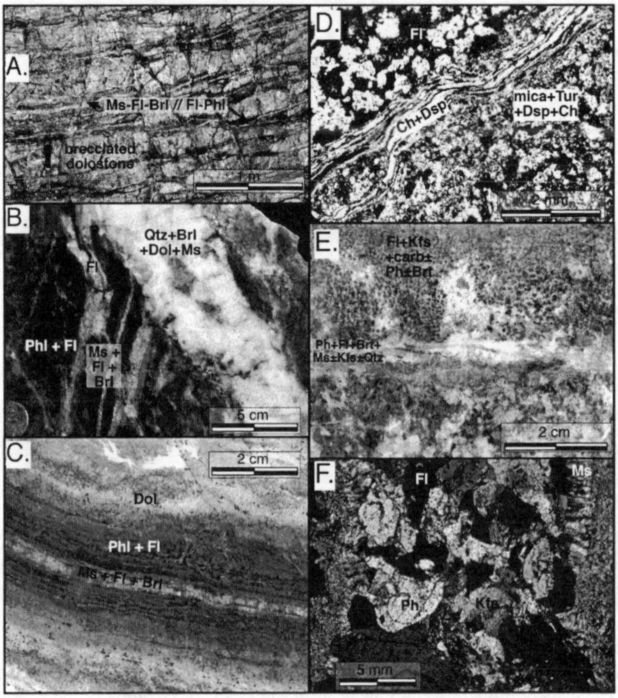

Figure 18. Textures of carbonate-hosted Be mineralization associated with strongly peraluminous granites showing characteristic F- and Al-rich veins and replacements. (A)-(C) Mica-fluorite-beryl (-quartz) veins from the deep W(-Mo) systems at McCullough Butte, Nevada (Appendix A). (A) Trench face with typical muscovite-fluorite(-beryl-pyrite-scheelite-sphalerite-bertrandite-quartz) veins and fluorite-phlogopite envelopes in brecciated dolostone. This exposure averages about 25% CaF_2 and 0.25% BeO. (B) Quartz-aquamarine-dolomite(±muscovite) vein cutting muscovite-fluorite-beryl(white) veins with inner fluorite to outer fluorite+phlogopite envelopes. (C) Rhythmically banded fluorite-phlogopite skarn envelope on muscovite-fluorite-beryl-pyrite vein. (D) Mottled chrysoberyl-rich replacement from the shallow Lost River Sn(-W) district, Alaska. Very fine-grained chrysoberyl+diaspore vein cuts mottled white mica (Li, Be-bearing)+tourmaline+diaspore+chrysoberyl replacement of limestone. (E)-(F) Phenakite-bertrandite-fluorite vein and replacement mineralization Mount Wheeler Mine, Nevada (Appendix A) which typical of distal fluorite-rich mineralization in carbonate rocks associated with many magma types (cf. Zabolotnaya 1977). (E) Phenakite(-bertrandite)-fluorite-muscovite-adularia vein-cutting fluorite-adularia-manganosiderite-phenakite replacement of limestone. Fluorite is dark and comprises more than half of these zones. (F) Micrograph of central vein from (F) showing phenakite-fluorite-adularia (Kfs) vein with muscovite border. All photos by M.D. Barton (except for A—modified from Sainsbury 1969).

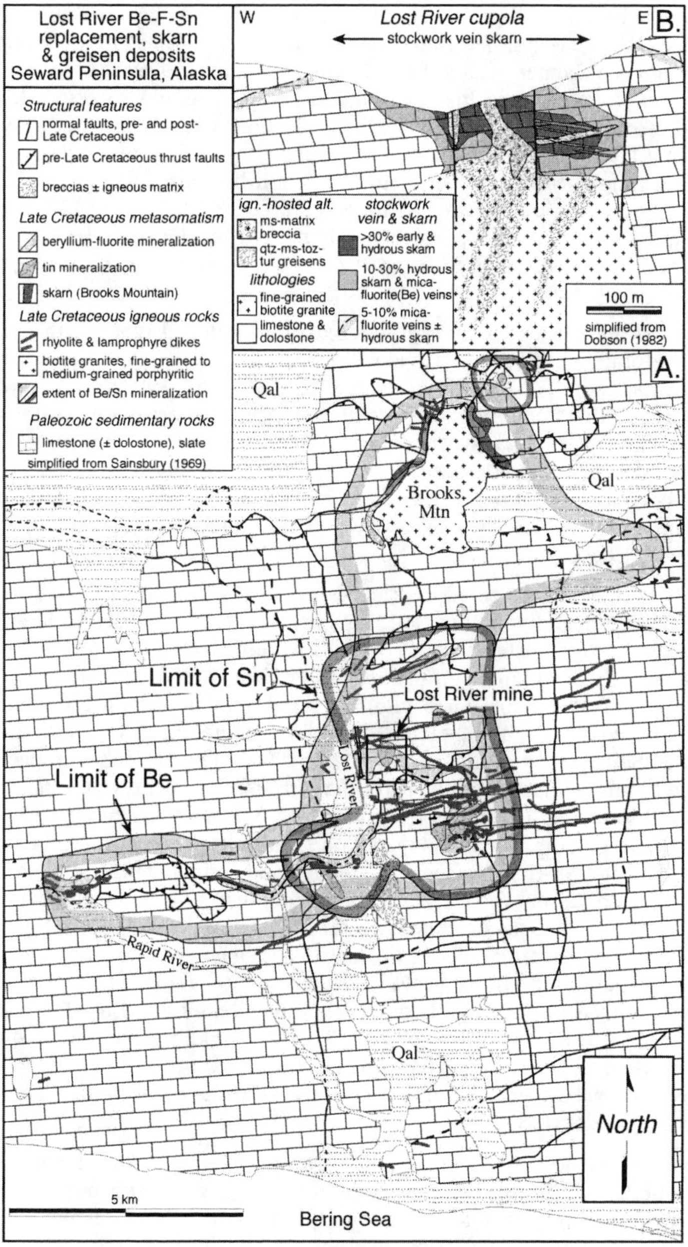

Figure 19. Geology of Be-F-Sn-W mineralization in the Lost River area, Seward Peninsula, Alaska. (A) Geologic map of the region showing the distribution of Be-F replacement mineralization, Sn and skarn mineralization, and igneous rocks. Note the strong structural control. Qal – Quaternary alluvium. Simplified from Sainsbury (1969). (B) Cross section of the Lost River mine area showing the generalized distribution of early and hydrous skarns and multiple stages (early and late) fluorite-mica-Be veins. Simplified from Dobson (1982).

High-grade Be mineralization in carbonate rocks typically consists of paragenetically complex fine- to coarse-grained fluorite±mica, K-feldspar, diaspore or tourmaline-rich open-space and replacement veins (Fig. 18; Ginzburg et al. 1979). These veins typically comprise stockwork systems that can extend up to several kilometers from known sources and proximal skarns (e.g., McCullough Butte, Nevada; Lost River, Alaska; Fig. 19). Rhythmic layering is common as it is in other Be-bearing systems (Fig. 18C). The ore minerals range from very fine-grained—"curdy" chrysoberyl is common and represents a metallurgical challenge for economic recovery (Fig. 18D; Apollonov 1967)—to quite coarse-grained, which is common with phenakite-bertrandite-fluorite replacements (Fig. 18E; Ginzburg et al. 1979). Chrysoberyl occurs principally in the lower pressure, better metallized districts.

Associated anhydrous calcic skarns range from reduced types with hedenbergitic pyroxene > aluminous garnet+vesuvianite to oxidized types with andraditic garnet± vesuvianite±magnetite > diopsidic pyroxene (cf. Newberry 1998). The latter are more common with shallow, Sn(-W) systems, like those associated with biotite granites in western Alaska or Tasmania, whereas the former are more common with W(-Mo-Sn) skarns associated with biotite±muscovite granites from southeastern China and the western USA (Fig. 4). Magnesian skarns of both types differ in having more abundant humite-group minerals, other Mg-silicates, sellaite, spinel and magnetite (cf. Einaudi et al. 1981; Kwak 1987).

One of the best-documented districts is at Lost River on the Seward Peninsula, western Alaska (Appendix A; Fig. 19). This area contains the second largest U.S. Be resource after the tuff-hosted ores at Spor Mountain, Utah. Biotite granites and late muscovite-bearing leucogranites in the Seward Peninsula are all peraluminous, but they differ from strongly peraluminous granitic suites in that they are commonly more oxidized (higher Fe^{+3}/Fe^{+2}) and they mostly lack Al-saturating phases such as muscovite or cordierite (Sainsbury 1969; Swanson et al. 1990). The Be deposits occur as fine-grained, commonly laminated chrysoberyl-diaspore-mica-fluorite veins and replacement bodies in limestones and dolomites (Fig. 18D). Also present are minor phenakite, beryl and euclase. These bodies are developed along faults and adjacent to dikes and extend kilometers from the Sn-rich greisenized granites and skarns. There is a strong vertical control on the distribution of hydrothermal alteration that reflects level of exposure in the systems. The Be-F-Al replacements are mainly distal and structurally high. Proximal iron-rich skarns can contain helvite with fluorite and magnetite, but more commonly are dominated by andraditic garnet, vesuvianite, magnetite, fluorite, scapolite and minor pyroxene overprinted by hydrous skarn with hornblende, biotite, fluorite, cassiterite and sulfides (Dobson 1982; Swanson et al. 1990). The skarns abut topaz-tourmaline-muscovite greisenized fine-grained biotite±muscovite granites (Fig. 19 inset). Carbonate-hosted greisen veins both predate and postdate the skarns and thus indicate multiple fluid release events (Dobson 1982).

Systems similar to Lost River occur elsewhere in Alaska and, notably, down the eastern side of Asia from Siberia into southern China (Fig. 2; Appendix A). Systems in southeastern China are associated with middle to late Mesozoic biotite±muscovite granites, typically with greisen mineralization within the intrusions. Unlike in North America, several of these complex polymetallic districts are economically important for other commodities. The Shizhuyuan, Hunan W(-Sn-Mo) district contains the world's premier W deposit and has Be-rich Mo-W-B-Sn±Cu±Pb skarns, greisens and replacement bodies (Mao et al. 1996b). At Xianghualing, Hunan, late chrysoberyl-fluorite-phlogopite ribbon rock and later fluoborite-chrysoberyl (-taaffeite group mineral) mica veins with spinel envelopes and minor sulfides overprint garnet-vesuvianite skarns and ribbon magnetite skarns. These magnesian skarns contain chondrodite, vesuvianite, diopside, amphibole, tourmaline and formed in dolomitic host rocks adjacent to beryl-bearing albitized and greisenized biotite granite and (Lin et al. 1995a, p. 238-242). Similar Be-F-rich mica-dominated veins overprint and are distal to complex polymetallic Sn skarns in the Geiju Sn district (Kwak 1987).

In districts with less compositionally evolved magmas, Be tends to be dispersed either in rock-forming silicates or as minor beryl. High Be concentrations occur only in particularly favorable traps. For example, many large hydrothermal systems are associated with Cretaceous two-mica granites in the Great Basin of the western United States (Appendix A). At McCullough Butte, Nevada two-mica granite contains 6-12 ppm Be and 0.03-0.2% F (Barton, unpubl. data), whereas contemporaneous two-mica granitoids emplaced at 20+ km depths in the Ruby Mountains have 0.5-4 ppm Be (Calvin Barnes, written comm., 2000). Around and above the shallower plutons (emplaced at 5-10 km depth), muscovite-phlogopite-fluorite-quartz-bearing veins and replacement zones overprint or are peripheral to reduced W(-Mo-Sn)-bearing pyroxene>garnet skarns (Fig. 18A-C). Some of these hydrothermal systems contain much Be (e.g., McCullough Butte, Fig. 3), but high concentrations (>0.1% BeO) only formed in distal locations where fluids traversed unreactive quartzites prior to encountering carbonate rocks. This is illustrated by McCullough Butte (see Fig. 18A) and, most extensively, around the Mount Wheeler mine, Mt. Washington district, Nevada (Fig. 20). In the Mt. Washington district, a substantial Be(-W-F) resource (Fig. 3) occurs in laterally extensive phenakite-bertrandite-beryl-scheelite-fluorite replacement bodies that formed in the basal carbonate unit above a thick clastic section overlying an unexposed granite (Fig. 20B). In place of skarn minerals, early mineral assemblages in this distal, cool hydrothermal system contain Fe-Mn carbonates and quartz, and were followed by deposition of progressively Al-, F-, and Be-enriched assemblages that culminate in muscovite-fluorite-beryl-quartz veins.

Metaluminous magma-related systems

Notable non-pegmatitic deposits of Be occur throughout the world with metaluminous or weakly peraluminous magmatism ("G" in Appendix A, Fig. 2). The preeminent region comprises Spor Mountain and other deposits associated with mid-Tertiary felsic magmatism in southwestern North America. In contrast to strongly peraluminous-related systems where Be can be economically concentrated in pegmatites and igneous-hosted vein deposits, metaluminous suites lack major intrusion-hosted deposits even though phenakite, beryl, helvite group and gadolinite occurrences are widespread. Instead, carbonate and volcanic rocks host the important deposits which mostly are fluorite-rich, bertrandite-, phenakite- or helvite-bearing veins and replacements. Beryl is normally subordinate to the other Be minerals or is completely absent.

Genetically related igneous rocks are highly felsic—typically biotite-bearing leucogranites and high-silica rhyolites (Fig. 5A,B). Coeval syenites and hornblende-bearing calc-alkaline granites are commonly present, whereas muscovite-bearing variants are scarce. They may occur in bimodal suites with mildly alkaline mafic rocks. Typical associated metals include Mo, Sn(>W), Ta, Nb and Zn. Most, but not all, of these systems are F-rich and many are sulfur-poor as evidenced mineralogically by widespread genthelvite-danalite solid solutions (Fig. 9) and other indicators of relatively low S such as Pb-enriched feldspar (amazonite). Another common feature of metaluminous-related systems, shared to some degree with other types, is an association with distal Mn-rich replacement or vein mineralization. In some areas, for example south-central New Mexico, helvite-bearing base-metal sulfide replacements formed near skarn and volcanic-hosted Be mineralization. A link to proximal Be-enrichments is not evident in other

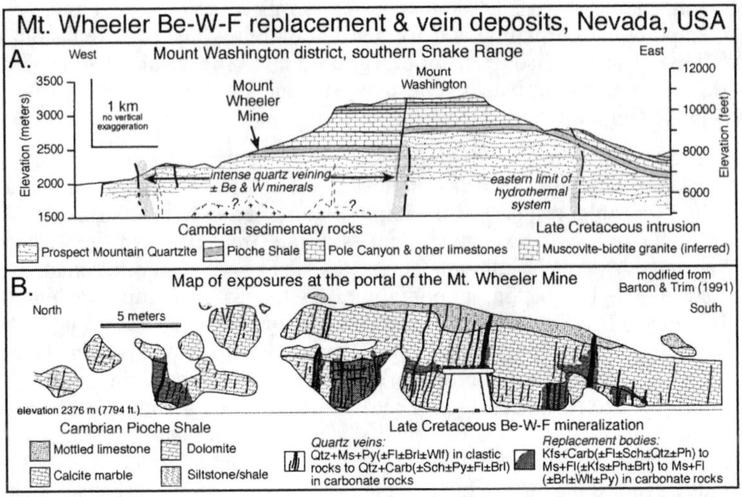

Figure 20. Geological relationships in the Mount Wheeler Mine area, Mount Washington district, Nevada. Beryllium mineralization here is associated with non-specialized strongly peraluminous granites. This district contains the most extensive and highest grade Be mineralization of the >20 occurrences associated with Late Cretaceous two-mica granites in the Great Basin (Barton 1987; Barton and Trim 1991). (A) Cross section through the district showing lateral extent of hydrothermal system. High-grade Be(-W-F) mineralization is laterally extensive but vertically restricted to carbonate rocks in the Pioche Shale, the lowest reactive beds in the sedimentary sequence. Only sparse granite porphyry dikes are exposed at the surface, but large two-mica granite bodies are exposed nearby in the southern Snake Range (Barton, unpubl.). (B) Sketch map showing the localization of high-grade (ca. 1% BeO & WO_3; 20% CaF_2) low-temperature (<300°C) replacement bodies that form where sheeted quartz veins intersect the basal carbonate rocks. See Figure 18E,F.

examples, including helvite in the epithermal Mn-rich mineralization at Silverton, Colorado and in the distal polymetallic Mn-mineralization at Butte, Montana district (Warner et al. 1959). These occurrences share features with distal rhodochrosite-bertrandite and Mn-silicate/carbonate occurrences in many districts in central and east Asia (e.g., East Qoynrat, Kazakhstan; Shizhuyuan, China).

Felsic host rocks in non-volcanic settings: In intrusive environments pegmatitic, miarolitic, albitized and greisenized bodies contain gadolinite, beryl, bazzite, phenakite, bertrandite, bavenite, and helvite group minerals. Although many features overlap with peraluminous systems, mineral assemblages are typically less acid (higher molar (K+Na/)Al). Cavity filling and alteration mineral associations are dominated by alkali feldspars, quartz and trioctahedral micas (siderophyllite, zinnwaldite, etc.). Muscovite is common in some greisens, but is generally less abundant than in strongly peraluminous examples. Beryl and (rare) euclase are largely restricted to muscovite-bearing greisens. In a number of regions, porphyry Mo style mineralization is associated with the Be-bearing intrusive complexes (e.g., in Norway, southwestern US, central Asia; cf. Geyti and Schoenwandt 1979; Burt et al. 1982; Burt and Sheridan 1986). Although few of these areas have detailed petrologic documentation, most appear to have formed at no more than 5 km depth from moderate salinity magmatic fluids. None of the granite-hosted deposits have been major sources of Be.

Biotite granites and monzonites with accessory Be minerals in miarolitic cavities and pegmatitic veins are well known in the southern Alps, Oslo Rift, Colorado (Pikes Peak batholith; Mt. Antero), and the Younger Granites Province of Nigeria (Appendix A). These occurrences are small and mainly of mineralogical interest. For instance, sparse gadolinite, bazzite, and bavenite are present in pegmatitic quartz-feldspar(±beryl) assemblages in biotite granite plutons in northern Italy and in miarolitic cavities in the peralkaline to metaluminous Oslo Rift intrusions. Mid-Tertiary calc-alkaline batholiths in Colorado (Mount Antero area, Mt. Princeton batholith) and Idaho (Sawtooth batholith) have biotite granites with miarolitic cavities and pegmatitic quartz veins that are noted for their aquamarine and other specimen material. At Mt. Antero, beryl, phenakite and bertrandite form sequentially in miarolitic cavities, pegmatites and muscovite-bearing quartz-molybdenite veins all associated with the apex of a small chemically specialized biotite leucogranite (Adams 1953; Jacobson 1993b). Geologic relationships and fluid inclusion data from Mt. Antero indicate that associated fluids were magmatic in origin had fairly low salinities (0.5-8 wt % $NaCl_{eq}$) and spanned a wide temperature range (~600 to 200°C; Kar 1991).

In metaluminous-related Be-rich hydrothermal systems, paragenetically early metasomatic K-feldspar±K-Fe mica assemblages are commonly mentioned; however, these assemblages are not well-described and they apparently lack Be minerals. This is a bit surprising given the common co-occurrence with beryl and other minerals in miarolitic cavities. Similarly, albitization with quartz loss and Li-Fe-Al mica growth in metaluminous systems mostly lacks Be minerals (e.g., Beus 1966; Charoy and Pollard 1989). An exception is at Sucuri, Brazil where helvite-group minerals occur with iron and base metal sulfides (Raimbault and Bilal 1993). Albitization of uncertain origin—magmatic or hydrothermal—is widely reported. Haapala (1997) argued that much of the albite is magmatic or due to local post-magmatic redistribution of components without significant sodium metasomatism. Albitization is widely reported in the Nigerian alkaline complexes in the same areas that contain Be-bearing greisens. According to Bowden et al. (1987) the albite has "snowball texture" (albite laths and zinnwaldite enclosed in large microcline and quartz crystals)—a texture that plausibly seems magmatic (e.g., Lin et al. 1995b, but cf. Kempe et al. 1999). Given the focus placed on albite-rich rocks as either an evolved magmatic source or a leached metasomatic source for Be this topic remains a fertile one for additional work.

In felsic host rocks hydrothermal Be mainly occurs in small greisen veins and pipes with or without associated albite-rich assemblages. Well-known regions are commonly Sn-rich and include Karelia, Brazil, Nigeria, Colorado and central Asia (Fig. 2, Appendix A). Hydrothermal quartz, Li-Fe-Al sheet silicates, and fluorite are ubiquitous. Topaz, cassiterite, wolframite and Mo-Zn(-Pb-As-Cu) sulfides are common. Cryolite can be present in fluid inclusions or as a separate phase. Sheet silicates include chlorite, Li-muscovite, siderophyllite and Li-Fe micas with chlorite being most common in the outer part of vein envelopes. Although muscovite is prominent in some deposits, greisens in many areas contain only trioctahedral micas (e.g., in Nigeria, Bowden and Jones 1978). Beryl is the most widely reported Be mineral and bertrandite, genthelvite and phenakite are common. These bodies typically zone from central quartz-rich bodies with topaz through inner mica-quartz envelopes to outer chlorite-mica-K-feldspar envelopes. A characteristic reddening of vein envelopes due to dispersed hematite is widely reported (e.g., Nigeria, Karelia). Beryllium minerals typically occur in the central part of greisen bodies along with other ore minerals. Petrological studies of several systems in Nigeria, Karelia and Mongolia show that mineralizing fluids are of magmatic origin and saline (10 to >40 wt % $NaCl_{eq}$), can have moderate CO_2 contents, commonly show evidence of phase separation, and were trapped at temperatures from 200 to 500°C (Haapala 1977a; Imeokparia 1992; Akande and Kinnaird 1993; Graupner et al. 1999).

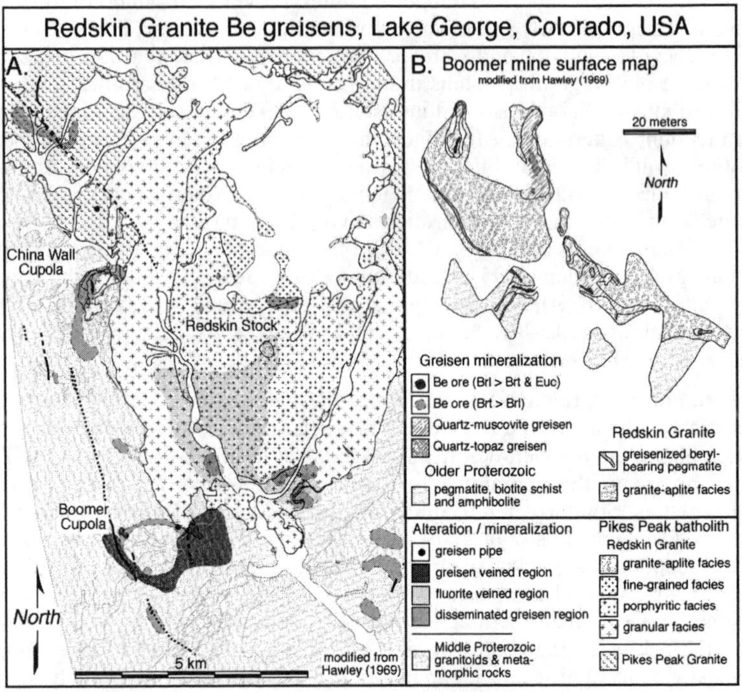

Figure 21. Geology of the greisen Be deposits associated with the Redskin biotite granite, Tarryall Mountains, Colorado (Hawley 1969). The Redskin granite is late, highly evolved phase of the 1 Ga Pikes Peak batholith (Desborough et al. 1980). Like some other fundamentally metaluminous systems, it has late muscovite-bearing phases which are associated with the Be-rich muscovite-quartz-fluorite-topaz greisens (inset). Miarolitic and other pegmatitic Be mineral occurrences occur widely in nearby parts of the Pikes Peak batholith (Eckel 1997). The distribution of rock types, hydrothermal features, and structures make it seem likely that the complex is tilted at moderate angles to the southwest giving an oblique section.

A well studied example of greisen-type mineralization is that associated with late units of the Redskin biotite granite in Colorado (Fig. 21; Hawley 1969). The Redskin granite is an evolved late phase of the variably alkaline 1.0 Ga Pikes Peak batholith, Colorado (Desborough et al. 1980). As a whole the batholith is known for Nb-Y-F-type miarolitic pegmatites that contain genthelvite, phenakite, bertrandite, gadolinite but lack beryl (Levasseur 1997; see next section). Quartz-lithian muscovite-topaz-fluorite (±wolframite-cassiterite-sulfide) greisens form small (1-20 m) pipes and veins developed within and above variably muscovite-bearing porphyritic and aplitic biotite granites. These occur along the southern and western phases of the intrusion which likely was the upper part of the now tilted intrusive system. Beryllium minerals (beryl+bertrandite±rare euclase) were quite localized within the greisen bodies and in places formed unusually high grade deposits (>5% BeO; Hawley 1969). The deposit at the Boomer Mine produced the first non-pegmatitic Be ores in the United States, operating between 1956 and 1965 (Meeves 1966; Hawley 1969).

Carbonate host rocks. As in the case of peraluminous-related systems, limestone and dolostone make excellent hosts for F- and Be-rich skarn and replacement deposits. These commonly accompany skarn, replacement and greisen Sn, W and base-metal mineralization. In some regions only carbonate-hosted mineralization comprises ore (e.g, Karelia, Haapala 1977b). Styles share some common features with strongly peraluminous systems, notably that anhydrous skarns with little Be are overprinted by hydrous, typically F-rich assemblages with higher Be contents. They differ in that the metaluminous-related skarns tend to be more oxidized (garnet/vesuvianite > pyroxene; cf. Einaudi et al. 1981) and have replacement assemblages that are typified by fluorite plus iron sulfides, oxides and silicates. Topaz- and muscovite-bearing assemblages can occur in greisenized areas, but they are minor compared to the abundant aluminous assemblages (micas+fluorite±plagioclase±diaspore±topaz) found with peraluminous granites. Tourmaline and other boron minerals can be abundant. Typical Be minerals are phenakite, bertrandite, danalite and helvite. Beryl and chrysoberyl are rare, but considerable Be can be bound in vesuvianite and other silicates (up to ~1% BeO, Beus 1966).

Notable examples of carbonate-hosted deposits include Pitkäranta, Russia, Iron Mountain, New Mexico, and a number of Sn skarn-greisen-replacement deposits in Tasmania and elsewhere in eastern Australia (e.g., Mt. Garnet, Queensland; see Appendix A and Fig. 2). This Sn-rich group has been thoroughly reviewed by Kwak (1987) and Newberry (1998). In comparison, a distinct group of Mo-W-bearing andraditic skarns contain only minor helvite (e.g., Oslo Rift, southeastern Arizona). In Tasmania, Be minerals occur in hydrous silicate assemblages that are developed after earlier calcic or magnesian skarns and they occur in iron sulfide/iron oxide-fluorite replacement bodies. Phenakite and bertrandite replace danalite in greisenized areas (Kwak and Jackson 1986). Kwak (1987) summarizes fluid inclusion data for the Tasmanian deposits which formed from high salinity (>30 wt % $NaCl_{eq}$) fluids and spanned a large temperature range. Lower temperature variants on these deposits are best expressed in volcanic-associated Be deposits, however Bulnayev (1996) links some moderate temperature (140-350°C) carbonate-hosted fluorite-bertrandite-phenakite deposits to subalkaline granitoids.

At Iron Mountain, New Mexico, 29 Ma porphyritic alkali granites intruded Paleozoic sedimentary rocks and formed oxidized Sn-, W-bearing andradite-magnetite skarns which are overprinted by helvite-bearing fluorite-rich assemblages (Fig. 22; Jahns 1944a). Skarns formed during several intrusive events (Robertson 1986) and the intrusive rocks contain abundant small veins consisting of quartz-K-feldspar, quartz-biotite-fluorite, or biotite. The Be ores form small bodies in skarn or marble and consist of rhythmically banded fluorite containing variable amounts of helvite, magnetite, vesuvianite, chlorite, and scheelite (Fig. 23). In his original description Jahns (1944b) coined the term "ribbon rock" for this texture while citing earlier descriptions at Lost River, Alaska and Pitkäranta, Karelia, Russia (formerly Finland). In New Mexico, contemporaneous volcanic rocks in adjacent fault blocks contain higher level bertrandite-rich mineralization (Meeves 1966), a few red beryl occurrences (Kimbler and Haynes 1980), and many volcanogenic cassiterite (wood tin) deposits (Rye et al. 1990). Southern New Mexico also hosts numerous other Be occurrences including helvite-bearing sulfide-carbonate replacement bodies at the Grandview Mine within 30 km of Iron Mountain (Warner et al. 1959).

Volcanic associated deposits: Shallow low-temperature (150-250°C) replacement and vein deposits of Be are linked to volcanic and hypabyssal high-silica rhyolites and granite porphyries. These "epithermal" deposits are the main source of non-pegmatitic Be ore (Spor Mountain, Utah) and the sole source of gem red beryl (Fig. 2; Appendix A). Although best known in the Basin and Range Province of the western United States, similar deposits are reported from a number of areas in Asia (Zabolotnaya 1977; Kovalenko and Yarmolyuk 1995). Genetically related volcanic and hypabyssal biotite-bearing topaz rhyolites (ongonites) are weakly peraluminous to metaluminous and are particularly rich in F, Be, and Li (Fig. 5C,D; Kovalenko et al. 1979; Burt et al. 1982; Christiansen et al. 1983). In the western United States they formed in an extensional tectonic setting and they belong to a compositionally expanded magmatic pattern that is bimodal in character and has felsic rocks that range from peraluminous to peralkaline.

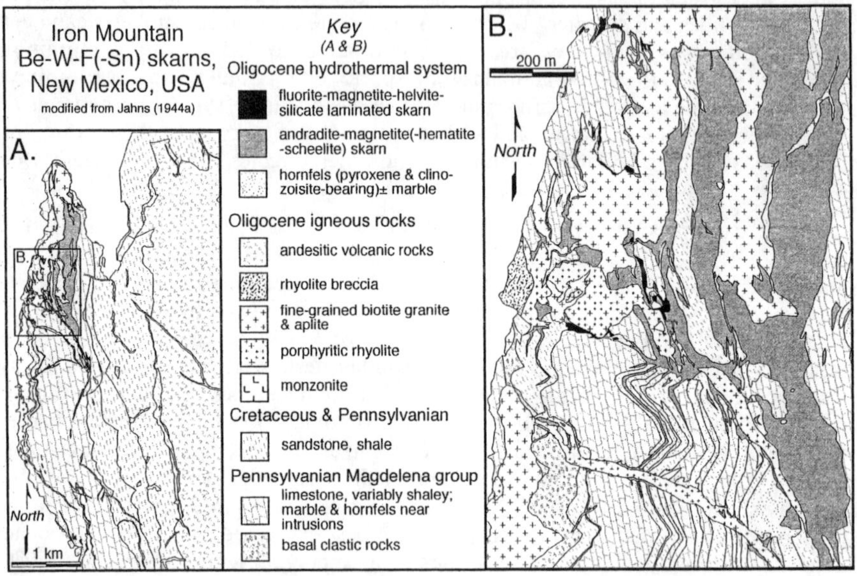

Figure 22. Geology of the Iron Mountain area, New Mexico (modified from Jahns 1946). (A) Simplified district geology showing extent of Fe-rich skarn and replacement deposits in Paleozoic carbonate rocks adjacent to Oligocene intrusions. The system is tilted about 30° to the east. (B) Detail of the central part of the district showing distribution of banded helvite-fluorite-iron oxide skarns at the marble contact or superimposed on earlier Sn-bearing andradite and W-bearing magnetite skarns. Be-F mineralization is related to the younger fine-grained granite (Robertson 1986).

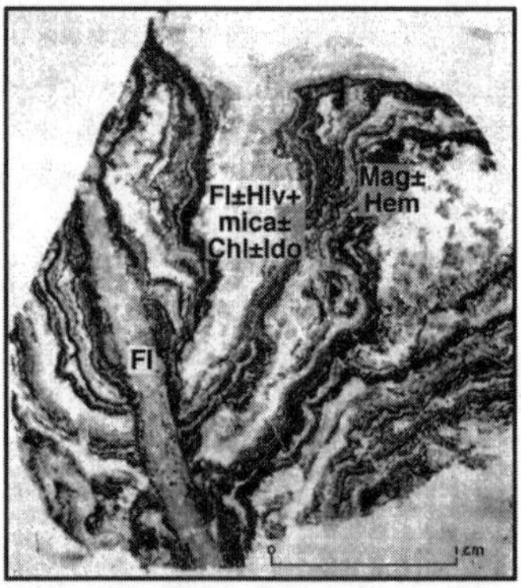

Figure 23. Rhythmically banded helvite-rich replacement from Iron Mountain, New Mexico. Magnetite±hematite (dark bands) alternating with fluorite-helvite-vesuvianite (light bands; photo modified from Jahns 1944a,b).

Fine-grained bertrandite and, rarely, beryl or behoite are associated with hydrothermal silica (quartz, chalcedony, opal), calcite; fluorite, carbonate, K-feldspar and Li clays in tuffs and breccias. Fluorite, although abundant in some of the better mineralized deposits, is not always present. Although sulfides are absent, Zn, Mo, Li and other metals can be concentrated in Mn oxides and clays (Lindsey 1975). BeO concentrations range

Fluorite-absent quartz/chalcedony veins with bertrandite-adularia-calcite±clays are described in Utah (Rodenhouse Wash, Griffitts 1965) and the former Soviet Union (Rozanov and Ontoeva 1987). Bertrandite-bearing (up to 2.5% BeO) clay-altered rhyolite tuff at Warm Springs, New Mexico also lacks fluorite (Hillard 1969). Most occurrences contain abundant fluorite in addition to silica minerals, bertrandite, calcite, K-feldspar and various clays. Zabolotnaya (1977) describes a quartz-fluorite-bertrandite±adularia stockwork with epithermal textures in subvolcanic Paleogene rhyolites. Similar deposits are present in Mongolia (Kovalenko and Yarmolyuk 1995).

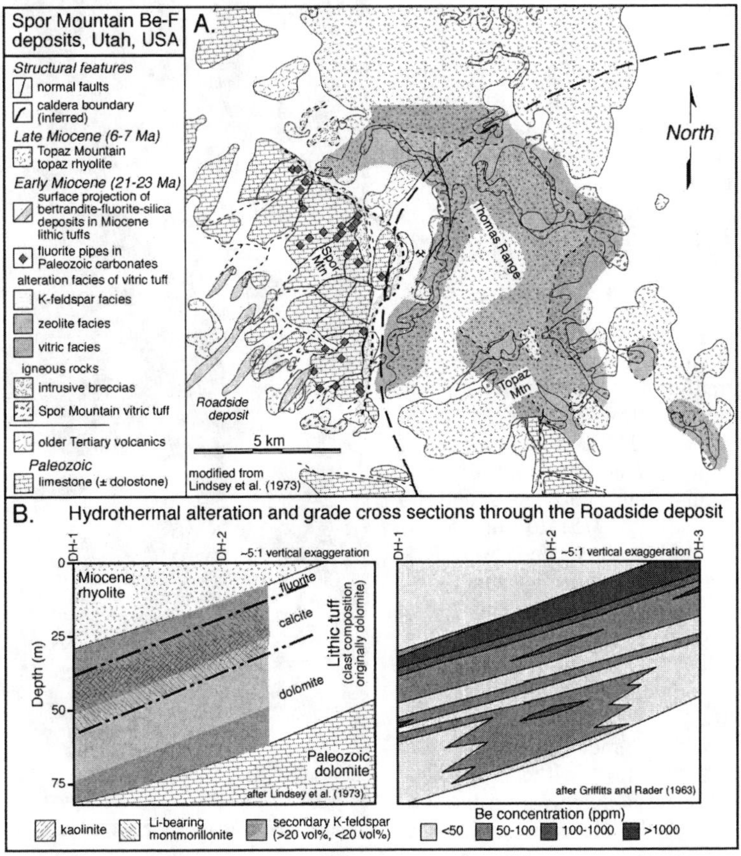

Figure 24. Generalized geological relationships and mineralization in the Spor Mountain and Thomas Range area, Utah, locus of the world's principal Be supply. (A) Geologic map of the district showing distribution of Be deposits in Miocene lithic tuff and regionally associated hydrothermal alteration (K-feldspar, fluorite and argillic types; adapted from Lindsey 1975). (B) Cross sections from the Roadside deposit (see A) showing types and distribution of hydrothermal alteration in early Miocene lithic tuffs and distribution of bertrandite mineralization in carbonate-clast-rich lithic tuff. The tuff is also enriched in Li, Zn and other elements. (Alteration from Lindsey et al. 1973. Be content from Griffitts and Rader 1963.)

The volcanic-hosted bertrandite-fluorite-silica ores of the Spor Mountain district are only one of several dozen Be occurrences in the region (Appendix A; Meeves 1966; Shawe 1966). At Spor Mountain and in the adjacent Thomas Range voluminous Cenozoic volcanic rocks overlie a carbonate-dominated Paleozoic sedimentary section (Fig. 24). Volcanism began with 39-38 Ma latites and andesites, followed by 30-32 Ma rhyolitic ash-flow tuffs. It culminated with early and late Miocene topaz rhyolite flows—the 21 Ma Spor Mountain Rhyolite, and the 6-7 Ma Topaz Mountain Rhyolite (Lindsey 1977). Interestingly, this bimodal distribution of topaz rhyolites occurs throughout western Utah. The topaz rhyolites have been intensively studied (e.g., Christiansen et al. 1984) and rocks of each episode have high concentrations of Be, F, Li and other lithophile elements (Fig. 5D). Uranium-lead dating of uraniferous silica yields an estimated oldest age of 20.8 Ma for Be mineralization but that younger thermal events were also likely (Ludwig et al. 1980).

Beryllium mineralization is localized in stratified tuff breccia immediately beneath rhyolite flows of Spor Mountain Formation (Figs. 24B and 25A). The tuff is extensively altered with a progression from regional diagenetic clays and K-feldspar (Lindsey 1975) to intense K-feldspathization with secondary sericite and smectite in the immediate vicinity of Spor Mountain (Lindsey et al. 1973). Dolostone clasts in the tuff breccia show corresponding alteration from original dolomite to calcite to silica to fluorite (Fig. 25B,C). Figures 24B and 25B,C show how Be grade increases with intensity of alteration in the tuff matrix and in carbonate nodules within the tuff. The source of the Be-bearing fluids is uncertain. Hydrogen and oxygen isotopic data (Johnson and Ripley 1998) are consistent with involvement of surface waters. Fluorite-rich, Be-poor (≤20 ppm) breccia pipes cut the Paleozoic carbonate rocks beneath the older rhyolites. These pipes lie along structures that also appear to control the Be orebodies. Lindsey et al. (1973) speculated that a connection to deeper Be mineralization exists; however, the deposits could reflect shallow

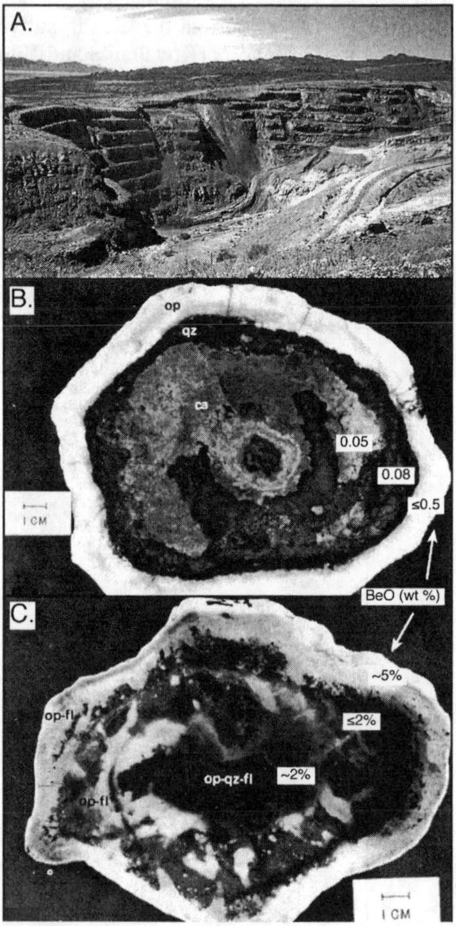

Figure 25. Photos of Spor Mountain mineralization. (A) Open pit with ore body at the bottom beneath unmineralized tuff (photo by Steve Young). (B) and (C) Two mineralized carbonate clasts from lithic tuff in the Roadside deposit illustrating the progression in alteration and BeO contents (cf. Fig. 7 in Lindsey et al. 1973). Mineralogy is labeled (qz = quartz, op = opal, fl = fluorite, ca = calcite), as are approximate BeO contents. (B) Partially replaced carbonate nodule from deeper part of the tuff with calcite core and quartz to opal outer zones with minor fluorite and bertrandite. (C) Carbonate clast fully replaced by fluorite-silica-bertrandite from the upper part of the tuff.

degassing of a magma without deeper mineralization, or they might have formed by leaching of the Be-rich Spor Mountain Rhyolite (cf. Wood 1992).

In the same region and of considerable gemological interest are occurrences of strongly colored, Mn-rich red beryl. These are restricted in occurrence to topaz rhyolites and were first described from the Thomas Range, Utah (Hillebrand 1905). In that area, small (<1 cm) red beryl occurs in gas-phase cavities along with topaz, bixbyite and quartz in 6-7 Ma rhyolites that overlie the Early Miocene rhyolites related to the Spor Mountain deposits. South of the Thomas Range, in the Wah Wah Mountains gem red beryl has been commercially produced from 22-23 Ma topaz rhyolite (Keith et al. 1994; Christiansen et al. 1997). In the Wah Wah Mountains, crystals exceeding 2 cm in length occur in cooling joints near the flow tops. Early bixbyite, topaz and silica polymorphs are followed by red beryl and ultimately kaolin plus mixed layer clays. The beryl is post-magmatic and is interpreted by Keith and coworkers to form by reaction of Be fluoride complexes released from the devitrifying rhyolite and subsequently react with feldspar and bixbyite along the joints. Red beryl also occurs the Sn-bearing rhyolites of the Black Range, New Mexico (Kimbler and Haynes 1980) which are close in time and space to the hypabyssal felsic intrusions that are associated with the Iron Mountain skarns and nearby volcanic-hosted bertrandite deposits (Meeves 1966). Given the small amounts of beryl present, it appears that these occurrences require no more than local redistribution of Be from the host topaz rhyolites.

Peralkaline magma-related systems

Sodic amphibole and sodic pyroxene-bearing granites, quartz syenites and nepheline syenites are associated with several large Be deposits and a number of mineralogically interesting occurrences (Appendix A). These are uncommon compared to deposits associated peraluminous and metaluminous igneous systems, likely reflecting the relative rarity of peralkaline magmas. This diverse group shares styles that range from magmatic pegmatitic assemblages to low-temperature (<200°C) hydrothermal systems hosted by a variety of rocks (Table 2; Fig. 4). As such they provide a useful mineralogical counterpoint to those deposits generated by more aluminous magmas. Consistent with the overall alkaline compositions, characteristic Be minerals include Na-Ca silicates (epididymite, chkalovite, leucophanite, milarite, leifite), gadolinite, and the Zn-Mn members of the helvite group (Table 1). Alteration assemblages in felsic rocks are dominated by Na(±K)-rich framework and chain silicates commonly with Li micas with or without quartz. Many of these assemblages can be termed fenites given that the key alteration minerals are K-feldspar or albite plus sodic pyroxenes and amphiboles. The local dissolution of quartz, the peralkaline silicates, and the rarity of acid assemblages provide a striking contrast with analogous post-magmatic metasomatism in peraluminous and most metaluminous systems (recall that Li-micas are not acid minerals; Eqn. 10). Carbonate-hosted systems differ less from the metaluminous environment. Fluorite-rich replacement bodies with or without skarns develop in carbonate rocks with characteristic Be minerals being phenakite, bertrandite, leucophanite and milarite among others.

In many regions, notably extension-linked alkaline felsic provinces like Nigeria, Norway and southwestern North America, metaluminous to weakly peraluminous biotite granites form concurrently with peralkaline aegirine- or riebeckite-bearing granites and syenites (Fig. 2). Where they coexist in the same igneous centers both peralkaline and metaluminous rocks may have rare element enrichments. Within the same intrusive suite, hydrothermal Be deposits more commonly form with the biotite granites, for example as in Nigeria, the Pikes Peak batholith, west Texas and nearby areas, and Norway (see Appendix A for references). Nevertheless, magmatic systems that are largely or entirely peralkaline do host major Be mineralization as described next.

Peralkaline granites. Beryllium mineralization generated by peralkaline granites ranges from pegmatitic to low temperature hydrothermal, paralleling the spectrum found in more aluminous systems. The hydrothermal systems have abundant K-feldspar and albite-rich alteration typically with quartz veining. Sodic amphibole, pyroxene and fluorides can be present as can be late hematitization and mica-rich greisens. Beryllium minerals tend to be late and are commonly distal. Zabolotnaya (1977) groups such deposits into helvite-group or leucophanite-bearing types with feldspathic metasomatism. Unfortunately, very few deposits are well described.

Feldspathic host rocks. At the high-temperature end of the spectrum, Nb-Y-F-type pegmatites and miarolitic cavities are common. For example, pegmatites in the locally peralkaline granites and quartz syenites from Pikes Peak batholith, Colorado have phenakite, bertrandite, genthelvite, barylite, and gadolinite in pegmatites containing quartz, albite, amazonite, and Li-mica (Levasseur 1997; Kile and Foord 1998; cf. Russian localities: Bazarov et al. 1972; Nedashkovskii 1983). More acid associations containing muscovite and beryl are relegated to greisen zones in the metaluminous Redskin biotite granite (last section). In the pegmatitic Strange Lake deposit, Labrador, the original Zr-Nb-Y-Be concentrations are likely magmatic (Miller 1996). Nonetheless, high-temperature sodic pyroxene hydrothermal assemblages and moderate-temperature calcic hydrothermal alteration culminate in hematite-fluorite-Be mineral (leifite, gadolinite, milarite) assemblages (Salvi and Williams-Jones 1996). This paragenesis parallels the shifts to calcic assemblages that occur in undoubted hydrothermal systems (see Novikova 1983 and carbonate-hosted systems below).

Even though proximal mineralized pegmatites are common, the best developed Be concentrations in most of these systems are hydrothermal, late, and distal. Permian-age fluorite-phenakite-helvite(-gadolinite-milarite-barylite-bertrandite) mineralization occurs on the outer fringes of a hydrothermal system that is associated with variably porphyritic riebeckite granites in the Verknee Espee district in Kazakhstan (Belov and Ermolov 1996). These granites have pegmatitic facies and are feldspathically altered adjacent to proximal Nb-Ta-Zr-REE mineralization. The latter is hosted by K-feldspar-riebeckite-aegirine-fluorite±quartz veins with albite-riebeckite-aegirine-biotite envelopes in clastic rocks and tuffs. Hingganite-rich quartz-fluorite-albite-aegirine stockworks with up to 0.15% BeO occur in apical portions of aegirine-riebeckite granites at Baerzhe, Inner Mongolia (Wu et al. 1996). Analogous albite, fluorite and hematite-bearing styles of hydrothermal alteration formed with the Be-enriched magmas at the Khaldzan-Buregtey Zr-Nb-REE deposit in Mongolia (Kovalenko et al. 1995b).

Variations on this style occur in the northern Ukrainian shield, where genthelvite- and phenakite-bearing Be-Ta-Sn deposits are associated with cryolite-bearing biotite and riebeckite granites of the Mesoproterozoic Perga complex (Appendix A; Esipchuk et al. 1993; Kremenetsky et al. 2000a). Quartz-K–feldspar metasomatic rocks are common but greisen-type alteration with late sulfides, siderite and cryolite is also present (Vynar and Razumeeva 1972). Muscovite-bearing greisens and other acid assemblages in these systems are rare. Zabolotnaya and Novikova (1983) describe a possible example of more acid alteration from an unnamed Mesozoic occurrence in Siberia where dickite+bertrandite occurs with alkaline granites and quartz syenites in a hydrothermal system that is otherwise characterized by alteration to typical K-feldspar, albite, hematite, and fluorite-rich facies.

Carbonate host rocks. Beryllium-bearing replacement bodies associated with peralkaline magmatic systems are fluorite-rich. Skarn alteration can be abundant or there may only be small amounts of calc-silicate minerals present. The calc-silicate-poor fluorite-bertrandite replacement bodies at Sierra Blanca, Texas and Aguachile, Coahuila belong to a continuum with this group because they formed in peralkaline intrusive centers, even though the most closely associated intrusions are not themselves peralkaline. Calcic-sodic Be silicates such as leucophanite, meliphanite and milarite are common in some deposits where they either post-date phenakite or bertrandite or are the main Be phase (e.g., Novikova 1984).

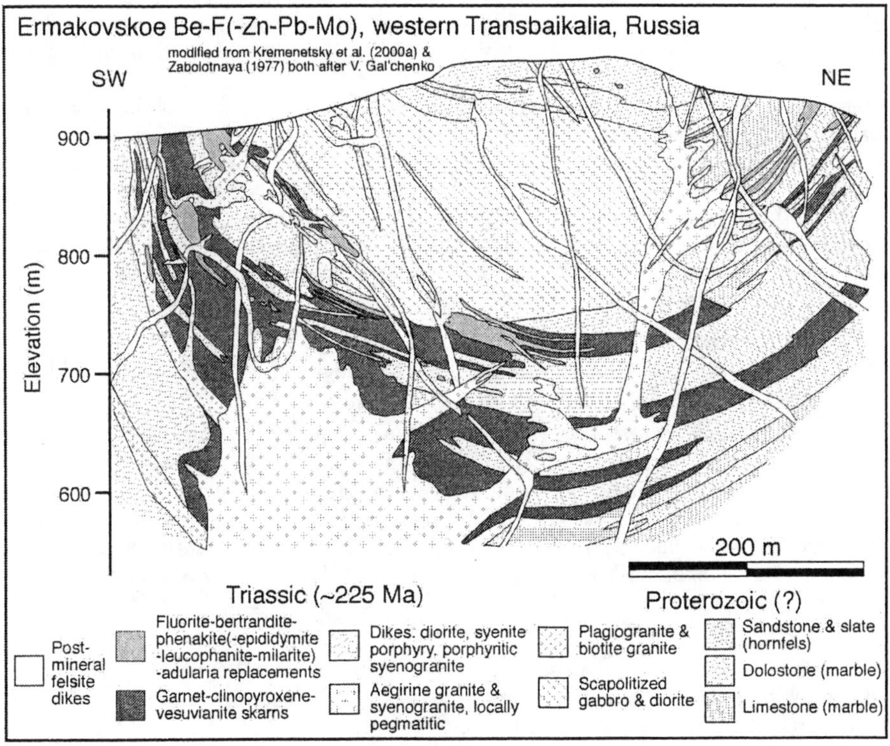

Figure 26. Cross section through the aegirine-riebeckite granite-related hydrothermal system in the Ermakovskoe district, western Transbaikal, Russia. Fluorite-bertrandite-phenakite-leucophane mineralization is distal to and superimposed on garnet-pyroxene-vesuvianite skarns surrounding Mo-mineralized feldspathically altered alkali granites and granite porphyries (modified from figures based on the work of V. Gal'chenko presented in Zabolotnaya 1977 and Kremenetsky et al. 2000a).This deposit produced Be in the past (Kremenetsky et al. 2000a).

The best-described skarn-related example is the Ermakovskoe deposit in Transbaikalia, Russia (Appendix A; Zabolotnaya 1977; Lykhin et al. 2001). This deposit produced Be ore for the Soviet Union (Kremenetsky et al. 2000a). At Ermakovskoe, proximal skarns and skarn-overprinting to distal Be-F replacement bodies formed around and over a Triassic aegirine granite and related syenitic and granitic dikes (Fig. 26). The intrusion is variably albitized and potassically altered and it contains minor quartz-feldspar-molybdenite mineralization in its upper portions. REE and Zr are also metasomatically enriched. Early metasomatism in the sedimentary and igneous host rocks created K-feldspar-rich assemblages in the aluminous rocks and vesuvianite(beryllian)-garnet-pyroxene skarns in the carbonate rocks. Beryllium occurs mainly as phenakite (deep) and bertrandite (shallow) in late dark fluorite-adularia-calcite-ankerite replacement zones (Novikova et al. 1994). Subsequently, more calcic phases of alteration overprint the system to form leucophanite, meliphanite, helvite, milarite and bavenite along with carbonates and sodic silicates. Fluid inclusion studies document a complex set of high-salinity, carbonate- and sulfate-rich fluids with Be contents approaching 1000 ppm (Reyf and Ishkov 1999). Kosals et al. (1974) report additional homogenization data from an unnamed deposit that closely resembles Ermakovskoe.

In another undisclosed Siberian location, leucophanite-fluorite-K–feldspar-albite veins and replacements occur in limestones along skarn-bearing contacts with a riebeckite granite (Kosals and Dmitriyeva 1973). Overprinting these is subordinate association containing fluorite, Li-mica, danalite, milarite, phenakite and bertrandite. Alteration in the granites and adjacent clastic rocks is dominated by riebeckite-albite-quartz and riebeckite-microcline-albite assemblages with Ta-Zr mineralization. Gadolinite joins leucophanite where rare fluorite veins cut the granite. The fluid inclusions studied by Kosals and Dmitriyeva (1973) indicate the leucophanite mineralization occurred between 380 and 490°C from variably saline fluids. In yet another unnamed deposit related to a aegirine-riebeckite granite, phenakite mineralization occurs in aegirine-alkali feldspar fenites and later fracture-controlled quartz-hematite-phenakite mineralization (Nedashkovskii 1970).

Fluorite in these systems is commonly dark in color and REE-rich (Hügi and Röwe 1970). Textures can be rhythmically banded or sieve-like with Be minerals and feldspars (Kosals et al. 1974; cf. Fig. 20).

Silica-saturated peralkaline volcanic settings. Deposits with peralkaline volcanic rocks are apparently rare. The best candidate is the Early Proterozoic Brockman Nb-Zr-Ta deposit, Western Australia (Appendix A) where Be is remobilized into bertrandite-quartz-carbonate veins in a weakly metamorphosed rare-metal rich volcanic tuff. Other possible examples include some of the volcanic-hosted bertrandite-fluorite occurrences in Mongolia (Kovalenko and Yarmolyuk 1995) and the helvite-bearing Shixi occurrence in China which is associated with hypabyssal dikes of Na-altered Nb-Ta-Zr-F-rich sodic rhyolite (Lin 1985; Appendix A).

Nepheline syenites and carbonatites. Hydrothermal Be enrichments may be more common with undersaturated igneous rocks than is generally appreciated. This is because the minerals are commonly dispersed in late igneous units or thin veins, they are easily weathered, and they are typically subtle in appearance. Chemically evolved units in nepheline syenite complexes commonly contain elevated Be in associated pegmatites, miarolitic cavities, hydrothermal veins and feldspathized rocks (Ilímaussaq, Mt. Saint-Hilaire, Lovozero, Khibiny, Oslo graben; Appendix A; Sørensen 1997). Where present (e.g., Mt. Saint-Hilaire, Lovozero), pegmatites can have complex internal structures and are dominated by albite, natrolite, sodalite with accessory chkalovite, leucophanite, and epidydimite. In addition, nearly all described systems contain multiple types of quartz-free metasomatic assemblages that contain albite, aegerine, analcime and other sodic minerals in both veins and wall rock alteration. Li micas are common whereas muscovite is absent. The absence of quartz and the scarcity of K-feldspar contrasts with the hydrothermal features associated with peralkaline granites. Genthelvite, bertrandite, epididymite, chkalovite, leucophanite are the more common of the large number of hydrothermal Be minerals present. Beryl forms only where quartz-bearing peraluminous rocks are cut by Be-bearing veins (Mt. Saint-Hilaire, Horváth and Gault 1990) and chrysoberyl has been reported in quartz-absent, aluminous(?) xenoliths and veins from the Khibiny massif (Men'shikov et al. 1999); euclase not reported. These mineral associations are consistent with the phase relationships presented in Figures 11 and 12.

Geological relationships are well defined in only a few areas. Engell et al. (1971) describe Be distribution in the Ilímaussaq nepheline syenite complex. Hydrothermal enrichments up to 0.1% BeO occur in zones with abundant veins. At Ilímaussaq more than a half dozen vein types contain combinations of aegirine, arfvedsonite, analcime, albite, natrolite, sodalite, Li mica, ussingite and other phases. These veins and associated albite-aegirine-arfvedsonite-natrolite fenites formed above late differentiates (lujavrites: eudialyte-acmite nepheline syenites) in the intrusion. The latter contain 10-30 ppm Be (30-80 ppm BeO). Hydrothermal Be minerals include chkalovite, tugtupite, bertrandite, beryllite among others (Engell et al. 1971; Sørensen et al. 1981; Markl 2001). Similar occurrences are known in the Lovozero and Khibiny complexes of the Kola Peninsula. Epididymite, leucophanite, chkalovite and many other Be minerals occur in hydrothermal albite-natrolite-polylithionite-bearing assemblages in pegmatites and veins (Men'shikov et al. 1999). As at Ilímaussaq, these are preferentially associated with late-crystallizing lujavrites. At Letitia Lake, Labrador (Appendix A; Fig. 2), epididymite and barylite constitute up to a few percent of zones with Nb-REE-Zn-bearing alkali-feldspar-rich veins and fenites in shallow syenitic and trachytic rocks. These hydrothermal zones contain about 0.4% BeO. In the Khibiny-Lovozero Complex, Kola, isotopic evidence points to involvement of meteoric waters in the later veins in these otherwise magmatic-fluid dominated systems (Borshchevskii et al. 1987). The atypical abundance of hydrothermal quartz in the syenite-related deposit at Thor Lake, Canada (Fig. 16; described with magmatic deposits) may reflect the local granite host or a hidden quartz-bearing intrusion or simply the cooling of magmatic fluids to low temperatures.

Hydrothermal occurrences in carbonate host rocks are apparently rare. A plausible candidate is at Hicks Dome in southern Illinois where Paleozoic carbonate rocks host bertrandite-bearing fluorite mineralization (Baxter and Bradbury 1980; Kogut et al. 1997). The latter centers on breccias are linked to alkaline magmatism, possibly a carbonatite. Mineralization is interpreted as due to mixing of F-rich magmatic fluids with basinal brines (Plumlee et al. 1995).

Overall, relatively few reported Be enrichments occur with carbonatite complexes, even those with fluorite mineralization. At Muambe, Mozambique massive Y-LREE bearing fluorite with up to 1% BeO replaces the marginal parts of a carbonatite complex adjacent to K_2O-rich fenitized shallow breccias (Appendix A). In the Magnet Cove, Arkansas, carbonatite complex, Erickson et al. (1963) found Be enrichments in late, thaumasite-bearing hydrothermal veins. At several other localities, Be enrichments occur in carbonatite-bearing alkaline complexes, but are most likely related to associated peralkaline syenites or granites. Barylite occurs in REE-Zr-Th-bearing fenitized rocks adjacent to nepheline syenites and carbonatites at Vishnevogorskii in the southern Urals (Zhabin and Kazakova 1960; Zhabin et al. 1960; Kogarko et al. 1995). Up to 0.6% BeO is reported with Zr-Y-Th-U-HREE breccia bodies associated with carbonatite dikes in the Coldwell alkalic complex, Canada, but are inferred by Smyk et al. (1993) on the basis of REE and rare metal enrichments to be derived from nearby felsic rocks.

Igneous-related emerald deposits

Although amagmatic origins for emeralds are clearly established in some deposits (e.g., see Giuliani et al. 1997, Franz and Morteani, this volume), most investigators link the majority of emerald deposits to metasomatism driven by granitic magmatism (Fig. 27, Appendix A; Kazmi and Snee 1989a; Sinkankas 1989). In the latter group, emerald forms where berylliferous granitic pegmatites or granite-derived quartz-mica-feldspar veins intersect Cr(±V) bearing host rocks. Most such host rocks are ultramafic or mafic in composition. The emerald-bearing veins and dikes typically have complex contact-parallel quartz-poor metasomatic zones which are dominated by biotite(-phlogopite), other Mg(-Ca) sheet silicates, amphiboles, and plagioclase (Figs. 28, 29). Phenakite and chrysoberyl can accompany emerald mineralization in ultramafic rocks (Fig. 14).

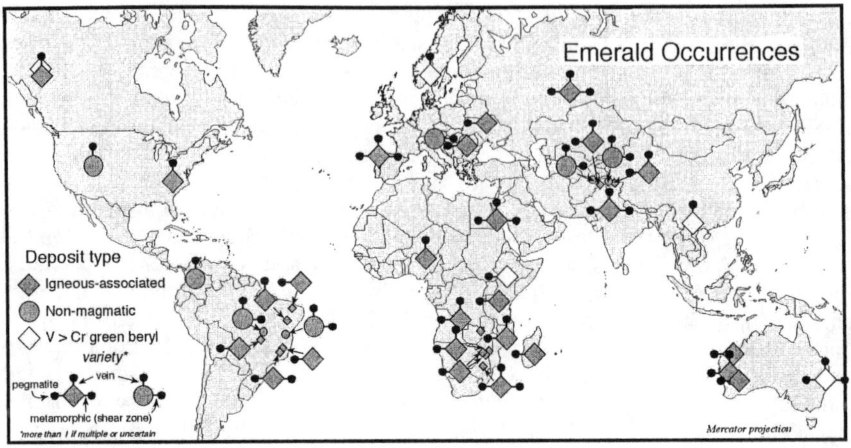

Figure 27. Global distribution of emerald and vanadian beryl deposits by geologic type. Compiled from Kazmi and Snee (1989a) and other sources. See text for discussion of and uncertainties in classification.

Mineralogically, these emerald-bearing systems have many parallels with apocarbonate greisens reflecting formation in analogous silica-poor, Mg-Ca rich hosts. In contrast to the ultramafic and mafic-hosted Cr-rich types, vanadian green beryls (V > Cr) typically form in pelitic rocks which have higher V/Cr than most mafic and ultramafic rock (Fig. 27). As many workers have pointed out, the rarity of emerald stems from the infrequent pairing of a beryl-forming environment with rocks that contain significant Cr. That said, many areas have this juxtaposition, yet lack reported emeralds and thus should be prospective for them.

Few emeralds form with metaluminous granitoids. Perhaps the only significant examples are the greisen-affiliated deposits related to the Nigerian Younger Granites (Abaa 1991; Schwarz et al. 1996). Syenitic occurrences are virtually non-existent because beryl does not form in these systems except where pelitic host rocks become involved (e.g., V-bearing "emerald" at Eidsvoll, Norway; ordinary beryl in hornfels at Mt. Saint-Hilaire, Quebec). Most deposits arise from peraluminous biotite and biotite-muscovite granites; this connection likely reflects the elevated Al_2O_3 activities that are required to make beryl (Fig. 12). Within this last group, emerald occurs with multiple magma and vein types. Genetically related granites and pegmatites can be chemically specialized as indicated by associated Li, Ta or Sn minerals (e.g., Hiddenite, North Carolina; Poona, Australia; some Brazilian and Egyptian deposits), however many of the better documented examples are associated with mineralogically simple granites and related veins (Reft River, Russia; Khaltaro, Pakistan; Egypt; Carnaiba and Socoto, Brazil; Crabtree, North Carolina; cf. Giuliani et al. 1990).

Figure 28 (opposite page). Zoning in igneous-affiliated emerald deposits showing the diversity of source units (quartz veins, aplitic and pegmatitic dikes. felsic schist) and host rocks (silica-rich and silica-poor ultramafic and mafic compositions). For several of these occurrences metamorphic origins that long post-date magmatism have been proposed (Habachtal, Gravelotte, Franqueira, Sekeit; also see discussion under metamorphic occurrences). Compiled from sources listed on the figure.

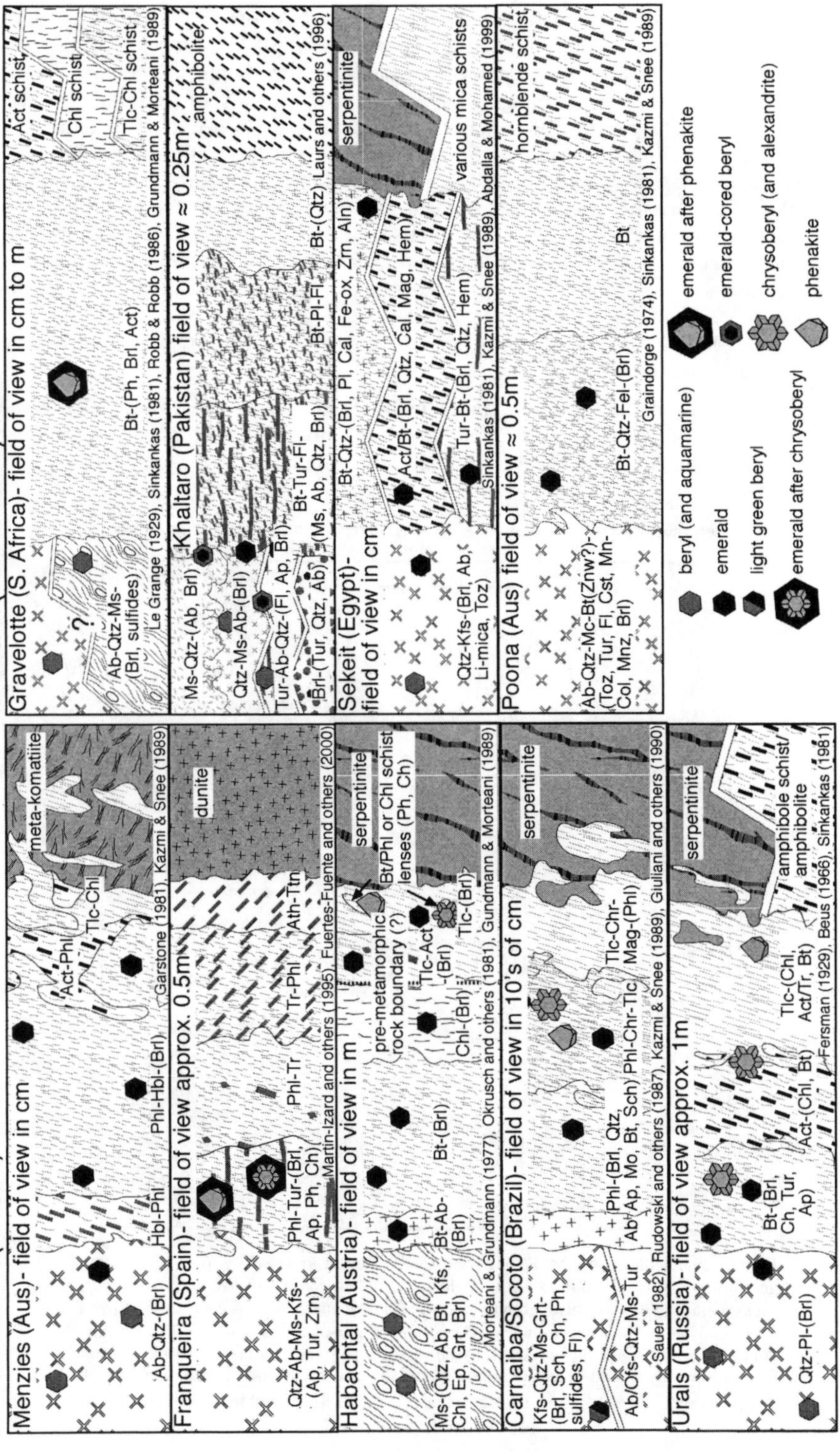
A. Ultramafic(-mafic) host
Menzies (Aus)- field of view in cm
Ab-Qtz-(Brl)
Hbl-Phl
Phl-Hbl-(Brl)
Act-Phl
Tlc-Chl
meta-komatiite
Garstone (1981), Kazmi & Snee (1989)
Franqueira (Spain)- field of view approx. 0.5m
Qtz-Ab-Ms-Kfs-(Ap, Tur, Zrn)
Phl-Tur-(Brl, Ap, Ph, Ch)
Phl-Tr
Tr-Phl
Ath-Ttn
dunite
Martin-Izard and others (1995), Fuertes-Fuente and others (2000)
Habachtal (Austria)- field of view in m
Ms-(Qtz, Ab, Bt, Kfs, Chl, Ep, Grt, Brl)
Bt-Ab-(Brl)
Bt-(Brl)
pre-metamorphic rock boundary (?)
Chl-(Brl)
Tlc-Act-(Brl)
Bt/Phl or Chl schist lenses (Ph, Ch)
Tlc-(Brl)
serpentinite
Morteani & Grundmann (1977), Okrusch and others (1981), Gundmann & Morteani (1989)
Carnaíba/Socotó (Brazil)- field of view in 10's of cm
Kfs-Qtz-Ms-Grt-(Brl, Sch, Ch, Ph, sulfides, Fl)
Ab/Ofs-Qtz-Ms-Tur
Phl-(Brl, Qtz, Ab, Ap, Mo, Bt, Sch)
Phl-Chr-Tlc
Tlc-Chr-Mag-(Phl)
serpentinite
Sauer (1982), Rudowski and others (1987), Kazmi & Snee (1989), Giuliani and others (1990)
Urals (Russia)- field of view approx. 1m
Qtz-Pl-(Brl)
Bt-(Brl, Ch, Tur, Ap)
Act-(Chl, Bt)
Tlc-(Chl, Act/Tr, Bt)
amphibole schist/amphibolite
serpentinite
Fersman (1929), Beus (1966), Sinkankas (1981)
B. Mafic(-ultramafic) host
Gravelotte (S. Africa)- field of view in cm to m
Ab-Qtz-Ms-(Brl, sulfides)
Bt-(Ph, Brl, Act)
Act schist
Chl schist
Tlc-Chl schist
Le Grange (1929), Sinkankas (1981), Robb & Robb (1986), Grundmann & Morteani (1989)
Khaltaro (Pakistan) field of view ≈ 0.25m
Ms-Qtz-(Ab, Brl)
Qtz-Ms-Ab-(Brl)
Tur-Ab-Qtz-(Fl, Ap, Brl)
Brl-(Tur, Qtz, Ab)
Bt-Tur-Fl-(Ms, Ab, Qtz, Brl)
Bt-Pl-Fl
Bt-(Qtz)
amphibolite
Laurs and others (1996)
Sekeit (Egypt)- field of view in cm
Qtz-Kfs-(Brl, Ab, Li-mica, Toz)
Bt-Qtz-(Brl, Pl, Cal, Fe-ox, Zrn, Aln)
Act/Bt-(Brl, Qtz, Cal, Mag, Hem)
Tur-Bt-(Brl, Qtz, Hem)
serpentinite
various mica schists
Sinkankas (1981), Kazmi & Snee (1989), Abdalla & Mohamed (1999)
Poona (Aus) field of view ≈ 0.5m
Ab-Qtz-Mc-Bt(Znw?)-(Toz, Tur, Fl, Cst, Mn-Col, Mnz, Brl)
Bt-Qtz-Fel-(Brl)
Bt
hornblende schist
Graindorge (1974), Sinkankas (1981), Kazmi & Snee (1989)
beryl (and aquamarine)
emerald
light green beryl
emerald after chrysoberyl
emerald after phenakite
emerald-cored beryl
chrysoberyl (and alexandrite)
phenakite

In both ultramafic and mafic rocks, emerald crystals form both within hybridized veins and dikes and in their biotite-rich envelopes (Fig. 28). Veins typically show evidence of reaction with the country rock and can be dominantly quartz or mica (phlogopite or muscovite) or plagioclase with a wide-variety of additional minerals including K-feldspar, tourmaline, fluorite, molybdenite, and scheelite. Pegmatitic dikes are variably metasomatically modified ("desilicated"), typically expressed by lack of K-feldspar, abundance of plagioclase and diminished amounts of quartz. This process may take place concurrently with emplacement or during later metamorphism, as can generation of the common (though not universal) mineral foliations. Where intense chemical exchange between the host and incoming fluids has affected the entire mass it may be impossible to tell if the causative fluids were magmatic or hydrothermal. Aquamarine and common beryl may be present in the less reacted interiors of pegmatites and veins, whereas emerald is most common in enveloping zones of metasomatic biotite±plagioclase±quartz.

In peridotites or serpentinites, outer metasomatic zones consist of chlorite, talc, actinolite plus other amphiboles and they commonly contain minor chrysoberyl and phenakite. These "blackwall" assemblages are similar to those found worldwide at

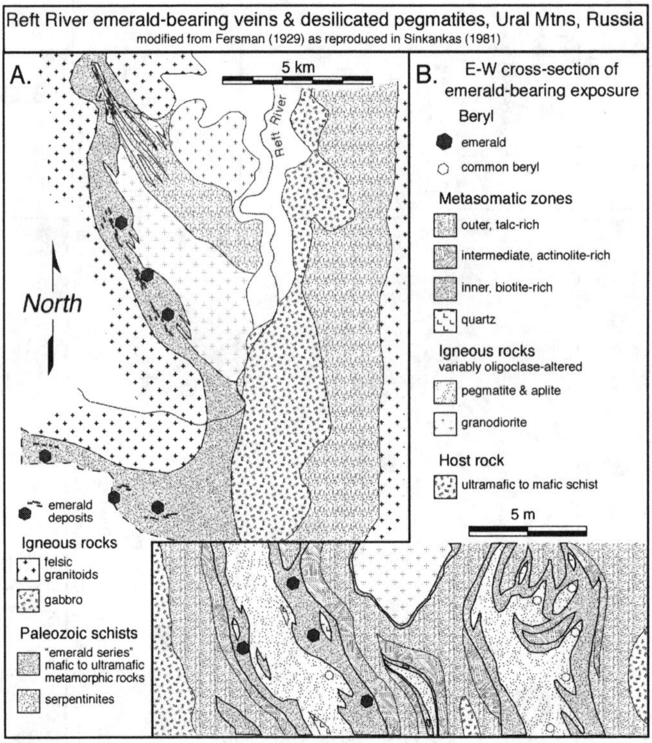

Figure 29. Geology of the Reft River, Urals, Russia emerald deposits. (A) Regional geology showing distibution of emerald deposits in mafic to ultramafic metamorphic rocks adjacent to two-mica granite plutons. The deposits themselves are localized by dikes and veins from the granites. (B) Cross-section of an exposure with emerald and common beryl occurring adjacent to desilicated pegmatites. These form an apparent continuum with emerald-bearing quartz and/or albite veins (Beus 1966, p. 236-244). Both A and B redrawn after figures from Fersman (1929) as illustrated in Sinkankas (1981).

contacts between felsic and ultramafic rocks (e.g., Coleman 1966). Outer zones in ultramafic hosts may be relatively siliceous, for example with talc-bearing assemblages (Gravelotte), however serpentine or olivine bearing silica-undersaturated assemblages are more common. These silica-poor rocks have the best-developed chrysoberyl and phenakite which form in the intermediate to outer metasomatic zones (Beus, 1966) and may in turn be overgrown by emerald (e.g., Martin-Izard et al. 1995; Fig. 28). Desilicated units can also contain rare Be minerals including bromellite (Klement'eva 1970), bavenite (Kutukova 1946), and epididymite and milarite (Černý 1963). Surinamite, which replaces chrysoberyl+talc at high-pressure (Hölscher et al. 1986), may represent the high pressure or metamorphosed equivalent of lower-pressure blackwall assemblages.

One of the better-studied ultramafic-hosted districts is along the Reft River in the Ural Mountains in Russia where pegmatites and veins derived from Devonian two-mica granites cut Paleozoic serpentinites (Fig. 29A). Emeralds are associated with both quartz-feldspar-mica pegmatites and mica (muscovite/phlogopite/margarite), plagioclase-mica, and quartz-albite veins (Beus, 1966; Vlasov, 1968). The mica-rich varieties have been termed "glimmerites." Metasomatic zoning is typical of the class and is similar around all varieties of veins and dikes—inner biotite to intermediate actinolite-chlorite to outer talc zones (Fig. 29B). Emeralds are concentrated in the biotite-rich rocks whereas phenakite and chrysoberyl are more distal. Some of the world's finest chrysoberyl (variety alexandrite) comes from the intermediate zone in these mines. Egyptian, Brazilian, South African and Australian occurrences (Appendix A; Figs. 27, 28) are similar. The common presence of fluorite and F-bearing micas suggests that the Mg-Ca-rich host rocks triggered F precipitation, which may have led to the precipitation of Be and Al from solution to form emerald, chrysoberyl and other minerals (e.g., Soboleva et al. 1972).

In contrast to the multiple metasomatic zones present in ultramafic host rocks, mafic rocks commonly develop only a biotite-rich envelope between the central veins or dikes and the host (Fig. 28). This likely reflects the higher activities of SiO_2 and CaO and a lower activity of MgO in mafic rocks (Fig. 14). For example, at Khaltaro, Pakistan emeralds formed at the contact between mafic amphibolites and young (~9 Ma) greisenized and albitized leucogranites (Fig. 30; Laurs et al. 1996). These include emeralds formed in the outer plagioclase-rich margins of aquamarine-bearing albite-quartz-muscovite veins and, rarely, in and around tourmaline-bearing simple pegmatites (Fig. 30). From whole rock analyses and modal mineralogy Laurs et al. (1996) deduced that the addition of F, alkalis and other elements to the amphibolite drove reactions making fluorite and biotite which in turn released the Cr and Fe needed for emerald formation.

Although magmatic-hydrothermal origins are well established in many areas, emerald paragenesis remains controversial in some regionally metamorphosed settings where igneous bodies are present. The dilemma stems from the fact that many occurrences are small bodies that could originate either (1) through local bimetasomatic exchange by intergranular diffusion over the relatively long times available during regional metamorphism or (2) by infiltration±diffusion metasomatism generated during shorter lived magmatic / hydrothermal events. In many areas a regional metamorphic overprint is absent and there is unambiguous evidence for magmatic conditions and geochemical signatures. Oxygen isotopic data can discriminate between some igneous-related systems and others of sedimentary or metamorphic origin (Fig. 31; Giuliani et al. 1997). These differences have been used with great success by Giuliani et al. (2000a) to deduce global sources and trading patterns in jewelry. For example, their results show that distinctive ^{18}O-enriched sedimentary-sourced Colombian emeralds spread throughout Europe within a decade of the discovery of the New World. In many cases, however, it is

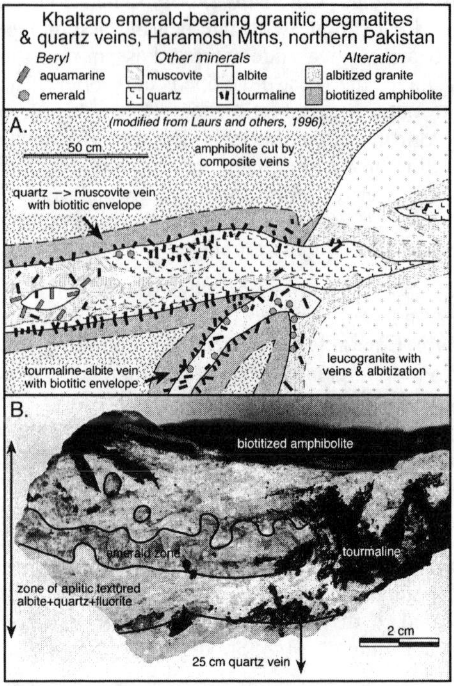

Figure 30. Khaltaro, Pakistan emerald deposit associated quartz veins and pegmatites from muscovite-biotite leucogranite intruding amphibolites (Laurs et al. 1996). (A) Sketch of relationships between quartz and albite vein. Modified from Laurs et al. (1996). (B) Photograph of emerald-rich sample from the margin of a quartz vein adjacent to the amphibolite. Note that the high abundance of beryl (emerald) far exceeds melt solubility and must be hydrothermal. Photo provided by Brendan Laurs.

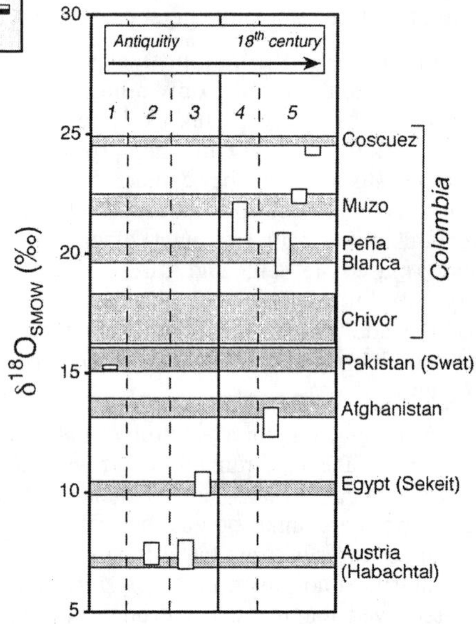

Figure 31. Oxygen isotopic compositions of emeralds for various deposits (vertical bars) and for different pieces of jewelry (numbered columns) as a function of fabrication age (redrawn from Giuliani et al. 2000a). This demonstrates the variability among in $\delta^{18}O$ in emeralds, which is a function of geologic environment (source, indicated by horizontal bands), and it shows that this systematic variation can be applied to the interpretation of trade patterns in gemstones. See Figure 27 for locations.

not clear if one mechanism or the other dominates or if both are involved (e.g., Gravelotte, Franqueira, Seikeit; Appendix A). Other tests might help resolve the mode of origin of some deposits, for example, determining the extent of mobility of Al or accurate dating of intrusions, metamorphism and emerald formation. Unfortunately, for many emerald occurrences, these are not well documented and clearly merit further investigation. In the next section we turn to those deposits where non-magmatic origins are clearly indicated.

NON-MAGMATIC OCCURRENCES

Non-magmatic processes are involved in many Be mineral occurrences, both in their genesis and in their subsequent modification. Local Be enrichments are noted in a varied group of Fe-Mn-oxide-rich rocks. Hydrothermal deposits related to sedimentary and metamorphic fluids comprise an important group of emerald deposits, notably the basin-related Colombian emerald deposits. Metamorphic and surficial processes redistribute Be through local concentration, mineralogical transformations, and placer processes. Some of these can be important gem sources—especially metamorphic emerald deposits and placer deposits of aquamarine and other materials.

Fe-Mn(-Zn) oxide-rich occurrences

A handful of Fe-Mn(±Zn)-oxide-rich rocks contain Be minerals or moderate Be enrichments. These occurrences clearly contrast with the obviously igneous-linked Fe / Mn associations discussed earlier which typically have magnetite-fluorite-helvite or sulfide-bearing helvite-Mn-carbonate/silicate±fluorite assemblages. The oxide-rich systems occur in supercrustal rocks of both sedimentary and volcanic origin; the role of igneous activity in their generation is problematic (Grew, this volume). The best-known example is the metamorphosed deposit at Långban, Sweden (Moore 1970) which contains twelve Be minerals (Introduction, this volume). Similar rocks at Franklin, New Jersey (Palache and Bauer 1930) may have a granitic source for their Be. Miscellaneous locations include helvite with Mn-rich jaspers in the Pyrenees Mountains (Ragu 1994b), bergslagite and other Be enrichments with Mn-Fe oxide layers in Switzerland (Graeser 1998; Brugger and Gieré 2000), and milarite with Mn oxide deposits in New South Wales (Kawachi et al. 1994). Mn-Fe-oxide-rich hot spring deposits in Nevada (Golconda, Sodaville) contain up to 60 ppm Be (160 ppm BeO) along with W and other metals (Warner et al. 1959). Zasedatelev (1973) suggested that similar materials occur elsewhere. Beyond a common oxide-rich, sulfide-poor geochemistry, it is not obvious that they share a common genesis.

Basinal (and metamorphic?) brine-related emerald deposits

Beryl, mainly emerald with minor aquamarine, is known from several areas around the world that share moderate temperatures of origin, saline fluids, high $\delta^{18}O$, and a lack of associated igneous rocks (Appendix A; Figs. 27, 31). Of these, the Colombian emerald deposits are seemingly unique in their geology as they are in the size, quality and number of emeralds produced. In all of these deposits, the concentration of beryl is low and most authors infer that Be is derived by local concentration from the wall rocks (see also Franz and Morteani, this volume).

Emerald deposits in Colombia are hosted by Cretaceous siltstones, sandstones and shales and occur in two belts, one west and one east of Bogota (Fig. 32). The clastic host rocks belong to a Mesozoic sedimentary basin that contains marine evaporites. Mineralization in the western belt formed during Oligocene compression whereas mineralization in the eastern belt formed in the latest Cretaceous during an extensional episode (Cheilletz et al. 1994; Branquet et al. 1999). Beryl (mainly emerald, rarely aquamarine; euclase occurs at Chivor) occurs in gash veins with carbonate, quartz, albite, pyrite and other minerals in fractured, brecciated, variably albitized rocks (Fig. 33; Beus 1979; Cheilletz and Giuliani 1996). Grades are extraordinarily low (Fig. 3): probably <10 ppb gem emerald, and well under 100 ppb emerald overall even though the rocks themselves contain 500-5000 ppb Be (Feininger 1970; Kozlowski et al. 1988).

Fluid inclusion studies show that the mineralizing fluids contained up to 40 wt % $NaCl_{eq}$ and were between 250 and 400°C (Kozlowski et al. 1988; Cheilletz et al. 1994; Giuliani et al. 1995, 1999; Banks et al. 2000). Stable isotopic studies of beryl and gangue

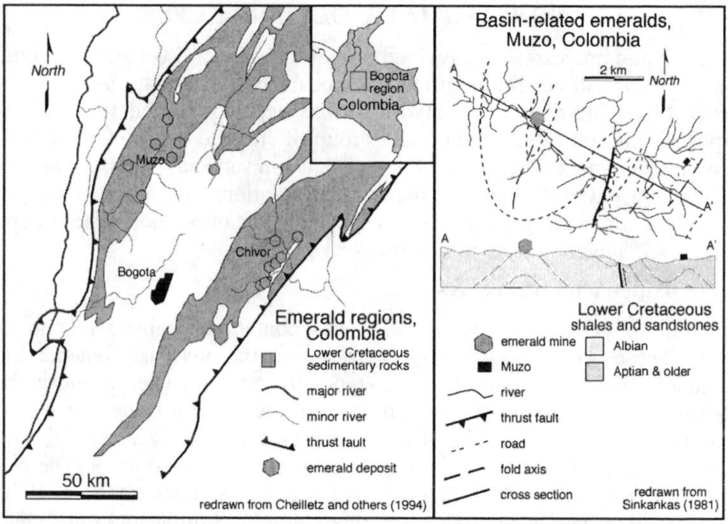

Figure 32. Geologic setting of the Colombian emerald districts. (A) Regional geology showing distribution of emerald districts in thrust belts surrounding inverted Cretaceous basin. Emerald deposits on the eastern side are latest Cretaceous in age and formed during extensional faulting, whereas on the western side they are localized by Eocene-Oligocene thrust and tear faults (Branquet et al. 1999). (B) Sketch of geologic relationships in the Muzo area showing correlation with compressional structures. Both figures redrawn from Sinkankas (1981).

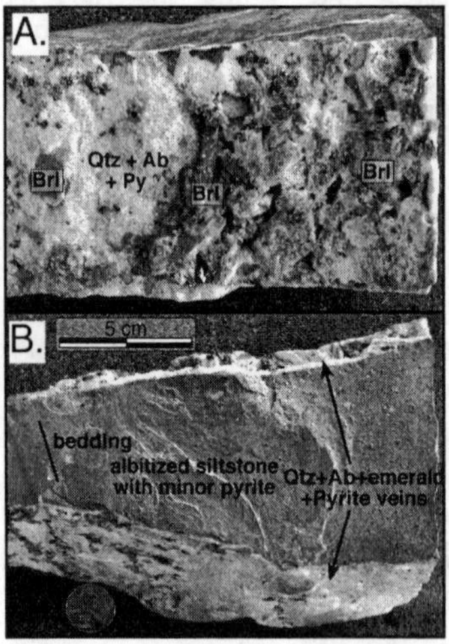

Figure 33. Front view (A) and side view (B) of emerald-rich quartz-albite-pyrite veins cutting at high angles across albitized siltstone from Chivor(?), Colombia. Albitization is common in the Colombian emerald districts with concomitant redistribution of Be, Cr and other components (Beus 1979; Cheilletz and Giuliani 1996). Photo and specimen courtesy of Frank Mazdab.

minerals indicate a basinal brine source with high $\delta^{18}O$ (16-25‰; Fig. 31) and sulfate reduction form sulfur to form pyrite. Hydrogen isotope ratios and fluid inclusion compositions indicate that two fluids were involved, a basinal brine and another, perhaps surface-derived fluid (Banks et al. 2000; Giuliani et al. 2000b). Basinal fluids are now generally accepted as the key ingredient in order to react with the host sedimentary rocks mobilizing the Be and Cr (±V) required to form emerald (Beus 1979; Ottaway et al. 1994; Giuliani et al. 1999) perhaps with deposition due to mixing with a second calcium-bearing fluid (Banks et al. 2000). Local redistribution of Be by breakdown of Fe-Mn hydroxides (Giuliani et al. 1999) or organic matter (Ottaway et al. 1994) is plausible, but not required.

Several other beryl occurrences have possibly analogous origins with basinal or metamorphic brines. In Uintah region of northeastern Utah, a single emerald crystal is reported from carbonate-rich veins that cut black shales and that have a basinal brine signature (Keith et al. 1996; Olcott et al. 1998). Gem beryl (aquamarine and emerald) occurs in the magnesite deposits of Brumado, Bahia, Brazil where it occurs with uvite, dolomite and topaz in quartz veins (Bodenlos 1954; Cassedanne and Cassedanne 1978). A basinal or metamorphic origin is conceivable given the association with bedded magnesite (which is commonly of basinal brine origin) and the lack of directly associated granitic bodies. A third candidate for basinal origin is the Mingora and nearby emerald deposits of the Swat district, Pakistan where beryl-carbonate-quartz veins cut ultramafic-bearing metamorphic rocks (Kazmi et al. 1989). Oxygen isotope ratios from Mingora are relatively high (Arif et al. 1996) and could indicate fluids either cooled from an S-type leucogranite or from a metamorphic or sedimentary source. Fluid inclusions reveal that temperatures ranged from 250° to 450°C and that the fluids contained ≤0.03 mole fraction of CO_2+CH_4 with salinities up to 20 wt % $NaCl_{eq}$ (Seal 1989). From these results Seal (1989) inferred that fluid mixing may have contributed to beryl deposition. Some Afghanistani deposits (e.g., Panjsher) that lie westward along the Indus suture share many of the characteristics that are compatible with a sedimentary or metamorphic brine origin (Kazmi et al. 1989; Giuliani et al. 1997; Fig. 31). Nwe and Morteani (1993) interpret similar brines that occur late in the paragenesis of the Gravelotte, South Africa emeralds to be of metamorphic origin and to be linked to the most Cr-rich crystals.

Metamorphic occurrences

A number of Be mineral localities, notably some emerald deposits and "alpine clefts," occur in regionally metamorphosed rocks and lack evidence for magmatic involvement (Figs. 2, 27). Their origins are attributed to local redistribution and in some cases by introduction of new material during metamorphism.

Shear zone or metamorphic emeralds. Based on studies in the Alps and other regions Morteani, Grundman and coworkers have recognized that a number of emeralds and related minerals show compelling evidence for growth during regional metamorphism (Morteani and Grundmann 1977; Okrusch et al. 1981; Grundmann and Morteani 1989; Appendix A, Fig. 27). These occurrences have been termed metamorphic or shear zone deposits (Grundmann and Morteani 1989) and in some reviews they are combined with other non-magmatic deposits (Giuliani et al. 1998). Shear zone deposits resemble many igneous-related emerald deposits. Both types have emerald±phenakite ±chrysoberyl in biotite-rich zones in metaultramafic and metamafic rocks. Although the biotite-rich zones commonly form at the contact with felsic lithologies, this is not always so. Textures show that beryl overgrows other metamorphic minerals, in some cases with curving inclusion trails clearly demonstrating synkinematic growth (Morteani and Grundmann 1977; Grundmann and Morteani 1989). As described above, these emeralds contain more Fe and Mg than most other types (Fig. 6).

At Habachtal, Austria emerald formed at the contact of metaperidotites and felsic gneisses. Pegmatites and quartz veins are absent, but the felsic gneisses contain ample Be to form the emeralds (Okrusch et al. 1981; Grundmann and Morteani 1989). There and elsewhere in the Tauern Window Mg-Fe-Na-rich beryl grew relatively late in the metamorphic history, commonly on preexisting phenakite (Franz et al. 1986). Fluid inclusions have moderate salinities and CO_2 contents consistent with a metamorphic fluid (Nwe and Grundmann 1990). Similar deposits that lack closely associated igneous rocks occur in Afganistan, Pakistan and Brazil (Kazmi and Snee 1989a; Giuliani et al. 1990).

These systems commonly have CO_2-bearing inclusions and non-magmatic oxygen isotopic values that are consistent with metamorphic origins (Giuliani et al. 1997, 1998; Fig. 31).

More controversial are deposits associated with pegmatites, veins and granites where Be was likely introduced as part of an igneous event, but emerald formation may significantly post-date magmatism. These include the deposits at Gravelotte, South Africa and Franqueira, Spain (Appendix A; e.g., Franz et al. 1996; Martin-Izard et al. 1996). Although textural and isotopic evidence can be compelling for metamorphic growth, the question of origin arises because available evidence does not necessarily preclude coeval magma emplacement. Synkinematic intrusion accompanied by progressive growth of metasomatic zones could lead to similar beryl overgrowth of phenakite during deformation. Conversely, metamorphism of pegmatitic or vein systems could drive additional local metasomatic exchange and emerald growth.

Alpine clefts. Of mineralogical interest only, alpine clefts comprise metamorphic gash veins and a variety of igneous-related cavities and veins, all of which can contain minor accessory Be phases (Appendix A; Stadler et al. 1973; Graeser 1998). Although the term "alpine cleft" has been widely applied to open-space mineral occurrences in other parts of the world (cf. Cook 1998), in order for it to be petrologically useful it should be restricted to types analogous to those seen in the Alps, especially the open-space metamorphic gash veins. The metamorphic occurrences represent a distinct environment for local concentration of Be where phenakite, milarite, bavenite and other minerals accompany quartz, chlorite, adularia, and hematite in open cavities that formed in moderate grade metamorphic rocks of felsic composition (Graeser 1998). Isotopic and fluid inclusion evidences indicate local metamorphic sources for Alpine-type veins (Luckscheiter and Morteani 1980; Mullis et al. 1994; Henry et al. 1996). In some ways this is analogous to the processes inferred for the Habachtal emerald deposits (above; Okrusch et al. 1981). In contrast to the phenakite, milarite assemblages of metamorphic clefts, the numerous Alpine igneous-hosted localities contain beryl, bazzite, bavenite, gadolinite and a variety of other minerals. The latter occur in aplites, miarolitic cavities, pegmatites and granite-related quartz veins of both Hercynian and Alpine age (Hügi and Röwe 1970; Pezzotta et al. 1999). Even among these occurrences, some may be primarily metamorphic in origin. The host felsic biotite granitoids that have modest Be enrichments in their differentiated phases (Hügi and Röwe 1970).

Weathering and placers

With the exception of some placer settings, Be minerals are typically dispersed during weathering and sedimentary processes (Grigor'yev 1986). Local increases in soils can occur where Be dispersed in rock-forming minerals concentrates in montmorillonite or, more rarely, oxides and hydroxides during weathering (Sukhorukov 1989). For the most part, however, Be concentrations are diluted except when particularly stable Be silicates are present.

Beryllium minerals in the BASH group typically resist weathering and thus they occur in placer deposits. Alluvial deposits, particularly in Sri Lanka and related Gondwanan occurrences, are major sources of gem aquamarine, chrysoberyl and taaffeite-group minerals (Menon and Santosh 1995; Dissanayake et al. 2000; Shigley et al. 2000). These materials can be of unusually high quality because alluvial processes preferentially remove damaged (flawed) parts of crystals (Sinkankas 1981). Even in the world class Sri Lanka gem placers Be contents estimated to be 1 to 13 ppm comparable to or only slightly enriched over the local crust (Rupasinghe et al. 1984; Rupasinghe and Dissanayake 1985).

Supergene concentration is unimportant (Grigor'yev 1986). Local concentrations in soils, stream sediments and plants mainly represent residual accumulation and subsequent dispersion during weathering of hypogene mineralization (Sainsbury et al. 1967; Grigor'yev 1997). Beryl, phenakite and chrysoberyl typically do not weather; however, several Russian studies (Ginzburg and Shatskaya 1964; Novikova 1967; Grigor'yev 1997) have shown that beryl and phenakite can break down during weathering of pyrite- and fluorite-rich rocks. In their examples, the products are Be-enriched clay minerals and secondary phosphates such as herderite and moraesite. The authors interpret the process to be one of acid attack by sulfate and fluoride bearing supergene fluids. Given that the association of sulfides, fluorite and Be minerals is common, this process should be fairly widespread. Bacterial enhancement of weathering might also be important. Experiments show that bacteria can increase dissolution rates of helvite, beryl and chrysoberyl by factors of 5 or more without apparent Be toxicity to the bacteria (Mel'nikova et al. 1990). Large supergene accumulations are not reported, but minor upgrading might be expected where F-bearing acid groundwaters flow from Be-bearing rocks into mafic or carbonate rocks.

In contrast to the BASH group, other Be minerals such as those in the helvite, gadolinite and Na-Ca-silicate groups weather readily (Grigor'yev 1986). They react under near-surface oxidizing and weakly acid conditions to produce clays, oxides and some secondary silicates. Supergene changes are more widely reported in alkaline rocks, for example on the Kola Peninsula where hydrothermal chkalovite, epidydimite and eudidymite weather to produce beryllite, bertrandite and other phases (Vlasov et al. 1966). These authors report no supergene enrichment.

SYNOPSIS OF DEPOSIT CHARACTERISTICS AND ORIGINS

In contrast to pegmatitic deposits of Be, which are widely distributed in time but of rather limited geological variability, non-pegmatitic deposits are mainly Phanerozoic in age and are geologically diverse (Figs. 2, 4; Appendix A; cf. Rundkvist 1977; Zabolotnaya 1977). This diversity is all the more remarkable given the ready substitution of Be into many rock-forming minerals at crustal abundances and the overall ineffective concentration of large amounts of Be by crustal processes. The largest districts contain no more Be than that in a few cubic kilometers of granite or rhyolite (cf. Fig. 3) in contrast to deposits of many other elements (e.g., Cu, Au, Mo) which require 10s to 100s of km^3 of crustal source materials. This section summarizes the key patterns in deposit characteristics at global and system scales. These patterns can readily interpreted in terms of experimentally determined Be mineral stabilities and well-known geologic processes. Ultimately, non-pegmatitic Be mineral occurrences form either (1) by concentration of Be during magmatic or hydrothermal processes, or (2) by local redistribution of Be during metamorphic or sedimentary processes.

Global-scale patterns

With geologic setting. As reviewed by others (e.g., Strong 1988), chemically evolved felsic magmas and related mineralization form in a wide variety of environments. Examples with prominent Be deposits (Fig. 2) include convergent margin compressional settings (e.g., the circum-Pacific Mesozoic), collisional to post-collisional transpressional to transtensional (e.g., the Indus suture, European Hercynian, Urals and central Asia), to extensional environments in rift or continental extensional environments (e.g., Norway, Niger-Nigeria, US-Mexico Cenozoic, Proterozoic alkaline systems across Laurentia). Amagmatic deposit types—whether metamorphic, basin-related or placer—require orogenesis to drive material transport via fluid flow (Colombian emeralds, Alpine clefts) or local diffusion and recrystallization (shear zone type emeralds, granulite-sourced placers). Thus, all these settings generally share the requirement of relatively thick continental crust, which is needed for generation and differentiation of the diverse felsic magmas required for igneous-related deposits or for the crustal thickening required to drive regional metamorphism.

Other contributing factors to regional patterns are less well understood. For example, felsic magmatism and lithophile element (Sn-W-Mo-F-Zn) systems are far more widespread than documented Be mineralization. Provincial differences are striking: Southwestern North America has many Be deposits ranging in age from Mesoproterozoic to Cenozoic, yet is devoid of economically important Sn deposits. Conversely, the major Bolivian and Thai-Malaysian Sn provinces lack major known Be occurrences. Could crustal characteristics (thus inheritance) be important as may be the case for elements like Sn? There seems little compelling reason to think so, given that ample Be is present in most felsic igneous rocks; however, other petrogenetic factors could be key to generating a favorable geologic environment (e.g., shallow F-rich hydrothermal systems or a distinctive differentiation path). Of course, worldwide patterns of Be mineralization are poorly known due to the lack of systematic Be exploration.

Over geologic time. Examination of Appendix A and Figure 2 shows that non-pegmatitic Be deposits are mainly Phanerozoic in age (cf. Rundkvist 1977; Zabolotnaya 1977) and in this respect resembles the temporal distribution of other types of intrusion-related and epithermal mineral deposits (Meyer 1981). Most major occurrences are younger than 350 Ma with the prominent exception of Proterozoic alkaline-related deposits in Canada and eastern Europe. This age distribution reflects the preservation potential of the generally shallow crust in which Be deposits form and the typically thick and thus elevated crust. Preservation of volcanic and subvolcanic levels is not expected for these environments any more than it is in other epizonal terrestrial mineralization (Meyer 1981). Thus volcanic and hypabyssal systems are most common in the late Mesozoic and Cenozoic, whereas deeper-seated varieties make up most older systems. The increased proportion of alkaline-related systems with age may reflect preservation by the crustal extension that typically accompanies alkaline magmatism (cf. Barton 1996).

Principal episodes of Be mineralization are in the Proterozoic (1.6-1.0 Ga; Laurentia, Brazil), Devonian to Carboniferous (western Europe, Ural Mountains through the Altaides into eastern Asia; Tasman belt), the late Mesozoic (180-65 Ma; northern circum-Pacific, east-central Asia; Nigerian rift; Colombia eastern belt), and the later part of the Cenozoic (35-5 Ma; western Colombian emeralds, southwestern North America, Indus suture, Alps). Other times can be regionally significant: for example, the Pan African (Late Proterozoic to early Paleozoic; Arabian-Nubian shield, eastern Brazil). Even within individual episodes the nature of the systems is diverse, as exemplified by the Cenozoic examples in southwestern North America. Thus, apart from the clear influence of preservation on the temporal distribution and the control exerted by magma types reflecting the pulse of orogenesis, the evidence shows that Be deposits have neither a compelling temporal progression nor a discernable global synchronicity.

System-scale patterns

Geological characteristics. As emphasized by the organization of this review, the most distinctive features of igneous-related Be systems are their felsic, F-rich character, and the systematic variation of mineralogical characteristics that reflect the variation in igneous compositions from peralkaline to strongly peraluminous and from quartz-rich to quartz-undersaturated (Table 3; Fig. 5). Non-magmatic deposits are much less common. Many of the latter have evidence for local, wall-rock sources of Be. Consequently, very few of them represent large mass accumulations of Be, even though some comprise major gem deposits (e.g., Giuliani et al. 1999).

Table 3. Synopsis of the paragenesis of Be minerals in non-pegmatitic occurrences.

Occurrence type	***Mineral associations & abundance***[1]	***Time-space distribution***
Magmatic		
Li-Cs-Ta (LCT) magmas	• minor Brl (or Hrd) with Qtz, alkali feldspar, Li micas	• late magmatic phase
Muscovite(-beryl) granites	• minor Brl	• late magmatic or post-magmatic in vugs and fractures
Nb-Y-F (NYF) miarolitic pegmatites	• minor Ph, Gad etc. with Li-Fe micas, alkali feldspars	• post-magmatic
Syenites	• dispersed in rock-forming minerals in late differentiates; rare Chk, Epd, etc. in pegmatites	• late magmatic phase or post-magmatic
Magmatic-Hydrothermal		
Strongly peraluminous		
Li-Cs-Ta-F-Sn (LCT magmas)	• rare Brl or Hrd in Qtz-Toz±Ms greisens or albitization	• early, proximal
Sn (-W) (Bt±Ms granites)	• feldspathic host: (1) Brl-Ab-Ms-Fl; (2) Brl(-Brt-Euc) in Qtz-Ms-Fl±Toz±Tur veins; (3) uncommon Brl-Qtz-feldspar • carbonate host: (1) Ch-Ph / Brt in Fl-mica±Dsp veins; (2) Hlv / Dn-Fl±Mag-silicate replacement	• (1) proximal (outer parts of intrusion), relatively early; (2) intermediate position and time; early and proximal • both groups typically post-date and can be distal to Fe-rich garnet skarns; sulfides later
W-Mo (Bt+Ms granites)	• feldspathic host: (1) Brl common in Qtz-feldspar(-mica-Wlf-Mo) veins; (2) Brl-Ab-Ms-Fl; (3) Brl(-Brt±Euc) in Qtz-Ms-Fl±Toz veins; sulfide-Brt- Fl-carbonate veins • carbonate host: (1) Brl-Ph or Brt in Fl-mica veins; (2) Ph-Brt-Fl-Kfs-Qtz-carbonate veins and replacements • ultramafic/mafic host: Brl(emerald)-Ph-Ch with Bt-Pl±Qtz	• (1) early, proximal veins with central Brl; (2) commonly at intrusion margins, intermediate timing; (3) intermediate in time & position, Brl commonly distal; (4) late or distal • (1) intermediate to late, can overprint in Grt-Px-Ido skarns; (2) distal may or may not be late • Brl is central, other Be minerals are distal
Metaluminous-weakly peraluminous		
Mo(-W-Sn) (Bt granites & rhyolites)	• feldspathic host: (1) Brl, Ph, Brt or Hlv in Qtz-Toz±Li-Fe mica or Kfs greisen veins; (2) Gnt-Brt-Qtz-Hem veins • carbonate host: (1) Hlv or Ph with Fl-Mag-K-silicates; (2) Brt-Fl±Qtz ±clay replacement • volcanic host: (1) Brt-Fl-silica-Kfs-clay-carbonate; (2) red Brl-Qtz-Mn-Fe oxides	• (1) Be intermediate to late & distal; (2) late, proximal? • (1) Be minerals late & typically distal after andradite-rich skarn; (2) distal, timing uncertain, skarn absent • (1) distal, overall timing uncertain; (2) intermediate timing during cooling of flows

Peralkaline		
Nb-REE-Y-F (aegirine-Rbk granites & rhyolites)	• feldspathic host: Gad-Ph-Hlv-alkali feldspar -Qtz±Li mica Rbk±Mag metasomatism • carbonate host: (1) Ph-Brt-Lph-Fl-Kfs-Qtz replacement; (2) Brt-Fl-clay±Kfs replacement	• proximal (pegmatitic) to distal (replacement) timing variable; proximal Nb-Ta-REE • (1) distal to / overprint Grt-Px Ido skarns; (2) distal in low-T shallow settings
Nb-REE-Y (undersaturated syenites)	• feldspathic host: (1) Eud-Bar-alkali feldspar-Na amphibole; (2) Epd-Lph-Chk etc.-Ab-Anc±Sdl (3) Fl-Brt replacement	• (1) position & timing unclear; (2) intermediate to late in host intrusion; (3) distal, timing unknown
Non-magmatic		
Fe-Mn-oxide	• rare, mainly non-silicates	• discrete Be minerals are late
Brines / basinal & metamorphic	• sparse Brl ±Euc) in Qtz-carbonate-feldspar-Py veins	• with main vein-forming event, temporal pattern uncertain
Metamorphic deposits	• shear zone: Brl(-Ph-Ch) with Bt(-Tlc-Act-Chl-Pl) • Alpine cleft: Ph-Mil-Qtz-Chl-Kfs-Hem	• near felsic contacts; syn-or post-peak metamorphism; some could be syn-magmatic • post-peak metamorphism
Surficial deposits	• minor Brl, Ch etc. in placers with other resistant minerals	• in alluvial systems downstream from pegmatite/granulite sources

[1] Mineral abbreviations after Kretz (1983) or as in Table 1. Feldspathic hosts include granitoids and clastic sedimentary rocks.

As shown in Figure 34, a salient feature of non-pegmatitic Be deposits is the wide range of formation conditions and fluid compositions. Depths range from the surface to >10 km and temperatures from magmatic conditions to surface temperatures, and commonly exhibit a broad range within a single deposit. Fluids vary from hypersaline to dilute; they may or may not have CO_2 and/or CH_4 and have diverse redox states and acidities. These fluid compositional characteristics generally correlate well with igneous compositions and depth. Sn, W and base metal-rich systems tend to be acid, saline and shallow. Oxidation and sulfidation states directly influence alteration and metallic mineralogies. These variables also govern helvite group stability (Fig. 9) and the transition metal contents of beryl. Hydrothermal alteration assemblages directly control the identity of stable Be minerals primarily by imposing values of $a_{Al_2O_3}$ and a_{SiO_2} (Figs. 11, 12).

The systematic progression in hydrothermal mineral associations resembles that in many other types of magmatic-hydrothermal ore deposits (Table 3). In feldspathic rocks early, high-temperature pegmatitic and miarolitic cavities form a geological continuum with feldspar-mica±quartz veins and sodic or potassic feldspathization. Later mica-rich and the latest-stage polymetallic assemblages form at somewhat lower temperatures and tend to be more extensive. Fluorine is typically abundant as evidenced by the presence of fluorite, topaz and/or F-bearing mica. Magmatic albite-rich rocks are common, however even more widespread is hydrothermal albitization which is clearly significant in many igneous-related systems and in the Colombian basin-related deposits.

Deposits hosted in carbonate-rich and ultramafic rocks have similar broad similarities, with differences in detail with host rock and magma type. Skarns form early. Vesuvianite, garnet and pyroxene typify calcareous hosts, whereas humite-group minerals and other magnesian silicates occur in magnesian carbonates (e.g., Kwak 1987).

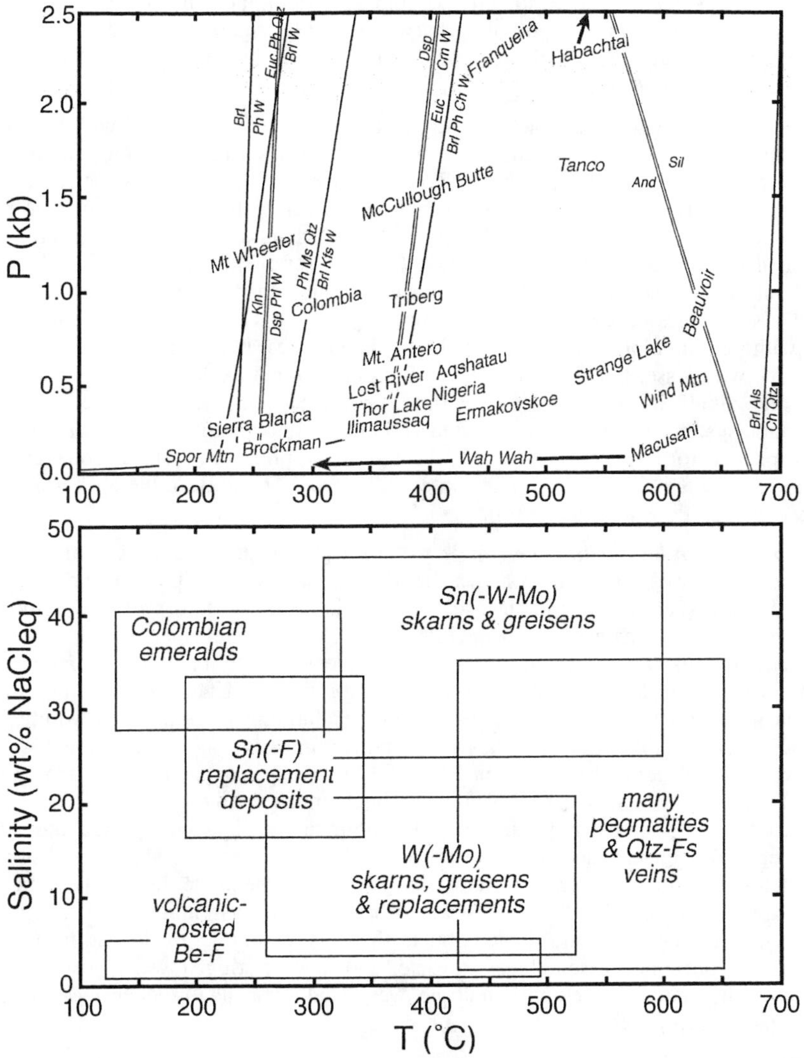

Figure 34. Synopsis of pressure-temperature-salinity conditions of formation for many types of Be-enriched deposits. Summarized from sources cited in the text and Appendix A.

Although there is a tendency for more reduced and pyroxene-rich skarns with increasing aluminum saturation index in associated intrusions, these patterns are complex as has long been recognized in other skarn systems (e.g., Einaudi et al. 1981; Newberry 1998). Skarn mineral assemblages may contain some Be but most ends up in later F-rich veins and replacements. As a class, low-T replacement deposits comprise the bulk of high-grade Be mineralization, whether as fluorite-silica-bertrandite after carbonate clasts at Spor Mountain, fluorite-diaspore-micas-chrysoberyl at Lost River, or fluorite-adularia-phenakite at Ermakovskoe and Mount Wheeler. Replacements of carbonate or skarn by fluorite and iron-rich oxides, sulfides and sheet silicates form another major group, typically with Sn-bearing hydrothermal systems. Reaction between veins and dikes in ultramafic rocks leads to metasomatic assemblages dominated by micas, plagioclase, other sheet silicates, and amphiboles that are broadly analogous to the skarns and apocarbonate greisens. The lack of garnet and pyroxene in ultramafic-hosted deposits reflects their high Mg/Ca.

Mineral associations are one of the most prominent correlations with magmatic compositions and the phase equilibrium reasons were described at some length under "Beryllium Mineral Stabilities" above. To summarize the key features: Micas change from muscovite-rich in peraluminous systems to Li-bearing trioctahedral micas in more alkaline systems. Mica-rich assemblages are not created equal—trioctahedral-mica dominated greisens are not acid (e.g., low K^+/H^+) assemblages (cf. Eqn.10). The latter thus should not be equated chemically with muscovite-bearing greisens or sericitic and related assemblages that occur in most porphyry Cu and Mo systems. BASH phases occur with peraluminous and some metaluminous granites; helvite group, phenakite and bertrandite with lesser beryl occur with metaluminous to peralkaline varieties and helvite group and alkali Be silicates in silica-undersaturated alkaline rocks (see Fig. 12). Amagmatic systems show a corresponding variation with alkalinity of the host materials: BASH phases form in aluminous rocks (e.g., Colombian and shear zone emerald deposits) whereas helvite group, phenakite and other Al-poor Be minerals form in Al-poor rocks (e.g., Fe-Mn oxide group).

Time-space relationships. No single pattern captures the time-space distribution of Be minerals in hydrothermal systems. Beryllium may be precipitated early or late, proximally or distally (cf. pegmatites; see Černý, this volume). Be is proximal in more F-poor systems both in granitic pegmatites and in quartz-feldspar veins, but more commonly it is distal and late. In volcanic-related deposits, Be minerals (bertrandite, red beryl) are generally late and low temperature (e.g., western Utah), whereas in systems with clear magmatic enrichments beryl (in peraluminous rocks), epidydimite (in peralkaline rocks) or other Be minerals are prominent in early stages (e.g., Mount Antero; Sherlova Gora, Sheeprock, Mt. Saint-Hilaire, Ilímaussaq, Lovozero), although a large suite of later Be minerals can crystallize through much of the sequence. The classic carbonate-hosted systems, regardless of the aluminousity of associated igneous rocks, tend to have Be concentrated in lower temperature assemblages, typically distal parts of the systems. Length scales can be meters to kilometers. Phenakite, bertrandite and, to a lesser extent chrysoberyl tend to be distal to beryl-bearing assemblages. Although one has the impression in many systems that there is a simple evolution from high-temperature, proximal assemblages to late and overprinting lower-T assemblages, patterns can be more complex with multiple events and reversals in sequence (Lost River, Shizhuyuan, Birch Creek).

Origins

Ultimately, these patterns in the non-pegmatitic occurrence of Be minerals reflect original rock compositions, controls on solubility, and magmatic evolution at the source of fluids. Magmatic concentrations are relatively low due to the limited solubility of Be minerals in silicate melts. Hydrothermal deposits derived from such melts would have proximal Be and might lack significant Be in distal or low-temperature assemblages if aqueous complexing is weak (e.g., in the F-poor quartz-feldspar vein association in Table 3). When Be is strongly complexed (e.g., by F) and is not near saturation at its source, aqueous fluids could travel some distance before Be precipitation occurs in a favorable physical or chemical setting. Precipitation mechanisms include cooling, mixing, neutralization, and removal of F from the fluid. Wall-rock reaction, particularly with carbonate or mafic mineral rich rocks, would be particularly effective for the last two depositional mechanisms; this is abundantly illustrated by fluorite-rich deposits of many types (Table 3). Differences in timing and zoning may also be due to evolution of the magma itself. Multiple magma batches, commonly of distinct origin, are well documented in many areas (e.g., Birch Creek, Cornwall, Beauvoir, Iron Mountain, Shizhuyuan).

The common features of Be-enriched igneous-related deposits are: (1) a magma that does not strongly remove Be in early formed minerals, and (2) a mechanism for generating and focusing F- and Be-bearing fluids. Melts with abundant Ca and Mg (i.e., most subalkaline types) will be unfavorable for magmatic enrichment because Be is partitioned strongly into sodic plagioclase (cf. London and Evensen, this volume). Likewise, mafic and calcic minerals sequester F in igneous systems either during crystallization (e.g., in biotite or apatite) or by formation of fluorite and other minerals on fluid release (cf. Barton 1987). For these reasons, igneous-related Be deposits are restricted to felsic, alkaline (low $CaO/[Na_2O+K_2O+CaO]$, cf. Fig. 5) igneous suites, but are not restricted in terms of alkalinity or silica saturation (in the sense of $a_{Al_2O_3}$ and a_{SiO_2}). Instead, the latter factors govern hydrothermal alteration types and Be mineral assemblages (Fig. 12), and are reflected in the suites of associated elements (Fig. 4; Table 3). Wall rocks as well as magmas may source Be in some of these systems, particularly where sodic plagioclase is destroyed and thus releases Be to the hydrothermal system (Beus et al. 1963). These processes deserve further investigation.

In contrast to igneous-related systems, in most non-magmatic deposits Be is only locally redistributed, Be concentrations are low, and the occurrences are mainly of mineralogical or gemological interest. Original enrichment in Be-bearing felsic rocks plays a role in some emerald deposits. Elsewhere, Be deposits are inferred to originate either by the action of aqueous fluids that release Be by mineralogical changes in the host rocks (e.g., Colombian emeralds, Alpine clefts) or by reworking by surficial processes (e.g., placer deposits).

ACKNOWLEDGMENTS

Any undertaking of this scope depends heavily on the work of others and we can only acknowledge a fraction of the underlying science through citation. We thank our reviewers, Gerhard Franz, David Lindsey and Steve Ludington for constructive and thorough comments; they bear no responsibility for errors and oversights. Ed Grew's editorial oversight and patience is especially appreciated. National Science Foundation grants to MDB partially supported preparation of this paper (EAR 98-15032) and earlier work on Be-mineral stabilities and granite-related mineralization in southwestern North America (most recently EAR 95-27009). Additional support for compilation and drafting was provided through the Center for Minerals Resources and cooperative work on with the U.S.G.S. Global Mineral Resource Assessment Project. MDB thanks Brendan Laurs and the Gemological Institute of America for help during research, particularly for the use of their library. Brendan Laurs, David Lindsey, and Frank Mazdab provided photographs. Meg Watt, Jared Bond and Steve Hubbs helped with preparation of the manuscript.

APPENDIX A: SUMMARY OF BERYLLIUM DEPOSITS BY REGION

(next eleven pages)

Appendix A. Summary of beryllium deposits by region

Province / Location	Type*	Age	References	Comments
NORTH AMERICA				
Cordillera — Cenozoic				
Southwest US-Mexico Tertiary	G, aG, S, pG (sk, rpl, vol, ma)	CenM	Warner et al. 1959; Meeves 1966; Shawe 1966; Burt and Sheridan 1988	Mainly replacement and vein Hlv, Brl and Brt mineralization with mid- and early Cenozoic rocks
Iron Mountain area: Iron Mountain district, Black Mountains, Apache Warm Springs, New Mexico, USA *(see Fig. 22)*	G(sk, volc)	CenM (29)	Jahns 1944a,b; Robertson 1986	Iron Mountain: Hlv-Fl-Ves skarns (0.2Mt @ 0.2-0.7% BeO) superimposed on Mag-Sch-(Sn)-Adr skarn; Black Range volcanics have red Brl and wood Sn; Warm Springs have volcanic-hosted Brt
Wind Mountain (Cornudas Mountain), New Mexico, USA	aS (m)	CenM (35)	Meeves 1966; McLemore et al. 1996; Nutt et al. 1997	BeO bearing (0.1%) pegmatitic Ne syenite in district of shallow alkaline laccoliths; Eud identified
Sierra Blanca, Texas, USA	aG (rpl)	CenM (37)	McAnulty 1980; Price et al. 1990; Rubin et al. 1990; Henry 1992	Fl-rich replacement bodies in limestone adjacent to Li-Be-Zn-Rb-Y-Nb-REE-Th rich alkaline Cyl-bearing rhyolites; 850 kT @ 1.5% BeO as Brt, Ph (Bht, Bb, Ch); minor grossularitic skarn; clays+Anc
Aquachile, Coahuila, Mexico	aS,aG (rpl)	CenM	McAnulty et al. 1963; Griffitts and Cooley 1978	Brt-Ad-bearing Fl replacement (17 Mt @ 0.1% BeO) adjacent to alkaline rhyolite & Rbk Qtz syenite
Mt. Antero, Colorado, USA	G (m,gr,mp)	CenM	Adams 1953; Sharp 1976; Kar 1991; Jacobson 1993b	Bt granite with Brl-rich aplite, miarolitic cavities with aquamarine to later Ph and Brt; greisen Qtz veins with Brl-Ms-Mo-Toz; gem source
Silverton, Colorado, USA	volc (vn-Mn)	CenM	Warner et al. 1959; Casadevall and Ohmoto 1977	Minor Hlv in Mn-rich epithermal Ag-Au veins
northern Basin & Range Province (Utah etc.)	rV, G, ±pG (rpl, gr, m, vn, sk)	CenM (MiocE)	Warner et al. 1959; Shawe 1966; Lindsey 1977; Burt and Sheridan 1986; Congdon and Nash 1991	Mainly western Utah Be belt, most occurrences in volcanic rocks, but also skarn (Hlv-Brl) and granite-hosted occurrences — dominant Be producer; magmas up to 80 ppm Be
Spor Mountain, Utah, USA *(see Fig. 24)*	rV (rpl), [rV (vn)]	CenM (21)	Shawe 1968; Lindsey et al. 1973; Lindsey 1977; Ludwig et al. 1980; Baker et al. 1998; Johnson and Ripley 1998	21 Ma Brt in Fl-silica replacement of carbonate clasts in lithic tuff (7Mt @ 0.72% BeO), extensive Li-Zn-bearing Kfs and clay alteration; overlying 6-7 Ma topaz rhyolites have red Brl
Wah-Wah, Utah, USA	rV (vn)	CenM (23)	Keith et al. 1994; Thompson et al. 1996	Topaz rhyolite with late red Brl+Kln in fractures with early Mn-Fe oxides; main gem red Brl source
Rodenhouse Wash, Gold Hill, Utah, USA	G(?) (vn)	CenM?	Griffitts 1965; Shawe 1966; Lindsey 1977	3 km long Qtz-Ad-Cal-Brt (low-F) fracture zone, (>1Mt 0.5% BeO)
Sheeprock Mtns, Utah, USA	G(m, p, gr)	CenM (22)	Cohenour 1963; Rogers 1990	Late magmatic, miarolitic and Qtz-greisen Brl in late phase of Bt(-Ms) granite (>1Mt, 0.01-0.1% BeO)

Sawtooth Batholith, Idaho, USA	G (p)	Cen (Eoc)	Sinkankas 1981; Kiilsgaard and Bennett 1985	Brl-bearing miarolitic pegmatites and Qtz veins in shallow Bt granite
Cordillera — Cretaceous				
Seward Peninsula, Alaska	pG (rpl, sk, gr)	MesL (Cret)	Sainsbury 1968, 1969; Swanson et al. 1990; Newberry 1998	Several Sn-W Bt granite centers with proximal Be-bearing skarns and extensive distal Be-F-Al replacement deposits; similar in NE-most Siberia
Lost River district, Alaska, USA *(see Fig. 19)*	pG (rpl, sk, gr)	MesL (CretL)	Sainsbury 1968, 1969, 1988; Dobson 1982	Extensive Ch-Dsp-Ms-Fl-replacement of carbonates other BASH minerals in Ms-Fl-Qtz veins (>3Mt @ 0.3-1.75% BeO) early and late relative to Sn-Fe-rich skarns (±Hlv) and greisen in Bt granite
Northern Cordillera, Yukon & Alaska	pG (sk, gr, vn) (E)	MesL (Cret)	Mulligan 1968, 1971; Dick and Hodgson 1982; Anderson 1988; Newberry et al. 1990; Layne and Spooner 1991	Early to Late Cretaceous granitoids with a wide range of commonly F-rich Mo-W, W-Mo-Cu and Sn-W systems; Brl and other Be minerals reported in a number of skarn and greisen/vein systems, also pegmatites
Crown (Finlayson Lake), Yukon, Canada	(E) (pG, gr)	MesL (CretE)	Groat et al. 2000a,b	Qtz-Tur±Sch veins associated with Ms Bt granite have emerald where they cut Chl-mica schists (green V-rich Brl reported with Lened stock to north)
Logtung, Yukon, Canada	pG (gr, sk)	MesL (CretE)	Noble et al. 1984; Anderson 1988	large hornfels and skarn hosted stockwork W-Mo(-Bi) deposit with Brl in early Qtz-Sch-Mo-Ms veins
Great Basin (Nevada)	pG (gr, sk)	MesL (CretL)	Olson and Hinrichs 1960; Shawe 1966; Barton 1987, 1990; Barton and Trim 1991; Van Averbeke 1996	>20 Late Cretaceous Bt-Ms granites exposed at different levels with greisen, skarn and replacement mineralization with F-Zn-W(-Mo)
Birch Creek, California, USA	pG (gr, sk)	MesL (82)	Barton 1987, 2000; Barton and Trim 1991	Ms-Phl-Fl-Brl veins (cut F-, Al-rich W(-Cu) skarn) and Brl-bearing greisen veins (cut albitization, Kfs-Qtz veins) in well-documented multi-event Bt-Ms granitic center
McCullough Butte, Nevada, USA	pG(sk, gr)	MesL (84)	Barton 1982a; Barton et al. 1982; Barton and Trim 1991	Brl(±Bav-Brt) Ms-Phl-Fl-Pl Zn-W(-Mo) skarns superimposed on earlier Grt-Px-Ves skarn (175Mt @ 0.027% BeO, 10% CaF_2) above greisenized Bt-Ms granite
Mt. Wheeler, Nevada, USA *(see Fig. 20)*	pG (rpl, vn)	MesL (85)	Stager 1960; Lee and Erd 1963; Barton and Trim 1991, unpubl. data; National Treasure Mines, unpubl. report	Ph-Brt(-Brl) Fl-rich-Ad-Mn Sid-Qtz-Ms-Sch replacements and Qtz veins in basal carbonate unit above hidden intrusion (>0.2Mt @ 0.75% BeO, 20% CaF_2, 0.2% WO_3)
Oreana, Nevada, USA	pG (sk, p, vn)	MesL (CretL)	Kerr 1938; Olson and Hinrichs 1960	Brl-Sch in pegmatitic Qtz-Pl-Fl-Ms veins with Bt±Ms granite stock; mined for W (0.1 to 10% Brl)
Uinta Mountains, Utah, USA	B (vn) (E)	PhanL	Keith et al. 1996; Olcott et al. 1998	Emerald in Cal veins in 950 Ma shale; formed at 200-300°C from Mes or Cen oil-field brine (no magmatism)

Appalachians

Quebec-New Hampshire Alkaline	S (p,vn), G (sk)	Mes	Burt 1980; Eby 1985; Foland and Allen 1991	Hot-spot-related Monteregian-White Mountain alkaline province has minor occurrences (notably Dn) plus spectacular locality at Mt. Saint-Hilaire
Mt. Saint-Hilaire, Quebec, Canada	S (p,vn)	MesL (123)	Currie et al. 1986; Gilbert and Foland 1986; Currie 1989; Mandarino and Anderson 1989; Horváth and Gault 1990	15 (possibly 17) Be minerals (Epd > Lph > Gnt > others; cf. Introduction, this volume) from pegmatites, miarolitic cavities & veins in shallow, composite Ne gabbro to sdl syenite intrusion
Canadian Maritime	pG (gr, p, vn)	PalM,L	Mulligan 1968; Currie et al. 1998	Scattered occurrences of Brl in granitic pegmatites, Qtz veins and greisens with or without W-Mo±Sn
North Carolina	pG (p, vn) (E)	PalM	Griffitts 1954; Sinkankas 1981, p. 565-572; Wood 1996	Simple and Li-Sn pegmatites and Brl-bearing veins, some associated with granodioritic stocks (122kT BeO estimated in Brl @ 0.02-0.1% BeO); emerald associated with desilicated pegmatites (Spruce Pine) and Qtz-Ms veins (Hdd)

Proterozoic North America and Greenland

Pikes Peak Batholith, Colorado, USA	(a)G, S (p, gr)	ProtL (1.0 Ga)	Gross and Heinrich 1966; Ludington 1981; Simmons et al. 1987; Levasseur 1997; Kile and Foord 1998; Smith et al. 1999	Variably rare-metal enriched, composite alkaline batholith; Gad, Ph, Hlv, Gnt, present in miarolitic pegmatites typically with Amz-Znw; local areas of F-rich veins and greisens with Brl-Brt
Redskin Granite & Boomer Mine, Colorado, USA *(see Fig. 21)*	G (gr)	ProtL (1.0 Ga)	Hawley et al. 1966; Hawley 1969; Desborough et al. 1980	Qtz-Ms-Toz greisens along margin of Redskin Bt granite (Pikes Peak derivative) were mined for Brl+Brt (>5Kt @ 2% BeO), minor Wlf-Cst-sulfides
Arizona-New Mexico Black Pearl & Tungstona mines	pG (gr, p)	ProtM (1.4 Ga)	Jahns and Wright 1944; Heinrich 1957; Olson and Hinrichs 1960; Dale 1961; London and Burt 1978; Schmitz and Burt 1990	Greisen W(-Mo-Be) veins (Boriana, Black Pearl), apogranites, and complex pegmatites with 1.4 Ga granites; similar in style & age to pegmatites in Colorado & New Mexico
Great Slave Thor Lake, NWT, Canada *(see Fig. 16)*	aG, S (gr, f, p)	ProtE (ArchL)	Davidson 1978; Trueman et al. 1988; Kretz et al. 1989; Sinclair and Richardson 1994; Richardson and Birkett 1996; Taylor and Pollard 1996	Thor Lake (T-zone) Ph-Brt(-Gad-Hlv) deposit (1.6 Mt @ 0.76% BeO) in Qtz-Fl-Pll "pegmatite" with Na- and Kfs alteration; nearby Nb-Ta-Zr mineralization in syenitic breccia; both with composite alkaline Blachford Lake complex (2.1 Ga); in same area are older, deeper Be(-Li-B) pegmatites & Qtz veins around 2.5 Ga granites
Nain	S, aG (m, p, f)	ProtM (1150-1340)	Richardson and Birkett 1996; Miller et al. 1997	Peralkaline volcanic and shallow intrusive rocks with widespread Be-Nb-Ta-Zr-Y-REE mineral occurrences

Strange Lake (Lac Brisson), Labrador, Canada	aG (m, p, f)	ProtM (1240)	Hill and Miller 1990; Birkett et al. 1991; Boily and Williams-Jones 1994; Miller 1996; Richardson and Birkett 1996; Salvi and Williams-Jones 1996, 1997; Miller et al. 1997	Lf-Gad(-Mil) and Zr-Nb-Y mineralization (52Mt @ 0.08% BeO) in lenticular zones in rare-metal-rich granite in larger alkali granite center; low-T F-rich hydrothermal overprint may have helped to concentrate rare elements
Letita Lake (Mann#1, Seal Lake), Labrador, Canada	S (f), aG (-)	ProtM (1327-1337)	Heinrich and Deane 1962; Miller 1989; Richardson and Birkett 1996; Miller et al. 1997	Eud-Bar mineralization (1.8Mt @ 0.375% BeO)with Nb(-REE-Zn) in alkali Fs veins & feldspathic replacement in hypabyssal Agt-Rbk syenite and trachyte; alkali granites lack deposits
Gardar Province, Greenland	S, aG (m, p, vn)	ProtM (1340-1168)	Upton and Emeleus 1987; Woolley 1987; Finch 1990; Steenfelt 1991; Macdonald and Upton 1993; Paslick et al. 1993; Goodenough et al. 2000	Alkaline granites (Ivigtut), agpaitic Ne syenite complexes (Ilímaussaq, Motzfeldt), and alkaline volcanics; magmatic and hydrothermal Be enrichments are common in later parts of agpaitic intrusions
Ilímaussaq, Greenland	S (m, p, vn)	ProtM (1168)	Engell et al. 1971; Bailey et al. 1981; Sørensen et al. 1981	Be enrichments in Agt-Ab-Anc etc. hydrothermal veins & fenites (Chk and 18 other Be minerals; cf. Introduction, this volume); plus modest late magmatic enrichment (in lujavrite; >20Kt BeO in Kjanefjeld area with Li, Zr, Nb, Zn, REE, Y, F)

SOUTH AMERICA

Andes — Cenozoic-Mesozoic

Andean Altiplano Macusani, Peru	rV (LCT, other?)	Cen (Mioc)	Pichavant et al. 1988a,b; Lehmann et al. 1990; Dietrich et al. 2000	Beryllium minerals apparently rare, but widespread Miocene rare metal enriched felsic intrusive and volcanic centers many with Sn-Ag(-B) ores, Fl is relatively uncommon; Macusani tuff is an end member
Colombia (Muzo, Chivor, Somondoco, Gachala, Peñas Blancas etc.) *(see Fig. 32)*	B (vn) (E)	MesL & CenM	Feininger 1970; Beus 1979; Cheilletz et al. 1994; Ottaway et al. 1994; Giuliani et al. 1995, 1997, 1999, 2000b; Cheilletz and Giuliani 1996; Zimmermann et al. 1997; Branquet et al. 1999; Banks et al. 2000	Emerald(@ 0.01-0.05 ppm BeO±rare aquamarine±Euc) in veins in variably albitized Cretaceous black shales and siltstones (2-10 ppm BeO); associated minerals include Qtz, Py, Cal, Na-plagioclase, Ms, Pas; two ages: Muzo & other western belt deposits generated during Oligocene compression; Chivor & eastern belt during Late Cretaceous extension; fluid inclusion, isotopic, and other evidence for basinal brine source with sulfate reduction by hydrocarbons; evidence for fluid mixing in both belts
Sierras Pampeanas, Argentina	pG (p, gr/vn?)	PalL (?)	Morteani et al. 1995	Hercynian Brl-rich pegmatites and Brl-bearing Qtz veins

Brazil — Proterozoic

Atlantic Shield, Brazil	M? B? pG (p, vn) (E), G (f)	ProtE, ProtL,PalE	Bilal and Fonteilles 1988; Kazmi and Snee 1989a; de Medeiros Delgado et al. 1994; Sinclair 1996; Giuliani et al. 1997; Ribeiro et al. 1997	Many Be deposits (gems, common Brl), mostly pegmatitic or vein related, possible shear zone and basin-related deposits (total Minas Gerais pegmatites: 106Mt @ 0.04% BeO); NYF, LCT and simple pegmatites; Hlv-Ph-Brl mineralization with albitization with Sucuri granite
Carnaiba & Socoto, Bahia, Brazil	pG (p, gr, vn) (E)	ProtE (1.9 Ga)	Kazmi and Snee 1989a; Giuliani et al. 1990; Cheilletz et al. 1993	Brl (-Mo-W) greisen, Qtz veins, and pegmatites associated with two-mica granites intrudes serpentinite-bearing metamorphic rocks locally forming emerald(±Ph)
Santa Terezinha de Goias, Goias, Brazil	M? G? (vn) (E)	PalE (520)	Kazmi and Snee 1989a; Giuliani et al. 1997; Ribeiro et al. 1997	Emeralds developed in intensely folded Bt-rich schists & Tlc±carbonate schists; pegmatite rare & unrelated(?)
Brumado, Brazil	pG? B/M? (vn) (E)	Prot?	Bodenlos 1954; Cassedanne and Cassedanne 1978; Kazmi and Snee 1989a	Emerald/aquamarine in Mgs unit in metamorphic sequence; Qtz, Drv, Toz, Mg-Ca silicates associated; uncertain origin, others suggest hidden granite, basinal / metamorphic origin possible
Rondonia	G (gr, f)	ProtM	Kloosterman 1974; Haapala 1995; Bettencourt et al. 1999	Ph ± other Be minerals with greisen-style Sn mineralization Bt granites; compared with Nigeria and broadly coeval rapakivi systems; cf. with Sucuri granite in Goias (above)

EUROPE

Alps — Cenozoic

Alpine clefts - metamorphic	M (vn)	Cen	Stadler et al. 1973 Hügi and Röwe 1970; Graeser 1998	Metamorphic Chl-Qtz-Ad fissures contain rare Be minerals (Ph, Brt, Mil, Bav etc., very rare Brl) typically in feldspathic host rocks
Alpine clefts - granitic	G (p, vn)	Cen	Stadler et al. 1973 Hügi and Röwe 1970; Graeser 1998; Pezzotta et al. 1999	Be-minerals (Brl, Bz, Bav, Gad) in pegmatites, aplites and veins associated with modestly Be-enriched granites
Habachtal, Austria	M (E)	Cen	Morteani and Grundmann 1977; Okrusch et al. 1981; Grundmann and Morteani 1989; Kazmi and Snee 1989a; Nwe and Grundmann 1990	Emerald, Ph & Ch form at contact along metamorphic shear zone where ultramafic rocks after serpentinite are in contact with felsic lithologies; textural evidence for syn-metamorphic growth of emerald after Ph

Western Europe — Paleozoic

West-central Europe	pG (g, m, p, sk, LCT)	PalL	many	
Cornwall, England	pG (gr, sk, p, vn)	PalL	Kingsbury 1961; Clark 1970; Embrey and Symes 1987; Manning and Hill 1990	Brl, Ph, Brt, Euc, Hlv group, etc. widespread in greisen, vein, pegmatite, skarn occurrences with Bt±Ms granites; LCT (Toz) granites not directly linked to mineralization

Massif Central, France e.g., Beauvoir *(see Fig. 15)*	pG (gr, mLCT)	PalL	Burnol 1978; Stussi 1989; Cuney et al. 1992; Raimbault et al. 1995; Raimbault and Burnol 1998	multi-stage LEM granite/veins; Be mainly in Li-mica; rare post-magmatic Brl and Hrd; multi-stage intrusion of progressively more evolved granitic magmas starting with main phase Ms-Bt granites
Erzgebirge, Germany / Krušné Hory, Czech Republic	pG (gr)	PalL	Baumann 1994	Brl & Hrd are rare; found early in Sn-W-Li-F greisens, Be enrichments are present in some of the specialized granites
Triberg, Schwarzwald, Germany	pG (gr)	PalL (315)	Markl and Schumacher 1996, 1997	Zoned system in Bt±Ms leucogranite with Brl-bearing miarolitic cavities overprinted by Brl-bearing albitization & Qtz-Ms(-Fl-Toz-Cst) greisens, late Ph & Brt
Western Iberia	pG (mLCT, gr, p) (E)	PalL	Kelly and Rye 1979; Voncken et al. 1986; van Gaans et al. 1995; Fuertes and Martin-Izard 1998; Charoy 1999	Present in Be-enriched magmas (Argemala), pegmatites (Franqueira), and W-Sn greisens (Panasqueira)
Franqueira, Galicia, Spain	pG (p) (E)	PalL (Carb 315)	Martin-Izard et al. 1995, 1996; Franz et al. 1996; Fuertes et al. 2000	Emerald, Ch, Ph formed around margin of granitic pegmatite in ultramafic rocks; alternative interpretations of syn-intrusion or metamorphic (post-intr.) origin
Oslo Graben	G S (sk, p, vn)	PalL (Perm)	Brinck and Hofmann 1964; Ihlen and Vokes 1978; Larsen et al. 1987; Larsen 1988; Jamtveit and Andersen 1993	Be-minerals associated with veins, skarns and pegmatites associated with Oslo Rift-related Bt granites, monzonites & syenites; multiple parageneses, minerals include Hlv group, Brm, Epd, Brl, Lph, Bav, Bar, Brt, Ph, Gad (22 in all)..
Eidsvoll (Byrud Minnesund), Oslo region, Norway	G(p) / S (p, vn) "(E)"	PalL (Perm)	Kazmi and Snee 1989a	Green V(0.9%)>Cr(<0.1%) Brl in Cambrian Alum Shale intruded by Qtz syenite
Kola Peninsula Lovozero & Khibiny complexes	aS (p, vn)	PalM (360-420)	Beus 1966, p. 244-255; Vlasov et al. 1966; Gerasimovsky et al. 1968; Borshchevskii et al. 1987; Kogarko et al. 1995; Men'shikov et al. 1999	Lovozero & Khibiny agpaitic syenitic intrusions have elevated Be overall reaching maximum in later units; complex pegmatites of various types follow and overlap with hydrothermal vein (e.g., Ntr-Fs-Pll); many Be minerals with varied paragenesis: Chk, Epd, Eud, Lph, Mil, Brm, Ttp, Lf, Bar, Brt, Byl, Gnt (15 in all)..
Northern Europe — Proterozoic				
Karelia — Finland, Russia	G (gr, sk, p)	ProtM (1.4-1.65)	Haapala 1977b; Haapala 1988; Haapala and Ramo 1990; Sviridenko and Ivashchenko 2000	Be-bearing Sn(-W-Zn) F-rich greisens (Brl, Hlv, Brt at Kymi, Finland; Eujaroki, below), polymetallic Sn-W skarns (Hlv, Brt, Ph at Pitkäranta, Russia)
Eurajoki, Finland	G (gr, vn)	ProtM (1.57)	Haapala 1977a; Haapala 1997	Multiple phase Bt(-Hbl-Fa) granite to Toz-bearing leucogranite with Be-bearing Sn(-W-Mo) greisens (Qtz-Sid-Toz-Chl etc.) with Brl and Hlv; albitization present; unusual Grt-Ghn greisens

Ukrainian Shield	aG S? (f, vn)	ProtM	Vynar and Razumeeva 1972; Zabolotnaya 1977; Sheremet and Panov 1988; Esipchuk et al. 1993; Kremenetsky et al. 2000a	Feldspathic metasomatic Be deposits with Gnt at Perga, Ukraine (>10Kt Be @ 0.54% BeO) and Diabazovoe, Belarus (>10Kt Be @ 0.3% BeO); in region of cyl-bearing rare-metal granites, syenites and Brl-bearing pegmatites
ASIA				
Central Asia — Cenozoic				
Indus Suture — Pakistan, Afghanistan	M(B?) (vn) (E); pG (p, vn) (E)	Cen	Kazmi and Snee 1989b; Seal 1989	Many emerald deposits of different styles along mafic/ultramafic-rich Indus suture; many other Brl-bearing pegmatites and granite-related veins
Mingora (Charbagh, Makhad, Gujar Kili), Swat, Pakistan	M (B?) (vn) (E)	Cen	Kazmi et al. 1986; Kazmi and Snee 1989a; Arif et al. 1996	Emerald in ultramafic schists with other lithologies nearby; Brl mainly with Qtz- or carbonate-rich zones with accessory Tur & Py; Mingora most important deposit
Khaltaro, Gilgit, Pakistan *(see Fig. 30)*	pG (vn,p) (E)	Cen (Mioc)	Kazmi et al. 1989; Laurs 1995	Emerald and aquamarine-bearing Qtz-Ms & Ab-Tur veins in Btitized mafic amphibolite plus miarolitic pegmatites all with variably albitized Bt leucogranite
Eastern Asia — mainly Mesozoic				
Transbaikalia / eastern Mongolia	aG, pG, G, S, rV (rpl, gr, m, f)	Mes	Kovalenko and Yarmolyuk 1995; Kremenetsky et al. 2000a	Many deposit types associated with mainly Mesozoic granitoids ranging from peraluminous to peralkaline, volcanic-hosted Be in Mongolia
Ermakovskoe, Buryatia, Russia *(see Fig. 26)*	aG (f, rpl, sk)	MesE (225)	Zabolotnaya 1977; Reyf and Ishkov 1999; Kremenetsky et al. 2000a	Fl-rich Ph (deep) to Brt (shallow) (± Lph, Eud, Mil) replacement deposits (>10Kt contained BeO @ 1.3% BeO) distal to variably porphyritic Agt granite stock in carbonate-clastic-igneous host; Grt-Px-Ves skarn; F-bearing, Na-Kfs alteration in intrusion
Sherlova Gora, Transbaikalia, Russian	pG (gr,mp)	MesM (Jur)	Beus 1966; Troshin and Segalevich 1977; Kosukhin 1980; Sinkankas 1981; Dukhovskiy 2000	Brl-rich Fl±Toz-Qtz greisen (Ms-Sid) veins ± miarolitic pegmatites associated with Bt leucogranites with modest Sn-W mineralization; gem locality
Baerzhe, Inner Mongolia, China	aG (f)	MesL (127)	Wu et al. 1996	Zr-Nb-HREE-Be deposit (0.02-0.15% BeO) with Hin ±Gnt in Ab-Fl-Qtz-Zrn altered Agt-Rbk granite
Far East / eastern Yakutia, Russia (e.g., Voznesenkoye)	pG (sk, gr, rpl)	MesL (PalM)	Bredikhina 1990; Kupriyanova and Shpanov 1997; Kremenetsky et al. 2000a; Rodinov 2000; Trunilina et al. 2000	Mostly Cretaceous (PalM to CenM) Sn(-W) systems (Bt granites) with scattered occurrences of Be mineralization; Vozneskoye (PalM) has well developed Be-rich Fl-mica replacements in skarn (2-10Kt @ 0.06% BeO)

Southeast China	pG (gr, sk, p, m?), V?	Mes (M,L)	Zhang 1992; Lin et al. 1995a; Kamitani and Naito 1998	Many skarn, vein greisen W-Mo-Sn systems with minor to abundant Be mineralization related to Jurassic and Cretaceous Bt±Ms granites
Shizhuyuan, Hunan, China	pG (sk,gr)	Mes (M?)	Moh 1988; Mao et al. 1996a,b; Newberry 1998	Polymetallic W(-Mo-Bi-Be-F-Sn) system with multi-stage greisen and F-rich skarn mineralization; Grt-Px-Ves skarn to Fl-Mag-rich hydrous skarns with W-Sn-Mo-Bi and distal Sn-Be-Cu; distal Hlv-bearing Mn base-metal
Xianghualing, Hunan, China	pG (sk, gr)	Mes	Lin et al. 1995a, p. 238-242	Albitized and greisenized Yanshanian polyphase Bt granite with Brl; Grt-Ves skarn, magnesian Mag-rich-chondrodite-Tur skarn in dolostone; Ch / Taf Fl(-Phl-fluoborite- etc.) banded skarns
Wangfengshan, Guangdong, China	pG (sk, gr)	Mes	Lin et al. 1995a, p. 242-244	Yanshanian volcanic and intrusive rocks, variably porphyritic intrusive rocks; intense albitization and fracture-controlled greisenization with Brl and Hlv-Dn early and Brt-Euc late
Malipo County, Yunnan, China	pG (p/vn) (E)	MesL (Cret L)	Kamitani and Naito 1998; Zhang et al. 1999	V > Cr "emerald" associated with shallow pegmatitic veins with Tur and Sch in area of coeval granite-related W-Sn mineralization
Shixi, Zhejiang, China	rV (f, gr?)	Mes?	Lin et al. 1995a, p. 254-255	Hypabyssal dikes of porphyritic sodic rhyolite, albitized and sercitized, high Be (as Hlv) and Nb, Ta, Zr, F
SE Asia	pG / G (p, sk, gr)	Mes	Manning et al. 1983; Pollard et al. 1995; Suwimonprecha et al. 1995; Linnen 1998; Zaw 1998	Major Sn(-W) province related to chemically .evolved Bt granites with pegmatites, greisens, skarns etc.; Be mineralization is rare except in pegmatites; generally low F

Central & Western Asia — Paleozoic

Altai	pG (gr, p, LCT), aG / S (m, p, f)	PalML,Mes E	Kartashov et al. 1993; Kovalenko and Yarmolyuk 1995; Graupner et al. 1996; Kremenetsky 1996	Be-bearing LCT and NYF pegmatites, greisen and feldspathic metasomatic rocks associated with peraluminous to peralkaline granites (± syenities)
Kyzyltau, Mongolia	pG (vn, gr)	PalL (Carb 340)	Graupner et al. 1996	Shallow W(-Y-Be-Mo) Qtz (-Fl-Wlf-Brl) vein system (+ albitization, greisenization) with multiple phase intrusive complex of Bt granites
Khaldzan-Buregtey, Altai, Mongolia	aS, aG (f, m)	PalM (Dev 385)	Kovalenko et al. 1995b; Kovalenko and Yarmolyuk 1995	Rare metal peralkaline granite phases in larger complex has Zr-Nb(-REE-Y-Be-Sn, Rb±Li etc) mineralization with Gnt & Bav; magmatic enrichment demonstrated by melt inclusions; post-magmatic metasomatism with formation of Pll, silicification, fluoritization, hematitization etc. but rare metals only in enriched protoliths

Verknee Espee, Kazakhstan	aG (f)	PalL (Perm)	Belov and Ermolov 1996; Kremenetsky et al. 2000a	Proximal Nb-Ta-Zr-REE mineralization with Bt-Ab-Kfs-Rbk veins associated with altered Rbk-Agt granites; Ph-Hlv-Gad-Bar-Mil-Brt occur in distal Be orebody
Central Kazakhstan	pG (gr,sk)	PalL (Carb L)	Beus 1966, p. 293-298; Zasedatelev 1973; Burshtein 1996; Malchenko and Ermolov 1996; Serykh 1996; Dukhovskiy 2000	Be-bearing greisen and skarn deposits associated with Hercynian Be-enriched leucogranites; Mo(-W) stockwork East Qonyrat (Kounrad), Mo-W-Be-F skarn and replacement Qatpar (sk), W(-Mo) greisen Aqshantau, Qaraoba (Karaoba); unnamed Mag-mica-Fl-Ch
Aqshatau, Kazakhstan *(see Fig. 17)*	pG (gr, vn)	PalL (Carb L)	Beskin et al. 1996; Matveev 1997; Ermolov 2000; Kremenetsky et al. 2000a	Large W(-Mo-Bi-Be) greisen system (16Kt contained BeO @ 0.03-0.07% BeO; Brl distal but early); complex zoned Qtz-Toz-Ms-Fl greisen in Bt granite and clastic rocks; late sulfides
Ural Mtns, Russia & Kokshetau, Kazakh. Boevskoe, Orlinogorski, Elenovskii	pG, G (gr, sk, p) (E)	PalM (Dev)	Zabolotnaya 1977; Sinkankas 1981; Odintsova 1996; Grigor'yev 1997; Kremenetsky et al. 2000a; Letnikov 2000	Mainly Devonian magmatism: Urals — Ms-Bt granites with Be-bearing pegmatites, greisens & mica-Fl metasomatites (Mo-W type) (Boevskoe: >10Kt Be @ 0.12% BeO); Kokshetau — Bt granites with Brl-Ph-Hlv Sn greisens etc. (up to 0.2% BeO) and alkaline granites with REE-Ta-Nb-Sn-Be
Reft River (Marinsky, Tro:tsky, etc.), Urals, Russia *(see Fig. 29)*	pG (p, vn, gr) (E)	PalM (Dev)	Fersman 1929; Zabolotnaya 1977; Sinkankas 1981; Kazmi and Snee 1989a	Two-mica granites in mafic-ultramafic metamorphic host rocks with extensive Brl(-Ch-Ph-Brt-Bav)-bearing Bt-Fl-rich desilicated veins & pegmatites; major emerald occurrence

Southern Asia — Proterozoic & Archean

Sri Lanka (placers & pegmatites) -- Gondwanan granulites	M, (p), P	Arch / Prot	Dahanayake 1980; Rupasinghe et al. 1984; Rupasinghe and Dissanayake 1985; Menon et al. 1994; Dissanayake and Rupasinghe 1995; Menon and Santosh 1995; Grew, this volume	Beryllium-bearing pegmatites and granulite-grade metamorphic assemblages (Brl, Taf, Ch); pegmatites may be syn- or post-metamorphic; major gem placers (1-13 ppm Be in placers is similar to bedrock); Pan-African age in adjacent southern India

AUSTRALIA

Tasman Orogen — Paleozoic

Tasman Orogen Tasmania, New South Wales, Queensland	pG (sk, gr, repl, p)	PalM / PalL	Taylor 1978; Kwak and Jackson 1986; Blevin and Chappell 1995; Walshe et al. 1995	Five regions of Sn-W mineralization with Bt granites has sporadically Be-bearing greisen, skarn and replacement mineralization; some Brl-bearing pegmatites

Tasmania, Australia — Mt. Lindsay, Mt. Bischoff, Moina, etc.	pG (sk, gr, rpl)	PalM / L (Dev-Carb)	Kwak and Askins 1981; Solomon 1981; Kwak 1983; Kwak and Jackson 1986; Bajwah et al. 1995; Halley and Walshe 1995	Be minerals (Hlv, Dn, Ph) in complex Fl-rich skarns and sulfide replacement bodies with Sn(-W-Zn) mineralization (up to 0.5% BeO); distal to greisenized Bt granites; boron common; greisen overprint converts Dn to Ph / Brt
Mole Granite (Emmaville, Torrington), NSW, Australia	pG (p, vn, gr) ("E")	PalL	Taylor 1978; Sun and Eadington 1987; Kazmi and Snee 1989a; Plimer and Kleeman 1991	Brl widespread with early pegmatites and Sn(-W-Mo) veins & greisens with Mole Bt granite; deep green Brl from deposits in host rocks have both V and Cu > Cr
Western Australia — Early Proterozoic & Archean				
Brockman, Western Australia	aVr m? rpl?	ProtE (1870)	Ramsden et al. 1993; Taylor et al. 1995a,b	Hydrothermally altered Fl-bearing alkali rhyolite with Nb-Zr-Ta-Be enrichment (4.3Mt @ 0.08% BeO); predates orogenesis
Western Australia emeralds (Menzies, Poona, Pilgangoora, Wodgina, Curlew)	pG (p, vn?, LCT) (E)	ArchL	Sinkankas 1981; Kazmi and Snee 1989a; Partington et al. 1995	Emeralds form in margin of variably desilicated simple and complex pegmatites in greenstones and ultramafic schists; other major LCT pegmatites in region (Greenbushes)
AFRICA / MIDDLE EAST				
Africa - Nigeria-Niger				
Nigerian-Niger Younger Granites	aG, S, G (gr, vn, p) (E)	MesM (Jur), ProtL,Pal	Bowden 1985; Bowden et al. 1987; Matheis 1987; Imeokparia 1988; Kuester 1990; Perez et al. 1990; Schwarz et al. 1996	Alkaline granites, syenites and volcanic rocks with Bt granites (Jurassic in Nigeria older to north); Brl-, Gnt-, and/or Ph-bearing greisens, veins & albitites with Sn, Zn, Ta-Nb deposits; Pan African Brl-bearing pegmatites in Nigeria
Nigerian ring complexes — Saiya-Shokobo, Ririwai (Liruei)	G (aG, S) (f, gr, vn, p) (E)	MesM (Jur)	Bowden and Jones 1978; Kinnaird 1985; Bowden et al. 1987; Imeokparia 1988; Abaa 1991; Schwarz et al. 1996	Brl (including emerald, aquamarine), Gnt, Brt, Ph associated with Sn-Zn-Ta greisens (Li micas), veins and miarolitic pegmatites mainly with Bt granites but also with peralkaline intrusive rocks
Taghouaji, Air Massif, Niger	aS, G (vn, gr)	PalE (430)	Bowden et al. 1987; Perez et al. 1990	Gnt-Brt-Hem with Zn-bearing silicates and Cu-Pb-Zn sulfides in Qtz veins associated with the only alkaline complex with Bt granite in Niger; Brl present in granite
Northern Africa - Arabia Pan-African				
Arabian-Nubian Shield	pG (gr, p) (E) / aG (m?, p) / M? (E)	ProtL (600-550)	Jackson and Ramsay 1986; du Bray et al. 1988; Binda et al. 1993; Grundmann and Morteani 1993; Mohamed 1993; Abdalla and Mohamed 1999	Late Proterozoic - early Paleozoic (Pan African) anorogenic granite province with multiple types of lithophile element suites commonly with Be enrichments; metamorphic emerald

Sekeit (Sikait) and nearby areas, Egypt	pG (gr, vn) (E), M? (E)	ProtL / PalE	Soliman 1982, 1984, 1986; Binda et al. 1993; Grundmann and Morteani 1993; Hassanen and Harraz 1996; Abdalla and Mohamed 1999	Brl (± emerald) occurs in greisens and Qtz veins associated with Pan African peraluminous variably Sn-Ta-W-Mo-enriched granites; emerald in veins etc. in mafic schists with control by regional shear zones; alkaline rare-metal granites & metasomatism also common in region
Baid al Jamalah, Saudi Arabia	pG (gr, vn)	ProtL (569)	Jackson and Ramsay 1986; du Bray et al. 1988; Kamilli et al. 1993	Proximal Brl in greisen W(-Mo-Sn) deposit associated with Be-enriched Bt granite; other systems (e.g., Silsilah) commonly lack Brl
Tuareg massif, Algeria	pG (p/vn?)	ProtL (680-610)	Anonymous 1987; Cottin et al. 1990; Goodwin 1996	Numerous Be occurrences in area of Pan-African magmatism superimposed on older basement
Southeastern Africa - Archean and Proterozoic				
SE Africa / Madagascar	G (p, vn), M (±E), P, carbonatite	Arch / Prot	Sinkankas 1981; Kazmi and Snee 1989a; Grew (this volume)	Abundant Brl-bearing pegmatites (LCT and other), also some vein systems & placers; some emeralds; granulite-related pegmatites
Gravelotte (Leysdorp), Transvaal, South Africa	pG / M? (p, vn) (E)	Arch	Sinkankas 1981, p. 492; Robb and Robb 1986; Grundmann and Morteani 1989; Kazmi and Snee 1989a; Nwe and Morteani 1993	Emeralds in Qtz veins and Bt-Tlc schist; desilicated pegmatites and Qtz veins in Act-Chl-Bt schists; evidence suggests metamorphic redistribution may be key for Ph and later emerald
Zimbabwe — Filabusi, Fort Victoria, Mweza	pG? (p,vn, gr) (E)	Arch	Sinkankas 1981; Kazmi and Snee 1989a	Emerald (± Ch) in ultramafic host adjacent to pegmatites, Qtz veins, greisens; many Be pegmatites including LCTs (e.g., Bikita)
Kafubu (Kitwe), Zambia	G? M? (vn, p?)	ProtL / PalE	Graziani et al. 1983; Sliwa and Nguluwe 1984; Kazmi and Snee 1989a; Kamona 1994; Mumba and Bartot 1998,M. Hitzman (pers. com.)	Brl(emerald) ± Ch in mafic schists in metamorphic complex of Lufilian arc near Copper Belt, with Qtz-Tur veins; spatially associated pegmatites but occurrences hydrothermal or metamorphic
Muambe, Mozambique	carbonatite (rpl)	Prot	Cilek 1989; Woolley 2001	Up to 1% BeO+Y-La enrichment in proximal Fl replacement in K_2O-rich fenite aureole around iron-rich calcic carbonatite

* Abbreviations for deposit types:

(1) genetic association: G = granite (Bt, Bt-Hbl or Fa), pG = peraluminous granite (Bt±Ms, Crd), aG = peralkaline granite (sodic pyroxene or amphibole present), S = syenitic rocks (commonly silica undersaturated and/or peralkaline), M = metamorphic origin (non-magmatic), B = basin-related hydrothermal, P = placer.

(2) mineralization styles: m = magmatic, LCT = Li-Cs-Ta-type enrichment, p = pegmatitic (miarolitic or otherwise), f = feldspathic replacement (albitization, K-feldspathization, can be peralkaline), gr = greisen (Toz, Ms, Li-Fe micas all included), vn = veins (lacking prominent envelopes), sk = skarn (metasomatic calc-silicates), rpl = replacement (typically in carbonates, F-rich, lacks calc-silicates)

(3) emerald occurrences are indicated by "(E)."

APPENDIX B: SELECTED STUDIES RELEVANT TO BERYLLIUM MINERAL STABILITIES

(see also Franz and Morteani, pp. 561ff, this volume)

Material	Synthesis and Reversal Experiments	Thermodynamic Data	Equation of State	Theoretical & Natural
Fluids				
Melt-mineral	Ganguli et al. 1975; Cemič et al. 1986; London et al. 1989; Icenhower and London 1995; Evensen et al. 1999; Evensen and London 2002; London and Evensen (this volume)			Bea et al. 1994
Melt-aqueous	London et al. 1988			
Aqueous species	Beus et al. 1963; Soboleva et al. 1977, 1984b; Samchuk and Mitskevich 1980; Barton 1986; Renders and Anderson 1987; Koz'menko et al. 1988; Prasad and Ghosh 1988; Clegg and Brimblecombe 1989	Samchuk and Mitskevich 1980; Barton 1986; Renders and Anderson 1987; Wood 1992		Wood 1992
BASH				
Behoite		Kostryukov et al. 1977; Barton 1986		
Bertrandite	Bukin 1968; Klyakhin et al. 1981; Hsu 1983; Lebedev and Ragozina 1984; Barton 1986	Kiseleva et al. 1985, 1986; Barton 1986; Hemingway et al. 1986	Hazen and Au 1986	Kosals et al. 1974; Burt 1978; Kupriyanova 1982; Wood 1992; Odintsova 1993
Beryl	Syromyatnikov et al. 1972; Franz and Morteani 1981; Klyakhin et al. 1981; Hsu 1983; Aines and Rossman 1984; Franz and Morteani 1984; Polupanova et al. 1985; Barton 1986; Cemič et al. 1986; Renders and Anderson 1987; Manier-Glavinaz et al. 1988,1989a; Wang et al. 1992; Evensen et al. 1999	Barton 1986; Hemingway et al. 1986; Kiseleva et al. 1986; Renders and Anderson 1987; Gurevich et al. 1989; Pilati et al. 1997	Morosin 1972; Deganello 1974; Schlenker et al. 1977; Hazen et al. 1986; Haussuehl 1993	Burt 1975, 1978; Kupriyanova 1982; Odintsova 1993; Kupriyanova and Shpanov 1997; Markl and Schumacher 1997; Evensen et al. 1999
Bromellite	Barton 1986	Barton 1986	Hazen and Finger 1987; Pilati et al. 1993	
Chrysoberyl	Franz and Morteani 1981; Hsu 1983; Ospanov 1983; Barton 1986; Cemič et al. 1986; Hölscher et al. 1986; Wang et al. 1992	Kiseleva et al. 1985; Barton 1986; Hemingway et al. 1986; Kiseleva et al. 1986; Hofmeister et al. 1987	Au and Hazen 1987; Hazen and Finger 1987	Burt 1978; Kupriyanova 1982; Odintsova 1993; Kupriyanova and Shpanov 1997
Euclase	Franz and Morteani 1981; Hsu 1983; Ospanov 1983; Barton 1986	Franz and Morteani 1981; Kiseleva et al. 1985; Barton 1986; Hemingway et al. 1986; Kiseleva et al. 1986	Hazen et al. 1986	Burt 1978
Phenakite	Bukin 1968; Syromyatnikov et al. 1972; Franz and Morteani 1981; Ospanov 1983; Lebedev and Ragozina 1984; Soboleva et al.	Matveev and Zhuravlev 1982; Topor and Mel'chakova 1982; Kiseleva and Shuriga	Hazen and Au 1986; Hazen and Finger 1987; Yeganeh-Haeri and Weidner	Kosals et al. 1974; Burt 1978; Odintsova 1993; Pilati et al. 1998

Appendix B, continued

Phenakite, cont.	1984a; Barton 1986; Wang et al. 1992	1983; Barton 1986; Hemingway et al. 1986	1989; Pilati et al. 1998	
Be-Mg-Al-Si	Hölscher et al. 1986		Schmetzer 1981	Grew 1981; Grew, this volume
Helvite group				
Danalite	Fursenko 1982	Ospanov 1983		Burt 1980, 1988
Genthelvite	Fursenko 1982	Ospanov 1983; Mel'chakova et al. 1991		Burt 1980, 1988
Helvite	Klyakhin et al. 1981; Fursenko 1982		Werner and Plech 1995 (also tugtupite)	Burt 1980, 1988
Others				
Chkalovite	Ganguli et al. 1975	Kiseleva et al. 1984a,b	Henderson and Taylor 1989	Markl 2001
Gadolinite	Ito and Hafner 1974		Demartin et al. 1993	
Phosphates			Klaska and Jarchow 1973; Henderson and Taylor 1984	Burt 1975

REFERENCES

Abaa SI (1991) Hydrothermal fluids responsible for the formation of precious minerals in the Nigerian Younger Granite Province. Mineral Dep 26:34-39

Abdalla HM, Mohamed FH (1999) Mineralogical and geochemical investigation of emerald and beryl mineralisation, Pan-African Belt of Egypt: genetic and exploration aspects. J Afr Earth Sci 28:581-598

Adams JW (1953) Beryllium deposits of the Mount Antero region, Chaffee County, Colorado. U S Geol Surv Bull 982-D:95-118

Aines RD, Rossman GR (1984) The high temperature behavior of water and carbon dioxide in cordierite and beryl. Am Mineral 69:319-327

Akande SO, Kinnaird J (1993) Characterization and origin of ore-forming fluids in the Nigeria mineral belts. Proc 8th Quadrenn IAGOD Symp, p 199-218

Anderson RG (1988) An overview of some Mesozoic and Tertiary plutonic suites and their associated mineralization in the northern Canadian Cordillera. *In* Taylor RP, Strong DF (eds) Recent advances in the geology of granite-related mineral deposits, Can Inst Min Metall Spec Paper 39:96-113

Anonymous (1987) Mineral deposit map of the Arab World. Arab Organization for Mineral Resources, Rabat, Morocco

Apollonov VN (1967) Morphologic features of chrysoberyl from Sargardon. Zapiski Vsesoyuz Mineral Obshch 96:329-332 (in Russian)

Arif M, Fallick AE, Moon CJ (1996) The genesis of emeralds and their host rocks from Swat, northwestern Pakistan: an investigative medium for stable isotopes. Mineral Dep 31:255-268

Au AY, Hazen RM (1987) High-pressure crystal chemistry of chrysoberyl, Al_2BeO_4: insights on the origin of olivine elastic anisotropy. Phys Chem Mineral 14:13-20

Aurisicchio C, Fioravanti G, Grubessi O, Zanazzi PF (1988) Reappraisal of the crystal chemistry of beryl. Am Mineral 73:826-837

Bailey JC, Larsen LM, Sørensen H (1981) The Ilímaussaq Intrusion, South Greenland: a progress report on geology, mineralogy, geochemistry and economic geology. Rapport Grønlands Geologiske Undersøgelse 103:1-130

Bajwah ZU, White AJR, Kwak TAP, Price RC (1995) The Renison Granite, northwestern Tasmania: a petrological, geochemical and fluid inclusion study of hydrothermal alteration. Econ Geol 90: 1663-1675

Baker JM, Keith JD, Christiansen EH, Griffen DT, Tingey DG, Dorais MJ (1998) Genesis of red beryl in topaz rhyolites, Thomas Range, Utah. Geol Soc Am Abstr Progr 30:370-371

Banks DA, Giuliani G, Yardley BWD, Cheilletz A (2000) Emerald mineralisation in Colombia: fluid chemistry and the role of brine mixing. Mineral Dep 35:699-713

Barton MD (1982a) Some aspects of the geology and mineralogy of the fluorine-rich skarn at McCullough Butte, Eureka Co., Nevada. Carnegie Institution of Washington, Year Book 81:324-328

Barton MD (1982b) The thermodynamic properties of topaz solid solutions and some petrologic applications. Am Mineral 67:956-974

Barton MD (1986) Phase equilibria and thermodynamic properties of minerals in the $BeO-Al_2O_3-SiO_2-H_2O$ (BASH) system, with petrologic applications. Am Mineral 71:277-300

Barton MD (1987) Lithophile-element mineralization associated with Late Cretaceous two-mica granites in the Great Basin, with suppl. data 87-16. Geology 15:337-340

Barton MD (1990) Cretaceous magmatism, mineralization and metamorphism in the east-central Great Basin. *In* Anderson JL (ed) The Nature and Origin of Cordilleran Magmatism. Geol Soc Am Mem 174:283-302

Barton MD (1996) Granitic magmatism and metallogeny of southwestern North America. Trans Roy Soc Edinburgh Earth Sci 87:261-280

Barton MD (2000) Overview of the lithophile-element-bearing magmatic-hydrothermal system at Birch Creek, White Mountains, California. *In* Dilles JH, Barton MD, Johnson DA, Proffett JM, Einaudi MT (eds) Contrasting Styles of Intrusion Associated Hydrothermal Systems: Society of Economic Geologists Guide Book Ser 32:9-26

Barton MD, Trim HE (1991) Late Cretaceous two-mica granites and lithophile-element mineralization in the Great Basin. *In* Raines GL, Lisle RE, Schafer RW, Wilkinson WH (eds) Geology and ore deposits of the Great Basin, p 529-538

Barton MD, Ruiz J, Ito E, Jones L (1982) Tracer studies of the fluorine-rich skarn at McCullough Butte, Eureka Co., Nevada. Carnegie Institution of Washington, Year Book 81:328-331

Barton PB, Jr., Skinner BJ (1979) Sulfide mineral stabilities. *In* Barnes HL (ed) Geochemistry of hydrothermal ore deposits. John Wiley & Sons. New York, p 278-403

Baumann L (1994) Ore parageneses of the Erzgebirge—history, results and problems. *In* von Gehlen K, Klemm DD (eds) Mineral deposits of the Erzgebirge/ Kru_né Hory (Germany/Czech Republic): reviews and results of recent investigations, 31. Gebrueder Borntraeger, Stuttgart, p 25-46

Baxter JW, Bradbury JC (1980) Bertrandite at Hicks Dome, Hardin County, Illinois. Trans Illinois State Acad Sci 73:1-13

Bazarov LS, Kosals YA, Senina VA (1972) Temperature conditions of the formation of zinnwaldite-amazonite-albite apogranites. Dokl Akad Nauk SSSR 203:685-688 [in Russian]

Bea F, Pereira MD, Stroh A (1994) Mineral/ leucosome trace-element partitioning in a peraluminous migmatite (a laser ablation-ICP-MS study). Chem Geol 117:291-312

Belov VA, Ermolov PV (1996) The Verkhnee Espe rare metal deposit in east Kazakhstan. *In* Shatov V, Seltmann R, Kremenetsky A, Lehmann B, Popov V, Ermolov P (eds) Granite-Related Ore Deposits of Central Kazakhstan and Adjacent Areas. Glagol, St. Petersburg, p 219-228

Beskin SM, Larin VN, Marin YB (1996) The greisen Mo-W deposit of Aqshatau, central Kazakhstan. *In* Shatov V, Seltmann R, Kremenetsky A, Lehmann B, Popov V, Ermolov P (eds) Granite-Related Ore Deposits of Central Kazakhstan and Adjacent Areas. Glagol, St. Petersburg, p 145-154

Bettencourt JS, Payolla BL, Leite WB, Jr., Tosdal RM, Spiro B (1999) Mesoproterozoic rapakivi granites of Rondonia tin province, SW Amazon Craton, Brazil: Nd, Sr, O, Pb isotopes and metallogenic implications. Geol Soc Am Abstr Progr 31(7):A-205

Beus AA (1966) Geochemistry of beryllium and genetic types of beryllium deposits. W. H. Freeman & Co., San Francisco–London

Beus AA (1979) Sodium: a geochemical indicator of emerald mineralization in the Cordillera Oriental, Colombia. J Geochem Explor 11:195-208

Beus AA, Sobolev BP, Dikov YP (1963) Geochemistry of beryllium in high temperature post-magmatic mineralization. Geokhimiya 3:316-323 (in Russian)

Bilal E, Fonteilles M (1988) A comparison of the development conditions of helvite, phenakite, and beryl in granitic surroundings: case of the Sucuri Massif (Brazil). C R Acad Sci Paris Série II 307:273-276 (in French)

Binda PL, Omenetto P, Warden AJ (1993) Mineral deposits and occurrences in the Precambrian of northeast Africa and Arabia: A review. Relaz e Monogr 113:429-516

Birkett TC, Miller RR, Salvi S, Williams-Jones AE (1991) The role of hydrothermal processes in the granite-hosted Zr, Y, REE deposit at Strange Lake, Quebec/ Labrador: evidence from fluid inclusions: Discussion and reply. Geochim Cosmochim Acta 55:3443-3449

Blevin PL, Chappell BW (1995) Chemistry, origin, and evolution of mineralized granites in the Lachlan fold belt, Australia: the metallogeny of I- and S-type granites. Econ Geol 90:1604-1619

Bodenlos AJ (1954) Magnesite deposits in the Serra das Eguas, Brumado, Bahia, Brazil. U S Geol Surv Bull 975-C:87-170

Boily M, Williams-Jones AE (1994) The role of magmatic and hydrothermal processes in the chemical evolution of the Strange Lake plutonic complex, Quebec-Labrador. Contrib Mineral Petrol 118:33-47

Borshchevskii YA, Borisova SL, Medvedovskaya NI, Amosova KB, Stepanova NA (1987) Isotope characteristics of minerals and rocks of the Khibiny-Lovozero Complex and some aspects of their origin. Zapiski Vsesoyuz Mineral Obshch 116:532-540 (in Russian)

Bowden P (1985) The geochemistry and mineralization of alkaline ring complexes in Africa (a review). J Afr Earth Sci 3:17-37

Bowden P, Jones JA (1978) Mineralization in the Younger Granite province of northern Nigeria. *In* Stemprok M, Burnol L, Tischendorf G (eds) Metallization associated with acid magmatism, 3. Geol Surv Prague, Prague, p 179-190

Bowden P, Black R, Martin RF, Ike EC, Kinnaird JA, Batchelor RA (1987) Niger-Nigerian alkaline ring complexes: a classic example of African Phanerozoic anorogenic mid-plate magmatism. Geol Soc Spec Pub 30:357-379

Branquet Y, Laumonier B, Cheilletz A, Giuliani G (1999) Emeralds in the Eastern Cordillera of Colombia: two tectonic settings for one mineralization. Geology 27:597-600

Bredikhina SA (1990) Physicochemical features of fluorite formation from deposits of the Voznesenskii ore field (Primor'e). Sov Geol Geophys 31:74-81

Brinck JW, Hofmann A (1964) The distribution of beryllium in the Oslo region, Norway: a geochemical stream sediment study. Econ Geol 59:79-96

Brownell GM (1959) A beryllium detector for field exploration. Econ Geol 54:1103-1114

Brugger J, Gieré R (2000) Origin and distribution of some trace elements in metamorphosed Fe-Mn deposits, Val Ferrera, eastern Swiss Alps. Can Mineral 38:1075-1101

Bukin GV (1968) Crystallization conditions of phenakite-bertrandite-quartz association (experimental data). Dokl Akad Nauk SSSR 176:121-123

Bulnayev KB (1996) Genesis of fluorite, bertrandite and phenakite deposits. Geol Rudn Mestorozhd 38:147-156 (in Russian)

Burnol L (1978) Different types of leucogranites and classification of the types of mineralization associated with acid magmatism in the north-western part of the French Massif Central. *In* Stemprok M, Burnol L, Tischendorf G (eds) Metallization associated with acid magmatism, 3. Geol Surv Prague, Prague, p 191-204

Burshtein EF (1996) Genetic types of granite-related mineral deposits and regular patterns of their distribution in central Kazakhstan. *In* Shatov V, Seltmann R, Kremenetsky A, Lehmann B, Popov V, Ermolov P (eds) Granite-Related Ore Deposits of Central Kazakhstan and Adjacent Areas. Glagol, St. Petersburg, p 83-92

Burt DM (1975) Beryllium mineral stabilities in the model system CaO-BeO-SiO_2-P_2O_5-F_2O_{-1} and the breakdown of beryl. Econ Geol 70:1279-1291

Burt DM (1978) Multisystems analysis of beryllium mineral stabilities: the system BeO-Al_2O_3-SiO_2-H_2O. Am Mineral 63:664-676

Burt DM (1980) The stability of danalite, $Fe_4Be_3(SiO_4)_3S$. Am Mineral 65:355-360

Burt DM (1981) Acidity-salinity diagrams: application to greisen and porphyry deposits. Econ Geol 76:832-843

Burt DM (1988) Stability of genthelvite, $Zn_4(BeSiO_4)_3S$: an exercise in chalcophilicity using exchange operators. Am Mineral 73:1384-1394

Burt DM, Sheridan MF (1986) Mineral deposits related to topaz rhyolites in the Southwest. Arizona Geol Soc Digest 16:170-178

Burt DM, Sheridan MF (1988) Mineralization associated with topaz rhyolites and related rocks in Mexico. *In* Taylor RP, Strong DF (eds) Recent advances in the geology of granite-related mineral deposits, Can Inst Min Metall Spec Paper 39:303-306

Burt DM, Sheridan MF, Bikun JV, Christiansen EH (1982) Topaz rhyolites: distribution, origin, and significance for exploration. Econ Geol 77:1818-1836

Calligaro T, Dran J-C, Poirot J-P, Querre G, Salomon J, Zwan JC (2000) PIXE/PIGE characterization of emeralds using an external micro-beam. Nucl Instr Meth Phys Res B 161-163:769-774

Carmichael ISE, Turner FJ, Verhoogen J (1974) Igneous Petrology. McGraw-Hill, New York

Casadevall T, Ohmoto H (1977) Sunnyside Mine, Eureka mining district, San Juan County, Colorado: geochemistry of gold and base metal ore deposition in a volcanic environment. Econ Geol 72: 1285-1320

Cassedanne JP, Cassedanne JO (1978) Famous mineral localities: the Brumado District, Bahia, Brazil. Mineral Rec 9:196-205

Cemič L, Franz G, Langer K (1986) Experimental determination of melting relationships of beryl in the system BeO-Al_2O_3-SiO_2-H_2O between 10 and 25 kbar. Mineral Mag 50:55-61

Černý P (1963) Epididymite and milarite, alteration products of beryl from Věžná, Czechoslovakia. Mineral Mag 33:450-457

Černý P (1991a) Rare-element granite pegmatites. Part I. Anatomy and internal evolution of pegmatitic deposits. Geos Can 18:49-67

Černý P (1991b) Rare-element granite pegmatites. Part II. Regional to global environments and petrogenesis. Geos Can 18:68-81

Černý P, Meintzer RE (1988) Fertile granites in the Archean and Proterozoic fields of rare-element pegmatites: crustal environment, geochemistry and petrogenic relationships. *In* Taylor RP, Strong DF (eds) Recent advances in the geology of granite-related mineral deposits. Can Inst Min Metall Spec Paper 39:170-207

Charoy B (1999) Beryllium speciation in evolved granitic magmas: phosphates versus silicates. Eur J Mineral 11:135-148

Charoy B, Pollard PJ (1989) Albite-rich, silica-depleted metasomatic rocks at Emuford, Northeast Queensland: mineralogical, geochemical, and fluid inclusion constraints on hydrothermal evolution and tin mineralization. Econ Geol 84:1850-1874

Cheilletz A (1998) La géologie des gisements d'émeraudes. Revue de Gemmologie A.F.G. 134-135:33-42

Cheilletz A, Giuliani G (1996) The genesis of Colombian emeralds: a restatement. Mineral Dep 31:359-364

Cheilletz A, Feraud G, Giuliani G, Ruffet G (1993) Emerald dating through ^{40}Ar/ ^{39}Ar step-heating and laser spot analysis of syngenetic phlogopite. Earth Planet Sci Lett 120:473-485

Cheilletz A, Feraud G, Giuliani G, Rodriguez CT (1994) Time-pressure and temperature constraints on the formation of Colombian emeralds: an ^{40}Ar/^{39}Ar laser microprobe and fluid inclusion study. Econ Geol 89:361-380

Christiansen EH, Burt DM, Sheridan MF, Wilson RT (1983) The petrogenesis of topaz rhyolites from the Western United States. Contrib Mineral Petrol 83:16-30

Christiansen EH, Bikun JV, Sheridan MF, Burt DM (1984) Geochemical evolution of topaz rhyolites from the Thomas Range and Spor Mountain, Utah. Am Mineral 69:223-236

Christiansen EH, Stuckless JS, Funkhouser MMJ, Howell KH (1988) Petrogenesis of rare-metal granites from depleted crustal sources: an example from the Cenozoic of western Utah, U.S.A. *In* Taylor RP, Strong DF (eds) Recent advances in the geology of granite-related mineral deposits. Can Inst Min Metall Spec Paper 39:307-321

Christiansen EH, Keith JD, Thompson TJ (1997) Origin of gem red beryl in Utah's Wah Wah Mountains. Min Engin 49:37-41

Cilek VG (1989) Industrial Minerals of Mozambique. Geological Survey, Prague

Clark AH (1970) Early beryllium-bearing veins, South Crofty Mine, Cornwall. Inst Min Metall Trans 79:B173-B175

Clegg SL, Brimblecombe P (1989) Estimated mean activity coefficients of aqueous beryllium chloride and properties of solution mixtures containing the beryllium(2+) ion. J Chem Soc, Faraday Trans 1 85: 157-162

Coats RR, Barnett PR, Conklin NM (1962) Distribution of beryllium in unaltered silicic volcanic rocks of the western conterminous United States. Econ Geol 57:963-968

Cohenour RE (1963) Beryllium and associated mineralization in the Sheeprock Mountains. Guide Geol UT 17:8-13

Coleman RG (1966) New Zealand serpentinites and associated metasomatic rocks. New Zealand Geol Surv Bull 76

Congdon RD, Nash WP (1991) Eruptive pegmatite magma: rhyolite of the Honeycomb Hills, Utah. Am Mineral 76:1261-1278

Cook RB (1998) FM-TGMS-MSA symposium on fluorite and related Alpine cleft minerals. Mineral Rec 29:4

Cottin JY, Guiraud M, Leterrier J, Lorand JP (1990) New geodynamic constraints on late Pan-African Orogeny, Laouni, south of central Ahaggar: stratified intrusion-granites-metamorphism relations. Publication Occasionnelle–Centre International Pour la Formation et les Echanges Géologiques 20: 255 (in French)

Cuney M, Marignac C, Weisbrod A (1992) The Beauvoir topaz-lepidolite albite granite (Massif Central, France): the disseminated magmatic Sn-Li-Ta-Nb-Be mineralization. Econ Geol 87:1766-1794

Cunningham LD (2000) Beryllium. Minerals Yearbook. U S Geol Surv, 12.1-12.5 [online]

Currie KL (1989) Geology and Composition of the Mont Saint-Hilaire. Geol Surv Canada Open-File Report 2031

Currie KL, Eby GN, Gittins J (1986) The petrology of the Mont Saint-Hilaire Complex, southern Quebec: an alkaline gabbro-peralkaline syenite association. Lithos 19:65-81

Currie KL, Whalen JB, Davis WJ, Longstaffe FJ, Cousens BL (1998) Geochemical evolution of peraluminous plutons in southern Nova Scotia, Canada: a pegmatite-poor suite. Lithos 44:117-140

Dahanayake K (1980) Modes of occurrence and provenance of gemstones of Sri Lanka. Mineral Dep 15:81-86

Dale VR (1961) Tungsten deposits of Gila, Yavapai, and Mohave Counties, Arizona. U S Bur Mines, Inform Circ 8078:104
Davidson A (1978) The Blachford Lake Intrusive Suite: An Aphebian alkaline plutonic complex in the Slave Province, Northwest Territories. Canadian Geol Surv Paper 78-1A:119-127
Davidson A (1982) Petrochemistry of the Blachford Lake Complex near Yellowknife, Northwest Territories. Geol Surv Canada Paper 81-23:71-79
Deer WA, Howie RA, Zussman J (1978) Beryl. Disilicates and ring silicates in Rock-Forming Minerals, 2nd edn, 1b. John Wiley & Sons, New York, N.Y., United States, 372-409
Deganello S (1974) Atomic vibrations and thermal expansion of some silicates at high temperatures. Z Kristallogr 139:297-316
Della Ventura G, Rossi P, Parodi GC, Mottana A, Raudsepp M, Prencipe M (2000) Stoppaniite $(Fe,Al,Mg)_4(Be_6Si_{12}O_{36})*(H_2O)_2(Na,\square)$ a new mineral of the beryl group from Latium (Italy). Eur J Mineral 12:121-127
Demartin F, Pilati T, Diella V, Gentile P, Gramaccioli CM (1993) A crystal-chemical investigation of Alpine gadolinite. Can Mineral 31:127-136
de Medeiros Delgado I, Pedreiha AJ, Thorman CH (1994) Geology and mineral resources of Brazil: a review. Intl Geol Rev 36:503-544
Dereppe JM, Moreaux C, Chauvaux B, Schwarz D (2000) Classification of emeralds by artificial neural networks. J Gemmol 27:93-105
Desborough GA, Ludington S, Sharp WN (1980) Redskin Granite: a rare-metal-rich Precambrian pluton, Colorado, USA. Mineral Mag 43:959-966
Dick LA, Hodgson CJ (1982) The MacTung W-Cu(Zn) contact metasomatic and related deposits of the northeastern Canadian Cordillera. Econ Geol 77:845-867
Dietrich A, Lehmann B, Wallianos A (2000) Bulk rock and melt inclusion geochemistry of Bolivian tin porphyry systems. Econ Geol 95:313-326
Dissanayake CB, Rupasinghe MS (1995) Classification of gem deposits of Sri Lanka. Geologie en Mijnbouw 74:79-88
Dissanayake CB, Chandrajith R, Tobschall HJ (2000) The geology, mineralogy and rare element geochemistry of the gem deposits of Sri Lanka. Bull Geol Soc Finland 72:5-20
Dobson DC (1982) Geology and alteration of the Lost River tin-tungsten-fluorine deposit, Alaska. Econ Geol 77:1033-1052
du Bray EA, Elliott JE, Stuckless JS (1988) Proterozoic peraluminous granites and associated Sn-W deposits, Kingdom of Saudi Arabia. *In* Taylor RP, Strong DF (eds) Recent advances in the geology of granite-related mineral deposits, Can Inst Min Metall Spec Paper 39:142-156
Dukhovskiy AA (2000) Granite-related Sn, W and Mo ore-magmatic systems: 3-D models and regularities of localization. *In* Kremenetsky AA, Lehmann B, Seltmann R (eds) Ore-bearing granites of Russia and adjacent countries—IGCP Project 373. IMGRE, Moscow, p 69-82
Dunn PJ (1976) Genthelvite and the helvine group. Mineral Mag 40:627-636
Dyachkov BA, Mairorova NP (1996) The rare metal deposits of the Kalba region in east Kazakhstan. *In* Shatov V, Seltmann R, Kremenetsky A, Lehmann B, Popov V, Ermolov P (eds) Granite-Related Ore Deposits of Central Kazakhstan and Adjacent Areas. Glagol, St. Petersburg, p 229-242
Eby GN (1985) Sr and Pb isotopes, U and Th chemistry of the alkaline Monteregian and White Mountain igneous provinces, eastern North America. Geochim Cosmochim Acta 49:1143-1153
Eckel EB (1997) Minerals of Colorado. Fulcrum Publishing, Golden, Colorado
Einaudi MT, Meinert LD, Newberry RJ (1981) Skarn deposits. *In* Skinner BJ (ed) Econ Geol, 75th Ann Vol. Economic Geology Publishing Company, p 317-391
Embrey PG, Symes RF (1987) Minerals of Cornwall and Devon. Mineralogical Record Inc., Tucson, Arizona
Engell J, Hansen J, Jensen M, Kunzendorf H, Loevborg L (1971) Beryllium mineralization in the Ilímaussaq intrusion, South Greenland, with description of a field beryllometer and chemical methods. Grønlands Geologiske Undersøgelse 33:40
Erickson RL, Blade LV (1963) Geochemistry and petrology of the alkalic igneous complex at Magnet Cove, Arkansas. U S Geol Surv Prof Paper 425
Ermolov PV (2000) Granite-related ore systems of Kazakhstan. *In* Kremenetsky AA, Lehmann B, Seltmann R (eds) Ore-bearing granites of Russia and adjacent countries—IGCP Project 373. IMGRE, Moscow, p 83-96
Esipchuk KY, Sheremet YM, Sveshnikov KI (1993) Rare metal granites and related rocks of the Ukrainian Shield. Bull Geol Soc Finland 65:131-141
Evans A (1982) Metallization associated with acid magmatism. John Wiley & Sons, Chichester, United Kingdom, 385

Evensen JM, London D (2002) Experimental silicate mineral/melt partition coefficients for beryllium and the beryllium cycle from migmatite to pegmatite. Geochim Cosmochim Acta (Yoder Spec Issue) 66:2239-2265

Evensen JM, London D, Wendlandt RF (1999) Solubility and stability of beryl in granitic melts. Am Mineral 84:733-745

Feininger T (1970) Emerald mining in Colombia: history and geology. Mineral Rec 1:142-149

Ferraris G, Prencipe M, Rossi P (1998) Stoppaniite, a new member of the beryl group: crystal structure and crystal-chemical implications. Eur J Mineral 10:491-496

Fersman AE (1929) Geochemische Migration der Elemente: III. Smaragdgruben im Uralgebirge. Abhandlungen praktische, geologische und Bergwirtschaftslehre 18:74-116

Finch AA (1990) Genthelvite and willemite, zinc minerals associated with alkaline magmatism from the Motzfeldt Centre, South Greenland. Mineral Mag 54:407-412

Foland KA, Allen JC (1991) Magma sources for Mesozoic anorogenic granites of the White Mountain magma series, New England, USA. Contrib Mineral Petrol 109:195-211

Franz G, Morteani G (1981) The system $BeO-Al_2O_3-SiO_2-H_2O$: hydrothermal investigation of the stability of beryl and euclase in the range from 1 to 6 kb and 400 to 800° C. N Jahrb Mineral Abh 140:273-299

Franz G, Morteani G (1984) The formation of chrysoberyl in metamorphosed pegmatites. J Petrol 25:27-52

Franz G, Grundmann G, Ackermand D (1986) Rock forming beryl from a regional metamorphic terrain (Tauern Window, Austria): parageneses and crystal chemistry. Tschermaks mineral petrogr Mitt 35:167-192

Franz G, Gilg HA, Grundmann G, Morteani G (1996) Metasomatism at a granitic pegmatite-dunite contact in Galicia: the Franqueira occurrence of chrysoberyl (alexandrite), emerald, and phenakite: discussion. Can Mineral 34:1329-1331

Fuertes FM, Martin-Izard A (1998) The Forcarei Sur rare-element granitic pegmatite field and associated mineralization, Galicia, Spain. Can Mineral 36:303-325

Fuertes FM, Martin IA, Boiron MC, Mangas VJ (2000) P-T path and fluid evolution of the Franqueira granitic pegmatite, central Galicia, northwestern Spain. Can Mineral 38:1163-1175

Fursenko DA (1982) Dependence of the composition of helvites on the acidity-basicity of mineralizing solutions. *In* Fisiko-khimicheskiye issledovaniya mineraloobrazuyuschikh system, Akad Nauk SSSR, Sibirskoye otdeleniye, Inst Geologii i Geofiziki 1982:104-107 (in Russian)

Ganguli D, De A, Saha P (1975) Synthesis and stability relations of chkalovite in the high-silica region of the system sodium oxide-bromellite (BeO)-silicon dioxide. Indian J Earth Sci 2:30-38

Ganzeeva LV, Bedrzhitskaya KV, Shumkova NG (1973) Leucophanite from the alkali metasomatites of the Russian Platform. Mineral Issled 3:25-28 (in Russian)

Garstone JD (1981) The geological setting and origin of emerald deposits at Menzies, Western Australia. J Royal Soc West Aust 64:53-64

Gerasimovsky VK, Volkov VP, Kogarko LN, Polykov AI, Saprykina TV, Balashov YA (1968) The Geochemistry of the Lovozero Massif: Part 2. Geochemistry. University of Toronto Press, Toronto

Geyti A, Schønwandt HK (1979) Bordvika: a possible porphyry molybdenum occurrence within the Oslo Rift, Norway. Econ Geol 74:1211-1220

Gilbert LA, Foland KA (1986) The Mont Saint-Hilaire plutonic complex: occurrence of excess ^{40}Ar and short intrusion history. Can J Earth Sci 23:948-958

Ginzburg AI, Shatskaya VT (1964) Migration of beryllium in the supergene zone of a fluorite-beryl deposit. Dokl Akad Nauk SSSR 159:1051-1054

Ginzburg AI, Zabolotnaya NP, Getmanskaya TI, Kupriyanova II, Novikova MI, Shatskaya VT (1979) The zoning of hydrothermal beryllium deposits. *In* Aleksandrov SM (ed) The zoning of hydrothermal ore deposits, Department of the Secretary of State, Translation Bureau, Ottawa, p 411-455

Giuliani G, Silva LJHD, Couto P (1990) Origin of emerald deposits of Brazil. Mineral Dep 25:57-64

Giuliani G, Cheilletz A, Arboleda C, Carrillo V, Rueda F, Baker JH (1995) An evaporitic origin of the parent brines of Colombian emeralds: fluid inclusion and sulphur isotope evidence. Eur J Mineral 7:151-165

Giuliani G, France LC, Zimmermann JL, Cheilletz A, Arboleda C, Charoy B, Coget P, Fontan F, Giard D (1997) Fluid composition, δD of channel H_2O, and $\delta^{18}O$ of lattice oxygen in beryls: genetic implications for Brazilian, Colombian, and Afghanistani emerald deposits. Intl Geol Rev 39:400-424

Giuliani G, France LC, Coget P, Schwarz D, Cheilletz A, Branquet Y, Giard D, Martin IA, Alexandrov P, Piat DH (1998) Oxygen isotope systematics of emerald: relevance for its origin and geological significance. Mineral Dep 33:513-519

Giuliani G, Bourles D, Massot J, Siame L (1999) Colombian emerald reserves inferred from leached beryllium of their host black shale. *In* Sangster AL, Zentilli M (eds) Latin American mineral deposits, 8. Canadian Institute of Mining, Metallurgy and Petroleum, Montreal, p 109-116

Giuliani G, Chaussidon M, Schubnel H-J, Piat DH, Rollion C, France-Lanord C, Giard D, de Narvaez D, Rondeau B (2000a) Oxygen isotopes and emerald trade routes since antiquity. Science 287:631-633

Giuliani G, France LC, Cheilletz A, Coget P, Branquet Y, Laumomnier B (2000b) Sulfate reduction by organic matter in Colombian emerald deposits: chemical and stable isotope (C, O, H) evidence. Econ Geol 95:1129-1153

Goodenough KM, Upton BGJ, Ellam RM (2000) Geochemical evolution of the Ivigtut Granite, South Greenland: a fluorine-rich "A-type" intrusion. Lithos 51:205-221

Goodwin AM (1996) Principles of Precambrian Geology. Academic Press, London

Graeser S (1998) Alpine minerals: a review of the most famous localities of the Central Swiss Alps. Rocks and Minerals 73:14-32

Graindorge JM (1974) A gemmological study of emerald from Poona, W.A. Austral Gemmol 12:75-80

Graupner T, Kempe U, Dombon E, Paetzold O, Leeder O (1996) The fluid regime in the tungsten (-yttrium) deposits of Kyzyltau (Mongolian Altai): comparison with fluid zonation in tungsten-tin ore systems. Progr Abstr Biennial Pan-American Conf Res Fluid Inclusions, Madison, Wisconsin, p 51-52

Graupner T, Kempe U, Dombon E, Paetzold O, Leeder O, Spooner ETC (1999) Fluid regime and ore formation in the tungsten(-yttrium) deposits of Kyzyltau (Mongolian Altai): evidence for fluid variability in tungsten-tin ore systems. Chem Geol 154:21-58

Graziani G, Guebelin E, Lucchesi S (1983) The genesis of an emerald from the Kitwe District, Zambia. N Jahrb Mineral Mon 1983:175-186

Grew ES (1981) Surinamite, taaffeite, and beryllian sapphirine from pegmatites in granulite-facies rocks of Casey Bay, Enderby Land, Antarctica. Am Mineral 66:1022-1033

Griffitts WR (1954) Beryllium resources of the tin-spodumene belt, North Carolina. U S Geol Surv Circ 309:1-12

Griffitts WR (1965) Recently discovered beryllium deposits near Gold Hill, Utah. Econ Geol 60:1298-1305

Griffitts WR, Cooley EF (1978) A beryllium-fluorite survey at Aguachile Mountain, Coahuila, Mexico. J Geochem Explor 9:137-147

Griffitts WR, Rader LF, Jr. (1963) Beryllium and fluorine in mineralized tuff, Spor Mountain, Juab County, Utah, in Geological Survey Research. U S Geol Surv Prof Paper 475-B:B16-B17

Grigor'yev NA (1986) Distribution of beryllium at the surface of the Earth. Intl Geol Rev 28:127-179, 327-371

Grigor'yev NA (1997) Unusual mineral balance of beryllium in the fluorite-hydromuscovite weathering crust, Boevskoe deposit, Urals. Geokhimiya:1066-1069 (in Russian)

Groat LA, Ercit TS, Marshall DD, Gault RA, Wise MA, Wengzynowski W, Eaton WD (2000a) Canadian emeralds: The Crown showing, southeastern Yukon. Newsletter Mineral Assoc Canada 63:1, 12-13

Groat LA, Ercit TS, Wise MA, Wengzynowski W, Eaton WD (2000b) The Crown emerald showing, southeastern Yukon. Geol Soc Am Abstr Progr 32:15

Gross EB, Heinrich EW (1966) Petrology and mineralogy of the Mount Rosa area, El Paso and Teller counties, Colorado: Part 2, Pegmatites. Am Mineral 51:299-323

Grundmann G, Morteani G (1989) Emerald mineralization during regional metamorphism: the Habachtal (Austria) and Leydsdorp (Transvaal, South Africa) deposits. Econ Geol 84:1835-1849

Grundmann G, Morteani G (1993) Emerald formation during regional metamorphism: the Zabara, Sikeit and Umm Kabo deposits (Eastern Desert, Egypt). *In* Thorweihe U, Schandelmeier H (eds) Geoscientific research in Northeast Africa: proceedings of the international conference. Balkema, Rotterdam, p 495-498

Gurevich VM, Gavrichev KS, Gorbunov VE, Fisheleva LI, Khodakovskii IL (1989) Low-temperature heat capacity of beryl. Geokhimiya:761-764 (in Russian)

Haapala IJ (1977a) Petrography and geochemistry of the Eurajoki Stock, a rapakivi-granite complex with greisen-type mineralization in Southwest Finland. Geol Surv Finland Bull 286

Haapala IJ (1977b) The controls of tin and related mineralizations in the rapakivi-granite areas of south-eastern Fennoscandia. Geologiska Föreningen i Stockholm Förhandlingar 99 Part 2:130-142

Haapala IJ (1988) Metallogeny of the Proterozoic rapakivi granites of Finland. *In* Taylor RP, Strong DF (eds) Recent advances in the geology of granite-related mineral deposits, Can Inst Min Metall Spec Paper 39:124-132

Haapala IJ (1995) Metallogeny of the Rapakivi Granites. Mineral Petrol 54:149-160

Haapala IJ (1997) Magmatic and postmagmatic processes in tin-mineralized granites: topaz-bearing leucogranite in the Eurajoki Rapakivi granite stock, Finland. J Petrol 38:1645-1659

Haapala IJ, Ramo OT (1990) Petrogenesis of the Proterozoic rapakivi granites of Finland. *In* Stein HJ, Hannah J (eds) Ore-bearing granite systems: petrogenesis and mineralizing processes, Geol Soc Am Spec Paper 246:275-286

Halley SW, Walshe JL (1995) A reexamination of the Mount Bischoff cassiterite sulfide skarn, western Tasmania. Econ Geol 90:1676-1693

Hassanen MA, Harraz HZ (1996) Geochemistry and Sr- and Nd-isotopic study on rare-metal-bearing granitic rocks, central Eastern Desert, Egypt. Precamb Res 80:1-22

Haussuehl S (1993) Thermoelastic properties of beryl, topaz, diaspore, sanidine and periclase. Z Kristallogr 204:67-76

Hawley CC (1969) Geology and beryllium deposits of the Lake George (or Badger Flats) beryllium area, Park and Jefferson counties, Colorado. U S Geol Surv Prof Paper 1044

Hawley CC, Huffman C, Jr., Hamilton JC, Rader LF, Jr. (1966) Geologic and geochemical features of the Redskin Granite and associated rocks, Lake George beryllium area, Colorado. U S Geol Surv Prof Paper 550-C

Hazen RM, Au AY (1986) High-pressure crystal chemistry of phenakite (Be_2SiO_4) and bertrandite ($Be_4Si_2O_7(OH)_2$). Phys Chem Mineral 13:69-78

Hazen RM, Finger LW (1987) High-temperature crystal chemistry of phenakite (Be_2SiO_4) and chrysoberyl ($BeAl_2O_4$). Phys Chem Mineral 14:426-434

Hazen RM, Au AY, Finger LW (1986) High-pressure crystal chemistry of beryl ($Be_3Al_2Si_6O_{18}$) and euclase ($BeAlSiO_4OH$). Am Mineral 71:977-984

Heinrich EW (1957) Pegmatite provinces of Colorado. Quarterly Colorado School Mines 52:1-22

Heinrich EW, Deane RW (1962) An occurrence of barylite near Seal Lake, Labrador. Am Mineral 47: 758-763

Hemingway BS, Barton MD, Robie RA, Haselton HT, Jr. (1986) Heat capacities and thermodynamic functions for beryl, $Be_3Al_2Si_6O_{18}$, phenakite, Be_2SiO_4, euclase, $BeAlSiO_4(OH)$, bertrandite, $Be_4Si_2O_7(OH)_2$, and chrysoberyl, $BeAl_2O_4$. Am Mineral 71:557-568

Henderson CMB, Taylor D (1984) Thermal expansion behaviour of beryllonite [$Na(BePO_4)$] and trimerite [$CaMn(BeSiO_4)_3$]. Mineral Mag 48:431-436

Henderson CMB, Taylor D (1989) Structural behaviour of chkalovite, $Na_2BeSi_2O_6$, a member of the cristobalite family. Mineral Mag 53:117-119

Henry C, Burkhard M, Goffé B (1996) Evolution of synmetamorphic veins and their wallrocks through a Western Alps transect: no evidence for large-scale fluid flow: stable isotope, major- and trace-element systematics. Chem Geol 127:81-109

Henry CD (1992) Beryllium and other rare metals in Trans-Pecos Texas. Bull West Texas Geol Soc 31: 1-15

Hill JD, Miller RR (1990) A review of Middle Proterozoic epigenic felsic magmatism in Labrador. *In* Gower CF, Rivers T, Ryan B (eds) Geol Assoc Can Spec Paper 38, p 417-431

Hillard PD (1969) Geology and beryllium mineralization near Apache Warm Springs, Socorro County, New Mexico. New Mex Bur Mines Min Res Circ 30:16

Hillebrand WF (1905) Red beryl from Utah. Am J Sci 19:330-331

Hofmeister AM, Hoering TC, Virgo D (1987) Vibrational spectroscopy of beryllium aluminosilicates: heat capacity calculations from band assignments. Phys Chem Mineral 14:205-224

Hölscher A, Schreyer W, Lattard D (1986) High-pressure, high-temperature stability of surinamite in the system $MgO-BeO-Al_2O_3-SiO_2-H_2O$. Contrib Mineral Petrol 92:113-127

Hörmann PK (1978) Beryllium. *In* Wedepohl KH (ed) Handbook of Geochemistry II/1. Springer-Verlag, Berlin, p 4-B-1 to 4-O-1, 1-6

Horváth L, Gault RA (1990) The mineralogy of Mont Saint-Hilaire, Quebec. Mineral Rec 21:284-359

Hsu LC (1983) Some phase relationships in the system $BeO-Al_2O_3-SiO_2-H_2O$ with comments on effects of HF. Mem Geol Soc China 5:33-46

Hügi T, Röwe D (1970) Beryllium minerals and beryllium-bearing granitic rocks of the Alps. Schweiz mineral petrogr Mitt 50:445-480 (in German)

Hutchison CS (1988) Geology of tin deposits in Asia and the Pacific: mineral concentrations and hydrocarbon accumulations in the ESCAP region. Springer-Verlag, New York

Icenhower J, London D (1995) An experimental study of element partitioning among biotite, muscovite, and coexisting peraluminous silicic melt at 200 MPa (H_2O). Am Mineral 80:1229-1251

Ihlen PM, Vokes FM (1978) Metallogeny (in the Oslo Paleorift). Norges Geol Undersøk Bull 45:75-90

Imeokparia EG (1988) Mesozoic granite magmatism and tin mineralization in Nigeria. *In* Taylor RP, Strong DF (eds) Recent advances in the geology of granite-related mineral deposits, Can Inst Min Metall Spec Paper 39:133-141

Imeokparia EG (1992) Geochemical and isotopic evidence for crystal melt + fluid phase equilibria and late stage fluid rock interaction in granitic rocks of the Ririwai Complex, northern Nigeria. African J Sci Tech, Ser B 6:60-79

Ito J, Hafner SS (1974) Synthesis and study of gadolinites. Am Mineral 59:700-708

Jackson NJ, Ramsay CR (1986) Post-orogenic felsic plutonism, mineralization and chemical specialization in the Arabian Shield. Inst Min Metall Trans 95:B83-B93

Jackson NJ, Willis RJ, Manning DAC, Sams MS (1989) Evolution of the Cornubian ore field, Southwest England: Part II, Mineral deposits and ore-forming processes. Econ Geol 84:1101-1133

Jacobson MI (1993a) Aquamarine in the United States: the search continues. Rocks and Minerals 68:306-319

Jacobson MI (1993b) Antero Aquamarines: Minerals from the Mount Antero – White Mountain region, Chaffee County, Colorado. L. R. Ream Publishing, Coeur d'Alene, Idaho

Jahns RH (1944a) Beryllium and tungsten deposits of the Iron Mountain District Sierra and Socorro counties, New Mexico. U S Geol Surv Bull 945-C:45-79

Jahns RH (1944b) "Ribbon rock", an unusual beryllium-bearing tactite. Econ Geol 39:173-205

Jahns RH, Wright LA (1944) The Harding beryllium-tantalum-lithium pegmatites, Taos County, New Mexico. Econ Geol 39:96-97

Jamtveit B, Andersen T (1993) Contact metamorphism of layered shale-carbonate sequences in the Oslo Rift: III, The nature of skarn-forming fluids. Econ Geol 88:1830-1849

Johnson TW, Ripley EM (1998) Hydrogen and oxygen isotopic systematics of beryllium mineralization, Spor Mountain, Utah. Geol Soc Am Abstr Progr 30:127

Kamilli RJ, Cole JC, Elliott JE, Criss RE (1993) Geology and genesis of the Baid Al Jimalah tungsten deposit, Kingdom of Saudi Arabia. Econ Geol 88:1743-1767

Kamitani M, Naito K (1998) Mineral resources map of Asia. Metal Mining Agency of Japan

Kamona AF (1994) Mineralization types in the Mozambique Belt of eastern Zambia. J Afr Earth Sci 19:237-243

Kar A (1991) Fluid inclusion and trace element studies of the gem pegmatites of Mt. Antero, Colorado. MS thesis, Virginia Polytechnic Institute and State University, Blacksburg, Virginia

Kartashov PM, Voloshin AV, Pakhomovsky YA (1993) Zoned crystalline gadolinite from the alkaline granitic pegmatites of Haldzan-Buragtag (Mongolian Altai). Zapiski Vseross Mineral Obshch 122: 65-79 (in Russian)

Kawachi Y, Ashley PM, Vince D, Goodwin M (1994) Sugilite in manganese silicate rocks from the Hoskins Mine and Woods Mine, New South Wales, Australia. Mineral Mag 58:671-677

Kazmi AH, Snee LW (1989a) Geology of the world emerald deposits: a brief review. *In* Kazmi A, Snee L (eds) Emeralds of Pakistan: geology, gemology and genesis. Van Nostrand Reinhold Co., New York, p 165-228

Kazmi AH, Snee LW (1989b) Emeralds of Pakistan: geology, gemology and genesis. Van Nostrand Reinhold Co., New York, NY, United States

Kazmi AH, Lawrence RD, Anwar J, Snee LW, Hussain S (1986) Mingora emerald deposits (Pakistan); suture-associated gem mineralization. Econ Geol 81:2022-2028

Kazmi AH, Anwar J, Hussain S, Khan T, Dawood H (1989) Emerald deposits of Pakistan. *In* Kazmi A, Snee L (eds) Emeralds of Pakistan; geology, gemology and genesis. Van Nostrand Reinhold, New York, p 39-74

Keith JD, Christiansen EH, Tingey DG (1994) Geological and chemical conditions of formation of red beryl, Wah Wah Mountains, Utah. Utah Geo Assoc Publ 23:155-170

Keith JD, Thompson TJ, Ivers S (1996) The Uinta emerald and the emerald-bearing potential of the Red Pine Shale, Uinta Mountains, Utah. Geol Soc Am Abstr Progr 28(7):85

Kelly WC, Rye RO (1979) Geologic, fluid inclusion, and stable isotope studies of the tin-tungsten deposits of Panasqueira, Portugal. Econ Geol 74:1721-1822

Kempe U, Goetze J, Dandar S, Habermann D (1999) Magmatic and metasomatic processes during formation of the Nb-Zr-REE deposits Khaldzan Buregte and Tsakhir (Mongolian Altai): indications from a combined CL-SEM study. *In* Finch (ed) Luminescence phenomena in minerals: thematic set 63. Mineralogical Society, London, p 165-177

Kerr PF (1938) Tungsten mineralization at Oreana, Nevada. Econ Geol 33:390-427

Kiilsgaard TH, Bennett EH (1985) Mineral deposits in the southern part of the Atlanta Lobe of the Idaho Batholith and their genetic relation to Tertiary intrusive rocks and to faults. U S Geol Surv Bull:153-165

Kile DE, Foord EE (1998) Micas from the Pikes Peak Batholith and its cogenetic granitic pegmatites, Colorado: optical properties, composition, and correlation with pegmatite evolution. Can Mineral 36 Part 2:463-482

Kimbler FS, Haynes PE (1980) An occurrence of red beryl in the Black Range, New Mexico. New Mex Geol 2:15-16

Kingsbury AWG (1961) Beryllium minerals in Cornwall and Devon: helvine, genthelvite, and danalite. Mineral Mag 32:921-940

Kinnaird JA (1985) Hydrothermal alteration and mineralization of the alkaline anorogenic ring complexes of Nigeria. J Afr Earth Sci 3:229-251

Kiseleva IA, Shuriga TN (1983) New data on thermodynamic properties of phenakite. Geokhimiya 1983:310-313 (in Russian)

Kiseleva IA, Mel'chakova LV, Ogorodova LP, Topor ND (1984a) Thermochemical study of chkalovite $Na_2[BeSi_2O_6]$. Probl Kalorim Khim Termodin 2:408-410 (in Russian)

Kiseleva IA, Mel'chakova LV, Ogorodova LP, Topor ND, Khomyakov AP (1984b) Thermodynamic properties of chkalovite $Na_2[BeSi_2O_6]$. Vestn Mosk Univ 25:278-81

Kiseleva IA, Mel'chakova LV, Ogorodova LP (1985) Heat of formation and heat capacity of bertrandite, euclase, and chrysoberyl. Moscow Univ Geol Bull 40:90-92

Kiseleva IA, Mel'chakova LV, Ogorodova LP, Kupriyanova IN (1986) Thermodynamic parameters of beryllium minerals: chrysoberyl, beryl euclase, and bertrandite. Geochem Intl 23:129-141

Klaska KH, Jarchow O (1973) The crystal chemistry of the beryllonite-trimerite region. Z Kristallogr 137:452 (abstr, in German)

Klement'eva LV (1970) A find of bromellite in the USSR. Doklady USSR Acad Sci (Earth Sci Sect) 188:152-154

Kloosterman JB (1974) Phenakite and nigerite in quartz-cassiterite veins in the upper Candeias, Rondonia. Mineracao Metalurgia 38:18-19

Klyakhin VA, Lebedev AS, Fursenko DA (1981) Equilibrium of beryllium minerals under hydrothermal conditions. Eksp Issled Sul'fidnykh Silik Sis 1981:55-61 (in Russian)

Kogarko LN, Kononova VA, Orlova MP, Woolley AR (1995) Alkaline Rocks and Carbonatites of the World. Part 2: Former USSR. Chapman and Hall, London

Kogut AI, Hagni RD, Schneider GIC (1997) The Okorusu, Namibia carbonatite-related fluorite deposit and a comparison with the Illinois-Kentucky fluorite district/ Hicks Dome. *In* Sangster D (ed) Carbonate-hosted lead-zinc deposits. Soc Econ Geol Spec Pub 4. Society of Economic Geologists, Littleton, Colorado, p 290-297

Kosals YA, Dmitriyeva AN (1973) Temperature of formation of leucophane-fluorite metasomatite. Tr USSR Acad Sci (Earth Sci Sect) 211:180-182

Kosals YA, Dmitriyeva AN, Arkhipchuk RZ, I. GV (1974) Temperature conditions and formation sequence in deposition of fluorite-phenacite-bertrandite ores. Intl Geol Rev 16:1027-1036

Kostryukov VN, Kostylev FA, Samorukov OP, Samorukova NK, Chesalina LA (1977) Thermodynamic properties of beryllium hydroxide. Zh Fiz Khim 51:1015 (in Russian)

Kosukhin ON (1980) Some distinctive features of the genesis of the granitoids in the Sherlova Gora and Adun-Chelon region (based on the study of inclusions in minerals). Sov Geol Geophys 21:8-14

Kotlyar BB, Ludington S, Mosier DL (1995) Descriptive, grade, and tonnage models for molydenum-tungsten greisen deposits. U S Geol Surv Open File Rep 95-584

Kovalenko VI (1978) The genesis of rare metal granitoids and related ore deposits. *In* Stemprok M, Burnol L, Tischendorf G (eds) Metallization associated with acid magmatism, 3. Geol Surv Prague, p 235-247

Kovalenko VI, Yarmolyuk VV (1995) Endogenous rare metal ore formations and rare metal metallogeny of Mongolia. Econ Geol 90:520-529

Kovalenko VI, Antipin VS, Petrov LL (1977) Distribution coefficients of beryllium in ongonites and some notes on its behavior in the rare metal lithium-fluorine-granites. Geochem Intl 14:129-141

Kovalenko VI, Samoylov VS, Goreglyad AV (1979) Volcanic ongonites enriched in rare elements. Tr USSR Acad Sci (Earth Sci Sect) 246:58-61

Kovalenko VI, Tsaryeva GM, Andreeva IA, Hervig RL, Newman S, Naumov VB, Cuney M, Rambault L (1995a) Melt inclusions of rare-metal magmas (granites, pantellerites, carbonatites, apatite rocks). EOS Trans Am Geophys Union 76:268

Kovalenko VI, Tsaryeva GM, Goreglyad AV, Yarmolyuk VV, Troitsky VA, Hervig RL, Farmer GL (1995b) The peralkaline granite-related Khaldzan-Buregtey rare metal (Zr, Nb, REE) deposit, western Mongolia. Econ Geol 90:530-547

Kozlowski A, Metz P, Jaramillo HAE (1988) Emeralds from Somondoco, Colombia: chemical composition, fluid inclusions and origin. N Jahrb Mineral Abh 159:23-49

Koz'menko OA, Belevantsev VI, Peshchevitskiy BI (1988) The solubility of BeO in aqueous HF at 250 and 350° C. Geochem Intl 25:135-138

Kremenetsky AA (1996) The rare metal pegmatite deposits of Mongolian Altai, eastern Kazakhstan, and China. *In* Shatov V, Seltmann R, Kremenetsky A, Lehmann B, Popov V, Ermolov P (eds) Granite-Related Ore Deposits of Central Kazakhstan and Adjacent Areas. Glagol, St. Petersburg, p 243-258

Kremenetsky AA, Beskin SM, Lehmann B, Seltmann R (2000a) Economic geology of granite-related ore deposits of Russia and other FSU countries: an overview. *In* Kremenetsky AA, Lehmann B, Seltmann R (eds) Ore-bearing granites of Russia and adjacent countries—IGCP Project 373. IMGRE, Moscow, p 3-60

Kremenetsky AA, Lehmann B, Seltmann R (2000b) Ore-bearing Granites of Russia and Adjacent Countries—IGCP Project 373. IMGRE, Moscow

Kretz R (1983) Symbols for rock-forming minerals. Am Mineral 68:277-279

Kretz R, Loop J, Hartree R (1989) Petrology and Li-Be-B geochemistry of muscovite-biotite granite and associated pegmatite near Yellowknife, Canada. Contrib Mineral Petrol 102:174-190

Kuester D (1990) Rare-metal pegmatites of Wamba, central Nigeria: their formations in relationship to late Pan-African granites. Mineral Dep 25:25-33

Kupriyanova II (1982) Dependence of beryllium mineral parageneses on the temperature and activity of some components. Dokl Akad Nauk SSSR 266:714-718 (in Russian)

Kupriyanova II, Shpanov EP (1997) Beryllium-fluorite ores of the Voznesensk Ore District (Primore, Russia). Geol Rudn Mestorozhd 39:442-455 (in Russian)

Kutukova EI (1946) Bavenite from the emerald mines. Dokl Akad Nauk SSSR 54:721-724

Kwak TAP (1983) The geology and geochemistry of the zoned, Sn-W-F-Be skarns at Mt. Lindsay, Tasmania, Australia. Econ Geol 78:1440-1465

Kwak TAP (1987) W-Sn skarn deposits and related metamorphic skarns and granitoids. Elsevier, Amsterdam

Kwak TAP, Askins PW (1981) The nomenclature of carbonate replacement deposits, with emphasis on Sn-F(-Be-Zn) 'wrigglite' skarns. J Geol Soc Aus 28:123-136

Kwak TAP, Jackson PG (1986) The compositional variation and genesis of danalite in Sn-F-W skarns, NW Tasmania, Australia. N Jahrb Mineral Mon 1986:452-462

Larsen AO (1988) Helvite group minerals from syenite pegmatites in the Oslo region, Norway. Norsk Geol Tidsskr 68:119-124

Larsen AO, Asheim A, Berge SA (1987) Bromellite from syenite pegmatite, southern Oslo region, Norway. Can Mineral 25:425-428

Laurs BM (1995) Emerald mineralization and amphibolite wall-rock alteration at the Khaltaro pegmatite-hydrothermal vein system, Haramosh Mountains, northern Pakistan. MS thesis, Oregon State University

Laurs BM, Dilles JH, Snee LW (1996) Emerald mineralization and metasomatism of amphibolite, Khaltaro granitic pegmatite-hydrothermal vein system, Haramosh mountains, northern Pakistan. Can Mineral 34:1253-1286

Layne GD, Spooner ETC (1991) The JC tin skarn deposit, southern Yukon Territory: I, Geology, paragenesis, and fluid inclusion microthermometry. Econ Geol 86:29-47

Lebedev AS, Ragozina TP (1984) Phase relations between phenakite and bertrandite in the BeO-SiO_2-HCl-HF-H_2O system at 400-600°C. Geokhimiya 19:59-65 (in Russian)

Lee DE, Erd RC (1963) Phenakite from the Mount Wheeler area, Snake Range, White Pine County, Nevada. Am Mineral 48:189-193

Le Grange JM (1929) The Barbara emeralds: A study of an occurrence of emeralds n the north-eastern Transvaal in the Murchison Range. Geol Soc South Africa Trans 32:1-25

Lehmann B, Ishihara S, Michel H, Miller J, Rapela CW, Sanchez A, Tistl M, Winkelmann L (1990) The Bolivian tin province and regional tin distribution in the Central Andes: a reassessment. Econ Geol 85:1044-1058

Letnikov FA (2000) Rare-metal granites of the Kokshetau block, northern Kazakhstan. *In* Kremenetsky AA, Lehmann B, Seltmann R (eds) Ore-bearing granites of Russia and adjacent countries—IGCP Project 373. IMGRE, Moscow, p 177-192

Levasseur R (1997) Fluid inclusion studies of rare element pegmatites, South Platte District, Colorado. MS thesis, University of Windsor

Lin C, LIu Y, Wang Z, Hong W (1995a) Rare element and rare-earth element deposits of China. *In* Mineral Deposits of China, 3, Beijing, p 226-279

Lin D (1985) Genesis of an altered volcanic type of beryllium deposit in southern China. Mineral Deposits = Kuangchuang Dizhi 4:19-30 (in Chinese)

Lin Y, Pollard PJ, Hu S, Taylor RG (1995b) Geologic and geochemical characteristics of the Yichun Ta-Nb-Li deposit, Jiangxi Province, South China. Econ Geol 90:577-585

Lindsey DA (1975) Mineralization halos and diagenesis in water-laid tuff of the Thomas Range, Utah. U S Geol Surv Prof Paper 818-B

Lindsey DA (1977) Epithermal beryllium deposits in water-laid tuff, western Utah. Econ Geol 72:219-232

Lindsey DA (1979) Geologic map and cross-sections of Tertiary rocks in the Thomas Range and northern Drum Mountains, Juab County, Utah. U S Geol Surv Map I-1176

Lindsey DA, Ganow H, Mountjoy W (1973) Hydrothermal alteration associated with beryllium deposits at Spor Mountain, Utah. U S Geol Surv Prof Paper 818-A:A1-A20

Linnen RL (1998) Depth of emplacement, fluid provenance and metallogeny in granitic terranes: A comparison of western Thailand with other tin belts. Mineral Dep 33:461-476

London D (1992) The application of experimental petrology to the genesis and crystallization of granitic pegmatites. Can Mineral 30:499-540

London D (1997) Estimating abundances of volatile and other mobile components in evolved silicic melts through mineral-melt equilibria. J Petrol 38:1691-1706

London D, Burt DM (1978) Lithium pegmatites of the White Picacho District, Maricopa and Yavapai counties, Arizona. *In* Burt DM, Pewe TL (eds) Guidebook to the geology of central Arizona, 2. University of Arizona, Tucson, p 61-72

London D, Hervig RL, Morgan GB (1988) Melt-vapor solubilities and elemental partitioning in peraluminous granite-pegmatite systems: experimental results with Macusani glass at 200 MPa. Contrib Mineral Petrol 99:360-373

London D, Morgan GB, Hervig RL (1989) Vapor-undersaturated experiments with Macusani glass + H_20 at 200 MPa, and the internal differentiation of granitic pegmatites. Contrib Mineral Petrol 102:1-17

Luckscheiter B, Morteani G (1980) Microthermometrical and chemical studies of fluid inclusions in minerals from Alpine veins from the Penninic rocks of the central and western Tauern Window (Austria/ Italy). Lithos 13:61-77

Ludington S (1981) The Redskin Granite: evidence for thermogravitational diffusion in a Precambrian granite batholith. J Geophys Res 86:B10423-B10430

Ludwig KR, Lindsey DA, Zielinski RA, Simmons KR (1980) U-Pb ages of uraniferous opals and implications for the history of beryllium, fluorine, and uranium mineralization at Spor Mountain, Utah. Earth Planet Sci Lett 46:221-232

Lykhin DA, Kostitsyn YA, Kovalenko VP, Yarmolyuk VV, Sal'nikova EB, Kotov AB, Kovach VP, Ripp GS (2001) Ore-spawning magmatism of the Ermakov beryllium deposit in western Transbaikalia: age, magma sources, and relation to mineralization. Geol Rudn Mestorozhd 43:52-70 (in Russian)

Macdonald R, Upton BGJ (1993) The Proterozoic Gardar rift zone, South Greenland: comparisons with the East African Rift system. Geol Soc Spec Pub 76:427-442

Macdonald R, Smith RL, Thomas JE (1992) Chemistry of subalkalic obsidians. U S Geol Surv Prof Paper 1523

Malchenko EG, Ermolov PV (1996) Metallogenic summary of central Kazakhstan and adjacent areas. *In* Shatov V, Seltmann R, Kremenetsky A, Lehmann B, Popov V, Ermolov P (eds) Granite-Related Ore Deposits of Central Kazakhstan and Adjacent Areas. Glagol, St. Petersburg, p 67-82

Mandarino JA (1999) Fleischer's glossary of mineral species 1999. The Mineralogical Record, Tucson, Arizona

Mandarino JA, Anderson V (1989) Monteregian treasures: the minerals of Mont Saint-Hilaire, Quebec. Cambridge Univ. Press, Cambridge, United Kingdom

Manier-Glavinaz V, Couty R, Lagache M (1988) Experimental study of the equilibrium between a natural beryl and hydrothermal fluids, geochemical inferences. Chem Geol 70:162 (abstr)

Manier-Glavinaz V, Couty R, Lagache M (1989a) The removal of alkalis from beryl: structural adjustments. Can Mineral 27:663-671

Manier-Glavinaz V, D'Arco P, Lagache M (1989b) Alkali partitioning between beryl and hydrothermal fluids: an experimental study at 600 degree C and 1.5 kbar. Eur J Mineral 1:645-655

Manning DAC (1982) An experimental study of the effects of fluorine on the crystallization of granitic melts. *In* Evans (ed) Metallization associated with acid magmatism. John Wiley & Sons, Chichester, UK, p 191-203

Manning DAC, Hill PI (1990) The petrogenetic and metallogenetic significance of topaz granite from the Southwest England orefield. *In* Stein HJ, Hannah JL (eds) Ore-bearing granite systems: petrogenesis and mineralizing processes, Geol Soc Am Spec Paper 246:51-70

Manning DAC, Putthapiban P, Suensilpong S (1983) An occurrence of bavenite, $Ca_4Be_2Al_2(SiO_3)_9{\cdot}xH_2O$, in Thailand. Mineral Mag 47:87-89

Mao J, Guy B, Raimbault L, Shimazaki H (1996a) Manganese skarn in the Shizhuyuan polymetallic tungsten deposit, Hunan, China. Shigen Chishitsu 46:1-11

Mao J, Li H, Shimazaki H, Raimbault L, Guy B (1996b) Geology and metallogeny of the Shizhuyuan skarn-greisen deposit, Hunan Province, China. Intl Geol Rev 38:1020-1039

Markl G (2001) Stability of Na-Be minerals in late-magmatic fluids of the Ilímaussaq alkaline complex, South Greenland. Geol Surv Denmark Bull 190:145-158

Markl G, Schumacher JC (1996) Spatial variations in temperature and composition of greisen-forming fluids: an example from the Variscan Triberg granite complex, Germany. Econ Geol 91:576-589

Markl G, Schumacher JC (1997) Beryl stability in local hydrothermal and chemical environments in a mineralized granite. Am Mineral 82:194-202

Martin-Izard A, Paniagua A, Moreiras D, Acevedo RD, Marcos-Pascual C (1995) Metasomatism at a granitic pegmatite-dunite contact in Galicia: the Franqueira occurrence of chrysoberyl (alexandrite), emerald, and phenakite. Can Mineral 33:775-792

Martin-Izard A, Paniagua A, Moreiras D, Acevedo RD, Marcos-Pascual C (1996) Metasomatism at a granitic pegmatite-dunite contact in Galicia: The Franqueira occurrence of chrysoberyl (alexandrite) emerald, and phenakite: reply. Can Mineral 34:1332-1336

Matheis G (1987) Nigerian rare-metal pegmatites and their lithological framework. Geol J 22:271-291

Matveev GM, Zhuravlev AK (1982) Thermodynamic properties of beryllium orthosilicate at high temperatures. Silikattechnik 33:203 (in German)

Matveev SS (1997) Evolution of the ore-formation process at the Akchatau greisen deposit according to the data of geochemical indicators. Petrologiya 5:326-336 (in Russian)

Mazurov AK (1996) The Koktenkol stockwork W-Mo deposit, central Kazakhstan. *In* Shatov V, Seltmann R, Kremenetsky A, Lehmann B, Popov V, Ermolov P (eds) Granite-Related Ore Deposits of Central Kazakhstan and Adjacent Areas. Glagol, St. Petersburg, p 155-166

McAnulty WN (1980) Geology and mineralization of the Sierra Blanca peaks, Hudspeth County, Texas. Guide NM Geol Soc 31:263-266

McAnulty WN, Sewell CR, Atkinson DR, Rasberry JM (1963) Aguachile beryllium-bearing fluorspar district, Coahuila, Mexico. Bull Geol Soc Am 74:735-744

McLemore VT, Lueth VW, Pease TC, Guilinger JR (1996) Petrology and mineral resources of the Wind River Laccolith, Cornudas Mountains, New Mexico and Texas. Can Mineral 34:335-347

Meeves HC (1966) Nonpegmatitic beryllium occurrences in Arizona, Colorado, New Mexico, Utah, and four adjacent states. U S Bur Mines Rpt Inves 6828:68

Mel'chakova LV, Kiseleva IA, Ogorodova LP, Fursenko DA (1991) Heat capacity and entropy of beryllium silicate-genthelvite. Moscow Univ Geol Bull 46:66-68

Mel'nikova EO, Avakyan ZA, Karavaiko GI, Krutsko VS (1990) Microbiological destruction of silicate minerals containing beryllium. Mikrobiologiya 59:63-69 (in Russian)

Menon RD, Santosh M (1995) The Pan-African gemstone province of East Gondwana. Geol Soc India Mem 34:357-371

Menon RD, Santosh M, Yoshida M (1994) Gemstone mineralization in southern Kerala, India. J Geol Soc India 44:241-52

Men'shikov YP, Pakhomovskii YA, Yakovenchuk VN (1999) Beryllium mineralization within veins in Khibiny massif. Zapiski Vseross Mineral Obshch 128:3-14 (in Russian)

Meyer C (1981) Ore-forming processes in geologic history. *In* Skinner BJ (ed) Econ Geol 75th Ann Vol, 1905-1980, p 6-41

Miller RR (1989) Rare-metal targets in insular Newfoundland. Report Activities—Mineral Devel Div Newfoundland Geol Surv 89-1:171-179

Miller RR (1996) Structural and textural evolution of the Strange Lake peralkaline rare-element (NYF) granitic pegmatite, Quebec-Labrador. Can Mineral 34:349-371

Miller RR, Heaman LM, Birkett TC (1997) U-Pb zircon age of the Strange Lake peralkaline complex: implications for Mesoproterozoic peralkaline magmatism in north-central Labrador. Precamb Res 81:67-82

Moeller P, Černý P, Saupe F, (eds) (1989) Lanthanides, tantalum and niobium: mineralogy, geochemistry, characteristics of primary ore deposits, prospecting, processing and application. Springer, Berlin

Moh GH (1988) Observations on complex tin ores of the West Pacific tin belt, Shizhuyuan, Hunan, China, with special attention to beryllium and tin minerals. Intl Symp Tin/Tungsten Granites in Southeast Asia and the Western Pacific, Extended Abstracts 5:91-95

Mohamed FH (1993) Rare metal-bearing and barren granites, Eastern Desert of Egypt; geochemical characterization and metallogenetic aspects. J Afr Earth Sci 17:525-539

Moore PB (1970) Mineralogy and chemistry of Långban-type deposits in Bergslagen, Sweden. Mineral Rec 1:154-172

Morosin B (1972) Structure and thermal expansion of beryl. Acta Crystallogr 28:1899-1903

Morteani G, Grundmann G (1977) The emerald porphyroblasts in the penninic rocks of the central Tauern Window. N Jahrb Mineral Mon 11:509-516

Morteani G, Preinfalk C, Spiegel W, Bonalumi A (1995) The Achala granitic complex with the pegmatites of the Sierras Pampeanas (Northwest Argentina): a study of differentiation. Econ Geol 90:636-647

Mulligan R (1968) Geology of Canadian beryllium deposits. Geol Surv Can Econ Geol Report 23:1-109Mulligan R (1971) Lithophile Metals and the Cordilleran Tin Belt. Can Min Met Bull 64:68-71

Mullis J, Dubessy Y, Poty B, O'Neill JR (1994) Fluid regimes during late stages of a continental collision: Physical, chemical and stable isotope measurements of fluid inclusions in fissure quartz from a geotraverse through the Central Alps, Switzerland. Geochim Cosmochim Acta 58:2239-2268

Mumba PACC, Bartot NR (1998) Les Mines d'Emeraudes de Kamakanga en Zambie: Vues sur l'Avenir. Revue de Gemmologie 134-35:165-167

Nechaev SV, Buchinskaya KM (1993) Greenockite in rare-metal metasomatites of the Ukrainian Shield. Mineral Zh 15:87-89 (in Russian)

Nedashkovskii PG (1970) The sources of beryllium in Be-deposits of different genetic types. Intl Union Geol Sci 2:16-18

Nedashkovskii PG (1983) Phenakite-bearing alkali pegmatites—a new genetic type of beryllium mineralization. Dokl Akad Nauk SSSR 271:157-158 (in Russian)

Newberry RJ (1998) W- and Sn-skarn deposits: a 1998 status report. *In* Lentz DR (ed) Mineralized intrusion-related skarn systems, Mineral Assoc Can Short Course Handbook 26:289-335

Newberry RJ, Burns LE, Swanson SE, Smith TE (1990) Comparative petrologic evolution of the Sn and W granites of the Fairbanks-Circle area, interior Alaska. *In* Stein HJ, Hannah JL (eds) Ore-bearing granite systems: petrogenesis and mineralizing processes, Geol Soc Am Spec Paper 246:121-142

Noble SR, Spooner ETC, Harris FR (1984) The Logtung large tonnage, low-grade tungsten(scheelite)-molybdenum porphyry deposit, south-central Yukon Territory. Econ Geol 79:848-68

Novikova MI (1967) Processes of phenakite alteration. Zapiski Vsesoyuz Mineral Obshch 96:418-424 (in Russian)

Novikova MI (1983) Evolution of beryllium mineralization in formations containing fluorite-phenakite-bertrandite ore deposits. Geol Rudn Mestorozhd 25: 30-37 (in Russian)

Novikova MI (1984) Leucophane, melinophanite, and eudidymite in a deposit of a fluorite-phenakite-bertrandite formation. Mineral Zh 6:84-90 (in Russian)

Novikova MI, Shpanov EP, Kupriyanova II (1994) Petrography of the Ermakovskoe beryllium deposit, western Transbaikal area. Petrologiya 2:114-127 (in Russian)

Nutt CJ, O'Neill JM, Kleinkopf MD, Klein DP, Miller WR, Rodriquez BD, McLemore VT (1997) Geology and mineral resources of the Cornudas Mountains, New Mexico. U S Geol Surv Open File Rep 97-0282

Nwe YY, Grundmann G (1990) Evolution of metamorphic fluids in shear zones: the record from the emeralds of Habachtal, Tauern Window, Austria. Lithos 25:281-304

Nwe YY, Morteani G (1993) Fluid evolution in the H_2O-CH_4-CO_2-NaCl system during emerald mineralization at Gravelotte, Murchison greenstone belt, Northeast Transvaal, South Africa. Geochim Cosmochim Acta 57:89-103

Odintsova YeA (1993) Beryllium mineral paragenesis in the beryllium oxide-water chemical-potential diagram. Dokl Akad Nauk SSSR 328:502-505 (in Russian)

Odintsova YeA (1996) Mineral facies of beryllium mineralization: example from Izumrudnykh Kopi Deposit, Northern Urals. Geol Rudn Mestorozhd 38:287-297 (in Russian)

Okrusch M, Richter P, Guerkan A (1981) Geochemistry of Blackwall Sequences in the Habachtal emerald deposit, Hohe Tauern, Austria: I, Presentation of geochemical data. Tschermaks mineral petrogr Mitt 29:9-31

Olcott JD, Duerichen E, Nelson ST, Keith JD, Tingey DG (1998) Genesis of economic fibrous calcite veins, Uinta Mountains, Utah. Geol Soc Am Abstr Progr 30:371

Olson DW (2000) Gemstones. Minerals Yearbook. U S Geol Surv, p 31.1-31.4

Olson JC, Hinrichs EN (1960) Beryl-bearing pegmatites in the Ruby Mountains and other areas in Nevada and northwestern Arizona. U S Geol Surv Bull 1082-D:135-200

Ospanov KK (1983) Dissolution sequence of beryllium minerals. Zh Neorg Khim 28:324-328 (in Russian)

Ottaway TL, Wicks FJ, Bryndzia LT, Kyser TK, Spooner ETC (1994) Formation of the Muzo hydrothermal emerald deposit in Colombia. Nature 369:552-554

Palache C, Bauer LH (1930) On the occurrence of beryllium in the zinc deposits of Franklin, New Jersey. Am Mineral 15:30-33

Partington GA, McNaughton NJ, Williams IS (1995) A review of the geology, mineralization, and geochronology of the Greenbushes Pegmatite, Western Australia. Econ Geol 90:616-635

Paslick CR, Halliday AN, Davies GR, Mezger K, Upton BGJ (1993) Timing of Proterozoic magmatism in the Gardar Province, southern Greenland. Bull Geol Soc Am 105:272-278

Perez JB, Dusausoy Y, Babkine J, Pagel M (1990) Mn zonation and fluid inclusions in genthelvite from the Taghouaji Complex (Air Mountains, Niger). Am Mineral 75:909-914

Pezzotta F, Diella V, Guastoni A (1999) Chemical and paragenetic data on gadolinite-group minerals from Baveno and Cuasso al Monte, Southern Alps, Italy. Am Mineral 84:782-789

Pichavant M, Kontak DJ, Briqueu L, Valencia HJ, Clark AH (1988a) The Miocene-Pliocene Macusani Volcanics, SE Peru: 2, Geochemistry and origin of a felsic peraluminous magma. Contrib Mineral Petrol 100:325-338

Pichavant M, Kontak DJ, Valencia HJ, Clark AH (1988b) The Miocene-Pliocene Macusani Volcanics, SE Peru: 1, Mineralogy and magmatic evolution of a two-mica aluminosilicate-bearing ignimbrite suite. Contrib Mineral Petrol 100:300-324

Pilati T, Demartin F, Gramaccioli CM (1993) Atomic thermal parameters and thermodynamic functions for corundum (α-Al_2O_3) and bromellite (BeO): a lattice-dynamical estimate. Acta Crystallogr 49:473-480

Pilati T, Demartin F, Gramaccioli CM (1997) Lattice-dynamical evaluation of thermodynamic properties and atomic displacement parameters for beryl using a transferable empirical force field. Am Mineral 82:1054-1062

Pilati T, Demartin F, Gramaccioli CM (1998) Lattice-dynamical evaluation of atomic displacement parameters and thermodynamic functions for phenakite Be_2SiO_4. Phys Chem Mineral 26:149-155

Plimer IR, Kleeman JD (1991) Geology, geochemistry and genesis of the Sn-W deposits associated with the Mole Granite, Australia. *In* Pagel M, Leroy J-L (eds) Source, transport and deposition of metals. A. A. Balkema. Rotterdam, Netherlands, p 785-788

Plumlee GS, Goldhaber MB, Rowan EL (1995) The potential role of magmatic gases in the genesis of Illinois-Kentucky fluorspar deposits: implications from chemical reaction path modeling. Econ Geol 90:999-1011

Pollard PJ (1989) Geochemistry of granites associated with tantalum and niobium mineralization. *In* Möller P, Černý P, Saupe F (eds) Lanthanides, Tantalum and Niobium. Springer-Verlag, Berlin, p 145-168

Pollard PJ (1995a) A special issue devoted to the geology of rare metal deposits: geology of rare metal deposits: an introduction and overview. Econ Geol 90:489-494

Pollard PJ (1995b) A special issue devoted to the geology of rare metal deposits. Econ Geol 90, 489-689

Pollard PJ, Nakapadungrat S, Taylor RG (1995) The Phuket Supersuite, Southwest Thailand: fractionated I-type granites associated with tin-tantalum mineralization. Econ Geol 90:586-602

Polupanova TI, Petrov VL, Kruzhalov AV, Laskovenkov AF, Nikitin VS (1985) The thermal stability of beryl. Geochem Intl 22:11-13

Prasad SN, Ghosh JC (1988) Thermodynamics of dissociation of beryllium sulfate in aqueous solution from 278.15 to 308.15 K. J Indian Chem Soc 65:831-833

Price JG, Rubin JN, Henry CD, Pinkston TL, Tweedy SW, Koppenaal DW (1990) Rare-metal enriched peraluminous rhyolites in a continental arc, Sierra Blanca area, Trans-Pecos Texas: chemical modification by vapor-phase crystallization. *In* Stein HJ, Hannah JL (eds) Ore-bearing granite systems: petrogenesis and mineralizing processes, Geol Soc Am Spec Paper 246:103-120

Ragu A (1994a) Helvite from the French Pyrenees as evidence for granite-related hydrothermal activity. Can Mineral 32:111-120

Ragu A (1994b) Helvite from the French Pyrenees as evidence for granite-related hydrothermal activity. Can Mineral 32:111-20

Raimbault L, Bilal E (1993) Trace-element contents of helvite-group minerals from metasomatic albitites and hydrothermal veins at Sucuri, Brazil and Dajishan, China. Can Mineral 31:119-127

Raimbault L, Burnol L (1998) The Richemont rhyolite dyke, Massif Central, France: a subvolcanic equivalent of rare-metal granites. Can Mineral 36:265-282

Raimbault L, Cuney M, Azencott C, Duthou JL, Joron JL (1995) Geochemical evidence for a multistage magmatic genesis of Ta-Sn-Li mineralization in the granite at Beauvoir, French Massif Central. Econ Geol 90:548-576

Ramsden AR, French DH, Chalmers DI (1993) Volcanic-hosted rare-metals deposit at Brockman, Western Australia: mineralogy and geochemistry of the Niobium Tuff. Mineral Dep 28:1-12

Renders PJ, Anderson GM (1987) Solubility of kaolinite and beryl to 573 K. Appl Geochem 2:193-203

Reyf FG, Ishkov YM (1999) Be-bearing sulfate-fluoride brines, a product of residual pegmatite's distallation within the alkaline granite intrusion (the Yermakova F-Be deposit, Transbaikalia). Geokhimiya 1999:1096-1111 (in Russian)

Ribeiro AAM, Cheilletz A, Giuliani G, Feraud G, Barbosa CG, Zimmermann JL (1997) $^{40}Ar/^{39}Ar$ and K-Ar geochronological evidence for two periods (differs from 2 Ga and 650 to 500 Ma) of emerald formation in Brazil. Intl Geol Rev 39:924-937

Richardson DG, Birkett TC (1996) Peralkaline rock-associated rare metals. *In* Eckstrand OR, Sinclair WD, Thorpe RI (eds) Geology of Canadian Mineral Deposit Types, P-1. Geological Society of America, Boulder, p 523-540

Robb LJ, Robb VM (1986) Archaean pegmatite deposits in the north-eastern Transvaal. *In* Anhaeusser CR, Maske S (eds) Mineral deposits of Southern Africa. Geol Soc South Africa, Johannesburg, p 437-449

Robertson DE (1986) Skarn mineralization at Iron Mountain, New Mexico. MS thesis, Arizona State University, Tempe, Arizona

Robie RA, Hemingway BS, Fisher JR (1978) Thermodynamic properties of minerals and related substances at 298.15 K and 1 bar (10^5 pascals) pressure and at higher temperatures. U S Geol Surv Bull 1452

Rodinov SM (2000) Tin metallogeny of the Russian Far East. *In* Kremenetsky AA, Lehmann B, Seltmann R (eds) Ore-bearing granites of Russia and adjacent countries—IGCP Project 373. IMGRE, Moscow, p 237-262

Rogers JR (1990) Origin of beryl in the Miocene Sheeprock Granite, west-central Utah. MS thesis, Brigham Young University, Provo, Utah

Rogers JR, Christiansen EH (1989) Magmatic beryl in the Sheeprock Granite, west central Utah. Geol Soc Am Abstr Progr 21(5):136

Rozanov KI, Ontoeva TD (1987) Low-temperature bertrandite mineralization. Geol Rudn Mestorozhd 29:97-103 (in Russian)

Rubin JN, Price JG, Henry CD, Kyle JR (1990) Geology of the beryllium-rare earth element deposits at Sierra Blanca, West Texas. *In* Kyle JR (ed) Industrial mineral resources of the Delaware Basin, Texas and New Mexico: Conference Guidebook, 8. Society of Economic Geologists, p 191-203

Rudowski L, Giuliani G, Sabate P (1987) The Proterozoic granite massifs of Campo Formoso and Carnaiba (Bahia, Brazil) and their Be, Mo, W mineralizations. *In* McReath I, Sabate P, Sial A-N (eds) International Symposium on Granites and Associated Mineralizations, Bahia, Brazil, p 253-257

Rundkvist DV (1977) The distribution of mineral zones associated with granitoid magmatism and rare-metal mineralization in space and time. *In* Stemprok M, Burnol L, Tischendorf G (eds) Metallization associated with acid magmatism, 2. Geol Surv Prague, Prague, p 11-19

Rupasinghe MS, Dissanayake CB (1985) Charnockites and the genesis of gem minerals. Chem Geol 53: 1-16

Rupasinghe MS, Banerjee A, Pense J, Dissanayake CB (1984) The geochemistry of beryllium and fluorine in the gem fields of Sri Lanka. Mineral Dep 19:86-93

Russkikh SS, Shatov VV (1996) The Verkhnee Qairaqty scheelite stockwork deposit in central Kazakhstan. *In* Shatov V, Seltmann R, Kremenetsky A, Lehmann B, Popov V, Ermolov P (eds) Granite-Related Ore Deposits of Central Kazakhstan and Adjacent Areas. Glagol, St. Petersburg, p 167-180

Rye RO, Lufkin JL, Wasserman MD (1990) Genesis of the rhyolite-hosted tin occurrences in the Black Range, New Mexico, as indicated by stable isotope studies. *In* Stein HJ, Hannah JL (eds) Ore-bearing granite systems: petrogenesis and mineralizing processes, Geol Soc Am Spec Paper 246:233-250

Sainsbury CL (1968) Tin and beryllium deposits of the central York Mountains, western Seward Peninsula, Alaska. *In* Ridge JD (ed) Ore deposits of the United States, 1933-1967 (Graton-Sales Volume), 2. American Institute of Mining and Metallurgy, Petroleum Engineers, New York, p 1555-1572

Sainsbury CL (1969) Geology and ore deposits of the Central York Mountains, western Seward Peninsula, Alaska. U S Geol Surv Bull 1287

Sainsbury CL (1988) Vertical and horizontal zoning from tin to beryllium deposits, Lost River District, Alaska. *In* Kisvarsanyi G, Grant SK (eds) North American conference on Tectonic control of ore deposits and the vertical and horizontal extent of ore systems, Proceedings. University of Missouri, Rolla, p 80-91

Sainsbury CL, Hamilton JC, Huffman C, Jr. (1967) Geochemical cycle of selected trace elements in the tin-tungsten-beryllium district, western Seward Peninsula, Alaska: a reconnaissance study. U S Geol Surv Bull 1242:F1-F42

Salvi S, Williams-Jones AE (1996) The role of hydrothermal processes in concentrating high-field strength elements in the Strange Lake peralkaline complex, northeastern Canada. Geochim Cosmochim Acta 60:1917-1932

Salvi S, Williams-Jones AE (1997) Fischer-Tropsch synthesis of hydrocarbons during sub-solidus alteration of the Strange Lake peralkaline granite, Quebec/Labrador, Canada. Geochim Cosmochim Acta 61:83-99

Samchuk AI, Mitskevich BF (1980) Complexing of beryllium in carbonate solutions. Geochem Intl 17: 62-66

Sauer DA (1982) Emeralds from Brazil. *In* Eash-DM (ed) Proceedings of the International Gemological Symposium. Gemol Inst Am, Santa Monica, p 357-377

Schlenker JL, Gibbs GV, Hill EG, Crews SS, Myers RH (1977) Thermal expansion coefficients for indialite, emerald, and beryl. Phys Chem Mineral 1:243-255

Schmetzer K (1981) The mineralogy of ternary oxides in BeO-MgO-Al_2O_3. Naturwissenschaften 68: 471-472 (in German)

Schmitz C, Burt DM (1990) The Black Pearl Mine, Arizona: wolframite veins and stockscheider pegmatite related to an albitic stock. *In* Stein HJ, Hannah JL (eds) Ore-bearing granite systems: petrogenesis and mineralizing processes, Geol Soc Am Spec Paper 246:221-232

Schwarz D, Kanis J, Kinnaird J (1996) Emerald and green beryl from central Nigeria. J Gemmol 25: 117-141

Seal RR, II (1989) A reconnaissance study of the fluid inclusion geochemistry of the emerald deposits of Pakistan and Afghanistan. *In* Kazmi A, Snee L (eds) Emeralds of Pakistan: geology, gemology and genesis. Van Nostrand Reinhold Co., New York, p 151-164

Seltmann R, Kaempf H, Moeller P (1994) Proceedings of the IAGOD Erzgebirge meeting: metallogeny of collisional orogens: focussed on the Erzgebirge and comparable metallogenic settings. Czech Geological Survey, Prague, Czech Republic

Serykh VI (1996) Granitic rocks of central Kazakhstan. *In* Shatov V, Seltmann R, Kremenetsky A, Lehmann B, Popov V, Ermolov P (eds) Granite-Related Ore Deposits of Central Kazakhstan and Adjacent Areas. Glagol, St. Petersburg, p 25-54

Shand SJ (1927) The Eruptive Rocks. Wiley, New York

Sharp WN (1976) Geologic map and details of the beryllium and molybdenum occurrences, Mount Antero, Chaffee County, Colorado. U S Geol Surv MF-810

Shaw AL, Guilbert JM (1990) Geochemistry and metallogeny of Arizona peraluminous granitoids with reference to Appalachian and European occurrences. *In* Stein HJ, Hannah JL (eds) Ore-bearing granite systems: petrogenesis and mineralizing processes, Geol Soc Am Spec Paper 246. Geological Society of America, Boulder, Colorado, p 317-356

Shawe DR (1966) Arizona-New Mexico and Nevada-Utah beryllium belts. U S Geol Surv Prof Paper 550-C:C206-C213

Shawe DR (1968) Geology of the Spor Mountain beryllium district, Utah. *In* Ridge JD (ed) Ore deposits of the United States, 1933-1967 (Graton-Sales Volume), 2. American Institute of Mining and Metallurgy, Petroleum Engineers, New York, p 1148-1161

Shawe DR, Bernold S (1966) Beryllium content of volcanic rocks. U S Geol Surv Bull 1214-C:C1-C11

Shcherba GN (1970) Greisens. Intl Geol Rev 12:114-150, 239-255

Sheremet YM, Panov BS (1988) Rare-metal leucogranites of the Ukrainian Shield. Intl Geol Rev 30: 275-283

Shigley JE, Dirlam DM, Laurs BM, Boehm EW, Bosshart G, Larson WF (2000) Gem localities of the 1990s. Gems Gemol 36:292-335

Shigley JE, Foord EE (1984) Gem-quality red beryl from the Wah Wah Mountains, Utah. Gems Gemol 20:208-221

Simmons WB, Lee MT, Brewster RH (1987) Geochemistry and evolution of the South Platte granite-pegmatite system, Jefferson County, Colorado. Geochim Cosmochim Acta 51:455-471

Sinclair WD (1996) Granitic Pegmatites. *In* Eckstrand OR, Sinclair WD, Thorpe RI (eds) Geology of Canadian Mineral Deposit Types, P-1. Geological Society of America, Boulder, p 503-512

Sinclair WD, Richardson DG (1994) Studies of rare-metal deposits in the Northwest Territories. Geol Surv Canada Bull 475

Sinkankas J (1981) Emerald and other beryls. Chilton Book Company, Radnor, Pennsylvania

Sinkankas J (1989) Emerald and other beryls. Geoscience Press, Prescott, Arizona

Sliwa AS, Nguluwe CA (1984) Geological setting of Zambian emerald deposits. Precamb Res 25:213-228

Smith DR, Noblett J, Wobus RA, Unruh DM, Douglass J, Beane R, Davis C, Goldman S, Kay G, Gustavson B, Saltoun B, Stewart J (1999) Petrology and geochemistry of late-stage intrusions of the A-type, mid-Proterozoic Pikes Peak Batholith (central Colorado, USA): implications for petrogenetic models. Precamb Res 98:271-305

Smyk MC, Taylor RP, Jones PC, Kingston DM (1993) Geology and geochemistry of the West Dead Horse Creek rare-metal occurrence, northwestern Ontario. Explor Min Geol 2:245-251

Snee LW, Kazmi AH (1989) Origin and classification of Pakistani and world emerald deposits. *In* Kazmi A, Snee L (eds) Emeralds of Pakistan: geology, gemology and genesis. Van Nostrand Reinhold Co., New York, p 229-230

Soboleva GI, Tugarinov IA, Khitarov DN (1972) Role of calcium and magnesium in country rocks during chrysoberyl mineralization. Geokhimiya:1392-1395 (in Russian)

Soboleva GI, Tugarinov IA, Kalinina VF, Khodakovskiy IL (1977) Investigation of equilibria in the system BeO-$NaOH$-HNO_3-H_2O in the temperature range 25-250° C. Geochem 7:1013-1024 (in Russian)

Soboleva GI, Tugarinov IA, Golitsina NS, Khodakovskii IL (1984a) Study of beryllium behavior in fluorine-containing hydrothermal solutions at 150-250°C. Geokhimiya:812-822 (in Russian)

Soboleva GI, Tugarinov IA, Golitsina NS, Khodakovskiy IL (1984b) The behavior of beryllium in fluorine-bearing hydrothermal solutions at 150-250° C. Geochem Intl 21:20-30

Soliman MM (1982) Chemical and physical properties of beryl and cassiterite ores, Southeastern Desert, Egypt. Inst Min Metall Trans 91:B17-B20

Soliman MM (1984) Geochemical exploration for Sn, Nb, Be, Mo and Bi mineralization, Homr Akarim area, Southeastern Desert, Egypt. J Afr Earth Sci 2:287-299

Soliman MM (1986) Ancient emerald mines and beryllium mineralization associated with Precambrian stanniferous granites in the Nugrus-Zabara area, Southeastern Desert, Egypt. Arab Gulf J Sci Res 4:529-548

Solodov NA (1977) Main features of the distribution of rare-metal mineralization in the Earth's crust. Sov Geol Geophys 18:7-14

Solomon M (1981) An introduction to the geology and metallic ore deposits of Tasmania. Econ Geol 76:194-208

Sørensen H (1992) Agpaitic nepheline syenites: a potential source of rare elements. Appl Geochem 7: 417-427

Sørensen H (1997) The agpaitic rocks: an overview. Mineral Mag 61:485-498

Sørensen H, Rose HJ, Peterson OV (1981) The mineralogy of the Ilímaussaq Intrusion. *In* Bailey JC, Larsen LM, Soerensen H (eds) The Ilímaussaq Intrusion, South Greenland: a progress report on geology, mineralogy, geochemistry and economic geology, 103. Grønlands Geologiske Undersøgelse, Copenhagen, p 19-24

Spendlove E (1992) "Red emeralds" of Utah. Rock & Gem 22:32-36, 38

Staatz MH, Griffitts WR, Barnett PR (1965) Differences in the minor element composition of beryl in various environments. Am Mineral 50:1783-1795

Stadler HA, De Quervain F, Niggli E, Graeser SA (1973) Die Mineralfunde der Schweiz. Neuarbeitung von R.L. Parker: Die Mineralfunde der Schweizer Alpen. Wepf & Co., Basel (in German)

Stager HK (1960) A new beryllium deposit at the Mount Wheeler Mine, White Pine County, Nevada. U S Geol Surv Prof Paper 400-B:B70-B71

Steenfelt A (1991) High-technology metals in alkaline and carbonatitic rocks in Greenland: recognition and exploration. J Geochem Explor 40:263-279

Stein HJ, Hannah JL, (eds) (1990) Ore-bearing granite systems: petrogenesis and mineralizing processes. Geol Soc Am Spec Paper 246. Geological Society of America, Boulder

Strong DF (1988) A review and model for granite-related mineral deposits. *In* Taylor RP, Strong DF (eds) Recent advances in the geology of granite-related mineral deposits, Can Inst Min Metall Spec Paper 39:424-445

Strunz H, Nickel EH (2001) Strunz mineralogical tables. Chemical-structural mineral classification system, 9th edn. Schweizerbart, Stuttgart

Stussi JM (1989) Granitoid chemistry and associated mineralization in the French Variscan. Econ Geol 84:1363-1381

Sukhorukov FV (1989) Zoning of the kaolin weathering profiles and distribution of rare alkaline elements, beryllium, and boron in them. Sov Geol Geophys 30:61-68

Sun SS, Eadington PJ (1987) Oxygen isotope evidence for the mixing of magmatic and meteoric waters during tin mineralization in the Mole Granite, New South Wales, Australia. Econ Geol 82:43-52

Suwimonprecha P, _ern_ P, Friedrich G (1995) Rare metal mineralization related to granites and pegmatites, Phuket, Thailand. Econ Geol 90:603-615

Sviridenko LP, Ivashchenko VI (2000) Ore-bearing grantites of Karelia, Russia. *In* Kremenetsky AA, Lehmann B, Seltmann R (eds) Ore-bearing granites of Russia and adjacent countries—IGCP Project 373. IMGRE, Moscow, p 281-294

Swanson SE, Newberry RJ, Coulter GA, Dyehouse TM (1990) Mineralogical variation as a guide to the petrogenesis of the tin granites and related skarns, Seward Peninsula, Alaska. *In* Stein HJ, Hannah JL (eds) Ore-bearing granite systems: petrogenesis and mineralizing processes, Geol Soc Am Spec Paper 246:143-159

Syromyatnikov FV, Makarova AP, Kupriyanova II (1972) Experimental studies of stability of beryl and phenacite in aqueous solutions. Intl Geol Rev 14:837-839

Tauson LV, Cambel B, Kozlov VD, Kamenicky L (1978) The composition of tin-bearing granites of the Kru_né Hory (Bohemian Massif), Spis-Gemer (Rudohorie) ore mts. (Western Carpathians) of Czechoslovakia and the Transbaikalian provinces of the U.S.S.R. *In* Stemprok M, Burnol L, Tischendorf G (eds) Metallization associated with acid magmatism, 3. Geol Surv Prague, Prague, p 297-304

Taylor RG (1978) Five tin provinces in eastern Australia: an approach to the classification and genesis of tin deposits. *In* Stemprok M, Burnol L, Tischendorf G (eds) Metallization associated with acid magmatism, 3. Geol Surv Prague, Prague, p 43-56

Taylor RP, Pollard PJ (1996) Rare earth mineralization in peralkaline systems: the T-Zone REE-Y-Be deposit, Thor Lake, Northwst Territories, Canada. *In* Jones AP, Wall F, Williams CT (eds) Rare Earth Minerals: Chemistry, origin and ore deposits. Chapman & Hall, London, p 167-192

Taylor RP, Strong DF, (eds) (1988) Recent advances in the geology of granite-related mineral deposits. Can Inst Min Metall Spec Paper 39:1-445

Taylor WR, Esslemont G, Sun SS (1995a) Geology of the volcanic-hosted Brockman rare-metals deposit, Halls Creek mobile zone, Northwest Australia. II. Geochemistry and petrogenesis of the Brockman Volcanics. Mineral Petrol 52:231-255

Taylor WR, Page RW, Esslemont G, Rock NMS, Chalmers DI (1995b) Geology of the volcanic-hosted Brockman rare-metals deposit, Halls Creek mobile zone, Northwest Australia. I. Volcanic environment, geochronology and petrography of the Brockman Volcanics. Mineral Petrol 52:209-230

Thompson TJ, Keith JD, Christiansen EH, Tingey DG, Heizler M (1996) Genesis of topaz rhyolite hosted red beryl in the Wah Wah Mountains, Utah. Geol Soc Am Abstr Progr 28:A-154

Tischendorf G (1977) Geochemical and petrographic characteristics of silicic magmatic rocks associated with rare-element mineralization. *In* Stemprok M, Burnol L, Tischendorf G (eds) Metallization associated with acid magmatism, 2. Geol Surv Prague, Prague, p 41-96

Topor ND, Mel'chakova LV (1982) Measurement of mineral heat capacities by scanning quantitative differential thermal analysis. Vestn Mosk Univ 4:50-58 (in Russian)

Troshin YP, Segalevich SF (1977) The H_2O^+ aureole around hypabyssal intrusions and hydrothermal mineralization. Sov Geol Geophys 18:106-110

Trueman DL, Pedersen JC, de St Jorre L, Smith DGW (1988) The Thor Lake rare-metal deposits, Northwest Territories. *In* Taylor RP, Strong DF (eds) Recent advances in the geology of granite-related mineral deposits, Can Inst Min Metall Spec Paper 39:280-290

Trunilina VA, Orlov YS, Roev SP, Ivanov PO (2000) Ore-bearing grantitic complexes of eastern Yakutia. *In* Kremenetsky AA, Lehmann B, Seltmann R (eds) Ore-bearing granites of Russia and adjacent countries—IGCP Project 373. IMGRE, Moscow, p 295-314

Upton BGJ, Emeleus CH (1987) Mid-Proterozoic alkaline magmatism in southern Greenland: the Gardar Province. *In* Fitton JG, Upton BGJ (eds) Alkaline igneous rocks, Geol Soc (London) Spec Pub 30: 449-471

Van Averbeke N (1996) Fluid inclusion studies of lithophile element mineralization in the Great Basin, Master's University of Arizona

van Gaans PFM, Vriend SP, Poorter RPE (1995) Hydrothermal processes and shifting element association patterns in the W-Sn enriched granite of Regoufe, Portugal. J Geochem Explor 55:203-222

Vlasov KA (ed) (1966a) Geochemistry of Rare Elements. Israel Program for Scientific Translations, Jerusalem

Vlasov KA (ed) (1966b) Mineralogy of Rare Elements. Israel Program for Scientific Translations, Jerusalem

Vlasov KA (ed) (1968) Genetic Types of Rare Element Deposits. Israel Program for Scientific Translations, Jerusalem

Vlasov KA, Kuzmenko MV, Eskova EM (1966) The Lovozero alkali massif. Hafner Publishing Company, New York

Voncken JHL, Vriend SP, Kocken JWM, Jansen JBH (1986) Determination of beryllium and its distribution in rocks of the Sn-W granite of Regoufe, northern Portugal. Chem Geol 56:93-103

Vynar ON, Razumeeva NN (1972) Hydrothermal mineralization in the Sushchano-Perzhanskaya Zone. Mineral Sb (Lvov) 26:197-205 (in Russian)

Walshe JL, Heithersay PS, Morrison GW (1995) Toward an understanding of the metallogeny of the Tasman fold belt system. Econ Geol 90:1382-1401

Wang Z, Tao K, Ye D, Liu W (1992) Phase transformations of beryl at high pressures and high temperatures. Dizhi Kexue 1992:161-167 (in Chinese)

Warner LA, Cameron EN, Holser WT, Wilmarth VR, Cameron EM (1959) Occurrence of non-pegmatite beryllium in the United States. U S Geol Surv Prof Paper 318

Werner S, Plech A (1995) Compressibility of tugtupite at high pressure. Z Kristallogr 210:418-420

Wilson M (1989) Igneous petrogenesis: a global tectonic approach. Unwin Hyman, London

Wood PA (1996) Petrogenesis of the Spruce Pine pegmatites, North Carolina. MS thesis, Virginia Polytechnic Institute and State University

Wood SA (1992) Theoretical prediction of speciation and solubility of beryllium in hydrothermal solution to 300°C at saturated varpor pressure: application to bertrandite/ phenakite deposits. Ore Geol Rev 7:249-278

Woolley AR (1987) Alkaline rocks and carbonatites of the world: Part 1, North and South America. British Museum (Natural History), London

Woolley AR (2001) Alkaline Rocks and Carbonatites of the World. Part 3: Africa. The Geological Society, London

Wu C, Yuan Z, Bai G (1996) Rare earth deposits in China. *In* Jones AP, Wall F, Williams CT (eds) Rare Earth Minerals: Chemistry, origin and ore deposits. Chapman & Hall, London, p 281-310

Yeganeh-Haeri A, Weidner DJ (1989) Elasticity of a beryllium silicate (phenacite: Be_2SiO_4). Phys Chem Mineral 16:360-364

Zabolotnaya NP (1977) Deposits of beryllium. *In* Smirnov V (ed) Ore deposits of the USSR, III. Pitman Publ., London, p 320-371

Zabolotnaya NP, Novikova MI (1983) Dickite-bertrandite mineralization in subeffusive rocks of the Mesozoic activation zone in a region of Siberia. Mineral Rudn Mestorozhd1983:13-18 (in Russian)

Zasedatelev AM (1973) The problem of genesis of berylliferous skarns. Intl Geol Rev 15:213-224

Zaw K (1998) Geological evolution of selected granitic pegmatites in Myanmar (Burma): constraints from regional setting, lithology, and fluid-inclusion studies. Intl Geol Rev 40:647-662

Zhabin AG, Kazakova ME (1960) Barylite, $BaBe_2(Si_2O_7)$, from the Vishnevye Gory, the first occurrence in U.S.S.R. Dokl Akad Nauk SSSR 134:419-421 (in Russian)

Zhabin AG, Mukhitdinov GN, Kazakova ME (1960) Paragenetic associations of rare elements of accessory minerals in exocontact fenitized rocks of the Vishnevye Gory miaskite intrusion. Trudy Inst Mineral Geokh Kristallokh Redk Elem 1960:51-72 (in Russian)

Zhang S (1992) Petrogenesis and mineralization of granite related rare-metal deposits in South China. J Southeast Asian Earth Sci 7:277-278

Zhang S, Feng M, Wang H, Lu W, Yang M (1999) Geological features and genesis of emerald deposit in Malipo County, Yunnan, China. Geol Sci Techn Inform 18:50-54 (in Chinese)

Zimmermann JL, Giuliani G, Cheilletz A, Arboleda C (1997) Mineralogical significance of fluids in channels of Colombian emeralds: a mass-spectrometric study. Intl Geol Rev 39:425-437